Energy and Fuels 2016

E3S Web of Conferences Volume 14 (2017)

Krakow, Poland
21 - 23 September 2016

Editors:

W. Suwala
M. Dudek
J. Leszczynski
S. Lopata

ISBN: 978-1-5108-3782-9

Printed from e-media with permission by:

Curran Associates, Inc.
57 Morehouse Lane
Red Hook, NY 12571

Some format issues inherent in the e-media version may also appear in this print version.

Printed by Curran Associates, Inc. (2017)

For additional information, please contact EDP Sciences – Web of Conferences at the address below.

EDP Sciences – Web of Conferences
17, Avenue du Hoggar
Parc d'Activité de Courtabœuf
BP 112
F-91944 Les Ulis Cedex A
France

Phone: +33 (0) 1 69 18 75 75
Fax: +33 (0) 1 69 28 84 91

contact-edps@webofconferences.org

Additional copies of this publication are available from:

Curran Associates, Inc.
57 Morehouse Lane
Red Hook, NY 12571 USA
Phone: 845-758-0400
Fax: 845-758-2633
Email: curran@proceedings.com
Web: www.proceedings.com

TABLE OF CONTENTS

E3S Web of Conferences
Open access conference proceedings in Environment, Energy & Earth Sciences
A flexible, fast and efficient publication service, for an open access recording of your event and a wide dissemination of your proceedings
www.e3s-conferences.org
edp open
Open Access Publishing from EDP Sciences

Committee of the Energy and Fuels 2016

Patronage

MINISTRY OF ENERGY

Minister of Energy
Krzysztof Tchórzewski

AGH
Akademia Górniczo-Hutnicza
im. Stanisława Staszica
w Krakowie

Rector of AGH University
of Science and Technology
Prof. Tadeusz Słomka

PK
Politechnika Krakowska
im. Tadeusza Kościuszki

Rector of Cracow
University of Technology
Prof. Jan Kazior

Former Rector of Cracow
University of Technology,
Prof. Kazimierz Furtak

Media patronage

CIRE.PL
Centrum Informacji o Rynku Energii

Diamond Sponsor

eit Knowledge & Innovation Community
KIC InnoEnergy

Platinum Sponsor

TAURON

MAGNETO®

edp sciences

E3S Web of Conferences

Statement of Peer review

In submitting conference proceedings to E3S Web of Conferences, I certify to the Publisher that I adhere to the **Policy on Publishing Integrity** of the journal in order to safeguard good scientific practice in publishing.

1. All articles have been subjected to peer review administered by the proceedings editors.
2. Reviews have been conducted by expert referees, who have been requested to provide unbiased and constructive comments aimed, whenever possible, at improving the work.
3. Proceedings editors have taken all reasonable steps to ensure the quality of the materials they publish and their decision to accept or reject a paper for publication has been based only on the merits of the work and the relevance to the journal.

Title, date and place of the conference

Energy and Fuels 2016
Kraków 21–23 September 2016

Proceedings editor(s):

Wojciech Suwała
Magdalena Dudek
Jacek Leszczyński
Stanisław Łopata

Date and editor's signature

Preface

The energy sector is now experiencing unprecedented changes and challenges of various magnitudes involving a range of fields. Limiting polluting emissions is a continuing process, especially in the European Union, where successive directives such as *Large Combustion Plants* and *Industrial Emissions* have set new and more stringent limits on emissions. Based on analyses of available techniques, plants are forced to invest in emissions abatement, which leads to higher costs for generated energy. Efforts to combat climate change are moving in the same direction. Over two decades after the adoption of the Kyoto Protocol, agreement on global efforts was reached at the 21st Conference of the Parties to the 1992 United Nations Framework Convention on Climate Change. Although the agreement does not impose decisive limits on greenhouse gas (GHG) emissions, it creates the obligation to undertake actions leading to a decrease in such emissions. Recently, the USA and China, which jointly are responsible for almost 40% of global GHG emissions, ratified the COP21 agreement. In terms of the energy sector, this means that fossil fuels should be gradually eliminated from the energy mix. The European goal is the domination of energy supplies by renewable sources by 2050. However, existing abundant reserves of cheap coal, a source which has no economic competitor, should not be forgotten. Despite being characterised by relatively high emissions and a negative impact on the environment and human health, coal will continue to play a role in energy supplies. However, since only clean coal technologies will be acceptable, improvements in efficiency and abatement technologies constitute a plausible path for coal use.

Air quality is becoming even more important in the case of Poland. While emissions from the energy sector are strictly controlled and have decreased substantially over the past two decades, so-called 'low emissions', mainly from domestic heating and transport, constitute a real threat, particularly in winter, when permissible concentrations are often exceeded. Polish society has become aware of possible adverse impacts on health and projects have been undertaken aimed at reducing the associated risks.

The introduction of support systems for renewable energy sources has led to an increase in their share of power generation. On one hand, this helps to mitigate the air quality problem, but on the other, RESs, which are still characterised by high capital costs and a low capacity factor in Polish conditions, contribute to increasing energy costs for the end consumer. Additionally, their intermittent nature requires completely new working conditions within the electricity supply system, mainly the more flexible operation of classic fossil generators to cover the residual load. Their unpredictability also requires the maintenance of backup capacity in a state of readiness to balance disappearing supplies. Development of high-capacity storage technologies could help to overcome the balancing problems.

Changes also involve energy-intensive sectors. Industry is continuously improving the efficiency of its processes, leading to lower energy consumption.

There is growing evidence that we will face a fuel changeover in the area of transport, from fossil-fuelled to electric vehicles (not, notably, for the first time in history), modifying demands for crude oil and electricity. All of these considerations barely scratch the surface of the problem.

The presentations, over one hundred in number, related to all important areas of the energy (energy policy, renewable energy sources, power systems and equipment, energy conversion and management, mathematical modelling in energy and industrial processes, energy storage technologies, and nuclear energy) and fuel (gasification of coal and solid fuels, biomass and biofuels, catalysis, methods of reducing pollutants during combustion of solid fuels, waste conversion and management, environmental protection topics, alternative fuel for fuel cells, electrochemical power sources) sectors were presented at the Energy and Fuels Conference 2016, held 21–23 September 2016 in Cracow, Poland and organised by the Faculty of Energy and Fuels, AGH University of Science and Technology, with the collaboration of the Institute of Thermal Power Engineering of the Cracow University of Technology.

Following a successful conference, we invited delegates to submit full-length manuscripts related to their conference presentations. The manuscripts were submitted for peer review prior to final acceptance for publication in the scientific journal *E3S Web of Conferences*. The editors address special warm thanks to the reviewers, who earnestly reviewed the papers and provided many valuable comments and remarks aimed at improving their quality.

We also thank the international scientific committee and members of the local organising committee of Energy and Fuels 2016 for their timely support.

We hope that the readers of the journal will find the presented papers interesting, useful, and stimulating for their future studies in the area of energy and fuels.

Kraków, Poland
January 2017

Editors:
Wojciech Suwała
Magdalena Dudek
Stanisław Łopata
Jacek Leszczyński

Near future of Polish energy system – COP21 and BAT conclusions

Wojciech Suwała[1,*], *Artur* Wyrwa[1] *and Marcin* Pluta[1]

[1]AGH University of Science and Technology, Faculty of Energy and Fuels, Al. Mickiewicza 30, 30-059 Krakow, Poland

Abstract. Next years will be very important for the development of the Polish energy system. Substantial part of the existing capacities are obsolete and need to be replaced or at least deeply modernised to fulfil expected emissions regulations. A hybrid of TIMES-PL and TIMES-MACRO energy-economic models was used to analyse the possible changes in the power generation mix and economic impact of different scenarios that combine the climate and air pollution regulations. The results show that to maintain the CO_2 emission reduction trend a combination of measures is needed including the use of renewables, nuclear, natural gas and clean coal technologies. However, such transition in the energy sector will lead to a cumulative GDP loss in the near future once the possible GDP gains could happen after 2045 in which the economic, technical and legislative conditions are very difficult to foresee.

1 Introduction

Next years will be very important for the development of the Polish energy system. There are problems inherited from the past. Due to insufficient investments in renewing the ageing generation assets during last two decades, nowadays the balancing of electricity demand becomes increasingly difficult, particularly during high demand periods. The constantly stiffening regulations on air pollution laid down by the EU Directives and Best Available Techniques (BAT) reference documents put burden on coal fired power plants, which dominate in the power generation mix. Another challenge is a reduction of GHG emissions in line with the Paris COP21 agreement and European Union declaration to decrease carbon dioxide emissions by 40% till 2030. One of EU instruments to reach such reduction is the EU Emission Trading System (ETS). In this system a limit on absolute amount of emissions from energy intensive industry, so called ETS sources (e.g. in power sector all generation units with total rated thermal input exceeding 20 MW) is set for a given period. At the same time the market of European Union Allowances (EUA) was established in which each emitter has to surrender sufficient number of EUAs to cover its emissions (1 EUA allows to emit 1 ton of CO_2). Number of EUAs available in a given period corresponds to the emission cap and EUAs are to be purchased by ETS sources in

* Corresponding author: wojciech.suwala@agh.edu.pl

auctioning system (for Polish electricity generators full auctioning will be obligatory after 2020). As a consequence of these climate and air pollution regulations a substantial part of existing obsolete capacities need to be replaced or at least deeply modernised. Taking into account that investments in the energy sector are irreversible all these aspects need careful consideration and discussion, involving decision makers, energy companies, experts and academics, in order to make the best possible transition of the energy system in near future. This discussion should be supported by the results of analytical tools, which are able to examine these diverse problems.

In this paper the hybrid of TIMES-PL and TIMES-MACRO energy-economic models was used to analyse the possible changes in the power generation mix and economic impact of different scenarios that combine the climate and air pollution regulations.

The paper starts with the discussion of circumstances and constraints forming scenarios of Polish energy system development, including time schedule of carbon dioxide emissions reduction and BAT conclusions. Nextly, the results of model calculations, mainly energy mix and costs of CO_2 emission reduction are presented. Capacities of TIMES model make it possible to estimate the values of main variables determining and describing the operation of the energy system. This information help to reduce the uncertainties of system development planning and preparation of regulations. Finally, the main conclusions are drawn.

2 Emissions and Regulations

Conference of the Parties in Paris (COP21) set new rules in establishing climate policies. Instead of a top-down approach proposed in Kyoto Protocol in which emission reduction targets were assigned to countries accordingly to meet the global reduction target, according to the Paris Agreement Parties declare their intended nationally determined contributions (INDC) which they commit to fulfil. The United States and China, the world's biggest emitters of greenhouse gases agreed to ratify the Paris climate change agreement. US intends to achieve 26-28% reduction below its 2005 level in 2025 whereas China aims at achieving peaking of its GHG emissions around 2030. The EU as a "bubble" has committed to a target of at least 40% domestic reduction in greenhouse gas emissions by 2030 compared to 1990 [1], but targets of individual Member States have not been declared. This declaration covers all greenhouse gases, and all sectors, i.e. these covered by the ETS but also others (non-ETS). In addition no international market based mechanisms will be included, but Land Use, Land Use Change and Forestry (LULUCF) contribution mechanism should be established before 2020.

In face of this declaration the problem of intensity of GHG emissions reduction arises, especially for Poland which is considered as coal dependent.

The data on greenhouse gases emissions for EU28 and Poland (Fig. 1) indicate that for both cases continuation of past trends give in 2030 emissions above the 40% reduction target, excess is about 10% and 5% for EU28 and Poland respectively. But EU trend for the last ten years (EU10Y) indicates much higher intensity of reduction - about 100 Mt per year, and continuation could give emissions 20% below the target.

The European Commission intends to reduce GHG emissions by 43% to 2030 comparing 2005 in ETS sector and 30% in non-ETS sectors [2]. Individual countries' targets of total reduction within non-ETS sectors has been established at a range from 40% (Luxembourg) to 0% (Bulgaria), for Poland this is 7%.

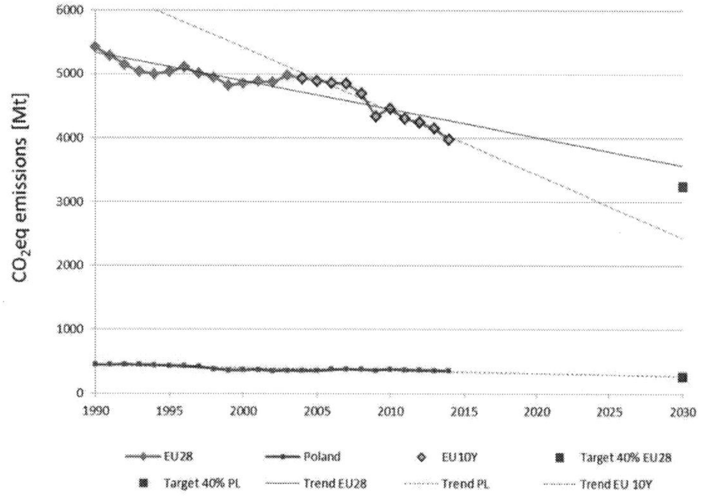

Fig. 1. GHG emissions in EU28 and Poland, trends and targets for 2030 (million metric tons). Source: own based on EUROSTAT data on greenhouse gas emissions by source sector

In case of Poland this target could be reached providing past trends will be continued (Fig. 2).

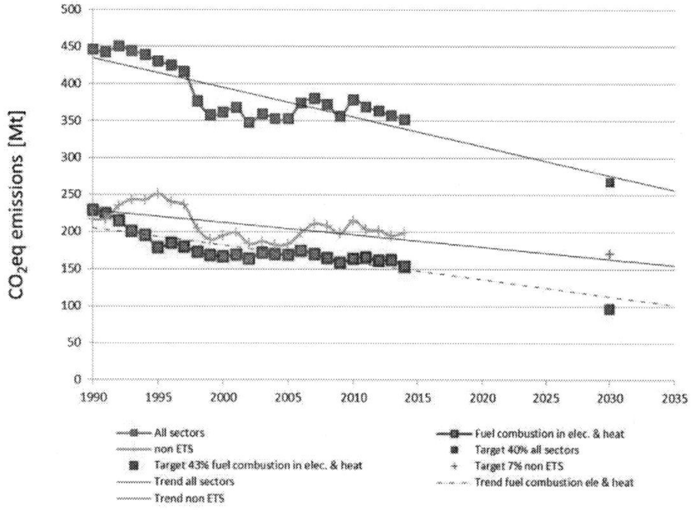

Fig. 2. GHG emissions in Poland, trends and targets for 2030 (million metric tons). Source: own based on EUROSTAT data on greenhouse gas emissions by source sector

For traditional air pollutants the emission standards for the power plants are determined in some documents, mainly the following directives:

- **LCP** - Large Combustion Plants Directive,
- **MCP** - Medium Combustion Plants Directive [3] referring generally to plants or their combination with a total rated thermal input equal to or greater than 1 MW and less than 50 MW,
- **IED** Industrial Emissions Directive,
- **BAT** new (conlucions) on attainable emissions standards, to be released in beginning 2017 and in force since 2021.

These documents limit the power station emissions of main pollutants such as sulphur and nitrogen oxides, particulates (dust). New BAT regulations introduce limits on mercury as well as a new factor i.e. the efficiency of transformation processes.

Table 1. Examples of emission limit values (mg/Nm3) for large combustion plants.

Year	Dust - particulates				SO_2				NO_x			
	LCP	IED	BAT existing	BAT new	LCP	IED	BAT existing	BAT new	LCP	IED	BAT existing	BAT new
2008	100				1200				600			
2016	100	20			1200	200			600	200		
2021		20	10	2*		200	130	10*		200	180	65*
2030			10	2*			130	10*			180	65*

Sources: [4, 5, 6] *The lower end of the range presented in [6].

The differences of allowed limits are impressive, especially for sulphur dioxide, for which limit is lowered by 120 times, however the restricted values are based of best available techniques analyses, thus attainable at rational cost.

3 Objective of the study

As presented in Fig. 2 CO_2 emissions from fuel combustion in public electricity and heat production have revealed the downward trend in the last two decades. The aim of the study was to (i) analyse possible future emission levels, (ii) estimate the CO_2 abatement potential within this sector for different EUA price scenarios and (iii) assess the impact of these scenarios as well as the expenditures that are necessary to adopt the generation units to BAT requirements on Gross Domestic Product.

4 Methodology

The analysis of possible future emissions for different energy mix and EUA prices was done with the use of TIMES-PL. TIMES-PL belongs to a family of bottom-up, partial equilibrium models developed with the use of IEA-ETSAP TIMES generator [7]. Analysis of the development of Polish energy system under various energy and environmental constraints with the use of TIMES-PL has been presented in the following papers [8-11]. The electricity supply side in TIMES-PL is reflected at the unit generation level or its homogenous representations, with individual technical and economic characteristics, including learning processes outcomes [10, 11]. The latest plans for decommissioning of existing capacities as well as for new investments are considered. New technologies such as Integrated Gasification Combined Cycle (IGCC) or small modular reactors are also considered and prospects of their deployment are evaluated. Additional data important for setting the optimal plan are prices of fuels, and demand for electricity and heat. The latter is derived based on econometric methods mainly considering the rate of economy development. The model indicates optimal capacity expansion plan and depicts the operation on generation units. The time horizon taken into consideration in the current analysis covers the period between 2011 and 2050. Each modelled year is split into 224 time slices. This approach allows to consider the temporal characteristics of supply and demand side especially behaviour of intermittent renewable technologies is taken into account in higher resolution. The marginal CO_2 emission abatement cost curve (MACC) has been built based on the results of consecutive TIMES-PL model runs with different EUAs prices. The first run started with the certificate price equal to 0 PLN/tCO_2 and in each subsequent run the price has been increased by 25 PLN/tCO_2 to finally reach 300 PLN/tCO_2. In each modelling year the abated CO_2 emissions (i.e. difference in emission level) resulting from an increase in the CO_2 certificate price were used to draw the abatement curve. Finally, the economic impact was assessed with the TIMES-MACRO model, which is also the results of the on-going IEA ETSAP project and aims at providing opportunity to analyse economic results of different policies. TIMES-PL and TIMES-MACRO have been coupled (Fig. 3) to allow estimation of changes in GDP level as results of BAT requirements introduction as well as of various levels of EUAs prices.

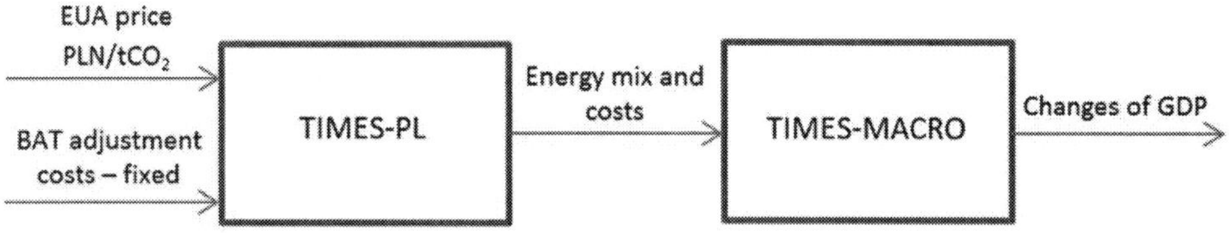

Fig. 3. The idea of the approach applied to estimate the economic results of different policies.

It should be noted, that the cost calculation of introduction of BAT conclusions is difficult without detailed data on possible emission controls that can be installed on each generation unit. Therefore the overall total costs of 10 Billion PLN were assumed as representing adjustment expenses, allowing estimation of impact on energy prices. The estimates of these costs, depending on scope and time distribution, range from 4 to 12 Billion PLN [12]. Therefore, the assumed amount is indicated as rational and in the line with the current policy oriented on ensuring energy security conditions.

5 Scenarios considered

The elements that form the scenario come from the main factors that will shape Polish energy system in the near future. These are mainly related to: (i) the EU and global climate policies resulting from COP21 agreement, and (ii) the restrictions on air pollutants' emissions which are expected to be determined based on the BAT conclusions [6]. Thus the following scenarios were formed:

- **REF** (reference) scenario based on the present regulations for emissions of traditional air pollutants and zero costs of carbon dioxide emissions allowances,
- **BAT** scenario assuming introduction of BAT conclusions, in which the costs necessary for modernisation and adjustments of generation assets to the BAT conclusions requirements were taken into account, based on the mentioned study [12],
- Different **EUA** price scenarios, in which the carbon price ranged from 150 to 300 PLN/tCO_2, as a measure of undertaken less/more ambitious EU policy targets to reduce carbon dioxide emissions in ETS sector, it is assumed that the technologies fulfil BAT requirements.

 DOI: 10.1051/e3sconf/20171401001

6 Results

One of the most informative tools for the GHG abatement policies evaluation are the aforementioned MACCs. Fig. 4 presents the emissions reduction that occurs for various levels of EUA price. This stepwise curve illustrates capacities of reduction, which as explained later, come from possibilities of replacing cheap but highly carbon intensive coal-based power generation with generation from other technologies. For instance, total CO_2 reduction potential as referred to the REF scenario in 2030 is nearly 100 Mt annually. This is about 2/3 of the present CO_2 emission level from fuel combustion for electricity and heat production. Such reduction could be achieved at high EUA price reaching 300 PLN/tCO_2 (approximately 70 €/tCO_2). Recent forecasts of EUA values [2] are lower; the 70 €/tCO_2 level is reached in 2045 linearly from present low values.

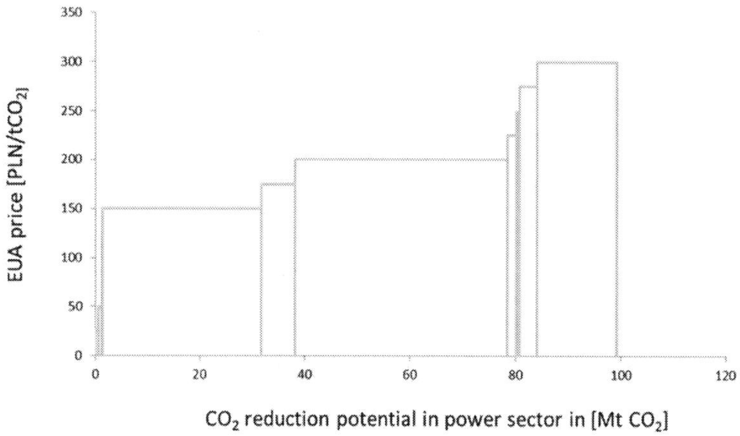

Fig. 4. Abatement cost curve for power sector in Poland for 2030 generated with the TIMES-PL model.

The details of technological changes to reach the emission reduction depicted in Fig. 4 are illustrated in Fig. 5. One can see that up to 2030 there is practically no change in electricity generation mix as compared to the REF scenario unless the price of EAU certificate exceeds 150 PLN/tCO_2. Above this price firstly the power generation in most inefficient hard coal fired power plants and CHPs is replaced with nuclear- and natural gas-based generation. The higher the EAU price the more natural gas is used but also new hard coal units equipped with CCS are built (although the future application of CCS in power sector is often questioned [10] in the current study the assumption has been made that CCS would be available starting from 2030). The stable generation from gas CHPs is the result of limited heat demand. Due to construction time constraint also limited generation is observed from new nuclear units.

DOI: 10.1051/e3sconf/20171401001

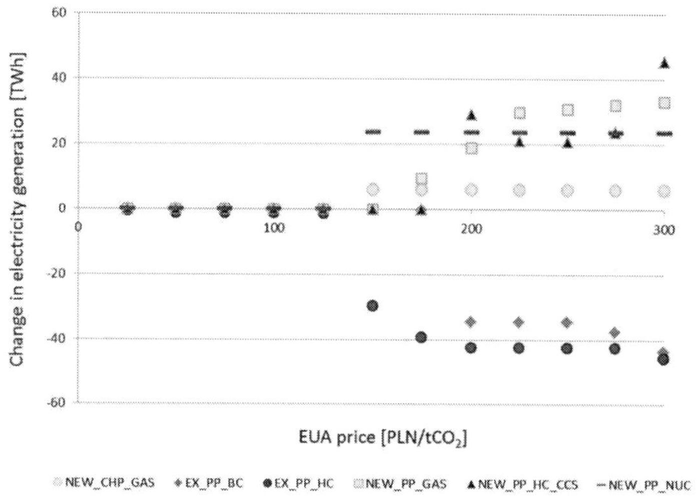

Fig. 5. Change in net electricity generation in 2030 as a result of EUA price change in relation to the Reference scenario (0 PLN/tCO_2) in most impacted power technologies.

The decreasing trend in CO_2 emissions with the increase in power demand is to be maintained till 2030 with the EAU certificate price around 150 PLN/tCO_2 (Fig. 6). This is mainly possible due to introduction of nuclear power generation and increase of gas use mainly in CHP units. One should note, that this EAU price is slightly lower that the current wholesale electricity price (average 180 PLN/MWh in 2015).

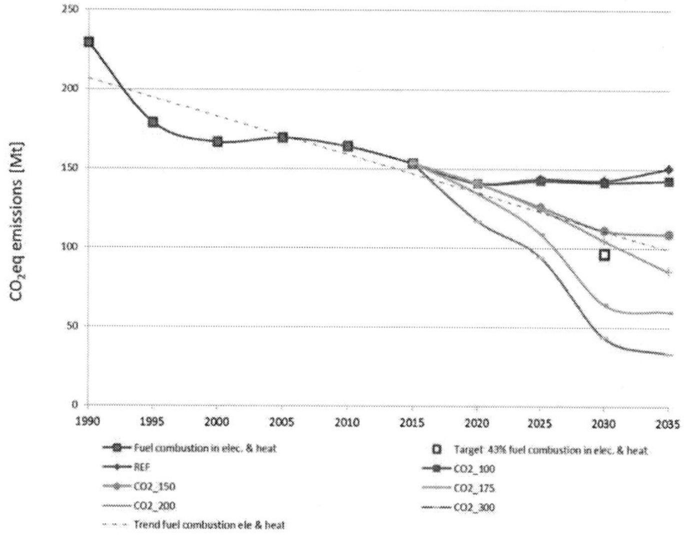

Fig. 6. Historical and estimated CO_2 emissions for different EUA price scenarios.

Data presented in Fig. 6 indicate that the EUA price of 175 PLN/tCO_2 (approximately 40 €/tCO_2) will lead the sector to the required level of emissions reduction after 2030. The electricity generation mix for different EUA price is presented in Fig. 7.

 DOI: 10.1051/e3sconf/20171401001

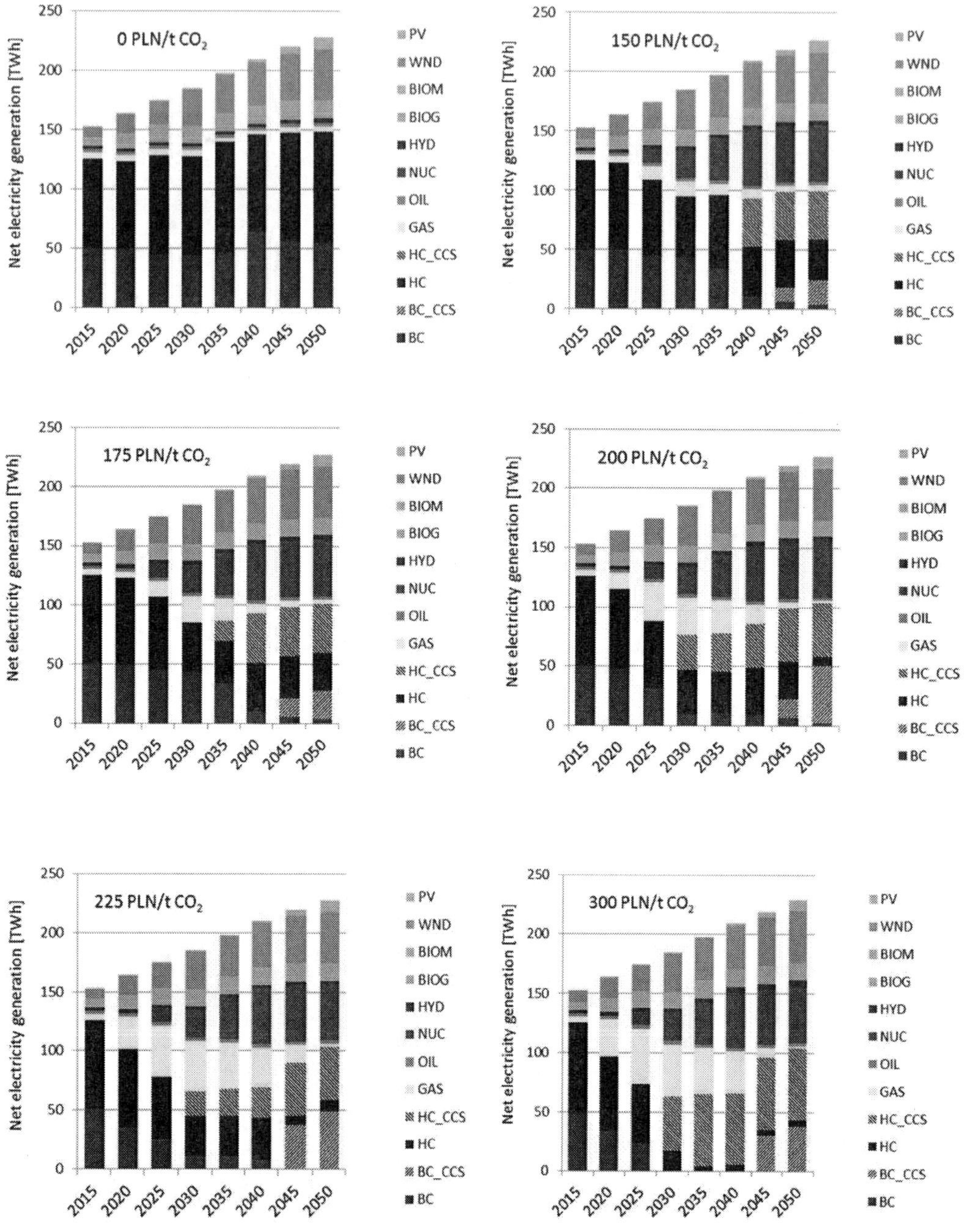

Fig. 7. Electricity generation mix for different EUA price.

Finally, the results of the impact of the energy scenarios on GDP are presented in Fig. 8. This figure presents the relative change in GDP compared to the REF scenario as the result of transitions of the energy system. One can see that in the mid-term adjusting of the power sector to the climate and air pollution regulations will result in a relative GDP loss in all scenarios. The increasing trend in the GDP loss is reversed due to lower energy costs and investments: (i) around 2030 for the BAT scenario and around (ii) 2040 for the other ones. In fact, in the case of first scenario group we can observe a GDP gains in the long term i.e. around 2050. Also for other two scenarios we see the rapid decline in GDP loss and if this trend was continued one could expect to have GDP gains before 2060. However, one should note that the cumulative GDP loss would not be compensated in any of these scenarios within the modelling time horizon.

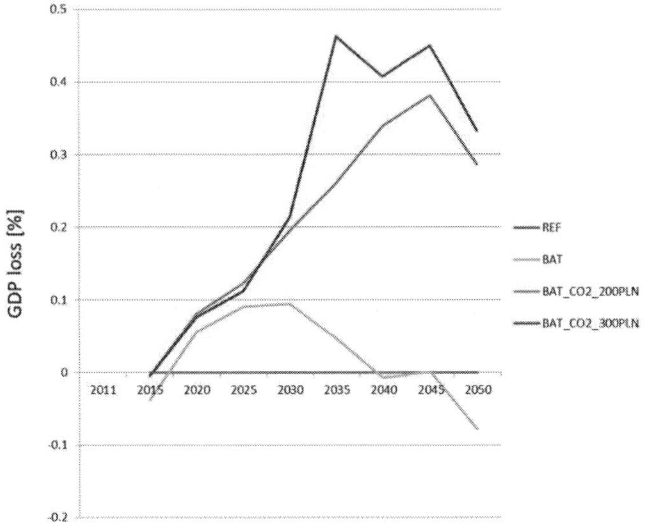

Fig. 8. The relative value of GDP loss for different energy scenarios referred to the REF scenario.

7 Conclusions

In this paper a hybrid of TIMES-PL and TIMES-MACRO energy-economic models was used to analyse the possible changes in the power generation mix and economic impact of different scenarios that combine the climate and air pollution regulations. The first observation is that the potential of renewable energy sources is fully exploited in all the scenarios even without stimulation of phasing out the intensive CO_2 generation (Fig. 7). This notwithstanding, still big CO_2 reduction potential exist in replacing inefficient hard coal fired power plants and CHPs with nuclear- and natural gas-based generation. Also clean-coal CCS technologies revealed to be competitive. In a short term i.e. till 2030 no significant change to the power mix could happen without very high EUA price signal. At the same time the study shows that cumulative GDP loss due to high EAUs prices and BAT-related expenditures would not be compensated within the modelling time horizon 2011-2050 in any of the scenarios considered. Moreover, the GDP loss will happen in the near future once the possible GDP gains could happen after 2045 in which the economic, technical and legislative conditions are very difficult to foresee.

References

1. Intended Nationally Determined Contribution of the EU and its Member States, Submission by Latvia And The European Commission On Behalf Of The European Union and its Member States Riga, 6 March 2015 http://www4.unfccc.int/Submissions/INDC/Published%20Documents/Latvia/1/LV-03-06-EU%20INDC.pdf (access 15.07.2016)
2. EU Reference Scenario 2016 Energy, transport and GHG emissions Trends to 2050, European Commission Directorate-General for Energy, Directorate-General for Climate Action and Directorate-General for Mobility and Transport, (July 2016)
3. DIRECTIVE (EU) 2015/2193 OF THE EUROPEAN PARLIAMENT AND OF THE COUNCIL of 25 November 2015 on the limitation of emissions of certain pollutants into the air from medium combustion plants
4. DIRECTIVE 2001/80/EC OF THE EUROPEAN PARLIAMENT AND OF THE COUNCIL of 23 October 2001 on the limitation of emissions of certain pollutants into the air from large combustion plants
5. DIRECTIVE 2010/75/EU OF THE EUROPEAN PARLIAMENT AND OF THE COUNCIL of 24 November 2010 on industrial emissions (integrated pollution prevention and control)
6. Best Available Techniques (BAT) Reference Document for Large Combustion Plants Industrial Emissions Directive 2010/75/EU (Integrated Pollution Prevention and Control) Joint Research Centre Institute for Prospective Technological Studies Sustainable Production and Consumption Unit European IPPC Bureau Final Draft (June 2016)
7. R. Loulou, U. Remme, A. Kanudia, A. Lehtila, G. Goldstein, Documentation for the TIMES Model – Part I, ETSAP, (2005)
8. M. Pluta, A. Wyrwa, T. Mirowski, J. Zyśk, *Wyniki wstępnych badań nad długookresowym rozwojem krajowego systemu wytwarzania energii elektrycznej w Polsce — Results of preliminary studies on the long-term development of the power system in Poland*, Polityka Energetyczna: ISSN 1429-6675, t. **15** z. 4, s. 85–96, (2012), (in Polish)
9. A. Wyrwa, M. Pluta, J. Zyśk, Modeling the mid-term development of the energy system in Poland with the use of TIMES-PL model, ENERDAY conference on energy economics and technology: energy policies and market design in Europe, Dresden, (2013)
10. L. Gawlik, [et al.], M. Pluta, W. Suwała, A. Szurlej, A. Wyrwa, J. Zyśk, Coal for Polish energy sector in 2050 perspective - scenario analyses, Mining Chamber of Industry and Commerce, Wydawnictwo Instytutu Gospodarki Surowcami Mineralnymi i Energią PAN, Katowice, (2013), (in Polish)
11. A. Wyrwa, M. Pluta, S. Skoneczny, T. Mirowski, Modelling the mid-term development of an energy system with the use of a technology explicit partial equilibrium model, E-science on distributed computing infrastructure: PLGrid plus. Springer-Verlag GmbH, LNCS 8500; p. 489-503, (2014)
12. K. Badyda, G. M. Niewiński, A. Patrycy, W. Orzeszek, *Próba oszacowania kosztów wdrożenia konkluzji BAT dla dużych źródeł spalania*, Nierówności Społeczne a Wzrost Gospodarczy, nr **46** (2/2016), (in Polish)

E3S Web of Conferences **14**, 01002 (2017) DOI: 10.1051/e3sconf/20171401002
Energy and Fuels 2016

A multicriteria approach for analysis of the impact of GHG limiting policies on economic growth in Poland

Jan Gadomski[1*], *Lech* Kruś[1], and *Zbigniew* Nahorski[1,2]

[1] Systems Research Institute, Polish Academy of Sciences
[2] Warsaw School of Information Technology

Abstract. The paper deals with the following questions: How to best proceed with a process of the macroeconomic transformation of a medium-size European country due to adjustment of its economy to the EU policy aimed at reducing CO_2 emission? What may be the consequences of the enforced emission limits for the economic development and future consumption? To answer these questions a computer-based system has been constructed including a macroeconomic model and a multicriteria optimization tool for two opposing objectives: maximization of the consumption and minimization of the greenhouse gases (GHG) emission. The model contains four production sectors with a special attention devoted to the energy sector, where four technologies are distinguished: the traditional, the modernized traditional and nuclear energy, and the renewable one. The model is the long-horizon one and describes equilibrium trajectories. The multicriteria optimization tool is developed using the reference point approach to support interactive analysis of Pareto optimal outcomes. Computational results are presented for the case of Poland. The welfare costs of the pursuing GHG limiting policy are assessed. Trajectories of the technological changes in time to achieve required limits of CO_2 emission are computed.

1 Introduction

Analysis presented in this paper merges two viewpoints: the technological and economic one. Limiting of the emissions of the greenhouse gases (GHG) aims at preventing the climate warming while the welfare/consumption of people is a common goal of the economic policy. Both goals are strongly intertwined. To a certain extent they are both complementary and substitutes. Decreased GHG emissions can be achieved by choosing less polluting technologies and/or lower output while larger consumption can be attained by increase of the output. Larger consumption can be also achieved by lowering the prices; however, we assume

* Corresponding author: jan.gadomski@ibspan.waw.pl

in the sequel that the modelling is performed with constant prices Therefore, these goals are contradictory. Aim of the analysis is to assess the consequences of curbing gaseous emissions for the choice of the production technologies and the level of welfare of the population.

Most analyses concerning these issues are performed using computable general equilibrium (CGE) models. In the Polish case the examples of such papers are: Antoszewski [1], Boratyński [2], Roberts [3]. Although very useful and popular, CGE models are based on the concept of equilibria of all markets with prices clearing these markets, and that all economic agents (producers and consumers) maximize their utility functions: profits and consumption respectively. Note that such an approach means microeconomic rationality in given policy framework (taxes, interest rates, etc.). These models often do not account for an impact of monopolies, common in some sectors, on the behaviour of the firms.

In our paper the approach is different. We analyse the effects of macroeconomic policy under the assumption that macroeconomic policy is performed optimally, while there are equilibria on all markets where the prices adjust with some inertia. In the analysis we use macroeconomic multicriteria optimization model [4], based on the earlier version [5], with two criteria: maximization of consumption and minimization of GHG emissions over the optimization period. Decisions concerning amounts of the output, choice of the production technologies and the trade exchange with abroad are found in the optimization process. Similarity in both approaches can be noticed. However, our approach is not considered as a substitute to CGE models but as their complementarity.

The paper is divided into four sections. In the next section the macroeconomic optimization model of Polish economy is presented and the two criteria methodology is outlined. The third section contains description of the simulation experiments. The final section concludes.

2 Method of analysis

In our analysis we use macroeconomic optimization model consisting of four production sectors, foreign exchange sector and the consuming sector containing both households as and the public sector. The novelty of the model used in this analysis is twofold: (i) time lags between investments and actual increment of the production capacities, and (ii) additional constraint requiring zeroing of the foreign debt/assets after predetermined time, are added.

The following production sectors are distinguished: the energy sector E (producing energy), the material sector M (producing non-energy intermediary goods), the consumption goods C (producing goods that are used up by the consuming sector consisting of households and the public sector), and the investment goods sector I (producing capital/investment goods used by all production sectors). Output of all production sectors can be exported or supported by imports. Output of the sector E is consumed by all sectors, output of the sector M is consumed by all production sectors, output of the sector C is consumed by the consuming sector, and the output of the sector I is used for building production capacities in all production sectors. All production sectors, except for the energy sector, can use two production technologies, and the energy sector has three technologies at disposal. These technologies are interpreted as the traditional fossil fuel, in Poland mainly coal generated energy, mix of the modernized coal generated and nuclear energy, and renewable energy respectively.

Parameters of production technologies in all sectors are distinguished using the following set of indices for the sectors, i= E, M, C, I; for the technologies, j=1, 2, 3; and for the years, t=1,..,T:

γ_{ijt} - productivity of fixed assets in year t in i-th sector and j-th technology, it is assumed that in the long term the technical progress increases the productivity of the fixed assets in each year by a constant ratio r_γ:

$$\gamma_{ijt} = \gamma_{ijt_0}(1+r_\gamma)^{t-t_0}; \tag{1}$$

where γ_{ijt_0} denotes productivity of the fixed assets in year t_0;

δ_{ij} - depreciation rate of the fixed assets in i-th sector and j-th technology;

α_{ij} - use of the goods produced in sector M in producing the unit of the gross product of i-th sector and j-th technology;

β_{ij} - use of goods produced in sector E in producing the unit of the gross product of i-th sector and j-th technology;

μ_{ijt} - emission per unit in producing the gross product of i-th sector and j-th technology in year t, it is assumed that in the long term the technical progress decreases the unit emission in each year by a constant ratio r_μ :

$$\mu_{ijt} = \mu_{ijt_0}(1+r_\mu)^{t-t_0} \tag{2}$$

where μ_{ijt_0} denotes unit emission in year t_0, while r_μ denotes the rate of the decrease of the emission unit;

τ_{ij} - duration of the investment process in years (natural number).

Production capacity defined as the potential gross output Q_{ijt} of sector i, i=E, M, C, I; using j-th technology, j=1, 2, 3; in year t , t=1,..,T; is described by the following one factor production function:

$$Q_{ijt} = \gamma_{ijt}K_{ijt-1}, \tag{3}$$

where K_{ijt} stands for the stock of the fixed assets in *i*-th sector and *j*-th technology at the beginning of year *t*. In this paper, the potential gross output (3) is also called the production capacity of *j*-th technology in sector E in year *t*.

Actual gross output X_{ijt} cannot exceed the production capacity

$$\varphi_{ijt}X_{ijt} = Q_{ijt}, \qquad \text{j=1, 2, 3; t = 1,..,T,} \tag{4}$$

where φ_{ijt} stands for the coefficient of the production capacity utilization in i-th sector, i=E, M, C, I; using j-th technology, j=1, 2, 3; in year t, assuming values from the range [0;1]. (In particular, $\varphi_{ijt} = 0$ indicates fully idle capital and $\varphi_{ijt} = 1$ represents full utilization of the production capacity of *j*-th technology in *i*-th sector in the year t).

Total actual output of i-th sector, i=E, M, C, I; is the sum of outputs produced using available technologies:

$$X_{it} = X_{i1t} + X_{i2t} + X_{i3t}, \qquad t=1,..,T. \tag{5}$$

Stock of the fixed assets K_{ijt} using *j*-th technology, *j*=1, 2, 3; in *i*-th sector, *i=E, M, C, I*; at the end of year *t* is given by the relationship:

$$K_{ijt} = K_{ijt-1}(1-\delta_{ij}) + I_{ij(t-\tau_{ij})}, \qquad j=1, 2, 3; t=1,..,T, \tag{6}$$

where I_{ijt} denotes investment in *j*-th technology, *j*=1, 2, 3; in *i*-th sector, *i*=E, M, C, I; in year *t*. (Note that the term $K_{ijt-1}\delta_{ij}$ denotes depreciation of the capital in *i*-th sector and j-th technology). For simplicity lag equal τ_{ij} years between the investment and its contribution to the stock of fixed assets is assumed.

Production of *i*-th sector using *j*-th technology in year *t* causes the emissions S_{ijt} of GHG:

DOI: 10.1051/e3sconf/20171401002

$$S_{ijt} = \mu_{ijt} X_{ijt} \qquad i=E, M, C, I; j=1, 2; t=1,..,T. \tag{7}$$

The total emission of GHG by i-th sector in year t equals:

$$S_{it} = S_{i1t} + S_{i2t} + S_{i3t}, \qquad i=E, M, C, I; t=1,..,T. \tag{8}$$

Gross income GI_t is defined as the sum of incomes generated in sectors E, M, C and I:

$$\begin{aligned} GI_t = & [1-(\alpha_{E1} + \beta_{E1})]X_{E1t} + [1-(\alpha_{E2} + \beta_{E2})]X_{E2t} + [1-(\alpha_{E3} + \beta_{E3})]X_{E3t} + \\ & + [1-(\alpha_{M1} + \beta_{M1})]X_{M1t} + [1-(\alpha_{M2} + \beta_{M2})]X_{M2t} + \\ & + [1-(\alpha_{C1} + \beta_{C1})]X_{C1t} + [1-(\alpha_{C2} + \beta_{C2})]X_{C2t} + [1-(\alpha_{I1} + \beta_{I1})]X_{I1t} + [1-(\alpha_{I2} + \beta_{I2})]X_{I2t}. \end{aligned} \tag{9}$$

Each year the country is endowed with certain number N_t of the emission allowances and its trajectory is determined by the following relationship:

$$N_t = f_N(t, N_{td}) \qquad t=1,..,T; \tag{10}$$

where N_{td} denotes the number of the emission allowances in the last considered period.

Disposable income DI_t equals the defined above gross income GI_t, decreased/increased by the debt servicing/income from the foreign assets:

$$DI_t = GI_t - r \cdot D_{t-1} + P_t(N_t - S_t) \tag{11}$$

where:

r - is the interest rate;

D_t - is the foreign debt (if positive) / foreign assets (if negative) at the end of year t:

$$D_t = D_{t-1} - (F_{Et} + F_{Mt} + F_{Ct} + F_{It}), \tag{12}$$

and P_t stands for the price of the emission allowance, N_t denotes the number of the emission allowances in year t, defined above, F_{it} i=E, M, C, I; t=1,..,T; denotes the foreign trade balance (export-import) in year t of i-th sector, and S_t denotes the actual total emission:

$$S_t = S_{Et} + S_{Mt} + S_{Ct} + S_{It}. \tag{13}$$

Equality constraints of the optimization model consist of the balance equations. The left hand sides of these equations denote domestic supply and the right hand sides represent domestic demand supplemented by the balances of foreign exchange of the given good.

The balance equation of E sector is expressed by the following equation:

$$\begin{aligned} X_{E1t} + X_{E2t} + X_{E3t} = & \beta_{E1t} X_{E1t} + \beta_{E2t} X_{E2t} + \beta_{E3t} X_{E3t} + \beta_{M1t} X_{M1t} + \beta_{M2t} X_{E2t} + \\ & + \beta_{C1t} X_{C1t} + \beta_{C2t} X_{C2t} + \beta_{I1t} X_{I1t} + \beta_{I2t} X_{I2t} + \rho_t \lambda DI_t + F_{Et}, \end{aligned} \tag{14}$$

where the term

$$\begin{aligned} & \beta_{E1t} X_{E1t} + \beta_{E2t} X_{E2t} + \beta_{E3t} X_{E3t} + \beta_{M1t} X_{M1t} + \beta_{M2t} X_{M2t} + \\ & + \beta_{C1t} X_{C1t} + \beta_{C2t} X_{C2t} + \beta_{I1t} X_{I1t} + \beta_{I2t} X_{I2t} \end{aligned} \tag{15}$$

denotes consumption of energy in year t in sectors M, E, C, I; using all technologies available in those sectors, and the term F_{Et} stands for the net balance of the foreign trade of the sector E (if $EXP_{Et} - IMP_{Et} = F_{Et} \geq 0$, then export EXP_{Et} exceeds import IMP_{Et} in the foreign trade of goods produced by the sector E; and if $F_{Et} < 0$, then import IMP_{Et} exceeds export EXP_{Et} in the foreign trade in energy). The term $\rho_t DI_t$, $0 < \rho_t \leq 1$, denotes consumption, or part of the disposable income DI_t in the year t designed for the purchases of the consumption goods, of which $\lambda \rho_t DI_t$ stands for the part of the total consumption expenditures directed for the purchases of energy. Note that the part $(1-\rho_t)DI_t$ of the disposable income equals the total investment expenditures. Coefficient ρ_t is not a constant as it depends on the propensity to invest. Constant coefficient λ, $0<\lambda\le 1$, denotes assumed constant share of the energy expenditures in the total consumption expenditures.

Supply of goods produced by sector M is supplemented by import, while some part of its output can be directed to export. The gross output of sector *M* is distributed in the way expressed by the following balance equation:

$$X_{Mt}=\alpha_{M1}X_{M1t}+\alpha_{M2}X_{M2t}+\alpha_{E1}X_{E1t}+\alpha_{E2}X_{E2t}+\alpha_{E3t}X_{E3t}+ \\ +\alpha_{C1}X_{C1t}+\alpha_{C2}X_{C2t}+\alpha_{I1}X_{I1t}+\alpha_{I2}X_{I2t}+F_{Mt}, \qquad t=1,..,T; \qquad (16)$$

where the term

$$\alpha_{M1}X_{M1t}+\alpha_{M2}X_{M2t}+\alpha_{E1}X_{E1t}+\alpha_{E2}X_{E2t}+\alpha_{E3t}X_{E3t}+\alpha_{C1}X_{C1t}+\alpha_{C2}X_{C2t}+\alpha_{I1}X_{I1t}+\alpha_{I2}X_{I2t} \qquad (17)$$

denotes consumption of the non-energy intermediate inputs in year *t* in sectors M, E, C, I, and F_{Mt} stands for the net balance of the foreign trade of sector M ($EXP_{Mt}-IMP_{Mt}=F_{Mt}\ge 0$ means that export EXP_{Mt} exceeds import IMP_{Mt} in the foreign trade of goods produced by the sector M, and opposite when $F_{Mt}<0$).

Supply of goods produced by sector I is supplemented by the import, while some part of its output can be directed to export. The gross output of sector I is distributed as described by the following balance equation:

$$X_{It}=I_t+F_{It}, \qquad t=1,..,T; \qquad (18)$$

where the term I_t

$$I_t=(1-\rho_t)DI_t=I_{M1t}+I_{M2t}+I_{E1t}+I_{E2t}+I_{E3t}+I_{C1t}+I_{C2t}+I_{I1t}+I_{I2t} \qquad (19)$$

denotes total investment in sectors M, E, C, I, and all technologies in year *t*, and F_{It} stands for the net balance of the foreign trade of sector I (if $EXP_{It}-IMP_{It}=F_{It}\ge 0$, the export EXP_{It} exceeds the import IMP_{It} in the foreign trade of goods produced by sector I, and opposite if $F_{It}<0$).

Supply of goods produced by sector *C* is supplemented by the import, while some part of its output can be directed to the export. The balance equation of sector *C* is as follows:

$$X_{Ct}=\rho_t\cdot(1-\lambda)\cdot DI_t+F_{Ct}, \qquad t=1,..,T; \qquad (20)$$

showing that the domestic supply (left-hand side of the above equation) of the non-energy consumption goods is equal to the demand generated by the part of the disposable income directed at purchasing non-energy consumption goods and the balance of the foreign trade in those goods (right hand side of the equation (20)). It is worth noting that the variable ρ_t can be interpreted as the propensity to consume. The term F_{Ct} stands for the net balance of the foreign trade of sector C (if $EXP_{Ct}-IMP_{Ct}=F_{Ct}\ge 0$, export EXP_{Ct} exceeds import IMP_{Ct} in the foreign trade of goods produced by sector C, and opposite if $F_{Ct}<0$).

Households and the public sector belong to the same sector called the consuming sector, where decisions being made concern the following factors: utilization of the production capacities in the sectors and technologies; distribution of the disposable income between the consumption and investment; technology choice; and the role of the foreign trade. Constant proportion between the household and public consumption is assumed.

Decision variables in the model include: the actual gross outputs in the sectors and technologies; investment in the capital assets in the sectors and technologies; and the foreign trade balances of all production sectors:

$$X_{E1t}, X_{E2t}, X_{E3t}, X_{M1t}, X_{M2t}\ X_{C1t}, X_{C2t}\ X_{I1t}, X_{I2t}, I_{E1t}, I_{E2t}, I_{E3t}, I_{M1t}, I_{M2t}\ I_{C1t}, I_{C2t}\ I_{I1t}, I_{I2t},$$
$$F_{Et}, F_{Mt}, F_{Ct}, F_{It}, N_{td}. \tag{21}$$

The inequality constraints are as follows.

Non-negative outputs and investments (foreign trade balances $F_{Et}, F_{Mt}, F_{Ct}, F_{It}$ can be either positive or negative):

$$X_{E1t}, X_{E2t}, X_{E3t}, X_{M1t}, X_{M2t}\ X_{C1t}, X_{C2t}\ X_{I1t}, X_{I2t}, I_{E1t}, I_{E2t}, I_{E3t}, I_{M1t}, I_{M2t}\ I_{C1t}, I_{C2t}\ I_{I1t}, I_{I2t} \geq 0. \tag{22}$$

Propensity to invest, defined as a ratio I_t / DI_t that cannot exceed the maximum propensity to invest:

$$I_t / DI_t \leq \sigma_{I/DI}, \tag{23}$$

where $\sigma_{I/DI}$ denotes the maximum value of the investment to the income ratio.

The above constraint reflects social resistance to the exceedingly high propensity to invest. The propensity to consume ρ_t is also constrained from beneath:

$$\rho_t \geq \sigma_{cons/DI} \tag{24}$$

where coefficient $\sigma_{cons/DI}$ denotes the minimum value of the consumption to the income ratio.

Another set of constraints deals with the feasible shares of the foreign trade in the sector outputs. The following constraints:

$$\sigma_{IMP/X} \leq \frac{IMP_t}{X_t} \leq \sigma_{IMP/X} \qquad j= \text{M, E, C, I}; \tag{25}$$

$$\sigma_{EXP/X} \leq \frac{EXP_t}{X_t} \leq \sigma_{EXP/X} \qquad j= \text{M, E, C, I}; \tag{26}$$

impose the maximum proportion of the import and export, respectively, in the national supply of the given product, where coefficients $\sigma_{IMP/X}$ and $\sigma_{EXP/X}$, j = M, E, C, I; denote, respectively, the maximum ratio of the import and export of a given product to its national gross output.

The following two constraints:

$$-r^{(-)}_{INV\ ij} \leq \frac{I_{ijt} - I_{ijt-1}}{I_{ijt-1}} \leq r^{(+)}_{INV\ ij}, \qquad j=1, 2, 3;\ j= \text{M, E, C, I}; \tag{27}$$

$$-r^{(-)}_{cons} \leq \frac{\rho_t DI_t - \rho_{t-1} DI_{t-1}}{\rho_{t-1} DI_{t-1}} \leq r^{(+)}_{cons}, \tag{28}$$

limit relative increases and decreases of investments in the sectors and the total consumption, respectively, where parameters $r^{(-)}_{INV\ ij}$ and $r^{(+)}_{INV\ ij}$ stand for the lowest and highest admissible rates of increase of the investment in technology j, j=1, 2, 3; i= M, E, C, I; while $r^{(-)}_{cons}$ and $r^{(+)}_{cons}$ denote the lowest and highest admissible rates of the consumption change, respectively. In particular, the constraint (28) reflects the social sensitivity to the changes in consumption and a possible resistance to them.

The following constraint reflects policy decisions concerning the desired share of a certain technology in the total output of a certain sector. In the current version of the model this constraint is the consequence of the requirement that in the energy sector the share of the renewable technology should be at least equal to 20%, beginning from the year 2030:

$$\frac{X_{E3t}}{X_{E1t} + X_{E2t} + X_{E3t}} \geq 20\%;\ t \geq 2030. \tag{29}$$

The following constraint limits the possibility of the excessive debt/credit as relative to the gross income

$$-0.60 \cdot GI_t \le D_t \le 0.60 \cdot GI_t \,. \tag{30}$$

The last constraint determines the policy time limit t_F of using the foreign debt / foreign assets as investment in of the means supporting technology conversion:

$$D_t = 0 \,, \qquad t > t_F \tag{31}$$

Two opposing criteria of macroeconomic development are considered:

maximization of the long-run consumption: $max \sum_{t=t_0}^{T} \rho_t DI_t (1+r_d)^{-(t-t_0)}$, (32)

minimization of the emission allowances in the final year t_d: N_{t_d}, (33)

The former is the present value of consumption (discounted consumption over the simulation period) and the latter is the cumulative emission over the simulation period.

Decrease of GHG emissions is implemented by allotting to a country a prenegotiated diminishing number of the emission allowances that should be met in a given time period. In this study, a linear-wise pathway of emission allowance limits is assumed. The pathway trajectories are formed by joining values of emission allowance limits in the initial year, in the intermediate years, and in the destination year, Figure 1. The actual GHG emission time trajectory in our study can however differ from the assumed pathway, due to allowed trade of the allowances.

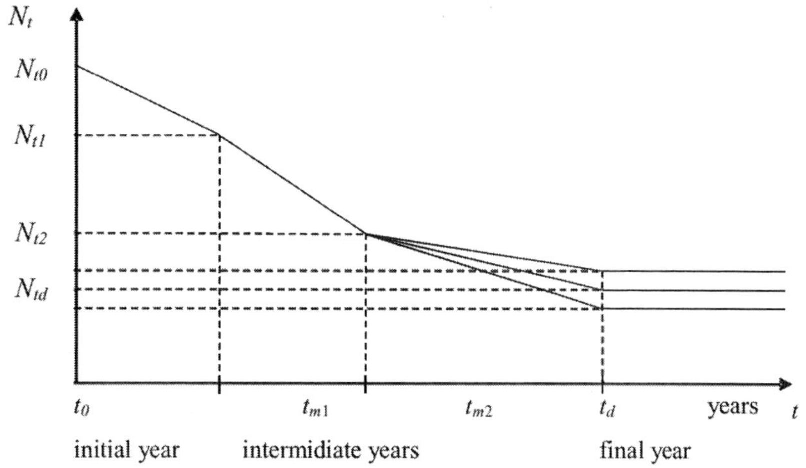

Fig. 1 Assumed emission allowance pathway trajectories. $t_0 = 2010$, $t_{m1} = 2020$, $t_{m2} = 2030$, $t_d = 2050$.

The initial year t_0 chosen in this study is 2010. The numbers of the emission allowances N_{t0} in the initial year and in the intermediate years are defined according to the EC agreements. The number of the allowances in the destination year 2050 is a variable that can be negotiated and is treated as a criterion considered together with the discounted consumption criterion. The multicriteria analysis is applied, with two criteria. The first, which is the discounted consumption that represents effects of the economic growth of the country, is maximized. The second, which is the number of the allowances in the destination year representing the emission curbing EC policy, is minimized.

As the model relations include only the affine expressions (1) – (31), they can be described in the form $Ax \le b$, where x is the vector of the decision variables, A is the matrix and b is the vector of the coefficients. The vector x – see expression (21), includes the decision variables, which are non-negative, according to the assumption (22).

Denote by $y(x)=[y_1(x),\ y_2(x)]$ the vector of the criteria, where y_1 is the discounted consumption (32), and y_2 is the number of the emission allowances in the destination year t_d. Due to the affine model relations, the criteria y_i, $i=1,2$, can be expressed as $y_i = c_i^T \cdot x + d_i$,where c_i, d_i, $i=1,2$ are vectors of coefficients. The criteria are conflicting. That is why the multicriteria optimization is applied, in which the decision variables that satisfy the constraints and jointly maximize y_1 and minimize y_2 are looked for. The problem is considered in two spaces, that of the decision variables, and that of the criteria. The model constraints define the set $\mathbf{X}_0$ of admissible values of the decision variables in the first space. In the second two-dimensional space there exists the set $\mathbf{Y}$ of attainable values of the criteria (outcomes). Decision variables leading to the nondominated (Pareto optimal) points in the set $\mathbf{Y}$ are looked for. We introducc thc following domination relation in the space $\mathbf{R}^2$ of criteria (y_1, y_2). A vector $y=(y_1, y_2)$ *dominates* a vector $v=(v_1, v_2)$, where $y, v \in \mathbf{R}^2$, if $y_1 \geq v_1 \wedge y_2 \leq v_2 \wedge y \neq v$. A vector $y=(y_1, y_2)$ *strictly dominates* a vector $v=(v_1, v_2)$, if $y_1 > v_1 \wedge y_2 < v_2$. The domination relation defines partial ordering in the criteria space, which is not a linear one. So, in this case the traditional optimality concept defined for one criterion is not valid. A vector y^p is *Pareto optimal* (*nondominated*) in the set $\mathbf{Y}_0$, if $y^p \in \mathbf{Y}$ and there is no vector $v \in \mathbf{Y}$ dominating the vector y^p. A vector y^{ps} is *weakly Pareto optimal* (*weakly nondominated*) in the set $\mathbf{Y}$, if $y^{ps} \in \mathbf{Y}$ and there is no vector $v \in \mathbf{Y}$ strictly dominating the vector y^{ps}. In the case analysed here the set $\mathbf{Y}$ is not given explicitly. Particular points of the set can be only found by computer simulations. The set of decision variables in $\mathbf{X}_0$ that corresponds to the set of the Pareto optimal points in $\mathbf{Y}$ is derived and analysed.

The multicriteria optimization problem is solved using the reference point approach with an order achievement function, developed by Wierzbicki [8], [9]. The elaborated computer system generates the Pareto optimal solutions in an interactive way. Assuming and assigning different reference values for the criteria and solving the resulting optimization problems, different Pareto optimal outcomes and decision variables are derived, compared and analyzed. For this, the order approximation achievement functions is used. In this approach, some reference points in the criteria space are given by a system analyst and then the computer-based system generates corresponding outcomes which are Pareto optimal in the set of attainable outcomes. This way a representation of the Pareto frontier can be obtained. Outcomes characterizing the Pareto frontier are derived by $\max_{x \in \mathrm{X}_0}[s(y(x), y^*)]$, where: $\mathbf{X}_0$ is a set of admissible decisions defined by the model relations, $y^* = (y_1^*, y_2^*)$ is a reference (aspiration) point assumed in the criteria space $\mathbf{R}^2$, and $s(y, y^*)$ is an order approximating achievement function.

The respective form of the achievement function has been applied and the optimization problem has been transformed to the linear optimization task solved sequentially, like in Gadomski at al. [5].

3 Simulations

Results described in this paper were obtained on the basis of the data from Input-output Tables of Poland in 2010 and from other data published in [6]. Unavailable data were estimated and, in some cases, based on the expert opinions. Simulations assumed 1.5% yearly increases of the productivity of capital in all sectors and technologies, and 1% yearly decreases of the unit emission in all sectors and technologies. All economic variables are expressed in constant prices. It is assumed that the international and national prices maintain fixed proportions.

As motivated earlier, the year 2010 is chosen as the initial year of the simulations. It is assumed that before the initial year the economy has grown along the equilibrium path with a steady growth rate which determined proportions between sectors. The imposed emission pathway disturbs the growth but, after a turbulent transition period of the technology conversion, the economy resumes to grow along the new equilibrium path. Our analysis is focused on the period where most macroeconomic adjustments is performed.

The emission of GHG in Poland in 2010 is related to the emission in 2005, which was equal 353.9 million ton of CO_2eq (Olecka et al. [7]). This value is adopted as a referential number of allowances allotted to Poland. The emission allowance pathway is assumed to have the shape presented in Fig. 1. The assumed numbers of emission allowances in the years 2020 and 2030 are 79% and 57%, respectively, as related to the their number in 2005. The number of allowances in the destination year is minimized. The discounted consumption in the full period of time is another criterion that is maximized. For different aspiration points assumed in the space of these two criteria, represented in Fig. 2 by small squares, the optimized nondominated points, represented by small circles, were obtained. Arrows indicate correspondence of the nondominated and the aspiration points, which form a representation of the set of the nondominated outcomes (Pareto frontier) in **Y** which is approximated in Fig. 2 by the dashed line. The outcomes located to the left of the Pareto frontier are unattainable, i.e. they do not belong to the set **Y**.

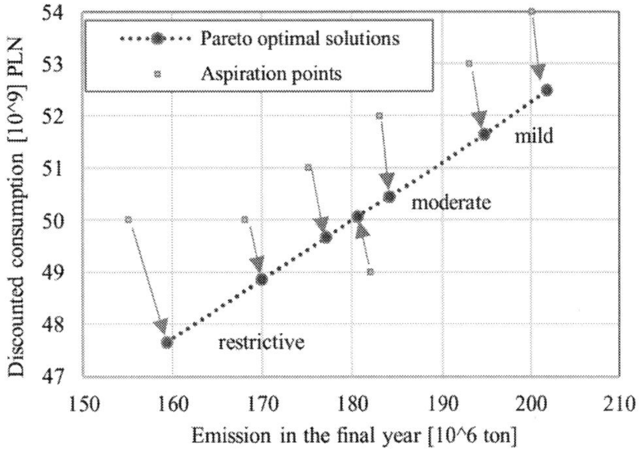

Fig. 2 Results of the interactive multicriteria analysis. The dashed line is an approximation of the Pareto set.

The numerical computed results are presented in Table 1. The case 8 relates to the maximum possible decrease of the number of the emission allowances in the destination year, for which the lowest feasible consumption constraint (29) is active. It is called the restrictive variant. It is the solution of the single criterion optimization problem with minimization of the emission allowances number in the destination year. It represents the greatest possible decrease of the emission allowances for the destination year within the assumed constraints, which is around 80% reduction of the Kyoto base emission for Poland. In the case 1, called the mild variant, there is no decrease of the number of emission allowances after the second intermediate year 2030. Among the intermediate points, the case 3 is chosen and called the moderate variant. It relates to a moderate decrease of the number of allowances in the destination year.

Two variants of the trajectory of numbers of the emission allowances were chosen for presentation: the mild and restrictive one. The former assumes that from year 2030 on the final yearly amount of allowances is equal to 57% of 2005 emission level and the latter assumes that from year 2050 on the final yearly amount of allowances is equal to 45% of 2005 emission level, see Fig. 3 (d). Main macroeconomic indicators that are GDP, consumption, investment, and emissions, are presented in Fig. 3.

Table 1. Selected results of the multicriteria analysis

	Aspiration points		Pareto optimal solutions						
Scenario	Emissin in the final year [10^6 ton]	Discounted consumption [10^9] PLN	Decrease of emission compared to the initial year	Emission in the final year [10^6 ton]	Discounted consumption [10^9] PLN	Decrease of consumption compared to the mild scenario [10^9] PLN	Decrease of consumption compared to the mild scenario in %	Cumulated consumption [10^9] PLN	Cumulated emission [10^6] ton
1 restrictive	155	50.0	45%	159.3	47.7	4.83	9.20%	112.6	21909
2	168	50.0	48%	169.9	48.9	3.62	6.90%	117.5	22869
3	175	51.0	50%	177.0	49.7	2.82	5.36%	121.0	23509
4 moderate	183	52.0	51%	180.5	50.1	2.41	4.60%	122.5	23829
5	182	49.0	52%	184.1	50.5	2.04	3.89%	124.2	24149
6	193	53.0	55%	194.7	51.7	0.84	1.60%	129.0	25130
7 mild	200	54.0	57%	201.7	52.5	0.00	0.00%	132.1	25748

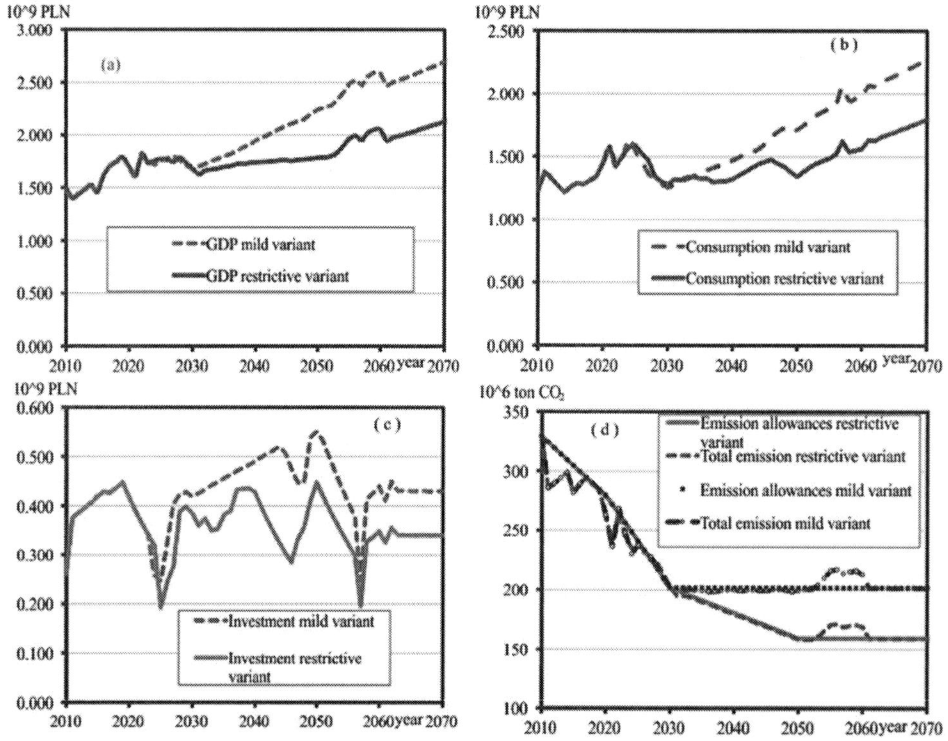

Fig. 3. (a) GDP, (b) consumption, (c) investment and (d) the amounts of the emission allowances and actual emissions in two variants (mild and restrictive) without technical progress.

All simulations have one element in common, namely successful technology conversion requires high market prices of the emission allowances. At low prices this conversion is considerably delayed or does not occur at all. In all sectors the cleanest technologies are employed. Moreover, in the energy sector employment of the modernized traditional technology derives from the volatility of the renewable energy sources, which would otherwise dominate.

After initial rise of GDP, consumption and investment, there follows long lasting recession. (Note that both mild and restrictive variants have similar characteristics).

In the initial phase technological conversion causes diminished utilization of the production capacities, the deepest in E sector, followed by M sector. This drop of the production capacities occurs the latest in sector I, as the burden of the technological change is concentrated there, see left hand panel of the Fig. 4, This phenomenon requires explanation, because it is accompanied by increase of the production capacities only in cleaner technologies. In modelling of the production system we have not taken into account the accelerated depreciation, decommissioning or scrapping of the idle capital assets. Abandoning of a technology results in gradual decrease of the capital assets accordingly with the equation (6). This result indicates that beside the cost of the technology conversion expressed in terms of consumption, there occur also losses related to idle capital assets.

The technological conversion has significant impact on the structure of the national economy. One should notice considerable increase of the share of sector E in the gross output. As there are negligible differences between the mild and restrictive variants, developments of the production capacities utilization and the shares of particular sectors in the national economy are presented in Fig. 4 using the results of the mild variant. One can conclude that the decreased rate of the production capacity utilization belongs to the cost of technological change. The greatest drop in the production capacities can be observed in the energy sector and the smallest in the investment sector.

Participation in the gross output changes considerably. One can note that the share of the energy sector significantly increases with the expansion of the newer technologies, see right hand panel of Fig. 4.

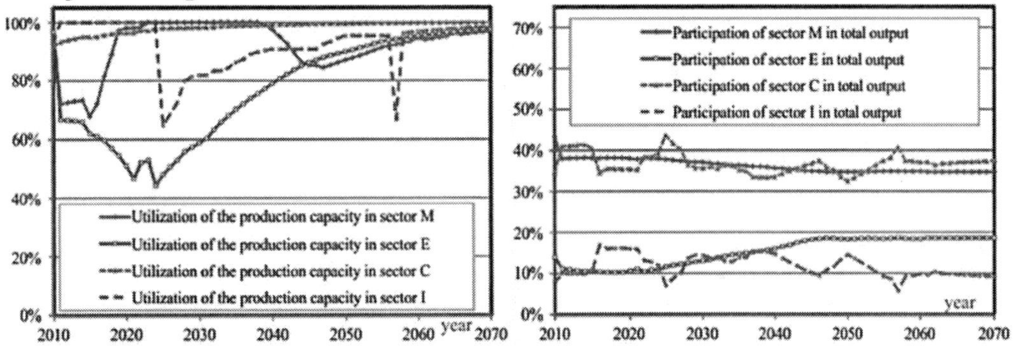

Fig. 4. Production capacities utilization, left hand panel, shares of particular sectors in the national economy, right hand panel.

Performed calculations allow an assessment of the cost of curbing emissions in terms of consumption lost as presented in Fig. 5.

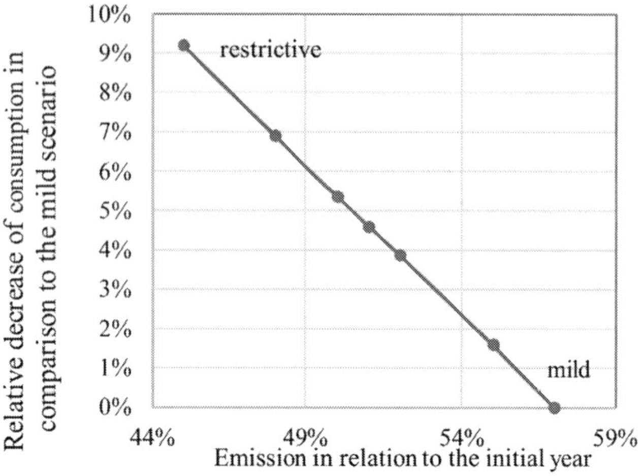

Fig. 5. Losses in consumption due to decreased emission.

4 Conclusions

The paper presents a macroeconomic model and multicriteria optimization tool for analysis of the consequences of the enforced emission limits for the economic development in Poland. A special attention is devoted to the energy sector. Different technologies are distinguished in the model, such as the modernized traditional and nuclear, as well as renewable, in comparison to the traditional one. The model determines equilibrium trajectories and makes it possible to analyze different scenarios of the technical progress development.

The applied multicriteria optimization focuses on two contradictory objectives: decreasing GHG emissions, and maintaining the highest possible growth rate. This enables analysis of the trade-off problem between two competing goals: reduction of the GHG emissions with the sustainable economic growth, as well as changes of the sectorial structure of investment and output. Also assessment of the cost of the GHG emission reduction in terms of the consumption lost is done.

The multicriteria optimization approach proved to be effective in analysis of the impacts of enforcing emission limits on the economic development process, and on the economic transformation caused by adjustment of the national economy to the emission decreasing policy. The technological conversion trajectory was derived for seven Pareto optimum solutions. Two solutions (mild and restrictive) are presented and discussed in the paper.

In all sectors but sector I (having negligibly low emissions in both technologies), new technologies replaced old ones in the investment outlays. It is necessary to note that in the energy sector the most expensive technology has been chosen (the one interpreted as the renewable). This can be explained by the severity of the end period emission constraints. However, very high volatility of supply from the energy sector makes problems in modelling and optimization, as its maximum capacity of production is practically never met, in particular in the case of the renewable technology. Considering year production, as in our analysis, the mean production can be predicted with good enough accuracy. But the short time production may vary considerably. Hence, high share of renewable energy requires keeping a large reserve of the traditional non-renewable technology power, which actually is the case in our modelling, or very advanced balancing of energy production and consumption, including energy storage. Therefore the minimum share of the second technology in total energy output has been included as a constraint in the model.

A necessity to adjust to the lowest emission levels at the end-period forces the economic system to cumulate consumption at the beginning period, with the similar impact on investment, and GDP, Fig. 3. As a result, the initial period of growth lasts only to 2013 and is followed by recession and then stagnation, both determined by the admissible level of emission.

Results performed by the proposed model confirm its applicability in the analysis of the impact of the policy of curbing the GHG emissions on the growth of the national economy. This model should not be treated as the substitute but as an alternative and a supplementary analytical tool to the more popularly used CGE models. It is important to remember that the results are presented in constant prices, and that an exogenous evolution in prices can be considered given credible scenario to obtain the results in actual prices.

References

1. M. Antoszewski, J. Boratyński et al., CGE model PLACE, MF Working Paper Series, Ministry of Finance, Republic of Poland, Warsaw (2015).
2. J. Boratyński, Historical simulations with a dynamic CGE model: Results for an emerging economy, Ecomod Conference, Seville. (2012)
3. B. M. Roberts, Calibration procedure and the robustness of CGE models: Simulations with a model for Poland, *Economics of Planning*, 189-210, Kluwer Academic Publishers (1994).
4. J. Gadomski, Assessment of the impact of the reduction of the gaseous emissions on growth in Poland. Assumptions and preliminary results. *Przegląd Statystyczny*, 3:237-287 (2016).
5. J. Gadomski, L. Kruś, Z. Nahorski, A multicriteria model for analysis of the impact of GHG limiting policies on economic growth. The case of Poland. Systems Research Institute of PAS, *Working Paper* No RB/37/2014. Warsaw (2014)
6. Central Statistical Office, Republic of Poland, Input-Output Table at Basic Prices in 2010, Warsaw (2011).
7. A, Olecka, K, Bebkiewicz, B. Dębski et al., National Inventory Report 2014. GHG Inventories for Poland for the years 1988-2012. The National Centre of Emission Management, Warsaw (2014) (in Polish)
8. A. P. Wierzbicki, On the completeness and constructiveness of parametric characterizations to vector optimization problems. OR Spectrum 8:73-87 (1986).
9. A. P Wierzbicki, M. Makowski, J. Wessels, Model-based Decision Support Methodology with Environmental Applications. Kluwer Academic Press, Dordrecht, Boston (2000).

Development perspectives of the Polish power generation sector according to the climate preservation conference COP21 policies

Radosław Szczerbowski[1,*], and *Bartosz* Ceran[1]

[1]Poznan University of Technology, Institute of Electric Power Engineering, Piotrowo 3A str., 60-965 Poznan, Poland

Abstract. Power generation policy equals the safety policy of a certain country. It should be reasonable and it should consider the access to available energy sources. The crucial issue revolves around minimizing the negative influence of power generation sector on the environment. At the same time, recent years have proven the united stand of world policies toward power generation. Poland has also attempted to determine a new model of power generation strategy. This strategy ought to consider the needs of both the recipients and climate challenges. But there is a crucial question of the shape of the new strategy for development of power system within the next years in the light of requirements of convention on climate change. During the Climate Conference in Paris in December 2015 – 195 countries accepted the first world agreement related to climate preservation. The agreement determines a world action plan, which is to prevent climate change as the result of climate warming. One of the goals is a quick reduction of emissions including power generation sector. This overview presents a current state of National Power System, availability of primary energy sources and various power technologies of future strategy for power system development. The technologies are described in view of their possible use for power generation and their applicability for the reduction of emissions of harmful substances to the atmosphere.

1 Introduction

The arrangements of 21 Conference of the Parties of UN Framework Convention on Climate Change COP21 which was held in 2015 in Paris, include several important goals which aim at the reduction of greenhouse gases and lowering global climate changes. The key points of the agreement, which was entered into by nearly 200 countries, include mainly:

- possibly quickest reduction of greenhouse gases emissions,
- sustaining the world temperature rise at the level of much less than 2°C and making attempts to limit the temperature to 1.5°C,
- the assessment of progress in terms of these goals shall be performed every five years.

* Corresponding author: radoslaw.szczerbowski@put.poznan.pl

As the consequence, it is assumed that, beginning with the year 2050, world industry may become nearly of zero emissions. Moreover, the agreement made at climate summit in Paris does not impose any defined commitments on its signees. In fact, it means the necessity to prepare new strategies for low emissions economic development in all the countries which have been sceptical toward the issue of gas emissions. The agreement will come into force when at least 55 parties of Climate Convention, responsible for 55% of world emission, decide to ratify it. It is expected that the Agreement will become effective at last at the end of 2020. Poland and other European countries, besides the new climate agreement concluded during the summit COP21 in Paris, must also abide by European climate regulations. Now, it is a climate and power generation package which expires in 2020. In the near future EU is going to discuss climate regulations for the years 2020 – 2030.

In the context of these changes, it is quite difficult and risky to predict the future of the power generation sector, especially for a longer perspective. Such forecasts are generally made on the basis of previous trends and future changes. Decisions are influenced by many factors, such as economic matters or difficult to predict – political agendas. Economic development of a country is determined by security of energy supply which, on the other hand, should be generated from easily available sources and at satisfactory prices. These indicators are the conditions of the choice of the sources which are going to shape the future composition of National Power System.

2 Current state of National Power System

National Power System is mainly based on power plants heated with coal; the system is one of the biggest in Europe; current installed capacity has exceeded 40 GW. Comparing to the year 2014, there has been an increase of installed capacity of more than 2.3 GW and it has been mostly caused by new investments into renewable energy sources (1.8 GW), mainly wind power plants [1]. Table 1 presents basic data concerning the current situation of National Power System. As it may be observed, the total installed capacity in coal power plants equals 28 638 MW, which constitutes more than 70% of installed capacity, however, the production of energy in these sources exceeds 83%.

Table 1. The present state of the National Power System (data on 31.12.2015).

List	Installed capacity	Share of installed capacity	Maximal capacity	Share of maximal capacity	Production	Share in production
	MW	%	MW	%	GWh	%
Power plants	31 927	78.9	32 069	80.6	141 901	87.7
Water power plants	2 290	5.7	2 330	5.9	2 261	1.4
Thermal power plants (hard coal, brown coal, gas)	29 637	73.3	29 739	74.8	139 640	86.3
Wind and RES	5 687	14.1	5 258	13.2	10 114	6.3
Industrial	2 831	7.0	2 451	6.2	9 757	6.0
Total	40 445	100	39 777	100	161 772	100.0

Source: Own elaboration on the basis of data [1]

The age structure of boilers and turbine sets working in Polish power plants indicates that more than 60% of them have operated for over 30 years. Thus, in the nearest 20 – 30 years they will be gradually removed from the energy system [2]. Despite such production structure, Poland has managed to reduce emissions of CO_2 to the atmosphere. Table 2 presents the data related to CO_2 emissions to the atmosphere for these recent years.

Table 2. Emissions of CO_2 from combustion of fuels in Poland in the years 1990-2013.

Years	1990	1995	2000	2005	2010	2012	2013	%change 90-13
CO_2 emissions from fuel combustion [million tonnes of CO_2]	344.8	333.4	289.7	296.3	310.4	296.8	292.4*	15.2%
CO_2 emissions from fuel combustion – coal [million tonnes of CO_2]	291.2	273.6	221.5	215.6	217.5	206.5	206.4	29.1%

Source: Own elaboration on the basis of data [3,4]
* Electricity and heat production 157.6 million t CO_2

Presently in Poland there are several ongoing investments related to the construction of new power capacity fuelled with hard coal, brown coal and gas. In the subsequent years there will be new power units started which will belong to JWCD [1]: unit 473 MW, gas and steam, in thermal plant Włocławek; unit 467 MW, gas and steam, in thermal plant Stalowa Wola (2016); unit 596 MW, gas and steam, in thermal plant Płock (2017 r.); unit 1075 MW, fuelled with hard coal in power plant Kozienice (2018); unit 496 MW, fuelled with brown coal, in power plant Turów (2018); 2 units each of 900 MW, fuelled with hard coal in power plant Opole (2019); unit 910 MW, fuelled with coal in power plant Jaworzno III (2019). In total the capacity joined to the system equals 5817 MW, which means that the lacking capacity, caused by removing old units, should be complemented by the year 2020.

3 Power generation with coal

The publication "Balance of mineral resources deposits in Poland" [5], published this year, presents the state of fossil fuels at the end of the year 2015. It also informs that Poland has presently got 91 deposits of brown coal, including 9 which are already in use, and 156 deposits of hard coal, including 51 which are already in use (table 3). Deposits of brown coal equal more than 23 516 million tons, and deposits of hard coal equal more than 56 220 million tons, including 71.6% of fossils which are energy-generating coals. (type 31–33).

Active mines of brown coal supply directly power plants in Bełchatów, Adamów, Konin and Turów. Coal mining for particular power plants is presented in Table 4. Additionally, there is the estimation of depletion of industrial and balance resources which are presently in use, assuming that the level of mining in a certain mine is not going to change. There is a presentation of balance resources which are not used and are stored in the vicinity of power plants; and there is also a calculation of time for depletion of recognised reserves, assuming that the demand of Polish power plants is not going to change and deposits are going to be used up completely.

On the basis of the data from the Table 4, it is reasonable to assume that none of the presently operating power plants is going to be cut off from brown coal supplies within life period of present energy units.

Now, there are no accepted plans for the construction of new open pits of brown coal in Poland. Thus, it is difficult to predict which deposits will be in use in the future. Table 5 presents theoretical reserves which may be used in the future for power generation. These reserves have been divided according to the extent of their recognition.

On the basis of data from Table 5, it may be observed that significant reserves of brown coal are located near Legnica, Gubin and Ścinawa. However, the construction of an open pit mine for brown coal in each of these locations may be difficult because of possible strong protests of local societies.

Resources of hard coal in Poland are located in: Górny Śląsk Mining Region, Lublin Mining Region and Dolny Śląsk Mining Region, which production was finished in the year 2000. Balance resources of active coalfields have been presented in Table 6. Additionally, the table shows the time of depletion of energy-generating coal resources at the present rate of production and also a recognised resources of energy-generating coal. Notice that average percentage of energy-generating coal in total production of hard coal has been recently at the level of around 85%, including around 60% which is consumed by power plants and heating plants. It can be also noted that the resources of coal deposits ,which have been already mined, are still of significant amounts, thus, in the future we may consider contribution of this fuel into the production of electricity. One of the fundamental obstacles to overcome for further production of hard coal, may be economic issues related mainly to the present and prospective coal prices at international markets and the price of production of domestic coal.

Table 3. Geological reserves of brown and hard coal in Poland in million tons (data on 31.12.2015).

Deposits	Number of deposits	Geological reserves		Industrial reserves
		balance	off-balance	
Brown coal				
Total	91	23 516.2	3 522.4	1 129.1
Active plants	9	1 418.7	48.3	1 112.2
Detailed recognition	35	5 838.7	872.6	16.8
Initial recognition	39	16 242.5	2 574.98	-
Discontinued operation	8	16.30	26.51	-
Hard coal				
Total	156	40 234.8	12 404.1	2 109.3
Active plants	49	12 071.2	4 789.0	2 097.1
Mines under construction	2	6.2	3.77	-
Detailed recognition	40	11 267.8	1 157.8	-
Initial recognition	19	13 828.1	6 194.5	-
Discontinued operation	47	3 061.5	259.0	12.22

Source: Own elaboration on the basis of [5]

Table 4. Brown coal mining, possible resources of brown coal in mining regions and the time of depletion of resources at the current rate mining (data on 31.12.2015).

Mine		Bełchatów	Adamów-Konin	Turów
Production [thousands of tons]		42 081	13 653	7 328
Resources [thousands of tons]	balance	946 566	105 949	364 815
	industrial	707 368	86 037	317 689
Time of depletion of balance resources [years]		22.5	7.8	49.8
Time of depletion of industrial resources [years]		16.8	6.3	43.4
Balance resources recognised [thousands of tons]	detailed	744 393	287 276	-
	initial	594 032	19 044	349 087
Time of depletion of resources recognised in details [years]		17.7	21.0	-
Time of depletion of resources recognised initially [years]		14.1	1.4	47.6
Theoretical time of operation for active mines of brown coal [years]		54.3	30.2	97.4

Source: Own elaboration on the basis of [5]

Because of the latest assumptions of climate policy, domestic energy system faces a great challenge. The proposal of significant reductions of emissions in power generating sector may exclude coal from power generating industry.

In the context of EU requirements the key challenges for Polish power generation industry are related to:
- lack of technological adjustment of power and heating plants in terms of parameters of emissions,
- significant dependence of power and heating plants on coal,
- poor technical condition of production sources and power networks,
- small scope of use of RES and lack of nuclear power production,
- low power effectiveness within the range of production and use of energy.

Table 5. Possible development of brown coal deposits in Poland (data on 31.12.2015).

Location	Geological reserves recognised in details [thousands of tons]	Geological reserves recognised initially [thousands of tons]
Legnica	1 702 950	1 723 049
Ścinawa	1 766 983	-
Czempiń-Krzywiń	-	1 701 085
Gubin-Brody-Torzym	1 613 500	2 862 849
Gostyń-Oczkowice	996 298	1 988 830
Szamotuły-Trzcianka	300 077	746 326

Source: Own elaboration on the basis of [5]

Table 6. Hard coal mining, possible resources of coal in mining regions and the time of depletion of resources at the current rate mining (data on 31.12.2015).

Mining Region		Górny Śląsk	Lublin
Total production [thousands of tons]		58 255	6 815
Resources of energy-generating coal in developed deposits [million of tons]	balance	11 402.6	674.7
	industrial	1 926.7	170.4
Time of depletion of balance resources [years]		195.7	99.0
Time of depletion of industrial resources [years]		33.1	25.0
Recognised balance resources of energy-generating coal [million of tons]	detailed	7 776.2	3 462.6
	initial	8 135.0	5 693.1

Source: Own elaboration on the basis of [5]

4 Power generation based on renewable resources

The alternative for power industry based on coal, and more and more important source of future energy production in all its forms, is undoubtedly renewable sources of energy. In Poland the biggest hopes are related to biomass, wind power and solar power. Total world wind power production has exceeded 430 GW. Especially significant dynamics is visible in investments into sea wind farms. Wind may be used in energy production if its speed falls into the range between 4 and 25 m/s. Wind conditions in Poland are of great changeability and there is a lack of high annual average wind speed. Wind power production also considers the second important factor which is the distribution of annual wind speed. Strong winds in Poland are present in winter months, i.e. from November to March. Almost 1/3 of Polish territory has proper wind conditions for the installation of wind plants. Polish wind power plants have the capacity of above 5.6 GW which still rises quickly. It is estimated that in 2020 it should reach 7–9 GW. National potential of wind power production equals around 7 GW on sea and 10-12 GW in land. However, annual coefficient of installed power use is estimated at 22–28% (1900–2400 hours depending on the location) [6]. The main problem of wide wind power use is the lack of possibility to store the energy, which solution would balance the vague energy production. July 1, 2016 was published the law on investments in wind power plants, which introduced significant restrictions on construction and location of wind turbines. According to many specialists in the wind energy industry, currently over 95% of new location projects of wind turbines were stopped [7].

Photovoltaic systems may also be expected to be more important in a long-term perspective. In 2015 world capacity of these systems reached 245 GW. In the next decades photovoltaic systems may become the important source of world power, with a domineering role in poor tropical countries, but also in countries of moderate climate, such as Poland. Presently in Poland photovoltaic systems have total installed capacity of over 90 MW, while in Germany their capacity exceeded 40 GW, which is the best illustration of prospective development of photovoltaic systems in Poland. Their further development mainly depends on unit costs of PV installation and possible supporting systems. According to the report of IRENA [6], installed capacity in 2030 in our country may reach 5 GW. The coefficient for installed capacity in use currently equals around 11–12%.

Poland has a small potential of water power plants which results from flat form of landscape and high soil permeability. Domestic production of energy from water is mainly focused on the Vistula river. Polish hydro-energetic reserves are estimated at 13.7 TWh a year, and the potential is used merely in around 12%, which relates to around 7% of installed capacity in a domestic power system. Water power production at the end of 2015 equalled installed capacity of around 1 GW. According to the report IRENA [6], the potential development of domestic water power generation is estimated at around 1.5 GW by the end of 2030.

The production of energy from wind power plants or photovoltaic power plants on a large scale in Poland will not be able to replace conventional sources of energy soon. It is caused by weather conditions, especially in terms of solar hours available and average wind speed a year. The development of wind power generation and photovoltaic systems will mainly depend on the possibility to store such energy. If the future allows for effective storage of energy for stabilization of network operation and better quality of energy, then these sources of energy will certainly thrive.

5 Nuclear power

Polish energy policy by the year 2030 includes the statement of development of nuclear power in our power generation system. A nuclear power plant of capacity of 3000 MW, which will possibly be joined to the system after the year 2030, is not going to improve energy crisis which is likely to be suffered earlier, i.e. in the years 2016–2019. Even after its introduction to the system, it will still not have a significant influence on Polish power generation system, which capacity in 2025 should exceed 40 GW. But there is a question of resigning from the construction of a nuclear unit and further development of this technology. Many countries, which base their system on nuclear power, schedule its further development and construction of new nuclear reactors. The examples are: Finland, Czech Republic or Great Britain. The construction of a nuclear power plant will definitely be an expensive investment for the period of 50-60 years as that is the time of new-generation power plants to operate. Moreover, a nuclear power plant operating at the base of load will improve the stability of the power system, which, in the perspective of significant development of unstable renewable energy sources, will be certainly of crucial importance.

6 Power generation based on gas fuels

Climate policy of zero emissions is integrated into plans of national energy groups which plan investments of new production capacities based on natural gas. Tauron works toward the construction of gas unit of capacity equalling 400 MW in Stalowa Wola, Energa works toward the construction of a gas unit of capacity equalling 860 MW in Grudziądz, and Polska Grupa Energetyczna intends to construct two gas units of capacity around 400-450 MW in the complex of power plants Dolna Odra; they also conduct preparatory works for the needs of construction of gas units in their heating plants in Bydgoszcz and Gorzów and consider the construction of a power plant in Puławy, including a gas unit of capacity equalling 400 MW. The construction of a gas unit of capacity equalling 500 MW in Płock is announced by PKN Orlen. Similarly, KGHM Polska Miedź has elaborated the concept of construction of two gas units of the capacity of 45 MW. The construction of own gas sources is also considered by PGNiG Termika; EC Żerań is going to construct a gas-steam unit of capacity equalling 450 MW [1]. The plans of energy investments include also a LNG port in Świnoujście. However, the operation of gas power plants in Poland and many European countries is presently unprofitable. The price of this material makes energy units based on gas fuel too expensive for use. The advantage is a low cost of investment and short time of construction.

7 Development plans for power generation system

Upon analysing available documents and other papers dealing with the attempt to determine the strategy for the development of power generation system, i.e. among others: Energy policy in Poland by 2030 [8]; Energy mix 2050, Analysis of scenarios for Poland, Warsaw 2011 [9]; Energy mix for Poland by 2060, elaborated for Chancellery of the Prime Minister by the Department of Strategic Analyses, [10]; Coal for Polish energy industry in the perspective of 2050 – scenario analyses, the document elaborated for the Polish Mining Chamber of Industry and Commerce [11]; The Strategy for Sustainable Development, Ministry of Development, draft law on the date 29 July 2016 [12]; it may be noticed that each of these strategies highlights the role of coal in the near future. Considering average values of production capacity given in the scenarios above, Table 7 presents the required production capacity by the year 2050.

Table 7. The forecast of power demand and a lack of installed capacity by 2050

Year	Installed capacity [GW]	Withdrawals [GW]	New production capacity [GW]	Forecast for power demand [GW]	Lack of installed capacity [GW]
2016	40.44	-	-	-	-
2020	-	4.97	5.82	46.00	4.70
2030	-	4.78	-	51.00	14.43
2040	-	8.19	-	56.00	27.63
2050	-	3.90	-	67.00	43.27

Source: Own elaboration on the basis of [1,8-12]

It will certainly be difficult to execute the construction of new open mining pits of brown coal, or to make a decision on the construction and starting of a new brown coal mine. The prices of coal on international markets may also influence the situation when it is more profitable to buy it from abroad. This may cause abandonment of investments into domestic production resources. However, the data above, in comparison to the plans of construction of new and conventional power units, show that, in theory, we have the potential to secure future supplies of power for the following years. If we consider the scenario of a nuclear unit in 2030 in NPS and in the subsequent year – the second one; and if we also consider the development of heating sources of gas technology and gas-steam technology, then one may think that our system is able to secure necessary power supplies. Of course, there are also renewable energy sources which are becoming more and more important. The development and contribution of renewable energy sources is determined not only by technological development, but mainly by all political decisions which impose a certain contribution of renewable resources into the total installed capacity. The political nature of changes in the recent months, mentioned above, has caused severe turbulence in the sector of renewable energy sources, mainly among the wind farms owners. Thus, it will be difficult to assess the right direction of the development of RES in our country. Following the opinions of different specialists, RES sector related to wind technology, as the result of necessity, will have to choose sea locations for their further investments. It is also worth-mentioning the potential of the development of small and dispersed production sources. The development of prosumer power generation in Denmark and Germany shows that also these sources may have a great share in national demand for energy supplies in the future. Dispersed sources which are supported by energy storage technologies will allow for joining them into virtual power plants of potential regulatory role in the whole system [13].

A very important aspect which is often omitted by various analyses is the flexibility of a power generating system. In the contemporary world flexibility gains special meaning as energy systems are more and more loaded as the result of rising demand for power supplies. Moreover, more significant contribution of renewable energy (wind power plants and photovoltaic systems) provides for the less stability of energy systems. The bigger share of renewable energy, the less predictable production of energy becomes. In order to take prompt actions in response to changeable situation of power generation, power plants must be flexible i.e. able to increase capacity in wide range in a short period of time required for the start-up of a power plant. Contrary to appearances, differences between technologies are sometimes enormous. Big steam power plants usually have a long period of start-up and are of low flexibility. It results from a slow process of steam production for the turbines to operate.

Modelling of energy system development has significant limitations. Altogether with the development of liberalisation processes of power sectors there are new factors which determine the choice of production technology. Because of the risk related to the competitive market the investors have preferred technologies with a short period of expense return and a short period of construction. A power sector is also more and more susceptible to the environmental policy, including the necessity to reduce CO_2 emissions. It requires a diametrical change into the directions of production sector development, especially for the benefit of clean coal technologies and nuclear plants, and also renewable energy sources (including decentralised systems). The selection of future technologies will depend on many factors. The comparison of the basic characteristics of production technologies are given in Tables 8a and 8b.

8 Conclusion

Energy policy constitutes a national security policy. It should be well-thought and consider the access to energy sources. Other important issue, is the necessity to be independent from imports of fuels. The identification of domestic power potential presented herein shows that in the perspective of 30 years, it is possible to base the power system on coal power plants. This time will certainly allow for the construction of a new energy mix, of greater and greater share of other power technologies. Decisions on the construction of new production sources must also consider their influence on the National Power System. Moreover, another aspect is also the access and price of the fuel, altogether with the costs of its future CO_2 emissions.

RES are clean energy sources and in the long-term perspective they will certainly be more important in the power industry [19]. However, having enormous amounts of coal we may expect that in the near future the energy system will be based on stable coal sources. It is sure that future will also bring the necessity to change the energy composition, as the result of, among others, shrinking fuel resources and the influence of environmental issues. Nevertheless, till these changes are made, there is the possibility to base the system on highly efficient and infallible coal sources. Reduction of CO_2 emissions should not be executed through rejecting coal from Polish energy mix, but it should take place through investments into new technologies supporting CO_2 reduction and the construction of energy units of higher efficiency. Significant share of solid fuels, coming mainly from domestic sources, should be the guarantee of energy security for the country in the next decades. The selection of technology for new generation sources in a long-term perspective must be based only on economic criterion, which considers the knowledge of expected total costs of energy, altogether with environmental costs. However, it is necessary to determine if iplementation of proposed processes of energy changes is possible and what will be its future cost. The agreement which was signed in Paris assumes that the future energy mix will engage all technologies. It is important, though, that this mix considers a balanced development.

Table 8a. Main characteristics of energy technologies [14-18].

Technology	Units power	Construction time	Lifetime [years]	Investments cost/kW	Fuel costs	CO_2 emissions
OCGT	Medium	Short	25	Low	High	Medium
CCGT	Medium	Short	30	Low	High	Medium
Coal power plant	Large	Long	45	High	Medium	High
NPP	Very large	Long	60	High	Low	N/a
Water power plant	Small - large	Long	100	Very high	N/a	N/a
Wind power plant	Small	Short	25	High	N/a	N/a
PV	Small	Short	25	Very high	N/a	N/a

Table 8b. Main characteristics of energy technologies [14-18].

Technology	Efficiency	Min load [%P_{max}]	Start up time [h]	Ramping speed [%P_{max}/min]	Black start	Quick start (10 min)
OCGT	30–40	20–50	0.2–0.5	10–20	Yes	Yes
CCGT	45–58	30–50	2–4	4–10	No	No
Coal power plant	35–48	30–60	4–8	2–6	No	No
NPP	30–35	40–50	25–48	0.5–5	No	No
Water power plant	80–90	15–20	0.1	5–40	Yes	Yes
Wind power plant	25–35	2–4	-	0.5–1	-	-
PV	15–20	-	-	-	Yes	-

Future domestic energy mix must be highly flexible as the share in renewable energy sources will increase. Optimal schedule for the energy system will allow for stability and effectiveness of use of available sources of energy. Power plants, which operate in a power system as the base, should be characterised mainly as highly-efficient and they should generate low emissions of harmful gases. Moreover, the system also requires power plants of high flexibility in order to cover changeable peak demands and cooperate with renewable energy sources to compensate momentary variations related to changeable wind strength or sun radiation. Planning the future development of the energy system it is necessary also to take account of possible risks. For example the lack of development of nuclear power in Poland, the cost of implementing RES and other economic determinants of implementation of different scenarios diversification of energy sources in Poland

References

1. Polish Transmision System Operator, data from : www.pse.pl (10.09.2016)
2. H. Kaproń, Z. Połecki, Rynek Energii, Eksploatacja podsystemu wytwórczego w Polsce w latach 1995–2010, **6**, 3–9, (2012)
3. International Energy Agency, CO_2 Emissions from Fuel Combustion-Highlights, (2015)
4. A. Olecka, K. Bebkiewicz, B. Dębski, M. Kanafa, J. Rutkowski, I. Kargulewicz, J. Skośkiewicz, D. Zasina, M. Zimakowska-Laskowska, M. Żaczek, Poland's National Inventory Report 2016, Greenhouse Gas Inventory for 1988-2014, Submission under the UN Framework Convention on Climate Change, National Centre for Emission Management (KOBiZE) at the Institute of Environmental Protection – National Research Institute, (2016)
5. M. Szuflicki, A. Malon, M. Tymiński, Bilans zasobów złóż kopalin w Polsce wg stanu na 31 XII 2015 r., Państwowy Instytut Badawczy, Państwowy Instytut Geologiczny, (2016)
6. D. Gielen, D. Saygin, N. Wagner, W. Budzianowski, REmap 2030 Perspektywy rozwoju energii odnawialnej w Polsce, IRENA, (www.irena.org/remap), Abu Dhabi, (2015)
7. Act of 20 May 2016 on Wind Energy Investments, Dz.U. z 2016, poz. 961 (*Journal of Laws from 2016, item 961*), Warsaw, (2016)
8. Energy policy in Poland by 2030, Warsaw, (2009)
9. Energy mix 2050, Analysis of scenarios for Poland, Warsaw, (2011)
10. Energy mix for Poland by 2060, elaborated for Chancellery of the Prime Minister by the Department of Strategic Analyses, Warsaw, (2015)
11. L. Gawlik (ed.), Coal for Polish energy industry in the perspective of 2050 – scenario analyses, the document elaborated for the Polish Mining Chamber of Industry and Commerce, Katowice, (2013)
12. The Strategy for Sustainable Development, Ministry of Development, draft law on the date 29 July 2016, (2016)
13. R. Szczerbowski, Możliwości zastosowania ogniw paliwowych w układach hybrydowych w celu eliminacji negatywnego wpływu chimerycznych źródeł odnawialnych, *IX Konferencja Naukowo-Techniczna „Optymalizacja w Elektroenergetyce"*, 189–199, (2015)
14. I. Pierre, F. Bauer, R. Blasko, N. Dahlback, M. Dumpelmann, K. Kainurinne, S. Luedge, P. Opdenacker, I. Pescador Chamorro, D. Romano, F. Schoonacker, G. Weisrock, Flexible generation: Backing up renewables, D/2011/12.105/47, (2011)
15. G. Papaefthymiou, K. Grave, K. Dragoon, Flexibility options in electricity systems, ECOFYS Germany, (2014)
16. P. Clerens, M. Farley, L. Jazbec, N. Kraus, K.D. Tigges, Thermal Power in 2030 Added Value For EU Energy Policy, European Power Plant Suppliers Association, (2015)
17. J.P. Deane, B.P. O′ Gallachoir, E.J. McKeogh, Renewable and Sustainable Energy Reviews,Techno-economic review of existing and new pumped hydro energy storage plant, **14**, 1293–1302, (2010)
18. Projected Costs of Generating Electricity, OECD Publications, (2010)
19. T. Olkuski, A. Szurlej, P. Janusz, Polityka Energetyczna, Realizacja polityki energetycznej w obszarze gazu ziemnego, **18/2**, 5–8, (2015)

Evaluation of the European Union-United States oil and petroleum-based fuels trade potential in the context of the negotiated TTIP agreement

Tadeusz Olkuski[1, *], *Andrzej Paweł* Sikora[2], *Mateusz Piotr* Sikora[3], *Adam* Szurlej[4]

[1] AGH University of Science and Technology, the Faculty of Energy and Fuels, Poland
[2] The Energy Studies Institute in Warsaw, Poland
[3] The Energy Studies Institute in Warsaw, Poland
[4] AGH University of Science and Technology, the Faculty of Drilling, Oil and Gas, Poland

Abstract. The article evaluates the European Union-United States oil and petroleum-based fuels trade potential. The planned trade structure and balance according to IEA (International Energy Agency) and IHS (IHS CERA www.ihs.com)) scenarios, the projected volume of imports and exports, and differences in price levels and costs are presented. The projected potential of the trade volume, taking into account the possible impact of the Transatlantic Trade and Investment Partnership (TTIP), is also presented. The analysis has shown that the elimination of trade barriers between the European Union and the United States would be more beneficial to US refineries. Due to the higher import tariffs to the EU, the potential benefits of US exporters are higher than those of the EU exporters to the US. This confirms the fears of European negotiators that some aspects of the agreement will have a negative impact on European businesses. However, in the case of petroleum products the TTIP agreement will have a negligible impact on increasing the export volume.

1 Introduction

The chemical composition of oil is variable depending on extraction site. There is a whole range of types of crude oil, while the most common classification of oils is based on two dimensions [1]:

- The density determining the proportions between light (more valuable) and heavy (less valuable) hydrocarbons in crude oil, measured by the API gravity. API Gravity is a specific scale developed by the American Petroleum Institute (API) for measuring the density of different types of oil:
    - Light crude oil – API higher than 33°,
    - Medium oil – API between 22° and 33°,
    - Heavy oil – API below 22°.
- Sulphur content of crude oil:

* Corresponding author: olkuski@agh.edu.pl

- Sweet crude oil – less than 0.5% sulphur,
- Medium sour crude oil – between 0.5 and 2.0% sulphur
- Sour crude oil – more than 2.0% sulfur.

In addition, the proportions of individual hydrocarbon groups in a given crude oil (paraffins, aromatics, and naphthenes), the shape of the distillation curve, vapor pressures, viscosities, pour points, the hydrogen sulfide, heavy metal, acidic compounds, chlorides and salt contents are all of a great importance. Conversion of almost every lighter and sweeter oil to heavier and sourer (sulfated) oil is technically possible. However, this raises the question whether such conversions are economically justified. The possibility of conversion in the opposite direction is limited due to the structure and processing capacities of refineries As a consequence, the mentioned relationships have to be taken into account when analyzing the potential oil trade between the United States and the European Union. An analysis of refinery products is easier, because these products (quality standards) are now very similar, while minimum differences should not have a significant impact on the potential oil trade. For example, the permissible sulfur content in diesel fuel in the United States and in the EU is 15 and 10 ppm, respectively.

2 The structure of the planned oil trade

The comparison of oil trade between the United States and the European Union (excluding Canadian and Norwegian deposits) shows that both areas remain importers of this raw material. It is worth noting that the level of dependence on external supplies of oil in the US is 2.5 times lower than in the EU (Fig. 1). When taking into account the production from Canadian and Norwegian deposits (Fig. 2) the situation of the EU slightly improves, while the economic situation of the United States changes dramatically. Canadian oil, along with unconventional deposits, can be potentially exported. The conditions prevailing in North America, favorable when compared to the EU conditions, result from, among others, the so-called shale revolution. It can be assumed that the shale oil revolution started in 2008, three years later than shale gas revolution [2].

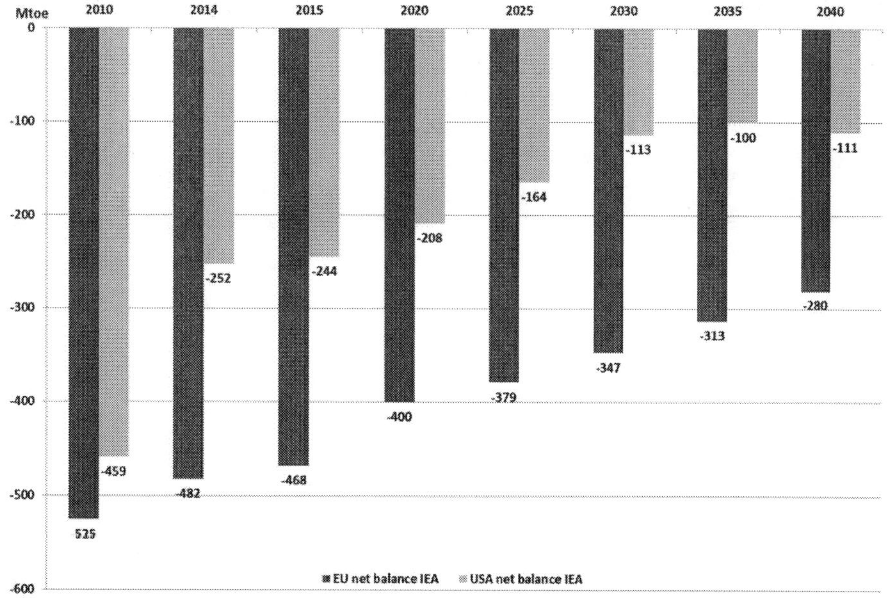

Fig. 1. The net trade balance of oil in the United States and European Union countries in the IEA baseline scenario – excluding Canadian and Norwegian deposits [3]

DOI: 10.1051/e3sconf/20171401004

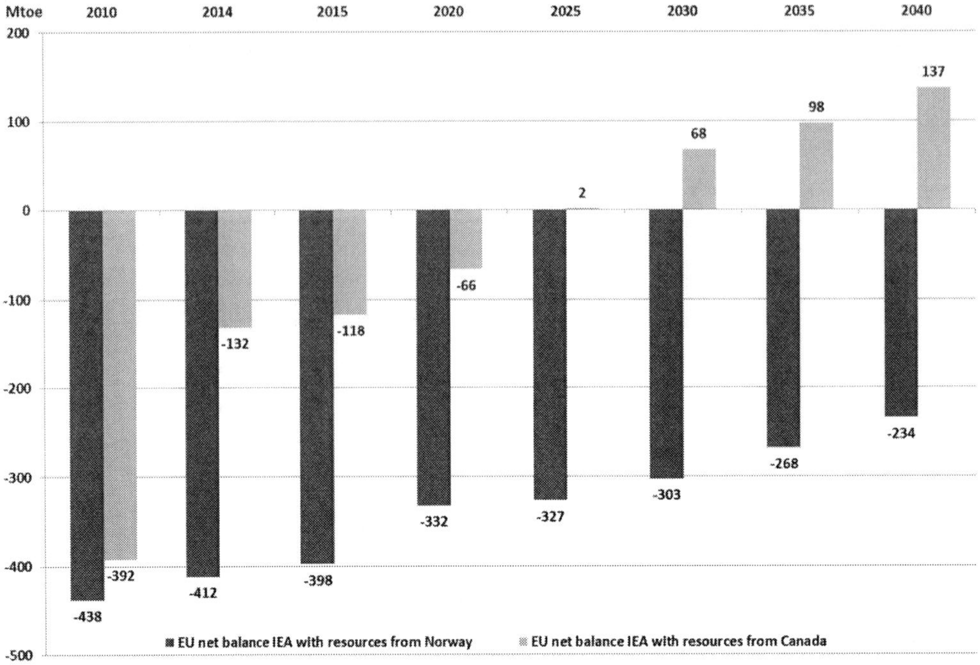

Fig. 2. The net trade balance of oil in the United States and the European Union according to the IEA baseline scenario – including Canadian and Norwegian deposits [3]

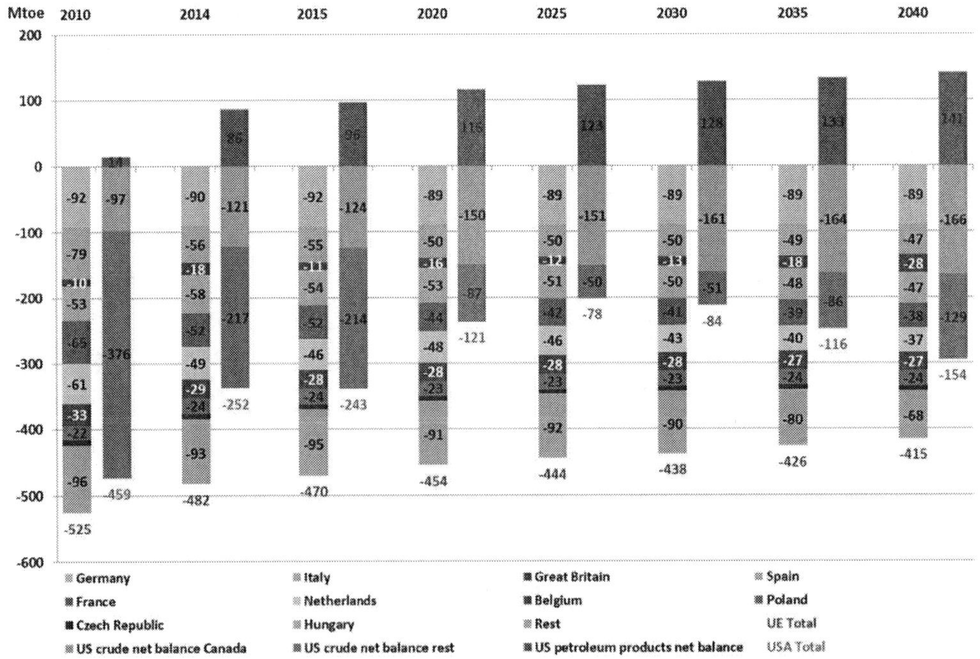

Fig. 3. The net trade balance of oil in the United States and European Union countries according to the available data compiled by the Energy Studies Institute in Warsaw [3]

The analysis carried out by the Energy Studies Institute in Warsaw, based on available market reports, has shown a similar trend (Fig. 3). Using the methodology of the International Energy Agency (IEA) [4], it is possible to take into account NGL (Natural Gas Liquids - ethane - LPG, sometimes natural gasoline from natural gas deposits) and the net trade balance of

petroleum products in addition to crude oil when estimating the oil trade on the basis of data compiled by the Energy Studies Institute in Warsaw. Experts from around the world are optimistic about the production increase in the United States, which in turn will lead to a surplus of products for export. The EU's dependence on imports decreases much more slowly than in the IEA forecast, while in the case of Germany, Italy, Belgium, and Poland, oil imports are not decreasing at all.
When analyzing the Energy Studies Institute in Warsaw data on the volume of imports and exports of oil between the United States and the European Union (Fig. 4), the increasing position of the United States in oil exports (the United States still import substantial quantities of oil) can be observed.

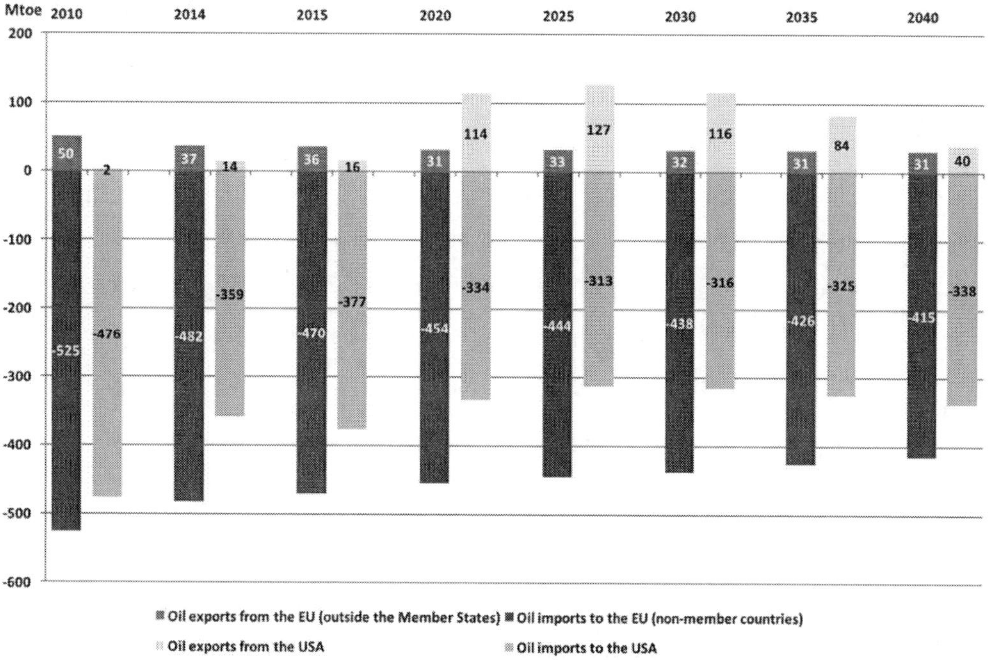

Fig.4. The forecasted volume of imports and exports of oil between the United States and the European Union on the basis of data compiled by the Energy Studies Institute in Warsaw [3]

In the recent past, the European Union (including Norway) has exported small (compared to the volumes shipped to the US from other countries) amounts of oil to the USA. However, according to the presented plans, the United States could begin to supply oil to the European Union already in 2017. On the basis of the available data, the Energy Studies Institute in Warsaw predicts initially low amounts in the range between 15-17 Mtoe/year (Fig. 5), which corresponds to the demand of the PKN Orlen refinery in Plock. Even if the TTIP agreement will fail to go beyond the stage of negotiations, and the US law will not lift the ban on imports, the obtained results have shown the economic viability of the discussed oil supplies.

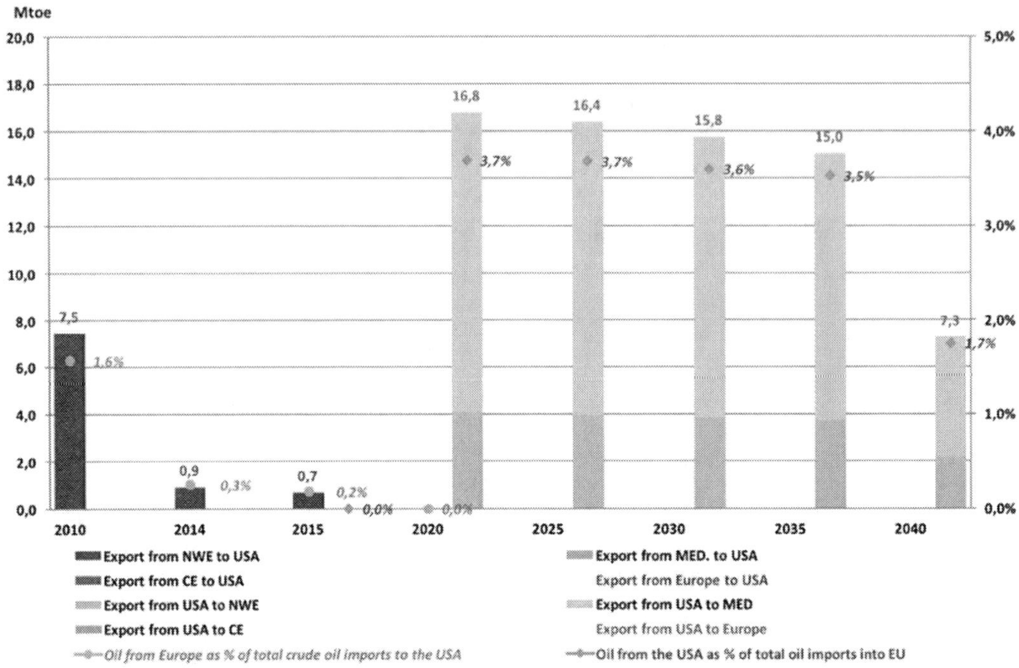

NWE – North-West Europe (The North Sea and Scandinavia)
MED – Southern Europe (the Mediterranean Basin)
CE – Central Europe (including Poland)

Fig. 5. The forecasted oil trade volume between the United States and the European Union (including Norway) according to the Energy Studies Institute in Warsaw (based on the available data) [3]

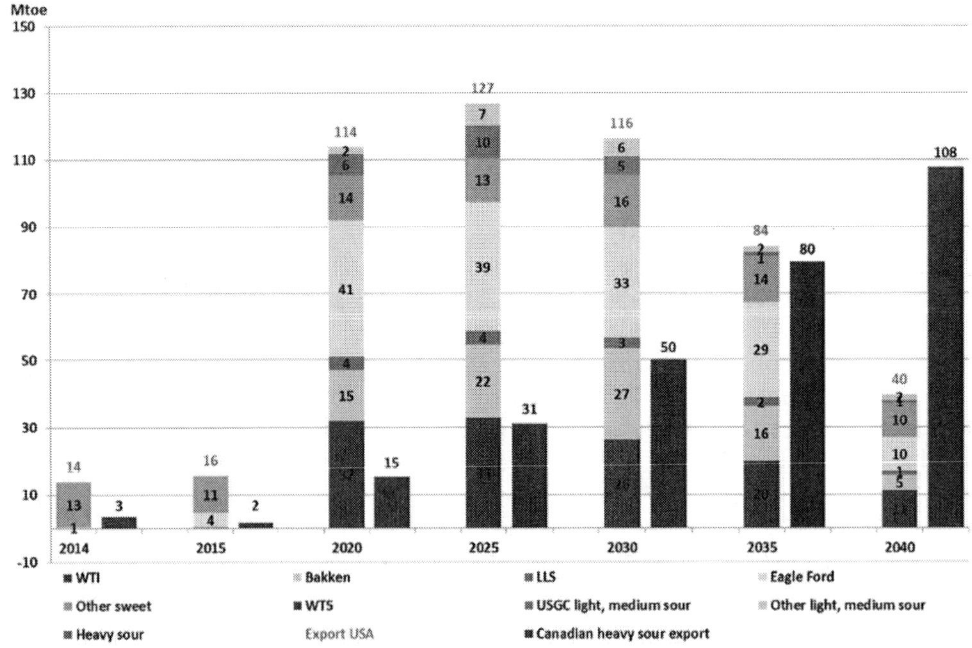

Fig. 6. The structure of the forecasted exports of crude oil from the United States and Canada broken down by oil types [3]

An analysis of the structure of the forecasted exports of crude oil from the United States has shown that 90-100% of the exported raw material can be classified as light and sweet crude oils (WTI, LLS) or very light and sweet (tight oil) (Bakken and Eagle Ford Shale). In this case, the export of heavy crude oils (e.g. Mars Blend), or even medium-heavy and sour crude oils, is not taken into account. Instead, the structure of the forecasted exports of crude oil takes into account Canadian heavy and sour crude oil from oil shale, while sour crude oils (light and medium), e.g. Bayou Choctaw Sour and Mississippi Sour enter the market from 2020 on (10-16 Mtoe) (Fig. 6).

Crude oil imports to the European Union amount to 250-280 Mtoe/year. This is mainly due to the dominant demand for middle distillates in the structure of yields, while the most economical way to obtain them is based on converting medium sour crude oils. Sour, medium and light crude oils are among main types of the imported crude oils. They account for about 60-65% of total imports and are imported from former members of the Commonwealth of Independent States (CIS) - mainly Russia (Urals) and Kazakhstan (CPC blend) and the Persian Gulf (Arab Light, Iran Light, Iran Heavy, Kirkuk). Light crude oil accounts for about 130-150 Mtoe/year and is usually imported from Africa (Libya, Algeria, Nigeria, and Angola) and the Caspian region (BTC blend, Azeri Light) Heavy oils, with API gravity below 22-23° are not refined in European petrochemical plants. This means that light and sweet crude oils from the United States and sour crude oil from Canada could form part of the imports to the EU (Fig. 7).

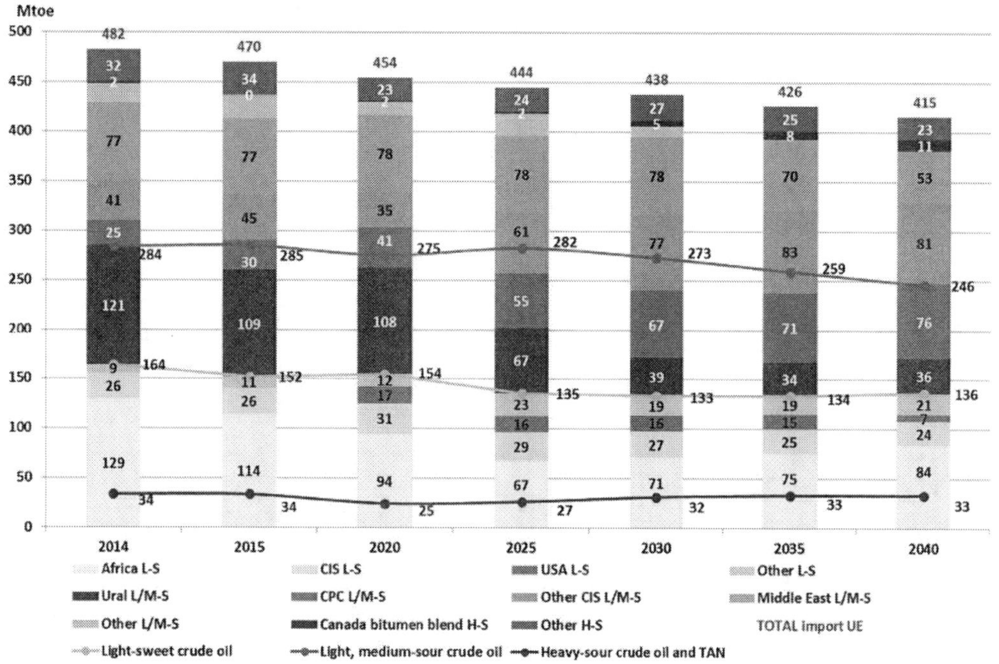

L-S light, sweet; L/S-K light/medium sour; C-K heavy/very heavy sour, TAN – „corrosive" crude oil.

Fig. 7. The structure of the forecasted imports of crude oil to the European Union broken down by oil types [3]

3 The structure of the planned petroleum products trade

The European Union will continue to import petroleum products (middle distillates, jet fuel and other refined products (mainly LPG)). Meanwhile, the United States is implementing the strategy to become the exporter of these products. On the basis of the available data (Fig. 8)

DOI: 10.1051/e3sconf/20171401004

it can be concluded that, the oil shortage in Europe accounts for 20-25% of the US oversupply. Firstly, it does not mean that the United States will be able to meet the European demand. However, the US can become a very interesting supplier. This, in turn, will increase the competition in the market. Secondly, other export destinations will also be of interest to the United States.

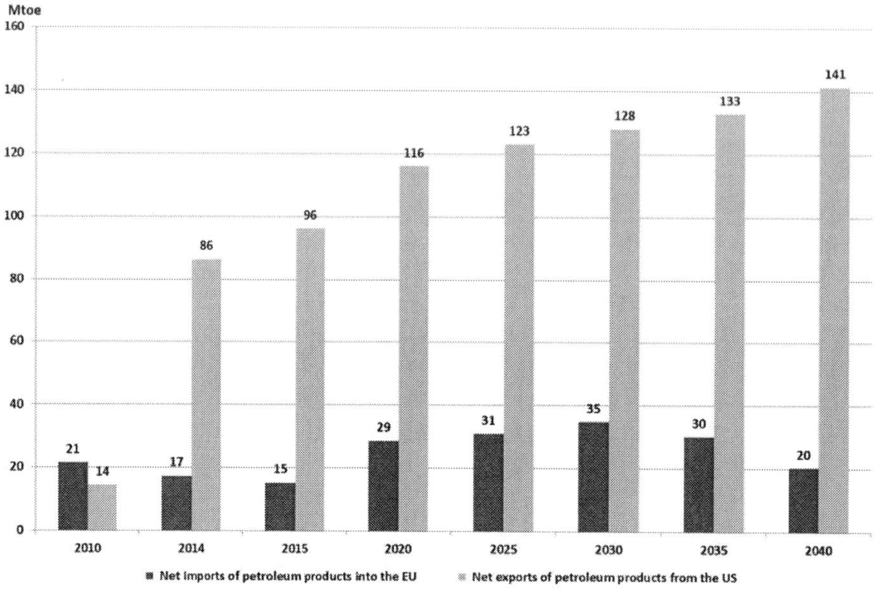

Fig. 8. Net imports of petroleum products to the EU and the net exports of petroleum products from the United States [3]

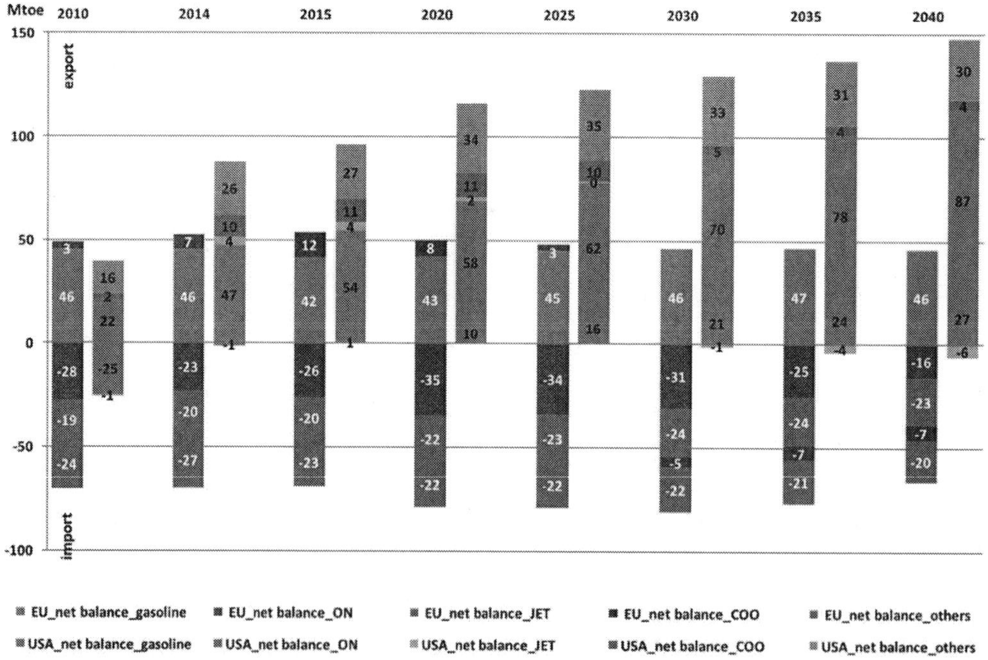

ON – refers to all middle distillates, JET – includes jet fuel and kerosene

Fig. 9. The net trade balance of individual petroleum products in the United States and European Union countries [3]

 DOI: 10.1051/e3sconf/20171401004

When broken down by individual types of petroleum products (Fig. 9), middle distillates, mainly diesel and, to a lesser extent, LPG and kerosene, are expected to be the leading export product of the United States. In the next 10-20 years, an increase in gasoline exports is also expected.

4 Differences in price levels, costs and logistics

4.1 Oil

As mentioned before, there is a whole range of types of crude oil. Therefore, the data compiled by the Energy Studies Institute in Warsaw on North American oils and crude oil imported to Europe [1] will be used for the purpose of further analysis (to compare products with similar properties and suitability) (Fig. 10).

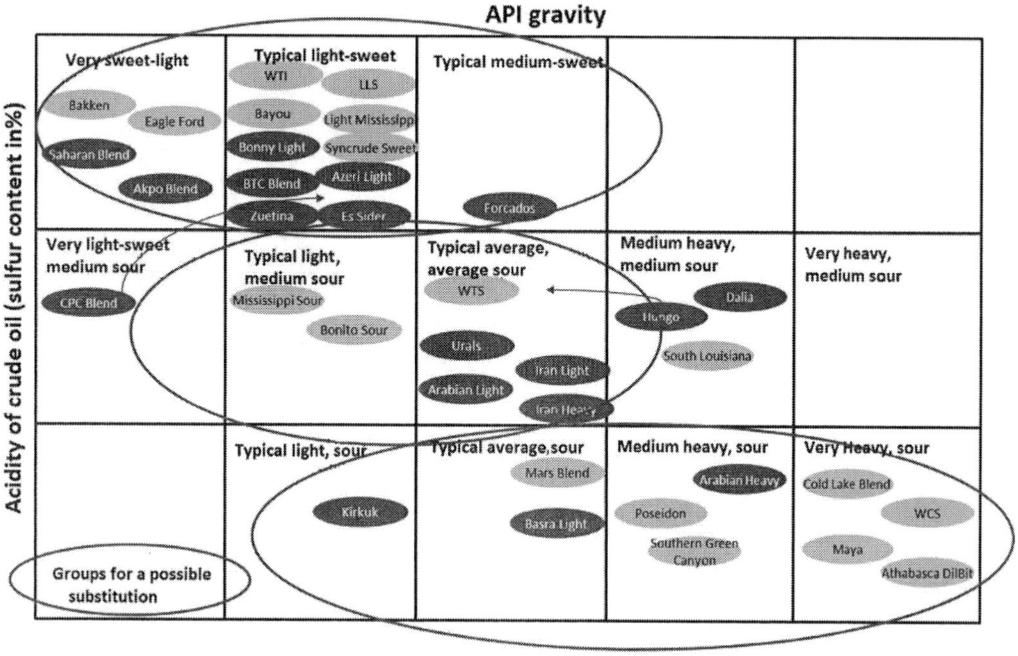

Fig. 10. The substitution matrix for individual types of crude oil used in North America and Europe [3]

It has been shown that light crude oils are the most suitable for substitution in the case of European imports. However their export potential is limited. On the other hand, the group of heavy and very sour crude oils, which could be exported, is not sought after by European refineries. The exception is the CPC Blend, which is very light (43-45° API), with sulphur content only slightly too high to consider it sweet (0.56%), which is why it is included in the group for a possible substitution with light, sweet crude oils.

Another important point is the cost of transport from the United States (mainly the Gulf of Mexico) to the European markets (NWE, MED, or CE). CIF prices - Cost, Insurance and Freight (named port of destination) - are known for oil supplies from the Middle East, Africa or former CIS countries, while the available quota and price forecasts for US crude oils (Loco prices) are provided by the USGC (the Gulf of Mexico). Therefore, the analysis should take into account these differences.

The analysis of the results shows that US crude oils:

- Light and sweet crude oils (Eagle Ford oil, WTI) are competitively priced and cheaper than their competitors for the entire period of the analysis (up to 2020 and 2025 for Bakken and LLS crude oil, respectively);
- Medium sour crude oils (WTS) are also attractively priced in relation to their substitutes. (up to 2025 and 2035 for Urals and Middle East crude oil, respectively);
- The analysis of medium heavy crude oils (API below 30°), heavy crude oils (API < 22°), and sour crude oils is more complicated, because only Mars Blend, Poseidon, and Southern Green Canyon are substitutes for crude oil from the Middle East. Other crude oils are considered to be much "heavier" and therefore are not of interest to European refineries. From these relations, it can be found that US crude oils are more attractively priced than crude oils from the Middle East.

To sum up, the price of imported oil should be cheaper by about 3-9 USD/bbl, while the trend should decrease to 0.5-1 USD/bbl. For petrochemical companies, processing huge amounts of this raw material, a price difference of $ 1 per barrel means millions of dollars in savings.

4.2 Petroleum products

It is clearly visible that higher (or lower) petroleum products prices are leading to a positive (or negative) net balance of trade (Fig.9). There is an oversupply of gasoline in both the United States and the EU, while the prices offered by European refineries are much more favorable.

As a consequence, the import of European products to the United States is expected. The price of jet and kerosene fuels is variable. Until 2019, the most profitable solution is to import oil from the United States, in the years 2020-2030 prices are unattractive for both parties, while after the year 2030 exports from the EU to the US seem to be the preferred option.

In the case of heavy fuel oil, changes in trade trends can be observed from the year 2030. When it comes to kerosene, an interesting paradox occurs: While a deficit of this product in the EU is accompanied by an oversupply in the US, the analysis has shown that it is more beneficial to export kerosene to the US. Another point worth mentioning is the planned elimination of customs duties. The elimination of import tariffs will raise the potential profits of exporters. This applies to all of the discussed products (except for medium distillates).

4.3 The forecast of the potential trade volume

4.3.1 Baseline scenario without changes resulting from the TTIP agreement

According to the available data, the volume of exports of sweet crude oil from the US to the European market (Fig. 5) during the peak period (2025-2030) will remain at a level of 15-17 Mtoe/year with a downward trend to 7 Mtoe (in the year 2040). Based on the expert analysis carried out at the Energy Studies Institute in Warsaw it has been shown that the volume of oil imports from the European Union to the United States (at current import tariffs) can reach up to 700 Mtoe (2015-2040) with the highest share for gasoline (65%), followed by heavy fuel oil (20%), jet fuel (12%), and kerosene (4%). In relation to the United States, the total trade volume may reach up to 850 Mtoe (with import tariffs on middle distillates) or 2520 Mtoe if no import tariffs are imposed. Without going into details, it is clear that the elimination of trade barriers between the European Union and the United States would be more beneficial to US refineries Due to the higher import tariffs to the EU, the potential benefits of US exporters are higher than those of the EU exporters to the US.

4.3.2 The potential impact of the TTIP agreement for the baseline scenario

Firstly, a possible change in the volume of oil imports from the United States to the European Union requires more favorable prices of comparable crude oils [1]. Secondly, an increase in the volume of oil imports can take place if European crude oils will be of equal interest to US customers.

It should be noted that the Transatlantic Trade and Investment Partnership and the resulting development of trade relations between the parties (e.g. shifting of export volumes from other "preferred" markets to the European market) would allow this scenario to occur. The realization of this scenario (Fig. 11) would allow the United States to export over 100 Mtoe of oil to the European Union (about 23% of the total demand of European refineries).

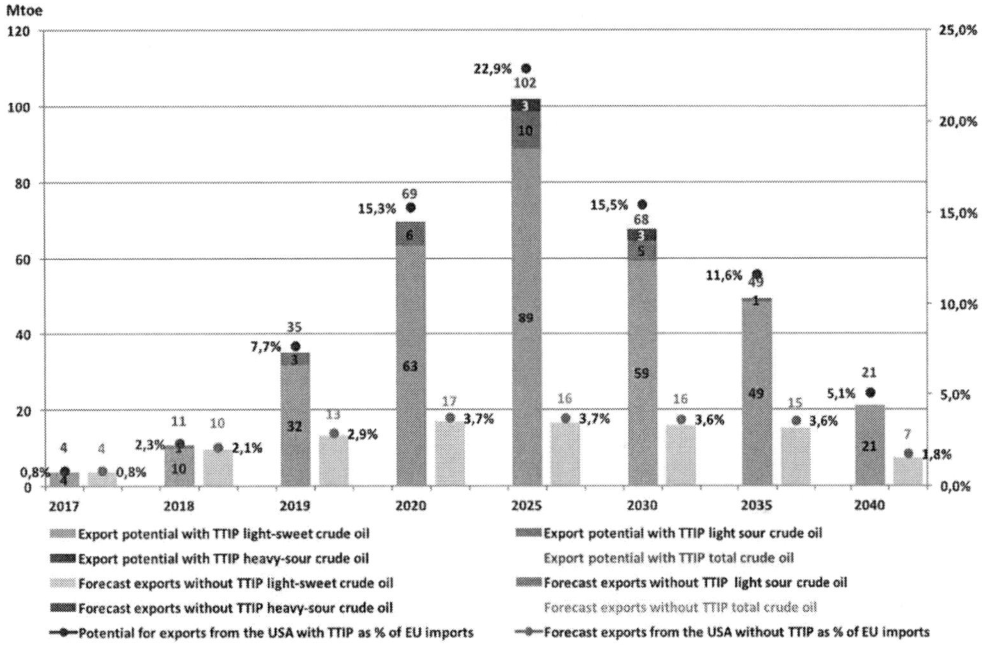

Fig. 11. Evaluation of the potential of oil exports from the United States and Canada to the European Union following the signing of the TTIP [3]

The low volume of medium sour crude oil and sour crude oil offered by the United States does not have a major impact on petrochemical plants in Poland. Sweet crude oil processed in Polish refineries in Płock and Gdańsk accounts for only about 10% of their processing capacity [5].

In the case of petroleum products, the TTIP agreement will have a negligible impact on increasing the export volume. Despite import tariffs, light and middle distillates including LPG, gasoline, and jet fuel already guarantee the profitability of exporters [6, 7]. The elimination of trade barriers would not increase volumes, though is expected to increase profitability (Fig. 12). It should be emphasized that the duties are much higher for goods imported to the European Union (3.5-4.7% ad valorem) compared to the tariffs in the US (0.052-0.52 USD/bbl or 0.1-0.5%).

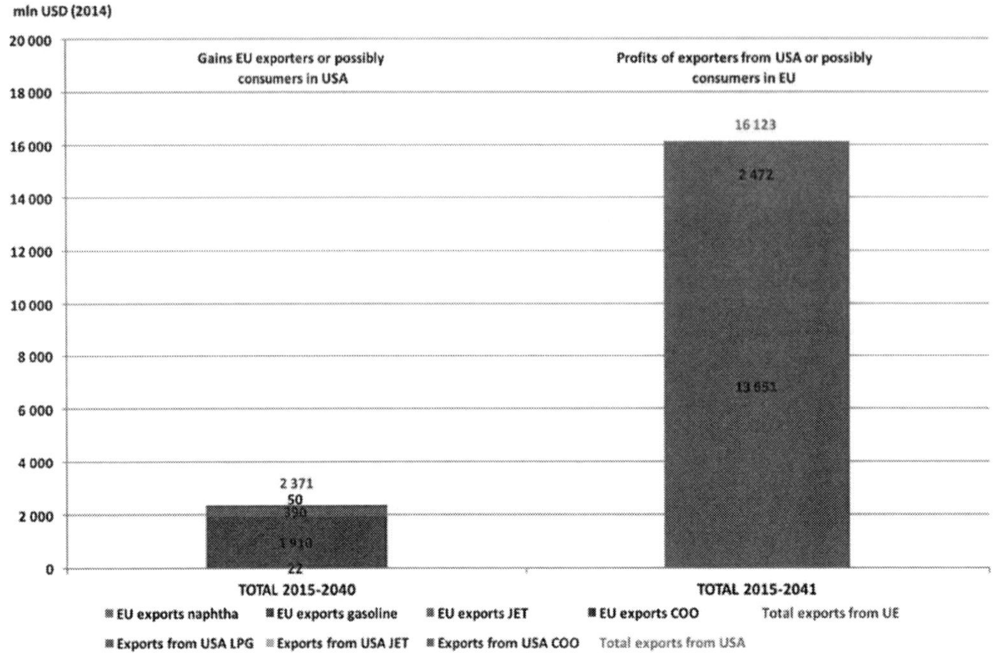

Fig. 12. The potential benefits resulting from the elimination of tariffs for exporters in the United States and the European Union [3]

5 Summary

The elimination of trade barriers between the European Union and the United States would be more beneficial to US refineries. Due to the higher import tariffs to the EU, the potential benefits of US exporters are higher than those of the EU exporters to the US.

The Transatlantic Trade and Investment Partnership and the resulting development of trade relations between the parties (e.g. shifting of export volumes from other "preferred" markets to the European market), would allow the United States to export over 100 Mtoe of oil to the European Union (about 23% of the total demand of European refineries).

The low volume of medium sour crude oil and sour crude oil offered by the United States does not have a major impact on petrochemical plants in Poland. Sweet crude oil processed in Polish refineries in Płock and Gdańsk accounts for only about 10% of their processing capacity.

In the case of petroleum products, the TTIP agreement will have a negligible impact on increasing the export volume.

The paper is an effect of the statutory project no. 11.11.210.217.

References

1. At a time when the article was prepared, TTIP negotiations and most of documents were confidential. TTIP text has not been published. Analysis in this article were based primarily on the analysis of documents and data selected by authors, primarily data from the European Commission - EU negotiating texts in TTIP - http://trade.ec.europa.eu/doclib/press/index.cfm?id=1230

2. BP Statistical Review of World Energy 2015
3. Costs and benefits for the Polish energy sector resulting from the planned trade agreement between the EU and the USA (TTIP) - The Energy Studies Institute in Warsaw. (unpublished material). (2015)
4. Janusz P., Kaliski M., Szurlej A., „The shale gas revolution' and changes on the LNG market. Mineral Resources Management", vol. 31, Issue 3, pp. 5–24. (2015)
5. Krupa M., Sikora A., Sikora M., „The Yankee has struck oil" (Jankesom uderzyła ropa) (The potential Transatlantic Trade and Investment Partnership (TTIP) impact on the European and Polish chemical sectors), Przemysł Chemiczny 94/10, October 2015, DOI: 10.15199/62.2015.10.2, pp. 1641-1645 (2015)
6. Olkuski T., Sikora A., Sikora M., Szurlej A., Tora B., „Rynek gazu ziemnego w kontekście umowy TTIP" (Natural gas market in context of Transatlantic Trade and Investment Partnership), Rynek Energii 5 (126), październik 2016 - http://www.cire.pl/item,137112,2,0,0,0,0,0,rynek-gazu-ziemnego-w-kontekscie-umowy-ttip.html
7. Sikora A., Sikora M., Krupa M., „Zniesienie zakazu eksportu ropy naftowej ze Stanów Zjednoczonych Ameryki Północnej oraz struktura prognozowanego eksportu" (Lifting the ban on oil exports from the United States of America and the structure of projected exports), CIRE, 30.12.2015 – http://www.cire.pl/item,122786,2,0,0,0,0,0,zniesienie-zakazu-eksportu-ropy-naftowej-ze-stanow-zjednoczonych-ameryki-polnocnej-oraz-struktura-prognozowanego-eksportu.html (2015)
8. Sikora M., „Sikora: polska dywersyfikacja dostaw ropy naftowej" (Polish diversification of crude oil supplies), CIRE, 9.05.2016 - http://www.cire.pl/item,128945,13,0,0,0,0,0,sikora-polska-dywersyfikacja-dostaw-ropy-naftowej.html (2016)
9. Sikora A., Sikora M., Szurlej A., Cios S., Jarosz R., „Analiza kosztów i korzyści Polski i UE związanych z planowaną umową handlową między UE i USA (TTIP) – podsumowanie chemia" (The potential Transatlantic Trade and Investment Partnership (TTIP) impact on the European and Polish chemical sectors), Przemysł Chemiczny, No. 5-2016, Vol. 95, May 2016, pp. 883-889, DOI: 10.15199/62.2016.5.1 (2016)
10. Sikora M., Sikora A., Szurlej A., Olkuski T., „The impact of TTIP agreement on the European Union-United States coal trade potential", International Scientific Conference on Sustainable Energy and Environment Development SEED'2016, AGH UST in Krakow, May 17th – 19th, 2016, DOI: 10.1051/e3sconf/20161000069 – http://www.e3s-conferences.org/articles/e3sconf/pdf/2016/05/e3sconf_seed2016_00069.pdf
11. Sikora A., Sikora M., „Rewolucja po amerykańsku" (American revolution), Chemia przemysłowa, 3/2016, str. 70
12. The World Energy Outlook 2014. IEA 2015. (2015)

E3S Web of Conferences **14**, 01005 (2017) DOI: 10.1051/e3sconf/20171401005
Energy and Fuels 2016

Energy security as a public good

Zdzisław Szalbierz [1,*], *Edyta* Ropuszyńska-Surma[2]

[1]Wrocław University of Science and Technology, Faculty of Computer Science and Management, Wyb. Wyspianskiego 27, Wroclaw, Poland
[2]Wrocław University of Science and Technology, Faculty of Computer Science and Management, Wyb. Wyspianskiego 27, Wroclaw, Poland

Abstract. Energy security is discussed as a multi-dimensional and complex category displaying features of a public good. Such an approach to energy security indicates that there are complex interdependencies between sector entities, consumers and public administration. The purpose of the paper is to identify and analyze relations between energy security and the policy of the state aiming at satisfying consumer demands. The key role here is played by legal and institutional regulations which have impact on all the defined dimensions of energy security.

1 Introduction

This article aims to identify and analyze the relations between energy security and the state policy in its responsibility for meeting the need for a variety of energy forms which households, businesses and public entities report at a particular time and place. The starting point for the discussion is the thesis according to which in Poland we are currently outside the realm of energy security, and we lack clear objectives for the protection of the environment (this relates especially to the targets for CO_2 emissions). The applied regulation instruments are becoming ineffective - among other things due to the lack of clearly defined objectives and the lack of a consistent long-term policy towards various stakeholders, e.g. prosumers.

In this context it is worthwhile to ask three questions:

1. What do we understand under the notion of energy security?
2. Is energy security a public good (common/free good) and in what sense is the state responsible for it?
3. What role does the legal and institutional regulation fulfill in reaching a higher level of energy security?

2 The nature of energy security

In recent years, after a relatively quiet period, there is again concern about the ability to satisfy energy needs by energy systems. This stems above all from the risk of interruptions of delivery (at determined times) in electricity and gas grids. The conviction that there is an

* Corresponding author: edyta.ropuszynska-surma@pwr.edu.pl

increasing risk of an energy deficit at specific times and places, and energy security has become the subject of increasingly intense discussions among scientists, economic activists and politicians. In Poland, two significant trends are observed. The first concerns the steady shrinkage in recent years of power reserves in relation to the maximum demand. The second consists in the increase of the peak demand, for example, on 23 June 2016 the record of the summer power demand was beaten, amounting to 22, 634 MW, while on 21 June 2013 the level was 1,033 MW less [1].

Energy security is becoming particularly important, however, it is defined in many different ways, taking into account various aspects.

A widely accepted general definition of energy security (security of electricity supply) has been "the ability to meet the demand for electricity, heat and gas in market conditions in terms of quantity and quality at the lowest possible price, resulting from the balance between the supply and the demand, while maintaining environment protection conditions [2].

The Union for the Co-ordination of Transmission of Electricity (UCTE) and *the European Transmission System Operators* (ETSO) see energy security from the angle of the technical operation of the power system. As a superior notion [3] the system reliability was adopted, understood as its capacity to supply customers with the necessary amount of energy in accordance with certain standards. At the level of the transmission and distribution grids the measures of reliability can be e.g.: the frequency, duration, and the likelihood of interruptions in deliveries, transmission and energy production. Within the reliability of the system two aspects are distinguished: adequacy as the system's ability to supply power and energy to cover the aggregate demand of consumers at any time, including planned or reasonably expected switch-offs of system components (standards of quality and reliability) and security - the ability of the power system to withstand sudden disturbances such as short circuits or unexpected loss of system components as a result of emergency shutdowns, natural (technical) disasters or terrorist attacks.

Polish Energy Policy until 2030, adopted by the Council of Ministers on 4 January 2009, defines energy security as "a state of the economy allowing to cover the current and prospective demand for fuels and energy, in a technically and economically justified way, while minimizing the negative impact of the energy sector on environment and social conditions" [4].

Finally, it is worthwhile to cite the definitions of energy security by T. Jamasb and M. Pollitt. According to these two eminent specialists from the MIT, since the energy market is undergoing liberalization and energy security has become multidimensional it is necessary to redefine the concept. They suggest in this context that in light of the liberalization of markets, the role of enterprises owned by the State Treasury, conditions of operation of the business environment (outside of the energy sector) and households the concept of energy security should be referred to the risk of a failure in supplying energy and the increased costs accompanying this risk [3]. As a result, the security of energy supply becomes a risk management strategy with strong references to cost effectiveness associated with both energy supply and energy demand. The question is how to interpret a situation in which the imbalance between the supply and the demand for important energy carriers (e.g. natural gas, electricity) is so big that it leads to a loss much larger than it would have cost if the equilibrium were taken care of in advance.

Thus, depending on the scope and scale of the analysis of energy security it is difficult to fully accept some definitions, because:

- on the one side the chain of the factors determining the level of energy security goes far beyond the area of the economy so that "energy security" is not a feature of the "state of the economy" only but also the result of a number of external conditions of its operation,
- on the other hand - to the recipient suddenly deprived of power it is, in fact, all the same whether the cause of this condition is outside the country or if it is the result of negligence

of his municipality or the company supplying fuel, electricity or heat to him; he is primarily interested in the question when the power is going to be restored, and what will be the possible financial consequences of the interruption, and, finally, what can you protect yourself against a repeated emergency like this.

It can be said that the elements determining the level of energy security constitute a chain whose individual parts are [3]:

1. access to primary energy carriers at a level that allows to meet the demand taking into consideration the transportation technology of primary carriers,
2. technologies of transformation of primary carriers into usable carriers,
3. long-distance transport technologies of usable carriers,
4. short-distance transport technologies of usable carriers,
5. mechanisms of financing the operations and development of energy sectors.

Each of these categories is subject of:

a) an economic analysis– ability of the economy to finance it, including the cost for limiting the environmental impact,
b) an environmental analysis – the level of the impact on the environment in the part that at present cannot be valued in economic terms,
c) the level of risk of interruptions including the interruptions stemming from political and military actions of other states,
d) an analysis of the threat of terrorist attacks,
e) extent of risks stemming from an interruption,
f) availability of procedures preventing interruptions.

Each category is also subject to an individual differentiation because of the specific nature of each carrier, both of primary energy and usable carriers, as well as the time horizon of the analysis.

It thus becomes clear that a deeper analysis of "energy security" leads to the conclusion that it is to far a simplification to ascribe to this notion the structure of a simple chain of relations. It has the structure of a multi-dimensional matrix, thus it is so tricky to attempt a sectional analysis.

Such many aspects of the notion of "energy security" mean that the array of risks that it is subject to, and of actions that need to be taken to protect it, is very differentiated, but also the responsibility for the state of energy security does not lie upon just one or a few entities. Rather, the statement seems substantiated that in Polish conditions energy security cannot be a matter dealt with only by entities of the energy sector since experience teaches us that they are not able to cope with the problem of the lack of energy security on their own.

The task of protecting energy security in its multifaceted aspects must thus be given to various institutions in the political, economic and social arena - in proportion to their role and competences: the highest authorities of public administration, its local authorities, energy sector companies, and the consumers. It is thus of particular importance to face the problem of the coordination of various actions at various levels since many aspects of energy security that are correlated with each other can be solved in an alternative way. In many cases, however, the adoption of one option excludes the rational application of another [3].

3 Energy security as a public/common good

There is no possibility of providing a clear definition of energy security as a good that would have a public/common character although intuitively we assume that being public/common is its important feature. In textbooks of classical economics energy security is not mentioned as a good, as opposed to national defense or the judicial system which are exhibited as model examples of purely public goods. Not everyone is able to perceive the category of public goods. An illustration of this statement can be the Polish PWN Encyclopedia of 2002 which does not contain the term of "public good". However, it contains the terms of a free good and a common good. Free is a good whose price equals zero. If we adopt this definition then free goods do not exist, thus they have become a historical category. A common good is a term for a collective value generated by communities to satisfy their individual interests [5]. A common good is interpreted differently depending on the approach. It seems that it also encompasses the interests of the state. The way how these interests are perceived depends on political convictions, and its content seems to be connected with its perception by an individual and the government. In this context it is worthwhile to fill this gap in relation to the notions of public good and energy security.

In the preceding bullet point the term of energy security is defined as a multi-aspect and complex category. For the sake of clear argumentation in our further analysis is has to underscored that the authors do not identify the notion of "energy security" with the service of supply of power (or any other carrier of energy) to the consumer. Energy security is - according to the economic nomenclature - one of thee many needs that is not satisfied by the supplier of the given energy carrier but by complex additional services, mostly invisible to the consumer. Such services, closely connected with the elements determining the level of energy security, include among others [6]:

- collection and monitoring of data on the demand for the generation capacity, the technical and economic state of the infrastructure,
- coordination of activities in the energy sector (including matching the supply and the demand in various periods),
- forecasting the directions of development of the technologies of production, storage, transport, distribution and use of energy,
- creation of the basis for an efficient use of energy.

From the point of view of the economics a public good is linked with a category of needs that can be satisfied by specific goods that are not subject to marketplace rules [7, p. 549; 8, p. 150; 9, p 131; 10, p. 59, p. 126]. In most economies the provision of public goods is the matter of the state, because - as stated earlier - they are not provided by the market. But the state does not have to be their supplier. The contemporary empirical and theoretical research indicates that both in the past[†] and in the present times communities, mostly local ones, have voluntarily taken initiatives of joint production and/or provision of public goods (e.g. combating crime) [10, pp. 86-87]. Such activities are associated with values such as social capital, or social ties. In the context of energy security, these initiatives may cover a relatively limited geographical area, and a key role in such projects will be played by local administration. But energy security should be seen in a much wider context than only through the prism of the local community. Therefore, the role of state bodies and its law regulations are crucial [11, p. 89-90, 95-96].

As noted earlier, the allocation of goods into various categories is not always clear. An important contribution to economics in this area was made by the classification of goods, proposed by Musegrave, Samuelson, and then developed by "club goods" of Buchanan. The division is based on two criteria: (1) the competition for the good among consumers (2) the possibility of being excluded from its consumption. According to this classification pure public goods are those where you can not exclude an entity from their consumption or such an exclusion is very expensive, and when consumers do not compete for the given good. From the definition of the public good results that "consumers (users) do not have to buy

[†] Hume noted in the first half of 18th century that neighbors may agree on the melioration of a meadow but that there is the risk that they will not agree on this because each of them will try to exploit the others. Therefore, the role of a judge, who would care for the common interest, was so important to Hume [Hume, 1739, p. 1911, 539; op. cit. 10, p. 37].

them"[‡] [8, p. 168]. This does not mean that the cost of their production is zero, because the public good is not for free. Public goods, especially those provided by the state[§], but not necessarily produced by it, are financed by taxes. In economics, the tax price determines the amount of additional tax corresponding to the costs which must be incurred in the event of increasing the production/delivery of a public good to different people. For this purpose a survey can be carried out among customers asking them "how much amount their expenditure on the public good would be, if there were a set price for each additional unit of the good" [8, p.168]. On this basis, you can determine the rate of substitution e.g. of a private good with a public good, and that informs you how much private goods or what number of units of a private good the consumer is willing to give up to increase the consumption of the public good by a unit [8, p.168-171]. In the same way research has been carried out that allows to calculate the so called "willingness to pay" (WTP) for the delivery/production of a unit of a specific good. An example would be to estimate the so-called willingness to pay for ensuring the continuity and reliability of electricity supply [12] or energy produced from renewable sources [13].

In addition to the lack of competition for the public good and the inability to exclude anyone from its consumption another feature of public goods are their positive external impacts, often constituting the main value of such goods [9, p. 132]. Ensuring energy security, as mentioned before, through the already mentioned package of services meets this criterion.

It follows that energy security on the one hand seemingly "escapes" the concept of the public good or the common good, on the other hand - especially on the international and national level, and not to a small extent also on the regional level - it does not bend to the rules of the market, and even to the principles of responsibility for security through companies that are public and are under state control. And the state control is implemented through legal and institutional regulations. We can at this point refer to the question that was repeatedly cited by Musgrave and referred to the availability of public goods - he believes that it is a political process "by which such goods are made available, rather than a process of public production " [10, p.37].

4 The Role of regulations

In the literature, it is emphasized that since 1990 regulation has been seen as a factor which plays a key role in the implementation of market reforms in the energy sector by improving the efficiency of investment in both the generation sub-sector and power grids, as a factor that ensures the effective operation and the possibility of competition via the TPA rule. The results of this regulation seem to be positive, especially in Great Britain [3].

From the point of view of the objectives of economic efficiency, proper regulation in the energy sector should play a particularly important role in providing incentives for energy supply security. However, in some countries or regions, it is the regulations that is pointed out as a reason for blackouts. The classical example is the energy crisis in California in 2000. The report of *The California Public Utilities Commission* (SPUC) pointed out that the cause of the crisis in California were too high prices and external errors of state authorities

‡ A similar approach is presented by Samuelson and Nordhaus for whom public goods are goods who benefit the whole community regardless of whether individual persons want to acquire them or not" [7, p. 549].

§ Throughout centuries two models of the decision-taking process on the delivery of public and private goods have developed, i.e.: (1) reaching a social consensus as a result of a vote (the concept by Wicksell of 1896 r.), (2) appointing a judge who on the basis of his preferences will determine the optimum allocation of resources between private and public goods (Samuelson, 1954) [10 p. 38-41].

inhibiting the development of new production capacities, which were the result of regulations introduced in 1996 and 1998. [14]. In this context, the aim of regulation is to ensure stimulators (incentives) for sufficient capacity and efficiency of both the grids and production capacities consuming different types of primary energy. The purpose of regulation is - from an economic point of view - apart from the security of energy supplies, a stability of prices charged to consumers, both in the high and low voltage range. At a higher level of generalization the aim of regulation will be energy security in the context of climate change, international exchange of energy carriers, as well as the proliferation of nuclear materials.

In the light of the foregoing it has to be stated that regulation, through the use of regulatory instruments that are effective and ensure efficiency, can and should play an important role in ensuring energy security in the short, medium and long term. The purpose of this text cannot be to offer instruments that can be used. The subject of this paper can merely be to identify the challenges and trends and to draw attention to the need for changes to the existing regulations in Poland.

From the perspective of regulation attention should primarily be paid to the classification of the factors affecting the level of energy security. However, it seems that so far, some of them are, in fact, have only been the means to achieve the goal, and the actual causative factors have not been defined.

The result of the evaluation of the level of energy security depends on three questions:

1. From which perspective is the evaluation done – that of a state, region, commune, industry or an individual consumer?
2. To what form of energy demand does it apply (what demand of the consumers does it relate to)?
3. What is the time horizon of the evaluation?

The concept of energy security used for the purpose of energy policy of the state must take into account - if it is to perform the desired function in the right way - both the global aspect as well as the local one, without neglecting any of the intermediate links, and extend over the longest possible time horizon. The USA may be a good example of the latter as it has developed its energy policy for the time horizon of 100 years. The EU is going in this direction, unfortunately lagging far behind the USA since its forecast horizon in this field is about 25 years [2].

As it was emphasized above the measure of energy security is the reliability of supplies at an economically justified level of prices and an impact on the natural environment limited to an economically and socially acceptable level. It is easy to observe that it is a measure constructed from the point of view of the both a local consumer, connected to the power system, in particular somewhere on the periphery, and the '*collective consumer*', which are individual fields of social life (such as communications, banking, health care, army, police, administration, etc.), and the national economy perceived as a whole. Not without significance in this case is the fact that the discussions on the international forum dedicated to the issues of energy security in the presented meaning are conducted under the banner of the '*security of supply*', regardless of whether they are focused on ensuring Europe's long-term access to oil and gas deposits, or on threats to the stable operation of transmission systems in the conditions of natural disasters, and the prevention of their effects.

The problem of energy security in Poland is becoming more urgent. Meanwhile, the entities in the energy sector, especially in the electricity sector can not find methods and tools for the efficient and effective investment. Decisions on new investment projects are postponed, and the capacities derived from units to be installed in the foreseeable future will not satisfy the growing demand - their example may be the unit with a capacity of 858 MW in Bełchatów and the long-debated two similar units in Opole. Due to the high regulatory and price risk, in the period 2010-2014 energy companies abandoned the construction of 10 new generation units. The condition of the present generating units is a reason for concern because

50% of the NPS capacity are units built in the 70s and before. In 2050 out of the current production capacity only 5 GW will still be in operation [15, p.8]. Energy companies have promised to build 11 new units until 2019, 10 of which will be units under the control of the State [15, p.11]. It means that other companies do not have sufficient incentive mechanisms to take up such activities. Other investment projects are at the stage of prolonged discussions, an example of which can be the construction of a nuclear power plant. Paradoxically, the new capacities are the much contested wind farms (4886 MW in 2015) [16, p. 46]. In Poland the RES capacity equilled 6 930 MW and most of it was in the form of wind turbines. Next part of this capacity was contained in hydro generators (964 MW in 2015) and biomass (740 MW in 2015).In 2015 we could observe the increase in solar energy capacity because the capacity of solar plants totalled 27 MW in 2014, but it equalled 108 MW in 2015 [16, p. 46]. In 2014 in Poland the share of the RES was 11.9% of the primary energy, while in Germany this value totalled 30%. For years from 2011 to 2014 in Poland the share of the RES increased by 1% (from 10.9% to 11.9%), while in Germany by 6 % (from 24% to 30%). [16, p. 24]. In 2015 363390 TJ came from the RES 363390 TJ which was 12.7% of total primary energy in Poland. Most of them came from biomass (72,22%) and from wind and bio-liquid fuels (respectively 10.76% and 10,78%) [16, p. 32]. We observe in Poland the increase in the RES, as measured by the capacity and generation. However, renewable energy generation (excluding wind turbines) has very little importance in Poland, and its development is very conservative, without clearly formulated goals, unlike in the Germany where power generation is developing in a multi-faceted way with the priority given to renewable energy. Nowadays, Germany realize policy called 'Energiewende', which is a complex program related to, among others the development of RES and the improvement in the efficiency of generation and consumption of all kinds of energy

This situation is confirmed by a report of the Supreme Control Chamber NIK [15], which found indeed that the current security, as guaranteed by companies of the sector, is provided in due degree, but efforts to ensure energy security in the long run are insufficient. Furthermore, alarming forecasts indicating a continuation of the upward trend in the maximum demand for power up to 40 GW in 2035 and 41-42 GW in 2050, and the demand for electricity from approx. 159 TW·h in 2015 to 230 TW·h in 2030 are emerging. [15, p. 8].

Energy grids, both transmission and distribution ones (for electricity and gas) are a natural monopoly and are subject to regulation which aims to guarantee the public interest (good). In this case, this interest (public good) manifests itself in the efficient and effective (in this article, these terms have praxeological meaning) work management of energy grids. It can be evaluated if, in practice, the idea of the public interest is realized or if, following J. Stiglitz, it should be assumed that in this case we have to do with the phenomenon of regulatory capture which means that a regulation is "captured" (perhaps in a big portion) by the regulated entities [17].

In this short article it is impossible to exhaust all topics of energy security and of the nature of regulation. We can only put forward certain hypotheses, cite some literature and practical examples to support them.

We allow ourselves to formulate the following theses:

1. The national energy generation, especially the energy grid is in the state of a substantial and increasing imbalance in terms of the security of energy supply. This applies to both power transmission and distribution systems as well as the generation sub-sector.
2. Regulations in the field of energy security are dispersed among international actors (within the EU) and numerous national ones. There is an apparent lack of co-ordination of regulatory tools used by individual entities. As a result, the existing regulations do not meet the requirements of efficiency and effectiveness, and they do not guarantee a reliable supply of energy.

In recent years in Poland there have been serious disruptions in the supply of electricity (Szczecin 2008, the area of Podkarpacie 2010, August 2015). Despite many investment projects into the grid, the state of the transmission and distribution networks is worsening. In addition, some urban areas are supplied basically via one transmission line, carrying enough power now but whose failure could cause a serious disruption in energy supply. According to the classical principle of risk protection (in this case it is the risk of a *blackout*), one should aim to diversify the sources of supply. This applies mainly to large urban agglomerations. Such a situation had for a relatively long period of time concerned Wroclaw, only since 2010 intensive effort has been made to diversify and expand connections. Currently, the security of supply of electricity for the city of Wrocław is rated to be at a good level [18]. An increase in the degree of diversification of power sources can also be achieved by expanding local dispersed energy sources, including renewable energy sources, which can be located directly at the destination. This idea is associated both with the concept of Smart Grids and the emergence - in Poland a still new - entity on the energy market - the so-called prosumer, as well as with the concept of ecological safety. Smaller towns, and mainly rural areas can choose from a catalog of more renewable energy technologies than large urban agglomerations.[**]
Table 1 shows the basic conditions for ensuring energy security on the national scale as well as - to a significant degree - in the region.

Table 1. Energy security conditions on the state and regional level

Basic conditions of energy security	Additional conditions
Sustained energy balance	Preserving economic security
Availability and sufficiency of infrastructure for processing, storage and distribution of fuels and energy	Preserving ecological and process safety
Diversification of energy carrier sources	Resistance to random and intentional threats

Source: [19]

The experience shows that periodically there may be serious discrepancies between the amount of offered energy (electricity, gas, heat from centralized sources) and the demand of the consumers that lead to losses on the part of the consumers (economic, financial, political and social) whose amount often significantly exceeds the expenditure for the production and transmission of energy. The above mentioned report by the NIK states that "the value of social losses caused by insufficient energy supplies in the years 2020-2030 can reach (....) 70 billion PLN, and the overall expenditure on the implementation of the above system solutions, e.g. power markets - is expected to be no more than 3-5 mln PLN" [15, p. 10]. It can be thus stated that from the economic point of view, if the necessary amount of energy is not delivered, it leads to losses whose height may significantly exceed the cost of production and delivery of the energy to consumers and be significantly higher than the expenditure necessary to implement new services ensuring the continuity of supply. Energy security therefore, also includes situations in which there are no physical shortages of energy but there are problems with its delivery, which in turn causes opportunity cost, the amount of which exceeds the costs incurred to produce and be prepared to make electricity available [17].

In recent years the importance of the guarantee of energy security has increased due to the occurrence of significant disruptions in the mechanism of energy supply. In real-life economy the activities of the entities of the energy sector in the scope of energy security supply are often not effective enough. Therefore there is an urgent need for harmonization

[**] See the results of the project ENERGYREGION (www.energy-region.eu).

with other entities, and above all with local government bodies, non-governmental organizations, as well as large, medium and small customers to guarantee a safe and effective matching of the amount of the produced energy with the demand for it [17].

5 Summary

In the context of the presented considerations there emerges an important question: how can the operation of the energy system and its regulation reduce the risk of energy deficit on a mass scale assuming that energy security is a public good? While trying to find an answer to this question we should formulate the basic objectives of regulation. The first is the construction of new power capacities, the second will be effectively functioning transmission and distribution networks; and the third - the relation between the operation of transmission and distribution networks and the "behavior of consumers."

In terms of the first objective, i.e. the construction of new generation capacities we have to distinguish among others the following factors: (1) identification of investment locations and obtaining permits for construction, (2) determination of the primary fuel and the effects of its use for the environment and economy (mainly the cost structure, keeping in mind that some renewable energy sources have variable costs close to zero, and therefore no marginal costs), (3) determination of the technical and economic parameters of the operation of the new power plant (4) transmission of power over the transmission system.

The second objective, i.e. an effective transmission and distribution grid should include: (1) investment projects necessary and sufficient to ensure the security, (2) building an effective structure of the sector and unbundling of the generation, transmission, distribution and marketing, (3) diversification of energy sources to help ensure competition, (4) legal, economic and financial responsibility for a breach of security both on the demand and supply sides, (5) international, national and regional coordination of regulatory tools.

The third objective, i.e. the relationship between the operation of the network and the "behavior of consumers" should include: (1) activation of processes of energy saving on the so-called demand side, (2) development of systemic mechanisms allowing for the participation of energy consumers, so that a flattening and/or a shift of the load curve occurs, (3) development of intelligent networks (Smart Grids) and flexible supply of energy taking into account micro-generation and dispersed generation, (4) active impact on the demand for energy by the public sector, (5) adaptation of legal regulations to incorporate the need for saving energy, (6) implementation of new - effective forms of management and ownership structure of the transmission and distribution grids.

It should be emphasized that energy security requires systemic instruments allowing for risk sharing among all energy market participants. The energy policy and legal create more or less comfortable conditions for modernization and making new investments.

References

1. *Roczne raporty PSE o Krajowym Systemie Elektroenergetycznym*, available www.pse.pl (18.08.2016)
2. *Doktryna zarządzania bezpieczeństwem energetycznym,* Ministerstwo Gospodarki i Pracy, Warszawa (maj 2004)
3. T. Jamasb, M. Pollitt, *Security of supply and regulation of energy networks,* Energy Policy **36** (2008)
4. *Polityka energetyczna Polski do 2030 roku. Załącznik do uchwały nr 202/2009* Rady Ministrów z 10 listopada 2009 r.
5. *Wielka Encyklopedia*, Wydawnictwo Naukowe PWN, Warszawa (2002)

6. Z. Szalbierz, E. Ropuszyńska-Surma, *Bezpieczeństwo energetyczne w układzie regionalnym*, Systems: Journal of Transdisciplinary Systems Science (2008)
7. W.D. Nordhaus, P.A. Samuelson, *Ekonomia t.1*, Wydawnictwo Naukowe PWN, Warszawa (2004)
8. J.E. Stiglitz, *Ekonomia sektora publicznego*, Wydawnictwo Naukowe PWN Warszawa (2013)
9. N. Acocella, *Zasady polityki gospodarczej*, Wydawnictwo Naukowe PWN, Warszawa (2002)
10. J.M. Buchanan, R.A. Musgrave, *Finanse publiczne a wybór publiczny. Dwie odmienne wizje państwa*, Wydawnictwo Sejmowe, Warszawa (2005)
11. Marszałek, M. Swoboda działalności gospodarczej wytwórcy – sprzedawcy energii elektrycznej, Wyd. C.H. Beck, Warszawa (2015)
12. D. Layten, K. Moeltner, *The Cost of Power Outages to Heterogenous House-holds*, Applications of Simulation Methods, Kluver Acd. Press (2005).
13. R. Scarpa, K. Willis, *Willingness-to-pay for renewable energy: Primary and discretionary choice of British households' for micro-generation technologies*, Energy Economics **32** (2010)
14. L. Steadman, *Energy Bills: Temporary Measures Versus Long-Term Relief*, Pacific Research Institute for Public Policy **57** (22.08.2000)
15. Najwyższa Izba Kontroli, *Zapewnienie mocy wytwórczych w elektroenergetyce konwencjonalnej. Informacja o wynikach kontroli.* KGP-4101-001-00/2014, Nr ewid. 17/2015/P/14/018/KGP (4.03.2015)
16. Główny Urząd Statystyczny, *Energia ze źródeł odnawialnych w 2015 r.*, Warszawa (2016).
17. A. Bieńkowska, E. Ropuszyńska-Surma, M. Bogrosz-Koczwara, Z. Szalbierz, *Kształtowanie sieci współpracy na rzecz bezpieczeństwa energetycznego Dolnego Śląska ze szczególnym uwzględnieniem aspektów ekonomiczno-społeczych*, Politechnika Wrocławska seria SPR **3**, Wrocław (2009)
18. *Aktualizacja założeń do planu zaopatrzenia w ciepło, energię elektryczną i paliwa gazowe dla obszaru Gminy Wrocław (wersja uzupełniona i poprawiona)* (27.12.2012)
19. G. Bartodziej, M. Tomaszewski, *Polityka energetyczna i bezpieczeństwo energetyczne,* Wyd. Energetyka i Środowisko, Warszawa (2008)

The pro-economical behaviour of households and their knowledge about changes in the energy market

Edyta Ropuszyńska-Surma[1,*], *Magdalena* Węglarz[2]

[1]Wrocław University of Science and Technology, Wybrzeże Wyspiańskiego 27, Wrocław, Poland
[2]Wrocław University of Science and Technology, Wybrzeże Wyspiańskiego 27, Wrocław, Poland

Abstract. The behaviour of energy and heat consumers is an important and interesting topic. For a long time the energy market was the classical example of a monopoly. In general, households are weak partners for energy companies, but energy markets are changing all over the world. This trend is noticeable in Poland too, but it is not as strong as in Germany or other countries where renewable energy sources (RES) are very popular and widely used. So, here we investigate why the development of RES in Poland is so weak. The main aim of the paper is to highlight the key factors, especially ecological and economic ones, determining households' decisions about installing RES in Poland. An additional aim is to present the level of energy awareness of households. We conducted a survey of Lower Silesia's households in November and December 2015. Economic factors are the best motivators for households to install RES. Among the economic aspects, the respondents indicated potential savings and potential profits.

1 Introduction – Polish context

Nowadays, the energy sector is developing on two levels in parallel. The first of them is the energy system at national level. It consists of, inter alia, a large number of conventional energy plants with big blocks. In Poland, most electricity (more than 80%) and heat are produced from coal and lignite. The second level is composed of dispersed energy systems, which are located near or at energy consumers' homes. Of course, this is typical for heating systems. In Poland, most heat comes from conventional fuels. Dispersed RES (renewable energy sources) intended for installation in households could be developed in Poland, but this process began later than in Germany and is progressing too slowly.

Statistical data are available about the capacity installed in Poland and the production of electricity and heat coming from bigger plants. However, the parameters of individual RES located in dispersed households are not identified. The most comprehensive source of information about RES is a periodical publication entitled "Energy from renewable sources" prepared by the Central Statistical Office (called GUS in Polish). Data are available for 2003-2015.The structure of RES in Poland is different from the EU-28. For example, in 2014 [1, pp. 26]:

* Corresponding author: edyta.ropuszynska-surma@pwr.edu.pl

- the proportion of energy obtained from bio-fuel totalled 43.8% in the EU and 76.6% in Poland,
- the proportion of energy obtained from liquid bio-fuel equalled 7.1% in the EU and 9.2% in Poland,
- wind energy accounted for 11.1% of energy production in the EU and 8.02% in Poland,
- solar energy is 6.1% of total energy production in the EU and 0.4% in Poland.

From an ecological point of view, it is important to reduce CO_2 emissions by reducing the consumption of energy (by the end-user). In Poland, the high level of such pollution is the result of using outdated heating systems based on conventional fuels. This is an important issue in villages and some cities (e.g. Cracow), where most citizens burn conventional fuel in individual heating systems, which causes smog. UE guidelines and legislation (e.g.: [2], [3]) impose an obligation to reduce pollution. In 2005, the share of RES in energy consumption in Poland totalled 7.2% and it should increase to 15% by 2020. In 2009, Polish households accounted for 20% of total energy consumption [4, p. 122], as did German households. However, only 7% of the energy consumed by Polish households was from renewable energy sources, while in Germany this figure was 17% [4, p. 124]. In 2011, these values were lower. Polish households used 19% of total energy consumption and renewable energy accounted for 7% of this, while German households used 17% of total energy consumption and renewable energy accounted for 18% of this, [5, p. 142] and [4, p. 144]. Authors often account for this by stating that Germany is a leading proponent of RES and its energy policy is effective and efficient. We should remember that Germany is the largest energy market in Europe. Since 2010 the share of RES in primary energy has dramatically increased in Germany, and recently the introduction the '*Energiewende*' programme has even strengthened this process, e.g. "from 1990 to 2012 the gross national product of Germany increased by 37%, while primary energy consumption overall decreased 7.7%" [6, p.230]. Recent publications have taken into consideration macroeconomic aspects of the '*Energiewende*' and its potential results related to Public Choice Theory and Interest Group Theory [6. 7].

Table 1. Consumption of energy from RES by households in Poland [TJ].

Kinds of energy	Category (class)	2008	2009	2010	2011	2011	2013	2014
Solid bio-fuel	Total final consumption	160630	162085	179138	187187	186927	198427	185807
	Including households	102500	102500	112746	115000	116850	116850	105450
Solar energy	Total final consumption	54	283	350	434	544	639	720
	Including households	-	200	250	300	380	460	520
Geothermal energy	Total final consumption	531	600	563	531	661	778	847
	Including households	402	460	440	430	510	561	608

Source: [1, pp. 52-53, 63], [8, pp. 31-32, 42]

Table 1 presents statistical data about the use of RES by households in Poland. There are no statistics showing how much electricity used by Polish households comes from wind and water, because these primary energy sources are used mainly to produce electricity. The energy balance sheet presents only primary energy sources with a division into kinds of users, e.g. households, transport, industry. We should remember that electricity is a derivative of energy production.

The development of RES at the consumer-end seems to be a key aim, so national and regional authorities should be interested in the factors influencing "ecologically friendly" behaviour in households. These observations are the background to asking about the main reasons for this situation. Therefore, our considerations focus on variables which describe households and could change their behaviour in a more ecological direction. So, the main aim of the paper is to highlight the key factors determining the decisions of Polish households about installing RES and ecologically friendly behaviour. From this point of view, we would like to answer the following important questions:

- do Polish households act pro-ecologically?
- do they possess basic knowledge about changes in the energy market which create favourable conditions for installing RES?
- are Polish households energy-aware consumers, especially in terms of electricity?
- is there any correlation between a household possessing RES and economic factors (e.g.: income, floor area) or demographic and social factors (e.g.: age, sex, number of persons in household).

Our research is conducted according to a microeconomics approach and we try to identify not only typical demographic and economic factors influencing households' behaviour, but social factors and, related psychological factors. It is not a popular and widely used approach to explore the behaviour of individual energy consumers, i.e. households, but recently some researchers have studied e.g. households' behaviour, their willingness-to-pay for RES micro-installations [9], green energy [10] or the factors determining prosumer behaviour [11] and they have taken into account the psychological features of individual energy consumers.

R. Scarpa and K. Willis dealt with similar aspects using a *choice experiment* related to *consumer theory*. Based on a sample of households in the UK, they identified the barriers to installing micro-generation technologies, such as solar photovoltaic cells, micro-wind turbines, solar thermal cells, heat pumps, biomass boilers and pellet stoves. The vast majority of British households had not installed these technologies, due to high capital costs. We should remember that the UK is richer than Poland [9], so we can imagine that Polish households are not interested in using these technologies

Admittedly, the Polish Statistical Office has been conducting research among households since 2006, every 2 years, including information on their energy and heat installations etc. However, these data do not enable us to make our analysis. However, based on it we can estimate the number of households who have RES installations (Table 2).

Table 2. The share of Polish households possessing solar collectors and heat pumps in 2009 and 2012.

The kind of RES installation	2009	2012
solar collectors	0.31%	0.41%
heat pumps	0.03%	0.05%

Source: [4, p. 116], [5, p. 120]

We can compare the results of our research to the outcomes of the Consumer Federation's survey conducted in 2015. This institution divided respondents into four segments: (1) prosumers, (2) energy-aware consumers, (3) consumers interested in RES and (4) consumers who do not know what renewable energy sources are. The results of this survey say that the most important barrier to installing RES is high cost (70%). Other obstacles are: the long period required for investment to become economically beneficial, lack of financial support (grants) and problems with finding appropriate advice and information [8].

This article is prepared on the basis of the results from a project called 'Modelling prosumers' behaviour on the energy market', which is funded by the NCN (National Centre of Science) by grant no. 2013/11/B/HS4/01070.

2 The research results – installing RES

We conducted a survey of households in Lower Silesia in November and December 2015. The sample was made up of 2000 households, but the number of correctly completed questionnaires was 949. Stratified random sampling was applied and the research was conducted by telephone. The stratification of the sample was based on the set of households in Lower Silesia according to the following strata: age, sex and place of residence according to poviats (local government units).

Based on respondents' answers, we identified the three following groups of respondents:

- prosumers – 45 households,
- potential prosumers – 281 households,
- unwilling to have RES (called 'not desiring RES') – 623 households.

In this paper, a prosumer is defined to be a household who has installed RES[†]. The second group of respondents did not have RES, but declared that they would like to install it in the future. The remainder of the respondents did not have RES and would not like to install it in the future. They are called 'unwilling to have RES'. Potential prosumers and those unwilling to have RES are together referred to as 'not possessing RES'.

We see that only 4.76% of the respondents have installed RES, but 31% of the households 'not having RES' would like to install it. Two interesting questions are why most households have not installed any RES and what causes the negative state of (lack of interest in) RES development in Polish households. We aimed to answer these issues through our research by asking the respondents about the barriers to installing RES (considered in this section) and we make conclusions based on statistical analysis (presented in section 2.2).

Only 41 of 45 prosumers answered a closed question linked to the reasons for installing RES. 68.3% of prosumers (28 households) stated that the reason for installing RES was long-term savings. Only a small number of households stated other reasons for installing RES: 'I am an ecologist' (17.1%), 'a professional persuaded me' (9.8%), 'neighbours or friends persuaded me' (4.9%).

Among economic aspects, the respondents indicated potential savings and potential profits from having RES[‡]. The most important advantage was 'smaller fees, savings'. This option was chosen by 39.6% of those in the groups of prosumers and potential prosumers[§]. 8.8% of these respondents answered 'own needs, free energy'. This answer is similar to 'smaller fees, savings' and so we can group them together. 'Financial profit' was another important issue, as mentioned by 18.7% of the respondents. Other answers connected with

[†] Here, our definition of prosumer is the same as in research by the Federation of Consumers [12].
[‡] This was an open question addressed to prosumers and potential prosumers. We asked about the advantages of being a prosumer.
[§] Not all prosumers and potential prosumers answered this question. We obtained 91 answers.

financial profit are: 'it is possible to sell surplus energy' - 3.3% of prosumers and potential prosumers chose this option; 13.2% chose the 'ecology and clean air' option; 8.8% selected 'financial and environmental benefits'; and 4.4% chose 'energy independence as a benefit from being a prosumer'. Therefore, saving and financial profit are the main advantages from having RES. Ecological aspects and independence are not such important benefits.

On the other hand, the respondents 'not possessing RES' specified the most important factors that influenced their reluctance to install RES. The most significant barriers to installing RES are: (1) high installation costs (27.0%)**, (2) the technical impossibility of installation (21.7%), (3) lack of interest in installing RES (15.98%), (4) unclear regulations (10.04%), (5) complicated process of connecting to the grid (9.3%), (6) lack of knowledge (9.2%).

Among the less important factors, the respondents also indicated the following opinions: unprofitable, no need to install RES, neighbours or friends dissuaded me, lack of time, lack of financial assets, long period required for the investment to become financially beneficial.

2.1 Economic and social behaviour of households

According to our research, the basic ecological knowledge of households and the associated behaviour are relatively advanced, but some ecologically friendly behaviour depends on the level of income. The psychological literature shows that human behaviour depends on, inter alia, emotional feelings, economic situation, education, knowledge and experience, see e.g. the model of pro-environmental behaviour according to Kollmus and Agyeman [13]. The marketing literature talks about ecological (green) consumers, e.g. in 1992 Meffert and Kirchgeorg made a typology of consumers according to their relation to the environment [14]. A similar approach was presented by Simmon, who divided American consumers into five groups [15, p.148].

Generally, the results obtained show that the average household does not know its tariff group or the meaning of word 'prosumer'. They do not even know what kind of electricity meter they have. We assume that these facts reflect the awareness of our respondents††. Only 12.3% of households know their electricity tariff group, but some of them (in the following question) gave the incorrect tariff group or made other mistakes, e.g. they did not give the electricity tariff group, but a gas tariff group, or they wrote the name of an electricity provider. 89.4% of respondents do not know what a prosumer is. Moreover, some respondents who declared that they knew what this word means, do not understand it completely. This understanding was checked by other questions. 27.8% of households have a smart meter.

In our research, we took three groups of respondents into consideration: (1) prosumers, (2) potential prosumers and (3) those unwilling to install RES. In all our statistical tests, we assumed a significance level of 5%, i.e. the p-value should be less than 0.05 to accept any hypothesis about the dependence between a given variable and a household's decision about installing RES. We divided potential factors that could influence a household's decision about installing RES into two groups: (1) demographic and social factors and (2) economic factors.

** The number in brackets indicates the percentage of households without RES who chose the given option. 807 households answered this question. This question was multiple choice.

†† The public awareness of solar water heaters has been investigated e.g. by Chinese researchers [16] and British researchers [17].

2.2 Identification of significant factors

2.2.1 The influence of demographic and social factors

In the first group of factors, we analysed the influence of such factors as: age, sex, number of people in the family. The most active group of prosumers are respondents in the age range of 30-49, and the second most active are respondents in the age range of 50-65. The group most reluctant to install renewable energy systems are respondents of age 65 or more. The result of the chi-square test of independence (Chi-square(6) = 19.799 with p-value = 0.003) clearly indicates a significant relationship between age and willingness to install RES. Additionally, we carried out Spearman's test of correlation, which also indicates that younger individuals are more willing to install RES (rho = 0.08, n=949 p-value = 0.014). Note that the lower the group number, the greater the willingness to install RES (those who have installed RES are in group 1, those who do not want to install RES are in group 3). Hence, a positive correlation between age and group number indicates a negative correlation between age and willingness to install RES.

We observed a significant association between sex and the three groups of respondents. Women more often do not have or want to install RES in the future (57.6%). Among prosumers, most are male (62.2%), as well as among potential prosumers (58.4%). The result of the corresponding test of independence (Chi-square(2) = 23.63 with p-value < 0.001) indicates a very significant relationship between sex and willingness to install RES. Additionally, we carried out Spearman's test of correlation which also indicated that males are more willing to install RES (rho = -0.158, n=949 p-value < 0.001).

Table 3. Influence of the number of people in a family on a household's decision about installing RES.

Number of people in a family	Prosumer	Potential prosumer	Unwilling to have RES	Total
1	0.8%	18%	81.2%	100%
2	2.4%	21.3%	76.3%	100%
3	6.7%	26.2%	67.1%	100%
4	7.3%	38.9%	53.8%	100%
5 or 6	6.6%	45.5%	47.9%	100%
7 and more	7.7%	57.7%	34.6%	100%

We now analyse the association between the number of people in a family and the three groups of respondents. The results are shown in Table 3. We can observe that households which consist of one or two people more often do not want to install RES (81% and 76%, respectively). There is a positive correlation between the number of individuals in a family and the willingness to install RES. One- and two-person households very rarely have installed RES (0.8% and 2.4%, respectively). Around 7% of larger households had installed RES (independently of the number of people in the household). On the other hand, the proportion of households wishing to install RES increases steadily as household size increases. The result of the chi-square test of independence (Chi-square(10) = 71.884 with p-value < 0.001) indicates a clear association between the number of individuals in a household and the willingness to install RES. Additionally, we carried out Spearman's test of correlation, which clearly indicates a positive association between the number of individuals in a household and willingness to install RES (rho = -0.263, n=949 p-value < 0.001).

2.2.2 The influence of economic factors

We analysed the relationship between installing RES and a number of factors, such as: income, expenditure on electricity, and floor or building area. For the purposes of presenting the results, we classified each economic factor into several categories. Table 4 presents the structure of all these categories in relation to the type of household (prosumer, potential prosumer or 'not desiring RES').

Table 4 The association between economic factors and willingness to install RES [%].

Economic factor	Category	Prosumer	Potential prosumer	Not desiring RES
income[‡‡]	< 3,000 PLN	2.0%	26.7%	71.3%
	3,001 – 4,000 PLN	4.9%	37.4%	57.7%
	4,001 – 5,000 PLN	3.6%	40.0%	56.4%
	5,001 – 6,000 PLN	11.1%	29.6%	59.3%
	6,001 – 8,000 PLN	35.3%	11.8%	52.9%
	8,001 – 10,000 PLN	25.0%	12.5%	62.5%
	> 10,000 PLN	16.7%	25.0%	58.3%
expenditure on electricity	< 50 PLN	2.8%	9.7%	87.5%
	51-100 PLN	1.9%	18.8%	79.3%
	101-200 PLN	5.5%	35.0%	59.6%
	201-300 PLN	8.5%	44.1%	47.5%
	>300 PLN	8.9%	44.3%	46.8%
floor or building area	< 40 m^2	0.0%	0.0%	100.0%
	40-79 m^2	0.0%	13.5%	86.5%
	80-149 m^2	6.7%	40.2%	53.0%
	>150 m^2	13.9%	57.6%	28.5%

Table 5. Results of statistical tests for an association between chosen economic factors and willingness to install RES.

Economic factor	Chi-square statistic	df	p-value	rho
income	68.414	12	0.000	-0.159
expenditure on electricity	78.690	8	0.000	-0.284
floor or building area	240.773	6	0.000	-0.499

Table 5 presents the results of statistical analysis related to each of the above mentioned variables. We tested the three following hypotheses:

- H1: household income is associated with the decision to install RES,
- H2: household expenditure on electricity is associated with the decision to install RES,
- H3: floor or building area is associated with the decision to install RES.

[‡‡] This question asked respondents to give their household's monthly net income. About 30% of the respondents did not answer this question. The number of households answering this question was 643 out of 949.

In each category of income, the majority of households belong to the 'not desiring RES' group. When household income is bigger than 5,000 PLN, then a relatively large proportion of respondents are prosumers and a smaller proportion of households are potential prosumers. We can hence see that as household income increases, potential prosumers become real prosumers. Statistical analysis (see Table 5) confirms that hypothesis H1 can be accepted (Chi-square(12) = 68.414 with p-value < 0.001).

We can also observe a tendency that households which have higher expenditure on electricity more often install RES or state that they would like to install it.

Looking at the association between willingness to install RES and 'floor or building area', we see that all the households possessing less than 40 m^2 floor area belong to the group 'not desiring RES'. This result should not seem surprising, because small apartments are located in cities, where the conditions for installing RES are unfavourable. Such households need a relatively small amount of energy mainly for heating and electricity. All the households which have installed RES possess a floor area of more than 80 m^2. When the floor or building area totals 80-149 m^2, 53% of households belong to the 'not desiring RES' group. We observed that when the floor or building area becomes larger, then the frequency of both prosumers and potential prosumers increases, e.g. for floor areas of 40-79 m^2, 13.5% of respondents are potential prosumers and for floor areas of 80-149 m^2, 40.2% of respondents are potential prosumers.

All of the selected factors are significantly associated with the decision to install RES, since the p-value is less than 0.05 in each case. Hence, we can reject the null hypotheses of independence and accept each of the hypotheses H1, H2 and H3. According to Spearman's correlation coefficient (Table 5 - rho), we see that 'the floor or building area' is most strongly associated with the willingness to install RES. The larger a property, the more likely it is that the household has installed RES or wishes to install RES. Expenditure on electricity and household income also have an important influence on the decision to install RES.

3 The research results – ecologically friendly behaviour of households

In this section we present factors related to ecologically friendly behaviour that are associated with households' decisions to install RES. We specified some examples of ecologically friendly behaviour that we called *factors*. Moreover, we noticed that the possession of certain home devices could be related to installing RES. Besides the factors presented in Table 6, we proposed the following factors: unplugging devices that are on standby, using eco-options on a washing machine, such as low water consumption or low energy options, measuring the amount of water to be boiled.

The respondents indicated the frequency of such behaviour on a six-level scale or on a four-level scale. We analysed the relation between each factor specified above and a household's decision to install RES. In this analysis we grouped respondents who have installed RES and those declaring that they wish to install RES in the future and compared them with those not desiring RES. In all our statistical tests, we assumed a significance level of 5%, i.e. the p-value should be less than 0.05 to accept any hypothesis about the dependence between a given variable and a household's decision about installing RES.

In the case of the factor described as 'unplugging devices that are on standby', the result of the test of association was (Chi-square(3) = 0.516 with p-value = 0.915). Hence, there is no evidence of an association between this factor and the willingness to install RES. Below we present the results of our research regarding key behaviour factors in the groups of respondents who have installed RES or wish to install RES.

In our research we specified eight key behavioural factors, each having a slightly different association with a household's decision to install RES (described below). The differences between the groups of 'prosumers' and 'potential prosumers' according to these factors were negligible, therefore, for the purposes of our analysis, we combined these two groups into one called 'pro-RES', and compared this combined group to the group called 'not desiring RES'.

In the case of the factor 'using eco-options in a washing machine', the answers were on a six-level scale, which had the following categories: never, very rarely, rarely, often, very often and always. The result obtained (Chi-square(5) = 12.113 with p-value = 0.033) allows us to reject the null hypothesis that willingness to install RES is independent of this factor. Spearman's test of correlation was also carried out to assess whether this relationship was monotonic (e.g. as the frequency of using eco-options increased, the willingness to install RES always tended to increase). There was no significant monotonic association between the frequency of using eco-options and the willingness to install RES (rho = -0.062, p-value = 0.101).

In the case of the other factors, the answers were on a four-level scale that had the following categories: never, rarely, often, and always. The results of the statistical analysis linked to these behavioural factors are shown in Table 6. We tested the seven following hypotheses:

- H4: using energy-saving lights (LED) is associated with the decision to install RES,
- H5: the utilization of electro-rubbish is associated with the decision to install RES,
- H6: installing energy-saving household equipment is associated with the decision to install RES,
- H7: sorting rubbish (garbage) is associated with the decision to install RES,
- H8: battery utilization is associated with the decision to install RES,
- H9: switching off the lights is associated with the decision to install RES,
- H10: washing or ironing at particular times is associated with the decision to install RES.

Table 6. The results of tests for the association of behavioural factors with the willingness to install RES.

Factor	Chi-square(3)	p-value
the utilization of electro-rubbish	15.687	0.001
using energy-saving lights (LED)	27.47	<0.001
installing energy-saving household equipment	13.881	0.003
sorting rubbish (garbage)	10.345	0.016
battery utilization	9.962	0.019
switching off the lights	8.819	0.032
washing or ironing at particular times	7.828	0.05

Six of these factors are significantly associated with the decision to install RES (we can accept hypotheses H4-H9 at a significance level of 5%). It is only unclear in the case of H10 whether this association is significant (the p-value is equal to 0.05). Three factors are significant at the 1% level: 'using energy-saving lights (LED)' – H4, 'the utilization of electro-rubbish' – H5, 'installing energy-saving household equipment' – H6. These relations are described below.

In the case of the 'utilization of electro-rubbish' factor, among pro-RES respondents 10.7% had never utilized electro-rubbish, whereas among respondents 'not desiring RES', this value is greater (equal to 18.9%). Among pro-RES respondents, 71.5% always utilized electro-rubbish, whereas among respondents 'not desiring RES', this value is smaller (equal to 60.7%). In the case of the 'installing energy-saving lights (LED)' factor, only 21.5% of the respondents from the pro-RES group did not have LED lights at all, whereas among the respondents 'not desiring RES', this number is equal to 36.5%. Among those 'not desiring RES', 16.4% of households have only such lights, whereas among the 'pro-RES' group, this percentage is 22.4%. In the case of the 'installing energy-saving household equipment' factor, 31.0% of the 'pro-RES' respondents do not have such devices at all, whereas among respondents 'not desiring RES', this proportion is greater (equal to 41.1%). Among those 'not desiring RES', 13.4% of households have only such devices, whereas among the 'pro-RES' this proportion is 20.6%. Statistical analysis (see Table 6) confirms that the hypotheses H4-H6 can be accepted at the 1% level.

4 Conclusions

The majority of households are not willing to install RES. The key economic reason is the excessive cost of installation. The next most important causes are the technical impossibility of installing RES and a lack of interest from respondents. The size of a house/flat is the main factor associated with the decision to install RES. Other significant factors are the expenditure on electricity and household income. Prosumers and potential prosumers emphasized savings and financial profit as the main advantages from installing RES. The same factors motivate them to install RES. The following are less important motivators: ecological aspects, neighbours and friends and energy independence (which means energy safety). Among demographic factors, the ones most significantly associated with willingness to install RES are: the number of people in a household and sex. In Poland, economic benefits give the strongest motivation to install RES, but we observed that some ecological friendly behaviours are positively associated with willingness to install RES. Three of these factors: installing energy-saving lights (LED), utilization of electro-rubbish, installing energy-saving household equipment, were significantly associated with the willingness to install RES at the 1% significance level.

Given these conclusions, we recommend the following:

- more intensive distribution of information, which should improve the awareness of society regarding energy,
- using economic tools in energy policy,
- organizing marketing action appropriate to the size of a household and the sex of the decision-maker.

References

1. Central Statistical Office. *Energy from renewable sources in 2014*. Series: Statistical Information and Elaborations. Warsaw (2015)
2. Directive 2009/28/EC of the European Parliament and of the Council of 23 April 2009 *on the promotion of the use of Energy from renewable sources and amending and subsequently repealing Directives 2001/77/EC and 2003/30/EC*, Official Journal of the EU L 140 (5.6.2009).
3. Directive 2001/81/EC of the European Parliament and of the Council of 23 October 2001 *on national emission ceilings for certain atmospheric pollutants*, Official Journal of the EU L 309 (27.11.2001).

4. Central Statistical Office. *Energy consumption in households in 2009*. Series: Statistical Information and Elaborations. Warsaw (2012).
5. Central Statistical Office. *Energy consumption in households in 2012*. Series: Statistical Information and Elaborations. Warsaw (2014).
6. O. Renn, J.P. Marshall. *Coal, nuclear and renewable energy policies in Germany: From the 1950s to the "Energiewende"*. Energy Policy **99** (2016) pp.224-232.
7. S. Strunz, E. Gawel, P. Lehmann. *The political economy of renewable energy policies in Germany and the EU*. Energy Policy **42** (2016) pp.33-41.
8. Central Statistical Office. *Energy from renewable sources in 2013*. Series: Statistical Information and Elaborations. Warsaw (2014).
9. R. Scarpa. K. Willis. *Willingness-to-pay for renewable energy: Primary and discretionary choice of British households' for micro-generation technologies*. Energy Economics **32** (2010) pp. 129-136.
10. *I. Diaz –Rainey, J. K. Ashton. Profiling potential green electricity tariff adopters: green consumerism as an environmental policy tool?* Business Strategy and the Environment **20** (2011) pp. 456-470.
11. L. Lampropoulos, G. Vanalme, W. Kling. *A methodology for modeling the behaviour of electricity prosumers within the smart grid*. Innovative Smart Grid Technologies Conference Europe (ISGT Europe), 2010 IEEE PES (October 2010) pp. 1-8.
12. Federacja Konsumentów. *Jak zostać prosumentem. Raport Federacji Konsumentów* (2015).
13. A. Kollmus, J. Agyeman. *Mind the Gap: Why do people act environmentally and what are the barriers to pro-environmental behaviour*, Environmental Education Research, **8** (2002) pp. 239-260.
14. H. Meffert, M Kirchgeorg. *Marktorientiertes Umweltmanagement*. Poeschel Verlag. Stuttgart (1992).
15. Ph. Kotler. *Marketing. Analiza, planowanie, wdrażanie i kontrola*. Gebethner i S-ka, Warszawa (1994).
16. X. Yuan, J. Zuo, C. Ma. Social acceptance of solar Energy technologies in China – End users' perspective. Energy Policy **39** (2011) pp. 1031-1036.
17. S. Taherian, M. Pias, G. Coulouris, J. Crowcroft. *Profiling energy use in Households and Office Space*. e-Energy'10 Proceedings of the 1st International Conference on Energy-Efficient Computing and Networking (2010) pp. 21-30.

The economic security of power plants

Dorota Niedziółka[1,*] and *Paweł* Czyżak[2]

[1] SGH Warsaw School of Economics Al. Niepodległości 164, 02-554 Warsaw, Poland
[2] SGH Warsaw School of Economics Al. Niepodległości 164, 02-554 Warsaw, Poland

Abstract. Currently, power plants in Poland have to work in a very uncomfortable situation. Unstable market conditions and frequent changes in the law may have serious adverse consequences for their economic security. Power plants play a very important role in the economy. The effectiveness of their performance affects the activity of all other businesses. Therefore, it is very important to provide a definition of economic security for the power plants' sector and the factors determining its level. Maintaining economic security will allow energy generation companies to grow in a sustainable way as well as limit operational risk. A precise definition can also be used to create analytical tools for economic security measurement and monitoring. Proper usage of such tools can help energy generation companies sustain their economic security and properly plan their capital expenditures. The article focuses on the definition of economic security in the "micro" context of a separate business unit (enterprise). We also present an analytical model that measures economic security of a company engaged in the production of energy - a company of strategic importance for the national economy. The model uses macroeconomic variables, variables describing prices of raw material and legal / political stability in the country, as well as selected financial indicators. The appliance of conclusions resulting from the model's implementation will help provide economic security for companies generating energy.

1 Introduction

The issue of economic security is a new research field and is drawing more and more attention because of dynamic changes in the environment of enterprises. It is equally important to define the term of enterprise economic security, as well as identify the factors that lead to its achievement.

Until now, the term of power plant economic security was unknown. However, the processes of privatization and liberalization led to major changes in the perception of power plant safety and the risk that it faces [1]. Power plants are very specific business entities. Their size, measured through the number of employees[2] and income[3], and their role in the economy, defined by the amount of generated energy, installed power, market share, are of

*Corresponding author: dniedz@sgh.waw.pl

[2] In the PGE Group, employment in 2015 reached over 39 900, 44% of which worked with conventional energy sources.

[3] The TAURON group had a sales income of 18 375 224 000 PLN in 2015 [4].

huge importance for the country and its government. At the same time, the possibility to invest in companies owning the power plants, the opening of the energy market, the realisation of new infrastructural investments, electricity trading, open the way to financial profit for potential investors and shareholders.

The electricity market differs greatly from other commodity markets, while also being very attractive. Therefore, it is essential to analyse the term of power plant economic security and the factors determining it. Providing a valid definition of economic security in the context of entities acting as energy producers is the main goal of this article. Also, it is equally important to point out factors that limit economic security and/or cause risk for the business entity. Such factors where proposed in the article and evaluated using a dedicated econometric model. The model was used to construct an analytical measure of economic security. Monitoring the value of this measure can help the business entity quickly respond to arising threats and sustain economic security at all times.

2 The definition of economic security

The state of security is equivalent to the lack of risk/threat. It is therefore a secondary term that evolved in a situation of immediate danger, or danger expected in the future [2]. Risk and fear became the foundations of security. Initially, security was defined in relation to social behaviour[4]. The security of an individual or a nation was of major interest for researchers, sociologists, historians. Security also had a military context and dimension. Political and military doctrines took into account the desirable state of political security, the achievement of which was based on satisfying several conditions: the integrity of the territory, the safety of the country's borders, the stability and durability of international agreements and military alliances.

The dynamics of political, economic and social changes caused an evolution in the meaning of security and broadened the research done in this field, especially in the second half of the 20th century. Globalization and the fusion of social and economic structures led to the increased significance of non-military security. Seeking the stability of an individual, new dimensions of security emerged: ecological security, energy security, information security and, finally, economic security. The need to define new areas of security is a result of increased and overwhelming uncertainty, as well as the complication and intensification of human relations with the environment, institutions and organizations. In the light of the modern World's accelerating evolution, questions arise regarding the means of providing peace and stability in all aspects of social and economic life.

The available research points out the many definitions of security, which are at the same time not precise and unclear. Such a broad spectrum of security perception introduces value for researchers. Defining security is used as a filter, enabling researchers to spot and analyse certain mechanisms, threats, risk or conditions of stability.

Economic security is a state, in which business entities have good conditions to realize their strategy and achieve their goals, e.g. increasing production, gaining market share, increasing stock prices, achieving financial liquidity, ensuring steady access to raw material [3]. The stability of the business entity's operations will be called the skill of limiting risk. The economic security of an enterprise should therefore be determined by the types of risk that the business entity will possibly be facing.

[4] That is why the term of security appears in the works of philosophers - A. Giddens, B. Russell, T. Hobbes.

3 Factors determining economic security of power plants

In the case of power plants, the factors shaping economic security are strongly correlated with the characteristics of the energy sector. Typically, the power plants market is a closed market with few players. The most popular forms of market structure are monopoly and oligopoly. However, due to globalization causing a strong pressure for deregulation and privatization, the recent decades have introduced some structural changes [5]. This forced the existing market players to be involved in competition and rivalry. This unknown pressure is also caused by the emergence of new players, especially renewable energy companies, who may not be direct competitors for conventional power plants, but introduce additional risk on the market.

Power plants can therefore achieve economic security by limiting the risk linked to their specific market. Companies owning the power plants have knowledge about the characteristics of the market and possess tools that allow them to minimize threats. The ability to control risk and adapt to changes occurring in the entity's direct and indirect environment is strictly bound with the development of economic security.

The economic security of a power plant depends on the amount of produced energy, the market development perspectives, as well as the current energy policy. The economic growth is a measure of stability – a lack of threats that could endanger the production process. Assuming the world economic growth equal to 2.8% annually (up to 2040) and an increase in energy usage effectiveness by 35% (in the same timeframe), the demand for electricity will rise by about 1.3% annually [6]. The electricity demand projections influence the future energy production plans, the amount of installed power and investment projects carried out by power plant owners, both the refurbishment of current plants and the opening of new ones.

The presence of new business entities in the energy production market should not necessarily be treated as a threat, because the location of the power plant and the strong entry barriers can significantly limit competition on the local market [5]. It is worth remembering that power plants are treated as strategic entities in the economy. Their role is determined by the nation's level of economic development, the structure of the economy, especially the share of industry and services in GDP, and the effectiveness of energy production. Up to the 1970s, energy markets were watched closely by the state – the energy market was treated not only as a determinant of social and economic development, but also as a strategic element of the country's security [5]. This type of market structure, with the domination of public ownership, limited the risk of a single entity's operations. Infrastructural investments were an essential part of the government's investment program and often had high priority. The energy prices were determined by administrative decrees. Energy markets were isolated, forming subsystems restricted to the country's borders. Only in the second half of the 20th century, changes of different magnitude arose, concentrating on the privatization and deregulation of the energy sector. The introduced reforms changed the structure of the market – energy production was isolated from energy distribution. New entities appeared, focusing on energy trading. The spot market was introduced, access to electro-energetic networks has become common and the energy prices were freed[5]. In many countries, the market reforms introduced social disturbance, caused mainly by the fear of rising electricity prices. This fear involved not only households, but also enterprises, for which the rising electricity costs would translate into rising prices of offered goods and services. The decreasing economic security was also a consequence of limited access to investment project funding and the high costs of market transformation.

[5] The character and level of introduced changes varied deeply – from widespread privatisation in the UK, to a public energy model in France and Germany [7]

Currently, the economic security of power plants means that the entity can operate in the short term without significant threats (or with limited risk) in the following areas: location, raw material supply, financial situation, legal and market environment.

The evaluation of economic security in energy sector companies that produce electricity can be ambiguous. On one hand, combining energy providers with mining companies gives interesting synergy possibilities – access to own raw material for the power plant and steady demand for the coalmine. On the other hand, merging energy companies with mining companies limits investment abilities of the former. Power plants have to cut their investment budgets and move this money to help mines that are deeply in debt. Such a scenario can be judged negatively by stockholders and investors. Messages about future mergers have a strong impact on the stock price of an entity. Analysts usually have a negative opinion about state ownership in companies – this ownership can be used by the government to realize its targets at the cost of smaller stockholders.

A significant role in shaping the economic security of energy companies plays their financial situation. This includes the asset structure (ratio between fixed assets and current assets), asset rotation, investment policy, funding sources, capital level, equity and liabilities structure, stability of funding sources, cost of funding, timely debt payoff, cash flow (ability to pay, coverage of fixed assets with capital) and net results, EBITDA, net debt to EBITDA ratio.

Some major causes of uncertainty include a change in electricity demand, the decrease of energy price or sales margin. Such events can trigger a drop in stock prices of companies from the energy sector. This can further influence their relationship with commercial banks that will notice an increased risk of credit granting. Therefore, obtaining funding for current operations and the realization of investment projects can become difficult and more expensive.

The risk of doing business in the energy sector increases also when changes in energy and/or climate policy are announced [8]. Countries that are members of the European Union are committed to implement the common law. Limits of CO_2 emission force changes in utilized energy sources or actions aimed at increasing production effectiveness and limiting emissions. These solutions compromise investment plans, increasing the expenses and causing pressure on energy prices. A change in energy prices is, however, negatively seen by the consumers and has a negative impact on the economy's competitiveness. The economic state of entities linked both vertically and horizontally is worsened.

At the same time, if the country does not meet its emission limits, funding from the so called 'renovation fund' – a support fund of the EU for countries that are making radical cuts in their CO_2 emissions, can be suspended. What's more, crediting from foreign financial institutions can be withdrawn or limited. This has a strong negative impact on the financial stability of companies and their economic security.

The development of renewable energy can also be seen as a threat for conventional energy companies, since renewable energy companies have priority in accessing the distribution network. A 'merit order' mechanism is in place – old and expensive coal power plants are displaced with renewable energy plants, which have almost no variable costs and, therefore, have a higher position in the price ranking [9]. Renewable energy plants can be conquered by modern hard coal plants, which have around 30% higher efficiency than currently used installations. The 'merit order' effect means that while more efficient installations (with lower variable costs – in Poland mainly costs of coal) enter the market, the energy prices decrease. The more energy is coming from 'cheaper' sources, the less energy from conventional plants is used.

An essential factor determining the power plant's economic security is access to raw material and its price. Energy groups having mines in their company portfolio are not affected by disturbances on the global supply market. They can also limit the risk of price changes.

DOI: 10.1051/e3sconf/20171401007

This aspect was strongly underlined in 2015, when very low oil prices have strongly impacted the financial condition of energy resellers. Upstream companies had record-breaking profits, while downstream entities recorded losses. Low material prices raised questions about the profitability of oil extraction, many downstream companies considered selling extraction licenses and limiting the search for new sources. Also, the price of raw material determines the structure of utilized primary energy sources. Thus, decisions related to building new power plants have to take into account this aspect.

A different factor having impact on the level of economic security is the location of the power plant. The location of the energy production facility is usually determined based on several factors – access to water, demand market size, access to raw material. Location influences the costs of energy production. Risk increases when new power plants are localized close to existing ones – competition arises. Due to the integration of energy markets, decisions of investors in neighbouring countries should also be considered – a new power plant located next to the country's border and connected to the energy system can change the structure of offered electricity and weaken the position of the local energy producer. Concerns can also arise when, due to the realization of cross-border investment projects and new energy transportation channels, energy trade increases. Because of the abovementioned observations, it is becoming more often to evaluate investment projects not only within the borders of one country, but also taking into account the energy strategy of other countries in the region.

In the light of broadening the electricity supply, a question about the power market, its structure and size, arises [10]. The economic security of a power plant can increase as a result of the power market introduction – such a market would act as a guarantee of the country's energy security.

The level of economic security is also influenced strongly by the stability of the law. A more stable and long-term energy policy allows energy companies to operate in a steady environment. This aspect gains importance taking into account the characteristics of the energy infrastructure – its immobility, huge costs of investment projects and the long lifetime of its components. These characteristics do not allow the law to change rapidly and deeply. Adapting to legal changes is very costly and can strongly affect the financial stability of energy companies.

A summary of the factors described above is provided on Fig. 1. These factors will be used to propose an analytical measure of power plant economic security.

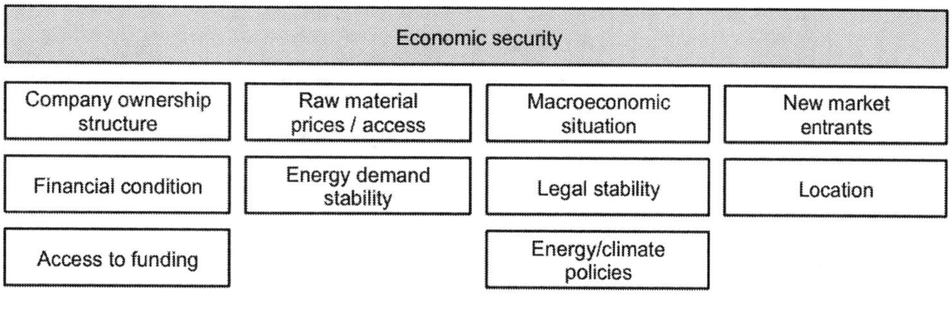

Fig. 1. Determinants of power plant economic security

4 Analytical evaluation of power plant economic security

In order to analyse factors determining the economic security of an energy production enterprise and measure the economic security objectively, a dedicated econometric model was prepared. The model utilized macroeconomic variables, financial indicators, coal price indices, political situation measures. The variables were chosen according to the factors outlined in the theoretical part of the article. As the dependent variable, the stock market price of Polska Grupa Energetyczna S. A. – the largest energy provider in Poland, was chosen. This is not an ideal measure of economic security, but the stock price indicates the moods of investors and their view on the risk a company is facing. The described model can be used to construct an objective measure of economic security of an energy production company.

Table 1. shows the explanatory variables included in the model. The variables were chosen from a larger pool of indicators from different groups, the choice was based on their statistical properties described further.

Table 1. Explanatory variables

Variable	Description	Expected dependency
	Macroeconomic variables	
WIG	The value of the Warsaw Stock Exchange Index	Positive (a general increase in the investors moods should result in the increase of stock prices of specific companies)
CPI	Consumer price index for Poland [last month = 100]	Negative in the long term, Positive in the short term (an increasing inflation can increase the company's profit, but in the long term it usually is a sign of macroeconomic problems in the country)
EURPLN	EUR/PLN exchange rate	Negative (an increase in the exchange rate can increase the costs of imported raw material, the rate change can also be an effect of macroeconomic problems)
	Coal prices	
RB_Index	Coal prices in Richard's Bay (South Africa) [USD]	Negative (declining coal prices have a positive effect on the margin of an energy producer)
	Financial indicators	
DEBT_EBIDTA_PGE	Net debt / EBITDA for PGE S.A.	Negative in the long term (debt increase is usually interpreted as a signal of financial problems of a company)
	Political / Legal stability indicators	
Legal_Acts	Number of signed legal acts	Negative (changes in regulations and bureaucracy are significant risks for an energy production company)

Monthly data from 12.2009 to 03.2016 was used, the data sources included: World Bank, GUS (Polish Statistical Office), NBP (Polish National Bank), PGE financial reports, biznesradar.pl and wnp.pl data aggregators. The DEBT_EBITDA_PGE variable was extracted from quarterly data using spline interpolation. All variables were transformed by a logarithmic function to enhance their statistical properties (exclude heteroscedasticity).

Based on visual analysis, most variables used in the model show a significant trend – such a situation is typical for non-stationary processes (processes which distribution changes in time) [11]. Stationarity can be measured analytically using the Augmented Dickey-Fuller procedure [12], in which a $y_t = \alpha y_{t-1} + \varepsilon_t$ model is estimated repeatedly for subsequent differences until the H_0: $\alpha < 1$ hypothesis is rejected. The so-called integration level marks the number of differentiations that have to be performed to transpose the process to stationarity. Testing for stationarity is important since non-stationary processes require special econometric treatment, such as cointegration methods. Variables are cointegrated, if a stationary linear combination of them exists. To test the existence of cointegration, the

 DOI: 10.1051/e3sconf/20171401007

Johansen procedure can be used [13]. The Johansen test is repeated for subsequent ranks of cointegration – accepting the H_0 for the given rank means that this rank is the rank of cointegration. If the H_0 is declined, the cointegration rank is higher than the current rank and the test should be repeated (up to the moment when H_0 is accepted). If the existence of cointegration is confirmed, a vector error correction mechanism (VECM) can be used. A VECM is an extension to a vector autoregression model (VAR) – contains an additional error term that can be used for analysing long-term dependencies between variables. The VECM has similar features as a VAR – its stability, residual distribution, autocorrelation should be tested to check if the model is correct. It is also important to properly choose the lag order – this can be done using information criteria such as BIC, AIC, HQ.

According to the methodology described above, the first step of the model's construction included ADF stationarity tests. The optimum lag was chosen using AIC criteria, testing down from 11. The results are presented in Table 2. All variables except EURPLN are clearly non-stationary and 1-st order integrated, EURPLN becomes 1-st order integrated if a 5% threshold is chosen, otherwise it seems to be stationary.

Table 2. ADF test results

Variable	p-value for model with constant value	p-value for model with constant value and linear trend
l_PGE	0.2	0.4757
l_WIG	0.2763	0.4442
l_CPI	0.8939	0.2827
l_EUR_PLN	0.06536	0.05917
l_RBIndex	0.9314	0.2229
l_DEBT_EBIDTA_PGE	0.3252	0.2734
l_LegalActs	0.9775	0.9853
Δl_PGE	4.731e-07	
Δl_WIG	1.666e-07	
Δl_CPI	2.094e-09	
Δl_EUR_PLN	1.548e-07	
Δl_RBIndex	1.031e-07	
Δl_DEBT_EBIDTA_PGE	0.03306	
Δl_LegalActs	0.001924	

Since the variables are non-stationary, their analysis can either be carried out through the use of cointegration methods, or by differentiating the variables to ensure their stationarity and then using a standard VAR model. Therefore, a Johansen cointegration test was performed with an unconstrained constant and linear cointegration trend setting. The optimum lag of 7 was chosen using AIC, BIC and HQC information criteria (the test is performed for a n-1 lag order). The results of the Johansen test indicate, that 3 cointegration relations exist in the model (p-value for rank 3 path test = 0,0528, L_{max} test = 0,1216).

Because a cointegration relation exists, a VECM could be created. The lag order was set to 7, the cointegration rank to 3, deterministic factors as in the Johansen test. The estimation results are correct, with most of the alpha values being between 1 and -1 (the system should return to equilibrium / the error correction mechanism should work properly). The model was tested for its correctness (Table. 3) – no residuals autocorrelation or ARCH effects were found, the normality of the residuals distribution was confirmed.

Table 3. VECM correctness tests

Variable	**p-value for residual autocorrelation test** (H0: autocorrelation does not exist)	**p-value for ARCH effects test** (H0: ARCH effects do not exist)	**p-value for residual normality** (H0: residuals have a normal distribution)
l_PGE	0.961	0.460088	0.9738
l_WIG	0.897	0.427474	
l_CPI	0.978	0.768965	
l_EURPLN	0.936	0.544592	
l_RBIndex	0.763	0.7929	
l_DEBT_EBITDA_PGE	0.819	0.357016	
l_LegalActs	0.224	0.826473	

The system is stable – all unit roots are inside the unity circle, with the roots related to cointegration relations being on the unity circle. The impulse response charts on Fig. 2 indicate the temporal behaviour of l_PKB in response to a peak change in the explanatory variables

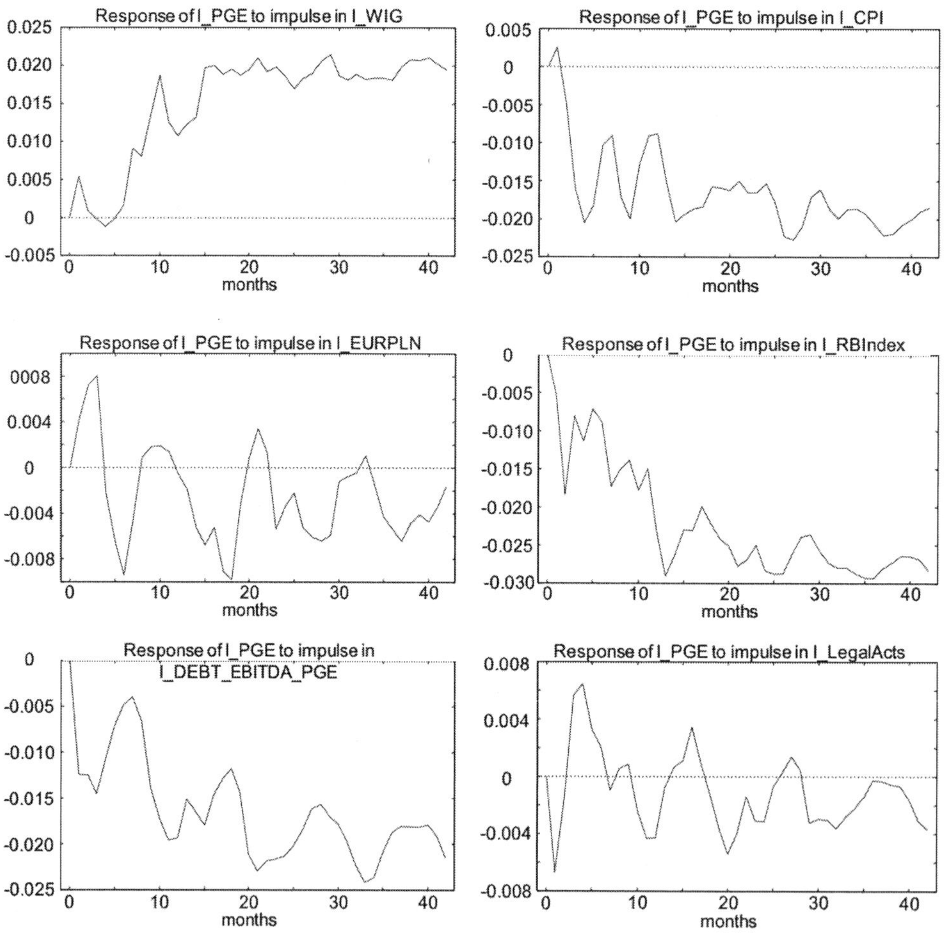

Fig. 2. Impulse response functions

Responding to an impulse in l_WIG, l_PGE increases in the first month and stabilizes on a positive value after around 15 months. This matches the expectations – the increase in

DOI: 10.1051/e3sconf/20171401007

general moods on the stock exchange has a positive impact on the prices of specific companies and their economic security. After an increase in l_CPI / inflation, l_PGE increases in the first month and stabilizes on a negative value after around 15 months. This also matches the expected behaviour – the increase in energy prices can temporarily increase the incomes and signify an increased economic growth. However, in the long term, the rising inflation has a negative impact on the economy and the economic security of companies. The change in EURPLN causes a significant increase in PGE in the first three months, but then the value oscillates around a negative number. The short-term increase can be caused by better conditions for energy or coal export from Poland to the EU. However, in the long term, the stronger EUR increases costs of coal import and can be a signal of macroeconomic problems in Poland, both factors having a negative effect on economic security. A positive impulse in RB_Index causes a systematic decrease in PGE, with a steady negative value reached after 20 months. This matches the expectations – the increase of coal prices increases costs and decreases margins, threatening the economic security. As indicated by the impulse response to a change in DEBT_EBITDA_PGE, increased debt also decreases the stock price and threatens the economic security – the impulse response function is clearly negative. The response to Legal_Acts is not clear, oscillating around a small negative value after significant (but declining in amplitude) increases and decreases. This behaviour could indicate a non-optimal choice of the explanatory variable – the increased number of introduced legal acts can be linked to increased burocracy, but also to positive changes in the current law. An optimum choice of a variable indicating the political and legal situation of a country, with a monthly frequency, is very difficult and will be the topic of further analysis in future publications. Generally, the impulse responses match the expectations, confirming the correct construction of the model.

Below, the variance decomposition chart is shown (Fig. 3a). The chart indicates that the error is determined mainly by lags of the dependent variable. This is correct for this model, since the explanatory variables were not chosen for forecasting purposes, but to isolate factors influencing economic security. According to Fig. 3a, in the short and mid-term, the company's debt and raw material prices have the biggest influence on its economic security. In a longer term, the inflation increases in significance, indicating the change in macroeconomic conditions. In the long term, coal prices remain the most significant factor of economic security. However, also l_WIG appears to increase in significance – stock behaviour is an important measure of the consumer trends and economic stability in the country and has an influence on the economic security of the company. Low significance can be attributed to exchange rates and legal situation. In Poland, most of the utilized coal is taken mainly from local sources so the exchange rates do not influence raw material costs. The low significance of legal factors results probably from a non-optimum choice of the explanatory variable – changes in energy law potentially can have a huge impact on the economic security of PGE.

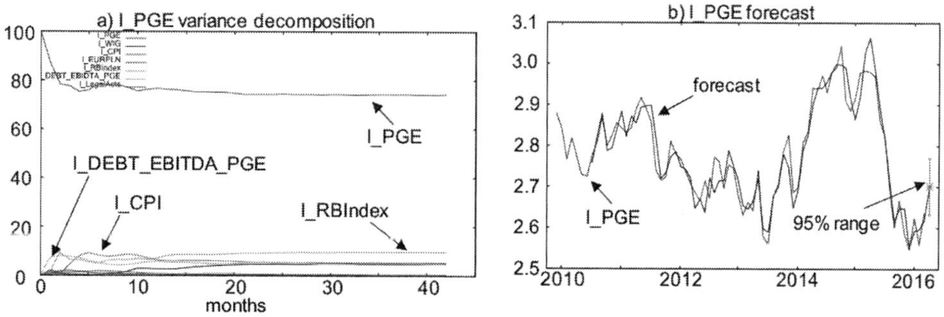

Fig. 3. Variance decomposition (a) and forecast (b) for l_PGE

Fig. 3b shows the original values for PGE stock prices and a forecast created using the VECM model. Even though the model was not optimized for stock price forecasting, the matching is quite good. Static in-sample forecasting was used (forecast for each observation is calculated from real values from previous periods, not from previous forecasts). Low error values (mean absolute percentage error around 1%) indicate that the model could be successfully used in forecasting, especially after further optimization for this purpose.

5 Conclusions

Energy companies are currently under huge pressure because of changes in the political, economical and legal setting, which strongly impacts their operations. This situation is a result of the energy market opening and the evolution of free market mechanisms, backed by the introduction of foreign capital. Strong vertical and horizontal connections and the national and international environment, push energy companies to seek new ways of achieving economic security. Though the state of economic security is not permanent, striving for its achievement gives business entities the possibility to limit risk. Sustaining economic security is possible using the analytical tools described in the paper, particularly monitoring the value of the proposed measures. Our research shows that the most significant factors shaping the economic security of an energy company are: raw material prices, the financial condition of the entity and macroeconomic conditions. Surely, the political and legal situation in the region is also a crucial aspect of economic security, however, it is very difficult to provide an adequate analytical measure in this field.

References

1. M. Zerka, Polskie Sieci Elektroenergetyczne S.A., web source: http://www.cire.pl/publikacje/Art_Zerka.pdf (November 2016)
2. U. Beck, *Społeczeństwo Ryzyka* (Wyd. Naukowe Scholar, Warsaw, 2004)
3. Y. Shutyak, O. Danylenko, D. Van Caillie, Conf. Proc. REDETE 2014, web source: https://ssrn.com/abstract=2640291 (November 2016)
4. web source: https://www.tauron.pl/tauron/grupa-tauron/tauron-w-liczbach/Strony/tauron-w-liczbach.aspx (August 2016)
5. D. Niedziółka, *Rynek Energii w Polsce* (Wyd. Difin, Warsaw, 2010)
6. Statoil's Energy Perspectives 2013, web source: http://www.statoil.com/en/NewsAndMedia/News/2013/Pages/14Jun_EnergyPerspectives.aspx (August 2016)
7. T. Jamab, M. Pollitt, Energy J., Spec. Issue on Euro. Elec. Liberal., p. 11-41, (2005)
8. B. Knopf, P. Nahmmacher, E. Schmid, Energy Policy, **85**, p. 50-60, (2015)
9. F. Sensfuß, M. Ragwitz, M. Genoese, Work. Pap. Sust. and Innov., **no. 7** (2007)
10. E. Bayer, W. Kukuła, J. Rączka, M. Stoczkiewicz, *Założenia rynku mocy w Polsce – analiza prawna i ekonomiczna*, (ClientEarth, Warsaw, 2016)
11. C.R. Nelson, C. R. Plosser, J. Monetary Econ., **10(2)**, p. 139–162, (1982)
12. D. A. Dickey, W. A. Fuller, J. Amer. Stat. Assoc., **74(366)**, p. 427–431, (1979)
13. S. Johansen, Econometrica, **59(6)**, p.1551–1580, (1991)
14. The IEA Model of Short-term Energy Security (MOSES) Primary Energy Sources and Secondary Fuels, International Energy Agency
15. Ch. Winzer Conceptualizing Energy Security, EPRG Working Paper 1123 Cambridge Working Paper in Economics 1151

E3S Web of Conferences **14**, 01008 (2017) DOI: 10.1051/e3sconf/20171401008
Energy and Fuels 2016

The economic value of the enterprise for stakeholders in the fuel and energy sector

Sylwia Lorenc[1,*]

[1]AGH University of Science and Technology, Department of Economy and Management in Industry, al. Mickiewicza 30, 30-059 Krakow, Poland

Abstract. The purpose of this article is to present certain trends in the distribution of value created by businesses for groups of stakeholders as part of a conception of sustainable growth. The subject of this analysis is the LW Bogdanka S.A. mining company, which was studied in terms of distribution of value for the period from 2009 to 2015. In this paper, the main stakeholders of the operation were identified as the shareholders, capital providers, employees, suppliers, local governments and the state. A method for measuring distributed value based on cash flow is demonstrated. Identification of the value added by the proposed methodology may assist of decisions related to the continuation or liquidation of operations.

1 Introduction

The primary aim of every business is to maximise its value in the long term. This effort to increase the business' value is inextricably linked with the attempt to answer the questions of what category of value is under consideration, for whom is this value meant to be created, and what methods are to be used in order to create it. Until recently, all models, methods, and tools used for assessment were based on an estimation of the value created for owners. Currently, as the aims and strategies of businesses are undergoing reorientation towards the creation of sustainable businesses, this value is increasingly coming to be understood as that created for all groups of stakeholders, including clients, employees, investors, and the immediate and further environment. As a result, modern business management practices involve the necessity to take a multi-dimensional approach to the value of the business, and to its fundamental role in creating mutual relations between the business and its environment. We can speak then of a bundle of aims which determines the common coordination of the interests of all groups of stakeholders in their effort to achieve economic, social, production and marketing, and environmental aims in order to meet the expectations of and deliver benefits to all the groups involved. In the case of fuel and energy businesses who depend on mutual relations with their surroundings, the building of sustainable value which meets the needs of all groups of stakeholders is essential.
The purpose of this article is to present current trends in the distribution of value generated by the business for its stakeholders, in accordance with the assumptions of sustainable growth. For the purposes of defining value for the individual groups of stakeholders in this

* Corresponding author: slorenc@agh.edu.pl

article, a research method using dynamic accounting based on free cash flow founded on value added has been used. The subject of this analysis is the LWB S.A. mining company.

2 Pricing model of company's sustainable value

The problem of assessing value for the owners of the business is relatively easy to solve by using common assessment methods such as EVA *(economic value added),* MVA (*market value added)*, TSR *(total shareholder return*) and others [1]. However, in what way should value be assessed for other groups of stakeholders? The identification and measurement of this value is a critical task for sustainable businesses. The development of an appropriate calculation model for all interested parties would be of great aid in solving this problem.

In our era of social and economic change, in which the stakeholders' need for information is constantly rising, it would seem essential not only to assess value, but above all to develop an efficient and transparent system for reporting the value created and the system of distribution of this value. It is after all on the basis of the quality and type of information presented that the various interest groups make their own assessments.

The distribution policy for the value created requires the definition of its structure. This structure illustrates the proportional distribution of value generated to the various groups of beneficiaries. In this paper, the distribution process for this value is presented based on the well-known concept of value added as it is earmarked for distribution among the various beneficiaries in the form of remuneration, interest, dividends, retained profits, and taxes [2]. However, methods of assessing value added which are well-known in the literature and in business make use of accrual accounting based on net profit, which supplies an insufficiently broad spectrum of information and is subject to numerous distortions[2, 3]. To eliminate the flaws of this method, the author of this paper decided to make use of a dynamic method of assessment of sustainable value, based on cash flows which have been used to present value generated in previous periods. These dynamic methods, based on DCF (*discounted cash flow*) would seem to be a more reliable tool for measurement than static tools. Income approaches compare the value of a business to its ability to generate income streams in the future, in accordance with the maxim that a business is worth as much as can be earned on it. Dynamic methods allow the structure of value distribution to be defined and give a realistic reflection of the financial benefits which are received by individual groups of beneficiaries. These benefits most often can be expressed in financial categories defined by cash flows. For the purposes of this analysis, two approaches to valuation have been used [4]:

- FCFF (*free cash flow to firm*)
- FCFE (*free cash flow to equity*)

The concept of valuation based on FCF (*free cash flow*) is based on the use of cash flows from operations, not accounting for liabilities, after taxation. A model based on FCFF permits the assessment of the total value of the business (equity E + debt D) for all parties which finance the operations (owners and lenders) [4]. FCFE permits the calculation of potential equity which is available to the owners.

The proposal to make use of cash flow makes it necessary to introduce the concept of FCF VA (*free cash flow from value added*). This is the sum of cash flow to various groups of beneficiaries, including owners FCFE (*free cash flow to equity*), employees FCFW (*free cash flow to workers*), providers of capital FCFD (*free cash flow to debts*), suppliers FCFS (*free cash flow to suppliers*), municipal administration FCFM (*free cash flow to municipal administration*) and the state FCFSt (*free cash flow to state*).

For the groups presented, the value distributed throughout the business takes the form of benefits from cash inflows, which for the company become an expense related to costs. For this reason, free cash flow for owners has been estimated as a derivative of profit, depreciation, investment expenses, demand for net working capital, and capital instalments.

Cash flow for banks are the result of paying interest and capital instalments. Cash flow for employees are a consequence of salaries, wages, and other social security benefits. Free cash flow for suppliers is the result of costs borne for materials, energy, and external services contracted for the benefit of the company. Cash flow for municipal administration and the state are a derivative of taxes and fees paid, as well as of income tax. The methodology accepted in this paper based on records of costs categorised by type in the operations sector is justified in the case of assessment of value added for all stakeholders by the need for a wider view on the aggregate costs of the operation of economic entities. Excessively generalised treatment of costs in this respect may result in incomplete information regarding the creation of real value for the interest groups [5].

3 Distribution of value added on the example LWB S.A.

The subject of this analysis is LWB S.A, a mining company which extracts coal in the Lublin Basin. The company is one of the leaders of the coal industry in Poland, exceptional due to its financial results, the efficiency of its extraction operations, and its investment plans for the development of new deposits [6]. LWB S.A. has an active policy regarding sustainable development including an implemented social responsibility strategy. LWB S.A reports the results of this sustainable development by means of CSR reports and integrated reports. Thanks to this, the company has been able to build a positive public image, creating value for shareholders and stakeholders alike. For these reasons, LWB S.A. would seem to be a benchmark in terms of best practice and constitutes a starting point for this analysis. The research methods, which combined accrual and cash flow accounting techniques, permit an analysis of the distribution of value created in the company between 2009 and 2015, from the company's IPO to the latest current data from financial reports.

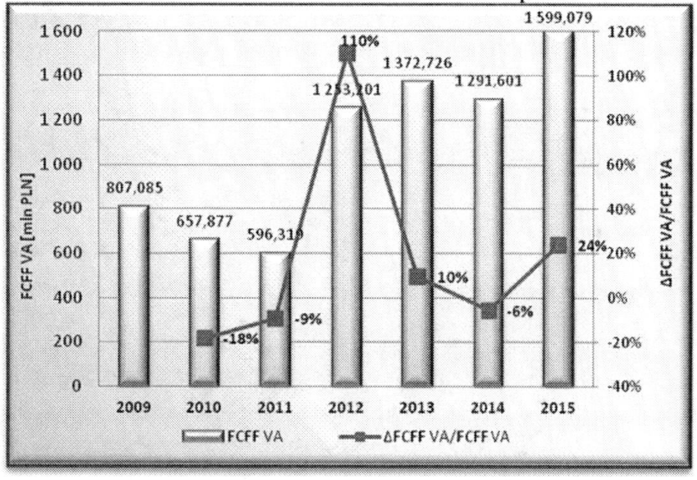

Fig. 1. Free cash flow from value added for all stakeholders generated by LWB S.A. in 2009-2015. Based on company data [7].

The benefits generated by the company which were estimated and are presented here have been assigned to individual groups of stakeholders. For specific years under study, the amounts are presented which reach different beneficiaries including shareholders, investors, employees, suppliers, local authorities, and the state. By using appropriate methodologies amounts of free cash flow have been calculated and analysed for each of these groups: free cash flow to equity, free cash flow to debts, cash flow based on costs by nature - free cash flow to workers, suppliers and municipal administration and free cash flow to state, in other words, income tax. The starting point for the analysis was to define FCFF VA- the total value

generated by LWB S.A. for its stakeholders in the years 2009-2015. The analysis was supplemented with an illustration of the dynamics of change, as shown in Fig. 1.
Fig. 1 illustrates the differentiated growth of value for specific periods. A drop in gross profits took place in the years 2009-2010 (18%), 2010-2011 (9%) and 2013-2014 (6%). The most dynamic change can be noted in the years 2011-2012 (growth of 110%) and 2014-2015 (growth of 24%). This greater than double increase in value (110%) was brought about by a change in the cash flow for owners (negative FCFE in 2011, followed by positive in 2012), and in turn a 24% increase by a change in cash flow for financing institutions FCFD).
In 2009-2015 total value added based on the cash flows which was generated by LWB S.A. for its stakeholders amounted 7 577 888 004 PLN. Tab. 1 and Fig. 2 show the way and proportions of distribution value for different stakeholders.

Table 1. The value added based on the cash flows generated by LWB S.A. for stakeholders in 2009-2015 [mln PLN]. Based on company data [7].

	2009	2010	2011	2012	2013	2014	2015
FCFE	149, 478	-209, 558	-321, 662	173, 015	262, 024	187, 537	188, 135
FCFD	-154, 925	-28, 108	-120, 679	-151, 191	-198,137	-247, 827	170, 506
FCFW	324, 755	317, 409	344, 950	403, 055	435, 376	484, 242	473, 271
FCFS	423, 501	501, 561	623, 159	728, 741	762, 165	764, 438	673, 879
FCFM	17, 281	18, 465	19, 815	26, 527	27, 297	29, 709	29, 863
FCFSt	46, 993	58, 107	50, 735	73, 055	84, 001	73, 502	63, 423
FCFF VA	807, 085	657, 877	596, 319	1 253, 201	1 372, 726	1 291, 601	1 599, 079
						SUM of FCF VA	**7 577, 888**

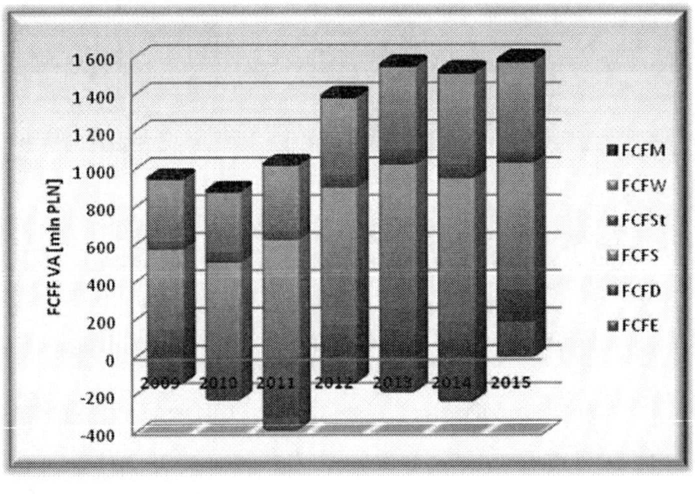

Fig. 2. Proportions of distribution value for different stakeholders in 2009-2015. Based on company data [7].

The largest share in the value generated by LWB S.A. is enjoyed by suppliers (59.1%). These noted especially high values in the year 2014, when the value created for this group totalled more than 764 million PLN. In the case of suppliers, there has been a general growth trend in share of value, with the exception of 2015, when a drop in value of 12% took place.

The second largest share in total value is enjoyed by employees (36.7%). For this group, as for the previous one, a steady increase in value with a slight drop in the last year (2%) can be observed in the period analysed.

Free cash flows for banks in the majority of years analysed were negative values. These negative values for this group indicate that in the years 2009-2014, capital instalments were negative in spite of positive interests. This fact results from the cash flow approach. In those years banks have not benefited from the mutual relations with the company.

The share of the state budget in the years analysed is in the area of 4-9%. In 2015, the taxes paid by the company to the state budget were the lowest to date in comparison with preceding years. The average mandatory tax burden which the company paid to the state treasury based on the mining operations of LWB S.A. hovers around 64 million PLN annually.

An interesting situation can be seen in the case of the owners, whose share in the consumption of value is highly differentiated. Their negative share in the distribution of value results primarily from high investment costs and from a negative of need for net working capital. Investments in fixed assets froze available cash resources, and thus lowered FCFE, however in the long term these investments are crucial in order to ensure the proper development of the company. Both investment in working capital and in fixed assets are necessary to maintain the current cash flows and to generate future revenues. The last years of the analysis indicate a stabilisation in cash flows, while recording positive values. The share of the owners in the consumption of value for the period 2009-2015 was on average approximately 5.7%, totalling around 61 million PLN annually.

The share of local government remained on average roughly 2-3%. This group is the smallest beneficiary of value. Payments to local governments in the form of taxes and fees paid by the company total roughly 25 million PLN annually.

The development of cash flow patterns for individual groups of stakeholders with an interest in the total value generated by the company between 2009 and 2015 is illustrated in Figs. 3-8.

The distribution of this value takes place in the following forms: payment of remunerations and benefits for employees, tax payments for local and national governments, interest for capital providers, dividends for owners, and profits for suppliers of goods and services.

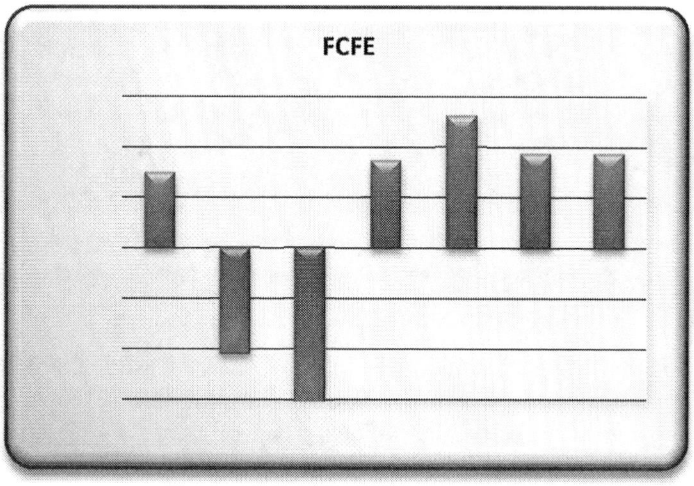

Fig. 3. Free cash flow to equity in 2009-2015 [mln PLN]. Based on company data [7].

The highest value which company has provided to owners can be observed in 2013 in the amount of 262 024 000 PLN. The lowest value has come at 2010-2011. The worst year for the owners in terms of the benefits they could receive was the year 2011. The lowest negative

 DOI: 10.1051/e3sconf/20171401008

value equal to - 321 662 000 PLN demonstrates of the need to provide the required cash for the company.

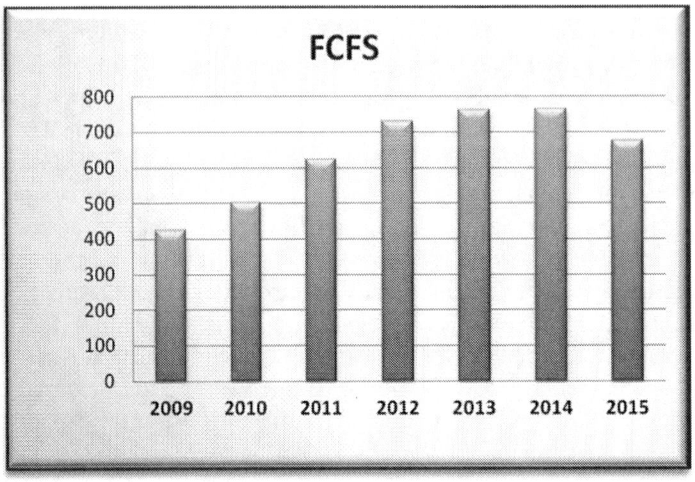

Fig. 4. Free cash flow to suppliers in 2009-2015 [mln PLN]. Based on company data [7].

In the case of suppliers who, according to the analysis, are the main beneficiary of the created value, the biggest gains for them observed in 2014 with amount equal 764 438 000 PLN. The year 2009 has proven to be the weakest in terms of distributed of benefits, only 423 501 000 PLN accounted profits for suppliers because of purchase by the LWB S.A. of materials, energy and external services.

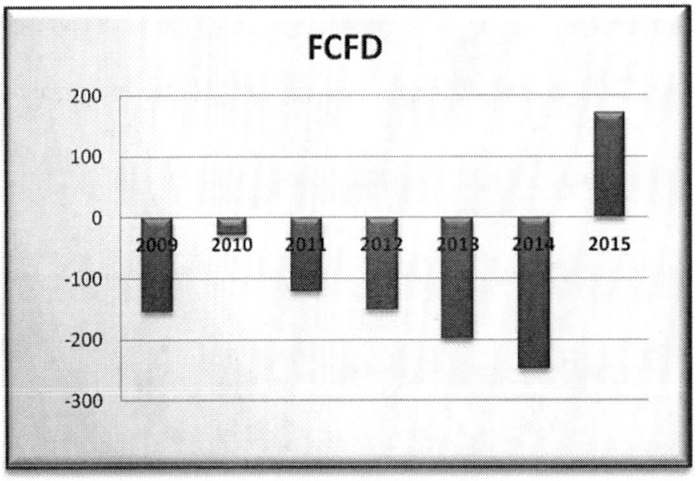

Fig. 5. Free cash flow to debts in 2009-2015[mln PLN]. Based on company data [7].

Providers of capital are the stakeholder, who have gained the least. In 2009-2014 banks have not benefited from relationships with LWB S.A. Only the last 2015 year, has brought a significant improvement in the results for this group of interests generating profits in the amount of 170 506 000 PLN.

DOI: 10.1051/e3sconf/20171401008

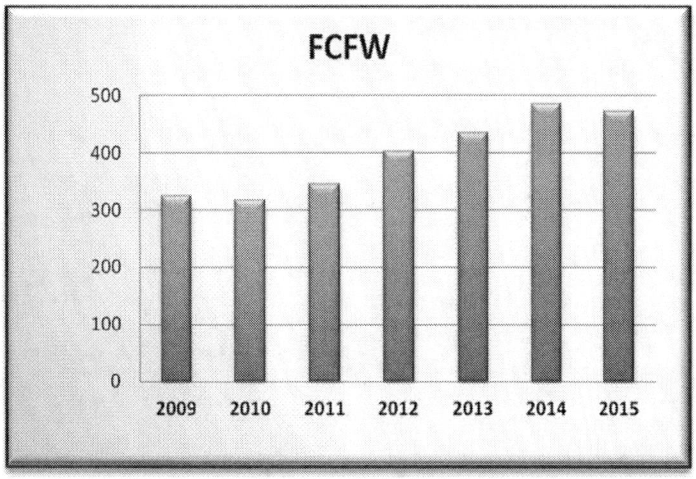

Fig. 6. Free cash flow to workers in 2009-2015 [mln PLN]. Based on company data [7].

For employees, who are the group strongly associated with the company, the highest value generated for them was in 2014 in the amount of 484 242 000 PLN. The lowest amount of salaries, wages and social security benefits for employees was observed in 2010 in the amount of 317 409 000 PLN.

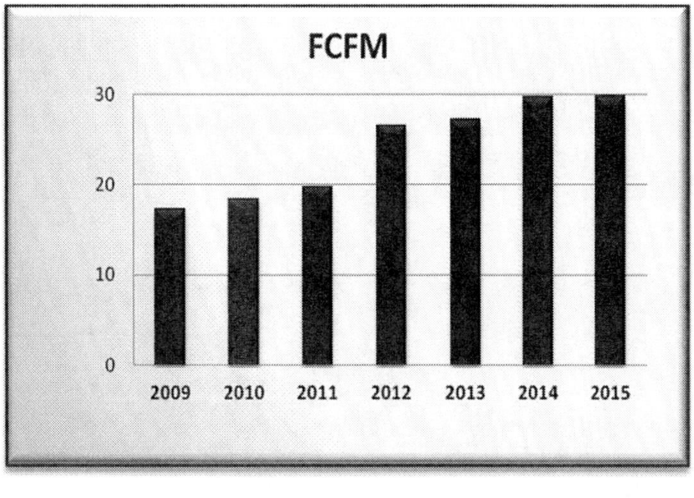

Fig. 7. Free cash flow to municipal administration in 2009-2015[mln PLN]. Based on company data [7].

The lowest value added the company generated for local governments. The highest benefits in the form of taxes and fees which went to this group of interest was recorded in 2015 (29 863 000 PLN). The lowest value was observed in 2009 in the amount of 17 281 000 PLN.

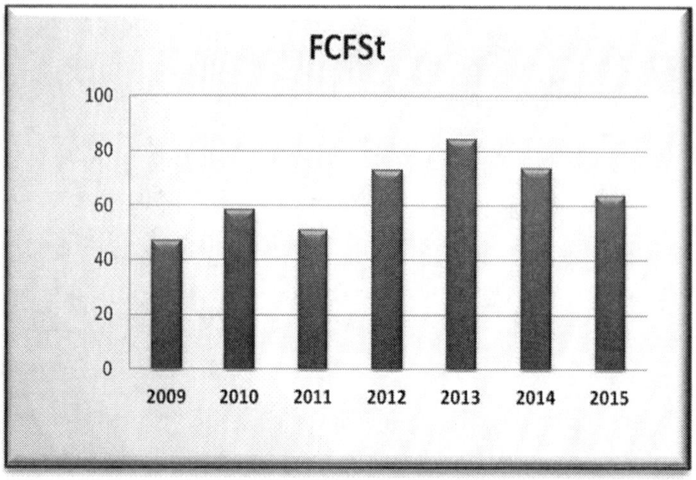

Fig. 8. Free cash flow to state in 2009-2015 [mln PLN].Based on company data [7].

Analysis of distribution of value added in the light of the benefits which goes to the state budget, shows that the highest income LWB S.A. created in 2013 in the amount equal 84 001 000 PLN. In turn, the lowest tax which was paid by the company amounted to 46 993 000 PLN and took place in 2009.

An analysis of charts 2-7 explains the role of LWB S.A. in the creation of private and public incomes, and shows how the operations of the company bring value to the community. Figure 9 shows the percentage of the various interest groups in the total value created in 2009-2015.

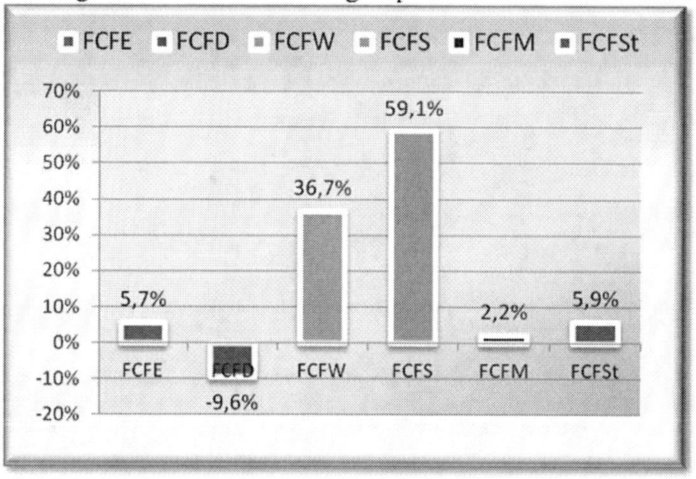

Fig. 9. The percentage of participation of individual stakeholders in the total value added in 2009-2015.Based on company data [7].

The above figure (9) shows that each of the groups in the analysis has its own larger or smaller share in the consumption of value, benefitting from mutual relations with the company.

The only group which did not participate in the benefits generated by the company in 2009-2015 are providers of capital. Over the last 7 years (except 2015) banks not only did not profit, it even they had to provide funds for the company (negative value added) (positive interest but negative capital instalments).

Increase on outside capital for the years 2014-2013, 2013-2012, 2012-2011, 2011-2010, 2010-2009 and 2009-2008 suggest that the company took out loans (positive capital instalments for the equity, negative for banks.) Banks obviously benefited from interest, but taking into account the free cash flow approach, capital instalments exceeded the amount of

interest from here we can observe the negative value for this group of interests. Only in 2015-2014 outside capital has decreased, (banks earned on the relationship with company) suggesting repay financial commitment to banks (negative capital instalments for the equity, a positive for banks), thus the total value of this year for the providers of capital is positive. The presented model emphasises the fact that the value created in the company is a derivative of the joint efforts of the various stakeholders working with the company. The analysis which was conducted shows in what way the effects of these common efforts are distributed among the various groups. The structure of distribution of value of course depends on the financial situation of the company, but not exclusively. In 2015, LWB S.A. recorded a loss, and yet this loss did not prevent the company from meeting the expectations of its stakeholders, for instance by paying salaries, debts and taxes, and by maintaining its suppliers. It is worth noting that despite a momentary lack of profitability, the company still delivered benefits to the community. The implementation of its social policy, based maintaining jobs and supporting local groups linked to the company, increases the trust and loyalty of the clients of the company, which being a publicly traded company must take care to cultivate transparency in its relations with stakeholders, and in doing so create a positive image for the entire energy, mining, and industrial sector.
Thanks to this type of presentation of results, LWB S.A. can accentuate in an accessible way the financial effects of its operations to build positive relations with stakeholders, which in turn influences the emphasis in the company on the implementation of a sustainable growth policy.

4 Conclusions

The efficient and transparent communication of a company with its stakeholders should be a priority in running any business whose operations significantly impact parties close to it and the further environment. This is especially true for heavy industries from the fuel and energy sector, for whom finical results cannot be the only indicator of success. The influence of fuel companies on the environment and the community is vast, and the usefulness of the information which is available in published reports is slight. For this reason, there is a need to generate additional information in the form of social reports which would indicate the global impact of a company on the environment and the community, and simultaneously represent an assessment of the value created for these groups. Such a report could include information about the specific value which the industry generates for the national economy in order to illustrate the rational for and usefulness of the operations of these companies.
The failure of contemporary mining and energy companies to generate profits makes it necessary to conduct an analysis accounting for the potential effects which would be felt if these companies were to be liquidated. The analysis conducted for the purposes of this article shows that such effects would be felt not only by the owners, but also by employees, suppliers, financing institutions, the state budget, and other stakeholders.
According to the analysis, the suppliers and employees are the groups who suffer the most in case of termination of activity LWB S.A.. They have the highest share in value which was created by the company. Then, the owners and financial institutions are the groups who suffer the least. The benefits which they receive are incomparably smaller than in the case of employees or suppliers. It is important to note, that employees and suppliers are a group of quantitatively greater than the groups of owners or financial institution. It not without significance is also the question of taxes paid by the company for the state budget and municipal administration. In the face of the present crisis in the mining and energy sector, there are many negative opinions which anchor in the social consciousness. It results mainly from the often false information about pay extra of billion PLN from the state budget to the mining and energy industry. Meanwhile, this analysis shows that it is quite the opposite. The

mining and energy industry is a powerful payer of all taxes and public law levy. From mining "live" not only municipalities in areas where the mines are located and coal is mined. Without an annual of cash injection from the mining companies, the state budget could have a huge problem. The situation is interesting because this analysis includes only one company which is LWB SA. In this case we can talk about tens of millions of PLN paid to the budget every year but much less in the case of the entire sector.

The presented analysis illustrates to what extent companies in strategic sectors have on their nearer and further environments, as well as on society as a whole. In this area, the presented methodology may assist of strategic decisions related to the continuation or liquidation of operations of mining and energy companies in terms of benefits which these companies provide for stakeholders.

For this reason, strategic decisions on the fate of these companies should be taken based not only on a financial analysis, but also on studies which permit the assessment of the company in terms of the range of benefits that it generates for the community, in accordance with the concept of sustainable development, in order to create a comprehensive and holistic view of the issue.

References

1. W.Skoczylas, *Determinanty i modele wartości przedsiębiorstw,* (PWE, Warszawa, 2007)
2. M.F. Morley, *The Value Added Statement in Britain*, Account Rev, **3** (1979)
3. https://www.isagen.com.co/informe-de-gestion/2011/archivos/informe_gestion2011_ingles.pdf
4. P. Fernandez, *Company valuation methods. The most common errors in valuations*, IESEP, **449** (2007)
5. S. Lorenc, A. Kustra, *Value added for stakeholders in mining operations based on cash flow*, IM, **2** (2016)
6. http://www.lw.com.pl/pl,2,s167,informacje_ogolne.html
7. http://www.ri.lw.com.pl/raporty-okresowe

Structure of financing investments in the energy sector

Barbara Kowal[1,*], *Robert* Ranosz[2] , *Wiktoria* Sobczyk[3]

[1] Faculty of Mining & Geoengineering, AGH University of Science & Technology, Al. Mickiewicza 30, 30-059 Krakow, Poland
[2] Faculty of Mining & Geoengineering, AGH University of Science & Technology, Al. Mickiewicza 30, 30-059 Krakow, Poland
[3] Faculty of Mining & Geoengineering, AGH University of Science & Technology, Al. Mickiewicza 30, 30-059 Krakow, Poland

Abstract. The purpose of this article is to discuss the issues of financing investments in the fuel and energy sector. The manner of financing business activities of every company depends on the decisions made by the management board, which need to take into consideration the effective striving for optimal level of the capital cost. The capital raised by the companies from the aforesaid sector may be in the form of equity or outside capital.

This study depicts such sources of capital as bank loans and corporate bonds considered outside capital as well as the issue of shares included in the equity. It is important for the managers making decisions on different types of financing investments in the energy sector to consider the capital cost. The outside capital cost shall be deemed to mean the interest rate on bank loans and corporate bonds, whereas the cost of equity shall be calculated on the basis of the Capital Asset Pricing Model (CAPM). The cost includes both the loss of profit represented by the risk-free rate (which is usually assumed at the level of 10-year old treasury bonds) and the risk to which the capital owner is exposed. The CAPM is usually used for determining the investment risk in a given company.

The article shows how the financing structure of the companies from the fuel and energy sector, listed on the Warsaw Stock Exchange, has evolved over the years. The authors also estimated the cost of equity. The results were compared with the chosen mining companies in Poland. Companies from the energy sector have lower investment risk than companies from the fuel sector. Looking at the profitability of investments it should be emphasized that the financing by outside capital is more advantageous than equity financing.

1 Introduction

Running business activities in the sector of companies operating on the energy and fuel market entails the necessity of making both financial and investment decisions. As part of

[*] Corresponding author: bkowal@agh.edu.pl

the investment decision-making process, a company needs to decide on the distribution of funds it managed to raise. On the other hand, the financial decisions make it necessary to choose appropriate sources of financing the planned investments. The selection of right sources of financing affects both the value of the implemented investment and the value of the company itself. A real impact on the choice of the source of financing have, among other things, such factors as: availability of capital and its cost. When considering the source of capital in the context of the investment efficiency, the capital structure determining the financial standing of each entity is not insignificant. Companies finance their business activities either from their equity (usually with limited access) or from third party equity, which is cheaper than their equity. An additional benefit derived from the financing of investment activities through outside capital is the fact that the interest paid on the obtained outside capital increases the operating costs of the business run and hence allows the company to pay lower income tax. However, it must be remembered that continuous accumulation of debts in a company is not beneficial and consequently may lead to the increase of costs of the raised capital or bankruptcy. Therefore, the choice of appropriate sources of capital for financing investments in the fuel and energy sector is an important factor, especially that it is characterized by capital intensity and long investment cycle. Their activities are subject to special social responsibility. On one hand, they must run their business activities efficiently and achieve their basic economic goals, but on the other - they need to remember (due to their nature) about the principles of sustainable development and social responsibility [7, 11, 8]. To efficiently implement the strategy and assumed goals, the management staff needs to invest financial resources in the business activity [3].

The authors of this article decided to investigate the capital structure of companies from the above-mentioned sectors and estimate the amount of the cost of equity held thereby. A number of methods, such as DGM, LEFAC and CAPM [9] may be used to estimate the cost of equity. In this article, one of the most popular methods, i.e. CAPM, was used to estimate the cost of equity.

2 Structure of financing

Equity is a capital good, which allows entities to run their business activities and to develop, which contributes to the increase of their goodwill and which allows them to achieve and maintain competitive advantage on the market. Equity is the capital brought in by the owners for an indefinite period of time. On the other hand, funds paid by creditors should be used by a company for a definite period of time (reimbursement on a specific date).

Business activities depend, among other things, on the structure of the sources of financing, which currently constitutes one of the most widely discussed issues in the field of finance [5]. The capital structure is reflected in maintaining appropriate proportions between equity and foreign equity. In the subject literature, the capital structure is not defined in an unambiguous manner. Some authors believe that it is the following [6, 12, 4, 2, 10, 1]:

- level of general indebtedness,
- structure of liabilities, which constitutes the structure of financing business of each entity,
- total share of long-term liabilities in the capital,
- structure of securities issued by a company (debt securities, proprietary shares),
- as ratio: (third party equity/equity) + 1 regardless of the due date.

Undoubtedly, it is the long-term capital that plays key role in the creation of the capital structure, which in turn has impact on the weighted average cost of capital of a company.

The authors analysed the financing structure in companies from the energy and fuel industry, listed on the Warsaw Stock Exchange in 2013-2015. Eight companies from the energy sector and three from the fuel sector were taken into account. At the beginning of the analysis, it was assumed that all sources of capital held by a given entity had to be included in equity, whereas the entirety of its liabilities constituted third party equity.

When analysing the financing structure, the share of outside capital in the entire financing capital for particular companies in the above-mentioned sectors was determined, which is shown in figure 1.

Table 1. Share of outside capital in financing for companies from the energy and fuel sector.

Companies from the energy sector	**2013**	**2014**	**2015**
ELEKTR. BĘDZIN S.A.	37.8%	30.0%	67.7%
ENEA S.A.	29.8%	33.7%	50.7%
ENERGA S.A.	53.0%	53.0%	52.5%
ZESP. ELEKTR. WR. KOG. S.A.	38.2%	40.1%	39.4%
POLENERGIA S.A.	45.8%	51.2%	56.3%
PGE S.A.	28.9%	32.4%	34.2%
TAURON S.A.	46.4%	48.0%	50.1%
ZESP. ELEKTROW. P-A-K S.A.	41.5%	44.4%	62.1%
Companies from the fuel sector	**2013**	**2014**	**2015**
GRUPA LOTOS S.A.	54.7%	56.4%	59.8%
PGNiG S.A.	41.2%	38.3%	38.3%
PKN ORLEN S.A.	49.5%	59.8%	53.9%

Source: own elaboration on the basis of data [13]

On the basis of percentage shares in the financing structure (table 1), it is evident that in the analysed period, the share of outside capital in the entire financing capital in case of the companies from the energy sector fluctuated between 28.9% and 67.7%. Two companies had a comparable share of this capital (Energa S.A., Zespół Elektr. Wrocławskich Kogeneracja S.A.), in other two - a slight increase of share of approx. 5 p.p. was noted (PGE S.A., Tauron Polska Energia S.A.), whereas a more significant increase of approx. 10 p.p. was observed in only one company (Polenergia S.A.). On the other hand, the remaining three companies experienced a considerable change in the analysed years of the outside capital share. The share fluctuated respectively: in Elektrociepłownia Będzin S.A. from 37.8% to 80.7%, in Enea S.A. from 29.8% to 50.7% and in Zespół Elektrowni Pątnów-Adamów-Konin S.A. from 41.5% to 62.1%. The financing structure shows that especially in the first two periods of analysis, the level of equity was higher than the level of debts (third party equity).

In the companies from the fuel sector, the share of outside capital is comparable to the level of the majority of companies from the energy sector. In the analysed period, it is maintained at the similar level. The increased share of the analysed capital was noted in case of two companies (Grupa Lotos S.A. from 54.7% to 59.8%, Polski Koncern Naftowy Orlen S.A. from 49.5% to 53.9%), whereas in case of one listed company - Polskie Górnictwo Naftowe i Gazownictwo S.A. - the share decreased from 41.2% to 38.3%.

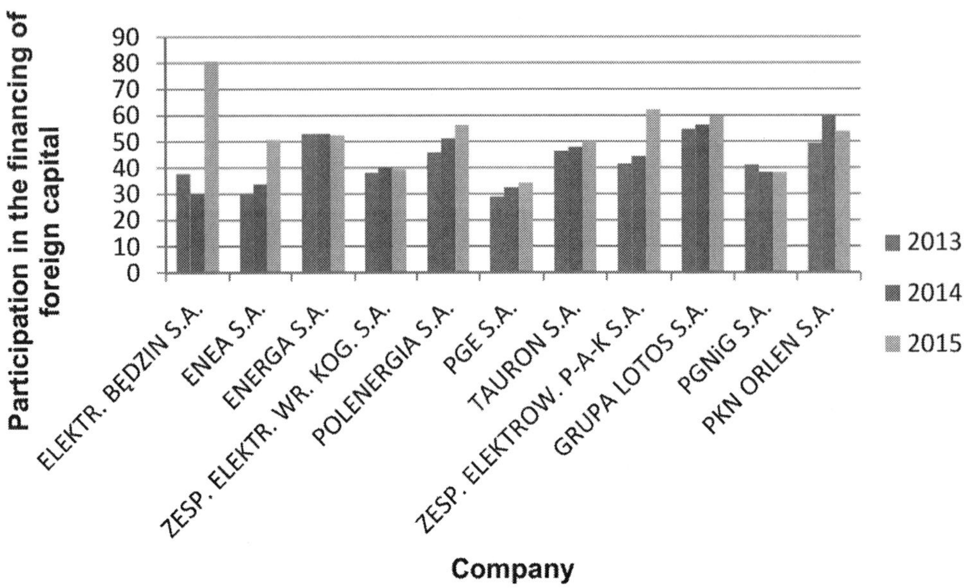

Fig. 1. Share of outside capital in the entire financing capital for companies from the energy and fuel sector, listed on the Warsaw Stock Exchange in 2013-2015.

Source: own elaboration on the basis of data [13]

The energy sector is less stable than fuel sector. The level of share of outside capital in the entire financing capital in six out of eight of the analysed companies significantly increased by comparison with the initial period (2013), exceeding 50%. In Elektrociepłownia Będzin S.A. - the level exceeded 67%. The companies from the fuel sector shape their financing structure in a more stable manner, thus, the level of share in the years 2013-2015 was not subject to significant fluctuations.

3 Cost of equity

The indebtedness of companies increases the investment risk, which in turn should cause an increase of the cost of equity. When calculating such cost, apart from the risk free interest rate, the risk of investment into a given company is also estimated. The Authors used one of the most popular methods of estimating cost of equity, namely the CAPM. The cost of equity for each company (table 3) was calculated according to the following formula (1):

$$\mathrm{CAPM} = r_e + (r_m - r_e) * \beta \quad (1)$$

where:

r_m - market rate (determined based on the WIG)

r_e - risk-free rate

β - beta factor

The β factor was calculated separately for each listed company from both the energy sector and fuel according (table 2) to the formula (2):

$$\beta = \frac{\mathrm{cov}\,(r_{tested}; r_m)}{s^2(r_m)} \quad (2)$$

Table 2. Beta factor for companies from the energy and fuel sector.

Companies from the energy sector	2013	2014	2015
ELEKTR. BĘDZIN S.A.	0.169	0.164	0.174
ENEA S.A.	0.554	0.583	0.629
ENERGA S.A.	0.613	0.681	0.730
ZESP. ELEKTR. WR. KOG. S.A.	0.448	0.447	0.434
POLENERGIA S.A.	0.566	0.547	0.540
PGE S.A.	0.955	0.944	0.978
TAURON S.A.	0.854	0.870	0.919
ZESP. ELEKTROW. P-A-K S.A.	0.714	0.735	0.771
Companies from the fuel sector	**2013**	**2014**	**2015**
GRUPA LOTOS S.A.	1.095	1.075	1.067
PGNiG S.A.	0.735	0.760	0.774
PKN ORLEN S.A.	1.141	1.147	1.136

Source: own elaboration on the basis of data [13]

The following assumptions were made: the risk free interest rate was adopted at the level of the 10-year old treasury bonds, i.e. 2.5%, whereas the market rate was adopted at the level of 7%. With such assumptions, the cost of equity for the companies from the energy sector and fuel sector was estimated. The results of the calculations are presented in table 3 and figure 2.

Table 3. Cost of equity calculated by using the CAPM.

Companies from the energy sector	2013	2014	2015
ELEKTR. BĘDZIN S.A.	3.26%	3.24%	3.28%
ENEA S.A.	4.99%	5.12%	5.33%
ENERGA S.A.	5.26%	5.57%	5.79%
ZESP. ELEKTR. WR. KOG. S.A.	4.52%	4.51%	4.45%
POLENERGIA S.A.	5.04%	4.96%	4.93%
PGE S.A.	6.80%	6.75%	6.90%
TAURON S.A.	6.34%	6.42%	6.64%
ZESP. ELEKTROW. P-A-K S.A.	5.71%	5.81%	5.97%
Companies from the fuel sector	**2013**	**2014**	**2015**
GRUPA LOTOS S.A.	7.43%	7.34%	7.30%
PGNiG S.A.	5.81%	5.92%	5.98%
PKN ORLEN S.A.	7.64%	7.66%	7.61%

Source: own elaboration on the basis of data [13]

Average cost of equity in case of companies from the energy sector oscillates in the range of 4 p.p. (from 3.26% to 6.90%). In six companies, the cost of capital increased in the analysed period from 0.02% to 0.53% (Elektrociepłownia Będzin S.A. by 0,02 p.p., Enea S.A., Energa S.A., PGE Polska Grupa Energetyczna S.A., Tauron Polska Energia S.A., Zespół Elektrowni Pątnów-Adamów-Konin S.A.), in the other two - the level of this cost decreased (Zespół Elektr. Wrocławskich Kogeneracja S.A. by 0.07 p.p., Polenergia S.A. by 0.11 p.p.).

On the other hand, in the fuel sector, the average cost of equity oscillated in the range of 2 p.p. (from 5.81% to 7.66%). In only one company, the cost of capital increased from 5.81% to 5.98% (PGNiG S.A.), in others - the cost slightly decreased in comparison with the analysed period (Grupa Lotos S.A. by 0.13 p.p., Polski Koncern Naftowy Orlen S.A. by 0.03 p.p.).

Table 4. Cost of equity for the selected mining companies.

Company	Cost of equity
KGHM S.A.	7.53%
JSW S.A.	8.45%
KHW S.A.	13.05%
Bogdanka S.A.	3.99%
KW S.A.	11.07%

Source: own elaboration on the basis of data [13]

To compare the amount of the cost of equity, its amount in 2015 was calculated for the selected mining companies, as shown in table 4. In case of similar financing structure of such companies (outside capital did not exceed 50%), compared to the financing structure of the companies from the energy and fuel sector, the cost was at a slightly higher level. Only in case of two companies (KGHM S.A. and Bogdanka S.A.), the cost was comparable to the cost of capital of the analysed energy and fuel companies.

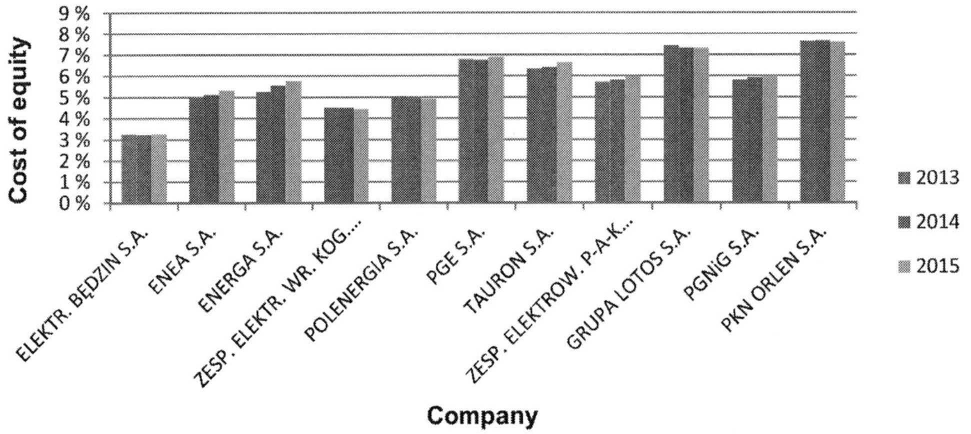

Fig. 2. Cost of equity for the companies from the energy and fuel sector in 2013-2015

Source: own elaboration on the basis of data [13]

According to the performed analysis, the cost of equity of the companies from the energy sector is lower than that of the companies from the fuel sector. Therefore, it may be concluded that according to investors, investment into the fuel sector is more risky than into the energy sector (higher cost of equity). When analysing the results obtained in the previous chapter (financing structure), it must be confirmed that there is a correlation between indebtedness of the companies from the above-mentioned sectors and the cost of equity. It should be also assumed that the cost of third party equity shall also increase.

4 Summary

Companies from the energy sector have lower investment risk than companies from the fuel sector. This difference may be due to the fact that enterprises of the fuel sector in the course of their business activities are exposed to more risks factors than companies from the energy sector. These main factors distinguishing these two sectors in terms of risk is that the susceptibility of financial performance (and therefore dividends) changes in foreign exchange rates and raw material (crude oil or natural gas).

In the case of companies in the energy sector there are no such threats as to changes in the price of energy, because mainly they are regulated. The article also shows the cost of equity for companies in the mining sector. The risk of investing in these companies is much higher than those analyzed in the article. The aim of showing that the risk in the mining sector is so high are the last "plans" for a merger of mining companies with companies from the energy sector. If such connections that occurred on a larger scale then it is expected that the cost of equity of energy companies would be falling growth.

Of course financing investments by foreign capital means a smaller tax base for each company. That causes the companies would pay less income tax in practice. Looking at the profitability of investments it should be emphasized that the financing by outside capital is more advantageous than equity financing.

The publication of the article was financed from statutory activities AGH agreement no 11.11.100.693

References

1. R. A. Brealey, S. C. Myers, *Principles of Corporate Finance*, McGraw Hill, New York (1991)
2. J. Downes, J. E. Goodman, *Dictionary of Finance and Investment Terms. Barron's Educational Series*, Inc., Hauppauge (1991)
3. W. Gabrusewicz, *Podstawy analizy finansowej*, PWE, Warszawa (2007)
4. E. A. Helfert, *Techniques of Financial Analysis. A Practical Guide to Managing and Measuring Business Performance*, Irwin, Burr Ridge (1994)
5. B. Jakubczyk, J. Lewandowska, *Teorie struktury kapitału w literaturze swiatowej*, Finanse i Prawo Finansowe, JOFFL, Kwartalnik, I(4): p. 65–81, **I**, No. 4 (December 2014)
6. M. Jasiniak, R. Pastusiak, W. Starzyński, *Struktura kapitału a efektywność funkcjonowania przedsiębiorstw sektora energetyki w krajach Europy*, W: *Finansowe i organizacyjne aspekty kooperacji nauki i lokalnej przedsiębiorczości*, ed. M. Wypych, P. Trippner, Przedsiębiorczość i Zarządzanie, Wydawnictwo SAN, Tom XV, Zeszyt 10, Część I, p. 53–63, Łódź-Garwolin (2014)
7. B. Kowal, D. Kowal, *The sustainable development management system of hard coal mining enterprises*, PJoES, ISSN 1230-1485, **20**, No. 4A, p. 145–150 (2011)
8. B. Kowal, A. Kustra, *Reporting of sustainable development strategy in energy industry*, E3S Web of Conferences **10**, 00129 (2016)

9. R. Ranosz, *Estimation of the cost of equity for mining companies using the P-LEFAC method*, ARCH MIN SCI, **59**, No 1, p. 257–269 (2014)
10. S. A. Ross, R.W. Westerfield, J. Jaffe, *Corporate Finance*, Irwin, Chicago (1996)
11. W. Sobczyk, *Sustainable development of rural areas*, PS, **9**, No. 1, p. 119-126 (2014)
12. M. B. Tudose, *Capital Structure and Firm Performance*, ETCognition, **15**, No. 2 (2012)
13. www.bankier.pl

The role of transport sector in CO_2 reduction in Poland

Sławomir Skwierz[1,*]

[1] Energy Market Agency (ARE SA), Statistics and Forecast Division, Warsaw University of Technology, The Faculty of Power and Aeronautical Engineering

Abstract. The need for greenhouse gases (GHG) reduction from the atmosphere, resulting from the International Treaties signed by Poland is a serious challenge that requires a complex action in all economy sectors. The UE action plan towards low carbon economy is in line with conclusions of the 21st UN Climate Change Conference (COP21), held in Paris in 2015. The current EU climate and energy objective is to reduce its GHG emissions by 40% below 1990 levels by 2030. The transportation must play a vital role in achieving the target. The transport sector in Poland now is responsible for 15% of the total domestic CO_2 emissions and this share is expected to grow as a consequence of economic development. The potential for CO_2 reduction in this sector is significant. Unlocking this potential can be beneficial for the Polish economy and rational in terms of cost. For example, it would help to stem dependence on oil import, contribute to air quality improvement and accelerate the process of transport infrastructure modernization. Defining the pathways of the future development of the transport sector in Poland under given environmental restrictions requires extensive analysis and appropriate tools to evaluate the effectiveness of various policy instruments, which affect the level of demand and consequently the GHG emissions. In the article, a new "bottom-up" energy model has been proposed to make such assessments. The model, named STEAM-PL (Set of Tools for Energy Demand Analysis and Modeling), has been elaborated to predict future energy demand and emissions of pollutants based on long-term economy and policy scenarios in all economy sectors in Poland. The paper presents the methodological basis for the construction of the model and shows the results obtained for different scenarios concerning the future development of the transport sector.

1 Introduction

Mobility is a fundamental need in modern society. Unfortunately, dynamic growth of transport leads to higher GHG emissions (including carbon dioxide being the main component of the GHG) as a result of increase in energy demand. The rise in fuel consumption also reflects negatively on trade balances and energy security of the oil

* Corresponding author: skwierz@are.waw.pl

importing countries. It exposes these countries to the risk of oil prices fluctuations, what represents a real menace to their economic growth prospects.

Poland is nearly fully dependent on oil supplies from abroad. Limiting energy consumption as a consequence of CO_2 abatement, coming from the commited obligations, can bring a lot of advantages for the economy and the environment, providing that the development strategy is implemented resolutely. In the EU action plan towards low carbon economy, it is indicated that countries being EU members need to reduce its GHG emissions by 40% to 2030 in comparison with 1990 [1]. The transport sector has to play a vital role in the process. Reduction target to 2030 for non-ETS sectors is a subject of ongoing work. Current proposals suggest necessity of 7% reduction level in reference to 2005 for Poland in the following sectors: transport, construction, waste management and agriculture [2]. Adoption of proposed solutions would mean a very serious challenge for the country, which economy is constantly growing and aspire to reach EU's average level of development on the horizon for 2030.

In 2014, the transport sector emitted 43.6 Mt of CO_2, what was about 15% of the total country emissions. In the future this share is expected to grow as it is observed in other European developed countries. In view of the structure and specific features of the transport sector in Poland, achieving any reduction in comparison with 2005 seems to be impossible. Since the 2005 up until now, considered emissions have already grown up by 26%.

In order to meet the CO_2 reduction goals in a secure and sustainable manner, governments must set in motion legislative initiatives which will provide citizens with incentives for following more sustainable patterns of consumption. Any policies to reduce emissions should be preceded by an impact study. In Poland there have not been almost any research on the development of the transport sector (in the context of the demand for fuels and energy carriers and emissions) with the application of advanced mathematical models that take into account current specific national conditions [3].

The main aim of the article is to present a methodology and a new "bottom-up" model STEAM-PL designed for energy demand and emission of pollutants evaluation in the medium and long term in Poland. Dedicated for the transport sector module STEAM_T, which is a part of the package STEAM-PL, allows to analyze the quantitative effects of implemented policies and measures, considering specific conditions of the Polish transport sector.

2 Energy consumption and CO_2 emissions in the transport sector in Poland

The transport sector is an area of activity, where the most significant energy demand growth took place among all other sectors, after political transformation in Poland.

Table 1. Final energy consumption in transport sector in Poland in years 1995-2014 [ktoe]

	1995	**2005**	**2014**
Gasoline	4 610	4 230	3 535
Diesel	2 796	5 657	9 154
LPG	193	1 702	1 744
CNG	5	15	24
Electricity	402	343	259
Jet fuels	269	319	584
Natural gas	0	236	362
Biofuels	0	49	705
TOTAL	**8 275**	**12 551**	**16 367**

Source: Energy Market Agency – ARE S.A.

 DOI: 10.1051/e3sconf/20171401010

Energy consumption in this sector nearly doubled since 1995 and reached the level of 16.4 Mtoe in 2014 [Fig. 1]. Growing share of energy consumption in transport is associated with the increasing role of the road transport, both passenger and freight.

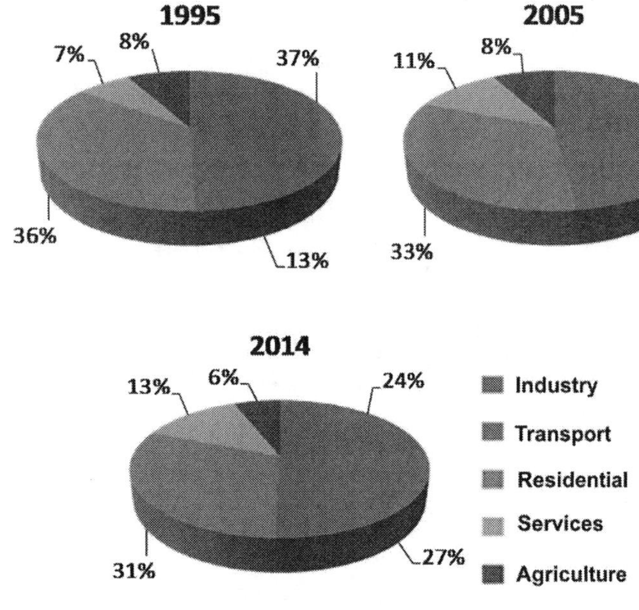

Fig. 1. Final energy consumption shares in particular economy sectors in Poland

In the non-ETS sectors the criteria of the measurement of CO_2 emissions are not defined. That is why they are estimated according to the amount of energy carriers consumption basing on CO_2 emission factors. Calculated by The National Centre for Emissions Management (KOBiZE) CO_2 emission volumes in transport in 2014 were at 43.6 Mt [4], what constituted 15% of the total country emission of the given component [Fig. 2].

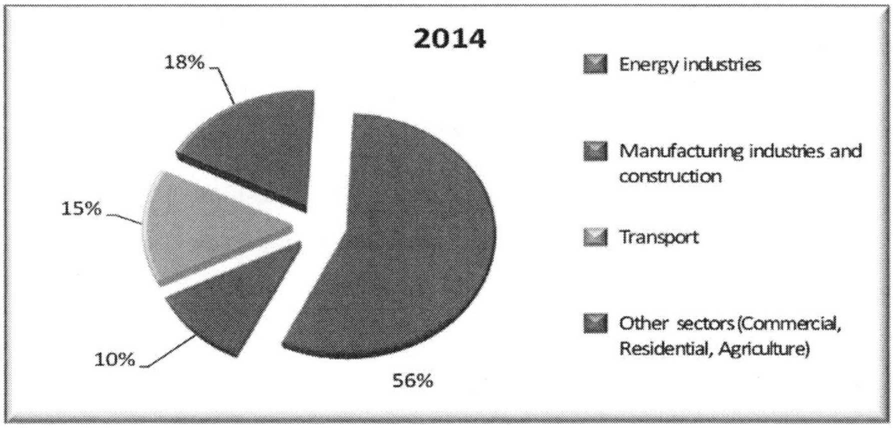

Fig. 2. CO_2 emission shares by economic sectors in 2014

Energy efficiency in transport clearly had an influence on the decrease in the rate of emissions growth. But this effect was not strong enough to balance the implications of the growing demand for the transport activity. One of the main reasons of this was a rebound effect (more fuel efficient car = higher its weight and more intensive usage). Among different modes of transport, road transport is a key one in terms of CO_2 reductions [Fig. 3].
In 2014 this type of mode accounted for 97% of the total CO_2 country emissions.

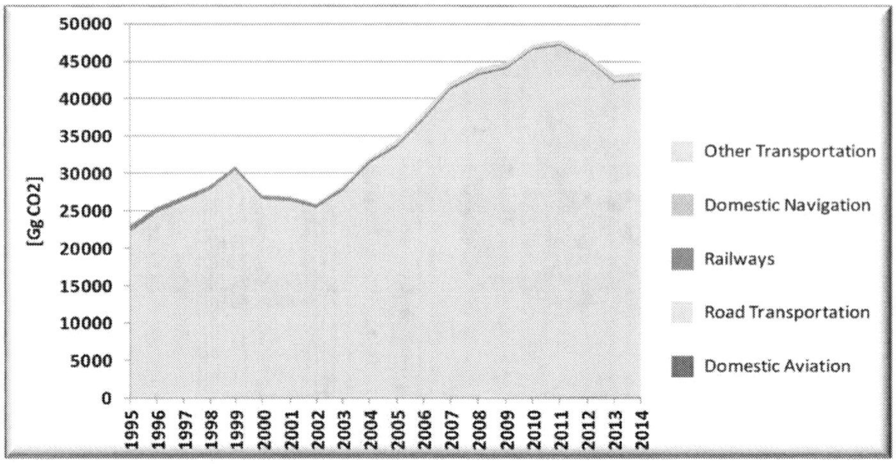

Fig. 3. CO_2 emission by transport branches in years 1995-2014

3 CO_2 reduction measures in transport

Technology improvements alone will not be sufficient to stop the growth in greenhouse gas emissions from transport, other policy measures will be necessary.
Paragraphs below provide a list of examples of CO_2 reduction measures that can be taken into account in the transport sector in Poland. Measures in this list can be divided into technical and other measures, that physically reduce the CO_2 emissions of transport.

1. The improvement of fuel efficiency of vehicles and reduction of emissions in road transport
    a. Improved combustion engines
    b. Improved aerodynamics
    c. Weight reduction
    d. Engine downsizing + turbo
    e. Low rolling resistance tyres
    f. Fuel efficient driving, lower driving speed
2. Alternative fuels (market-orientated measures to influence users choice towards more fuel-efficient cars):
    a. LPG, CNG, Liquid biofuels and biogas, Electricity, Hydrogen
    b. Hybrid vehicles
3. Modal shift in passenger transport
    a. Prioritization public transport and non motorized modes
    b. High speed rail system development
4. Modal shift in freight transport
    a. Promoting intermodal transport
    b. Development of inland transport
5. Transport infrastructure modernization
    a. Modernizing infrastructure and the service of conventional rail transport
    b. Inteligent Transportation Systems
6. Improved logistics and goods distribution process [5].

DOI: 10.1051/e3sconf/20171401010

4 Methodological approach

Proposed in the article model, named STEAM (**S**et of **T**ools for **E**nergy Demand **A**nalysis and **M**odelling) consists of five independent modules, dedicated for analysis of particular economy sectors (Fig. 4).

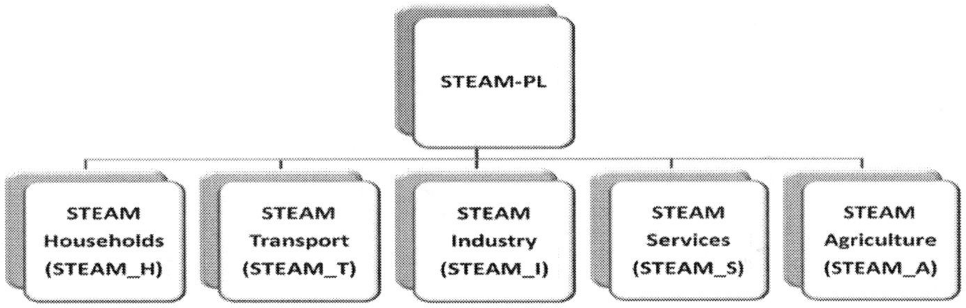

Fig. 4. STEAM-PL modular structure

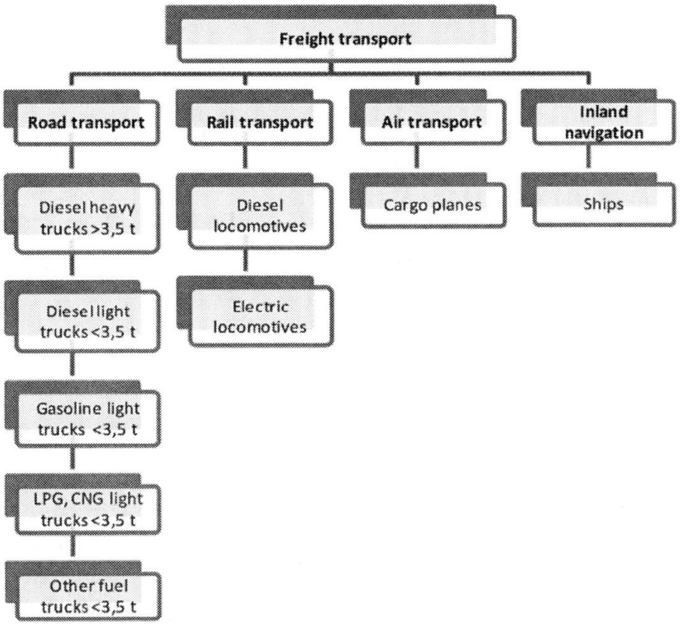

Fig. 5. Model STEAM_T structure – Freight transport

 DOI: 10.1051/e3sconf/20171401010

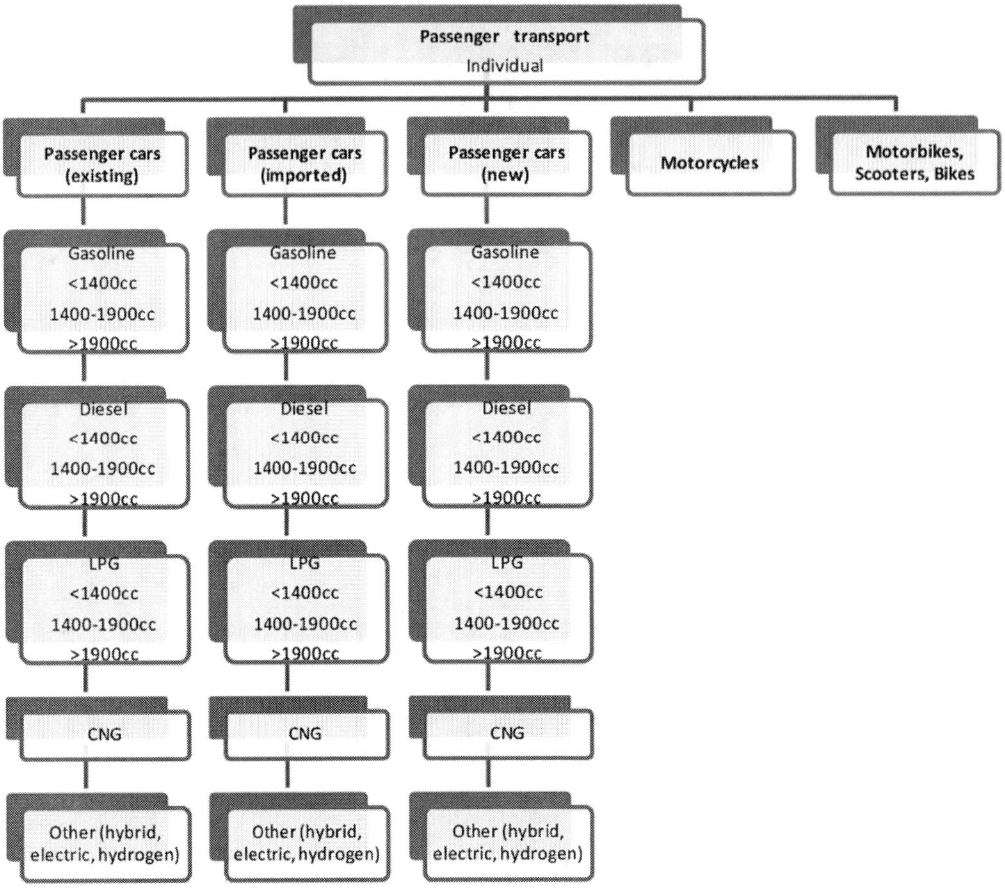

Fig. 6. Model STEAM_T structure – Individual passenger transport

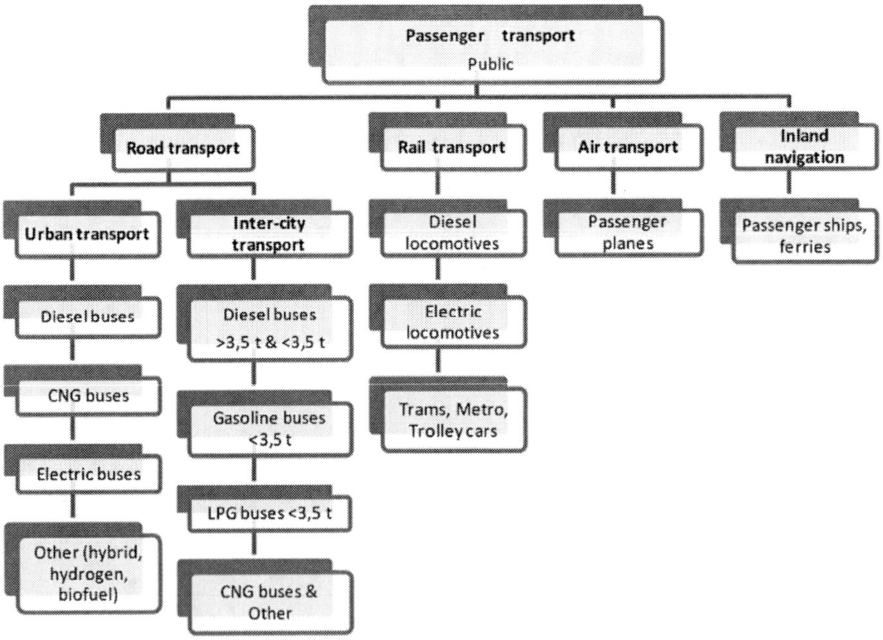

Fig. 7. Model STEAM_T structure – Public passenger transport

STEAM-PL is an integrated „bottom-up" and "top-down" model, which reflects all important technical aspects of energy utilization, including technological development. An engineering approach, has been supplemented with econometric modeling of market shares, employed to estimate the penetration of supply alternatives. A market share algorithm is used in the model to determine the response of various energy demand segments to changes in energy prices and costs of particular technologies. A market share algorithm has been adopted from the ENPEP/BALANCE methodology [6]. This solution makes possible to analyze the effects of various economic instruments e.g. fuel tax, cars subsidizing. STEAM-T is a module dedicated for analysis of the transport sector in Poland. Very detailed structure basing on the most accurate statistical data verified by different sources allows to include various specific conditions of functioning of the transport sector in Poland, like: average age of vehicles and their technical condition, number of vehicles really in use, which is totally different from the number of vehicles registered and high share of imported cars. STEAM-T module generates energy demand and emissions of pollutants projections broken into passenger (individual, public) and freight transport modes as well as different types of vehicles specified within each group. The demand for a transport fuel is equal to the product of vehicle utilization rate and total stock of vehicles:

$$D_t = S_t * R_t * U_t \tag{1}$$

Where D_t – the demand for fuel at time t
S_t – the vehicle stock
R_t – the utilization rate (kilometers per year)
U_t – the unit energy consumption (litre per kilometer)

The level of vehicle stock is linked with the economic, social and technological growth. The pace of a new car additions is limited by vehicles per 1000 population ratio, estimated by using comparisons with other European countries. The model employs a market share algorithm using a logit function to estimate the share of different fuel and vehicle types (2):

$$MS_i = \frac{(1/(P_i \times Pm_i))^{\gamma}}{\sum_i^n (1/(P_i \times Pm_i))^{\gamma}} \tag{2}$$

where MS – market share of a given type of vehicle
i – given type of vehicle
P – cost of operation
PM – premium multiplier
γ – price sensitivity coefficient

STEAM_T module has a very detailed structure, reflecting all key elements and aspects of the functioning of the transport sector in Poland (Fig. 5, 6 & 7).

Table 2. Main assumptions

<table>
<tr><td>Time horizon</td><td>2030</td></tr>
<tr><td>GDP growth</td><td>2016-2020 3.9% AAGR, 2021-2030 3.3% AAGR Source: [7]</td></tr>
<tr><td>Population</td><td>2020 – 38.2 mln, 2030 – 37.2 mln Source: [8]</td></tr>
<tr><td>Fuel prices

ON1 – diesel for non-commercial use

ON2 – diesel for commercial use</td><td>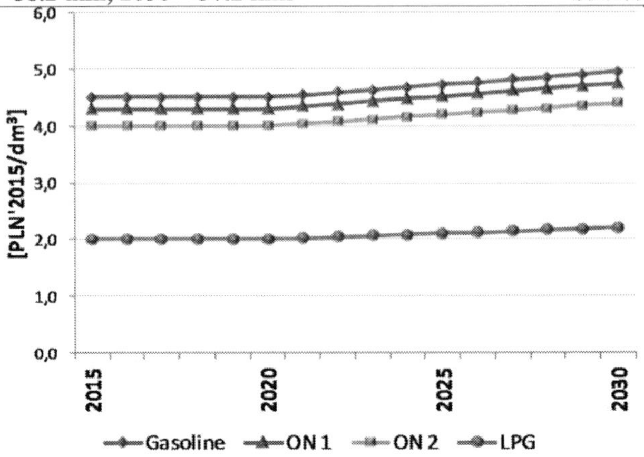

Source: [9], own assumptions</td></tr>
<tr><td>Number of passenger cars (in use)</td><td>2020 – 16.0 mln
2030 – 17.5 mln Source: Econometric model</td></tr>
<tr><td>Passenger transport activity</td><td>2020 – 410 [Gpkm]
2030 – 461 [Gpkm] Source: Econometric model</td></tr>
<tr><td>Passenger cars average mileage (annual) [km]</td><td>
<table>
<tr><th></th><th>2020</th><th>2030</th></tr>
<tr><td>Gasoline cars</td><td>7 000-8 500</td><td>7 500-9 000</td></tr>
<tr><td>Diesel cars</td><td>10 560-11 000</td><td>11 000-11 320</td></tr>
<tr><td>LPG cars</td><td>10 200-10 670</td><td>10 500-11 440</td></tr>
<tr><td>CNG cars</td><td>12 000-12 400</td><td>13 000-13 900</td></tr>
<tr><td>Electric cars</td><td>7 900-8 760</td><td>8 300-9 710</td></tr>
</table>
Source: [10], own assumptions</td></tr>
<tr><td>Passenger cars fuel consumption [dm³/100km]</td><td>
<table>
<tr><th></th><th>2020</th><th>2030</th></tr>
<tr><td>Gasoline cars <1400cc</td><td>5.3-6.6</td><td>5.1-6.3</td></tr>
<tr><td>Gasoline cars 1400-1900cc</td><td>6.3-7.8</td><td>6.1-7.6</td></tr>
<tr><td>Gasoline cars >1900cc</td><td>7.9-9.8</td><td>7.7-9.6</td></tr>
<tr><td>Diesel cars <1400cc</td><td>4.6-5.7</td><td>4.4-5.5</td></tr>
<tr><td>Diesel cars 1400-1900cc</td><td>5.9-7.3</td><td>5.7-7.1</td></tr>
<tr><td>Diesel cars >1900</td><td>6.8-8.5</td><td>6.6-8.3</td></tr>
<tr><td>LPG cars <1400cc</td><td>6.0-7.6</td><td>5.8-7.3</td></tr>
<tr><td>LPG cars 1400-1900cc</td><td>10.9-13.7</td><td>10.6-13.2</td></tr>
<tr><td>LPG cars >1900cc</td><td>12.6-15.7</td><td>12.2-15.3</td></tr>
<tr><td>Hybrid cars</td><td>4.0-5.0</td><td>3.5-4.5</td></tr>
<tr><td>CNG cars [Nm3/100 km]</td><td>9.0-11.3</td><td>8.8-11.0</td></tr>
<tr><td>Electric cars [kWh/100km]</td><td>12.8-13.5</td><td>11.0-12.5</td></tr>
</table>
Source: [10], car manufacturers data, own assumptions</td></tr>
<tr><td>Transport of goods</td><td>2020 – 2 096 519 [1000 t]
2030 – 2 301 670 [1000 t] Source: Econometric model</td></tr>
<tr><td>Freight transport activity</td><td>2020 – 410 [Gtkm]
2030 – 473 [Gtkm] Source: Econometric model</td></tr>
<tr><td>Number of trucks (in use)</td><td>Heavy trucks >3,5 t 2020 – 0,7 mln, 2030 – 0,72 mln
Light trucks <3,5 t 2020 – 2,1 mln, 2030 – 2,1 mln
Source: Econometric model</td></tr>
<tr><td>Trucks average mileage (annual) [km]</td><td>
<table>
<tr><th></th><th>2020</th><th>2030</th></tr>
<tr><td>Diesel heavy trucks >3.5t</td><td>28 700</td><td>31 900</td></tr>
<tr><td>Diesel light trucks <3.5t</td><td>7 600</td><td>7 700</td></tr>
<tr><td>Gasoline light trucks <3.5t</td><td>15 300</td><td>17 000</td></tr>
<tr><td>LPG light trucks <3.5t</td><td>19 500</td><td>19 000</td></tr>
</table></td></tr>
</table>

	CNG light trucks <3.5t	17 000	19 000
	Source: [10], car manufacturers data, own assumptions		
Trucks fuel consumption [dm^3/100km and Nm3/100km for CNG]]		**2020**	**2030**
	Diesel heavy trucks >3.5t	38.0	35.0
	Diesel light trucks <3.5t	9.2	8.5
	Gasoline light trucks <3.5t	8.8	7.0
	LPG light trucks <3.5t	11.6	10.6
	CNG light trucks <3.5t	10.9	8.7
	Source: [9], own assumptions		
CO_2 emission factors	LPG – 63.1 kg/GJ Gasoline – 69.3 kg/GJ Jet fuel – 70.0 kg/GJ Diesel – 74.1 kg/GJ		*Source: [4]*

5 Main assumptions

Two scenarios have been prepared to analyze the future energy demand and possibility of CO_2 reduction in the transport sector. The first one (Business as usual scenario – **BAU**) assumes no active government policy in this field. Energy efficiency progress is made mainly basing on market rules. No additional solutions are taken into account, except those that have already been accepted.

The second one (Modernization scenario – **MOD**) assumes implementation of a sets of policy instruments, which require active role of government and public funds involvement, including:

1. Support the use of electric cars (approx. 1.4 million cars in use in 2030).
2. Support the use of hybrid cars (approx. 1.6 million cars in use in 2030).
3. Support the development of rail passenger and freight transport, including intermodal (increase the number of passengers transported by rail from 260 mln in 2014 to 460 mln in 2030, increase the weight of goods transported by rail in the total transported weight from approx. 12% to 20 % in 2030).
4. Support the process of introducing in the bus fleet of public transport the electric, CNG, hybrid and bio fuel combustion vehicles (the share of electric and hybrid buses in the total number of buses increases to 15% in 2030, while the CNG to 7%).
5. Support the development of the CNG technology in cars and trucks.
6. Support the development of water transport (increase the weight of cargo transported by inland waterways in the total weight of 0.7% to 1.3% in 2030)
7. Implementation of higher biofuels share in fuels sold (up to 10%).

Main assumptions taken into the calculations common for the prepared scenarios have been presented in the table 2:

6 Results

Results of elaborated scenarios in reference to energy demand and CO_2 emissions in the domestic transport sector have been presented below.

In the BAU scenario fuel consumption increases from 16.9 Mtoe to 19.7 Mtoe in years 2014-2030 [Fig. 8], while CO_2 emissions rises from 43.6 Mt to 50.4 Mt at the given time [Fig. 10]. In comparison with the year 2005, which constitutes the reference year set by ETS and non-ETS system, emissions increase by 46%. Implementation of the policy instruments listed in the section 5, allows to reduce the growth to 35% only. Energy demand in the MOD scenario increases to 18.6 Mtoe in 2030 (1.1 Mtoe less than in the BAU scenario) [Fig. 9]. CO_2 emissions reach the level of 46.6 Mt. This result is far from the expectations coming from

 DOI: 10.1051/e3sconf/20171401010

EU proposals and shows that CO_2 reduction in the Polish transport sector will be very challenging. More actions are needed exceeding those assumed in the MOD scenario in order to achieve deeper reductions.

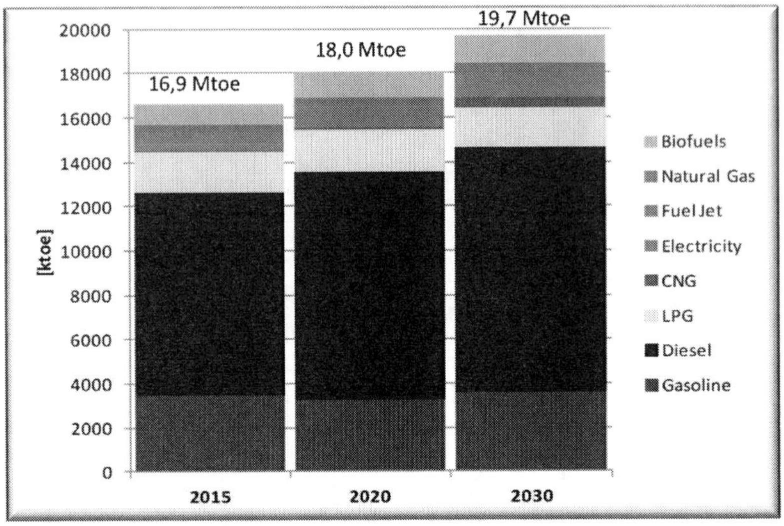

Fig. 8. Energy demand projection in the transport sector in Poland – BAU scenario

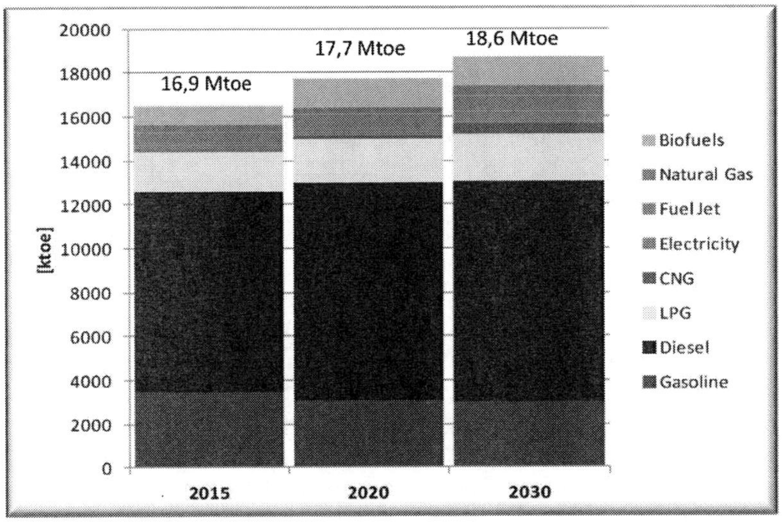

Fig. 9. Energy demand projection in the transport sector in Poland – MOD scenario

DOI: 10.1051/e3sconf/20171401010

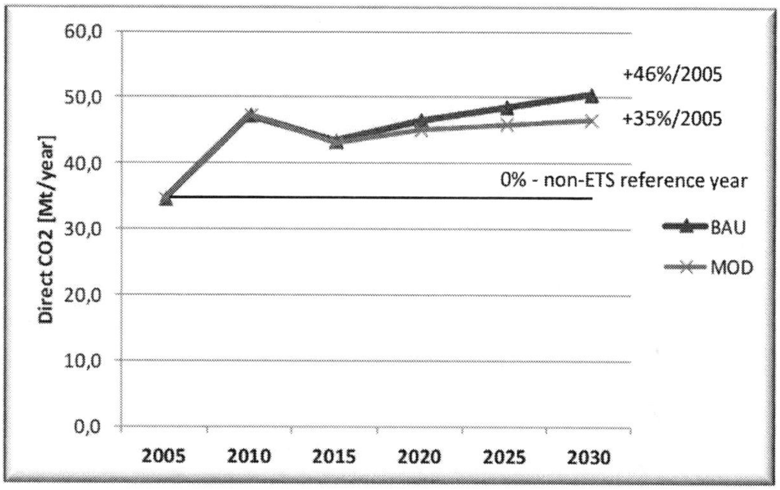

Fig. 10. CO_2 reduction pathways in the transport sector in Poland

7. Summary

Poland as a EU member and signatory to the Paris Agreement must face the challenge of CO_2 reduction up to 2030 and afterwards. The transport sector has to play an important role in this process. This sector accounts for approx. 15% of the total domestic emissions and its share is expecting to grow as a consequence of the economic growth. Actions taken towards low-carbon mobility can be beneficial for the Polish economy, providing that respective solutions are implemented resolutely with preventing excessive costs as one of the objectives. The key issue in the process of the strategy formulation is the correct assessment of the effectiveness of regulatory instruments. In the article a new energy model STEAM-PL and a dedicated to transport sector analysis module STEAM_T have been proposed to capture the effects on energy demand and emisson of pollutants.

The module is based on a "bottom-up" approach, supplemented with econometric modeling of market shares, employed to estimate the penetration of supply alternatives. A market share algorithm is used to determine the response of various energy demand segments to changes in energy and technology prices. Described functionalities and very detailed structure of the module considering specific conditions of the transport sector in Poland makes it a very useful tool to carry out quantitative analysis. Obtained results show that, without active role of the government CO_2 emissions will rise in the transport sector by approx. 46% in comparison with 2005. Implementation of a set of reduction measures listed in the article will allow to constrain the increase up to 35%, but still it is far from lastly proposed on the EU forum obligations. Further reduction requires additional measures of CO_2 abatement.

References

1. European Council – Conclusions on 2030 Climate and Energy Policy Framework. Brussels, 23 and 24 October 2014
2. European Commission, Press Release Database. Energy Union and Climate Action: Driving Europe's transition to a low carbon economy. Brussels, July 2016
3. P. Kaszyński, J. Kamiński, Mirowski T. – A proposal for an implementation of the transport sector model under Polish conditions. *Energy policy Journal* (2013)

4. National Centre for Emissions Management (KOBiZE) - Poland's Informative Inventory Report 2016. Warsaw, February 2016
5. R. Smokers, A. De Buck, M. Van Valkengoed – Marginal abatement costs for greenhouse gasemission reduction in transport compared with other sectors. Delft, May 2009.
6. Argonne National Laboratory. Energy and Power Evaluation Program (ENPEP-BALANCE). Brief Model Overview – Version 2.25. Chicago Illinois, January 2008
7. Ministry of Finance – Macroeconomic assumptions for the local government unit forecasting purposes. Warsaw, May 2016.
8. Central Statistical Office – Projections of population in Poland in years 2014 -2050. Warsaw, October 2014.
9. International Energy Agency – World Energy Outlook 2015. Paris, November 2015.
10. J. Waśkiewicz, Z. Chłopek, Pawlak P. – Professional forecasts of the road transport activity changes. Motor Transport Institute, Warsaw October 2012.

Ecological and economical aspects of solar energy use

Wiktoria Sobczyk[1]*, *Patrycja* Pelc[2], *Barbara* Kowal[3] *Robert* Ranosz[4]

[1] Faculty of Mining & Geoengineering, AGH University of Science & Technology, Al. Mickiewicza 30, 30-059 Krakow, Poland
[2] Regional Directorate of Environmental Protection in Warsaw, Faculty of Environmental Impact Assessment, Poland
[3] Faculty of Mining & Geoengineering, AGH University of Science & Technology, Al. Mickiewicza 30, 30-059 Krakow, Poland
[4] Faculty of Mining & Geoengineering, AGH University of Science & Technology, Al. Mickiewicza 30, 30-059 Krakow, Poland

Abstract. The natural environment is devastated by the industry and by households, which use fuels such as coal, oil and natural gas for heating water and buildings as well as for generating electricity. One of the ways to limit the degradation of the environment and the usage of natural resources is to use unconventional and renewable resources and to implement energy-saving technologies in construction, industry and households. A dynamic development of systems based on renewable energy sources such as biomass, water, wind, heat from inside the Earth and sunbeams has recently occurred in the whole world. This thesis is related to the purest and the least onerous for environment source of energy which is undoubtedly the energy that comes directly from the solar radiation. In this article the solar energy application is presented, taking into account its impact on the environment and financial costs of developing solar techniques. The installation of alternative energy sources are entitled to grants and credits, which reduces the investment costs. Ecological effects, although immeasurable, are significant and definitely testify in favor of the solar system.

1 Introduction

As of the beginning of the 19th century and - in some parts of the world - even until the beginning of the 20th century, almost all energy used for human consumption came from wind, water and wood, which stored the Sun's energy. At a later point in time, people started exploiting fossil energy sources from below the Earth's surface. Their use contributed to the rapid evolution of transport and building technologies. As at the end of the 20th century, people began to notice various disadvantages of fossil fuels: limited resources, unequal distribution and environmental pollution caused by their combustion.
Future developments in science taught people how to use solar energy for both heating the dwelling houses and water as well as for producing electricity. The most important parameter

* Corresponding author: sobczyk@agh.edu.pl

in solar power technology is the total annual incident radiation, which reaches the ground surface at a given point (in Poland - approx. 950 kWh/m^2) [1].

Solar power is used for human consumption both passively and actively. Active solar techniques related to the process of converting solar radiation energy into other types energy mainly include the photothermal conversion (conversion of solar radiation into thermal energy in solar collectors), photovoltaics (conversion of solar radiation into electricity using photovoltaic panels) and photochemistry (chemical effects of light).

2 Sustainable energy and energy security

Sustainable development of energy is mainly based on its durability, which should be understood as the long-term availability. Renewable sources of energy are considered sustainable; however, their application must include such factors as source access, generation capacity, demand and dispersion in a given area.

Energy security is understood as the ability to access and use inexpensive, secure and environmentally friendly energy. One of the basic obligations of a state energy policy is to ensure the adequate level of energy security. In the 1970s, the term "security" was mostly associated with the area of military security. Now, economy and industrial economics in particular are becoming increasingly important. Security is construed not only as a lack of specific threats, but also as having opportunities to develop. The energy security of a state depends on the condition of its domestic infrastructure, including the technical condition and efficiency of the facilities and systems used in transport and distribution of fuels and energy. Moreover, it depends on the size of reserves. Another functional aspect of energy supply security is the system capacity to counteract sudden disturbances. These may be breakdowns of transport systems or loss of a supply sources.

The negative impact of supply reliability of any specific energy carriers becomes particularly sharp when raw materials cannot be obtained, when their prices increase, but also when the society is reluctant to use a specific fuel. And so, energy security as defined above, is a highly complex category. Energy security indicators, as well as its measuring units and standards should be analyzed in terms of technical (probability of supply disruption indicator),economic (costs of energy acquisition and costs of losses related to the lack of energy), and environmental (emission to the atmosphere plus the resultant threats) perspectives. In each of the areas highlighted above, threats may occur randomly or through an intentional effort of a state or organization, but they can also result from changes in the economic, political or natural environment (climate changes) [2, 3].

In terms of reduction of GHGs emissions, the following goals were formulated:
- reduction of domestic GHGs emissions by 80% until 2050 when compared with emissions in 1990 (in all EU Member States);
- program of obligatory reduction of GHGs emissions in the subsequent years: 25% in 2020, 40% in 2030, 60% in 2040, 80% in 2050 [4].

More widespread utilization of low-emission technologies is emphasized in particular. In terms of energy generation, using energy from renewable sources and other low-emission solutions will be promoted. Improving the efficiency of transport will be an important priority, while another important task will be reducing energy consumption in residential construction. Strong support for renewable energy sources is reflected in its enormous share in the gross final energy consumption: approx. 75% in 2050, with a 97% share of renewable energy sources in electricity production.

3 Economic analysis of the use of solar power

The cost of using solar panels is very low. It includes the cost of driving the circulating pumps and the cost of periodical replacement (at intervals of several years) of the fluid used in the thermal circuit of the system. The most beneficial is the installation of photovoltaic cells, which are also characterized by the longest service life. The most expensive are the systems using vacuum tube collectors and flat-plate collectors. On the other hand, organic solar cells entail lower operational costs. In Poland, the cost of a complete solar set for a single-family house, where 4 people live, is PLN 16,000 (EUR 3500) depending on the type of the installed collectors, their number and execution technology.

Passive solar systems can use some elements of the building (e.g. glazing) to accumulate heat. No intermediate medium is used for heat transport and transfer. Sun rays in the form of visible and infra-red waves which penetrate into the rooms, being accumulated in walls, floors and ceilings. The elements heat up and emit thermal radiation, producing the greenhouse effect. Apart from passive solar systems, installation of photovoltaic cells, solar collectors and use of other alternative sources of energy, the construction of a low-energy house, which constitutes a method for reducing the cost of using such building, seems the most beneficial from the point of view of environmental protection and potential return on investment [5].

When building a single-storey house with a converted loft of a total floor area of 115 m^2 and a pitched roof, the works begin with laying the foundations [6]. The cost of foundation slab and thermal insulation is PLN 180/m^2 (EUR 40/m^2).

Table 1. Cost of building a traditional house and low-energy house

Operation	Traditional house PLN/EUR	Low-energy house PLN/EUR
raw state (laying foundations and building foundation walls, creating anti-damp insulation and thermal insulation)	15000/3330	21000/4660
building shell	105000/23300	105000/23300
windows, window sills, entrance door	24000/5300	33000/7300
total cost of installation	57100/12700	114300/25400
cost of insulation	38800/8600	53500/11900
cost of building	240000/53300	327000/72200

The difference between expenses incurred in case of a traditional house and low-energy house (table 1) of the same size is approx. PLN 90,000 (EUR 20,000) and may seem major. Nonetheless, any additional investments involved will be fully recouped thanks to efficient thermal and electric energy consumption. If we decide to purchase a prefabricated frame house and dispense with a couple of technical elements, such as a heat pump, we will reduce the costs of building the low-energy house, hence its price will be comparable to the price of an average brick house [7]. Exploitation of an eco-friendly building should ensure maximum use of sunshine to light rooms, low energy consumption thanks to alternative energy sources and appropriate location, recycling of the so-called greywater, and compounding into the surrounding [8].

4 Ecological analysis of the use of solar power

Ecological effect was calculated on the basis of the simulation performed in the CASAnova program [9], using the selected building permit design [10].
The following formula was used to calculate the amount of the fuel used (B):

$$B = \frac{E}{Qu}[kg], [m^3] \tag{1}$$

where: E - energy demand in the building, MJ
Q_u - calorific value of the fuel (the amount of heat which is emitted during the complete combustion), MJ/kg, MJ/m^3.

The amount of gaseous emission of polluting substances emitted into the atmosphere as a result of energy-releasing fuel combustion (table 2) was calculated on the basis of the following formula:

$$E = B \times w \; [kg] \tag{2}$$

where: w - emission factor, kg/Mg, kg/mln m^3
B - amount of fuel burned, Mg.
For dust:

$$E = B \times w \times \frac{100-\eta}{100-k} \; [kg] \tag{3}$$

where: B - amount of fuel burned, Mg;
w - emission factor, kg/Mg;
η - operability of the dust collecting device, %;
k - amount of flammable parts in dust, %.

Table 2. Data for calculating ecological effect of combustion of various fuels [11,12, own elaboration]

Fuel	Lower Heating Value	Energy demand in the building for domestic hot water	Amount of fuel	Ash content A, %	Amount of flammable parts in dust, %	Amount of total sulphur S%, mg/m^3
Bituminous coal	28 [MJ/kg]	24334,54 [MJ]	869.091 [kg]	5	10	0.41
Coke	27 [MJ/kg]		901.279 [kg]	9		0.6
Natural gas	39,5 [MJ/m^3]		616.06 [m^3]	0		40

The rates of reduction of the SO_2, NO_x, CO, CO_2 and dust emission in case of meeting the energy demand in the building for domestic hot water through the solar collector system are showed on the figures 1-5.

 DOI: 10.1051/e3sconf/20171401011

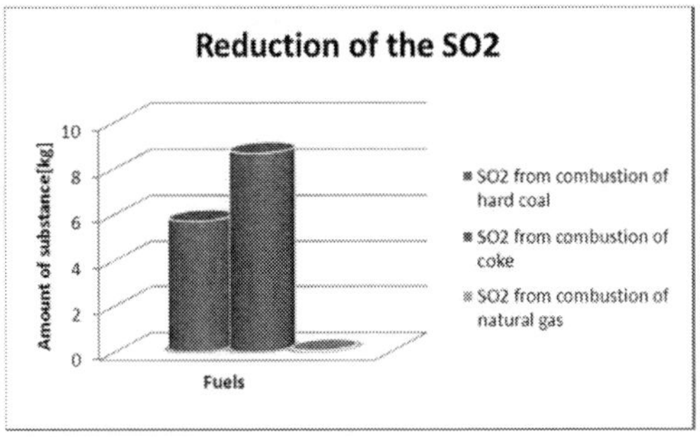

Fig. 1. The rate of reduction of the SO_2 emission in case of meeting the energy demand in the building for domestic hot water through the solar collector system [own elaboration]

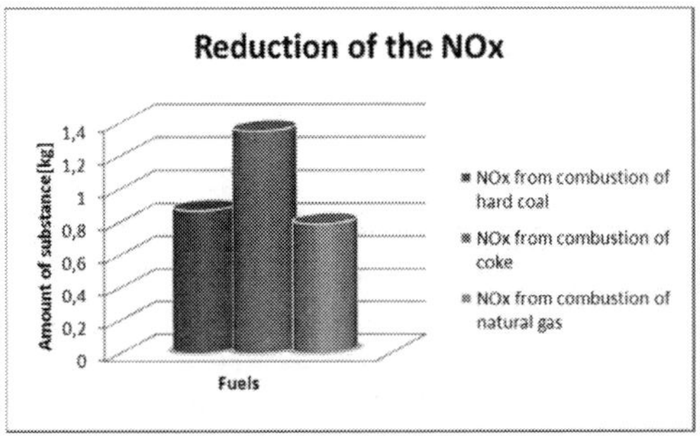

Fig. 2. The rate of reduction of the NOx emission in case of meeting the energy demand in the building for domestic hot water through the solar collector system [own elaboration]

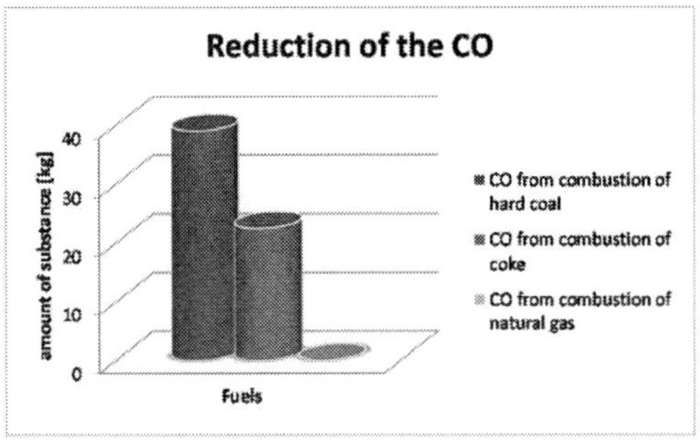

Fig. 3. The rate of reduction of the CO emission in case of meeting the energy demand in the building for domestic hot water through the solar collector system [own elaboration]

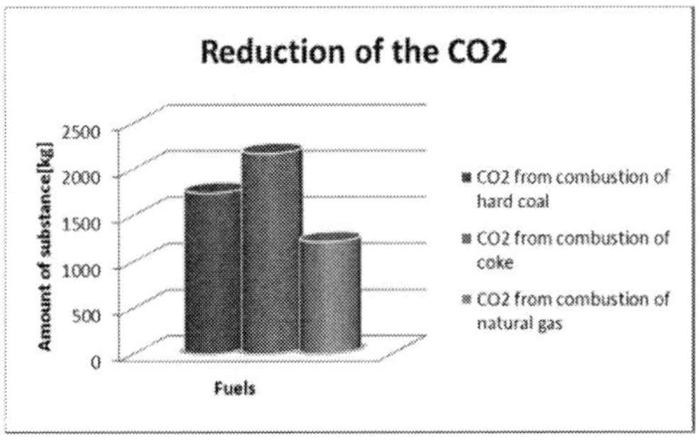

Fig. 4. The rate of reduction of the CO_2 emission in case of meeting the energy demand in the building for domestic hot water through the solar collector system [own elaboration]

**CO_2/100 – rate of reduction of the CO_2 emission divided by 100 to increase readability of the graph*

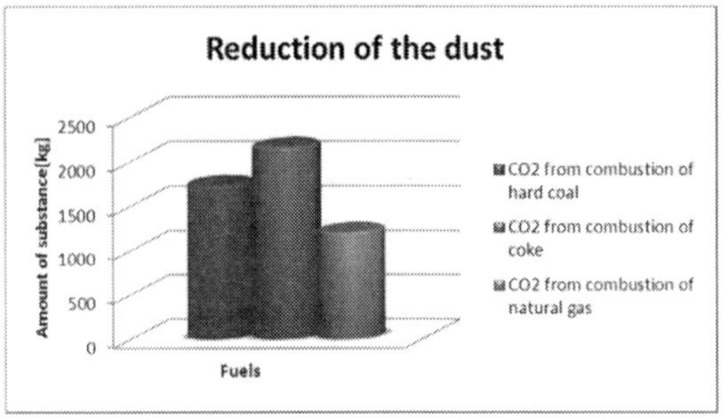

Fig. 5. The rate of reduction of the dust emission in case of meeting the energy demand in the building for domestic hot water through the solar collector system [own elaboration]

According to the calculations, it is evident that burning coke causes the greatest emission of pollutants. Combustion of coke causes greater emission only in case of CO. The smallest amount of pollution occurs due to combustion of natural gas. Changing the heat source into solar power will prevent emission of hazardous substances in quantity given in the table 3. It must be remembered that the aforesaid pollutants are emitted by one single-family house only. On a global scale, the change of the energy source allows to avoid emission of huge amounts of pollutants.

Investment outlays on the installation of the solar power systems differ depending on the implemented technology and are much higher in case of systems used for producing electricity than in case of devices used for the production of heat energy. However, the conversion of solar radiation also has some negative environmental impact, since the production of collectors and cells consumes energy and requires natural resources, whose excavation and processing causes emission of greenhouse gases.

Table 3. The results of calculation of reduction of hazardous substances emission in case of meeting the energy demand in the building for domestic hot water through the solar collector system [own elaboration]

Fuel	**SO_2**	**NOx**	**CO**	**CO_2**	**Dust**	**Energy demand in the building for domestic hot water**	**Amounts of hazardous substances**
	[kg]						
Bituminous coal	5.701	0.869	39.109	1738.182	7.242	24334.54 [MJ]	869.091 [kg]
Coke	8.8652	1.352	22.532	2163.07	13.519		901.279 [kg]
Natural gas	0.049	0.789	0.222	1209.95	0.01		616.1 [m^3]

5 Environmental education of the society – a chance for environmental protection

The environmental education of society in the field of utilizing alternative solutions in building thermoregulation should therefore be the top priority for state environmental and energy policy. Raising awareness of the benefits of the implementing non-conventional solutions in residential buildings may bring about huge environmental and social gains. Thoughtless discharge of heated air into the atmosphere through building chimneys is a needless wastefulness [13-17].

So far 40.000 passive houses have been built in the world [18]. Inhabitants of these buildings confirm that the costs of energy have significantly dropped, while the comfort indoors remains high. Thermomodernization undertakings are a great opportunity for environmental protection, development of the labour market and cities in the future.

In terms of energy use, Poland cannot demonstrate required efficiency. This means that the depletion of natural resources, fossil fuels and environmental pollution are gradually weakening the national economy [19-20]. All the buildings in Poland completed in the past years must undergo thermomodernization and the construction of new passive buildings should commence. The cost of building a passive house is slightly higher than the cost of a traditional building (an increase by 15% of the investment outlays), however, passive constructions require no financial expenditure for the heating system. Therefore additional costs are compensated for fairly quickly. Furthermore, the construction of a passive house does not require testing new building materials as well-known materials are used, such as are also used in traditional construction. Erecting passive buildings is based on technologies which have already been tried and tested. This means that passive technologies may be used in a wide range of applications.

Educating the society about the possibilities offered by the modern construction industry is necessary for passive buildings to become more popular. Investors usually concentrate on standard solutions – cheap and easy to implement, without making the effort to find out what else is available. The designer's task is to present all solutions and to spread knowledge about innovations, so that we can enjoy living in healthy, energy-efficient, green houses.

6 Conclusions

Sustainable development of energy requires reduction of pressure on the environment by dispensing with large amounts of goods, whose production is related to the demand for natural resources. The supply of energy raw materials must be also secured, which is possible thanks to the reduced energy consumption and replacement of traditional energy sources by renewable sources.

To solve the problem of excessive emission of greenhouse gases caused by the combustion of fossil fuels, the return to the renewable energy sources must be implemented. The total resources of energy from renewable sources far exceed the demand for both current and future generations.

An economy must be built on the basis of sustainable development, where the correlations between economic growth, concern for the natural environment and quality of life are deliberately shaped.

Acknowledgements: The paper has been prepared within the AGH University of Science and Technology – statutory research work No 11.11.100.482.

References

1. M. Nowicki, (Nadchodzi era Słońca), Państw. Wyd. Nauk., Warszawa (2012)
2. K. Stala-Szlugaj, Solid Waste from Hard Coal Combustion in the Municipal and Housing Sector, Ann. Set The Env. Prot., **14**, 909-918 (2012)
3. K. Stala-Szlugaj, PM Emission from Hard Coal Combustion in Heating Plants of Thermal Power Under 50 MW According to Obligatory Standards Emission , Ann. Set The Env. Prot., **15**, 1689-1704 (2013)
4. http://ec.europa.eu/clima/citizens/eu/index_pl.htm

5. A. Ostrowska, W. Sobczyk, M. Pawul, Evaluation of economic and ecological effects of solar energy on the example of a single-family house, Ann. Set The Env. Prot., 15 (2013)
6. Ustawa – Prawo budowlane. Dz.U. 2016, nr 0, poz. 290 Rzeczypospolitej Polskiej z dnia 9 lutego 2016 r. w sprawie ogłoszenia jednolitego tekstu ustawy – Prawo budowlane (isap.sejm.gov.pl)
7. www.kb.pl
8. S. Belniak, M. Głuszak, M. Zięba M., (Budownictwo Ekologiczne - Aspekty ekonomiczne). Wyd. Nauk. PWN, Warszawa (2013)
9. http://nesa1.uni-siegen.de/index.htm/softlab/casanova_e.htm
10. H. Koczyk, B. Antoniewicz, M. Basińska, (Ogrzewnictwo praktyczne. Projektowanie. Montaż. Certyfikacja energetyczna. Eksploatacja), Wyd. SYSTHERM D. Gazińska s.j., Poznań (2009)
11. www.kwsa.pl
12. http://egesa.pl/gaz_ziemny
13. A. Malada, W. Sobczyk, (Uprawa roślin energetycznych jako forma aktywizacji środowisk wiejskich), Zesz. Nauk. Kat. Inż. Proc. Uniwersytetu Opolskiego, II, 92-98 (2005)
14. W. Sobczyk, The energetic basket willow's harvest – in the light of investigations, Polityka Energetyczna, 10, 2, 547-556 (2007)
15. W. Sobczyk, Evaluation of harvest of energetic basket willow, TEKA Kom. Mot. i Energ. Roln. PAN, XI, 343-352 (2011)
16. B. Klojzy-Karczmarczyk, Variability of Mercury Content in Various Fractions of Soils from the Vicinity of Krakow Ring Road Section, Ann. Set The Env. Prot. **16**, 1, 363-375 (2014)
17. B. Klojzy-Karczmarczyk, Analysis of Long-term Research on Mercury Content in the Soils in the Immediate Surroundings of the Southern Ring Road of Krakow, Ann. Set The Env. Prot. **15**, 2, 1053-1069 (2013)
18. gazetaprawna.pl/nieruchomosci/artykuly/domy-pasywne 2014
19. B. Kowal, D. Kowal, M. Karkula, Sustainable development in the model of mining company strategy. Pol. Journ.of Env. Stud., 21, 5A, 222-226 (2012)
20. R. Ranosz, Mining and its role in the global economy. Mineral Res. Manag., 30, 1, 5-20 (2014).

Forecast of power generation and heat production from renewable energy sources

Tadeusz Pydych[1], *Mateusz* Szydłowski[1], and *Janusz* Sowiński[1,*]

[1]Czestochowa University of Technology, Faculty of Electrical Engineering, 42-200 Czestochowa, Poland

Abstract. The share of renewable energy sources (RES) in the end use of energy in the UE will increase from the present level of about 25% to 50 % in 2030 according to the assumptions of the European Commission. In Poland the RES Act was passed in 2015. The act defines mechanisms and instruments for supporting the production of electricity and heat from renewable energy sources. Statistics (2003-2014) of electricity generation and heat production from RES in Poland were used in the research. Because of amendments to regulations connected with promoting RES and the emissions trading system (ETS) as well as the uncertainty associated with further directions of the energy and environmental policy, generation of electricity and heat based on the use of RES must be modelled while taking risk into account. A number of dynamic processes incorporating random events may be modelled by stochastic equations using Ito calculus. By applying Euler's method to solve stochastic differential equations (SDE), it is possible to simulate the development of the use of renewable energy carriers in electricity generation and heat production in the future.

1 Introduction

Current interest in renewable energy sources (RES) is largely due to the targets of the EU policy on energy generation and environment [1], but it also results from the fact that conventional energy sources, such as coal, oil and natural gas are running out [8, 10]. Besides, economic growth typically causes increase in demand for energy, including renewable energy, which has a significantly less detrimental impact on the environment, especially with respect to the greenhouse effect, than energy obtained from fossil fuels [9]. By using locally available sources, renewable energy boosts development of distributed power generation. The fact that resources are located at a short distance from consumers reduces the cost of energy transmission in two ways – it eliminates costly investment in transmission lines and prevents energy losses. On the other hand, it requires the strengthening and development of distribution lines, including investment in the smart grid. A dynamic development of innovative technologies has to be accompanied by stable operation of the power system.

The changing energy market calls for a strategy that will ensure energy security in Poland [5, 6]. Measures are taken in order to determine as precisely as possible the role of

* Corresponding author: jansow@el.pcz.czest.pl

Polish coal in the future and to predict a development of RES. This in turn implies that the energy market, including the power market, will be undergoing transformation.

Utilizing RES is at the moment the key point of all the activities aimed to reduce emissions in electrical energy generation and consumption of fossil fuels, especially coal and lignite. It is believed that further technological progress will increase efficiency of energy generation from RES and its cost will be lower [12]. Due to EU policies promoting RES [1] and imposing the emission trading system (ETS), renewable energy is becoming increasingly popular as a target of future investments. The growth of RES is not however completely unobstructed – the area of land where biomass can be grown is limited, there are environmental constraints (other than pollutant emission), and the infrastructure is insufficient [3, 4].

Until recently, the most important mechanism for supporting renewable energy producers in Poland was the system of energy certificates and sales tax exemption. The RES Act has modified this system by introducing auctions and solutions for prosumers.

According to the government's announcement, the amount of renewable energy purchased obligatorily in 2017 will not be 20 % of total purchased energy as was stipulated in the Act, but only 16 %, including 15.5 % of green certificates energy and 0.5 % of blue certificates energy (i.e. energy sold by agricultural biogas plants). Producers of non-renewable electrical energy are obliged to buy green certificates at the stock exchange and submit them to Energy Regulatory Office, or alternatively, they have to pay a fee. Excessive supply of green certificates negatively impacts their price at the stock exchange: at the beginning of 2014 the price of a green certificate was 250 PLN/MWh and in August 2016 it was 56 PLN/MWh, which is almost an 80 % drop. Withdrawing the support for renewable energy by diminishing the value of green certificates and at the same time decrease in energy prices puts investors into a disadvantage.

In July 2016 the so called distance act was passed, on which the minimal distance between a wind plant and houses or protected areas is to be ten times greater than the height of a wind turbine. The new regulation threatens both new projects, rendering most of them impossible to implement as well as existing wind plants, increasing taxes on them.

The amendment of the act on renewable energy passed on the 1st of July 2016 introduced new rules of settling accounts with prosumers by means of discounts, taking into account the balance between energy input into the grid, energy consumed and energy stored by a prosumer in the period of one year.

2 Statistics of electricity generation and heat production from RES in Poland

The present research uses the statistical report compiled by the Polish Central Statistical Office (GUS) "Energy from renewable sources", published in yearbooks [2]. The statistical data published by GUS [2] cover chronological data sets from the years 2003-2014. Sources regarded as RES include those exploiting the force of wind, sunlight, geothermal energy, sea waves, currents, tides and energy produced from solid biofuels, biogas and liquid biofuels as well as energy collected from natural environment by heat pumps.

For the calculations, statistics were used regarding electricity generation in GWh and heat production in TJ from renewable energy carriers in the years 2003÷2014. The statistics include chronological data sets regarding total production and production broken by the following carriers: water (power plants with capacity below 1MW, between 1 MW and 10 MW, and above 10 MW), wind, solid biofuels, biogas (including biogas from landfill sites, from sewage treatment plants and other biogas), bioliquids and solar energy.

DOI: 10.1051/e3sconf/20171401012

2.1 Obtaining primary energy from RES

The share of renewable energy in total primary energy has been increasing in the recent years in Poland, with the variation in the amount of total primary energy obtained being relatively small. The share of renewable energy in total primary energy in the years 2007-2014 is presented in Table 1, and the structure of renewable energy sources in 2007 and in 2014 is presented in Fig.1. Of all the RES, liquid and solid biofuels by far exceed other carriers, wind energy is rapidly increasing and hydropower is slightly decreasing.

Table 1. Share of renewable energy in total primary energy in the years 2007-2014 (Source: GUS [2])

Item	2007	2008	2009	2010	2011	2012	2013	2014
Total primary energy [TJ]	3040160	2985356	2816880	2824028	2882363	3038921	3006461	2853825
RES primary energy [TJ]	203141	226788	253352	287313	312148	355259	356693	337659
Share of RES energy in total primary energy	6.7%	7.6%	9.0%	10.2%	10.8%	11.7%	11.9%	11.8%

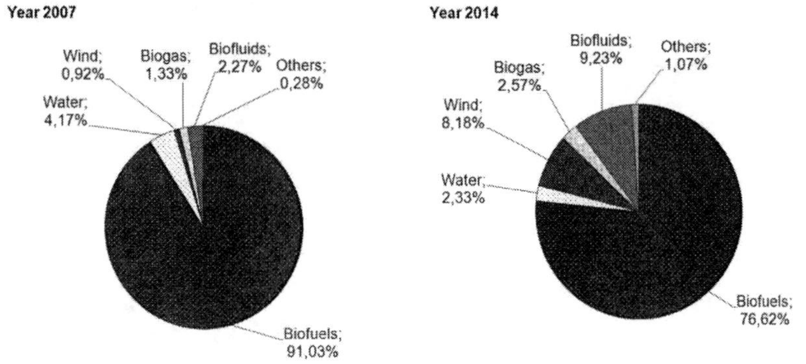

Fig. 1. Structure of renewable energy sources in 2007 and 2014 [2]

2.2 Generation of electrical energy from renewable sources

The calculations presented below are based on the statistics on electrical energy generation (in GWh) and heat generation (in TJ) from RES in the years 2003-2014. The statistics include chronological series of data on total generation and are divided into the following sources: water (hydroelectric plants of power below 1MW, from 1 MW to 10 MW and above 10 MW), wind, solid biofuels, biogas (including biogas from landfill, from sewage treatment plants and other), bioliquids and solar energy. Generation of electrical energy in the period 2003-2014 is presented in Table 2 and in Fig. 2, and its structure in 2003 and in 2014 is shown in Fig. 3.

DOI: 10.1051/e3sconf/20171401012

Table 2. Electricity generation from renewable energy sources in the years 2003-2014 [GWh] (Source: GUS [2])

Item	2003	2004	2005	2006	2007	2008	2009	2010	2011	2012	2013	2014
Total	2,250	3,074	3,848	4,291	5,429	6,606	8,679	10,889	13,137	16,879	17,066	19,842
Water of which power plants	1,671	2,082	2,201	2,043	2,352	2,152	2,375	2,920	2,331	2,037	2,439	2,182
below 1 MW	242	273.5	358.2	247.9	306.3	290.2	292.2	516	307	320.7	351.9	322
between 1 and 10 MW	431	616.9	504.2	566.6	658.1	605.4	627.9	667.2	636.1	619.5	645.3	565
above 10 MW	998	1,191	1,339	1,228	1,388	1,257	1,455	1,738	1,388	1,097	1,442	1,296
Wind	124	142.3	135.5	256.1	521.6	836.8	1,077	1,664	3,205	4,747	6,004	7,676
Solid biofuels	399	768.2	1,400	1,833	2,360	3,365	4,904	5,905	7,148	9,529	7,932	9,160
of which co-firing	-	620.5	1,236	1,645	2,126	2,963	4,661	5,593	6,389	7,239	3,929	4,510
Biogas of which	56	82.2	111.3	160.1	195.2	251.6	319.2	398.4	451.1	565.4	689.7	816
biogas from landfill sites	53	63.3	75.3	92	113.6	148.4	174.8	219.9	233.7	236.5	240.7	225
biogas from sewage treatment plants	2	18.1	35.4	66.7	79.5	94.9	122.7	132.4	149.8	193.7	233.5	253
other biogas	1	0.8	0.6	1.5	2.1	8.3	21.7	46.1	67.7	135.1	215.5	388
Bioliquids	-	-	-	-	-	-	3	0.9	1.4	0.2	0.6	0
Solar energy	-	-	-	-	-	-	-	-	0.2	1.1	1.5	7

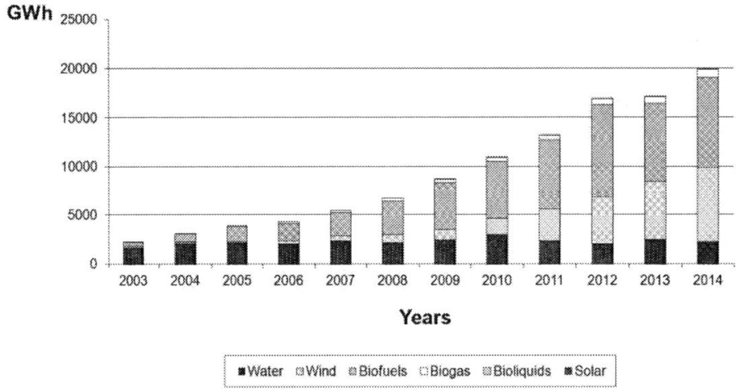

Fig.2. Electricity generation from renewable energy sources in the years 2003-2014 [2]

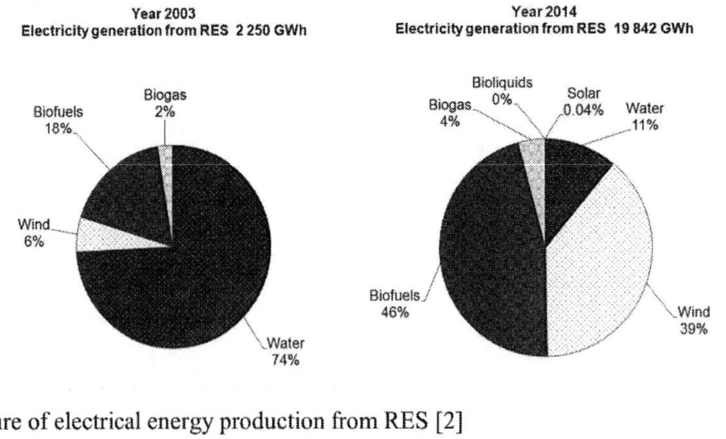

Fig. 3. Structure of electrical energy production from RES [2]

DOI: 10.1051/e3sconf/20171401012

A significant increase in renewable energy generation took place in the period 2003 – 2014. A comparison between the RES structure in 2003 and in 2014 indicates considerable changes as well: renewable energy obtained from solid biomass and wind significantly increased with the share of hydropower decreasing, despite the fact that the total volume of energy generated by hydroelectric plants increased by over 20%. The share of bioliquids and solar energy in 2014 was much below 1%.

2.3 Production of heat from renewable sources

The statistical data on heat production from renewable sources in the period 2003-2014 used in the calculations are presented in Table 3 and in Fig. 4.

Table 3. Heat production from RES in the period 2003-2014 [TJ] (Source: GUS [2])

Item	2003	2004	2005	2006	2007	2008	2009	2010	2011	2012	2013	2014
Total	2766	2791	3589	3748	4706	6340	11270	12231	13452	19052	15988	14272
Solid biofuels	2465	2242	2768	3049	4008	5414	10448	11479	13320	18840	15611	13973
Biogas of which	301	549	821	699	698	926	817	751	131	212	377	299
biogas from landfill sites	230	136	91	109	30	148	112	113	62	69	74	69
biogas from sewage treatment plants	61	411	727	583	658	734	624	617	23	40	128	86
other biogas	10	2	3	7	10	44	80	21	46	103	174	144
Bioliquids	-	-	-	-	-	-	5	1	1	0	0	-

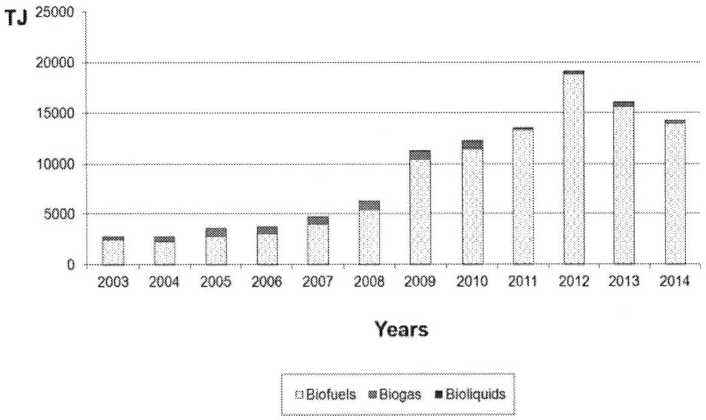

Fig.4. Heat production from RES in the period 2003-2014 [2]

In the recent years heat production from RES rose dynamically. The greatest increase occurred in 2012 – by 38.7 % with respect to 2011.

The structure of renewable sources used in heat production in two selected years is presented in Fig.5. The most popular renewable energy carrier is solid biofuel.

Directive 2009/28/EC stipulates that Poland obtains a 15% share of renewable energy in total gross end-use energy consumption in 2020. It is a realistic target, considering the fact that in 2014 the share of renewable energy in total gross energy end-use consumption was 11.8 %.

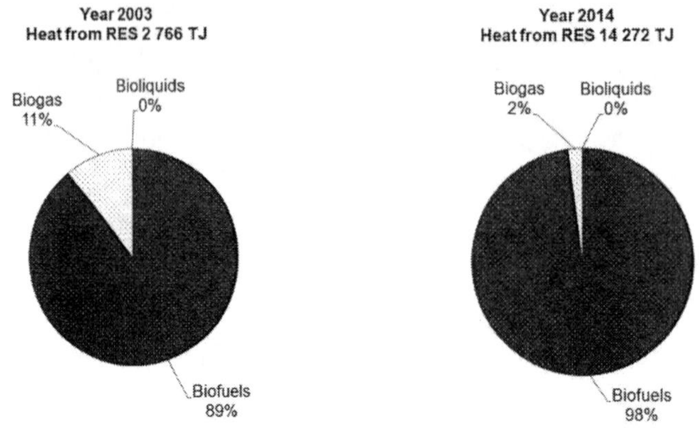

Fig. 5. Structure of heat production from RES in 2003 and in 2014 [2]

3 Dynamic processes modelling

A number of dynamic processes with random errors in economy and in power industry can be modelled by means of stochastic equations, such as the Ito equations [11]. They provide a powerful tool for the analysis of real economic phenomena and they have been fruitfully applied for modelling processes in power industry as well. When the stochastic differential equations are solved by means of Euler's method [7], it is possible to simulate utilization of RES for electrical energy and heat generation in the future.

A stochastic process with a number of continuous-time variables can be represented by means of stochastic differential equations (SDE) of the following general form:

$$dX_t = F(t, X_t)dt + G(t, X_t)dW_t, \tag{1}$$

where: X – the state variable, W – the Wiener process (Brownian motion) variable, F – the trend-determining function, G – the diffusion function.

It is possible to define special variants of the model on the basis of the general formula (1). The simplest variant is a linear trend model with a random Brownian motion (BM) component. This variant includes two components: deterministic and random:

$$dX_t = A(t)dt + V(t)dW_t, \tag{2}$$

Another variant is the model of relative variation in the state variable with a random component represented by geometric Brownian motion (GBM):

$$dX_t = B(t)X_t dt + V(t)X_t dW_t, \tag{3}$$

The variability of the random component can be modelled as constant elasticity of variance (CEV):

$$dX_t = B(t)X_t dt + V(t)X_t^{\alpha(t)}dW_t, \tag{4}$$

A fundamental property of the CEV stochastic model is the possibility of determining the relationship between the state variable and its variability through the value of the exponent $\alpha(t)$.

Another variant of model (1) is known as SDE from Mean-Reverting Drift (SDEMRD):

$$dX_t = S(t)\left(L(t) - X_t\right)dt + D\left(t, X_t^{\alpha(t)}\right)V(t)dW_t, \tag{5}$$

where: $S(t)$ – the mean reversion speed, $L(t)$ – the mean reversion level (i.e. long run average level).

If the function $D\left(t, X_t^{\alpha(t)}\right) = 1$ (from Eq. 5), then it is possible to transform the formula into the Hull-White/Vasicek (HWV) model:

$$dX_t = S(t)\left(L(t) - X_t\right)dt + V(t)dW_t, \tag{6}$$

The simulation of RES utilization treated as the state variables X_t consists in obtaining a numerical solution of the SDEs, for which Euler's method is proposed. Since the renewable sources are a group of correlated variables, the most suitable simulation model is GBM including a multidimensional vector of correlated random variables, as in the formula below:

$$dX_t = \mu X_t dt + \sigma dW_t, \tag{7}$$

where: μ – the mean value of the variable X_t, σ – the standard deviation of the variable X_t.

Correlations between variables are taken into account for the multidimensional vector of random variables in Wiener processes. The GBM model adequately represents phenomena following the commodity-exchange rules. In the model under scrutiny, the risk measure corresponds to the value of standard deviation σ. The model includes a deterministic component, representing the trend of the variable and a random component, in which the uncertainty measure is the standard deviation of the variable. Uncertainty is to be taken in its economic sense, whereas in the mathematical sense it would be the risk measure since the probability distribution of the variable is known.

4 Forecast of electrical energy and heat generation from RES under risk

Selected simulations of electricity generation and heat production from renewable carriers are presented in Fig. 6 and 7 and the structures of electricity generation and heat production in 2020 are presented in Fig. 8 and 9. The simulations for the future period 2015-2020 can be considered as the forecasts. The average values and standard deviations of variables X_t in formula (7) have been estimated on the basis of historical sequences. Estimates of trend coefficients were limited in such a way that the particular simulations allow for achieving the basic objective for Poland, which is a 15% share of renewable energy carriers in the end demand.

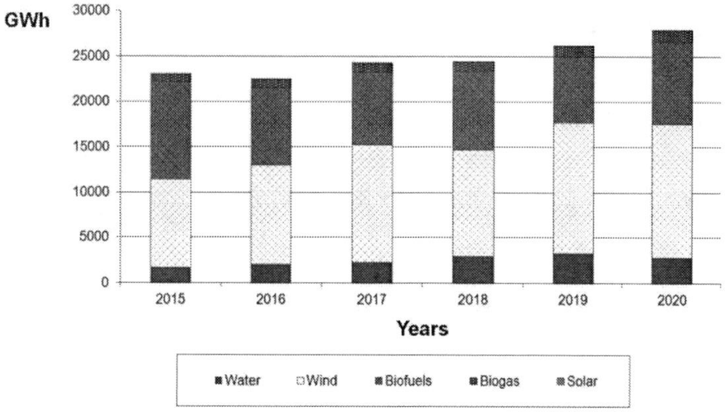

Fig.6. Selected simulation of electricity generation from RES in years 2015-2020

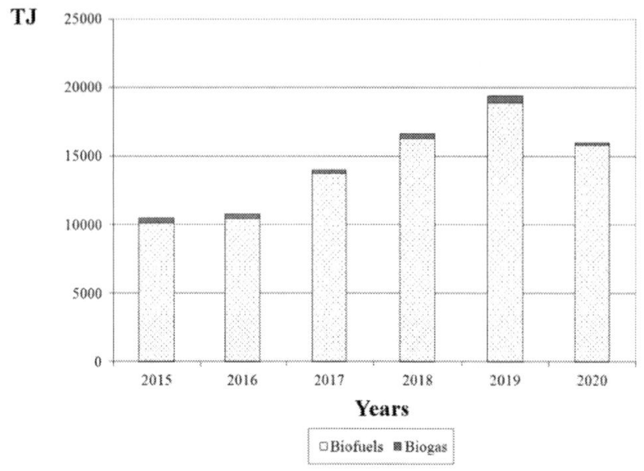

Fig.7. Selected simulation of heat production from RES in the years 2015-2020

Since the value of the standard deviation was estimated on the basis of RES utilization in the past period, the risk level was also assumed to be the same as in the past. If a greater value of the standard deviation is assumed, the risk level is also taken to be greater. A detailed discussion on risk quantification can be found e.g. in [11].

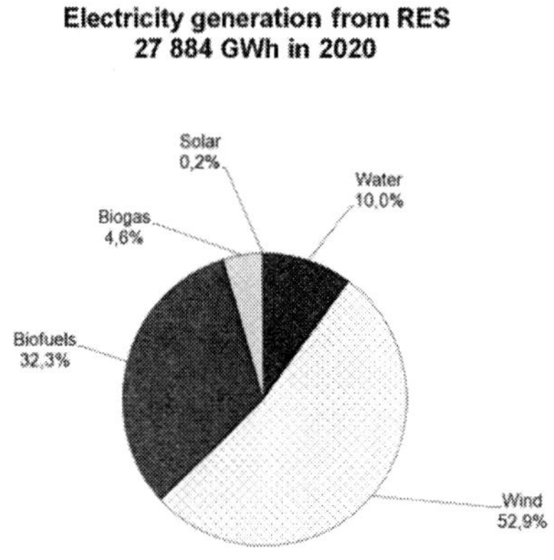

Fig.8. Selected simulation of the structure of RES utilized in electrical energy generation in 2020

5 Conclusions

Simulations of electricity generation and heat production from renewable carriers until 2020 were performed using Euler's method. The derived sequences should be treated as possible scenarios on condition that the trends observed on the RES market to date continue. Simulations are based on approximate solutions obtained from the GBM model including correlation among the variables (i.e. renewable energy carriers), represented by the Wiener processes.

The simulation results presented in Fig. 6 and 8 confirm the dynamic growth of the electricity production from RES. The increase in electricity production in 2020 is about 40% (27 884 GWh) as compared with 2014 (19 842 GWh). The largest share in 2020 in the structure of electricity production is noted for wind power plants (approx. 53 %), followed by biofuels power plants (approx. 32 %) and hydropower plants (approx. 10 %).

The simulation results of heat production from RES are presented in Fig. 7 and 9. Increase in heat production in 2020 is about 12 % (16 002 TJ) as compared with 2014 (14 272 TJ). The share of biofuels in the structure of heat production is about 99 %.

Fig.9. Selected simulation of the RES structure utilized in heat production in 2020

Different simulation sequences may be obtained by changing the assumptions about the development of variables in the future. For example, different results are obtained if the correlation among variables in the GBM model is not taken into account. Other scenarios will be produced if it is assumed that the behaviour of variables is determined by market rather than commodity-exchange mechanisms. Then, simulations should be performed by means of the mean-reverting drift model (SDEMRD) or the Hull-White/Vasicek model (HWV) [7].

A crude analysis has been carried out with regard to the impact of uncertainty about the environment of the model on the simulation results. Uncertainty connected with a number of factors – such as: the uncertainty concerning legal regulations regarding RES, energy and environmental policy of the EU and Poland, opportunities for obtaining renewable carriers, their prices, etc. – was modelled by introducing into equation (7) a random component, taking into account total risk quantified with the value of the standard deviation σ. The presented methodology of modelling processes in power engineering in the presence of risk seems to be a promising analytical tool, undoubtedly requiring further research. The risk measure corresponds to the value of standard deviation σ in the models. The values of the standard deviation were estimated on the basis of RES utilization in the past period, that's why the risk level was also assumed to be the same as in the past. If a greater value of the standard deviation is assumed in the models, the risk level is also taken to be greater.

The world power industry has already changed a lot from the traditional model of monopoly towards competitive markets. The changes in the technology of energy and heat generation as well as continuous progress in lowering the investment cost of renewable energy legitimize the conclusion that in the future renewable energy technology should be competitive with respect to other technologies with high variable costs.

When it comes to evaluation of power generation technologies, their impact on the environment is becoming increasingly important. It can be suspected that the power industry will remain tightly connected to environmental issues in the future as well, which will inevitably lead to a new paradigm in power industry, advantageous both for economy and society.

References

1. Directive 2009/28/EC of the European Parliament and of the Council of 23 April 2009 on the promotion of the use of energy from renewable sources.
2. Central Statistical Office of Poland (GUS), *Energia ze źródeł odnawialnych. Yearbooks 2003-2013* (Warsaw)
3. A. Gawlak, L. Poniatowski, Main directions of investment in the sector of electrical energy distribution, EPE 2014, DOI: 10.1109/EPE.2014.6839418
4. A. Gawlak, M. Kornatka, Comparative analysis of operating conditions in Polish medium-voltage and 110 kV networks, Proceedings of The 8th International Scientific Symposium ELEKTROENERGETIKA 2015, Stará Lesná, Slovak Republic, pp.53-56
5. Polish Ministry of the Economy, *Energy policy of Poland until 2030* (Warsaw 2009).
6. Polish Ministry of the Economy, *National action plan for energy from renewable sources* (Warsaw 2010)
7. B.K. Øksendal, *Stochastic Differential Equations: An Introduction with Applications* (Berlin Springer, 2003)
8. I. Marciniak, Światowe i unijne normatywne dokumenty a rozwój OZE w Polsce, Przegląd Elektrotechniczny, **90**, 7 (2014)
9. J. Paska, T. Surma, Wyzwania dla Polski w świetle nowej polityki energetycznej Unii Europejskiej, Rynek Energii, **113,** No.4 (2014)
10. J. Popczyk, Synergetyka, Przegląd Elektrotechniczny, No. 6 (2011)
11. J. Sowiński, *Inwestowanie w źródła wytwarzania energii elektrycznej w warunkach rynkowych* (WPCz, Częstochowa, 2008)
12. J. Sowiński, Koszty energii elektrycznej z odnawialnych źródeł energii, Przegląd Elektrotechniczny, **90**, No. 8 (2014)

E3S Web of Conferences **14**, 01013 (2017) DOI: 10.1051/e3sconf/20171401013
Energy and Fuels 2016

Wind Energy in Poland - economic analysis of wind farm

Renata Gnatowska[1,*], *Agnieszka* Wąs[1]

[1]Czestochowa University of Technology, Institute of Thermal Machinery, Poland

Abstract. Wind power generation is the most dynamically developing branch of renewable energy sector in Poland. The high price of coal and its negative environmental impact result in ever growing support for the renewable energy in the energy sector. Wind power industry produces clean energy, but costs of investments are very high. Process of creating a wind farm is very long and complicated. It requires the knowledge of many disciplines such as law, economics, energetic. The aim of this paper was study the state and the development of the wind power sector, especially wind farm in Poland. The economic analysis on investment in wind farm in Poland according to a new act on renewable energy sources was presented. The subject of this analysis was to estimate profitability of wind farm project 40 MW capacity. The analysis includes the comparison of various support systems - the system of green certificates and the new - auction model on the profitability of investment in wind farm. This aims to show how important is appropriate government support for producers of green energy words.

1 Wind Energy in Poland - present state and prospects of development

Poland, as of the European Union member is obliged to execute the decisions that were taken by the Council of the EU. Given RES talking about the Europe 2020 strategy, which talks about the growth of the share of RES in total energy production to 20% compared to 1995. The constant development of renewable energy sources also indicate the provisions of the Paris Climate Change Conference 2015 (COP21) - (ang. United Nations Framework Convention on Climate Change, the 21st Conference of the Parties, COP21), the extent of which is global, not just European. For the development of renewable energy sources it is mainly the speed limit climate change caused by emissions of carbon dioxide, is produced by the combustion of fuels [1-3].

Wind power generation is the most dynamically developing branch of renewable energy sector in Poland (see figure 1). According to the data of the Energy Regulatory Office (URE), the total power of wind turbines in Poland was 2497 MW, at the end of March 2012. In the end of March 2016 there were 1129 wind installations (single turbines and farms) operating in Poland, of the total power of 5660 MW, which results in a nearly

* Corresponding author: gnatowska@imc.pcz.czest.pl

double increase in power [4]. There are a few dozens of operating wind farms, mainly located in the northern part, the central part of the country (Fig. 2).

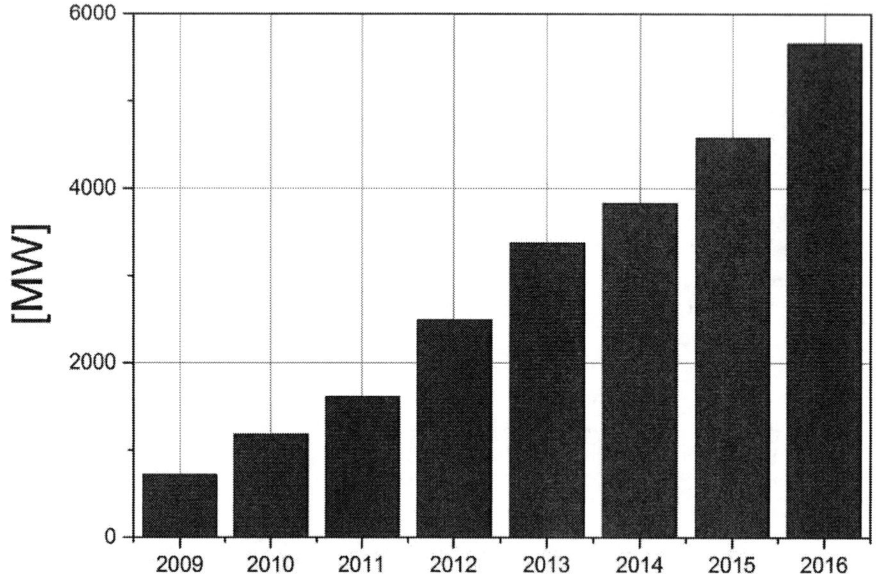

Fig. 1. Capacity installed in Wind farm in Poland [MW] [4].

Reforms introduced by the Renewable Energy Sources Act 2015 [5], marked a significant step forward, however, subsequent amendments to the RES Act have illustrated how the Polish government is in a difficult position of striking a balance between developing RES for energy diversification and rescuing coal industry.

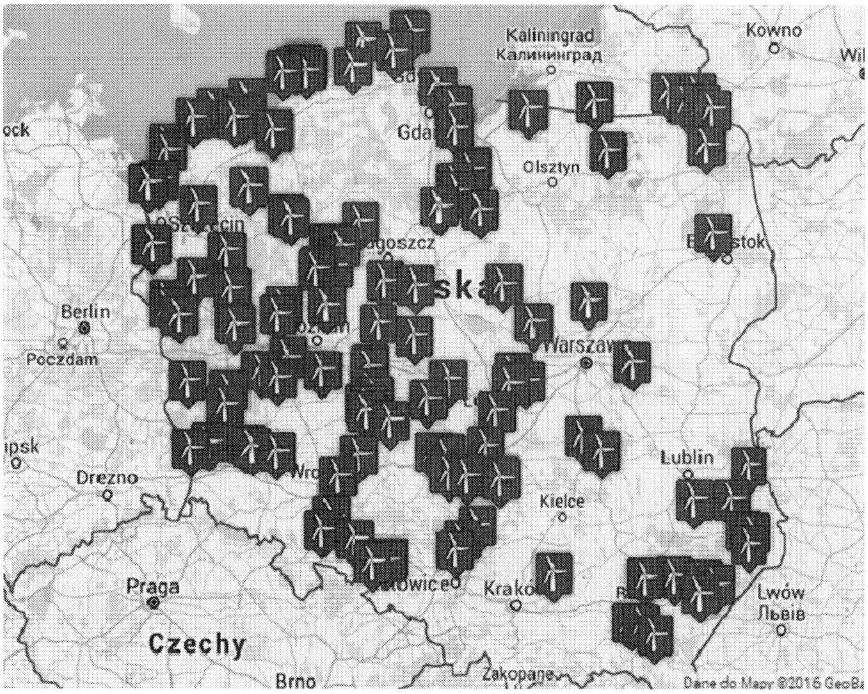

Fig. 2. Map of wind energy installations in Poland [6].

The RES Act, (which is the implementation of the provisions of Directive 2009/28/EC into Polish law), sets a new regulatory and economic framework for investments in renewable projects in Poland [7]. The Auction System replaces the previous renewable energy investment incentive system, under which installations were required to obtain tradable/transferable certificates of origin from RES ('Green Certificates') and to sell property rights arising from those one. The property rights arise once the Green Certificates are registered on a register of certificates operated by the Polish Power Exchange 'Towarowa Gielda Energii S.A.' ('TGE'). The Green Certificates can be traded on a regulated market. The value of Green Certificates is set by the market. The Green Certificate System had worked well during 2000–2010, due to the development of 'co-firing'. The rapid increase in biomass co-firing created an enormous oversupply of Green Certificates. In March 2016, the oversupply was over 20 TWh [8]. The oversupply of Green Certificates caused their value to drop significantly (from around PLN 280/MWh in early 2012 to less than PLN 100/MWh by February 2013 [9]). The outcome was a reduction of profitability of RES projects. Poland has also severely limited growth in wind generation by passing the Act, aimed at restricting wind power development. The purpose of Act on Windfarms [10] is to determine places where wind power plants can be built and sets out the minimum distance required between a wind power plant and residential buildings, forests or national parks. The distance is set at 10 times the height of a wind power plant (in practice approximately between 1.5 and 2 km). This requirement seems very strict. It will be a challenge to find plots of land which meet the statutory distance requirement, in many cases it will be impossible to build new or expand already existing plants. In addition, the Act quadruples the rate of tax payable on existing turbines, making them unprofitable [10].

The aim of this paper was study the state and the development of the wind power sector, especially wind farm in Poland. The economic analysis on investment in wind farm in Poland according to a new act on renewable energy sources was presented. The subject of this analysis was to estimate profitability of wind farm project 40 MW capacity. The analysis includes the comparison of various support systems - the system of green certificates and the new - auction model on the profitability of investment in wind farm. This aims to show how important is appropriate government support for producers of green energy. The new formal and legal framework for the wind power sector in Poland was described in relation to the new Renewable Energy Sources Acts [5, 10-11]. The economic analysis of three scenarios of the development of the wind power sector in Poland was presented.

2 Wind farm case study

In order to estimate the economic efficiency of wind farm examines the case of a wind farm with a capacity of 40 MW, located on Polish territory. Where the average annual wind velocity is 6 m / s. The analyzed wind farm consists of 20 wind turbines of the Danish company VESTAS - model V80 2.0 MW. On the basis of this chart below (Figure 3), estimated annual energy production by the wind farm.

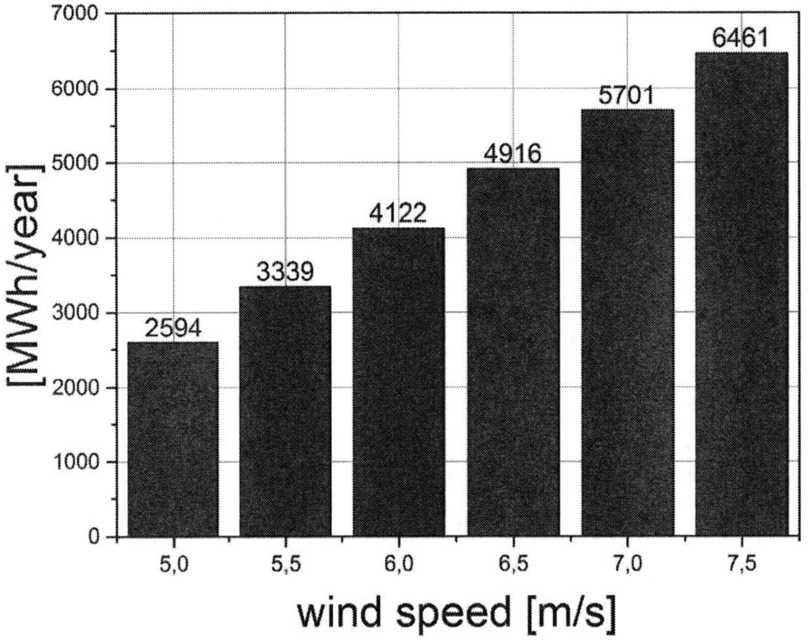

Fig. 3. The average annual production of electricity by turbine VESTAS V80 2.0MW [12].

The investment costs of a wind farm was estimated at 252.5 million PLN, of which the largest part was the purchase of wind turbines, representing 81% of investment costs (see Tab. 1). In order to analyze the efficiency of investment in a wind farm was estimated operating costs (Tab. 2). [3, 13, 14] and sources of financing.

Table 1. Summary of investment costs of analysed wind farm.

COMPONENTS	COST [PLN]	[%]
Wind turbines	205 787 500	81,5
Project (with costs of research)	11 362 500	4,5
Construction work (with work infrastructure)	17 675 000	7
Connecting to network	13 887 500	5,5
Internal energy network	3 787 500	1,5
Total costs	252 500 000	100,00

Table 2. Summary of the annual operating costs of a wind farm.

COMPONENTS	COST [PLN]	[%]
Property tax	2 000 000	27,36
Service and maintenance	3 300 000	45,14
Balancing the energy	1 500 000	20,52
Insurance	360 000	4,92
Land rent	150 000	2,05
Total costs	7 310 000	100,00

Case Study is funded by the following sources: own contribution, financing and credit for 10 years, the interest rate is 2 % per annum. The annual installment of the credit is a fixed amount. The structure of the funding sources are presented in table 3.

Table 3. Summary of the annual operating costs of a wind farm.

COMPONENTS	VALUE [PLN]	PERCENTAGE [%]
OWN CONTRIBUTION	152 500 000	60,40
FINANCING	40 000 000	15,84
CREDIT	60 000 000	23,76

3 Analysis of economic efficiency of wind farm

Return on investment in a wind farm in the Polish conditions is the subject of very extensive. The profitability of such an investment is influenced not only the amount of energy produced, which results directly from the wind speed occurring in the area. Attention should be paid to several other factors. Among other things, the volatility of prices of electricity sales (sales of "green energy" is subject to special rules). Construction of a wind farm requires huge financial investment, so you should carefully examine all factors that could affect its economic efficiency and identify risks.

3.1 Definition of parameters

In order to assess the economic efficiency will be used three indicators, which will determine not only whether the investment is profitable, but also determine in which year of the wind farm will start to make a profit. The cash flow (*CF*), which specifies the aggregate for a given period cash flows. It is the difference of income and expenditure [13]. This ratio is determined using the expression:

$$CF = \Sigma^{n}_{t=0}\ (earinings - costs)_t \qquad (1)$$

were: t - years of investment.

The net present value (*NPV*) of the flow of profits over simulation time is calculated as the aggregated value of discounted annual profits and is given by:

$$NPV = \Sigma^{n}_{t=0}\ [CF_t\,/(1+r)_t] - I_0 \qquad (2)$$

where: CF - net cash inflow during the period t, r - discount rate, I_0 -total initial investment costs, t – number of time periods, n - the terminal year of the simulation.

The internal rate of return (*IRR*) is the discount rate for which the *NPV* is equal to 0. The investment is profitable when the *IRR* assumes a greater value of the discount rate, and the difference between them is of greater value, the investment is more profitable ($IRR > r$). This indicator can be represented by the formula [13-14]:

$$\Sigma^{n}_{t=0}\ [CF_t\,/(1+r)_t] - I_0 = 0 \qquad (3)$$

3.2 Analysis

Economic analysis includes three cases of investment in a wind farm. For each, the discount rate is 5% and the duration of the investment 25 years. Case studies vary in way of the sale of energy:

• **CASE A** - the sale of energy takes place at auction, and the earnings of the wind farm is estimated at PLN 385/MWh [15],

• **CASE B** - the sale of energy takes place in a system of "green certificates", where the total purchase price consists of the amount from the sale of certificates, which the average value in 2015 amounted to PLN 123.6/MWh , and the price of the average quarterly price of electricity determined in July 2016 year by the URE to 173.5 PLN/MWh [16],

• **CASE C** - the sale of energy takes place without the support system.

The comparison was made for different cases of energy sale from the wind farm(various support schemes). This analysis seems to be interesting from the point of view of investors on Poland energy market. In frame of this work was analyzed the new auction system, the existing system of green certificates and the lack of support energy sales for the producers of energy from renewable sources. The risk of erroneous estimation of investment profitability is high, because the auction system is a new system that has not yet started its operation and provide a stable selling price of energy for the producers of "green energy".

The premise of the auction system are stable prices in the perspective of 15 years, while the above analysis applies to the operation of a wind farm in 25 years. For the purposes of analysis it was assumed that the selling price of electricity for the described event will be fixed in 25-year time horizon and will amount to 385PLN per MWh.

Analysis of the profitability of the investment functioning in the system of green certificates is a bit more complicated, what is the cause of instability in their prices. Average price of the certificate was in 2009 - 250.03 PLN/MWh, while in later years has experienced drastic declines - for example in February 2013 and June 2015 amounted to approx. 100 PLN/MWh, and now it is no longer even 40 PLN/MWh [16]. Also note that if the sale is not their only source of income of a wind farm. The wind farm sells certificates are merely certificates of origin, and the electricity is sold at standard prices prevailing on the market. In this case it will be PLN 173.5/MWh. This price will also be taken into account for the calculation for the wind farm, which does not exist in any of these support systems.

DOI: 10.1051/e3sconf/20171401013

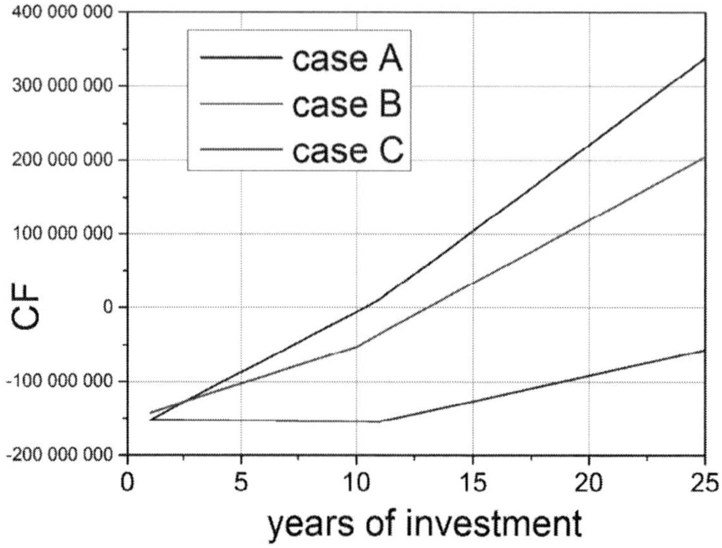

Fig. 4. Values of CF for the cases A, B and C.

The graph on Fig. 4 summarizes the value of the paramerer *CF* for all three cases. It shows that the best profitability shows for case A operating in the auction system, slightly worse results are achieved system of green certificates. Economic viability for 25 years operation of the farm does not show the wind farm, which does not receive any support. *CF* ratio does not take into account the changing value of money over time, so in order to determine a more accurate economic viability should take into account the value of the index *NPV*, the results of which are summarized on Fig. 5.

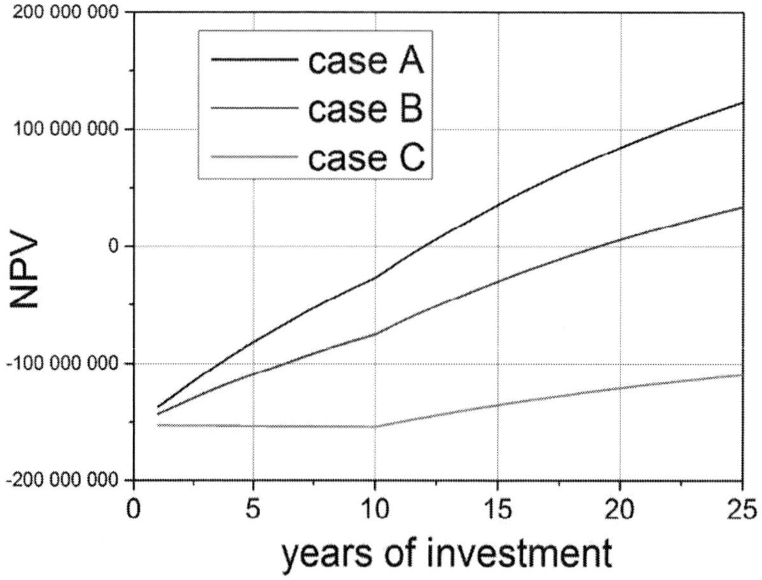

Fig. 5. Values of NPV for the cases A, B and C.

According to the above graph (Fig. 5) can be identified in which year each of the cases will begin to be profitable. Both indicators *CF* and *NPV* best results achieved for case A. According to the parameter *NPV* wind farm operating in the auction system will start to be profitable in the 12th year of the investment. In the case of green certificates will be the 20th year of operation of the investment. In the case of a system that does not receive support investment in the wind farm will not be profitable.

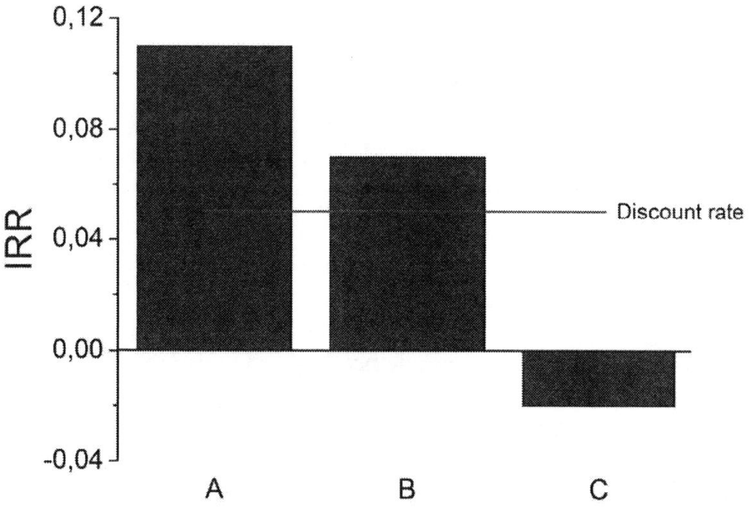

Fig. 6. Values of IRR for the cases A, B and C in the 25. year of investment.

The graph above (Fig. 6) showing the results of *IRR* in the 25th year of the investment confirms the results obtained for indicators of *CF* and *NPV*. In cases A and B the value of IRR is higher than the accepted level of 5% discount rate. For the case C *IRR* in the 25th year has a value -2%, which indicates a failure of the investment.

The obtained results allow to become conscious of what the most important are costs and minimize them, which is one of the most important contributions. It will be useful to improve the competitiveness of floating wind farms in the future. The authors are aware that the presented results do not contain a comprehensive analysis. The further study is needed to determine additional questions regarding the profitability changes according to wind resource, wind farm's lifetime or carbon emissions. Obtain results can be enhanced also by feedback from relevant people related to the operation and maintenance activities taking place at the wind farm.

4 Summary

Made in the work of analysis clearly showed that the operation of a wind farm in conditions where it has no support is undoubtedly highly unprofitable. In addition, there is no certainty that for every reported investment auctions will be allocated. This situation indicates that the development of RES installations, including of course the wind energy influenced by the state authorities. Support in the form of the auction will receive probably investors who will be needed to fulfill the minimum targets of the strategy for 2020. In the case of our country should consider, therefore, that certainly tends to the production of "green energy" because of the benefits that entails the production of or only from a sense of obligation to fulfill international agreements. Favor of the government to promote the development of renewable energy over the required level, resulting with international obligations is therefore a question mark, which increases the risk of the economic efficiency of this type of investment and discourages potential investors. It appears that the government tends to reduce support for the RES energy and will most likely treat biomass installations especially as opposed to wind farms.

The new Act on Windfarms might suggest an occasion to invest in offshore wind farms on Baltic Sea. The productivity of the sea farms are twice better than the onshore and the location is not so controversial [17]. The first Polish offshore wind farm is projected to be operational in 2021.

References

1. M.M. Al-Maghalseh, E.M. Maharmeh, Procedia Computer Science 83, 790-798 (2016)
2. P.J. Simons, W.M. Cheung, Journal of Cleaner Production. (2016).
3. M. Ligus, *Efektywność inwestycji w odnawialne źródła energii. Analiza kosztów i korzyści* (CeDeWu, Warszawa, 2012)
4. The Energy Regulatory Office website, http://www.ure.gov.pl/pl/rynki-energii/energia-elektryczna/odnawialne-zrodla-ener/potencjal-krajowy-oze/5753,Moc-zainstalowana-MW.html [01.09.2016 r.]
5. The Act on renewable energy sources, Ustawa z dnia 20 lutego 2015 r. o odnawialnych źródłach energii (Dz.U. 2015 poz. 478)
6. The Energy Regulatory Office website, http://www.ure.gov.pl/uremapoze/mapa.html [01.09.2016 r.]
7. Report: Renevable Investments in Poland 2016 - Onshore Wind & Photovoltaic, http://wysokienapiecie.pl/energy-poland/1229-report-renevable-investments-in-poland-2016-onshore-wind-photovoltaic [01.09.2016 r.]
8. Association of Energy Trading Report, "status on 31 March, 2016", http://webcache.googleusercontent.com/search?q=cache:tIqKJfjKZ8EJ:www.toe.pl/pl/wybrane-dokumenty/rok-2016%3Fdownload%3D1462:electricity-and-gas-market-in-poland-status-on-31-march-2016-toe-report+&cd=1&hl=en&ct=clnk&gl=uk, [01.09.2016 r.]
9. Argus Media. Polish renewable subsidy crash hits biomass, http://www.argusmedia.com/pages/NewsBody.aspx?id=836517&menu=yes [01.09.2016 r.]
10. The Act on Windfarms, Ustawa z dnia 20 maja 2016 r. o inwestycjach w zakresie elektrowni wiatrowych (Dz.U. 2016 poz. 961)
11. Ustawa z dnia 22 czerwca 2016 r. o zmianie ustawy o odnawialnych źródłach energii oraz niektórych innych ustaw (Dz.U. 2016 poz. 925)
12. Polish Wind Energy Consulting webside, http://www.polishwindenergy.com/index.php/elektrownie-wiatrowe-w-polsce/item/39-vestas-v80-2-mw.html [01.09.2016 r.]
13. Z. Lubośny, *Farmy wiatrowe w systemie elektroenergetycznym* (Wydawnictwo Naukowo-Techniczne, Warszawa, 2009)
14. R. Gnatowska, A. Wąs, Inż. Ochr. Środow. **18**, 1 (2015)
15. The Energy Regulatory Office website, http://www.ure.gov.pl/pl/rynki-energii/energia-elektryczna/aukcje-oze/dokumenty/6539,Ceny-referencyjne.html [01.09.2016 r.]
16. The Energy Regulatory Office website, http://www.ure.gov.pl/pl/tagi/21,srednie-ceny.html [01.09.2016 r.]
17. S.O. Effiom, B.N. Nwankwojike, F.I. Abam, Energy Reports 2, 48-53 (2016)

DOI: 10.1051/e3sconf/20171401014

The analysis of the wind potential in selected locations in the southeastern Poland

Krzysztof Sornek[1,*], *Mariusz Filipowicz*[1], *Wojciech Goryl*[1], *Eugeniusz Mokrzycki*[2], *Tomasz Mirowski*[2] and *Maciej Duraczyński*[1]

[1]AGH University, Faculty of Energy and Fuels, Department of Sustainable Energy Development, Mickiewicza Av. 30, 30-059 Krakow, Poland
[2] Mineral and Energy Economy Research Institute of the Polish Academy of Sciences, Wybickiego Str. 7C, 31-261 Krakow, Poland

Abstract. This paper shows the results of analysis of the wind potential in selected locations in the southern Poland (placed in the Małopolskie, Świętokrzyskie and Podkarpackie Voivodships). The measurements and analysis of the wind resources in potential locations of the wind turbines are important part of the investment process. The statistical analysis involves the creation of histograms (e.g. histogram of the wind speed and direction) and fitting those histograms to theoretical distributions (e.g. Weilbull distributions of wind speed). Such analysis has been described and conducted using measurement data for four selected locations. Basis on the conducted analysis, the economy efficiency and environmental impact of wind turbine operation has been estimated. Three market available wind turbines have been included to calculate NPV, IRR and SPBT indicators. Then, the avoided emissions of CO_2, NO_x, SO_2 and dust have been calculated. There were also conducted some calculation using TRNSYS simulation software. The results of simulations have been compared with measurement data and the level of convergence have been found.

1 Introduction

As regards the number of wind power plants, south-eastern part of Poland is the area where relatively small number of wind turbines have been installed so far. Due to the lack of detailed data concerning the potential, the observed interest of possible investors is not as high as in other parts of the country (in the north-west and in central Poland). Since the development of wind power sector depends primarily on proper location of wind farms, it is necessary to acquire and process suitable data, first of all including estimated wind power potential in certain area. Known wind power resources make it possible to carry out financial and ecological analysis of undertakings, which could be completed in selected locations.

This study involves an analysis of measurement data from four selected measurement masts located in Małopolskie, Świętokrzyskie and Podkarpackie Voivodeships. The acquired data provided the grounds for carrying out statistical analysis and identifying basic parameters characterizing the wind. Then, the following were carried out: the assessment of

* Corresponding author: krzysztof.sornek@agh.edu.pl

wind power resources, and financial and ecological analysis of the investment in selected types of wind turbines. Obtained results have been compared to the results of dynamic simulations performed using the TRNSYS software.

2 Assessment of wind parameters

The measurements and analysis of wind resources in general can be divided into two stages: long-term data acquisition from measurement masts (primarily wind speed and direction), and their further statistical analysis.

Measurement masts are provided with devices, which generally can be divided into: sensors, recorder and auxiliary equipment. The most frequently used sensors are: wind speed sensors (e.g. cup anemometers, vane anemometers, and ultrasonic anemometers), wind direction sensors (potentiometer-type transducers), temperature sensors, and – in some cases – pressure and humidity sensors. The measurements are most often carried out at a height below planned height of wind turbine rotor axis (installation of so high mast would be very expensive and difficult due to technological reasons). As a result of this, two or more anemometers are used in order to obtain information on wind speed at different heights. In this way, it is possible to determine wind speed profile, which allows its extrapolation at required heights. Specific configuration of sensor arrangements is selected i.a. depending on terrain roughness class.

Wind speed v affects the parameter called air stream power density P according to the formula (1):

$$P = \frac{1}{2}\rho v^3, \tag{1}$$

where ρ is air density. Power density may be given in watts per square meter.

The first stage of statistical analysis involves developing histograms of data - wind speed and direction, depending on the duration of particular measurements (monthly, daily, etc.), and wind speed and direction depending on height. These histograms are then being fitted by theoretical distributions: statistical wind speed distributions, daily speed histograms and models of wind speed on measurement height. Wind speed histograms are most often represented using a two-parameter Weibull distribution or sometimes one-parameter Rayleigh distribution (which is a special case of the Weibull distribution). This does not illustrate precisely the frequency of occurrence of certain wind speeds. In fact, it may differ even by 30% from true distribution determined empirically [1].

3 Results

3.1 Description of selected locations

Measurement data used in the analysis of wind power potential in selected areas of Małopolskie, Świętokrzyskie and Podkarpackie Voivodeships originate from four measurement masts installed in the following locations:

- location in eastern part of Świętokrzyskie Voivodeship (north-west of Sandomierz) – wind speed and direction measurement at the height of 50 and 100 m;
- location in eastern part of Małopolskie Voivodeship (south of Gromnik) – wind speed measurement at the height of 10 and 40 m and wind direction measurement at the height of 40 m;
- location in south-western part of Podkarpackie Voivodeship (south of Jasło) – wind speed and direction measurement at the height of 50 and 100 m;

- location in northern part of Podkarpackie Voivodeship (north of Leżajsk) – wind speed and direction measurement at the height of 50 and 100 m;

The selected locations are shown on the map (see Figure 1).

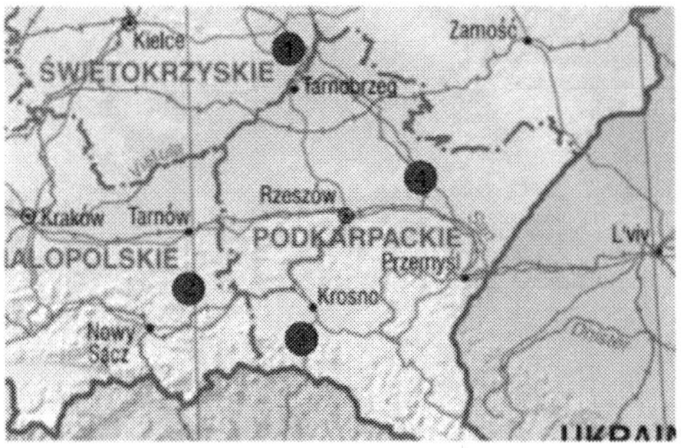

1 - location in the vicinity of Sandomierz (Świętokrzyskie Voivodhip)
2 - location in the vicinity of Gromnik (Małopolskie Voivodhip)
3 - location in the vicinity of Jaslo (Podkarpackie Voivodhip)
4 - location in the vicinity of Lezajsk (Podkarpackie Voivodhip)

Fig. 1. Location of data gathering masts.

The comparison presented further in this work includes data obtained directly for the height of 50 m in locations no. 1, 3 and 4, and data calculated for the height of 50 m in case of location 2 (recalculation has been carried out from the level of 40 m using a power model – formula 2). Moreover, the comparison of the impact of height h above ground level on available wind resources *v(h)* has been made for a given location:

$$v(h) = v_0 \cdot \left(\frac{h}{h_0}\right)^{\alpha}, \tag{2}$$

where:

v_0 – wind speed at the height of h_0,
h_0 – reference height,
α – coefficient dependent on terrain roughness.

3.2 Comparison of wind speeds for selected locations

The following comparisons have been made based on measured wind speeds in the selected locations: comparison of average monthly wind speed fluctuations, comparison of daily wind speed fluctuations and comparison of Weibull distributions.

3.2.1 Comparison of average monthly wind speed fluctuations

One of the basic aspects of wind power engineering is uneven wind speed distribution over a year. In Polish conditions, higher wind speeds appear in transition periods (spring, autumn) and in winter season, and lower speeds – in summertime. This relation is also observed in case of the discussed locations, which is shown in Fig. 2.

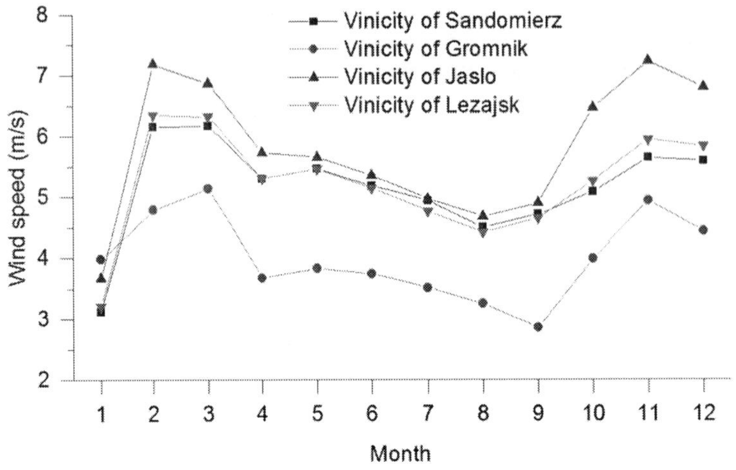

Fig. 2. Monthly wind speed profiles for selected locations.

As shown on fig. 2, the trend of changes in the average monthly wind speeds is similar. Highest wind speeds occur in February, March, November and December, and in the other months they are visibly lower. We can observe that highest average wind speeds over an entire year have been measured near Jasło, and lowest – near Gromnik. The comparison of highest and lowest average wind speeds in respective months of a year, and average wind speed over a year is presented in Table 1.

Table 1. Yearly highest, lowest and average wind speeds in selected locations

Location	Highest average monthly wind speed in a year [m/s]	Lowest average monthly wind speed in a year [m/s]	Average annual value of monthly wind speed [m/s]
Vicinity of Sandomierz	6.2	3.1	5.2
Vicinity of Gromnik	5.1	2.9	4.0
Vicinity of Jaslo	7.2	3.7	5.8
Vicinity of Lezajsk	6.4	3.2	5.2

3.2.2 Comparison of daily wind speed fluctuations

Apart from known average monthly wind speeds, it is also possible to compare daily fluctuation of wind speeds. Wind speed profiles have been fitted using the theoretical relations adapted from works [2-3]. The formulas are based on the daily wind speed characterised by low values until early morning hours, wind speed increase from that time until late afternoon, and wind speed drop to its minimum value in the evening. Since daily wind profiles being an average of many days has been considered, fitting has been carried out with the formula 3.

$$V = \begin{cases} V_{min} & t \in [0, T_1) \\ V_{min} + V_{max} \cdot sin\left(\frac{t-T_1}{A_1}\right) & t \in [T_1, T_2) \\ V_{min} + V_{max} \cdot sin\left(\frac{T_3-t}{A_2}\right) & t \in [T_2, T_3) \\ V_{min} & t \in [T_3, 24h) \end{cases} \quad (3)$$

where:
T_1, T_2, T_3 – individual stages of wind speed pattern variability (fitting parameters),
A_1, A_2 – fitting parameters,
V_{min}, V_{max} – minimum and maximum wind speed in a daily histogram,
t – current time,
V – value of speed from theoretical fitting.

Comparison of daily wind speed fluctuations for the considered locations is provided in Fig. 3

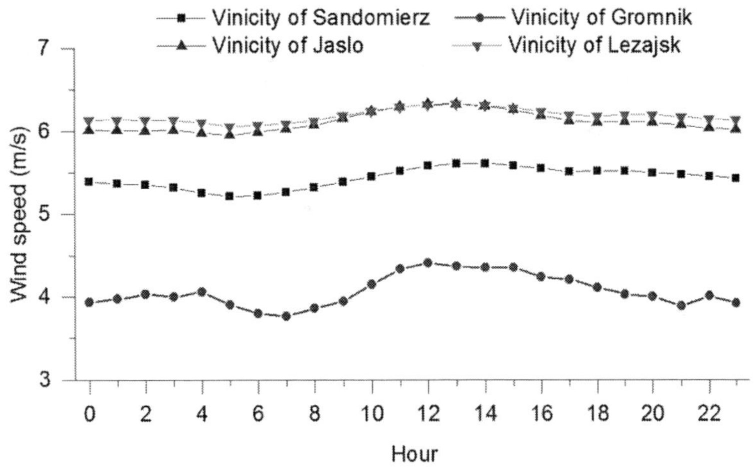

Fig. 3. Daily wind speed profiles

According to Fig. 3, daily wind speed profile is very stable for the vicinity of Sandomierz, Jasło and Leżajsk. Visibly higher daily wind speed fluctuations appear near Gromnik, where the speed changes from 3.8 to 4.4 m/s over 24 hours.

3.2.3 Weibull distributions

Based on completed measurements, it is also possible to determine probability density function f(v) for the occurrence of a certain wind speed over a year. For that purpose, Weibull distributions have been specified for individual locations using the formula 4.

$$f(v) = \left(\frac{k}{c}\right)\left(\frac{v}{c}\right)^{k-1} \cdot exp\left[\left(-\frac{v}{c}\right)^{k}\right] \quad (4)$$

where:
c, k – fitting parameters. Parameter c has the function of the so-called scale parameter, which means in general that the higher value of that parameter, the wider is a given distribution. In this case, it means greater range of wind speed fluctuations at a given location. Parameter k is the so-called shape parameter and it affects distribution shape. Changing the shape parameter from k=1 (and less) to higher values, we move from the category of exponential distributions to the distributions having a maximum. Parameter c determines with some approximation the position of distribution maximum (especially for distributions with small k).

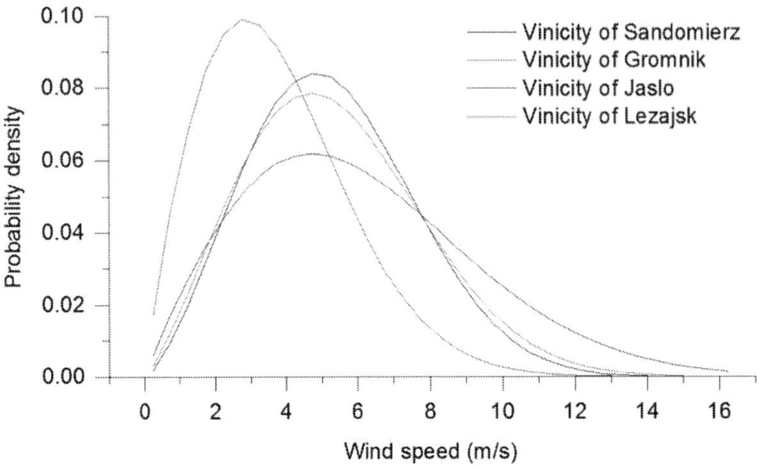

Fig. 4. Fitted Weilbull distribution of wind speed for selected location

Comparison of Weibull distributions for the selected locations shows evidently higher share of speeds ranging from 9 to 16 m/s near Jasło than in the other considered locations. On the other hand, speeds from range 0-5 m/s prevail near Gromnik. Then, in the vicinity of Sandomierz and Leżajsk the conditions are much the same – speeds from range 4-7 m/s prevail in these regions. Thus, one may conclude that the vicinity of Jasło is characterised by the most favourable wind conditions among all examined locations.

3.3 Determination of wind speeds for the considered heights

Because wind speed is strongly dependent on the height above ground level, an additional analysis of changes in monthly and daily wind speed at the height of 100 m was carried out for the location near Jasło. The results obtained for the vicinity of Jasło (the location with the best wind conditions determined at the height of 50 m) are shown in Figs. 5, 6 and 7.

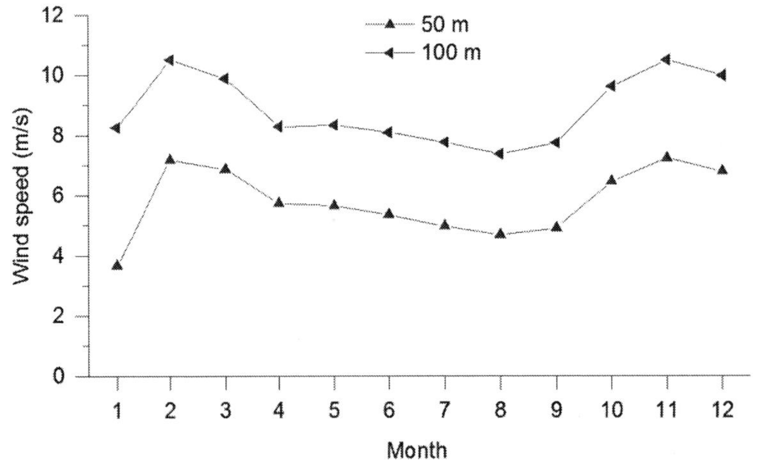

Fig. 5. Monthly wind speed profiles at two heights for Jasło.

Comparison of fluctuations of average monthly wind speeds at the heights of 50 m and 100 m shows big similarity (Fig. 6). The average difference between the values measured at both heights is ca. 3 m/s – highest in January (ca. 4.6 m/s) and lowest in April (ca. 2.6 m/s). The situation is slightly different when we analyse daily wind speed distribution. Analysis of the data from Fig. 6 allows observing that the trajectory of changes in wind speed at the height of 100 m is less even than in the case of 50 m height. Moreover, average daily wind speed in the case of 100 m is ca. 48% higher than the speed measured at the height of 50 m.

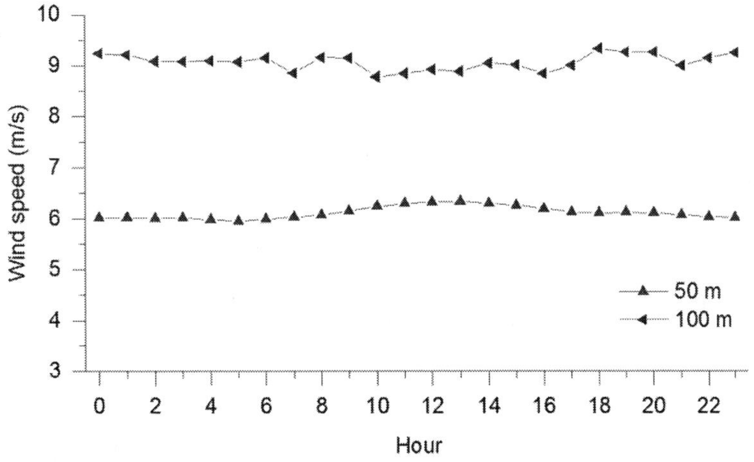

Fig. 6. Daily wind speed profiles at two heights for Jasło

Also, the probability density for the occurrence of higher wind speeds (ranging from 8 to 16 m/s) is considerably higher in case of 100 m. Then, high share of speeds ranging from 0 to 6 m/s is visible for the height of 50 m. Consequently, in both cases Weibull distribution has the shape as shown in Fig. 7.

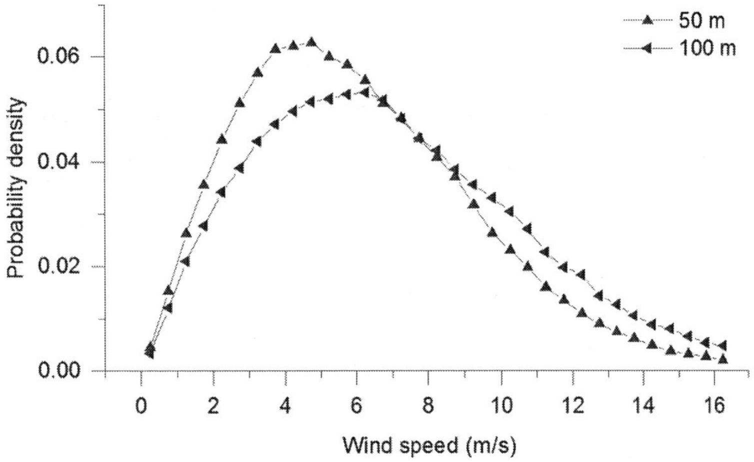

Fig. 7. Weibull wind speed distribution for heights 50 and 100 m for Jasło.

3.4 Estimated energy generation in wind turbines

Based on the analysis of wind resources at different locations, the area in the vicinity of Jasło has been chosen for further considerations. This area is characterised by greatest potential regarding the investment in wind turbines among all discussed locations. The use of four market available wind turbines with parameters specified in Tab. 2 has been discussed for this location. In case of the two turbines, for which tower height remains within range from 50 m to 100 m, wind energy E(h) has been recalculated (eq. 5) for the appropriate height h.

$$E(h) = E_0 \cdot \left(\frac{h}{h_0}\right)^{3\alpha}, \quad (5)$$

where:
E_0 – wind energy at the height of h_0,
h_0 – reference height,
α – coefficient dependent on terrain roughness.

Table 2. Energy generation for selected wind turbines

Turbine	Nominal power [kW]	Tower height [m]	Rotor diameter [m]
WT1	2,000	85	71
WT2	1,500	80	77
WT3	2,050	100	92.5
WT4	2,500	100	90

The amount of possible electric energy yield when using wind turbines according Table 2 is shown in Fig. 8. As we can see, the most favourable solution is to use the wind turbine no. 4. It allows production of ca. 6.75 GWh of electric energy per year.

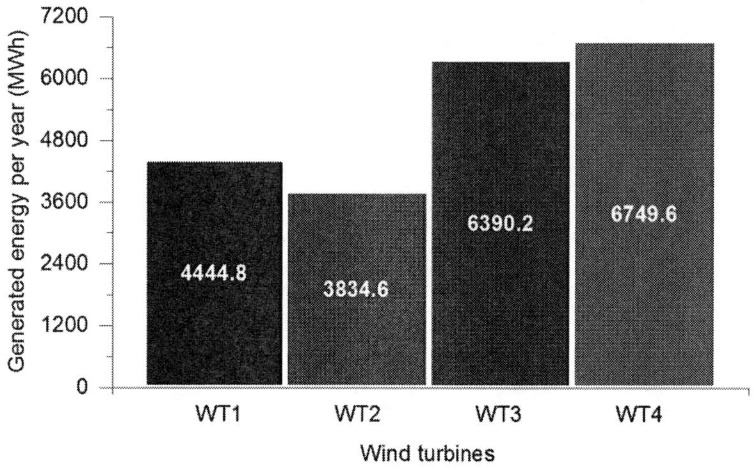

Fig. 8. Amount of energy generation for selected turbines

Assuming that the cost of wind farm construction in Poland is ca. 7 million PLN per 1 MW of installed power, and the price per a unit of electric energy produced by wind power plant (with installed power exceeding 1 MW) is 385 PLN/MWh, it is possible to approximate the values of the indexes: SPBT (Simply Pay Back Time), IRR (Internal Rate of Return) and NPV (Net Present Value). These calculations also take into account operating costs (land lease costs, wind farm servicing and management, energy balancing costs, insurance and the costs of energy for own purposes), which are assumed to reach the level of ca. 10% of the revenues. The values of economic indexes calculated for the assumptions specified above are as follows: SPBT = 7.4 years, NPV = 0.746 million PLN (for considered period of 10 years), and IRR = 7.1 %.

Calculations regarding ecological effect can be carried out parallel to the economic calculations. In general, ecological effect can be described as the volume of pollution not discharged into the environment due to operation of new equipment being the subject of the

DOI: 10.1051/e3sconf/20171401014

investment. The volume of emission avoided *e* per one year can be expressed in the most general way using the formula (6):

$$e = \sum E \cdot w_{e,i} \tag{6}$$

where:

E – the volume of produced electric energy, MWh

$w_{e,i}$ – emission factor of "i" pollution, kg/MWh

In case of construction of wind power plant with turbine no. 4, emission volumes computed in this way are respectively (compared to a typical coal power plant): e_{CO} = ca. 28 Mg, e_{CO2} = ca. 60 181 Mg, e_{NOx} = ca. 9 Mg, e_{SO2} = ca. 46 Mg and e_{dust} = ca. 84 Mg.

3.5 Simulations of the hybrid wind-diesel power plant in TRNSYS

Due to the variability in the amount of the energy generated in wind turbines, it is necessary to preserve the capacity e.g. in gas and water peaking power plants. During conducted simulations, diesel electricity generator and DEGS dispatch controller have been used to adapt the number of running engines to a given electrical load which is partly met by Wind Turbines (see Fig. 8).

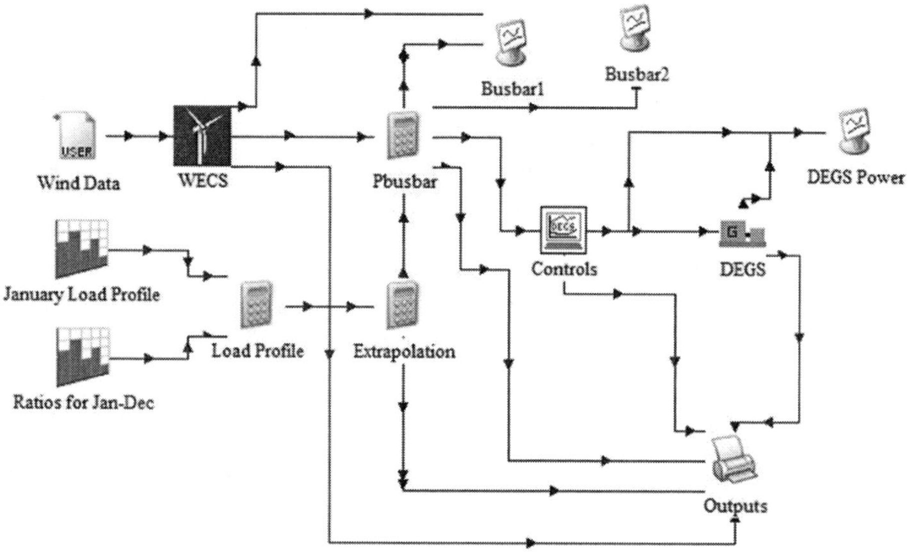

Fig. 9. TRNSYS simulation diagram

The simulations were conducted using wind data for the vicinity of Jasło (wind speed measured at a level of 100 m) as an input, 500 kW wind turbine(s) and 300 kW diesel generator. The simulation results, shown in Fig. 10, show that in case of using the off-shore systems it is necessary to store energy and use peaking energy source (e.g. diesel generator). The same situation is in the case of on-shore systems connected to the grid, where shortage of energy is met by conventional power plants. The width of peaks characteristic for diesel generator working time changes depending on installed power of the turbines.

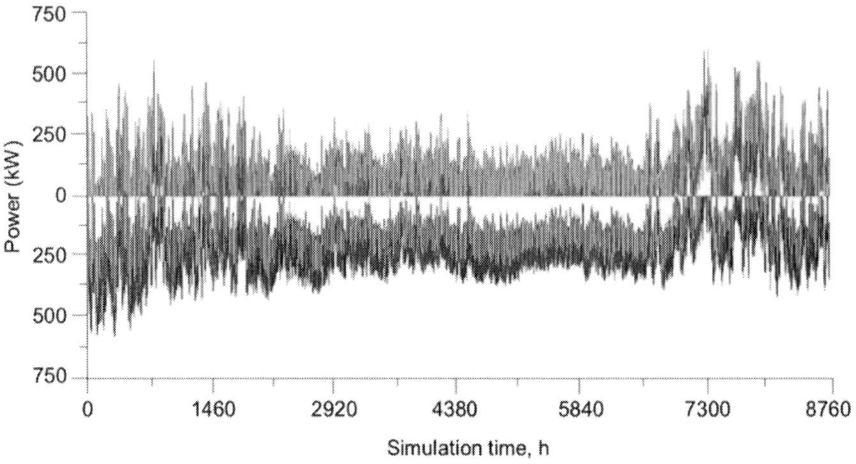

Fig. 10. Results of simulations conducted in TRNSYS.

4 Conclusions

Results of measurements conducted in four selected locations show that the selected regions of Małopolskie, Świętokrzyskie and Podkarpackie Voivodeships mostly have quite good wind conditions. The average monthly wind speed is about 5-6 m/s at the height of 50 m in case of three of four considered locations (the vicinity of Sandomierz, Jasło and Leżajsk). Lower wind speed has been measured for location near Gromnik. Taking into account the most promising location (close to Jaslo) we can observe that average wind speed in case of the height of 100 m is ca. 48% higher than the speed measured at the height of 50 m. It ranges within 8-11 m/s (considering average monthly speed values). Depending on the type of considered wind turbine, it is possible to generate up to 6.8 GWh of electricity in the selected location. Due to high instability of energy production in wind turbines, it is necessary to reserve the capacity in gas and water peaking power plants.

Acknowledgements

The work has been completed as part of the statutory activities of the Faculty of Energy and Fuels at the AGH University "Studies concerning the conditions of sustainable energy development"

References

1. T. Mirowski, E. Mokrzycki, R. Ney, *Energetyka wiatrowa - stan obecny i szanse rozwoju* (Wydawnictwo IGSMiE PAN, Krakow, 2015)
2. M. Donatelli, G. Bellocchi, E. Habyarimana, R. Confalonieri, F. Micale, Comput. Electron. Agr. **69**, 165-170 (2009)
3. J. Ephrath, J. Goudrian, A. Marani, Agr. Syst. **51**, 377-293 (1996)

Damage assessment in wind turbine technology

Wiesław Ostachowicz[1,2*], *Paweł H.* Malinowski[1], *Rohan* Soman[1] and *Tomasz* Wandowski[1]

[1] Institute of Fluid-Flow Machinery, Polish Academy of Sciences, 80-231, Gdansk, Poland
[2] Warsaw University of Technology, Faculty of Automotive and Construction Machinery Engineering, 02-524, Warsaw, Poland

Abstract. The economic efficiency of wind turbines WT is strictly dependent on their availability and reliability. This problem is most important in the case of offshore wind turbines OWT. There is a search for new materials, new manufacturing process, and new rules of structural designs. The aim of it, is to achieve far tighter safety margins than previously developed wind turbine solutions. Attention of many researchers is focused on the problem of WT/OWT reliability.
In this paper Structural Health Monitoring (SHM) approach for wind turbines is presented. This approach is based on such topics as piezoelectric transducers, elastic waves propagation phenomenon, fibre Bragg gratings sensors, structural vibrations analysis. The combination of mentioned techniques allows to perform efficient both local and global SHM of the OWT.
Research results presented in this paper are mainly related to fibre reinforced polymer materials FRPs due to fact that they are widely utilized in wind turbine blades manufacturing. However, certain results for metallic structures (tower structure) were also presented.
Considered above investigations are intended to develop future SHM techniques that include sensors, hardware, software and methodologies of WT/OWT structural assessment.

1 Introduction

Wind Energy is seen as one of the most promising solutions to man's ever increasing demands of a clean source of energy. The use of wind energy has received an impetus due to the advancements in the field of materials engineering. Newer, bigger wind turbines are now possible which are more robust, and lighter in weight. The main drawback of the wind energy is the high initial costs for setting up and high maintenance costs. These high initial cost make the energy more expensive than the conventional energy sources like fossil fuels and nuclear and hence has not been widely accepted. In order to reduce the cost of generation, there is effort to increase the life-time of the wind turbines, reduce maintenance costs and ensure high availability [1].
The lifetime may be increased by ensuring a more robust design while the maintenance cost may be lowered and the high availability ensured through the use of condition monitoring (CM) and structural health monitoring (SHM). SHM allows early detection of damage and

* Corresponding author: wieslaw@imp.gda.pl

allows maintenance planning which reduces the cost. Furthermore, it can allow us to avoid unnecessary downtime, hence increasing the availability of the system.

The wind turbine is a complex machine consisting of electrical and structural components. The structural components include the gear box, tower, rotor and several auxiliary mechanisms like the yaw system, pitch system etc. Considerable amount of research has been carried out in the area of SHM of rotating machinery which specifically deals with the CM of the gearbox system and can be found in [2, 3]. The auxiliary systems although important will not cause catastrophic failure of the system and hence are not significant. The SHM of the structural components on the other hand is essential and a big challenge due to the specific problems associated with them. The study in this area can be classified as the one dedicated to the metallic structures including the support structure and the composite parts including the rotor blades. Due to the very different operating conditions and the dominant failure modes of these systems different SHM techniques need to be used for the monitoring based on the advantages and disadvantages of the individual techniques.

2 Metallic Structures

As can be seen in Fig. 1, the support & housing sub-assembly of the wind turbine has the highest downtime per failure. This is attributed to the need of specialized equipment needed for the maintenance or the replacement of the sub-assembly. The numbers indicated are for on-shore wind turbines and are expected to go even higher for off-shore structures. Thus the support structure foundation is one of the key components of the wind turbine that need to be closely monitored. The foundation of the wind turbine is exposed to the most extreme environmental conditions including, salinity, extreme temperatures, scouring, drifting sea ice and higher possibility of collision with passing vessels and as such has to be carefully monitored.

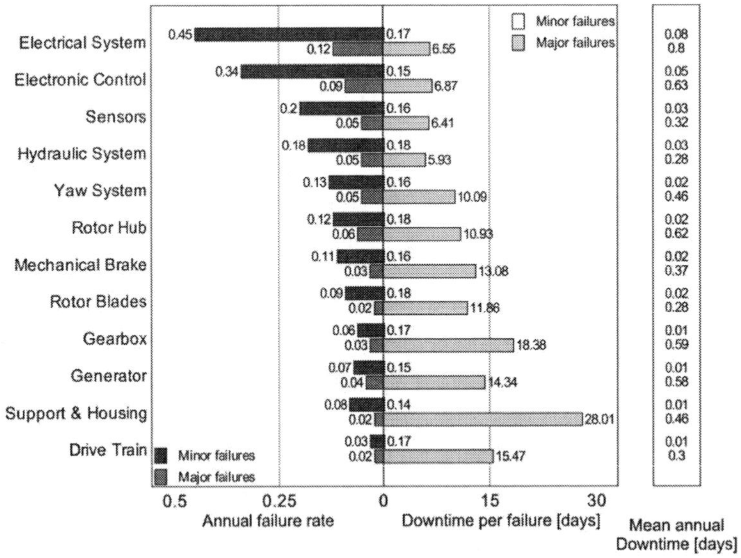

Fig. 1. Reliability Characteristics of Sub-Assemblies [4]

The tripod structure is expected to be the future of the offshore wind turbine foundation structures. This is due to the trend of setting up wind farms in deeper waters where the traditional monopole foundations will not be able to fulfill the structural requirements. The major modes of failure of the tripod structure are fatigue at the main node and scouring at the support where the tripod is connected to the sea bed. A frequency based damage detection

DOI: 10.1051/e3sconf/20171401015

technique for fatigue induced cracks has been proposed in [5]. Similar methodology may also be used for the detection and localization of the deterioration of the support condition due to scouring. The method quantifies the change in the frequency response function (FRF) of the healthy structure and deteriorated one using the root mean square difference (RMSD). But any other tool may also be used in order to monitor the change in the FRF. The Fig. 2 shows the placement of the fiber Bragg grating (FBG) strain sensors on the scaled tripod used for the validation of the methodology.

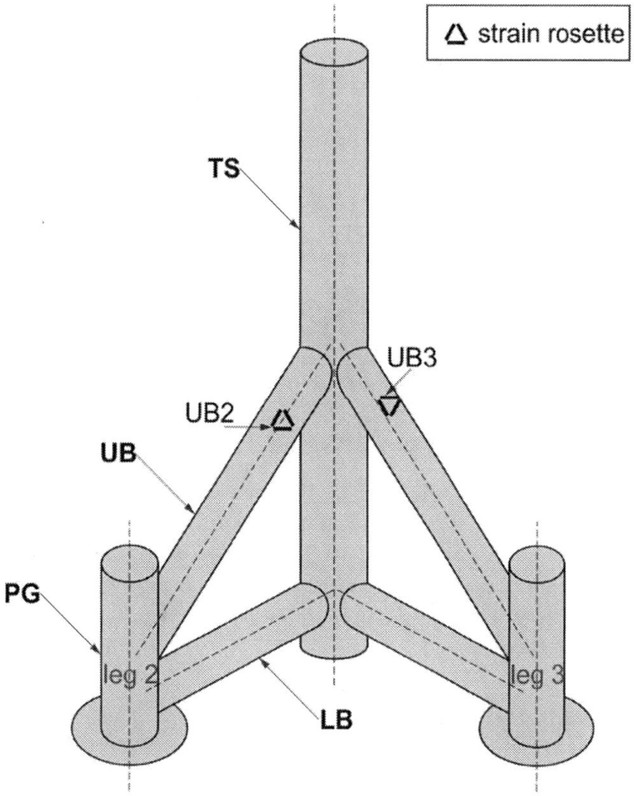

Fig. 2. Nomenclature of the components of the scaled tripod [5]; UB-Upper Brace, LB- Lower Brace, TS- Tower Support, PG- Pile Guide and UB2 and UB3 refer to the strain rosettes on leg 2 and leg 3 respectively

The FBG sensors were used due to their corrosion resistance and capability of working underwater in harsh environments. The working principle and the details about FBG sensors may be found in [6].

The strains were measured under random hammer excitation (output only) in order to simulate real conditions where the excitation of the support structure due to waves cannot be measured. The legs of the tripod were secured to the anti-shaker table with bolts and the deterioration was simulated by removing the support of one leg. The experimental validation of the methodology was limited to the extreme case while the efficacy of the method for intermediate conditions was validated using the finite element model.

The

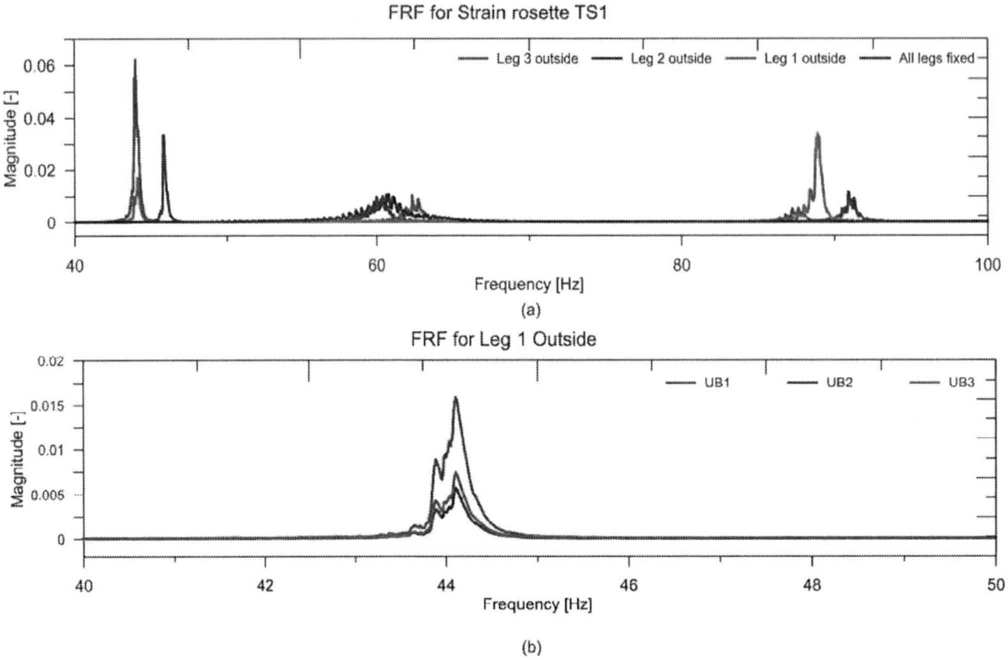

Fig. 3 (a) shows the experimental results for the four cases namely all legs bolted and each of the leg unsupported. As can be seen, the removal of the support leads to the new peaks in the FRF spectrum. If the RMSD value of the peaks is higher than a certain threshold determined based on engineering judgement the damage is said to be detected. The peaks for the damaged condition are at the same frequencies owing to the symmetry of the structure. By analyzing the new peaks in the FRF, the damage can be further localized.

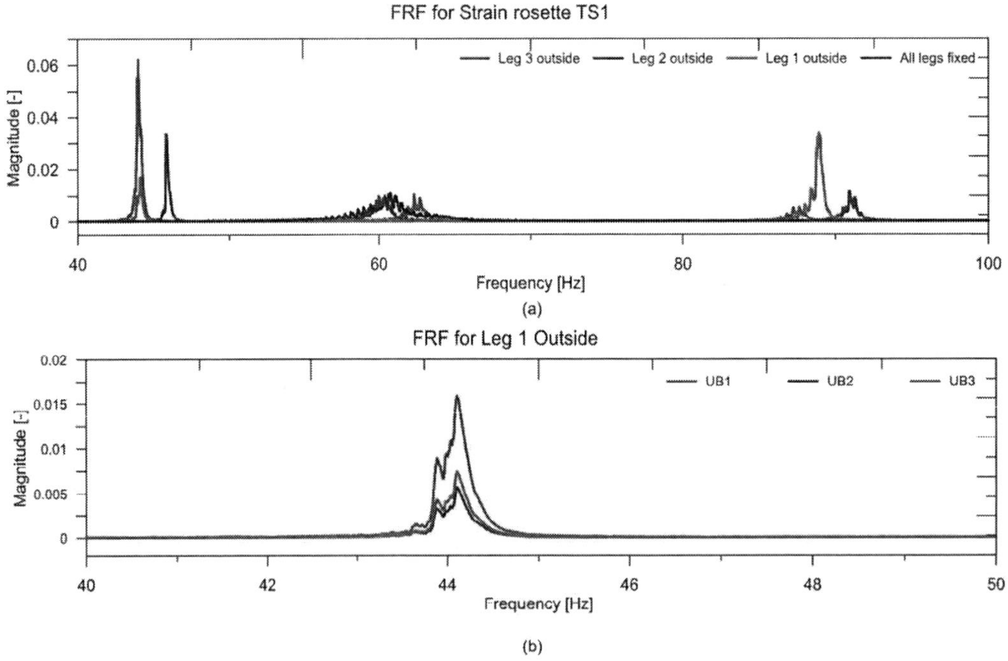

Fig. 3. FRF Spectrum for: a) each of the scenarios healthy and each leg support removed, b) each sensor on UB for the scenario Leg1 outside

DOI: 10.1051/e3sconf/20171401015

As can be seen in

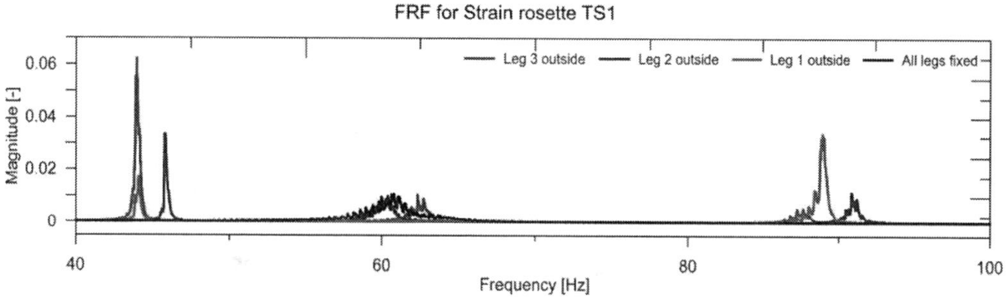

(a)

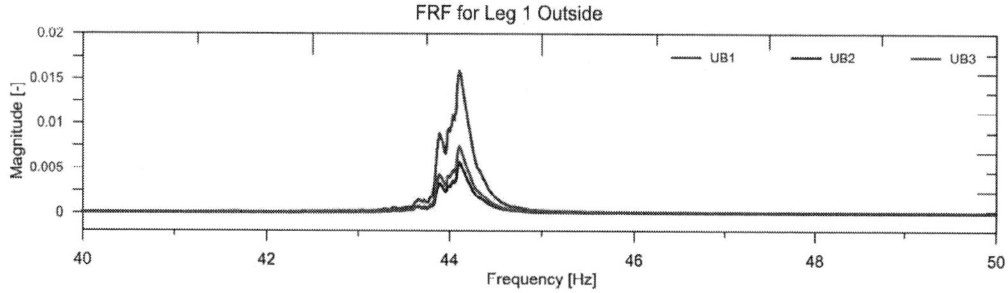

(b)

Fig. 3 (b), the peak related to the deteriorated support is highest. Although this might seem a trivial damage localization criterion, it is only valid for the vibration modes where the UB of the tripod has significant deflection. Thus a validated FE model is necessary for supporting the damage localization process. Although a development of a validated FE model is time consuming and difficult, the present practice in design of the support structures the entire wind farm has only a few standardized designs of the support based on the sea bed depth and condition. Hence, the proposed methodology is still cost effective.

3 Composite Structures

In this section selected non-destructive testing method for assessment of composite structures are presented. Composite structures during exploitation are exposed to mechanical damage (delamination) and physico-chemical degradation (moisture, thermal degradation, aggressive agents). Therefore, there is a need of development of assessment methods that are able to detect mentioned typed of material degradation.

3.1 Active thermography

First method was based on active thermography. In this approach pulse excitation based on halogen lamps (2 x 150 W, 7s heating time) was utilized. Temperature increase due to heating was equal to 30°C. Sensing process was performed using FLIR SC5600 infrared camera. Experimental stand was presented in Fig. 4. Research was conducted for GFRP panel with dimensions 500 mm x 500 mm x 1.5 mm. In this panel 20 mm x 20 mm teflon insert was located in order to simulate delamination. Result of measurement is presented on differential thermogram (Fig. 5). The location of teflon insert is clearly visible.

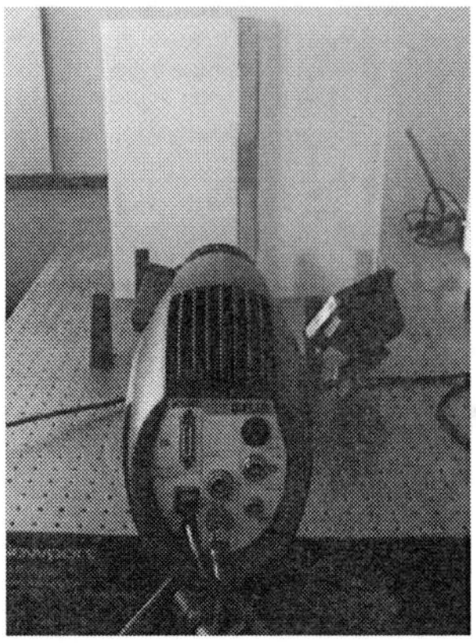

Fig. 4. Active thermography stand

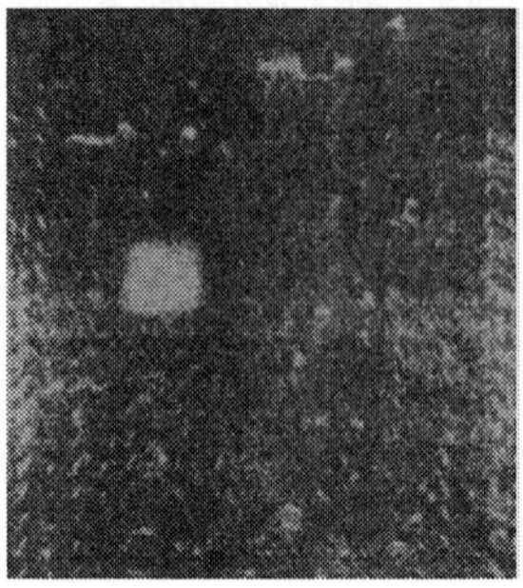

Fig. 5. Thermogram for GFRP sample with teflon insert (20 x 20 mm)

3.2 Scanning Laser Doppler Vibrometry SLDV

In this section results of experimental investigations based on Polytec PSV-400 Scanning Laser Doppler Vibrometer (SLDV) are presented. In the conducted research only one scanning laser head was utilized (measurements along laser beam approximating out-of-plane velocities). Measurements were performed for dense mesh of points spanned over the area of investigated composite panels.

A simple and widely used full wavefield signal processing method is based on root mean square (RMS) of the time domain signals. As a result energy distribution related to propagation of wave in the structure is obtained. The location and size of damage are visualized by much higher amplitudes in areas of their occurrence than in undamaged areas. Weighted root mean square (WRMS) is a modification of approach based on RMS energy. In WRMS a weight factor is added that decreases the importance of the time samples at the beginning (large wave amplitudes due to excitation) and increases the importance of the samples closer to the end of signal where waves amplitudes are small due to damping. The Weighted Root Mean Square (WRMS) can be calculated as:

$$WRMS = \sqrt{\frac{1}{N}\sum_{k=1}^{N} w_k {s_k}^2} \quad (1)$$

where coefficient w_k is defined as follows: $w_k = k^m \quad m > 0$. In the case of $w_k = k^0$ we obtain classical RMS. For case of $w_k = k^1$ linear amplification of signal amplitude with time will be obtained. Utilization of $w_k = k^2$ will give square function of amplification with time. Weighting factor $w_k = k^{1.8}$ was utilized in this research.

 DOI: 10.1051/e3sconf/20171401015

In this research three CFRP panel with planar dimensions 500 mm x 500 mm x 3.5 mm and 5 mm were used. In these panels teflon inserts were located in order to simulate delamination. Three sizes of teflon inserts were investigated: 10 x 10 mm, 15 x 15 mm and 20 x 20 mm. Each panel has one simulated delamination. Results in the form of WRMS energy maps for investigated delamination sizes are presented in Fig 6. Strong energy concentration in the delaminated areas can be observed. These maps show location, shape and size of delaminations.

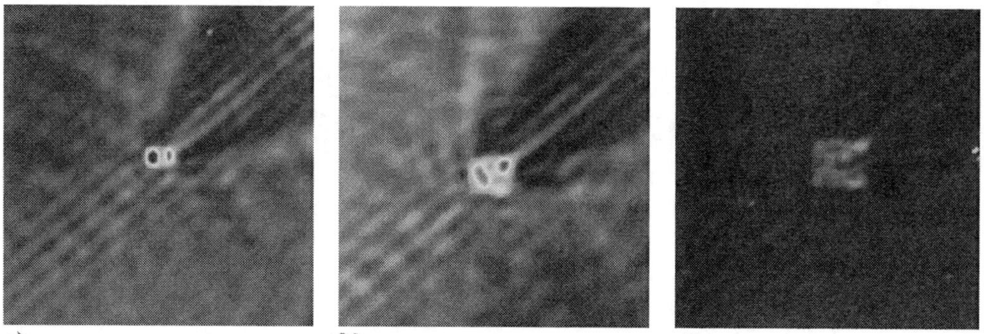

a) b) c)

Fig. 6. WRMS damage map for CFRP sample with teflon insert: a) 10 x 10 mm, b) 15 x 15 mm, c) 20 x 20 mm; excitation frequency 100 kHz

The next investigated case was aerospace CFRP panel with omega stiffeners. Panel consists of two CFRP laminates bonded together (one laminate includes formed stiffeners). In this panel circular delamination simulated by teflon insert with diameter 30 mm was located (**Fig. 7**). The result in the form of WRMS energy map for excitation frequency 50 kHz was presented in **Fig. 8**. In the WRMS energy map stiffeners locations can be clearly observed. Moreover, location of delamination with diameter 30 mm can be also noticed.

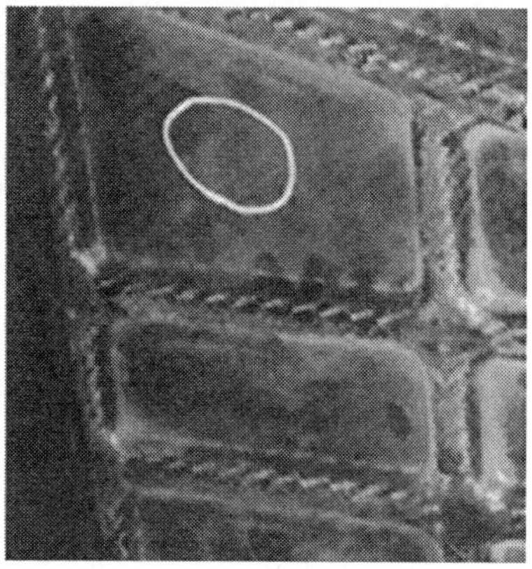

Fig. 7. CFRP stiffened aerospace panel with teflon insert (diameter 30 mm)

Fig. 8. WRMS damage map for CFRP stiffened aerospace panel with teflon insert; excitation frequency 50 kHz

3.3 Laser Induced Fluorescence LIF

In this section research was focus on a spectroscopic method that can be applied for thermal degradation detection in composite elements. In this technique spatial distribution of fluorescence intensity can be monitored with CCD imaging camera equipped with bandpass optical filters. Laser Induced Fluorescence LIF spectra were recorded using laboratory system equipped with laser excitation sources: DPSS cw Nd:YAG 532 nm, 0.2 – 2 W output power (Spectra Physics) and pulsed Nd:YAG 532 nm, 6 ns pulse diration (Briliant B, Quantel). The 532 nm excitation was chosen based on previous research. Excitation at this wavelength showed maximal sensitivity to surface state condition. Fluorescence detection system was based on 0.3 m Czerny-Turner type monochromator (SR-303i, Andor) and spectroscopic ICCD camera (DH-740 Istar, Andor). In the detection path excitation laser radiation was blocked by a band pass filter (OG550, Shott). Fluorescence spectra were recorded in 10 randomly chosen points on sample in order to avoid local surface inhomogeneity. Additionally, the spatial distribution of laser induced fluorescence was recorded with digital camera equipped with cut-off bandpass filter (OG570, Shott). Thermal degradation of samples was carried out by means of heat gun producing stream of hot air 5 mm in diameter and temperature 450 ^{0}C. The influence of time duration of heat source exposure was investigated. The research was conducted on CFRP. Four times of exposure were tested 5, 10, 20 and 30 s in this research. The photographs of the heat exposed area under the LIF laser illumination are presented in Fig. 9. The woven structure of the composite is clearly visible. The increase in the intensity in the visible light is also clearly observed. With increasing time of heating a clear circular shape forms indicating the overheated area can be observed.

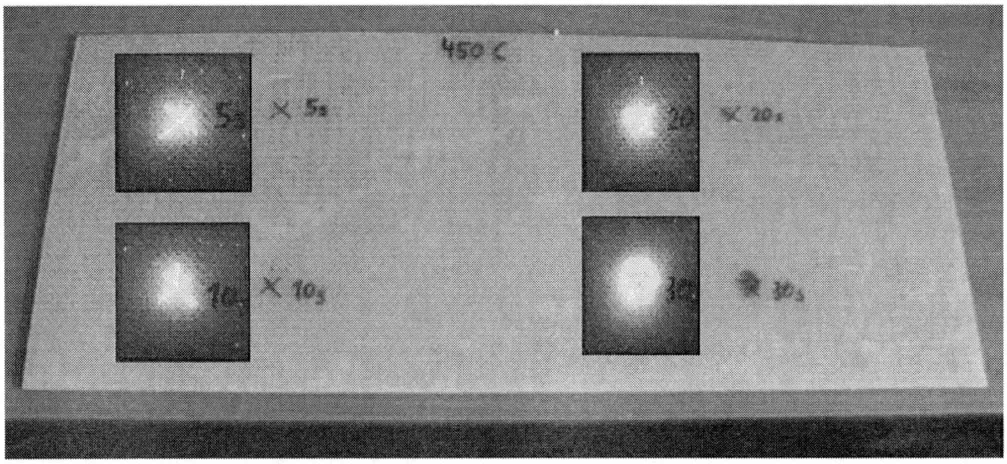

Fig. 9. Fluorescence intensity distribution registered at GFRP sample exposed to heat source at 450°C for 5, 10, 20and 30 seconds - pictures made with photo camera

4 Conclusions

In this paper different assessment method for metallic and composite parts of structures were presented. The investigation of metallic structure with the FBG sensors showed promising results. Strain tracking allowed to distinguish the damage case after analysis of the FRF response. In the case of composite elements three non-contact nondestructive testing NDT measurement techniques were discussed. Active thermography and laser vibrometry were utilized for delamination detection in composite samples. Delaminations were simulated by teflon inserts. Both method gave very good results of delamination localization. In the case of LIF method thermal degradation in samples was detected. Thermal degradation was induced by a heat gun. The LIF method also gave good results of thermal degradation localization.

Research was partially supported by the project entitled: Non-invasive Methods for Assessment of Physicochemical and Mechanical Degradation (PBS1/B6/8/2012) granted by National Centre for Research and Development in Poland.
This research was also supported by the project titled: Development of the selection method of the offshore wind turbine support structure for polish maritime areas (PBS1/A6/8/2012, Aquilo) granted by National Center for Research and Development in Poland.
Authors would like to acknowledge the European Commission for their research grant under the project FP7-PEOPLE-2012 ITN 309395 MARE-WINT (new Materials and REliablity in offshore WINd Turbines technology).
Authors would like to give special thanks to Academic Computer Centre in Gdansk (CI TASK) for access to ABAQUS software.
The authors would like to thank Mirosław Sawczak for the research on LIF technique.

References

1. R. N. Soman, P. H. Malinowski and W. M. Ostachowicz, 2015. Bi-axial neutral axis tracking for damage detection in wind-turbine towers. Wind Energy, vol. **19**, 4, pp. 639–650.
2. Y. Feng, Y. Qiu, C. J. Crabtree, H. Long and P. J. Tavner, Monitoring wind turbine gearboxes, Wind Energy, vol. **16**, no. 5, pp. 728–740, 2013.
3. S. Sheng, Wind turbine gearbox condition monitoring round robin study–vibration analysis, Contract, vol. **303**, pp. 275–3000, 2012.
4. S. Faulstich, B. Hahn and P. J. Tavner, Wind turbine downtime and its importance for offshore deployment, Wind Energy, vol. **14**, no. 3, pp. 327–337, 2011.
5. S. Opoka, R. Soman, M. Mieloszyk and W. Ostachowicz, 2016. Damage detection and localization method based on a frequency spectrum change in a scaled tripod model with strain rosettes. Marine Structures, **49**, pp.163-179.
6. E. Udd and W. B. Spillman, *Fiber optic sensors: an introduction for engineers and scientists*. (John Wiley & Sons, 2011).

Evaluation of the impact of adjusting the angle of the axis of a wind turbine rotor relative to the flow of air stream on operating parameters of a wind turbine model

Stanisław Gumuła[1*], *Małgorzata* Piaskowska – Silarska[2], *Krzysztof* Pytel[2], *Henryk* Noga[2], *Wojciech* Kulinowski[2]

[1]AGH University of Science and Technology, al. Mickiewicza 30, 30-059 Kraków, Poland
[2] PU of Cracow, ul. Podchorążych 2, 30-084 Kraków, Poland

Abstract. The aim of this study was to determine the effect of regulation of an axis of a wind turbine rotor to the direction of wind on the volume of energy produced by wind turbines. A role of an optimal setting of the blades of the wind turbine rotor was specified, as well. According to the measurements, changes in the tilt angle of the axis of the wind turbine rotor in relation to the air stream flow direction cause changes in the use of wind energy. The publication explores the effects of the operating conditions of wind turbines on the possibility of using wind energy. A range of factors affect the operation of the wind turbine, and thus the volume of energy produced by the plant. The impact of design parameters of wind power plant, climatic factors or associated with the location seismic challenges can be shown from among them. One of the parameters has proved to be change settings of the rotor axis in relation to direction of flow of the air stream. Studies have shown that the accurate determination of the optimum angle of the axis of the rotor with respect to flow of air stream strongly influences the characteristics of the wind turbine.

1 Introduction

All over the world there are areas where wind energy can be used to produce energy in a commercially reasonable manner. Therefore, the power of newly built wind power plants is constantly increasing, there are more of them and they are bigger. Grouping of the plants in ecologically justified locations of wind farms results in the need to reduce the distance between objects. [1]. Power plants that are located too close to one another may disturb the direction of the wind flow and the changing wind direction can affect the amount of energy obtained by the wind power plant. [2]. Essential elements of the wind power plant are the tower, nacelle and rotor. The tower holds the weight of the rotor and nacelle while being resistant to vibrations and the effects of fatigue resulting from wind pressure and cyclic rotation of the rotor blades. Nacelle includes gears, generator and control systems managing

* Corresponding author: stangum@gmail.com

DOI: 10.1051/e3sconf/20171401016

the work and setting of the rotor of power plant. At the rotor wheel there is the conversion of wind energy into mechanical energy. Moving away the rotor axis of wind turbine from the main wind direction causes change in the effective area of the rotor work and consequently the differences in the amount of obtained energy. Positioning mechanism for the optimum direction enabling maximum use of wind energy plays a role in the efficient use of wind energy. A mechanism tracking the wind direction and the position of a nacelle, managed by an electronic system, turns the nacelle slightly during the changes in wind direction. Right positioning of the rotor of wind power plant is an extremely important element of the aero energy system. [3].

Wind energy is a rapidly growing global economic sector. [4]. In comparison with the solar energy, it is characterized by lower costs per 1 kWh and comparing it with geothermal energy less technical obstacles will be encountered. Use of wind energy is convenient as the wind blows in virtually every place on the globe, near the land, the ocean or high above the surface of our planet.

The efficiency of energy production from wind varies. This is related to constantly changing climatic conditions and different regions of the world. A good complement to energy production is usually between wind and solar energy, which are the two most suited renewable sources of energy. A good solution is the creation and operation of hybrid wind and solar energy systems, optimally matching energy installations for the effective and economical use of energy resources. In order to optimize the operation of such plants, it is important to determine the operating parameters of both systems, i.e. solar and wind power installations [5–10].

The aim of this study was to determine the effect of regulation of an axis of a wind turbine rotor to the direction of wind on the volume of energy produced by wind turbines.

2 Power coefficient and tip speed ratio

Wind turbine may be designed as a device with a horizontal or vertical axis of rotation of rotor. Horizontal axis wind turbines have a rotor mounted to a nacelle. The rotors have not the same amount of blades, but often there are three blades in rotor. Rotor blades are crucial elements of wind turbine because of their ability to capture wind energy. Design and dimensions of the rotor blades have a substantial influence on an energy output. Rotor blades drive a generator, that produce planned amount of energy. A produced energy can be shown as a power output in Watts and in non-dimensional form as a power coefficient C_p [Fig. 1].

$$C_p = 2\frac{P}{\rho V_0^3 A} \tag{1}$$

where:
C_p – power coefficient;
P –power, W;
ρ — density of air, kg/m^3;
V_0 – wind speed, m/s;
A – area of a rotor, m^2, given by: $A=\pi R^2$.

A maximum value of a theoretical C_p stand for C_{Pmax} = 16/27 = 0.593 and is defined as the so-called Betz limit. Contemporary wind turbines work with C_p up to about 0.5. Figure 1 shows that presented wind turbine has a maximum efficiency of approximately C_p = 0.3 for a wind speed between 6 and 8 [m/s].

Tip speed ratio refers to the wind speed and the speed of the tips of the blades of wind turbine. It is calculated as:

$$\lambda = \omega R/V_0 \tag{2}$$

where:
λ – tip speed ratio;
ω – rotational speed, rpm;
R – radius of the rotor, m;
V_0 – wind speed, m/s.

Power coefficient can be presented as a function of the tip speed ratio λ (TSR). [Fig. 2]. TSR vary with rotational and wind speed. Generally, the greater number of blades, the lower is the tip speed ratio. [11-13].

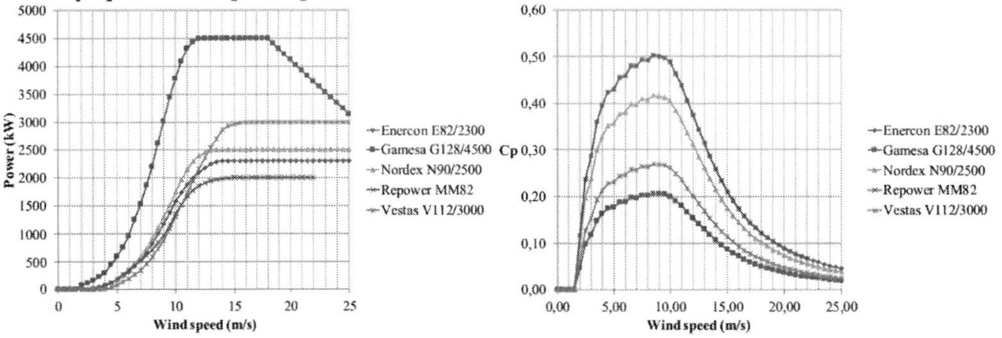

Fig. 1. Comparison between sample power curve (left) and power coefficient (right) as a function of the wind speed, [14].

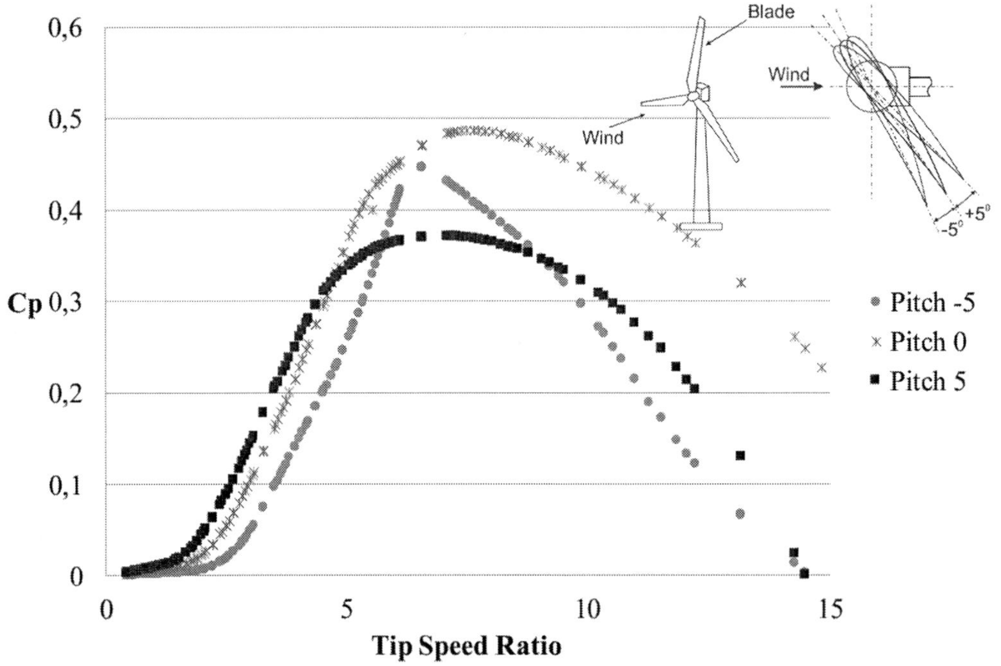

Fig. 2. Comparison between sample C_p curve and TSR for pitch values: –5, 0 and 5. Insert shows a coordinate system applied in research.

The angle of the axis of the rotor in relation to the direction of flow of the air stream is one of the factors that make the system work efficiently.

3 Experimental procedure

The aim of the research is to determine the impact of the setting of the rotor axis of the power plant towards the wind direction on the amount of energy produced by the wind power plant. Moreover, the role of optimal setting of the blades of the power plant in the rotor towards the wind direction was specified.

The measurements were carried out on the test position, which allows research on models of wind power plants working in conditions close to real. The drive system of the tunnel is composed of sixteen axial fans with capacity of 3,800 m^3/h at a rotational speed of 2300 rpm. controlled independently by inverter. Before entering the measurement space the air stream passes through the system consisting of the diffuser and bundle–pipe straightener. In order to increase the speed of the air flow and align its velocity profile, the tunnel was tightly connected with Witoszyński type shape nozzle.

In the test space the model of wind power plant with a rotor of 0.5 m was placed. The power plant was connected with the control and register unit, which measured the torque and rotational speed of the rotor and therefore the power on the power plant shaft. The recording system performs acquisition of torque and rotational speed. The AT2016 used for measurements is a multifunction analog and digital I/O card that features the measurement and control functions for PC/AT systems: D/A and A/D conversion, digital input/output and vounter/timer. It has 16 analogue and 8 digital channels, the sampling rate from 0.002 [Hz] to 200 [kHz], the input signal is ± 5 [V]. The maximum measuring error does not exceed ±0.07%. A measuring computer was equipped with an electronic signal acquisition module E.S.A.M. that recorded more than 20,000 samples of each measured values during the tests. The measuring system in a power plant was composed of two sets. The first one measured the speed of the fan using the optical method. Torque measurement was made using a strain gauge torquemeter mounted in such a configuration to change a torque into a beam bending [Fig. 3].

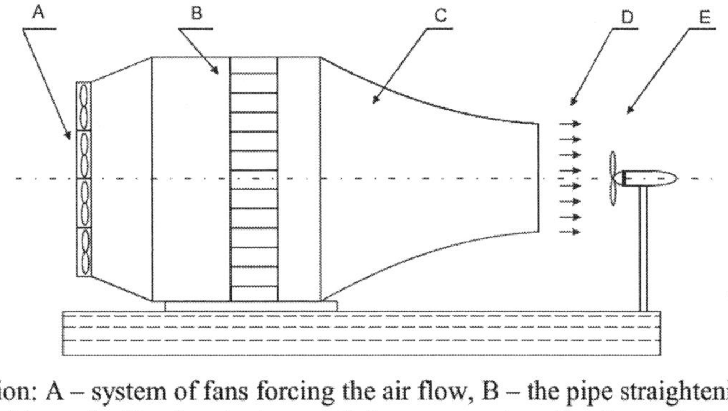

Fig. 3. Test position: A – system of fans forcing the air flow, B – the pipe straightening the air flow, C – the Witoszyński nozzle, D – the air stream of given parameters, E – the model wind power plant.

4 Results and discussion

Analyses of the impact of specific working conditions on the produced energy have been conducted for the model wind power plant. Measurements of the impact of changes in the direction of the rotor axis towards the flow direction of the air stream on the operating of a model wind power plant was carried out for the power plant with the three-bladed rotor with profiled blades. The measurements were carried out for five air stream velocities 4.3 m/s, 5.3 m/s, 7.1 m/s, 7.9 m/s and 11.5 m/s. The angle of deviation of the rotor axis from the direction of the air stream was 0°, 10°, 20°, 30° and 45°. Results of chosen measurements are presented in the charts showing the changes of power coefficient (C_p), depending on the tip speed ratio (λ) for selected values of the angle between the direction of flow of the air stream and the direction of the setting of the rotor axis and the selected flow rate of the air stream (V_0).

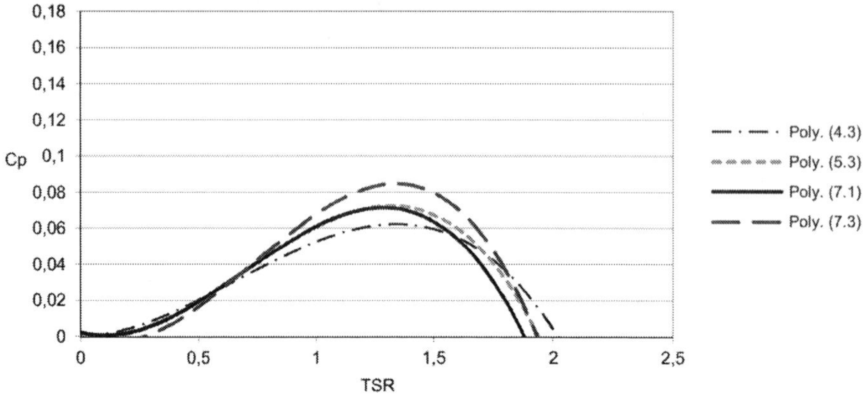

Fig. 4. Changes in Cp = f (λ) for the angle of 0° between the direction of flow of the air stream and the direction of the rotor axis for the selected wind speed values 4.3 – 7.3 m/s.

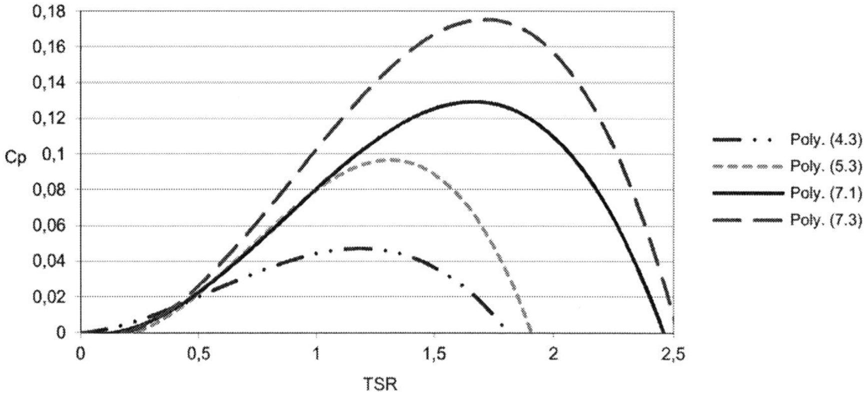

Fig. 5. Changes in Cp = f (λ) for the angle of 10° between the direction of flow of the air stream and the direction of the rotor axis for the selected wind speed values 4.3 – 7.3 m/s.

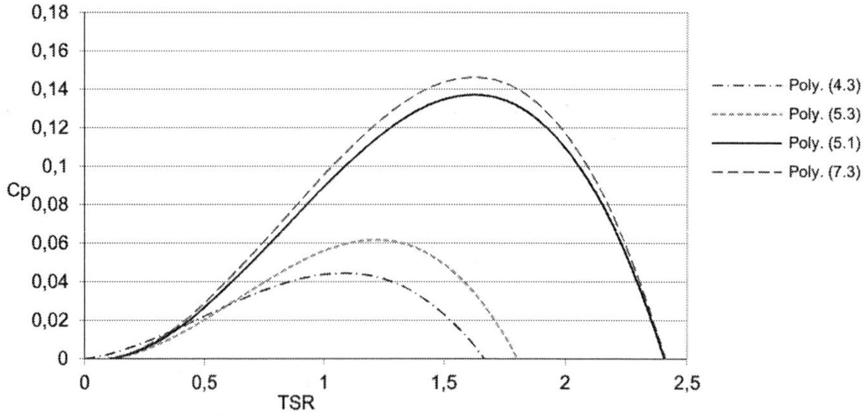

Fig. 6. Changes in Cp = f (λ) for the angle of 20° between the direction of flow of the air stream and the direction of the rotor axis for the selected wind speed values 4.3 – 7.3 m/s.

DOI: 10.1051/e3sconf/20171401016

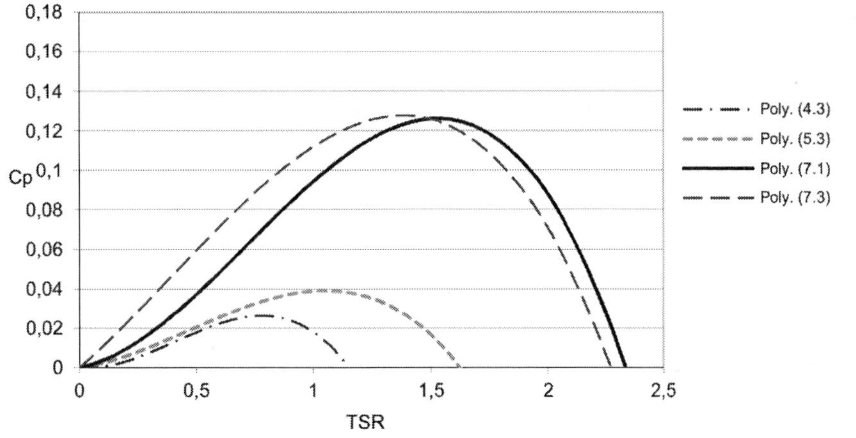

Fig. 7. Changes in Cp = f (λ) for the angle of 30° between the direction of flow of the air stream and the direction of the rotor axis for the selected wind speed values 4.3 – 7.3 m/s.

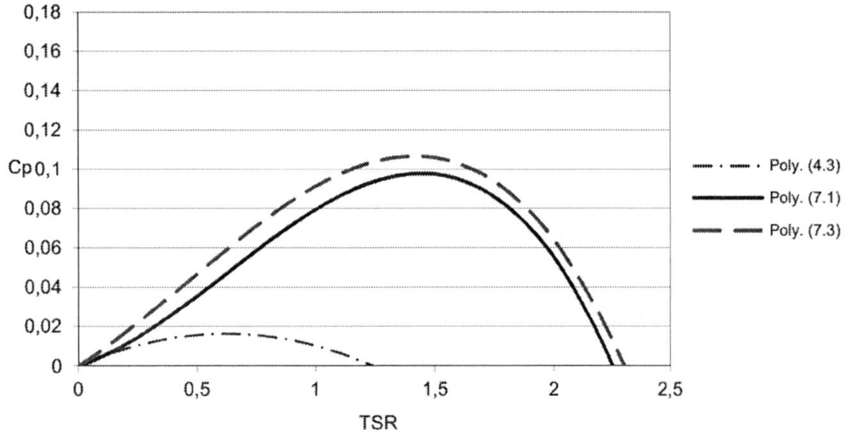

Fig. 8. Changes in Cp = f (λ) for the angle of 45° between the direction of flow of the air stream and the direction of the rotor axis for the selected wind speed values 4.3 – 7.3 m/s.

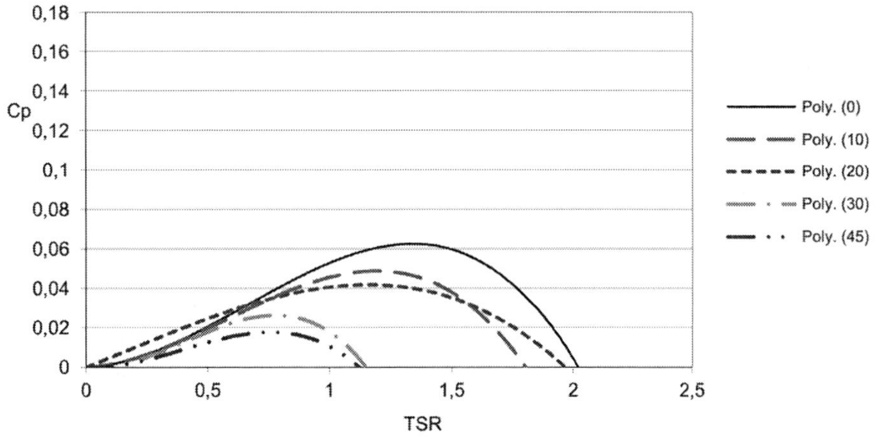

Fig. 9. Changes in Cp = f (λ) for the speed of air stream flow v = 4.3 m/s for the selected angles between the direction of flow of the air stream and the direction of the rotor axis (0 to 45°).

 DOI: 10.1051/e3sconf/20171401016

Figures from 4 to 8 describe the changes in $C_p = f(\lambda)$. The results were approximated with polynomial of degree 3. It has been observed that for the angle of 0° the C_p maximum values for the test model wind power plant, regardless of the set speeds, are similar and in the range of 0,06-0,085. The maximum λ value is also comparable for the C_{pmax}. At the same time it was observed that for an angle of 10° the C_p values are the maximum out of the analysed variants, hence the conclusion is that the amount of energy from the wind is the greatest. It was observed that with increasing the speed of moving air mass the value of C_p raises as well. Simultaneously, change in the value of angle between the direction of flow of the air stream and the direction of setting of the rotor axis affects the change in value of tip speed ratio. Increase in speed means raise of the maximum value of tip speed ratio.

Figures from 9 to 12 describe the changes in $C_p = f(\lambda)$ for permanent values of flow velocity of the air stream for a variable angle between the direction of flow of the air stream and the direction of the axis of the rotor.

Increasing the value of the angle of deviation of the axis of a power plant from the flow direction of the air stream reduces the use of wind energy. For the tested speeds of 4.3 and 5.3 m/s the use of wind energy decreases along with increase in the angle of deviation of the axis of power plant from the direction of the air inflow. For the air stream speed of 7.1 and 7.3 m/s, the use of wind energy increases at the beginning and then decreases along with increasing the yaw angle of the axis of power plant. The greatest use is for the angle of 10°, the smallest for the angle of 0°. Power plant at an angle of 10° uses pressure of wind stream on the blades of power plant. The increase in speed means that for the angle of 0° the C_p value increases slightly, whereas for 10° it grows rapidly. The power plant uses the power of pressure of the air stream.

The analysed wind power plant model is characterized by low efficiency. A diameter of a measurement space is 1.4 [m]. A model wind turbine with a rotor diameter no more than half of the test section was placed in the measurement space. For small wind turbines like the wind tunnel model, the Low-Reynolds-Number effects will diminish the performance significantly. The real wind power plants reach Cp up to about 50 %. However, the actual relationship between the deviations of the rotor axis towards the free airflow can be estimated.

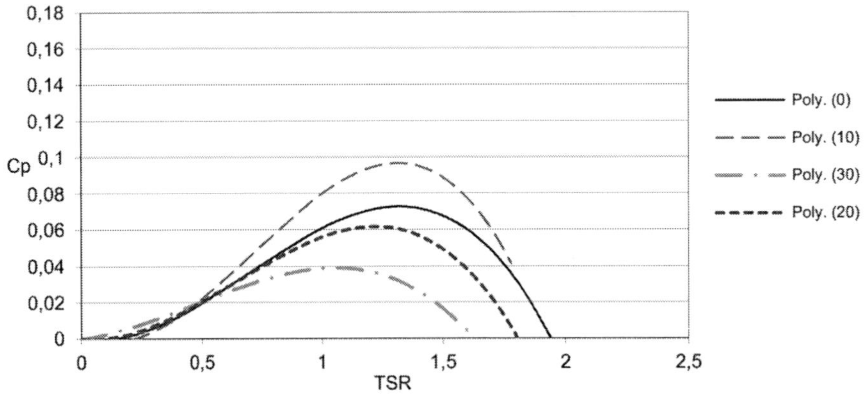

Fig. 10. Changes in Cp = f (λ) for the speed of air stream flow v = 5.3 m/s for the selected angles between the direction of flow of the air stream and the direction of the rotor axis (0 to 45°).

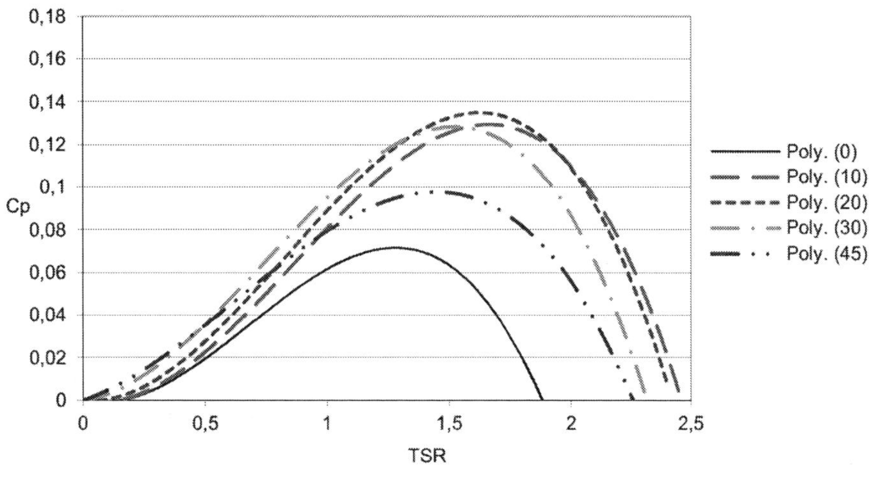

Fig. 11. Changes in Cp = f (λ) for the speed of air stream flow v = 7.1 m/s for the selected angles between the direction of flow of the air stream and the direction of the rotor axis (0 to 45°).

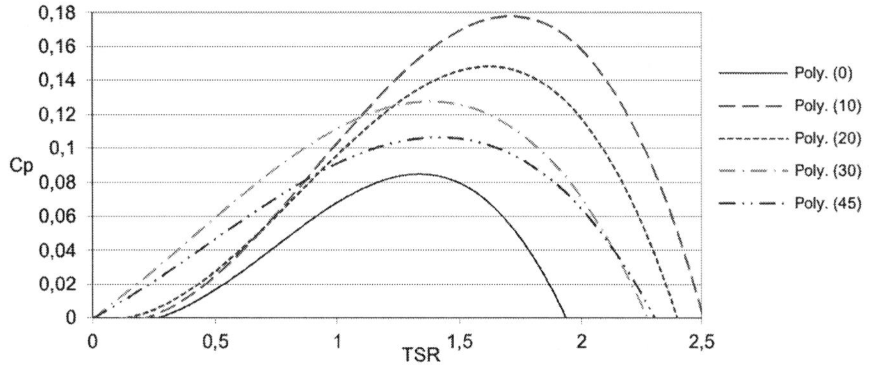

Fig. 12. Changes in Cp = f (λ) for the speed of air stream flow v = 7.3 m/s for the selected angles between the direction of flow of the air stream and the direction of the rotor axis (0 to 45°).

5 Summary

The performances of chosen wind turbines have been analysed. The publication explores the issues of the operating conditions of wind turbines on the possibility of using wind energy. A number of factors affect the operation of the wind power plant, and thus the amount of the energy produced by the plant [13–15]. Among them, we can isolate the impact of the structural parameters of the wind power plant [16–20]. wind turbine drive and control systems [21–26], materials of components of the wind turbines [27–35], climatic factors or seismic events associated with the location [36–39].

Wind is considered as relatively well-known carrier of a low carbon source of energy. Throughout the world, there are areas where wind energy can be used in power plants in a commercially reasonable manner. Therefore, power of constructed wind power plants steadily increases. Number of them is increasing and they are getting larger, as well. Grouping of wind turbines in ecologically justified wind farms results in a need to reduce a distance between individual objects. Wind turbines located too close to each other can interfere with the flow direction of wind and wind direction changes can affect the amount of energy generated by wind turbine.

 DOI: 10.1051/e3sconf/20171401016

One of the parameters proved to be the change of setting or the rotor axis towards the direction of the air stream flow. [40–45]. Studies showed that accurate determination of the optimum angle of the axis of the rotor towards the flow of the air stream very strongly influences the characteristics of the wind power plant.

References

1. S. Gumula, K. Pytel, M. Piaskowska-Silarska, Polish Journal of Environmental Studies, **23** (2014) 2315–2320.
2. W. Hudy, K. Jaracz. K. Pytel, International Carpathian Control Conference **16** (2015) 81–185.
3. S. Gumuła, T. Knap, *Determination of the optimal geometry and movement parameters of the wind power plants depending on the average annual wind speed.* National Forum "Small Energy - '97." Zakopane 1997.
4. S. Gumula, W. Hudy, M. Piaskowska–Silarska, K. Pytel, Chemical Industry, **94** (2015) 1000–1003.
5. S. Gumula, K. Pytel, M. Piaskowska–Silarska, Polish Journal of Environmental Studies, **23** (2014) 2321–2325.
6. H. Noga, Information technologies and technical education **1** (2009) 165–169.
7. I. Turekova, J. Depesova, A. Haskova, T. Bagalova, EDULEARN **15** (2015) 1913–1921.
8. K. Pytel, W. Hudy, H. Noga, W. Kulinowski, International Carpathian Control Conference **17** (2016) 401–406.
9. K. Pytel, W. Hudy, H. Noga, W. Kulinowski, International Carpathian Control Conference **17** (2016) 407–412.
10. K. Pytel, K. Jaracz, S. Gumula, International Carpathian Control Conference **13** (2012) 592–595.
11. T. Chmielniak, Basic theory of profiles and palisades vane. Warsaw PAN 1989.
12. J. Elsner: Aerodynamics of the palisades vane. Warsaw PAN 1989.
13. J. Madejski: The treaty of propellers. Wroclaw Ossolineum 1991.
14. http://www.thewindpower.net, access Jan 1st, 2017
15. R. Carriveau, Fundamental and Advanced Topics in Wind Power, InTech, 2011.
16. G. Ronsten, Journal of Wind Engineering and Industrial Aerodynamics **39**.1 (1992) 105–118.
17. Chen, Jin, et al. Applied Mathematical Modelling **40**.3 (2016): 2002–2014.
18. P. Krajewski, Ł. Flaga, A. Flaga, Technical Transactions Civil Engineering (2016) 81–91.
19. Fischer, Gunter Reinald, Timoleon Kipouros, and Anthony Mark Savill, Renewable Energy **62** (2014) 506—515.
20. Spinato, F., et al. "Reliability of wind turbine subassemblies." IET Renewable Power Generation **3**.4 (2009) 387–401.
21. R. Jablonski, M. Turkowski, R. Szewczykn (Eds.). *Recent Advances in Mechatronics.* Proceedings of 7th International Symposium Mechatronics held Sept. 19–21 in Warsaw, Poland. 2007.
22. F.D. Bianchi, H. de Battista, R.J. Hernán, Wind Turbine Control Systems. Principles, Modelling and Gain Scheduling Design, Springer, 2007.
23. L. Borowik, P. Ptak, Przeglad Elektrotechniczny **88** (2012) 142–145.
24. T. Prauzner, Electrotechnical overview **88** (2012) 205–208.
25. Keung, Ping-Kwan, et al. IEEE Transactions on Power Systems **24** (2009) 279–287.
26. A. Björn, J. Birk. *A high power density converter system for the Gamesa G10x 4, 5 MW wind turbine.* European Conference on Power Electronics and Applications (2007) 1–8.

27. K. Pytel, J. Suchanicz. M. Livinsh, A. Sternberg, Phase Transitions. **87** (2014) 1181-1190.
28. J. Suchanicz, B. Garbarz-Glos, G. Stopa, T. Kruzina, J. Kusz, M. Zubko, W. Hofmeister, I. Jankowska-Sumara, D. Wcisło, K. Konieczny, R. Rosiek, A. Finder, K. Pytel, M. Dambekalne, A. Sternberg, Integrated Ferroelectrics **108** (2009) 98-105.
29. K. Pytel, J. Suchanicz, M. Livinsh, A. Sternberg, Condens. Matter Phys., **16** (2013) 31706.
30. Lavassas, I., et al. Engineering structures **25**.8 (2003) 1097–1106.
31. J. Suchanicz, A. Finder, I. Smeltere, K. Konieczny, I. Jankowska-Sumara, B. Garbarz-Glos, M. Sokolowski, R. Bujakiewicz-Korońska, K. Pytel, M. Antonova, A. Sternberg, Integrated Ferroelectrics, **123** (2011) 102–107.
32. J. Suchanicz, G. Stopa, J. Kusz, M. Zubko, W. Hofmeister, M. Antonova, A. Kalvane, M. Dambekalne, A. Sternberg, I. Jankowska-Sumara, B. Garbarz-Glos, D Wcisło, K. Konieczny, K. Pytel, A. Finder, Journal of Materials Science **45** (2010) 1453–1458.
33. P. Dulian, B. Garbarz-Glos, W. Bąk, M. Antonova, C. Kajtoch, K. Wieczorek-Ciurowa, H. Noga, Ferroelectrics **497** (2016) 62–68.
34. B. Garbarz-Glos, W. Bąk, H. Noga, M. Antonova, A. Kalvane, W. Śmiga, Integrated Ferroelectrics **173** (2016) 12–18.
35. W. Śmiga, B. Garbarz-Glos, W. Piekarczyk, H. Noga, D. Sitko, M. Karpierz, M. Livinsh, Integrated Ferroelectrics **173** (2016) 46–52.
36. W. Sobczyk, K. Sternik, E.J. Sobczyk, H. Noga, Envoronmental protection. **17** (2015) 1113–1124.
37. Jonkman, Jason Mark. Dynamics modeling and loads analysis of an offshore floating wind turbine. ProQuest, 2007.
38. A. Bergek, S. Jacobsson. *The emergence of a growth industry: a comparative analysis of the German, Dutch and Swedish wind turbine industries.* Change, transformation and development. Physica-Verlag HD (2003) 197–227.
39. M. Wolsink, Renewable and sustainable energy reviews **11**.6 (2007) 1188–1207.
40. G. Bir, *Multiblade coordinate transformation and its application to wind turbine analysis.* ASME Wind Energy Symposium (2008) 1–17.
41. M. Korczyński, R. Głowski, S. Opaliński, S. Czaban, *The utilization of wind-tunnel to estimate the dust content of chosen bedding materials.* ISAH–2007 Tartu, Estonia. 2007.
42. Howell, Robert, et al. Renewable energy **35**.2 (2010) 412–422.
43. G. Ahmad, Ghulam, anA. Uzma. Renewable Energy **101** (2017) 286–292.
44. W. Yang, et al. Wind Energy **17**.5 (2014) 673–693.
45. C. Kaidis, U. Bahri F. Amoiralis. IET Renewable Power Generation **9**.8 (2015) 892–899.

E3S Web of Conferences **14**, 01017 (2017) DOI: 10.1051/e3sconf/20171401017
Energy and Fuels 2016

A simulation and simple optimization of a wind-solar-hydro micro power source with a battery bank as an energy storage device

Jakub Jurasz[1,*], *Adam* Piasecki[1]

[1]AGH University of Science and Technology, Faculty of Management, Cracow, Poland

Abstract. Renewable energy sources (RES) are not the backbone of the Polish electricity generation sector. Even though the use of such resources is beneficial in terms of, e.g., CO_2 emissions, current policy seems to create more and more obstacles hindering their further development on an industrial scale. The present paper proposes a simulation model of a hybrid micro power source coupled with a battery bank supplying a small group of households with an annual energy demand of 30 MWh. Results indicate that, for the selected site, a power source consisting of a wind turbine – 8kW, photovoltaic array – 9kW, water turbine – 2kW and 256 kWh energy storage capacity of a battery bank can be a reliable energy source. However, due to the intermittent nature of the selected energy sources there is still a need to remain on-grid in order to avoid excessive energy surpluses (in the case of an oversized system) and deficits. This work opens several interesting directions for future studies, which will be discussed in later sections.

1 Introduction

Some RESs such as solar, wind and hydro exhibit significant temporal and/or spatial variability [1]. Therefore, a significant amount of research has been conducted in order to quantify, describe and overcome those impediments resulting from the stochastic nature of wind speed, irradiation and flow rate [2, 3]. One of the most frequently cited – and questioned [4] – works of Delucchi & Jacobson [5, 6] points to several solutions which may ease and facilitate the process of integrating variable renewable energy (VRE) into the national power system. Their proposed solutions include, *inter alia*: the creation of hybrid power sources based on complementary RESs; usage of energy storage devices such as battery banks or hydro-pumped storage, etc.. Research has so far concentrated mainly on the combinations of: PV–battery bank [7], PV–wind turbine–battery bank [8, 9], PV–diesel generator [10] and various, numerous others. In this paper we build upon those concepts and on the research conducted to date on hybrid power sources [11]. In the following section we introduce our vision of a power source consisting of water and wind turbines, an array of photovoltaic modules and a battery bank. The third section describes the mathematical model which will be further used to simulate the behaviour of the proposed power source and enable a simple

* Corresponding author: jakubkamiljurasz@gmail.com

DOI: 10.1051/e3sconf/20171401017

optimization. The second-to-last section describes the obtained results and presents some observations. The conducted research is then summarized in the last section, entitled *Conclusions*.

2 RES hybrid energy sources

The driving force behind the increasing popularity of micro hybrid RES energy sources is the need to overcome the weaknesses of one source (e.g. the diurnal nature of irradiation) with the strengths of another (e.g. the relatively stable energy output from a water turbine). The variable energy output from PV, wind turbine (WT) and, to a lesser degree, from water turbine must meet an energy demand which itself fluctuates (see charts included in Fig. 1). It has been already mentioned that solar, wind and hydro energy to some extent exhibit temporal and spatial complementarity [12, 13, 14]. Therefore, it has been assumed that this is a good premise for investigating the concept of a hybrid RES energy source consisting of the above-indicated energy sources. Furthermore, it is important to note that energy demand can be divided basically into two parts: the so-called 'base load', which remains at much the same level over the course of the year; and local morning and late afternoon demand peaks resulting from residents' behaviour patterns (e.g. preparing meals after work). Hence, it seems reasonable to use a relatively stable energy source in the form of hydropower, which will cover the bulk of baseload demand. Meanwhile, optimally-sized and more variable energy sources (namely PV and WT) coupled with a battery bank will cover the remaining demand.

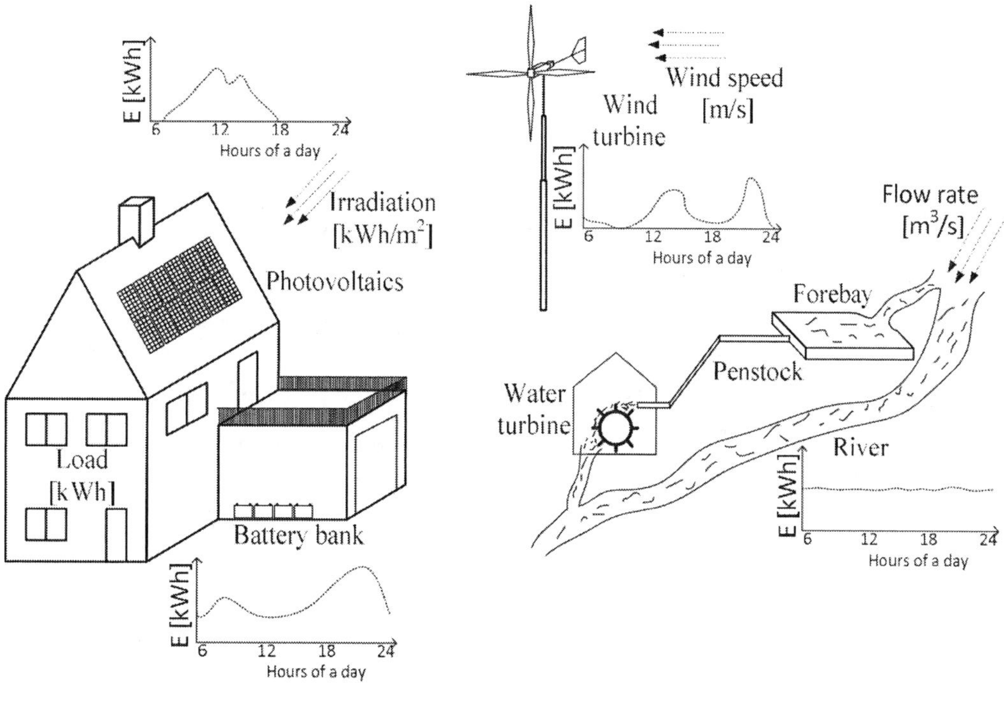

Fig. 1. Conceptual design of an off-grid PV–WT–Hydro hybrid power source. Charts show exemplary energy consumption/yield patterns.

 DOI: 10.1051/e3sconf/20171401017

3 Simulation model

After Karkula [15] simulation models are usually applied when it is impossible or very hard to obtain an analytical solution to the investigated problem – such situations occur in the case of dynamic systems. Without simulation it is very hard to predict the impact of individual variables on a system's behaviour. Therefore, by applying various simulation tools (e.g. mathematical models) it is possible to conduct various experiments on a system, to test different scenarios, or, in other words, to change the reality in which the simulated system will operate. In this paper we introduce a binary mathematical model which enables the simulation of a micro power source consisting of a wind turbine and a water turbine, a photovoltaic system and a battery bank. It has been assumed that such a system should be able to cover the energy demand of several households. Theoretically, such a system can be coupled to the power grid in order to ensure that occasional energy surpluses can be transferred via grid to other consumers. Conversely, energy deficits which will inevitably appear in the case of an improperly sized power source can be covered by energy from the power grid. Nonetheless, the impact of VRE energy sources on the power grid should be limited – meaning that unexpected transfer of energy from and to the grid should be as low as absolutely possible. Therefore, in the objective function presented below (see Eq. 12) an assumption has been made to limit the total sum of energy deficits and surpluses. For the sake of simulation model clarity all used variables, coefficients and indices have been summarized in Table 1.

Table 1. Variables, coefficients and indices used in the simulation model

Name	Variable	Unit
Day/Hour	$i=1...365/j=1..24$	[-]
Time interval for calculations	t	[1 h]
Standard Testing Conditions	STC	[kW/m^2]
Overall PV installation efficiency	η^{PV}	[%]
Battery bank charging/discharging efficiency	η^{cha}/η^{dis}	
Irradiation	H_{ij}	[kWh/m^2]
Flow-rate	Q_{ij}	[m^3/s]
Minimal working flow-rate	Q^{min}	
Maximal working flow-rate	Q^{max}	
Total head of the hydro turbines used	h	[m]
Gravity acceleration	g	[m/s^2]
Water density	ρ	[kg/m^3]
Wind speed	v_{ij}	[m/s]
Cut-in wind speed	v^{cut-in}	
Cut-out wind speed	$v^{cut-out}$	
Rated output wind speed	v^{rated}	
PV/WT/water turbine nameplate capacity	$P^{PV}/P^{WT}/P^{Hydro}$	[kW]
Polynomial approximating wind turbine power curve	$f(v)$	[-]
Polynomial approximating water turbine efficiency	$f(Q)$	[-]
Energy yield: PV, WT and water turbine	$E_{ij}^{PV}/E_{ij}^{WT}/E_{ij}^{Hydro}$	[kWh]
Energy demand	E_{ij}^{D}	
Energy stored in battery bank	E_{ij}^{BB}	
Battery bank storage capacity	E^{BB_Max}	
Energy surplus	E_{ij}^{S}	
Energy deficit	E_{ij}^{Def}	
Binary variables	$x_{ij}^{I...III}$	[0,1]

For the parameters described in Table 1, the following values have been used in simulation: STC = 1000 kW/m², η^{PV} = 80%, η^{cha} and η^{dis} = 80%, h = 1 m, g = 9.81 m/s², ρ = 1000 kg/m³. Irradiation (H_{ij}) – (www.soda-pro.com), wind speed (v_{ij})/flow rate (Q_{ij}) – IMGW NRI (www.imgw.pl), and energy demand (E^D) – [9] time series covered the year 2015 and had a one-hour time step. Those time series were representative for the area near Ostrowiec Świętokrzyski, located in south-central Poland. The polynomial approximating and water turbine efficiency have been calculated based on data presented by Kaldellis [16] for a small Kaplan turbine, and the other, describing the wind turbine power curve, has been calculated for a small, 1-kW, horizontal-axis wind turbine manufactured by Kastrel (www.kestrelwind.co.za). The following subsections introduce formulas and equations used to determine the values of individual variables. Note that in formulas describing energy generation, indices have been omitted.

Energy yield from PV – after [14]

$$E^{PV} = \frac{HP^{PV}}{STC}\eta^{PV}t \tag{1}$$

Energy yield from wind turbine – after [14]

$$E^{WT} = \begin{cases} 0, for\ v < v^{cut-in} \\ f(v)t, for\ v^{cut-in} \leq v < v^{rated} \\ P^{WT}t, for\ v^{rated} \leq v \leq v^{cut-out} \end{cases} \tag{2}$$

Energy yield from water turbine – after [16]

$$E(Q) = f(Q)\rho ghQt \tag{3}$$

$$E^{Hydro} = \begin{cases} 0, for\ Q \leq Q^{min} \\ E(Q), for\ Q^{min} \leq Q \leq Q^{max} \\ P^{Hydro}t, for\ Q^{max} \leq Q \end{cases} \tag{4}$$

Energy flow

$$E_{ij}^{B1} = E_{ij}^{PV} + E_{ij}^{TW} + E_{ij}^{Hydro} - E_{ij}^{D} \tag{5}$$

$$x_{ij}^{I} = \begin{cases} 1, if\ E_{ij}^{B1} \geq 0 \\ 0, otherwise \end{cases} \tag{6}$$

Battery bank – modified after [17]

$$E_{i,j+1}^{BB} = \min\left[(E_{i,j}^{BB} + \eta^{cha}E_{ij}^{B1})x_{ij}^{1}; E^{BB_{Max}}\right] + \max\left[(E_{i,j}^{BB} + \frac{E_{ij}^{B1}}{\eta^{dis}})(1 - x_{ij}^{1}); 0\right] \tag{7}$$

Energy deficit

$$x_{ij}^{II} = \begin{cases} 1, if\ E_{ij}^{BB} = 0 \\ 0, otherwise \end{cases} \tag{8}$$

$$E_{ij}^{Def} = \left| E_{i,j-1}^{BB} + \frac{E_{ij}^{B1}}{\eta^{dis}} \right| x_{ij}^{II} \tag{9}$$

Energy surplus

$$x_{ij}^{III} = \begin{cases} 1, if\ E_{ij}^{BB} = E^{BB_{Max}} \\ 0, otherwise \end{cases} \tag{10}$$

$$E_{ij}^{S} = \min(\left(E_{ij}^{B1} - \frac{E^{BB_{Max}} - E_{ij-1}^{BB}}{\eta^{cha}} \right) x_{ij}^{III}; 0) \tag{11}$$

Optimization – objective function

$$\min Z = \sum_{i=1}^{365} \sum_{j=1}^{24} E_{ij}^{Def} + \sum_{i=1}^{365} \sum_{j=1}^{24} E_{ij}^{S} \tag{12}$$

The above-described model assumes that energy flow is possible only between selected components (see Fig. 2). However, the precise values of energy used from, e.g., wind turbines, to cover the energy demand, or how much energy from PV will be stored in the battery bank remained beyond the scope of this study. Therefore, formulas for their calculation will not be presented.

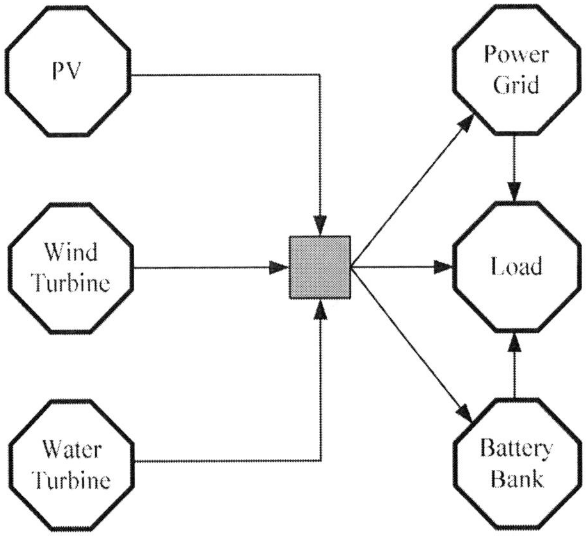

Fig. 2. Energy flow in the investigated hybrid power source which in theory has been connected to the power gird

4 Results and discussion

The creation of hybrid RES power sources is mainly based on the assumption that the selected energy sources will complement each other, thereby providing a stable power source. Such systems, in which the failure of one component will not impede 100% effectiveness, are known as parallel reliability structures [19]. For the purpose of this study a one-year-long hourly time series of wind speed, flow rate and irradiation covering the year 2015 was

DOI: 10.1051/e3sconf/20171401017

selected. Based on formulas (1, 2, 3 and 4), the energy yield for a photovoltaic array, wind and water turbine (each rated 2 kW) have been calculated. Fig. 3 depicts the energy output from hydropower as a function of energy yield from photovoltaics and wind turbine. The nominal capacity of the water turbine for the given flow time series has been estimated based on [18]. As can be observed, in the case that there is no energy generation from PV one can statistically expect a significant amount of energy from hydropower, but only when there is also energy available from wind turbines (above 30% of WT rated capacity). On the other hand, when there is plenty of energy from PV and moderate energy yield from WT, the observed energy generation from hydropower will be relatively low (below 75% of the rated capacity of the water turbine). Of course the dependencies presented above are valid only for the selected parameters of the water turbine and might change significantly if another were selected. Please note that this chart (Fig. 3) refers to energy generated by devices such as wind turbines etc., not the potential energy availability resulting from wind speed or irradiation.

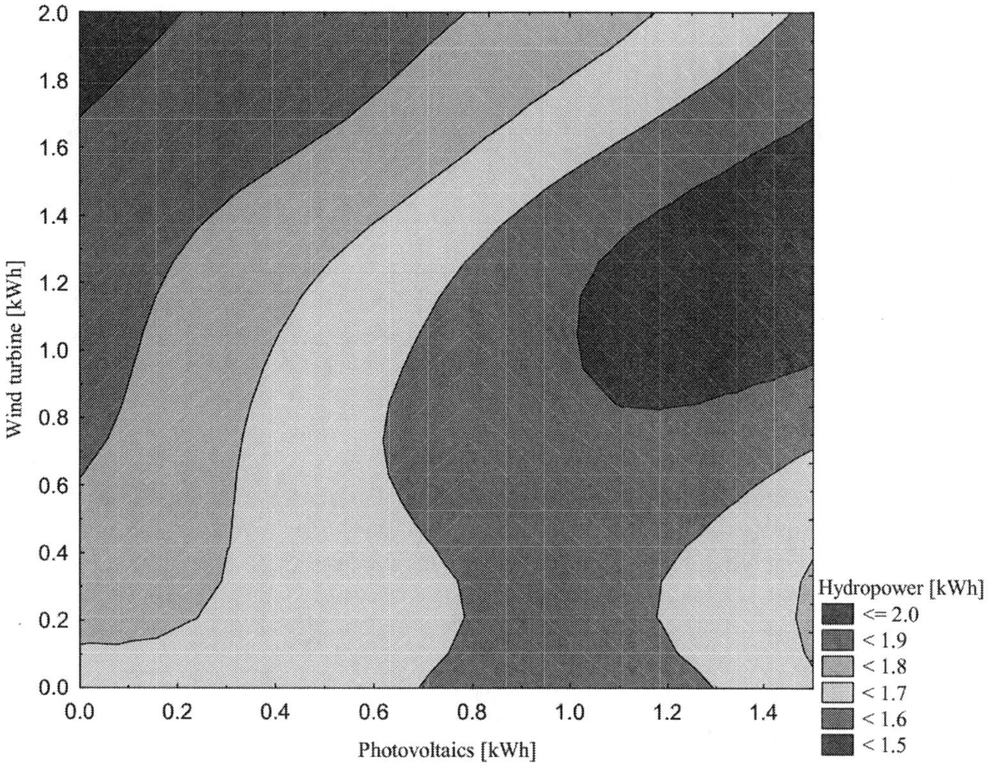

Fig. 3. Energy generation from photovoltaics, wind turbine and water turbine, each rated 2 kW

For the purpose of this study a time series depicting the energy demand of a group of several household has been selected. The total annual energy consumption amounted to 30 MWh – which translates to a mean daily energy demand of 82.2 kWh. Based on Eq. 1–11 the behaviour of such an energy source has been simulated. The optimal nameplate capacity of wind turbines, PV arrays and battery bank has be estimated by an Excel Non-Linear solver. Prior to the optimization, the following constraints were imposed: the nameplate capacity of PV, WT and battery bank must be an integer value; additionally, the capacity of the battery bank cannot exceed the average two-day energy consumption plus theoretical energy losses. Therefore, the maximal battery bank capacity should not exceed 256 kWh. The results obtained indicate that the optimal value of the objective function will be achieved for the following input variable values: P^{PV}=9 kW, P^{WT}=8kW, E^{BB_Max}=256 kWh. As a result, the proposed system will cover over 92% of the initial energy demand, whilst energy surpluses will be slightly over 1.8 MWh. The energy deficit will be equal to 2.25 MWh. This means that by adding a little bit to the energy generating capacity of the proposed hybrid energy source the overall energy balance might be equal to zero. The exemplar energy generation and demand values are presented in Fig. 4. Note the significant energy surpluses from the wind generation and smooth energy generation from water turbine.

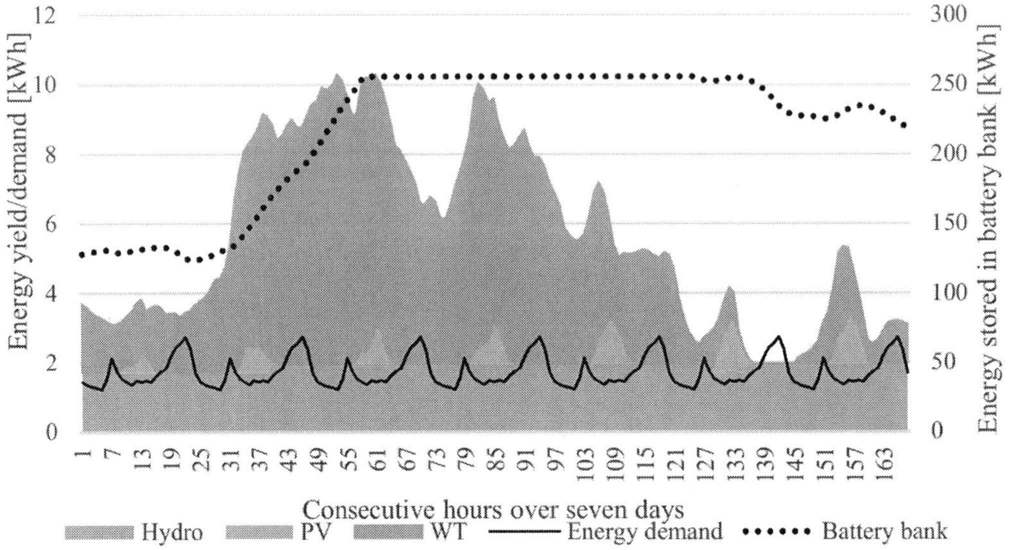

Fig. 4. Energy generation, demand and storage over seven consecutive days in January

The seasonal and diurnal patterns of energy generation and demand make the sizing procedure very complicated. The conducted simulation enabled us to grasp those patterns and estimate the optimal nameplate capacities. In Fig. 5 we have presented the cumulative mean energy generation values for the three investigated energy sources. Please note the impact of the relatively low energy yield values in February and October on the mean hourly volume of energy stored in battery bank, which is depicted in Fig. 6.

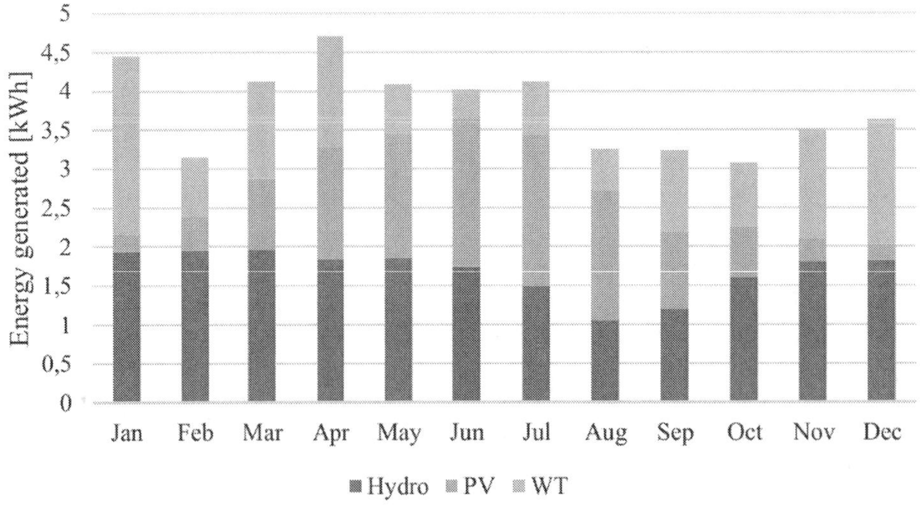

Fig. 5. Mean hourly energy generation from selected sources within individual months

DOI: 10.1051/e3sconf/20171401017

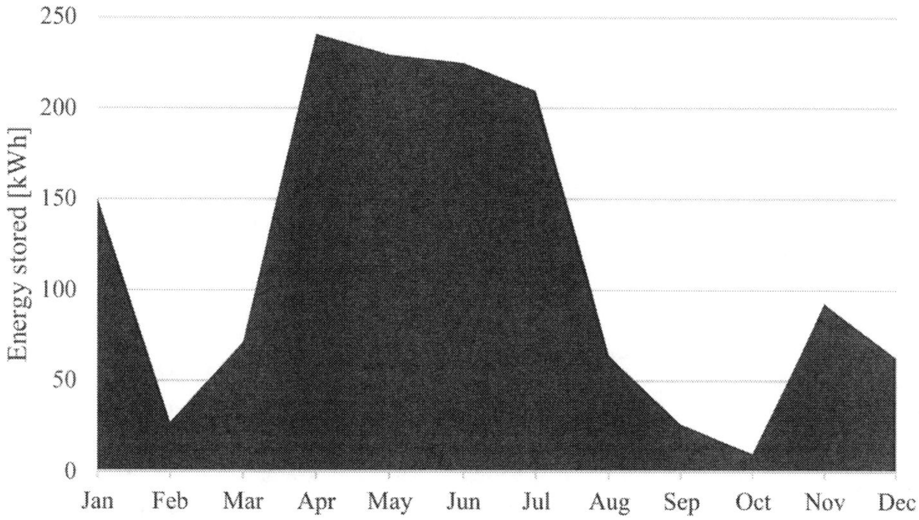

Fig. 6. Mean hourly volume of energy stored in battery bank during each month

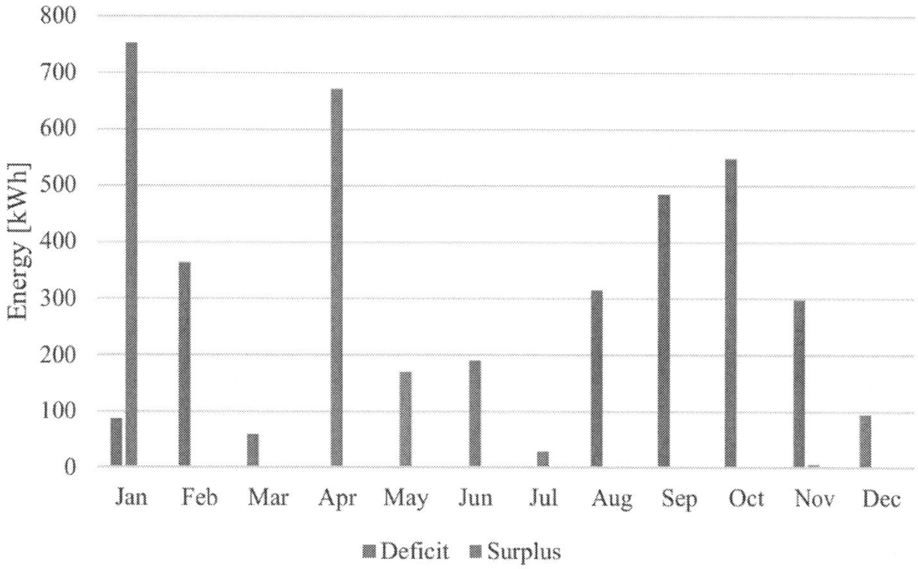

Fig. 7. Sum of energy deficits and surpluses occurring over the course of the investigated year

A decrease in energy availability from PV, WT and hydropower results in increased use of the battery bank and the more frequent occurrence of energy deficits (see Fig. 7). The opposite can be observe in months such as January and April, when the energy generation from wind turbines leads to a significantly higher overall energy generation than in the remaining months. Note that, in those two months alone, the energy surpluses accounted for almost 80% of the total energy surplus observed over the simulation period. However, the variable nature of wind and irradiation values and the limited capacity of the battery bank lead to a situation where some energy deficit occurred even in January. The question therefore arises of what nameplate capacities for individual components of this hybrid energy source would ensure energy self-sufficiency. To partially investigate this issue we decided to gradually increase the capacity of the battery bank. The obtained results are depicted in Fig. 8.

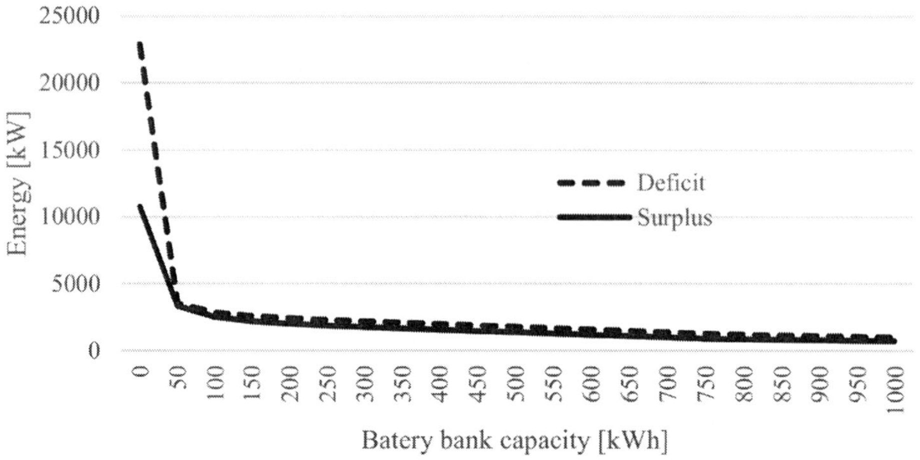

Fig. 8. Energy deficit and surplus as a function of battery bank capacity

As can be seen in Fig. 8, the most significant decrease in energy deficit and surplus can be observed for battery capacities ranging from 0 to 50 kWh. This highlights the role and importance of energy storage devices when using energy sources which rely on variable renewable energy sources. The data presented in Fig. 8 reveals that, for the selected nameplate capacities of energy sources and patterns of energy demand, each kWh added to the battery bank capacity (for capacity ranging from 200 to 1000 kWh) will result in a 1.6 kWh decrease in energy deficit.

5 Conclusions

An analysis of the available literature indicates that hybrid renewable energy power sources are gaining importance due to the increasing role of RESs in power systems. Greater capacity installed in VRE requires actions to enable their smooth and cost-effective integration. One of the viable solutions is to introduce to the power system new RESs which are complementary in time, thereby limiting and curtailing the probability of instances of concurrent zero energy availability from all VREs. Referring back to the current state of knowledge, the majority of models focuses on realizing given objective functions, such as meeting a given energy demand or minimizing total energy generation costs. In this paper we have introduced a simple yet effective mathematical model which enables the simulation of a micro power source which generates electric energy from three variable renewable energy sources. The model has been tested on a potential hybrid power source located in one Polish voivodeship. The single objective optimization function enabled us to determine the nameplate capacity of wind turbines, photovoltaics and battery bank. Due to the volumetric constraints of this paper, several interesting issues have not been raised. These include: estimation of nameplate capacities which will enable energy self-sufficiency; and the impact of adding more capacity to the wind turbines and photovoltaics on energy surpluses and deficits. What is more, the proposed hybrid power source should be juxtaposed with an energy source which uses only wind and solar energy – because it tends to exhibit positive correlation with energy demand over certain periods [20], or only one of those sources. Finally, it is worth underlining that the presented model can easily be enhanced with the economic and environmental aspects of the proposed hybrid energy source, and this will be the scope of our future research.

References

1. E. L. Jones, *Renewable energy integration: practical management of variability, uncertainty, and flexibility in power grids*, Academic Press (2014)
2. J. Kleissl, *Solar energy forecasting and resource assessment*, Academic Press (2013)
3. C. E. Hoicka, I. H. Rowlands, *Solar and wind resource complementarity: Advancing options for renewable electricity integration in Ontario, Canada*, Renewable Energy, **26(1)**, 97–107, (2011)
4. T. Trainer, *A critique of Jacobson and Delucchi's proposals for a world renewable energy supply*, Energy Policy, **44**, 476–481, (2012)
5. M. A. Delucchi, M. Z. Jacobson, *Providing all global energy with wind, water, and solar power, Part I: Technologies, energy resources, quantities and areas of infrastructure, and materials*, Energy Policy, **39(3)**, 1154–1169 (2011)
6. M. Z. Jacobson, M. A. Delucchi, *Providing all global energy with wind, water, and solar power, Part II: Reliability, system and transmission costs, and policies*, Energy policy, **39(3)**, 1170–1190 (2011)
7. M. Mehrabankhomartash, et al., *Practical battery size optimization of a PV system by considering individual customer damage function*, Renewable and Sustainable Energy Reviews, **67**, 36–50 (2017)
8. S. K. Nandi, H. R. Ghosh, *Prospect of wind–PV–battery hybrid power system as an alternative to grid extension in Bangladesh,* Energy, **35(7)**, 3040–3047, (2010)
9. J. Jurasz, J. Mikulik, *Solar & wind hybrid power source for residential building mathematical model approach*, Architecture Civil Engineering Environment, **8(4)**, 5–10 (2015)
10. R. Dufo-López, J. L. Bernal-Agustín, *Design and control strategies of PV–Diesel systems using genetic algorithms,* Solar energy, **79(1)**, 33–46 (2015)
11. O. Bajpai, V. Dash, Hybrid renewable energy systems for power generation in stand-alone applications: a review, Renewable and Sustainable Energy Reviews, **16(5)**, 2926–2939 (2012)
12. J. Jurasz, A. Piasecki, *Evaluation of the Complementarity of Wind Energy Resources, Solar Radiation and Flowing Water–a Case Study of Piła*, Acta Energetica, **2** (2016)
13. F. Monforti, et al., *Assessing complementarity of wind and solar resources for energy production in Italy. A Monte Carlo approach,* Renewable Energy, **63,** 576–586 (2014)
14. S. Jerez, et al., *Spatio-temporal complementarity between solar and wind power in the Iberian Peninsula*, Energy Procedia, **40**, 48–57 (2013)
15. M. Karkula, Weryfikacja i walidacja dynamicznych modeli symulacyjnych procesów logistycznych, Logistyka, **2**, 717–726, (2012)
16. J. K. Kaldellis, et al., *Techno-economic evaluation of small hydro power plants in Greece: a complete sensitivity analysis*, Energy Policy **33(15)**, 1969–1985 (2005)
17. H. Ren, et al., *Optimal operation of a grid-connected hybrid PV/fuel cell/battery energy system for residential applications*, Energy, **113**, 702–712 (2016)
18. B. Karolewski, P. Ligocki, *Wyznaczanie parametrów małej elektrowni wodnej*, Praca Naukowe Instytutu Maszyn, Napędów i Pomiarów Elektrycznych Politechniki Wrocławskiej, **56** (2004)
19. B. Tchórzewska-Cieślak, *Niezawodność i bezpieczeństwo systemów komunalnych: na przykładzie systemu zaopatrzenia w wodę*, Oficyna Wydawnicza Politechniki Rzeszowskiej (2008)
20. J. Jurasz, J. *Mikulik, Investigating Theoretical PV Energy Generation Patterns with Their Relation to the Power Load Curve in Poland*, International Journal of Photoenergy, 7 pages, **2016**. doi:10.1155/2016/3789840 (2016)

DOI: 10.1051/e3sconf/20171401018

The analysis of the hydropower potential in Podkarpackie Voivodship

Wojciech Goryl*, *Mariusz* Filipowicz, and *Krzysztof* Sornek

AGH University of Science and Technology, Faculty of Energy and Fuels, Department of Sustainable Energy Development, al. Mickiewicza 30, 30-059 Krakow, Poland

Abstract. The paper presents an analysis of the use and potential of hydropower in Podkarpackie (Subcarpathian) Voivodship. The current usage of hydropower in the Region is presented, divided for range of generated power, geographical location and given rivers. Existing and planned artificial water reservoirs are taken into consideration in frame of hydropower development in the Region. Selected case studies related to varied hydropower plants are discussed. Problems connected with development of the hydropower were identified and considered. For each county in the Region theoretical available power possible to install in the main rivers was indicated, taking into account environmental aspects, technical problems with connection to the electrical grid and overall possibility of the facility installation. For each county communities with the best favorable situation are indicated. It is discussed the analysis of the evaluation of the use of hydropower in Podkarpackie Voivodship and provided information on planned and considered investments in hydropower plants. In addition, an analysis of the energy potential of the main rivers of the Podkarpackie Region have been procced to indicating the most favorable areas for location of hydropower plants. Moreover, given the environmental conditions, an analysis of the power estimation of the hydropower plants were made.

1 Introduction

Podkarpackie Voivodship is located in south-eastern Poland and it borders with Slovakia and the Ukraine. In terms of size, it is the eleventh region in Poland, and in terms of the population ninth. Population is over 2 million, the population density is 119 people/km^2. The biggest city in the Region is Rzeszow with a population over 180 thousand [1].

Climate has transitional character between lowland and mountain. Average rainfall is 700 - 800 mm and 800-1000 mm in the northern part and in the southern part, respectively. In some parts of the Bieszczady Mountains rainfall increases up to 1200 mm [2].

Podkarpackie Voivodship has three distinct physiographic regions. The northern part of the Region is Sandomierz Basin Plain, the central part the Carpathian Foothills, while the southern part the Low Beskids and Bieszczady Mountains.

* Corresponding author: wgoryl@agh.edu.pl

Generaly, Poland has a small technical potential of hydropower of approx. 8.3 TWh/year [3] compared to other European countries: Switzerland 41 TWh/year, Austria 56 TWh/year [4], France 100 TWh/year [5], Norway 240 TWh/year [6]. Moreover, in Poland annual electricity generation is 2.1 TWh, Switzerland 34.5 TWh, Austria 37.6 TWh, France 69.8 TWh and Norway 116.3 TWh [7]. This is caused mainly by the fact that Poland is a lowland country. The greater the difference in height between the upper and lower reservoir then the power of the hydropower plant is higher. Unfortunately, in order to obtain optimum operating conditions for water turbines it is required to build a very expensive dam, which are designed not only to accumulate water to yield more energy, but also to regulate the flow of the river and serve as a flood protection. The largest number of hydropower plants in Poland are located in the southern part, mainly in the region: Malopolska, Lower Silesia, Silesia and Podkarpackie. Hydropower plants over 5 MWe are usually a pumped storage hydroelectric plants, while small hydropower plants are mainly the run-of-river hydroplants.

In the Podkarpackie Voivodship it was able to identify 16 operating hydroelectric power plants of different sizes [8,9], it is planned to build five more. There exist 26 potential locations for next hydropower plants. Figure 1 presents the number of hydropower plants in case of the installed power.

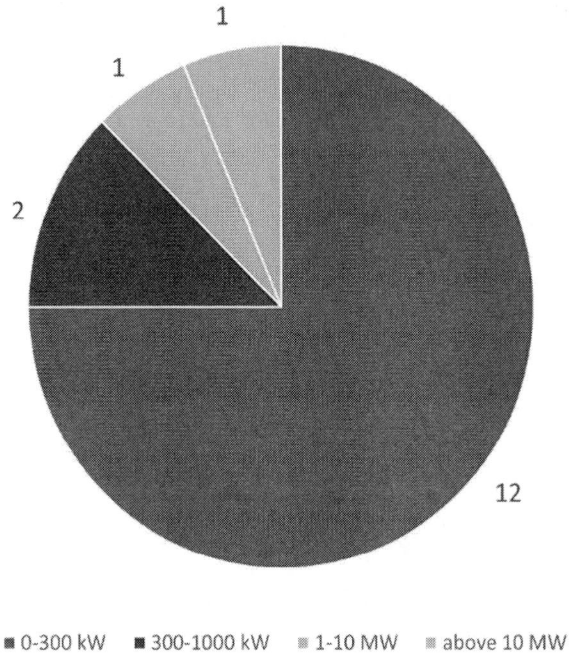

Fig. 1. Number of hydropower plants in Podkarpackie Voivodship in case of installed power

Some of the existing hydropower plants are a few dozen kW, where the produced energy is consumed nearby. Due to the installed power, the vast majority plants are so-called small hydropower plants which has less then 5 MWe installed power. These include hydropower plants in Wilcza Wola, Żołynia, Krempna, Sieniawa, Radawie and Nienowicach which are connected to the medium voltage distribution network. The total installed power of small hydro power plants is estimated several MWe.

Two large hydropower plants located in in the Podkarpackie Voivodship, Solina are Myczkowce hydropower plant on the San river (Solina – 200 MWe, Myczkowce – 8.3 MWe).

1.1 Hydro turbines types

The most the most commonly used hydro turbines in the Podkarpackie Voivodship are Kaplan and Fracis turbines. In some small hydropower plants also propeller turbines are used. Below, short description of the particular turbine type [10,11]:

1.1.1 Kaplan turbine

Rotor of the Kaplan turbine is constructed as a propeller with 2 to 8 adjustable (rotated on shaft neck) blades. Pin blades are mounted in the hub of the rotor. This is the vertical type of turbine. Recently it is spread a new type of Kaplan turbine - tubular turbine which is placed in the tubular housing. Water supply to the stator blades take place parallel to the turbine axis.

In the bulb turbine water flows through the support blades, keeping the turbine in the axis of the pipe and then through the stator blades and rotor blades, where water transfer its energy and flows straight to the axial suction pipe.

One of the variants – the Staflo turbine is a turbine with the external generator which is installed outside the water channel and possible transmission in the engine room, where they are easy to maintenance.

1.1.2 Francis turbine

The water flows to the rotor through a spiral casing and then, radially through the variable blades placed on the outer perimeter of the rotor. Francis turbine's rotor consists of two rims and blades. They are used for creation of variable cross-section channels in the direction of water flow. The blades in the Francis turbine are not adjustable. The most common configuration is a horizontal single-rotor. This version is relatively easy to install and maintain. The version with double rotor is used to larger flows. The vertical version is used to the larger generated powers.

1.1.3 Propeller turbine

Can be considered as a Kaplan turbine with non-adjustable blades, in particular, is very sensitive to changes in flow rate, because the slower flow cannot strikes the blade at the correct angle

1.2 Hydropower potential in Podkarpackie Voivodship

The main rivers of the Podkarpackie Region are: right-bank tributaries of the Vistula River (San and Wisłoka), left tributary of the San (Wisłok) and left-bank tributary of the Wisłoka (Ropa).

Energy resources of major rivers are determined by tributaries of the following rivers:
- San: Hoczewka, Osława, Sanoczek, Magierówka, Baryczka, Łubienka, Wiar, Wisznia, Szkło, Lubaczówka, Wisłok, Trzebośnica, Tanew i Bukowka;
- Wisłok: Pielnica, Morwawa, Lubatówka, Stobnica, Strug, Świerkowica, Mleczka;
- Wisłoka: Krempna, Wilsznia, Iwielka, Kłopotnica, Żółków, Ropa, Jasiołka, Czarna, Wielopolka, Tuszynka i Breń;
- Ropa: Sękówka, Moszczanka, Lubuszanka, Olszynka, Bednarka and 7 unmarked, on available maps, small rivers and torrents.

 DOI: 10.1051/e3sconf/20171401018

2 Estimation of available hydropower generation

For each county in the Region theoretical available power possible to install in the main rivers was indicated, taking into account environmental aspects, technical problems with connection to the electrical grid and overall possibility of the facility installation.

Table 1. Theoretical and available hydropower in Podkarpackie Voivodship (only rivers with theoretical generation power larger than 0.5 MW are listed)

County	River with P≥0,5MW	Teorethical potential in river flow [MWh/a]	Avarage usage hydropower [MW]	Possible electricity generation [MWh/a]
brzozowski	San, Wisłok	91 156.18	2.60	21 877
dębicki	Wisłoka with the Wielopolka river estuary	83 770.02	2.30	20 105
jarosławski	San with the rivers estuary: Lubaczówka, Szkło, Wisznia	97 528.31	2.74	23 407
jasielski	Wisłoka with the rivers estuary: Jesiołka, Ropa	70 523.31	1.93	16 925
krośnieński and Krosno city	Wisłok	22 497.94	0.62	4 760
leski	San with the Hoczewka river estuary and Myczkowce Dam	165 340.09	5.30	39 682
leżajski	San, Wisłok	21 355.00	0.59	5 125
mielecki	Wisłoka with the Tuszynka river estuary	70 338.29	4.93	16 881
niżański	San with the Tanew river estuary	195 761.30	5.50	46 983
przemyski	San with the Wiar river estuary	196 878.46	5.53	47 251
przeworski	San, Wisłok with the Mleczka river estuary	16 576.98	0.46	3 979
rzeszowski and Rzeszów city	San, Wisłok with the Strug river estuary	59 836.96	1.65	14 361
sanocki	San with the Osława river estuary, Wisłok	173 203.21	4.85	41 569
stalowski	San with the Bukowa river estuary	117 087.26	3.29	28 101
strzyżowski	Wisłok with the Stobnica river estuary	18 367.02	0.50	4 408
	Sum	**1 400 220.33**	**39.83**	**335 414**

The available power was estimated based on assumption that in natural flow of the river the hydraulic losses are compensated by potential energy and kinetic energy decrease of the water flow. According to the assumption of the model, the theoretical construction of the densely located weirs will transform the hydraulic losses into the estimated, useful energy resource. Model estimated the resource of useful energy and the expected electricity generation from hydropower. The model simplified calculations, assuming that the 50% of the energy potential of the river is possible to use for electricity generation. The model indicators were adopted and tested for the average flow of the Dunajec river, between the water dams: Rożnów – Czchów [12].

Analyzing the potential of tributaries of major rivers in the Voivodship, such as San Wisłok, Wisłoka, Ropa it should be noted that the total theoretical potential in the river stream is 1 400 220.33 MWh/year, which provides a principle for the small hydropower plants construction with a capacity of 0, 8-1,0 MW (data for each rivers in the Table 1).

Nowadays, in Podkarpackie Voivodship over a dozen hydropower plants are in operation which are located in i.a. Wilcza Wola, Żołynia, Krempna, Sieniawa, Radawa, Nienowice.

The special importance has the Solina Dam, which is the biggest hydraulic engineering structure in Poland. Moreover, hydropower plant located in the Solina Dam is the biggest storage hydroelectric plant working on natural water flow. Among the working small hydropower plants in Podkarpackie Voivodship it is worth to mention about the Klimówka Hydropower Plant with 1.1 MWe (generated electricity is putted to electric grid Gorlice – Wysowa of 15kV), small hydropower plant Pilzno with 825kWe and Tabor with 30kWe. The river which energy potential is used in greatest extent is San – Water Power Station Solina – Myczkowce and to a much lesser extent Wisłoka i Wisłok. The Table 1, presented the theoretical and useage potential for each county in Podkarpackie Voivodship.

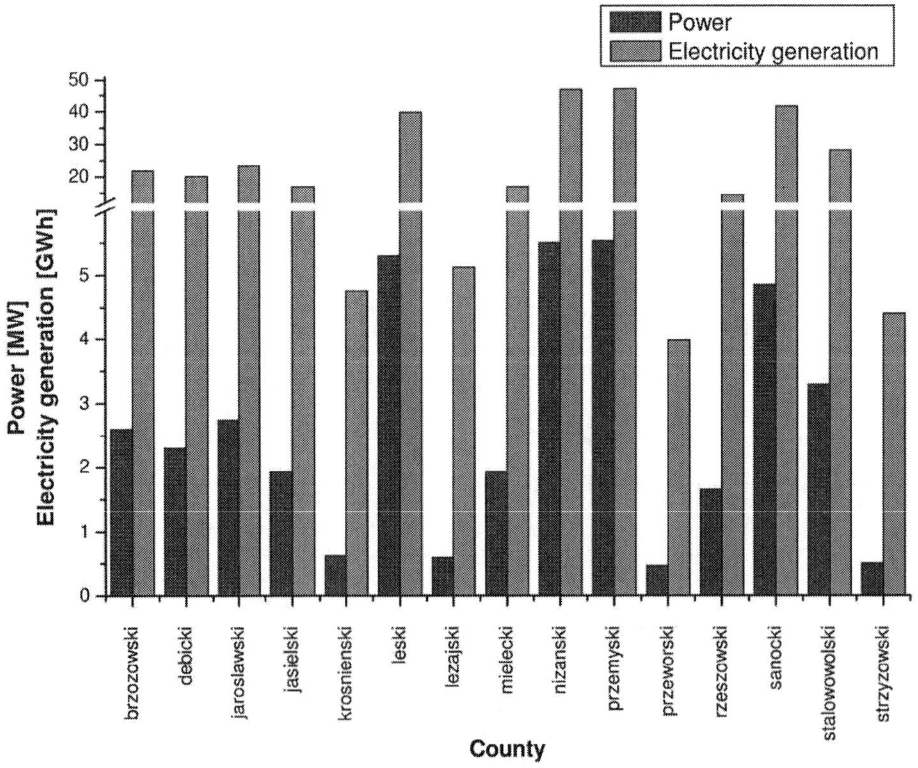

Fig. 2. Power and possible generation in each conty of the Podkarpackie Voivodship

Figure 2 shows the results of the estimation of the potential hydropower in a given county in Podkarpackie Voivodship. There were estimated the power potential of hydroelectric power plants and electricity generation. Plants of the highest installed power could be built in the county of Przemysl, Nisko, Sanok and Lesko. Those plants may have separately up to approx. 5 MW. The situation is similar with electricity generation. Annually it could be generate approx. 40-50 GWh per county. The worst potential is characterized by counties Przeworsk, Strzyżów, Leżajsk and Krosno with total powers possible to install up to 500 kW per each and a total electricity generation of 4-5 GWh. In total, in Podkarpackie Region it could be installed approx. 40 MW of new power where electricity generation could reach up to 335.5 GWh.

The above data show the average possible power to achieve from hydro in a given section of the river and the possible electricity generation annually. For investments, the construction of hydroelectric power plants are more promising where those values are much larger.

The development of hydropower in the Podkarpackie Voivodship is mainly driven by two main factors. The first factor is the hydro potential. The second factor is the flood protection of an areas located close to rivers by building water dams. Next are environmental concerns and possible connection to the elctrical grid. Whereas, generaly, the possibility of the location of water dams depend largely on local plans and the actual land use nearby the river, the possibility of legal land acquisition, geological conditions and the required number of arrangements, for example opinions of the respective Water Management Authority.

The development of hydropower in a given place usually depends on the following factors: natural, technical, legal and economic.

Regarding environmental factor, it is important to assess the water resources in the given area, the possibility of water damming up, eg. based on topographic maps and geological conditions.

The technical factor is very important to determine the mode of operation of the hydro plant, type and size of the turbines and other auxiliary equipment.

Legal, it is necessary to obtain all necessary permits for the execution of investment in accordance with applicable law. This is one of the most difficult point in the realization of hydroelectric power plant.

Economic, it is needed to perform calculations, eg. the economic efficiency of investment for the various technical options, sources of financing, the dynamics of electricity prices change in the long term.

3 Examples of existing hydropower plants

The total annual generation of electricity from hydropower in Poland is 2 261 GWh [13]. In Podkarpackie Voivodship it is estimated at about 245 GWh: in the Solina-Myczkowce Hydro Power Plant 230 GWh, Mokrzec 5 GWh and other small hydropower plants – 10 GWh. Most of the hydroelectric power plants in Podkarpackie Regtion has low power of several hundred kilowatts. Only two plants have larger capacities, plants belonging to the ZEW Solina-Myczkowce Those plants are used three types of turbines: the Kaplan, Francis and Propeller [14]. The list of ten hydropower plants is presented in the Table 2.

Table 2. Summary of basic information about existing or planned hydropower plants in Podkarpackie Voivodship [14]

Place/River	**County**	**Owner/grid connection**	**Number of turbines x electrical power [kW]**	**Type of turbine**	**Total power [kW]**	**Built in**
Mokrzec/ Wisłoka	Dębica	MEW Pilzno	3x275	Kaplan	825	2002
Nienowice/Wisznia	Jarosław	ZKE	3x50+1x100		250	1998
Radawa/ Lubaczówka	Jarosław	ZKE	1x300	Propeller	300	1993
Wróblik Szlachecki/ Tabori	Krosno	RZE	4x7.5	Propeller	30	2000
Solina/ San	Lesko	ZEW Solina-Myczkowce	4x50 000	Francis	200 000	1968
Myczkowce/San	Lesko	ZEW Solina-Myczkowce	2x4100	Kaplan	8 200	1961
Dębiaki/ Babulówka	Mielec	Individual person	1x23	Francis	23	
Przeworsk/Mleczka	Przeworsk	Individual person	3x21	Propeller	63	2011
Zaklików/ Sanna	Stalowa Wola	Individual person	1x18	Francis	18	1996
Rzeszów/ Wisłok	Rzeszów	RZE	1x660	Kaplan	660	2012

It is visible that capacity of the power plants started from 18 kW up to 200 MW. Four installation have power less than 100 kW. Each type of the turbines is commonly popular, however for small installations propeller and Francis turbines are used. Kaplan turbines are installed in medium capacity units and Francis for big installations. All installations are build after 60th of the XX centaury and the most of the small installations – in last 30 years. Special importance has Solina hydropower plant –third largest pumped-storage electricity in Poland.

4 Conclusions

The Podkarpackie Voivodship has large potential of hydropower, however its usage is limited due to significant variation of flow [15]. It is caused by varied hydrologic conditions and hilly type of rivers.However, comparing to wind energy, the water flow is definitelly more stable [16], the power usage is 2-3 times larger comparing wind power plants.

The counties przemyski and niżański has the largest potentian of power generation – more than 5 MW. The one of the most important factors influencing the development of hydropower are environmental conditions. However, well recognized potential for installation of small hydro-power exists.

The work has been completed as part of the statutory activities of the Faculty of Energy and Fuels at the AGH University "Studies concerning the conditions of sustainable energy development"

References

1. http://demografia.stat.gov.pl/bazademografia/Tables.aspx [access: 26.01.2017]
2. http://maps.igipz.pan.pl/aims/AIMS.dll?REQUEST=GetPage&PAGE=sg&MAP=fiz%5Cklimat%5CSuma%20opadow%20roczna&width=480&height=480 [access: 26.01.2017]
3. Report of the Ministry of Environmental Protection, "Strategy of the Development of Renewable Energies", Warsaw (2000)
4. B. Wagner, Ch. Hauer, A. Schoder, H. Habersack, Renew. Sust. Energy. Rev., **20** (2015)
5. T. Hubert, E. Vidalenc, Energy Proced., **20** (2012)
6. International Renewable Energy Agency, Vol. 1: power sector Issue 3/5 (2012)
7. W. Matuszek, Elektroener., **52** (2005)
8. http://monitoruj.podkarpackie.pl [access: 15.12.2016]
9. http://www.baza-oze.pl/bazadanych.php [access: 15.12.2016]
10. J. Twidell and T. Weir, Renewable Energy Resources, Second edition, Taylor & Francis, (2006)
11. T. Price, D. Probert, Applied Energy, **57** (1997)
12. J. Sumera, Dane z analizy zasobów energii wody na obszarze województwa podkarpackiego, Baza danych odnawialnych źródeł energii województwa podkarpackiego
13. Report of the KSE (2015)
14. G. Kalda, JCEEA, 61 (2014)
15. M. Blok, B. Tomaszewska, TPG, **2** (2015)
16. M. Duraczyński, M. Filipowicz, Polityka Energetyczna, **14**, 1 (2011)

Forecasting the Nysa Kłodzka flow rate in order to predict the available flow for a run-off-river (ROR) power plant

Jakub Jurasz[1,*], *Marcin* Wdowikowski[2]

[1]AGH University, Department of Management Engineering, 30-067, Gramatyka 10 St., Cracow, Poland
[2]Institute of Meteorology and Water Management - NRI, 01-673 Warsaw, Podleśna 61, Poland

Abstract. Hydroelectricity is generally perceived as a stable and predictable power source. However ROR power plant without reservoir energy output is mainly driven by changing flow rate. This study applies artificial neural networks to create flow rate forecasts with one hour lead time. Forecasting models were built for Nysa Kłodzka catchment which possesses significant potential for new hydropower plants development as well as leads to frequent floods. The best of the obtained model gives satisfactory results both in terms of root mean square error (0.6379 m^3/s) as well as Nash-Sutcliffe performance indicator (0.9978). Obtained results were compared with currently used forecasting models and were proven to be superior.

1 Introduction

Development of the so-called renewable energy sources (RES) in Poland is driven mainly by the European Union regulations and directives. To a much lesser degree an increasing installed capacity in RESs can be attributed to their superior economic performance and social awareness of the environmental protection. However in future it is the economics that may play a crucial role due to dwindling non-renewable resources and/or their picking up prices. Additionally foreseeable is a reduction in RES costs as well as their increasing efficiency which will undoubtedly lead to their greater role in national energy markets. Unfortunately three the most important representatives of RESs namely: solar, wind and hydropower in some circumstances (e.g. hydroelectricity dependent on current flow rate, photovoltaics and wind energy without energy storage facilities) introduce undesirable variability to the operation of energy system (by some they are denoted as VRE – Variable Renewable Energy). Their non-dispatchable nature results from dependence on naturally varying phenomena such as: wind speed, irradiation and flow rate. Therefore there is not only a need to overcome fluctuating demand side of energy market but also its supply side. Hence the increasing research interest in areas such as: energy yield forecasting from solar [1,2] wind [3] power sources; temporal and spatial complementarity assessment of certain RES

* Corresponding author: jakubkamiljurasz@gmail.com

[4,5]; energy storage [6]; hybrid RES power sources [7]; RES integration to the national power systems [8].

The Directive 2009/28/EC of the European Parliament on the promotion of the use of energy from renewable sources required from Poland to 2020 an achievement of 15% share of renewables in the total of primary generated energy. The national authorities concerns indicate that the goals of the Directive rather won't be fulfilled. These fears were confirmed by the computation of renewable energy (RE) share in the national electricity consumption for the last several years. In 2009 the share of RE was 5.7% while in 2011 it rose to 8% which it is still not enough [9,10].

According to the Polish Energy Regulatory Office (http://www.ure.gov.pl/) as on 30.06.2016 the installed capacity in RES amounted to almost 8.25 GW (excluding co-firing biomass in commercial power plants). This was however unevenly distributed among: wind power (68.7%), biomass power plants (15.4%), hydroelectricity (12%), biogas power plants (2.8%) and slightly above 1% in photovoltaics. As can be seen, hydroelectricity amounts to almost 1 GW of the installed capacity in Polish RES. The biggest contribution to this capacity is made by hydro-pumped storage located on rivers (382.7 MW) and run-off-river power plants with capacity greater than 10 MW (311 MW). However, Polish hydroelectricity is dominated by relatively a small run-off-river units. From total of 752 hydropower plants 98.7% has nominal power smaller than 10 MW, from which 76.7% is not greater than 300 kW. According to the RESTOR HYDRO (http://www.restor-hydro.eu/pl/) project there is an opportunity to develop over 8 500 pico, micro and even bigger hydro installations in Poland. In order to minimize the environmental impact and financial investments those hydro generators may be built on existing weirs, dams and other lateral structures and become a run-off-the-river power plants. Therefore in future not only the role of small scale hydroelectricity may increase in Poland but in some cases it will play crucial role in smoothing and balancing inherently variable energy generation coming from photovoltaics and wind turbines [11]. Hence it will be a necessity for a power grid operator to possess knowledge on future VRE energy generation capabilities.

Above described situation in Polish renewable energy market and current problems in area of RES enable formulation of a following goal of this study: "Assessment of the possibility of applying artificial neural networks (ANN) to create a flow rate forecast with a one hour lead time for Nysa Kłodzka river and juxtapose obtained results with those coming from current models used by HydroProg [12].

2 Artificial neural networks

The concept of artificial neural networks has been inspired by biological neural networks which operate in human and animal brains to estimate approximate functions based on usually large sets of often unknown input variables. Typical ANN consists of three layers, namely: input, hidden and output each of which performs its idiosyncratic tasks. Additionally the process of ANN model creation can be divided into three phases, namely: teaching, validation and testing each of which is performed on previously distinguished subsets. Thanks to the ANN structure and applicability for non-linear systems they have been widely applied in many areas of research such as: medicine [13]; energy demand [14]; waterworks operation/water demand [15]; drought forecasting [16]. Comprehensive description of ANN and their performance as forecasting tools can be found for example in [17].

Due to the rapid development of computing capabilities of modern computers the process of ANN creation has been in large portion automated and the analysis of various ANN architectures even when the data set is very large tend not to be a big problem. This study applied the Statistica software developed by StatSoft for creating multilayer perceptron (MLP) ANNs. Statistica has Broyden-Fletcher-Goldfarb-Shanno teaching algorithm built in

 DOI: 10.1051/e3sconf/20171401019

along with tool dividing data into teaching, validation and testing subsets which correspond to accordingly 70%, 15% and 15% of the initial set. However the area which has not been automated so far is the selection of appropriate explanatory variables - this problem is usually encountered in area of building forecasting models. According to [18] following approaches to selecting most appropriate explanatory variables subsets may be implemented in area water resources modelling:

- based on expertise and former knowledge on investigated system,
- correlation analysis between dependent and independent variables,
- heuristic approaches based on creating various ANN models and adding or subtracting subsequent explanatory variables,
- extracting data from newly created ANN and using it to perform sensitivity analysis,
- combination of tools used in various methods.

On the basis of the previous experiences with ANN models creation [14, 15, 19]] it has been assumed that in this study the explanatory variables subset will be selected based on correlation analysis combined with expertise and reports from previous studies on Nysa Kłodzka catchment properties. Obtained results were then compared with those available at HydroProg hydrological forecasting system (free access: http://www.klodzko.hydroprog.uni.wroc.pl/) which is a well-known and established forecasts source for the Nysa Kłodzka catchment [12, 20, 21]. Because the forecasting performance of the models available there is assessed based on two criteria, namely: root-mean squared error (RMSE) and Nash–Sutcliffe [25] model efficiency coefficient (E) they were also applied in this study. They were calculated based on formulas (1) and (2). In case of the RMSE criterion the smaller its value the better is the forecasting model – therefore it is a useful tool to compare various models tested on the same data sets. The second criterion values can range from -∞ to 1 where E = 1 indicates the perfect match between modelled and observed values, whereas E = 0 informs that the model accuracy is as good as using a mean of the set as a forecast. The Nash-Sutcliffe values less than 0 show that created model can be easily outperformed by simply using the mean of the set to predict system behaviour.

$$RMSE = \sqrt{\frac{1}{n}\sum_{i=1}^{n} e_i^2} \tag{1}$$

where: n – number of observations in testing subset, e – forecast error

$$E = 1 - \frac{\sum_{t=1}^{T}(Q_o^t - Q_m^t)^2}{\sum_{t=1}^{T}(Q_o^t - \bar{Q}_o)^2} \tag{2}$$

where: $\bar{Q}_o$ - mean value of observed discharges, Q_m^t– modelled discharge at time t, Q_o^t – is observed discharge at time t.

3 Case study – Nysa Kłodzka catchment

The Nysa Kłodzka River is a left tributary of the Odra River, into which it flows at km 181.3 in the Opolskie voivodeship – Southern West Poland. The river's total length is 181.7 km, while the catchment area reaches 4565.7 km^2. The Nysa Kłodzka is a mountain river, located in Kłodzko Valley. It has an average gradient of 9.05‰, and starts in the Lower Silesia voivodeship, through which it flows nearly 50% of its total length – 89.4 km. The highest elevation within the basin is equal to 1425 m a.s.l. The river supplies several important for the flood protection purposes reservoirs: the Topola, Kozielno, Otmuchow and

Nysa. Its key tributaries are: the Bystrzyca Kłodzka, Biała Lądecka, Bystrzyca Dusznicka, Ścinawka and Buszówka. The Nysa Kłodzka catchment was developed intensively by hydropower facilities in recent years. There are 24 hydroelectric power plants (of either run-of-the-river (ROR) or conventional and dam (CON) type) with capacities from 0,002 to 5 MW (detailed table [9]). The Otmuchow and Nysa-Glebinow power plants stands as the highest capacities (with 4.8 MW and 3.28 MW respectively), while the lowest capacities (several kW each) goes to Nowa Morawa on the Biala Ladecka, Klodzko on the Bystrzyca Dusznicka and Rynarcice on the Scinawka Niemodlinska. Most of the functioning power plants belongs to the local or national authorities, but there are also commercial companies like (among others) Hydroelektrownie Dolnego Slask, TAURON Ekoenergia or Hydroenergia as well as private investors. Total capacity of all actually existing hydropower plants on the Nysa Klodzka River covers nearly 17.6 MW. [9, 12, 20, 21]

Hydrometeorological data were derived from the national hydrological and meteorological measurement and observational network held by Institute of Meteorology and Water Management – National Research Institute (IMWM-NRI). In the Nysa Kłodzka river catchment there are monitoring structures there are 12 water level gauges including hydrological station in Bardo, and about 12 meteorological stations including synoptic station in Kłodzko (Fig. 1). Water level is measured on all mentioned stations and the flow rate is computed from the relationships based on flow measurements made by IMWM-NRI. All stations in the Nysa Kłodzka river basin are automatically gathered. Data series used in this study becomes from standard measurement system, that can be seen from (http:/www.pogodynka.pl). Basic time period for water level measurement is 10 minutes.

Because of the very frequent flooding occurrence in the Kłodzko Valley, local authorities decided to build early warning system based on the 22 automatic hydrologic gauging stations where water level is measured every 15 minutes, and 18 automatic meteorological stations (with various sets of measured parameters) recording observations every 15 minutes. That hydrometeorological system is owned by Crisis Management Center in Kłodzko and it is called LSOP (Lokalny System Osłony Przeciwpowodziowej in polish, free access at http://lsop.powiat.klodzko.pl). Besides measuring intervals the main difference between LSOP and IMWM-NRI monitoring is that LSOP gives only water level in case of the hydrological data.

Using LSOP hydrological data the HydroProg forecasting system was applied in the recent years [12, 20, 21].

DOI: 10.1051/e3sconf/20171401019

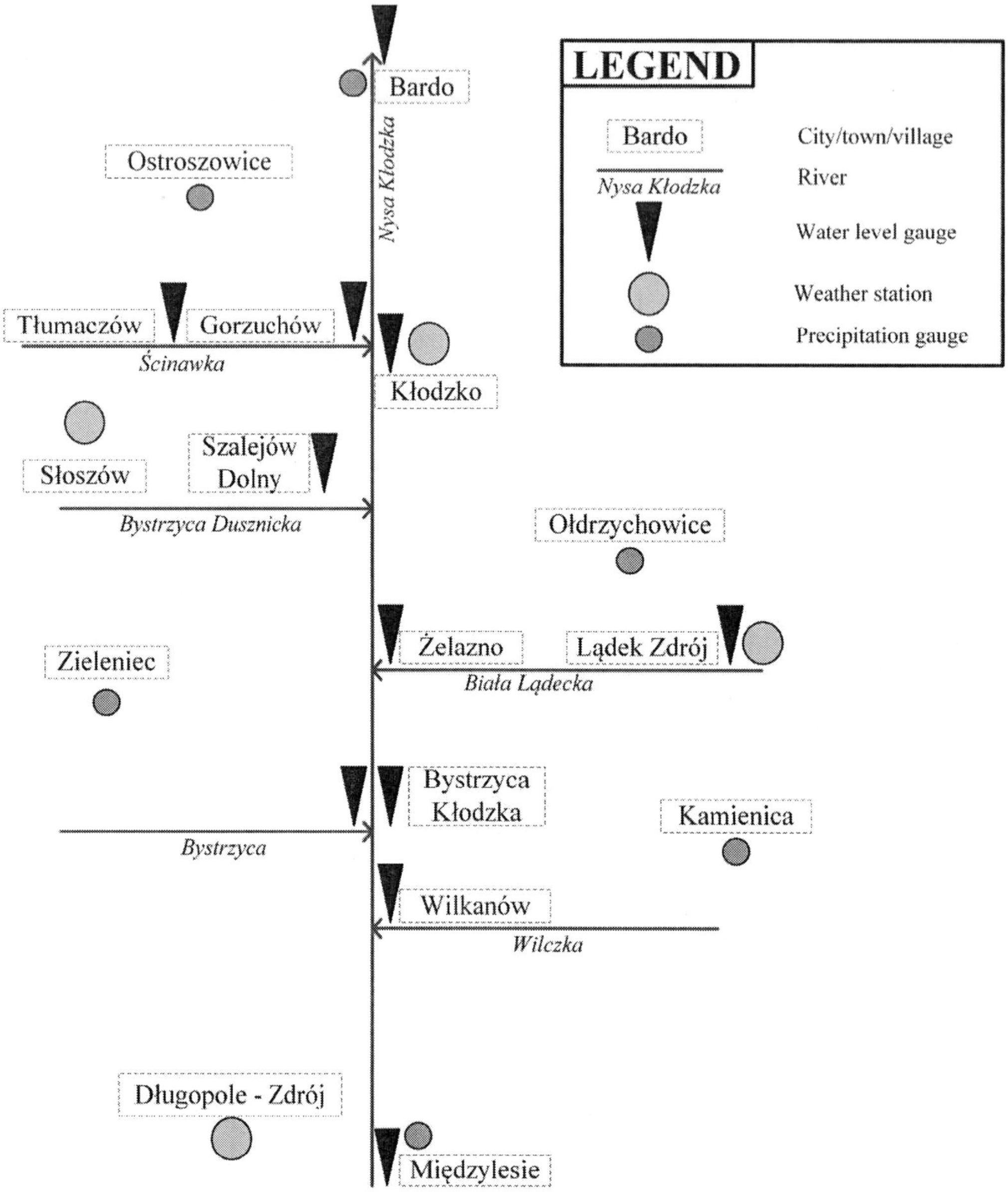

Fig. 1. The Nysa Kłodzka catchment along with location of weather stations and water level gauges from which data were obtained for the purpose of this study.

Over the considered time period (2010-2013) the hourly values of Nysa Kłodzka flow rate exhibited following statistical properties: mean – 20.87 m^3/s; standard deviation – 19.75 m^3/s; coefficient of variation – 94.65 %; maximal value – 240.78 m^3/s; minimal value – 5.22 m^3/s. Example of flow rate variation over twelve month period is given in Fig. (2) notice rapidly occurring flooding flow rate peaks and how much time it takes for them to smooth out. Random swells of waters may not only lead to loss of human life, destruction of infrastructure but also have impact on the operation of hydroelectricity.

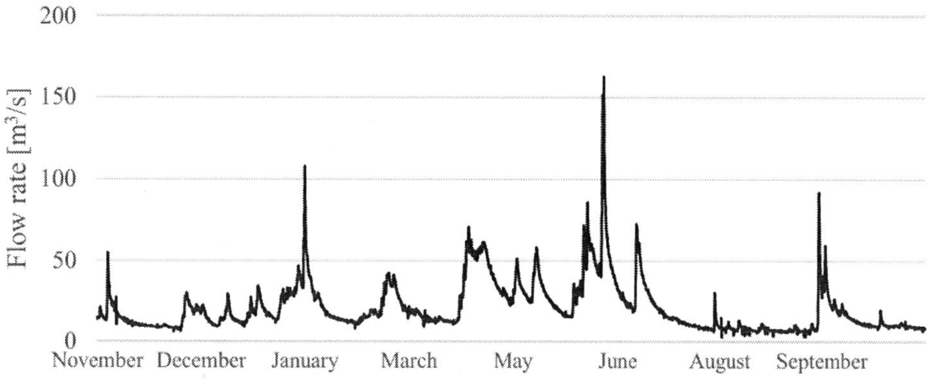

Fig. 2. Hourly values of Nysa Kłodzka flow rate over the year 2013 measured in Bardo.

4 Forecasting models - results

The ANN forecasting models have been created to predict the flow rate in Bardo (which is located in the upper part of the scheme depicted in Fig. (1). Based on described in the second section approach eight sites with water gauges were selected to use collected there data sets as explanatory variables. Additionally historical hourly flow rates measured in Bardo were also considered as input variables. Expertise indicate that in the case of Nysa Kłodzka catchment the lag time ranging from 1 to 12 hours is sufficient (denoted in this paper as e.g. *t-1*). It has been assumed that forecast will be made with one hour lead time. Table 1 contains calculated correlation coefficients, all of which were significant with **p – value < 0.05**, therefore all those subsets were used as explanatory variables.

Table 1. Matrix of correlation coefficients between set of explanatory variables and a dependent variable (flow rate in Bardo in time t).

		Location of hydrological water level gauges							
		Bardo	**Kłodzko**	**Szalejów**	**Międzylesie**	**Bystrzyca Kłodzka**	**Wilkanów**	**Żelazno**	**Gorzuchów**
Time lag in a reference to Bardo	**t-1**	0.998	0.937	0.851	0.744	0.832	0.750	0.849	0.828
	t-2	0.993	0.940	0.854	0.754	0.835	0.758	0.849	0.829
	t-3	0.987	0.940	0.855	0.764	0.836	0.766	0.849	0.827
	t-4	0.979	0.937	0.854	0.772	0.836	0.772	0.846	0.823
	t-5	0.971	0.932	0.851	0.780	0.833	0.776	0.841	0.817
	t-6	0.962	0.925	0.846	0.785	0.830	0.779	0.835	0.811
	t-7	0.953	0.917	0.840	0.789	0.824	0.779	0.829	0.803
	t-8	0.943	0.907	0.833	0.790	0.817	0.777	0.821	0.795
	t-9	0.934	0.898	0.824	0.790	0.809	0.772	0.813	0.787
	t-10	0.924	0.888	0.815	0.788	0.800	0.767	0.805	0.779
	t-11	0.914	0.878	0.806	0.785	0.790	0.761	0.796	0.771
	t-12	0.904	0.867	0.796	0.780	0.780	0.754	0.787	0.762

According to the literature [22, 23] using meteorological parameters such as precipitation, temperature or humidity in hydrological forecasting models, based on flow hydrodynamic transformation between neighbouring gauges does not lead to an improvement in accuracy.

The so called *"Automatic network search"* in-built functionality in Statistica was used to construct ANN models. Although it initially suggests to investigate only 20 various ANN structures it has been calculated that exactly 500 different architectures of MLP ANN should be tested. This number results from the following assumptions:

- the minimal and maximal number of neurons in so called hidden layer is a problematic issue which is often very controversial. Therefore an approach presented in [24] has been applied. It states that the minimal number of neurons in hidden layer should be equal to the square root of the product of the number of neurons in input and output layers. Whereas the maximal number should not be greater than three times the minimal number of neurons in the hidden layer plus one,
- five various (namely: *sine, linear, logistic, hyperbolic tangent* and *exponential*) activation functions have been considered in the hidden and output layers.

Therefore assuming 20 various numbers of neurons in hidden layer, 5 different activation functions in both hidden and output layers it is possible to create as much as 500 unique ANN architectures. Created models have been assessed based on criteria previously mentioned (Eq. 1 and 2) but only with a reference to the testing subset which does not take part in the process of ANN creation. Table 2 summarizes the most important characteristics of selected five ANN which outperformed the reaming ones in terms of forecasting accuracy measured as a correlation between observed and models flow rate values within the testing subset. Those five models were further investigated in the next section.

Table 2. Architecture of five ANN models selected for a further investigation (BFGS – stands for Broyden-Fletcher-Goldfarb-Shanno teaching algorithm).

No.	Number of neurons in the hidden layer	Teching alogrithm	Activation function	
			Hidden layer	Output layer
ANN1	23	BFGS 59	Expotential	Sine
ANN2	9	BFGS 97	Hyperbolic tangent	Linear
ANN3	14	BFGS 58	Expotential	Sine
ANN4	22	BFGS 80	Logistic	Linear
ANN5	28	BFGS 124	Hyperbolic tangent	Linear

The original dataset was a 26294 hours long time series. However some missing values in flow rate due to the equipment failure or malfunctions were observed. Therefore this set has been reduced to 15856 hours long subset in order to make sure that each forecast will be created based on a complete set of explanatory variables. Obtained complete set was further divided into three smaller ones, namely:

- teaching subset 11000 cases,
- validation subset 2378 cases,
- testing subset 2378 cases.

Classification of the cases to the individual subsets was performed based on an in-built in the Statistica software algorithm. But in order to avoid type III errors [26] a k-fold cross-validation with k = 10 was conducted. It basically means that the procedure of dividing the original set into three subsets was done ten times and the obtained performance indicators values (RMSE and E) were averaged.

Table 3 summarises the RMSE and E for each of the five ANN models. Observed values are quite similar and differentiable at two decimal places. In terms of Nash-Sutcliffe criterion all models exhibited equal performance and the values of E suggests that created forecasting models explain almost completely the behaviour of modelled phenomena. In case of forecasts generated by models using one hour prediction time period e.g used by HydroProg, exhibited values of E equal to 0.93 what means that all five proposed in this paper models came out to be superior. However it is clear that model ANN2 which had only nine neurons in the hidden layer and used hyperbolic tangent and liner functions as activations ones respectively in hidden and output layer outperformed the remaining ones in terms of RMSE criterion. Although the greatest observed absolute error in case of the ANN2 model was 12.48 m^3/s, the mean of all absolute errors was 0.2836 m^3/s. Majority (95%) of absolute errors was smaller than 1 m^3/s and almost 86% of them was not greater than 0.5 m^3/s. Figure (3) depicts the 30 observed and forecasted by the ANN2 model flow rates for 30 subsequent observations.

Table 3. RMSE and Nash-Sutcliffe forecasting metrics values for five ANN models.

	Artificial neural network model				
	ANN1	**ANN2**	**ANN3**	**ANN4**	**ANN5**
RMSE	0.6498	0.6397	0.6616	0.6456	0.6454
E	0.9987	0.9987	0.9986	0.9987	0.9987

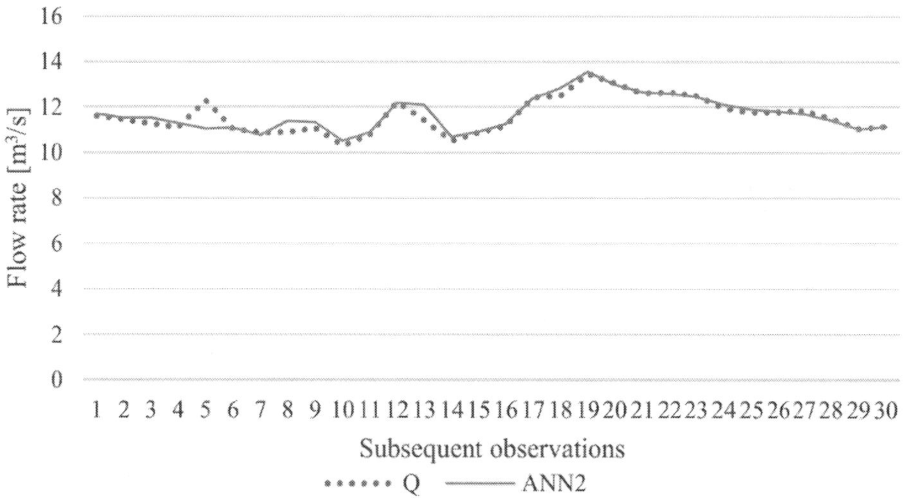

Fig. 3. Relation between modelled (ANN2) and observed flow rate for 30 subsequent observations for The Nysa Kłodzka dataset were obtained for the purpose of this study.

5 Conclusions

The results and observations coming from this study may be applied for the future analysis of the operation of run-off-river power plants with or without reservoir. Information about the flow rate which will be available an hour ahead may give the operator/owner of the ROR type hydropower plant an opportunity to perform better (financially) on the energy market but also to adjust energy yield to the current market needs. It is important to note that flow rate forecasts are currently of a great importance for civilian safety services but in future the variable renewable energy may start such a huge role in energy system that it will be necessary to put in service all possible measures which ease their inherent stochastic nature.

There seems to emerge several interesting directions for future studies, which should focus on:

- investigating the impact of smaller explanatory variables subsets and longer lag times on the forecast accuracy;
- building hybrid forecasting models based on other machine learning tools and data pre-processing techniques such as wavelet transform;
- scrutinizing the impact of forecast accuracy on the real ROR hydropower energy output;
- creating a simulation and optimization model (which will incorporate data from flow rate forecasts) of a ROR power plant with a reservoir operating on the energy market.

References

1. A. G. R. Vaz, B. Elsinga, W. G. J. H. M. van Sark, M. C. Brito, *An artificial neural network to assess the impact of neighbouring photovoltaic systems in power forecasting in Utrecht, the Netherlands* Renewable Energy, **85**, 631-641 (2016)
2. S. Kim, Y. Seo, V. P. Singh, *Estimating Global Solar Irradiance for Optimal Photovoltaic System* Procedia Engineering, **154**, 1237-1242 (2016)
3. S. S. Soman, H. Zareipour, O. Malik, P. Mandal, *A review of wind power and wind speed forecasting methods with different time horizons* In North American Power Symposium (NAPS), 2010 (pp. 1-8). IEEE. (2010)
4. J. Jurasz, A. Piasecki, *Evaluation of the Complementarity of Wind Energy Resources, Solar Radiation and Flowing Water–a Case Study of Piła* Acta Energetica **2** (2016).
5. C. E. Hoicka, I. H. Rowlands *Solar and wind resource complementarity: Advancing options for renewable electricity integration in Ontario,* Canada Renewable Energy, **36(1)**, 97-107 (2011)
6. X. Luo, J. Wang, M. Dooner, J. Clarke, *Overview of current development in electrical energy storage technologies and the application potential in power system operation* Applied Energy, **137**, 511-536, (2015)
7. M. A. Ramli, A. Hiendro, Y. A. Al-Turki, *Techno-economic energy analysis of wind/solar hybrid system: Case study for western coastal area of Saudi Arabia* Renewable Energy, 91, 374-385, (2016)
8. S. Weitemeyer, D. Kleinhans, T. Vogt, C. Agert, *Integration of Renewable Energy Sources in future power systems: The role of storage* Renewable Energy, **75**, pp. 14-20 (2015)
9. R. Kasperek, M. Wiatkowski: *Hydropower generation on the Nysa Kłodzka river* Ecological Chemistry and Engineering, **21(2)**, pp. 327-336 (2014)
10. Directive 2009/28/EC of the European Parliament and of the Council of 23 April 2009 on the promotion of the use of energy from renewable sources and amending and subsequently repealing Directives 2001/77/EC and 2003/30/EC.
11. I. Graabak, M. Korpås, *Balancing of variable wind and solar production in Continental Europe with Nordic hydropower–A review of simulation studies* Energy Procedia, **87**, 91-99 (2016)
12. T. Niedzielski, B. Miziński, M. Kryza, P. Netzel, M. Wieczorek, M. Kasprzak, P. Migoń, M. Szymanowski, J. Jeziorska, M. Witek, *HydroProg: a system for hydraulic forecasting in real time, based on the multimodelling approach,* Meteorology Hydrology and Water Management vol. **2** issue 2. IMGW-PIB, Warszawa, pp. 65-72. (2014)

13. U. Smyczyńska, J. Smyczyńska, & R. Tadeusiewicz. *Neural modelling of growth hormone therapy for the prediction of therapy results* Bio-Algorithms and Med-Systems. **11(1)**: 33-45. doi: 10.1515/bams-2014-0021 (2015)
14. J. Jurasz, J. Mikulik, *Day ahead electric power load forecasting by WT-ANN,* Przegląd Elektrotechniczny, **4**, pp. 152-154, (2016)
15. A. Piasecki, J. Jurasz, W. Marszelewski. *Wykorzystanie wielowarstwowych sztucznych sieci neuronowych do średnioterminowego prognozowania poboru wody-studium przypadku.* Ochrona Srodowiska **38.2** (2016): 17.
16. J. Adamowski, H. F. *Chan A wavelet neural network conjunction model for groundwater level forecasting.* Journal of Hydrology, **407(1-4)**, 28-40. doi:10.1016/j.jhydrol.2011.06.013 (2011)
17. D. J. MacKay, *Information theory, inference and learning algorithms.* Cambridge university press (2003)
18. G. J. Bowden, G. C. Dandy, H. R. Maier: *Input determination for neural network models in water resources applications. Part 1 – background and methodology.* Journal of Hydrology, Vol. **301**, No. 1–4, pp. 75–92 (2005)
19. A. Piasecki, J. Jurasz, and Skowron, R. *Application of artificial neural networks (ANN) in Lake Drwęckie water level modelling.* Limnological Review, **15(1)**, 21-30. (2015)
20. T. Niedzielski, B. Miziński, D. Yu, *Hydrological forecasting in real time: an experimental integrated approach*, et. al.: Jasiewicz, J., Zwoliński, Z., Mitasova, H., Hengl, T. (Eds.), Geomorphometry for Geosciences, Bogucki Wydawnictwo Naukowe, Adam Mickiewicz University in Poznań – Institute of Geoecology and Geoinformation, Poznań, pp. 97–101 (2015)
21. T. Niedzielski, B. Miziński *Real-time hydrograph modelling in the upper Nysa Kłodzka river basin (SW Poland): a two-model hydrologic ensemble prediction approach.* Stochastic Environmental Research and Risk Assessment, Online first: 28 April 2016 DOI: 10.1007/s00477-016-1251-5 (2016)
22. O. Bozorg-Haddad, M. Zarezadeh-Mehrizi, M. Abdi-Dehkordi, H. A. Loáiciga, M. A. Mariño, *A self-tuning ANN model for simulation and forecasting of surface flows,* Water Resources Management, **30(9)**, pp 2907–2929. DOI:10.1007/s11269-016-1301-2 (2016)
23. B. Kang, Y.H. Ku, Y.D Kim, *A case study for ANN-based rainfall–runoff model considering antecedent soil moisture conditions in Imha Dam watershed,* Korea. Environmental Earth Sciences, **74(2)**, pp 1261–1272, DOI:10.1007/s12665-015-4117-0 (2015)
24. K. Shibata, Y. Ikeda, *Effect of number of hidden neurons on learning in large-scale layered neural networks.* Proc. of the ICROS-SICE International Joint Conference, pp. 5008–5013, (2009
25. J. Nash, and J. V. Sutcliffe, *River flow forecasting through conceptual models part I—A discussion of principles,* Journal of hydrology **10.3** (1970): 282-290
26. F. Mosteller, *A k-sample slippage test for an extreme population.* In: Selected Papers of Frederick Mosteller, Springer, New York pp. 101–109, (2006)

An analysis of hybrid power generation systems for a residential load

Bartosz Ceran[1,*], *Qusay* Hassan[2,3], *Marek* Jaszczur[2] and *Krzysztof* Sroka[1]

[1]Poznań University of Technology, Faculty of Electrical Engineering, Institute of Electrical Power Engineering, Poland
[2]AGH University of Science and Technology, Faculty of Energy and Fuels, Department of Fundamental Research in Energy Engineering, Poland
[3] University of Diyala, Departement of Mechanical Engineering, Iraq

Abstract This paper presents the results of an energetic and economical analysis of a hybrid power generation system (HPGS) which utilises photovoltaic modules, wind turbines, fuel cells and an electrolyzer with hydrogen tank working as the energy storage. The analysis was carried out for three different residential loads, local solar radiation and local wind speed, based on the real measurement values. The analysis shows the optimal solution and the limits of the investment costs required for the system construction. The presented results confirm the effectiveness of the proposed approach, which could be assumed as a very useful tool in the design and analysis of a hybrid power generation system.

1 Introduction

The global challenge of providing reliable, clean and cost-effective energy remains one of the major challenges which the world is facing with this century [1]. The renewable energy sources are considered as the most important and most effective solution for the future. The wind energy and the solar energy are considered as the most effective among the various rentable energy resources, they have made a rapid and substantial breakthrough in last year's [2].

Solar energy is an inexhaustible source used depending on various techniques such as solar systems [3]. The power generated by the photovoltaic systems is a proven method with a large number of applications throughout the world [4]. The photovoltaic system can be integrated with the grid connection (grid-on) or can be autonomous systems (grid-off). Usually photovoltaic system consists of the photovoltaic modules, storage batteries, inverters and power control components sometimes with mechanical or electronic tracking systems. From the local point of view the photovoltaic systems are zero greenhouse gases emissive which make energy generation environmentally friendly. In addition to the flexible process ability to expanded for meeting the growing energy needs furthermore cost effective and minimal maintenance requirements.

Wind energy is an important energy source, environmentally and economically competitive and expected to have a significant role to meet the global challenges of clean energy and sustainable development under climate change [5-7]. After the first oil crisis was

* Corresponding author: bartosz.ceran@put.poznan.pl

introduced wind energy is considered as a medium that uses local resources and promotes energy independence, security, and decentralization of power system [8].

1.1 Hybrid power system configuration and optimization

The hybrid power generation system (HPGS) optimal (or possible) configuration depends on three main factors: energy resource (conventional and/or renewable), demanded load, and the cost (operating expenditure and capital expenditure) [8]. The main problem which face power generation using a hybrid power system is the variation in renewable resources (wind speed and solar radiation) and the variation in load demand. Therefore, the majority concern in the design of the hybrid power system that uses renewable energy sources is the best choice of components for the system that can meet the load demand and economically pregnancy. Based on the components operating cost, fuel, labor and transportation, maintenance and highly desirable to assess the most cost-effective scaling of all components to meet peak loads.

The system size optimization of hybrid power system has to reduce the net present costs while meeting demand as reliable and cost effective. In this work the computer simulation model has been developed to optimize hybrid power generation system which utilises photovoltaic modules (PV), wind turbines (WT), PEM fuel cells (FC), electrolyzer (EL) and hydrogen tank (HT) as the energy storage. The main purpose of this project is to study the possibility of interbreeding renewable energy sources through the design and optimization of the hybrid PV/WT/FC system used based on the optimal computer design.

In order to utilize in the economical and efficient way renewable energy resources, it is important to optimise the size of system components. The optimal system components sizes can help to ensure minimum investment cost with the fully use of the system components. In such cases the hybrid system can operate in the optimal conditions in terms of the investment as well as in terms of the energy system reliability requirements. Solar and wind energy systems are renewable energy systems resources, with and without the grid connection widely used among the most developed ones. The sizing tool determines the optimum size of the system components. In hybrid energy power systems around 35% of the total energy is lost due to the non-optimum size of the system [9]. The optimal configurations can be found by using the computer simulation. The Matlab simulation has been used in order to obtain the optimum configuration and the sizes of the system components.

In the last decades large number of the research projects, technologies and configurations which concentrate on the optimal design of the hybrid power system components have been done. Juhari et al. [10] optimized energy systems in the context of the minimizing cost of energy and excess energy. Kellog et al. [11], optimized the unit size of the hybrid photovoltaic/wind power system. Kamaruzzaman et al. [12] optimize the operational strategy of the components cost. The calculation using Net Present Cost (NPC) as the key function to minimise in the case of hybrid energy systems design was done by Lambert et al. [13]. Kamel et al. [14] used HOMER software in order to find the optimum components size at lowest cost for the hybrid power system in stand-alone applications. Fahmy et al [15] design an optimal economic renewable hybrid power system depending on wind & solar radiation and showed these two sources of energy can have a high effect on reducing COE more than other renewable energy sources. Khare et al [16] presented the methodology for components size calculation as well as for stand-alone PV/diesel/battery hybrid system optimization. Shivrath et al. [17] shows methodology for optimally designed the solar and wind hybrid system for a remote drip irrigation system. Rashidi et al.[18] was focused on the optimal design of a standalone solar-hydrogen system.

1.2 Study Area

Poznań is a city located on the Warta River in western Poland with renewable energy resources such as the wind and solar radiation that can be exploited for alternative energy solutions. Poznań is situated in latitude 52'4064° N and longitude 16'9252° E with abundant wind speed most the year round. According to Poznań weather stations this area is endowed with an annual wind speed average 6 m/s and has an annual average daily solar radiation of about 2.81 kW/m^2/day.

2 Energy analysis of hybrid power generation system

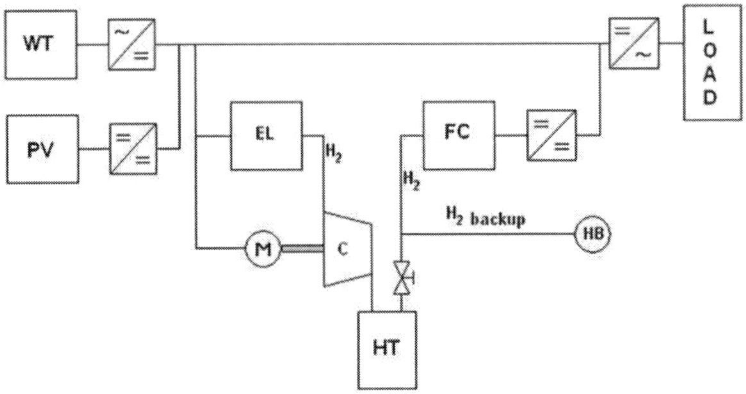

Fig. 1. Flowchart-hybrid power generation system.

Hybrid power generation system flowsheet analysed in this paper is shown in Fig. 1.

Temporary power shortages associated with stochastic nature of the generation of energy by renewable sources is compensated by fuel cell operation. This mode of operation is enabled by operating characteristics of fuel cells, which are perfectly suited for varying loads and show high efficiency in a wide power range. A very valuable feature of fuel cells is their high efficiency at low load at 45-55 % for PEM fuel cells. Primary, hydrogen is supplied from the hydrogen tank HT. In the case of running out hydrogen in the tank, the fuel cell is powered by the hydrogen backup HB. The use of additional hydrogen in the backup purposes may be subjected to small emissions of pollution associated with its production, depending on the used technology (steam reforming of hydrocarbons, coal gasification and biomass gasification).

The balance of the power in the system, depends on the ratio of power generated by renewable energy sources RES to the demand of the energy load. For the cases where the demand is lower than the energy power balance equation takes the following form:

$$P_{load} = P_{PV} + P_{WT} - P_{EL} - P_{comp} \quad (1)$$

where: P_{load} [kW] – power consumed by the recipient, P_{PV} [kW] – power generated by photovoltaic installation, P_{WT} [kW] – power generated by wind power plants, P_{EL} [kW] – power consumed by the electrolyzer, P_{comp} – power used for compression of hydrogen.

If the demand exceeds the power generated by the sources, the energy power balance equation takes the form:

$$P_{load} = P_{PV} + P_{WT} + P_{FC} \quad (2)$$

where: P_{FC} [kW] – power generated by the fuel cell.

DOI: 10.1051/e3sconf/20171401020

Energy analysis was conducted on the basis of the balance equations (eq.1 and eq.2.) produced and received from hybrid power generation system. The analysis was carried out for three different residential loads, local solar radiation and local wind speed, all based on the real measurement values. The Fig. 2 and 3 shows the annual timetables of solar energy distribution and wind speed for which distribution technical-economic analysis be adopted.

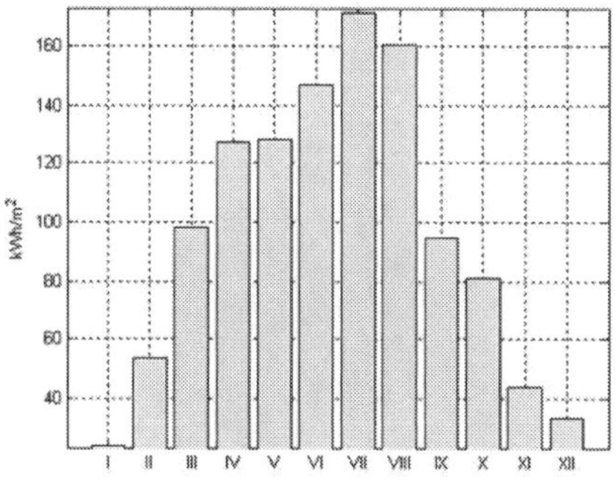

Fig. 2. The annual distribution of solar energy.

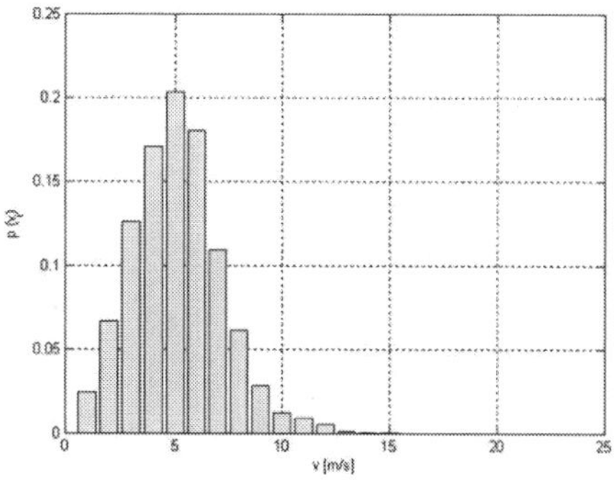

Fig. 3. The annual distribution of wind speeds.

The Fig.4, 5 and 6 show the weekly energy profiles of recipient 1,2 and 3. Profile 1 for is a two-person apartment where a gas oven and gas cooker are installed. The second profile is a family of four where all devices are powered by electrical energy (no gas installation). The third profile is a student apartment. In this case, the students occupy the apartment for four days a week. The rest of the time they spend in their hometowns.

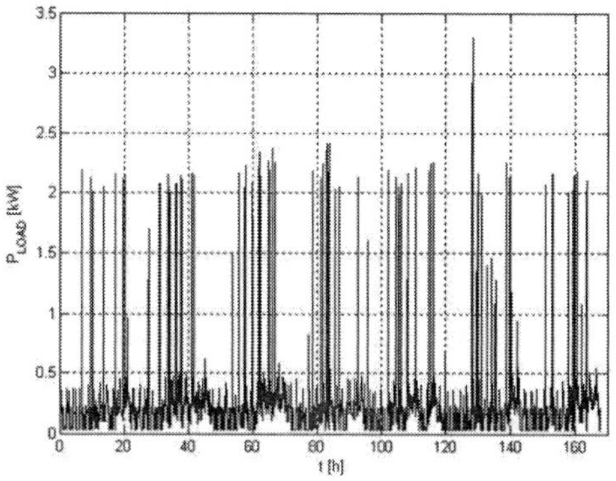

Fig. 5. The energy profile of the recipient 2.

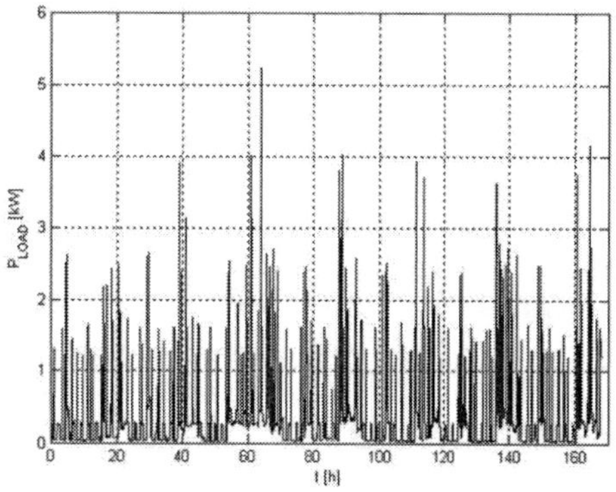

Fig. 4. The energy profile of the recipient 1.

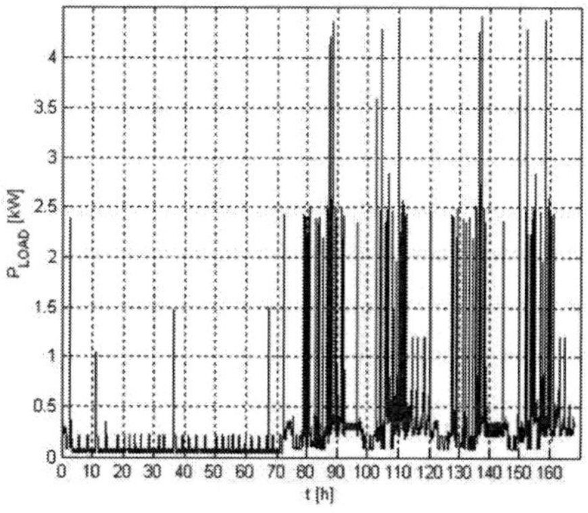

Fig. 6. The energy profile of the recipient 3.

In order to compare the profiles, annual electricity consumption by individual customers the following values has been designated: the instantaneous value of the maximum power consumption by the given receiver P_{max}, average power value P_{av_load} and an annual energy consumption A (see Table 1).

Table 1. Comparison of recipients.

	Recipient 1	**Recipient 2**	**Recipient 3**
A [MWh/a]	1.62	3.04	2.35
P_{av_load} [kW]	0.19	0.35	0.27
P_{max} [kW]	3.37	5.22	4.4

With a view to the selection of the composition of the generating units to meet the needs of the customer, we used the following equations [19]:

$$\Delta P = P_{av\_gen} - P_{av\_load} \rightarrow \min \tag{3}$$

$$P_{av\_gen} = n_{cf\_PV} \cdot P_{PV} + n_{cf\_WT} \cdot P_{WT} \tag{4}$$

where: P_{av_gen} – power generated by renewable sources, P_{av_load} – average input power by the recipient, n_{cf} – capacity factor of each renewable energy sources.

The value under the average for the period T_a is designated with the dependencies:

$$P_{av\_PV} = \frac{A_{PV}}{T_a} \tag{5}$$

$$A_{PV} = W_k \cdot W_w \cdot \int_0^T P_{PV} dt \tag{6}$$

where: A_{PV} – energy produced by photovoltaic system at time t, W_k – correction factor that takes into account the angle of installation referral to the south and the angle of deviation of the modules from the level, W_w - coefficient of performance.

For the determination of the amount of electricity produced by photovoltaic modules there were used the characteristics of the P_{PV} = f (E) developed on the basis of external characteristics family U = f (I) (see Fig. 7). Photovoltaic modules system is equipped with an MPPT maximum power point tracking ($P_{PV} = P_{MMP}$) [20].

In order to determine the average power produced by wind turbines P_{av_WT} operational characteristics was implemented P = f(v) and the distribution of wind speeds was presented in the form of a continuous function f(v) using the Weibull distribution (see Fig. 8):

$$f(v) = \frac{k}{\lambda}\left(\frac{v}{\lambda}\right)^{k-1} e^{-(v/\lambda)^k} \tag{7}$$

where: v - wind speed [m/s], k - shape parameter (k > 0), λ - the scale parameter (λ > 1)

$$P_{av\_WT} = \int_0^\infty P(v) f(v) dv \tag{8}$$

Energy storage has been selected basing on the following assumptions:

- fuel cell is able to meet the instantaneous maximum demand:

$$P_{FC} = P_{max\_load} \tag{9}$$

- electrolyzer is able to accept temporary power generated by photovoltaic system and wind sources:

$$P_{EL} = m \cdot (P_{max\_WT} + P_{max\_PV}) - P_{min\_load} \quad (10)$$

where: P_{EL} – rated electrolyzer power, P_{max_WT} – the maximum instantaneous wind turbines, P_{max_PV} – instantaneous maximum power photovoltaic installation, P_{min_load} – minimum instantaneous customer demand, m – coefficient of simultaneity of maximum power generation by the photovoltaic system and the wind turbine.

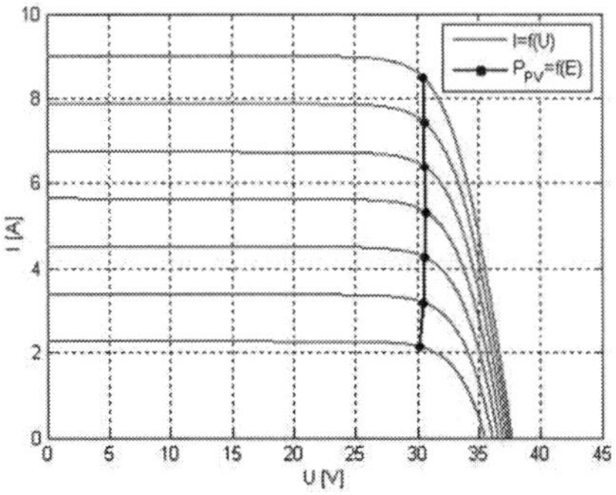

Fig. 7. External characteristics of PV modules family.

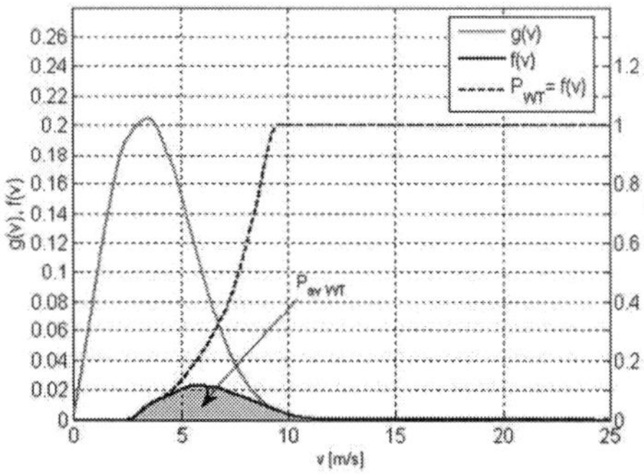

Fig. 8. Determination of annual average wind power capacity.

To determine the amount of the hydrogen produced by the electrolyzer and the quantities of hydrogen consumed by fuel cell, the characteristics P= f(V_{H2}) (see Fig. 9) developed on the basis of external characteristics of the fuel cell and electrolyzer U = f(I) (see Fig. 10) was used.

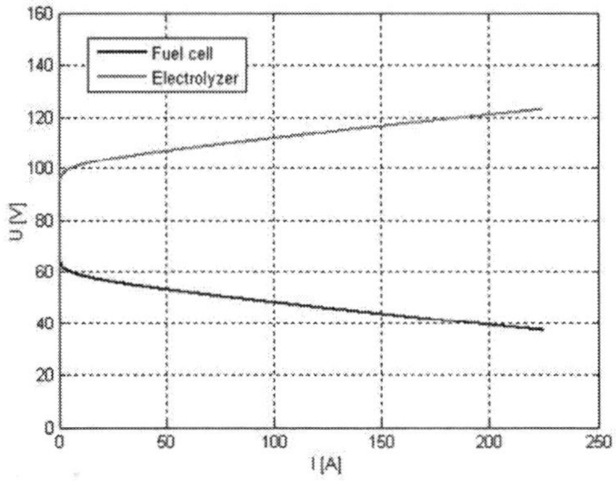

Fig. 9. The characteristics U = f (I) of fuel cells and the electrolyzer.

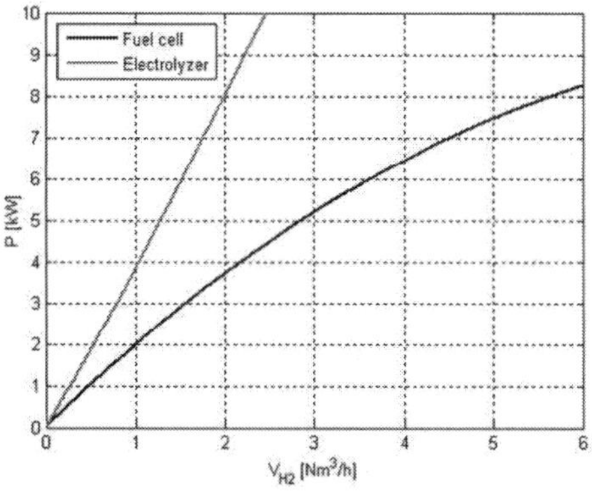

Fig. 10. The characteristics P = f (V_{H2}) of fuel cells and the electrolyzer.

Basing on the equations 1 - 10 and the authorial code written in MATLAB it has been chosen a set of devices of a hybrid system and the annual demand for additional hydrogen for each customer has been specified in order to ensure uninterrupted power supply. The results for the hybrid system are shown in Table 2.

In order to determine the optimal composition of the generating units of the hybrid power generation system for each of the profiles, two additional variants were considered, except for the basic variant (PV and WT):

PV variant – the source of renewable energy as only photovoltaic installations, hybrid system is not equipped with wind power plants.

WT variant – the source of renewable energy are the only wind power plants, hybrid system is not equipped with photovoltaic installations.

Table 2. Comparison of recipients

	Basic Variant PV + WT			Variant PV			Variant WT		
Recipient	**1**	**2**	**3**	**1**	**2**	**3**	**1**	**2**	**3**
RES nominal power [W]	1050	1850	1350	1500	2750	2250	600	1200	900
Photovoltaic modules [W_p]	3x250	5x250	3x250	6x250	11x250	9x250	-	-	-
Wind turbines [W]	1x300	2x300	2x300	-	-	-	2x300	4x300	3x300
Fuel cell [kW]	3	5	4.2	3	5	4.2	3	5	4.2
Elecktrolyzer [kW]	1	1.5	1.2	1.5	2.5	2	1	1.5	1
H_2 backup [Nm^3]	**459.4**	**812.6**	**464.9**	**706.6**	**1130.3**	**774.64**	**443.73**	**606.2**	**476.73**

By analyzing the results of the calculations, it can be concluded that the differences between the demand for hydrogen for the variant "WT" and "PV + WT" are marginal. On the other hand, in the case of the variant "PV", resignation from wind turbines must be compensated by more frequent work of the fuel cell and much larger hydrogen consumption from back-up HB. This is related to the fact that the photovoltaic installation works only during daylight and its power strongly depends on the solar power radiation. During the cloudy days or at night, the recipient will always be powered by fuel cells. This relationship applies to all the profiles of the concerned recipients.

The smallest value of nominal power P_N was obtained for the variant excluding PV system. It should be noted that the analysis assumes the wind profile obtained on the basis of annual measurements.

Conducted energy analysis is the basis for economic analysis, which gives a full comparison of the variants considered.

3 Hybrid power system components with an economical constraints and analysis

Three different configurations are considered in this analysis. The each configuration has different size and components illustrated in Table. 2. The main components are wind turbine (WT), Photovoltaic modules (PV), Converter, Fuel cells (FC), Electrolyzer (EL) and Hydrogen tank (HT), The cost, number of units to be used, operating hours, etc. need to be specified in MATLAB for all equipment's. The description of all these components is given in the following sections. In the present simulation, the main component include two major renewable energy which photovoltaic panel & wind turbine with a fuel cell as a storage unit. The photovoltaic panel & fuel cell involved inverter for AC/DC and an electrolyzer used with hydrogen tank to feed the fuel cell. The system schematic diagram is shown in Fig. 1.

3.1 Wind turbine unit

The wind is the source of energy in wind turbine unit. In general the production depends on the wind variation, and in this work annually average (6 m/s) is considered as in the selected site. The wind turbine unit size considered (0.3 kW AC), the installation and replacement costs of this size taken as ($1000) with operation & maintenance cost ($100/yr) for (20 years) lifetime.

3.2 Photovoltaic array and power converter model

The photovoltaic array is the second energy producer unit, affected by solar radiation variation. The installation and replacement costs of (1 kW) solar energy systems are taken as ($1000) with operation & maintenance cost ($300/yr) and the lifetime was taken (20 years) without sun tracking system.

An electronics power converter is needed to maintain energy flow between the (DC) and (AC) components. The installation and replacement costs of (1 kW) are taken as ($500) with operation & maintenance cost ($200/yr) and a lifetime of a unit are considered (15 years) with an efficiency 85 %.

3.3 Fuel cell, electrolyzer and hydrogen tank unit

A fuel cell is the energy storage unit in this system. A commercial type model with size (3 kW) for all configurations was considered in this simulation. The estimated lifetime is (30000 hr) and the installation with replacement costs are taken ($500/kW) while the maintenance cost is expected at ($ 0.1/hr).

The electrolyzer is the device that produces hydrogen by using electrical load and at the same time consider one of two main loads on the system. The installation and replacement costs are taken is ($500/kW) with operation & maintenance cost ($500/yr) and the lifetime (20 years).

The hydrogen tank is a part of this system that is the storage of hydrogen produced from electrolyzer which used to feed the fuel cell. The installation with replacement costs is taken ($500/kg) with operation & maintenance cost ($300/yr) and the lifetime (25 years).

4 Results

The project lifetime considered (20 years) with an annual discount rate of 1% for all configurations. After simulation and optimization system components for each configuration found the optimal contribution of each component to each combination, the annual electricity production for each configuration is illustrated in Table 3.

Table 3. Electricity production of hybrid system components

Configuration 1			Configuration 2		Configuration 3	
Production component	**kWh/year**	**%**	**kWh/year**	**%**	**kWh/year**	**%**
PV array	827	5	1.756	49	0	0
WT	12.362	81	0	0	24.724	94
FC	2.156	14	1.812	51	1.676	6
Total	15.345	100	3.567	100	26.399	100

Table 4 and Figure 11 show the cash flow summary for the optimal system for each configuration for all components, capital cost, replacement cost, operation and maintenance cost and salvage value. For the configuration 1, the total capital cost is considered to be ($6500) and the configuration 2, considered ($6500) and configurations 3, considered ($6000) for whole project lifetime.

The cost of electricity (COE $/kWh) and total net present cost (NPC $) for each configuration of the hybrid power system is shown in Figure 12 where the configuration 1 (COE is $0.287 and NPC $7962), configuration 2 (COF is $0.301 and NPC $8359) and configuration 3 (COE is $0.243 and NPC $6739) are summarized.

Table 4. Overall cost for each component of all combinations for hybrid power system

	Configuration 1			Configuration 2			Configuration 3		
	Capital ($)	Replace ($)	O&M ($)	Capital ($)	Replace ($)	O&M ($)	Capital ($)	Replace ($)	O&M ($)
PV	750	0	41	1.500	0	81	0	0	0
WT	1.000	0	180	0	0	0	2.000	0	0
FC	1.500	0	776	1.500	0	74	1.500	0	604
Conv	1.250	1.077	90	1.250	1.077	90	1.000	861	72
EL	1.000	0	90	1.500	0	135	1.000	0	90
HT	1.000	0	108	750	0	81	500	0	54
Total	**6.500**	**1.077**	**1.286**	**6.500**	**1.077**	**462**	**6.000**	**861**	**821**

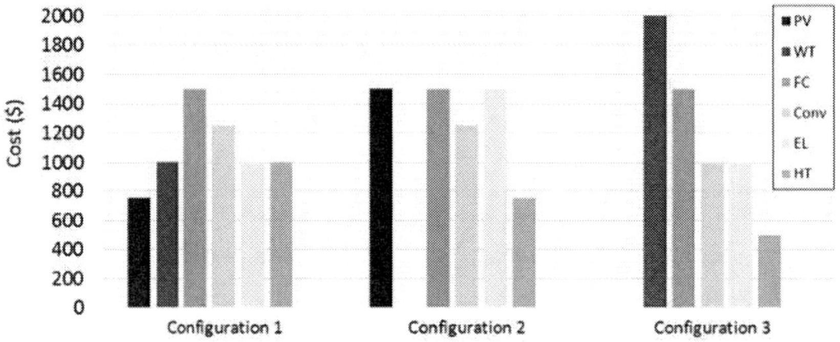

Fig. 11. Cash flow summary for components cost of all combinations for hybrid power system.

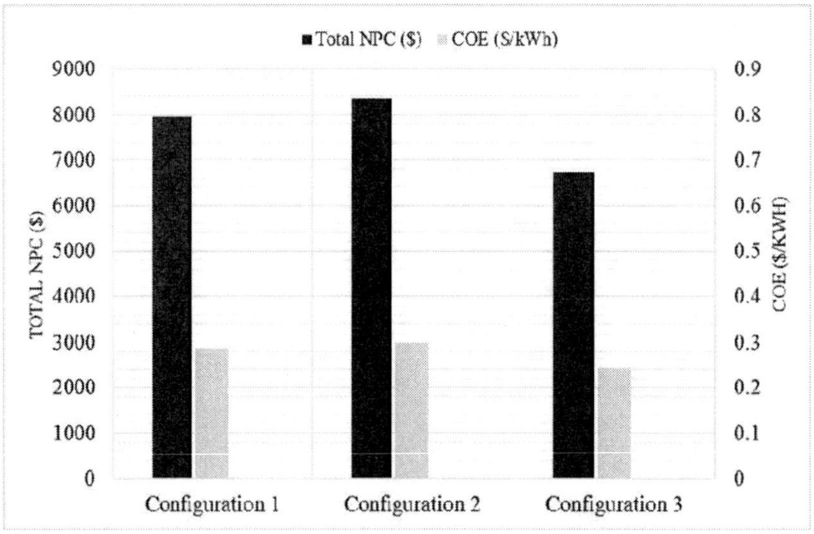

Fig. 12. Cost of electricity and total net percent cost for hybrid power system configurations.

5 Conclusions

The electrification using hybrid power systems are worldwide very promising in the recent years. Solar and wind power are considered as the renewable energy resources due to their high reliability and safety as well they are zero emissive. This study presents a complete

DOI: 10.1051/e3sconf/20171401020

analytical design of three configurations of the hybrid power systems with optimizing units for different loads presented in figures 4, 5, 6. The simulation and optimization results for each configuration has been prepared and the electricity demand can be met with a hybrid system comprising.

Resulting, the systems can be profitable with a reasonable investment for production renewable energy. It utilizes all renewable energy, component PV, WT and storage unit FC as showed in Table 3. However, the reliability of the system is ensured due to used high-reliability storage unit. The cost of energy generated from the above configurations has been found to be \$0.287, \$0.301, and \$0.243 for the configuration 1, 2 and 3 respectively and the price can be very beneficial and suitable for long-term investments particularly with partial governmental support. Also the nature of the hybrid power system is environmentally-friendly that can be depicted from the annual emissions.

References

1. S. C. Bhattacharyya, Energy for Sustainable Development: a critical review and analysis, Energy for sustainable Development, **16(3)**, 260-71, (2012)
2. Alliance for Rural Electrification (ARE). Hybrid power systems based on renewable energies. ARE Position Paper (2008)
3. M. Thirugnanasambandam, S. Iniyan, R. Goic, Renewable and Sustainable Energy Reviews **14(1)**, 312–22, (2010)
4. I. P. Panapakidis, D. N. Sarafianos, M. C. Alexiadis, Renewable and Sustainable Energy Reviews, **16(1)**, 551–563, (2012)
5. V. R. Vanajaa, Int. Journal of Applied Engineering Research ,**11(4)**,2579–2586 (2016)
6. G. M. Joselin, S. Iniyan, E. Sreevalsan, S. Rajapandian, Renewable and Sustainable Energy Reviews; **11(6)**:1117–45 (2007)
7. R. Saidur, M. R. Islam, N. A. Rahim, K. H.Solangi, A review on global wind energy policy, Renewable and Sustainable Energy Reviews, **14(7)**,1744–1762, (2010)
8. A. Bataineh, A. Alqudah, A. Athamneh, Jordan. Energy and Environment Research, **4(3)**, 9–20, (2014)
9. A. J. Peterson, R. Perez, B. Bailey, K. Elsholz, Solar Energy, **65**, 227-235, (1999)
10. J. A. Razak, S. Kamaruzzaman, A. Yusoff, International Journal of Energy, **3(1)**, 77 – 81, (2007)
11. W. Kellog, M. H. Nehrir, G. Venkataramanan, V. Gerez, Electric Power Systems Research, **39**, 35-38, (1996)
12. S. Kamaruzzaman, Z. Azami, A. Yusoff , M. N. Zulkifli, A. R. Juhari, S. M. Nor, Transactions on Mathematics **4 (7)**,130-140 (2008)
13. T. Lambert, HOMER: The Hybrid Optimization Model for Electrical Renewables.
14. S. Kamel, C. Dahl, Agriculture in Egypt, Energy, **30**, 1271-1281 (2005)
15. F. H. Fahmy, N. M. Ahmed, H. M. Farghally, Smart Grid and Renewable Energy, **3**, 43-50 (2012)
16. A. Khare, S. Rangnekar, International Journal of Renewable Energy research, **3(3)**, 673-681 (2013)
17. Y. Shivrath, P. B. Narayana, S. Thirumalasetty, E. L. Narsaiah, International Journal of Modern Engineering Research (IJMER), **2(4)**, 2947-2950 (2012)
18. H. Rashidi, S. Niazi, J. Khorshidi, Australian Journal of Basic and Applied Sciences, **6(10)**, 249-256 (2012)
19. C. Wang, M. H. Nehrir, Power management of a stand-alone wind/photovoltaic/fuel cell energy system, IEEE Transactions, 957-967, (2008)
20. B. Ceran, K. Sroka, Performance Analysis of a Hybrid Generation System of Wind Turbines, Photovoltaic Modules, and a Fuel Cell, Acta Energetica, **2/23**, 36–42, (2015)

DOI: 10.1051/e3sconf/20171401021

Cogeneration – development and prospect in Polish energy sector

Dominika Matuszewska[1,*], *Marta* Kuta[1] , and *Jan* Górski[1]

[1]AGH University of Science and Technology, Faculty of Energy and Fuels, Department of Thermal and Fluid Flow Machines, al. A. Mickiewicza 30, 30-059 Krakow

Abstract. Next 10-15 years are crucial for condition of Polish energy sector in light of challenges arising mainly from increasing demand for electric energy, need of reducing greenhouse gases emissions and shutdowns of old units. In this situation cogeneration can be one of the most rational way to meet those circumstances. This paper analyzes present development of cogeneration in Poland and its prospect for future.

1 Introduction

In the next 10 - 15 years the energy sector in Poland is facing important and serious challenges arising mainly from increasing demand for electricity. In a situation where the demand for heat in the country over the last few years is falling, and at the same heat production with over 90% efficiency is not a big technological problem, the biggest issues of the heating sector seem to be modernization old heating plants, thereby reducing fuel consumption, lowering maintenance costs and reducing overall emissions [1]. From this perspective, the problems in the national power system seem at this point far more pressing and urgent. In August 2015, PSE SA restricted the supply and consumption of electricity in our country. It is worth noting that these restrictions have been introduced due to the decrease in the production capacity of Polish power plants and though they had no direct influence on the individual recipient, they significantly affected on large customers (industrial plants, factories, etc.). The introduction of 20 level of power supply resulted in that the major electricity consumers in Poland could only receive the minimum power - which was required to ensure the safety of people and to prevent possible damage of equipment and technological facilities [2].

A few years earlier, it was said about future problems with uncontrolled supply of electric energy and year 2015 showed that the problem is real and with the increasing demand for electricity growing even more serious [3,4,5].

It is estimated that in 2030, Poland will need 46 - 54 GW (depending on available forecasts) of electricity, which now puts before us the question about ways to meet the demand for power in 2030 [6]. Figure 1 presents an estimated installed capacity in 2030 [GW] based on various reports.

* Corresponding author: dommat@agh.edu.pl

 DOI: 10.1051/e3sconf/20171401021

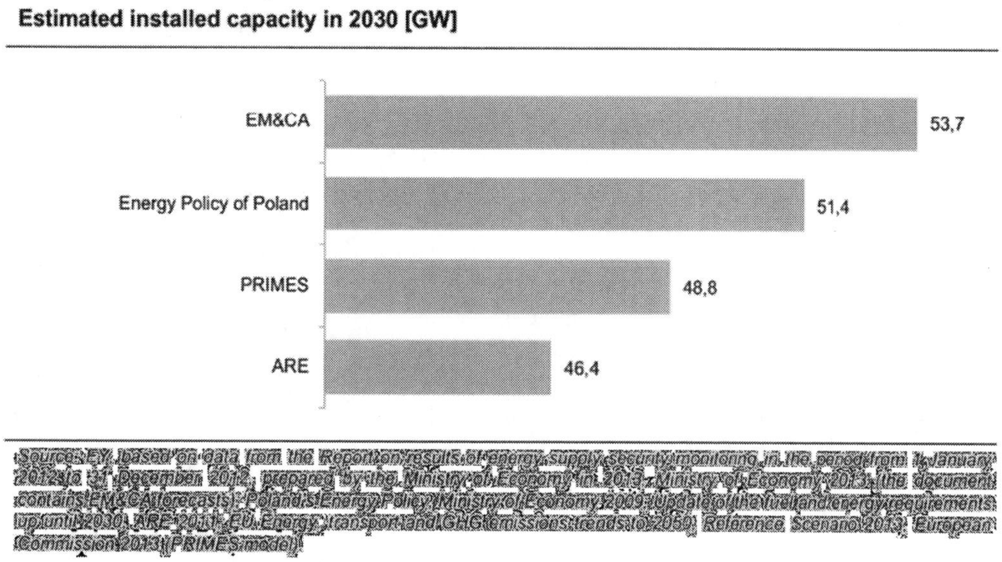

Fig. 1. Estimated installed capacity in 2030 [GW] [6]

The energy sector to meet these requirements must take into account a number of factors including the internal politics of the country, as well as international obligations or technologies available on the market. The development of the energy sector in Poland is closely connected with EU policies, the priorities of which include the creation of an internal, unified and competitive energy market, ensuring security of energy supply, increasing the share of renewable in energy production, reduction of GHG and increase energy efficiency [6].

During the electricity production the energy efficiency is very important and crucial. In case when the efficiency of heat production often hovers around 90%, the efficiency of generating electricity using heat engines is much lower and so in the steam power plants is approx. 46%, in power stations with gas turbines - more than 40%, in power station with combustion engines - around 48%. The value of the electricity generation efficiency is largely limited by thermodynamics, therefore, combined heat and power production seems to be much more rational - both in terms of thermodynamic and economical - way of use the chemical energy of fuels. CHP (i.e., simultaneous generation of different energy) can be successfully implemented in high power systems (eg., industrial and power plants) as well as in distributed systems low power. Both in Poland and in the EU can be observed an increased amount of investments in the construction of cogeneration systems (despite higher capital expenditures). The positives flowing from the construction of such facilities (including the reduction of fuel consumption, operating costs and total emissions) influence significantly on increased interest in this type of installations [1, 7].

2 Development of cogeneration in Poland

Poland is one of the member countries of the EU, where the tradition of combined heat and power is long, and awareness of the benefits arising from the use of these technologies quite broad. In the 1970 - 1980 years, the share of the production of combined heat in the country grew steadily in industrial and power plants. In the 90s, it can be observed a decrease in the amount of heat generated in industrial CHP plants, while in the professional CHP plants growth has been maintained, on which undoubtedly were affected by economic change and restructuring in the country. Overall, it can be said that the years of the 90th brought in Poland stabilization of heat production from CHP units. The biggest jump of heat production from CHP dates back to the years 1970 - 1995, when it reached a 3-fold increase of the production (approx. 164 PJ). [8, 9].

As for the electricity production in cogeneration, from 70s it can be observed its continuous growth with varying intensity, wherein the years 1970-1995 are characterized by the highest growth rates [10].

Table 1 presents installed electrical capacity, total heat supplied, total electricity generation and total share of electricity production in Poland in the period 2006 – 2012. The installed capacity of high-efficiency cogeneration in this period was approximately 8.7 GW, while electricity production fluctuated around 27 TWh. As for the heat production from cogeneration, it can be seen minimal decline over the years 2006 - 2012, while the electricity production cogeneration has grown steadily and in 2012 amounted to almost 17% of total electricity production [11].

Table 1. Cogeneration in Poland in the period 200-2012 [11]

CHP	2006	2007	2008	2009	2010	2011	2012
Installed electrical capacity [GW]	8.47	9.02	8.78	8.63	8.69	8.84	8.33
Total heat supplied [TWh]	76.5	72.5	72.1	71.8	77.0	73.2	72.4
Total electricity generated [TWh]	26.0	27.6	26.4	26.1	27.7	27.1	27.0
Total % of gross electricity production	16.05	17.30	16.91	17.19	17.57	16.58	16.68

The date from the table 1 are presented in graphic form below in Figure 2.

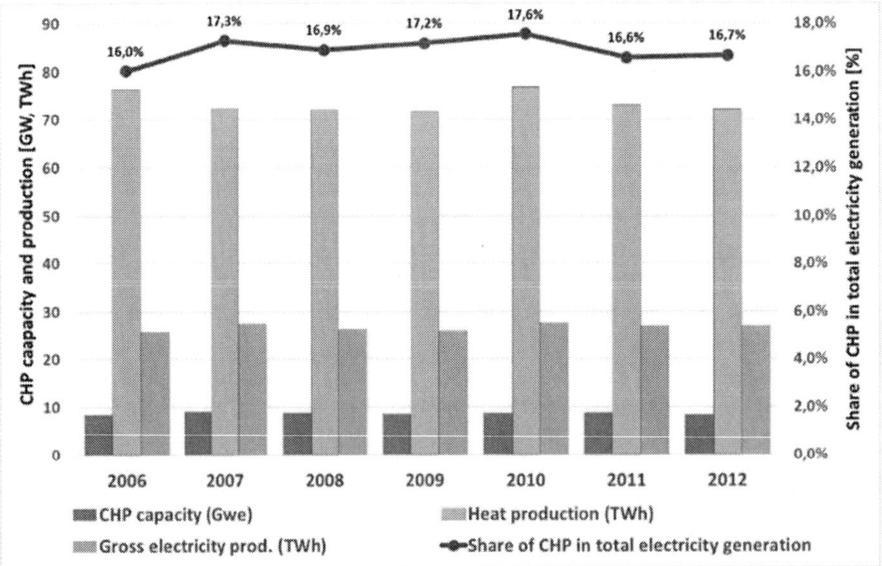

Fig. 2. Electricity and heat production in high-efficiency CHP and share in total electricity production in the period 2006 – 2012 [11]

In 2014, heat production in Poland was about 393.2 thousand of TJ, of which a significant amount was used for their own needs, while 60% (taking into account transmission losses) went to customers connected to the network. It is worth mentioning that more than 64% of the heat production was produced in cogeneration with electric power (Figure 3) [12].

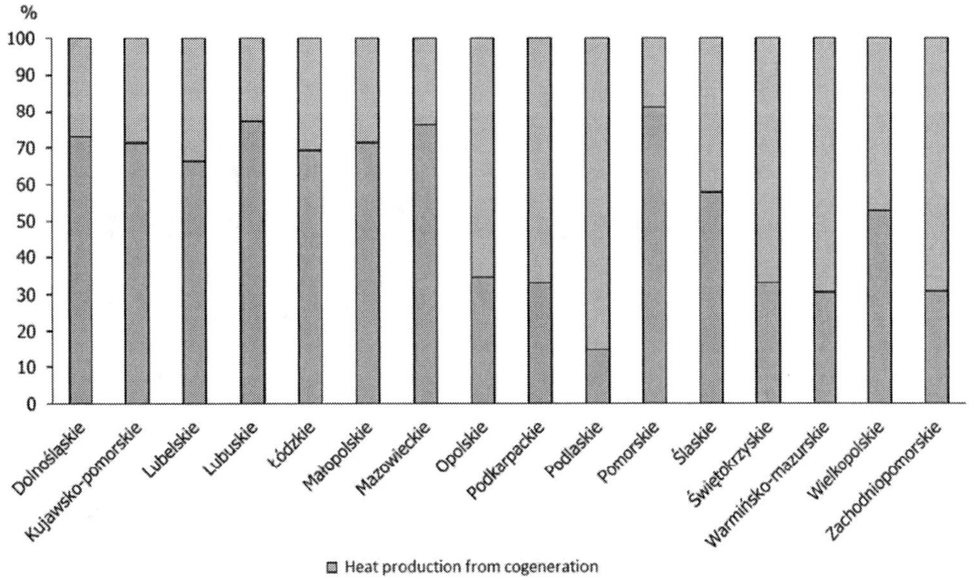

Fig. 3. The share of cogeneration in heat production for different regions in Poland (2014) [12]

The share of cogeneration in heat production varies for different regions in Poland and so - was the highest in the regions of Pomerania (over 80%), Lubuskie, Mazowieckie (over 75%) and the lowest in the region of Podlasie (below 20%).

The electricity production in cogeneration (with use of steam turbines) in Poland is mainly based on the coal technologies (about 88%). The share of electricity generated in gas cogeneration units (combined cycle with a gas turbine CCGT) and low power units of less than 1 MW (as internal combustion engine) was in 2012 less than 12% of the electricity generated in cogeneration [11]. Production of electricity from cogeneration is shown in the following figures (Figure 4).

 DOI: 10.1051/e3sconf/20171401021

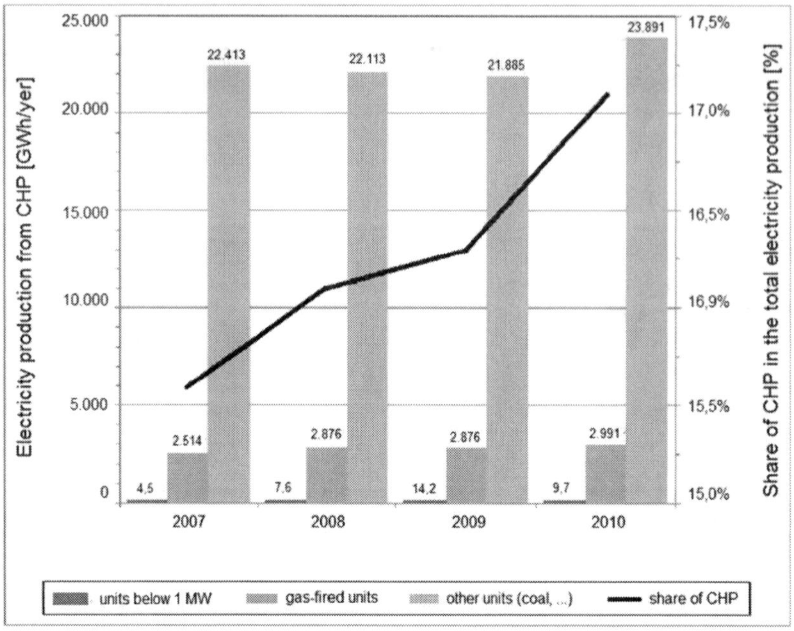

Fig. 4. Electricity production in high-efficiency CHP and share in total electricity production [11]

3 Prospect of cogeneration in Poland

Cogeneration is considered to be one of the key technologies that can help meet the main goals of energy and climate policy in Poland, which include, among others, improving energy efficiency, increasing the use of renewable energy sources and reduce greenhouse gas emissions and dust [1,6, 11]. It seeks to increase the efficiency of energy conversion of fuel used to produce electricity and heat, which is very desirable from both energy, environmental and economic points of view. [1, 13]. The figure 5 illustrated mechanism of obtaining a fuel savings by using cogeneration, and thus indirectly influence the emission of harmful substances into the environment. Government estimates show that the use of cogeneration potential reduce primary energy consumption and emissions (32 Mt CO_2) and thus help achieve the objectives set out in the Climate and Energy Package 2020 [6, 11].

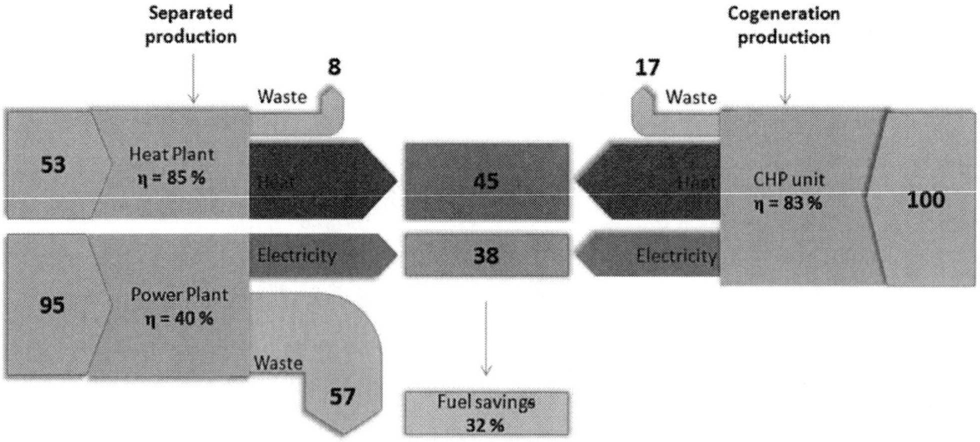

Fig. 5. The mechanism of obtaining a fuel savings in cogeneration

DOI: 10.1051/e3sconf/20171401021

In near future, the energy system in Poland will face with problems of capacities shutdowns, on what will be affected - in addition to the EU climate policy - the age and condition of existing units. In 2013 the vast majority of units generating electricity were power plants working over 30 years - having a total installed capacity of 18 GW which represents 58.65% of the total installed capacity (Figure 6).

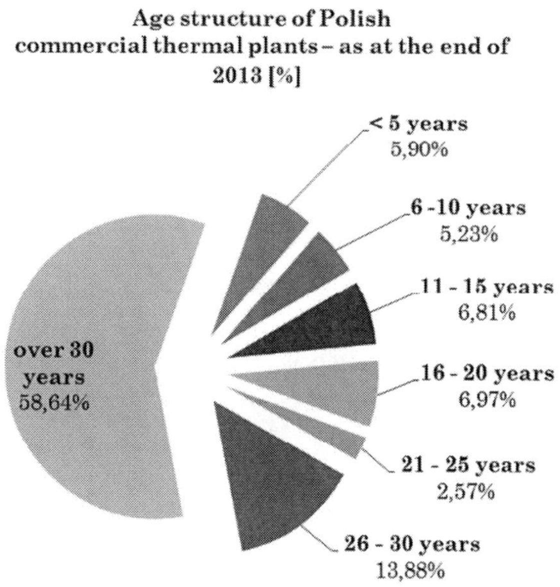

Fig. 6. Age structure of Polish comercial thermal plants – as at the end of 2013 [%] [6]

The scale of the planned shutdowns of currently operating power plants and the estimation of installed capacity in 2030 (depending on the forecast: 46-54 GW) open up huge investment opportunities in new generation capacities. Taking into account the assumptions of Climate and Energy Package (about greenhouse gases emissions and renewable energy), cogeneration and hybrid systems based on renewable energy sources [14] may be a good option for power plants based on fossil energy sources.

The technical potential of heat supply from cogeneration varies depending on the forecast - for 2020 was estimated at approx. 206 PJ. New projections show that Poland has the potential to achieve 12.051 MW, which would increase by more than 4,000 MW compared to 2010 (Figure 7). It is estimated that the level of production of electricity in cogeneration by 2030 will increase from 21 TWh to 48 TWh or 22% of total demand for electricity. Forecasts also indicate that approx. 20% of installed capacity in cogeneration will be from renewable sources [11].

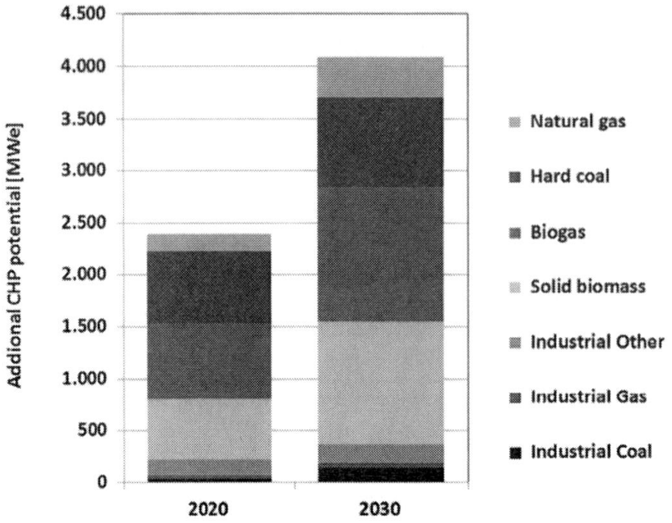

Fig. 7. Additional CHP potential till the year 2030 [11]

4 Conclusion

Cogeneration is a very efficient way to produce electricity and heat. CHP systems to produce a certain amount of electricity and heat need less fuel than if production is from separated units. This influence not only on fuel savings but also on lowering operating cost of such unit and on reduce of greenhouse gases emissions and environmental pollution.

In summary, Poland has a long tradition in the combined production of heat and electricity. Despite this, there is still great potential for new cogeneration units especially considering the future of capacity demand forecast and the necessity of withdrawal of the old, outdated units. The forecasts predict that the share of cogeneration in heat and electricity will be growing. However, greater emphasis will be put on unit cooperating with renewable energy sources (approximately 20% of installed capacity in cogeneration in 2030 will be from renewables).

This work was supported by AGH – University of Science and Technology (Project 11.11.210.216).

References

1. J. Skorek, J. Kalina, *Gazowe układy kogeneracyjne* (Wydawnictwo Naukowo-Techniczne) (2005)
2. PSE SA, *Statement from 10th August 2015 about a threat to the security of electricity supply, taken actions and measures to remove the threat and prevent its negative effects, and the introduction of restrictions on the supply and consumption of electricity*. (2015)
3. *Poland Energy Report*, Enerdata (2012)
4. Ministry of Economy, *Report on results of energy supply security monitoring in the period from 1 January 2011 to 31 December 2012* (2013)
5. Ministry of Economy, *Report on results of energy supply security monitoring in the period from 1 January 2013 to 31 December 2014* (2015)
6. Report of EDF, *Is any divresification needed ? poland's potenial electricity mix in 2030 in the light of internal and external deteminants* (2014)
7. J. Szargut, A. Ziębik, *Podstawy energetyki cieplnej* (PWN Warszawa) (1998)

8. J. Marecki J., Wokół Energetyki (2005)
9. J. Kalina, *Small-scale cogeneration in Poland. 2003 market report.* The work package WP2 (title *Cogeneration and eco-buildings*) of the OPTI_Energy Centre of Excellence, Fifth Framework Programme of the European Commission (2003)
10. J. Lewandowski J., et al., *Program rozwoju kogeneracji w Polsce.* (Uczelniane Centrum Badawcze Energetyki i Ochrony Środowiska) (2010)
11. *CODE2 Cogeneration Observatory and Dissemination Europe D5.1 Final Cogeneration roadmap for Poland* (Jozef Stefan Institute) (2014)
12. A. Buńczyk A., Thermal Power in numbers – 2014 (URE) (2015)
13. M. Sołtysik, K. Mucha-Kuś, Acta Energetica **3/24** (2015), 97-102
14. J. Kiciński, P. Lampart, Acta Energetica 2/2 (2009)

Shortening start-up and an extension of the power unit load range

Jan Taler[1,*], *Marcin* Trojan[1], *Dawid* Taler[2], *Piotr* Dzierwa[1] and *Karol* Kaczmarski[1]

[1]Institute of Thermal Power Engineering, Cracow University of Technology, Cracow, Poland
[2]Institute of Thermal Engineering and Air Protection, Faculty of Environmental Engineering, Cracow, Poland

Abstract. A power plant with additional water pressure tanks was proposed. The maximum rise in the block electric power resulting from the shut-off of low-pressure regenerative heaters was determined. At that time, the boiler is fed with hot water from water pressure tanks acting as heat accumulators. Accumulation of hot water in water tanks is also proposed in the periods of the power unit small load. In order to lower the plant electric power in the off-peak night hours, water heated to the nominal temperature in low-pressure regenerative heaters is directed from the feed water tank to pressure tanks. The water accumulated during the night is used to feed the boiler during the period of peak demand for electricity. Pressure accumulators are proposed to be used for the rapid start-up of the boiler from a cold state. The evaporator of the boiler is filled at the beginning of start-up with hot water from the accumulators. Drops in the power block electric power were determined for different capacities of the tanks and periods when they are charged. The tanks may also be used to ensure a sudden increase in the electric power of the unit that is operating in the automatic system of frequency and power control (in Polish: ARCM).

1 Introduction

Large oscillations occur in electric power generation due to the development of photovoltaic cells, wind farms and other distributed energy sources. Electricity should be supplied as soon as possible to the power grid by the thermal power units, during the power shortage that occurs in the electric power system. A requirement is also imposed on state-of-the-art power units to make it possible to raise or lower the power capacity at the rate of 2÷8% of installed power per minute in the full range of control, i.e. from the minimum to the maximum load. The power unit start-up from different states should also proceed fast.

The manufacturer of modern power units is obliged to carry out the following tests:

- Start-up from the cold state (after 8 hours of unit standstill)
- Start-up from the hot state (after 8 to 50 hours of unit standstill)
- Start-up from the cold state (after more than 50 hours of unit standstill)

The aim this article is an analysis of the possibility of using hot water pressure tanks to improve a 200 MW power unit flexibility.

* Corresponding author: taler@mech.pk.edu.pl

 DOI: 10.1051/e3sconf/20171401022

The water pressure tanks may be used to reduce the power unit power capacity in the off-peak night hours, due to low demand for electricity. In this, the period of low demand for electricity the boiler may operate at the minimum permissible load (at the level of the boiler technical minimum) then the steam is used both to generate electrical energy and to heat up hot water accumulated in the tanks.

In the period of peak demand for electricity, which in Poland is between 4 p.m. and 10 p.m., the power unit power capacity may be raised. Low-pressure regenerative bleeds are then shut off, which involves a rise in the electric power of the block due to a bigger mass flow rate of steam through the turbine. Hot water accumulated in the tanks is used for feed the boiler.

The water pressure tanks may also be used to fill the boiler with hot water at the beginning of the boiler start-up. According to the boiler drum optimum heating method proposed in [1-3], the boiler evaporator together with the drum may be filled with hot water with a temperature much higher than the initial one. Hot water accumulated in the pressure tank may be used to feed the bottom headers of the boiler furnace chamber walls, and the water displaced from the boiler evaporator may be directed to the bottom part of the pressure tank. Owing to such heat accumulators, the boiler start-up from the cold state becomes faster and safer.

Hot water pressure tanks may also be used in the automatic system of frequency and power control (ARMC) to ensure a sudden increase in the electric power of the power unit. It is possible to raise the maximum power capacity of the power unit by more than 7% for several hours feeding the boiler with hot water from the tanks. In order to rise rapidly maximum power capacity, low-pressure regenerative bleeds are shut off and the boiler is fed with hot water from the tanks.

Diagram of the power unit with a hot water storage tank used in the period of peak demand for electricity and low demand for electricity is shown in Figure 1.

a)

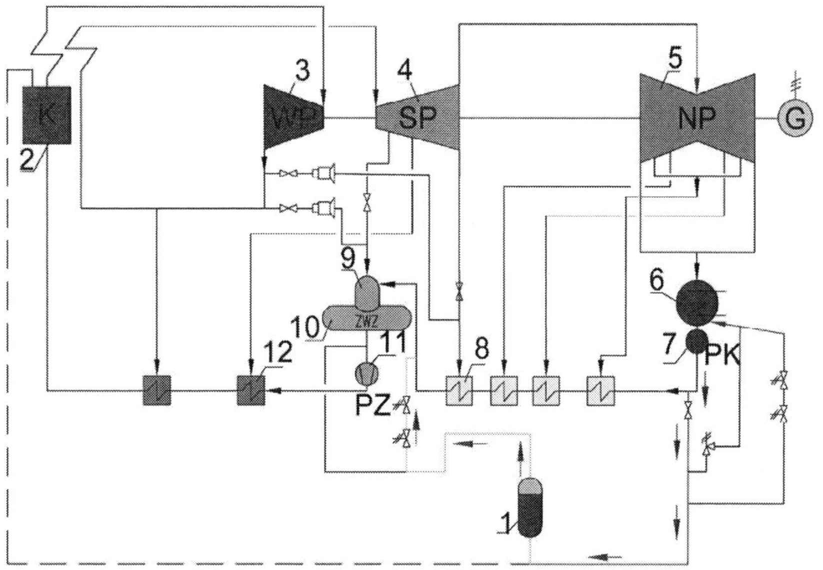

b)

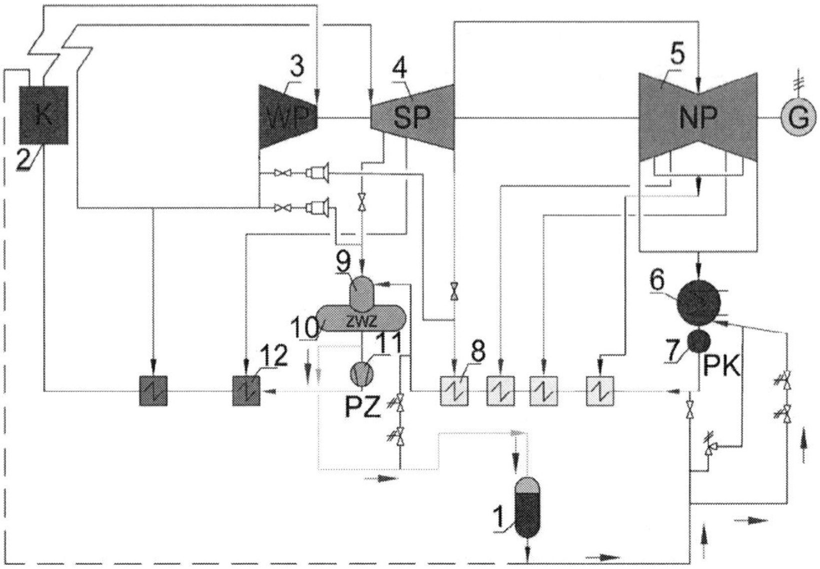

c)

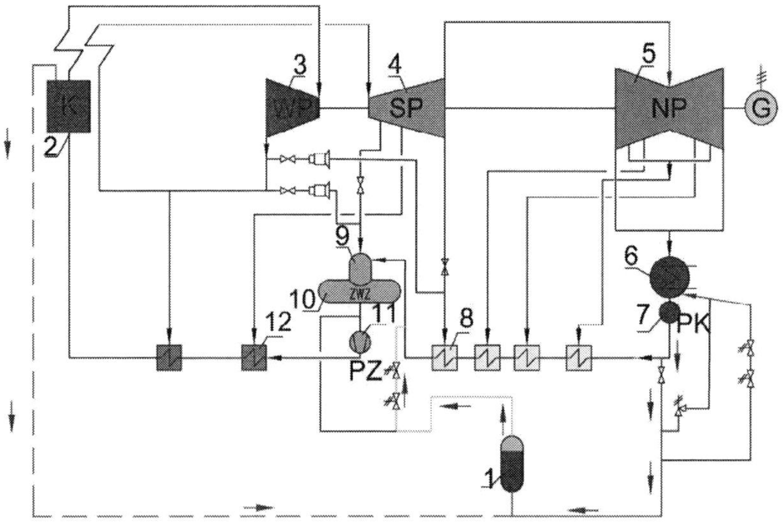

Fig. 1. Diagram of the thermal power plant with storage of hot water; a) rapid increase in the power block, b) reduction of the power block by filling the storage tank with hot water, c) rapid start-up of the boiler from the cold state by filling the boiler with hot water; 1 - heat storage tank, 2 - boiler, 3 - high pressure turbine, 4 - medium pressure turbine, 5 - low pressure turbine, 6 - condenser, 7 - condensate pump, 8 - low-pressure regenerative heaters, 9 - deaerator, 10 - feedwater tank, 11 - feedwater pump,12 - high-pressure regenerative heaters.

In this paper, a mathematical model for an optimum heating the boiler evaporator with hot water is also presented. The power increase in the peak load using the water pressure tank and power reduction during the night by closing turbine bleeds are also calculated.

2 Analysis of using the water tanks to fill the evaporator with hot water

Pressure tanks may be used to flood the boiler evaporator with hot water during the power unit start-up from the cold state (Fig. 1a). Filling the evaporator and then the drum with the initial temperature T_0 with hot water having temperature $T_0+\Delta T_{cz}$ (Fig. 2), the drum heating rate may be reduced in further stages of the start-up. The start-up time is only a bit shorter compared to the time obtained using the Standard PN-EN 12952-3, but in contrast to this standard, circumferential stress on the edge of the hole in the drum-downcomer junction does not exceed the permissible value [1-3].

If the boiler evaporator is heated at rates determined according to Standard PN-EN 12952-3 [4], permissible stress values are exceeded on a part of the hole perimeter [2]. This may quickly give rise to cracks on the downcomer tube hole edge. The method of the boiler start-up proposed in [1-3], where the boiler evaporator is filled with hot water, is safe. Unlike the Standard PN-EN 12952-3, the method allows an abrupt rise in the water temperature in the drum at the beginning of the drum heating process. The water temperature in the boiler drum may be raised even up to 100°C, which makes it possible to reduce the consumption of heavy oil fired in the boiler. In the start-up first phase, the temperature of the water is raised to the nominal temperature of about 315°C from 100°C but not from 20°C to 315°C.

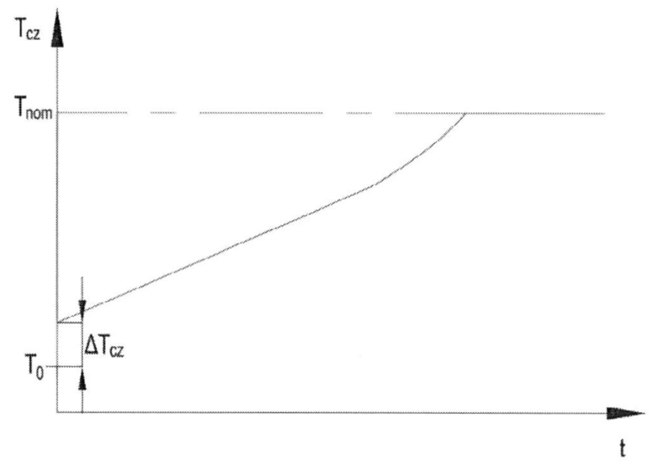

Fig. 2. Proposed optimum change in the temperature of the medium in the drum.

A mathematical model of the boiler evaporator was developed to determine the optimum value of the sudden temperature increase ΔT_{cz} in the feed water supplied to the boiler throguh lower headers of the evaporator so that the circumferential stress on the edge of the hole in the drum does not exceed the allowable stress.

Also, a steady- state mathematical model of the entire power block equipped with hot water storage tanks was developed to determine maximum power capacity increase during the peak loads and power decrease during nights when the demand for electrical energy is low.

3 Mathematical model of the boiler evaporator heating

A scheme of water cooling and metal heating in the evaporator after it is suddenly filled with hot water in power unit with the capacity of 200MW is depicted in Fig. 3.

In order to simulate heating boiler evaporator, mathematical models of water walls and downcomers were developed. The cold air flows through the combustion chamber during evaporator heating. The outer surfaces of the downcomers are thermally insulated.

A differential form of the energy conservation equation for water flowing through the evaporator tubes and the energy conservation equation for the evaporator tube wall were derived first. After the formulation of all balance equations, a system of ordinary differential equations was solved using the finite difference method.

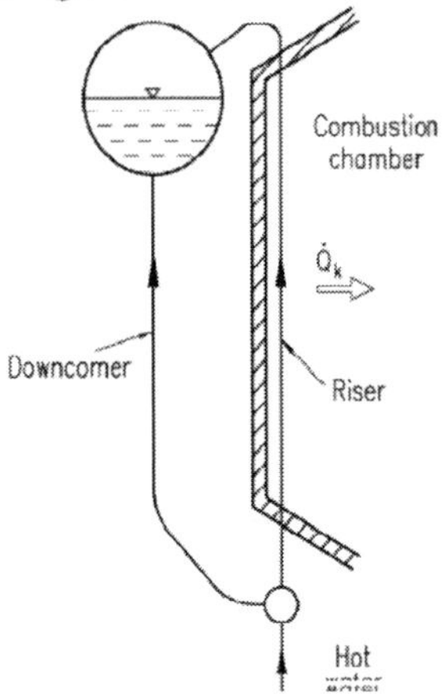

Fig. 3. Scheme of water cooling and metal heating in the evaporator after it is suddenly filled with hot water.

3.1 Energy balance equation for water flowing through the evaporator and the riser tubes

The energy balance equation for water flowing in the evaporator tube is

$$\tau_c \frac{\partial T_c}{\partial t} + \frac{1}{N_c} \frac{\partial T_c}{\partial s^+} = -(T_c - T_w) \tag{1}$$

where the number of transfer units for water N_c and time constant of water τ_c are

$$N_c = \frac{h_w \pi d_w L_s}{\dot{m}_c c_c} \text{ and } \tau_c = \frac{d_w}{4h_w} \rho_c c_c \tag{2}$$

It was assumed that at the time $t = 0$ the temperature of water in the evaporator tubes is constant and equal to ambient temperature T_{ot}, i.e. the initial condition is

$$T_c\big|_{t=0} = T_{ot} \tag{3}$$

Hot water with temperature T_{cg} flows into the riser tubes through lower headers to the boiler waterwalls and the boundary condition is of the following form

$$T_c\big|_{s^+=0} = T_{cg} \tag{4}$$

3.2 Energy balance equation for waterwall tubes

The energy balance equation derived for a wall was derived assuming that the wall is an element with a lumped thermal capacity

$$\tau_{we}\frac{dT_w}{dt}+T_w=\frac{2r_w h_w T_c + r_z h_{zr} T_{pow}}{2r_w h_w + r_z h_{zr}} \tag{5}$$

where the time constant is given by

$$\tau_{we}=\frac{\left(r_z^2-r_w^2\right)\rho_w c_w}{2r_w h_w + r_z h_{zr}} \tag{6}$$

The value of the waterwall tube time constant (6) is highly dependent on the tube wall thickness and on the weighted heat transfer coefficient h_{zr} on the air side, which is much lower than the heat transfer coefficient h_w on the water side.
The initial condition for Equation (5) has the following form

$$T_w\big|_{t=0}=T_{ot} \tag{7}$$

3.3 Energy balance equation for air in the combustion chamber

The energy balance equation for the air flowing upwards through the combustion chamber is

$$\tau_p\frac{\partial T_p}{\partial t}+\frac{1}{N_p}\frac{\partial T_p}{\partial s^+}=\left(T_w-T_p\right) \tag{8}$$

where the number of transfer units for air N_p and time constant of air τ_p are

$$N_p=\frac{n_r \pi\, r_z h_{zr} L_s}{\dot{m}_p\, c_p} \text{ and } \tau_p=\frac{A_k\, \rho_p\, c_p}{n_r \pi\, r_z h_{zr}} \tag{9}$$

The initial and the boundary conditions are, respectively

$$T_p\big|_{t=0}=T_{ot} \text{ and } T_p\big|_{s^+=0}=T_{ot} \tag{10}$$

Similar energy conservation equation with appropriate boundry and initial conditions were formulated for downcomer tubes. The finite difference method was used to determine the water and air temperature as well as the temperature of the tube walls.

4 Determination of the power capacity for the power unit while using heat storage tanks

Mass flow rates and temperature of the water and the steam were determined from solving a system of equations of mass and energy balance set for all components of the thermal power plant. An example the control region for the boiler is shown in Fig. 4.
Figure 4 presents simplified diagram of the boiler with intake of injection water from outside the control region boundary used to write the mass and energy balance equations.

DOI: 10.1051/e3sconf/20171401022

The expressions for calculating the power of the turbine set in the night off-peak are given below:

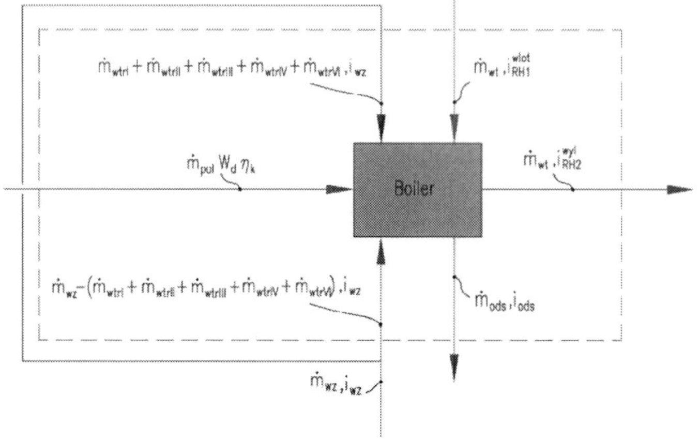

Fig. 4. Simplified diagram of the boiler

- internal power generated in a high-pressure part of the turbine set

$$\dot{Q}_{wp} = \dot{m}_{ps}\left(i_{SH5}^{wyl} - i_I\right) + \left(\dot{m}_{ps} - \dot{m}_I\right)\left(i_I - i_{II}\right) \tag{11}$$

- internal power generated in an intermediate-pressure part of the turbine set

$$\begin{aligned}\dot{Q}_{sp} &= \dot{m}_{sp}\left(i_{sp} - i_{III}\right) + \left(\dot{m}_{sp} - \dot{m}_{III}\right)\left(i_{III} - i_{IV}\right) + \left(\dot{m}_{sp} - \dot{m}_{III} - \dot{m}_{IV}\right)\left(i_{IV} - i_V\right) + \\ &+ \left(\dot{m}_{sp} - \dot{m}_{III} - \dot{m}_{IV} - \dot{m}_V\right)\left(i_V - i_{VI}\right)\end{aligned} \tag{12}$$

- internal power generated in a low-pressure part of the turbine set

$$\dot{Q}_{np} = \dot{m}_{np}\left(i_{np} - i_{VII}\right) + \left(\dot{m}_{np} - \dot{m}_{VII}\right)\left(i_{VII} - i_{znp}\right) \tag{13}$$

- the total internal power of the turbine set

$$\dot{Q}_t = \dot{Q}_{wp} + \dot{Q}_{sp} + \dot{Q}_{np} \tag{14}$$

- the relative drop in the power unit related to the design power $\dot{Q}$

$$\dot{Q}_{proc} = \frac{100\dot{Q}_t}{\dot{Q}} \tag{15}$$

- drop in power during the night off-peak hours

$$S_M = 100 - \dot{Q}_{proc} \tag{16}$$

The drop in the boiler output and in the power unit power in the night off-peak hours duringthe filling of the tank is shown Table 1.

Table 1. List of results – the drop in the boiler output and in the power unit power in the night off-peak hours during the filling of the tank (power capacity 206 MW).

Item	Number of tanks	Time of filling the tanks during the night off-peak hours, h	Water mass flow to tanks, t/h	Boiler output during the filling of tanks, t/h	Relative drop in power, %
1	2	3	108.46	607.38	2.29
2		4	81.34	610.51	1.79
3		5	65.07	612.51	1.46
4		6	54.23	613.89	1.24
5	4	3	216.91	596.99	3.96
6		4	162.67	601.80	3.19
7		5	130.15	605.05	2.66
8		6	108.46	607.38	2.29
9	8	3	433.83	583.03	6.21
10		4	325.37	589.13	5.22
11		5	260.30	593.60	4.51
12		6	216.91	596.99	3.96
13	12	3	650.75	574.21	7.63
14		4	488.06	580.46	6.62
15		5	390.45	585.30	5.84
16		6	325.37	589.13	5.22

5 Summary

The results of the performed calculations and analyses indicate that installation of water pressure tanks is well justified. The investment is profitable. Water pressure tanks may not only be used to reduce the power unit power during the off-peak night hours and raise it in the periods of peak demand, but also to increase the power capacity fast at any time. Hot water tanks may also be used to fill the boiler with hot water at the beginning of the boiler start-up. According to the boiler drum optimum heating method proposed in [1-3], the boiler evaporator together with the drum may be filled with hot water at a temperature much higher than the initial one. The pressure hot water tank may be used to feed the bottom headers of the boiler furnace water walls. The cold water displaced from the boiler evaporator may be directed to the bottom part of the pressure tank. Owing to such heat accumulators, the boiler start-up from the cold state becomes faster and safer.

The results of the analysis of the 200 MW power unit thermal system and of the thermal calculations lead to the following conclusions:

1. The power unit maximum power capacity of 206 MW may be raised in the period of peak demand for electricity by 7.32% (0.0732*206= 15 MW)
2. The period of the power unit operation with the raised maximum power capacity may last:
   - 30 minutes at two tanks with the total capacity of 360.5 m^3
   - 60 minutes at four tanks with the total capacity of 721 m^3
   - 120 minutes at eight tanks with the total capacity of 1442 m^3
   - 180 minutes at twelve tanks with the total capacity of 2163 m^3

3. The maximum reduction in the power unit power capacity (which is 135 MW before the activation of the heat storage tanks) during the night off-peak hours is 16.27% (0.1627*135=21.96 MW) at eight tanks with the total capacity of 1442 m^3. The period in which the tanks are charged, and consequently the time of the power unit operation under a reduced load, is 3 hours.
4. If eight water tanks with the total capacity of 1442 m^3 are charged for 6 hours, the reduction in the power unit power capacity is 8.15% (0.0815*135=11MW). If the water pressure tanks are discharged during 2 hours in the period of peak demand for electricity, the increment in the power unit power capacity totals 7.32% (15 MW). This solution, which anticipates installation of 8 tanks charged for 6 hours during the night off-peak period and discharged for 2 hours in the peak period, is recommended to be used in practice.
5. The tanks may be made as vertical structures with stratification of cold and hot water and operating in a parallel system. The hot water tanks may also be horizontal. In this case, they operate as a series-parallel system.
6. Hot water pressure tanks may also be used in power units participating in the automatic system of frequency and power control (ARMC) to ensure a sudden increase in the power unit power at any time of the day.

In order to shorten the start-up from the cold state, the boiler evaporator may be flooded with hot water from one of the pressure water tanks.

Nomenclature

c	– specific heat capacity at constant pressure, J/(kg·K)
d	– diameter, m,
D	– diameter, m,
d_w	– inner diameter of the tube, m
G	– generator,
h	– heat transfer coefficient ,W/(m^2·K)
i	– specific fluid enthalpy, kJ/kg,
i_I	– enthalpy of bleed I steam, kJ/kg
i_{II}	– enthalpy of bleed II steam, kJ/kg
i_{III}	– enthalpy of bleed III steam, kJ/kg
i_{IV}	– enthalpy of bleed IV steam, kJ/kg
i_V	– enthalpy of bleed V steam, kJ/kg
i_{VI}	– enthalpy of bleed VI steam, kJ/kg
i_{VII}	– enthalpy of bleed VII steam, kJ/kg
$i_{RH_1}^{wlot}$	– steam specific enthalpy upstream the steam reheater stage I, kJ/kg,
$i_{RH_2}^{wyl}$	– steam specific enthalpy downstream the steam reheater stage II, kJ/kg,
i_{np}	– enthalpy of the steam flow to the turbine low-pressure part, kJ/kg
i_{ods}	– specific enthalpy of brine, kJ/kg,
i_{SH5}^{wyl}	– specific enthalpy of the boiler live steam, kJ/kg,
i_{sp}	– enthalpy of the steam flow to the turbine intermediate-pressure part, kJ/kg

 DOI: 10.1051/e3sconf/20171401022

i_{wz} – specific enthalpy of feed water, kJ/kg,
i_{znp} – specific enthalpy of the steam flow from the low-pressure turbine part, kJ/kg,
L_s – tube length, m
n – number of tanks,
N_c, N_p – number of transfer units for water and air, respectively
r_w, r_z – inner and outer tube radius, m
s^+ – dimensionless curvilinear coordinate, m
t – time, s
T_c, T_p, T_w – temperature of water, air, and riser tube wall, respectively, °C,
$\dot{Q}$ – heat flow rate, MW,
W_d – fuel calorific value, kJ/kg,
$\dot{m}$ – fluid mass flow rate, kg/s,
$\dot{m}_I, \dot{m}_{II}, \dot{m}_{III}, \dot{m}_{IV}, \dot{m}_V, \dot{m}_{VI}, \dot{m}_{VII}$ – steam mass flow from bleed: I, II, III, IV,V, VI, VII,kg/s
$\dot{m}_{np}$ – steam mass flow to the turbine low-pressure part, kg/s
$\dot{m}_{ps}$ – boiler output (live steam mass flow at the boiler outlet), kg/s
$\dot{m}_{sp}$ – steam mass flow to the turbine intermediate-pressure part, kg/s

Greek symbols

η_k – boiler efficiency.
ρ – density , kg/m^3
τ_c, τ_p, τ_{we} – time constant of water, air, and riser tube, respectively, s

References

1. J. Taler, P. Dzierwa, D. Taler, P. Harchut, Energy, **92**, 1 (2015) 160–170
2. P. Dzierwa, D. Taler, J. Taler, M. Trojan, "Optimum heating of thick wall pressure components of steam boilers", ASME 2014 Power Conference POWER2014, July 28-31, 2014, Baltimore, Maryland, USA, p. 1-9
3. P. Dzierwa, J. Taler, J Press Vess-T ASME, **137** (2014) 011202-1-8
4. Polish Standard PN-EN 12952-3, "Kotły wodnorurowe i urządzenia pomocnicze. Część 3: Konstrukcja i obliczenia części ciśnieniowych [Water-tube boilers and auxiliary installations. Part 3: Design and calculation for pressure parts.], PKN [Polish Standardization Committee], Warszawa 2004

Mathematical modeling of unsteady response of plate and fin heat exchanger to sudden change in liquid flow rate

Anna Korzeń[1,*], and *Dawid* Taler[2]

[1]Cracow University of Technology, Faculty of Mechanical Engineering, Institute of Thermal Power Engineering, Al. Jana Pawła II 37, 31-864 Cracow, Poland
[2]Cracow University of Technology, Faculty of Environmental Engineering, Institute of Thermal Engineering and Air Protection, Ul. Warszawska 24, 31-155 Cracow, Poland

Abstract. A mathematical simulation of plate fin and tube heat exchanger will be presented. The transient operation of the heat exchanger was carried out using general numerical model developed in [1]. Reynolds number of the water flowing inside the tubes varied in the range from 4000 to 12000. A detailed analysis of transient response was modeled for growth in the water volume flow in time.
At first, heat transfer correlations for air and water were determined based on the experimental data. Unknown parameters appearing in the relationships for the Nusselt numbers on the air and water-sides were estimated using the least squares method. The power-type form of the relationship for the air-side Nusselt number was used while two correlations of different form were chosen for the water-side Nusselt number. The first is the exponential correlation, and the second form is similar to the relationships of Petukhov-Kirillov and Gnielinski. These correlations were used in the mathematical model of the heat exchanger for the simulation of its transient operation.
The results of the numerical simulations of a heat exchanger using experimentally determined air and water-side heat transfer formulas for calculation of heat transfer coefficient were compared with the experimental data. Very good agreement of computation results (i.e. air and water temperature at the outlet of the heat exchanger) with the experimental data was obtained.

1 Introduction

Publications on the dynamics on the plate fin and tube heat exchangers (PFTHE) are not numerous. Mathematical models of heat exchangers to simulate the non-steady state operation are needed to analyze heat exchangers startup and shutdown, and in PID control systems as well as in control systems using mathematical models of the controlled heat exchanger.

However, most of the works refers to a steady state of these exchangers [2].

* Corresponding author: korzen@mech.pk.edu.pl

The transient response of PFTHE was modeled in [1,3-6]. Usually, fins are modeled as elements with lumped thermal capacity while the equivalent heat transfer coefficient on the air side is calculated as in the steady-state [1,3-6]. The steady-state temperature distribution in fins is used to calculate the equivalent heat transfer. This simplification can lead to some errors, especially for higher fins.

The system of the governing differential equations is solved by the Laplace transform or the finite difference method. The Laplace transform method is widely used by Roetzel and Xuan [3] to model the transient operation of various heat exchangers. Taler [5] modeled the transient response of the PFTHE for a stepwise increase in liquid or gas temperature using the Laplace transform method. The transient temperature of the fluids was compared with temperature obtained by the finite volume method.

Korzeń and Taler [1] developed a new mathematical model of the PFTHE to simulate its transient operation. In contrast to the existing methods for the modeling transient response of heat exchangers with extended surfaces, in which the weighted steady-state heat transfer coefficient on the finned tube side is used, the transient temperature distribution will be calculated in each fin. This allows for computing more exactly the heat flow rate from the fins to the flowing gas.

Usually, modeling PFTHEs assumed that the fluid flow inside the tubes is turbulent. However, many low-duty heat exchangers operate in the transitional region when the Reynolds number varies in the range of 2 300 to about 12 000.

In this paper, numerical modeling of transient work of a PFTHE for low Reynolds numbers on the liquid side will be carried out.

2 Mathematical formulation of the problem

The system of partial differential equations describing the space and time changes of liquid T_1, tube wall T_w, and air T_2 temperatures in one-row plate-fin and tube heat exchanger are [1]:

- liquid

$$\frac{1}{N_1}\frac{\partial T_1}{\partial x^+}+\tau_1\frac{\partial T_1}{\partial t}=-\left(T_1-T_w\right) \tag{1}$$

- tube wall

$$U_m\delta_w\rho_w c_w\frac{\partial T_w}{\partial t}+\frac{m_f c_f}{s}\frac{\partial \overline{T}_f}{\partial t}=\frac{k_w U_m\delta_w}{L_{ch}^2}\frac{\partial^2 T_w}{\partial\left(x^+\right)^2}+h_w U_{in}\left(T_1-T_w\right)+h_o U_{bo}\left(\overline{T}_2-T_w\right) \tag{2}$$

- air

$$\frac{1}{N_2}\frac{\partial T_2}{\partial y^+}+\tau_2\frac{\partial T_2}{\partial t}=T_w-T_2 \tag{3}$$

where $\overline{T_2}$ denotes the mean air temperature over the row thickness, defined as

$$\overline{T_2}\left(x^+,t\right)=\int_0^1 T_2\left(x^+,y^+,t\right)dy^+ \tag{4}$$

The symbols $x^+ = x/L_{ch}$ and $y^+ = y/p_2$ in equations (1-3) stand for dimensionless coordinates. The numbers of heat transfer units N_1 and N_2 are given by

$$N_1 = \frac{h_1 A_{in}}{\dot{m}_1 c_{p1}} , \qquad N_2 = \frac{h_2 A_{bo}}{\dot{m}_2 c_{p2}} \tag{5}$$

where: $s = L_{ch}/n_f$, $A_{in} = n_r U_{in} L_{ch}$, $A_{bo} = n_r U_{bo} L_{ch}$.

The time constants τ_1 and τ_2 are

$$\tau_1 = \frac{m_1 c_{p1}}{h_1 A_{in}} , \qquad \tau_2 = \frac{m_2 c_{p2}}{h_2 A_{bo}} . \tag{6}$$

The symbols in equations (1-6) denote:

$$m_1 = n_r A_{in} L_{ch} \rho_1 , \qquad m_2 = n_r \left(p_1 p_2 - A_{oval} \right)\left(s - \delta_f \right) n_f \rho_2 , \qquad m_w = n_r U_m \delta_w L_{ch} \rho_w$$

$$m_f = \left(p_1 p_2 - A_{oval} \right) \delta_f \rho_w , \quad A_{in} = \pi ab , \; A_{oval} = \pi \left(a + \delta_w \right)\left(b + \delta_w \right), U_m = \left(U_{in} + U_{bo} \right)/2 .$$

The subscript w refers to the wall, f to the fin, and m to the mean value.
The weighted heat transfer coefficient h_o is defined by

$$h_o\left(t\right) = h_2\left(t\right)\left[\frac{A_{bbf}}{A_{bo}} + \frac{A_f}{A_{bo}} \cdot \eta_f\left(h_2, t\right) \right] \tag{7}$$

The initial temperatures of both fluids are equal and amount to T_0. The initial conditions are

$$T_1\left(x^+, t\right)\Big|_{t=0} = T_{1,0}\left(x^+\right) \tag{8}$$

$$T_w\left(x^+, t\right)\Big|_{t=0} = T_{w,0}\left(x^+\right) \tag{9}$$

$$T_2\left(x^+, y^+, t\right)\Big|_{t=0} = T_{2,0}\left(x^+, y^+\right) \tag{10}$$

The boundary conditions have the following form

$$T_1\left(x^+, t\right)\Big|_{x^+=0} = f_1\left(t\right) \tag{11}$$

$$T_2\left(x^+, y^+, t\right)\Big|_{y^+=0} = f_2\left(t\right) \tag{12}$$

$$\left.\frac{\partial T_w}{\partial x}\right|_{x=0} = 0 \tag{13}$$

$$\left.\frac{\partial T_w}{\partial x}\right|_{x=L_{ch}} = 0 \tag{14}$$

where $f_1(t)$ and $f_2(t)$ are functions of time describing the variation of the liquid and air temperature at the inlets to the exchanger.

 DOI: 10.1051/e3sconf/20171401023

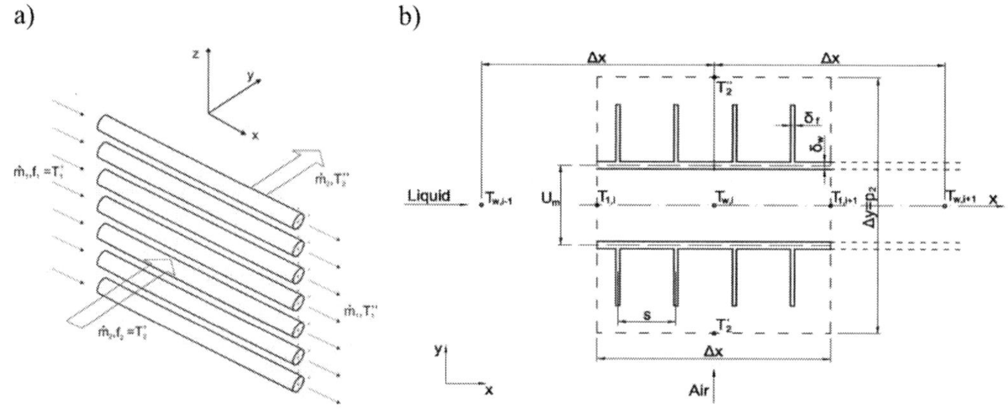

Fig. 1. Scheme of the one-row heat exchanger (a) and control volume (b).

The fin efficiency appearing in the relationship (7) was calculated using the Finite Volume Method based on the Finite Element Method. The division of the fin into finite volumes with the number of nodes is shown in Fig. 2.

The initial-boundary value problem defined by equations (1-14) applies to heat exchangers made of bare or finned tubes. For bare tubes m_f is equal to zero because there are no fins.

The transient temperature of the fluids and tube wall in the one-row heat exchanger (Fig. 1) were determined by the explicit finite difference method. To calculate time-dependent efficiency η_f of the rectangular fin attached to an oval tube the finite volume – finite element method (FVM-FEM) was used. Based on the mathematical model of the one-row heat exchanger, a mathematical model of the whole heat exchanger with a complex flow system was built.

The automotive radiator for the spark-ignition combustion engine with a capacity of 1580 cm^3 is a double-row, two-pass plate-finned heat exchanger. The radiator consists of aluminium tubes of oval cross-section. The cooling liquid flows in parallel through both tube rows.

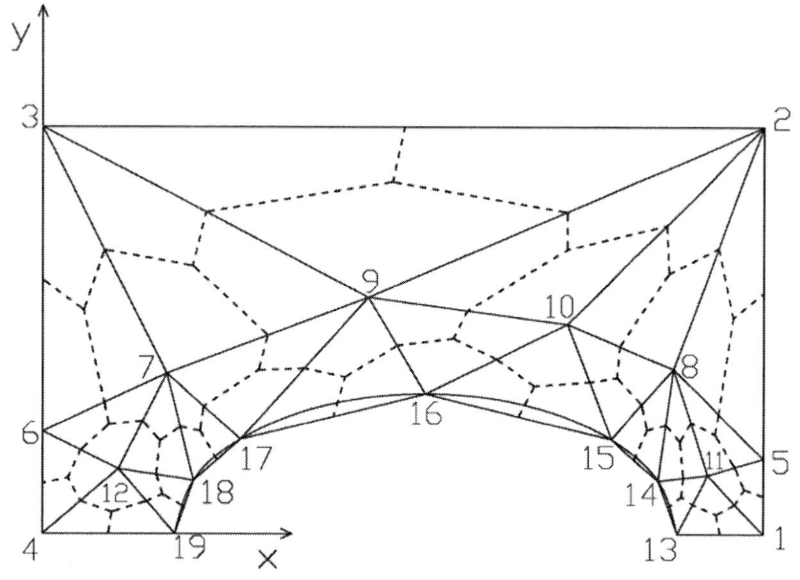

Fig. 2. Division of the fin model into finite volumes with node numbers.

3 Experimental correlations for Nusselt numbers at water and air side

In the developed mathematical model of the exchanger, the heat transfer coefficients on the air and water side are calculated from experimental correlations. The correlations for the heat transfer coefficient were derived assuming a fully developed turbulent flow in straight tubes. One of the most common correlations is the Dittus-Boelter formula [7-9]:

$$\mathrm{Nu}_w = 0{,}023\,\mathrm{Re}_w^{0,8}\,\mathrm{Pr}_w^n\left[1+\left(\frac{d_r}{L_{ch}}\right)^{2/3}\right]. \tag{15}$$

where $n = 0.4$ when the fluid is heated and $n = 0.3$ when it is cooled. In recent years, the Gnielinski correlation is becoming more and more popular [10-11].

$$\mathrm{Nu}_w = \frac{(\xi/8)(\mathrm{Re}_w - 1000)\mathrm{Pr}_w}{(1+12{,}7\sqrt{(\xi/8)}\left(\mathrm{Pr}_w^{2/3}-1\right)}\left[1+\left(\frac{d_r}{L_{ch}}\right)^{2/3}\right], \tag{16}$$

where the friction loss coefficient ξ is defined as:

$$\xi = \frac{1}{(1{,}82\log \mathrm{Re}_w - 1{,}64)^2} = \frac{1}{(0{,}79\ln \mathrm{Re}_w - 1{,}64)^2}. \tag{17}$$

In Formula (16), symbols d_r and L_{ch} denote the tube hydraulic diameter and length, respectively. The reason for the growing popularity of the Gnielinski formula (16) is its greater accuracy in the transition area of $2300 < \mathrm{Re}_w < 10000$. Despite the fact that the Gnielinski proposal gives smaller values of the heat transfer coefficient in the transition area compared to the Dittus-Boelter formula, these values are still inflated [12-13]. If $\mathrm{Re}_w < 2300$, the flow is laminar; the Nusselt number is $\mathrm{Nu}_w = 4.364$ and does not depend on the Reynolds number [7-8]. It can be checked easily that both the Dittus-Boelter and the Gnielinski formulae give values of Nuc much higher than 4.364. For this reason, the correlations for the Nusselt numbers on the water and air side were determined based on experimental testing.

The following form of the correlation for the Nusselt number at the air side was taken:

$$\mathrm{Nu}_a = x_1\,\mathrm{Re}_a^{x_2}\,\mathrm{Pr}_a^{1/3}. \tag{18}$$

Considering that the range of changes in the Reynolds number at the air side is not too big, it is possible to approximate experimental data well using the exponential equation written above (18).

The correlation for the Nusselt number at the water side was taken in a form similar to the formulae developed by Petukhov and Kirillov [14] and Gnielinski [10].

$$\mathrm{Nu}_w = \frac{(\xi/8)(\mathrm{Re}_w - x_3)\mathrm{Pr}_w}{(1+x_4\sqrt{(\xi/8)}\left(\mathrm{Pr}_w^{2/3}-1\right)}\left[1+\left(\frac{d_r}{L_{ch}}\right)^{2/3}\right] \tag{19}$$

The values of coefficients $x_1, x_2, ..., x_m$ will be selected so that the sum of the squares of differences in temperature

$$S=\sum_{i=1}^{n}\left[f''_{w,i}-T''_{w,i}\left(x_1,...,x_m\right)\right]^2=\min \tag{20}$$

$$f''_{w,i}=T'_{w,i}-\varDelta T_{w,i} \tag{21}$$

should reach the minimum.

The real values of the determined parameters $\tilde{x}_1,...,\tilde{x}_m$ are included with probability $P=(1-\alpha)\cdot 100$ [%] in the following intervals [15-16]:

$$x_i-t_{n-m}^{\alpha/2}\cdot s_t\sqrt{c_{ii}}\le\tilde{x}_i\le x_i+t_{n-m}^{\alpha/2}\cdot s_t\sqrt{c_{ii}}\ , \tag{22}$$

where

x_i – parameter value determined by means of the least squares method,

$t_{n-m}^{\alpha/2}$ – t-Student distribution quantile for the confidence level $1-\alpha$ and $n-m$ degrees of freedom.

Quantity s_t^2 is the variance estimate calculated from:

$$s_t^2=\frac{\sum_{i=1}^{n}\left[T''_{w,i}\left(x_1,...,x_m\right)-f''_{w,i}\right]^2}{n-m}=\frac{S\left(\tilde{x}_1,...,\tilde{x}_m\right)}{n-m}. \tag{23}$$

Where n – number of measuring points and m – number of sought parameters.

Quantities c_{ii} are diagonal elements c_{ii} of the covariance matrix.

In this case, the number of measuring points is n = 47, and the number of unknown parameters is m = 4.

3.1 Experimental correlations

The thermal and flow measurements of the exchanger were performed for different air and water flow velocities. The measurement results are presented in [17].

The Reynolds number values at the air side: $\mathrm{Re}_a=w_{\max}d_h/\nu_a$ and at the water side: $\mathrm{Re}_{wg}=w_{wg}d_r/\nu_w$. The air physical properties were calculated at the air average temperature. Similarly, the thermophysical properties of water were calculated at the average temperature. The average temperature is understood as the arithmetic mean of the average temperature at the exchanger inlet and outlet. The water velocity in the upper pass w_{wg} is smaller than the water velocity in the lower pass w_{wd} because in the upper pass water flows in parallel through 20 tubes whereas in the lower one – through 18.

Based on the data presented in Table 1, the following values of individual coefficients were obtained:

$x_1=0{,}0713\pm0{,}0053$; $x_2=0{,}7055\pm0{,}0136$; $x_3=0{,}4624\pm0{,}3194$; $x_4=22{,}2273\pm0{,}3163$

The coefficients are present in Formulae (18) and (19).

The value of the sum of squares: S = 1.07041 K^2.

The correlations for the Nusselt numbers on the water and air side have the following form:

$$\mathrm{Nu}_a=0{,}0713\,\mathrm{Re}_a^{0{,}7055}\,\mathrm{Pr}_a^{1/3}\ ,\qquad 60\le\mathrm{Re}_a\le351 \tag{24}$$

 DOI: 10.1051/e3sconf/20171401023

$$Nu_w = \frac{(\xi/8)(Re_w - 0,4624)Pr_w}{(1+22,2273\sqrt{(\xi/8)}(Pr_w^{2/3}-1))}\left[1+\left(\frac{d_r}{L_{ch}}\right)^{2/3}\right], \qquad 3850 \le Re_w \le 11317 \quad (25)$$

In the mathematical model of the radiator operation under unsteady-state conditions the relationships (24) and (25) were used.

3.2 Increase in water mass flow in time

The comparison of the results obtained from measurements and calculations is presented in Fig. 3.

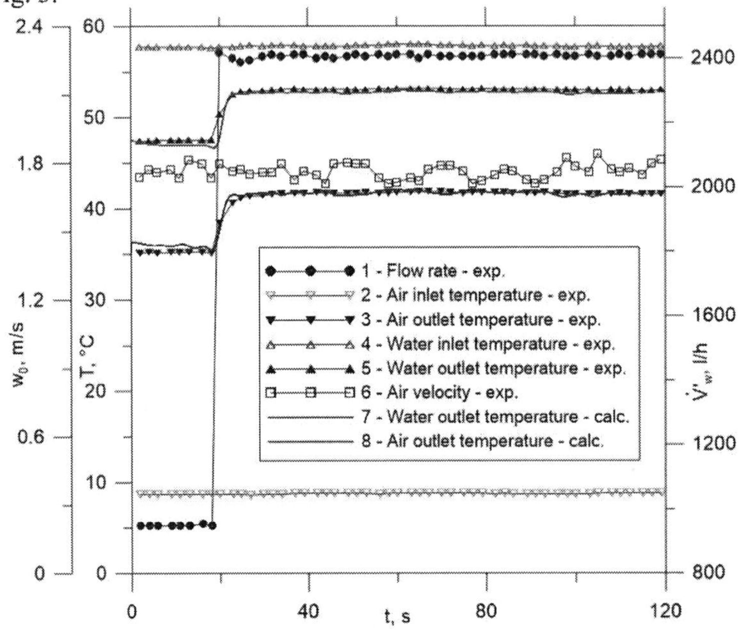

Fig. 3. The exchanger unsteady-state operation caused by a sudden reduction in the water volume flow;

1 – $\dot{V}'_w$ – water volume flow at the exchanger inlet, l/s

2 – T'_{am} – average air temperature before the exchanger, °C

3 – T''_{am} – average air temperature after the exchanger, °C

4 – T'_w – water temperature before the exchanger, °C

5 – T''_w – water temperature after the exchanger, °C

6 – w_0 – air velocity before the exchanger, m/s

7 – T''_w – calculated water temperature after the exchanger, °C

8 – T''_{am} – calculated average air temperature after the exchanger, °C.

The increase in the water volume flow causes a rise in the water and air temperatures at the exchanger outlet (Fig. 3). The increment in the air temperature is bigger than the increment in the water temperature. This results from the smaller flux of the air heat capacity $\dot{m}_a c_{pa}$ compared to the heat capacity flux of water $\dot{m}_w c_w$. If the heat flux given up by water is the same as the one absorbed by air, the increment in the air temperature is higher.

Fig. 4 presents changes in the water and the tube wall temperatures on the exchanger length in time $t = 119$ s (Fig. 3).

 DOI: 10.1051/e3sconf/20171401023

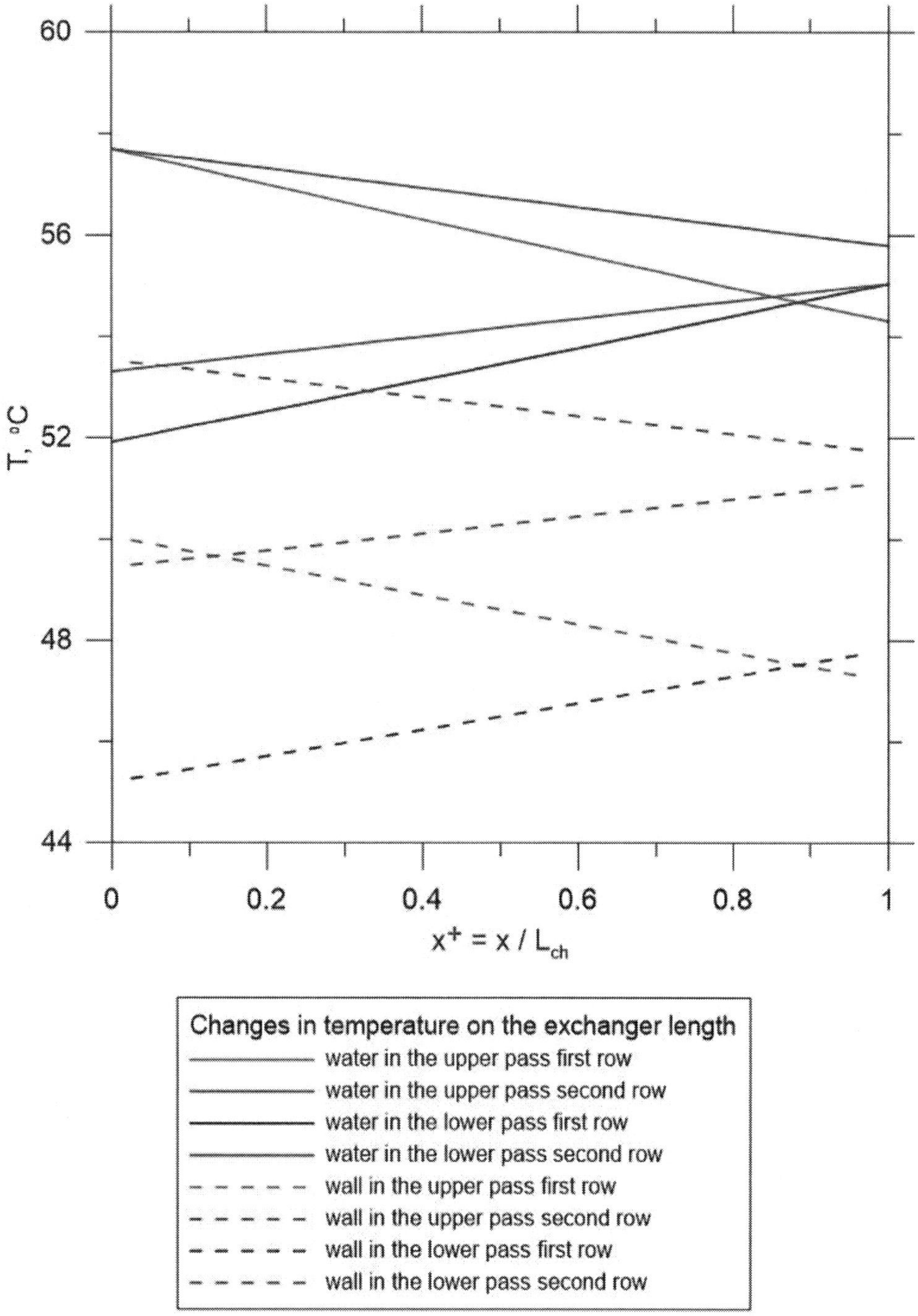

Fig. 4. Distribution of the water and the tube wall temperatures on the exchanger length for time t = 119 s in the steady state.

Analysing the results presented in Fig. 4, it can be seen that water in the first row tubes cools down more than in the tubes placed in the second row. This concerns both the upper and lower pass. It can be noticed that the tube wall temperature is lower than the water temperature due to the fact that heat is absorbed by cold air.

DOI: 10.1051/e3sconf/20171401023

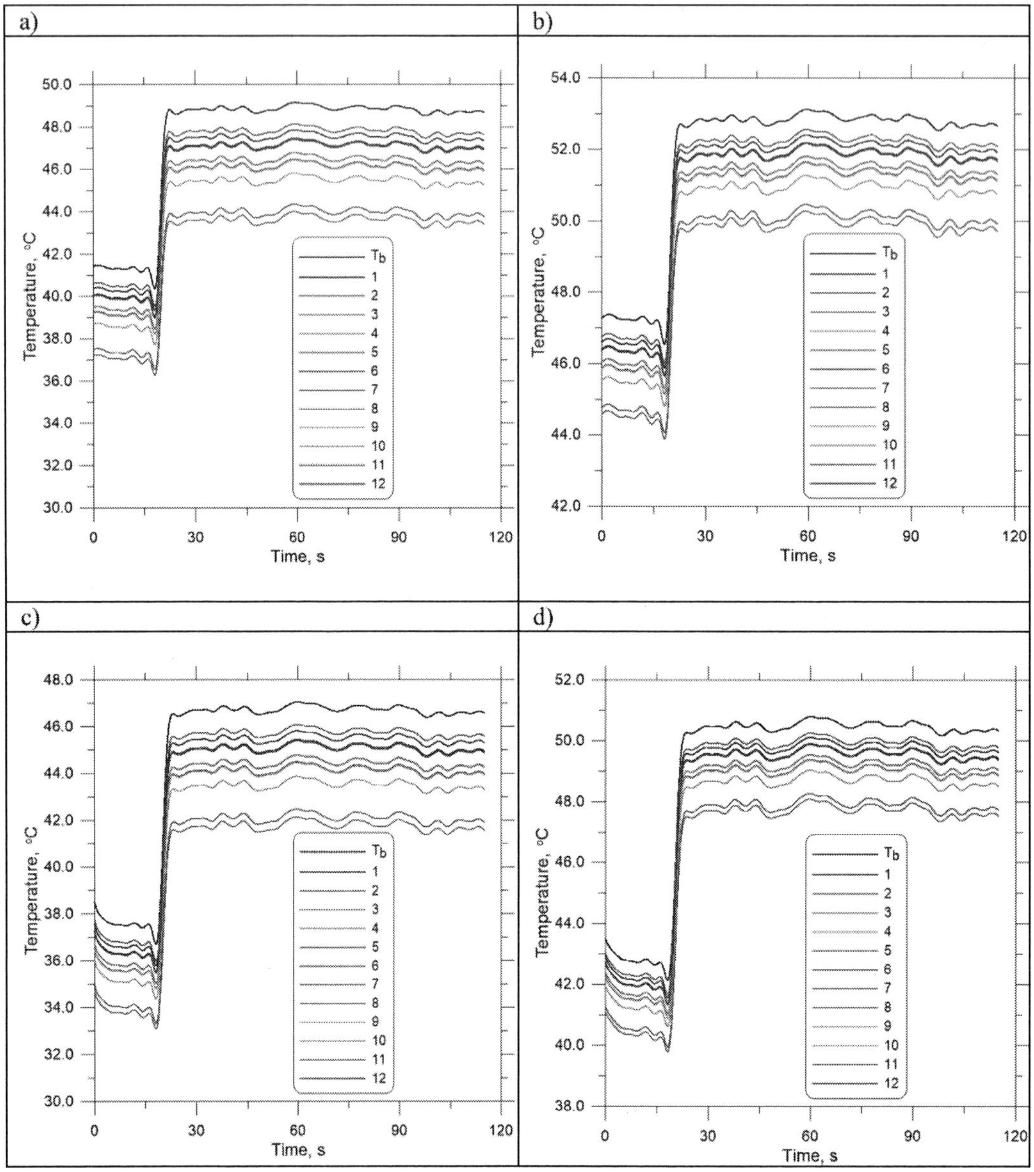

Fig. 5. Changes in the fin base temperature T_b and the fin temperature in 12 nodes, that are marked in Fig. 2, at a distance of 247 mm (9.5 cell lengths) from each pass inlet:
(a) – first tube row in the first (upper) pass, (b) – second tube row in the first (upper) pass,
(c) – first tube row in the second (lower) pass, (d) – second tube row in the second (lower) pass.

The curves illustrating time-dependent changes in the temperature of selected fins are presented in Fig. 5a – d. An analysis of the results indicates that the highest temperature is at the fin base, and the lowest in node 3 (Fig.2), which is the most distant from the tube axis. It can be seen that the temperatures of the fins in the first tube row (Fig. 5a and 5c) are lower than in the second row (Fig. 5b and 5d). Cold inflowing air cools the fins and tubes located in the first row better. The air temperature after the first tube row is much higher compared to the inlet air temperature. Due to a reduction in the difference between the temperatures of the flowing air and the tubes located in the second row, the temperature of the water and fins in the second tube row is higher than in the first row.

4 Conclusions

The calculations and measurements carried out in the paper confirm that the developed numerical model of the plate fin and tube heat exchanger has a high accuracy. This model is particularly useful to simulate exchanger, where the temperature of the fins can be very high. Thanks to the experimental determination of heat transfer correlations on the air and water sides a very good agreement of calculations and measurements was reached. The developed model can be used to simulate the heating and cooling heat exchanger and in automatic control systems.

References

1. Korzeń A., Taler D.: Modeling of transient response of a plate fin and tube heat exchanger, *Int. J. Therm. Sci.*, **92** (2015), pp.188-198.
2. Taler D.: Mathematical modeling and control of plate fin and tube heat, *Energy conversion and management*, **96** (2015), pp. 452-462.
3. Roetzel W., Xuan Y.: *Dynamic behaviour of heat exchangers, Computational Mechanics Publications Vol.3*, WIT Press, Southampton 1998.
4. Ataer Ö.E.: An approximate method for transient behavior of finned-tube cross-flow heat exchangers, *Int. J. Refrig.*, **27** (2004), pp. 529-539.
5. Taler D.: *Direct and inverse heat transfer problems in dynamics of plate fin and tube heat exchangers*. in: Belmiloudi A. (ed.) *Heat transfer, mathematical modelling, numerical methods and information technology*, InTech, Rijeka 2011, pp 77–100, free online edition: www.intechopen.com.
6. Vaisi A., Talebi S., Esmaeilpour M.:*International Communications in Heat and Mass Transfer*, **38** (2011), pp. 951-957.
7. Arpaci V. S., Kao S. H., Selamet A.: *Introduction to Heat Transfer*, Prentice Hall, Upper Saddle River, NJ, USA 1999.
8. Bejan A., Kraus A. D., Ed.: *Heat Transfer Handbook*, John Wiley & Sons, Inc., Hoboken, New Jersey 2003.
9. Wiśniewski S, Wiśniewski T. S.: *Wymiana ciepła* (*Heat exchange*), Wydanie 6, WNT, Warszawa 2009.
10. Gnielinski V.: Neue Gleichungen für den Wärme- und den Stoffübergang in turbulent durchströmten Rohren und Kanälen, *Forschung im Ingenieurwesen*, 41 (1975) Nr.1, S. 8-16
11. Webb R. L.: *Principles of enhanced heat transfer*, John Wiley & Sons, Inc., Hoboken, New Jersey 1994.
12. Mirth D. R., Ramadhyani S., Hittle D. C.: Thermal performance of chilled-water cooling coils operating at low water velocities. *ASHRAE Transactions*, Vol. 99 (1993), Part 1, pp. 43-53.
13. Mirth D. R., Ramadhyani S.: Correlations for predicting the air side Nusselt numbers and friction factors in chilled-water cooling coils. *Experimental Heat Transfer*, Vol. 7 (1994), pp. 143-162.
14. Petukhov B.S., Genin A.G., Kovalev S.A.: *Heat transfer in nuclear power plants*. Atomizdat, Moscow 1974 (in Russian).
15. Brandt S.: *Data Analysis. Statistical and computational methods for scientists and engineers, 3rd edn*. Springer, Berlin 1999.
16. Coleman H.W., Steele W.G.: *Experimentation, validation, and uncertainty analysis for engineers, 3rd edn*. Wiley, Hoboken 2009.
17. Korzeń A.: *Mathematical modelling of the operation of cross-flow finned-tube heat exchangers*, Wydawnictwo Politechniki Krakowskiej, Kraków 2015.

Micro cogeneration - rich-methane gasifier and micro gas turbine

Danuta Król[1,*], *Sławomir* Poskrobko[2], and *Józef* Gościk[3]

[1]Silesian University of Technology, Konarskiego 18, 44–101 Gliwice, Poland
[2]Technical University of Białystok, Wiejska 45A, 15–351 Białystok, Poland
[3]Technical University of Białystok, Wiejska 45C, 15–351 Białystok, Poland

Abstract. The study presents a concept of integration a gasifier system with distributed generation of electricity and heat cogeneration system based on a gas microturbine. The gas generator is supplied by the RDF fuel from waste and biomass fuel Bio-CONOx. In the scale considered, the cogeneration system is designed to produce 30kWe of electricity and approx. 50kW of heat. Important perspective directions of technology development are: (i) the possibility of gas microturbine to cooperate with the gasifier (up to date, in such systems were used, and continue to apply only piston engines), (ii) the production of syngas in the gasifier (for efficient cogeneration) in the composition of which there is a high content of methane (CH_4 = 18%-22%). In the first step of possible commercialization a mathematical model to simulate single shaft gas turbine cogeneration plant has been developed. In conceptual design is application of microturbine as the prime mover of Combined Heat and Power (CHP) system but with especial emphasis on possible use of a low calorific gas attainable from presented in details a gasifier unit. To support the calculations for preliminary design analysis, a computer program is developed in EES software environment.

1 Introduction

The use of gasifiers in the energy sector is limited by too low efficiency of chemical energy conversion in the fuel, into chemical energy derived from post-processing gas. When the process is carried out in a fixed bed of solid fuel under low pressure conditions (atmospheric), the efficiency of fuel conversion into gas (chemical effectiveness) reaches up to approx. 60-70%. The chemical effectiveness is defined by the ratio of the calorific value of the charge to the calorific value of the resulting syngas [1]. Paper [1] describes the technology of gasification in fluidized compact bed, which has successfully been implemented in the 50s-60s. Later research aimed at improving the chemical efficiency by improving kinetic conditions of the gasification process were carried out using catalysts [2,3]. The increase in the conversion rate was achieved by improving the construction of gasification chambers. Gasification in cyclone chambers [4], with fluidized grate in bladder and circulation bed [5],

* Corresponding author: dankrol@wp.pl; danuta.j.krol@polsl.pl

double layer gasification [6], gasification in plasma [7]- all these were implemented. The effect of process improvements has been achieved by reactive additives to fuels, that affect the increase in the content of flammable components in the generator gas, which translates into an increase in the calorific value of the syngas. For example, various types of biomass containing resin pine biomass, biomass or coal with plastics [8] etc. are gasified. The additives to fuels do not play the catalyst role, they are additional fuel- a substrate taking part in the thermal change process. Their participation does not significantly increase the calorific value of the fuel. Their share in the fuel directs the process towards obtaining combustible gases i.e. in particular CO and H_2. The best results, taking into account the technology used, were obtained during gasification in a fluidized bed [9]. Fluid technologies were successfully implemented in Finland, Austria and the USA, where syngas with a calorific value of approx. 6–8 MJ/Nm^3 [10] was obtained.

The paper presents possibilities of generating the syngas in the gasification process of pellets fuel from waste (RDF), with methane forming addition of Bio-CONOx [11]. The resulting gaseous fuel with high calorific value, achieved through significantly increased content of CH_4. Such obtained and tested gasifier system can be treated as a very interesting source of fuel which is applicable to supply a gas microturbine and consequently to create a concept of cogeneration plant designed for supply industrial process or residential needs with pressurized, hot water.

2 Materials and methods

2.1 Properties of RDF and Bio-CONOx

The RDF pellets fuel used in these studies is generated from the combustible fraction of municipal waste. The preparation Bio-CONOX is composed of combustible biomass waste materials. Tables 1 and 2 show RDF and Bio-CONOX preparation fuel properties.

Table 1. Pellets fuel properties of RDF on a dry weight basis and elemental composition of flammable substance expressed on a dry ash-free basis.

Properties	Symbol	Value
Moisture content	[%]	5.4
Flammable fraction	[%]	87.1
Nonflammable fraction	[%]	12.9
Lower heating value LHV	[kJ/kg]	21798
carbon c^{daf}	[%]	66.7
hydrogen h^{daf}	[%]	7.5
nitrogen n^{daf}	[%]	4.1
sulphur s^{daf}	[%]	1.8
chlorine cl^{daf}	[%]	0.3
oxygen o^{daf}	[%]	19.6

Table 2. Physical, chemical, and fuel properties of Bio-CONOx expressed on a dry basis and elemental composition of flammable substance expressed on a dry ash-free basis.

Properties	Symbol	Value acc. [16]	Value
Moisture content	[%]	<12	7.3
Flammable fraction	[%]	>90	93.2
Nonflammable fraction	[%]	<10	6.8
Lower heating value LHV	[kJ/kg]	>17000	17745
Higher heating value HHV	[kJ/kg]	>15500	18942
Protein content	[%]	24-40	31.0
Plant fats content	[%]	2.5-5.0	4.4
carbon c^{daf}	[%]	49.00-52,00	50.2
hydrogen h^{daf}	[%]	6.00-8,00	6.3
nitrogen n^{daf}	[%]	6.00-8.00	5.9
sulphur s^{daf}	[%]	1.00-1.5	1.2
chlorine cl^{daf}	[%]	<0.05	0.02
oxygen o^{daf}	[%]	30.45-38.00	36.38

RDF fuel and mixtures of the fuel with Bio-CONOX 13%, 23%, 33% were applied to the gasification process.

2.2 Test stand

Gasification was carried out on a laboratory scale, in the tubular gasifier with fluidized bed (Fig. 1). The gasifying agent was air (0.64 g/s) that was that was feed to the bed by a perforated grate from below the gasifier. The diameter of the perforation-hole in the grate d=1mm. During the experiment the air was not heated – the air temperature was 22°C. The maximum thermal power of the gasifier was 5kW. The temperature of the process was defined in the range of 380-410°C, and above the layer of fuel approx. 900°C. In a fluid bed (in the layer of fuel) - at a temperature of 380-410°C – the gasification process occurred autothermally. In the gaseous phase (above the bed), the temperature reached 880-930°C. An intense evaporation of high-calorific condensable substances, including tar, occurred. It is suggested that these high-calorific compounds were partially burned over the fuel layer in an atmosphere of unreacted oxygen, which resulted in the temperature increase up to the value of ~ 900°C. That is evidenced by the lack of high-calorific hydrocarbons CnHm in the generator gas.

The fuel was fed into the reaction chamber continuously with a feed screw driven by a stepper motor, with a precise speed control, i.e. 0,41 g/s fuel flow. The ash was removed in the process exhausting from the gasification chamber by revolving (vacuum and performance) of the exhaust fan rotor. The resultant gas was cooled by a stream air sucked from the environment to a temperature of approx. 180^{0}C. The gas generator and blowing air

 DOI: 10.1051/e3sconf/20171401024

channel are fitted with electric heating system, which is used to heat the reaction zone, where spontaneous combustion of fuel took place. The stabilization of the process was carried out by regulating the flow of fuel and the gasification air flow.

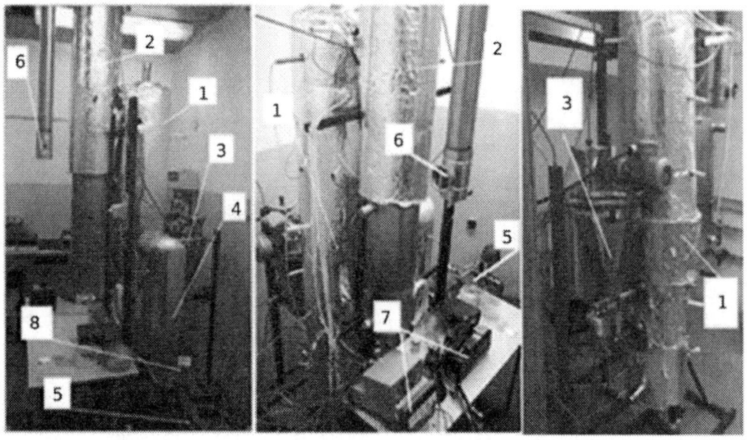

Fig. 1. Laboratory test stand – fluidized bed gasifier, where: 1 – gasifier, 2 – combustion chamber and chimney channel, 3 – fuel tank and dosing system, 4 – air heater, 5 – air flow channel and fan, 6 control chimney draft, 7 – control panel, 8 – measurement of airflow.

3 Results and discussion

3.1 Formatting the text Gasification

The test results of RDF fuel gasification containing formulation Bio-CONOx are shown in Fig. 2 and 3. Fig. 2 shows the dependence of the of generator gas composition CH_4, CO, H_2, CO_2 from the amount of oxygen. Taking into account that the air flow during the gasification process was maintained at a constant level, the demand of the oxygen is significantly reduced, with increasing participation of the additive Bio-CONOx in the fuel (Figure 2). This is expressed in increasing participation of oxygen in the generator gas. These results show, that the addition of Bio-CONOx clearly influences the formation of strongly reducing zone in the reaction environment. Under these conditions the temperature in the bed was maintained in the range of 380-410°C, the temperature over the bed gradually grew from 880°C (gasification without the addition of Bio-CONOx) to 930°C (gasification with 33% addition of Bio-CONOx). Changes in the temperature suggest the increased participation of condensing vapour cracking (including tars), which favoured methane forming processes according to the scheme: (TAR)gas $\rightarrow V_{CO}CO + V_{CH4}CH_4 + V_{H2}H_2 + V_{CO2}CO_2$. The high temperature above the bed was assured by the process of partial combustion, mainly high-calorific tars, according to Brydon kinetic model: $CH_{1,522}O_{0,0228} + 0,867O_2 \rightarrow CO + 0,761H_2O$. Gasification of RDF fuel with air made it possible to obtain a synthesis gas, having a composition of combustible gases up to 10,3%CH_4, 11,6%CO and 7,3%H_2, and calorific value max. at ~ 6000kJ/Nm^3.

Gasification kinetics is directed by a complex interdependence of chemical and physical phenomena. The speed of heterogeneous gasification reactions depends both on the speed of chemical reactions, and the speed of diffusion of the gasifying agent into the pores of fuel:

$$Vr = - dnc/dt = f(k_{ef}, O_{,mr}, C, w, p, d) \qquad (1)$$

 DOI: 10.1051/e3sconf/20171401024

where: k_{ef} - effective reaction speed constant; O - specific surface area of the fuel; m_r - fuel mass in the reaction space; C - the concentration of the gasification agent in the reaction space; w- the flow rate of the reactants; p - pressure; d - diameter of the fuel grains.

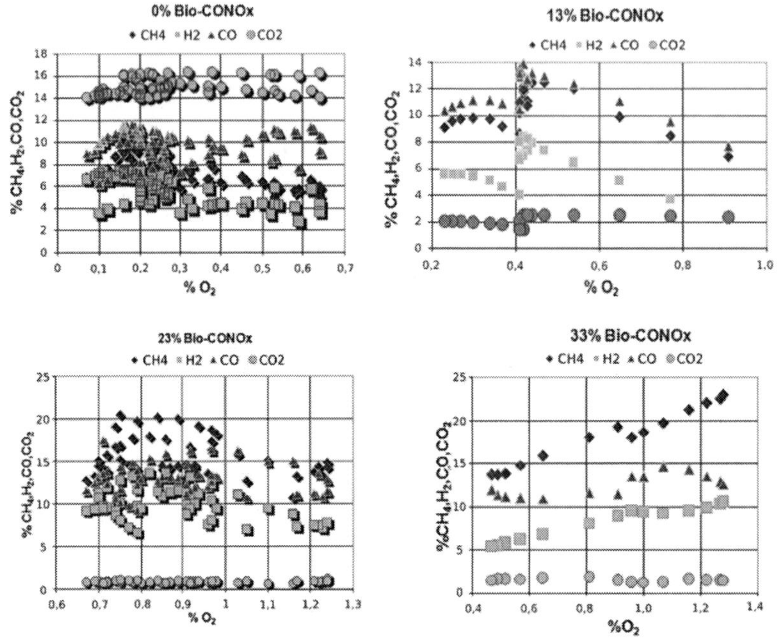

Fig. 2. %CH_4, H_2, CO, CO_2 = f (O_2) for different shares of Bio-CONOx in the fuel.

The results shown in Figure 2 indicate that use of the preparation Bio-CONOx in amounts of 13, 23 and 33% of the fuel mixture, causes expansion of the reducing zone. The result is a continuous increase in the proportion of methane CH_4 (even double with the 33% share of Bio-CONOx in the fuel) and hydrogen H_2 in the syngas. Attention is drawn to a significant drop in carbon dioxide content from 16% (for the gasification of RDF fuel only) to a very low level of 1÷2.5% when Bio-CONOx was added. In addition, the amount of CO_2 in the process remains unchanged (Figure 2 - points are arranged parallely to the axis X). This reflects the dominant reducing zone in the area of the gasification reaction, favouring the increase of CH_4 share. The primary objective of the processes of the second and third generation is, among others, the preparation of higher calorific gas consisting mainly of methane. The obtained results clearly fit in with this trend. They show that in the area of the fuel layer, in favorable temperature conditions 380-410°C [15], exothermic methanation reactions occurred. Basic relations oriented towards methanation of synthesis gases are:

$$CO + 3H_2 \leftrightarrow CH_4 + H_2O \quad \Delta H = -206 kJ/mol \tag{2}$$

The water formed during the methanation reaction also reacts with carbon monoxide according to the equation:

$$CO + H_2O \leftrightarrow CO_2 + H_2 \quad \Delta H = -41,5\ kJ/mol \tag{3}$$

When in the reaction system the concentration of carbon monoxide decreases, the balance of the reaction (3) moves. This reaction changes direction. It begins to occur from the right side to the left.

Carbon monoxide which is being formed undergoes the methanation according to equation (2). By adding reaction (2) to reaction (3) – with its course from right to left – we get the sum equation of the carbon dioxide methanation reaction:

$$CO_2+4H_2 \rightarrow CH_4 +2H_2O \quad \Delta H = -164{,}9\ kJ/mol \tag{4}$$

Hydrogen in the low-temperature thermal conditions of deposit, was created in the radical reactions $H^* + H^* \rightarrow H_2$ and that hydrogen was involved in the methanation reaction of CO and CO_2. Methanation reactions of CO and CO_2 are exothermic reactions taking place with the reduction of volume, so reducing the temperature favours the shift of the equilibrium in the direction to increase methane concentration in the equilibrium gas mixture. The methane forming process is more widely described in[12. Intensification of methanation process for the gasification of fuel with 23% and 33% shares of Bio-CONOx, where the highest share of CH_4 and calorific value of LHV_{syngas} ~10MJ/Nm^3 were recorded. The calorific value of the generator gas was calculated on the basis of the following formula (its source is the instruction manual of the analyzer):

$$LHV_{(syngas)} = 126\ [\%CO] + 108\ [\%H_2] + 359\ [\%CH_4] + 665\ [\%C_nH_m] \quad [kJ/Nm^3] \tag{5}$$

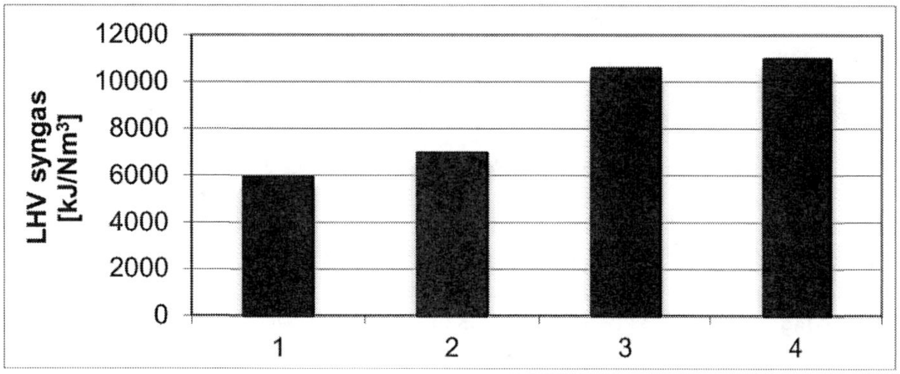

Fig. 3. Calorific value of the synthesis gas depending on the share of Bio-CONOx in the fuel: 1 – 0%Bio-CONOx; 2 –13%Bio-CONOx; 3 - 23%Bio-CONOx; 4 - 33%Bio-CONOx

The LHV values of obtained generator gases indicate an upward trend with increasing addition of Bio-CONOx fuel.

3.2 Modeling of a syngas microturbine based cogeneration plant

The paper presents a conceptual, small scale cogeneration plant (30 kW) that combines a gasifier, a single shaft microturbine supplied by syngas and a flue gas heat exchanger (FGHEX). So, it is assumed that worked out units of the plant (gasifier and syngas conditioning parts) can supply a gas microturbine (GM) with so called pure syngas which enable conversion chemical energy to electrical energy and heat without known potential problems. The basic scheme of a cogeneration plant is shown in Fig. 4.

The part of installation responsible for producing syngas were described elsewhere in details. So shortly it is only necessary to mention that for next investigation some idealized Bryton cycle is considered (we assumed as negligible such effects as – to named just a few: air intake and exhaust duct losses and pressure losses in a combustion chamber and in the FGHEX gas side).

 DOI: 10.1051/e3sconf/20171401024

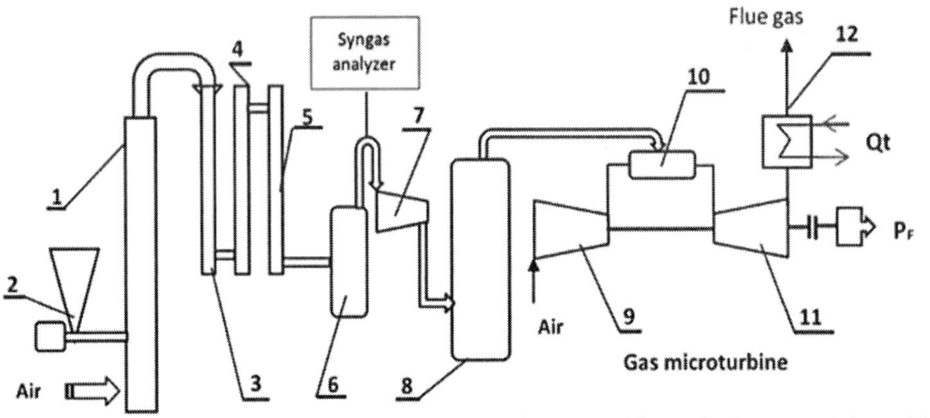

Fig. 4. Simplified scheme of a GM based cogeneration plant: 1-gasifier, 2-feed system, 3,4-scrubbing unit,5-carbon (or ceramic) filter, 6-compensatory syngas container, 7-compressor for syngas, 8-container for compressed syngas, 9-GM (air compressor), 10-GM (combustion chamber), 11-GM (expansion turbine), 12-FGHEX, PE-electric power, Qt-heat power.

3.2.1 Gas microturbine model

Computational part of the work undertaken concerns mainly on thermodynamic aspects of a microturbine work. The GM consists of four components: compressor, combustion chamber, turbine and generator. At presented paper it is assumed that thermodynamic cycle should be equivalent of classical Brayton cycle (simple, open without regeneration). In the cycle considered atmospheric air (*state 1*-inlet GM) is continuously drawn into the compressor and then after compression (*state 2*-outlet GM) is deliver to the combustion chamber. Syngas fuel is used to increase the temperature of compressed air through a combustion process. Hot gases (combustion products) leaving the combustion chamber (*state 3*) expand in the turbine which produces work to drive the compressor, generator and finally discharges flue gas (*state 4* for flue gas also subscript *g1*).

Air compressor model

Knowing the air inlet temperature T_1 to compressor, pressure ratio π_c and isentropic efficiency $\eta_{i(c)}$ for compressor, we can determine the following parameters:

$$\pi_c = p_2 / p_1 \tag{6}$$

where p_1 and p_2 are compressor inlet and outlet air pressure respectively;

$$\eta_{i(c)} = (T_{2s} - T_1)/(T_2 - T_1) \tag{7}$$

where T_1 and T_2 are compressor inlet and outlet air temperature respectively, and T_{2s} compressor isentropic outlet temperature, so the final temperature of the compressor can be obtained from equation

$$T_2 = T_1 \cdot \left(1 + \left(\pi_{i(c)}^{pc} - 1\right) \Big/ \eta_c\right); \quad pc = (k_c - 1)/k_c \tag{8}$$

Assuming that blade cooling is not taken into account, the specific work of the compressor w_c and consequently compressor power N_C can be calculated by

$$w_c = c_{p(a)} \cdot T_1 \cdot \left(\pi_c^{pc} - 1\right) \Big/ \left(\eta_{m(c)} \cdot \eta_{i(c)}\right) \;\rightarrow\; N_C = \dot{m}_a \cdot w_c \tag{9}$$

where $c_{p(a)}$ is the specific heat of air and $\eta_{m(c)}$ is the mechanical efficiency of the compressor.

Combustion chamber model

From energy balance in the combustion chamber:

$$\dot{m}_a \cdot c_{p(a)} \cdot T_2 + \dot{m}_f \cdot W_d + \dot{m}_f \cdot c_{p(f)} \cdot T_f = \left(\dot{m}_a + \dot{m}_f\right) \cdot c_{p(g)} \cdot T_3 \quad (10)$$

where, $\dot{m}_f$ is fuel mass flow rate (kg/s), $\dot{m}_a$ is air mass flow rate (kg/s), W_d is low heating value, T_3 turbine inlet temperature, T_f is temperature of fuel, $c_{p(f)}$ is specific heat of fuel and $c_{p(g)}$ specific heat of flue gas. After manipulating (5), the fuel air ratio can be expressed as

$$FAR = \dot{m}_f / \dot{m}_a = \left(c_{p(g)} \cdot T_3 - c_{p(a)} \cdot T_3 \cdot \left(1/\pi_t^{pg}\right)\right) / \left(W_d + c_{p(f)} \cdot T_f - c_{p(g)} \cdot T_3\right) \quad (11)$$

Gas turbine model

The exhaust gases temperature from gas turbine is given by

$$T_4 = T_3 \cdot \left(1 - \eta_{i(t)} \cdot \left(1 - 1/\pi_t^{pg}\right)\right); \quad pg = \left(k_g - 1\right)/k_g \quad (12)$$

where T_4 is also the inlet gases temperature ($=T_{g1}$) to FGHEX.
The specific shaft work w_t of the turbine and turbine output power are given by

$$w_t = c_{p(g)} \cdot T_3 \cdot \eta_{i(t)} \cdot \left(1 - \left(1/\pi_t^{pg}\right)\right) / \eta_{m(t)} \rightarrow N_T = \left(\dot{m}_a + \dot{m}_f\right) \cdot w_t \quad (13)$$

The MG powers ((9), (12)) allow to express the net output power from the turbine as

$$N_{GT} = \left(N_T - N_C\right) \cdot \eta_g \quad (14)$$

If the heat supplied is expressed as

$$E_{ch} = \dot{m}_f \cdot W_d \quad , \quad (15)$$

then the gas turbine efficiency η_{th} can be determined by equation

$$\eta_{th} = N_{GT} / E_{ch} \quad . \quad (16)$$

3.2.2 Gas microturbine model Flue gas heat exchanger

For recover waste heat from gas turbine, exhaust gas flows into FGHEX (it is assumed with T_{g1}, T_{g2} inlet and outlet flue gas temperature, respectively). Usually FGHEX represents a single pressure heat recovery steam generator (HRSG) or water-gas heat recuperator.

3.2.3 Thermodynamic performance

There are known and commonly used three thermodynamic methodologies for the performance evaluation of cogeneration plant including of course gas fired systems. Two of them are fundamentally energetic providing to calculate fuel utilization efficiency and power to heat ratio efficiency. Third methodology operates on exergetic approach and allow to system evaluation on the base of second low, exergetic efficiency. In the paper for the evaluation overall efficiency we restricted our attention to apply the first approach (named also as Energy Utilization Factor) and defined as

$$\eta_{ov} = EUF = \frac{N_T + Q_{ex}}{E_{ch}} \quad where \quad Q_{ex} = (\dot{m}_a + \dot{m}_f) \cdot \tilde{c}_{p(g)} \cdot (T_{g1} - T_{g2}) \quad (17)$$

3.2.4 Results

All calculations have been performed using EES software. For better reliability of the GM results are compared with the technical data available for commercial microturbine Capstone C30 (in fact data partially are mixed also with those available for model 330).

Table 3. Microturbine performance data (Capstone Model C30).

Electrical power N_T [kW]	30
Total heat extraction Q_{ex} [kW]	56
Heat added E_{ch} [kW]	115
Flue gas temperature T_4 [K]	271[°C] / 544.15[K]
Electrical efficiency η_{el} [-]	26
Overall efficiency *EUF* [%]	77
Rotational speed [r/min]	96000
Air flow rate at compressor inlet [kg/s]	0.31

Comment: designed for ambient conditions (ISO-standard reference conditions): temperature 288.15 [K], pressure 101.325 [kPa], relative humidity 60%.

Table 4. Design parameters for simulation of the model microturbine.

Compressor	$\pi_c = 4{,}5$, $\eta_{i(c)} = 0{,}8$, $\eta_{m(c)} = 0{,}95$, $k_a = 1{,}4$
Turbine:	$\pi_{i(t)} = 4.5$, $\eta_{i(t)} = 0.88$, $\eta_{m(t} = 0.95$, $k_g = 1.326$
Combustion chamber:	$\eta_{cc} = 1{,}0$
Generator: generator efficiency	$\eta_g = 1{,}0$
Mass flow rate of air [kg/s]	0.31
Outlet temperature [K]	544.15
Output power N_{GT} [kW]	30.0

Table 5. Operating parameters of the model GM results of calculation using EES.

State of cycle	Pressure [kPa]	Temperature [K]
1 – inlet air to the compressor	101.3	288.15
2 – outlet air from the compressor	456	481.53
3 – inlet gas to the turbine	456	(see Table 6)
4 – gas exhaust from the turbine	101.3	544.15
Compressor input power N_C [kW]	63.8	
Turbine output power N_T [kW]	94.4	
Net output power N_{GT} [kW]	30.0	

Table 6. Calculated cases for different fuels including design parameters[1/].

	Fuel I	Fuel II	Fuel III	Fuel IV	Methane
Temperature of inlet gas to turbine T_3 [K]	811.2	815.3	821.1	821.3	828.3
Mass flow rate of fuel [kg/s]	0.02216	0.0172	0.01027	0.01007	0.002182
Ratio air/fuel AFR [-]	13.99	18.03	30.18	30.78	142.1
Heat extracted [kW][2/]	58.229	59.112	56.996	56.961	55.557
Heat added [kW]	107.395	107.827	108.431	108.449	109.136
Efficiency η_{th} [%]	28.5	28.39	28.23	28.23	28.05
Efficiency EUF [%]	82.98	81.82	80.23	80.19	78.39

Comments:[1/] calorific values W_d : 4.847 MJ/kg - Fuel I, 6.270 MJ/kg - Fuel II, 10.558 MJ/kg - Fuel III, 10.769 MJ/kg - Fuel IV and for pure methane 50 MJ/kg (all W_d based on Lower Heating Values);[2/] heat extracted rates have been calculated taking reference temperature (T_{g2} = 368,15[K]/95[°C] – the temperature of the exhaust gases exit from the FGHEX).

4 Conclusions

The studies discussed in the paper showed that co-gasifying with air (in compact bed, in the laboratory tubular reactor with a capacity of 5 kW) the RDF fuel (from waste) containing the biomass additive Bio-CONOx, a generator gas with a high methane content can be obtained. Bio-CONOx share in the fuel was subsequently: 0%, 13%, 23%, 33%. Temperature distribution in the reaction chamber of the gasifier (in the fuel layer 380-410°C and above the fuel layer 880-930°C) allowed to obtain syngas - produced during the gasification of RDF without the addition of Bio-CONOx – with methane content of even approx. 10%. In such conditions, in the layer of fuel, exothermic methanation process took place, i.e. hydrogenation of CO and CO_2, according to the mechanism of Fischer-Tropsch and Sabatier Sanders. Over the fuel at high temperatures, process conditions conducive to cracking tars were created (in line with Brydon kinetic concept). The product of cracking was, among others, methane. The results showed that in the co-gasification process with air of RDF fuel with the addition of Bio-CONOx, the addition significantly affects the methane forming transformations in the gasification process, making the obtained synthesis gas has reach higher calorific value of over 10 MJ/Nm^3.

The obtained syngas we used as an interesting and commercially attractive fuel for conceptual design a microturbine combined heat and power a small scale system. At this stage, we firstly presented some important part of the design namely analytically performed sizing of main feeding system. Creation of reliable set of results concerning fuel flow rate (see Table 6) was possible on the base fuel characterization (syngas detailed analysis) and finally combustion concept taken.

In order to develop the cogeneration model and to perform systematic calculations it has been assumed that a commercially available microturbine (Capstone C30) works at nominal, steady state mode of operation corresponding precisely of design point. The very important fixed parameters were: air flow rate at compressor inlet, net output power N_{GT} as well as the temperature of gas exhausting the turbine.

Based on the represented findings we can conclude that developed model can be used as a kind of computer tool giving a valuable reference level for necessary modification in concept design and for expediency estimation of improvements in moving a small scale cogeneration systems toward thermodynamic and/or technology perfection.

References

1. P.D. Lebiediew, A.A. Szczukin. Przemysłowa technika cieplna, WNT Warszawa (1962)
2. W.Zhu, W.Song, W.Lin. Fuel Proc. Techn., 890.(2008).
3. R.Brown, Q.Liu, G.Norto. Biomass and Bioenergy,**18**, 499,(2008)
4. X.Guo, B.Ciao, S.Liu, Z.Hu, S.Luo, M.He. J of Hydr. En., **34**, 1265 (2009)
5. K.Li, R.Zhang, J.Bi. Energy, **35** 2722 (2010).
6. S.Poskrobko S. XII Symp. Wym. Ciepła i Masy, Darłówko, 841 (2007)
7. M. Materazzi, P. Lettieri, R. Taylor, C. Chapman. Waste Manag. **47** 256(2016)
8. P.McKendry. Biores Techn. **83**, 55 (2002)
9. D.L.Granatstein. Lahti Finland. Natural Resources Canada/CANMET Energy Technology Centre (CETC), IEA BIOENERGY AGREEMENT—TASK 36, (2002).
10. Ch.Hanping, L.Bin, Y.Haiping, Y.Guolai. Z.Shihong. En & Fuels, **22**, 3493,(2008)
11. S.Poskrobko, D.Król, J.Łach. Fuel Proc. Tech.,.**101**,. 58, (2012)
12. D. Król, S. Poskrobko. Energy **116.** 592, (2016).

DOI: 10.1051/e3sconf/20171401025

Computer monitoring and optimization of the steam boiler performance

Tomasz Sobota[1,*]

[1]Cracow University of Technology, Institute of Thermal Power Engineering, Al. Jana Pawła II 37, 31-864 Cracow, Poland

Abstract. The paper presents a method for determination of thermo-flow parameters for steam boilers. This method allows to perform the calculations of the boiler furnace chamber and heat flow rates absorbed by superheater stages. These parameters are important for monitoring the performance of the power unit. Knowledge of these parameters allows determining the degree of the furnace chamber slagging. The calculation can be performed in online mode and use to monitoring of steam boiler. The presented method allows to the operation of steam boiler with high efficiency.

1 Introduction

The fouling processes caused by slagging and ash deposits are strongly affected by the kind of coal [1-6]. They become more intense if coal is co-fired with biomass. In the case of the furnace chamber, the fouling processes are different on its individual walls. Depending on the furnace chamber aerodynamics, local slag overhangs may be formed in the burner area or at other locations on the chamber walls. In places where temperature is high, such as the platen superheater area, the superheater surface slagging may also be caused by melted drops of ash forming deposits on the surface of tubes [7]. If biomass is co-fired, due to the low melting point of ash, intense processes of ash deposit formation and slagging may occur on the superheater stages located in zones with a lower flue gas temperature.

Ash fouling processes also affect the lifetime of the boiler superheaters [8-9]. It should be noted that about 40% of the boiler failures are caused by damage to boiler superheaters. For this reason, there is a need to develop an effective and efficient method of determination of thermo-flow parameters for steam boilers. The presented in this paper method allows to the operation of steam boiler with high efficiency.

2 Calculations of the boiler furnace chamber

The degree of fouling of the boiler furnace chamber walls may be assessed by an online mode determination of the flue gas temperature at the furnace chamber outlet T_{fe}, the heat flux absorbed by the furnace chamber walls $\dot{Q}_r$ and the heat effectiveness ψ of the furnace

* Corresponding author: tsobota@mech.pk.edu.pl

chamber walls. The outlet of the combustion chamber is located directly beneath of the platen superheater at a height of 29 m (Fig. 1a). Adopted position of the outlet section of the furnace is consistent with the industrial definition, which says that the outlet from the furnace is where the spacing of the platens is less than 2.0 m. The spacing of the platens is 0.7 m in the boiler analyzed in the paper. The heat effectiveness ψ of the furnace chamber is defined as an average ratio of the heat flux absorbed by the water-wall to the heat flux falling on the water-wall of the combustion chamber.

Based on experimental data obtained from the testing of large power boilers, a new formula containing parameter M, which characterizes the location where the maximum flame temperature occurs in the boiler, was proposed [11]:

$$\frac{T_{fe}}{T_{ad}} = \frac{Bo^{0,6}}{M\,\varepsilon_{pal}^{0,6} + Bo^{0,6}} \tag{1}$$

Parameter M in Eq. (1) is a function of the kind of fuel (oil, gas or coal) and of the relative height at which burners are placed [10-12]. After temperature T_{fe} is calculated from (1), the average heat flux $\dot{Q}_r$ absorbed by the furnace chamber walls may be determined. The degree of slagging of the furnace chamber walls may be assessed determining the waterwall heat efficiency coefficient ψ. The ψ coefficient gets smaller in proportion to slagging of the furnace chamber walls. If the furnace chamber walls are covered with a layer of slag, the temperature of the slag surface is high and the density of the heat flux $\dot{q}$ absorbed by the wall (i.e. the wall thermal load) is much smaller than in the case of a clean wall. Slagging of the furnace chamber walls results in an increase in the flue gas outlet temperature T_{fe} and a decrease in the heat flux absorbed by the furnace chamber walls

3 Determination of the average degree of the furnace chamber fouling

The waterwall heat efficiency coefficient ψ will be determined in an online mode from the following nonlinear equation:

$$\dot{m}_{par}^{zm} = \dot{m}_{par}^{obl}(\psi), \tag{2}$$

where: $\dot{m}_{par}^{zm}$ and $\dot{m}_{par}^{obl}$ denote properly measured and calculated mass flows of steam. The steam mass flow $\dot{m}_{par}^{obl}$ is a function of the waterwall heat efficiency ψ and it is calculated from the boiler evaporator energy balance. The symbol $\dot{m}_{par}^{zm}$ denotes the steam mass flow at the boiler drum outlet determined using the steam mass flow $\dot{m}_p$ measured at the boiler outlet and the water mass flows $\dot{m}_{w1}$ and $\dot{m}_{w2}$ fed into steam attemperator No. 1 and No. 2, respectively.

$$\dot{m}_{par}^{zm} = \dot{m}_p - \dot{m}_{w1} - \dot{m}_{w2}. \tag{3}$$

The condition of equality of the calculated and measured values of the steam mass flow allows an assessment of the degree of the furnace chamber fouling ζ.

The waterwall heat efficiency coefficient ψ is equal to the product of the waterwall shape factor x, which for tight walls in the OP-380 boiler is equal to unity, multiplied by the fouling coefficient ζ of the chamber furnace heating surfaces:

$$\psi = x\zeta, \tag{4}$$

where: $x = 1$ for membrane water-walls used in the boiler studied.

Determining ζ so that the calculated and measured values of the boiler efficiency are the same, it is possible to assess the average degree of the fouling of the walls of the furnace chamber.

Owing to this way of determining the value of ζ the temperature of flue gases at the outlet from the furnace chamber T_{fe} may be calculated much more accurately.

If the fouling of the chamber walls is higher, the waterwall efficiency coefficient decreases: $\psi = \zeta$. The heat flux absorbed by the boiler evaporator $\dot{Q}_{par}$ gets smaller and the flue gas temperature at the furnace chamber outlet gets higher. As a result, there is a reduction in the steam mass flow generated in the evaporator and directed to the superheaters, which, at a higher flue gas temperature T_{fe}, causes an increment in the reheated steam temperature (at the same mass flow of fuel: $\dot{Q}_{pal}$). In order to maintain the reheated steam temperature at the set level, bigger mass flows of water fed into the reheated steam attemperators are necessary.

Also the flue gas temperature after individual superheater stages rises due to the fouling of the furnace chamber walls.

The following quantities:

- the waterwall heat efficiency ψ,
- the mass flow rates of water $\dot{m}_{w1}$ and $\dot{m}_{w2}$ injected into attemperators,
- the flue gas temperature after individual superheater stages

are strongly dependent on the degree of the fouling of the furnace chamber walls. If ψ is reduced, and $\dot{m}_{w1}$ and $\dot{m}_{w2}$, as well as T_{sp}, after individual stages reach permissible set values, the steam slag and ash blowers in the furnace chamber should be activated.

Apart from the above-mentioned indices characterizing the fouling of the boiler heating surfaces in terms of heat exchange, the increase in the flue gas subatmospheric pressure after the superheater stages located further down the convective duct (at a higher distance from the furnace chamber) is another significant symptom of the fouling of the boiler – mainly of the surfaces of superheaters. Increased aerodynamic resistance caused by ash fouling of superheaters leads to a higher power consumption of the fans. The measurement of the flue gas pressure and of the engine power capacity may also be used to assess the degree of ash fouling of superheaters.

4 Mathematical model of the boiler evaporator

The live steam mass flow is calculated from the boiler evaporator mass and energy balance (Fig. 1b):

$$\dot{m}_{wz} = \dot{m}_p - \dot{m}_{w1} - \dot{m}_{w2} + \dot{m}_{ods}, \tag{5}$$

$$\dot{m}_{wz} h_{wz-2} + \dot{Q}_{par} = \dot{m}_{ods} \cdot h'(p_n) + (\dot{m}_p - \dot{m}_{w1} - \dot{m}_{w2}) \cdot h''(p_n), \tag{6}$$

Substitution of (5) in (6) results in:

$$\dot{m}_p = \frac{\dot{Q}_{par}}{h''(p_n) - h_{wz-2}} - \dot{m}_{ods} \frac{h'(p_n) - h_{wz-2}}{h''(p_n) - h_{wz-2}} + \dot{m}_{w1} + \dot{m}_{w2}, \tag{7}$$

where $\dot{Q}_{par}$ is the heat flux transferred from flue gases to the evaporator via radiation and convection.

The heat flux $\dot{Q}_{par}$ may be calculated from the following expression:

$$\dot{Q}_{par} = \dot{Q} - \dot{m}_g \cdot c_{p,g}\Big|_0^{T_{fe}-273.15} (T_{fe} - 273.15), \tag{8}$$

where $\dot{Q}$ is the heat flux brought to the furnace chamber with fuel and air, and which is defined as:

$$\dot{Q} = \dot{m}_g \cdot c_{p,g}\Big|_0^{T_{ad}-273.15} \cdot (T_{ad} - 273.15), \tag{9}$$

a)

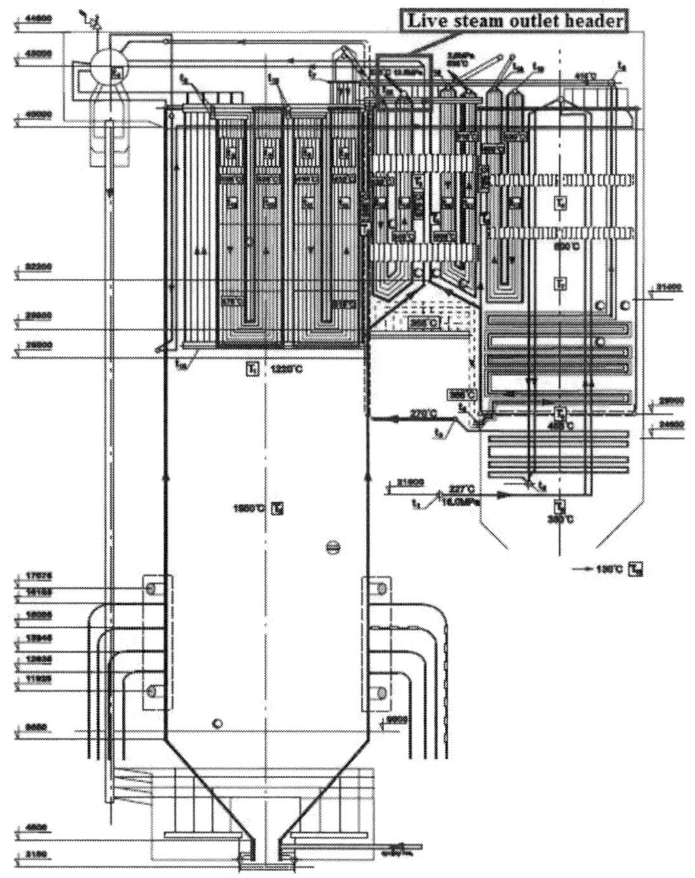

b)

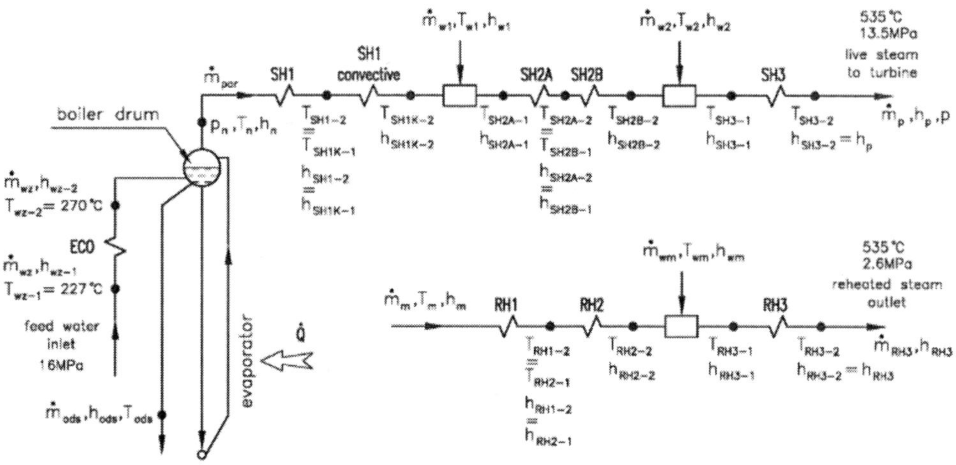

Fig. 1. Natural circulation steam boiler OP-380 with capacity of $380 \cdot 10^3$ kg/h: a) longitudinal cross-section of the boiler, b) diagram of control volume for the boiler evaporator mass and energy balance

The symbol T_{fe} denotes the flue gas temperature at the furnace chamber outlet and The adiabatic combustion temperature T_{ad} is calculated according to the [10].

It should be added that both efficiency and thermal calculations require the development of many functions and programs used for the calculation of thermal properties of water, saturated and superheated steam, and flue gases.

5 Determination of fuel mass flow rate

The calculation of the fuel mass flow rate is based on the fuel chemical composition or on the coal calorific value. The second approach is more appropriate for practical applications since the chemical composition of the coal is not determined continuously. Boiler efficiency is determined online by the indirect method [6, 13]. According to the ASME Performance Test Codes: PTC 4 - Fired Steam Generators [14] the stack gas temperature is corrected for air heater leakage based on O_2 difference between flue gas into air heater and flue gas O_2 leaving air heater. After correcting the temperature of the flue gas exiting the boiler the thermal efficiency of the boiler is calculated.

Based on the boiler efficiency η determined in an online mode, the coal mass flow is calculated from the definition of the boiler heat efficiency

$$\begin{aligned}\eta = \frac{\dot{Q}_n}{\dot{Q}_h} = & \frac{\left(\dot{m}_p - \dot{m}_{w1} - \dot{m}_{w2}\right)\left(h_p - h_{wz-2}\right)}{\dot{m}_{pal}\, W_d} + \frac{\left[\left(\dot{m}_{w1} + \dot{m}_{w2}\right)h_p - \dot{m}_{w1}h_{w1} - \dot{m}_{w2}h_{w2}\right]}{\dot{m}_{pal}\, W_d} + \\ & + \frac{\dot{m}_{ods}\left(h_n - h_{wz-2}\right)}{\dot{m}_{pal}\, W_d} + + \frac{\dot{m}_m\left(h_{RH1-2} - h_{RH1-1}\right)}{\dot{m}_{pal}\, W_d} + \frac{\dot{m}_m\left(h_{RH2-2} - h_{RH2-1}\right)}{\dot{m}_{pal}\, W_d} + \\ & + \left(\dot{m}_m + \dot{m}_{wm}\right)\left(h_{RH3-2} - h_{RH3-1}\right)/\dot{m}_{pal}\, W_d - \dot{m}_{wm}h_{wm}/\dot{m}_{pal}\, W_d ,\end{aligned} \tag{10}$$

Heat balance across the superheater attemporators is used to determine spray water mass flows. The fuel mass flow rate at various boiler loads is determined from Eq. (10) taking into account that the boiler efficiency η is known,

$$\begin{aligned}\dot{m}_{pal} = & \frac{\left(\dot{m}_p - \dot{m}_{w1} - \dot{m}_{w2}\right)\left(h_p - h_{wz-2}\right)}{\eta\, W_d} + \frac{\left[\left(\dot{m}_{w1} + \dot{m}_{w2}\right)h_p - \dot{m}_{w1}h_{w1} - \dot{m}_{w2}h_{w2}\right]}{\eta\, W_d} + \\ & + \frac{\dot{m}_{ods}\left(h_n - h_{wz-2}\right)}{\eta\, W_d} + \frac{\dot{m}_m\left(h_{RH1-2} - h_{RH1-1}\right)}{\eta\, W_d} + \frac{\dot{m}_m\left(h_{RH2-2} - h_{RH2-1}\right) - \dot{m}_{wm}h_{wm}}{\eta\, W_d} + \\ & + \left(\dot{m}_m + \dot{m}_{wm}\right)\left(h_{RH3-2} - h_{RH3-1}\right)/\eta\, W_d .\end{aligned} \tag{11}$$

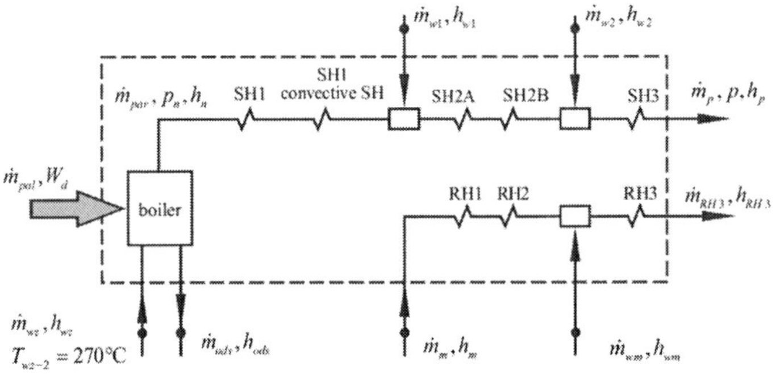

Fig. 2. Diagram of control volume for the boiler mass and energy balance

rocessing Techn. 88, pp. 1082alid only for steady state conditions. The symbols: h_{wz-2}, h_n, h_{w1}, h_{w2}, h_{ods}, h_p, h_{RH3} in equations (10) and (11) denote, respectively, the enthalpy of: feed water after the economizer, steam saturated at the boiler drum pressure, injection water in superheated steam attemperators, brine, live steam at the boiler outlet, and reheated steam at the boiler outlet (Fig. 2).

The calculations of the volume and mass flow rates of the flue gases are based on the determined fuel mass flow rate. Based on the determined excess air number λ using the expression: $\lambda = 21/(21–O_2)$, defined as the ratio of the air to the air flow rate required theoretically, the volume and mass flows of moist flue gases are found. The symbol O_2 denotes the volume fraction of oxygen in the flue gas.

6 Calculation of heat flow rates absorbed by superheater stages and monitoring the degree of fouling of superheaters and reheaters

The degree of fouling of the *i*-th superheater stage with ash ξ_i is calculated continuously from the following expression:

$$\xi_i = \dot{Q}_{i,z} / \dot{Q}_{i,c} \tag{12}$$

where: $\dot{Q}_{i,z}$, in W, denotes the heat flow rate absorbed by the superheater stage affected by fouling, and $\dot{Q}_{i,c}$, in W, is the heat flow rate absorbed by a clean superheater stage. The heat flow rate $\dot{Q}_{i,c}$ at a given value of the excess air number λ is a function of the boiler load. After the boiler is thoroughly cleaned, the measurements of $\dot{Q}_{i,c}$ are made depending on the steam mass flow rate $\dot{m}_i$ flowing through a given superheater stage. The advantage of formula (12) is no need for measuring the temperature of the flue gas after the each superheater stage, which is especially difficult in the regions of high flue gas temperatures.

If the excess air ratio in the boiler or the slagging of the waterwalls of the combustion chamber vary greatly over time, the fouling factor calculated from the formula (12) can lead to less accurate results. An alternative way to assess the degree of ash fouling of a superheater stage is the ratio of the current overall heat transfer coefficient $U_{i,z}$ versus overall heat transfer coefficient for the clean surface $U_{i,c}$.

$$\xi_{U,i} = U_{i,z} / U_{i,c} \tag{13}$$

The overall heat transfer coefficient U_i related to the surface area A_i is calculated in the on-line mode using the following relationship:

$$U_i = \dot{Q}_i / \left(A_i \cdot \Delta T_{m,i} \right) \tag{14}$$

where the symbol $\Delta T_{m,i}$ denotes the mean logarithmic temperature difference (LMTD) between the flue gas and steam. The flue gas temperature at the outlet of the combustion chamber and before each superheater stage was calculated based on the flue gas temperature measured after the last superheater stage in the flue gas duct. The flue gas temperature is then low and easier to measure.

Assuming that the measured value of the flue gas temperature after the entire superheater in known, the flue gas temperature before each stage of the superheater including flue gas temperature at the outlet of the combustion chamber can be determined sequentially. Based on the energy balance equation for the i-th superheater stage

$$\dot{m}_{s,i} \bar{c}_{ps,i} \left(T_{s,i}^{out} - T_{s,i}^{in} \right) = \dot{m}_{g,i} \bar{c}_{pg,i} \left(T_{g,i}^{in} - T_{g,i}^{out} \right) \tag{15}$$

the flue gas temperature $T_{g,i}^{in}$ at the inlet of the *i*-th stage of the superheater can be obtained

$$T_{g,i}^{in} = T_{g,i}^{out} + \left(T_{s,i}^{out} - T_{s,i}^{in} \right) \dot{m}_{s,i} \bar{c}_{ps,i} / \dot{m}_{g,i} \bar{c}_{pg,i} \tag{16}$$

The steam temperature at the inlet $T_{s,i}^{in}$ and outlet of the superheater stage $T_{s,i}^{out}$ are known from measurements. The flue gas temperature $T_{g,i}^{out}$ after the *i*-th superheater stage is known from the measurement or from the calculation of the preceding superheater stage. The steam mass flow rate $\dot{m}_{s,i}$ was also measured, and the mass flow rate of the flue gas $\dot{m}_{g,i}$ was determined based on the known fuel mass flow rate $\dot{m}_F$.

Knowing the temperature of the inlet and outlet exhaust gas and vapor for each stage of the superheater can then be calculated $\Delta T_{m,i}$. When the superheater stage is fouling with ash, then the value of the parameter $\zeta_{U,i}$ is decreasing. The heat flow rate absorbed by the

 DOI: 10.1051/e3sconf/20171401025

superheater is determined based on the energy conservation equation. The control area and the method of the superheater energy balance determination is shown in Fig. 3.

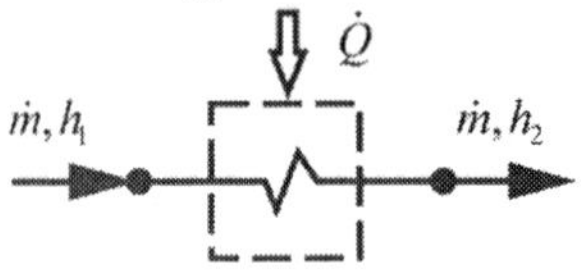

$$\dot{Q} + \dot{m} \cdot h_1 = \dot{m} \cdot h_2$$
$$\dot{Q} = \dot{m}(h_2 - h_1)$$

Fig. 3. The control area for the superheater energy balance

The heat flow rates absorbed by individual stages of the live steam superheater in the OP-380 boiler may be determined from the following expressions (Fig. 1):

- platen superheater, which includes wall superheaters which absorb heat flow rate $\dot{Q}_{SH1}$ and hanging platens which absorb heat flow rate $\dot{Q}_{SH1K}$:

$$\dot{Q}_{GR} = \dot{Q}_{SH1} + \dot{Q}_{SH1K} \tag{17}$$

where

$$\dot{Q}_{SH1} = \dot{m}_{par}\left(h_{SH1-2} - h_n\right) \tag{18}$$

and

$$\dot{Q}_{SH1K} = \dot{m}_{par}\left(h_{SH1K-2} - h_{SH1K-1}\right) \tag{19}$$

Considering that $h_{SH1-2} = h_{SH1K-1}$ (Figs. 1-2) and substituting equations (18) and (19) into (17), the following expression is obtained:

$$\dot{Q}_{GR} = \dot{m}_{par}\left(h_{SH1K-2} - h_n\right) \tag{20}$$

- second stage platen superheater

$$\dot{Q}_{SH2} = \left(\dot{m}_{par} + \dot{m}_{w1}\right)\left(h_{SH2B-2} - h_{SH2A-1}\right) \tag{21}$$

- third stage (final) superheater

$$\dot{Q}_{SH3} = \left(\dot{m}_{par} + \dot{m}_{w1} + \dot{m}_{w2}\right)\left(h_{SH3-2} - h_{SH3-1}\right) \tag{22}$$

Considering that $\dot{m}_{par} + \dot{m}_{w1} + \dot{m}_{w2} = \dot{m}_p$ and $h_{SH3-2} = h_p$, equation (22) assumes the following form:

$$\dot{Q}_{SH3} = \dot{m}_p\left(h_p - h_{SH3-1}\right) \tag{23}$$

The heat flow rates absorbed by individual stages of steam reheaters in the OP-380 boiler may be determined from the following expressions (Figs. 1-2):

- wall reheater

$$\dot{Q}_{SH3} = \dot{m}_p\left(h_p - h_{SH3-1}\right) \tag{24}$$

- intermediate reheater

$$\dot{Q}_{RH2} = \dot{m}_m\left(h_{RH2-2} - h_{RH2-1}\right) \tag{25}$$

- final reheater

$$\dot{Q}_{RH3} = \left(\dot{m}_m + \dot{m}_{wm}\right)\left(h_{RH3-2} - h_{RH3-1}\right) \tag{26}$$

Considering that $\dot{m}_m + \dot{m}_{wm} = \dot{m}_{RH3}$ and $h_{RH3-2} = h_{RH3}$, equation (26) assumes the following form:

$$\dot{Q}_{SH3} = \dot{m}_{RH3}\left(h_{RH3} - h_{SH3-1}\right) \tag{27}$$

Knowing the heat flow rate $\dot{Q}_{i,c}$ absorbed by the clean i-th stage of the superheat at a given boiler load and heat flow rate $\dot{Q}_{i,z}$ absorbed by the fouled i-th stage at the same load of the boiler the fouling degree of the i-th superheater stage can be evaluated in the on-line mode.

Pre-setting the limiting value of the fouling degree ξ_i for i-th superheater stage, the frequency of sootblowers activating can be controlled. If the fouling degree ξ_{is} lowers below the prescribed limit then the sootblower is activated. Steam and water enthalpies were determined using International Steam Tables [15]. The heat flux measurements using different methods are presented in [16-18].

7 On-line monitoring of the steam boiler thermo-flow parameters

The ability to monitor the thermo-hydraulic parameters of steam power boiler is shown on the example of OP-380 unit used in CHP Krakow S.A. RAFAKO design OP-380 steam boiler is drum type, radiant, pulverized coal fired boiler, with natural circulation. Its characteristic is shown in Table 1.

Table 1. Characteristic of the OP-380 power boiler

Boiler type	With natural circ.	Reheated steam pressure	2.65 MPa
Firing system	Tangential	Feed water temperature	227°C
Output capacity	105.6 kg/s	Boiler efficiency	91%
Live steam temperature	540°C	Fuel type	Hard coal
Live steam pressure	13.9 MPa	Fuel low heating value	21.0 MJ/kg
Reheated steam temperature	540°C		

a)

b)

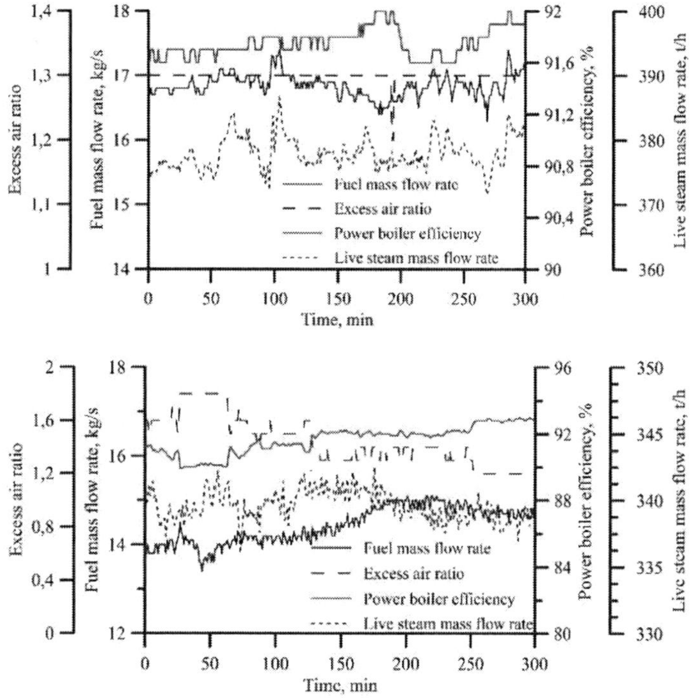

Fig. 4. Live steam and fuel mass flow rates, excess air ratio and efficiency of the power boiler time distributions

The Figs. 4a and 4b shows the time changes of the fuel and live steam mass flow rates and also changes in the excess air ratio and the boiler efficiency. Fig. 4a shows the stable operation of the boiler close to nominal capacity. Throughout the time period the value of the excess air ratio does not change, and the boiler efficiency is high - in the range of 91.6%-92.0%.

A different situation is presented in Fig. 4b. This figure shows large time variation of the excess air ratio. In the first part of the analyzed interval excess air ratio reaches a value of approx. 1.6 which is accompanied by a decrease of the boiler efficiency up to 90%. It is associated with high temperature in the combustion chamber and hence a large heat losses in the flue gases. In the second part of the diagram 4b excess air ratio decreases to 1.2, which is a typical value for this type of boiler. This situation is reflected into an increase in boiler efficiency above 92%.

Heat flow rates received by the power boiler evaporator (Fig. 5), live steam superheater (Fig. 6) and steam reheater are also monitored.

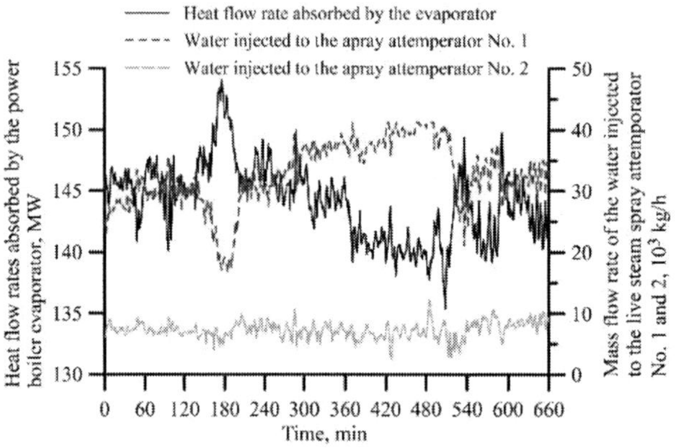

Fig. 5. Time distribution of the heat flow rate absorbed by the boilers evaporator

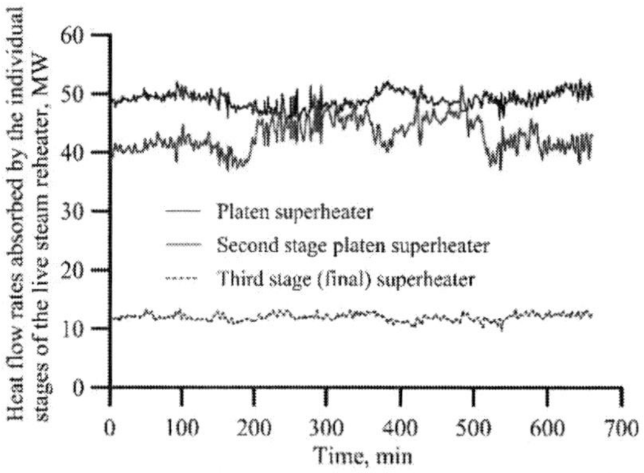

Fig. 6. Time distribution of the heat flow rates absorbed by the live steam superheater stages

On-line monitoring of the heat flux received by the steam boiler heating surfaces can be used to assess the fouling and slagging degree of the combustion chamber and live steam superheater. The increase in fouling of the combustion chamber is reflected by a decrease in live steam flow. This information can be a signal for the operator to start the boiler water or steam soot blowers.

8 Conclusion

The ability of the developed system to monitor thermal and flow parameters, thermal stresses and lifetime of critical pressure components has been shown on example of a steam boiler with natural circulation.

The system of power boilers operation monitoring allows to determine following parameters: heat losses and boiler efficiency, coal mass flow rates, fouling and slagging degrees of the combustion chamber and all superheater stages. The monitoring systems performs the on-line heat transfer calculations of the boiler combustion chamber and heat flow rates absorbed by superheater stages.

An important task of the computer system developed is also determining thermal stresses in critical pressure parts of the boiler and to estimate their remnant life. The proposed computer system ensures economical and safe operation of the boiler in steady states and during load changes. Due to the development of wind farms blocks in thermal power plants should be capable to rapid starts. By installing developed system start-up time of the boiler can be greatly reduced without the risk of overheating superheater tubes and shortening service life of the boiler structural elements.

References

1. H. Bilirgen, *Slagging in PC boilers and developing mitigation strategies.* Fuel **115**, pp. 618-624 (2014)
2. N. S. Harding, D. C. O'Connor, *Ash deposition impact in the power industry*, Fuel Processing Techn. **88**, pp. 1082-1093 (2007)
3. L. M. Romeo, R. Gareta, *Hybrid System for Fouling Control in Biomass Boiler*, Eng. Apps. of Artificial Intelligence **19**, pp. 915-925 (2006)
4. L. M. Romeo, R. Gareta, *Fouling Control in Biomass Boiler.* Biomass and Bioenergy **33**, pp. 854-861 (2009)
5. A. U. Syed, N. J. Simms, J. E. Oakey, *Fireside corrosion of superheaters: Effect of air and oxy-firing of coal and biomass.* Fuel **101**, pp. 62-73 (2012)
6. J. Taler, M. Trojan, D. Taler, *Monitoring of Ash Fouling and Internal Scale Deposits in Pulverized Coal Fired Boilers*, Nova Science Publishers, New York (2011)
7. M. Poniewski, A. Salij, J. Stępień, *Impact of biomass co-firing on selected parameters of a 225MW power unit*, J. of Power Techn. **95**, pp. 84-90 (2015)
8. H. Othman, J. Purbolaksono, B. Ahmad, *Failure investigation on deformed superheater tubes.* Eng. Failure Analysis **16**, pp. 329-339 (2009)
9. A. K. Ray, Y. N. Tiwari, R. K. Sinha, P. K. Roy, S. K. Sinha, R. Singh, S. Chaudhuri, *Remnant life assessment of service-exposed pendent superheater tubes.* Eng. Failure Analysis **9**, pp. 83-92 (2002)
10. N. V. Kuznetsov, V. V. Mitor, I. E. Dubovskij, and E. S. Karasina, *Thermal Calculations of Steam Boilers. Standard Method*, Energy, Moscow, Russia (1973)
11. A. G. Blokh, *Heat Transfer in Steam Boiler Furnaces*, Hemisphere, Washington, (1988)
12. S. Kakaç, *Boilers, Evaporators, and Condensers*, Wiley, New York (1991).
13. J. Taler (Ed.), *Thermal and Flow Processes in Large Steam Boilers. Modeling and Monitoring*, WNT, Warsaw (2011)
14. *ASME PTC 4. Fired Steam Generators. Performance Test Codes*, The ASME, New York (2013)
15. W. Wagner, H. J. Kretzschmar, *International Steam Tables Properties of Water and Steam Based on the Industrial Formulation IAPWS-IF97*, Springer-Verlag, Berlin (2008)
16. J. Taler, D. Taler, P. Ludowski,. *Measurements of local heat flux to membrane water walls of combustion chambers*, Fuel **115**, pp. 70–83 (2014)
17. J. Taler, D. Taler,. Measurement of Heat Flux and Heat Transfer Coefficient. Chapter 1, in: Heat Flux: *Processes, Measurement Techniques and Applications*, Editors: G. Cirimele and M. D'elia, Nova Science Publishers New York, pp. 1-104 (2012)
18. J. Taler, D. Taler, Measurements of Local Heat Flux and Water-Side Heat Transfer Coefficient in Water Wall Tubes, Chapter 1, in *Heat Transfer*, Editor: D. M. Salim Newaz Kazi, InTech, Rijeka – Shanghai, pp. 3-34 (2012)

E3S Web of Conferences 14, 01026 (2017) DOI: 10.1051/e3sconf/20171401026
Energy and Fuels 2016

Visualization and research of gas-liquid two phase flow structures in cylindrical channel

Sebastian Stefański[1*], *Wojciech* Kalawa[1], *Kaja* Mirek[1] and *Maciej* Stępień[1]

[1] AGH University of Science and Technology, Faculty of Energy and Fuels, Department of Thermal and Fluid Flow Machines, al. Mickiewicza 30, 30-059 Krakow

Abstract. Two-phase flows are commonly found in many industries, especially in systems, where efficient and correct functioning depend on specific values of flow parameters. In thermal engineering and chemical technology the most popular types of two-phase mixture are gas-liquid or liquid-vapour mixtures. Bubbles can create in flow different structures and determine diverse properties of flow (velocity of phase, void fraction, fluctuations of pressure, pipe vibrations, etc.). That type of flow is difficult to observe, especially in liquid-vapour mixture, where vapour is being made by heating the medium. Production of vapour and nucleation process are very complicated issues, which are important part of two-phase flow phenomenon. Gas-liquid flow structures were observed and described with figures, but type of structure depends on many parameters. Authors of this paper made an attempt to simulate gas-liquid flow with air and water. In the paper there was presented specific test stand built to observe two-phase flow structures, methodology of experiment and conditions which were maintained during observation. The paper presents also the structures which were observed and the analysis of results with reference to theoretical models and diagrams available in literature.

1 Introduction

Two-phase flows are commonly found in many industries, especially in systems, where efficiency is determined by specific parameters of mixture flow. In thermal and chemical engineering the most popular types of two-phase flows are gas-liquid and liquid-vapour mixtures. In heat and power engineering two phase flows can be found in boiler pipes producing steam and in evaporators of refrigerating systems. Structure of two-phase flow has an important impact on heat and mass transfer, so it is important to observe and specify characteristics of structures. Productions of vapour and nucleation process are very complicated issues, which are important part of two-phase flow phenomenon. Gas-liquid flow structures were observed and described with figures, but type of structure depend on many parameters, such as diameter of the pipe, volume flow rate of each phase, void fraction and heat flow value [1].

* Corresponding author: stefanski@agh.edu.pl

DOI: 10.1051/e3sconf/20171401026

2 Fundamentals

Two-phase flow structures are determined by position (angle) of pipe and flow direction of the mixture (upwards or downwards). For vertical orientation and upwards flow structures from figure 1 can be observed. In case of downwards flows the observed structures are different. There is a lack of a common opinion about their systematics in the literature. The exemplary classification is given below, in the figure 2.

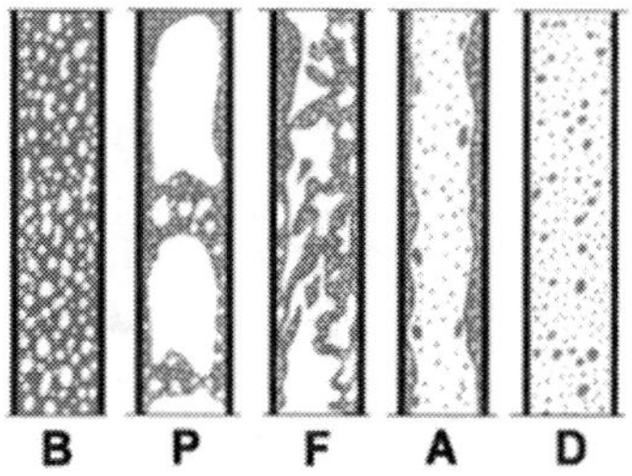

Fig. 1. Two-phase flow structures for vertically oriented channel (upwards flow) [2]

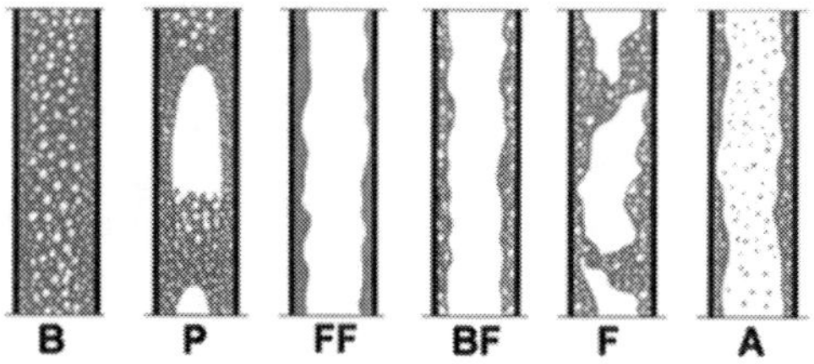

Fig. 2. Two-phase flow structures for vertically situated channel (downwards flow) [2].
B –bubble flow, P – plug flow, F – floating film flow, BF – bubble-foam flow, F – foam flow, A – annular flow

The structures presented in figure 1 are described as follows:
- bubble flow (B) – gas phase constitute dispersed bubbles in continuous liquid phase, with similar velocity;
- plug flow (P) – gas phase constitute plugs occupies almost the whole diameter of the channel;
- foam flow (F) – continuity of liquid phase is disturbed by high volume flow of gas phase;
- annular flow (A) – liquid phase flows on inner surface of the channel with high velocity of gas phase in the middle;
- dispersed flow (D) – high volumetric flow rate of the gas phase causes formation of small drops of liquid [2].

Due to the fact that the boundary between structures is difficult to observe, it is common to apply simpler classification including below mentioned structures:
- bubble flow (B),
- transitory flows (P, F),
- annular flows (A, BF, FF).

Radically different structures can be observed in horizontally situated channels. The classification of two-phase flow structures suggested by Baker is presented in the figure 3 [2].

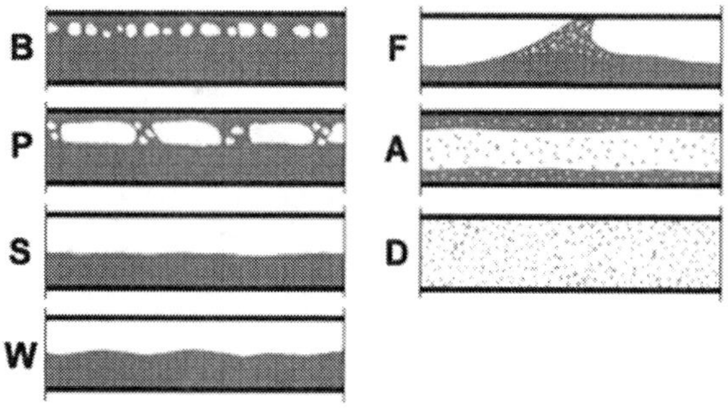

Fig. 3. Two-phase flow structures for horizontally oriented channel [2].

 DOI: 10.1051/e3sconf/20171401026

The structures presented in the figure above are defined as:

- bubble flow (B) – in the continuous liquid phase there are bubbles cumulating in the upper part of the channel;
- plug flow (P) – gas phase flows in continuous liquid phase in round shape plugs cumulated in upper part of a channel;
- stratified flow (S) – continuous gas phase flows in the upper part of the channel, phases separation surface is smooth;
- wavy flow (W) – the structure which is formed as a result of volumetric flow rates increase;
- flash flow (F) – phases separation surface is wavy and wave picks touches upper wall of the channel;
- annular flow (A) – liquid phase constitute thin layer on the inner channel surface, gas phase flows in the middle of the pipe, asymmetry of the liquid layer thickness there may occur and the surface between phases is wavy;
- disperse flow (D) – the structure found when the gas phase velocity is significantly higher than liquid, drops are entrained from wave picks; for extreme conditions liquid layer fades and the whole medium flows as small drops [2].

Extent of structure occurrence can be presented on chart known as regime map depending on particular parameters, where structures are represented by areas separated with lines (fig. 4). The regime maps can be based on different coordinate systems, channel geometry and the direction of mixture flow. Usually on the axes can be values of velocities, volumetric flow rates, mass flow rates or phases' physical properties such as viscosity, density, surface tension. Lines separating the areas of structures cannot be considered as the strict boundary of two phenomena. It is the result of the complexity of two-phase flow issue reflected in combining two different types of flow structure [3].

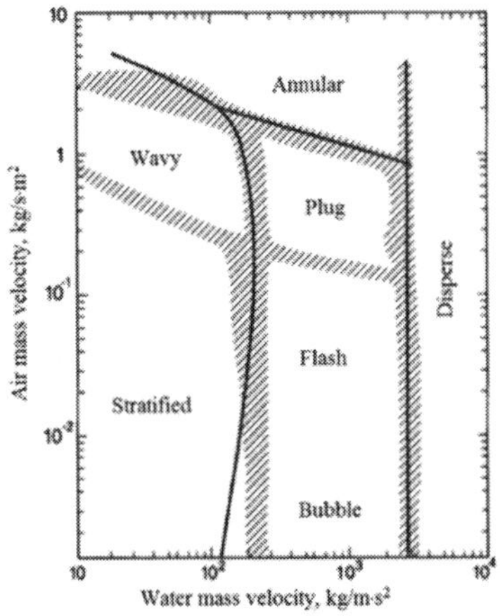

Fig. 4. Weisman chart. The regime map of air-water mixture flow in cylindrical channel (int. diameter 51mm) depending on air and water mass flow rates [3].

3 Methodology of experiment

3.1 Test stand

Two-phase flow structures were observed using the test stand consisting of several autonomic parts (fig. 5, 6). The basic element of this stand is pumping system including impeller pump (10), suction and delivery pipelines with flow regulation valve (1), pressure transducer (2),

DOI: 10.1051/e3sconf/20171401026

water temperature sensor (3), flowmeter (4). Pump operation parameters and pipelines diameters determined technical characteristics of farther part of the stand. Pumping system was equipped with control unit which enables to control volumetric flow rate of water and display values from sensors. Cylindrical channel used for observations of two-phase flow structures is made of acrylic, an 1800 mm long pipe with internal diameter 56 mm (7). The pipe was permanently connected to steel mountings. This element was connected to steel rack using flange coupling. To the flanges there are connected normalized hydraulic pipes and hoses. In the flange installed on the delivery pipeline there were two 5 mm channels used for supply air to transparent channel (5, 6).

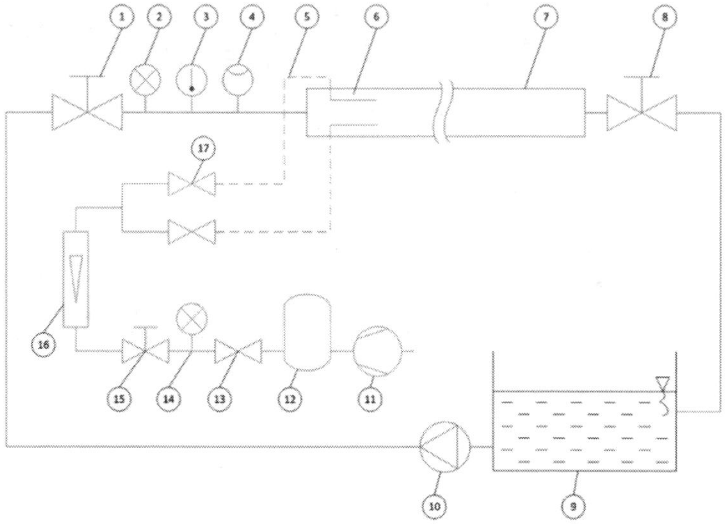

Fig. 5. Test stand diagram. 1 – flow regulation valve, 2 – pressure transmitter, 3 – thermometer, 4 – flowmeter, 5 – air flexible hose, 6 – air supply system, 7 – acrylic channel, 8 – pressure regulation valve, 9 – opened water container, 10 – circulation pump, 11 – compressor, 12 – gas cylinder, 13 – pressure regulation valve, 14 – manometer, 15 – flow regulation valve, 16 – rotameter, 17 – valve

The aim of the project was to build a compact test stand which enables observing two-phase flow structures in the horizontally and vertically oriented pipe with downwards and upwards direction of flow. For this reason the steel rack is connected to basic frame with bolt. That connection allows changing the angle between the transparent channel and ground.

Water is pumped (10) to the flow regulation valve (1) and then it flows through the pressure transducer (2), temperature sensor (3), and flowmeter (4). With the flexible hose water is directed to transparent acrylic channel (7). Behind the channel there is the DN65 valve (8) used for controlling the mixture pressure. With the second flexible hose water is directed to the opened container (9) where the phases of mixture are separated. From that container water transported to the suction port of the pump. The other phase in the mixture is air supplied with the compressor (11) to the gas cylinder (12). Pressure of air can be regulated with the valve (13). Compressed air flows through the flexible 8mm air hose to the manometer (14), flow regulation valve (15), rotameter (16) and then to T-pipe splitting the flow between two ports in flange connected to the channel. Air flow in each port can be regulated with the valves (17) installed on flexible hoses (5). The air is distributed to the transparent channel with two 10 cm long hoses with a diameter 6 mm (6).

Two-phase flow structures are observed with high speed video camera (300 FPS) (5, fig. 6) and graphically edited with computer software. For better lightning of the channel there is LED band installed on steel rack under the channel, and the matt screen placed behind the channel (fig. 7).

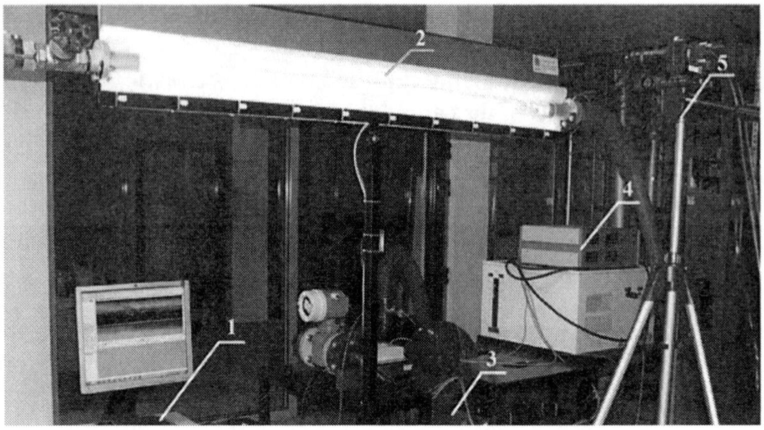

Fig. 6. Photo of test stand. 1 – data canvassing station, 2 – acrylic channel, 3 – pumping system, 4 – control unit, 5 – high speed video camera

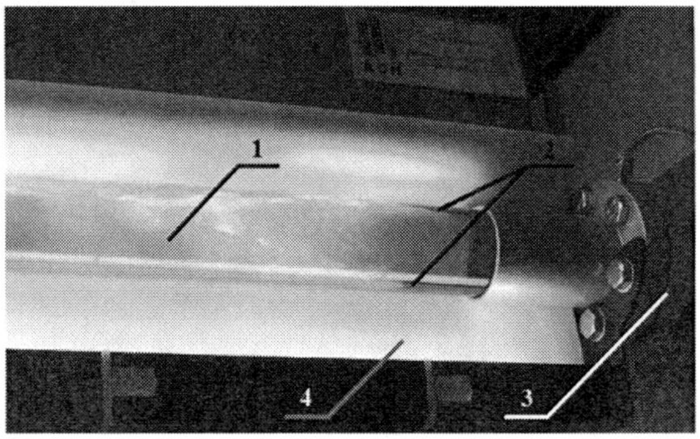

Fig. 7. Photo of inlet of acrylic channel and air supply system. 1 - acrylic pipe, 2 - air supply hoses, 3 - flange coupling, 4 - screen

3.2 Experiment procedure

For the purpose of making the analysis of the results easier and enabling the reference to studies available in literature, there were recreated conditions similar to the air-water flow structures studies by Weismann [3] (horizontal and vertical downwards direction of the flow) and Barnea [4] (vertical upwards direction of the flow). The geometry of pipe applied in those studies (int. diameter 51 mm) was the most similar to acrylic channel in authors' test stand (int. diameter 56 mm). Conditions set in Weismann's and Barnea's experiments were following: net pressure of the mixture 1 bar, temperature 25°C. To achieve similar conditions of experiment, air was being supplied with one orifice located in the inlet of the channel.

During the experiment two-phase flow structures in horizontal and vertical orientation were observed. For each configuration of the channel there was carried out series of measurements for volumetric air and water flows given in the table 1.

Table 1. Test point flow values.

Water volumetric flow rate, $\dot{V}_w$, l/min	10		30		100		200		400	
Air volumetric flow rate, $\dot{V}_a$, l/min	10	50	10	50	10	50	10	50	10	50

The test stand was set in particular configuration before each series of experiment. Water volumetric flow rate was set on value given in table 1 and after stabilization of flow parameters; air was supplied with the net pressure of 1 bar. When volumetric flow rate

DOI: 10.1051/e3sconf/20171401026

fluctuations were less than 1 l/min, the net pressure of the mixture were set with the regulation valve on 1 bar. Observed structures were registered with high speed video camera on the length of 0.4 m. This procedure was repeated for each test point. Canvassed sequences of frames were subjected to graphic processing and compared to studies and diagrams available in literature.

4 Results of experiment

Tests taken according to described methodology enabled to observe and canvass two-phase flow structures for three stand configurations. The results of experiment, that is type of observed structure and volumetric flow rates of phases, are collected in table 2.

Table 2. The results of experiment for three configurations of test stand.

Direction of the flow	Vertical, downwards		Vertical, upwards		Horizontal	
Water volumetric flow rate [l/min]	Air volumetric flow rate [l/min]					
	10	50	10	50	10	50
10	Annular	Annular	Plug	Plug	Stratified	Stratified
30	Annular	Annular	Foam	Foam	Stratified	Wavy
100	Annular	Combined	Bubble	Foam	Flash	Wavy
200	Bubble	Bubble	Bubble	Bubble	Flash	Flash
400	Bubble	Bubble	Bubble	Bubble	Bubble	Bubble

For horizontally oriented channel structures of two-phase flow are radically different from the vertically ones. The reason of those differences is fact that buoyant and gravity force act perpendicularly to the direction of the flow. Structures which were observed for that configuration are consistent with theoretical classification [2]. They are presented in figure 8.

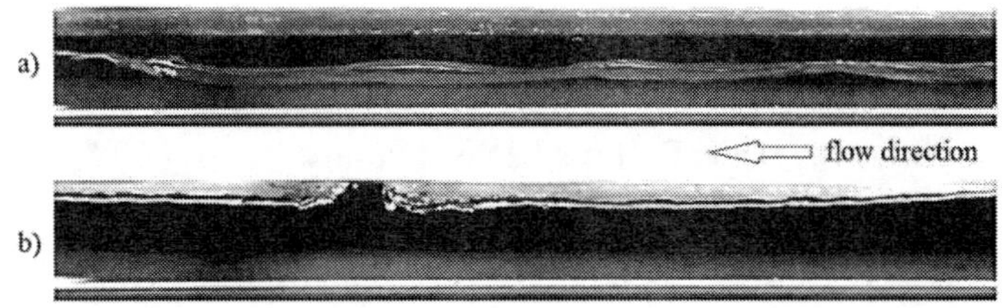

Fig. 8. Structures of two-phase flow for horizontally directed flow. a – wavy flow (water volumetric flow rate 30 l/min, air volumetric flow rate 50 l/min), b – flash flow (water volumetric flow rate 100 l/min, air volumetric flow rate 10 l/min). Flow direction from right to left.

Characteristic feature of downwards flow in vertical orientation is deceleration of air bubbles in relation to water velocity. It is caused by buoyant force acting on bubbles and oriented upwards. For considered range of test points there were two structures included in classification described by Oschinovo and Charles [5]. In the figure 9 there are presented photos of two structures which were observed during experiment. For upwards flow the buoyant force caused acceleration of the bubbles in relation to liquid phase velocity. For that reason the structures of flow presented above appear for different test points than for downwards flow configuration. Exemplary structures are presented in figure 10.

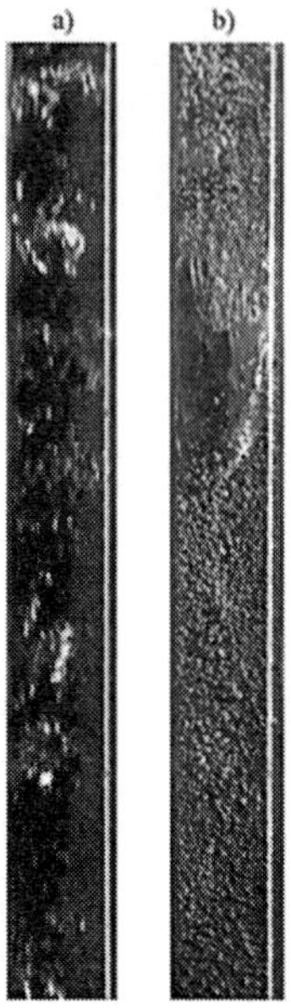

Fig. 9. Structures of two-phase flow for upwards flow. a – bubble flow (water volumetric flow rate 200 l/min, air volumetric flow rate 10 l/min), b – plug flow (water volumetric flow 10 l/min, air volumetric flow rate 10 l/min)

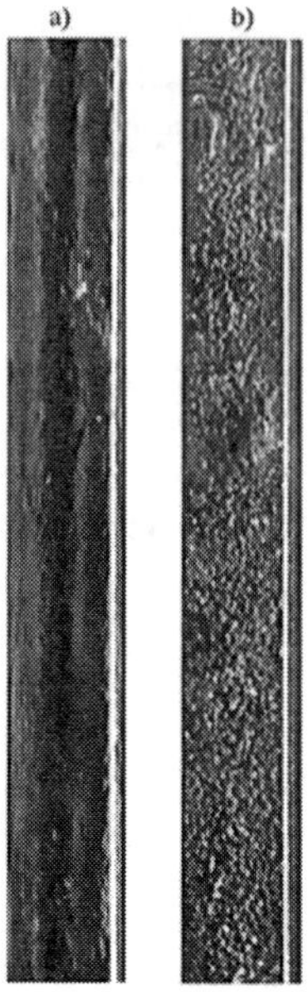

Fig. 10. Structures of two-phase flow for downwards flow. a – annular flow (water volumetric flow 30 l/min, air volumetric flow rate 10 l/min), b – bubbles flow (water volumetric flow rate 100 l/min, air volumetric flow rate 10 l/min)

4.1 Experimental uncertainties

For verification of the experiment results with regime maps, there is essential to calculate other parameters characterising two-phase flow, such as apparent velocities of phases and apparent mass velocities in flow.

The test stand was equipped with following measuring devices:

- pressure transmitter Huba Control 691 -1÷9 bar, max. measurement error 0.3% f.s.,
- RTD thermometer Endress+Hauser Omnigrad T TST487, max. measurement error ±0.25Ω (25°C),
- electromagnetic flowmeter Endress+Hauser Proline Promag 10W, max. measurement error ±0.5% o.r.,
- spiral pressure gauge -1÷10 bar, max. measurement error ±0.2 bar,
- rotameter Dwyer MMA 5÷50 l/min, max. measurement error 4% f.s.

On the basis of measurement errors of apparatus the calculation of apparent velocities and mass velocities measurement uncertainties were carried out. For that calculation best-worst case analysis were used. Values of air flow rate read from rotameter were calculated for each test point because of nonstandard operating conditions as follows:

$$\dot{V}_a = \dot{V}_{ar} \cdot \sqrt{\frac{p \cdot T_N}{p_N \cdot T}} \tag{1}$$

where:

$\dot{V}_a$ - standard air flow rate corrected for pressure and temperature
$\dot{V}_{ar}$ - actual air flow rate reading
p - actual absolute air pressure, bar
T_N - normal temperature NTP, 293.15 K

p_N - normal pressure NTP, 1 bar
T - actual air temperature, K

For temperature of air T=(25±1)°C and pressure p_a=(1±0.2) bar there was estimated air density ρ_a=(2.34±0.24) kg/m^3 on the basis of ideal gas law [6]. Density of water was estimated to be (997.1±0.1) kg/m^3 [6]. The next step was calculation of apparent velocities of phases.

The apparent velocity is defined as the linear velocity on the assumption that the whole volumetric flow rate passes through full cross-section area of the channel. It can be calculated using formula (2) [7].

$$u_i = \frac{\dot{V}_i}{A} \tag{2}$$

where:
u_i - apparent velocity of phase i, m/s
$\dot{V}_i$ - volumetric flow rate of phase i, m^3/s
A - channel cross-section area, 24.64 ·10^{-4} m^2

Apparent velocities uncertainties were estimated with best-worst case analysis using flow rate measurement error. The results are presented in table 3.

By analogy, apparent mass flow rate of the phase is defined as mass flow rate in relation to cross-section of the channel like in the formula (3) [7] given below.

$$w_i = \frac{\dot{m}_i}{A} = \frac{\rho_i \cdot \dot{V}_i}{A} = \rho_i \cdot u_i \tag{3}$$

where:
w_i - apparent mass velocity of phase i, kg/(s·m^2)
$\dot{m}_i$ - mass flow rate of phase i, kg/(s·m^2)
ρ_i - density of phase i, m^3/kg
A - cross-section area, m^2
u_i - apparent velocity of phase i, m/s

Apparent mass velocities uncertainties were estimated by analogy to previous analyses, using flow rate measurement error and density uncertainties. The results are presented in table 4.

Table 3. Apparent velocities and mass velocities of phases in channel for different flow rates.

Water volumetric flow rate, $\dot{V}_w, \frac{l}{min}$	10	30	100	200	400
Water volumetric flow rate, $\dot{V}_w, \frac{dm^3}{s}$	0.167 ±0.001	0.500 ±0.003	1.667 ±0.008	3.333 ±0.017	6.667 ±0.033
Apparent water velocity, $u_w, \frac{m}{s}$	**0.0677 ±0.0003**	**0.2030 ±0.0010**	**0.6767 ±0.0034**	**1.3534 ±0.0068**	**2.7070 ±0.0135**
Apparent mass water velocity, $w_w, \frac{kg}{s \cdot m^2}$	**68 ±1**	**202 ±1**	**675 ±3**	**1349 ±7**	**2699 ±14**
Air volumetric flow rate, $\dot{V}_a, \frac{l}{min}$	10		50		
Air volumetric flow rate, $\dot{V}_a, \frac{dm^3}{s}$	0.167 ±0.033		0.833 ±0.033		
Apparent air velocity, $u_a, \frac{m}{s}$	**0.0677 ±0.0135**		**0.3383 ±0.0135**		
Apparent mass air velocity, $w_a, \frac{kg}{s \cdot m^2}$	**0.158 ±0.048**		**0.792 ±0.113**		

 DOI: 10.1051/e3sconf/20171401026

The calculations described with formulas (1) and (2) enabled to give results on regime maps available in literature [3] [4]. The results of experiment given on the abovementioned regime maps are presented in the figures 11-13 below.

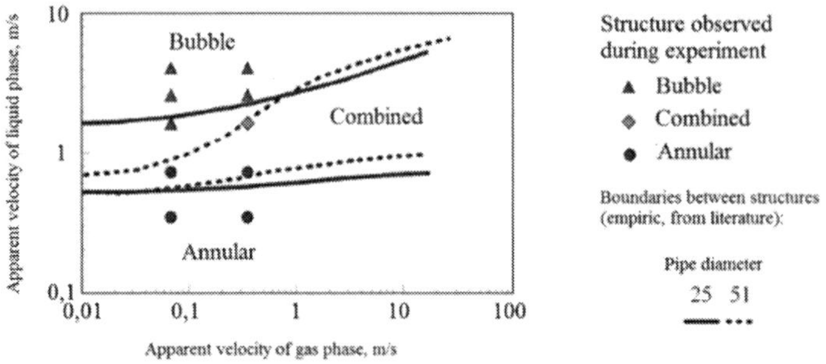

Fig. 11. Experimental results in comparison to Barnea's regime map [4] for downwards vertical gas-liquid flow in pipes with diameter of 25 mm and 51 mm, net pressure 0,1 MPa and temperature of the mixture 25°C.

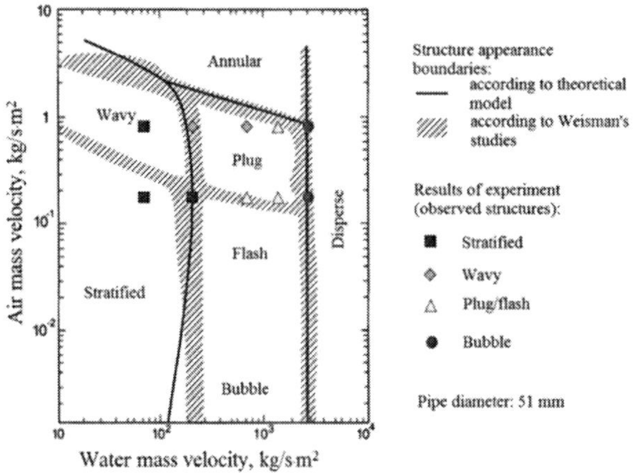

Fig. 12. Experimental results in comparison to Weisman's regime map [3] for horizontally directed gas-liquid flow in pipe with diameter of 51 mm, net pressure 0,1 MPa and temperature of mixture 25°C.

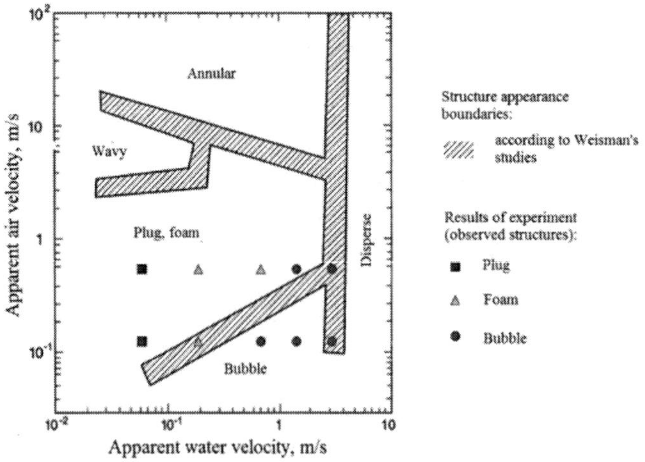

Fig. 13. Experimental results in comparison to Weisman's regime map [3] for upwards vertical gas-liquid flow in pipe with diameter of 25 mm, net pressure 0,1 MPa and temperature of mixture 25°C.

5 Conclusions

Comparing the experimental results to regime maps developed for geometrically similar systems working under close net pressure lead to the conclusion that the test stand correctly fulfills its role. A great majority of observed two-phase flow structures can be adjusted to the areas from literature regime maps. Slight deviations from empirically developed areas in regime maps are caused by gradual change of the structure when changing the parameters of the flow and geometry of considered channel. The boundaries defined on the maps describe the probability of changing from one structure to the other, not the exact conditions for its appearance [1]. The performed tests proved that there is possibility of existence of different two-phase flow structures and strong differences in interfacial turbulence. Disturbances in the two-phase flow in industrial systems have direct influence on heat transfer coefficient value, which determine the intensity of heat transfer, for example during boiling process.

Making the comparison of results obtained in experiment to those found in literature shows that the published regime maps cannot be taken as universal ones. Possibility of appearance of the particular structure is determined by many parameters, for instance temperature, net pressure, geometry of the channel and thermodynamic characteristic of fluids. For that reason adjusting the map for comparing one experiment to the other must consider the compatibility of mentioned parameters.

Test stand does not enable achieving as high phases velocities as presented on regime maps available in literature. It is cause by volumetric flow rate limit of pump and compressor installed in the test stand. Disperse flow could not be observed during the experiment for any flow direction. Despite described limits and problems, the test stand provides scientific and didactic value, and with some improvements can be used for further analysis or observations.

References

1. A. Kuchczyńska, *Warunki pracy aparatu ze wznoszącym i opadającym przepływem dwufazowym gaz-ciecz*, doctor's thesis, Opole (2010)
2. Didactic materials for subject *Operacje Mechaniczne w Inżynierii Procesowej*, Opole University of Technology (2015)
3. J. Weisman, *Two-Phase Flow Patterns. Handbook of Fluids in Motion*, Ann Arbor Science Publ., (1983)
4. D. Barnea, *A unified model for predicting flow-pattern transitions for the whole range of pipe inclinations*, International Journal of Multiphase Flow (1987)
5. R. Koch, A. Noworyta, *Procesy mechaniczne w inżynierii chemicznej*, WNT (1992)
6. K. Ražnjević, *Handbook of Thermodynamic Tables and Charts,* Hemisphere Publishing Corporation (1976)
7. Z. Orzechowski, *Mechanika płynów w inżynierii środowiska*, WNT (1997)

 DOI: 10.1051/e3sconf/20171401027

Polyhedral meshing as an innovative approach to computational domain discretization of a cyclone in a fluidized bed CLC unit

Marcin Sosnowski[1,*], *Jaroslaw* Krzywanski[1], and *Renata* Gnatowska[2]

[1]Jan Dlugosz University in Czestochowa, Faculty of Mathematics and Natural Sciences, 13/15 Armii Krajowej Av, 42-200 Czestochowa, Poland
[2]Czestochowa University of Technology, Faculty of Mechanical Engineering and Computer Science, 21 Armii Krajowej Av, 42-200 Czestochowa, Poland

Abstract. Chemical Looping Combustion (CLC) is a technology that allows the separation of CO_2, which is generated by the combustion of fossil fuels. The majority of process designs currently under investigation are systems of coupled fluidized beds. Advances in the development of power generation system using CLC cannot be introduced without using numerical modelling as a research tool. The primary and critical activity in numerical modelling is the computational domain discretization. It influences the numerical diffusion as well as convergence of the model and therefore the overall accuracy of the obtained results. Hence an innovative approach of computational domain discretization using polyhedral (POLY) mesh is proposed in the paper. This method reduces both the numerical diffusion of the mesh as well as the time cost of preparing the model for subsequent calculation. The major advantage of POLY mesh is that each individual cell has many neighbours, so gradients can be much better approximated in comparison to commonly-used tetrahedral (TET) mesh. POLYs are also less sensitive to stretching than TETs which results in better numerical stability of the model. Therefore detailed comparison of numerical modelling results concerning subsection of CLC system using tetrahedral and polyhedral mesh is covered in the paper.

1 Introduction

1.1 Simulation of fluidized bed CLC unit

The combustion of fossil fuels in energy production and transportation is the main source of climate change. Therefore the development of new combustion technologies characterized by low pollutant emission rates is in the spotlight of contemporary research. Therefore the Circulating Fluidized Bed (CFB) boilers are popular alternatives to the traditional pulverized coal boilers mostly due to decreased emission, low combustion temperature and

[*] Corresponding author: m.sosnowski@ajd.czest.pl

high fuel flexibility. In order to segregate the CO_2, the Chemical Looping Combustion (CLC) was introduced to CFB designs. The principle of CLC was developed and patented in 1954 by Lewis and Gilliland [1]. A key factor in the CLC performance is the Oxygen Carrier (OC), which circulates between the air reactor and fuel reactor (Fig. 1). Apart from the OC also the process design is important. Generally it consists of two reactors: the air reactor in which the OC is loaded with oxygen and the fuel reactor where the OC supplies the oxygen for fuel combustion. The majority of process designs currently under investigation are systems of coupled fluidized beds. The system analysed within the confines of the research is based on a design presented in [2] with bubbling fluidized bed as fuel reactor and a fast fluidized bed riser as air reactor.

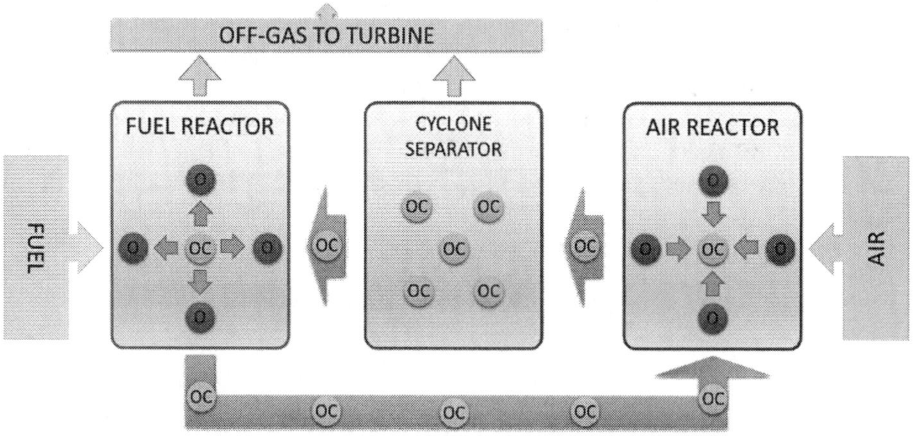

Fig. 1. Schematic diagram of CLC (O – oxygen, OC – oxygen carrier).

The heat and mass transfer between the phases, the chemical reactions and complex hydrodynamics of CFBs are still not completely understood and difficult to predict despite its widespread applications. The knowledge concerning CFB systems relied mostly on experimental studies but in recent years researchers have focused on improving CFB efficiency by utilizing the capabilities of numerical methods. With the rapid development of high performance computers, various techniques such as artificial intelligence including neurocomputing and fuzzy logic approach [3-7] were used in CFB modelling [8, 9] but Computational Fluid Dynamics (CFD) turned out to be one of the best tools in such research and plays an important role in the design and optimization of industrial systems [10]. A common challenge for CFD is the expensive computational cost in terms of time required to reach a converged solution. The very first issue influencing the CPU-time is the computational domain discretization, therefore the research carried out within the confines of this paper focus on the innovative approach to discretization of a fluidized bed CLC unit.

1.2 Polyhedral meshing

Advances in aerospace [11], automotive [12], civil engineering [13], medicine [14] and other numerous branches of industry could not have been achieved without applying numerical modelling as a research tool. The primary and critical activity in numerical modelling is the computational domain discretization. This process consists in dividing the analysed geometry into numerous small control volumes and it is commonly called "meshing". In fluid dynamics applications it is the most desirable to divide the domain into hexahedral (HEX) control volumes denoted red in Fig. 2 as the resulting mesh is characterized by low numerical diffusion particularly in case of flow perpendicular to the faces of control volumes. Unfortunately it is not always possible to construct structured HEX mesh for complex geometries or it can be particularly time-consuming part of CFD analysis. Therefore semi-automatic tetrahedral (TET) mesh generation algorithms have been implemented into commercial pre-processors. TETs denoted blue in Fig. 2 are the simplest volume elements as they are built out of four faces. The faces are plane segments, so both face and volume centroid locations are well defined. The unquestionable advantage of TET mesh is the ease of its generation even in case of complicated geometry. On the other hand TETs cannot be stretched excessively, so significantly larger number of elements has to be used in comparison to structured HEX mesh in order to achieve a reasonable accuracy in boundary layers, long channels or small gaps. Moreover TETs have only four neighbours, therefore computing gradients can be problematic due to spatial position of neighbour nodes, which may lie in nearly one plane. In consequence, the numerical diffusion of TET mesh is significantly higher than HEX mesh and low quality TETs result in convergence errors and significantly reduce the accuracy of the solution.

The polyhedral (POLY) mesh denoted green in Fig. 2 was introduced in STAR-CCM+ solver and ANSYS Fluent in order to combine the advantages of HEX (low numerical diffusion resulting in accurate solution) and TET mesh (rapid semi-automatic generation) as well as to overcome the disadvantages of both the above mentioned types of control volumes.

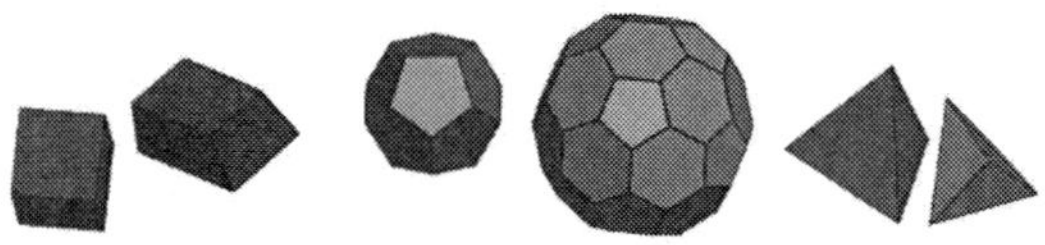

Fig. 2. Hexahedral (red), polyhedral (green) and tetrahedral (blue) control volumes.

The major benefit of POLYs is that each individual cell has many neighbours, so gradients can be much better approximated in comparison to TETs. POLYs are also less sensitive to stretching than TETs which results in better numerical stability of the model. In some cases, POLYs can even achieve better accuracy than HEXs due to larger number of neighbouring elements. They allow the exchange of mass over a larger number of faces, reducing numerical diffusion effects caused by flows not perpendicular to any of the cell faces. This proves to be advantageous in situations where no prevailing flow direction can be identified and leads to a more accurate solution achieved with a lower cell count.

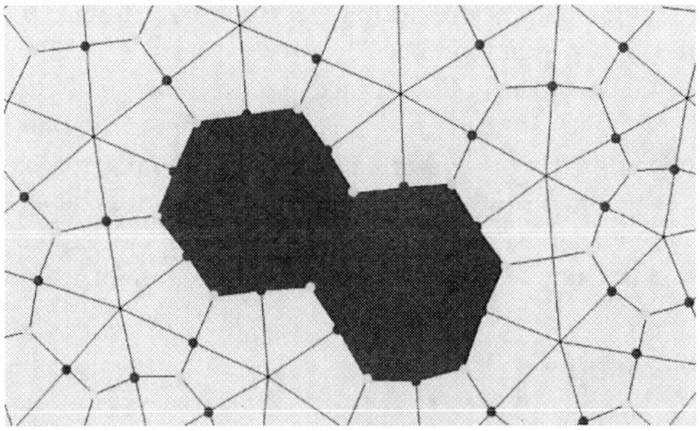

Fig. 3. TET to POLY conversion: edge centroids (red dots) and face centroids (yellow dots) of TETs.

POLY mesh generation in commercial CFD code like ANSYS Fluent consists in simple conversion of TET elements to POLYs by decomposition of cell into multiple sub-volumes depicted in Fig. 3. In order to do so, new edges are created on each face between the face centroid (yellow dots in Fig. 3) and the centroids of the edges of that face (red dots in Fig. 3). Subsequently new faces are created within the cell by connecting the cell centroid to the new edges on each face. The newly-created faces may be adjusted and merged with neighbouring faces during the agglomeration process in order to minimize the number of faces on the resultant polyhedral cell.

 DOI: 10.1051/e3sconf/20171401027

2 Methods

2.1 Research object

The lab-scale fluidized bed CLC unit was the object of the carried out research. The unit was designed on the basis of the design presented by Thon et al. [2] and it was divided into three main parts: fuel reactor, air reactor and cyclone separator (Fig. 4). The last mentioned is the most demanding part of the unit in terms of computational fluid dynamics therefore it was selected for detailed analysis in the preliminary stage of the research concerning the evaluation of polyhedral mesh application in discretization of the computational domain.

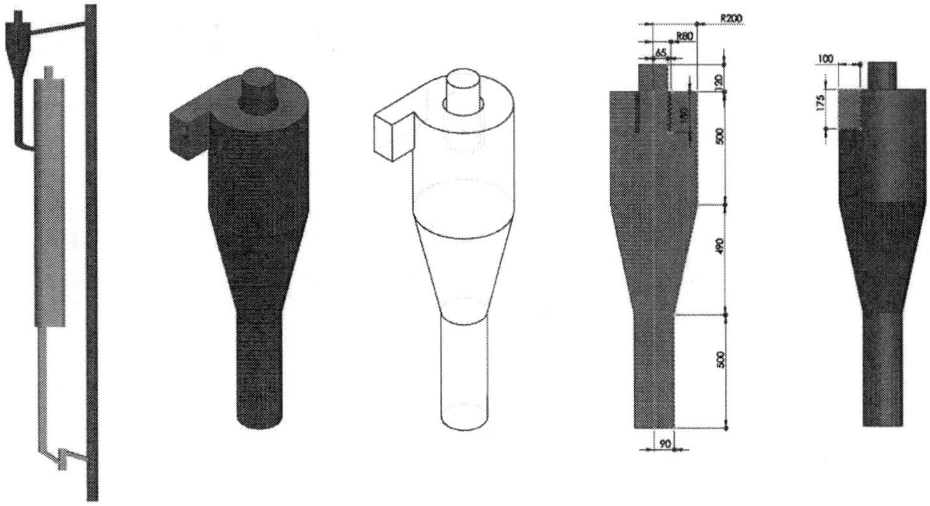

Fig. 4. The analysed fluidized bed CLC unit: air reactor (red), fuel reactor (green), cyclone separator (blue) and detailed view of the cyclone.

2.2 Analysed configurations of computational domain discretization

The computational domain was discretized with TET cells (Fig. 5) using ANSYS Meshing. Six configurations of input parameters were used in order to generate TET meshes of number of cells ranging from approx. 35k to 516k.

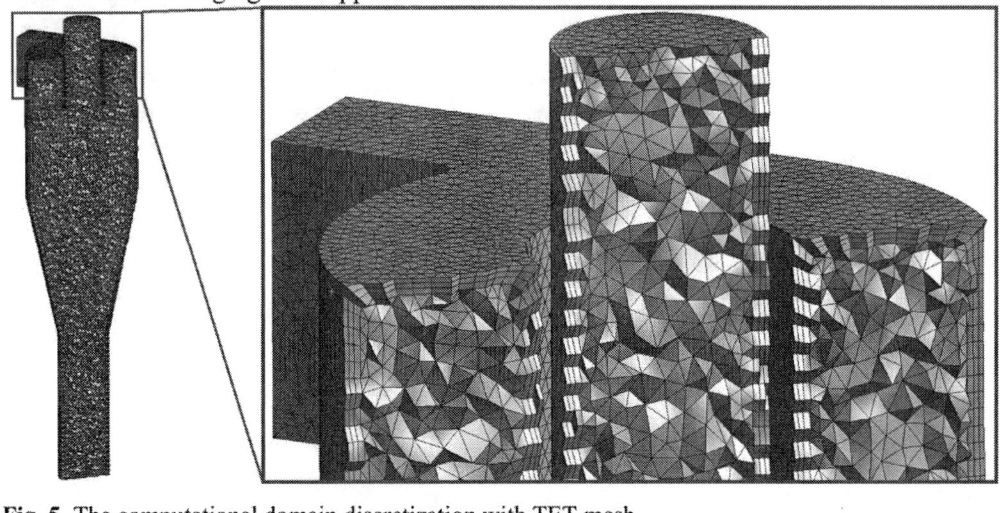

Fig. 5. The computational domain discretization with TET mesh.

The uniform size function with specified max face size and max TET size according to Fig. 6 was used. Moreover inflation of 3 layers, growth rate equal 1.2 and maximum thickness as specified in Fig. 6 was generated by post algorithm on all surfaces except for inlet and outlet. Each TET mesh was converted to POLY mesh in ANSYS Fluent.

tetrahedral mesh						
Max Face Size	0.03	0.025	0.02	0.017	0.015	0.01
Max TET Size	0.045	0.0375	0.03	0.026	0.0225	0.015
Inflation	0.03	0.025	0.02	0.017	0.015	0.01
No. of cells	**34 393**	**49 195**	**88 520**	**137 036**	**193 117**	**515 859**
polyhedral mesh (conversion)						
No. of cells	**18 913**	**23 345**	**35 260**	**50 311**	**65 862**	**153 422**

Fig. 6. Parameters of investigated meshes.

The minimum orthogonal quality for both the TET mesh and POLY mesh did not decrease below 0.11 (orthogonal quality ranges from 0 to 1, where values close to 0 correspond to low quality) and maximum ortho skew did not exceed 0.89 (ortho skew ranges from 0 to 1, where values close to 1 correspond to low quality). These quality factors indicate good quality of the generated meshes.

2.3 Reference mesh

The structured HEX reference mesh of approx. one million cells was generated in order to validate the results obtained using TET and POLY meshes of various number of cells. The HEX elements are considered as the most desirable in CFD analysis and the mesh-dependency studies carried out within the research proved, that the solution obtained with approx. one million cells is mesh-independent. Therefore the results obtained by calculations using the reference mesh are considered as the baseline for subsequent analysis of evaluation of POLY mesh application in discretization of the computational domain.

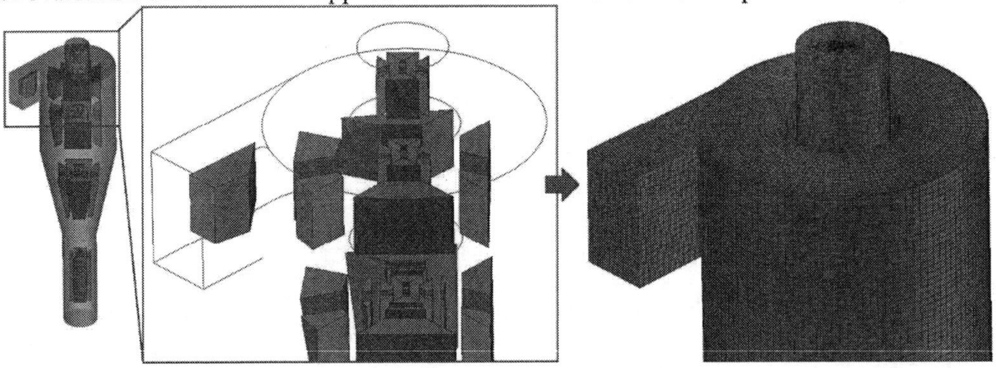

Fig. 7. Reference mesh: blocking and output mesh.

The structured HEX mesh was constructed in ANSYS ICEM CFD by applying the blocking technique. The final blocking topology consisted of 54 blocks created by splitting the initial block in order to describe the main part of the cyclone separator plus one additional block created by extruding the face representing the inlet duct. Triple o-grid was applied to arrange mesh lines into an "O" shape to reduce skew in cases of a block corner located on a continuous curve as depicted in Fig. 7.

2.4 Computational model

The exactly identical computational model (except for the mesh) was used for all thirteen cases (reference HEX mesh, six configurations of TET mesh and six configurations of POLY mesh obtained by conversion of TET mesh in ANSYS Fluent) analysed within the research. The solver was configured as pressure-based and the analysis were performed for steady state. Standard k-epsilon viscous model was used along with standard wall functions. The viscous model constants are listed in Table 1.

Table 1. k-epsilon viscous model constants.

Model Constant	Value
Cmu	0.09
C1-Epsilon	1.44
C2-Epsilon	1.92
TKE Prandtl Number	1
TDR Prandtl Number	1.3

Velocity-inlet boundary condition type was assigned to the inlet (green surface in Fig. 8) with velocity magnitude normal to the boundary equal 5 m/s. The turbulence parameters were specified as intensity and hydraulic diameter equal 5% and 0.127m respectively.

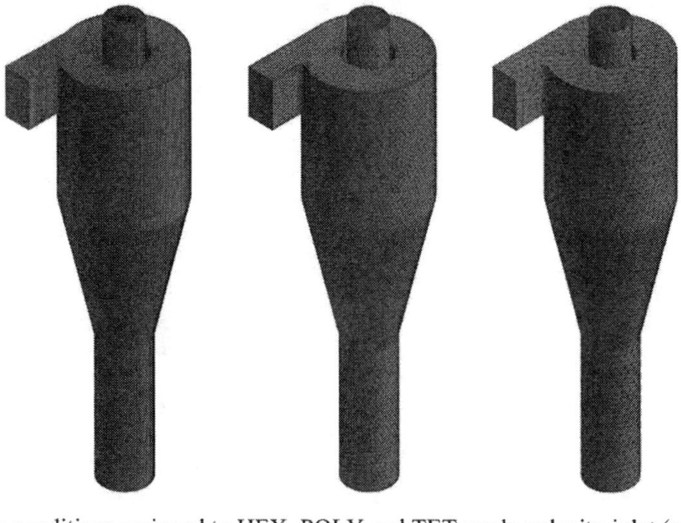

Fig. 8. Boundary conditions assigned to HEX, POLY and TET mesh: velocity-inlet (green), pressure-outlet (red), wall (blue).

The outlet of the computational domain (red surface in Fig. 8) was defined as pressure-outlet with gauge pressure of 0Pa (operating pressure – 101 325Pa), backflow turbulent intensity of 5% and backflow hydraulic diameter of 0.15m.

Wall type condition was assigned to the remaining surfaces (blue surfaces in Fig. 8). The roughness height was set to 0m and roughness constant to 0.5.

The solution was assumed as converged when residuals of x-velocity, y-velocity, z-velocity, k, epsilon decreased below 1e-5 and continuity residual decreased below 1e-3. Moreover three additional surface monitors were defined as integral of velocity magnitudes on outlet surface, XY plane and YZ plane.

Pressure-velocity coupling by SIMPLE algorithm was used as a solution method. This algorithm uses a relationship between velocity and pressure corrections to enforce mass conservation and to obtain the pressure field.

 DOI: 10.1051/e3sconf/20171401027

3 Results

The results showed in Fig. 9 are presented as velocity contour plots on XY and YZ planes for the reference HEX mesh, POLY meshes and TET meshes for all configurations. It allows detailed qualitative and rough quantitative comparison of the results in respect of mesh type and number of elements.

Integrated velocities on the outlet surface and XY plane were normalized to the values obtained for the reference HEX mesh and are shown in Fig. 10. Each one of the six cases is presented for the reference HEX mesh, TET mesh and POLY mesh obtained by conversion of the respective TET mesh in ANSYS Fluent according to the data listed in Fig. 6.

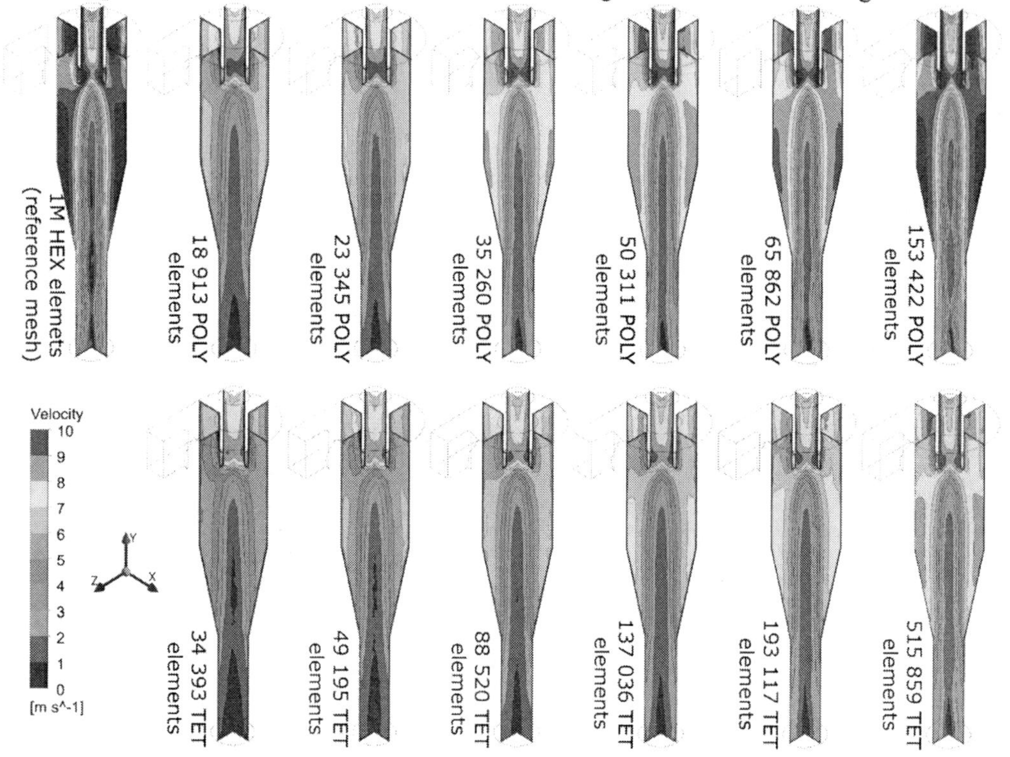

Fig. 9. Velocity on XY and YZ plane for all analysed meshes.

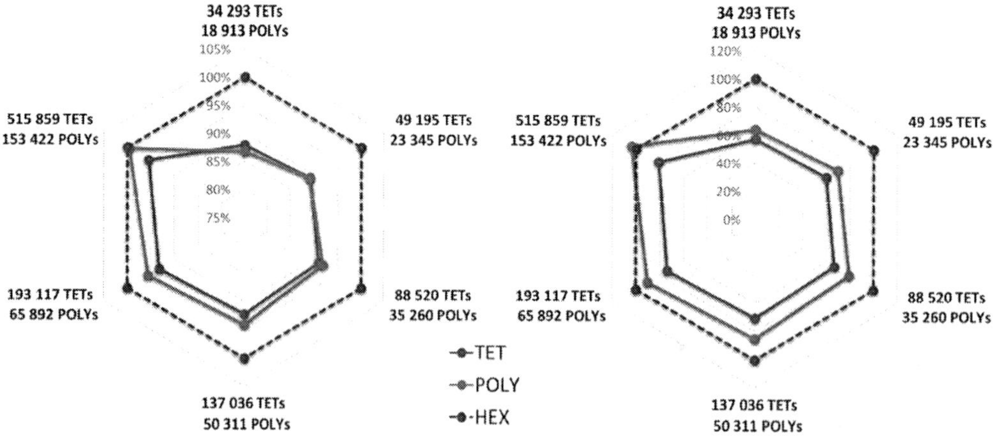

Fig. 10. Integrated velocity on outlet surface (left) and XY plane (right) for analysed POLY and TET meshes normalized to reference HEX mesh.

 DOI: 10.1051/e3sconf/20171401027

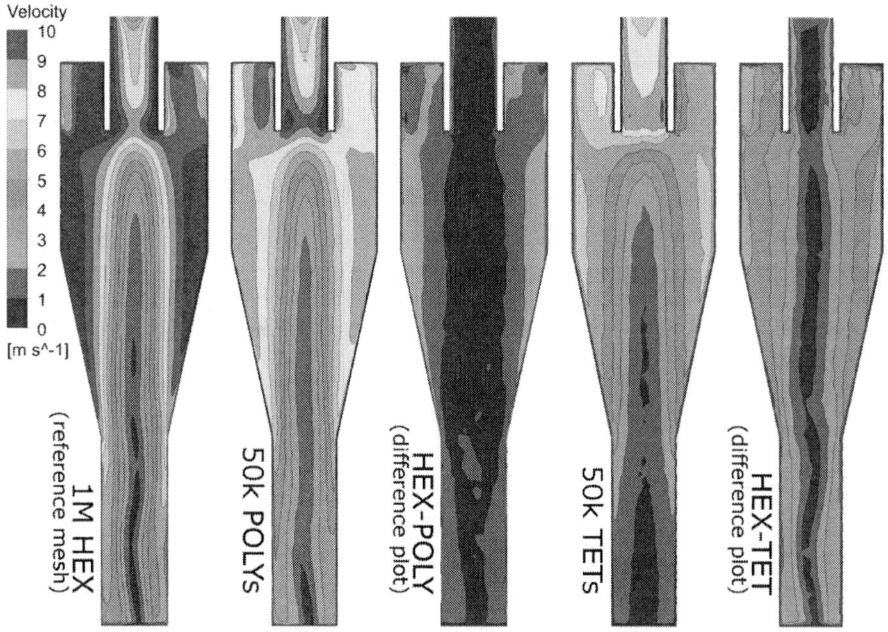

Fig. 11. Velocity on XY plane for analysed meshes with difference plots of HEX-POLY and HEX-TET.

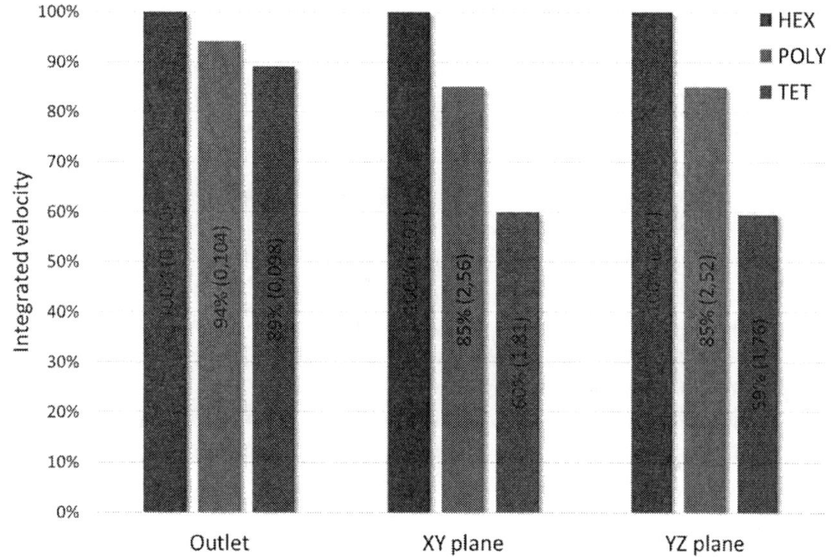

Fig. 12. Integrated velocity on outlet surface, XY plane and YZ plane for POLY, TET and the reference HEX mesh.

Fig. 11 is presented in order to compare the results obtained using TET and POLY meshes of comparable number of elements (50 311 POLYs and 49 195 TETs). The figure shows the velocity in XY plane for the reference HEX mesh, POLY and TET meshes as well as difference plots generated by subtracting the local velocity for reference mesh and POLY or TET mesh respectively.

The quantitative outlook for the solution accuracy is presented in Fig. 12, which shows comparison of the integrated velocity on outlet surface, XY and YZ planes obtained with the analysed mesh types. The integrated velocity is shown as absolute values as well as normalized to the values generated with the reference HEX mesh.

Despite the results accuracy, the calculation time is also a very important factor in terms of the overall cost of the analysis. The POLY mesh turned out to be beneficial in this field as the overall calculation time in case of POLY mesh is approx. three to five times shorter in comparison to the respective TET mesh depending on the analysed cases.

4 Discussion

The aim of the carried out research was to investigate the possibility of applying POLY meshing as an innovative approach to computational domain discretization of a fluidized bed CLC unit. Six TET meshes of number of elements ranging from approx. 34k to 516k were prepared as well as six POLY meshes obtained by conversion of the respective TET meshes in ANSYS Fluent. Reference HEX mesh of approx. 1M elements was used as a baseline and results obtained for all TET and POLY meshes were referenced to it.

As can be concluded on the basis of Fig. 9 and Fig. 10, the results obtained with POLY mesh are significantly more accurate than TET mesh even though the element count for POLY meshes was from 1.8 to 3.4 times lower. The velocity distribution obtained for the TET mesh of 516k elements is comparable to POLY mesh of 50k elements.

The significant difference between POLY and TET meshes of comparable number of elements (approx. 50k) can be determined on the basis of Fig. 11. The velocity contour plot for TET is not even comparable to the reference HEX mesh, while POLY is in a good qualitative agreement and quantitative difference of approx. 10%-15%. This thesis is proved by Fig. 12, where the integrated velocity on XY plane for TET mesh is only 60% of the reference value while in the case of POLY mesh it is 85%. It has to be mentioned that discretizing the cyclone separator domain by 50k elements generates rather coarse mesh and 85% agreement obtained for POLY mesh is a very good result.

Moreover the POLY mesh of 153k elements indicates 100% agreement with the reference mesh in terms of integrated velocity on outlet surface and almost 100% agreement in terms of integrated velocity on XY plane. These factors for almost three times finer TET mesh (516k elements) are respectively in 95% and 81% agreement.

Therefore the overall conclusion from the carried out research is that the major benefit of POLY mesh consists in the more accurate solution obtained with fewer cells, which in consequence significantly reduces the computational time.

The carried out research revealed a great potential in discretization of computational domain using POLY mesh but further investigation is necessary in order to validate this innovative technique with transient multiphase flow, advanced turbulence models and different oxygen carriers. Moreover the influence of the under-relaxation factors as well as different algorithms of pressure-velocity coupling on the convergence time has to be analysed.

References

1. Gilliard, E. and W. Lewis, *Production of pure carbon dioxide.* 1954.
2. Thon, A., et al., *Operational experience with a system of coupled fluidized beds for chemical looping combustion of solid fuels using ilmenite as oxygen carrier.* Applied Energy, 2014. **118**: p. 309-317.
3. Krzywanski, J. and W. Nowak, *Modeling of bed-to-wall heat transfer coefficient in a large-scale CFBC by fuzzy logic approach.* International Journal of Heat and Mass Transfer, 2016. **94**: p. 327-334.

4. Krzywanski, J., et al., *A generalized model of SO2 emissions from large- and small-scale CFB boilers by artificial neural network approach Part 1. The mathematical model of SO2 emissions in air-firing, oxygen-enriched and oxycombustion CFB conditions.* Fuel Processing Technology, 2015. **137**: p. 66-74.
5. Krzywanski, J., et al., *A generalized model of SO2 emissions from large- and small-scale CFB boilers by artificial neural network approach Part 2. SO2 emissions from large- and pilot-scale CFB boilers in O-2/N-2, O-2/CO2 and O-2/RFG combustion atmospheres.* Fuel Processing Technology, 2015. **139**: p. 73-85.
6. Krzywanski, J. and W. Nowak, *Artificial Intelligence Treatment of SO2 Emissions from CFBC in Air and Oxygen-Enriched Conditions.* Journal of Energy Engineering, 2016. **142**(1).
7. Krzywanski J., et al., *The Non-Iterative Estimation of Bed-to-Wall Heat Transfer Coefficient in a CFBC by Fuzzy Logic Methods.* Procedia Engineering, 2016, 157, p. 66-71.
8. Krzywanski, J., R. Rajczyk, and W. Nowak, *Model research of gas emissions from lignite and biomass co-combustion in a large scale CFB boiler.* Chemical and Process Engineering-Inzynieria Chemiczna I Procesowa, 2014. **35**(2): p. 217-231.
9. Krzywanski, J., et al., *A 1.5D model of a complex geometry laboratory scale fuidized bed CLC. Equipment.* Powder Technology, 2016.
10. Li, T.W., et al., *CFD simulations of circulating fluidized bed risers, part I: Grid study.* Powder Technology, 2014. **254**: p. 170-180.
11. Kroll, N., *The ADIGMA Project*, in *Adigma-a European Initiative on the Development of Adaptive Higher-Order Variational Methods for Aerospace Applications: Results of a Collaborative Research Project Funded by the European Union, 2006-2009*, N. Kroll, et al., Editors. 2010, Springer-Verlag Berlin: Berlin. p. 1-9.
12. Jamrozik, A., et al., *Numerical simulation of two-stage combustion in SI engine with prechamber.* Applied Mathematical Modelling, 2013. **37**(5): p. 2961-2982.
13. Gnatowska, R., *A Study of Downwash Effects on Flow and Dispersion Processes around Buildings in Tandem Arrangement.* Polish Journal of Environmental Studies, 2015. **24**(4): p. 1571-1577.
14. Spiegel, M., et al., *Tetrahedral vs. polyhedral mesh size evaluation on flow velocity and wall shear stress for cerebral hemodynamic simulation.* Computer Methods in Biomechanics and Biomedical Engineering, 2011. **14**(1): p. 9-22.

Mathematical model for the power generation from arbitrarily oriented photovoltaic panel

Qusay Hassan[1,3], *Marek* Jaszczur[1,*], *Estera* Przenzak[2]

[1]Department of Fundamental Research in Energy Engineering, Faculty of Energy and Fuels, AGH University of Science and Technology, Poland
[2]Department of Sustainable Energy Development, Faculty of Energy and Fuels, AGH University of Science and Technology, Poland
[3]Department of Mechanical Engineering, University of Diyala, Iraq

Abstract. In this paper, a mathematical model for modelling the solar radiation components and photovoltaic arrays power outputs from arbitrarily oriented photovoltaic panel has been presented. Base on the model electrical power prediction of the photovoltaic system in realistic local condition has been presented and compared with experimental measurement. The results show the effectiveness of the proposed model, which provides tools to better understand the performance and reliability as well as decision-making tool in designing of a hybrid renewable energy base power generation system. It has been shown that base on the model prediction, the efficiency and possible failures of the system can be found which are important from the technical and economical point of view.

1 Introduction

Among of the renewable energy sources, solar energy is one of the most alternative and sustainable energy resources in the world. It's ubiquitous, safe, plentiful, available for free, and environmentally friendly [1-3]. The key drawback of solar systems is unpredictable nature and in consequences the unpredictable system power output. The electrical energy production depends on the local weather and climatic conditions [4-6]. Despite the great progress in weather forecast still accurately solar radiation prediction it is not possible. Additionally, due to still relatively low solar energy to electrical energy conversion [3], the research and development efforts should be made to optimize the overall system performance of the solar systems and create new ways to predict the exact production from the environmental and climatic conditions available [7]. This is particularly important for the hybrid energy system design and analysis. Selected site data of the solar radiation are the best source of information for the proper design and investigation of solar energy conversion systems assessment [8]. In the literature [2, 13, 14] a large number of models have been proposed in order to calculate energy production from solar radiation and they can be used for a variety of purposes in the design and development of solar power systems. But most of the models has based on the daily, monthly or yearly averaged statistical data and it is not able to estimate instantaneous power production. From production-

[*] Corresponding author: jaszczur@agh.edu.pl

consumption point of view, instantaneous power production and not averaged are highly desirable and key in complex system analysis of hybrid systems which utilities wind turbines, heat pumps, fuel cells and energy storage.

The designing and evolution of solar power systems depend primarily on the solar radiation data in particular on the total solar radiation solar radiation incident on (normal component) the PV module surface. But the most of the weather stations measure intensities of total (diffusive and global) solar radiation on the horizontal surfaces [9]. This generates significant difficulty in the direct utilization of measured data to evaluate the incident solar irradiation on an arbitrarily oriented photovoltaic panel. Incoming radiation on tilted surface is the sum of the sunbeam radiation, the three components of diffuse radiation and the radiation reflected from the various surrounding surfaces for example ground. In order to reconstruct direct beam and diffuse solar radiation components from the global horizontal irradiation (measurement) mathematical models have been used [10, 11].

In the presented paper, a mathematical model for modelling the solar radiation components as well as electrical power outputs from arbitrarily oriented photovoltaic panel has been shown. Base on the realistic local condition the electrical power prediction of the PV system has been presented and compared with the experimental measurement from Faculty test system.

2 Theory and models

The weather data (solar radiation, air temperature) and PV system instantaneous power were acquired from the weather meters (Pyranometers & Thermometer) and solar panels located at AGH University of Science and Technology campus, building D4. The global solar radiation and air temperature data were taken for the year 2015, from 1st January to 31st December with the sampling 5 minutes. The PV model and solar model was used for the determination of the power generation from the photovoltaic panel for the location 50.066354N, 19.918191E, Krakow, Poland. In Fig.1a the local monthly average ambient temperature T_{air} base on the measurement and in Fig.1b the monthly average daily solar radiation together with clearness index are presented. The PV power output directly depends on the amount of radiation incident the surface of the PV array and array temperature.

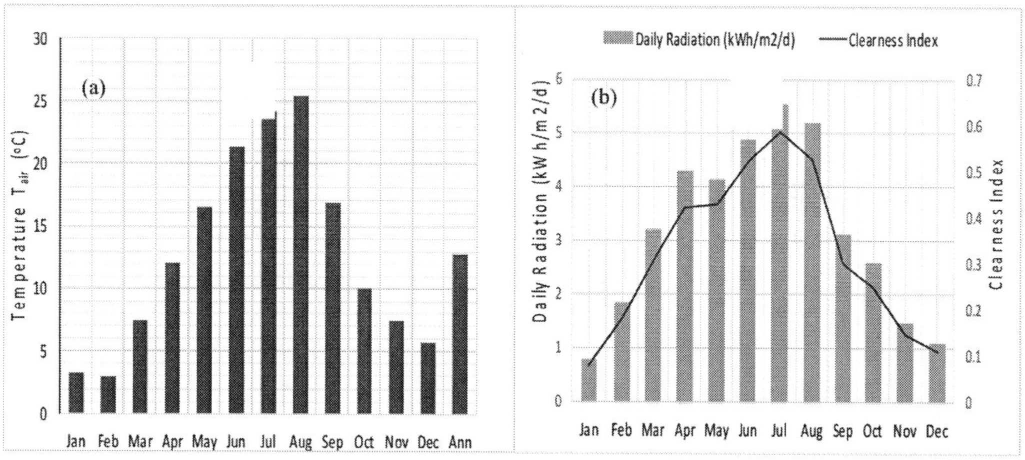

Fig. 1. (a) The ambient temperature T_{air} and (b) the total solar radiation/clearness index.

The global solar radiation incident on the arbitrary oriented the PV array surface was calculated using Hay and Davies, Klucher and Reindl (HDKR) Model [16].

2.1 Determination solar radiation components

In general, the time of the day affects the location of the sun in the sky, which we can describe by an hour angle ω (in degree). This value is equal zero at solar noon, negative and positive before and after solar noon respectively and can be calculated as follows [13].

$$\omega = (ts - 12hr) \bullet 15° / hr \tag{1}$$

where t_S is the solar time calculates from the civil time as:

$$t_S = t_C + \frac{\lambda}{15° / hr} - Z_C + E \tag{2}$$

where t_C is the civil time in hours corresponding to the midpoint of the time step, λ is the longitude [°], Z_C is the time zone in hours east of GMT and E is the equation of time.

The equation of time E accounts for the effects of obliquity and the eccentricity of the earth's orbit, and can be calculated as follows [13]:

$$E = 3.82\begin{pmatrix} 0.000075 + 0.001868\cos B - 0.032077\sin B \\ -0.014615\cos 2B - 0.04089\sin 2B \end{pmatrix} \tag{3}$$

where B is calculated using the day of the year n:

$$B = 360\frac{(n - 1)}{365.2425}$$

2.2 Determination of Extraterrestrial Solar Radiation on Horizontal Surfaces

The extraterrestrial normal solar radiation G_{on} is a function of the distance between the earth and the sun and is given as [12,13,19]:

$$G_{on} = G_{sc}\left(1 + 0.033412\cos\frac{360n}{365}\right) \tag{4}$$

where G_{sc} is the solar constant 1.367 (kW/m^2).

The extraterrestrial horizontal radiation G_o (kW/m^2) incident on the horizontal surface at the outer layers of the atmosphere is equal:

$$G_o = G_{on}\cos\theta_z \tag{5}$$

where θ_z [°] is the zenith angle, value between a vertical line and the line to the sun and this value is zero when the sun is directly overhead, and 90° when the sun is on the horizon which is expressed as:

$$\cos\theta_z = \cos\phi\cos\delta\cos\omega + \sin\phi\sin\delta \tag{6}$$

where ϕ [°] is the latitude and δ [°] is the solar declination position of the solar noon with respect to the plane of the equator, can be calculated using the formula proposed by Cooper [14] as follows:

$$\delta = 23.45°\sin\left(360°\frac{284 + n}{365}\right) \tag{7}$$

The model designed to compute each value every time step so by integrating equation (5) over one time step the average extraterrestrial horizontal radiation $\overline{G_O}$ (kW/m^2) over the time step [13] expressed as:

$$\overline{Go} = \frac{12}{\pi} G_{on} \left[\cos\phi \cos\delta \left(\sin\omega_2 - \sin\omega_1 \right) + \frac{\pi\left(\omega_2 - \omega_1\right)}{180°} \sin\phi \sin\delta \right] \quad (8)$$

the period between hour angles ω_1 and ω_2 which define an hour (where ω_2 is the larger)

2.3 Determination the Diffuse Radiation

The solar radiation coming from the sun is attenuated by the atmosphere and the clouds before it is reaching the surface of the Earth. The ratio of global horizontal radiation entering the Earth's surface to the extraterrestrial horizontal radiation has been defined as:

$$k_T = \frac{\overline{G}}{\overline{Go}} \quad (9)$$

where $\overline{G}$ is the global radiation on the earth's surface averaged over the time step.

The global horizontal radiation incident on the earth's surface consists two main parts, beam radiation (direct radiation) $\overline{Gb}$ and diffuse radiation $\overline{Gd}$. The diffusive radiation component comes almost from all the directions and does not cast a shadow. It consists up to three components (isotropic, circumsolar and horizon brightening). The sum of the beam and diffuse radiation is called global solar radiation $\overline{G}$ which is expressed as:

$$\overline{G} = \overline{Gb} + \overline{Gd} \quad (10)$$

The surface orientation has a high effect on the beam radiation (it comes directly from the sun direction) than it does on the diffuse radiation (comes from many directions, on some models is even the direction independent). In the most of the cases the only global horizontal (total) solar radiation and not it's direct and diffuse components. The diffusive radiation component can be evaluated $\overline{Gd}$ calculated using the clearness index k_T [15]:

$$\frac{\overline{Gd}}{\overline{G}} = \begin{cases} 1.0-0.09k_T & \text{for } k_T \le 0.22 \\ 0.9511-0.1604k_T+4.388k_T^2-16.638k_T^3+12.336k_T^4 & \text{for } 0.22<k_T \le 0.80 \\ 0.165 & \text{for } k_T \le 0.80 \end{cases} \quad (11)$$

When the sky is clear and the sun is at the highest position on the sky the direct radiation is about 85% and diffuse radiation is about 15% (when neglecting reflections) while the percent of diffuse radiation keeps going up to 40% when the sun is 10° above the horizon [16].

2.4 Determination of Incident Solar Radiation on PV array

The total radiation received by the PV module surface consists of a direct radiation, diffuse radiation and reflected radiation. The diffused radiation is a combination of three components (isotropic, circumsolar and horizon brightening) and the reflected radiation is the radiation reflected from the surrounding ground or any other surfaces if exists. For this reason, different models had been proposed in the literature for estimating the total solar radiation incident the PV module surfaces.

One of the models has been proposed by Hay and Davies, Klucher and Reindl (HDKR model) [16]. This model takes into consideration all key radiation components and the incident total solar radiation $\overline{G_T}$ can be calculated from the following formula:

DOI: 10.1051/e3sconf/20171401028

$$\overline{G_T} = \left(\overline{Gb} + \overline{Gd}\,A_i\right)R_b + \overline{Gd}\left(1 - A_i\right)\left(\frac{1 + \cos\beta}{2}\right)\left[1 + f\sin^3\left(\frac{\beta}{2}\right)\right] + \overline{G}\rho\left(\frac{1 - \cos\beta}{2}\right) \quad (12)$$

where β is the slope of the PV module, ρ is the ground reflectance, which is also called the albedo, f is the horizon brightening factor, A_i is the anisotropy index and R_b is the ratio of beam radiation on the tilted surface to beam radiation on the horizontal surface all calculated as follows:

$$f = \sqrt{\frac{\overline{Gb}}{\overline{G}}} \quad , \quad A_i = \frac{\overline{Gb}}{\overline{Go}} \quad , \quad R_b = \frac{\cos\theta}{\cos\theta_z} \quad (13)$$

The incidence angle θ is, the angle between the direction of direct beam radiation and the normal to that surface, and it is given by:

$$\begin{aligned}\cos\theta &= \sin\delta\sin\phi\cos\beta - \sin\delta\cos\phi\sin\beta\cos\gamma \\ &+ \cos\delta\cos\phi\cos\beta\cos\omega + \cos\delta\sin\phi\sin\beta\cos\gamma\cos\omega \\ &+ \cos\delta\sin\beta\sin\gamma\sin\omega\end{aligned} \quad (14)$$

where γ is the surface azimuth angle defined as the skew of the projection on a horizontal plane of the normal to the surface from the local meridian, with zero due south, east negative, and west positive [18].

2.5 Determination of PV array power

The PV module generating electrical power depends on the total solar radiation $\overline{G_T}$ incident on the PV surface, which in the most cases is not horizontal but inclined in order to have higher energy production, better condition for module cooling. In order to predict the PV generating power in each computational time step the model must also calculate the global solar radiation incident on the surface of the PV module every time step [13, 15, 19]. The PV module generating power P_{PV} can be calculated according to the following equation:

$$P_{PV} = Y_{PV} f_{PV}\left(\frac{\overline{G_T}}{\overline{G_{T,STC}}}\right) \quad (15)$$

where Y_{PV} is the rated capacity of the PV module (power output under STC conditions), f_{PV} is the PV derating factor introduced taking into account additional factors affecting PV (soiling, losses, shading, snow cover) and the $\overline{G_{T,STC}}$ is the incident radiation at standard test conditions.

When the information about PV module temperature is known, it is possible to take into account PV efficiency variation due to thermal effects, in that case, the PV module generating power P_{PV} can be calculated as follows:

$$P_{PV} = Y_{PV} f_{PV}\left[1 + \alpha_P\left(T_c - T_{c,STC}\right)\right]\left(\frac{\overline{G_T}}{\overline{G_{T,STC}}}\right) \quad (16)$$

Where α_P is the PV temperature coefficient for the power (%/°C), T_C is the PV module instantaneous temperature and $T_{C,STC}$ is the PV module temperature (°C) at STC conditions. Standard Test Conditions (STC) properties (solar radiation of 1 kW/m^2, PV cell temperature of 25°C, and no wind).

3 Experimental and numerical results

Mathematical model eq. (1)-(16) have been used to calculate energy production from PV modules. The polycrystalline silicon, monocrystalline silicon and thin film amorphous silicon PV modules of the size 1000W and the specifications shown in Table 1. where considered PV derating factor 85%, ground reflection 0% or 25% and azimuth angle $\gamma = 0°$ and three different angles β were used (0°, 25°, 45°) and the angles γ, β which follow the sun beam direction (i.e. two-directional tracking systems method was implemented). For the cases where the module temperature was taken into consideration, the cell temperature was calculated base on the air temperature and the solar radiation using the energy balance proposed by Antonio *et al.* [13].

Table 1. PV modules specifications [13].

PV type	Average value of efficiency at STC [%]	Temperature coefficient of power [%/°C]
Polycrystalline silicon	13.0	-0.48
Monocrystalline silicon	13.5	-0.46
Thin film amorphous silicon	5.5	-0.20

The Fig. 2 and 3 show the PV power generating (for 1 kW PV array) for a different angle β and for the tracking system. Results are presented without module temperature consideration and with module temperature taken into account (results denoted by _T).

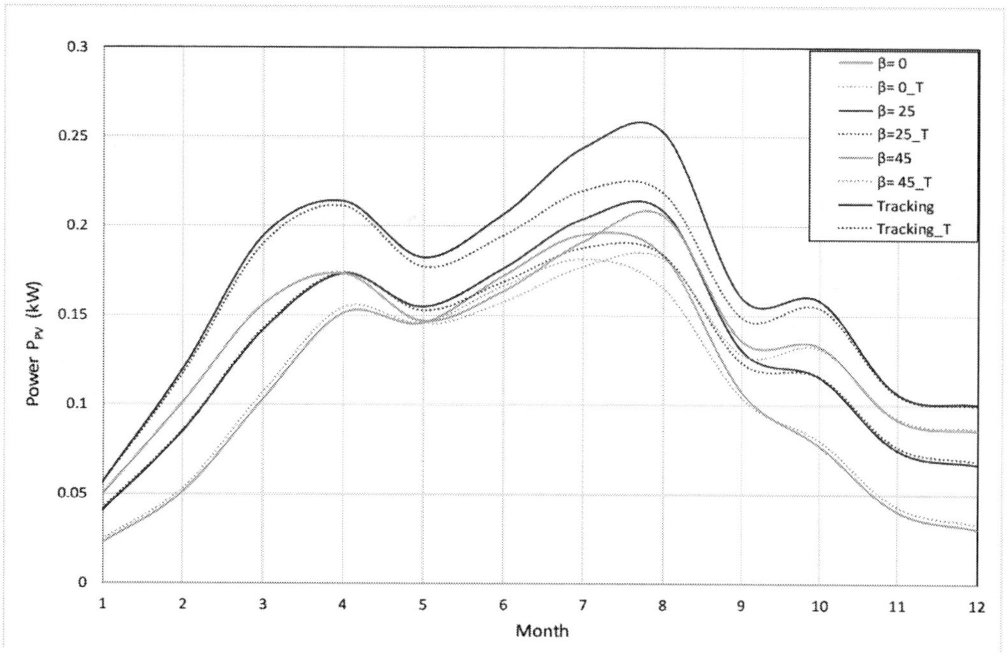

Fig. 2. The monthly average of PV power generation P_{PV} for Polycrystalline silicon.

DOI: 10.1051/e3sconf/20171401028

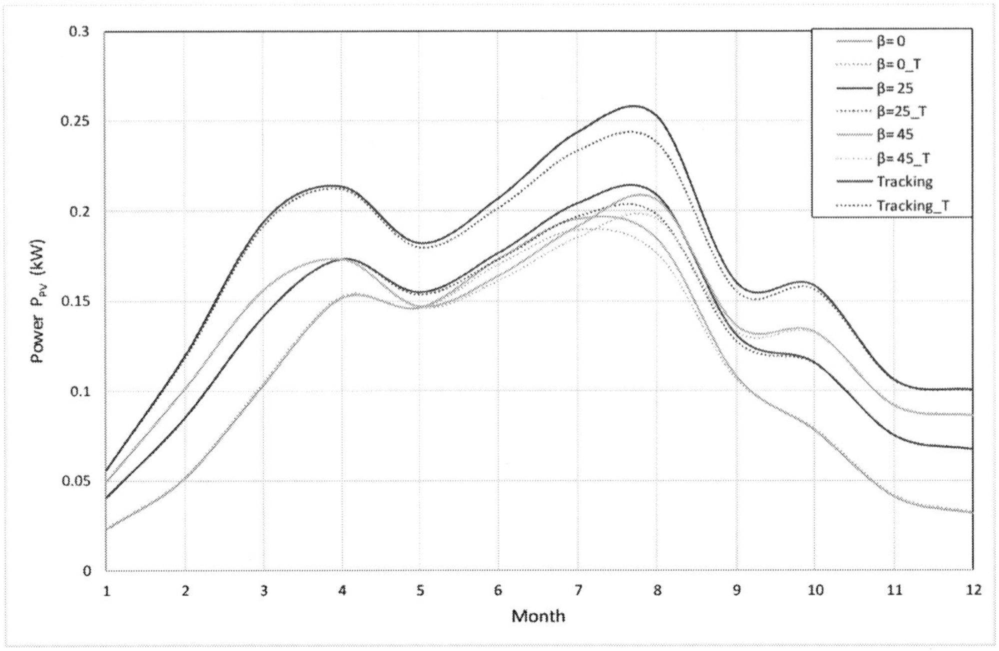

Fig. 3. The monthly average of PV power generation P_{PV} for thin-film amorphous silicon.

It can be seen from Fig.2 and, 3 that the maximum power generation is for the modules with the tracking system and its degrees by decreasing module slope angle. The total temperature effect is not large and in general it cause power decreasing during the summer and power increasing for other seasons. The results from Fig.2 and 3 and also for the Monocrystalline silicon are summarized in Table 2 where the annual PV energy production E_{PV} is presented. Modules type temperature affect the annual electrical energy generation and the difference in energy production is about 5%. When the tracking system is used energy production for all PV types is about 60% higher in reference to the case β =0°.

Table 2. Annual energy production in (kWh/yr).

PV type	$\beta = 0°$		$\beta = 25°$		$\beta = 45°$		Tracking	
	T_C non eff.	T_C eff.	T_C non eff.	T_C eff.	T_C non eff.	T_C eff.	T_C non eff.	T_C eff.
Polycrystalline silicon	940	921	1149	1115	1196	1157	1459	1384
Monocrystalline silicon	940	922	1149	1117	1196	1159	1459	1388
Thin film amorphous silicon	940	930	1149	1131	1196	1176	1459	1432

Fig. 4 illustrate the ground albedo effect on the PV power generation for the case with the module slope angle β =45° and for Polycrystalline silicon. Three different values for albedo are considered ρ=0%, 20% and 40%. The PV module temperature effect is also analyzed. It can be seen that increasing ground albedo power generation is also increasing. At the same time for the cases with module temperature consideration, PV power generation is significantly lower than for the summer time and is increasing for other seasons.

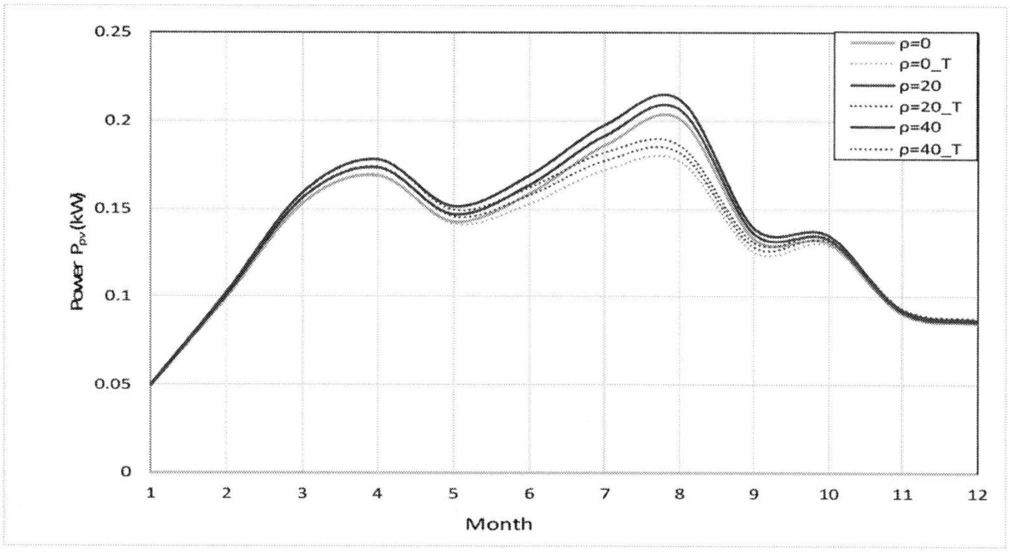

Fig. 4. Albedo of effect on the PV power generation P_{PV} for β =45° for Polycrystalline silicon.

4 Model experimental set-up

In this section, the results from the experimental set-up are compared with the model prediction in order to evaluate system efficiency. The PV system was located on the building D4, AGH University of Science and Technology and is shown in Fig 5(left). The system size was 1.62 kW (12 panels x 135W) the panels were located with the slope angle β=7.5° from the horizontal plane and azimuth degree γ=20° west. The cells have been made from thin-film amorphous a-Si/μc-Si (Sharp model NA-E135L5), the temperature coefficient of power was -0,24 %/°C, the nominal operation temperature NOCT 46°C with efficiency 9.6 % at STC. The PV modules were connected to the 1.61 kW converter working with the efficiency 94.2 %. It was assumed for the simulations that, the albedo value is 25% and derating factor 85% (which include energy losses of 5% and 10% for the panel surface pollution). Fig. 5 (left) shows the array composed of polycrystalline silicon type modules.

Fig. 5. PV modules located on the roof of D4 building, AGH University of Science and Technology: (left) thin film amorphous silicon type and (right) polycrystalline silicon type.

From the Fig. 5 it can be seen shadowing as well as additional reflection from the ventilation devices located on the roof. This may significantly influence the PV system. In Fig. 6a experimental results obtained from measurement are compared with the numerical prediction for two different days (at winter and summer). On the 30th of July 2015 at morning and afternoon hours due to shadow effect power production is much smaller that

 DOI: 10.1051/e3sconf/20171401028

model prediction. At winter time when solar radiation is lower and additionally due to a large number of cloudy days, the solar radiation diffusive component is dominated the difference is smaller. But in general, for most of the time, real power production is much lower than theoretical prediction. The difference is caused by significant system shadowing and the modules overheating.

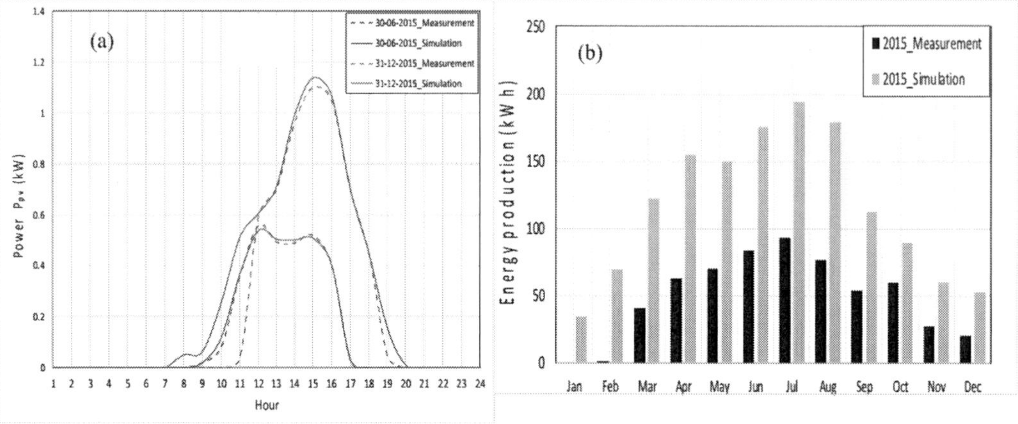

Fig. 6. (a) The PV modules power generation: experimental measurement vs model prediction, (b) The energy generation E_{PV} by the PV system: for the test system and for model prediction.

In the Fig. 6b the monthly energy production obtained from the measurement has been compared with the model prediction for the whole year 2015. It can be seen that system generate much lower electrical energy than was obtained from the computer simulations. The total energy production predicted with the presented model was E_{PV}=1371 kWh which agree well with theoretical prediction for Poland (for 1.61kW P_{PV_max} modules up to 1610kWh electrical energy should be produced), while according to experimental measurement only 589 kWh was produced. This shows that the system efficiency is about 57% of the nominal conditions. This also shows that only about 3% of the primary solar energy is converted to the electrical energy.

Conclusions

In the present paper mathematical model for photovoltaic modules power estimation together with the model for total solar radiation decomposition have been presented. Presented model can be used for the electrical energy production calculation and provides tools to better understanding the performance of the systems and can be considered as a decision-making tool in designing of a hybrid renewable energy base on solar energy. In the present work three different most popular PV module types have been analyzed. It has been showing that the modules angle, as well as PV module temperature or ground albedo, may influence the system instantaneous power generation. Results show that tracking system may increase energy generation about 60%. Ground reflectance has also a significant effect on the energy production, particularly when the slope angle β is relatively high.

It has been detected that PV experimental set-up has very high losses due to shadow effective from ventilation devices (for example almost no energy production during January and February). The total electrical energy production for 2015 estimated using proposed model and local solar radiation measurement is 1371 kWh while the real energy production obtained from the measurement is only 589 kWh. This shows that the actual system efficiency is about 57% of nominal conditions and the system performance has to be improved.

Through these results, we can explain that the actual system lows efficiency due high effect of the shadow and non-optimum slope angle of the panels β=7.5° and this lead losing solar radiation and increase panel temp during the summer for very low ventilation as well as increase panel surface pollution for non-effect rain angle cleaning.

The present work was supported by the Polish Ministry of Science (Grant AGH No.11.11.210.312).

References

1. H. Li, W. Ma, Y. Lian and X.Wang, *"Estimating daily global solar radiation by day of year in China"*. Applied Energy, **87**: pp. 3011-3017. (2010)
2. P. Hersch and K. Zweibel, *"Basic photovoltaic principles and methods"*. Antimicrob. Agents Chemother., vol. **58**, no. 12, pp. 7250–7, 1982.
3. A.Q. Jakhrani, A.K. Othman, A.R.H. Rigit and S.R. Samo, *"A simple method for the estimation of global solar radiation from sunshine hours and other meteorological parameters"* IEEE ICSET 2010, pp. 6-9, (2010)
4. P. Nema, R.K. Nema and S. Rangnekar, *"A current and future state of art development of hybrid energy system using wind and PV-solar: A review"* Renewable and Sustainable Energy Reviews, **13(8)**: pp. 2096-2103, (2009)
5. E.M. Nfah, J.M. Ngundam and R. Tchinda, *"Modeling of solar/diesel/battery hybrid power systems for far-north Cameroon"* ,Renewable Energy, **32**: pp. 832–844, (2007)
6. M. T. Y. Tadros, *"Uses of sunshine duration to estimate the global solar radiation over eight meteorological stations in Egypt,"* Renewable Energy, **21 (2)**, pp. 231–246, (2000)
7. M.K. Deshmukh, S.S. Deshmukh, *"Modeling of hybrid renewable energy systems"* Renewable and Sustainable Energy Reviews, **12(1)**: pp. 235-249, (2008)
8. A.A. Sabziparvar, *"A simple formula for estimating global solar radiation in central arid deserts of Iran"*. Renewable Energy, **33**: pp. 1002–1010 (2008)
9. E.G. Evseev and A.I. Kudish, *"The assessment of different models to predict the global solar radiation on a surface tilted to the south"* ,Solar Energy, **83**: pp. 377–388, (2009)
10. K. Soga, H. Akasaka and H. Nimiya, *"A comparison of methods to estimate hourly total irradiation on tilted surfaces from hourly global irradiation on a horizontal surface"*, Kagoshima University, 1-4-7, Korimoto, Kagoshima, 890-0065, (1999)
11. A.Q. Jakhrani, A.K. Othman, A.R.H. Rigit, S.R. Samo and S.A. Kamboh, *"A novel analytical model for optimal sizing of standalone photovoltaic systems"* Energy, **46**, pp. 675-682, (2012)
12. J. Duffie, W. Beckman "Solar Engineering of Thermal Processes" 4th edition, Wiley, New York, NY, (1991)
13. H. Antonio, S. Hegedus, "Handbook of photovoltaic science and engineering", (2003)
14. D. Erbs, S. Klein, J. Duffie ,"Estimation of the diffuse radiation fraction for hourly, daily, and monthly-average global radiation", Solar Energy,(1982)
15. N. Mahammad, Encyclopaedia of Geomorphology , pp. 66, (2006)
16. K. N. Shukla, S. Rangnekar, and K. Sudhakar, *"Comparative study of isotropic and anisotropic sky models to estimate solar radiation incident on tilted surface: A case study for Bhopal, India"*, Energy Reports, **1**, pp. 96–103, (2015)
17. B. Liu, R. Jordan, *"The long term average performance of flat plate solar energy collectors"*, Solar Energy, **7**, pp. 53–7, (1963)
18. VA. Graham, KGT. Hollands *"A method to generate synthetic hourly solar radiation globally"*, Solar Energy, **44 (6)**, pp. 333-341, (1990)
19. C. D. Whiteman and K. J. Allwine, "Extraterrestrial solar radiation on inclined surfaces",**1**, 3 (1986)

Energy consumption of buildings depends on the daylight

Ewa Piotrowska[1,*], *Adam* Borchert [2]

[1] Bialystok University of Technology, Department of Electrical Power Engineering, Photonics and Lighting Technology

[2] Bialystok University of Technology, Department of Power Electronics and Electrical Drives The Bialystok University of Technology is located in Poland

Abstract. In order to reduce energy consumption in heated buildings and thus fossil fuels, there is a need for proper modernization of existing buildings and new construction with reduced energy demand. The size and the insulation of windows have a decisive influence on the amount of heat loss. The study looks into the impact of decreased power consumption through optimal use of sunlight through the selection of the size of windows, type of windows, and adjusting the light intensity using an automatic control of lighting according to the amount of sunlight reaching the room. The research related to the differences between the six types of windows in relations to the energy consumption of the building.

1 Introduction

It is estimated that the department of building consumes half the energy produced. The most important thing during the heating season is to provide the building with heat gains form the sun. The amount of solar radiation in Poland is low, but in the winter per square meter south oriented facades of the building located in Warsaw falls about 385 kWh of energy. This is the amount of heat corresponding to the quantity of energy obtained combustion of 38 liters of fuel oil.

In addition to obtaining solar energy by using special devices such as solar panels, it can be used it in its pure, unaltered form also, setting the appropriate building in relation to parts of the world and improving participation of glazed surfaces in stalls external walls.

It is possible to obtain in this way increase heat gains from the sun and a larger share of solar energy in the lighting of rooms. This affects the final energy balance of the building by reducing the the demand for heat which is usually obtainable from the heat source or situated in a building, which receives heat using traditional fuels such coal, gas or fuel oil. The use of daylight systems or intelligent lighting control by reducing artificial lighting decreases energy use of the building for electricity by about 15%. This is important for the evaluation of the effective use of solar energy in the construction industry, in particular in passive houses, or a zero-energy houses, [5].

* Corresponding author: e.piotrkowska@onet.pl

2 The energy balance of buildings and daylight

Energy balance is the difference between the profits of heat energy in the building and its losses. The energy efficiency of the house is a ratio of the amount of energy saved as compared to the amount of energy consumed or forecast. Efficient use of energy means reducing the energy consumption of production, operation and running of services. The building refers to the energy needed for heating, cooling, ventilation and air-conditioning or electricity consumed by the device. In order to eliminate or at least reduce consumption of non-renewable fuels in Poland in 2020 in line with the recommendations of the Climate Package of 2015 years, it is planned to reduce greenhouse gas emissions by 20% and improving energy efficiency by 20%. Public buildings also define the energy needs for lighting purposes, [2].

2.1 Design of daylight lighting in the rooms

While designing the lighting in the room, it is crucial to include factors that affect the intensity of solar radiation, as shading of the building, type of glazing, inclination of windows, location in relation to parts of the world, location, time of year and time of day. The size of the average intensity of solar radiation, depends on the the location of the building relative to parts of the world for individual months. Location of the room on the south side positively affects the heat balance of the building, especially during the coldest months: January, February and November compared to the rest of the orientation of the building.

2.1.1 The windows in the building

The flow of power through the window has the biggest impact on the balance of energy in building. The most important matter in designing daylight rooms is the selection of the size and the type of windows. The characteristics of energy-efficient window frames are determined not only by the heat transfer coefficients, but also by the parameters describing the properties of the glazing, such as the ratio of the total solar energy transmittance and light transmittance. It is important to remember that this parameter is different for the window frame, window, and the entire window. Insulation window improves increasing the number of windows in the profile and filling the space between the noble gases, eg. argon or krypton or coating glass coatings selective. It shall also apply to the so-called. warm frame, or inter-frame polymer which improve the thermals of the window (reduction in the linear coefficient of heat transfer at the connection to the frame of the glass package).

2.1.2 Heat losses through the windows

In terms of energy (heat loss by penetration), the window is characterized by the heat transfer coefficient U_w; $W/(m^2K)$, depending on the insulation frame, glazing and heat transfer coefficient of linear thermal bridge at the junction of the glass from the window frame. Heat losses through the opacified partition (like wall) during winter are approximately 25%. In winter, a small area of vertical windows (less than that due to daylight) in which the share exchange heat through the windows of the south is less than 30% in the monthly balance, and in the windows on the north side is about 35%. When the surfaces of the windows is above the applicable minimum surface due to the daylight part of the heat demand for space heating due to heat loss through the window significantly increased to over 60% in the case of the southern windows, and over 70% in the case of the northern windows, [5].

In conclusion if surface of a window is big, specially if this window is on the roof, it affects the need for cold or heat in this room which grows along with the surface of windows and tilting from the vertical (at site level).

3 Energy consumption for heating and consumption for the lighting of buildings

3.1 The final indicators of consumption for energy of building

3.1.1 Indicator of the primary energy - EP

Indicator of the primary energy determined for the annual consumption of the building on non-renewable primary energy falling on 1m^2 rooms with controlled temperature (cooled or heated). It is determined by formula, [1]:

$$EP = Q_P / A_f \quad (1)$$

$$Q_p = Q_{p,H} + Q_{p,W} + Q_{p,C} + Q_{P,L} \quad (2)$$

Q_p- non-renewable primary energy to heat annual consumption for heating and ventilation ($Q_{p,H}$) energy consumption for cooling rooms ($Q_{p,C}$) energy consumption for heating water ($Q_{p,W}$) energy consumption for the electricity power needed for operation of technical systems which are associated with those purposes. The public buildings are also referred to energy consumption for lighting ($Q_{p,L}$).

3.1.2 Index of usable energy - EU

Index of usable energy describes the energy required for different needs associated with the use of the building; this is a monthly balance of gains and losses of heat energy for heating, cooling and heating water to appoint by formula, [1] (building energy consumption for heating building):

$$EU = Q_u / A_f \quad (3)$$

$$Q_u = Q_{H,nd} + Q_{W,nd} + Q_{C,nd} \quad (4)$$

$$Q_{H,nd} = \sum_i (Q_{ht,i} - \eta_{H,i} \cdot Q_{gn,i}) \quad (5)$$

Af – surface of room with controlled temperature, $Q_{H,nd}$ – annual energy consumption for usable energy for heating and ventilation $Q_{H,i}$ - dimensionless utilization coefficient of heat gains in the coming months, $Q_{ht,i}$ - heat losses in the following months, kWh/month, $Q_{gn,i}$- heat gains in the following months, kWh/month.

DOI: 10.1051/e3sconf/20171401029

3.1.3 Index of the final energy- EK

The indicator for the final energy is the annual energy per m^2, needed to provide cooling, heating, ventilation and lighting facilities, hot water preparation and taking into account losses in technical systems. Calculated by, [1]:

$$EK = Q_k / A_f \tag{6}$$

$$Q_k = Q_{k,H} + Q_{k,W} + Q_{k,C} + Q_{k,L} \tag{7}$$

$$Q_{k,H} = Q_{H,nd} / \eta_{H,tot} \tag{8}$$

$Q_{k,H}$ - annual energy consumption for the final energy for heating and ventilation,

$\eta_{H,tot}$ - average seasonal total efficiency of the heating system of the building (from the production of heat transfer in the room).

$$\eta_{H,tot} = \eta_{H,g} \cdot \eta_{H,s} \cdot \eta_{H,d} \cdot \eta_{H,e} \tag{9}$$

$\eta_{H,g}$ - the average seasonal efficiency of producing heat transfer medium of energy supplied to the borders of the building (the final energy). The value depends on the type of heat source (boiler, pumps, heaters or other). It falls generally in the range {0.5-1.0}, $\eta_{H,s}$ - the average seasonal efficiency of the heat accumulation in the capacitive elements of the heating system of the building. This value is assumed to be depending on the presence and the type of the installed buffer. These values are in the range {0.91-1.00}, $\eta_{H,d}$ - the average seasonal efficiency of the transport of heat transfer medium within the building, $\eta_{H,e}$ - he average seasonal efficiency of regulation and the use of heat in the building.

3.1.4 Energy consumption for lighting

The energy consumption for lighting evaluated in public building is calculated by:

$$Q_{pL} = w_{el} \cdot E_{kl} \tag{10}$$

w_{el} - correction factor for the media electricity, E_{kl} - aggregate annual energy consumption for lighting building [kWh/year]:

$$E_{kL} = E_{Lj} \cdot A_f \tag{11}$$

A_f - the surface area of each room of the building; m^2, E_{Lj} - annual unit energy consumption for every room of a building; kWh / m^{2}year, according to PN-EN 15193: 2010 is an indicator of LENI, [1, 5].

$$E_{Kj} = \{F_c \cdot \frac{P_j}{1000} \cdot [(t_D \cdot F_0 \cdot F_D) + (t_N \cdot F_0)]\} \tag{12}$$

F_c - factor related to the reduction of light intensity to the required level, F_O - coefficient taking into account the absence of members in the workplace, F_D - coefficient taking into account the use of daylight, P_j - power unit lighting luminaires base, W/m^2, t_D - time of use light per day, h/r, t_N - time of use of the lights during the night; h/r.

DOI: 10.1051/e3sconf/20171401029

4 Intelligent lighting control systems

Building Management System (BMS) contributes to the increase in the comfort of usage, safety and rational use of energy, with the economic aspect. The use of daylight and modern electrical installations, provided with systems for building automation (BMS) can lead to a reduction of electricity consumption for lighting purposes up to 80%, and in terms of the total consumption of electricity and heat, can bring savings of 30% and more, [3].

4.1 Intelligent lighting control system with daylight

The first level of integration can be distinguished for the indoor lighting. The control takes into account the artificial lighting alone or the takes into account artificial lighting and its control by external information like day lighting, occupancy, etc. This system complements the system of electric light sources, so that electric lighting is complementarity with the intensity of daylight to the required level. This is done through controls of the luminous flux "dimmer". They may be devices controlled manually or automatically. In the second case, the control system controlling electric lighting must interact with the sensor measures the illuminance level. The "intelligent system" of this type provides automatic adjustment of the size of the luminous flux of electric light sources to the changes in the share (level) of daylight, while maintaining a given level of brightness contributing to the reduction of electricity consumption of the building.

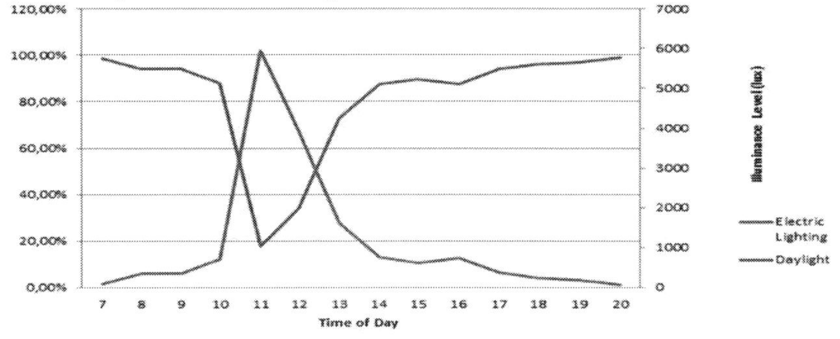

Fig. 1. The result of daylight automatic control is energy savings for 23 July 2016 (Clear sky).

The energy savings were calculated by taking the highest point of electricity used for the day, and then subtracting anytime the actual value was lower to find the difference. The differences are then totaled and averaged to find the overall saving percentage for each day. The measurements were made for 15 (fig.2) and 23 (fig.1) July 2016.

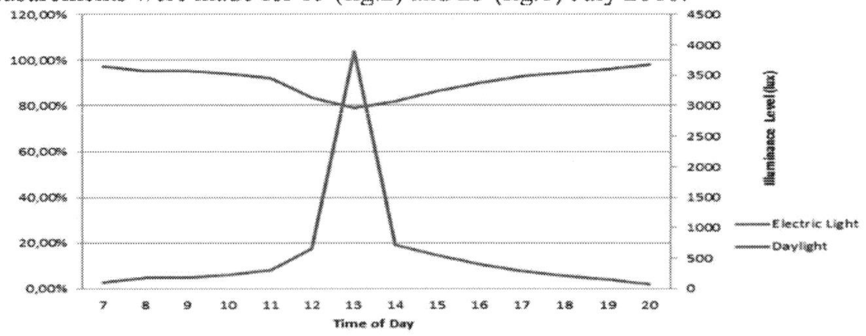

Fig. 2. The result of daylight automtic control is energy savings for 15 July 2016 (Overcast sky).

 DOI: 10.1051/e3sconf/20171401029

4.2 HDR Images as an innovation of the use of the daylight for systems of automatic lighting control

In order to map the sky luminance of the light used HDR images. The process involves taking the picture LDR (Fig. 3) of the sky through objective "fisheye" and transforming it into the HDR image. Then, on the HDR image is applied to the mesh distribution of the sky equal to 145 Tregenza`s pieces. At the end, the average together luminance values part of the mesh like that show Figure 4.

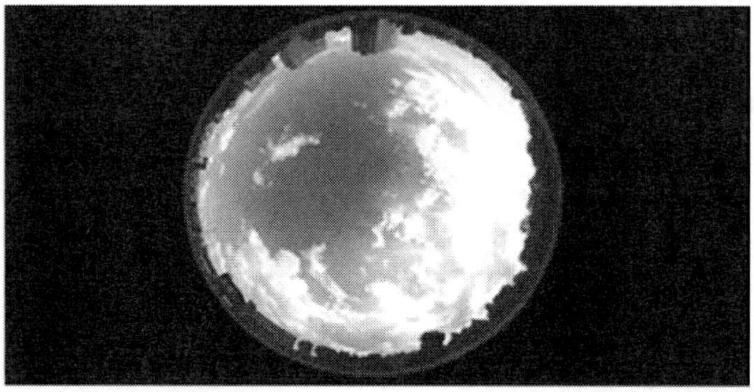

Fig. 3. Example of LDR sky image

Fig. 4. Luminance distribution in Bialystok 11-th of July a) HDR sky image, b) with Tragenza mesh, c) with mesh and averaged luminance.

Through the analysis of HDR images from the test room, it can be concluded that it affects the electricity consumption through the daylight. Obtained results allow determining the appropriate size and number of window openings depending on the location of the building, which, in turn, reduces the electricity consumption for artificial lighting, as well as allows analyzing the energy balance of the building. Optimizing these two issues significantly reduces the used electricity and thermal energy for heating and cooling the building.

5 Research and analysis of the results

The effect of the selection of characteristics of window and its effect on the energy balance of the building. The assumption was a room with dimensions of 3.27 x 3.04 x 2.77 m with one window with dimensions of 1.74 x 1.45 m. Wall with a window set on the south side of the best sunshine in a year. The window was tested in variants according to the Polish standards in Figure. 3:1. Window box triple, 2. windows loom single glazed window 3. double-glazed unit in single-chamber 4. single-glazed window with glass composite twin-chamber 5. the window composite double glazed window 6. composite triple glazed (one single-chamber glass unit).

 DOI: 10.1051/e3sconf/20171401029

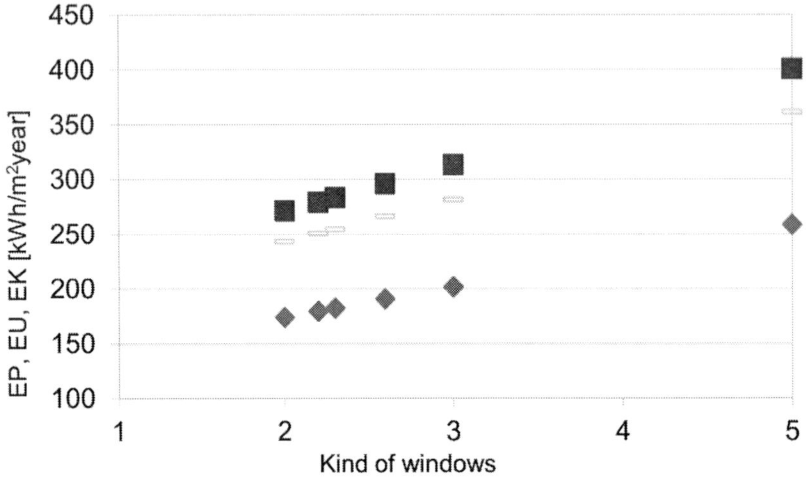

Fig. 5. Indicators energy consumption for EP, EK and EU depend on kind of window.

The lowest value of the energy consumption of the room for every kind of window in the case of energy the usable, the largest and the primary energy. Graphs EP, EK, EU and have the same shape and in each case the largest energy consumption (approximately 35% more than other kinds) in the case of a window 2 which is a single glazed. The lowest power consumption is in the case of windows box 1 triple and 6 composite triple glazed (one single-chamber glass unit).

5.1 Energy losses and gains, depending on the selection of a window

The effect of energy gains, for this room from sunlight is shown in Fig.6.

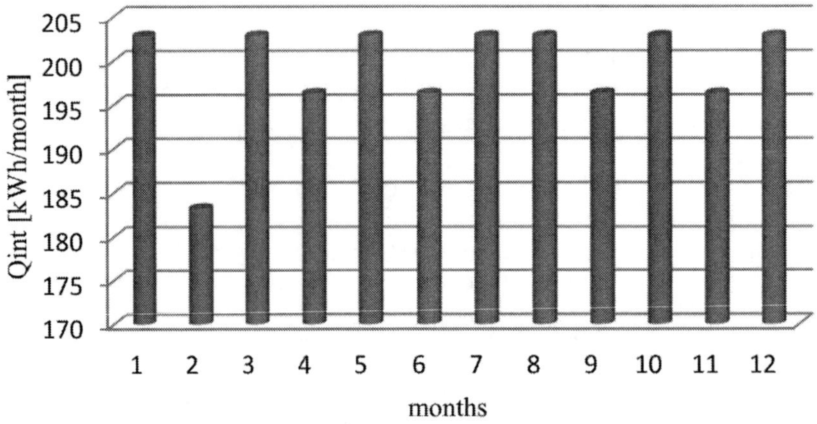

Fig. 6. Heat gains from sunlight (daylight).

It is easy to notice that the heat gains coming from sunlight are rather high and steady throughout the year. Only in February the value of heat gains is low. High values of more than 200 kWh / month are received not only in the summer months like July and August,

 DOI: 10.1051/e3sconf/20171401029

but also during the winter months like December and January. The effect of heat energy losses for this kind of room depending window system is shown in Fig.7.

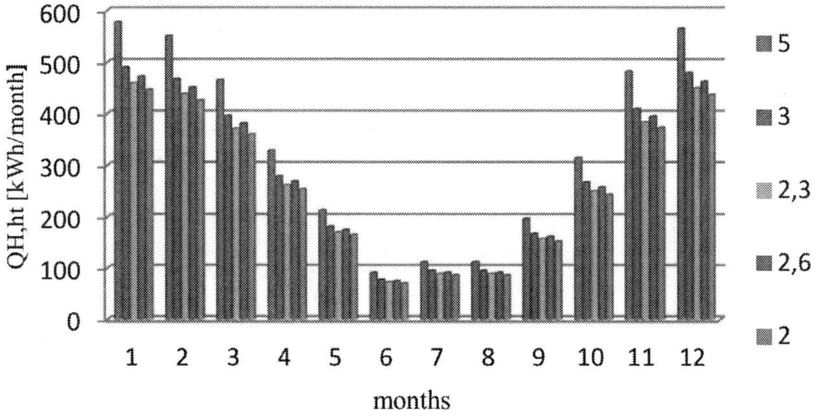

Fig. 7. Losses of heat energy depend on the kind of a window.

The greatest loss of heat coming from the windows appears in the winter months (January-March and November-December) and the smallest in the summer months from June to August. In this case, the worst results were obtained, regardless of the month, in window 2, while the other windows have values closer to each other. Fig.8 analyzed the annual heat demand for the test windows.

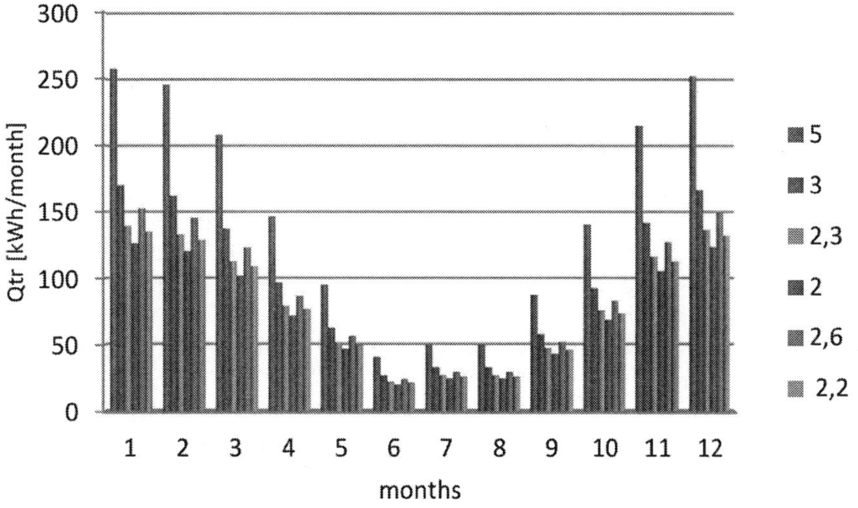

Fig. 8. The annual energy consumption for heating the test room depends on the kind of a window.

The graph shows the results of calculation of the heat demand of the room depending on the change in the type of windows, [1]. Single-windows greatly increase the heat demand of the building (as 50% greater than the other) which is very disadvantageous. The demand for heat is the lowest in the case of the window in the system 1, and comparable to much larger window, type 4 single-glazed window with glass composite twin-chamber and 6. Windows type 3 double-glazed units in single-chamber and 5 composite double glazed windows to higher demand by about 10% higher than the windows kind 1, 4 and 6.

5.2 Energy consumption for lighting of a room

Table 1 shows the results of calculations of demand for the lighting of a room depending on the applied controls: manual control and automatic (as in the case of an intelligent lighting management system).

Table 1. Energy consumption demand for lighting control, in depends on manual or automatic.

	manual control	automatic control
	(kWh/m²year)	(kWh/m²year)
E_l	1,51	1,24
E_{kl}	135,87	111,28
E_{pl}	407,61	333,83

In the test room it adopted normative duration of lighting for classrooms with a value of 1800h / year. The use of light for manual control is 10% higher than in the case of automatic control. Usage of a light sensor daylight using HDR declined the operation time of lighting during the day by adapting accordingly their intensity and reducing electricity consumption for lighting, [5]

6 Summary

In order to optimize the energy balance of the building and the use of the phenomenon of free heat gain, we should take into account the impact of lighting sunlight at the design stage.

The energy consumption during the year is changing: it is the highest during the summer while the lowest during winter. Heat gains through windows are relatively similar throughout the year, hence a greater impact on energy losses through the buildings which have windows. Loss of heat and energy through the window openings is the greatest in the winter.

The size and the insulation of windows have a decisive influence on the amount of heat loss. The study of cubic kind of windows depends on their design. It noted the influence of selection of appropriate properties joinery building on the gains and losses of heat during the year.

The most economical use of energy in the building is the type of windows Type 1. window box triple or Type 4. single-glazed window with glass composite twin-chamber or Type 6. composite triple glazed (one single-chamber glass unit). The least favorable in increasing energy demand is Type 2. single glazed loom window. Type 3. windows double-glazed unit in single-chamber and Type 5. window composite double glazed window have on average about 10% larger energy consumption compared to windows Type 1, 4 and 6.

Considering the construction of windows and their function (partition transparent), windows are ones of the main elements of the building, which has a significant impact on its thermal characteristics. As a result, it can be concluded that the selection of windows in the building during the upgrade to a better and more efficient performance affects to a fairly large extend the overall energy demand, which in large residential buildings and offices reduces the overall consumption of thermal energy of the building. The windows are very important when planning energy-efficient refurbishment. The windows largely determine the heat balance of the building, not only the loss but also the profits (sourced in a passive solar energy). Transparent partitions are responsible for an average of 15-25% of heat loss. Therefore, the important thing is the imposition of appropriate standards for the selection of windows for the building renovation and new buildings, not only in terms of the efficiency of heat transfer through the window but also for the whole structure. Further cost reductions can be achieved (due to maximize of use a daylight to illuminate the room) by using automatic lighting control systems.

By controlling the light intensity during the on and off switching, the lighting system allows for optimal use of natural light, which not only reduces the consumption of electricity for lighting by 10-15% but also increases user's comfort. The use of daylight sensors captures real luminance sky dome using HDR images which contributes significantly to the energy consumption for artificial lighting on intelligent systems.

All these measures result in reduced demand for heat in buildings, and the use of non-renewable fuels and heating oil. Reducing the energy consumption of buildings affects not only the building itself, but the whole environment.

References

1. CIE Standard General Sky Guide: Technical Report (2015).
2. D. Chwieduk, *Wytyczne Tworzenia Koncepcji Energetycznej Budynku*, Budownictwo, Rok 109, Zeszyt **3**, 2-B/2012
3. S. Darula, R. Kittler, H. D. Kambezidis, A. Bartzokas, *Guidelines for more realistic daylight exterior conditions in energy conscious designs*, National Observatory of Athens Institute of Environmental Research and Sustainable Development and Institute of Construction and Architecture Slovak Academy of Sciences Bratislava, Athens (2000)
4. E. Klugmann-Radziszewska, *Odnawialne Źródła Energii Przykłady Obliczeniowe*. Gdańsk, (2009)
5. E. Piotrowska, M. Zajkowski: Modernizacja infrastruktury budynków a ich właściwości środowiskowe i energetyczne, V Międzynarodowa Studencka Konferencja Naukowa : Inżynieria Środowiska Młodym Okiem, Białystok, 2016
6. E. Piotrowska: Modelowanie natężenia oświetlenia dziennego w programie DIALux, Zastosowania Komputerów w Elektrotechnice : ZKwE'2016 : XXI Konferencja Naukowa, Poznań, 2016

Energy savings due to building insulation of different thickness

Tadeusz Orzechowski[1,*], and *Mateusz* Orzechowski[2]

[1]Kielce University of Technology, Al. Tysiąclecia PP 7,25-314 Kielce, Poland
[2]Statistical Office in Kielce, Wróblewskiego 2, 25-369 Kielce, Poland

Abstract. In the plans of thermo-modernisation of historic buildings, strict requirements on energy performance are often relaxed. Detailed analyses are performed to select thermal upgrading technology that would ensure maximum environmental benefits while preserving the historic value of the building. The analysis of the costs of thermal upgrading with the use of heat-preserving plasters having different insulation properties is made for a coal-fired boiler plant. Optimal and advantageous ranges of insulation materials application are shown in reference to coal savings in boilers with variable and constant efficiency. Climatic conditions and environmental benefits are indicated.

1 Introduction

Big cities have to cope with poor quality of ambient air, which is related to emissions from transport and heating. It is a particularly serious problem in urban areas with densely arranged old buildings, where heat is supplied by local coal-fired facilities. The actions of local governments aimed at substituting coal with less harmful fuels are not always effective. The reason for users' preference for coal are lower prices when compared with other energy carriers. Additionally, old buildings under supervision of conservation officers cannot be thermally upgraded using typical technologies that involve the external walls insulation retrofit by over-cladding. That would ruin the historic and architectural value of the building and the environment. When thermal upgrades of such buildings are prepared, strict requirements concerning the energy performance are often relaxed. Detailed analysis are made to choose such renovation solutions that would ensure maximum environmental protection and preserve historic heritage [1]. As a result, technologies selected to fit specific buildings generate higher costs. Therefore, it is necessary to seek funds that would allow the conservation of the possible largest historic environment and, at the same time, be acceptable to the public [2].

As regards materials for construction industry, an extremely wide rage of options are available. The descriptions of the materials can be found in literature, e.g. [3]. In study [4], a systematic approach to proper selection and identification of the best upgrading solutions for existing buildings is discussed, including those intended for reduction in energy consumption and greenhouse gases emissions. Historic buildings that are in various

* Corresponding author: todek@tu.kielce.pl

condition, were constructed using different technologies and materials used at the times a given structure was built. Every time the building envelope upgrading is planned, it is necessary to check the chemical composition, or conduct an analysis of the materials that will be used [5], [6]. Investigations conducted in France indicate that energy consumption in old buildings is affected by their ambient surroundings: microclimate, location in the city and the area, shading, and others [7]. The envelope insulation has influence not only on energy consumption but also, indirectly, on CO_2 emissions into the atmosphere. Authors of study [8] discuss multi-parameter optimisation analysis carried out with respect to economic and environmental criteria applied jointly. Heating costs also depend on external conditions of the building envelope, including the orientation with respect to the cardinal directions [9].

Taking into account all the constraints mentioned above, it is possible to make improvements in thermal properties of the envelope of historic buildings by installing insulation on the internal side. That reduces emissions into the atmosphere, even if the fuel for heating is not changed. Costs calculations and listing the optimal parameters of the prospective investment could become an encouragement for the users to choose thermal upgrading of that kind.

2 Problem formulation

In Poland, the material most frequently used to improve the insulation properties of the building envelope is foamed polystyrene. The reason is low price and popularity of this technology [10]. In some cases, the material can be used for internal insulation, but every time it is necessary to make detailed analysis of moisture passing through the building wall, and to check the ventilation quality with respect to moisture excess removal. To avoid risks mentioned above, it is necessary to apply different technologies that, as a rule, are more expensive. That affects return on investment, the accessibility of loan, etc. The application of a coat of plaster, which can be perlite-based, having enhanced insulation properties is one of currently recommended technologies. If plaster with the addition of pre-sorted granules is prepared on the site, thermal conductivity is approx. 0.15 W/mK. Perlite plaster mortars with thermal conductivity coefficient of 0.067 W/mK are also available, but their price is almost twice higher. The same refers to other insulation materials.

When balancing thermo-modernisation, to seek for optimal solutions, it is necessary to achieve balance of total costs $C_{A,N}$, which contain heating costs and insulation placing costs, with the costs of necessary materials, which can be written as follows:

$$C_{A,N} = \left(\frac{f_1}{R_w + d_{ins}/\lambda_{ins}} + U_2 f_2 \right) C_E \sum_{j=1}^{Nb} \left(T_i - T_{ej} \right) \frac{\tau_j}{\eta_{1j}\eta_{2j}} + C_{Ni}\left(d_{ins} C_{ins} + C_d \right) \tag{1}$$

where f_1 and f_2 represent the shares of insulated and uninsulated (e.g. window openings) exterior closures, respectively, R_w and d_{ins}/λ_{ins} is heat conductivity resistance of uninsulated enclosure and the insulation material with thickness d_{ins} and heat conductivity λ_{ins}. U_2 is heat transfer coefficient of the uninsulated part. C_E, C_{Ni}, C_{ins} and C_d denote, respectively: energy costs, coefficient that refers to the cost of credit in the balanced period, insulation costs and insulation installing costs.

In old buildings that have not been modernised for various reasons, heating systems are often based on coal boilers, usually hopper-fed, operating in the systems without buffer storage tank. Boiler capacity is adjusted in accordance with the demand for heating power, which results from actual weather conditions. At such operational arrangement of the boiler, its efficiency η_1 changes with load at transmission efficiency η_2. The amount of heat necessary to maintain the designed temperature T_i is proportional to the appropriate temperature difference, and heat consumption is proportional to the time of temperature difference occurrence τ_j. In formula (1), it is assumed that the heating of the building starts when the external temperature T_e is lower than the base temperature T_b (hereafter adopted T_b=12 °C). Its value allows to determine the number of intervals summation N_b in Eq. (1).

If no external constraints or requirements (e.g. standard-specific ones) are found, the investment should be realised in accordance with criteria imposed to ensure optimal handling of the investment. Those criteria include the minimum total costs. From Eq. (1), it is necessary to calculate the minimum total costs with reference to the insulation thickness. The condition necessary for the occurrence of the extremum is the zero value of the total costs first derivative $C_{A,N}$ with respect to insulation thickness d_{ins}, which leads to the following expression:

$$d_{insOpt} = \sqrt{\frac{\lambda_{iz} f_1 C_E \sum_{j=1}^{Nb} \left(T_i - T_j\right) \frac{\tau_j}{\eta_{1j}\eta_{2j}}}{C_{Ni} C_{ins}}} - \lambda_{ins} R_w \tag{2}$$

Adopting a constant index of the annual fuel price increase r_E, the total costs of the required combustion of fuel C_E after N successive years can be calculated acc. the formula:

$$C_E = \frac{\left(1 + r_E\right)^N - 1}{r_E} C_{E1} \tag{3}$$

C_{E1} is the per-unit price of the fuel in the first year.
According to forecast given in [11], annual increase in coal prices for private consumers up to the year 2030 will be approx. 3%. In study [12], over 2025 time horizon, this value is only about 1.35 %. For further calculations, the value is assumed to be r_E=2 %.

If the thermal upgrading investment is funded from a bank loan with fixed interest rate i_r, the total repayment over N years is:

$$C_{A,insN} = C_{A,ins} C_{Ni} = C_{A,ins} N i_r \frac{\left(1 + \frac{i_r}{12}\right)^{12N}}{\left(1 + \frac{i_r}{12}\right)^{12N} - 1} \tag{4}$$

where $C_{A,ins}=d_{ins}C_{ins}+C_d$ (see Eq. 1).

3 Results

To calculate the total costs involved in thermo-modernisation investment, it is necessary to have a forecast of weather conditions for the term of the loan, or the investment amortisation period. Unfortunately, weather forecasts that cover time lengths longer than a month are highly unreliable. To be on the safe side, it is necessary to take into account average values obtained from long–term weather records, provided by Central Statistical Office of Poland [13], which are presented in Fig. 1.

 DOI: 10.1051/e3sconf/20171401030

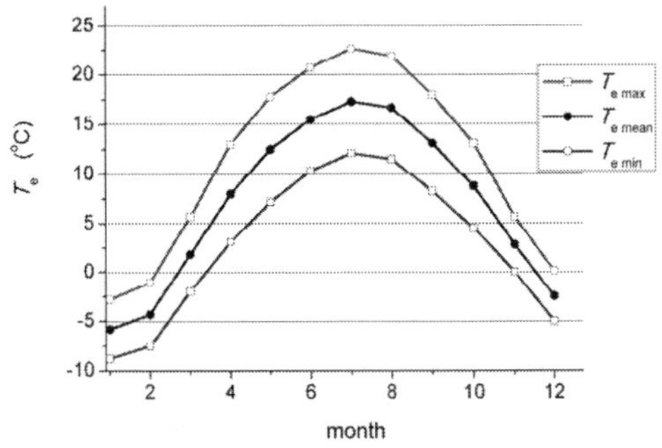

Fig. 1. Mean temperatures in Poland [13].

To account for the actual conditions during the heating season, the number of heating degree days (HDD) for a selected base temperature was determined from the Hitchin's formula [14]. Calculations were conducted for mean temperature values, shown in Fig. 2 while taking necessary constants into consideration [15].

Exemplary computations were conducted for the building wall of thermal resistance 0.5 W/mK and two types of insulation materials. The first is the blend of separately purchased components mixed at the construction site ($\lambda \approx 0.15$ W/mK). The other, manufactured at specialised production facilities, contains additives that enhance insulation properties ($\lambda = 0.076$ W/mK). Figure 2 shows changes in the optimal thickness of heat-preserving, perlite-based plaster as the term of the loan increases acc. Eq. 2.

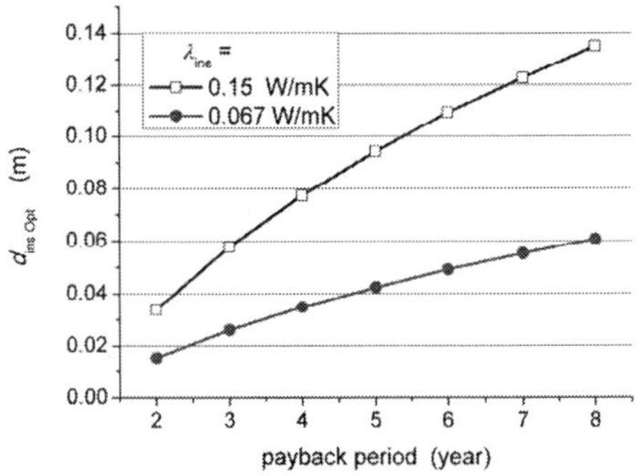

Fig. 2. Optimal thickness of the heat-preserving plaster over the term of the loan at 5 % interest rate.

Calculations were made for the heating system with a coal-fired boiler, the efficiency of which depends on the load. Linear dependence is assumed, so at the minimal load, boiler efficiency is 20% and it grows until the 80% maximum value is reached at full load.

The thickness of the plaster with lower insulation properties, optimised with respect to total costs, is over twice as much as that of the other type of plaster, and the ratio of appropriate values shows, with good approximation, price relationship between those two. Also, it should be noted that a longer term of the loan makes it possible to apply a thicker insulation layer.

Changes in both components (i.e. costs of heating and insulation) of the total costs as a function of the plaster thickness is shown in Fig. 3. For a five-year term of the loan and fixed interest rate i_r=5 %, the optimal insulation thickness is 9.4 cm, and the overall cost of thermal upgrading amounts to almost 142.5 PLN/m^2 (1 USD ≈ 4 PLN).

DOI: 10.1051/e3sconf/20171401030

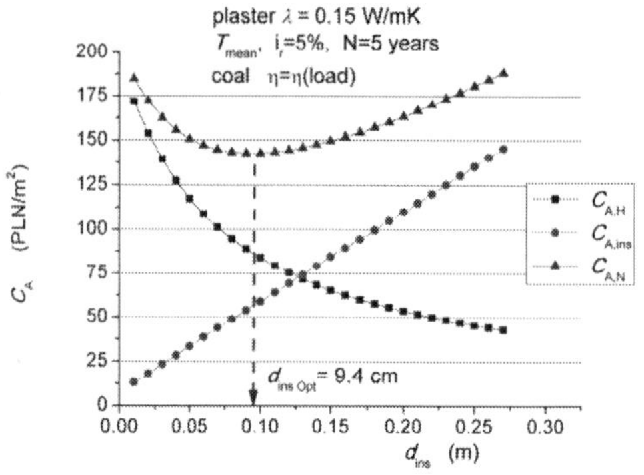

Fig. 3. Heating, insulation and total costs for coal heating system (λ=0.15 W/mK).

Figure 4 shows total costs for the mean distribution of monthly temperatures in the heating season and also for extremes of temperatures: minimum and maximum ones in each month of the heating season. The dependences shown in Fig. 4 indicate considerable changes in costs caused by weather cycles and they can be used to assess the investment risk. The calculations show that as much as 30% deviation from the optimum thickness leads to only a few percent increase in total costs.

Poorly predictable weather conditions are a main risk factor for thermal upgrading investment. Figure 4 shows total costs of the investment for the assumption that average monthly temperatures take the extreme values as shown in Fig. 2. It is obvious that at milder temperatures during the heating season, costs are the lowest, and the highest at low temperatures. In the case under consideration, differences of a few percent are found.

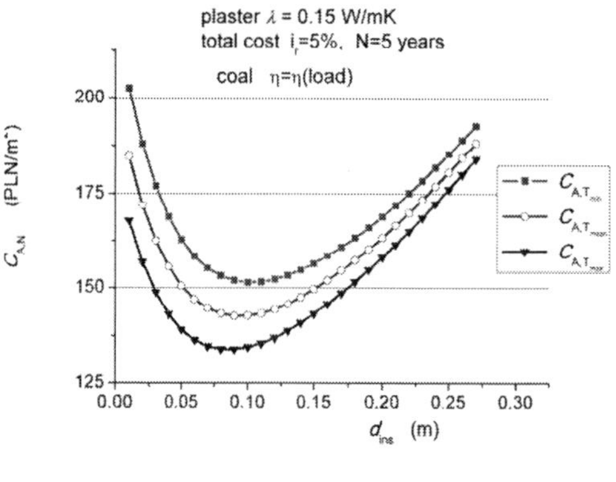

Fig. 4. Change in total costs for different plaster thicknesses under various climatic conditions.

The comparison of the total costs of thermo-modernisation with the use of a plaster coat having different insulation properties over a five-year term of the loan and fixed interest rate of 5 % shows lower costs of thermal upgrading with plaster of lower coefficient of thermal conductivity for thinner layers. For a 6 cm thick layer, the costs are the same. As the layer thickness of the worse quality plaster grows, the costs are clearly lower, which is shown in Fig. 5.

 DOI: 10.1051/e3sconf/20171401030

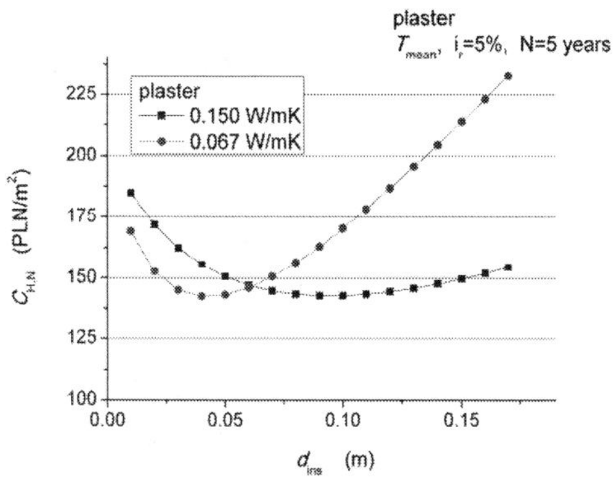

Fig. 5. Total costs of thermal upgrading with plaster having different insulation properties, depending on the clad layer thickness.

The most important environmental effect of insulation installing are savings in fuel consumption. That directly reduces carbon dioxide emissions into the atmosphere, equivalent particulates emissions, and also that of SO_2 and N_2. Figure 6 presents the amount of coal saved in a year for a plaster coat having different insulation properties. Heating is provided by a boiler, the heating efficiency of which grows linearly with an increase in load, and for the sake of comparison, by a constant 80% efficiency boiler. The results refer there only to heat loss by conduction.

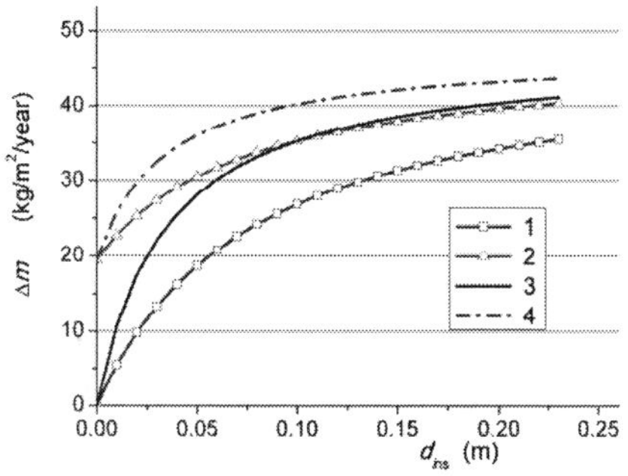

Fig. 6. Coal savings for plaster layers of different thickness 1 and 2 - λ=0.15 W/mK, 3 and 4 - λ=0.067 W/mK, heating by boiler with load-regulated variable efficiency (1, 3), and with high, constant efficiency (2, 4).

The costs involved in the investment are mostly affected by the heat source that uses much fuel and has low efficiency. When thermo-modernisation of the building is planned, it is advisable to upgrade the heating system so that the boiler should have constant, and possibly high efficiency regardless of the actual demand for heat. If possible, the heating system upgrades should include hybrid or co-generation installations [16].

4 Conclusions

Thermo-modernisation of historic buildings requires particular diligence and generates high costs for the property owner. While seeking for possible solutions, it is essential to aim at minimum costs and maximum environmental effects that are of key importance for large, old cities. The application of heat-preserving plaster always leads to improvement in conditions and a reduction in operating costs. Depending on the plaster coat thickness that is possible to apply, more expensive material with better insulation parameters should be used for thinner coatings, whereas for thicker coatings, a cheaper material is acceptable (see Fig. 5). Thermo-modernisation of the building envelope offers an incentive to invest in heat source modernisation that always results in considerable costs reduction and enhancement of the environmental effect (see Fig. 6). Balancing the costs and benefits of the investment of that type is also related to weather risks. Depending on climatic conditions, weather risks can affect the profits from fuel savings, thus the expected return on the investment, as shown in Fig. 4.

Interference in the structure of the historic buildings is subject to special oversight under supervision of conservation officers. For this reason, the final selection of the thermo-modernisation parameters should also take into account - if any - local and government regulation.

References

1. Ü. Alev, L. Eskola, E. Arumägi, J. Jokisalo, A. Donarelli, K. Siren, T. Broström, T. Kalamees, Energ. Buildings **77**, 58 (2014)
2. K. Fabbri, Cult. Heritage **14S**, 25 (2013)
3. S. Schiavoni, F. D'Alessandro, F. Bianchi, F. Asdrubali, Renew. Sust. Energ. Rev. **62**, 988 (2016)
4. Z. Ma, P. Cooper, D. Daly, L. Ledo, Energ. Buildings **55**, 889 (2012)
5. Z. Owsiak, W. Grzmil, J .Civil Eng. **KSCE** 1 (2014)
6. Z. Owsiak, J. Zapała-Sławeta, P. Czapik, B. Pol. Acad. Sci. Tech. Ser. **63-1**, 23 (2015)
7. R. Cantin, J. Burgholzer, G. Guarracino, B. Moujalled, S. Tamelikecht, B.G. Royet, Build. Environ. **45**, 473 (2010)
8. R. Dylewski, J. Adamczyk, Build. Environ. **46**, 2615 (2011)
9. M. Ozel, Appl. Energ. **88**, 2429 (2011)
10. T. Orzechowski, COW **6/47**, 229 (2016)
11. *https://bip.um.**katowice**.pl/dokumenty/2015/3/20/1426859144.pdf* (in Polish)
12. Collective work ed. by Gawlik, L., *Coal for Poland's energy production over 2050 time horizon– analysis of scenarios* (in Polish), Katowice (2013)
13. Local Data Bank. Warsaw: *www.stat.gov.pl*, last accessed June 2016
14. *Degree-days: theory and application TM41:2006*, (Chartered Inst. Build. Services Engineers -London)
15. J. Dopke, COW **5/45**, 171 (2014)
16. W. Stanek, W. Gazda, W. Kostowski, Energy **92**, 279 (2015)

 DOI: 10.1051/e3sconf/20171401031

The thermoelectric generators use for waste heat utilization from cement plant

Karol Sztekler[1,*], *Krzysztof* Wojciechowski[2], *Maciej* Komorowski[1], *Milena* Tarnowska[1]

[1]AGH University, Faculty of Energy and Fuels, Mickiewicza Av. 30, 30-059 Krakow, Poland
[2]AGH University, Faculty of Materials Science and Ceramics, Mickiewicza Av. 30, 30-059 Krakow, Poland

Abstract. Nowadays, one of the major economic problems is the increasing energy consumption and the long-term forecasts electricity demand by 2050 will increase several times as compared to 1990. Hence also to reduce the emissions of harmful combustion products there are investments undertaken for inter alia renewable energy sources and seeks to make the most efficient manufacturing system by levelling eg. heat losses at various stages of production, eg. electricity or also another product.
Production often entails the formation of by-product which is waste heat. One of the equipment processing heat into electricity is a thermoelectric generator. Its operation is based on the principle of thermoelectric phenomenon, which is known as a Seebeck phenomenon. The simplicity of thermoelectric phenomena allows its use in various industries, in which the main waste product is in the form of heat with the temperature of several hundred degrees. The study analyses the possibility of the thermoelectric systems use for the waste heat utilization resulting in the cement production at the cement plant. The location and design of the thermoelectric system that could be implemented in cement plant is chosen. The analysis has been prepared in the IPSEpro software.

1 Introduction

The cement industry in Poland, in terms of production volume is on seventh place among the European cement producers. Deep technical modernisation, which took place in the industry in the last decade, caused the industry in terms of technology is at the European and world forefront. 13 cement plants operating in the country is one of the most modern in Europe. These plants in the past few years have been modernised at the cost of 10 billion PLN. The cement industry in Poland directly and indirectly employs more than 24,000 people. In 2013, cement sales in Poland amounted to 14.3 million tonnes and was lower by 8% compared to the 2012 year. The 2014 annual sale of cement in Poland was close to the level of 15.2 million tons.

Favourable trend continued in 2015, when cement sales amounted to approximately 16.4 million tons [9, 11]. Cement plant average consumes approx. 4 GJ energy to produce 1

* Corresponding author: sztekler@agh.edu.pl

tonne of cement (in China up to 5.4 GJ/t) [2,3,4,6]. In table 1, the energy consumption (heat and electricity) for various types of rotary kilns used in the cement industry is presented.

Table. 1. Energy consumption for different methods of clinker in rotary kilns [6]

Methods production cement	Heat energy [GJ/t]	Electricity [MWh/t]
wet method furnaces long	5,0-7,5	0,025
half wet method - furnaces long / Lepol	3,4-4,0	0,03
half dry method - furnaces long / Lepol	3-3,9	0,03
dry method – furnaces with precalciner	3,1-3,2	0,022
shaft furnaces	3,7-6,6	0,03

Based on the gathered information, it can be concluded that the cement industry due to high consumption of heat and electricity (which transfer into CO_2 emissions) and emissions associated with the clinker production process adversely affect environment and all actions aimed at improvement of cement production efficiency will benefit to reduce energy consumption and reduce emissions of harmful gases into the atmosphere. The aim of this study is to present possible directions of waste heat utilisation in the cement plant with a strong focus on the use of thermoelectric generators for the utilisation of waste heat from rotary kilns. Calculations were made for the rotary kiln installed in Rudniki cement plant and the results of the analysis will help to show the potential use of thermoelectric generators in large scale industry [2, 3, 4].

2 Reduction of the cement plant negative impact on the environment

The cement industry is a highly energy-intensive and high consumption of raw materials and natural non-renewable fuels, joined the realisation of the policy of effective management of raw materials and energy, in line with the policy of sustainable development leads to: Increase in the production of the share of renewable and alternative energy sources; Promote energy efficiency; Reducing the negative impact of manufacturing processes and use of fuel and energy to the environment; Improvement (further development) of production processes; Valorisation of wastes in production processes; Replacement of fuels with a high content of C element with less C element fuels (eg. switching from coal to gas); The capture of carbon dioxide from the flue gases; Reduce the amount of carbon dioxide emitted by vehicles [4, 5]. Emission of carbon dioxide from the cement industry accounts for about 5% of global emissions resulting from human activity. In Poland in 2002 year the emissions from the cement industry was around 7,824 thousand tons, equivalent to the emissions per unit of approximately 0.69 kg/tonne of produced cement [4]. Manufacturing of 1 ton of clinker is accompanied by 0.5 tonnes of emitted technology CO_2. Over 50% of the total emissions from the cement industry is technology CO_2 [2,3,4,6]. There are several methods reducing the amount of carbon dioxide emitted per tonne of produced cement until 2020 year [6] for example: change electricity production (Increasing the efficiency of energy production and electricity from the fuels with low carbon content) -reduction carbon dioxide less than 5%; use of alternative fuels in cement production process- reduction carbon dioxide less than 5%; emission from the transport (change the type of fuel)- reduction carbon dioxide less than 20%. Despite a significant reduction in carbon dioxide emissions (in Poland, the amount of emitted carbon dioxide per unit of clinker decreased of about 40% compared to the beginning of the nineties of the last century), cement industry carries out activities aimed at further reducing its emissions. Another way to reduce the negative impacts on the environment in the cement industry is the use of waste energy. Even approx. 15% of the waste energy can be recovered in the cement plant through the use of appropriate technical solutions which are discussed below [8]. The overall efficiency of the manufacturing process of cement clinker with dry process in rotary kiln is approx. 50%. Some rotary kilns operating at full power declare the efficiency of approx. 55%. The use of heat recovery can apparently increase the efficiency of rotary kilns which will help to reduce the consumption of fuel, energy and CO_2 emissions. Manufacture of the clinker in the rotary kiln is the most energy intensive step in the cement production process (consumes approx. 90% of the total heat energy consumption). While to produce each tonne of cement an average of 90 - 130 kWh of electricity is consumed. The share of energy costs in the form of fuel and electricity is on average about 50% of the total costs associated with the production of one tonne of cement [4, 5]. As is apparent from (Fig. 1) the thermal balance of modern rotary kiln on the dry method shown in the Sanke graph, it is energy intensive process with an efficiency of 50%. Significant, more than 40%, the impact of the high energy consumption of the clinker burning process in a rotary kiln have heat losses: In the exhaust gases; With excess air of the clinker cooler grate; Radiation and convection of hot surfaces and rotary kilns. The enthalpy of the exhaust gases from the rotary kiln and the enthalpy of excess air from the clinker cooler is about 30% of the heat losses of the clinker burning process. In the Fig. 1 the sources of waste energy from the rotary kiln and possible ways of its use are shown. Till now common in the cement industry process of waste heat use (particular enthalpy of flue gases and excess air) is to use it for raw materials, coal and slag drying. Therefore, it is advisable to search for new technologies for waste heat utilisation (production of electricity) which might arise from the exhaust gases and radiant heat of the rotary kiln. Therefore, the electricity generation in a cement plant from waste heat from the rotary kiln can be achieved by: Classic steam cycle (Clausius-Rankine cycle); An organic Rankine cycle (ORC); Kalina cycle; Installation of a steam engine block. Practically all of these methods originate from the classic Clausiusa-Rankine cycle. Both ORC and Kalina cycles have found practical use in the cement industry. Modern installations of electricity generation, so-called WHR (Waste Heat Recovery) are used in the cement industry.

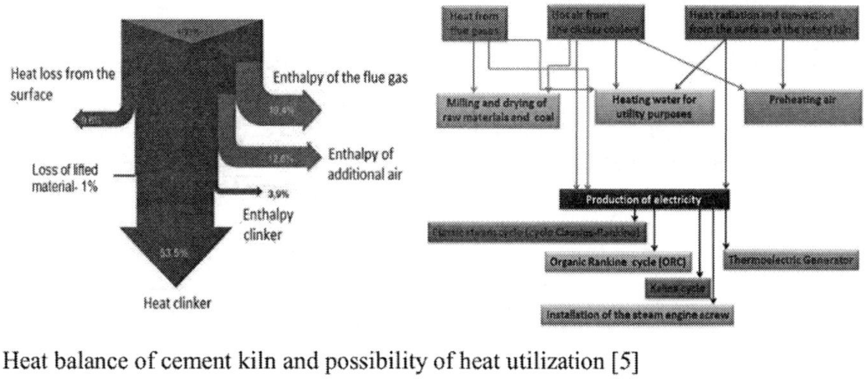

Fig. 1. Heat balance of cement kiln and possibility of heat utilization [5]

A global leader in the implementation of WHR cogeneration systems in the cement industry is the Japanese company Kawasaki Heavy Industries (KHI), which in 1980 in the Sumitomo Osaka Cement Plant has built and launched the first WHR installation. But already in 1982, KHI launched Taiheiyo Cement plant with an electrical capacity of 15 MW, securing more than 30% of the electric power demand by the cement plant. In recent years, to 2009 Kawasaki company launched 106 such plants with a total electric capacity of 1413 MW. The average energy recovery in new cogeneration systems in cement plants is approx. 30-45 kWh/Mg cement [5,7,8].

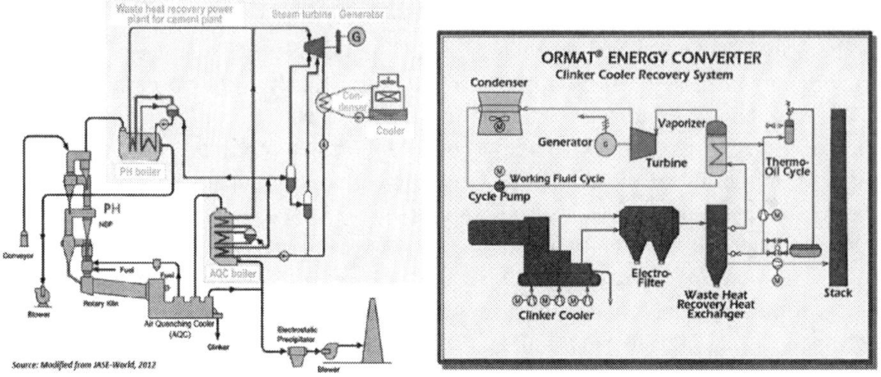

Fig.2. Electricity production from waste heat a) steam cycle; b) ORC cycle [12]

Accordingly, the duplication of world WHR solutions based on a classical Rankine cycle is very limited for offered terms. This causes a relatively low temperature of the flue gases derived from the rotary kiln (approx. 620 K), high enthalpy variation of excess air from the coolers (large temperature fluctuations) and the required heat in the drying process of fuel and other products require a different technology than the classic Rankine way generating electricity from the waste enthalpy. In Poland, in practice the implementation of WHR technology is limited to the two techniques - ORC and Kalina cycle. In 2010-12-08 - The Swiss company Holcim increased the efficiency of their cement plant because in Untervaz use of ABB modern heat recovery system based on ORC cycle. Such solution may be implemented in many cement plants in Poland [12, 6]. Another possible technical solution for the utilisation of waste heat emitted by radiation from the rotary kiln shell is the use of the thermoelectric generator and the direct electricity production. Device transforming heat into electricity is a thermoelectric generator. Its operation is based on the principle of thermoelectric phenomenon, a phenomenon known as the Seebeck effect. Seebeck Effect (Fig .3) depends on the formation of electromotive force in a circuit consisting of two conductors connected to the two ends of which the contacts are located at different temperatures is thus possible to convert thermal energy into electricity, an essential element of executing this process is a thermocouple. For the system to work with an assumed efficiency suitable large temperature difference between the hot and cold side is required. Hence thermoelectric generator is nothing else like a heat exchanger, which in addition to electricity production can produce hot water for utility purposes or any other medium used for the thermoelectric generator cooling. The investigations were conducted at the AGH University of Science and Technology.

The tested system (Fig. 3) in optimal conditions generate a power of 200 W, acting with efficiency in the range of 1.1-1.9 %. Maximum efficiency of thermoelectric modules was amounted to 3%. At the AGH Laboratory thermoelectric modules were developed that can be used at higher temperatures. Their highest possible efficiency is 9%. This type of solution is based on thermoelectric generators is examined in this article [1].

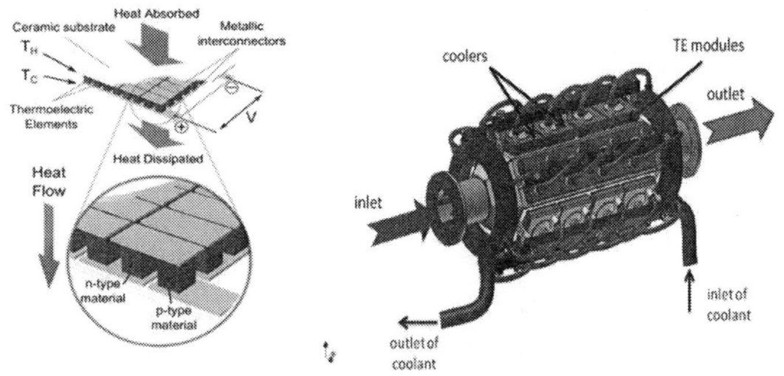

Fig.3. Thermoelectric generator a) The principle of thermoelectric generator b) thermoelectric generator prototype made by AGH [1]

3 Waste heat utilisation from the rotary kiln in Rudniki Cement Plant

Rudniki Cement Plant near Czestochowa is engaged in production: cements (slag, mixed or composite, limestone, steel), road binders (from the year 2009) and limestone. In Rudniki annual production capacity of cement is 1200 kt, while the clinker is 650 kt. [10,13]. Currently in the Rudniki Cement Plant are 2 ways of waste heat utilisation which is emitted by the rotary kiln installation. In 2009 is run pipeline of the hot gases received from the furnace, which are used for drying of the raw material fed to the raw meal preparation in two ball mills. In 2014 it was built another second pipeline of exhaust gases supplied to the coal department, where they are used for coal drying in a ball mill. There is no heat recovery from the rotary kilns radiation in cement plant [10, 13].

3.1 Design assumptions

Hot surface of the rotary kiln, which can reach temperatures of over 400 °C is another significant source of heat loss. Convection and radiation losses may constitute approx. 8-15% of the energy input. Rotary kilns by radiation emit considerable amounts of heat that may be used by thermoelectric generator to produce electricity and heat (in the form of hot water and for heating of the cement plant buildings) The heat dissipated by the rotary kiln is done by radiation and convection. However, in the zone of the highest temperature heat transfer is dominated by radiation (approx. 4 times higher coefficient of heat transfer by radiation than convective coefficient) [14]. Thus, in the paper it is contemplated the use of thermoelectric generator operating as a radiation heat exchanger for a rotary kiln no. 4. Diagram of the system operation for waste heat utilisation from the rotary kiln is presented in Fig no.4. Thermoelectric generator (TEG) must be placed in the zone of the rotary kiln no. 4, where is the highest temperature. In the section of 16-28 m initial length of the cement kiln, where there was on a selected range of Ar 1 maximum temperature of 348.7 °C, this part of the furnace was considered as a potential place to install thermoelectric generator. In view of the operational and technological considerations exchanger must not obstruct the operation, repair work or maintenance, so it is expected to placement in suitable distance from the rotary kiln. Distance of thermoelectric generator to the cement kiln should be as small as possible, but the above reasons, considering the application of the mechanism governing this distance. Most preferred would be to place the heat exchanger enclosing the rotary kiln in figure 4, but for the reasons of TEG implementing it will be located under the furnace and the generator is formed as a cylinder surface comprising a sixth circuit of the rotary kiln, which

 DOI: 10.1051/e3sconf/20171401031

is a solid angle of 60° within a distance of 1m from the furnace, provided with a frame on wheels, which can be easily moved.

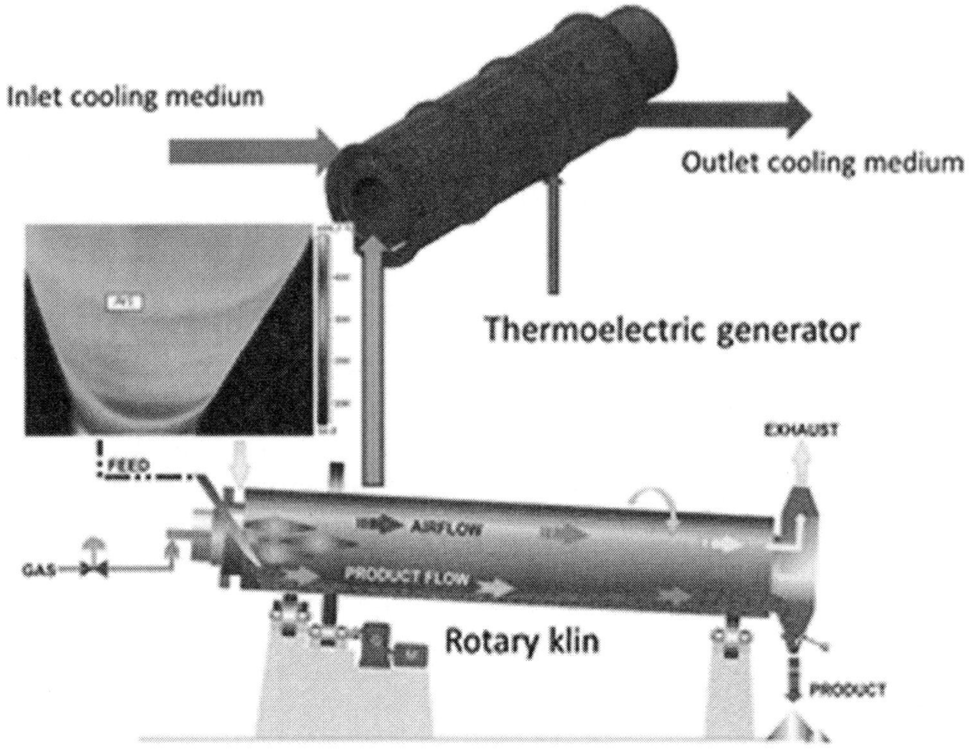

Fig.4. Schem instalation thermelectic generation in rotary klin

In addition, it can be in the area of the furnace throat, wherein the diameter is changed from 3.75 m to 3.45 m, in areas where supports are provided, and where run frequently used road transport. Other assumptions for the calculation: averaged temperature of the rotary kiln shell in section 16-28 m: t_l = 316 °C; the length of the exchanger installation l_l = 12 m; for the cooling of the thermoelectric generator is used the water at input temperature of 12 °C and output temperature of 65 °C; Calculations are performed for the temperature T_2 of the wall surface of the thermoelectric generator (TEG) in the range of 37.6 to 40.5 °C; efficiency of electricity generation using the thermoelectric generator in this temperature range is assumed to be 7%, based on the literature [1].
For the thermal - flow calculation of the thermoelectric generator heat exchanger should be designated [14]:

The configuration factor

$$\varphi_{1-2} = \frac{1}{A_1} \int_{A_1} \int_{A_2} \frac{cos\varphi_1 cos\varphi_2}{\pi r^2} dA_1 dA_2 \quad (1)$$

A_1 - surface discharge heat (rotary klin) [m^2]
A_2- area of absorbing heat (thermoelectric generator) [m^2]
φ_1- configuration factor for surface discharge heat
φ_2- configuration factor for surfaces absorbing heat
r - the distance between the furnace and the heat exchanger the thermoelectric generator [m]

- Calculation of heat flux absorbed by the TEG (thermoelectric generator) walls from the rotary kiln

$$q_{1-2} = \varepsilon_1 \cdot \varepsilon_2 \cdot \varphi_c \cdot C_0 \left[\left(\frac{T_1}{100}\right)^4 - \left(\frac{T_2}{100}\right)^4\right] \left[\frac{W}{m^2}\right] \quad (2)$$

ε_1 – emissivity of the surface area of the furnace
ε_2 – emissivity of the surface of steel pipes
C_0 – technical constant radiation of a black body [$W \cdot K^4/m^2$]
T_1– temperature surface rotary kiln [K]
T_2 – temperature surface thermoelectric generator [K]

- Calculation of the thermal power absorbed by the TEG [28]

$$Q_{1-2} = q_{1-2} \cdot \pi \cdot d_z \cdot L \ [W] \ (3)$$

L – the length of the pipe in the thermoelectric generator [m]

- Calculation of water flow velocity in the TEG tubes

$$v = \frac{Q_{1-2}}{\frac{\pi \cdot d_w{}^2}{4} \cdot \rho \cdot c_p \cdot \Delta t_f} \left[\frac{m}{s}\right] \quad (4)$$

$$\Delta t_f = t_{wy} - t_{we}$$

t_{we} – inlet temperature of the cooling water TEG [^0C]
t_{wy} – outlet temperature of the cooling water TEG [^0C]
ρ – density of water [kg/m^3]
C_p– specific heat at p = const [J/(kg·K)]
d_w , d_z –inner/ outer diameter pipe in the thermoelectric generator [m]

- Calculation of the Reynolds number

$$\mathrm{Re} = \frac{v \cdot d_w}{v_f} \quad (5)$$

v – kinematic viscosity [m^2/s]
v_f– kinematic viscosity of water at a reference temperature [m^2/s]

- Calculation of Prandtl number for water at reference temperature t_{fm} and the wall temperature t_2

$$\mathrm{Pr} = -3{,}379 \ln(t) + 17 \ (6)$$

- Calculation of heat transfer coefficient

$$\alpha = \frac{Nu \cdot \lambda}{d_w} \left[\frac{W}{m^2 \cdot K}\right] (7)$$

λ – thermal conductivity [W/(m·K)]
Nu - Nusselt number

Used Michiejew correlation for Nusselt number, depending on the type of flow
• laminar flow:

$$\mathrm{Nu} = 0{,}15 \cdot Re^{0{,}33} \cdot Pr_{fm}^{\ 0{,}43} \cdot Gr^{0{,}1} \left(\frac{Pr_{fm}}{Pr_2}\right)^{0{,}25} \cdot \varepsilon_L (8)$$

Prfm - Prandtl number for reference parameters
Pr2- Prandtl number for surfaces absorbing heat

- Calculation of heat flux acquired by the exchanger tubes of the rotary kiln

$$q_{2-1} = \alpha(t_2 - t_{fm}) \left[\frac{W}{m^2}\right] (9)$$

t_{fm} - reference temperature, t_{fm} = 37.5 °C
α- heat transfer coefficient [W / (m^2K)]

4 Calculation results and conclusions

The results of the calculations for the TEG with a length of 12 m at a distance from the rotary kiln s = 1-4 m have been summarised in the table. Below, the calculation for the interval L = 12 m (16-28 m) are at different distances from the furnace shell, wherein the shell is the average temperature t_1 = 316 °C.

Table 2. Results of the calculations for the TEG installed on rotary kiln

s	t_2	φ_c	q_{1-2}	Q_{1-2}	P_{elTEG}	v	Re	Pr_2 for t_2	Nu	α	q_{2-1}
m	°C	-	W/m²	kW	Kw	m/s	-	-	-	W/(m²K)	W/m²
1	39.4	1.38	4232.6	163.34	11.43	0.459	29782	4.59	157.19	2199.22	4178.52
1.4	39.2	0.802	2460.48	94.95	6.65	0.267	17313	4.6	101.75	1423.62	2420.15
1.8	39.1	0.544	1670.42	64.46	4.51	0.181	11754	4.61	74.61	1043.84	1670.14
2.2	39.1	0.404	1239.92	47.85	3.35	0.134	8725	4.61	56.2	786.29	1258.06
2.6	39.1	0.318	975	37.63	2.63	0.106	6861	4.62	45.23	632.81	949.22
3	39.1	0.26	797.71	30.78	2.15	0.086	5613	4.61	36.39	509.16	814.65
3.4	39.1	0.219	672.01	25.93	1.82	0.073	4729	4.61	29.37	410.89	657.43
3.8	39.2	0.189	578.71	22.33	1.56	0.063	4072	4.6	23.77	332.59	565.4
4	39.3	0.176	540.64	20.86	1.46	0.059	3804	4.6	21.4	299.35	538.84

For the case under consideration the use thermoelectric generator for waste heat utilisation emitted from the surface of the rotary kiln in the radiation path is proposed. This paper proposes a thermoelectric generator with a length of 12 m located on the length of 16-28 m of no. 4 rotary kiln. The rotary kiln will not be completely built with thermoelectric generator in this section, but it will be placed under the furnace in the form of cylinder surface covering 1/6 of the circumference of the rotary kiln, which is the solid angle 60°. It will receive heat from approx. 1.35% of the area of the side of the furnace. TEG can produce from P_{elTEG} = 1.46 kWe to about P_{elTEG} = 11 kWe of electricity which translates to receive by radiation of Q1-2=10kWth to Q1-2 = 163 kWth depending on the distance to the surface of the TEG furnace. The highest achieved power TEG is possible at a generator distance of 1 m from the rotary kiln surface. Since the essence of the action TEG receives heat from the hot side, this allows for production of hot water for hot water and heating plant. With the TEG with a maximum power of approx. 163 kWth and 11 kWe Cement Plant reaches saving of burned fuel. The plant will be able to afford to completely disable the boiler room in summer (May-September), while in winter there will be aided working exchanger. A very important element is to equip TEG in the continuous measurement of the temperature of input and output water, which will give an indication to the approximation, or the distance from the kiln shell heat exchanger depending on the current needs of the Plant for hot tap water. In addition, heat has no influence on the quality of the final material produced in the Plant and the whole technological process.

The article was funded from government money Faculty of Energy and Fuels number 11.11.210.216 and Faculty of Materials Science and Ceramics number 11.11.160.438

References

1. K. Wojciechowski, J. Merkisz, P. Fuć, J. Tmankiewicz, R. Zybała, J. Leszczyński, P. Lijewski, P. Nieroda, *Prototypical thermoelectric generator for waste heat conversion from combustion engines* Combustion Engines ; ISSN 0138-0346. — R. 52 nr 3 (2013)
2. A. Uliasz Bocheńczyk A., E. Mokrzycki, *Emisja dwutlenku węgla w przemyśle cementowym.* Polityka Energetyczna, tom 20, zeszyt 2, s. 129-143 (2003)
3. B. Vanderborght, B. Brodmann, *The Cement CO2 Protocol: CO2 Emissions Monitoring and Reporting Protocol for the Cement Industry.* Guide to the Protocol, version 1.6 www.wbcsdcement.org (2001)
4. A. Uliasz Bocheńczyk A., E. Mokrzycki, *Możliwości ograniczenia emisji CO2 w przemyśle cementowym.* Polityka energetyczna Tom 7 Zeszyt specjalny Wyd. Instytut GSMiE PAN Kraków s. 555-564 PL ISSN 1429.6675 (2004)
5. A. Duczkowska –Kądziel A., J. Duda, *Methods of increasing the electric power generated from waste heat in association with a rotary kiln,*.Heat Transfer and Renewable Sources of Energy HTRSE-2012
6. E. Worell, L. Price, N. Martin, Ch. Hendriks, M.L. Ozawa, *Carbon dioxide emission from cement industry.* Annual Review of Energy and the Environment,26, pp. 303-329 (2001)
7. J. Duda, K. JesionekJ, *Zagospodarowanie procesowego ciepła odpadowego na przykładzie Cementowni.* „Prace Naukowe", Politechnika Warszawska, Warszawa , s. 47–53 (2011)
8. J. H a r d e r, *Trends in power generation from waste heat in cement plants,* „Zement–Kalk–Gips" s. 36–47 (2011)
9. .http://www.polskicement.pl/aktualnosci/W_2015_roku_wzrost_sprzedazy_cementu_w_Polsce_do_16_mln_ton-308
10. Cementownia Rudniki. Dostępny: http://www.cemex.pl/cementownia-rudniki.aspx

DOI: 10.1051/e3sconf/20171401031

11. http://www.polskicement.pl/files/Pages/225/uploaded/informator%202014%20-%204%20strony%20-%20low.pdf
12. H. Legmann, ORMAT Industries, *Recovery of industrial heat in the cement industry by means of the ORC process.* papers\IEEE 2002\4066\July 11, 2002
13. Tarnowska M., *Projekt wymiennika do odzysku ciepła wydzielanego przez piece obrotowe"* Praca dyplomowa, 2016
14. Poniewski M., Orzechowski T., Cesarz Cz., Chajewski R., Gołaszewski K., Lubowicz M., et al.: *Utylizacja ciepła odpadowego cz. II,* Zlec. nr 1839/13, Kielce, Politechnika Świętokrzyska, Samodzielny Zakład Termodynamiki i Mechaniki Płynów, 1993

The thermoelectric generators use for waste heat utilization from conventional power plant

Karol Sztekler[1,*], *Krzysztof* Wojciechowski[2], *Maciek* Komorowski[3]

[1]AGH University, Faculty of Energy and Fuels, Mickiewicza Av. 30, 30-059 Krakow, Poland
[2]AGH University, Faculty of Materials Science and Ceramics, Mickiewicza Av. 30, 30-059 Krakow, Poland
[3]AGH University, Faculty of Energy and Fuels, Mickiewicza Av. 30, 30-059 Krakow, Poland

Abstract. On the base of available data, it is estimated that the industrial approx. 20-50% of the energy is removed into the atmosphere as waste heat include in the form of hot flue gases, cooling water, the heat losses from the equipment hot surfaces or heated products. However, according to the data from the US market in 2010, in the form of waste heat is emitted more than 96 · 106 TJ annually (2.7 · 1010 MWh), means more than 57% of the produced energy. According to statistics, currently the energy production in the US amounts to approx. 26% of the world's energy production. Assuming the same indicators, the total annual amount of waste heat in the scale of the world equals 370 · 106 TJ (10.4 · 1010 MWh). One of the ways to increase the energy efficiency of manufacturing processes and reducing energy consumption and negative impacts to the environment is the use of waste energy [1,2,3] In this work it was investigated the possibilities of the waste heat utilization from conventional thermal power plant using thermoelectric generators, the operation of which is based on the Seebeck effect.

1 Introduction

Nowadays, one of the major economic problems is the increasing energy consumption. This results in a rapid depletion of fossil fuel resources. To reduce the consumption level of these resources, a lot of investment in renewable energy and development investment in these sources. However, there is also the possibility of improving the efficiency of energy production from non-renewable sources. Energy production often involves the formation of by-product which is waste heat. However the waste heat is a certain volume of energy carrier which can still be utilized. One of the equipment processing heat into electricity is a thermoelectric generator. Its operation is based on the principle of thermoelectric phenomenon, which is known as a Seebeck phenomenon. The simplicity of thermoelectric phenomena allows its use in various industries, in which the main waste product is in the form of heat with the temperature of several hundred degrees. The most beneficial seemingly waste heat with the highest temperature. Thermoelectricity was discovered in the nineteenth century. The phenomenon of heat conversion into electricity and vice versa are called

* Corresponding author: sztekler@agh.edu.pl

thermoelectric effects. Distinguishes the 3 main types of these phenomena: Seebecka; Peltiera; Thomson [1, 2, 3]. Thermoelectric effect in the 20th century was used in various fields, eg. power the spacecraft, drawing heat for the heating of a single connector on the nuclear reactions. Thermoelectric materials were also used to power the pacemaker. Thermoelectric generator can also be used in the exhaust systems of cars. According to estimates, in modern engines with spark ignition and compression ignition heat contained in the exhaust gases is approx. 30% of the total heat from burning fuel. This heat instead of giving ineffectively to the environment, can be though to some extent re-use. This will increase the efficiency of the drive system. The use of thermoelectric generators to convert about 6% derived waste heat will reduce fuel consumption by up to 10%.

1.1 Construction and operation of thermoelectric generator

Thermoelectric generators are built of thermoelectric modules. A single module is made with a thermocouple. A single thermocouple generates a small thermoelectric force, and therefore to the formation of one module is used even hundreds of thermocouples. The module is encased ceramic layer which also acts as a housing and the insulator. Under the housing elements are arranged semiconductor . The thermal energy is supplied to the heat exchanger. Then, the heat is directed to the thermoelectric modules in which a partial conversion of thermal energy into electricity. The rest of the heat is discharged to the cooler. The process of Thermoelectric (Figure 1) begins at the start of one side of the generator heating. Absorption of a certain amount of heat causes diffusion of the mobile charge carriers along the temperature gradient. Negative particles flow to the "cold" side by an semiconductor n- type, and the particles of the positive p-type semiconductor. As a result, between the two parties to the generator potential difference is formed [2, 3].

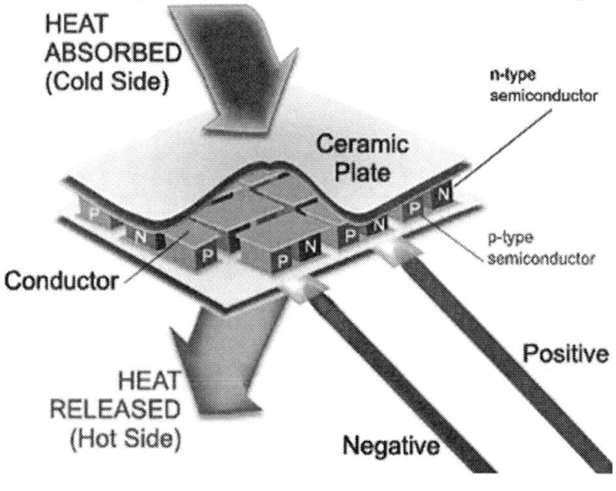

Fig.1. Scheme and operation of thermoelectric module [3]

The benefit would be getting the greatest power in the process. It is possible to increase the value of received power. For this purpose semiconductors are connected in the cell. This increases the voltage of the cell, and thus also the power.

1.2 The use of thermoelectric systems

Practical use of thermoelectricity phenomenon began in the twentieth century. The beginnings of this field is inter alia powering radio or creation of the first thermoelectric generator with an efficiency of 5% in the first half of the twentieth century. Later, one of the most significant initiatives was the American SNAP project, which was used by NASA when

DOI: 10.1051/e3sconf/20171401032

powering space probes, including expeditions to the moon. These facilities have used radioactive reactions as a heat source. Also, Voyager 1 space probe (sent into space in 1977), now the farthest located object sent by man, is powered by thermoelectric generators. The universality of such systems application due especially to the lack of moving parts and the possibility of long-term work. The success of the project on the Voyager probes led to further projects the power of objects sent into space by utilizing the thermoelectricity phenomenon. General-Purpose Heat Source Radioisotope Thermoelectric Generator (GPHS-RTG) was designed. These generators used as power of Cassini probes, which has been sent to Saturn, New Horizons was sent on Pluto, Galileo was directed to Jupiter and Ulysses investigating regions of the sun. In France Thermoelectricity used to power a pacemaker, and in the 90s the company producing watches named Seiko has introduced a thermal watch, which used the heat of the human body. In recent years, a pioneer in the use of thermoelectricity in the industry is an American company Alphabet Energy [6].

Research in this area is also GMZ Energy cooperating inter alia with the company car Honda. The company has developed an advanced generator for Honda Accord. There has generator power at 1 kW. This result was achieved through a combination of five generators with a capacity of 200 W. The electrical energy will support the operation of the automotive alternator. The company's activities are co-financed by the US army. Tests were also conducted for the possibility of generator using by the military. GMZ used to this materials of half-HEUSLER type. By using nanostructures have improved mechanical integrity or stability in areas of high temperatures . Also, the Fiat Group in collaboration with companies trying to develop the technology and bring it to their vehicles. Model, for which the study was conducted was the Fiat Iveco Daily fitted with a diesel engine with a capacity of 2.3 liters. As the thermoelectric material bismuth telluride Bi_2Te_3 was used (exhaust gas temperature of 450 °C, gas flow of 70 g/s (at maximum load), 140 g/s (at full load) and the parameters of the coolant temperature of 60 °C, flow rate 0.33 l/s). Reached limit the amount of consumed fuel by 4%. Tests have shown that the thermoelectric materials allowed to work at a maximum temperature of 270 °C, so it is necessary to use materials for higher temperature ranges. [8] In Poland, research on thermoelectric generator installed in the exhaust system of the ignition engine 1.3 JTD (Figure 2) were conducted. They were conducted at the AGH University of Science and Technology under the direction of Professor Krzysztof Wojciechowski. Due to the use of thermoelectric modules built on the compounds of tellurium, bismuth and antimony generator provided at the place where the temperature of exhaust gas is about 300 ° C to ensure the best use of them. The tested system in optimal conditions generate a power of 200 W, acting with efficiency in the range of 1.1-1.9%. Maximum efficiency of thermoelectric modules amounted to 3% and has been achieved for temperatures: the cold side $T_c = 25$ °C, the hot side $T_h = 200$ °C [2,8, 4].

Fig. 2. Picture of the tested at the AGH University of Science and Technology thermoelectric generator [8]

Thermoelectric Laboratory at AGH developed modules that can be used at higher temperatures. Their highest possible efficiency is 9%. Thanks to their use the electric power of the thermoelectric generator can be increased to 1 kW [8].

2 Thermoelectric generator use in conventional thermoelectric power station

The electric power installed in Polish power plants in 2015 was 40.45 GW . 82% of the power was supplied by the combined heating plants - CHP (ie. producing energy for distribution and sale in the domestic energy system), use of traditional fuels (coal: 54.5%, lignite: 25.2%, gas 2.3 %). Another 6.3% provide wind turbines. As a result of the manufacturing process, the release of huge amounts of waste heat, which is not in any way disposed of and used further for energy purposes. One way to use waste heat of a sufficiently high temperature is to use thermoelectric generators which on the one hand, will produce electricity on the other hand will heat the cooling medium of thermoelectric generator, which can later be used for other industrial processes. Thermoelectric generator is possible for an application in any installation emits waste heat. The problem of waste heat is not limited only to the extraction and processing. It also causes huge losses, estimated at 20-50% of total energy consumption in the industry [Energetics, Energy Use, Loss, and Opportunities Analysis: U.S. Manufacturing & Mining, p. 17, 2004]. Almost 15 GW is lost in the US manufacturing sector. Also transport generates huge amounts of waste heat. Approximately 33% of the fuel energy is wasted in the exhaust system of the vehicle, a 29% loss in the cooling system [Steven Chu & Arun Majumdar, Opportunities and challenges for a sustainable energy future, Nature vol 488]. According to data from the US market in 2010, in the form of waste heat emitted annually is more than $96 \cdot 10^6$ TJ ($2.7 \cdot 10^{10}$ MWh), or more than 57% of the produced.energy According to statistics, currently the production of energy in the US amounts to approx. 26% of the world's energy. Assuming the same indicators, the total annual amount of waste heat in the scale of the World is amounted of $370 \cdot 10^6$ TJ ($10.4 \cdot 10^{10}$ MWh). If the heat will be converted using thermoelectric generators with an average efficiency of 5%, you can get $5.2 \cdot 10^9$ MWh of electricity, which (with the price of energy 165 PLN/MWh), would annually generate electricity worth 865 billion PLN [3, 5.6]. Conventional steam power plants use the chemical energy of fuel to produce electricity. Electricity can be produced in power plants or in combined heat and power plants CHP (combined with thermal energy). The combustion of fuel in the boiler generates heat, which is transmitted working medium - water, which evaporates. The steam drives a turbine blade, allowing it to work, produced power is transmitted to the electrical generator that produces electricity. This paper is focused on analyzing the possibilities of waste heat utilization for a small industrial power plants which are in Poland a dozen and the most popular boiler used in the steam production is OR-32/2 boiler . In combined heat and power plants one of the used boilers types are pulverized coal and grate boilers tab.1.

 DOI: 10.1051/e3sconf/20171401032

Table.1. Main parameters of power plant installation

Electrical power P_G	1,5	MW
Steam pressure	36	bar
Steam temperature	425	°C
Steam mass flow	32	t/h
Total efficiency of cycle	37	%
Boiler efficiency	76	%
The temperature of the boiler feed water	105	°C
Flue gas mas flow	56900	m^3/h

The calculations take into account the parameters of one such OR-32 boiler. Temperature of water supplied the boiler is 105 °C. After leaving of this boiler type the steam parameters are as follows: pressure 3.6 MPa, temperature of 425 °C, maximum continued capacity 32 t/h, rated thermal output approx. 25.2 MW and a guaranteed efficiency of 72%. The boiler is supplied with coal. Flue gas parameters relevant for the thermoelectric generators application, which is temperature of the air preheater outlet and a stream at the boiler outlet are respectively 200 °C and 56900 m^3/h . Achievable power depends on whether the steam from the boiler is used for other processes of plant or also a whole may be directed to the turbine. Below is a sample steam – water cycle of combined heat and power plant (Fig.3).

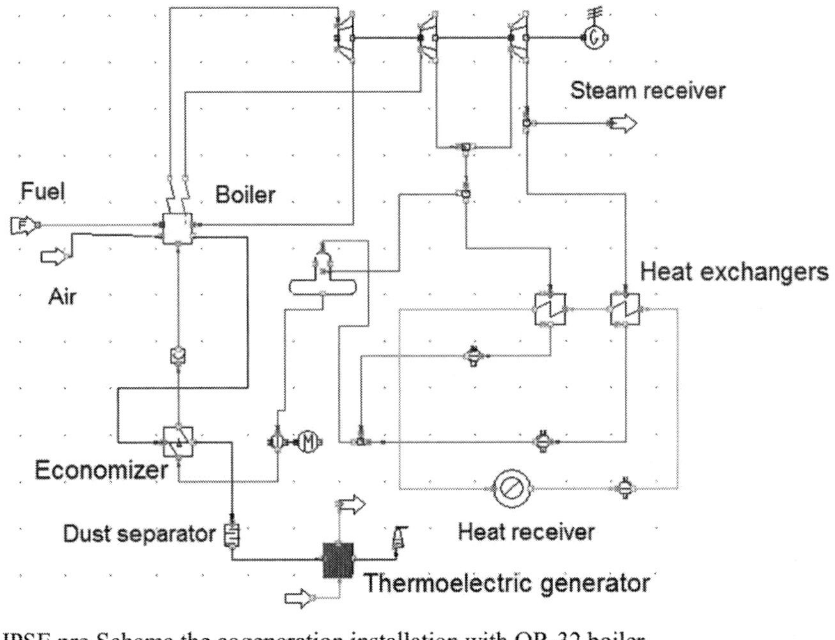

Fig.3. IPSE pro Scheme the cogeneration installation with OR-32 boiler

After analyzing the steam – water cycle of small heat and power plant for potential local use of waste heat by thermoelectric generators. The case analysis shows that from the operating of heat and power plants point of view is important to thermoelectric generator the least interfered with the structure of the plant's water steam power cycle because any change could result in unplanned downtime and additional costs of installation. Therefore it determined that the best place for the waste heat utilization from combined heat and power plants is flue gas duct after the flue gas cleaning system where the flue gas reaches a temperature of about 200 °C. In the flue gas duct thermoelectric genrators are arranged in such a way as not to affect the flow of flue gas through the electrostatic precipitator or other flue gas cleaning equipment . In most cases, for the flue gas cleaning process at low gas flow used are small-sized electrostatic precipitators. Small-sized electrostatic precipitators are used for extraction of gas from boilers with power being in the range of 5-50 MW coal-fired or biomass. These are called. "Small sources of dust." They clean gases from water grate and steam boilers. Small-sized electrostatic precipitators produced by Elwo allow for flue gas cleaning which flow rate is in the range of 10000 to 65000 Nm^3/h and the concentration of flue gas dust is from 1 to 10 g/Nm^3. On the basis of the manufacturer's data specified that the drop of flue gas temperature in the flue gas cleaning unit is approx. 5 °C, so when the enter flue gas temperature of 200 °C will be the cooling to the temperature of approx. 195 °C [9,10,11,12].

3 Results of analysis

The calculation considers the energy conversion for the three cases. They differ in the temperature to which the flue gases are cooled (respectively 150, 145 and 130°C). The flue gas temperature after extraction before the process of energy conversion is 195 °C. The valuesof the flue gas stream is contemplated in the range of from 30000 to 80000 m^3/h by adjusting its value at 5000 m^3/h. Adopted cooling temperatures of the flue gas are selected on the basis of a graph representing the content of sulfur in the operational state coal. The content of this element in the fuel affects the dew point of the flue gas. Low sulfur content results in a lower dew point temperature, which makes it possible to cool the flue gas to a lower temperature. Considered the grate furnace due to the characteristics of the OR-32 boiler.

For the simulation calculations in IpsePro software adopted the following chemical composition of flue gases: [15] –CO_2 0.18659 kg/kg; - H_2O 0.18044 kg; - N_2 0.62693 kg; 0.00016641 kg O_2/kg - SO_2 0.0058755 kg/kg. Considered the use of thermoelectric generators studied at the AGH University of Science and Technology [3,8]. Maximum generators efficiency investigated at the AGH University of Science and Technology was 1.9%. Calculations of realizable power were also conducted for generators wit efficiencies of 3 and 7%.

Table. 2. Production of electricity by thermoelectric generators with the efficiency of 7%

Exhaust gas flow [m^3/h]	Exhaust gas temperature Flow out from of TGE		
	150°C	145°C	130°C
30000	15.32	17.19	23.04
40000	20.43	22.92	30.72
50000	25.54	28.65	38.41
60000	30.65	34.39	46.09
70000	35.76	40.12	53.77
80000	40.87	45.85	`61.45

Table. 3. Production of electricity by thermoelectric generators with the efficiency of 3%

Exhaust gas flow $[m^3/h]$	Exhaust gas temperature Flow out from of TGE		
	150°C	145°C	130°C
30000	6.56	7.36	9.87
40000	8.75	9.82	13.16
50000	10.94	12.28	16.46
60000	13.13	14.73	19.75
70000	15.32	17.19	23.04
80000	17.51	19.65	26.33

Table. 4. Production of electric energy by thermoelectric generators with the efficiency of 1.9 %

Exhaust gas flow $[m^3/h]$	Exhaust gas temperature Flow out from of TGE		
	150°C	145°C	130°C
30000	4.16	4.66	6.25
40000	5.54	6.22	8.34
50000	6.93	7.77	10.42
60000	8.32	9.33	12.51
70000	9.70	10.89	14.59
80000	11.09	12.44	16.68

For the flue gas flow of 60,000 m^3/h can be achieved up to 45 kWe for AGH modules (efficiency 7%) and 12 kWe of electricity with the use of commercial embedded thermocouples in the flue gas duct for which efficiency is at least 1.9%, assuming that the drop in temperature is 650 °C.

Application of this case involves the reconstruction of the flue gas duct and flue gas cleaning installation. Such a solution would be the most beneficial for combination of flue gas cleaning system with the fabric filter, where the temperature reduction is indicated for technological and safety reasons where local flue gas temperature is much higher.

4 Conclusions

The aim of the study was to analyze the possibility of the thermo elements use to convert waste heat into electricity. Simulations were performed using the IPSEpro software. It allows to simulate all kinds of thermal cycles, power units and cooling cycles. The study analyzes the possibility of the thermoelectric generators use in a steam, OR-32 grate boiler, behind which is placed the small electrostatic precipitator for flue gas cleaning. Considering the obtained results, for systems with thermoelectric efficiency of 1.9%, or such as that achieved at the AGH University of Science and Technology it is noted that the parameters characterizing the OR-32 boiler (flue gas stream of 56900 m^3/h) electric power will be from about 7 to about 13 kW, depending on the temperature to which the flue gas are cooled. For generators with an efficiency of 3% can be achieved several kilowatts of electric power. Using the thermoelectric chips on the efficiency of 7% can be obtained the power of 30 kW, and the temperature was reduced to 130°C even at the level of 40 kW. Increased value of the flue gas stream would result in higher values obtained power. They are not too high value as compared with the losses in the boiler, due to the relatively low values of thermoelectric generators, but the resulting power may be useful as a partial meet the needs of their own power. Examples of auxiliary OR-32 boiler, such as primary and secondary air fans show demand for power respectively 22 and 7.5 kW (based on the technical documentation of Fakop), these are values that can be covered with energy obtained as a result of thermoelectric systems use behind the electrostatic precipitator. In the next stages of work out economic analysis allows to determine the payback period for the construction of thermoelectric generator will be carried. The use of thermocouples which is a very innovative solution possible to implement not only in the power systems but also in other installations where there is a problem with the waste heat utilization and switch it to a useful product.

The article was funded from government money Faculty of Energy and Fuels number 11.11.210.216 and Faculty of Materials Science and Ceramics number 11.11.160.438

References

1. J. Vazaquez, M. A. Zanz-Bobi, R. Palacios, A. Arenas, *State of the art of thermoelectric generators based on heat recovered from the exhaust gases of automobiles*. Proc. of 7th European Workshop on Thermoelectrics, Pamplona, Spain (2002)
2. U. Birkholz, *Conversion of waste exhaust heat in automobile using FeSi2 thermoelements*. Proc. 7th International Conference on Thermoelectric Energy Conversion, Arlington, USA, pp.124-128 (1988)
3. K. Wojciechowski, M. Schmidt, R. Zybala, J. Merkisz, P. Fuć, P. Lijewski, *Comparison of waste heat recovery from the exhaust of a spark ignition and a diesel engine*. Journal of Electronic Materials, Volume 39, Issue 9, pp.2034 (2010)
4. J. Bass, N.B. Elsner, A. Leavitt, *Performance of the 1 kW thermoelectric generator for diesel engines*. Proc. 13th Int. Conf. Thermoelectrics B, Mathiprakisam, edn., AIP Conf. Proc., New York, No. 295 (1995)
5. K. Sztekler, M. Komorowski, D. Kot ,Modelling of the energy use of ventilation air from the mines SEED 2016 : the international conference on the Sustainable Energy and Environment Development : Kraków, *Poland, May 17* Wydawnictwo Instytutu Zrównoważonej Energetyki,— ISBN: 978-83-944254-0-1. — S. 143(2016)
6. K. Sztekler, T.M. Wójcik , *Wykorzystanie metanu z powietrza wentylacyjnego z kopalń na cele energetyczne.* Bezpieczeństwo Pracy i Ochrona Środowiska w Górnictwie *; ISSN 2081-4224. nr 6, s. 15–21 (2015)*
7. A. Królicka, A. Hruban, A. Mirowska, *Nowoczesne materiały termolektryczne – Przegląd Literaturowy*. Electronic Materials. Volume 40, Issue 9 (2012)
8. K. Wojciechowski, J. Merkisz, P. Fuć, J. Tmankiewicz, R. Zybała, J. Leszczyński, P. Lijewski, P. Nieroda, Prototypical thermoelectric generator for waste heat conversion from combustion engines Combustion Engines ; ISSN 0138-0346. — R. 52 nr 3 (2013)
9. K. Stanisław, *Kotły: konstrukcje i obliczenia* Oficyna Wydawnicza Politechniki Wrocławskiej, Wrocław (2001)
10. Badania parametrów pracy małogabarytowego elektrofiltru do usuwania aerozoli higroskopijnych soli amonowych z gazów odlotowych Instytut chemii i techniki jądrowej, Warszawa (2002)
11. Dokumentacja techniczna kotła OR-32 firmy Fakop
12. M. Pronobis. ,*Modernizacja kotłów energetycznych* Wydawnictwo Naukowo-Techniczne,Warszawa (2002)

E3S Web of Conferences **14**, 01033 (2017) DOI: 10.1051/e3sconf/20171401033
Energy and Fuels 2016

Reasonableness of phase change materials use for air conditioning – a short review

Marta Kuta[1*], *Dominika* Matuszewska[1] and *Tadeusz M.* Wójcik[1]

[1]AGH University of Science and Technology, Faculty of Energy and Fuels, Department of Thermal and Fluid Flow Machines, al. A. Mickiewicza 30, 30-059 Krakow

Abstract. PCM-based systems for cooling of building spaces are a developing area with great potential. Especially since cooling systems are becoming more popular resulting in increased demand for energy. The aim of this work is to analyse how application of phase change materials can support and improve cooling efficiency of air conditioning. Authors discuss and compare different types of PCM-based solutions.

1 Introduction

It can be observed over last few years that with an improvement of quality of life, usage of air conditioning becomes more frequent [1]. The consequence of this situation is the increase of the demand for energy not only for heating, but also for cooling, what causes general increase of energy consumption for thermal stabilization of buildings. It is important to improve existing and design new systems for controlling temperature, ventilation and humidity of buildings. Especially if we take into consideration how growing use of air conditioning influences the energy consumption. One of the possibilities for improving cooling efficiency of air conditioning is use of phase change materials (PCMs). It can be one of efficient method for improving the process as well as reducing the size of the system. Phase change materials are able to absorb, store and release large amount of heat. It is promising storage medium for a large number of applications such as waste-heat recovery, cooling and heating of buildings or heat pump-systems [2]. It is also considered for air conditioning systems. Phase change materials can be design to change the phase at the suitable temperature range, phase change temperature can also be designed for air conditioning requirements [3].

2 PCM for cooling applications in buildings (including air conditioning)

Phase change materials are substances which the most important attribute is the ability to absorb, store and release the heat under certain circumstances. General principle of operation is the same, but properties (phase change transition temperature, latent heat of fusion, specific heat, thermal conductivity, aspirations for gas generation and changes in

* Corresponding author: marta.kuta@agh.edu.pl

volume during the transition, life span, tendency to corrosion, flammability or explosiveness, subcooling, recyclability, environmental friendliness, availability or cost) are individual for every material. Figure 1 presents the schematic work of phase change materials on the example of solid-liquid PCMs. For this kind of transition it can be observed that when the ambient temperature rises, the temperature of PCM also rises. When temperature reaches the level of phase transition PCM starts to melt and at the same time absorb heat without changes in PCM temperature. When the temperature decreases the opposite process occurs. PCM releases the heat and returns to original solid state.

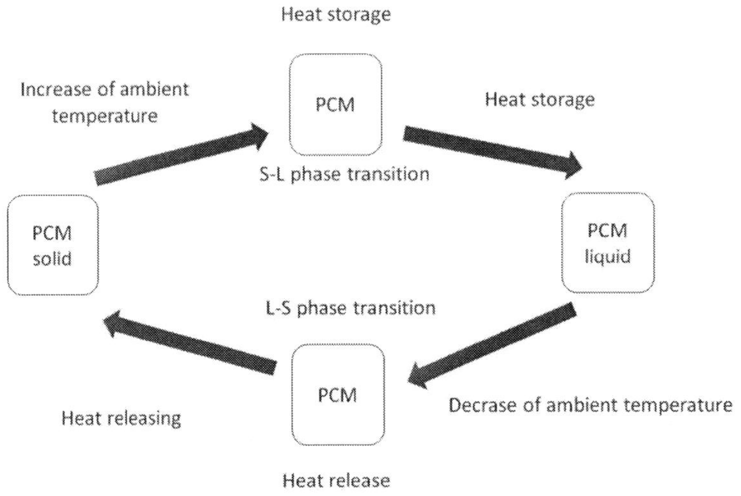

Fig. 1. The scheme of transition for solid – liquid phase change materials

Phase change materials can be used to store the thermal energy or to stabilise the temperature in the specific range. This is the reason why they have a big potential for building applications. In order to maximize the impact of PCM use, detailed analysis of the properties in order to select the most advantageous solution is necessary. The most important properties have been included in table 1. They refer to every type of PCM: organic (paraffin, non-paraffin), inorganic (salt hydrate, metallic) and eutectics (organic-organic, inorganic-inorganic, inorganic-organic)

Table 1. Important properties and requirements of phase change materials[4]

Important properties and requirements of phase change materials			
Thermo – physical	Kinetic	Chemical	Another (economical, environmental)
- suitable temperature of phase transition, - high latent heat of transition, - high specific heat, - high thermal conductivity, - high density, - small or no volume change, - minimal gas generation during the transition	- minimal subcooling, - high rate of crystallization	- no degradation and stability of properties after many cycles, - complete reversibility - non-corrosive - compatibility with the system/construction, - non-flammable, - non-explosive, - easy to recycle	- cost-effective, - environment friendly, - available

Number of phase change materials suitable for cooling applications in buildings are commercially available and numbers of new PCMs are investigated. Some of them have been selected and included in table 2.

Table 2. Selected phase change materials suitable for cooling building applications

Phase change material	Melting point, °C	Latent heat of fusion, kJ/kg	Reference
Emerest 2325	17–21	138–140	[5]
Glycerin	17.9	198.7	[4]
Hexadecane $CH_3–(CH_2)_{14}–CH_3$	18	236	[6]
Paraffin C16-C18	19	152	[7]
$Na_2SO_4·10H_2O$	21	198	[8]
Heptadecane $CH_3–(CH_2)_{15}–CH_3$	22	214	[6]
RT25 Paraffin	24	164	[9]
$Mn(NO_3)_2·6H_2O$	25.8	125.9	[10]
Vinyl stearate	27	122	[10]
Octadecane $CH_3–(CH_2)_{16}–CH_3$	28	244	[6]

Phase change materials from table 2 are only examples. There is wide range of phase change materials with different properties. The most important stage with PCMs applications is to choose material characterised by adequate properties for an application.

3 PCM based systems

Number of researchers conducted their reserch in the area of PCM based building thermal stabilisation, including air conditioning. Selected examples are shown and discussed in this paragraph.

Rahdar et al. [11] analyzed vapor compression air conditioning system via two strategies of hybrid systems: ice thermal energy storage system and PCM tank. Schematic configuration of two thermal storage systems are shown in figure 2. Authors analysed systems in order to check and compare capabilities of shifting the load from on-peak to off-peak mode from exergetic, economic and environmental point of views.

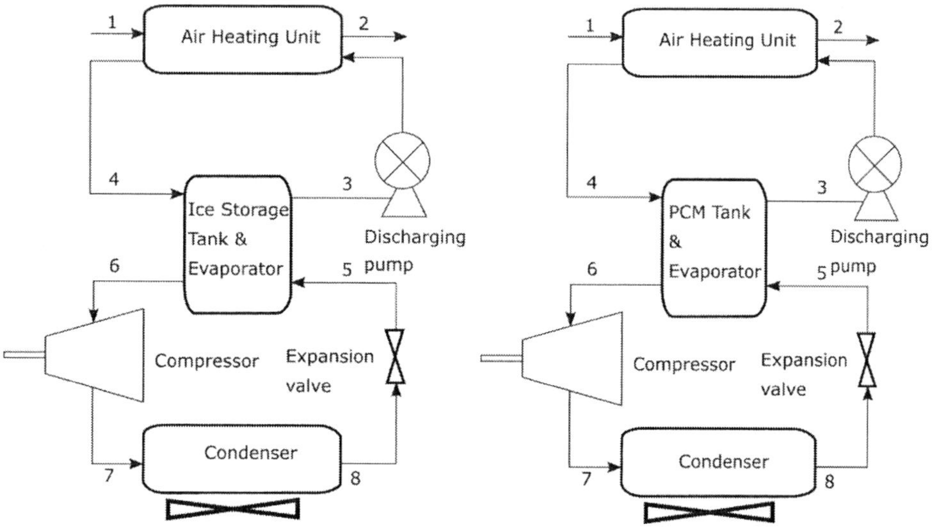

Fig. 2. Schematic configuration of two thermal storage systems[11]

DOI: 10.1051/e3sconf/20171401033

The summary of results shows that :
- power consumption in comparison to conventional system is 4.59% lower for ice system and 7.58% lower for PCM system
- CO_2 emission production for ice system in comparison to conventional system is 17.8% lower for ice system and 27.2% lower for PCM system
calculated pay back period amounts : 3.16 years for ice system and 5.56 years for PCM system.

Navidbakhsh et al. [12] analysed in their work energy, exergy, economic, and environmental aspects of the ice thermal energy storage system incorporating a phase change material module for air conditioning applications. Schematic diagram of the modeled ice thermal energy storage system incorporating PCM as the partial cold storage is presents in the figure 3. Authors compared system containing PCM module with the original ice system and with conventional air-conditioning system.
The summary of results shows that :
- power consumption for PCM system in comparison to ice system and conventional system was respectively 6.7% and 17.1% lower
- CO_2 emission production for PCM system in comparison to ice system and conventional system decreased respectively 7.2 and 17.5%
- calculated pay-back period amount : 3.97 years for PCM system and 3.39 years for ice system

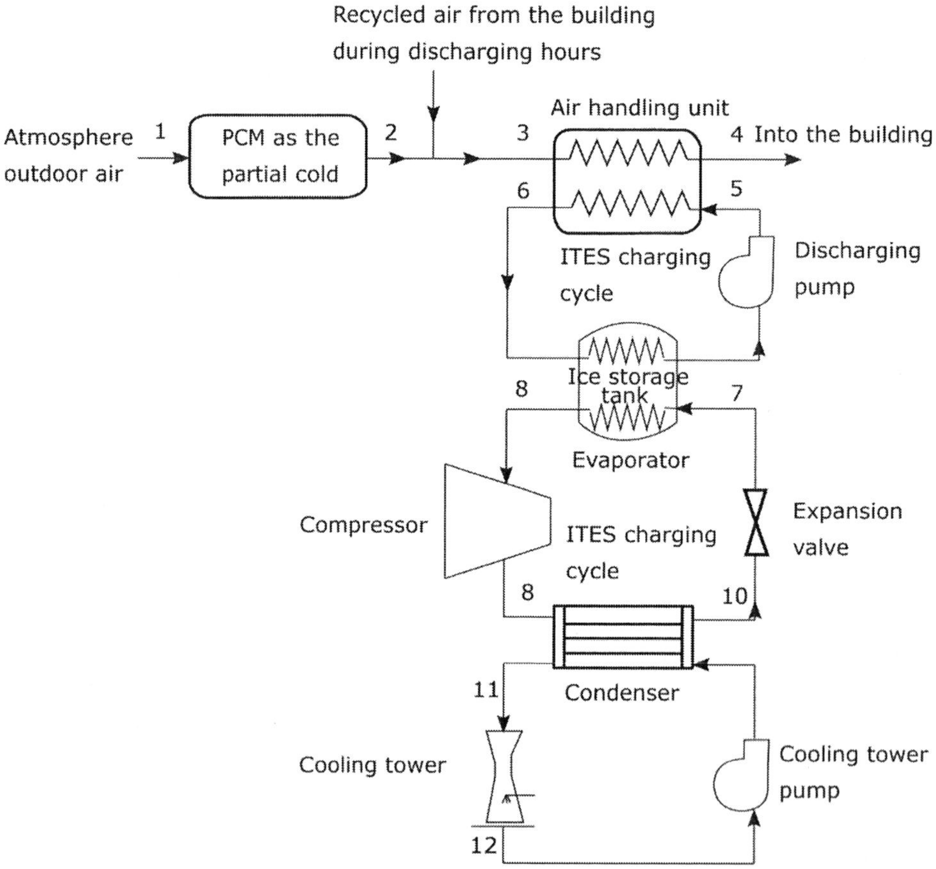

Fig. 3. Schematic diagram of the modeled ice thermal energy storage system incorporating PCM as the partial cold storage [12]

Mostaffa et al. [13] present air conditioning system in the form of combination of a latent heat thermal storage unit and a refrigeration system. Figure 4 shows the schematic of the system.
Authors analyse three cases with different size (l – length, d- thickness) of PCM slabs and different type of PCM. Description of all analysed by Author cases has been presened in table 3.

Table 3. Description of parameters of PCM slab for three analysed by Authors cases[13]

Case no.	Length, l	Thickness, d	PCM
case 1	1.8 m	15 mm	RT27
case 2	1.7 m	10 mm	S27
case 3	1.4 m	15 mm	SP25

The analysis is conducted from exergoeconomic and environmental points of view. The analysis resulted with conclusions that considering : highest cooling load of 11.8 kW, highest overall product exergy rate of 0.084 kW, lowest investment cost rate of 0.026 M$ and lowest amount of CO_2 emission respectively : case 3 has the highest coefficient of performance (COP) of 3.434, case 1 has the highest exergy efficiency of 2.93%, case 2 has the lowest total cost rate of 4094 $/year.

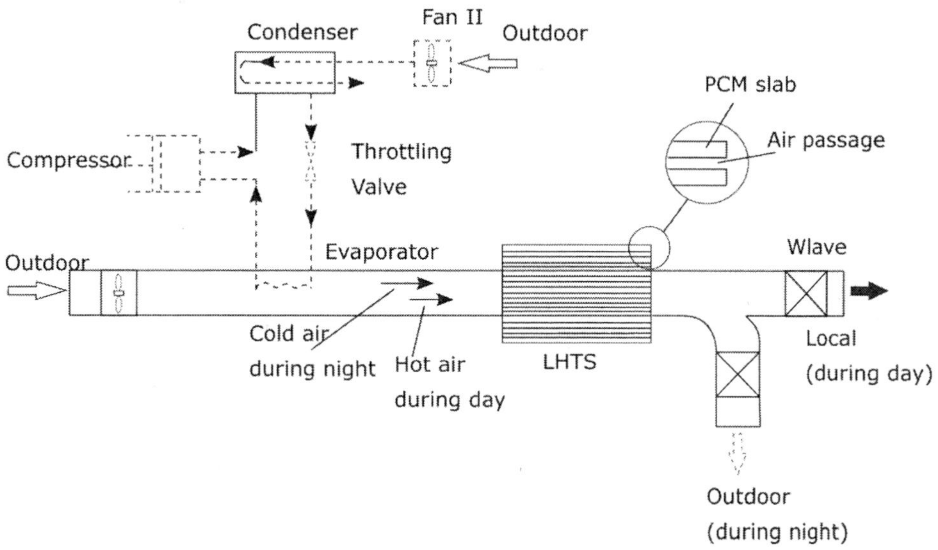

Fig. 4. Schematic of the air cooling system incorporating a latent heat thermal storage unit[13]

Zhao and Tank [14] present in their work possibility of integration of shell-and-tube based phase change material (PCM) thermal storage system with conventional air-conditioner. The proposed system use water (for charging loop) and air (for discharging loop) as a media for the heat transfer fluid. Schematic diagram of the air-conditioner integrated with PCM thermal storage is shown in the figure 5.

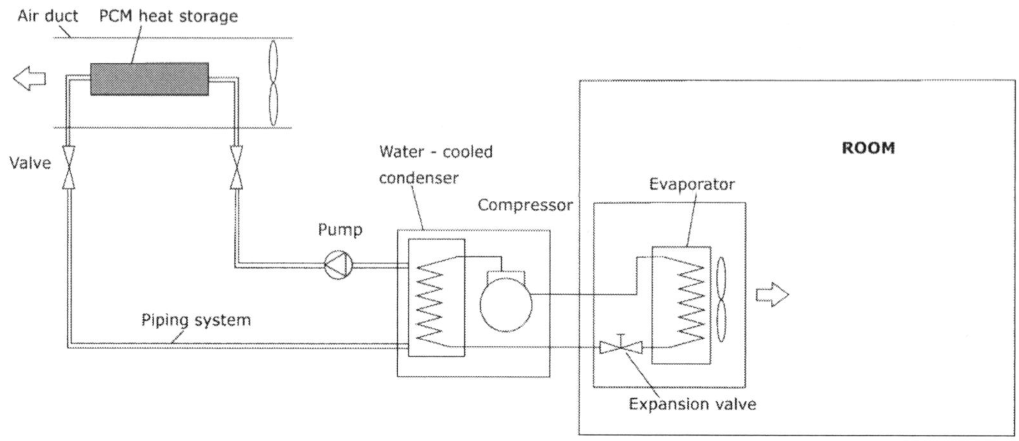

Fig. 5. Conceived schematic diagram of the air-conditioner integrated with PCM thermal storage[14]

Authors developed a numerical model for the PCM thermal storage unit and evaluated the influence on PCM TES system performance of heat transfer fluid inlet temperature, mass flow rate and conductive fin height. Authors give suggestions related to optimal parameters of both – the system and the system operation. They observed that charging time of the system depend on inlet temperature, mass flow rate and fin height: higher parameters – schorter charging time. They also concluded that PCM based thermal energy storage system proposed in this work is easy to scale-up and is characterised by effectiveness higher than 0.5.

Chaiyat[15] presents a concept of using phase change material for improving cooling efficiency of an airconditioner. They test modified air-conditioning system. Modification consists of addition of PCM ball pack bed. Research is conducted for Thailand climate. Experiment results are compared and verified with the mathematical model giving high consistency. Details of the testing room schematic are shown in figure 6 and some additional information are presented in table 4.

Author concluded that results show potential for using the packed ball bed of PCM (RT-20) to reduce energy consumption of the system under discussed circumstances and observes advantages:

- power consumption in comparison to conventional system is 9% (3.94 kW h/d) lower
- calculated pay-back period amounts: 3.16 years

Author selected for the purpose of the experiment phase change material manufactured by Rubitherm: RT-20, which is characterised by melting temperature: 22 °C, freezing temperature: 20 °C and heat of fusion: 160-180 kJ/kg. PCM ball was made from celluloid, diameter 4 cm. PCM bed is filled with balls at arroud 70% volume.

Table 4. Charging and discharging performances of the PCM bed depends on its thickness[15]

Thickness of the PCM bed, cm	40	30	20
Number of PCM ball	1100	933	619
Volume of PCM, liter	57.60	48.85	32.41
Charging time, minute	110	100	70
Discharging time, minute	180	140	80

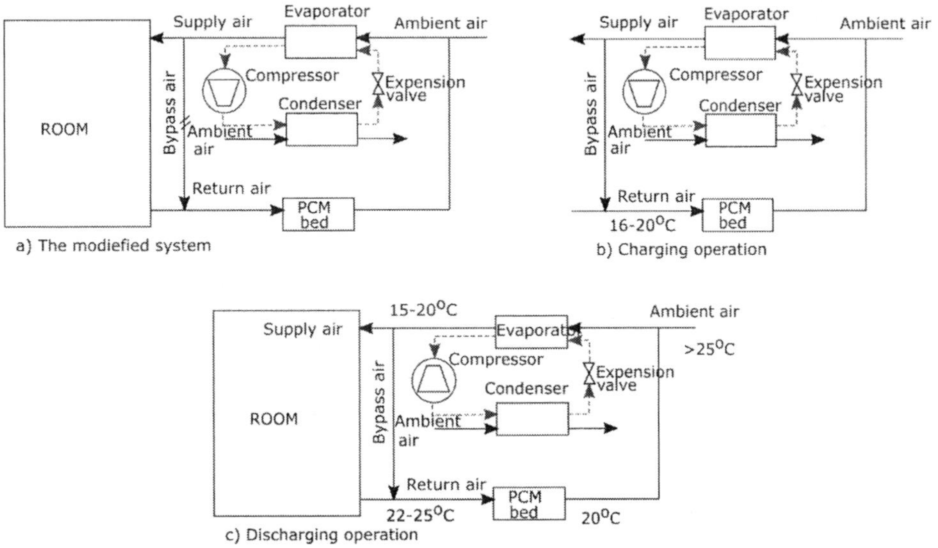

Fig. 6. Schematic of the testing room[15]

4 Summary

Current researches and achievements in the field of building thermal stabilisation supported by the use of phase change materials show that there is a great potential, but also number of difficulties which have to be faced in the further works.

Wide range of different phase change materials are available on the market. They are group into three main groups (organic-organic, inorganic-inorganic, inorganic-organic) and it can be observed that the biggest attention is around organic PCMs, especially paraffins. Despite many advantages, there is also one main disadvantages for the paraffin use: low thermal conductivity. Many researches are conducted to reduce this problem. It is important to correctly choose the phase change material and construction of the thermal energy storage system in order to effective use of the system.

Phase change materials enable short and long therm thermal energy storage. For cooling building applications they are especialy useful fo the electrical energy peak load shifting. Proper use of phase change materials for air conditioning gives possibility of reducing the power consumption and CO_2 emission.

Acknowledgement This work was supported by AGH – University of Science and Technology (Project 11.11.210.216).

References

1. L. W. Davisa and P. J. Gertlera, Proceedings of the National Academy of Science of the United States of America 112 (2015)
2. A. Sharma, V.V. Tyagi , C.R. Chen , D. Buddhi, Renewable and Sustainable Energy Reviews **13** (2009) 318–345
3. G. Li, Y. Hwang, R. Radermacher, International journal of refrigeration **35** (2012) 2053-2077
4. A. Sharma, V.V Tyagi, C.R. Chen, D. Buddhi, Renewable and Sustainable Energy Reviews **13**, (2009) 318-345

5. S. Scalat, D. Banu, D. Hawes, J. Paris, F. Haghighata, D. Feldman, Solar Energy Materials and Solar Cells **44** (1996) 49-61
6. M. Koschenz, B. Lehmann, Energy and Buildings **36** (2004) 567–578
7. B. Zalba, J.M. Marıin, L.F. Cabeza, et al., International Journal of Refrigeration **27** (8) (2004) 839–849
8. J. Turnpenny, D. Etheridge, D. Reay, Applied Thermal Engineering **20** (2000) 1019-1037
9. J. Marin, B. Zalba, F. Cabeza, H. Mehling, International Journal of Refrigeration, **27** (2004), 839–849
10. P. Tatsidjodoung, N. Le Pierrès, L. Luo, Renew Sustain Energy Rev, **18** (2013), pp. 327–349
11. M. H. Rahdar, A. Emamzadeh , A.Ataei, Applied Thermal Engineering **96** (2016) 391–399
12. M. Navidbakhsh, A. Shirazi, S. Sanaye, Applied Thermal Engineering **58** (2013) 30-41
13. A.H. Mosaffa, L. Garousi Farshi, Applied Energy **162** (2016) 515–526
14. D. Zhao, G. Tan, Applied Energy **138** (2015) 381–392
15. N. Chaiyat, Energy Conversion and Management **94** (2015) 150–158

CFD modelling and PIV experimental validation of flow fields in urban environments

Renata Gnatowska[1,*], *Marcin* Sosnowski[2] , and *Václav* Uruba[3]

[1]Czestochowa University of Technology, Faculty of Mechanical Engineering and Computer Science, 21 Armii Krajowej Av, 42-200 Czestochowa, Poland
[2]Jan Dlugosz University in Czestochowa, Faculty of Mathematics and Natural Sciences, 13/15 Armii Krajowej Av, 42-200 Czestochowa, Poland
[3] Academy of Sciences of the Czech Republic, Institute of Thermomechanics, v.v.i.; Dolejškova 1402/5, 182 00 Praha 8, Czech Republic
University of West Bohemia, Department of Power System Engineering, Univerzitní 22, 306 14 Plzeň Czech Republic

Abstract. The problem of flow field in the urban boundary-layer (UBL) in aspects of wind comfort around buildings and pollutant dispersion has grown in importance since human activity has become so intense that it started to have considerable impact on environment. The issue of wind comfort in urban areas is the result of complex interactions of many flow phenomena and for a long time it arouses a great interest of the research centres. The aim of article is to study urban atmospheric flow at the local scale, which allows for both a detailed reproduction of the flow phenomena and the development of wind comfort criteria. The proposed methodology involves the use of PIV wind tunnel experiments as well as numerical simulations (Computational Fluid Dynamics, CFD) in order to enhance understanding of the flow phenomena at this particular scale in urban environments. The analysis has been performed for the 3D case of two surface-mounted buildings arranged in tandem, which were placed with one face normal to the oncoming flow. The local characteristics of flow were obtained by the use of commercial CFD code (ANSYS Fluent). The validation was carried out with reference to the PIV results.

1 Introduction

Increased interest in the problem of flow fields in urban environments has been observed over the past few decades. It is governed by potential applications of the results in the design process of new buildings as well as optimization of the existing urban areas and city centres. Nowadays, this issue plays an important role in environmental protection, it concerns in particular the so called wind comfort around and between buildings, the life quality in urban areas as well as specific economic aspects of utilization of the defined zones. The studies of the wind environment around buildings are generally performed as wind tunnel experiments since it is the most well-established way to simulate a natural

[*] Corresponding author: gnatowska@imc.pcz.czest.pl

wind. In the past decades, numerous studies showed that the prediction of flow field in the surroundings of high-rise buildings in urban environments is very effective by the use of Computational Fluid Dynamics (CFD) [1-4]. Most of the studies have dealt with complicated, real configurations so it is uncertain whether the prediction of CFD is accurate. However, one of the most important aspects of this type of simulations is that the methods and results need to be validated before such predictions can be relied on. Therefore, PIV verification and validation of CFD studies are imperative, as well as detailed sensitivity studies that can provide guidance in the selection of computational parameters for future CFD analysis [5].

There is relatively little verification of CFD which simulate the air flows around objects in different configurations. For a real situation, the accuracy of CFD cannot be tested because of the lack of related experimental data.

CFD simulation of wind flow around 3D buildings started with fundamental studies for isolated buildings, often with a cubical shape [6-8]. Other often investigated configurations are cubic obstacles in a tandem arrangement, where the flow field is further complicated by mutual interference [8-10]. The flow structure around surface mounted obstacles depends on various factors: the grid resolution, the influence of the boundary conditions, the oncoming boundary layer thickness δ related to the height of obstacles H and by comparing the performance of various types of turbulence models in steady RANS simulations [7-8]. Other important factors are also the oncoming flow turbulence, the non-dimensional distance ratio (S/H) and the height ratio of the two consecutive bodies. Tall buildings, defined as those which protrude above their neighbours, act as scoops to collect the wind over much of their height and deflect it to the ground level [11]. In front of a building exposed to the wind, a vortex forms below the stagnation point. This vortex is shed on each side of the building and it can produce excessively high winds at ground level, causing discomfort or even danger.

The present paper provides a validation study of influence of the geometry of buildings arrangement on flow fields. The CFD simulation results are compared with Particle Image Velocimetry (PIV) wind tunnel measurements. The analysis has been performed for the 3D case of two surface-mounted buildings arranged in tandem, which were placed with one face normal to the oncoming flow. The local characteristics of flow were obtained by the use of commercial CFD code (ANSYS Fluent).

2 The applied research methods

The program of the study consists of wind-tunnel experiments and numerical simulation. In the frame of the numerical study, the three-dimensional steady RANS simulations have been carried out using a commercial CFD code, FLUENT v.17, with the realizable version of k-epsilon turbulence model. According to the literature ([3-5, 12]) this model is widely used for analysis of flows for different configurations of buildings.

The geometries of the analysed cases of two rectangular in-line surface-mounted square cylinders immersed in a boundary layer are sketched in figure 1. The measurements were carried out for configurations of two elements with different height, aligned in one line and the distance between them was S/D=1.5. The results presented in this work relate to a fixed ratio of object height $H_1/H_2 = 0.6$ and value of their "immersion" in boundary layer $H_2/\delta = 1$.

2.1 CFD simulations: settings and parameters

The dimensions of the computational domain were determined based on the best practice guidelines ([13, 14]). The upstream distance between inlet plane and windward

 DOI: 10.1051/e3sconf/20171401034

building facade is 7.5D, while the downstream distance between leeward building facade and outlet plane is 29D. The distance between the side walls of the building and the side planes of the domain is 4.5D on both sides. For computation purposes, the flow domain is divided into a number of hexahedral cells. The grid is clustered near the object and the spacing is increased to a growth ratio of 1.2 in the surroundings of the building model. The grid is obtained based on grid-sensitivity analysis. Three grids were made by increasing the resolution by the factor of 2, from Grid A (182388 cells) thru Grid B (341571 cells) to Grid C (523634 cells). The results obtained on these three grids are compared in terms of the pressure on the surface of first and second object. These tests indicate that the resolution of the finer grid (Grid C) is suitable.

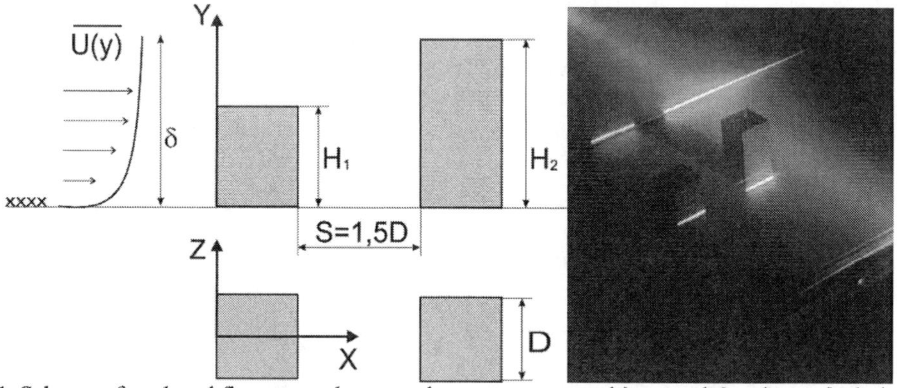

Fig. 1. Scheme of analysed flow around two tandem arrangement objects and the photo of wind tunnel setup.

The inlet mean velocity profile has been approximated after wind tunnel experiment using the power distribution:

$$U(y) = U_\delta \, (y/\delta)^\alpha \tag{1}$$

where, α characterizes the terrain type (α=0.15; suburban boundary layer). In addition to the inlet velocity profile, the k-ε realizable turbulence model requires an appropriate distribution of the turbulent kinetic energy k and the dissipation rate ε which together define the velocity and length scale of the turbulent motion. One of the best methods to describe the inlet boundary conditions is the method proposed by Richards and Hoxey [15]. The turbulent kinetic energy profile is estimated from the u_{rms} profile measured in wind tunnel. The turbulence dissipation rate is given by Eq. (2) [14]:

$$\varepsilon(y) = u_*^3 / \kappa y \tag{2}$$

where, κ is the von Kármán constant equal 0.41, and u_* is the friction velocity u_* = 0.04**U** [16]. This, method has been applied for the calculations described in this paper.
The standard wall functions in FLUENT is based on the work of Launder and Spalding [17], and have been used in this calculations. The governing equations are solved using the control volume method. Pressure-velocity coupling by SIMPLE algorithm was used as a solution method with second-order pressure interpolation. Second-order discretisation schemes are used for both the convection terms and the viscous terms of the governing equations. In order to assess the convergence of the solution, both the scaled residuals and the streamwise wind speed in selected points of the domain were monitored as a function of the number of iterations, in accordance to the guidelines by Franke et al. [13] and Tominaga et al. [14]. Convergence was obtained when the scaled residuals levelled off and reached a

minimum of 10^{-6} for x-, y-, z-velocity, k and ε and 10^{-4} for continuity. The static pressure on objects surfaces was also monitored.

2.2 Wind tunnel experiments

The experiment with Particle Image Velocimetry (PIV) was carried out in an open-circuit wind tunnel at the Institute of Thermomechanics of Academy of Sciences of the Czech Republic. The analysed configuration of buildings model was located within boundary layer generated by the facility output from the contraction nozzle of a cross-section of 250x250 mm^2. The test section of wind tunnel is 3*m* long, but the measurements were performed in the central part of tunnel. For seeding particles, a standard SAFEX fog generator has been used. The laser beam from a dual-head Nd:YLF laser with a pulse energy of 10mJ at 2kHz was made into a thin light-sheet by optics to illuminate tracer particles. A Dantec CCD camera (SpeedSense 611) was mounted to capture pairs of particle images at a frame rate of 100Hz. In order to evaluate the flow around the tandem cubes, two-component particle image velocimetry (PIV) is employed. The PIV measurements provide a velocity vector field with 2 velocity components in horizontal streamwise and spanwise x- and z-direction (U_x and U_z) at different section characterised by z/D parameter with step z=0.005m from the axis system. The farthest measuring section from the axis of building system was located at z/D = 2. For both numerical and experimental cases the Reynolds number was set to Re_D = 9 000 based on the free-stream velocity U = 5.5 *m/s* and the cube width D = 0.025 m. Figure 2 shows the measured mean wind velocity and velocity fluctuations, as well as the fitted numerical distributions, which will be used as inlet boundary conditions in the carried out CFD simulation. The vertical turbulence structure (u_{rms}) have not got the ideal shape of "traditional" boundary layer, because it have been measured at the inlet to the test section in wind tunnel at the Institute of Thermomechanics, Academy of Sciences of the Czech Republic.

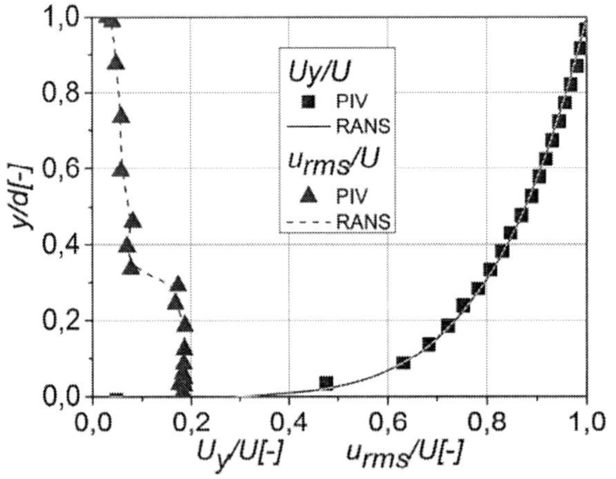

Fig. 2. The inlet profiles of mean wind speed (*Uy/U*) and velocity fluctuations u_{rms}/U measured in the wind tunnel at the building position (PIV) and the inlet profiles used in the CFD simulations (RANS).

3 CFD results and validation with wind tunnel measurements

For multiple bluff bodies arrays, the flow field is more complex than the flow around a single object due to mutual interactions and interference phenomena. The flow around the three-dimensional bluff-body is composed of a range of separation and adherence points classified as singular nodal and saddle points. One can distinguish here a horseshoe vortex and post-edge vorties located a close distance from the rear side of the object. The flow changes to strongly complicated if another object is placed in the aerodynamic wake, because of both the periodicity of vortex shedding and the cyclic oscillatory air flow between the obstacles. The results from the CFD simulation are compared with the experimental PIV data in terms of wind velocity ratio, velocity-vector fields and streamlines around two surface-mounted buildings arranged in tandem along the vertical planes.

The CFD simulations that are shown in the validation study are the ones considered the most accurate. These are the results obtained with Grid C (see section 2). Note that the experimental velocity-vector fields in the immediate vicinity of the ground are considered less reliable due to effects of shadows or reflections that might have produced uncertainties in the PIV measurements at these positions. Also the very high wind speed gradients in this area might have caused additional uncertainties in the PIV results.

The CFD validation results are shown in Figs. 3 and 4. These figures show the flow through selected sections in transverse direction of flow in analysed tandem configuration. The flow field is symmetric.

The figure 3 presents the time-averaged flow structure around two surface-mounted buildings in the tandem array. Streamlines are plotted in vertical planes at four different values of z/D of 0, 0.2, 0.4 and 1. The arrangement characterized by H_1/H_2=0.6 parameter, which is conducive to the occurrence of the so-called "downwash" effect [18]. This effect consists of washing the front side of the leeward object with large air masses, which results in strong air circulation in the area between cuboids, which determines flow structure between them. This situation is presented in Fig. 3a-c and Fig. 3e-g, which shows the result of numerical modeling (CFD) and experimental measurements (PIV) in the gap between the objects. The streamlines within this region, the structure of the recirculation, changes with height, indicating that the building wake has a three-dimensional structure. Fig. 3d and 3h present results for section located at z/D=1, where the flow tends to an undisturbed state by the presence of buildings arrangement. Similar conclusions result from Fig. 4, which presents the x-component of mean velocity field (U_x/U) and vectors magnitude around two surface-mounted buildings in the tandem array.

The results presented in this work show a good agreement between the PIV and the CFD methods. Note that some discrepancies are present near the ground, but as mentioned before, the PIV results are less reliable here. Fig. 4a and Fig. 4e show a very good agreement between the PIV and the CFD streamlines fields in the vertical axis plane. Both the PIV and CFD results show a standing vortex with similar size and position, in a gap between buildings and downstream of the leeward building.

The present study has focused on validation of CFD simulations of outdoor wind flow around configuration of simple system of buildings, and on the analysis of the flow phenomena at this particular scale in urban environments. The following limitations of the study are mentioned:

• The validation has only been performed for 3D steady RANS with the realizable k-epsilon model for tandem configuration of buildings. Future studies should focus on the evaluation of different turbulence modelling approaches (RANS, DES and LES), different turbulence models, and also on different building configurations, including multi-zone buildings.

• The present study has only focused on flow parameters. Future work should focus on the analysis of physical and numerical simulation of pollutant dispersion in flow field in aspects of wind comfort in urban areas.

 DOI: 10.1051/e3sconf/20171401034

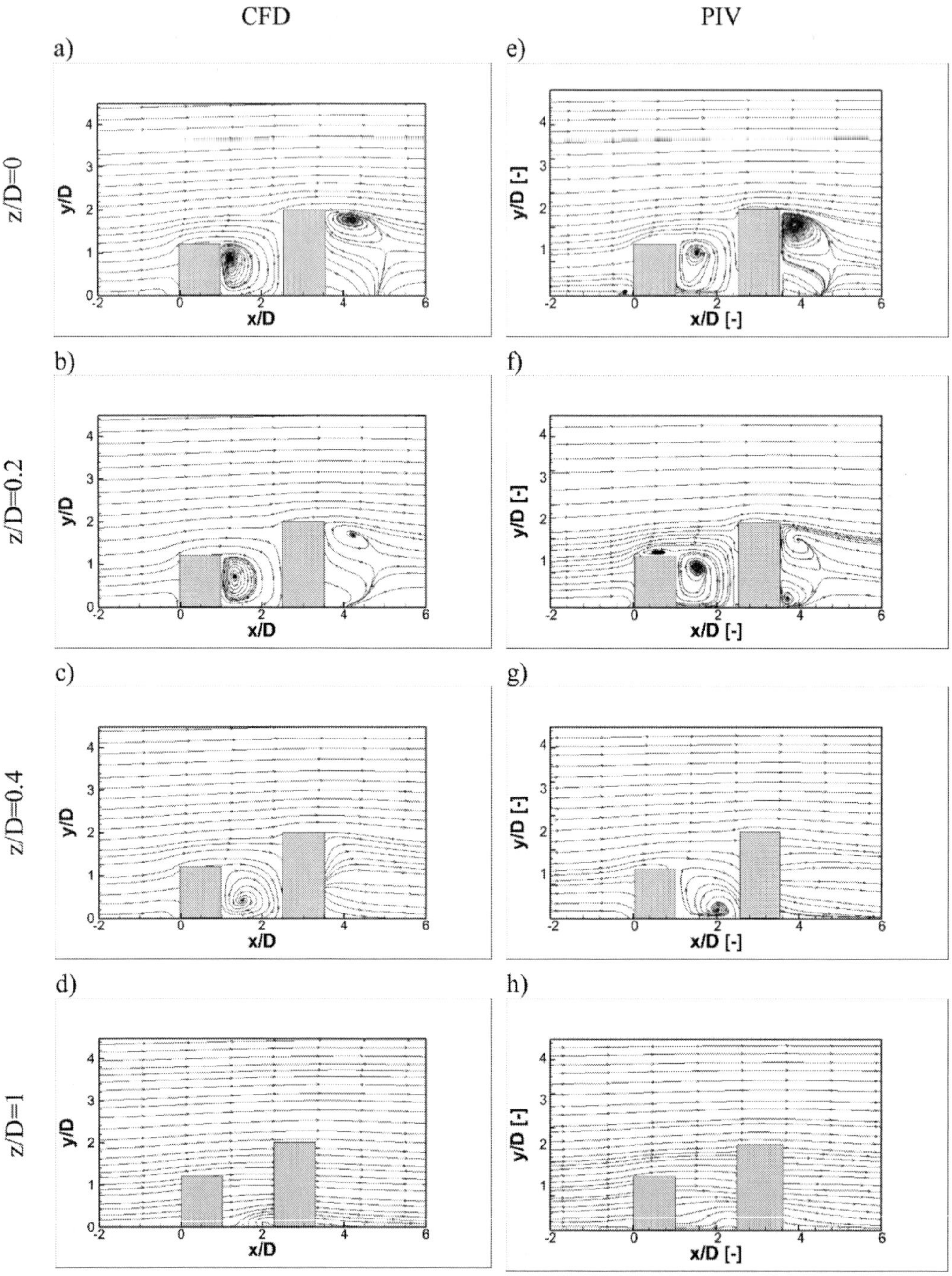

Fig. 3. Validation results: comparison of the numerical (CFD) results and the experimental (PIV) measurements in terms of streamlines in vertical planes at *z/D*=0, *z/D*=0.2, *z/D*=0.4 and *z/D*=1 around two surface-mounted buildings arranged in tandem.

DOI: 10.1051/e3sconf/20171401034

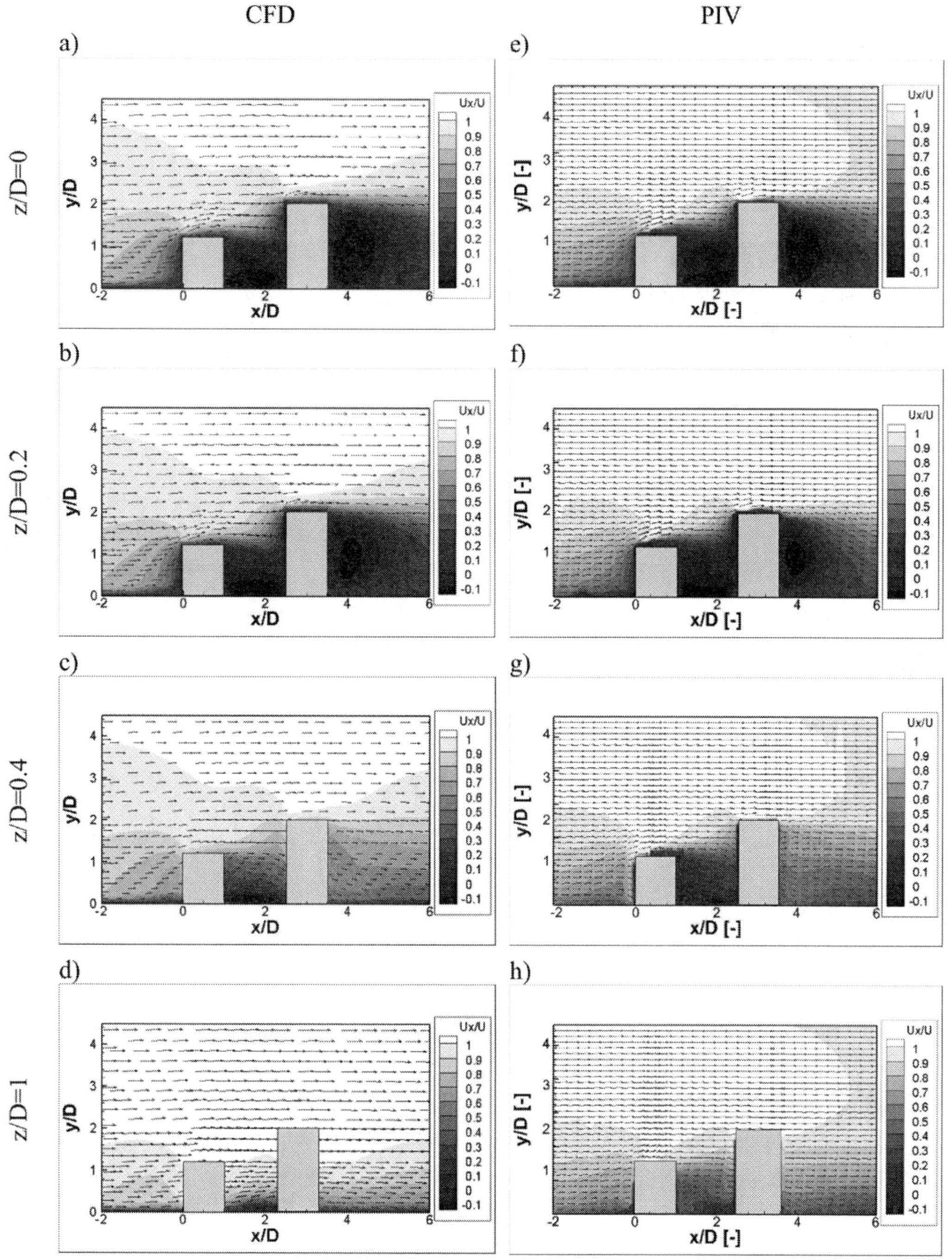

Fig. 4. Validation results: comparison of the numerical (CFD) results and the experimental (PIV) measurements in terms of x-component of mean velocity U_x/U and vector magnitude velocity field in vertical planes at z/D=0, z/D=0.2, z/D=0.4 and z/D=1 around two surface-mounted buildings arranged in tandem.

In spite of these limitations, it is important to mention that PIV experimental data will be beneficial to CFD modeling. This work will help to analyse the flow around the building in urban and suburban boundary layer and improve the understanding of the mechanism that influence the interference effects.

4 Summary and conclusions

Wind flow in built-up areas may be affected by a large range of parameters, mainly by wind velocity around buildings and adequate air quality. Those criteria are often in contradiction, because ensuring proper air quality requires adequate ventilation of built-up areas. Its effectiveness depends on wind direction, buildings configuration, etc. This paper has presented CFD simulations of wind-flow around buildings. The 3D steady Reynolds-Averaged Navier-Stokes (RANS) approach was applied with the realizable k-epsilon model. The intention of the paper was twofold: (1) to provide a validation study of outdoor wind flow around building arrangement by comparison with detailed Particle Image Velocimetry (PIV) measurements in an wind tunnel; (2) to provide information on the flow field and wind climate in urban areas. It is one of the important factors that must be taken into account in the process of architectural designing in urban organizations. Experimental and numerical analysis of the physical phenomena and mechanisms associated with airflow in the vicinity of objects in tandem arrangement immersed in a boundary layer made it possible to obtain a detailed description of the velocity field in the built-up environment.

The results of the validation study showed a good to very good agreement for analyzed configuration. The influence of the geometry on the fluid flow structure varies with height and is indicated mostly in the space between objects and also in the immediate vicinity of the ground. The application of both CFD numerical simulation of wind flow in a built environment with wind tunnel PIV experiment allowed to carry out verification and validation of results and also seems to be appropriate approach to proper planning of buildings configuration in order to achieve local wind comfort.

References

1. B. Blocken, J. Wind Eng. Ind. Aerodyn. **129** (2014)
2. B. Blocken, W.D. Janssen, T. van Hooff, Environ Model Softw. **30** (2012)
3. B. Blocken, T. Stathopoulos, J.P.A.J. van Beeck, Build Environ. **100** (2016)
4. Y. Hui, A. Yoshida, Y. Tamura, Y. Journal of Fluids and Structures, **37** (2013)
5. B. Blocken, Build Environ **91** (2015)
6. D.A. Paterson, C.J. Apelt, J. Wind Eng. Ind. Aerodyn. **35** (1990)
7. S. Murakami, A. Mochida, Y. Hayashi, S. Sakamoto, J. Wind Eng. Ind. Aerodyn. **44,** 1-3 (1992)
8. T. Stathopoulos, J. Wind Eng. Ind. Aerodyn. **94,** 11 (2006)
9. H. Sakamoto, H. Haniu, J. Wind Eng. and Ind. Aerodyn. **31,** 41 (1988)
10. R.J. Martinuzzi, B. Havel, J. Fluid Eng. **122,** 24 (2000)
11. T. Lawson, T.V. Lawson. Building aerodynamics. London: Imperial College Press, (2001)
12. T.H. Shih, W.W. Liou, A. Shabbir, Z. Yang, J. Zhu, Comput. Fluids. **24** (1995)
13. J. Franke, A. Hellsten, H. Schlünzen, B. Carissimo, Int J Environ Pollut. **44**, 1-4 (2011)
14. Y. Tominaga, T. Okaze, A. Mochida, Build Environ. **46**, 4 (2011)
15. P.J. Richards, R.P. Hoxey, J. Wind Eng. Ind. Aerodyn. **46&47** (1993)
16. C.-H. Hu, F. Wang, Build Environ. **40** (2005)
17. B.E. Launder, D.B. Spalding, Comput Method Appl M. **3** (1974)
18. R. Gnatowska, Pol J Environ Stud. **24**,4 (2015)

Heat losses in power boilers caused by thermal bridges

Monika Kocot[1,*], *Stanisław* Łopata[2,**] and *Zbigniew* Plutecki[3]

[1]Heat not Lost Sp. z o.o. ul. Górnych Wałów 7, 44-100 Gliwice, Poland
[2]Cracow University of Technology, Institute of Thermal Power Engineering, al. Jana Pawła II 37, 31-864 Kraków, Poland
[3]Opole University of Technology, Department of Innovative Technological Processes, ul. Ozimska 75, 45-370 Opole, Poland

Abstract. In this article the analysis of heat losses caused by thermal bridges that occur in the steam boiler OP-140 is presented. Identification of these bridges were conducted with use of thermographic camera. Heat losses were evaluated based on methodology of VDI 4610 standard, but instead of its simplified equations, criterial equations based on Nusselt number were used. Obtained values of annual heat losses and heat flux density corresponding to the fully insulated boiler surfaces were compared to heat losses generated by thermal bridges located in the same areas. The emphasis is put on the role of industrial insulation in heat losses reduction.

1 Introduction

The amount of heat losses in power boilers has a significant influence on their capability to process energy. The improvement of energy processing efficiency is achievable through:

- optimization of coal incineration processes,
- obtaining higher steam parameters,
- minimization of static heat losses.

Among the main causes of heat losses in power boilers are bad technical condition, improper installation or a complete lack of thermal insulation. Previously installed thermal insulations were performed according to the now withdrawn ordinance of the Minister of Economy dated September 17 1999 concerning safety and hygiene of work with power equipment and installations, which stated that the external surface temperature of the insulation layer should not exceed 60°C – thus protecting workers from heat-related injuries [1]. In most cases, such insulation had nothing to do with heat losses reduction nor with reaching the maximum return on investment at potentially lowest losses of energy. Nowadays growing energy prices, regulations targeted at reduction of carbon dioxide emissions as well as growing awareness of energy manufacturers create favourable environment for:

* Corresponding author: monika.kocot@hnl.pl
** Corresponding author: lopata@mech.pk.edu.pl

- selection of the economic thickness of insulation,
- modernizations, e.g. isolating of installations and elements that were not isolated before,
- replacement of damaged insulation.

It is commonly known, that insulation of industrial installations and machines is performed to achieve certain goals, such as:

- minimization of heat transfer to the environment (reduction of heat losses),
- prevention of thermal deformations,
- reduction of temperature loss of the operating medium flowing through the vessels,
- control over the temperature of an operating medium,
- prevention of steam condensation on the vessel's surface,
- protection of health and providing safety for the workers (e.g. by heat-related injuries prevention).

Currently in Poland isolation of industrial installations is also regarded as one of the most important and efficient tools of energy efficiency improvement [2].

Additionally, on May 20 2016 the Energy Efficiency act was adopted by the Polish parliament, which regulates activities of enterprises from the energy sector by setting a number of obligations and restrictions, such as [3]:

- obligation to perform actions targeted at energy efficiency improvement resulting in 1,5% savings of electric, heat or natural gas energy measured in toe (tonne of oil equivalent) sold to end-users on the territory of Poland per annum,
- gradual restriction of compensation payment opportunities – certain payment thresholds will be introduced,
- obligation to contract out or conduct energy audits every four years (concerns enterprises defined by the Freedom of Economic Activity Act as non-SMEs).

Obligations defined by legislation as well as benefits resulting from energy efficiency improvement, boost popularity of thermal modernizations.

2 Technical characteristics of the analysed power boiler

The analysis has been conducted on the OP-140 power boiler operating at one of the national power plants. This boiler is coal-fired and the biomass in the participation 15%. It cooperates with a similar boiler and each of them generates 145 tons of steam per hour. For steam characteristics are: temperature of 540 °C and pressure of 13,5 MPa. It is channelled by pipelines of an integrated collector system to a pass-out condensing turbine used for heating purposes. The first launch of the analysed power boiler took place in 1976. The boiler has a typical two-pass configuration: radiant surfaces are located above the furnace in the first pass (I – absorb heat by thermal radiation), and convective surfaces are arranged in the second pass (II – absorb heat via a fluid i.e. gas). These flue gas ducting are connected by a horizontal gas duct (the fire-bridge) at the top of boiler.

The internal surface of the analysed boiler is covered with isolation that consists of specific insulation layer and protective coating. To the present this isolation had been in use for 39 years [4]. It was made of mineral wool mats. Those mats were made of several layers of thickness 100 mm + 2×50 mm, with an average density of 100 kg/m^3. Additionally the isolation contained: bearing structure, galvanized mesh, steel sheet coating of thickness 1.0 mm, in some places flat and in some places trapezoidal. Condition of the isolation presented on the attached images (Figure 1) had not been verified for a long time and a visual inspection revealed a number of irregularities, including noticeable chipping, discontinuities and corrosion damages.

DOI: 10.1051/e3sconf/20171401035

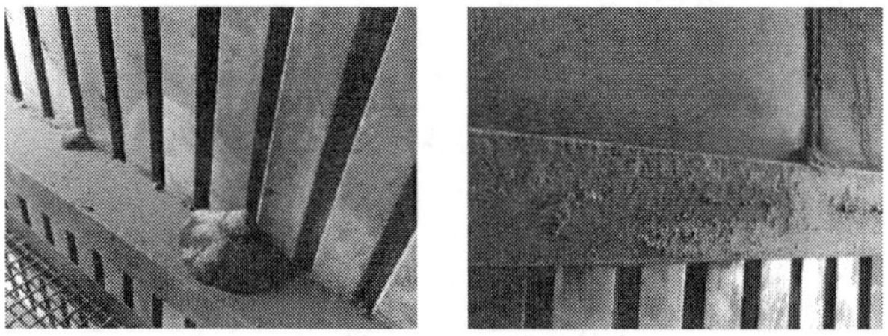

Figure 1. Damage of the isolation layer spotted on the analysed power boiler OP-140.

3 Analysis of generated heat losses

3.1 Thermographic analysis of the boiler

Thermographic examination of the boiler was conducted in November 2015 between 8 a.m. and 3 p.m. The temperature of the external environment of the boiler varied at different altitudes from +12 °C (level +0,00 m) up to around 40 °C (level +35,00 m). The examination was carried out while the boiler was operating at nearly nominal heat output (nominal power). The boiler's surface temperature and thermal bridges surface temperature were studied with the use of Flir E60 thermographic camera and a multifunctional device with sensors (for measurements temperatures of the surfaces and liquid environments). This allowed to diagnose the places of occurrence of thermal bridges simultaneously.

Due to the boiler's construction and the varying environmental temperature, the installation was divided in zones (for the needs of calculation and analysis). Numbers and descriptions are given (and altitudes in brackets):

- zone 1 - fragment includes the 1st pass (I) of the boiler (0,00 m ÷ 8,10 m),
- zone 2 - fragment includes the 1st pass (I) of the boiler (8,10 m ÷ 15,85 m),
- zone 3 - fragment includes the 1st pass (I) of the boiler (15,85 m ÷ 20,70 m),
- zone 4 - fragment includes the 1st and the 2nd pass (I and II) of the boiler (20,70 m ÷ 27,00 m),
- zone 5 - fragment includes the 2nd pass (II) of the boiler (15,85 m ÷ 20,70 m),
- zone 6 - fragment includes the 2nd pass (II) of the boiler (9,30 m ÷ 15,85 m),
- zone 7 – fragment includes the structural ceiling of the boiler space (27,00 m ÷ 33,00 m),
- zone 8 – steam drum of boiler (29 m).

Locations of the given zones are presented on figure 2 (along with the overall scheme of the boiler [5]).

The only area which was hard to reach and examine was around the structural ceiling space above the boiler's duct. The temperature between structural beams reached 300 °C which accompanied by intensive air flows, causing this area physically inaccessible. In addition, this space was isolated only from the outside. Therefore, in order to assure the most optimal isolation in this area, it would be recommended to apply additional of injection isolation or loose mineral wool.

 DOI: 10.1051/e3sconf/20171401035

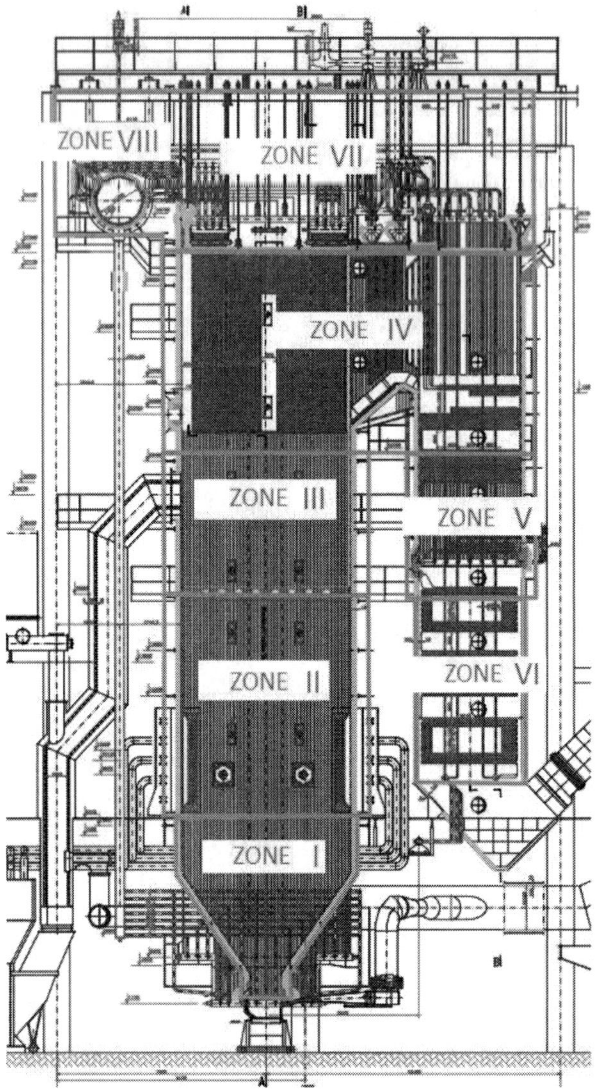

Figure 2. Scheme of the power boiler OP-140 and locations of zones of thermographic analysis.

3.2 Identification of thermal bridges

Thermographic examination of the power boiler revealed its thermal bridges and provided an opportunity to further analyse them by classification into relevant categories based on their key features. The following categories of thermal bridges were distinguished [6]:

- operational, resulting from chipping and displacement of specific insulation (occurred mainly on straight surfaces, covering the heat transfer surfaces of the boiler, e.g. in zones of waterwalls panels, reheaters and superheaters),
- structural (constructional), originating from many integrated elements which provide the operational stability of the boiler, such as: built-in sealing rings located on its circumference (on levels: +7,20 m, +10,45 m, +12,33 m, +13,90 m, 15,93 m, +18,10 m, +20,45 m, +22,80 m, +24,65 m, +26,50 m and +27,95 m) as well as e.g. inspection hatches, valves, thermowell of the measuring equipments (these elements usually remain uninsulated),

 DOI: 10.1051/e3sconf/20171401035

- assembly-related, caused by the improper isolation assembly (e.g. excluding bypassing the rules, rushed work, lack of supervision, which might lead to improper thickness, uneven distribution or even a complete lack of insulation).

Figures 3-5 demonstrate the abovementioned types of thermal bridges identified on the analysed boiler (they are showing their thermographic images and their real-life photographs). White dashed lines create boundaries for which Flir Tools + has calculated average temperatures that were later used for heat loss calculation.

a

b

c

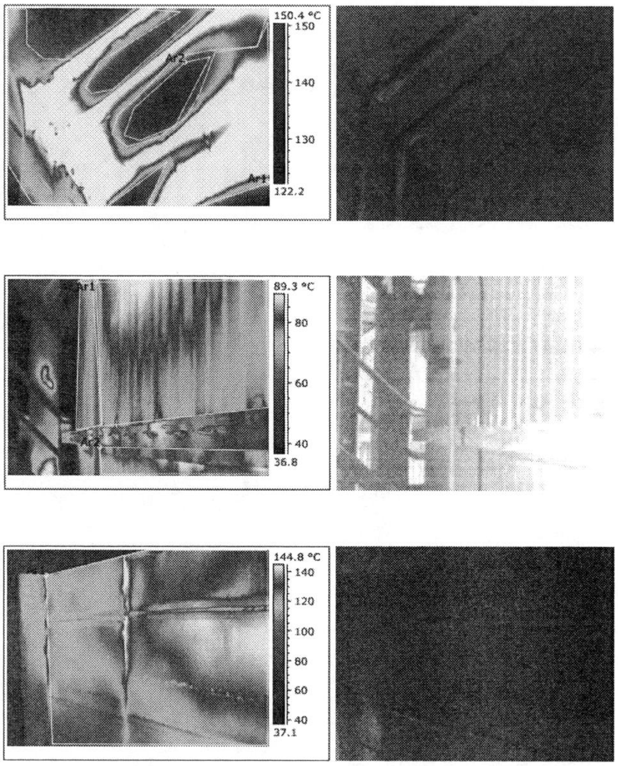

Figure 3. Operational thermal bridges: a – boiler's duct, b – zone IV, front wall, c – zone VIII, left wall.

a

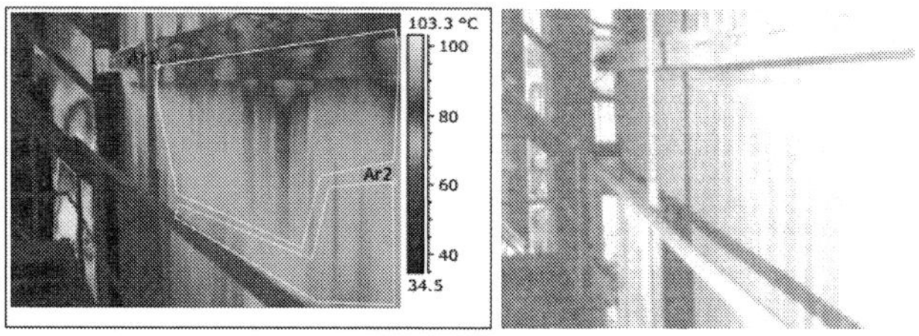

 DOI: 10.1051/e3sconf/20171401035

b

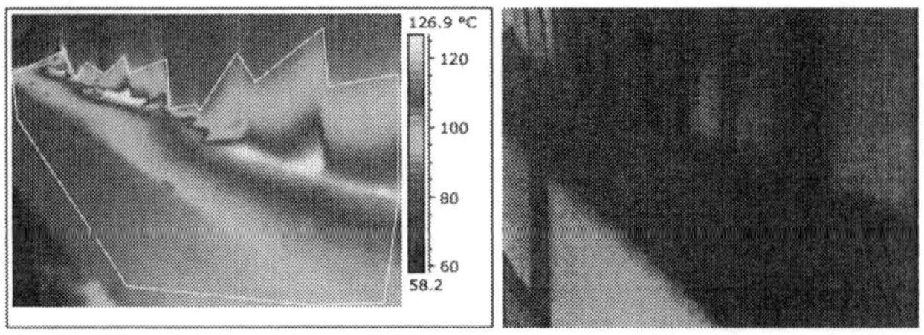

c

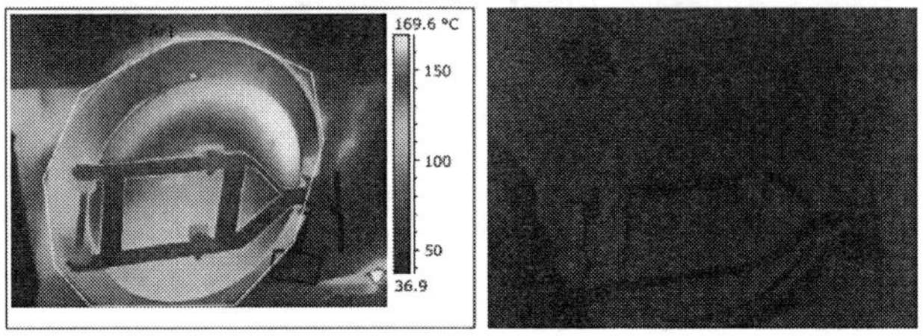

d

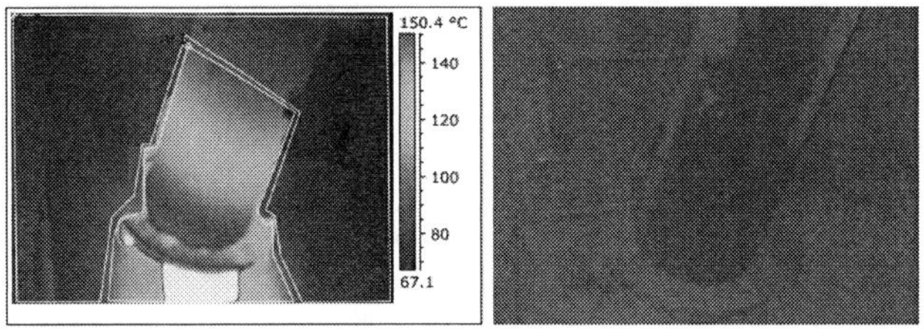

Figure 4. Structural thermal bridges: a – zone IV, front wall, b – zone IV, left wall, c – zone VI, right wall (inspection hatch), d – zone III (OFA nozzle).

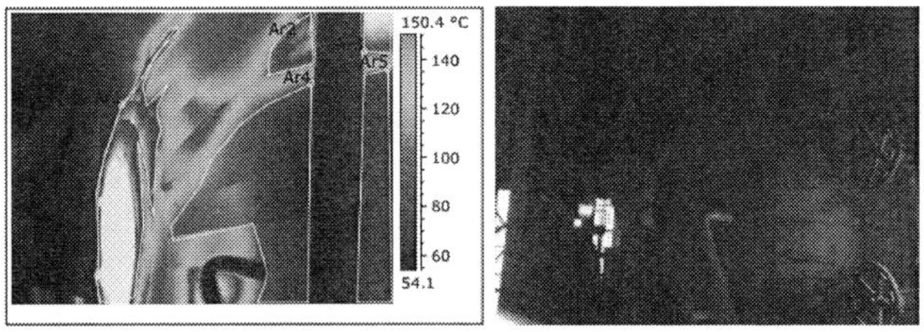

Figure 5. Assembly-related thermal bridge: the bottom of the boiler drum.

3.3 Heat losses calculation

3.3.1 Defining average temperatures of the analysed surfaces of the boiler

A thorough analysis of the thermographic images enabled us to define average temperatures of insulation coatings and at chosen locations of thermal bridges. The used equipment (Flir Tools + software) allowed us to select a region and calculate the average temperatures for it. The program also provides an opportunity to introduce parameters that might have influence on the interpretation of thermographic images. These are e.g.: emissivity coefficient, reflected temperature RTC, environmental temperature, air flow velocity, as well as distance between the camera and a measured object. All mentioned parameters were measured during the examination with the use of a specified tape Introl HB, that has a constant emissivity coefficient, Lambert radiator and a set of thermometers accompanied by sensors designed to measure temperatures of liquids and surfaces. All surfaces where heat exchange occurs, including parts of thermal bridges, along with their obtained average temperatures are listed in Table 1 [7].

Table 1. Heat exchange areas and corresponding to them average temperatures (of the insulation coating in selected boiler zones and thermal bridges).

Name	A_1, m²	T_1, °C	A_2, m²	T_2, °C
Zone 1	155	57	35	100
Zone 2	204	77	42	82
Zone 3	131	76	43	78
Zone 4	331	84	62	95
Zone 5	70	69	25	79
Zone 6	115	60	25	75
Zone 7	290	84	–	–
Zone 8	63	93	–	–
Other thermal bridges (measuring equipment, inspection hatch, bottoms etc.)	–	–	15	120

Notions used in Table 1: A_1 – are of the isolation surface, m², A_2 – area of the thermal bridge, m², T_1 – insulation coating temperature, °C, T_2 – thermal bridge temperature, °C.

3.3.2 Methodology of heat losses calculation

Thermographic images of heat distribution in each zone of the boiler were made, on which temperature maps on the surfaces of the boiler and thermal bridges are clearly distinguishable (e.g. on figures 3-5). Then calculations of heat losses were made:

- through the isolated surface of the boiler taking into account measured temperatures of insulation coatings,
- through identified structural, assembly-related and operational thermal bridges.

Heat losses calculations were done based on the norm VDI 4610 [8]. It enables the evaluation of heat losses generated by the isolation and thermal bridges. Used quantitative dependencies were simplified with regard to the analysed object. The overall flow of heat losses generated by an installation was calculated by using the following equation:

$$Q = \sum_{i=1}^{n} Q_i \tag{1}$$

where:
Q - the overall heat flow, W,
Q_i - the heat loss flow generated by an i-th component of the system (in a given case – by the zone of a boiler), W.

Heat losses of a chosen zone of a boiler:

$$Q_i = Q_{WD,i} + Q_{WB,i} \tag{2}$$

where:
$Q_{WD,i}$ - heat loss flow generated by the i-th standard element of a component, W,
$Q_{WB,I}$ - heat loss flow generated by the i-th thermal bridge or by the i-th uninsulated part of the component, W.

In case of the boiler the heat losses flow through the chosen zone is calculated with taking into account of heat losses originated by thermal insulation as well as through regular and irregular thermal bridges. The following equation was used:

$$Q_i = q_p \cdot A_m + \sum_{j=1}^{m} Q_{WBd,i,j} + \sum_{y=1}^{m} Q_{WD-BT,i,y} =$$

$$\left[\frac{1}{R_{si}+R_{sc}+R_{se}} \cdot A_m + \sum_{j=1}^{m} (k \cdot A_m)_{WBd,i,j} + \sum_{y=1}^{m} (k \cdot A_m)_{WD-BT,i,y}\right] \cdot \Delta\vartheta_i =$$

$$\left[\frac{1}{\frac{1}{\alpha_w}+\frac{Gr_{sc}}{\lambda_{sc}}+\frac{Gr_{iz}}{\lambda_{iz}\cdot a}+\frac{Gr_{pl}}{\lambda_{pl}}+\frac{1}{\alpha_k+\alpha_r}} \cdot A_m + \right.$$

$$\sum_{j=1}^{m}\left(\frac{1}{\frac{1}{\alpha_w}+\frac{Gr_{sc}}{\lambda_{sc}}+\frac{1}{\alpha_k+\alpha_r}} \cdot A_m\right)_{WBdi,j} + \tag{3}$$

$$\left.\sum_{y=1}^{m}\left(\frac{1}{\frac{1}{\alpha_w}+\frac{Gr_{sc}}{\lambda_{sc}\cdot a}+\frac{Gr_{iz}}{\lambda_{iz}}+\frac{Gr_{pl}}{\lambda_{pl}}+\frac{1}{\alpha_k+\alpha_r}} \cdot A_m\right)_{WD-BTi,y}\right] \cdot \Delta\vartheta$$

where:
q_p - heat loss flow generated by thermal insulation, per square meter (heat flux density), W/m^2,
A_m - heat exchange area, m^2,
$Q_{WBd,i,j}$ - heat loss flow generated by irregular thermal bridges occurring in insulation (structural and assembly-related thermal bridges), W
$Q_{WD\text{-}BT,i,y}$ - heat loss flow generated by the isolated component of the object (operational thermal bridges), W,
R_{si} - heat resistance of the internal side of the object, (m^{2}K)/W,
R_{sc} - heat resistance of the object's wall, (m^{2}K)/W,
R_{se} - heat resistance of the external side of the object, (m^{2}K)/W,
k - heat transfer coefficient, W/(m^{2}K),
$\Delta\vartheta_i$ - difference of temperatures between the operating medium and the external environment, °C,
α_w - convective heat transfer coefficient of the internal side of the object, W/(m^{2}K),
α_k - convective heat transfer coefficient, W/(m^{2}K),
α_r - radiative heat transfer coefficient, W/(m^{2}K),

λ - thermal conductivity coefficient (in subscripts, correspondingly: sc – the wall of the object, iz – isolation, pl – isolation coating), W/(mK),
Gr – thickness (in subscripts, correspondingly: sc – the wall of the object, iz – isolation, pl – isolation coating), m,
a – deterioration factor of insulation, –.

All convectional heat transfer coefficients were calculated with regard to the heat exchange process between the external surface of the boiler and the surrounding air using criterial equations, allowing to determinate the Nusselt number. In case of the analysed power boiler OP-140 a free convection takes place. Air flows upwards along the the boiler's surface. This phenomenon is caused by different air densities on different altitudes. No horizontal air currents were registered at the walls of the object. Depending on the geometry of the object the Nusselt number may be derived from the empirical equations, e.g. from Pchelkin's model (examples are presented in Table 2) [9].

Table 2. Empirical equations for the chosen types of convectional heat transfer.

Convection model	Type of convection, Nusselt criterion, Range of applicability, Calculations temperature, ^{o}C
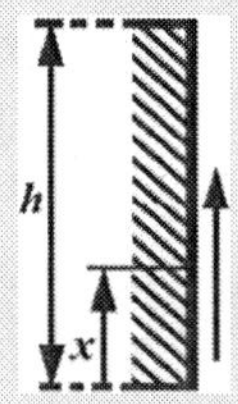	Free, laminar $Nu_x = 0{,}676\cdot Pr^{0,5}\cdot(0{,}861+Pr_p)^{-0,25}\cdot(0{,}25\cdot Gr_{xp})^{0,25}$ $10^4 < Gr_{xp}Pr_p < 6\cdot 10^{10}$ $t = t_p = t_\infty$; $Pr_p = Pr(t_p)$; $Gr_p = Gr(t_p)$
	Free, laminar $Nu_h = 0{,}902\cdot Pr^{0,5}\cdot(0{,}861+Pr_p)^{-0,25}\cdot(0{,}25\cdot Gr_{hp})^{0,25}$ $10^4 < Gr_{hp}Pr_p < 6\cdot 10^{10}$ $t = t_p = t_\infty$; $Pr_p = Pr(t_p)$; $Gr_p = Gr(t_p)$
	Free, turbulent $Nu_x = 0{,}030\cdot Gr_{xt}^{0,4}\cdot Pr_t^{0,47}\cdot\left(1+0{,}494\cdot Pr_t^{0,66}\right)^{-0,4}$ $6\cdot 10^{10} < Gr_{xt}\cdot Pr_t$ $t = t_t = 0{,}5(t_p+t_g)$; $Pr_t = Pr(t_t)$; $Gr_t = Gr(t_t)$
	Free, turbulent $Nu_h = 0{,}025\cdot Gr_{ht}^{0,4}\cdot Pr_t^{0,47}\cdot\left(1+0{,}494\cdot Pr_t^{0,66}\right)^{-0,4}$ $6\cdot 10^{10} < Gr_{ht}\cdot Pr_t$ $t = t_t = 0{,}5(t_p+t_g)$; $Pr_t = Pr(t_t)$; $Gr_t = Gr(t_t)$

Notions used in Table 2: Nu - Nusselt number, Pr - Prandtl number, Gr - Grashoff number, h, x - according to the drawing in column 1, t_g - medium temperature (e.g. gas), ^{o}C, t_p - surface temperature, ^{o}C.

3.3.3 Results of calculations

Total heat losses are calculated as the sum of heat losses that occur regularly on the boiler insulation surface and heat losses generated by thermal bridges. Calculated values of annual heat losses, expressed in GJ/year, for the selected zones of the analysed boiler OP-140 are presented on Figure 6. Heat losses generated without the influence of thermal bridges identified in those areas are presented on Figure 6 simultaneously. Moreover, Figure 6 also illustrates values of heat flow density, expressed in W/m^2 (which are also total and without taking into account thermal bridges).

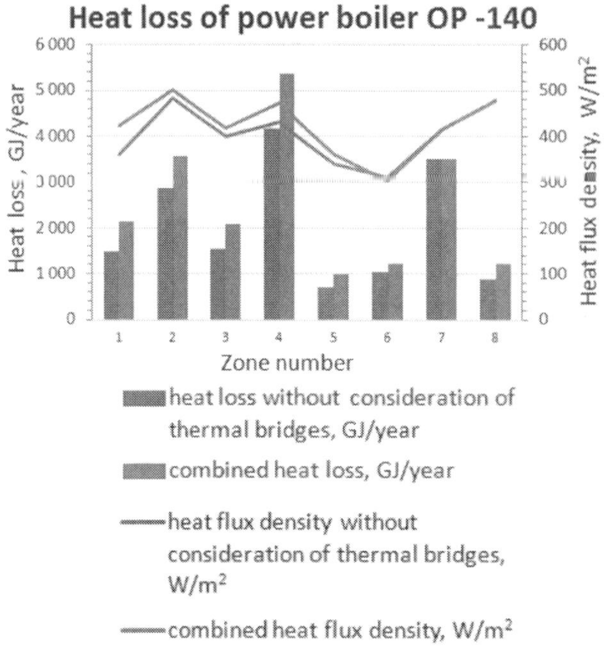

Figure 6. Heat losses and heat flow density (with and without taking into account thermal bridges) calculated for the selected zones of the OP-140 power boiler (at an operating with nominal power).

Analysis of Figure 6 leads to the assumption that thermal bridges located in different parts of power boilers might share up to 30% of the total heat losses occurring in the installation. Therefore, reducing of the number and influence of thermal bridges will provide a significant energy efficiency improvement. At the same time it will contribute to reduce the negative influence on the natural environment.

4 Summary and final remarks

Fuel consumption is directly related to heat losses. The latter also include the thermal bridges, which can be classified as operational, structural and assembly-related. They occur in thermal power equipments and in industrial installations. Heat losses generated by thermal bridges are not always properly evaluated and sometimes they are completely neglected. Such is often the case of structural elements of power boilers. Little attention was paid to them at the end of the previous century. An attitude which has radically changed since then and national environmental policies represent just one of many examples of these changes.

In case of the analysed power boiler OP-140 the most significant heat losses were caused by insulation chipping and improper insulation mounting, that are in fact thermal bridges (ring seals, inspection hatches, tubes and other uninsulated elements etc.). Thermal calculations revealed that by eliminating thermal bridges one may reach energy savings of 15-30 %.

References

1. Regulation of the Minister of Economy of September 17 1999 on safety and hygiene of work with equipment and power installations (in polish), Dz. U. 1999 r., Nr **80,** poz. 912

DOI: 10.1051/e3sconf/20171401035

2. Announcement of the Ministry of Economy dated December 12 2012 concerning the detailed list of actions targeted at energy efficiency improvement (in polish), M. P. 2013 r., poz. 15
3. Act of May 20 2016 about Energy Efficiency (in polish), Dz. U. 2016 r., poz. 831
4. Operating instructions of the energy boiler OP-140 (unpublished)
5. Material companies RAFAKO (Figure concept of OP-140)
6. S. Łopata, Heat not Lost-2015, Energy Efficiency, thermal insulation industry, Ed. Z. Plutecki, Studies and Monographs, 30-36 (Heat not Lost Sp. z o.o., Cracow 2015)
7. M. Kocot, Classification and Heat Loss Analysis of Thermal Bridges in Power Boiler OP-140, Thesis engineering, Cracow University of Technology (2016)
8. German standard VDI 4610: VDI 4610 Blatt 1:2012-12, Energieeffizienz betriebstechnischer Anlagen Wärme - und Kälteschutz (2012)
9. R. Sobociński, Z. Nagórski, T. Kośmicki, Collection of tasks of technical thermodynamics (Publishing Warsaw University of Technology, Warsaw 1996)

Some harvesting system transforming energy wastes of compressed air to electricity

Jacek S. Leszczynski[1*], *Krzysztof* Kastelik[1], *Ryszard* Kaminski[1], *Bartłomiej* Tomasiak[1], *Dominik* Grybos[1], *Milosz* Olszewski[1], *Anna* Plewa[1], *Pawel* Mermer[2], *Roman* Rygal[3], *Marian* Soinski[3]

[1]AGH University of Science and Technology, Faculty of Energy and Fuels, Department of Hydrogen Energy, Al. Mickiewicza 30, 30-059 Krakow, Poland
[2]Hafner Sp. J./O Czestochowa, ul. Główna 14 42-200 Czestochowa, Poland
[3]Magneto Sp z o.o., ul. Odlewników 43, 42-200 Czestochowa, Poland

Abstract. Energy harvesting devices are self-powered systems which are associated with input ambient energy and convert such energy to electrical one. In this study, we focus on input ambient energy deriving from compressed air. Using arbitrary pneumatic machine, the compressed air after work - with its typical parameters as overpressure 6 bar or higher and variable volume flow - is exhausted to atmosphere. This generates energy losses. We have constructed some device which transforms the loss of compressed air energy to electricity. It illustrates the principle of action very well and does not perturb operation of pneumatic machines being source of energy losses.

1 Introduction

Alternative energy generation systems are rapidly expanding and being interconnected to distributed transmission networks (grid-on) or supply internal networks (grid-off). At present, one observes development in electrical energy storage technologies [1] and their application potential in power systems. However, it should be noted huge amount of energy losses at global scale. Thus, there is a plenty of room to recover the energy loss and to reproduce electricity. Within this background, energy harvesters [2] are a booming business at the level of watts to kilowatts. By name, energy harvesting devices [3] are self-powered systems which are associated with input ambient energy and its conversion to electrical energy. However, classical input ambient energies include heat (thermoelectric modules), light (photovoltaic cells), radiation (rectifying antennas) and vibration (piezoelectric, electro-magnetic, electro-static transducers).

In this study, we focus on input ambient energy deriving from compressed air. There is huge amount of systems, where primary air makes work due to compression, is exhausted from the system. Such process one observes in food industry - when packing or other machines are used, automotive industry – where pneumatic robots are applied, mechanical industry – where pneumatic presses or pneumatic motors are used, transport industry –

* Corresponding author: jale@agh.edu.pl

where compressed air is necessary to open or close doors, etc. Moreover, one can produce compressed air from other phenomena, such as: frequently inhibiting/stopping vehicles, elevators which are falling down gravitationally or doors which are opening and closing very frequently. Generally speaking, production of compressed air costs input energy. After work, the compressed air, with typical parameters as overpressure 6 bar or higher and variable volume flow, is exhausted to atmosphere. This generates energy loss. Here we focus on such energy losses. We have constructed some device, which transforms the loss of compressed air energy - exhausted to atmosphere - to electricity.

2 State of the art and test rig

Our idea is restricted and patented in Polish Patent Office [4]. Fig. 1 illustrates scheme of installation which is able to transform any energy losses of compressed air to electrical energy.

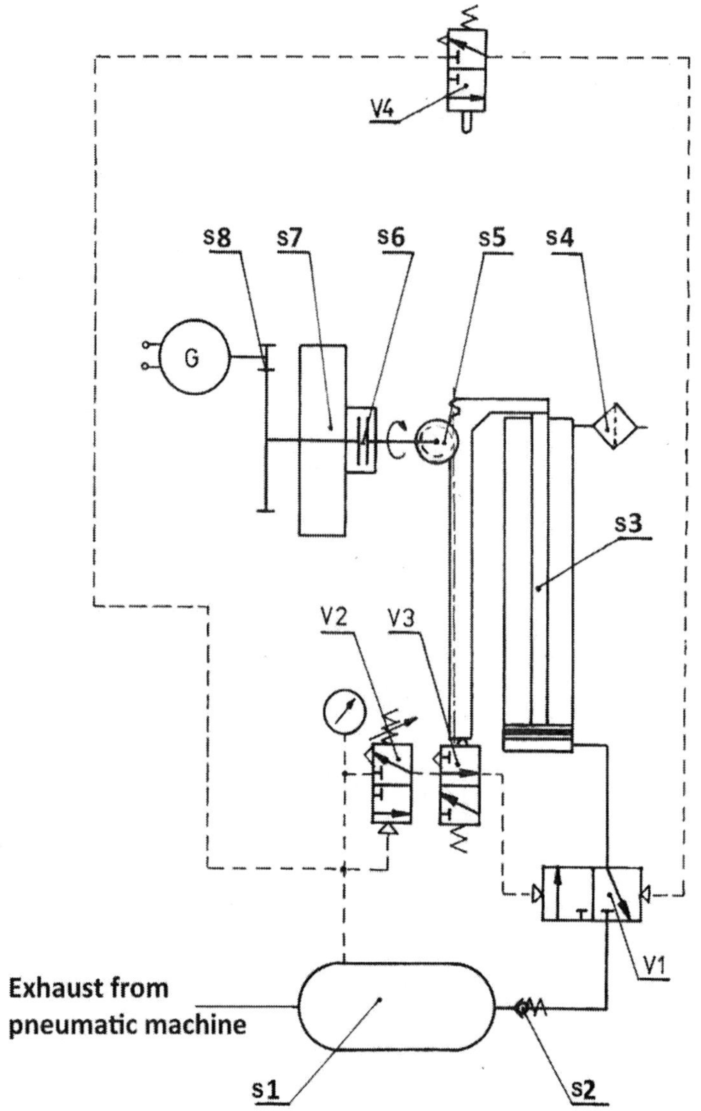

Fig. 1. Scheme illustrates energy recovery, according to our pending patent [4].

We look into energy of compressed air exhausting to atmosphere which is storing in some air tank (S1). Control valve (V1) is double controlled by valve (V2) and control valve (V4). When overpressure in tank (S1) riches control overpressure manually controlled from range 0.5 bar to 1.5 bar then, through one directional valve (S2), control valve (V1) is opening and low-compressed air drives cylinder (S3) automatically. Cylinder (S3) goes up running a kinematic pair - rack gear (S5). The kinematic pair drives power generator (G) through system of gears (S7) and (S8) with unidirectional clutch (S6). When cylinder (S3) hits high position then control valve (V4) switches automatically control valve (V1) and cylinder (S3) is falling down due to gravity till to lower positon determined by control valve (V3). Above description illustrates one cycle of operation. Next cycle starts and stops automatically being dependent on overall volume of tank (S3), the speed of compressed air filling tank (S3) and manually adjustable the air overpressure in the tank.

According to idea illustrated by Fig. 1 we are able to construct a test rig. Using CAD/CAM technology we have elaborated the virtual model of our harvester. The model is presented by Fig. 2.

Fig. 2. Virtual construction of energy harvester: 1) overpressure tank; 2) pressure management system; 3) gears; 4) low speed power generator; 5) battery management system.

 DOI: 10.1051/e3sconf/20171401036

One may distinguish main components of the energy harvester as: overpressure tank (1), pressure management system (2), system of friction and rack gears (3), low speed power generator (4) and battery management system (5).

In order to demonstrate how our idea operates in practice we have constructed the test ring in semi-industrial scale. Fig. 3 shows our construction.

Fig. 3. Test rig in semi-industrial scale.

We mounted air tank (1) – with volume V_z=10 dm^3, low-speed power generator (2) which generates 500 W in 400 obr/min, battery management system (3), system of rack - gearwheel (4) with d_1=0.5 m and length of rack l_r= 1 m, system of friction gear d_2=0.5 m, d_3=0.02 m. We applied the cylinder with the following technical data: d_s=0,08 m, l_c=1 m. Figures 4 and 5 present overpressure control system and battery management system in more details.

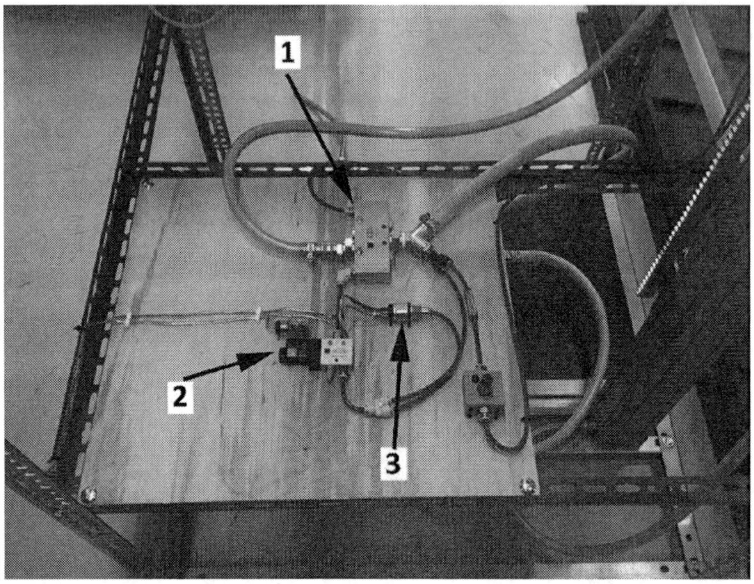

Fig. 4. Photograph of overpressure control unit.

The main components of the overpressure control unit are the following:
- 3/2 control valve (1) of 1/4 inch which is normally open and has spring return cylinder,
- control valve (2) of 1/8 inch which is normally close and is controlled by electromagnetic coil,
- pressure sensor (3) controls the valve (1) by using the valve (2).

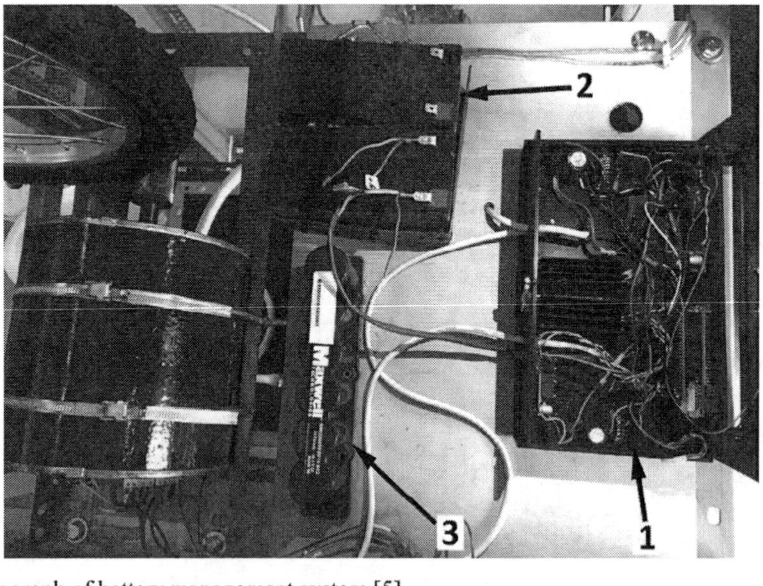

Fig. 5. Photograph of battery management system [5].

The main components of the battery management system are the following:

- electronic control unit composes from one AC/DC and two DC/DC transformers which transform arbitrary voltage level to 12 V DC,
- bank of two batteries with rechargeable battery capacity C_a=7.2 Ah and operating voltage U_a=12 V DC,
- bank of supercapacitors with maximum U_c=16 V DC operating voltage, rechargeable supercapacitors capacity C_c=58 F.

Using the technology demonstrator, as presented by Fig. 1 completely, we are able to perform several tests in order to assess how our system operates in real working environment.

3 Technological tests

Before we begin full technological tests, we try to investigate how our power generator works. In this case we use FEMM code [6] that to simulate the generator behaviour in order to assess its mechanical constructions. We assume idle operating state of the generator that to observe magnetic flux density. Fig. 6 shows distribution of magnetic flux density in the kinematic pair rotor-stator.

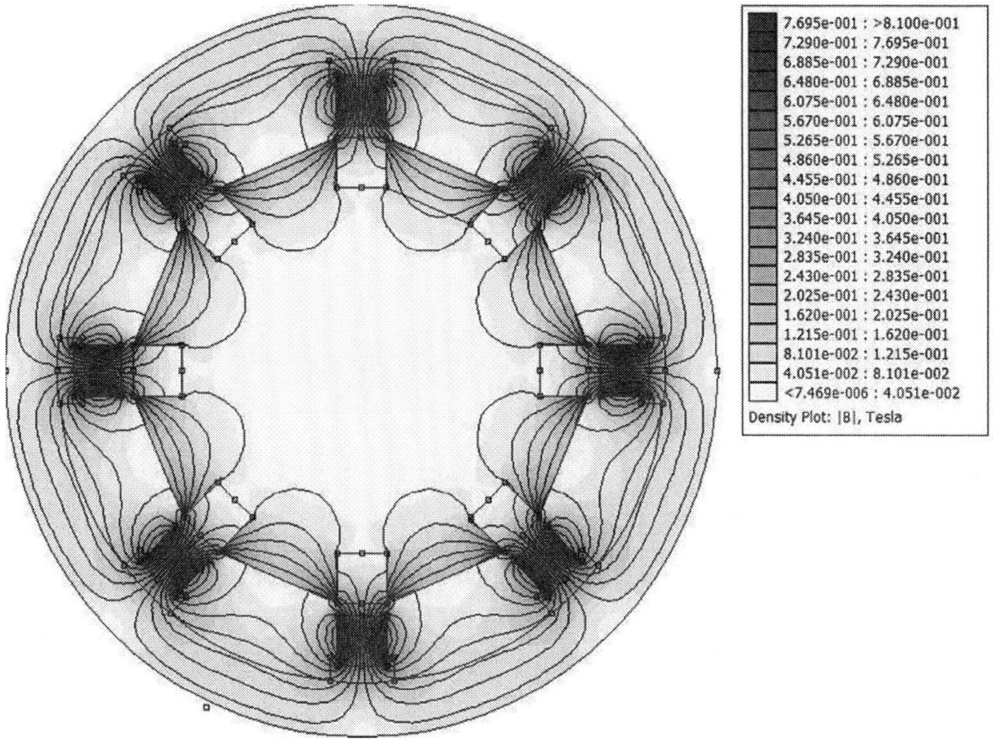

Fig. 6. Simulation of magnetic flux density has been done by FEMM [6], cross section, idle state of the power generator.

In simulating conditions the generator rotates with the rotational speed 300 r/min. We observe concentration of the magnetic flux near neodymium magnets and magnetic field lines closes. This means that there is no big dissipation of magnetic energy. Next we would like to estimate how the magnetic field is shielded. Fig. 7 presents magnetic flux density distribution in horizontal section.

 DOI: 10.1051/e3sconf/20171401036

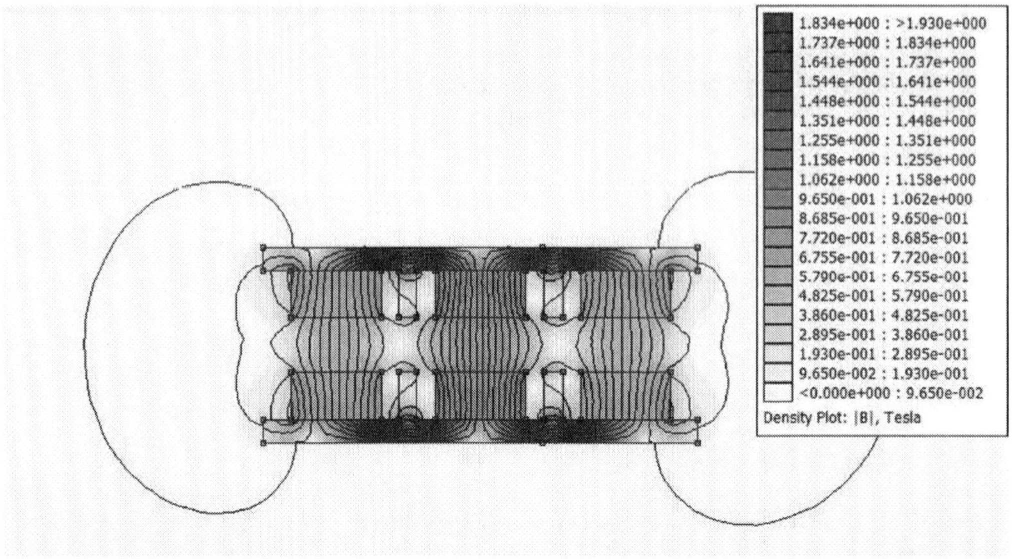

Fig. 7. Simulation of magnetic flux density has been done by FEMM [6], horizontal section, idle state of the power generator.

We are able to assess the thickness of rotating plate in order to find minimal thickness being robust for magnetic saturation. Here we establish h=5 mm thickness of rotating steel plate. One notices closed lines of magnetic flux. This indicates no magnetic dispersion and therefore the minimal value of plate thickness fulfils its function, i.e. the magnetic shield.

Next we analyse spatial distribution of the magnetic flux density in the piece of low speed generator. Fig. 8 shows such situation.

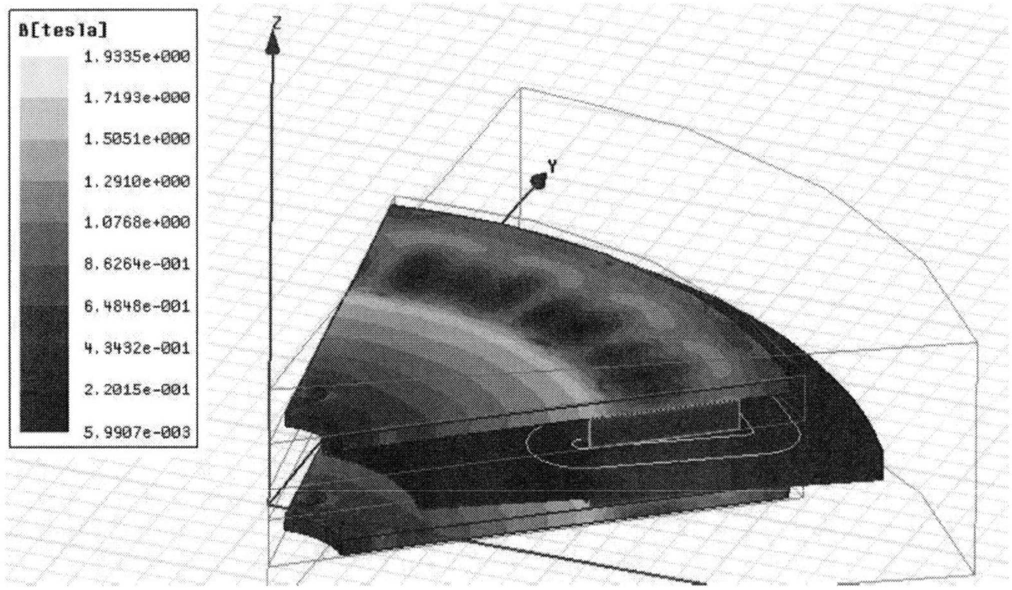

Fig. 8. Simulation of magnetic flux density has been done by FEMM [6], spatial-view, idle state of the power generator.

We observe right distribution of the magnetic flux density in quarter of the kinematic pair rotor-stator. Values of magnetic fluxes are not high and not exceed saturation. This means the distribution of magnetic fluxes guarantees right operation of the power generator. To summarise, we have analysed the power generator computationally. There is no obstacle in the proposed construction and therefore the power generator is ready for operation.

In order to proof how our harvester operates, we performed several industrial tests. We connected our unit throughout the tank to a pneumatic machine, to the valve where compressed air was exhausted to atmosphere. Notice oscillations of overpressure of compressed air in the pneumatic machine around 6 bar. After work, the compressed air through a system of valves is exhausted to atmosphere. Our tank with volume V_z= 10 dm^3 fills quite fast to the control overpressure 1.5 bar. After the control overpressure is reached, the control valve opens and then the stored compressed air in the tank runs the cylinder. In the tank we notice decompression till to 0.75 bar. The cylinder uses 2250 dm^3/h volumetric flow of the compressed air in one cycle of its operation. Through the system of gears, the cylinder runs the power generator. Table 1 shows values of the power generator registered in one cycle of cylinder operation – from the bottom position to the top position.

Table 1. Values of power generator registered under idle state.

Measurement no.	U_o [V]	n_o [r/min]
1	70	32
2	98	202
3	100	214
4	80	93
5	99	208

Above table presents voltage and rotating speed values which were registered under the idle state of power generator. This means no external load has been noted. Mean values are the following: $\overline{U}_0 = 89\ \text{V}$ and $\overline{n}_0 = 150\ \text{r/min}$.

Next we would like to observe what happens when we upload the generator. Fig. 9 and Fig. 10 present two scenarios of the operating generator under different loads.

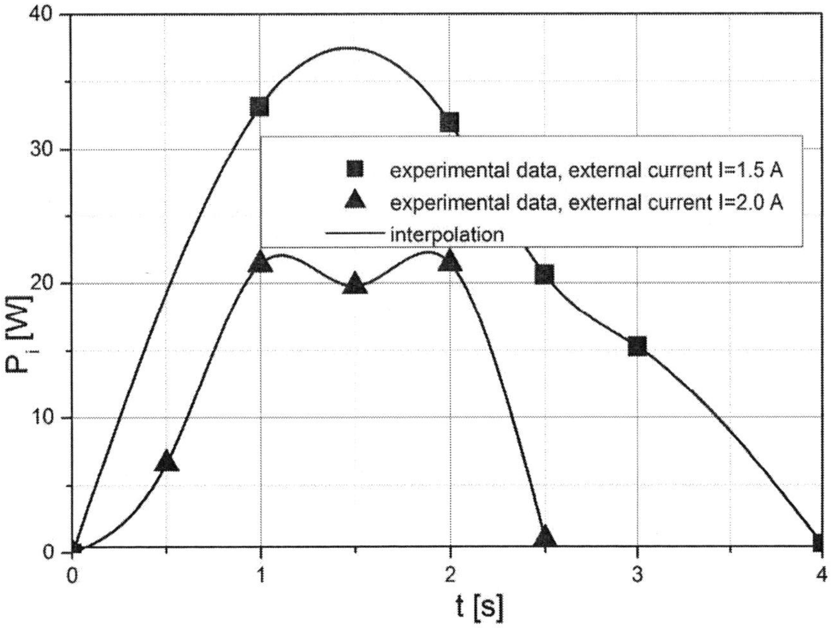

Fig. 9. Power produced in generator over time registered under one motion of cylinder in two load states.

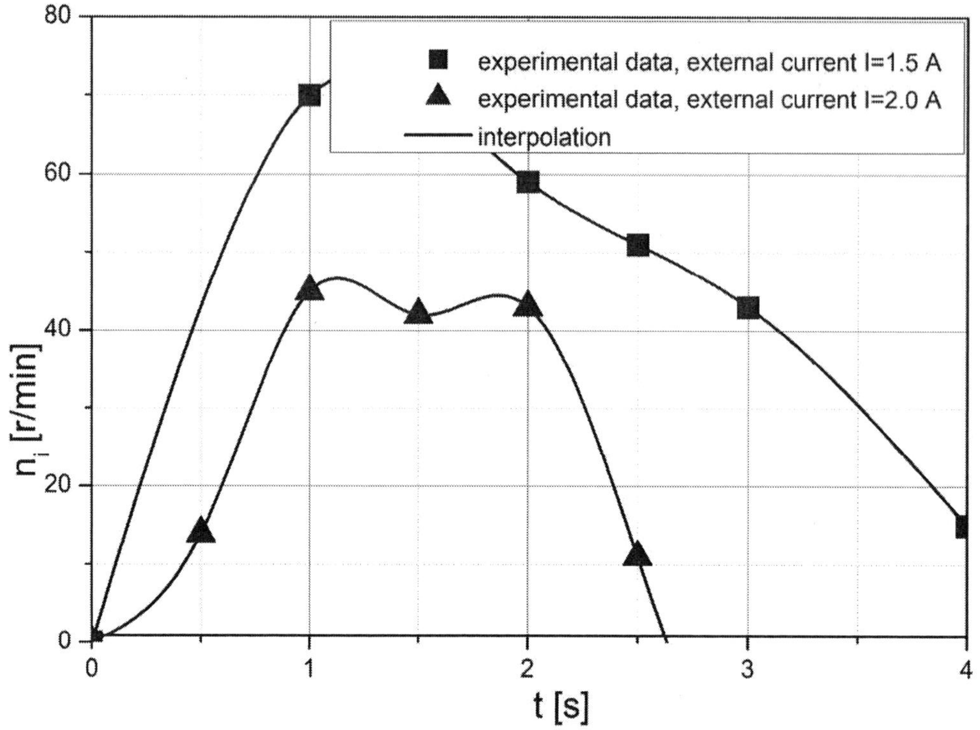

Fig. 10. Rotational speed of generator over time registered under one motion of cylinder in two load states.

In one cycle of cylinder operation we use 2250 dm^3/h volumetric flow of the compressed air. The compressed air during its work into the cylinder is decompressed dynamically. The cylinder starts go up from overpressure 1.5 bar and ends the cycle to 0.75 bar. Duration over time the adiabatic decompression process varies from equals around 1.5 s, when we load the power generator. Time for the idle return of the cylinder, from the upper position to the lower one under gravity only, equals to 2.5 s. Here we noted average velocities of the piston motion around w_1=0.67 m/s in the cylinder operating state and velocity w_2=0,4 m/s in the cylinder idle return. In order to assess how effective is our energy harvester we perform two technological tests when the power generator was loaded in some resistance way and we noted the following currents 1.5 A for test no. 1 and 2.0 A for test no. 2. Fig. 9 shows changes of power over time which was registered on the generator ends. One notes more power registered over time for low load resistance, when current was 1.5 A. This situation was explained in Fig. 10 where rotational speed over time for the two above cases was registered. Here we observe higher values of the rotational speed when resistance load was 1.5 A. To conclude we have noted that the power generator is not sufficient to work in low rotational speeds. This means we need to elaborate a novel model of the power generator featuring low internal resistance and therefore being more efficient in the range of low rotational speeds. However, our industrial test does not perturb normal operation of the pneumatic machine being involved that to recover the loss of energy of compressed air.

4 Concluding remarks

In this study, we focus on input ambient energy deriving from compressed air. Using arbitrary pneumatic machine we add our harvester to values where the compressed air, with its typical overpressure 6 bar or higher, after work is exhausted to atmosphere. Here we constructed the device which transforms the loss of compressed air energy to electricity.

Additional process of energy transformation does not perturb the primary operation of pneumatic machine under condition that maximal value of overpressure in the tank does not exceed 1.5 bar. Our energy harvester is not efficient and not perfect device. At present the harvester illustrates the principle of action very well and does not perturb operation of pneumatic machines being source of energy losses. In further researches we need to elaborate a mathematical model which will illustrate the dynamics of compressed air and the dynamics of all devices involved in the energy harvester. Simple technological tests confirm that the power generator used in the harvester features by high internal resistance and therefore is not adequate to operate in its low rotational speeds.

References

1. X. Luo, J. Wang, M. Dooner, J. Clarke, Overview of current development in electrical energy storage technologies and the application potential in power system operation, Applied Energy **137**, 511–536 (2015)
2. P. Harrop, *An Introduction to Energy Harvesting* (IDTechEx, www.IDTechEx.com, 2009)
3. P. Spices, L. Mateu, M. Pollak, *Handbook of Energy Harvesting Power Supplies and Applications* (Pan Stanford Publishing Pte Ltd., Singapore, 2015)
4. J.S. Leszczynski, K. Kastelik, R. Kaminski, et al., A device for processing pulses of gravitational energy into electricity, patent pending No. P.414139, Polish Patent Office, Warsaw, Poland (2016)
5. B. Tomasik, A. Plewa, J.S. Leszczynski, Electronic system for pulse energy storage (Bulletin of the Polish Hydrogen and Fuel Cells Association 9, 14th Symposium on Fast ionic conductors : 22–25.09.2015, Bialka Tatrzanska, Poland, 58, 2015)
6. D. Meeker, *Finite Element Method Magnetics* (Version 4.2, User's Manual, October 25, 2015)

The influence of control parameters on energy efficiency of switched reluctance generator for vehicle applications

Mariusz Korkosz[1], *Adam* Powrózek[1*]

[1]Rzeszów University of Technology, The Faculty of Electrical and Computer Engineering, Rzeszow, Poland

Abstract. In the paper, the possibilities of application of switched reluctance generators in modern variable-speed drives were discussed. The process of mechanical energy conversion into electric energy was discussed. Based on simulation and laboratory tests, output power profiles, efficiency profiles and profiles of ratio of output power to phase current were determined in the single-pulse mode. The authors showed that proper selection of control parameters allows the operation of the SRG with maximum power, maximum efficiency or maximum ratio of output power to phase current. Finally, conclusions concerning advantages of using SRGs in variable-speed applications were drawn.

1 Introduction

Due to the development of variable-speed applications and hence increasing demand for electrical energy, research on efficient, fault tolerant, compact and reliable generation systems are being conducted. In recent years, the intensified development of vehicles with hybrid and electric drives can be seen. Electric machines, which operate depending on the situation, among others, as a generator, are used in both of these types of drives. The development of these types of vehicles is a part of a global tendency to decrease negative impact of combustion engines on the environment. Despite strict emission standards, combustion engines will still have negative impact on the environment. Stricter emissions standards can only decrease but not eliminate a problem of exhaust emissions into the atmosphere. The vehicles with hybrid drives do not solve this problem, but they allow covering short distances only with electric drive. Only vehicles with electric drive alone allow eliminating exhaust emissions into the atmosphere. However, we should be aware that vehicles with hybrid or electric drives are not completely harmless for the surrounding environment. In hybrid vehicles, combustion engine drives electric machine which operates as a generator and charges batteries. In vehicles with electric drive, batteries are charged from the mains. In both electric and hybrid drives, part of accumulated kinetic energy of moving vehicle is converted into electric energy. It is possible when electric machine operates as a generator. Recently, systems for converting kinetic energy into electric energy

* Corresponding author: apow@prz.edu.pl

during vehicle braking are used in many vehicles with combustion engines which are equipped with Start-Stop systems. Such vehicles are very often called semi-hybrid recently.

In vehicles with electric and hybrid drives and with Start-Stop systems, energy recovery during vehicle braking depends largely on the road conditions. During smooth deceleration or braking, energy recovery should occur at as high efficiency of kinetic energy conversion into electric energy as possible. Braking distance is very important while sudden stoppage of the vehicle is necessary. In this case, braking with classic mechanical brakes is very important. However, part of kinetic energy can be converted into electric energy in this case. Emergency braking process should be assisted as much as it is possible by electric machine. It means that converting of kinetic energy into electric energy has to be directed into obtaining as high braking torque as possible.

Various types of electric machines are used in above-mentioned solutions [2-4]. However, brushless motors with permanent magnets are currently most widely used. They are characterized by the highest efficiency of energy conversion and very small dimensions. High efficiency can be obtained by using high-energy permanent magnets but this solution has two main disadvantages. The first one is high price of permanent magnets and the second one is the source of origin of permanent magnets. Most of the well-known resources of rare-earth elements which are used in manufacturing of high-energy permanent magnets are in the possession of one of the Asian countries. Therefore, studies on alternative technologies of electric machines for above-mentioned applications should be conducted and switched reluctance machines could be one of them [5-7].

In the paper, generating operation of switched reluctance machine is analysed. The process of mechanical energy conversion into electric energy is discussed. Based on simulation and laboratory tests, the authors show that proper selection of control parameters allows operation of SRG with maximum efficiency, maximum power or maximum ratio of output power to phase current (*P/I*). Finally, conclusions concerning advantages of using SRGs in variable-speed applications are presented.

2 SRG-based energy conversion

Switched reluctance machines are categorized among machines with electronic commutation where classic commutator is replaced by dedicated power converter with control system. Fig.1a shows a structure of 8/6 switched reluctance machine (Fig.1a). The switched reluctance machines have a rotor without windings; they are concentrated only in a stator. The electromagnetic torque is produced by the tendency of alignment of stator and rotor. The rotor tends to the position where reluctance is the smallest.

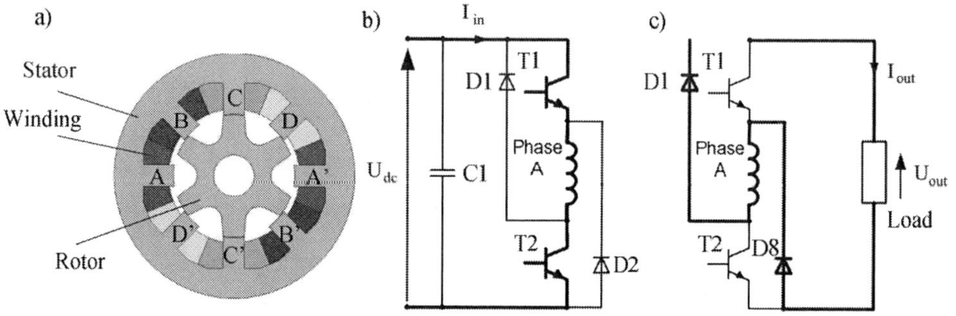

Fig. 1. a) A structure of the 8/6 switched reluctance machine, b) one leg of the power converter during excitation stage and c) during generation stage

Figs.1b and 1c show one leg of the power converter with marked directions of phase current flow at the excitation stage (Fig.1b) and at the generation stage (Fig.1c) [1]. In the single-pulse mode, both switches are turned-on at the turn-on angle θ_{on}, current flows from the dc source and energy is stored in the magnetic field of the machine. Beyond the turn-off angle θ_{off}, both switches are turned-off and the energy provided by the prime mover through a shaft is converted into electric energy. Fig.2 shows idealised waveforms of phase back-EMF e_{ph} (Fig.2a), phase electromagnetic torque T_{eph} (Fig.2b), phase current i_{ph} (Fig.2c) and phase power P_{ph} (Fig.2d) in the single-pulse mode. The phase BEMF e_{ph} changes its direction after aligned position of rotor θ_a (Fig.2a). The sign of e_{ph} is determined by $dL_{ph}/d\theta$ and its amplitude varies with the rotor speed, so behaviour of phase current i_{ph} (Fig.1b) is determined by the relationship between e_{ph} and U_{dc}. The change of phase BEMF sign to negative also causes a change of a sign of the generated electromagnetic torque T_{eph} (Fig.1c). Switched reluctance machine starts to operate in a braking mode or in a generating mode (Fig.1c). It can be seen in Fig.1c that when turn-on angle is earlier than aligned position, electromagnetic torque is positive and machine operates in a motoring mode for a while. Output power in the switched reluctance generator is a function of supply voltage U_{dc}, speed n and control variables (θ_{on} and θ_{off}).

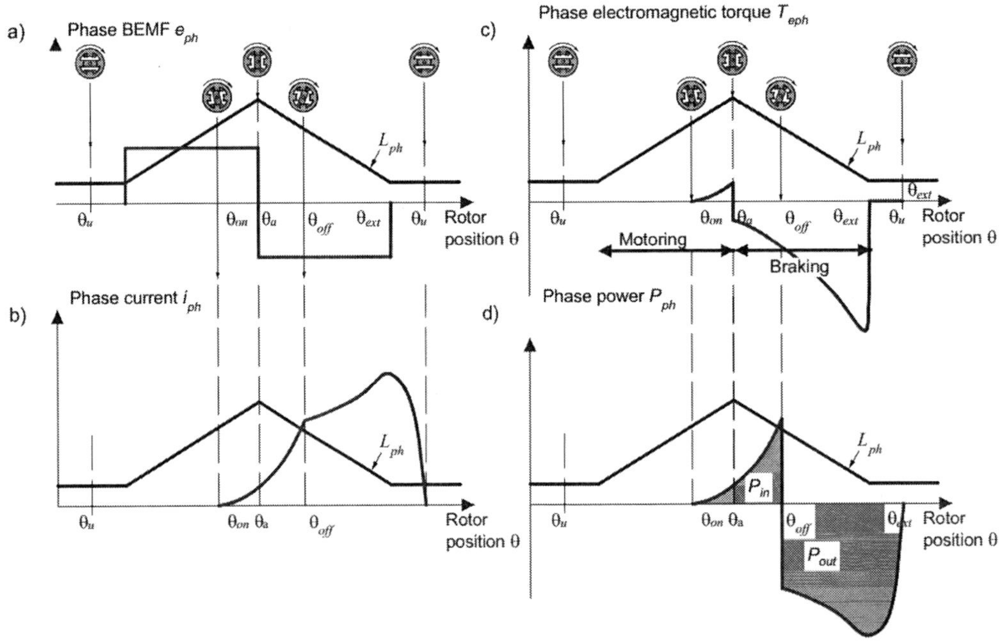

Fig. 2. Idealised waveforms of a) phase back-EMF e_{ph}, b) phase electromagnetic torque T_{eph}, c) phase current i_{ph} and d) phase power P_{ph} in the single pulse-mode

3 Results of simulations

In table 1, basic parameters of tested switched reluctance machine are presented.

Table 1. Basic parameters of the tested SRM.

Parameter	Value
Number of stator phases N_s	8
Number of rotor poles N_r	4
Rated voltage U_N [V]	24
Rated power P_N [W]	750
Rated speed n_N [rev/min]	3000

The simulation tests of the SRG were conducted based on nonlinear simulation model built in the Matlab/Simulink environment [8]. The influence of control parameters on properties of switched reluctance generator was determined based on the simulation model. Figs.3-5 show exemplary simulation results. Fig.3 shows dependence of average output power P_{outav} in the function of control parameters (turn-on angle θ_{on} and turn-off angle θ_{off}) at constant speed n=3000 rev/min and constant dc-link voltage U_{dc}=24V and with marked points of maximum power.

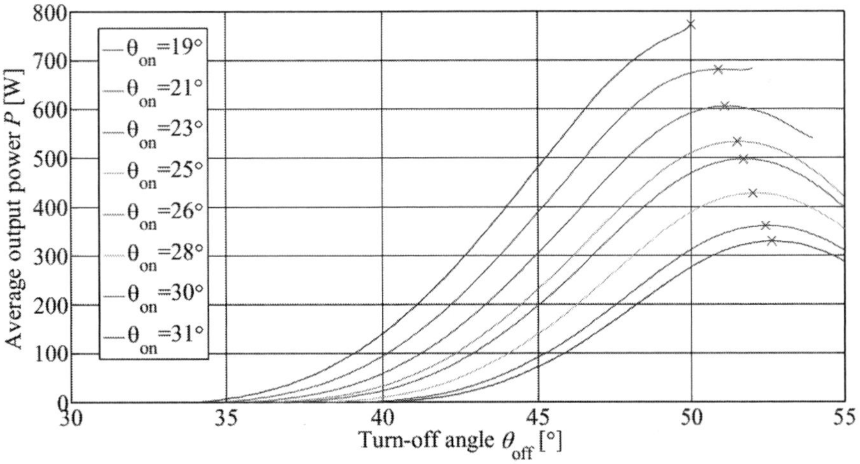

Fig. 3. The average output power P_{outav} vs. turn-off angle θ_{off} at θ_{on}=const, U_{dc}=24 V and n=3000 rev/min

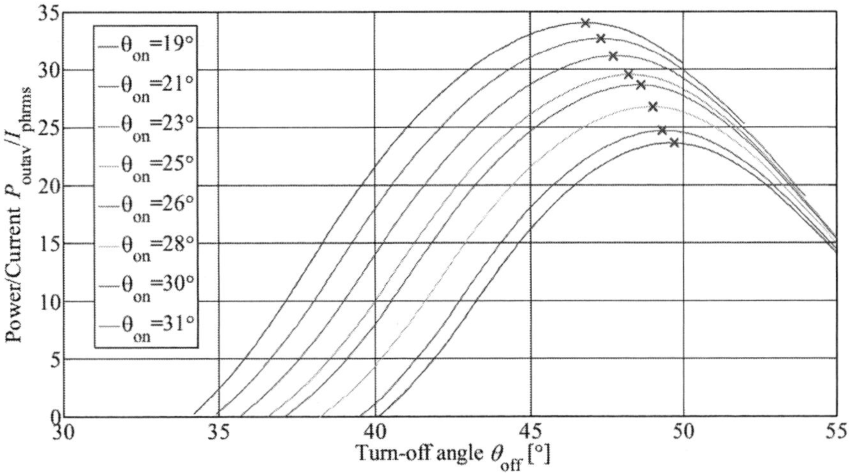

Fig. 4. The power/current ratio P_{outav}/I_{phrms} vs. turn-off angle θ_{off} at θ_{on}=const, U_{dc}=24 V and n=3000 rev/min

It can be seen from Fig.3 that selection of control parameters has to be dependent on value of required output power. However, other criteria should be taken into account when control parameters are selected. One of them is criterion of maximum efficiency. The second one is criterion connected with ratio of average output power to phase RMS current P_{outav}/I_{rms}. Fig.4 shows dependence of ratio P_{outav}/I_{rms} in the function of control parameters at n=3000 rev/min and U_{dc}=24V. When control parameters are selected to obtain maximum efficiency, output power is significantly limited. However, when control parameters are selected to obtain maximum output power, efficiency is noticeably lower. The selection of control parameters to obtain maximum ratio P_{outav}/I_{rms} allows obtaining higher output power than for criterion of maximum efficiency. At the same time, efficiency of energy conversion is higher than for criterion of maximum output power. Fig. 5 shows dependence of average output power P_{outav} in the function of speed n. Control parameters were selected to obtain maximum values of average output power P_{outav} and ratio P_{outav}/I_{rms}.

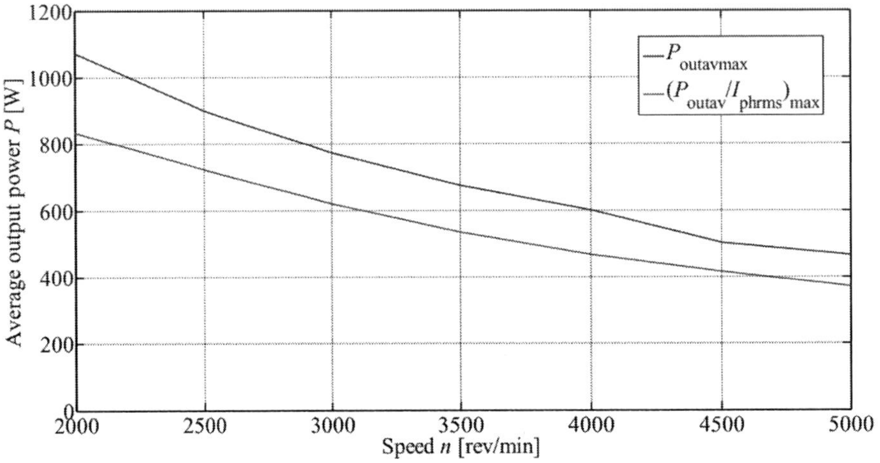

Fig. 5. The output power P_{outav} vs. speed n – simulation tests

The control with maximum ratio P_{outav}/I_{rms} gives 23% decrease of average output power.

4 Results of laboratory test

Fig.6 shows a schematic diagram of a laboratory setup. Yokogawa WT1600 digital power meter was used to measure all electric (current, voltage and power) and mechanical (speed, torque) parameters of the tested machine.

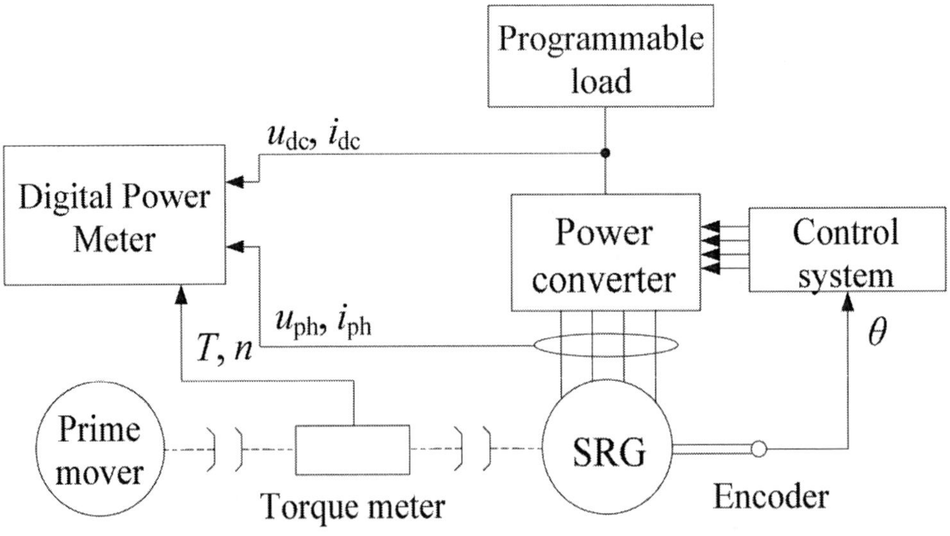

Fig. 6 A schematic diagram of the laboratory setup

In laboratory conditions, practical tests of control parameters influence on properties of switched reluctance generator were conducted. Fig.7-10 show results of laboratory tests. Fig.7 shows dependence of average output power P_{outav} in the function of control

DOI: 10.1051/e3sconf/20171401037

parameters (turn-on angle θ_{on} and turn-off angle θ_{off}) at constant speed n=3000 rev/min and constant dc-link voltage U_{dc}=24V and with marked points of maximum power.

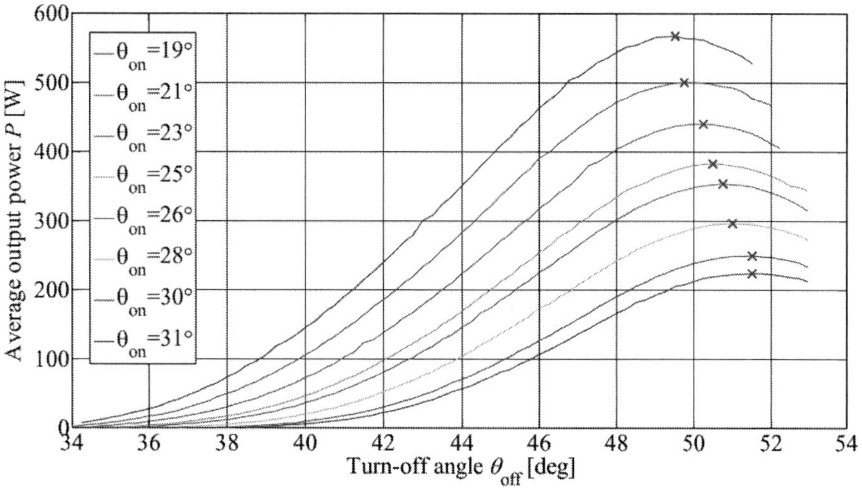

Fig. 7. The average output power P_{outav} vs. turn-off angle θ_{off} at θ_{on}=const, U_{dc}=24 V and n=3000 rev/min

Efficiency of energy conversion η in the function of control parameters (turn-on angle θ_{on} and turn-off angle θ_{off}) at constant speed n=3000 rev/min and constant dc-link voltage U_{dc}=24V was shown in Fig. 8.

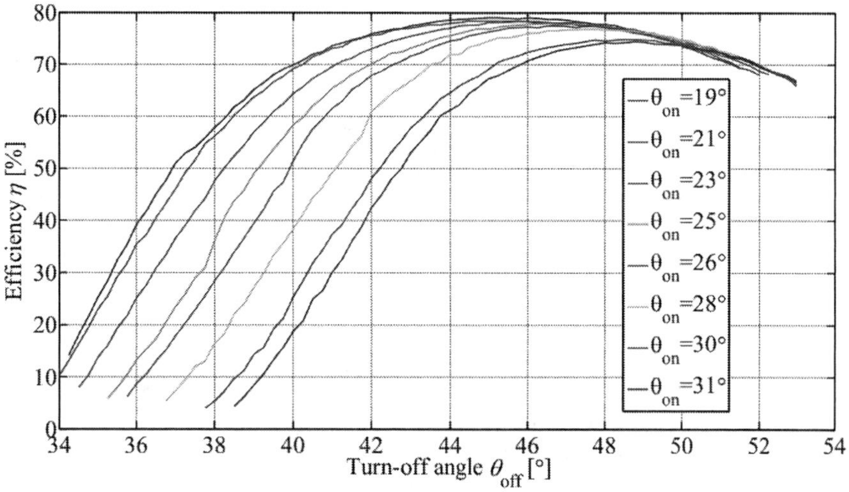

Fig. 8. The efficiency η vs. turn-off angle θ_{off} at θ_{on}=const, U_{dc}=24 V and n=3000 rev/min

Points of maximum efficiency was obtained at completely different control angles than points of maximum output power. Fig.9a shows dependence of average output power P_{outav} in the function of speed n where control angles were selected to obtain:

- maximum output power P_{outav},
- maximum ratio P_{outav}/I_{rms},
- maximum efficiency η.

 DOI: 10.1051/e3sconf/20171401037

Fig. 9b shows dependence of control angles (θ_{on} and θ_{off}) in the function of speed at which above mentioned criteria were obtained.

a)

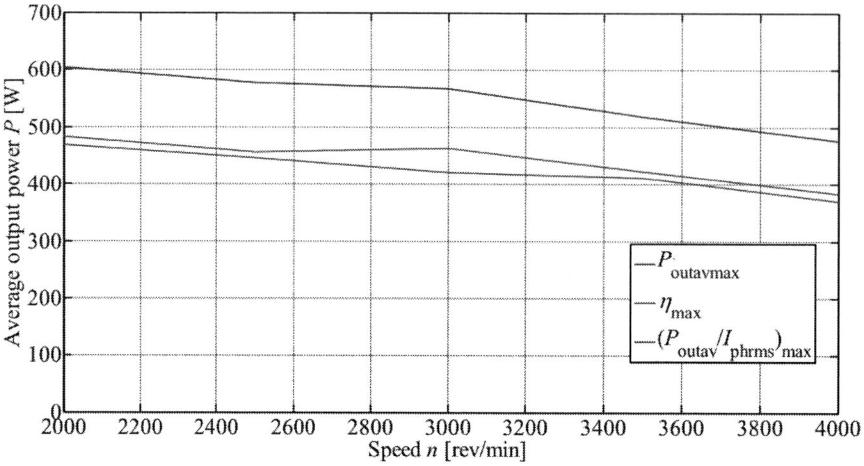

b)

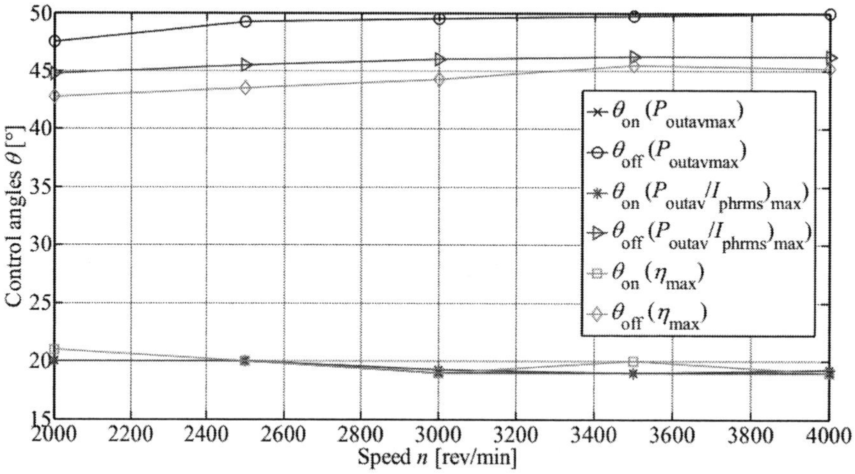

Fig. 9. a) Control angles θ_{on}, θ_{off} vs. speed n, b) the output power P_{outav} vs. speed n – laboratory tests

Results of laboratory tests confirmed that control angles at which maximum power was obtained give lower efficiency of energy conversion, so this is not optimal control. Increase of efficiency is connected with decrease of generator output power.

5 Conclusions

The switched reluctance machines are an alternative technology for brushless machines with permanent magnets. The lack of permanent magnets allows their manufacturing in European market. Despite simple structure, control of SRG compared to other types of machines is more complicated. The control directed to obtain maximum output power causes decrease of energy conversion efficiency. However, this state can be used for emergency braking of vehicle when as high braking torque as possible is required. The control with maximum efficiency or maximum ratio P_{outav}/I_{rms} can be used when a vehicle brakes smoothly or during typical generating operation i.e. wind turbine. During generating or braking operation, the control with maximum efficiency of energy conversion should be used for small powers and the control with maximum value of ratio P_{outav}/I_{rms} should be used for medium powers.

References

1. D. A. Torrey, *IEEE Transactions on Industrial Electronics*, Vol. **49**, pp:3-14 (2002)
2. Z. Q. Zhu, D. Howe, *Proceedings of the IEEE*, Vol. **95**, No. 4, pp: 746-765 (2007)
3. I. Boldea, L. N. Tutelea, L. Parsa, D. Dorrell, *IEEE Transactions on Industrial Electronics*, Vol. **61**, No. 10, pp: 5696-5711 (2014)
4. V. Nasirian, Sh. Kaboli, A. Davoudi, *IEEE Transactions on Industry Applications*, Vol. **49**, No. 3, pp: 1031-1042 (2013)
5. B. Fahimi, A. Emadi, R.B. Sepe, *IEEE Transactions on Energy conversion*, Vol. **19**, No. 1, pp: 116-124 (2004)
6. N. Schofield, S. Long, *IEEE Transactions of Vehicular Technology*, Vol. **58**, No. 1, pp: 48-56 (2009)
7. K. M. Rahman, B. Fahimi, G. Suresh, A. V. Rajarathnam, M. Ehsani, *IEEE Transactions on Industry Applications*, Vol. **36**, No. 1, pp: 111-121 (2000)
8. A. Powrózek, *Scientific letters of Rzeszow University of Technology, Electrical engineering,* No. **289**, pp: 193-203 (2013)

The assessment of energy efficiency of electric machines for domestic appliances drive

Piotr Bogusz[1], *Mariusz* Korkosz[1,*], *Jan* Prokop[1]

[1]Rzeszow University of Technology, The Faculty of Electrical and Computer Engineering, Rzeszow, Poland

Abstract. In the paper, the division of domestic appliances into categories connected with input power was presented. The authors discussed issues connected with energy efficiency of these devices. An example of such a device power consumption of which affects considerably on overall power consumption from the mains is a vacuum cleaner. Vacuum cleaners are used in almost all households and they were covered by the EU regulations which introduced limitations in power consumption from the mains. In the paper, the assessment of energy efficiency of classic electric motors for vacuum cleaners drive was presented. Results of practical tests of chosen vacuum cleaners were presented. The digital power meter was used to measure electric parameters of tested vacuum cleaners and the PC was used to collect measuring data. The assessment of input power influence on energy consumption and energy efficiency was conducted based on tests results. It was shown in conclusions that the one of development directions of domestic appliances, which can cause improvement of energy efficiency, are alternative technologies of electric machines with much higher efficiency i.e. energy-saving electric machines with electronic commutation.

1 Introduction

Nowadays, domestic appliances are used on a mass scale in households. In Poland, there were over 13 million flats in 2013 according to the statistical data [1]. Domestic appliances in flat equipment include: washing machines, dishwashers, ovens, refrigerators, cookers, hoods, multifunctional food processors or vacuum cleaners. Some of them require one (i.e. a vacuum cleaner) or even several electric motors (i.e. a washing machine). Very small or small power electric motors (i.e. dishwashers) are generally used in most of domestic appliances. Thus, efficiency of energy conversion has negligible influence on energy consumption and in consequence on energy efficiency class. The situation changes when motor which is used in the appliance consumes energy which considerably affects overall power consumption from the mains. An example of such domestic appliance can be a vacuum cleaner which is used in almost all households. In Poland, there are over 13 million vacuum cleaners. The introduction of limitations on power consumption from the mains concerns this device [2]. Working time of a vacuum cleaner depends on a flat area.

* Corresponding author: mkosz@prz.edu.pl

The average usable area of flats equals about 73.1 m^2 [1]. Manufacturers of electric machines for vacuum cleaners assume that these machines operate 50 hours during the year. Additionally, it is assumed that they operate twice a week. Therefore, average working time of the vacuum cleaner equals about 28 minutes. In practice, working time of this device equals about 15-50 minutes during a single vacuuming.

In the paper, the assessment of energy efficiency of classic electric motors for vacuum cleaners drive was presented. Results of practical tests of chosen vacuum cleaners were presented. The digital power meter was used to measure electric parameters of tested vacuum cleaners and the PC was used to collect measuring data. The assessment of input power influence on energy consumption and energy efficiency was made based on tests results. It was shown in conclusions that one of development directions of domestic appliances, which can cause improvement of energy efficiency, are alternative technologies of electric machines with much higher efficiency i.e. energy-saving electric machines with electronic commutation.

2 Energy efficiency of domestic appliances

To improve energy efficiency in the EU, many activities were introduced to reduce energy consumption by 20% by 2020. One of these activities is improvement of energy efficiency of the household equipment. Additional information as labelling of most household equipment was introduced to allow the customer to make more conscious choice of a device. The energy label of a device is defined individually for each device, but it describes energy efficiency class of the device and its basic parameters. Energy efficiency classes are marked with letters from A to G, with A being the most efficient. Energy efficiency class is determined based on energy consumption of a given device to standard energy consumption of this type of devices. Separate regulations describe the standard energy consumption of each type of devices which are labelled. Three additional classes A^+, A^{++} and A^{+++} have been introduced for several devices i.e. washing machines, refrigerators, dishwashers and TV sets since December 2010. The parameter which must be also placed on the label of each device is average annual energy consumption (kWh per year). Thereby, the consumer is able to compare energy consumptions of different devices of the same type.

The one of the most crucial parameters which influences energy efficiency is input power from mains with its efficiency of conversion. Domestic appliances can be divided into three categories according to their purposes:

- I converting electric energy into thermal energy,
- II converting electric energy into thermal energy and a little bit into mechanical energy,
- III converting electric energy into thermal energy and significantly into mechanical energy,
- IV converting electric energy into mechanical energy.

The first category includes devices such as: a toaster, an electric kettle, etc. These devices are not labelled so far due to minimal possibilities of the energy efficiency improvement. These devices generally contain heating elements. The second category includes devices which contain both heating elements and low-power electric machines e.g. dishwashers. The power consumption of these electric machines has negligible influence on the overall power consumption of the device. The third category of devices are domestic appliances which contain instead of heating elements also electric machines whose power consumption has significant influence on the overall power consumption, e.g. a washing machine. The fourth category contains devices where an electric machine is the only energy receiver and it converts electric energy into mechanical energy, e.g. mixers, blenders, coffee grinders or vacuum cleaners. Most of these devices use electric machines with rated powers which do not exceed several hundred watts, except for vacuum cleaners. Before 01.09.2014, manufacturers of vacuum cleaners were offering, to individual consumers, devices whose power exceeded 2 kW. Therefore, vacuum cleaners were covered by EU regulations [2] which imposed limitations on input power. Since 01.09.2014, every new vacuum cleaner should not consume more than 1600 Watts of power from the mains. After 01.09.2017, power consumed by vacuum cleaners should be less than 900 Watts. In table 1, acceptable values of average energy consumption per year for particular energy classes of vacuum cleaners are presented [2].

Table 1. The average energy consumption for particular classes of vacuum cleaners

Energy efficiency class	Annual energy consumption AE [kWh/annum]	
	After 01.09.2014r.	After 01.09.2017r
A+++	-	AE ≤ 10,0
A++	-	10,0 < AE ≤ 16,0
A+	-	16,0 < AE ≤ 22,0
A	AE ≤ 28,0	22,0 < AE ≤ 28,0
B	28,0 < AE ≤ 34,0	28,0 < AE ≤ 34,0
C	34,0 < AE ≤ 40,0	34,0 < AE ≤ 40,0
D	40,0 < AE ≤ 46,0	AE > 40,0
E	46,0 < AE ≤ 52,0	-
F	52,0 < AE ≤ 58,0	-
G	AE > 58,0	-

As can be seen in table 1, after 01.09.2017, additional energy classes will be introduced for vacuum cleaners, i.e. A+, A++ and A+++. Other parameters of the device, which are important in terms of device operation, are also placed on the energy label. Fig. 1 shows currently used energy label for vacuum cleaners.

DOI: 10.1051/e3sconf/20171401038

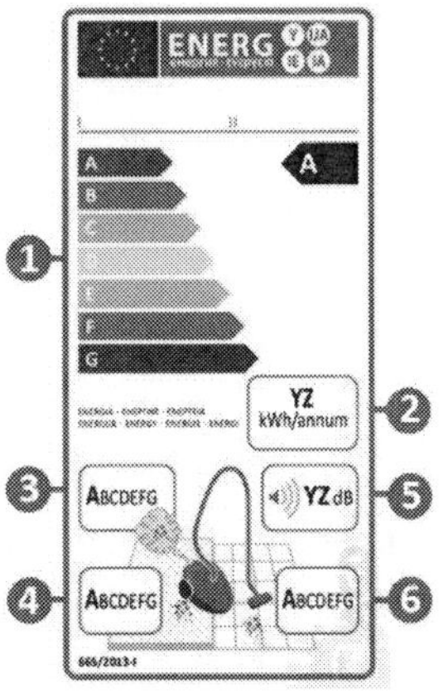

Fig. 1. The energy label used for vacuum cleaners

Besides energy efficiency class of the vacuum cleaner marked with (1) in Fig. 1, the energy label contains also:

- average annual energy consumption (2),

- dust re-emission class (ABCDEFG) (3),

- soft floor cleaning performance class (ABCDEFG) (4),

- sound power level in dB (5),

- hard floor cleaning performance class (ABCDEFG) (6).

The annual energy consumption AE in kWh/year does not result directly from input power. It is calculated separately for soft floors (AE_c) and hard floors (AE_{hf}). Equations 1 and 2 allow calculating the annual energy consumption for soft floors and hard floors, respectively [2].

$$AE_c = 4 \times 87 \times 50 \times 0{,}001 \times ASE_c \times \left(\frac{1 - 0{,}20}{dpu_c - 0{,}2} \right) \quad (1)$$

$$AE_{hf} = 4 \times 87 \times 50 \times 0{,}001 \times ASE_{hf} \times \left(\frac{1 - 0{,}20}{dpu_{hf} - 0{,}2} \right) \quad (2)$$

whereby:

ASE_c – is the average specific energy consumption in Wh/m^2 during soft floor test,

ASE_{hf} – is the average specific energy consumption in Wh/m^2 during hard floor test,

dpu_c – is the dust pick-up on model soft floor,

dpu_{hf} – is the dust pick-up on model hard floor.

For general-purpose vacuum cleaners, the average annual energy consumption can be calculated from:

$$AE_{gp} = 0{,}5 \times AE_c + 0{,}5 \times AE_{hf} \quad (3)$$

as a sum of half of energy which is required for hard floor vacuuming and half of energy which is required for soft floor vacuuming – which results from laboratory tests.

DOI: 10.1051/e3sconf/20171401038

Additionally, it is assumed that the standard surface to be cleaned equals 87 m^2 and this surface is cleaned 50 times per year. In laboratory tests, it is also assumed that the vacuum cleaner passes over each point on the floor four times (two double strokes). Laboratory tests are very often different compared to real conditions, which will be shown in part 3 of the paper.

3 Results of practical tests of chosen domestic appliances

Practical tests of chosen domestic appliances were conducted in real conditions. Yokogawa WT1600 digital power meter was used to measure electric parameters of tested devices (voltage, current, power, energy) and a PC was used to collect measuring data. Fig.2 shows a schematic diagram of a laboratory setup.

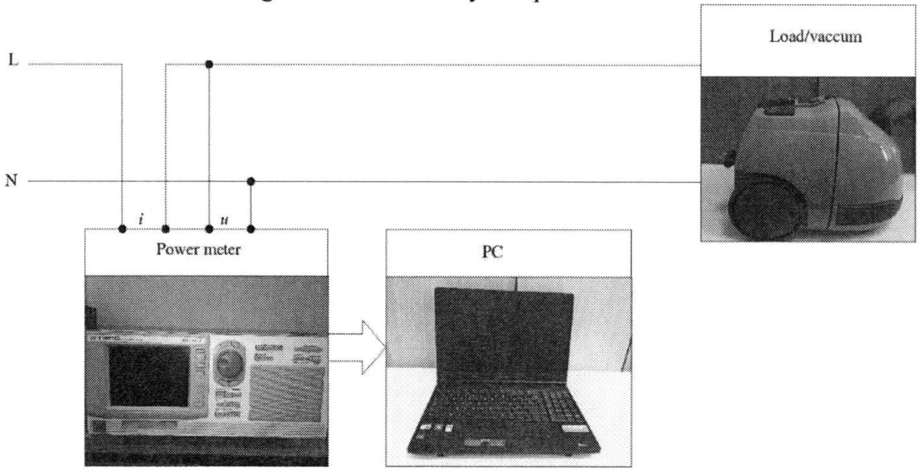

Fig. 2. A schematic diagram of the laboratory setup

Besides input power, power factor was also taken into account during the tests. Waveforms of currents drawn by tested domestic appliances were also registered.

Mixers, blenders, food processors, coffee grinders or vacuum cleaners are typical devices included in the category IV. In vacuum cleaners, electric motors with power exceeding 2 kW were generally used, but introduction of EU's restrictions of power consumption which should not exceed 1.6 kW caused that these motors cannot be used in new devices. At the same time, they are still in a practical use but their contribution in the market will be systematically decreasing and now they constitute over 50% of all vacuum cleaners. Several general-purpose vacuum cleaners with different powers were selected for the tests: 850 W (marked as I), 1000 W (marked as II), 1100 W (marked as III), 1650 W (marked as IV), 2000 W (marked as V) and a central vacuum cleaner with power 1500 W (marked as VI). Practical tests were conducted during vacuuming of a duplex flat with surface to be cleaned of 106 m^2. The flat was equipped with a central vacuum cleaner (the device marked as VI). Both hard floors (75%) and soft floors (25%) were in the tested flat. Practical tests were conducted in conditions which were different from conditions described with equation 3 (larger flat area, more hard floors than soft floors, higher supply voltage). Figs. 3-5 show waveforms of input active power P and power factor $\cos\phi$ of tested devices while vacuuming (I – Fig.3, V – Fig.4 and VI – Fig.5).

 DOI: 10.1051/e3sconf/20171401038

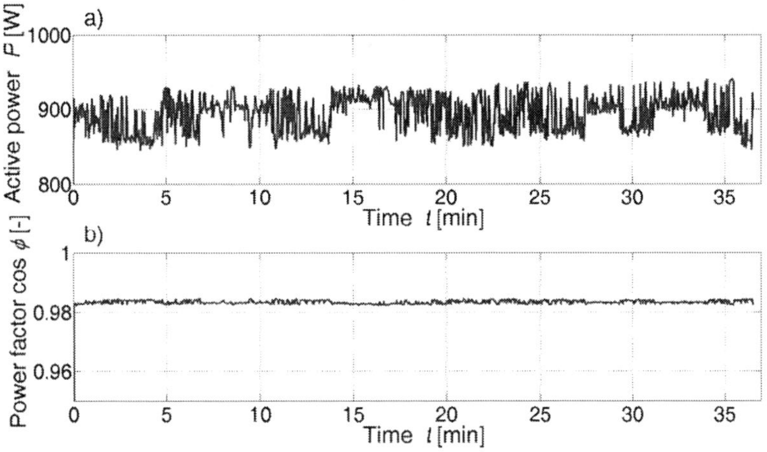

Fig. 3. Waveforms of a) input active power P and b) power factor ϕ for device I

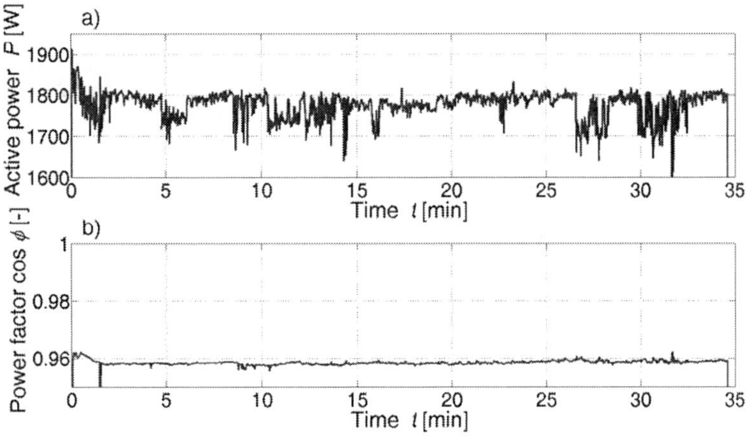

Fig. 4. Waveforms of a) input active power P and b) power factor ϕ for device V

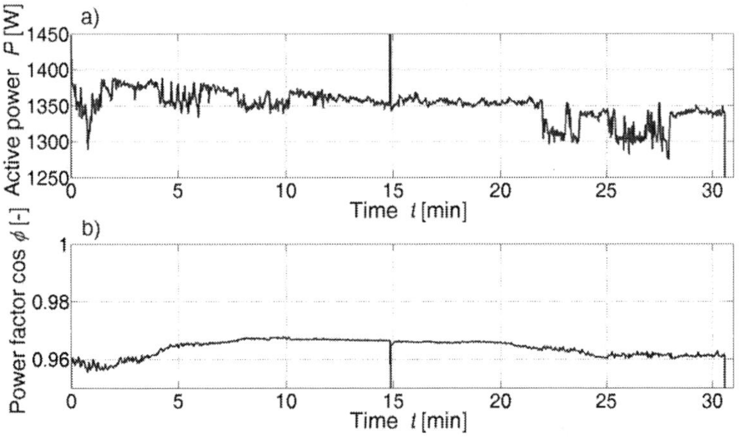

Fig. 5. Waveforms of a) input active power P and b) power factor ϕ for device VI

In table 2, chosen results of the tests are presented.

Table 2. Chosen results of practical tests of vacuum cleaners

Parameter/Vacuum cleaner	I	II	III	IV	V	VI
Input active power P_{inav} [W]	889	1065	1321	1472	1770	1347
RMS voltage U [V]	240,7	241,5	238,2	242,4	239	239,2
Power factor $\cos\phi_{av}$ [-]	0,983	0,991	0,974	0,977	0,96	0,963
Time t [min]	36,5	34,8	35,7	35,9	34,6	30,6
Time $t_{av\,(I\text{-}V)}$ [min]	35,5					30,6
Energy consumption E [kWh]	0,526	0,6301	0,7816	0,8709	1,02	0,688
Estimated annual energy consumption AE [kWh/annum]	26,3	31,51	39,08	43,55	52,36	34,4

The device with the lowest power has had the smallest cleaning head area. In particular, cleaning process with devices III-VI involved much more effort. What can be seen in Fig. 3 are numerous instantaneous changes of input power due to necessity of pulling cleaning head from the surface in order to change point on the floor to be cleaned. Vacuuming of soft floors was dubious, especially on a carpet with long hair. The tested vacuum cleaner with the lowest input power had serious problems with vacuuming such surfaces. It is obviously a subjective opinion which arises from testing of each vacuum cleaner. In the case of the central vacuum cleaner, vacuuming time was noticeably shorter. In this particular case, a significant influence on shorter vacuuming time had vacuuming of stairway which was much more comfortable and less troublesome.

A working point of the motor changes during operation of a vacuum cleaner due to adhesion way of cleaning head to the surface and instantaneous fluctuations of the supply voltage. During tests, the supply voltage was exceeding 230 V. The supply voltage has a significant influence on parameters of a commutator motor. Higher supply voltage about 10 V caused increase in the input power in devices I (about 39 W), II (about 69 W) and III (about 221 W) compared to power given in the data plate. In vacuum cleaners marked as IV, V and VI, which were produced before 01.09.2014, it can be seen that the input power was smaller in tests conditions despite higher supply voltage. These differences were due to the fact that manufacturers give information of power which is consumed only by mere suction unit in a laboratory setup. Older vacuum cleaners have also a little bit lower power factor. The supply voltage in real conditions can vary in the range of $\pm 10\ \%U_N$, so at power voltage of 230 V it means changes in the range of (207 V÷253 V). In laboratory conditions, some vacuum cleaners were supplied with voltages: 207 V, 230 V and 253 V. Tests of vacuum cleaners marked as I, III, IV and V were conducted in two variants. In the first one, input power was measured while vacuuming hard floor and in the second one while vacuuming soft floor. The test was connected with regulation of input power. All tested devices despite central vacuum cleaner allow infinitely, partly infinitely or step regulation of input power. Fig. 6 shows dependencies of active input power P and reactive input power Q in the function of power factor $\cos\phi$ during regulation of suction power at supply voltages: 207 V, 230 V and 253 V (for unit V). In table 3, chosen test results are presented.

 DOI: 10.1051/e3sconf/20171401038

The change of supply voltage in the range of ±10 %U_N has great impact on the value of input power and in consequence on suction power. In the case of unit I, a 10 % increase of the supply voltage caused a 15 % increase of the input power and in other units it was a 17 % increase of the input power. When the supply voltage decreased about 10 %, the input power decreased by about 19 % in unit I. The input power in other units decreased by about 18 %. It is not a good situation because suction power decreases noticeably with decreasing of the supply voltage. The supply voltage can equal about 207 V when there is a long feeder or when there are other high-power loads on the same phase (an oven, an induction hob etc.). It is a very problematic situation especially for low-power vacuum cleaners and such devices will be used after September 2017. In general, it should be assumed that when the supply voltage decreases to about 0,9U_N the vacuum cleaner will lose about 17 % of the power output. Results of laboratory tests of an exemplary suction unit with a motor with a classic commutator at various supply voltages confirm this statement (Fig. 7).

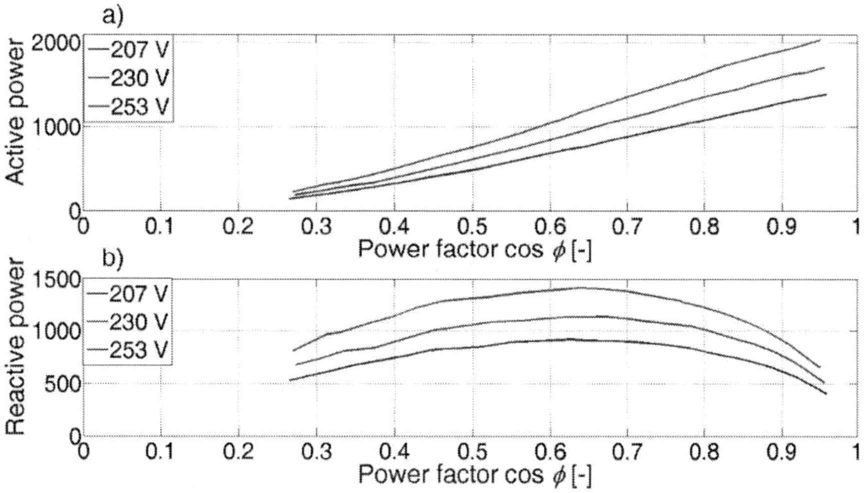

Fig. 6. A dependence of a) active input power P, b) reactive input power Q in the function of power factor cosϕ at U=var

Table 3. Chosen tests results

Voltage	Floor	I	III	IV	V
253 V	Hard	946 W	1507 W	1646 W	2031 W
	Soft	961 W	1418 W	1598 W	1913 W
230 V	Hard	804 W	1256 W	1375 W	1709 W
	Soft	848 W	1187 W	1350 W	1617 W
207 V	Hard	651 W	1077 W	1122 W	1388 W
	Soft	687 W	986 W	1099 W	1310 W

DOI: 10.1051/e3sconf/20171401038

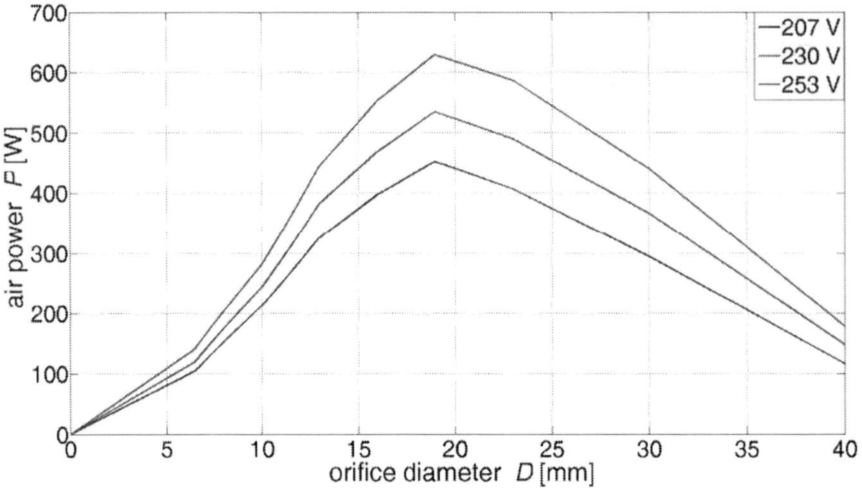

Fig. 7. A dependence of air power *P* of the suction unit in the function of an orifice diameter *D* at *U*=var

The change of the supply voltage has a significant influence on the suction power of the vacuum cleaner. Moreover, the points of maximum air power were obtained at the same orifice diameter regardless of the supply voltage value.

Classic commutator motors were used in all tested units. During operation with decreased supply voltage, there is no possibility to maintain set suction power (performance classes of vacuuming soft and hard floors), as long as manufacturers do not allow such possibility. Among tested vacuum cleaners, only unit III almost disposes declared power at the decreased supply voltage. When new energy efficiency classes appear, used vacuum cleaners should have input power much less than 900 W to satisfy restrictions of highest classes, i.e. A+, A++, A+++. Alternative electric machines with higher efficiency of energy conversion can be one of development directions of improving energy efficiency of vacuum cleaners and also partial compensation of voltage drop by changing control parameters of motor [3-7].

4 Conclusions

The energy efficiency improvement of electric machines used in modern domestic appliances is one of development directions of reducing global energy consumption. It is very important especially in domestic appliances from category IV where electric motor with input power equalling several hundred watts is only one energy receiver. Decreasing of energy consumption by using vacuum cleaners with class A instead of e.g. class D is unnoticeable in terms of one household (it is barely several kilowatts per year). Practice tests show that in real conditions these differences are lower. However, it has significant meaning on a country or world scale because these devices are currently mass-used. Customers will pay all costs which are connected with the introduction of domestic appliances with improved energy efficiency. A purchase of the domestic appliance with A class or higher could be profitable only when lifetime of this device will be extended. Currently, manufacturers of domestic appliances tend to reduce lifetime of their products so the purchase of the energy efficient device is not economically justified. An application of brushless DC motors with electronic commutation on mass-scale in vacuum cleaners is surely one of directions of vacuum cleaners development. However, the purchase of the vacuum cleaner with this kind of motors should guarantee at least over a decade of failure-free operation. The customer should choose the moment of disusing the domestic appliance and not the device itself through limitations introduced by manufacturers which reduce its durability.

Acknowledgements The authors would like to thank the Zelmotor Ltd. Company for providing the test results of the suction unit and chosen devices for practical tests.

References

1. Central Statistical Office of Poland, Poland in Figures, Warsaw (2015)
2. Official Journal of the European Union, Commission regulation (EU) No 666/2013 of 8 July 2013 implementing Directive 2009/125/EC of the European Parliament and of the Council with regard to ecodesing requirements for vacuum cleaners (2013)
3. J. F. Gieras, M. Wing, *Permanent Magnet Motor Technology - Design and Applications, Second Edition, Revised and Expanded*, (Marcel Dekker Inc., New York/ 2002)
4. R. Krishnan, *Permanent Magnet synchronous and Brushless DC Motor Drives*, (CRC Press, New York, 2009)
5. M. A. Rahman, A. Chiba, T. Fukao, IEEE Power Engineering Society General Meeting, Vol. 2, pp.1272-1275 (2004)
6. Ziyuan Huang, Jiancheng Fang, IEEE Transactions on Industrial Electronics, Vol. **63**, Issue:5 (2016)
7. Dong-Hee Lee, Huynh Khac Minh Khoi, Jin-Woo Ahn, International Conference on Electrical Machines and Systems (ICEMS), pp: 1595-1598 (2010)

Labview based testing system for the aim of construction of energy efficient magnetic cores

Wojciech Pluta[1,*], *Roman* Rygał[2], *Marian* Soiński[2], *Jacek* Leszczyński[3]

[1]Faculty of Electrical Engineering, Czestochowa University of Technology, Czestochowa, Poland
[2]Industrial Magnetic Research Center, Magneto Ltd., Czestochowa, Poland
[3]Department of Hydrogen Energy, AGH University of Science and Technology, Krakow, Poland

Abstract. There is increasing demand for more and more energy efficient magnetic cores to be used in modern electrical devices like inverters or smart grid instrumentation (high and medium voltage instrument transformers). The most important magnetic parameters are as high as possible magnetization characteristics ($B_m = f(H_m)$ curve) associated with lowest specific total core losses. As magnetic properties of magnetic cores become worse during production processes it is extremely important to test them in order to improve the production process. In the case of large magnetic cores they cannot be tested in generally used computer systems. Therefore, there is a need for quality testing system for cores of large mass, in the order of several hundred kilogramms.

The paper presents an example of an application for testing the system based on Labview platform for quality checking of large magnetic cores in the order of 1000 kg. Example of testing data for one and three phase core are presented.

1 Introduction

There is increasing demand for more and more energy efficient magnetic cores to be used in modern electrical devices like inverters or smart grid instrumentation (high and medium voltage instrument transformers). The most important magnetic properties of magnetic cores, when assessing their quality - should include characteristics of magnetization ($B_m = f(H_m)$), relative permeability ($\mu_r = f(H_m)$) and specific total loss ($P_S = f(B_m)$).

There are many methods for quality checking of magnetic materials like electrical steel sheets, amorphous or nanocrystalline tapes as for example and they are toroidal sample [1], Epstein frame [2], SST [3]. However, technical challenge is to check properties of readymade magnetic cores. Large mass of cores in the order of dozen of kilograms or several hundred kilograms make impossible to test such cores in generally used universal computerized systems for testing of magnetic materials as for example [4-9]. The systems posses limitation of range of flux density measurements at which measurements are performed with specified accuracy at exactly specified waveform of magnetic flux distortion. In a case of relatively large masses in the order of dozens and more kilograms commercially available computer test systems are not usually used. This is due to the need to provide high-power necessary to magnetization if the core which has big amount of a ferromagnetic material. The

* Corresponding author: w.pluta@gmail.com

DOI: 10.1051/e3sconf/20171401039

measurement range depends on the size of the core, electrical steel tape grade, as well as the power of amplifier of the measuring system. For example, for 2000 W power amplifier and the core weight approx. 25 kg achieved flux density is equal to 1.6 T for core made of HiB tape, or only 1.1 T in a case of core made of non-oriented tape.

Production of cores from electrical steel tapes requires the use of such technological processes as, for example: cutting, packaging of, squeezing, twisting or welding. These processes, as well as undesirable air gaps cause further deterioration in the properties of the final product. Additionally, magnetic properties of cores become more deteriorated as they are made of high quality ferromagnetics as for example from laser scribed electrical steel. Hence, the correct determination of the magnetic properties of cores gives the ability to maintain the high quality of the final product. Therefore it is necessary to test such large magnetic cores in other testing systems. In the case of magnetic cores producers there is an increasing need for quality testing system working in full or half automatic mode. Labview [10] platform gives possibility to complete this setup it in relatively short time and is open for modernizations.

In this paper the example of an application for testing system is presented to quality checking of one or three phases cores with weights up to 1000 kg. The testing system under consideration is made as portable system and enables test at power frequency 50 Hz/60 Hz. The system allows for precise and first of all repeatable determination of magnetic properties of tested magnetic cores.

a)

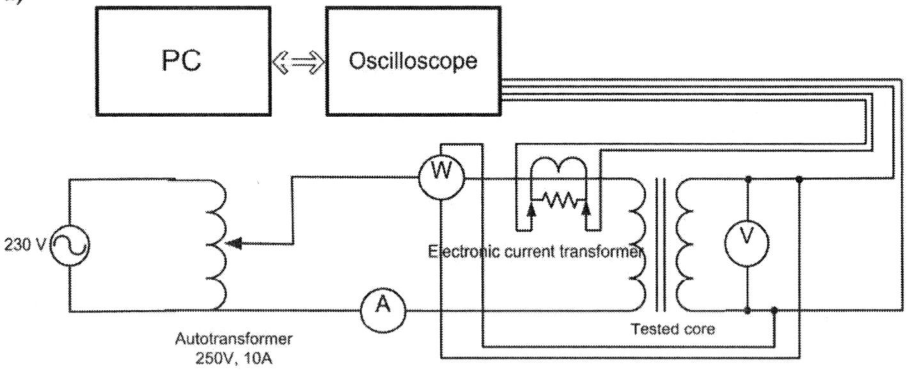

b)

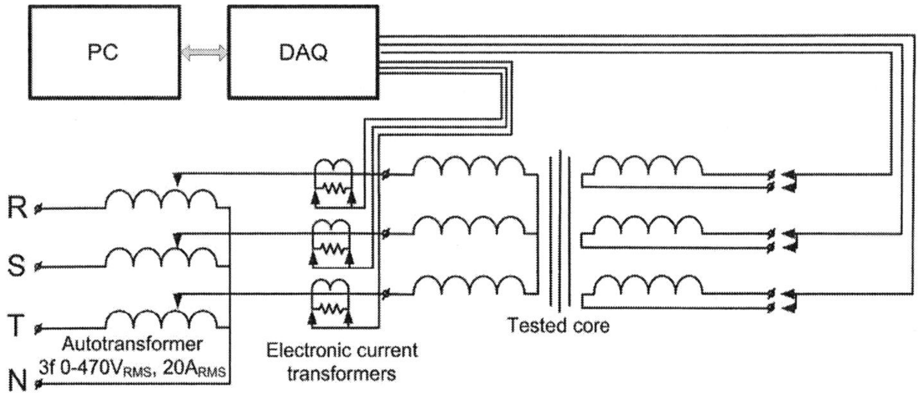

Fig. 1. Schematic diagrams of testing systems using wattmeter method for quality checking of: a) one phase magnetic cores with oscilloscope, b) three phase cores with DAQ card and Labview

2 Measurement system

Study the properties of soft magnetic materials is to determine, among other such parameters as specific total loss P_S [W/kg] measured at different flux density B_m levels and corresponding, effective flux density B_{eff} [T], maximum field strength H_m [A/m], effective field strength H_{eff} [A/m], remanence B_r [T], coercive field strength H_c [A/m], relative permeability (μ_r) or specific apparent loss S_S [VA/kg] and ac hysteresis loops as well. All the parameters are calculated on a base of measured waveforms: primary current $i(t)$ and secondary voltage $u_2(t)$ whose corresponds to flux density waveform $b(t)$ and magnetic field strength $h(t)$ calculated form (1) and (2).

For determination of magnetic properties of large cores most often wattmeter method is used (Fig. 1a) [11, 12]. This method was considered until recently as more labor intensive and less accurate. The paper presents the application of this method, by using Labview. Also some examples of test results for single and three-phase core are presented later in the paper. The schematic diagram of this system is shown in Fig. 1b).

The measuring system shown in Fig. 1a) is composed of wattmeter, ammeter A, voltmeter V, electronic current transformer and oscilloscope used to control the waveform of magnetic flux density. In that measurement process is also useful to use a computer for data acquisition, which reduces the time of acquisition data and processing results. Power supply of system from Fig. 1a) directly from the network preserve the waveform factor to be within ± 1% as required by standards [1-3] and only at high magnetic field strength it goes to higher values. However, as shown in [12] discrepancies due to waveform factor are not high and can be omitted when core are tested for industrial quality checking purpose.

The system presented in Fig. 1 b) is very similar to the one presented above in Fig. 1a). The main difference is the use of DAQ card instead of many devices such as three wattmeters, ammeters and voltmeters plus minimum three inputs oscilloscope. In presented testing system there was used NI USB-6251 BNC M-series [13]. The DAQ card posses eight differential BNC analogue inputs each of 16 bits resolution at sampling rate of 166 kS/s when used six channels. This card has also twenty four digital I/O channels and some of them are planned to use for measurement automation purposes as for example to change shunt resistors or voltage divider.

The measurement system is equipped with a device that prevents engagement of the connection to the test core when the autotransformer is not in the zero position. It enables to supply one phase or three phase test core.

An ac sinusoidal current inject into the excitation winding which induces a magnetic field H in the sample. The value of current is obtained from electronic current transformer. The value of applied magnetic field H is directly proportional to the current according to line integral form of Ampere's law.

$$h(t) = \frac{u_H(t) z_{ct}\, z_1}{R_b l_{av}} \tag{1}$$

where: l_{av} is the effective magnetic path length of the test core, u_H is the voltage on R_b resistance, R_b is the burn resistance of an electronic current transformer, z_1 is the total number of turns in the excitation winding, z_{ct} is the number of turns of an electronic current transformer.

As autotransformer does not provide isolation from electrical network the use of an electronic current transformer allows electric separation from electrical network. To obtain the induced flux density B the induced voltage is integrated over time. This results from induction law and is described by (2):

DOI: 10.1051/e3sconf/20171401039

$$b(t) = \frac{1}{s_{Fe} z_2} \int u_B(t) dt \tag{2}$$

where: s_{Fe} is the cross sectional area of the test specimen on which is wound a secondary winding (test winding), u_B is the induced voltage in measurement winding (secondary site), z_2 is the total number of turns in the measurement winding.

Set out by measurement way instantaneous values of magnetic field intensity (1) and of magnetic flux density (2) allow for determination of a number of parameters characterizing the magnetic properties of the tested core. These two waveforms contain all the information about the magnetic properties of interest. If they are plotted in an X-Y plane a hysteresis loop is obtained (Fig. 2a). The area of the loop is directly proportional to the active power loss dissipated in the material and the specific total loss can be calculated from (3).

$$P = \frac{1}{\gamma s_{Fe} z_2 T} \int_0^T h(t) u_B(t) dt \tag{3}$$

where: t is the time, T is the cycle of measured waveform, γ is the mass density of material the tested core is made of.

The apparent power loss also is calculated directly from current and voltage waveforms but taking into account the turn ratio. The apparent power loss is a measure of total power, which needs to be delivered to the core in order to magnetise it to a required level.

Using the system as in Fig. 1b) can be determined such dependences as the dynamic magnetization $B_m = f(H_m)$ or the relative permeability characteristics $\mu_r = f(H_m)$ calculated from the peak values of $b(t)$ and $h(t)$ waveforms. Examples of hysteresis loops for stacked and welded core, weighing approx. 23 kg, made of grainoriented tape shown in Fig. 2.

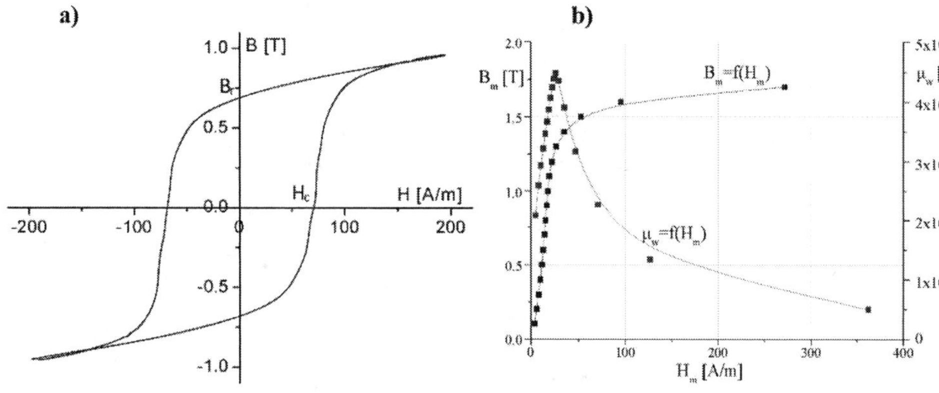

Fig. 2. Specific experimental examples of the characteristics measured at frequency of 50 Hz: a) the hysteresis loop for the of magnetic flux density approx. 1.0 T determined experimentally for the core stacked welded with a mass of approx. 23 kg, made of grainoriented tape grade M130-27S and b) the characteristics of $B_m = f(H_m)$ and $\mu_r = f(H_m)$ grainoriented tape grade M130-27S tested in SST (own measurements)

As it is shown later wattmeter method based on Labview and with use of DAQ card, allows easier and more accurate measurement of magnetic properties of the cores of relatively large masses.

3 Labview implementation of the measuring system

Labview is a programming environment based on a graphical user interface [10, 14] and is a set of programs created by National Instruments, serves as a development environment, containing an extensive set of routines optimized for fast processing of collecting and analyzing data. What distinguishes the programming of Labview in other programming environments is a way to write code. Labview use graphical interface, wherein the program code is created similarly to build electronic circuits as in Pspice [15] by combining elements in such a way as to implement a predetermined function. There is also the possibility of inserting routines written in other programming languages such as Assembler, however, the concern is usually very specialized calculations. The program is executed in accordance with the concept of Data Flow, wherein the lead wires to the function data, and then receive results that have to carry out the following functions [10]. This approach to programming is very intuitive for electronics engineers and thanks to a well-developed system of aid does not need to remember the names of functions and the order and types of their arguments. National Instruments also provides a dedicated measuring devices connected to the computer (controller) allow for automated data acquisition (usually measurements of electrical signals) and outputting output signals.

The application created in Labview consists of two main components. The first is the front panel constituting the interface input and output. While the second element is a window Block Diagram, in which is stored the source code (in graphical form) program. The user interface (front panel) for the measurement, analysis and storage of electromagnetic is shown in Fig. 3.

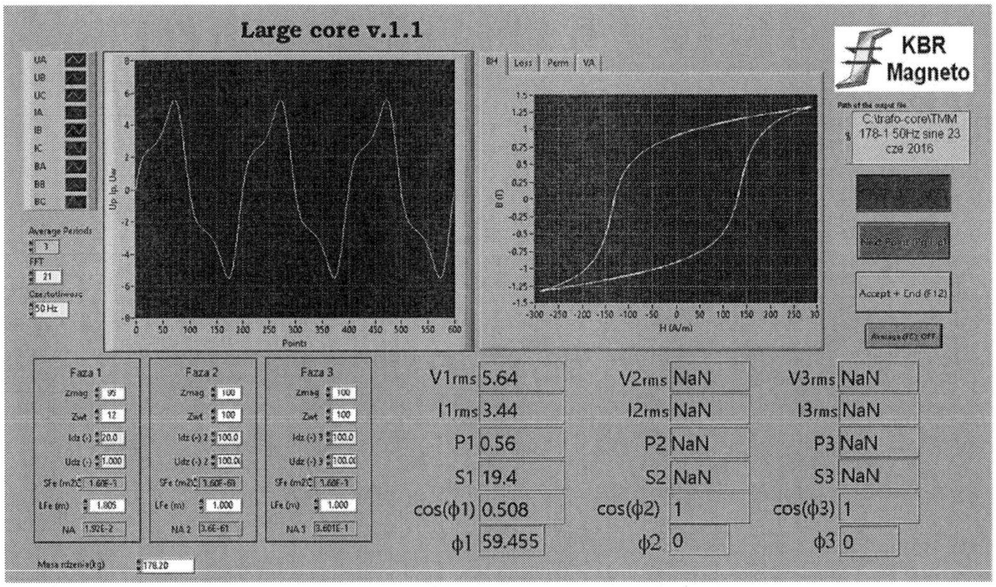

Fig. 3. Front panel of the application

The block diagram and example of code is presented in Fig. 4.

DOI: 10.1051/e3sconf/20171401039

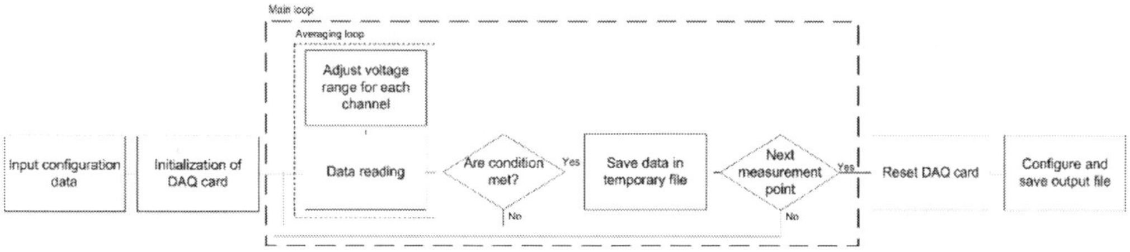

Fig. 4. Block diagram of presented system

Source code of developed in this environment measurement application is organized usually in following sections of program. They are configuration section of communication with a used device, section of data acquisition and on-line calculation and last section of data saving. The configuration part consists of creation of metrics of measurements and DAQ card initialization. The file of metrics of measurements consists of date and time of measurement, user ID and path the file saved at the first time. Also at part of the code is loaded data of a magnetic core to be tested including such basic data as core mass, average magnetic flux length, numbers of turns of magnetizing and secondary coils, core cross section on which the secondary coil was wounded and frequency. In core configuration file also are included scaling data of current transformers and voltage dividers. The initialization of a DAQ card consists of choosing a proper card, channels definition and timing setup. In the next section there are two loops which one consists of data reading and adjusting voltage range for each channel which is also used for averaging purposes. The second consists of mentioned averaging loop and calculation and presentation of chosen quantities. In this main loop also temporary data saving is included. The last section resets DAQ card and configure and in the long run save output file.

For precise measurement can be used data acquisition card (DAC) with simultaneous sampling of measured signals. Such DAQ cards are rather very expensive if taken into account accuracy, sampling rate or number of voltage input ranges (increasing significantly accuracy of measurement). In cheaper cards commonly one analogue-to-digital converter is used. As an effect the signals seen by the computer are artificially delayed. This delay can be significant for accuracy in loss calculation. The necessity of precise measurement results from a definition of active power loss P and inductive load that shows magnetic core when tested. The active power loss is equal to:

$$P = UI\cos(\varphi) \tag{4}$$

where: U is the RMS value of voltage induced in secondary winding (proportional to the rate of change of flux density), I is the RMS value of magnetising current supplied to the primary winding (proportional to the magnetic field) and φ is the phase shift between I and U.

The angle φ takes higher values as grows magnetizing current (magnetic field strength) and at the same time increases error in active power loss calculation (due to the trigonometric function cosine). Taking into account that when the system is used to test a magnetic core for three phase transformer there is a need to use minimum 6 channels for measurement and the error in measurement can be significant. For this reason software trigerring was used. The software decides by comparison to one channel at which point each channel should be triggered. The software trigerring applied in this system was succesfully used in [16]. The precize trigerring is particullarly important as averaging of input signal and the aim of the use is to improve accuracy of the measurements [17].

RMS and average voltages are calculated in the conventional manner, the Fourier transform is used for harmonic analysis and THD calculation and values for coercivity and remanence are calculated by linear interpolation of the *BH* loop appropriate neighbouring points. Data processing can only be carried out once and complete cycle of the magnetizing

 DOI: 10.1051/e3sconf/20171401039

waveform is acquired. Therefore, the time of each iteration and thus the convergence time depend on the the fact whether are measured 3-phases or 1-phase magnetic core. The resulting file consists of metricsof the file, configuration data, calculated results such as maximum, average and RMS values of flux density and magnetic field strength, relative permeability, coercive field, remanence, specific total loss and apparent loss. Also collected waveforms of magnetizing currents, secondary voltages, flux densities and magnetic field strengths for each phase are stored in the resulting files. Name of the resulting file consists of core name, operators name and measurements time. Such name allow easy finding of the files in the future.

4 Results

Three cores for test were chosen: two stacked one phase, welded and bolted cores made of non-oriented electrical steel tape of grade M400-50A one of 28 kg in mass (Fig. 5 a) and one of 220 kg (Fig. 5 b). It is important to mention that the core is designed for choke and it has an air-gap of 2 mm. The third core was made as three phase core of 165 kg in mass and was made as cut core and of grain-oriented electrical steel tape of grade M130-30S (Fig. 5 c).

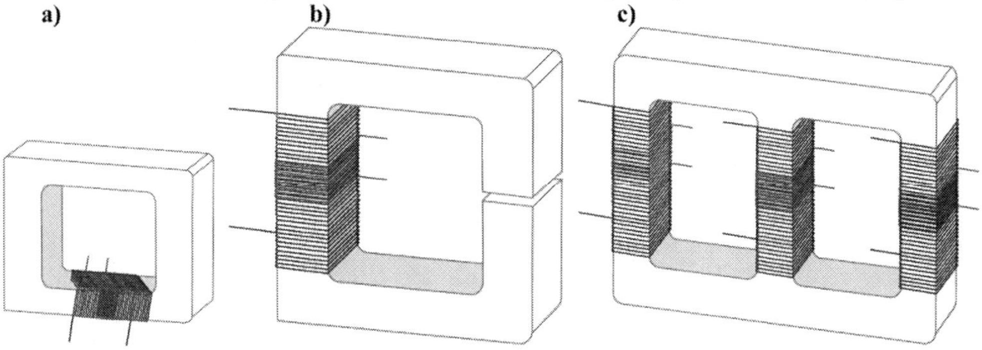

Fig. 5. Schematic drawing of tested cores: a) special transformer core(28 kg), b) choke core with air-gap (220 kg), c) three phase core (165 kg)

As presented in the paper [12] the computer system with power amplifier of 2000 W was able to measure properties of a stacked core with distributed air gap of 15 kg (similar to the one presented in Fig. 5a) only up to flux density of 1.6 T. By using wattmeter method was possible to determine the magnetic properties of 28 kg core (Fig. 5a) up to flux density of 1.9 T and the larger flux density was not measured due to weak control on very fast increase of magnetizing current. Thus was improved by the system described in the in the paper (Fig. 6 and 7).

Non-linear characteristics of the magnetization of the ferromagnetic core introduces a significant distortion of the magnetizing current and of the induced voltage on the secondary side even for sinusoidal voltage source. It has been found that with the increase of magnetic flux density from 1.0 T to 1.5 T, there is more than 3-fold increase in THD factor of the secondary voltage [12]. The accuracy of the measurement of the power loss is strongly dependent on the shape of the magnetic flux, in turn, proportional to the secondary voltage [1-3, 5, 11].

It is worth to mention that in general measurements using wattmeter method can be exposed to considerable error. This is mainly due to measure of the maximum magnetic field strength by ammeter and then multiply it simply by $\sqrt{2}$. The error of determining maximum magnetic field strength by ammeter can be as high as 50% in relation to measured waveform of the magnetic field strength. The error will be different for different construction and different material of magnetic core. It was also shown that the method of using the measurement of waveform *h(t)* allows more precise measurement of the magnetic properties of cores through more precise definition of H_m.

It should be noted that during the magnetization deforms not only the magnetizing current (as discussed above), but also the induced magnetic flux and induced the secondary voltage. Changes waveform factor FF corresponds to the nature of changes in harmonic distortion THD of tested cores and the increase in THD is translating into an increase in loss of higher harmonics in the total loss [5, 12]. In the existing standards [1-3], it is assumed that the aspect ratio of the magnetic flux not to deviate from 1.1107 (i.e. for a sinusoid) more than ± 1%. As presented [12] it is possible for stacked core of about 30 kg to take measurements to the value of 1.2 T within the standard requirement.

Fig. 6 shows the characteristics of magnetization $B_m = f(H_m)$, and relative permeability $\mu_r = f(H_m)$ determined by computer system [9] for sample core (Fig. 5b) V made of electrical steel tape grade M400-50A and by the presented wattmeter method with DAQ card (Fig. 1 b) of made choke core (Fig. 5 b).

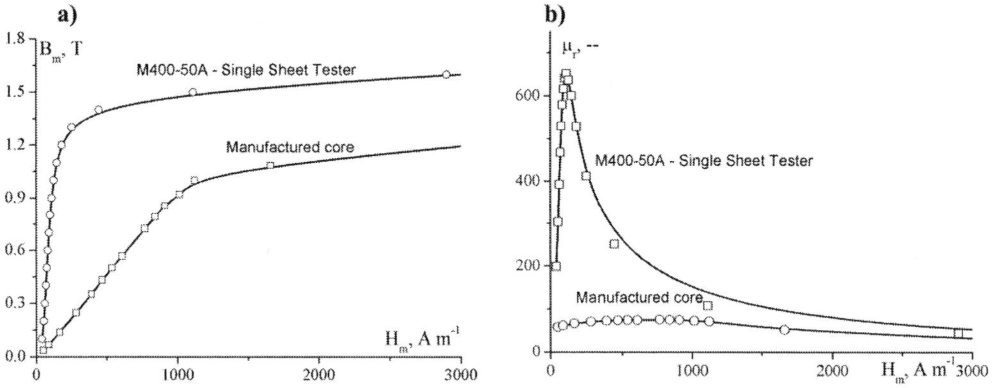

Fig. 6. Research results of sample core of 220 kg made of electrical steel tape grade M400-50A determined by computer system and wattmeter method with DAQ card: a) magnetisation characteristic $B_m = f(H_m)$, b) relative permeability $\mu_r = f(H_m)$

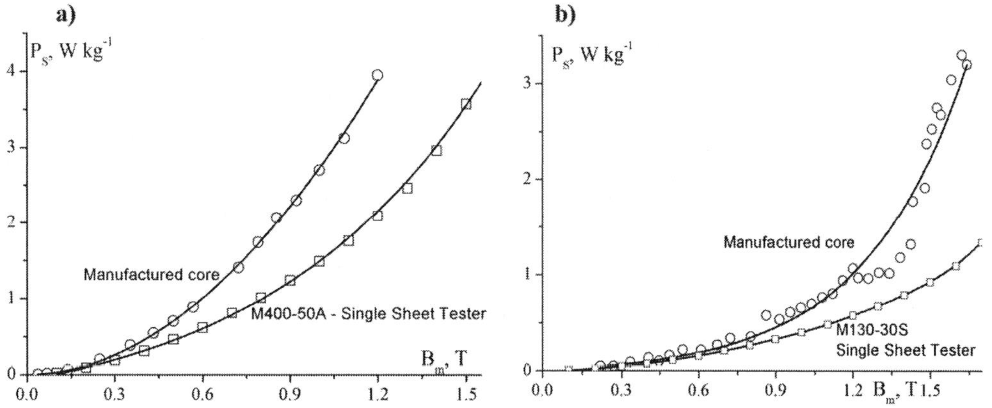

Fig. 7. Comparison of specific total loss $P_S = f(B_m)$ measured using computer system with SST [3] and wattmeter method with DAQ card for: a) one phase choke core of 220 kg made of electrical steel tape grade M400-50A and b) special construction of three phase transformer core of 165 kg made of electrical steel tape grade M130-30S

The presented in Fig. 6 difference between characteristics of electrical steel core sample results from airgap of choke core (Fig. 5b). Hence, first part of magnetisation characteristic of choke core is straight line and relative permeability is very low. The test of 220 kg choke core was performed up to magnetic field strength of 3000 A/m and relatively low flux density is considered as satisfactory. Using this advantage Fig. 7 shows the results of measurements of the core loss of one and three phase cores (Fig. 5 b), to a value above 1.5 T.

The data presented in Fig. 7 confirms usefulness of wattmeter method with DAQ card for measuring relatively large magnetic cores. Determination of loss is possible for such core at flux density values 1.4 or 1.6 T (respectively core screw or welded) and up to 1.9 T for smaller cores (like in Fig. 5 a).

Wattmeter method with DAQ card is relatively simple, but is deprived of advantage of forming shape of magnetic flux. The computer system is equipped with an adaptive feedback loop that allows the formation of magnetic flux in the core ensuring that it is sinusoidal in nature - in accordance with the requirements of the standard [1-3]. Wattmeter method with DAQ card is simpler, but not possessing this advantage. In this method, the magnetizing current flows directly from the network via an autotransformer - thus has no possibility to influence the induced magnetic flux in the core. Therefore, it should be count the possibility that for higher values of the flux density B its shape will differ from the standard. However, presented results confirm the high compatibility of the two measurement methods and hence the usefulness of the wattmeter method based on Labview, to assess the quality of cores with very large masses of the order of 1000 kg.

5 Conclusion

There is increasing demand for low cost and efficient transformers. At the same time the magnetic cores of transformers are subjected to ever more demanding magnetisation conditions. The intermediate testing allows to gather knowledge about the production process and hence to improve the quality of magnetic cores and finally to increase the efficiency of end product that is transformers.

Development in recent years of environments to support measurement allows relatively easy to create new applications of testing systems. In this work is presented one of such testing systems based on Labview. Selecting the Labview environment was mainly due to an extensive database routines and drivers dedicated acquisition and analysis of measurement data as well as considerations of programming easy for improvement and modernization. Thus it provides a relatively easy adjustment of the measuring system for current needs.

There are presented also test result of some chosen magnetic cores made of electrical steel tapes of masses in the order of 30 kg to 250 kg. It is believed that the system is capable to test magnetic cores and it is estimated that the mass could be so large as 1000 kg. The cores differ not only by mass but also by different construction as stacked and welded and stacked and bolted design for use in one and three phase systems and also made of non- and grain-oriented electrical steel tapes. There were measured basic magnetic properties and every test point is recorded and the control criteria are calculated. It allows real time recording of $b(t)$ and $h(t)$ waveforms which allow to calculate all parameters. The measurement is low cost and can be performed in semiautomatic mode of a series of complex measurements. Presented results of magnetic characteristics of magnetizing $B_m = f(H_m)$, relative permeability $\mu_r = f(H_m)$ and specific total loss $P_S = f(B_m)$ indicate the usefulness of the proposed method for industrial testing of large magnetic cores.

It is acknowledged that this work has been carried out within the project grant "Industrial research a new type of magnetic cores made of amorphous and nanocrystalline strips, thin magnetic sheets and composite materials operating in higher frequencies", and was supported by the National Centre for Research and Development under European Regional Development Fund in the frame of European Smart Growth Funds, under contact No. POIR.01.01.01-00-0306/15-00. The support is gratefully acknowledged.

References

1. IEC 60404-6:2003 Magnetic materials - Part 6: Methods of measurement of the magnetic properties of magnetically soft metallic and powder materials at frequencies in the range 20 Hz to 200 kHz by the use of ring specimens
2. PN-EN 60404-2:2003/A1:2008 Materiały magnetyczne – Część 2: Metody pomiaru własności magnetycznych stalowych blach i taśm elektrotechnicznych przy użyciu aparatu Epsteina
3. IEC 404-3:1999; Magnetic materials. Methods of measurements of soft magnetic properties of electrical steel and type with the use of single sheet tester
4. Product Catalogue, Electrical Steel Tester C 510, http://www.brockhaus.com Brockhaus Messtechnik
5. W. Pluta, Measurements of magnetic properties of electrical steel sheets for the aim of loss separation, JEE, **12**, (2011), pp. 122-127
6. F. Fiorillo, *Characterization and measurement of magnetic materials*, Academic Press (2005)
7. S. Tumanski *Handbook of magnetic measurements*, CRC Press, 2011
8. W. Pluta, R. Rygał, M. Soiński: Nowoczesne techniki określania własności materiałów magnetycznie miękkich. WE, nr **8**, (1999), pp.406-412 (in Polish)
9. J. Chojancki, P. Liszewski, M. Soinski, R. Rygał, W. Pluta, S. Zurek, P. Pinkosz, Harmonic distortion of magnetizing current in combined wound toroidal cores, *IEEE Trans. on Magnetics*, vol. 40, 2008 nr 11, pp. 3819-3819
10. National Instruments, LabVIEW, http://www.ni.com/labview/
11. M. Nałęcz, J. Jaworski, *Pomiary magnetyczne*, PWN Warszawa (1965) (in Polish)
12. W.A. Pluta, M. Soiński: Warunki pomiaru własności magnetycznych rdzeni ferromagnetycznych metoda watomierzową. ER **R. 90** nr 1 (2014), pp.217-220 (in Polish)
13. DAQ M Series User Manual NI 622x, NI 625x, and NI 628x Devices, www.ni.com
14. Nesimi Ertugrul: *Electrical Circuits and Machines Laboratory with LabVIEW* Prentice Hall Professional (2002)
15. A. Król, J. Moczko, *Spice symulacja i optymalizacja układów elektronicznych*, Wydawnictwo NAKOM (1998)
16. S. Zurek, T. Meydan: Precise post-triggering and correction of interchannel delay for alternating and two-dimensional measurements of magnetic properties. 8th International Workshop on 1&2-Dimensional Magnetic Measurement and testing 27-28.09.2004, Ghent, Belgium
17. V. Gorican, et al., "The measurement of power losses at high magnetic field densities or at small cross-section of test specimen using the averaging", *JMMM*, vol. **215-216** (2000), pp. 393-695

Cost, resources, and energy efficiency of additive manufacturing

Piotr Dudek[1,*] and *Krzysztof* Zagórski[1]

[1]AGH University of Science and Technology, Faculty of Mechanical Engineering and Robotics, al. A. Mickiewicza 30, 30-059 Kraków, Poland

Abstract Additive manufacturing (AM) is the process of joining materials to make objects from Computer Aided Design (CAD) model data, usually layer upon layer, as opposed to using subtractive manufacturing methods. The use of rapid prototyping technologies has increased significantly in recent years. These new techniques, while still evolving, are projected to exert a profound impact on manufacturing. They can reduce energy use and time to market and offer industry new design flexibility. We include a brief study on the cost and energy efficiency of selected methods of additive manufacturing compared to traditional methods of manufacturing parts. One common claim is that 3D printers are more energy-efficient than other manufacturing technologies. We present energy efficiency and time requirements for producing a typical mechanical part and a very complicated element, using both traditional manufacturing and rapid prototyping methods. This paper represents an attempt to answer the questions of when 3D printing can be used efficiently and of choosing the appropriate technology on the basis of batch size, element size, complexity, and material requirements.

1 Introduction

Rapid prototyping represents a relatively novel technology in manufacturing, associated with potentially strong stimuli for sustainable development. As part of digital fabrication technology, Additive Manufacturing (AM) systems are able to manufacture three-dimensional components and products directly from raw material and 3D design data. Synonyms used to refer to this technology include *additive techniques*, *layered manufacturing*, *rapid prototyping*, *digital manufacturing*, and *solid freeform fabrication* [1]. The layer-by-layer operating process of these systems does not require the use of tools, moulds or dies. The main feature of AM methods is their flexibility in terms of the design of a product, making these methods responsive to almost any shape. Adoption of these technologies announces a future in which value chains are shorter, more localised, and more collaborative, while offering significant sustainability benefits. Observers of technology speculate that Additive Manufacturing will have a profound economic impact on the manufacturing sector and, indeed, on the society as a whole [2, 3].

* Corresponding author: pdudek@agh.edu.pl

Compared to conventional manufacturing, such as machining, forging, injection moulding, casting, and other processes, AM has certain advantages. Firstly, AM can reduce environmental impact due to its avoidance of the tools, dies, moulds, and material waste associated with conventional manufacturing. Secondly, and most importantly for designers, this technology enables the manufacture of parts with novel geometric designs that would be difficult or impossible to achieve using conventional manufacturing processes. Thirdly, novel geometries enabled by AM technologies can also lead to environmental benefits or improved performance in a final product containing elements produced by AM. For example, AM is used by aircraft companies to produce parts that weigh less, which can lead in turn to greater aircraft fuel efficiency. General Electric uses 3D printing technology to produce fuel nozzles for the next-generation LEAP jet engine which are five times more durable than the previous model. 3D printing enabled engineers to design them as a single part rather than 20 individual parts, reducing the number of brazes and welds that would have been necessary using traditional methods [4, 5].

The first environmental studies on AM processes proposed the potential for gains in terms of environmental impact compared with more traditional processes, such as machining [6-8].

Table 1. Classification of AM technology [9]

Technology	Description	Related technologies
Vat photopolymerisation	Liquid photopolymer in a vat is selectively cured by light-activated polymerisation.	Stereolithography (SLA), Digital Light Processing (DLP)
Binder jetting	The binder jetting process uses two materials: a powder-based material and a binder. A liquid bonding agent is selectively deposited to join powder materials.	3DP (3D Printing)
Material jetting	Material jetting creates objects in a method similar to a two-dimensional inkjet printer. Droplets of build material are selectively deposited.	Multi-jet modelling (MJM), PolyJet, Droplet on Demand (DoD)
Material extrusion	Material is drawn through a nozzle, where it is heated and then deposited layer by layer. The nozzle can move horizontally, and a platform moves up and down vertically after each new layer is deposited.	Fused deposition modelling (FDM)
Sheet lamination	Sheets of material are bonded to form an object.	Laminated object manufacturing (LOM), ultrasonic consolidation (UC)
Powder bed fusion	Thermal energy selectively fuses regions of a powder bed.	Electron beam melting (EBM), selective laser sintering (SLS), selective metal melting (SLM), direct metal laser sintering (DMLS)
Directed energy deposition	Focussed thermal energy is used to fuse materials by melting as the material is being deposited.	Laser Engineering Net Shaping (LENS)

2 Additive Manufacturing technologies

Although the term '3D Printing' is used by the media as a synonym for all additive manufacturing processes, there are actually many different processes which vary in their method of layer manufacturing. These processes differ depending on the material and technology used, and can be classified in a few categories (Table 1).

The AM processes can also be classified based on the state of the starting material used [1].

3 Environmental implications

Since the introduction of AM in the 1980s, the benefits of producing small quantities of complex parts rapidly have been well understood among manufacturing circles. Despite this, the industry is just beginning to understand exactly how transformative the technology will be within the future of manufacturing [10 - 12].

Today, the most established use of additive manufacturing is making prototypes. 3D printers can quickly produce complex pieces in small quantities for the purpose of evaluating a design. Mould makers and plastics processors frequently use additive equipment in precisely this way – to test a part before the tooling is finalised and before full-scale production begins. Additive manufacturing and traditional manufacturing face different trade-offs, with each process likely to play a role in the deployment of manufacturing capabilities. AM has the potential to vastly accelerate innovation, compress supply chains, minimise materials and energy usage, and reduce waste [5, 13, 14].

The following are the some of the most significant benefits of additive manufacturing, highlighting where the technology is taking us today:

- Less waste: building objects up layer by layer, instead of using traditional machining processes that cut away material, can reduce material needs and costs by up to 90%. AM can also reduce the 'cradle-to-grave' environmental footprints of component manufacturing through avoidance of the tools, dies, and material scraps associated with conventional machining processes.
- Lower energy consumption: AM saves energy by eliminating production steps, using substantially less material, enabling reuse of by-products, and producing lighter products.
- Innovation and design freedom: in traditional machining, every complex detail has to be produced through additional tool paths or other extra steps. By comparison, the complexity of a part in additive manufacturing adds very few extra considerations, or none. A part can be made purely for functionality, without manufacturing-related constraints. Design engineers have only started to explore the implications of this.
- Short lead times: Items can be fabricated as soon as the 3D digital description of the part has been created, eliminating the need for expensive and time-consuming part tooling and prototype fabrication. As soon as the part is printed, engineers can begin testing its properties, instead of waiting weeks or months for a prototype or part to come in.
- Light weight: With the elimination of tooling and the ability to create complex shapes, AM enables the design of parts that can often be made to the same functional specifications as conventional parts, but with less material.

Compared to other methods, additive manufacturing can produce parts quickly, because it is not necessary to perform all tasks with all the necessary technology for preparing for manufacture, and it is possible to produce parts with complicated shapes. The costs of parts are readily predictable. However, it must be mentioned that in these technologies, though

DOI: 10.1051/e3sconf/20171401040

they produce less waste, raw material is more expensive than material used in conventional types of manufacturing.

4 Energy usage in the SLS process

The Department of Manufacturing Systems, Faculty of Mechanical Engineering and Robotics, AGH, conducted a study to determine power usage in the production of parts using selective laser sintering (SLS) technology. To do this, a platform was prepared, including 75 parts for a total height of 92 mm for an EOS Formiga P100 (Fig. 1). Parts were printed using polyamide PA2200. The measured build time was 12.5 hours plus an additional 3 hours of preheating, with the temperature set at 172°C. Cooling time was 12.5 hours.

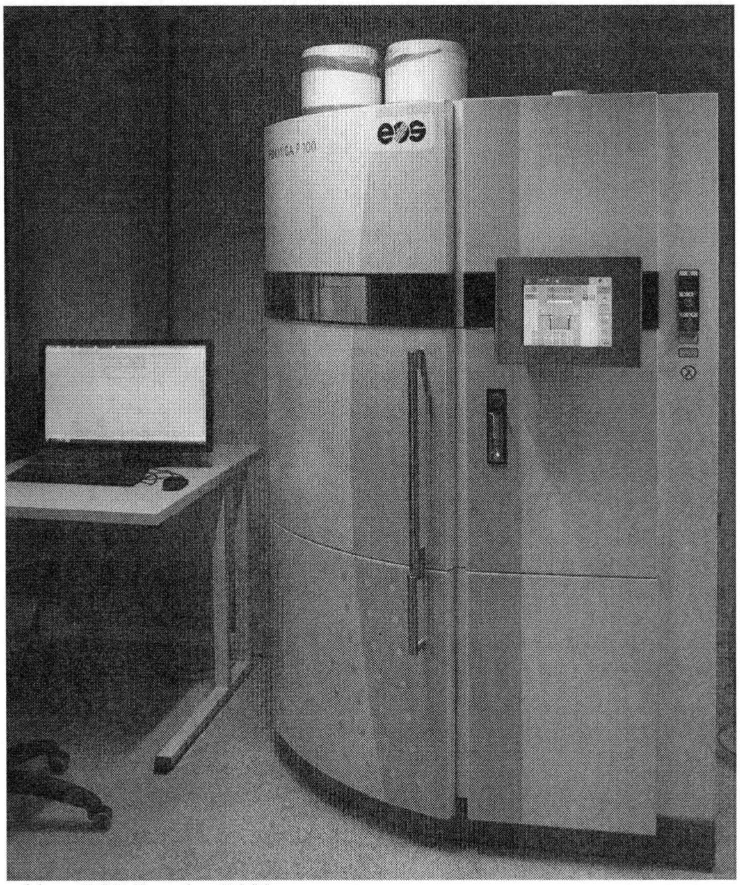

Fig 1. SLS machine: EOS Formiga P100

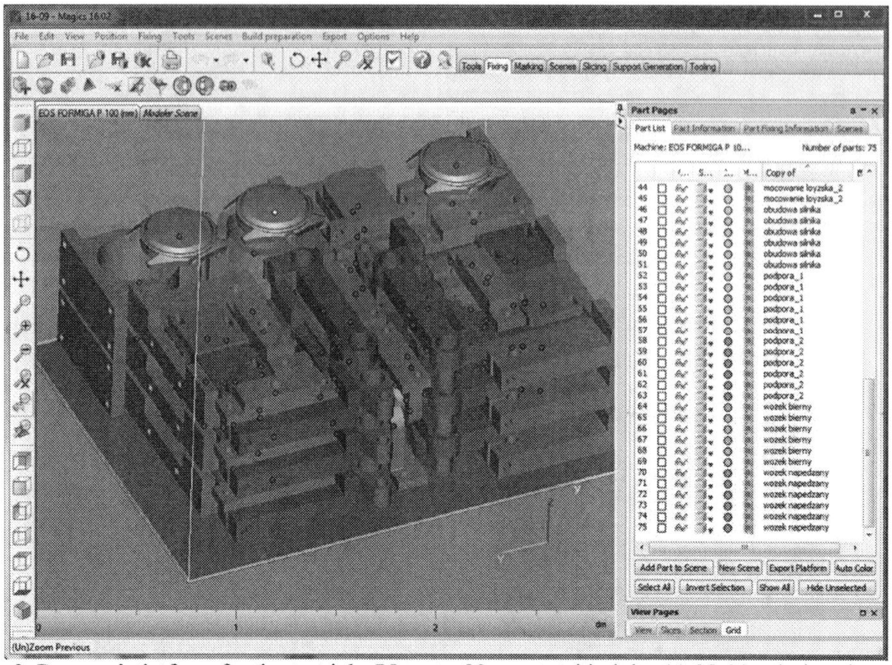

Fig. 2. Prepared platform for the test job: 75 parts, 92 mm total height, 11:38:00 printing time

Before printing was started on the test job, the work and removal chambers, heaters and air channels were cleaned and the optics and pyrometer were checked. Parts were positioned using Magics software (Fig. 2), and, after the arrangement was completed, the parts were merged and exported as one STL file. The volume of this part was 1,196,500 mm^3; it was located in a bin with the dimensions 200×250×92 mm. After some calculation it can be determined that the useful portion of this volume is approximately 26%.

After printing and waiting 12.5 hours for cooling of the printed parts, the machine was cleaned and the building bin removed for post-processing of the parts. The parts were removed from the bin and cleaned with a sandblasting machine. The rest of the powder was sieved and stored for reuse in future jobs. It should be noted that this material should be mixed (new powder with used; the best ratio is 50/50) for stabilising the printing process and to minimise waste. Some types of material cannot be reused, for example polyamide PA12 filled with aluminium (Alumide); others, for example PrimeCast or powder with a polystyrene base, can be reused 100% after sieving. Because of this, the cost of the material for printing depends not only on the volume of printing and cost of material, but also on the ability to reuse powder. The best ratio is associated, of course, with material based on polystyrene, but this material has poor mechanical and thermal properties, and is used only to create patterns for investment casting, master patterns for vacuum casting, or production of lost patterns for the plaster and ceramic shell casting processes. After certain printing processes, leftover material cannot be reused, and thus becomes waste. In analysing actual printing jobs, it can be observed that over a long time period, waste for SLS averages 20–40% for each job. The amount of material waste is affected by a number of factors, including the size of the printed piece, the preparation of the machines, packing density when preparing to print, etc. On every job, some material is removed by a vacuum cleaner when cleaning and preparing the machine for the next job; this quantity is the same for large and small jobs. For less waste, a good solution is to pack dense parts, and, if possible, to reduce the height of the printed piece. This is an arduous process, even with the use of automatic placement tools, but it can significantly reduce waste. Since the SLS process for plastic parts needs no support structure parts, parts can be placed one over another. This process is characterised by relatively little waste compared to other AM technologies.

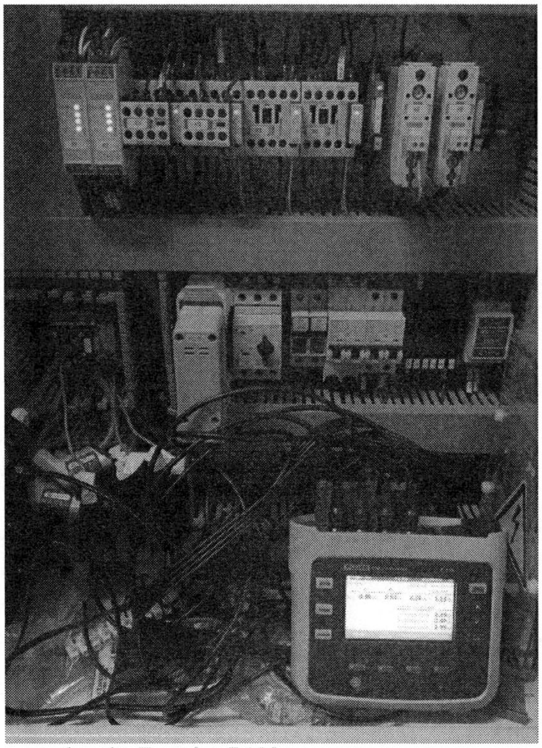

Fig. 3. Fluke logger connected to the Formiga P100

In observing the print process, it would seem that the process is energy-intensive. The machine must heat the working and removal chambers and provide power to the motors, laser, and electronic circuits, and the printing process itself is time-consuming. To examine the power usage of the Formiga P100, the Fluke 1736 energy logger was used (Fig. 3). Because the laser circuit, electronics, and heaters use different phases, it is possible to measure energy usage for these different elements.

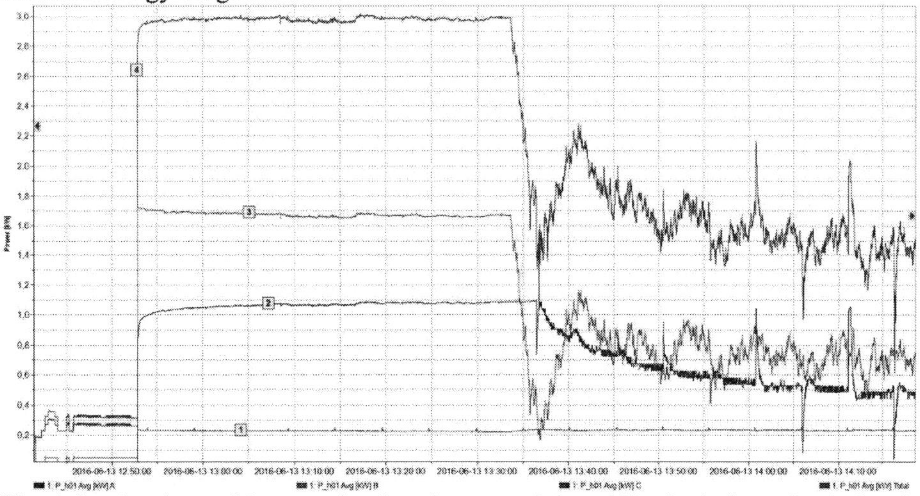

Fig. 4. Starting the machine to preheating phase – total time, 180 min. 1– laser circuit; 2 – electronics, motors, and removal chamber; 3 – heater circuit; 4 – total power

Fig. 4 shows the energy usage of the Formiga P100 from starting the machine to the preheating stage. The machine was switched on after it had cooled to room temperature. After the machine starts and become operable the process of distributing the powder on the working bed begins. Next, the process of preheating is initiated, seen as a rise in the power during phases 2 (removal chamber) and 3 (working chamber). Total power for preheating does not exceed 3 kW. After 40 minutes of heating, the power needed for this operation decreases, and for 140 min the power is regulated as the temperature in the chambers slowly increases.

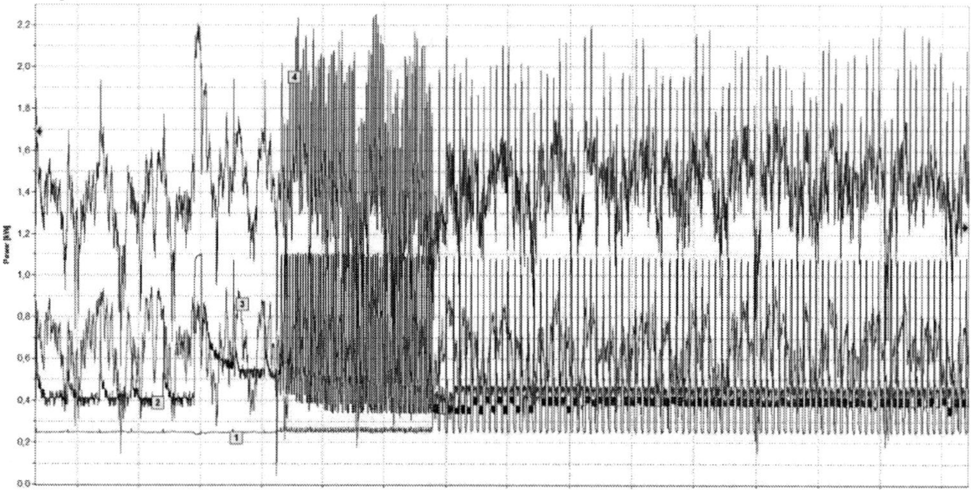

Fig. 5. Preheating phase – start of job. 1– laser circuit; 2 – electronics, motors, and removal chamber; 3 – heater circuit; 4 – total power

After the preheating phase, the process and removal chambers are heated up to working temperature (approximately 12 minutes); then, printing is started. The first 60 layers are empty, so the laser is idle; after that, the laser starts working (the peaks in line 1, Fig. 5), and the chambers are also heated to the desired working temperature; due to the new applied layer of powder, the temperature must be raised. Printing time is 12.5 hours, and the process uses an average of 1.4 kW of power.

The last 40 layers are also empty, and after this the cooling phase begins. The laser, heaters, and motors are switched off, and only the electronics and the nitrogen generator (which needs no power) work; here, an average of 0.5 kW was used. It can be seen that heating is the most energy-intensive part of the SLS process, and uses 50% of the required power. The total energy used for the production of 75 parts was approximately 29 kWh, or 104.4 MJ. It must be mentioned that this represents the energy of the printing machine; the power used for the compressor or machines for post-processing was not measured.

5 Conclusions

The RP industry continues to expand. Current research on additive manufacturing costs reveals that this technology is cost-effective for manufacturing small batches. Due to the complexities of measuring additive manufacturing costs, the current study is limited to the scope of energy effectiveness and efficiency.

Currently, research also reveals that material costs constitute a major proportion of the cost of a product produced using additive manufacturing, but a number of factors, including build orientation, envelope utilisation, build time, energy consumption, product design, and labour, complicate attempts to minimise the cost of additive manufacturing. Even a simple change in part orientation and the part's position in the building chamber can significantly reduce energy and material consumption. These issues must be considered when analysing the cost of additive manufacturing, making it difficult and complicated to minimise costs.

The results and analysis support some major recommendations for future work to reduce SLS energy consumption:

- Build volumes should be packed as densely as possible to maximise the ratio of part output per build height;
- Time and energy consumption for the preheating phase, preparation, and laser scanning of each layer should be reduced;
- Material-related energy consumption should be improved by engineering long-term recyclable powder and by reducing powder loss during handling.

The scope of future energy studies could also be extended to include factors such as distribution, waste handling, and infrastructure. AM technology introduces greater freedom and expanded possibilities to the design stage, which may further improve the sustainability of a designed product and reduce its environmental impact during the service stage, factors that were not considered in this study.

References

1. I. Gibson, D. Rosen, B. Stucker, Additive Manufacturing Technologies (Springer, 2015)
2. R. A. D'Aveni, 2013. Harv. Bus. Rev. **91**, 34-35 (2013)
3. S.S. Muthu, M. M. Savalani, *Handbook of Sustainability in Additive Manufacturing* (Springer 2016)
4. Quadrennial Technology Review 2015, *Innovating Clean Energy Technologies in Advanced Manufacturing*. U.S. Department of Energy
5. Quadrennial Technology Review 2015, *Additive Manufacturing Technology Assessment*, U.S. Department of Energy
6. R. Huang, M. Riddle, D. Graziano, J. Warren, S. Das, S. Nimbalkar, J. Cresko, E. Masenet, J. Clean. Prod. **135**, 1559-1570 (2016)
7. M. Gebler, A. J.M. Schoot Uiterkamp, C. Visser, Energ. Policy **74**, 158-167 (2014)
8. Wohlers Report 2014, *3D Printing and Additive Manufacturing State of the Industry*, Annual Worldwide Progress Report, Wohlers Associates, 2014
9. The ASTM International Committee F42 on Additive Manufacturing Technologies
10. *3-D Printing: Sustainability Opportunities and Challenges*, BSR Report 2015
11. C. McAlister, J Wood, *eceee Industrial Summer Study* (2014),
12. T. S. Dougles, S.W. Gilbert, *Costs and Cost Effectiveness of Additive Manufacturing*, NIST Special Publication 1176 (2014)
13. M. Despeisse, S. Ford, Centre for Technology Management working paper series, University of Cambridge, **3**, 2015
14. S. Huang, P. Liu, A. Mokasdar, L. Hou, Int. J. Adv. Manuf. Tech. **67**, 1191-1203 (2013)

Modelling and simulation of lead-acid battery pack powering electric vehicle

Damian Burzyński[1*], *Leszek* Kasprzyk[1]

[1]Poznan University of Technology, Institute of Electrical Engineering and Electronics, Poland

Abstract. This paper presents aspects of modelling and simulation of energy storages based on the example of a lead-acid battery pack for powering an electric vehicle. The most frequently used energy storages, with particular emphasis on the difficulties in their proper selection for mobile equipment, was discussed. The mathematical model of the lead-acid battery and the relations describing its respective parameters was also presented. A selected energy storage was subjected to detailed analysis with regards to the electrical and thermal parameters while powering an electric vehicle during an standard driving cycle in an urban area. The simulation was performed in the Matlab Simulink environment.

1 Introduction

Energy storage has been an important element of man's activities for a long time. Initially, the issue concerned mainly the heating of the place of living or the cooling of the stored food, however, as the civilization kept developing, the need for accumulation of energy in various forms was apparently becoming more and more important. At present, energy storage constitutes an essential element of the global energy economy while balancing the energy demand and surplus and ensuring the energy security. This is particularly important in the case of energy systems cooperating with renewable sources, which are characterised by high instability of performance depending on weather conditions.

For this reason, a large number of various types of energy storages, especially those accumulating electric energy, has appeared recently. The type of the used storages depends on many factors of which those most important include mobility and ability to accumulate energy as well as easiness of its transfer – i.e. the energy density and power density which a given storage may accumulate. These parameters are the main factors, which determine the application of various energy storages [1-2].

The sources of electric energy also become more and more popular because of a great increase in the number of portable electrical and electronic equipment such as laptops, mobile phones, etc. However, the manufacturers of this equipment more and more often expect real methods of anticipation of the behaviours of these sources, especially since their behaviour under various operating and environmental conditions depends on many factors, in particular, in systems with high dynamics of load changes such as electrical vehicles [3].

*Corresponding author: damian.burzynski@put.poznan.pl

DOI: 10.1051/e3sconf/20171401041

Sometimes such analyses are very complex and time-consuming, therefore, they require the application of advanced calculation methods [4-5]. For this reason, this paper deals with the problems related to the analysis of the electric energy sources which can power an electrical vehicle based on the example of lead-acid batteries.

2 Energy storages used in electrical vehicles

Some of the mobile electric energy loads, which require the source of high power for their operation are electric vehicles, which have become more and more popular in the recent years. One example of a country belonging to the European Union, where an increase in the number of electrical vehicles is the greatest is Norway. In this country, in the year 2015, the electrical and hybrid plug-in vehicles constituted over 22% of the newly registered passenger cars [6]. An increase in the popularity of these vehicles results, to a great extent, from the appearance of more and more effective energy sources.

Depending on the possibility of energy conversion as well as the possibility of its accumulation in large quantities, the energy is accumulated using the following storages: mechanical (pneumatic, pumped hydroelectric energy storages, kinetic energy storages – mainly as flywheels), electrochemical (flow batteries and popular cells/lead-acid batteries, lithium-ion and nickel-cadmium batteries etc.), chemical (fuel cells, electrolysers), electric (accumulating energy in the electrical field – supercapacitors, or in the magnetic field – superconducting coils) and thermal [7].Popular examples of their application are presented in figure 1.

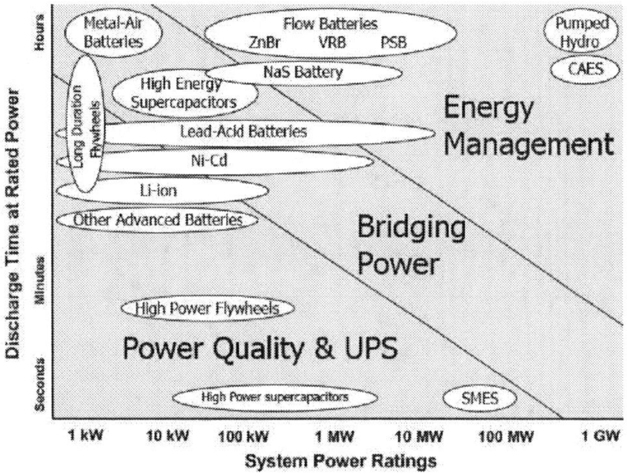

Fig. 1. Energy storages and their functions [2].

From among the energy storages presented in the figure above, the electrochemical energy sources are used most frequently to power electric vehicles. This follows from the fact that the vehicles require high energy and power densities from the storages – the ability of accumulating large amount of energy in a small and light storage which, at the same time, is characterised by high power handling capacity, and in the case of the regenerative braking, also high power consumption (charging with high current).

For this reason, a significant share of the energy storages used in vehicles includes lithium batteries and lead-acid batteries. Among them, the lithium batteries are characterised by better technical parameters, but are also much more expensive and dangerous (their damage poses a risk of very intense fire). Therefore, in many solutions, the lead-acid batteries are still popular. However, it must be remembered that these batteries are sensitive to high currents, which cause their accelerated degradation. While designing a drive powered by these batteries, it is necessary to analyse their behaviour in detail at different loads and in different environmental conditions, using professional mathematical models.

3 Mathematical model of the lead-acid battery

In systems, where the low power load is present, the power supply battery may often be treated as an ideal or real voltage source. In such cases, the mathematical model constitutes a circuit diagram which comprises an ideal voltage source or a voltage source with serially connected resistor that represents the internal resistance of the battery. However, while considering the dynamic systems, in which load is distributed unevenly and which are fast-changing, a more accurate model that takes into consideration the behaviour of the battery and the parasitic and thermal phenomena taking place in it must be used. One of the more accurate mathematical models used to analyse the battery performance is the electric circuit consisting of the voltage source E_m and the pairs of capacitors and resistors joined in parallel R_nC_n [8-10]. The electric charge losses caused by the self-discharge of the battery are represented by the occurrence of the P–N branch, through which the parasitic current I_p flows. Depending on the expected model accuracy, the appropriate number of pairs of RC dynamic branches must be selected. The number as such also depends on the dynamics of changes in the load of the analysed battery. The electric diagram of the discussed n-order model of a single cell of the lead-acid battery is presented in figure 2 (with the n-number of the connected RC branches) [8,11].

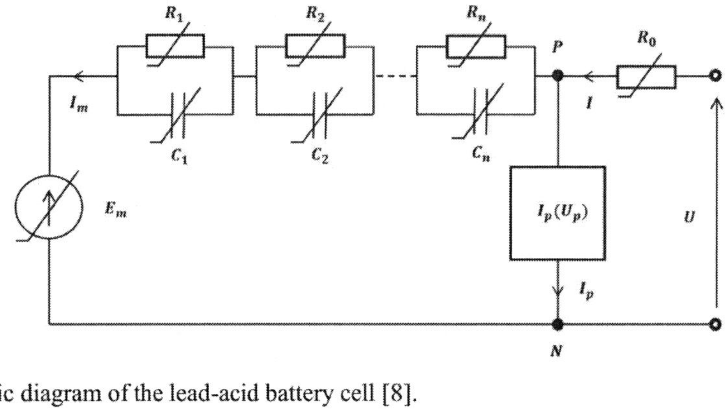

Fig. 2. Electric diagram of the lead-acid battery cell [8].

The value of the majority of parameters of the cell's equivalent circuit depends, above all, on the battery state of charge *SOC*, the load and the electrolyte temperature T_e. The parameters, which affect the operational properties of the battery are described in detail in publications[8,11], and their values for the second-order model can be determined from following equations:

$$E_m = E_{m0} - K_E(1 - SOC)(273.15 + T_e) \quad (1)$$

$$SOC = 1 - \frac{Q_e}{C(0,T)} \quad (2)$$

$$R_0 = R_{00}(1 + A_0(1 - SOC)) \quad (3)$$

 DOI: 10.1051/e3sconf/20171401041

$$R_1 = -R_{10} \ln(DOC) \tag{4}$$

$$C_1 = \frac{\tau_1}{R_1} \tag{5}$$

$$R_2 = R_{20} \frac{\exp[A_{21}(1-SOC)]}{1+\exp(\frac{A_{22}I_m}{I_n})} \tag{6}$$

where: E_m – electromotive force of the cell,E_{m0} – open-circuit voltage of fully charged battery at 0°C, SOC – battery state of charge, T_e – electrolyte temperature, Q_e – electric charge drawn from the battery (described by formula7), $C(0,T)$ – unloaded battery capacity at T, R_0, R_1, R_2 – main branch resistances, K_E, R_{00}, R_{10}, R_{20},A_0, A_{21}, A_{22} – constants depended on battery parameters, DOC – battery depth of charge, τ_1 – time constant of the dynamic branch, I_m– current flowing in the main branch, I_n – rated battery current.

The charge drawn from a single cell and the battery energy capacity are given by equations (7–8) [8,11]:

$$Q_e(t) = \int_0^t -I_m(t)dt \tag{7}$$

$$C(I,T) = \frac{K_C C_{0*}(1+\frac{T_e}{-T_f})^{\varepsilon}}{1+(K_C-1)(\frac{I}{I_n})^{\delta}} \tag{8}$$

where: K_C, δ, ε – constants set on the basis of the manufacturer's data, C_{0*} – battery capacity at 0°C, T_f – electrolyte freezing temperature, I – battery load current.

4 Simulation of operation of the lead-acid battery pack powering an electric vehicle

For the purposes of simulation of the performance of the lead-acid battery pack during the movement of the electric vehicle, the second-order mathematical model presented in figure 3 was assumed.

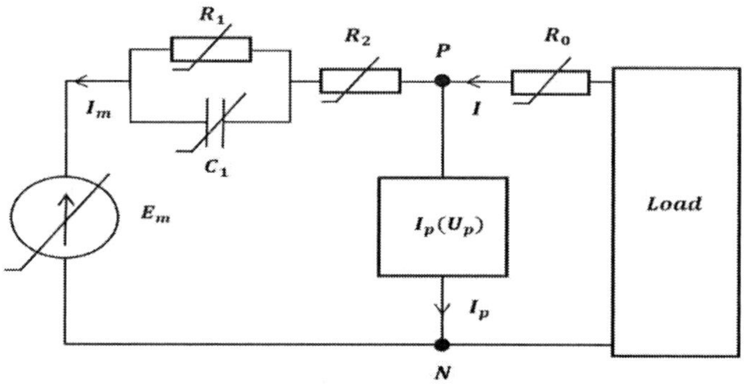

Fig. 3. Circuit diagram for the lead-acid battery cell selected for the simulation.

The model is considered to be sufficiently accurate to carry out the simulation in dynamic states [8-9]. Parameters of the equivalent circuit were determined on the basis of equations (1)-(8) presented in chapter 3 and constants listed in table 1. The parameter which forces the operation of the system was the electric power with which the lead-acid battery pack was loaded. The power was determined based on the construction data of the electric vehicle [1,12] (with the assumed weight equal to 1150 kg and frontal surface of the vehicle equal to 2.3 m^2) and vehicle characteristics in the urban area.

Table 1. Electric parameters of the equivalent circuit of one battery cell.

Description	Value
Unloaded battery cell voltage	E_{m0} = 2.18 V
Ambient temperature	T_a = 22°C
Constants which enable the determining the battery capacity	K_C =1.19 C_n = 50 Ah δ = 0.71 ε =0,00692 T_f =–35°C
Constants which enable the determining the values for respective situations	R_{00} = 0.0025 R_{10} = 0.03 R_{20} = 0.143 A_0 = –0.37 A_{21} = –7.48 A_{22} = –3.07

The standard EPA–FTP–75 driving cycle [13], whose speed profile is presented in figure 4 was used for this purpose.

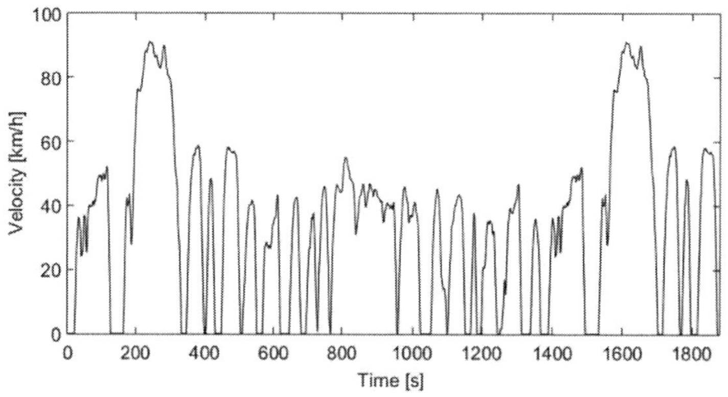

Fig. 4. Velocity profile of the standard EPA–FTP–75 driving cycle [13].

The energy storage subject to modelling comprised a lead-acid battery pack, whose parameters were selected based on publications [1,7].It consisted of 4 battery packs including 8 batteries of the nominal voltage of 12 V connected in series and capacity of 50 Ah. In order to achieve the desired capacity respective battery packs were connected in parallel. In accordance with the methodology presented in publications [8,11] and based on the manufacturer's data, the electric parameters of the equivalent circuit of the battery cell, necessary to carry out the computer simulation, were assigned.

In order to map the behaviour of the battery pack, the equivalent circuit presented in figure 2 was implemented in the MATLAB Simulink environment. The lead-acid battery pack which was loaded with power in accordance with the characteristics presented in figure 5,

 DOI: 10.1051/e3sconf/20171401041

was selected to the performance simulation. The simulation that was carried out enabled the determining all important electric parameters of the battery pack under consideration.

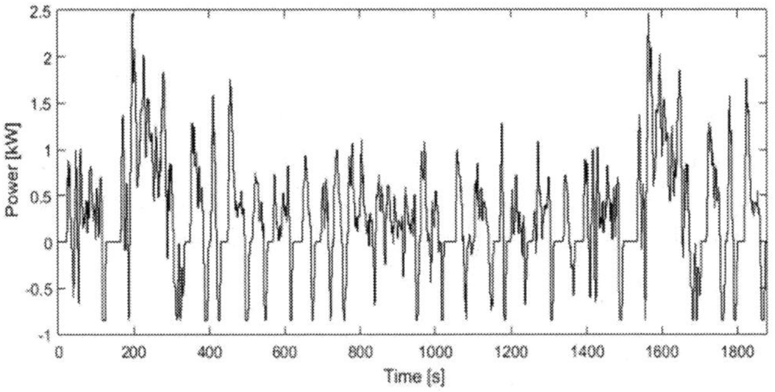

Fig. 5. Power demand of electric vehicle(positive during speeding up, negative during braking).

Characteristics of the selected battery parameters in the function of duration of one driving cycle are presented in figures 6–9.State of charge (Fig. 6) was determined from equations (7)–(8), andcurrentsandvoltageswere calculatedon the basis ofconventional methodsfor circuit analysisin transient states. Due to the possibility of appearance of the overvoltages in the moments of sudden braking in the case of occurrence of high voltages on the battery terminals (over 19 V) use of mechanical brakes was assumed. The result was also a limitation in the value of the charging current (negative values of current in Fig. 7) – equal to 0.3 C (maximum value recommended by the manufacturer). Because of marginal values of parasitic current (Fig. 9) their impact on the operation of battery pack can be omitted.

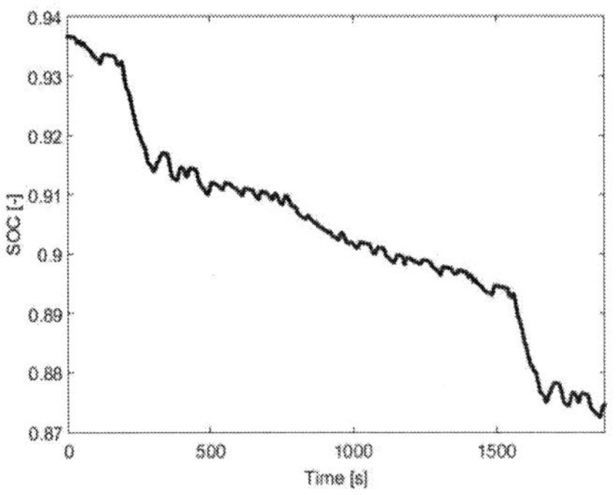

Fig. 6. Battery state of charge.

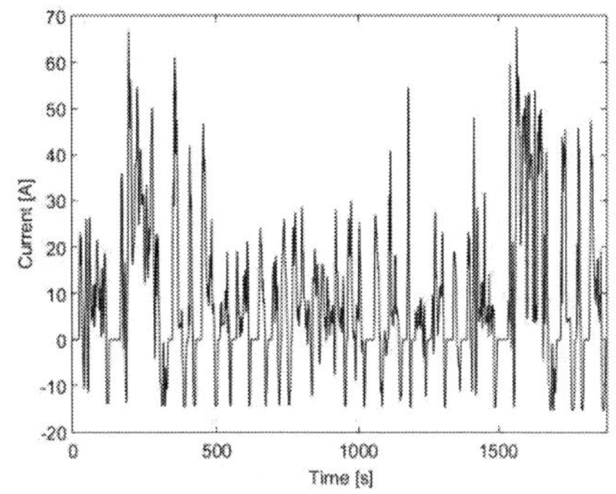

Fig. 7. Current flowing through the main branch of the cell.

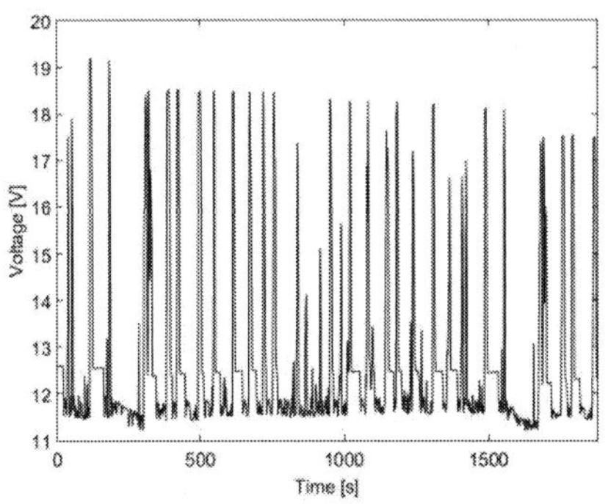

Fig. 8. Voltage on the battery terminals.

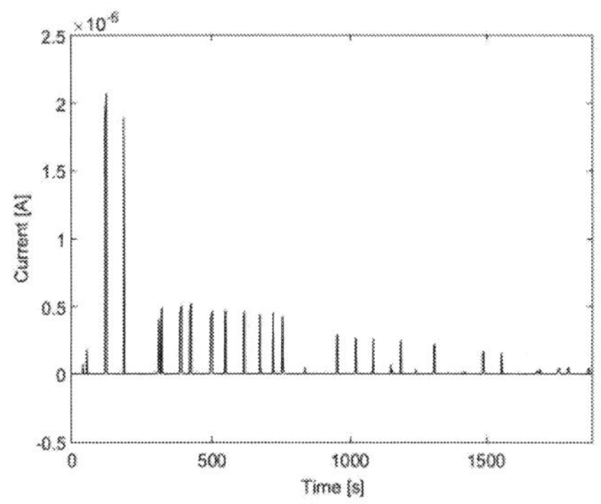

Fig. 9. Current flowing through the parasitic branch.

5 Summary

This paper deals with the issue of modelling of performance of lead-acid battery pack, which powers an electric vehicle. The analysis of operation of electrochemical sources is a complex problem, therefore, the second-tier mathematical model of the battery was used for its needs. On the basis of the calculated parameters of the energy storage and the assumed parameters of the electric vehicle, the computer simulation for the standard EPA–FTP–75 driving cycle were performed. It represents the performance of the battery pack while driving the electrical vehicle. Because of marginal values of (just a few microamperes) parasitic current flowing through the battery during the movement of the electrical vehicle under consideration, its presence may be omitted.

The application of the mathematical model of the battery pack may be very advantageous, e.g. at the stage of designing and selecting of energy storages for electric vehicles, in terms of tests of their behaviours in the states which deviate significantly from the nominal state. Not only does it entail benefits in terms of the finances and time (tests carried out using real systems are very costly and time-consuming), but also it constitutes a response to the current issues related to ecology (degradation, and as a consequence of this, the necessity to dispose of waste equipment used during the tests).

Issues related to the modelling of the performance of energy storages with non-linear loads are interdisciplinary by nature. The used mathematical model may be successfully applied in systems cooperating with renewable energy sources.

References

1. L. Kasprzyk, K. Bednarek, *Selection of the hybrid electric energy storage for the electric vehicle*. Przegląd Elektrotechniczny, Vol. **91**, No. 12, pp.129-132, (2015)*(in polish)*
2. A. A. Jamali, N. M. Nor, T. Ibrahim, *Energy storage systems and their sizing techniques in power systems - a review*. IEEE Conference on Energy Conversion, pp. 215-220, (2015)
3. X. Wu, D. Freese, A. Cabrera, W. A. Kitch, *Electric vehicles energy consumption measurement and estimation*. Transportation Research Part D 34, pp. 52-67, (2015)
4. L. Kasprzyk, A. Tomczewski, K. Bednarek, *Efficiency and economics aspects in electromagnetic and optimization calculations of electrical systems*. Przegląd Elektrotechniczny, Vol. **88**, No. 7B, pp. 57-60, (2012)
5. L. Kasprzyk, *Optimization of Lighting Systems with the use of the Parallelized Genetic Algorithm on Multi-Core Processors using the .NET Technology*. Przegląd Elektrotechniczny, Vol. **88**, No. 7B, pp. 131-133, (2012)
6. Statistical pocketbook 2015. The European Council On Clean Transport. http://www.theicct.org. [access: 01.09.2016]
7. L. Kasprzyk, *Electric vehicles and the issues related to the selection of the electric energy storage in the aspect of environment protection*. Europejski Wymiar Bezpieczeństwa Energetycznego a Ochrona Środowiska, pp. 691-708, (2015) *(in polish)*
8. M. Ceraolo, *New dynamical models of lead-acid batteries*. IEEE Transactions On Power Systems, Vol.15, No.4, (November 2000)
9. H. He, R. Xiong, Y. Chang, *Dynamic modeling and simulation on a hybrid power system for electric vehicle applications*. Energies**3**, pp. 1821-1830, (2010)
10. R. Carter, A. Cruden, P. J. Hall, A. S. Zaher, *An Improved Lead-Acid Battery Pack Model for Use in Power Simulations of Electric Vehicles*. IEEE transactions on Energy Conversion, No. 1 (**27**), pp. 21-28, (March 2012)

DOI: 10.1051/e3sconf/20171401041

11. S. Barsali, M.Ceraolo, *Dynamical Models of Lead-Acid Batteries: Implementation Issues.* IEEE Transactions On Energy Conversion, Vol. **17**, No. 1, (March 2002)
12. C. De Cauwer, J. Van Mierlo, T. Coosemans, *Energy Consumption Prediction for Electric Vehicles based On Real-World Data.* Energies **8**, pp. 8573-8593, (2015)
13. T.J.Barlow, S. Latham, I.S. McCrae, P.G. Boulter, *A reference book of driving cycles for use in the measurement of road vehicle emissions.* Published Project Report PPR354, Version 3, TRL Limited, (June 2009)

Power sources involving ~ 300W PEMFC fuel cell stacks cooled by different media

Magdalena Dudek[1,*], *Andrzej* Raźniak[1], *Bartłomiej* Lis[1], *Michał* Kawalec[2], *Mariusz* Krauz[2], *Tadeusz* Wójcik[1]

[1]AGH University of Science and Technology, Faculty of Energy and Fuels, Krakow, Poland
[2]Institute of Power Engineering Ceramic Department CEREL, Boguchwała, Poland

Abstract Two constructions of ~300W PEMFC stacks, cooled by different media, were analysed. An open-cathode ~300W PEMFC stack cooled by air (Horizon, Singapore) and a PEMFC F-42 stack cooled by a liquid medium (Schunk, Germany) were chosen for all of the investigations described in this paper. The potential for the design and construction of power sources involving fuel cells, as well as of a hybrid system (fuel cell-lithium battery) for mobile and stationary applications, is presented and discussed. The impact of certain experimental parameters on PEMFC stack performance is analysed and discussed.

1 Introduction

Fuel cells (FCs) are electrochemical devices that directly convert chemical energy from fuel into electricity and heat. These electrochemical devices can operate as long as fuel (mainly hydrogen) and an oxidant (usually oxygen taken from the air) are supplied. Fuel cells are generally classified into two basic categories, according either to the type of electrolyte used or to the temperature of operation. In terms of the electrolyte used, five main types of fuel cells are distinguished: a) proton exchange membrane fuel cells, or PEMFCs, temperature range 30–80°C; b) alkaline fuel cells, or AFCs, temperature range 50–200°C; c) phosphoric acid fuel cells, or PAFCs, operating temperature ~200°C; d) molten carbonate fuel cells, or MCFCs, operating temperature ~650°C; e) solid oxide fuel cells (SOFCs, operating temperature 500–1000°C [1]. PEMFCs, due to their high power density, low operating temperature, and ability to start up rapidly when cold, are considered to be the most promising candidates for next-generation power sources for transportation, stationary, auxiliary, and portable applications [2,3]. Nafion-based polymer membranes are still used in PEMFCs as solid electrolytes. The main drawback of these membranes is the strong dependence of their protonic conductivity on water content and temperature. This requires continuous monitoring of temperature and humidity during single-PEMFC and PEMFC-stack performance [4]. However, In platinum catalysts of electrode materials, which become passivated in catalytic activity there is a strong need for cleaning in the reformation process of kerosene in order to achieve a tolerably low CO content (in the ppm range) for the fuel cell. Therefore, one of the

* Corresponding author: potoczek@agh.edu.pl

main tasks in the future should be to develop compact reformers and ways to reduce fuel poisons such as CO and sulphur. Until these reforming problems are solved, PEMFCs are suited only for applications where the use of pure hydrogen as fuel is feasible and acceptable. The high price of platinum and limited resources make it necessary to find new, cheaper catalysts with high tolerance for impurities [5,6].

A PEMFC fuel cell stack consists of many single cells, stacked so that the cathode of one cell is electrically connected to the anode of the adjacent cell. In this way, exactly the same current passes through each cell. The key aspects of a fuel cell stack design are: uniform distribution of reactants to and inside each cell; maintenance of the required temperature in each cell; minimum resistive losses made possible through choice of materials, configuration; uniform contact pressure; mechanical sturdiness (ability to withstand internal pressure, including thermal expansion, and external forces during handling and operation, including shocks and vibrations) [7,8].

A PEMFC fuel cell is very sensitive to the flow rate of the reactants. Each cell in the stack must receive approximately the same amount of reactant gases. Uneven flow distribution would result in uneven performance of the cells. Water and heat are the by-products of fuel cell operation, and the supporting system must include the means for their removal. It is possible to re-use water or heat, at least partially – for example, for humidification of the reactant gases. The PEMFC stack is the main part of a fuel cell system. In designing power sources involving PEMFC stacks, supporting equipment is necessary for operation. A fuel cell system typically involves the following subsystems: oxidant supply (oxygen or air), fuel supply (hydrogen or hydrogen-rich gas), heat management, water management, power conditioning and instrumentation, and controls [9,10]. During PEMFC stack operation, heat, as a by-product occurring in the course of electricity production, must be effectively removed from the fuel cell system. The most favourable operating temperature range for existing PEMFCs is usually 60–80°C. A higher temperature can significantly exacerbate the degradation of the membrane and the catalyst and reduce the stack's performance. On the other hand, a lower temperature is not favourable for the reaction kinetics and may also cause flooding due to lower water saturation pressures at low temperatures, which is a major concern from the water management perspective. Effective cooling is critical for safe and efficient operation of PEMFC stacks [11-13]

Despite advanced interdisciplinary research and development work on various designs of fuel cell power generators, certain issues concerning their operation, including the potential for and barriers to their application in integrated electrical systems have not been fully recognised or explained, even in regard to commercially available PEMFC stacks [14,15]. It must be emphasised that the fuel cell stack market in EU countries is thriving; these products can be easily bought and delivered to their intended recipients. Moreover, any increase in the use of fuel cell technology in Poland requires acquisition of necessary knowledge concerning electrochemical sources of energy (fuel cells, batteries and accumulators, supercapacitors), their advantages, and operating capacities and applications. Therefore, training of specialised staff should cover interdisciplinary procedures, enabling staff members to learn about the design and operating principles of fuel cells, the potential for constructing fuel cell power generators, and their research methodology, as well as specifying the potential for and barriers to their use within specific technical applications [16].

In this study, the performance of power sources involving a 300-W PEMFC stack cooled by different media was considered. Special focus was placed on the effect of experimental conditions on the performance of designed PEMFC stacks. The potential for modification of electrical parameters, as well as for reduction of the mass and cost of fuel cell systems, was also analysed in this study.

 DOI: 10.1051/e3sconf/20171401042

2 Experimental

2.1 Selection of types of PEMFC stacks with ~300 W electric power

Two commercially available PEMFC stacks were selected for the construction of an electric generator involving fuel cell technology.

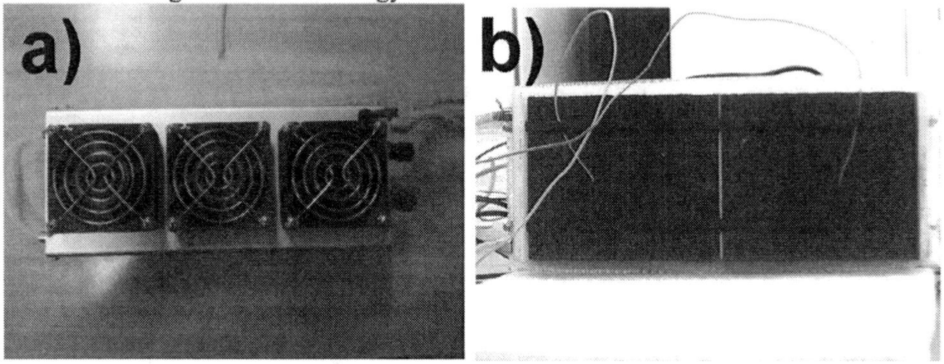

Fig. 1. Photo: a) the initial Horizon H-300 PEM Fuel Cell without modification and b) H-300 PEM Fuel Cell with modification - divided into two modules of 35 cells each, using a gold-plated board.

The first was the Horizon H-300 PEM Fuel Cell, a 300-watt (Fig. 1a), air-fed/air-cooled, self-humidified hydrogen fuel cell suitable for many medium-scale projects and uses. The open-cathode compact PEMFC stack permitted the performance of some necessary modifications aimed at adapting the electrical parameters of the stack to the requirements of the electrical system. The first modification was aimed at lowering the starting voltage of the PEMFC stack. The initial H-300 PEMFC stack was divided into two modules of 35 cells each, using a gold-plated board (Fig.1b) with channels for hydrogen supply as an electrical separator. The modules operated in parallel mode. The gas tightness of the new elaborated stack was checked and all hydrogen leakage was eliminated. The potential for the construction of a hybrid power system (fuel cell battery) was also considered. The new elaborated self-humidifying PEMFC stack was connected with a 6S lithium ion battery pack for peak energy demand and energy storage for safety and stack start-up purposes. The element connecting the cell stack and the LiPo battery is a charge-discharge control system, which consists of two subsystems. The charge subsystem enables the collection of excess electricity during lower demand periods (horizontal flight or descending especially in UAV applications). This system was built on the basis of a high-performance DC-DC converter limiting the charging voltage to the LiPo battery pack to a safe range (4.2V per cell, CV mode) and a charging current. The discharge subsystem can use energy stored in the battery. In this mode, the current flows through the diode in the reverse direction. In addition, the LiPo battery pack can be equipped with an active balancer evening the charge level of individual cells in a package. PEMFC stacks were tested with a laboratory electronic load (Dynaload XBL 50-150-800). Dry hydrogen was supplied from a laboratory installation under a pressure of 0.4 bar.

During the operation of two series-connected PEM fuel cell stacks, the distribution of temperatures at the surface of each stack was recorded without contact, using a ThermoTracer H2640 thermographic camera (manufactured by NEC) with a thermographic detector (also an NEC product) with dimensions of 640×480 pixels to record the temperature with a thermal resolution of 0.03°C. During the measurements, the temperature of the air in the laboratory was T_a=20.1°C; the emissivity ratio of the camera was set to ε=0.91. Thermal images recorded as the result of measurements were processed and analysed using Thermography Studio Professional, a software program dedicated to this camera.

The second type was a 360-W PEMFC stack with a closed cathode, cooled by a liquid medium (Schunk, Germany).

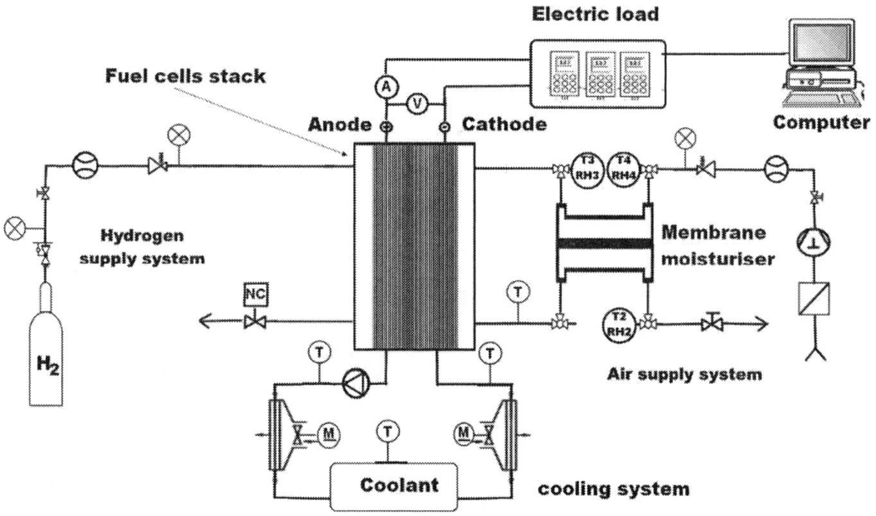

Fig. 2. Schema of an experimental setup designed to characterise an FC-42 PEMFC stack cooled by liquid media.

Fig.2 present a schema of an experimental setup designed to characterise an FC-42 PEMFC stack cooled by liquid media. The fuel cell stack was powered by high-purity hydrogen. The hydrogen dosing system was equipped with a pressure regulator with a control manometer, which lowered the pressure to 0.3 bar, as required by a fuel cell stack. The amount of hydrogen fed to the PEMFC stack was monitored using a Bronkhorst flow measurement valve. At the hydrogen outlet of the PEMFC stack, there is an electrically controlled normally – closed valve functioning as a purge valve whose primary function is to rinse the anode component of the cell during the start, finish, and operation of the pile through periodic opening and introduction of hydrogen into the surroundings. The fuel cell stack's cathode was powered with air introduced by a dry-running air compressor (model ACO-006, 75 W, efficiency 110 l/min) through a system of filters, pressure regulators, and rotameters to measure the flow.

3 Results

Electrically powered small unmanned aerial vehicles (UAVs) in particular are already suited for the application of PEMFC fuel cells. By replacing their batteries either partially or totally with a fuel cell system, flight times can be significantly extended [17].

Each galvanic cell (including a single fuel cell) and fuel cell stack is characterised by maximum power, represented by a pair of (voltage-current) max parameters. For some types of galvanic cells, this point may not be observed in power-current dependence curves if its position is far outside the range of useful voltages or currents. In the case of fuel cells, the position of the (voltage-current) max point is usually clearly marked and plays an important role in the performance of the generator. Gradual loading of a fuel cell stack leads to a self-regulating change in work conditions (voltage drop, current increase, increase in the amount of hydrogen fuel used), which leads to an appropriate increase in the power drawn from the stack according to rising need. This will remain in effect until the maximum power point is achieved, after which a rapid fall in the power supplied by the cell stack is observed despite

 DOI: 10.1051/e3sconf/20171401042

the increasing load current. The shortage created in the power balance of the drive unit may lead to a loss of aerodynamic lift, with dramatic consequences for the plane [18].

3.1 Performance measurements PEMFC stack air-cooled.

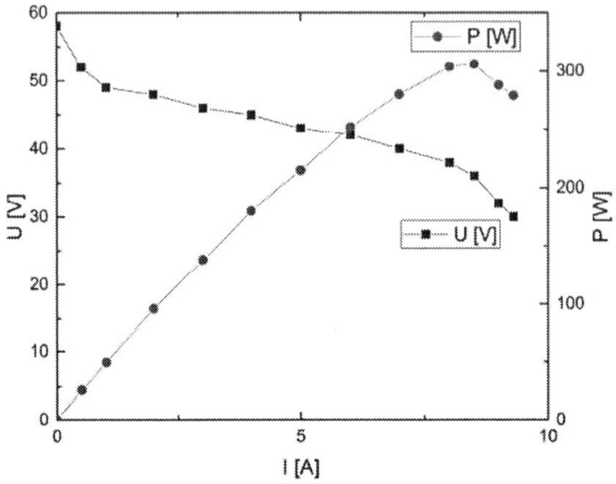

Fig. 3. Current voltage and current-power curves of unmodified H300 fuel cell stack.

In Fig.3, the) current (I) - voltage (U and current (I) – power (P) relationships of the original 300-W H-stack are presented. It was found that the maximum of power output (P_{max})~300 W, was recorded for the following pair of electrical parameters: I=8.4A, U=36.5V. Based on recorded characteristics, the performance of a PEMFC stack is recommended to be close to the electrical parameters recorded for P_{max}. Unfortunately, the open circuit voltage U for I= 0A was close to 60V; all recorded voltage values under load are higher than 35V. In the case of the application of the original H-300 PEMFC stack as a main power source for supplying small electrical brushless (AXI, Czech Republic or Hacker Germany) in small unmanned aerial vehicle (UAV) construction, these values are higher than the requirements of the electric motor management controller.

The first electrical modification was aimed at lowering the starting voltage of the PEMFC stack. The initial H-300 PEMFC stack was divided into two modules (35 cells in each module) using an electrical separator. Gold-plated board (Fig. 1b) with channels for hydrogen supply was used as an electrical separator. The modules operated in parallel mode. As can be seen in Fig. 4, reductions in open circuit voltage (OCV) were obtained, as well as in the values of voltage under load. In this case, the voltage and current values are suitable for most electric brushless motors applied in UAV technology.

DOI: 10.1051/e3sconf/20171401042

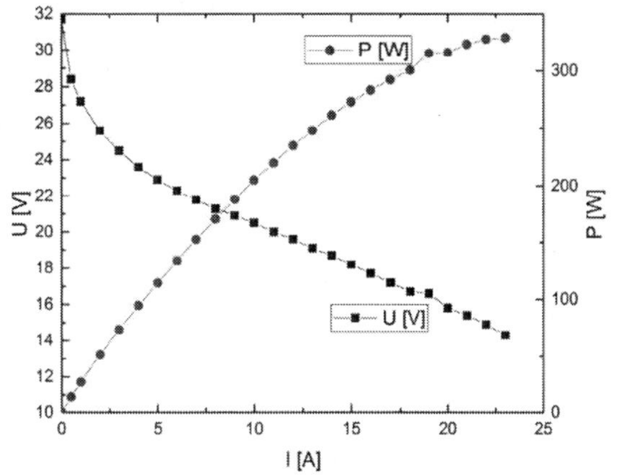

Fig. 4. Current voltage and current-power curves of modified 300W fuel cell stack.

In the case of instantaneous power deficit, one frequently used solution is the employment of an auxiliary unit consisting of a storage battery pack or super-capacitor system. In this study, a PEMFC stack was connected in parallel with a lithium polymer battery pack and tested. It is extremely important to determine the voltage-current and power-current characteristics of the complete system to be sure that the above-mentioned power deficit does not occur, even in unusual conditions of aircraft operation.

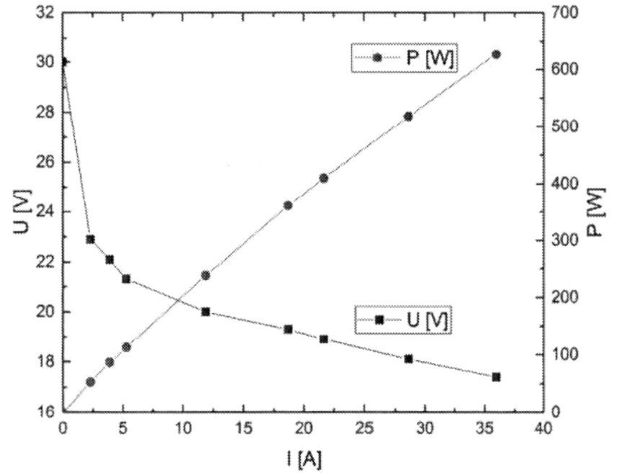

Fig. 5. Current voltage and power curve of hybrid fuel cell – lithium battery system

In Fig. 5, the U-I and P-I dependences for a hybrid system (PEMFC stack + battery system) were determined. An increase of power density up to 600 W was recorded. In the case of a hybrid power system, the P_{max} of this source was not visible.

The results of the electrical investigation of a PEMFC stack with an open cathode indicated that, in this type of construction, it is relatively easy to modify a PEMFC stack to fit the fuel cell as a power source to the requirements of electrical motors (24–32 V) typical of UAV construction. The reduction of the total mass of the stack, which is also a crucial parameter, from 2100 to 1300 g was achieved as well in the newly modified construction.
Fig. 4a presents the thermographic image of temperature distribution on the surface of the PEMFC stack on the air inlet side, recorded on the bench during measurement of the stack while operating under electricity load.

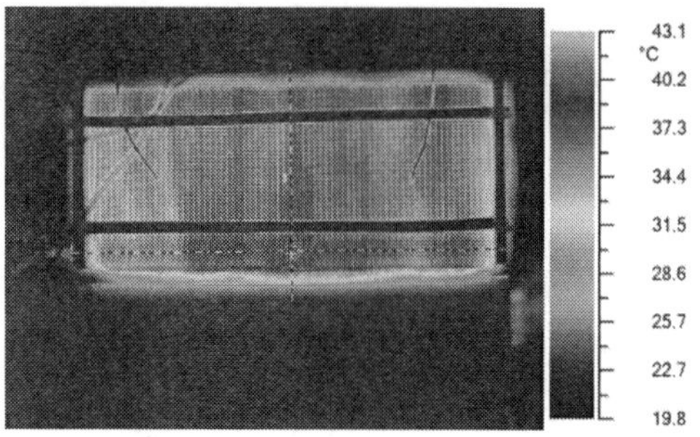

Fig. 6a. Thermographic image of temperature distribution on the PEMFC stack on the air inlet side

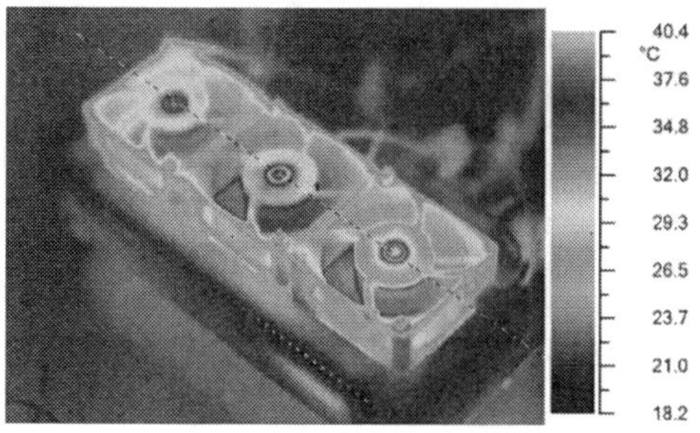

Fig. 6b. Thermographic image of temperature distribution on the PEMFC stack on the fans side

Based upon analysis of the thermographic image (Fig. 6a), two horizontal dark blue stripes are visible in the foreground (relative to the stack of PEM cells heated by their operation). These are two metal rods, extending beyond the stack, which serve to compress the individual fuel cells making up the stack, and thin thermocouple wires used to control the temperature of individual FC stacks. During PEMFC operation under load, the temperature, which varies within a range of approximately 35 to 40°C, is lower on the left side. This is because on this side, located at the bottom, was a supply (inlet) of fresh gaseous hydrogen at ambient temperature, with an outlet at the top from the anode space where reaction products were discharged during periodic purges ('purge'). In Fig. 6b, temperature is distributed along a line passing through the centres of the three fans used to cool the FC stack. It can be seen that the highest temperature (approximately 40°C) is attained by the centres of the fan rotors, under which both bearings (on which they are grounded) are located, and by the windings of the DC electric motors they are driven by. These results clearly indicate that all modified PEMFC stacks operate within a safe temperature range (30-40°C). No unexpected local increases (> 60°C) of temperature were observed in the modified PEMFC stack.

3.2 Performance measurements PEMFC stack liquid-cooled supplied with dry or humidified air.

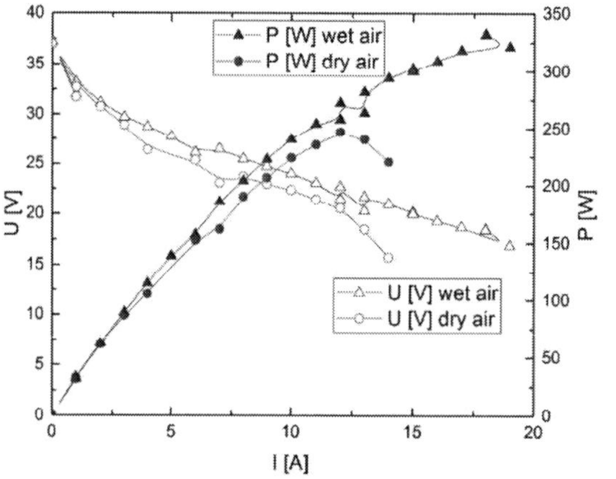

Fig. 7. The family of U-I and P-I curves recorded for an FC-42 PEMFC stack supplied with dry or humidified air

In Fig. 7, the (U)-current (I) and power (P)-current (I) curves are recorded for an FC-42 stack supplied with dry or humidified air.

Direct comparative analysis of U-I and P-I curves obtained for an FC-42 stack fed by dry or humidified air reflects the difference in electrical performance of the investigated power sources. Analysis of U-I and P-I dependencies enabled us to distinguish three areas of electrical response.

(1) In the first area of applied current load, the dominance of activation losses is visible. Although it was noted previously that practically the same value of open circuit voltage U1=U2 was obtained (U=37V, I=0A), progressive switch on electrical loading of a PEMFC stack with small currents (from 1 to 2 A) leads to a gradual decrease in voltage values. In these measurement conditions, no significant differences in the electrical response of the PEMFC stack were observed.

(2) In the second area of switch on current load to the PEMFC stack, where the current I varies from 3 to 12A, the difference in the electrical power output of the energy generator involving the PEMFC stack becomes more visible. Initially, when the current load increased to more than 5A, slightly lower values of voltage were observed for a PEMFC stack fed with dry air than for one fed with humidified air. In these test conditions of applied current load, near-linear dependency of U-I was obtained. In this area, the main source of energy losses is ohmic polarisation. An increase of resistance due to the polymer membrane is responsible for the occurrence of these losses. The presence of humidity in air can improve protonic conductivity in a Nafion membrane, due to humidification of the polymer. These electrolytes are very sensitive to H_2O content in the cell.

(3) The third area of curves U=f(I) corresponds to current load I>12A for a PEMFC stack supplied with air without humidification and I > 18 A for a stack supplied with humidified air. In these experimental conditions, the maximum power (P_{max}) output of the power sources was obtained. It was found that in the case of an FC-42 supplied with dry air, the maximum power output P_{max} was observed for the pair of values I=12A and U=27V (P_{max}=246W), but in the case of an FC-42 stack supplied with humidified air, improved parameters were recorded (P_{max}=332W, I=18A). Above the P_{max} a rapid voltage drop was recorded due to an increase in the current load. Switch–on of higher values of current load is not recommended for PEMFC stacks, due to not only the consequent reduction of electrical power but also the potential introduction of irreversible changes in the PEMFC structure and destruction of the cells.

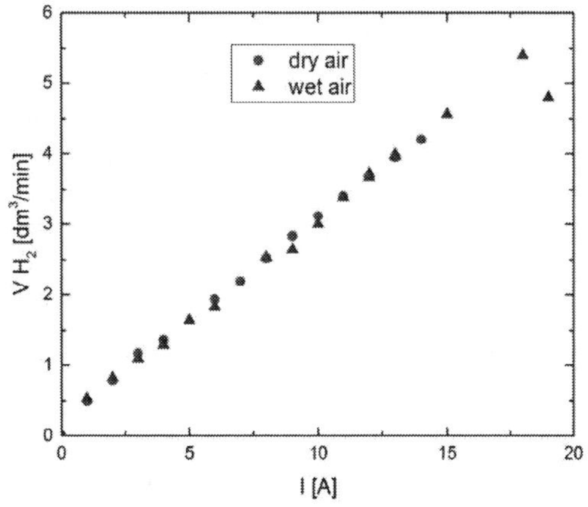

Fig. 8. Dependency of the amount of hydrogen consumed by the Schunk FC-42 fuel cell stack vs the current generated by the investigated power source. The PEMFC stack was supplied with humidified air (A) or dry air (B).

In order to determine the amount of hydrogen consumed during operation of the PEMFC FC-42 stack, the flow rate of hydrogen was measured using a mass flow meter. Fig. 8. shows the dependency of the amount of hydrogen consumed by the Schunk FC-42 fuel cell stack vs the current generated by the investigated power source. The measurements were

 DOI: 10.1051/e3sconf/20171401042

made in analogous conditions. The PEMFC stack was supplied with humidified air (A) or dry air (B). Based on this analysis, it can be concluded that the amount of hydrogen utilised as a fuel for supplying a PEMFC stack rises with increasing current produced by the stack. For I=18A, maximum hydrogen consumption was found to be 5.5 dm^3/min.

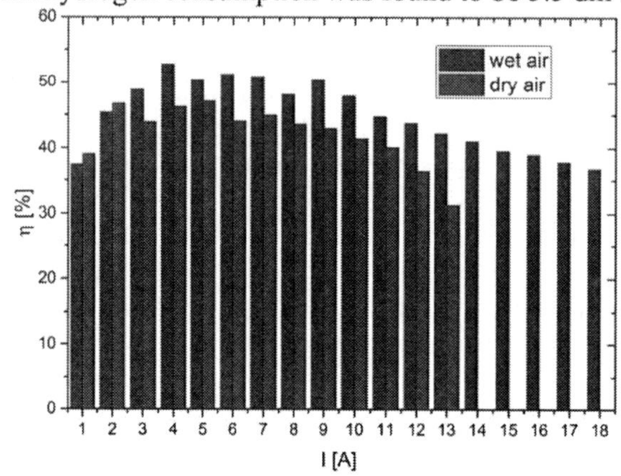

Fig. 9. Comparison of the electrical performance obtained from the PEMFC FC-42 stack supplied with dry or humidified air.

However, a direct comparison of the levels of power output (P_{max}) obtained from the PEMFC FC-42 stack supplied with dry or humidified air, respectively, indicated a difference in the electrical performance of the investigated power sources (Fig. 9.). The highest power output of the PEMFC FC-42 stack was recorded when humidified air was introduced into the cathode chamber. The observed increase in the electrical power of the PEMFC stack was caused by appropriate moisturisation of the polymer electrolyte with water introduced along with air into the cathode chamber.

4 Conclusion

Two electrical generators involving ~300W PEMFC stacks, cooled by air or by a liquid medium, were designed and investigated. Their performance, along with the potential for modification of their electrical parameters and simplification of the construction of power sources, is presented. It was found that in the case of an open-cathode stack, the modification of electrical parameters and construction aimed at the reduction of mass is simpler when compared to a similar electrical generator involving a PEMFC stack cooled by liquid media. The potential for the construction of a hybrid system (battery-fuel cell stack) is also demonstrated. The elaborated hybrid system, thanks to its customised electrical parameters as well as its compact light construction, seems suitable for transport applications (small UAVs, bikes, etc). On the other hand, power sources involving PEMFC stacks cooled with a liquid medium, due to their additional cooling system, seem to be better suited for stationary applications (UPS systems).

Acknowledgement: This work is supported by National Centre of Research and Development, as a Sector Program INNOLOT, project HYBRIDRIVE "Technologies of hybrid power drive for lightweight or unmanned aerial vehicles."

References

1. K. Kordesch, G. Simader, *Fuel Cells and their applications*, (VCH Publisher, Inc, New York, USA, 2001)
2. M. Oszcipok, M. Zedda, D. Riemann, D. Geckeler, J. Power Sources **154**, 404 (2006)

3. Y. Wang, K.S. Chen, J. Mishler, S.Ch. Cho, X.C. Adroher, Applied Energy **88,** 981 (2011)
4. P.P. Lopes, K.S. Freitas, E.A.Ticianelli, Electrocatalysis **1,** 200 (2010)
5. S.B. Fabry, L.Dubau, Y. Ahmad, K. Guerin, M. Chatenet, Electrocatalysis **6**, 521 (2010)
6. X. Wu, B. Zhou, J. Power Sources **324**, 804 (2010)
7. Z.M. Huang, A. Su, Y.Ch. Liu, Energy **72,** 547 (2014)
8. S.W. Jeon, D. Cha, H.S. Kim, Y. Kim, Applied Energy **166,** 165 (2016)
9. D. Chu, R. Jiang, J. Power Sources **83,** 128 (1999)
10. N. Sammes, *Fuel Cell Technology: Reaching Towards and Commercialization*, (Springer –Verlag London 2006)
11. Ch.Y. Chen, K.P. Huang, W.M. Yan, M.P. Lai, Ch.Ch. Yang, Int. J. Hydrogen Energ. **41,** 11784 (2016)
12. M.G. Santarelli, M.F. Torchio, M. Calı, V. Giaretto, Int. J. Hydrogen Energ. **32**, 710 (2007)
13. G. Zhang, S. G. Kandlikar, Int. J. Hydrogen Energ. **37**, 2412 (2012)
14. Ó. Gonzalez-Espasandin, T.J. Leo, E. Navarro-Arévalo, The Scientific World Journal **2014**, 497642 (2014), Review Article Fuel Cells: A Real Option for Unmanned Aerial Vehicles Propulsion http://dx.doi.org/10.1155/2014/497642
15. R. Agarwal, F. Collier, A. Schaefer, A. Seabridge, *Green Aviation* (John Wiley & Sons, 2016)
16 C. Harms, F. Köhrmann, Al. Dyck, Solid State Ionics **275**, 75 (2015)
17 J.M. Krawczyk, A.M. Mazur, T. Sasin, A.W. Stokłosa, Transactions of the Institute of Aviation **4**, 49 (2014)
18. M. Dudek, P. Tomczyk, P. Wygonik, M. Korkosz, P. Bogusz, B. Lis, Int. J. Electrochem. Sci., **8,** 8442 (2013)

 DOI: 10.1051/e3sconf/20171401043

Numerical model for evaluation of the effects of carbon deposition on the performance of 1 kW SOFC stack – a proposal

Konrad Motylinski[1,*] and *Yevgeniy* Naumovich[1]

[1]Thermal Processes Department, Institute of Power Engineering, Warsaw, Poland

Abstract. Solid oxide fuel cells are high-temperature electrochemical energy conversion devices which operate at elevated temperature (600-900°C). As a result it possible to internally reform the incoming fuel, thus except hydrogen and carbon monoxide, SOFCs can be fuelled with various hydrocarbonaceous gases. The presence of carbon-containing compounds in the fuel might result in the formation and of carbon in a form of a thin layer on the SOFC anode. The carbon deposition process depends on the thermodynamic conditions, such as temperature and steam to carbon ratio. The higher the temperature, the longer period of time is required for the solid carbon particles to deposit on the porous surface. The correlation used for this study is based on creating the ternary diagrams or Gibb's diagrams. The presented results cover a first stage of the analysis of the carbon deposition processes in SOFCs, focusing mainly on the numerical study of the changes of the fuel cell performance due to degradation of anode performance. A dedicated model of SOFC was proposed. It accounts for the diminution of the active area and/or deactivation by the increase of the resistance of the anode. The article presents the proposed methodology and the numerical approach.

1 Introduction

The carbon deposition process is an important issue, which can occur in various chemical and power systems. Fuels or generally gas mixtures containing hydrocarbons such as methane, ethane, propane, etc., and optionally other volatile organic compounds *i.e.* alcohols or ethers such as dimethyl ether (DMR) can be a subject of such a reaction. The iron and steel industry is highly dependent on metallurgical coke, which is produced from high-grade bituminous coal, and has crucial roles as an energy source, reducing agent, and maintaining bed permeability in blast furnace operation. A pyrolysis process is used to convert high-grade coal to coke, tar, and gas in a coke oven unit. The main product, coke, is transported to the top of the blast furnace and used as a reducing agent. Tar vapor and gases are the by-products and contain high amounts of carbon and energy. The tar material is mainly composed of condensable organic materials and may cause operational problems such as pipe plugging, condensation, and tar aerosol formation [1]. For the gasification

* Corresponding author: Konrad.Motylinski@ien.com.pl

processes it is crucial to achieve a long-life catalyst for biomass conversion. One of the major mechanisms in the deactivation of gasification catalysts is carbon deposition. This process results in a decrease in the active surface area of the catalytic metal particles. The strategy to prepare catalysts with a tolerance for carbon deposition is to use a basic support material, achieve a high dispersion state of metal particles, and introduce promoters to the catalysts [2]. Oxy-fuel IGCC is a concept of new coal fueled power generation with CO_2 capture capability without large penalty on thermal efficiency. An imperative issue to be solved to realize this concept is to reconcile pre-combustion sulfur removal and retaining the inorganic gas composition of the coal derived gas. Although introduction of dry gas sulfur removal is desired to protect gas turbine from corrosion and to preserve the gas composition, the sulfur removal sorbent containing zinc ferrite installed in the process may suffer from the severe side reaction of carbon deposition [3]. The process of carbon solid particles deposition also on the porous surface might result in changing of the flow parameters. The deposit, also called soot, alters the structure which later affects thermal management and flow pattern of channel. Additionally, in reactors and electrochemical systems, the carbon layer affects the geometry of the functional layer and leads to a significant deterioration of the physical-chemical properties of the material and can considerably suppress the ability to withdraw the heat from the process in case of high temperature systems. An example of the porous surface where carbon deposition can lead to severe disorders and deterioration of the parameters are the electrodes of Solid Oxide Fuel Cells (SOFC). The problem can be significant in fuel cell systems which are fed by fuel which composition and flow can fluctuate, for example SOFCs combined with gasifiers [4].

Solid oxide fuel cells are the high-temperature electrochemical energy conversion devices which operate at elevated temperature, typically in range 600-900°C. The high operating temperature allows for internal reforming process of hydrocarbonaceous fuels, thus SOFCs can be fueled with various gases containing hydrocarbons and their mixtures as well alternative energy carriers including dimethyl ether and ammonia [5]. As a result, under proper thermodynamic conditions, the deposition of solid carbon particles on the SOFC anode might occur. An example of the massive carbon deposition on the surface of a planar SOFC is shown in Fig. 1. Consequences of the soot formation are clearly observable on parts, exposed to anodic subspace (A and B). In this case high extent of deposits caused a complete cell disintegration.

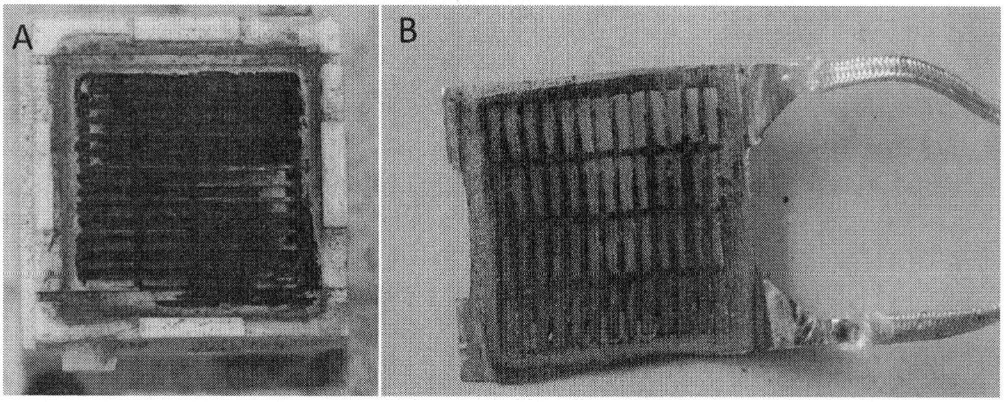

Fig. 1. Massive carbon deposition on the surface of a planar SOFC, where: A - manifold after removing the SOFC cell, B - golden current collector [6]

The formation and deposition of carbon on the porous surface is a well-known problem, which was studied and analyzed by many research groups. Significant number of articles were published, approaching the carbon deposition problem in three different yet correlated aspects.

The first aspect concerns studying the influence of anode porous material type on the carbon deposition process [7]. In the field of fuel cells with solid oxide electrolytes, two major problems were identified when fuelling with gases composed of carbon and hydrogen. One problem is high sensitivity to impurities such as sulfide and chloride, which poisons the SOFC anode and leads to the rapid performance degradation [8]. The other problem is carbon deposition which can be accelerated by catalysts present in the materials making the SOFCs. Nickel/yttrium-stabilized zirconia (Ni/YSZ) cermet is the most commonly used SOFC anode material due to its catalytic effect on the electrochemical oxidation of fuel. At the same time, nickel is an excellent catalyst for the carbon deposition reactions, such as methane cracking (Eq. (1)), reduction of carbon monoxide (Eq. (2)) and disproportionation of monoxide (Eq. (3)) [7, 9-11].

$$CH_4 \rightarrow C + 2H_2 \tag{1}$$

$$CO + H_2 \rightarrow C + H_2O \tag{2}$$

$$2CO \rightarrow C + CO_2 \tag{3}$$

The deposited carbon can deactivate and eventually passivate the Ni-catalyst and can cause rapid degradation of the cell's performance and its mechanical stability through worsening of the thermal management and the distribution of reagents. Because of these, several research groups proposed alternative dopants to the anodic materials, which hinder the carbon deposition process, without compromising the fuel cell performance [12-14].

The second area of research is oriented at studying the effects of various SOFC operation parameters, such as the temperature, fuel type or the steam-to-carbon ratio (S/C) which describes the ratio of molar amount of water steam to molar amount of carbon elements in the anode gas and SOFC current load. Subotic et al. [10] analysed the SOFC behavior under various operating conditions as well as the carbon formation process in regimes of different fuel compositions. They have prepared a set of experimental data showing the correlations between specific fuel parameters, affecting the carbon deposition mechanisms. Koh et al. [15] found that carbon deposited with humidified methane on Ni/YSZ porous surface is reversible when it is oxidized at a current load of thermodynamically carbon-free condition. This reversible carbon deposition does not affect the performance of a cell. The results published by Chen et al. [9] show that the morphology of the deposits and the degradation of the cell performance were significantly influenced by the operating temperature of the SOFC. They stated that for the fuel cell with Ni/YSZ anode fueled with syngas, the cell degraded faster once the carbon was formed and deposited. The damages at the microstructural level in the proximity of anode surface induced a cell's conductivity decrease. Alzate-Restrepo et al. [16] studied the electrolyte-supported cell with Ni/YSZ anodes exposed to CO/H_2 mixtures at the temperature of 1073 K operating at the current density of 10 mA/cm^2. During the tests the carbon formation occurred after a longer period of time (in the worst case after 6 h of operation) and led to deactivation of the anode by increasing the polarization resistance, reducing the porosity and blocking the active sites. Singh et al. [17] investigated the rate and speed of carbon deposition process in a SOFC fueled by tar biogas and operating in wide range of parameters. The thermodynamic analysis indicated that the amount of carbon decreased as the steam content of the fuel stream was increased. Carbon deposition found to be negligible as the operating conditions were varied to achieve the fuel mixture with higher water content (from 5% to 15% of H_2O in feed gas).

Kupecki et al. [18] discussed the effects of flow parameters on the stack performance in different flow configurations which can affect the thermodynamic parameters in the anodic compartments for different fuels. The problem of carbon deposition in micro-cogenerative power systems was discussed with respect to serially connected fuel cell stacks [19].
In such a configuration the role of a fuel processor is played by the first stack, therefore the threat of carbon formation inside the anodic compartments of the second or third stacks in series is eliminated. Depending of the fuel type and the expected system functionality, the fuel processing unit has to be adjusted to enable high electrical and overall efficiency as well eliminate the potentially dangerous issues, such as the discussed carbon formation [20].
The third aspect is the usage of the different mathematical models for describing and simulating the mechanisms and conditions required for the carbon formation and deposition process to take place. The common method to simulate the coke formation and deposition process is the Gibbs Free Energy Minimization (GFEM) method, which allows to compute the field of thermodynamic parameters which favor the formation of solid carbon particles [21-23]. The key parameter affecting the carbon formation reaction is the ratio between the steam and carbon (S/C) which is defined according to Eq. 4.

$$S/C = \textit{molar flow of steam / molar flow of carbon-containing gases (except } CO_2) \quad (4)$$

Additionally, the temperature of the cell and the gases affects the process. The higher is the temperature, the longer period of time is required for the solid carbon particles to deposit on the porous surface. The correlation used for this study is based on creating ternary diagrams, commonly called the Gibb's diagrams (see Fig. 2).

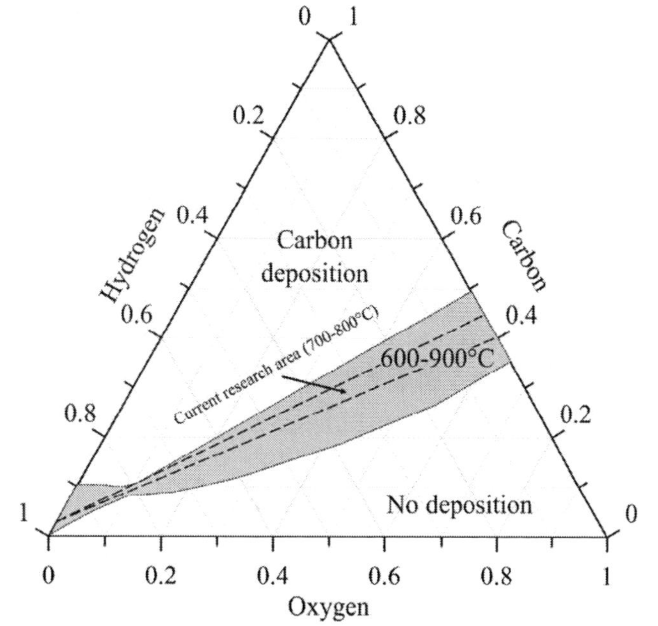

Fig. 2. Gibb's diagram at 1 atm. (data based on [24]). Gray area is a domain where carbon deposits may have transient morphology, from fine soot at lower boundary to graphite on upper.

The theoretical boundaries of the carbon deposition process are defined by the carbon formation isotherms, which depict the equilibrium composition for different gas mixtures, which contain carbon, hydrogen and oxygen atoms. The Gibb's diagrams aid in estimating the regions of parameters and compositions which should be generally avoided [24]. However, considered model are targeted on boundary region, where carbon formation is expected, its morphology is not settled (Fig. 2).

Despite the existence of numerous publications reporting the studies of the influence of SOFC operation parameters on the coke deposition process, authors clearly point the need for *in situ* analysis combined with numerical modeling. The two combined methods can enable detailed multi-variant study of selected parameters and their effect on carbon formation and deposition. Additionally, in the current literature lacks the reports on the verisimilar impact of the anode gas velocity and degree of flow turbulence (Reynolds number) on the speed and rate of coke concretion. For that reason a new modeling approach was proposed.

2 Methodology

The proposed model of a solid oxide fuel cell is based on zero-dimensional approach, which is a commonly used method. Solid Oxide Fuel Cell voltage is defined by the following function [25,26]:

$$E_{SOFC} = (\mathcal{E} - i_{max} \cdot r_1 \cdot \eta_f) / (r_1 \cdot (1 - \eta_f)/r_2 + 1) \tag{5}$$

where:

$\mathcal{E}$ – e.m.f, the maximum voltage defined by the work of an isothermal process; η_f - fuel utilization factor defined by the current working conditions of the cell; i_{max} – the maximum current density limited by amount of delivered fuel or oxidant; r_1 – the area specific resistance defined by ionic conductivity of the solid electrolyte and r_2 – the area specific resistance defined by electronic conductance [27].

The fuel cell generates power in processes at constant temperature, i.e. isothermal process. Thus, the maximum voltage is defined by the maximum work of the isothermal process. Mixtures of various components occur in the case of the analyzed fuels. Given these circumstances, the general form of Nernst's equation is used to estimate the e.m.f. of the integral electrochemical process in SOFC:

$$\mathcal{E} = R \cdot T \cdot ln(p_{O2,cathode}/p_{O2,anode}) / 4F \tag{6}$$

where: T – the absolute temperature; R – the universal gas constant; F – the Faraday constant; $p_{O2,cathode}$ - oxygen partial pressure at the cathode outlet; $p_{O2,anode}$ - oxygen partial pressure at the anode outlet. Taking in account negligible electronic transference numbers of YSZ in conditions, typical for SOFC (see, for example [28,29]), voltage losses associated with electrolytic permeability of the solid electrolyte membrane may be ignored. The open circuit voltage E_{OCV} formula was written as follows:

$$E_{OCV} = \mathcal{E} / (r_1/r_2 + 1) \tag{7}$$

The maximum cell current presented in Equation 5 depends on the fuel flow and Faraday constant:

$$I_{max} = 2F \cdot n_{H2,eq} \tag{8}$$

Factor $n_{H2,eq}$ was introduced to account for variable composition of incoming fuel. Under assumption that fuel undergoes complete external or internal reforming reaction, the hydrogen equivalent molar flow to the anode compartments of a cell writes:

$$n_{H2,eq} = n_{H2} + n_{CO} + 3n_{CH3OH} + 4n_{CH4} + 6n_{C2H5OH} + 7n_{C2H6} + 10\, n_{C3H8} \tag{9}$$

Hydrogen equivalent can be calculated for gases containing different hydrocarbons and for other hydrogen carriers. By this mean different fuels can be compared on the same basis. Hydrogen equivalent becomes of high attention when performance of SOFC fed by different fuels is evaluated. Electric current in a fuel cell depends on a gas flow, therefore fuel utilization factor η_f was introduced to correlate these. The factor is basically defined as a ratio of fuel consumed by the cell to generate electrical current I, to the maximal current I_{max} dependent on total amount of fuel supplied to a cell:

$$\eta_f = I_{SOFC} / I_{max} \tag{10}$$

Model was implemented in Aspen HYSYS 8.0 modeling software. Full description of the methodology was presented in [18]. The numerical model was also preliminary validated, simulating SOFC stack operation, fueled by the gas mixture of hydrogen and nitrogen [18]. For the current study, the presented SOFC mathematical model is combined with the model which computes the equations of carbon formation isotherms [23]. Theoretical boundary between non-deposition and deposition can be defined by finding equilibrium composition for different gas mixtures, based on three elements: carbon, hydrogen and oxygen. The set of mathematical equations refers to water, carbon monoxide, carbon dioxide, methane and hydrogen. Boundaries can be found for different temperatures as so-called carbon formation isotherms. Gases with different compositions can be located on the Gibb ternary diagram (see Fig. 2) and evaluation of the possible threat of solid carbon formation and deposition can be performed.
The equilibrium can be described by the following reactions:

$$CH_4 \leftrightarrow <C_{solid}> + 2H_2 \tag{11}$$

$$<C_{solid}> + H_2O \leftrightarrow CO + H_2 \tag{12}$$

$$CO_2 + H_2 \leftrightarrow CO + H_2O \tag{13}$$

$$2H_2O \leftrightarrow O_2 + 2H_2 \tag{14}$$

The following equilibrium constants correspond to reactions 11-14 respectively: K_1, K_2, K_3 K_4. The equilibrium in the system can be then excepted after simple rearrangement as four equations [30].

$$K_3 = n_{CO} \cdot n_{H2O} / n_{CO2} \cdot n_{H2} \tag{15}$$

$$K_1 \cdot K_2 = p^2 \cdot n^3{}_{H2} \cdot n_{CO} / n_{CH4} \cdot n_{H2O} \tag{16}$$

$$4n_{CH4} + 2n_{H2} + 2n_{H2O} - n_H \cdot S = 0 \tag{17}$$

$$n_{H2} = const \tag{18}$$

with the normalization condition:

$$n_{CH4} + n_{CO} + n_{CO2} + n_{H2} + n_{H2O} = 1 \tag{19}$$

where factor S of equation 17:

$$S = 5n_{CH4} + 2n_{CO} + 3n_{CO2} + 2n_{H2} + 3n_{H2O} \tag{20}$$

Equation for the content of water in the system can be found by putting n_{H2}=A and n_H as a parameter, and eliminating variables from the above equations:

$$b_1 n_{3H2O} + b_2 \cdot 2n_{2H2O} + b_3 \cdot n_{H2O} + b_4 = 0 \tag{21}$$

where:

$b_1 = 2g_2 \cdot (1 + g_3)$

$b_2 = g_1 \cdot (2 + g_3) - 3g_2 \cdot g_3 + Ag_2 \cdot (2 + 3g_3)$

$b_3 = 2Ag_1 \cdot (1 + g_3) - 2(1 + g_1 \cdot g_3)$

$b_4 = A \cdot (g_3 - 2) + 4 - g_3$

$g_1 = K_1 \cdot K_2 / p^2 \cdot A^3$

$g_2 = K_1 \cdot K_2 / p^4 \cdot A^4 \cdot K_3$

$g_3 = n_H / (1 - n_H)$

The content of methane, carbon mono- and dioxide in the system can be found using equations 22-24, respectively.

$$n_{CH4} = (1 - n_{H2O} - A) / (1 + g_1 \cdot n_{H2O} + g_2 \cdot n_{H2O}^2) \qquad (22)$$

$$n_{CO} = g_1 \cdot n_{CH4} \cdot n_{H2O} \qquad (23)$$

$$n_{CO2} = g_2 \cdot n_{CH4} \cdot n_{H2O}^2 \qquad (24)$$

Amount of oxygen in the system is fund as:

$$n_{O2} = n_{H2O}^2 / n_{H2}^2 \cdot K_{pw} \qquad (25)$$

where K_{pw} is the equilibrium constant of water formation reaction. The positions of the equilibrium concentrations in the Gibbs triangle (corresponding to different content of carbon, hydrogen and oxygen) can be found using reactions 17, 26 and the normalization condition 27.

$$n_{CH4} + n_{CO} + n_{CO2} - n_C S = 0 \qquad (26)$$

$$n_H + n_C + n_O = 1 \qquad (27)$$

The algebraic method can be used for determining the equilibrium compositions of the gas mixture, however parameters n_{H2} and n_H have to be manipulated to do so. It was previously reported [31] that these methods are especially precise in the regions of a Gibbs triangle where equilibrium is highly sensitive to changes of n_C, n_H and n_O.

According to the calculation scheme, calculations can be performed for different temperatures in order to define carbon formation isotherms. Precise definition of the carbon formation boundary allows taking countermeasures to prevent this dangerous process from occurring.

For this study, complex experiments will be performed in order to analyze the formation and deposition of coke process on the SOFCs anode for different operating points of the fuel cell. Based on the realized experiments, a proper numerical model will be made, describing this process, and taking into consideration SOFCs operation temperature, fuel cell current load, steam-to-carbon ratio, including the process kinetics, which influence anode gas flow turbulence rate. Modeling of an SOFC using zero-dimensional dynamic models as well as modeling of static fuel cells on fixed operating point are presented in [23,25,27,32]. The prepared numerical method will be adopted and extended to enable use of mathematical tools for describing the carbon deposition mechanisms and later to allow the validation and tuning of the models using experimental data. The proposed methodology targets the determination of the operating parameters of SOFC which eliminate the risk of formation of the carbon layer on the porous surface of the anode. In addition, the consideration of flow parameters on the process kinetics allows to uniquely figure out impact of the flow turbulence on secretion and deposition of carbon on the surface of the surface of dense ceramic and on a porous layers. The usage of dedicated mathematical model will allow to generalize the mechanisms description in a parametric way, allowing the determination of process boundaries in units and processes using gases composed of hydrogen and carbon elements.

3 Conclusion

The proposed numerical tool combines the electrochemical model which is based on the electric circuit equivalent, and the calculation algorithm which serves as a computational model for determination of the carbon formation isotherms under given working conditions. The concept for correlating the coke formation and deposition in the anodic compartments relies on the foreseen reduction of the active area of the SOFC and/or deterioration of the ionic transport due to pore clogging. This can be included in the dynamic model by correlation of the area specific internal ionic and electronic resistances of the cell. In order to identify the process, correlate the rate of formation, thickness of the deposit layer and the cell's performance electrochemical degradation, the series of experimental investigations is needed. Further work will be related to the experimental analysis of the process in order to provide data for the model calibration and validation. The objective is to qualitatively and quantitatively analyze selected factors and their influence on the electrochemical fuel oxidation in regimes in which the soot formation can be expected. This will be achieved using a test bench which is currently under construction

Acknowledgment This work was financially supported by the National Science Centre, Poland, Grant No. 2015/19/N/ST8/01876.

References

1. R. Cahyono, N. Yasuda, T. Nomura, T. Akiyama, Fuel Proc. Tech. **119**, 272-277 (2014)
2. J. Ozaki, M. Takei, K. Takakusagi, N. Takahashi, Fuel Proc. Tech. **102**, 30-34 (2012)
3. M. Kobayashi, Y. Nakao, Fuel **152**, 19-28 (2015)
4. K. Badyda, J. Kupecki , J. Milewski, Rynek Energii **88**(3), 74-79 (2010)
5. J. Kupecki, Int. J. Hyd. Energy **40**(35), 12009-12022 (2015)
6. J. Jewulski, M. Wierzbicki, A. Zieleniak, *Wykorzystanie gazu ze zgazowania w ogniwach paliwowych [in] Nowoczesne Technologie Pozyskiwania i Wykorzystywania Biomasy* (Institute of Power Engineering, Warsaw, p. 320-335, 2010)
7. J. Macek, B. Novosel, M. Marinsek, J. Eur. Ceram. Soc. **27**, 487-491 (2007)
8. F. Cayan, M. Zhi, S. Pakalapati, et al., J. Power Sources **185**, 595-602 (2008)
9. T. Chen, W. Wang, H. Miao, et al., J. Power Sources **196**, 2461-2468 (2011)
10. V. Subotic, C. Schluckner, C. Hochenauer, J. Energy Inst. **89**, 121-137 (2016)
11. Y. Zhang, Z. Yang, M. Wang, J. Power Sources **185**, 595-602 (2008)
12. T. Takeguchi, R. Kikuchi, T. Yano, et al., Catal. Today **84**, 217-222 (2003)
13. T. Takeguchi, Y. Kani, T. Yano, et al., J. Power Sources **112**, 588-595 (2005)
14. T. Borowiecki, A. Gotebiowski, B. Stasifiska, App. Catal. A: General **153**, 141-156 (1997)
15. J.H. Koh, Y.S. Yoo, J.W. Park, et al., Solid State Ionics **149**, 157-166 (2002)
16. V. Alzate-Restrepo, J. Hill, J. Power Sources **195**, 1344-1351 (2010)

17. D. Singh, E. Hernandez-Pacheco, P. Hutton, et al., J. Power Sources **142**, 194-199 (2005)
18. J. Kupecki, J. Milewski, A. Szczesniak, R. Bernat, K. Motylinski, Int. J. Hyd. Energy **40**(45), 15834–15844 (2015)
19. J. Kupecki, M. Skrzypkiewicz, M. Wierzbicki, M. Stepien, Int. J. Hyd. Energy (2016), In press, dx.doi.org/10.1016/j.ijhydene.2016.07.222
20. J. Kupecki, J. Jewulski, K. Badyda, Rynek Energii, **97**(6), 157-162
21. K. Girona, J. Laurencin, J. Fouletier, F. Lefebvre-Joud, J. Power Sources **210**, 381-391 (2012)
22. K. Eguchi, H. Kojo, T. Takeguchi, et al., Solid State Ionics **152**, 411-416 (2002)
23. J. Kupecki, *Analysis of Micro-Combined Heat and Power Unit with Solid Oxide Fuel Cells* (doctoral thesis, Warsaw University of Technology, 2013)
24. Z. Jaworski, B. Zakrzewska, P. Pianko-Oprych, Rev. Chem. Eng. (2016)
25. J. Milewski, K. Swirski, M. Santarelli, P. Leone. *Advanced Methods of Solide Oxide Fuel Cell Modeling* (Springer, 2011)
26. J. Kupecki, App. Mech. Mat. **607,** 205-208 (2014)
27. J. Kupecki, J. Milewski, J. Jewulski., Cen. Eur. J. Chem. **11**(5), 664-671 (2013)
28. D. Bray, U. Merten, J. Electrochem. Soc. **111**(4), 447-452 (1964)
29. L. Zhang, L. Zhu, A. Virkar, J. Power Sources **302**, 98-106 (2016)
30. C. Huang, S. Shy, C. Chien, et al., J. Power Sources **195**, 2260-2265 (2010)
31. J. Jewulski, L. Suski L. *Thermodynamic Data For Technology, Series B: Chemical equilibrium diagrams relevant to the molten carbonate fuel cell* (Polish Scientific Publishers, Cracow, I edition, 1985)
32. J. Milewski, Fuel Cells **12**, 709-72 (2012)

Optimization of proton conductors for application in solid oxide fuel cell technology

Konrad Świerczek[1,*], *Wojciech* Skubida

[1]AGH University of Science and Technology, Faculty of Energy and Fuels, Department of Hydrogen Energy, al. Mickiewicza 30, 30-059 Krakow, Poland

Abstract. Application of proton-conducting ceramic membranes in various electrochemical devices (e.g. SOFCs, SOECs) is considered as highly beneficial. Because of a unique transport mechanism associated with proton movement in oxide materials, designing and optimization, as well as practical application of such proton conductors is a challenge. In this work, apart from discussion of literature data, authors' own results are provided, which are focused on material-related issues, including development of electrolyte and electrode materials exhibiting high proton conductivity.

1 Introduction

Among contemporary issues of energy sector, emission of carbon dioxide and low efficiency of conventional electricity generation systems are of crucial importance. Solid Oxide Fuel Cells (SOFC), which are electrochemical devices that convert chemical energy of a fuel (typically hydrogen) and an oxidizer directly into electricity and heat, and may exhibit high efficiency, are of special interest in an emerging hydrogen economy, i.e. usage of H_2 as the energy carrier, but also in so called clean coal technologies, in which traditional, carbon-containing fuel are utilized more effectively and in more environmental friendly way [1, 2]. Interestingly, if SOFC is supplied with electrical energy, it can also be used as Solid Oxide Electrolyzer Cell (SOEC), in which water can be split to produce hydrogen and oxygen. SOFC and SOEC technology relies on purely ion-conducting electrolytes (membranes), which are typically ceramic oxides conducting oxygen anions, e.g. yttria-stabilized zirconia (YSZ). Recently, in order to improve parameters of SOFCs and lower their operating temperature, a replacement of the oxygen ion-conducting electrolyte with material exhibiting proton conductivity was proposed and shown to be effective [3]. The same beneficial impact can be expected if proton-conducting membrane is used in the high-temperature electrolyzer.

In this work selected issues concerning application of proton-conducting oxides, having perovskite structure, are discussed. It seems that apart from controlled B-site substitution (acceptor-type), a simultaneous doping at the A-site can be also done, affecting considerably physicochemical properties of the materials. Consequently, structural parameters, protonic conductivity, as well as unwanted reactivity of the compounds in CO_2-containing atmospheres may be adjusted. Additionally, possible new trend in development of SOFCs with proton-conducting membrane, symmetrical cell concept, is presented.

* Corresponding author: xi@agh.edu.pl

2 Application of SOFC and SOEC technology

In a standard design, a single fuel cell comprises dense, gas-tight and purely ionically-conducting electrolyte, porous electrodes (anode and cathode), as well as interconnects serving as electrical connectors. Typically, high-temperature Solid Oxide Fuel Cell is supplied by hydrogen or another gaseous fuel, such as syngas or methane, on the anode side, while an oxidant (e.g. air, pure oxygen) is supplied on the cathode side. This is due to the nature of ionic conduction in the electrolyte (conduction of oxygen ions via oxygen vacancy mechanism) and mechanism of electrochemical reactions taking place on both electrodes [4]: at the anode: $H_2 + O^{2-} \rightarrow H_2O + 2e^-$; $CO + O^{2-} \rightarrow CO_2 + 2e^-$ or $CH_4 + 4O^{2-} \rightarrow 2H_2O + CO_2 + 8e^-$ (with internal reforming); at the cathode: $1/2O_2 + 2e^- \rightarrow O^{2-}$; giving overall: $H_2 + 1/2O_2 \rightarrow H_2O$; $CO + 1/2O_2 \rightarrow CO_2$ or $CH_4 + 2O_2 \rightarrow 2H_2O + CO_2$, respectively. However, relatively low power density obtained per square centimeter of the cell (on the order of single watts), as well as voltage of a single cell being in a range up to 1 V, force to combine cells into packets and stacks, in order to construct a generator capable for delivering desired electrical power, as well as heat (up to hundreds of kW_e) [5, 6]. Such possibility of assembling large-scale systems creates vast potential concerning application of SOFCs in many areas, from household to industrial systems. In practical application, residential Combined Heat and Power (CHP) units operating on natural gas can provide additional electrical power for home usage, as well as heat for central heating and domestic hot water (Fig. 1a). However, nowadays the biggest disadvantage of such devices is their limited power output (approximately up to 2 kW_e and 1 kW of thermal power). Consequently, so far they are rather considered as auxiliary units, not the main power source. Industrial-scale SOFC systems having higher power range are suitable for utilization of the waste heat energy, for example as shown in Fig. 1b, where gas turbine is also used for electricity generation. This kind of solution allows to improve overall efficiency up to ca. 60% [7].

Despite of many material-related problems (corrosion, thermal degradation, etc.), high operating temperature of SOFCs, up to 1000 °C, has some advantages. The most important one is the mentioned above (last overall reaction) possibility of usage of carbohydrates as a fuel, without a need for a complicated and expensive external reforming. Inside the cell fuel, methane can be catalytically converted to hydrogen and carbon oxide, forming synthesis gas, which then can be electrochemically oxidized on the anode to carbon dioxide and steam, generating electricity and heat. The main reason limiting usage of methane as a fuel is its contamination with sulfur compounds, which cause cell poisoning, and results in a lowering of efficiency of the system. Higher concentrations of sulfur compounds can even cause mechanical destruction of the cells [8].

Another reason hindering usage of CH_4 is carbon deposition, which can occur on the anode as a result of uncontrolled fuel reforming. This can result not only in an efficiency decrease, but also in mechanical destruction of the unit. There are several ways to prevent such the effect, for example, steam excess or some oxygen can be applied, but it is always connected with lowering of the efficiency, as well as a decrease of lifetime of the device [10, 11]. Search for alternative materials that will be resistant to the abovementioned sulfur and carbon deposition-related problems is currently ongoing in many laboratories. In the case of SOECs, apart from production of hydrogen and oxygen from a splitting of water, an interesting other application can be also considered. If SOEC unit with the oxygen-conducting electrolyte is supplied with H_2O, CO_2, as well as with electrical energy, H_2 and CO (as well as O_2) can be produced, which can be used for production of synthetic fuels [12].

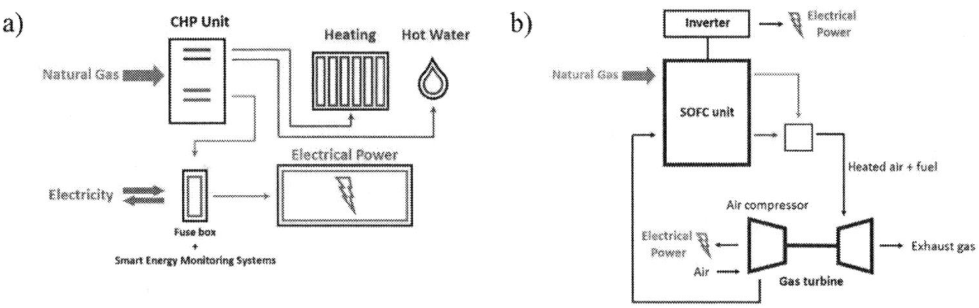

Fig. 1. a) CHP generator for domestic use, scheme [9]; b) SOFC system with gas turbine [7].

Replacement of conventional electrolyte transporting oxygen anions by proton-conducting compounds in various electrochemical devices is a relatively new idea. Such cells are often abbreviated as PCFCs, Proton Ceramic Fuel Cells, and are described in more details in the following chapter.

3 Proton Ceramic Fuel Cells and Electrolyzer Cells

Usage of proton conductor as electrolyte brings important advantages, concerning operation of SOFCs and SOECs. Apart from a change of direction of flow of ions in the electrolyte, the most important ones are related to changes of the reactions taking place at the respective electrodes. If supplied with hydrogen and oxygen, water, being a product of the reaction in the fuel cell, is formed on the cathode side of the cell (Fig. 2a), rather than at the anode. This enables a more efficient use of the fuel supply downstream [13]. The same effect concerns water vapor electrolyzer, which produces hydrogen from water vapor (Fig. 2b). If proton-conducting electrolyte is used, the efficiency of the process increases, and the obtained product (i.e. hydrogen) is undiluted. Other advantages are related to a fact that the Nernst voltage is maintained at a high level, and there are no problems with reduced stability of the anode, resulting from a presence of the water vapor [4]. Proton-conducting electrolytes show also higher values of the conductivity in the intermediate temperature range (400-600 °C), which is especially interesting concerning application [13]. Unfortunately, it is not possible to fuel PCFC stack directly with hydrocarbon fuels.

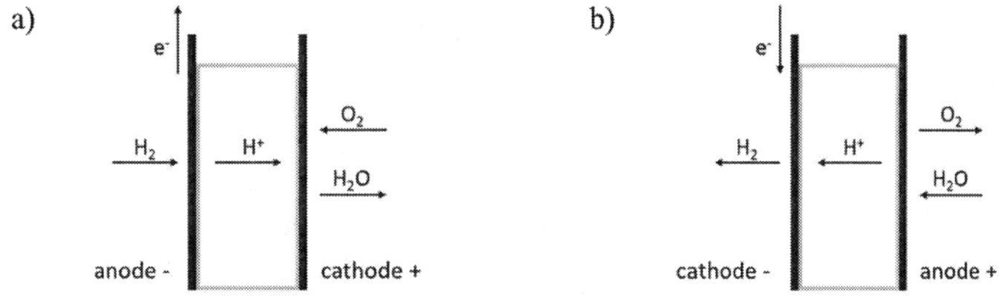

Fig. 2. Diagram of a fuel cell (a), and water vapor electrolyzer (b), with proton-conducting electrolyte.

The basis for design of Protonic Ceramic Fuel Cells and also Solid Oxide Electrolyzer Cells with proton-conducting electrolyte are materials from a group of so called High Temperature Proton Conductors (HTPC) [14]. This group of oxides consists mostly of acceptor-type substituted perovskite-type oxides, and is discussed below.

While all the above indicates that proton conductors are promising electrolytes for the intermediate or low temperature PCFCs and SOECs, the reported so far performance is still inadequate for practical application, primarily due to a lack of proper electrode materials, which would be compatible with such electrolytes, as well as fundamental issues concerning

DOI: 10.1051/e3sconf/20171401044

presence of mixed, proton, oxygen ion and electron conduction in a particular material at the same time [15]. In most cases, when it comes to testing of the proton- conducting electrolytes in the cells, mixture of electrolyte material and metal oxide (which during operation is reduced to metal) is used as the anode. Typically NiO is utilized, and its cermets with the electrolyte powder display excellent catalytic properties for hydrogen oxidation, as well as allow for a good current collection. Unfortunately, there is no "ideal anode material" known, i.e. exhibiting mixed protonic and electronic conductivity at the same time, which would allow to use a single phase-type anode material (Fig. 3a). With usage of cermets, the electrochemical reaction at the anode is limited to the triple phase boundary (electrolyte, anode material and gas phase), marked as a circle in Fig. 3b. Unfortunately, nickel metal in the cermets tends to agglomerate after prolonged operation, leading to a reduced three-phase boundary and increased cell's resistance [4]. That indicates the importance of searching for new materials, also oxides, which can replace Ni.

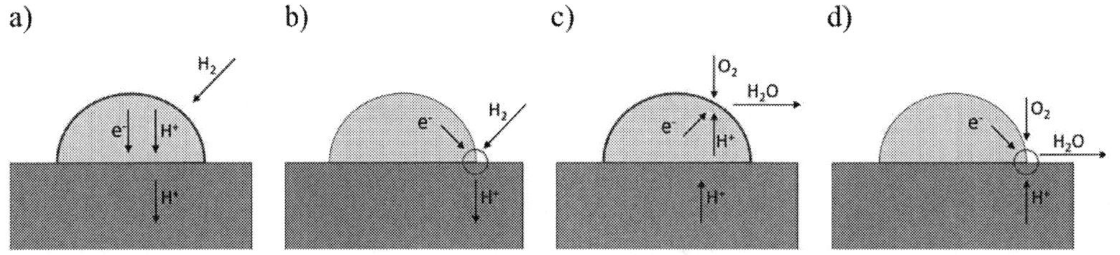

Fig. 3. Anode and cathode reactions in PCFC: a, c) mixed protonic-electronic "ideal" electrode, b, d) only electronically-conducting electrode (in this case usage of cermet-type electrode is necessary).

The electrochemical reaction at the cathode in PCFCs is shown in Fig. 3c and d. Due to similar issues, no "ideal" cathode material is available, and also cermets are utilized, in which, apart for the electrolyte phase, secondary material having good electrical conductivity is needed. Typically, such materials are $ABO_{3-\delta}$-type perovskites, which show high catalytic activity towards oxygen reduction, e.g. $La_{1-x}Sr_xCo_{1-y}Fe_yO_{3-\delta}$. Interestingly, such materials are actually used in SOFCs with the oxygen-conducting electrolyte [15, 16]. As can be concluded, in the case of cells with proton-conducting electrolyte, composite-type electrodes must be used, in which the overall efficiency is related the triple phase boundary length and density in the electrode volume. Therefore, optimization of the composites (from the point of view of selection of materials and also microstructure) is necessary in order to obtain functional electrodes. Illustration of PCFC with a mixed ionic-electronic conducting composite-type electrodes is shown in Fig. 4.

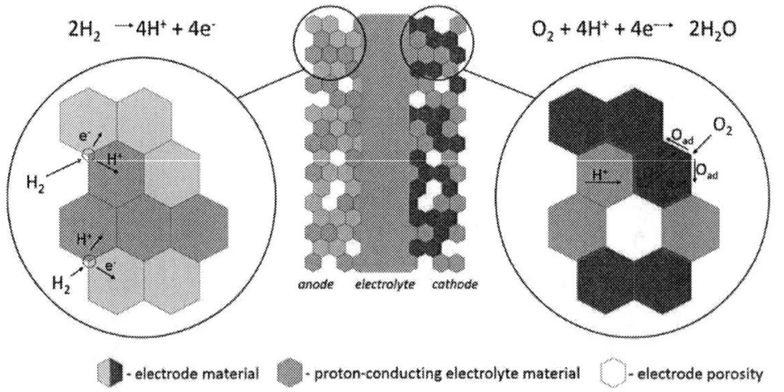

Fig. 4. Schematic illustration of PCFC with mixed ionic-electronic conducting composite electrodes. Zoomed areas represent model of expected charge transport mechanisms in the anode and cathode. Blue regions marked on the sides correspond to the areas where electrode reactions are expected to occur.

4 Proton transport in perovskite-type oxides

The mentioned HTPC oxides do not have protons in their structure, and for the proton conduction occurrence, humidified atmosphere is needed. Typically, these compounds possess perovskite-type crystal structure, however other material are also known, as briefly summarized in the chapter 5. In the dry atmosphere such perovskite-type oxides are characterized by a rather insulating properties, with a very low value of the *p*-type electronic conductivity. In the presence of water vapor and at high temperatures, protons ca be built into crystal structure of the material. In most cases it is easily achieved due to high negative enthalpy of the process [4]. Proton incorporation causes lowering of electronic component of the conductivity and an appearance of the high protonic conductivity with low activation energy, which occurs at a relatively wide range of temperatures and oxygen partial pressures [17, 18]. Protons are absorbed into the structure of HTPC from the surrounding atmosphere that contains water vapor and/or hydrogen, according to the following chemical reactions $H_2O + V_O^{\bullet\bullet} + O_O^x \leftrightarrow 2OH_O^{\bullet}$ and $H_2 + 2O_O^x \leftrightarrow 2OH_O^{\bullet} + 2e'$, respectively.

In the case of perovskite oxides, the second reaction is considered to be dominant. The water vapor dissociates at the surface of the material, forming hydroxide ion and the proton, then, the OH^- ion takes place of the oxygen vacancy site, while proton forms a bond within the oxygen lattice [19]. The adsorption of water is an exothermic reaction, which explains the dominant nature of the proton conductivity at low temperatures, while usually high-temperature conductivity is dominated by the vacancy mechanism (conduction of oxygen) [20]. The above-described mechanism of the appearance of mobile protons in the structure of the oxide requires presence of the oxygen vacancies. This can be realized by an appropriate introduction of the acceptor-type dopant. To form oxygen vacancies in perovskite oxides, in which the A cation is at the +2 oxidation state and the B cation possesses +4 charge, a partial substitution by the trivalent cation (M) at the B site is used. The reaction of formation of the oxygen vacancies in dry atmosphere can be written as [19]: $2B_B^x + O_O^x + M_2O_3 \leftrightarrow 2M_B' + V_O^{\bullet\bullet} + 2BO_2$. In the steam-containing atmosphere protons can be incorporated directly as charge compensating defects during synthesis, which takes place according to the equation [4]: $M_2O_3 + 2AO + H_2O_{(g)} \leftrightarrow 2A_A^x + 2M_B' + 4O_O^x + 2OH_O^{\bullet}$. The solubility of the acceptor addition increases with the partial pressure of water, and the protons can be the dominant defects ensuring electroneutrality condition. However, practical use of the above reaction is difficult to implement, because the synthesis (annealing, sintering) would have to be carried at high temperatures, in which protons are only a small part of the existing defects [4].

Generally, there are two ways of proton transport described in the literature: vehicular mechanism and Grotthuss mechanism, shown schematically in Fig. 5. In the first one, the diffusion of protons takes place through movement of the carrier accompanied by the H^+ ion. Most often this is OH^-, H_3O^+ or NH_4^+ group [4]. Simulations and isotopic studies indicate that in perovskite oxides high temperature proton transport occurs within the structure according to the Grotthuss mechanism. In this case, H^+ ions move between successive oxygen anions, and not with OH^- medium, as in vehicular mechanism [21]. It can be also stated that the mechanism of H^+ ions transport is similar in different families of oxides, due to the high degree of self-localization and a strong coupling between protons moving in the same crystal structure. This means that the hydrogen cations are located in the electron cloud of the oxygen anions, to form hydroxide ions having a positive charge $OH_O^{\bullet}$ in respect to the (oxygen) lattice [4]. While proton interacts strongly in the created O-H bond, it is affected by other, surrounding oxygen anions. Depending on the distance between two oxygens (O-O), protons can form the hydrogen bond between the two oxygen anions, which can be symmetrical (i.e. of the same length), if O-O distance is less than 0.24 nm. However, if the distance is in the 0.25-0.28 nm range, proton is more strongly associated with one of the adjacent oxygens, and the binding sites are of different length [14]. The activation energy of the proton hopping from one oxygen to the next one depends on the O-O distance, due to a decrease of the electron density and increase of the potential between oxygens. Regardless of a class of oxide, it is approximately 2/3 of the value of the activation energy of oxygen conduction (through vacancy mechanism). This observation can be explained by a fact that migration of protons is dependent on the oxygen network vibrations. Both, proton and oxygen conductivity require to overcome the same energy barrier, but while O^{2-} anion must overcome energy barrier of the migration process between the energy minima, the proton can jump when ion on which it is located overcomes 2/3 of the energy barrier [4].

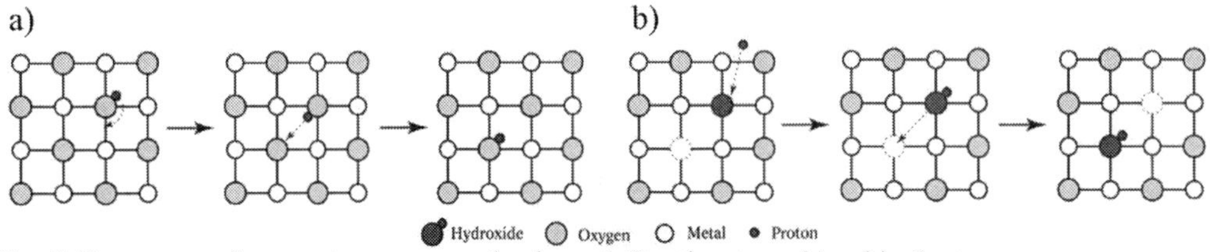

Fig. 5. Two types of proton transport mechanism: a) Grotthus type, b) vehicular type

Proton conductivity shows an important phenomenon called isotopic effect. This effect stems from a large difference in mass of H^+ and D^+. Due to a higher mass, deuterium isotope moves slower through structure of the material (oxide) than hydrogen. There are two models describing such the effect: a classic model, which only takes into account change between frequency of vibration O-H and O-D, and a pseudo-classic one that additionally takes into account quantum effects connected with the isotope change. From the first model, which says that mobility of protons μ in the HTPC is proportional to the frequency of hydrogen bond vibration, it can be stated that $\mu_H/\mu_D \approx \sqrt{2}$. Assuming that isotope change does not affect concentration of charge carriers, conductivity ratio can be given as $\sigma_H/\sigma_D \approx \sqrt{2}$. The pseudo-classic model allows to compute change of the energy of the basic state ΔE_0 between proton and deuterium, which is 0.054 eV, and such difference of the activation energy of conductivity can be expected in the temperature range at which proton conductivity is dominant [22, 23].

5 Overview of proton-conducting oxides

In 1980s and 1990s Iwahara et al. discovered proton conductivity in $(Ba,Sr)CeO_{3-\delta}$ oxides at high temperatures [24, 25], which triggered great interest and intensive research concerning HTPCs in many laboratories worldwide [26-29]. Initially, the B-site substituted $(Ba,Sr)(Zr,Ce)O_{3-\delta}$ perovskites were studied [30]. It was shown that choice of cations with different ionic radii has a profound effect on the degree of distortion from the cubic structure in these oxides. Considering preference of the cubic structure, B-site doped $BaIn_{1-y}M_yO_{3-\delta}$ (M^{4+} = Zr, Sn) materials, having high content of In and a disordered oxygen vacancy sublattice, were also found to be of great interest [31, 32]. Analysis of results of the available studies indicates that good proton conductivity is connected to high alkalinity of A cations and a high deviation from stoichiometry in the oxygen sublattice, but without oxygen vacancy ordering present [4, 33]. Chemical composition of the oxide is strongly affecting mobility of protons inside the structure of perovskite-type oxide. The low value of the enthalpy of mobility of protons can be attributed to a presence of large cations in the A-sublattice, which allows for a better dynamics of movement of H^+. The enthalpy of proton migration is also affected by the repulsion between H and B cations. In regular perovskites, where the B cation is on +5 oxidation state, the energy is much higher than for materials, in which the cations are at the +4 oxidation state [18]. Due to a small value of the energy migration and low defect formation energy, perovskite oxides can be generally considered as good proton conductors.

Perovskites with a cubic structure (with Goldschmidt's tolerance fact $t \approx 1$), exhibit higher conductivity than materials exhibiting lower symmetry, because positions of oxygen ions are energetically equal, so there are no places in the structure that would require higher amount of energy, and H^+ jump is possible in any direction with equal probability [4]. In a group of perovskite-type HTPCs, one of the highest proton conductivity (on the order of 10^{-2} S·cm^{-1} at 600 °C show oxides based on the substituted $BaCeO_3$. Unfortunately, these materials exhibit a significant decrease of the transfer of protons at elevated temperatures (proton transference number decreases) [34]. Although $SrCeO_3$-based oxides have a lower conductivity of H^+ ions, their high temperature proton transference number remains high. However, problem associated with practical use of the abovementioned compounds is their chemical instability and relatively high reactivity in atmospheres containing CO_2 or SO_3 and water vapor [11]. The alternative candidates may be strontium or barium zirconates, having excellent chemical stability over a wide temperature range. However, they are characterized by a significantly lower conductivity (influence of grain boundaries), comparing to $BaCeO_3$ [4]. Among high temperature proton conductors there are materials with more complex structure or a derivative from the perovskite one, such as for example $Sr_{6-2x}Nb_{2+2x}O_{11+3x}$, $Sr_{6-2x}Ta_{2+2x}O_{11+3x}$ or $(Ba_{1-y}Ca_y)_6Nb_2O_{11}$. Those material can incorporate one molecule of water per one molecule of compound, and show pure protonic conductivity (10^{-3} S·cm^{-1}) at 600 °C [35]. Other type of the structure that shows proton conductivity is brownmillerite one, observed for $Ba_2(In_{1-x}Ti_x)_2O_{5+\delta}$ materials, which show H^+ conductivity of 10^{-3} S·cm^{-1} at 450 °C [36]. Recently, also La_6WO_{12}, $La_2Zr_2O_7$ or $LaNbO_4$ materials are of interest, which stems from a lack of Ba in their structure [37, 38].

6 Structure and transport in $Ba_{0.9}La_{0.1}Zr_{0.25}Sn_{0.25}In_{0.5}O_{2.8}$

Choice of the chemical composition of perovskite-type HTPC has profound influence on the physicochemical properties of the material. While acceptor-type chemical substitution on the B-site is commonly used to modify properties (mainly, increase concentration of the intrinsic oxygen vacancies), introduction of +3 cations (Ln^{3+}) at the A-site is relatively less studied, as it would act in opposite way, concerning concentration of $V_O^{\bullet\bullet}$. However, for a small degree of substitution, such doping may be of interest, especially concerning studies of relationship between crystal structure and transport properties of the materials.

For the particular example of $Ba_{0.9}La_{0.1}Zr_{0.25}Sn_{0.25}In_{0.5}O_{2.8}$ composition, which was obtained by a solid state reaction at 1600 °C, crystal structure was determined as cubic, with *Pm-3m* symmetry, and unit cell parameter a = 4,1676(1) Å. The recorded in different atmospheres (synthetic dry air, wet air with ca. 3 vol.% H_2O or D_2O) impedance data for a dense sinter, presented in Nyquist-type plots, are shown in Fig. 6. As can be seen in Fig. 6a, at 400 °C the impedance curve comprises of three semi-circles, first of which (at the highest frequency) can be attributed to the bulk ionic conductivity, second one to the grain boundaries, and the one visible at lowest frequencies to the influence of the electrode. A visible decrease of size of first two, when the atmosphere was switched from dry air to the D_2O- and H_2O-containig one proves an existence of proton (or D^+) component of the conductivity [39, 40]. At the same time, impedance data registered at 600 °C suggests an appearance of electronic component of the conductivity, as the recorded spectra are shifted in relation to the origin. Summary of the transport data in $Ba_{0.9}La_{0.1}Zr_{0.25}Sn_{0.25}In_{0.5}O_{2.8}$ is presented in Tab. 1. While the protonic (or deuterium-related) conductivity is relatively high at 550 °C, the respective transference numbers are low in the considered oxide. This indicates that in temperatures below 600 °C the oxygen vacancy-related conductivity is considerable. The mentioned appearance of the electronic component of the conductivity at 600 °C and above, while detrimental for the electrolyte behavior, may be utilized considering application of this material for construction of the electrodes in PCFCs (e.g. Fig. 3c).

 DOI: 10.1051/e3sconf/20171401044

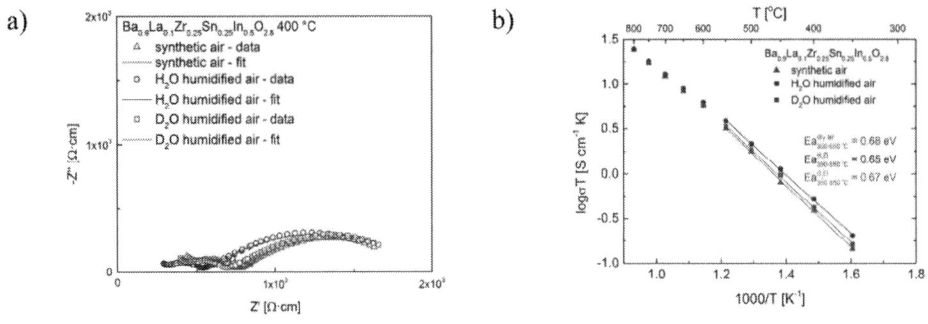

Fig. 6. a) Nyquist-type plot of impedance data recorded at 400 °C, b) temperature dependence of electrical conductivity for $Ba_{0.9}La_{0.1}Zr_{0.25}Sn_{0.25}In_{0.5}O_{2.8}$ in different atmospheres.

Table 1. Transport properties of $Ba_{0.9}La_{0.1}Zr_{0.25}Sn_{0.25}In_{0.5}O_{2.8}$ at different temperatures.

property	T = 350 °C	T = 550 °C
Protonic conductivity σ_H (σ_{H2O} - σ_{dry}) [$S\cdot cm^{-1}$]	$0.9\cdot10^{-4}$	$0.8\cdot10^{-3}$
Deuterium conductivity σ_D (σ_{D2O} - σ_{dry}) [$S\cdot cm^{-1}$]	$0.3\cdot10^{-4}$	$0.2\cdot10^{-3}$
Protonic transference number t_H (σ_H/σ_{H2O})	0.29	0.18
Deuterium transference number t_D (σ_D/σ_{D2O})	0.11	0.05

7 Concept of Symmetrical Fuel Cell

A novel idea of a symmetrical cell, built using a redox-stable electrode material as both, the anode and cathode, was shown recently [41]. In the classic cell, anode and cathode are made of different materials, which is directly connected with very strict requirements (e.g. chemical stability and sufficient transport properties under reducing and oxidizing atmospheres). In the symmetrical construction, both electrodes are made of the same oxide material or composite (Fig. 7a). Initially, this approach was utilized alleviate problems with possible sulfur poisoning or coke formation on the surface of the SOFC anode operating with hydrocarbon-type fuels, e.g. methane, but also synthesis gas. It was shown that anodes made of Mo- and W-containing materials (e.g. $Ba_{2-x}Sr_xM_{2-x}Mo_xO_{6-\delta}$ and $Ba_{2-x}Sr_xM_{2-x}W_xO_{6-\delta}$, M: Mg, Fe, Co, Ni, etc.) are more tolerant to carbon deposition than Ni-based cermets [41, 42]. Interestingly, these compounds are also tolerant to a changing oxygen partial pressure, which allows for example to flush the anode with air, in order to remove any sulfur or carbon species adsorbed on the electrode, and thereby to regenerate the electrode after sulfur or coke deactivation. As some of the mentioned oxides are also usable in air at elevated temperatures and show suitable electrical conductivity, they can be successfully applied on the cathode side. This new configuration with identical materials (or composites) used on both electrodes leads to a number of advantages. One is related to a fact that manufacturing process of such SOFCs can be more simple and cheaper (only one sintering process is needed to prepare both electrodes). In addition, a redox stable cathode is also expected to enhance the cathode durability, since oxygen partial pressure at the cathode triple phase boundary region can be quite low, especially when SOFCs are operated at a low cell voltage.

Fig. 7b shows results of an exemplary cell (obtained by authors) built with ca. 0.5 mm think $Ce_{0.8}Gd_{0.2}O_{1.9}$ oxygen ions-conducting electrolyte, and composite electrodes (anode and cathode) consisting of cation-disordered $SrFe_{0.75}Mo_{0.25}O_{3-\delta}$ oxide mixed in 3:2 wt. ratio with electrolyte powder. The obtained power density was about 80 $mW\cdot cm^{-2}$ at 900 °C when fueled with hydrogen. In the case of methane or propanol fuels used, the registered power density is such symmetrical configuration did not exceed 10 $mW\cdot cm^{-2}$, but resistance of the used material to carbon deposition and improved properties concerning sulfur poisoning are a basis for further studies. Recently, a significant number of emerging literature reports on the subject can be noticed [43-45].

DOI: 10.1051/e3sconf/20171401044

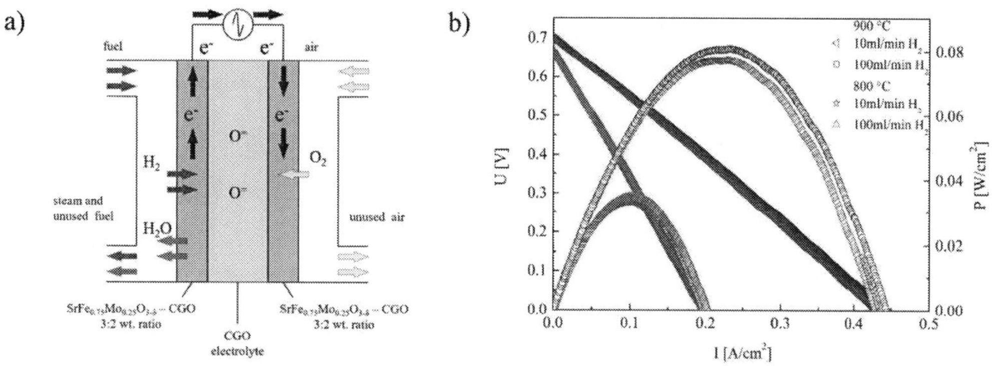

Fig. 7. a) Concept of symmetrical SOFC construction, based on [41] with exemplary electrolyte and electrode materials shown, b) performance data measured for symmetrical cell (see text for details).

Symmetrical cell construction in the case of SOFCs with oxygen-conducting electrolyte seems justified. Similar approach can be also considered in the case of PCFCs and SOECs with proton-conducting electrolyte. Initial results concerning the abovementioned group obtained by the authors indicate that a relatively high resistance of the electrode/electrolyte is among the most important issues, which need to be solved. Also, it seems that interfusion of cations between such electrode materials and commonly used proton-conducting electrolytes at the temperatures needed for manufacturing of the cells will be a hindrance. More studies are needed in order to design effectively working electrode material for such cells, but symmetrical cell approach also seems to be justified for PCFCs and SOECs.

8 Summary

Proton-conducting oxides seem as an interesting choice for the solid electrolyte for high temperature fuel cells. However, there are still material-related issues, which needs to be solved before efficient PCFCs can be built. This includes development of functional electrolytes, but also electrode materials exhibiting high proton conductivity. It seems that it is achievable by a proper selection of composition of the compounds and composites.

Acknowledgements The project was funded by the National Science Centre Poland (NCN) on the basis of the decision number DEC-2012/05/E/ST5/03772.

References

1. A. Gugliuzza, A. Basile, *Membranes For Clean And Renewable Power Applications*, Woodhead Publishing Limited (2014)
2. V.V. Kharton, Solid State Electrochemistry II: *Electrodes, Interfaces And Ceramic Membranes*, Wiley-Vch Verlag & Co. Kgaa (2011)
3. N. Ito, M. Iijima, K. Kimura, S. Iguchi, *J. Power Sources*, **152** (2005) 200
4. T. Ishihara, *Perovskite Oxide for Solid Oxide Fuel Cells*, Springer (2009) NY,
5. J. W. Fergus, R. Hui, X. Li, D. P. Wilkinson, J. Zhang: *Solid Oxide Fuel Cells Materials Properties and Performance*, CRC Press (2009) Boca Raton,
6. S.C. Singhal, K. Kendall: *High Temperature Solid Oxide Fuel Cells: Fundamentals, Design and Applications*, Elsevier Science (2003) Oxford
7. Development of SOFC for Products, *Mitsubishi Heavy Industries*, Ltd.
8. M.L. Toebes, J.H. Bitter, A.J. van Dillen, K.P. de Jong, *Catal. Today*, **76** (2002) 33
9. *Fuel Cell Today*, 13 Nov. 2016, http://www.fuelcelltoday.com/applications/stacionary
10. A.L. Dickcs, *J. Power Sources*, **61** (1996) 113

11. R.H. Cunningham, C.M. Finnerty, R.M. Ormerod: *Solid Oxide Fuel Cells V. The Electrochemical Society Proceedings*, Pennington (1997) NJ
12. S.H. Jensen, X. Sun, S.D. Ebbesen, R. Knibbe, M. Mogensen, *Int. J. Hydrogen Energy*, **35** (2010) 9544
13. W. Zając, D. Rusinek, K. Zheng, J. Molenda, *Cent. Eur. J. Chem.*, **11** (2013) 471
14. A. Mielewczyk-Gryń, *Właściwości strukturalne i transportowe ceramicznego przewodnika protonowego – domieszkowanego niobanu lantanu*, Dissertation for the degree of Philosophiae Doctor
15. L. Yang, Z. Liu, S. Wang, Y.M. Choi, C. Zuo, M. Liu, *J. Power Sources*, **195** (2010) 471
16. T. Kariya, H. Tanaka, T. Hirono, T. Kuse, K. Yanagimoto, K. Uchiyama, M. Henmi, M. Hirose, I. Kimura, K. Suu, H. Funakubo, *J. Alloys Compd.*, **654** (2016) 171
17. K.D. Kreuer, *Solid State Ionics*, 136 (2000) 149
18. Y. Wang, A. Chesnaud, E. Bevillon, J. Yang, G. Dezanneau, *Int. J. Hydrogen Energy*, **36** (2011) 7688
19. E. Fabbri, D. Pergolesi, E. Traversa, *Chem. Soc. Rev.*, **39** (2010) 4355
20. K.D. Kreuer, Annu. *Rev. Mater. Res.*, **33** (2003) 333
21. N. Agmon, *Chem. Phys. Lett.* **244** (1995) 456
22. N. Bonanos, A. Huijser, F.W. Poulsen, *Solid State Ionics* **275** (2015) 9
23. N. Bonanos, *Solid State Ionics*, **145** (2001) 265
24. H. Iwahara, T. Esaka, H. Uchida, N. Maeda, Solid State Ionics **3/4** (1981) 359
25. H. Iwahara, H. Uchida, K. Ono, K. Ogaki, J. Electrochem. Soc. **135** (1988) 529
26. T. Norby, Solid State Ionics, **40/41** (1990) 857-862.
27. E. Matsushita, T. Sasaki, Solid State Ionics **125** (1999) 31-37.
28. W. Zając, E. Hanc, A. Gorzkowska-Sobas, K. Świerczek, J. Molenda, Solid State Ionics **225** (2012) 297-303.
29. W. Zając, E. Hanc, Funct. Mater. Lett. **7** (2014) 1440014
30. P. Berastegui, S. Hull, F. J. García-García, S.-G. Eriksson, J. Solid State Chem. **164** (2002) 119-130.
31. P. Colomban, Ann. Chim. Sci. Mat, **24** (1999) 1-18.
32. J. Jankovic, D.P. Wilkinson, R. Hui, J. Power Sources **201** (2012) 49-58.
33. K. Świerczek, W. Zając, A. Klimkowicz, K. Zheng, N. Malikova, B. Dabrowski, *Solid State Ionics*, **275** (2015) 58
34. H. Iwahara, T. Yajima, T. Hibino, H. Ushida, *J. Electrochem. Soc.*, **140** (1993) 1687
35. S.M. Haile, D.L. West, J. Campbell, *J. Mater. Res.*, **13** (1998) 1576
36. K.-D. Kreuer, A. Fuchs, J. Maier, *Solid State Ionics*, **77** (1995) 157
37. R. Haugsrud, *Solid State Ionics*, **178** (2007) 555
38. T. Omata, K. Ikeda, R. Tokashiki, S. Otsuka-Yao-Matsuo, *Solid State Ionics*, **167** (2004) 389
39. A.S. Nowick, A.V. Vaysleyb, Solid State Ionics **97** (1997) 17-26
40. Y. Aoki, S. Kobayashi, T. Yamaguchi, E. Tsuji, H. Habazaki, K. Yashiro, T. Kawada, T. Ohtsuka, *J. Phys. Chem.*, 2016 (29) 15976
41. Q. Li, V. Thangadurai, *J. Power Sources*, **196** (2011) 169
42. S. Vasala, M. Lehtimäki, S.C. Haw, J.M. Chen, R.S. Liu, H. Yamauchi, M. Karppinen, *Solid State Ionics*, **181** (2010) 754
43. D.M. Bastidas, S. Tao, J.T.S. Irvine, *J. Mater. Chem.*, **17** (2006) 1603
44. X. Zhu, Z. Lü, B. Wei, X. Huang, Y. Zhang, W. Su, *J. Power Sources*, **196**(2) (2011) 729
45. T.L. Cable, S.W. Sofie, *J. Power Sources*, **174**(1) (2007) 221

The perspectives for the use of hydrogen for electricity storage considering the foreign experience

Tomasz Blacharski[1], *Krzysztof* Kogut[2,*] and *Adam* Szurlej[3]

[1]Polish Gas Company, Kasprzak Avenue 25, 01-224 Warszawa, Poland
[2]AGH University of Science and Technology, Faculty of Energy and Fuels, Mickiewicz Avenue 30, 30-059 Kraków, Poland
[3]AGH University of Science and Technology, Faculty of Drilling, Oil and Gas, Mickiewicz Avenue 30, 30-059 Kraków, Poland

Abstract. Over the last years, the European Union has seen a rapid increase in installed capacity of generating units based on renewable energy sources (RES). The most significant increase in installed capacity was recorded in 2015, in wind farms and solar PV installations. One of the most serious is the volatile character of RES on a time basis. Therefore, for a further expected increase in the use of RES and their effectiveness improvements, investments are needed allowing for electricity to be stored. One of the electricity storage options is to use excess electricity in order to produce hydrogen by electrolysis of water. Although this process plays a marginal role in obtaining hydrogen on a worldwide basis due to high costs, experience in recent years has shown that periodically low (negative) electricity prices, developing on the power exchanges in the situation where there is surplus electricity available, affect economic requirements for hydrogen production technologies. The paper shows activities undertaken by European countries (mainly Germany) aiming at making it possible for hydrogen to be stored in the natural gas grids. A particular attention is given to material resource issues and possible operational problems that might arise while blending natural gas with hydrogen into the grid. The experiences of selected European countries are of particular interest from the Polish perspective, having regard to significant increase of RES in electricity generation during the last few years and adopted objectives for the growing importance of RES in the Poland's energy balance.

1 Introduction

The realization of EU energy policy relies on investing in RES-based energy technologies. The dynamics with which RES technologies are used can be traced on the example of the increasing installed power from RES. According to The European Wind Energy Association (EWEA) data of 2015, the highest growth of installed power capacity was observed in wind farms – 12 800 MW (i.e. 44.2% new installed power in EU) and solar parks – 8500 MW (29.4%). It is worth noting that in the previous year the power of gas-fed power stations increased only by 6.4%, as

* Corresponding author: kogut@agh.edu.pl

DOI: 10.1051/e3sconf/20171401045

compared to the situation ten or so years ago when the investments in gas-based units dominated. It should be also observed that not only West European countries specialize in RES technologies. In 2015 the increase of installed power in wind farms was highest in Germany (6013 MW), i.e. in a country of highest installed power in EU wind farms. Poland occupies the second position in the ranking with its 1266 MW but in Poland 85% of electricity is still generated from conventional sources, based on hard coal and lignite. A further increase in installed wind farm capacity can lead to a situation in which coal-fired units will reach the limits of their regulatory capabilities, and the power system will become unstable [1-3]. The high dynamics of RES increase can be seen on the example of the indices of annual increase for particular RES technologies used for energy production in the years 1990–2015: photovoltaic cells – 44.1%, wind farms – 22.1%. Total participation of RES in the World's electrical energy production in 2014 equaled to 22.3% and was second of magnitude after coal with its 40.7%. In the years 1990–2014 the average annual increase of electrical energy generated by RES installations remained on the level of 3.6%, with the global increase of electrical energy production of 2.9% [4]. In 2015 the World's highest increase of RES was observed in the wind farms (from 370 GW at the close of 2014 to 433 GW), and in photovoltaics from 177 GW to 227 GW [5].

Figure 1 illustrates changes of the electricity generation in Poland and in Germany in the years 2005 and 2015.

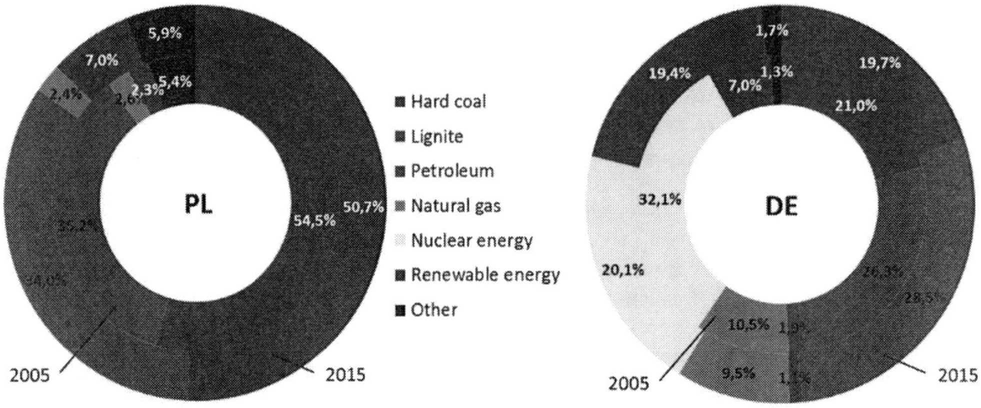

Fig. 1. Structure of electricity generation in Poland and Germany [own analyses based on 6–9]

The analysis of figure 1 reveals that the participation of RES is systematically increasing. The growing share of RES is connected with the fact that electricity is generated very unevenly because of the varying atmospheric conditions [10]. Frequently the period of low energy demand (e.g. holiday time) coincides with a high productivity time, and in the high demand periods (e.g. evening hours) the production drops down. Solving of this issue is very important in view of the planned development of RES applications.

2 Storing of electrical energy

Storing of electrical energy is frequently associated with batteries, accumulators or energy of water. Among other available methods of energy storing are [11]:

- Electrical energy – making use of condensers, super condensers (DLC) or Superconducting Magnetic Energy Storage (SMES),

 DOI: 10.1051/e3sconf/20171401045

- Mechanical energy – starting from pumped-storage hydroelectricity (PSH) and flywheels (FES), through springs and Compressed Air Energy Storage (CAES), to hydraulic accumulators,
- Chemical energy – biofuels, synthetic fuels, synthetic methane (SNG), methanol or hydrogen (H2),
- Electrochemical energy – lead accumulators (LA), Li-Ion accumulators, sodium accumulators (NaS, NaNiCl), nickel accumulators (BEV); Redox Flow Batteries (RFB) and fuel cells,
- Thermal energy – latent heat of melted salt

Figure 2 gives a comparison of rated electrical power, time of use of stored energy and amount of collected energy in various technologies of storing.

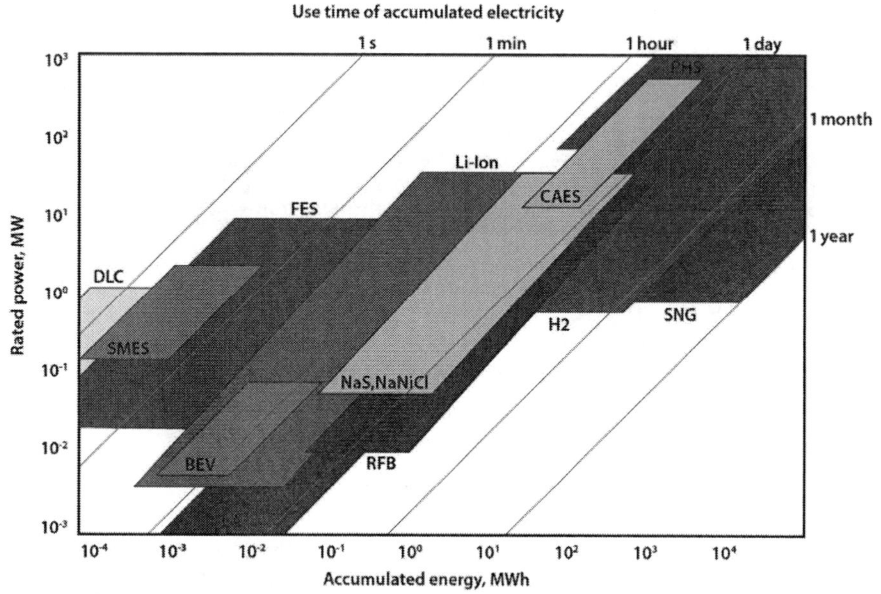

Fig. 2. Comparison of various technologies of storing electrical energy [11, 12]

3 Storing of hydrogen

As presented in figure 2 the longest time of energy storing allows one to use hydrogen in the pure molecular form, and bounded in natural gas (SNG) after prior methanization (1, 2):

$$CO_2 + 4H_2 \rightarrow CH_4 + 2H_2O \quad (1)$$

$$CO + 3H_2 \rightarrow CH_4 + H_2O \quad (2)$$

Additionally, the use of hydrogen allows for storing largest quantities of energy and obtaining highest electrical power. Technically, this technology is most advantageous.

One of the newest such technologies is *Power to Gas*.

3.1 Power to Gas

In this technology the electrical energy can be converted into gaseous fuel. The *Power to Gas* concept has been presented in figure 3.

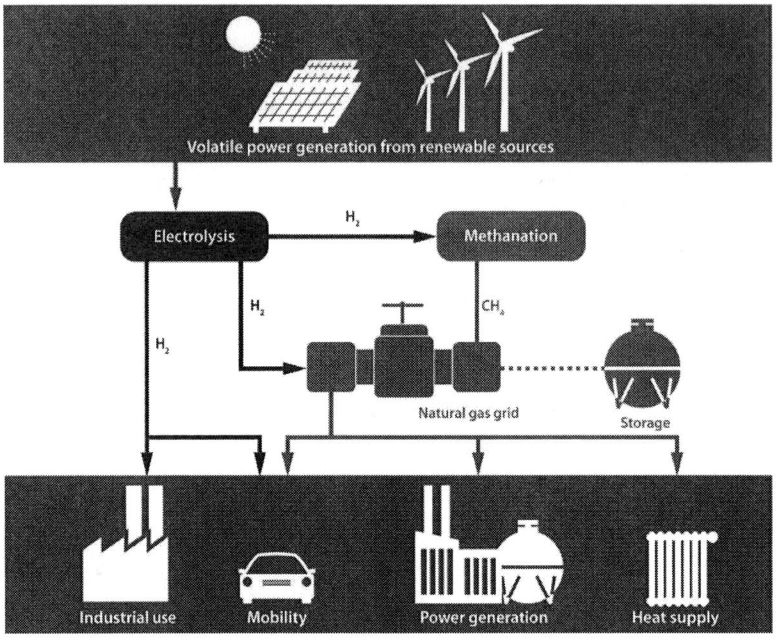

Fig. 3. Power to Gas solution [own preparation based on 13, 14]

The excess of RES electrical energy, most frequently produced in periods of lower energy demand, is used for hydrolyzing water with of one of the presently available methods (alkaline electrolysis, proton exchange diaphragm electrolysis, solid oxide electrolysis cell), which will be briefly discussed further in this paper.

3.2 Technical problems – storing of hydrogen

Owing to its properties, the storing, transmission and ultimate use of hydrogen create considerable problems. These are purely material problems caused by the smaller size of the hydrogen particle as compared to, e.g. methane, which is a dominating component of natural gas. The diverse combustible properties of hydrogen have to be also accounted for. For this reason the risk analysis should be performed for transmission of hydrogen [15, 16].

When the gas interacts with the pipes or container, there is a risk that hydrogen may locate inside the crystalline structure of steel. In the case of steel pipes the hydrogen corrosion is also dangerous as it leads to the formation of cracks and diffusion of hydrogen particles outside the tank. Synthetic pipes (PE, PA11) also create problem as they are more liable to leakages than their steel counterparts [17], though their mechanical strength remains on the same level [18, 19].

Apart from material and strength issues, one should also account for the impact of hydrogen on parameters, which are important for:

- transport – viscosity, density, specific heat, Joule-Thomson coefficient,
- combustion – burning heat, Wobbe index, flammability, regular combustion rate, minimum ignition energy, etc.,
- measurement – for gas meters and recalculation devices or gas analyzers.

The degree of preparation of particular groups of elements working together (transmission, storing, use) in a function of hydrogen content is presented in figure 4.

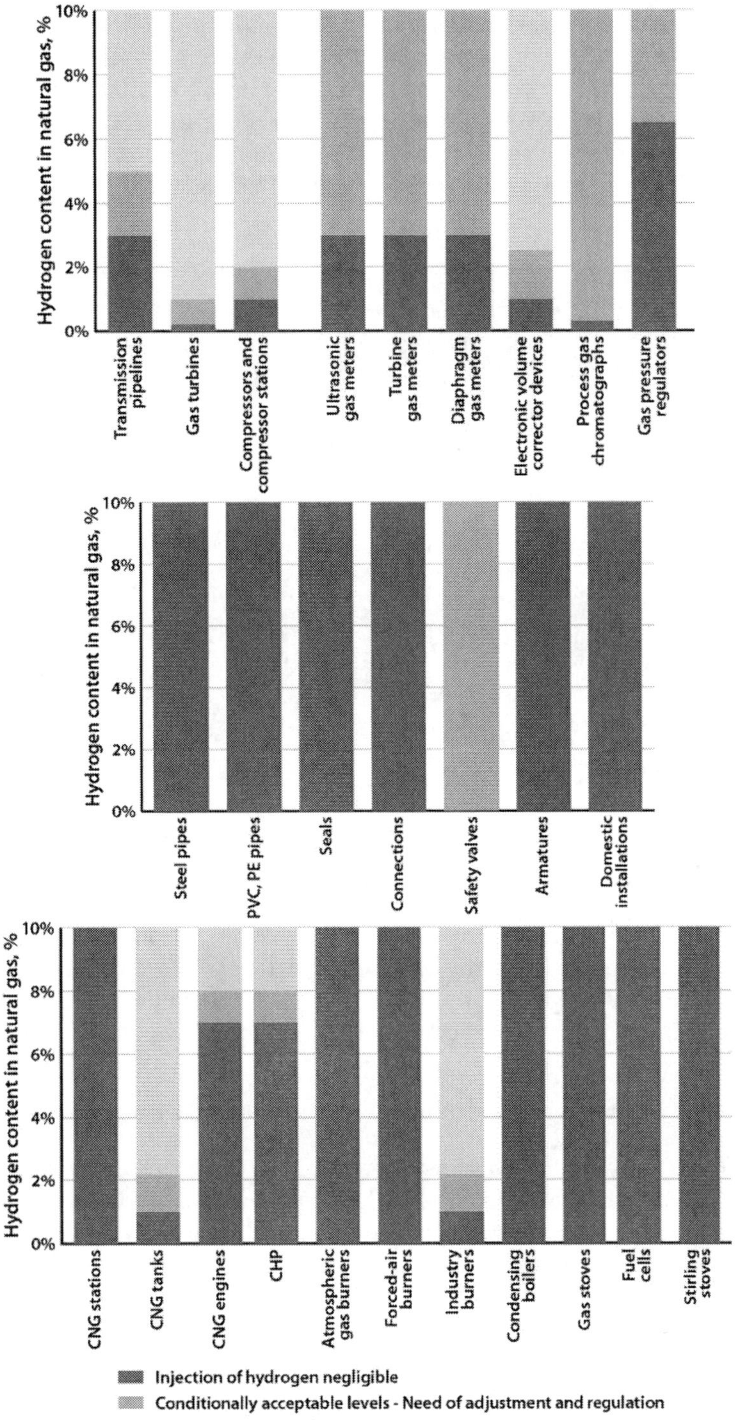

Fig. 4. Adjustment of devices in particular gas delivery segments (transmission, distribution, use) in a function of hydrogen content in natural gas [20]

Most of the elements of the gas transmission-use system can transport natural gas with small quantities of nitrogen. The most difficult situation can be encountered in the case of gas turbines, process chromatographs used for determining gas composition, or gas leakage detection systems [21].

The influence of the increasing hydrogen content in high-calorific natural gases used in Poland on the change of major parameters determining the possibility of transmitting gas through the gas pipeline network is presented in figure 5. The influence of a hydrogen concentration reaching up to 50% was analyzed. 'Domestic gas' (E) is produced in Poland and belongs to the E group of gases. 'East gas' has parameters of gas imported from Russia, whereas 'mean gas' has features of a mixture of imported and E gas.

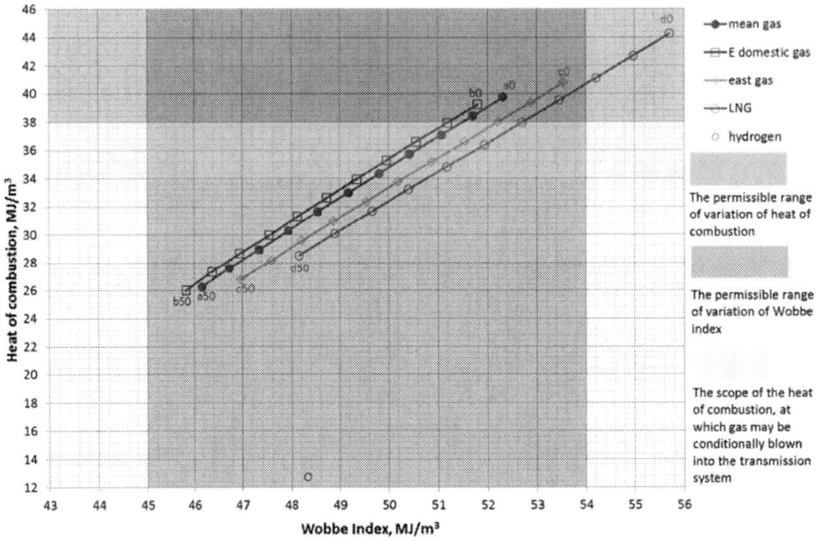

Fig. 5. Change of burning heat value and Wobbe index under the influence of increasing participation of hydrogen, depending on type of gas. "0" – clean gas, "50" – 50% hydrogen content in gas, change of hydrogen content every 5% [own analysis based on 22, 23]

There is a danger that after adding ca. 10% of hydrogen to natural gas the latter will not meet the requirements of the Economy Minister regulation [MG_2010]. In the case of the gas pipelines, where the transmitted natural gas flows in considerably bigger streams, the risk of making the gas unfit is much lower than in the distribution networks [21, 22].

3.3 Production of hydrogen – selected technical and economic aspects

At present most of the World's hydrogen is produced with the use of the methane reforming with steam. The popularity of natural gas in hydrogen production stems from, e.g. low cost of production (low cost of investment and exploitation), lowest CO_2 coproduction per mole of obtained hydrogen, and also natural gas resources [24]. Considering the dynamic development of RES, observed mainly in the EU countries, one should expect an increased production of hydrogen through electrolysis – a way of managing excess of electrical energy from RES. Typically for the RES technologies, the work does not go uniform. The electrolysis itself is a very energy-consuming process. The following types of professional electrolyzers are used [25–27]:

- alkaline – water solution of hydroxide, most frequently KOH or rarely NaOH, is the electrolyte, electrodes are separated with a porous diaphragm to stop mixing of generated hydrogen and oxygen. The process is conducted at a temperature of 70–80°C and atmospheric pressure, and at a temperature of 90–100°C in pressure electrolyzers; the efficiency is of 70%,
- polymeric PEM – polymeric diaphragm of high ionic conductance is used in the process with hydrogen ions (protons) as charge carriers; hydrogen is produced on the cathode

while the electrolyzer is operating. The process is conducted at a temperature of 70–80°C, the efficiency equals to 70–80%, and the obtained hydrogen has purity of the level of 99.999%.

Table 1 summarizes the main characteristics of alkaline and PEM electrolysis systems.

Table 1. Parameters of typical water electrolysis cells [26]

Electrolysis process	Alakline	PEM
Ions transport	Hydroxyl OH^-	Protons H^+
Anode reaction	$4OH^-_{(aq)} \rightarrow O_{2(g)} + 2H_2O_{(l)}$	$2H_2O_{(l)} \rightarrow 4H^+_{(aq)} + O_{2(g)} + 4e^-$
Cathode reaction	$2H_2O_{(l)} + 2e^- \rightarrow H_{2(g)} + 2OH^-_{(aq)}$	$4H^+_{(aq)} + 4e^- \rightarrow 2H_{2(g)}$

At present the commercial solutions bring about over 80% energy efficiency [28]. Advantageously for this technology, the process can be performed economically if the cost of the provided electrical energy is lower than the marker price. The analysis of table 2 reveals that the cost of electrical energy dominates in the structure of hydrogen production through electrolysis. Considering the high dynamics of RES investments, characterized by uneven production of electrical energy, the electrical energy prices may drastically drop down to even reach negative values in a situation of high energy sales from RES installations (mainly wind farms and photovoltaic parks), at moderate demand for energy (e.g. holiday time). This can be evidenced by a detailed analysis of electrical energy prices on the German market as of 8 May 2016 (Sunday), and visualized in figure 6. On this day the RES-based energy technologies covered the afternoon demand for electricity (only solar parks generated about 26 GW of power supplied at 1 p.m., so higher than maximum demand of Poland in 2015, i.e. 25.1 GW of 7 January 2015) [8]. At a low demand for electrical energy on 8 May 2016, the RES installations covered over 90% of demand for electrical energy in Germany (exceptionally high participation) in particular hours of the day.

Table 2. Structure of hydrogen production [%] through steam reforming and electrolysis (polymeric electrolyzer) [after 29, 30]

Production cost category	**Steam reforming**	**Electrolysis**
Capital charges	29	15
Fuel and electricity	61	78
Operation and maintenance	10	7

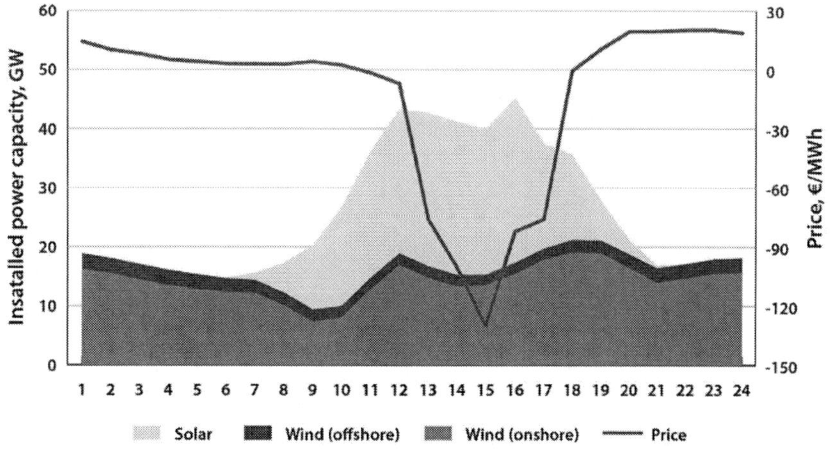

Fig. 6. Prices and production of electrical energy from RES in Germany on 8 May 2016 [31, 32]

Judging from German experience, the development of the use of RES and varying electrical energy prices, in that negative prices, may have impact on the economic attractiveness of hydrogen production through electrolysis. The analysis of hydrogen production through electrolysis presented in [33] shows that the lowest values were observed in nuclear plants – 4.15 USD/kg, higher for wind farms – about 6 USD/kg and highest for solar parks – 7 to 10.5 USD/kg. The comparison of hydrogen production through electrolysis and other technologies based on coal (gasification), or natural gas (steam reforming) reveals the dominance of technologies based of fossil fuels. In the case of electrolysis-based technology the cost of electrical energy dominates in the hydrogen production budget, whereas in the case of reforming the cost of natural gas prevails, totaling to about 60%. Thus a drop of natural gas prices observed over the last years, especially on the American market as a consequence of the shale gas revolution, made the World's popular hydrogen technology even more attractive [32–35].

It should be emphasized that thanks to the use of innovative solutions in the production of electrolyzers the capital costs relating to their production were significantly lowered from over 2500 USD/kW$_e$ (2001) to less than 400 USD/kW$_e$ (2012) [29].

Hydrogenics and Proton Motor are typical examples of companies producing electrolyzers on a commercial scale. In April 2013 Hydrogenics Corporation announced that for the first time on the World's scale they used a PEM electrolyzer 1 MW for storing energy, as ordered by E.ON in Hamburg. The installation will be used for hydrogen production with the use of excessive electrical energy from RES, mainly from wind farms. On 15 October 2015 a 1.5 MW unit started its operation in Hamburg (also employing PEM electrolyzer) [36]. Tables 3 and 4 contains selected technical data of various electrolyzers – alkaline and PEM respectively.

Table 3. Selected technical data of illustrative alkaline electrolyzers [after data provided by 36–41]

Parameter	Unit	Model				
		HyProvide™ A60-	**S-556**	**HySTAT-60-10**	**A-300**	**TITAN EL-N**
Manufacturer		GreenHydrogen	IHT	Hydrogenics	NEL Hydrogen	Teledyne
Pressure	barg	30	32	10	250	9
Number of piles		1	NA	4	NA	NA
Max H_2 flow rate	m^3/h	60	760	60	300	500
Installed power	kVA	270	ca. 3420	2×240 + 35	1300	NA
Purity of H_2	%	99.995	NA	99.998	99.9	99.999
H_2O consumption	l/m^3 H_2	1.0	NA	1.5-2	0.9	1.0
Weight	Mg	ca. 3.5	NA	ca. 16	NA	NA

Table 4. Selected technical data of illustrative PEM electrolyzers [after data provided by 42–44]

Parameter	Unit	Model		
		C 30	**McLyzer S 18 MP**	**Mercury G192**
Manufacturer		Proton OnSite	McPhy	ErreDUE
Pressure	barg	30	12	30
Number of piles		1	1	1
H_2 flow rate	m^3/h	30	12	128
Installed power	kVA	275	63	684
Purity of H_2	%	99.9998	99.999	99,9995
H_2O consumption	l/m^3 H_2	0.90	0.83	0.84
Weight	Mg	3	1.5	11.4

In the case of the most advanced model of the yield 60 m^3/h and a relatively small stream of gas in the pipeline (ca. 10 000 m^3/h), the obtained hydrogen concentration in natural gas remains on a level of 6‰ at most. Only when some electrolyzers are used at the same time the hydrogen concentration equals to a few percent (the flow rate of 10 000 m^3/h).

Obviously, if the same model is applied and hydrogen is pumped to the distribution gas pipeline, where the flowing stream is much smaller, the hydrogen concentration will be respectively higher.

4 Conclusions

The realization of goals of the EU energy policy, sustained development and reduced resources of fossil fuels will have influence on the direction of changes in the energy sector. In the times of an intense increase of installed power capacity in RES-based technologies, the storing of electrical energy is the most important problem to be solved. As shown in the paper, there are many methods of storing electrical energy and numerous research works nowadays concentrate on increasing their efficiency. One of the most interesting energy storage solutions is the *Power to Gas* concept. German experience shows to the increasing significance of this technology manifesting itself in a higher capacity of the electrolyzers, higher number of installations and hydrogen production through electrolyzers. A broader use of the *Power to Gas* technology is interesting to companies representing the natural gas sector. Over the last few years the demand for natural gas has lowered in many European countries, except Poland. Thus the development in of *Power to Gas* projects will contribute to the increase of use of the existing gas infrastructure, which shall have influence on the higher efficiency of work of the energy sector. However, considering the specific properties of hydrogen, mainly before it is introduced to the gaseous networks, a polydimensional analysis should be performed in view of the risk accompanying transport of a mixture of natural gas and hydrogen. The plans of the Polish government regarding the more intense use of electrical vehicles and planned sales of hydrogen-fed vehicles by one of the World's leading concerns may be important factors giving spur to the use of the *Power to Gas* technology in Poland.

References

1. EWEA, *Wind in Power: 2015 European statistics* (2016), www.ewea.org
2. Ł. Bartela, J. Kotowicz, K. Dubiel, Journal of Power Technologies, **96**, 2, 92–100 (2016)
3. J. Kotowicz, Ł. Bartela, D. Węcel, K. Dubiel, Energy, **118**, 156-171 (2017)
4. IEA, *Key Renewables Trends Excerpt from: Renewables information* (Pairs, 2016)
5. REN21, *Renewables 2016 Global status report* (2016)
6. www.mg.gov.pl
7. Raport Prezesa URE (2016); www.ure.gov.pl
8. Polskie Sieci Elektroenergetyczne; www.pse.pl
9. DIV Berlin, EEFA, *Evaluation tables of the energy balance for Germany* (2016)
10. J. Kotowicz, M. Jurczyk, D. Węcel, W. Ogulewicz, Journal of Power Technologies, 96, 3, 149-156 (2016)
11. K. Hajdrowski, Energia elektryczna, **11**, (2012)
12. http://forschung-energiespeicher.info/en/wind-to-hydrogen/project-list/project-details//Stacks_auf_Herz_und_Nieren_testen/, (2016)
13. e-on, Technology & Innovation, *Energy Storage, Power to Gas* (2015)
14. Hydrogenics, *Power-to-Gas Solution* (2012)
15. A. Rusin, K. Stolecka, Rynek Energii, **3**, 87-93 (2011)

16. T. Włodek, M. Łaciak, K. Kurowska, Ł. Węgrzyn, AGH Drilling, Oil, Gas, **33**, 2, 379-396 (2016)
17. M.H. Klopffer, P. Berne, S. Castagnet, M. Weber, G. Hochstetter, E. Espuche, *Polymer Pipes for Distributing Mixtures of Hydrogen and Natural Gas: Evolution of their Transport and Mechanical Properties after an Ageing under an Hydrogen Environment* (18th World Hydrogen Energy Conference, Essen, 2010)
18. S. Castagnet, J.C. Grandidier, M. Comyn, G. Benoit, Int. J. Pres. Ves. Pip. **89,** 203-209 (2012)
19. M. Ball, A. Basile and T. N. Veziroglu (ed.), *Compendium of Hydrogen Energy* (Elsevier) (2015)
20. G. Müller-Syring, M. Henel, *Wasserstofftoleranz der Erdgasinfrastruktur inclusive aller assoziierten Anlagen* (2014)
21. T. Blacharski, P. Janusz, M. Kaliski, Ł. Zabrzeski, AGH Drilling, Oil, Gas, **33,** 2, 515-528 (2016)
22. *Rozporządzenie Ministra Gospodarki z dnia 2 lipca 2010 r. w sprawie szczegółowych warunków funkcjonowania systemu gazowego*, Warszawa (2010)
23. https://swi.gaz-system.pl/swi/public/, (5 IX 2016)
24. T. Borowiecki, A. Gołębiowski, P. Kowalik, J. Ryczkowski, Przem Chem, **8,** 1306-1315 (2015)
25. A. Małek, Ł. Grabowski, K. Pietrykowski, R. Sochaczewski, G. Barański, M. Szlachetka, M. Gęca, Autobusy, Technika, Eksploatacja, Systemy transportowe, **10**, 305-310 (2011)
26. I. Dincer, Int J Hydrogen Energ **37,** 1954-1971 (2012)
27. D.M.F. Santos, C.A.C. Sequeira, J.L. Figueiredo, Quim. Nova, **36**, 1176-1193 (2013)
28. S.A. Sherif, D. Yogi Goswami, E.K. Stefanakos, A. Steinfeld, *Handbook of Hydrogen Energy*, (Taylor & Francis Group, 2015)
29. D.R. Peterson, P. Zelenay, *Overview of U.S. Department of Energy Efforts on Hydrogen Production from Water Electrolysis.* (ElectroHyPEM Workshop, Taormina, 2014)
30. K. Schoots, F. Ferioli, G.J.Kramer, B.C.C.van der Zwaan, Int J Hydrogen Energ **33,** 2630-2645 (2008).
31. http://www.epexspot.com/en/market-data/dayaheadauction/chart/auction-chart/2016-05-08/DE (2016)
32. https://www.eex-transparency.com/homepage/power/germany/production/usage/solar-wind-power-production/solar-wind-power-production-chart (2016)
33. P. Nikolaidis, A. Poullikkas, Renew Sust Energ Rev **67**, 597-611 (2017)
34. E. Gross-Gołacka, W. Lubiewa-Wieleżyński, A.P. Sikora, A. Szurlej, R. Biały, Przem Chem, **92**, 1393-1398 (2013)
35. A. Szurlej, P. Janusz, Gospod Surowcami Min, **29**, 4, 77-94 (2013)
36. Hydrogenics, *Hydrogenics to Install World's First Megawatt PEM Electrolyser for E.ON Power-to-Gas Facility* (2013), http://www.fuelcelltoday.com
37. *HyProvide™A60*, (2016), http://www.greenhydrogen.dk
38. *S-556*, (2016), http://www.iht.ch
39. M. Bodner, A. Hofer, V. Hacker, WIREs Energy Environ, , 365-381 (2015)
40. *A-300*, http://www.nelhydrogen.com
41. *Titan™ EL-N Series*, (2016), http://www.teledynees.com
42. Proton OnSite, *Hydrogen Generation Systems. C Series. Technical Specifications.* (2016), http://www.protononsite.com
43. http://www.mcphy.com
44. ErreDUE, *Mercury – Hydrogen generators*, (2016), http://www.erreduegas.com

An analysis of the thermodynamic cycles with high-temperature nuclear reactor for power generation and hydrogen co-production

Michał Dudek[1,*] and *Marek* Jaszczur[1]

[1]AGH University of Science and Technology, Faculty of Energy and Fuels, Department of Fundamental Research in Energy Engineering, Krakow, Poland

Abstract. In the present paper, numerical analysis of the thermodynamic cycle with the high-temperature nuclear gas reactor (HTGR) for electricity and hydrogen production have been done. The analysed system consists of two independent loops. The first loop is for HTGR and consists of a nuclear reactor, heat exchangers, and blower. The second loop (Rankine cycle) consist of up-to four steam turbines, that operate in heat recovery system. The analysis of the system shows that it is possible to achieve significantly higher efficiency than could be offered by traditional nuclear reactor technology (PWR and BWR). It is shown that the thermal efficiency about 52.5% it is possible to achieve when reactor works at standard conditions and steam is superheated up to 530°C. For the cases when the steam has supercritical conditions the value of thermal efficiency is still very high and equal about 50%.

1 Introduction

The growing demand for electricity consumption is one of the most important key indicators showing the development stages of countries. The most important challenges for world energy industry based nowadays on three main problems: how to reduce the impact of the industry on the environment, how to ensure the low cost of the energy production and how to improve the system overall performance. For companies operating in the Polish energy market, the number of challenges is much greater than companies operating in different countries where Nuclear Energy is used. Still open and very important for the Polish energy sector remain the issue of the construction of the first, Nuclear Power Plant. After the accident at the Fukushima Daiichi Nuclear Power Station, many countries have changed point of view to nuclear energy. More safety technologies with lower capital cost and higher thermal efficiency have been searched to ensure more attractive solutions in the future development of nuclear energy in industrial application. The choice of the first Polish nuclear power technology used to produce electricity should be determined to take into account nuclear safety, computer technology, and cyber security as well as materials science, and technical analysis, which allow today design and implement new solutions for nuclear energy. The new solution with high-efficiency energy conversion systems should provide a lower unit cost of

* Corresponding author: jaszczur@agh.edu.pl

 DOI: 10.1051/e3sconf/20171401046

energy conversion and give new possibilities in energy and fuel production [1]. A new program for future nuclear energy systems, Generation IV has been created in an effort to provide next - generation technologies that will compete in all markets with the most cost - effective technologies expected to be available over the next three decades [4]. Gas-cooled reactor systems have several fundamental features that distinguish them from the other types of reactors and provide significant advantages. In particular, the fuel is in the form of small ceramic - coated particles capable of operating at very high temperatures, the moderator is solid graphite and the coolant is helium or carbon dioxide inert for neutrons. Gas-Cooled Nuclear Reactor provides the potential for increased energy conversion efficiency and can be treated as a high-temperature heat source to different process heat applications, such as coal gasification or thermochemical hydrogen production. In contrary to the fossil fuels, hydrogen is cleaner and less harmful to the environment. By using this fuel we avoid most kinds of pollution, greenhouse gases, and acid rains. Several methods of hydrogen production are known and some of them required high-temperature heat.

A new generation of high-temperature nuclear reactors (HTR) is the most innovative concept among current advanced nuclear reactor technologies [2-4] and give us the opportunity to achieve a higher value of thermal efficiency and the possibility to the increase industrial application compared to another type of nuclear reactor and also give us the opportunity to realize large number heat consuming industrial, chemical and thermochemical process.

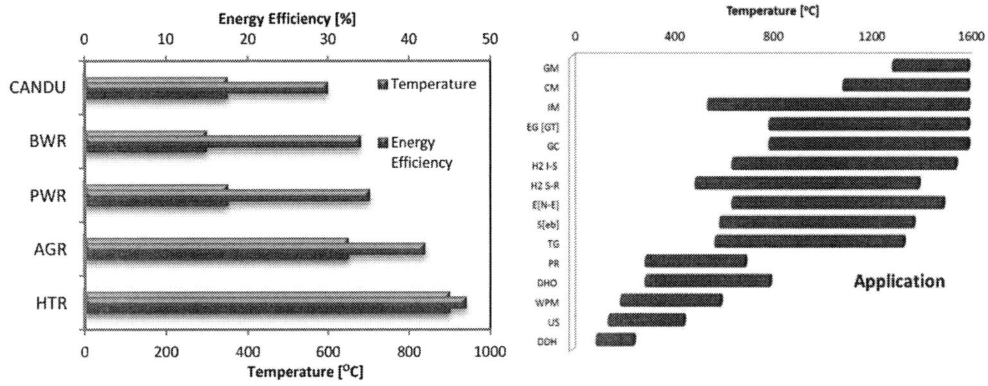

Fig. 1. (a) Operational temperatures and energy efficiency for various nuclear technologies, (b) industrial application operational temperatures

2 High-Temperature Nuclear Reactor

High-Temperature Nuclear Reactor is a very promising concept for using high outlet temperature for industrial purposes, at present, there is only one demonstration site in the world where such reactors are currently in construction. In Shidao Bay NPP in China the two 250 MW_t HTR-PM (High-Temperature Reactors-Pebble-bed Module) reactors, are able to generate power of 210 MW_e. The construction has begun in 2012 and the operation is predicted for 2017. The outlet temperature will be about 750°C and is in the low-temperature range required for efficient hydrogen production [5]. These type of reactors but designed to supply temperatures, up to 1000°C are called VHTRs but they are still at the research stage with no industrial use. One project launched by the United States Department of Energy (DOE) is a Next Generation Nuclear Power Plant (NGNP) which assumes designing a VHTR similar to Areva's SC-HTGR (Steam-cycle high-temperature gas-cooled reactor) with the aim of using up to 30% of a process heat to produce hydrogen. The prototype of NGNP reactor is likely to be deployed by approximately 2030 [6,7]. Another program dedicated to

VHTRs' development was submitted by Korean Atomic Energy Research Institute (KAERI). In cooperation with China and Japan, they plan to start operation of their first VHTR, coupled with hydrogen production in 2020 [8]. With the outlet temperature of 900-1000°C supplied by a Very High-Temperature Reactor, the hydrogen is to be produced at efficient processes of the sulfur-iodine cycle or the electrolysis. However, with such high temperatures the problem of safe and reliable heat exchangers occurs [9]. The aim of the present work is to analyses the possibility of hydrogen production with the lower temperatures and using the nuclear systems that will be introduced to the industry in the near future. General parameters for the different type of Nuclear Reactor are described in Table 1.

Table 1. General parameters for the different type of Nuclear Reactor [10].

	PWR	BWR	AGR	CANDU	RMBK	HTR
Power [MW]	1200	600	600	600	1200	1300
Fuel [% ^{235}U]	3,2	2,6	2,3	0,7	2	10
Dimensions of Core	3,0x3,7	3,7x3,7	9,1x8,5	7,1x5,9	11,8x7	9.8x6
Pressure [bar]	150	70	40	86	47	48
Outlet Temperature [°C]	324	288	650	305	284	720
Steam Pressure [bar]	4-6	7	16	4,7	6,5	17,5
Superheated Steam Temperature [°C]	270-290	280	565	260	280	530
Thermal Efficiency [%]	32-35	32-34	42	30	31	40

Actually, numerous studies have been conducted on the implementation of high-Temperature Nuclear Reactors in the United States, Russia, Japan, France [6] At present on HTGR's advanced research programs are in progress in many countries. Table 2 includes basic parameters for High-Temperature Nuclear Reactor which can be used in various type of cycle [7].

Table 2. Main thermodynamic parameters for High-Temperature Nuclear Reactor [11].

Parameter	HTR Cycle General Concepts			
	GTMHR	GTHTR300	ANTARES	VHTR NGTCC
Power [MWth]	600	600	600	350
Thermodynamic cycle	Inter-Cooled Recuperated Brayton Cycle	Recuperated Brayton Cycle	Combined Cycle	Combined Cycle
Working fluid	He	He	He/N_2	He
Reactor inlet/outlet temperature [°C]	491/850	587/850	355/850	400/950
Turbine inlet temperature [°C]	850	850	800	950
Reactor gas pressure [Bar]	71	70	55	71
Compression ratio	2.86	2	2	1.94
Plant Net Power [MWe]	286	274	280 (80 GT, 200 ST)	180(50 GT, 130 ST)
Thermal efficiency [%]	47.6	45.6	47.0	51.5
Number of compressor stages	9	6	4	6
Turbine blade cooling	Uncooled	Uncooled	Uncooled	First, two stages cooled

 DOI: 10.1051/e3sconf/20171401046

3 High-Temperature Steam-Cycle Model Description

In the present paper, numerical analysis of high-temperature gas nuclear reactor (HTGR) thermodynamic cycle for electricity and hydrogen production have been done. The analysed system consists of two independent loops. First loop is a helium loop and is equipped with nuclear reactor, heat exchanger, and blower. The second loop is steam loop and is equipped with up to four steam turbines, that operate in heat recovery mode. The analysis of proposed system shows that is possible to achieve much higher efficiency than could be offered by traditional nuclear reactor technology (PWR 32-35% and BWR 32-24%).

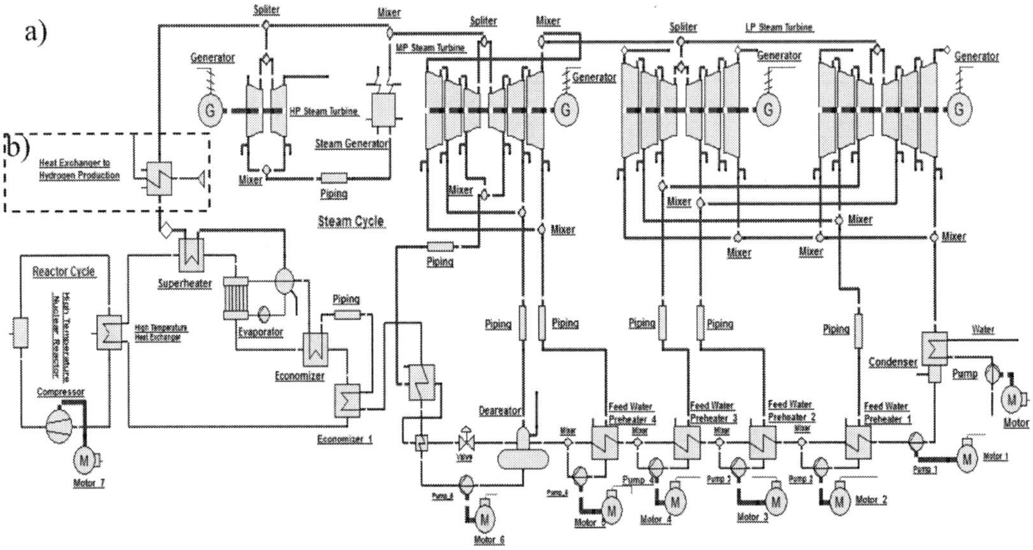

Fig. 2. The High-Temperature Nuclear Reactor thermodynamic cycle, coupled with heat recovery steam cycle: without hydrogen production (a) and with hydrogen production (b).

4 Mathematical Model of High-Temperature Nuclear Reactor with heat recovery steam cycle

Following equations are used to describe a mathematical model of the High-Temperature Nuclear Reactor coupled with the heat recovery steam cycle.

Reactor thermal power is calculated as follow:

$$N_R = q_j \Sigma_f \phi V = q_j \sigma_f \phi m \frac{N_A}{A} \tag{1}$$

q_j - heat generated in one act cleavage, σ_f - microscopic cross section [cm^2], ϕ - the average neutron flux [1/cm^{2}s], m – mass of fissile material [kg], N_A – Avogadro Number, A- mass number.

Heat recived from Nuclear Reactor:

$$\dot{Q}_R = \dot{m}_{He} c_{pHe} \left(T_{out} - T_{in} \right) \tag{2}$$

$\dot{m}_{He}$ - helium mass flow rate [kg/s], c_{pHe}- helium specific heat [kJ/(kgK)], T_{out} - reactor outlet temperature [K], T_{in} – reactor inlet temperature [K].

 DOI: 10.1051/e3sconf/20171401046

Heat received from high-temperature heat exchanger nuclear reactor:

$$\dot{Q}_{R1} = \dot{m}_{He1} c_{pHe} \left(T_{out1} - T_{in1} \right) \tag{3}$$

T_{out1} - heat exchanger outlet temperature [K], T_{in1} – heat exchanger inlet temperature [K].

Compressor work:

$$W_{comp} = \dot{m}_{He} \left(h_2 - h_1 \right) / \eta_{comp} \eta_{isent} \tag{4}$$

h_2 -compressor outlet enthalpy [kJ/kg], h_1 -inlet enthalpy to compressor [kJ/kg], η_{comp} - compressor mechanical efficiency, η_{isent} - isentropic efficiency

Superheater energy balance:

$$\dot{m}_{He1} c_{pHe} \left[T_{out1} - T_{ot} \right] + \dot{m}_{pp} c_{pspTot}^{Tinp} [T_{inp} - T_{ot}] = \\ = \dot{m}_{He1} c_{pHe} \left[T_{out2} - T_{ot} \right] + \dot{m}_{pp} c_{pppTot}^{Toutp} [T_{outp} - T_{ot}] \tag{5}$$

$\dot{m}_{He1}$ - helium mass flow rate at inlet to superheater [kg/s], $\dot{m}_{pp}$ - mass flow rate of saturated steam/superheated steam [kg/s], c_{pspTot}^{Tinp} - saturated steam specific heat [kJ/(kgK)], c_{pHe} -helium specific heat [kJ/(kgK)], c_{pppTot}^{Toutp} -superheated steam specific heat [kJ/(kgK)]] T_{inp} - saturated steam inlet temperature [K], T_{out2} - superheater helium outlet temperature [K], T_{outp} - superheated steam outlet temperature [K], T_{ot} -ambient temperature [K].

Evaporator energy balance:

$$\dot{m}_{He1} c_{pHe} \left[T_{out2} - T_{ot} \right] + \dot{m}_{w} c_{pw1Tot}^{Tinw} [T_{inw} - T_{ot}] = \\ = \dot{m}_{He1} c_{pHe} \left[T_{out3} - T_{ot} \right] + \dot{m}_{pp} c_{pspTot}^{Tinp} [T_{inp} - T_{ot}] \tag{6}$$

$\dot{m}_{w}$ - mass flow of water at inlet to evaporator [kg/s], c_{pw1} - water specific heat [kJ/(kgK)], T_{inw} - evaporator water inlet temperature [K], T_{out3} - evaporator helium outlet temperature [K], T_{inp} - saturated steam outlet temperature [K], c_{pspTot}^{Tinp} - saturated steam specific heat [kJ/(kgK)].

Economizer energy balance:

$$\dot{m}_{He1} c_{pHe} \left[T_{out3} - T_{ot} \right] + \dot{m}_{w} c_{pwTot}^{Tinw} [T_{inw} - T_{ot}] = \\ = \dot{m}_{He1} c_{pHe} \left[T_{out4} - T_{ot} \right] + \dot{m}_{w} c_{pw1Tot}^{Tinw} [T_{inw} - T_{ot}] \tag{7}$$

T_{out4} - economizer helium outlet temperature [K], c_{pw} - water specific heat at inlet to economizer [kJ/(kgK)]. c_{pw1} - water specific heat at outlet from economizer [kJ/(kgK)].

Economiser one energy balance:

$$\dot{m}_{He1}c_{pHe}\left[T_{out4}-T_{ot}\right]+\dot{m}_{w}c_{pw2Tot}^{Tinw1}[T_{inw1}-T_{ot}]= \\ =\dot{m}_{He1}c_{pHe}\left[T_{in1}-T_{ot}\right]+\dot{m}_{w}c_{pw1Tot}^{Tinw}[T_{inw}-T_{ot}] \tag{8}$$

T_{inw1} -economiser water inlet temperature [K], T_{in1} - heat exchanger helium inlet temperature [K]. c_{pw2} - water specific heat at inlet to economizer one [kJ/(kgK)], c_{pw1} - water specific heat at outlet from economizer one [kJ/(kgK)].

HP steam turbine power:

$$P_{HP}=\eta_{gen}\eta_{isent}\eta_{mech}0.5\dot{m}_{pp}[h_{ppi}-h_{pout}] \tag{9}$$

η_{gen} - electrical efficiency, η_{isent} - isentropic efficiency, η_{mech} - mechanical efficiency, $\dot{m}_{pp}$ - mass flow of superheated steam [kg/s], h_{ppi} - steam turbine specific enthalpy at the inlet to the high-pressure [kJ/kg], h_{pout} - steam turbine specific enthalpy at the outlet [kJ/kg].

Mixer after steam generator energy balance:

$$0.5\dot{m}_{pp}h_{ppi}+0.5\dot{m}_{pp}h_{asg}=\dot{m}_{pp}h_{ppim} \tag{10}$$

h_{ppi} -specific enthalpy for superheated steam after the splitter [kJ/kg], h_{asg} - specific enthalpy for superheated steam after steam generator [kJ/kg], h_{ppim} -enthalpy of superheated steam before MP Steam Turbine [kJ/kg]

MP steam turbine power:

$$P_{MP}=\eta_{gen}\eta_{isentMP}\eta_{mechMP}[\dot{m}_{pp}[h_{ppim}-h_{ppoim1}] \\ +[\dot{m}_{pp}-\dot{m}_{pu1-2}][h_{ppom1}-h_{ppoim2}]+[\dot{m}_{pp}-\dot{m}_{\dot{m}_{pu1-2}}-\dot{m}_{pd}][h_{ppom2}-h_{ppoim3}]] \tag{11}$$

η_{gen} -electrical efficiency, $\eta_{isentMP}$ -isentropic efficiency of steam turbine, η_{mechMP} - mechanical efficiency of steam turbine $\dot{m}_{pp}$ - mass flow at inlet to steam turbine [kg/s], $\dot{m}_{pp5}$ - mass flow at first sluice inlet [kg/s], $\dot{m}_{p1}$ - mass flow of steam at second sluice inlet, h_{ppim} - specific enthalpy of steam at inlet to MP steam turbine [kJ/kg], h_{ppoim1} -specific enthalpy of steam at inlet to first sluice [kJ/kg], h_{ppoim2} -specific enthalpy of steam at inlet to second sluice [kJ/kg], h_{ppoim3} - specific enthalpy of steam at outlet from MP steam turbine [kJ/kg], $\dot{m}_{pp}$ - mass flow of superheated steam at inlet to MP steam turbine [kg/s] $\dot{m}_{pu1-2}$ - mass flow of saturated steam at inlet to preheater 5 [kg/s], $\dot{m}_{pd}$ - mass flow of saturated steam at inlet to the deareator [kg/s], $\dot{m}_{p4}$ - saturated steam mass flow at inlet to feed water preheater 4 [kg/s].

MP steam turbine mass and energy balance:

$$\dot{m}_{pp}h_{ppim} = \dot{m}_{pu1-2}h_{ppim1} + \dot{m}_{pd}h_{ppim2} + \dot{m}_{pf4}h_{ppim3} + \dot{m}_{ppo}h_{ppim4} \tag{12}$$

$$\dot{m}_{pp} = \dot{m}_{pu1-2} + \dot{m}_{pd} + \dot{m}_{pf4} + \dot{m}_{ppo} \tag{13}$$

Mixers after MP Steam Turbine

Mixer_1

$$\dot{m}_{pu1} + \dot{m}_{pu2} = \dot{m}_{pu1-2} \tag{14}$$

Mixer_2

$$\dot{m}_{pu3} + \dot{m}_{pu4} = \dot{m}_{pd} \tag{15}$$

Mixer_3

$$\dot{m}_{pu5} + \dot{m}_{pu6} = \dot{m}_{pf4} \tag{16}$$

Mixers after first and second LP Steam Turbines

Mixer_4

$$\dot{m}_{pu7} + \dot{m}_{pu13} = \dot{m}_{pf3} \tag{17}$$

Mixer_5

$$\dot{m}_{pu10} + \dot{m}_{pu16} = \dot{m}_{pf2} \tag{18}$$

Mixer before condenser

$$\dot{m}_{pu9} + \dot{m}_{pu12} + \dot{m}_{pu15} + \dot{m}_{pu18} = \dot{m}_{wskpo} \tag{19}$$

$\dot{m}_{pu1}$ - mass flow of steam from first sluice [kg/s], $\dot{m}_{pu2}$ - mass flow of steam from second sluice [kg/s], $\dot{m}_{pu3}$ - mass flow of steam from third sluice [kg/s], $\dot{m}_{pu4}$ - mass flow of steam from fourth sluice [kg/s], $\dot{m}_{pu5}$ - mass flow of steam from fifth sluice [kg/s], $\dot{m}_{pu6}$ - mass flow of steam from sixth sluice [kg/s], $\dot{m}_{pu7}$ -mass flow of steam from seventh sluice[kg/s], $\dot{m}_{pu13}$ -mass flow of steam from the thirteenth sluice [kg/s], $\dot{m}_{pu10}$ -mass flow of steam from on the tenth sluice [kg/s], $\dot{m}_{pu16}$ - mass flow of steam from the sixteenth sluice [kg/s], $\dot{m}_{pu9}$ -mass flow of steam from ninth sluice [kg/s], $\dot{m}_{pu12}$ - mass flow of steam from the twelfth sluice [kg/s] $\dot{m}_{pu15}$ -mass flow of steam from fifteenth sluice [kg/s], $\dot{m}_{pu18}$ -mass flow of steam from fourth sluice [kg/s].

Water pump before Feed water preheater_1

$$W_{p1} = \dot{m}_{wskpo}[h_{wpo} - h_{wpin}] \tag{20}$$

h_{wpin} - specific enthalpy of water at the inlet to first water pump [kJ/kg].

Feed water preheater_1 energy balance:

$$\dot{m}_{wskpo}h_{wpo} + m_{pf1}h_{pf1} = \dot{m}_{pf1}h_{pwf1} + \dot{m}_{wskro}h_{waf1} \tag{21}$$

$\dot{m}_{pf1}$ - mass flow of steam at inlet to feed water preheater 1 [kg/s], h_{pf1} - specific enthalpy of steam at inlet to feed water preheater 1[kJ/kg], $\dot{m}_{wskpo}$ - mass flow of water at inlet/outlet to/from feed water preheater 1 [kg/s], h_{wpo} - specific enthalpy of water at inlet to feed water preheater 1 [kJ/kg], h_{pwf1} - specific enthalpy of water at outlet from feed water preheater 1 [kJ/kg], h_{waf1} - specific enthalpy of water after feed water preheater 1 [kJ/kg].

Water pump before Feed water preheater_2

$$W_{p2} = \dot{m}_{pf1}[h_{wpf2} - h_{wpf1}] \tag{22}$$

Mixer before Feed water preheater_2 energy balance.

$$\dot{m}_{pf1}h_{pwf2} + \dot{m}_{wskpo}h_{waf1} = \dot{m}_{wf2}h_{wf2'} \quad (23)$$

Feed Water Preheater _2 energy balance:

$$\dot{m}_{pf2}h_{pf2} + \dot{m}_{wf2}h_{wf2'} = \dot{m}_{pf2}h_{wf2} + \dot{m}_{wf2}h_{waf2} \quad (24)$$

$\dot{m}_{pf2}$ - mass flow of steam at inlet to feed water preheater 2 [kg/s], h_{pf2} - specific enthalpy of steam at inlet to feed water preheater 2[kJ/kg], $\dot{m}_{wf2}$ - mass flow of water at inlet/outlet to/from feed water preheater 2 [kg/s], $h_{wf2'}$ - specific enthalpy of water at inlet to feed water preheater 2 [kJ/kg], h_{wf2} - specific enthalpy of water at outlet from feed water preheater 2 [kJ/kg], h_{waf2} - specific enthalpy of water after feed water preheater 2 [kJ/kg].

Water pump before Feed water preheater_3

$$W_{p3} = \dot{m}_{pf2}[h_{wpf3} - h_{wpf2}] \quad (25)$$

Mixer before Feed water preheater_3 energy balance.

$$\dot{m}_{pf2}h_{wf3} + \dot{m}_{wf2}h_{waf2} = \dot{m}_{wf3}h_{wf3'} \quad (26)$$

(26)**Feed Water Preheater _3 energy balance:**

$$\dot{m}_{pf3}h_{pf3} + \dot{m}_{wf3}h_{wf3'} = \dot{m}_{pf3}h_{wf4} + \dot{m}_{wf3}h_{waf3} \quad (27)$$

$\dot{m}_{pf3}$ - mass flow of steam at inlet to feed water preheater 3 [kg/s], h_{pf3} - specific enthalpy of steam at inlet to feed water preheater 3[kJ/kg], $\dot{m}_{wf3}$ - mass flow of water at inlet/outlet to/from feed water preheater 3 [kg/s], $h_{wf3'}$ - specific enthalpy of water at inlet to feed water preheater 3 [kJ/kg], h_{wf4} - specific enthalpy of water at outlet from feed water preheater 3 [kJ/kg], h_{waf3} - specific enthalpy of water after feed water preheater 3 [kJ/kg].

Water pump before Feed water preheater_4

$$W_{p4} = \dot{m}_{pf4}[h_{wpf5} - h_{wpf4}] \quad (28)$$

Mixer before Feed water preheater_4 energy balance.

$$\dot{m}_{pf3}h_{wf5} + \dot{m}_{wf3}h_{waf3} = \dot{m}_{wf4}h_{wf4'} \quad (29)$$

Feed Water Preheater _4 energy balance:

$$\dot{m}_{pf4}h_{pf4} + \dot{m}_{wf4}h_{wf4'} = \dot{m}_{pf4}h_{wf4} + \dot{m}_{wf4}h_{waf4} \quad (30)$$

$\dot{m}_{pf4}$ - mass flow of steam at inlet to feed water preheater 4 [kg/s], h_{pf4} - specific enthalpy of steam at inlet to feed water preheater 4 [kJ/kg], $\dot{m}_{wf4}$ - mass flow of water at inlet/outlet to/from feed water preheater 4 [kg/s], $h_{wf4'}$ - specific enthalpy of water at inlet to feed water preheater 4 [kJ/kg], h_{wf4} - specific enthalpy of water at outlet from feed water preheater 4 [kJ/kg], h_{waf4} - specific enthalpy of water after feed water preheater 4 [kJ/kg].

Water pump before deareator

$$W_{p4} = \dot{m}_{pf5}[h_{wpf5} - h_{wpf4}] \quad (31)$$

Mixer before deareator energy balance.

$$\dot{m}_{pf4}h_{wf5} + \dot{m}_{wf4}h_{waf4} = \dot{m}_{wf5}h_{wf5'} \quad (32)$$

Deareator energy balance

$$\dot{m}_{pd}h_{pd} + \dot{m}_{wf5}h_{wf5'} + \dot{m}_{pp}h_{wpp} = [\dot{m}_{pp} + \dot{m}_{wf5} + \dot{m}_{pd}]h_{wpd} \quad (33)$$

$\dot{m}_{pd}$ - mass flow of steam at inlet to deareator [kg/s], h_{pd} - specific enthalpy of steam at inlet to deareator [kJ/kg], $\dot{m}_{wf5}$ - mass flow of water at inlet to deareator from mixer before deareator [kg/s], $h_{wf5'}$ - specific enthalpy of water at inlet to deareator from mixer before deareator [kJ/kg], $\dot{m}_{pp}$ - mass flow of water at inlet to deareator from control valve [kg/s] h_{wpp} - specific enthalpy of water at inlet from control valve [kJ/kg], h_{wpd} - specific enthalpy of water deareator [kJ/kg].

Water pump after deareator

$$W_{p5} = [\dot{m}_{pp} + \dot{m}_{wf5} + \dot{m}_{pd}][h_{wpdo} - h_{wpd}] \tag{34}$$

h_{wpdo} - specific enthalpy of water at the outlet from pump [kJ/kg]

Total Efficiency

$$\eta_{TH2} = \frac{P_{HP} + P_{MP} + P_{LP} - \sum W_{pi} + \eta HHV_{H2} - Qu_{H2}}{N_R} \tag{35}$$

P_{HP} / P_{MP} / P_{LP} High/Medium/Low-Pressure Steam Turbine Power [MW], N_R - Reactor thermal power [MW], Q_{uH2} - Heat lost to heat up helium to hydrogen production [MW], η - thermal efficiency of hydrogen production, HHV - higher heating value [kJ] W_{pi} - work of water pump [MW]

5 Model assumption

High-temperature Nuclear Reactor coupled with heat recovery steam cycle is the main topic of present research for hydrogen and electricity production. Model assumptions for all simulation at this work and main thermodynamic parameters are described in Table 3.

Table 3. Model assumption and thermodynamic parameters.

Main Parameter	**Value**	**Unit**
High-Temperature Nuclear Reactor Power	600	MW_{th}
Helium Pressure	70	Bar
Helium Inlet Temperature	235.9	°C
Helium Outlet Temperature	800-940	°C
HP/MP/LP Steam Turbine Isentropic Efficiency	0,88	-
HP/MP/LP Steam Turbine Mechanical Efficiency	0.998	-
Electrical Efficiency	0,9856	-
MP Turbine first sink pressure	29.4	Bar
MP Turbine second sink pressure	10,5	Bar
MP Turbine third sink pressure	4.8	Bar
LP Turbine first sink pressure	1,92	Bar
LP Turbine second sink pressure	0.24	Bar
LP Turbine third sink pressure	0.08	Bar
Dearator inlet Pressure	10.5	Bar
Condenser inlet Pressure	0.08	Bar
Mass Flow for Water Cooling	4707	kg/s
Heat Exchanger to Hydrogen Production		
Helium Mas Flow Rate	10	kg/s
Helium Inlet Pressure	1	Bar
Helium Inlet Temperature	20	°C

6 Results

Results of the numerical simulations for HTGR system coupled with heat recovery steam cycle are presented in Figures 3-5. In the Fig.4(a) steam cycle in the h-s graph and in Fig. 3(b) superheated steam mass flow for different the reactor outlet temperature are presented.

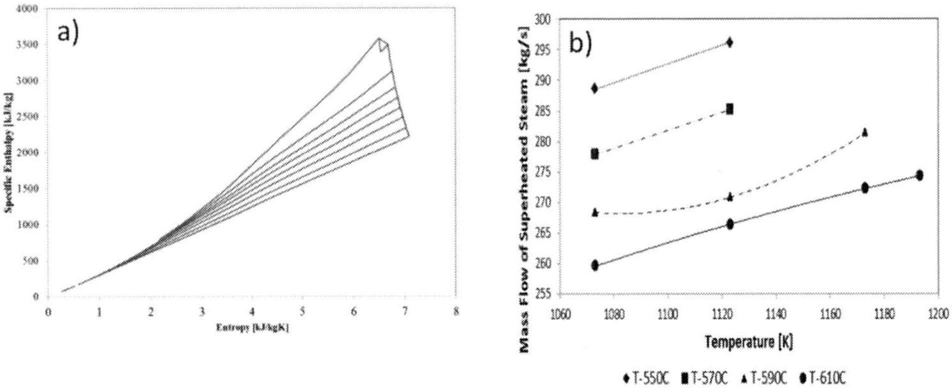

Fig. 3. h-s diagram (a) and superheated steam mass flow rate for different reactor outlet temperature (b).

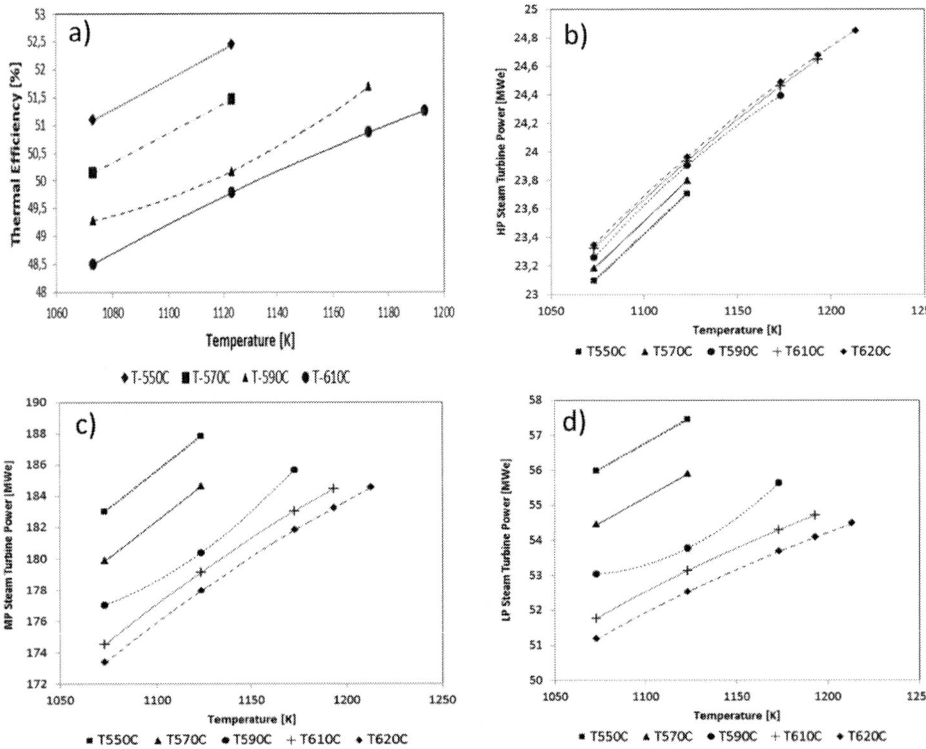

Fig.4. The thermal efficiency for heat recovery steam cycle without hydrogen production(a), HP-Steam turbine power at the various point of superheated steam (b), MP-Steam turbine power (c), LP-steam turbine power (d).

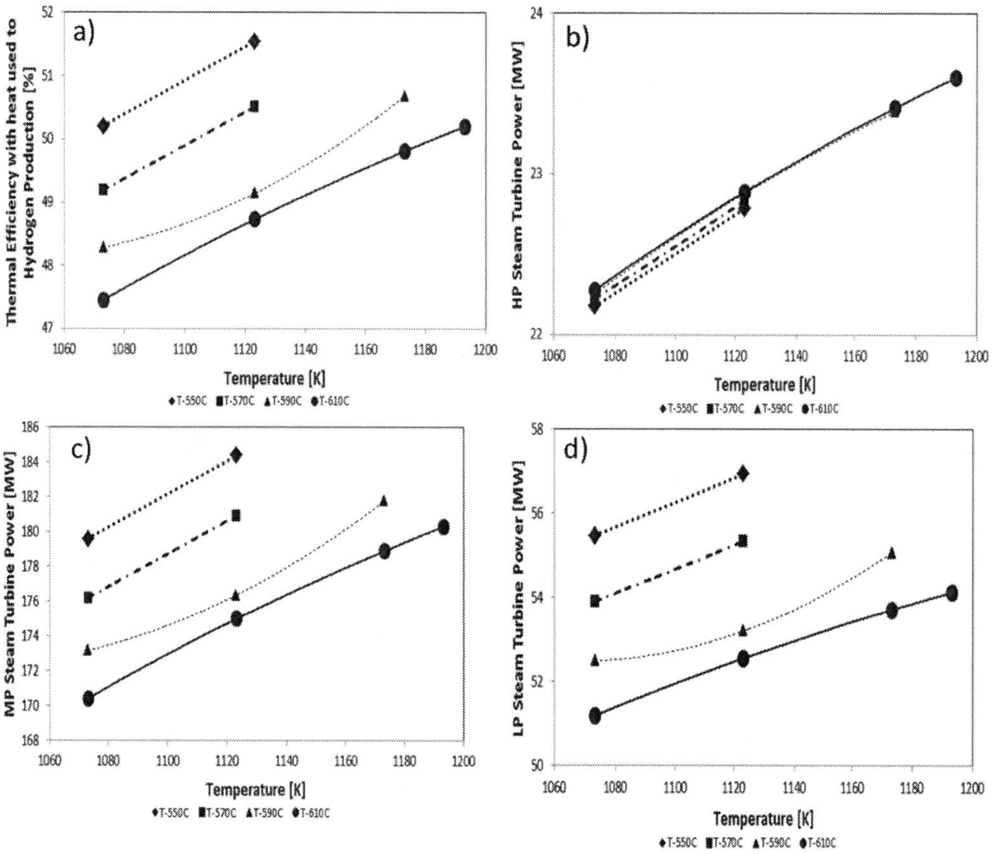

Fig.5. The thermal efficiency for heat recovery steam cycle with hydrogen production(a), HP-Steam turbine power at the various point of superheated steam (b), MP-Steam turbine power (c), LP-steam turbine power (d).

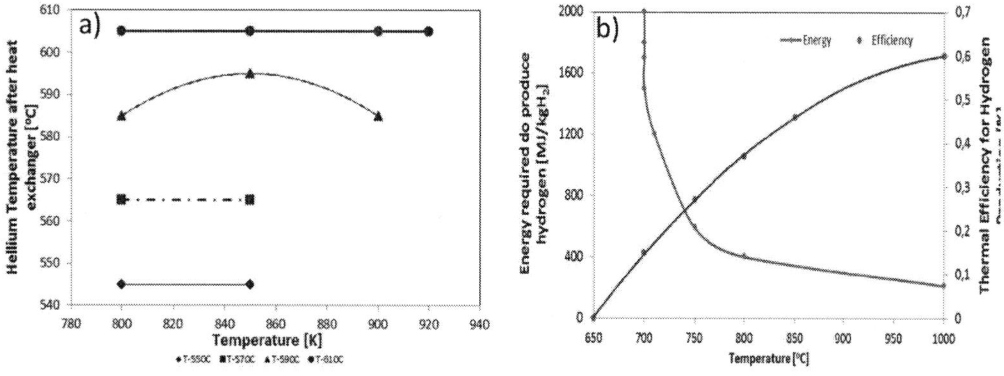

Fig.6. The graphs present (a) helium temperature to hydrogen process (b) thermal efficiency and thermal energy required to produce 1 kg of hydrogen using the MHR–SI [12].

7 Conclusions

High or very high-temperature nuclear reactors can be used to produce electricity with the high value of thermal efficiency. The power conversion system presented in Fig.1. and Fig. 2. has been analysed with and without hydrogen production in order to maximize the cycle efficiency. From a thermodynamic point of view, the most appropriate value of the thermal efficiency 52.5% is possible to achieve when cycle work in normal conditions for High-Temperature Nuclear Reactor and superheated steam should be superheated up to 530°C but when High-Temperature Nuclear Reactor coupled with heat recovery steam cycle works in supercritical conditions the value of thermal efficiency will be reduced by 2 to 3%. The lowest value of thermal efficiency for cycle working in normal and supercritical conditions is respectively equal 51% and 48.5% . When the outlet temperature from a nuclear reactor is increased up to 940°C it is possible to work in supercritical conditions what gives the opportunity to increase the value of the mass flow of superheated steam and increase a power from HP steam turbine, but the same time we can observe decrease the power of MP and LP steam turbine. The analysis shows that High-temperature Nuclear Reactor allows us effective realized much of industry process, give us the opportunity to produce hydrogen in a thermochemical cycle of coupled electricity generation and thermochemical hydrogen production process. What is more the analysis show that it is possible to achieve a high value of thermal efficiency with and without hydrogen production. The temperatures are not only one parameter to improve the thermal efficiency for High-Temperature. Another most important parameters are isentropic and mechanical efficiencies of the turbines and the compressor which has significant influence. High-temperature reactors used for electricity and hydrogen production have significant potential to improve efficiency by raising the reactor outlet temperature or steam temperature conditions.

The present work was supported by the Polish Ministry of Science (Grant AGH No.11.11.210.312).

References

1. S.J. Bae, J. Lee, Y. Ahn, J.I. Lee, Annals of Nuclear Energy **75** (2015)
2. U. Given, G. Velidi, Proceedings of International Congress on Advances in Nuclear Power Plants, ICAPP (2011)
3. B.J. Marsden, S.L. Fok, G. Hall, International Conference on Global Environment and Advanced Nuclear Power Plants (2003)
4. *Generation IV Roadmap: R&D Scope Report for Gas-Cooled Reactor Systems*, Gen IV IF (2002)
5. Z. Zhang, Y. Dong, F. Li, Z. Zhang, H. Wang, X. Huang, Hong Li, B. Liu, X. Wu, H Wang, X. Diao, H. Zhang, J. Wang, 2 (2016)
6. *Next Generation Nuclear Plant Research & Development Program Plan*, Idaho Nat Lab (2005)
7. *International Prismatic Block HTGR Commercial Deployment Meeting Summary in Washington DC*, March 8-10 (2016).
8. S.K. Mylavarapu, X. Sun, R. E. Glosup, R. N. Christensen, M.W. Patterson Applied Thermal Engineering **65** (2014)
9. Y. Shin, T.Lee, K. Lee, M. Kim Applied Energy **173** (2016)
10. G. Jezierski, Conventional and Nuclear Power [in Polish], Energetyka Cieplna i Zawodowa **10** (2009)
11. C.F. McDonald, Appl Thermal Eng **73** (2014)
12. B.Yildiz, M. S. Kazimi, International Journal of Hydrogen Energy **31** (2006)

E3S Web of Conferences **14**, 01047 (2017) DOI: 10.1051/e3sconf/20171401047
Energy and Fuels 2016

Development of numerical models for Monte Carlo simulations of Th-Pb fuel assembly

Mikołaj Oettingen[1*]

[1]AGH University of Science and Technology, Faculty of Energy and Fuels, Department of Nuclear Energy, al. Mickiewicza 30, 30-059 Krakow, Poland

Abstract. The thorium-uranium fuel cycle is a promising alternative against uranium-plutonium fuel cycle, but it demands many advanced research before starting its industrial application in commercial nuclear reactors. The paper presents the development of the thorium-lead (Th-Pb) fuel assembly numerical models for the integral irradiation experiments. The Th-Pb assembly consists of a hexagonal array of ThO_2 fuel rods and metallic Pb rods. The design of the assembly allows different combinations of rods for various types of irradiations and experimental measurements. The numerical model of the Th-Pb assembly was designed for the numerical simulations with the continuous energy Monte Carlo Burnup code (MCB) implemented on the supercomputer Prometheus of the Academic Computer Centre Cyfronet AGH.

1 Introduction

From one side, the development and design of any research or commercial nuclear system need numerous precise neutron transport and burnup calculations for the prediction of its behaviour under radiation field. From the other side, the numerical tools, nuclear data and models applied in these calculations, should be validated experimentally so that the results will be credible. Such validation must be done on a small scale compared to the real nuclear power systems, for instance by means of assemblies designed for dedicated integral experiments.

This paper presents the first step of a detailed calculations being performed before to conduct an integral irradiation experiment on a thorium-lead (Th-Pb) nuclear system, i.e. it shows the procedure applied for the development of numerical models for the High Performance Computing (HPC) using Monte Carlo methods. The presented numerical model reproduces the Th-Pb fuel assembly available at the AGH University, Faculty of Energy and Fuels, Department of Nuclear Energy, Krakow, Poland. The Th-Pb assembly consists of ThO_2 fuel rods manufactured in Bhabha Atomic Research Centre (BARC), India [1] and metallic Pb rods manufactured at AGH University. The design of the assembly allows different combinations of rods for various types of irradiations and experimental measurements. As neutron source the ^{252}Cf isotope has been foreseen since it is easily accessible [2]. Moreover, its energy spectrum, comparing to the other radioisotope sources,

* Corresponding author: moettin@agh.edu.pl

is the most similar to the fission energy spectrum of main fissionable isotopes and may be approximated by the Watt fission spectrum [3]. The lead was selected as a neutron reflector because of its numerous advantages, such as: low neutron absorption cross section, high atomic mass and thus little thermalisation, neutron multiplication capability in the (n,2n) reaction and perspectives of application as a coolant in the future fast nuclear energy systems [4]. The numerical model of the Th-Pb assembly was designed for the numerical simulations with the Monte Carlo continuous energy Burnup code (MCB) [5] implemented on the supercomputer Prometheus of the ACK Cyfronet AGH [6]. The design of a detailed numerical model of the Th-Pb fuel assembly is necessary to perform reliable simulations in the following scientific areas preceding experimental measurements:

- neutron spectrum and flux density distribution for different configurations of assembly, showing the influence of material composition on those characteristics,
- expected activity of thorium rods and other samples, for the selection of proper irradiation conditions,
- neutron induced transmutations and formation of new isotopes, especially fissile ^{233}U,
- fission and radiative capture reaction rates on various isotopes presented in the initial and irradiated fuel.

Section 2 briefly introduces the background of the thorium fuel cycle. Section 3 presents the MCB code used for numerical simulations. Section 4 shows the engineering geometry of the Th-Pb assembly while Section 5 the designed numerical model. The study is concluded in Section 6.

2 Thorium fuel cycle

The existing nuclear reactors operate on the uranium-plutonium fuel cycle (U-Pu), where the fission reaction mostly occurs on the naturally occurring ^{235}U and bred ^{239}Pu. However, the fuel cycle can be also based on the thorium (^{232}Th), which natural abundance is about 3 to 4 times larger than uranium [7]. The thorium itself cannot undergo nuclear fission but serves as a fertile isotope for the production of fissile ^{233}U. Thus, thorium fuel has to be combined with the fissile isotopes like ^{233}U, ^{235}U or ^{239}Pu. The thorium-uranium (Th-U) fuel cycle is a promising alternative against U-Pu fuel cycle, but it demands many advanced research before starting its industrial application in commercial nuclear reactors [8]. The leading country conducting research on the thorium fuel cycle is India due to the large resources of thorium and limited resources of uranium. India postulates implementation of a three stage nuclear power programme linking the closed fuel cycles of Pressurized Heavy Water Reactors (PHWR), Liquid Metal-cooled Fast Breeder Reactors (LMFBR) and Advanced Heavy Water Reactors (AHWR) [7]. In the first stage PHWR will utilize natural uranium fuel and thus produce plutonium and depleted uranium. Next, the depleted uranium and plutonium will fueled LMFBR with thorium blanket. In this way ^{233}U will be bred from ^{232}Th and ^{239}Pu from ^{238}U. The bred ^{239}Pu will continuously fueled LMFBR while ^{233}U will be used in the last step to fuel AHWR with Th component for continuous breeding of ^{233}U. Therefore in equilibrium state AHWR will operate on ^{233}U and will be fuel just with Th. The main advantages and drawbacks of the Th-U fuel cycle comparing to the traditional U-Pu fuel cycle are presented below:

Advantages:

- Good chemical and physical properties,
- Lower inventory of minor actinides and plutonium in spent fuel,
- Proliferation resistance (^{232}U contamination of irradiated fuel),
- Possibility of breeding fuel in thermal reactors and fast reactors,
- Thorium is more abundant, safer and more effective in mining.

Disadvantages:
• Neutron source (other fuel) required to produce fissile ^{233}U from ^{232}Th,
• Difficult and costly fuel fabrication (high γ radiation from ^{232}U),
• Relatively long time of ^{232}Th breeding to ^{233}U,
• Likelihood of using ^{233}U for military purposes,
• R&D and licensing needed before commercial implementation.

3 MCB code

Generally, in Monte Carlo methods a physical system is described by the probability density functions, from which the sequences of random numbers are sampled and used for numerical simulation. The sampling process is repeated many times and events of interest are recorded by scoring functions, called tallies. The final results of the simulation are the averages over events of interest simulated in many trials and associated variations. The Monte Carlo methods in nuclear engineering allow simulation of the advanced nuclear systems with a heterogeneous material composition and complex geometry. The presented paper shows the procedure of design numerical models for the Monte Carlo Continuous Energy Burnup Code (MCB) developed at the AGH University, Faculty of Energy and Fuels, Department of nuclear Energy. The MCB code integrates the commercial Monte Carlo transport code A General Monte Carlo N-Particle Transport Code (MCNP) [3] and the novel Transmutation Trajectory Analysis Code (TTA) [9] at the level of the FORTRAN source code. The MCNP subroutines use Monte Carlo methods for neutron transport simulations while the TTA forms and analyzes transmutation and decay chains for nuclide density evolution in time function. All MCNP features are available in MCB, and some new ones were implemented. In the MCB calculations one could obtain the following parameters: neutron multiplication factor, nuclear reaction rates, potential doses, material activities, neutron flux and power distributions, evolution of material densities and many others. The MCB code was already applied in many national and international projects related to the design of innovative nuclear systems [10].

4 Engineering Geometry

Fig. 1A. presents the Th-Pb fuel assembly located in the radiometric laboratory at the AGH University. The assembly consists of three main components, i.e. support structure, lead reflector and core. The assembly is mounted on a bottom support plate with four wheels, which can move the set-up between chosen position in the laboratory, e.g.: position for preparation of assembly arrangement and insertion of neutron source, position for short-time irradiations and position for long-time irradiations with an additional shielding. The upper support consists of a cylindrical top plate, top bars and grid. The bars and grid determine the rectangular shape of the assembly core. The top and bottom support plates are connected by four side bars. All structural element were fabricated using standard construction steel. The assembly core is surrounded by the reflector mounted using rectangular Pb bricks designed especially for radiation shielding purposes. The bricks are jointed together by means of "V" shape wedges located on each side of them. The standard dimensions of the brick equal 100x100x50 mm (LWH). However, some bricks were cut into smaller pieces to fix them into current assembly arrangement.

Fig. 1B shows the assembly core composed of Pb and ThO_2 rods arranged in the hexagonal pattern with pitch of 13 mm and length of 1200 mm. The current pattern consist of 246 cylindrical ThO_2 rods and surrounding hexagonal Pb rods. The hexagonal shape of the Pb rods allows their compact package while cylindrical shape of the ThO_2 rods, easy

DOI: 10.1051/e3sconf/20171401047

replacement of Pb rods without dismantling the whole assembly. The pins on both ends of the rods allow for their precise placement in the bottom plate and fixing bars. The fixing bars made of Al (see Fig. 1A), serve as a support structure for the pins. The fixing bars freely settle on pins and can be easily removed if necessary. The adjacent rods and outer Pb reflector can be vertically extracted after having removed the Al bars. A gap among rods of about 0.4 mm allows for their easy removal and the modification of the set-up configuration. Seven most inner ThO_2 rods were replaced Six of them were replaced by six Pb rods for the intensification of a neutron flux due to (n,2n) reactions on lead and by a space in the center of the assembly used to insert the Cf neutron source. The use of hexagonal pins surrounded by plane made of Pb bricks will form some gaps at the interface between the bricks and Pb rods (see Fig. 1B). The influence of the gap on the measurements will be checked in the numerical simulations preceding the conduction of the integral experiments.

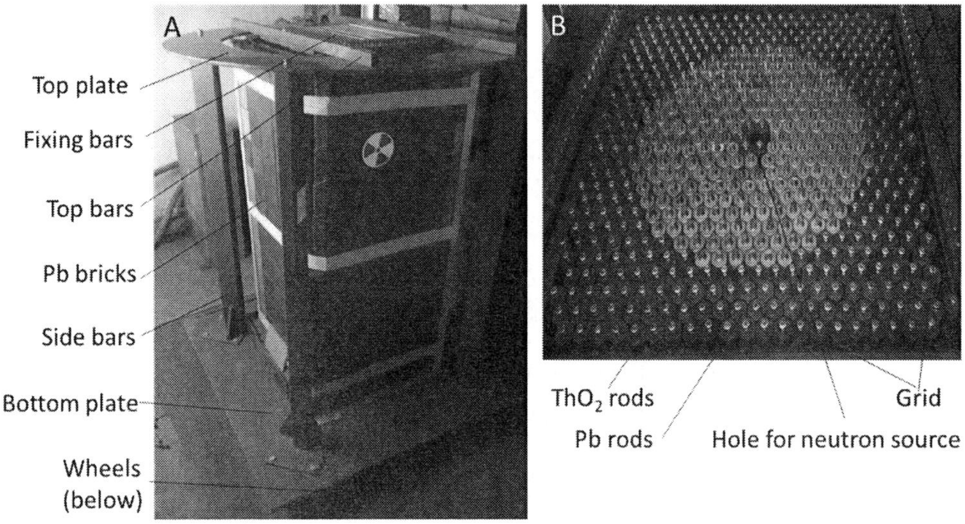

Fig. 1. Th-Pb assembly (A). The core of the assembly without aluminium fixing bars (B).

Fig. 2 shows both types of rods while Fig. 3 schema of the ThO_2 fuel rods manufactured in BARC. Unfortunately, the AGH University does not possess the mock-up of the thorium rod, which would allow the detailed geometrical measurements for the numerical modelling. The dismantling of the rods and extraction of theirs inner components would be troublesome due to the potential risk of alpha contamination as well as legal issues related to radiological protection. Thus, for the numerical modelling the specification of the rods from the available documentation was applied [11]. Table 1 present the main characteristic of the ThO_2 rods.

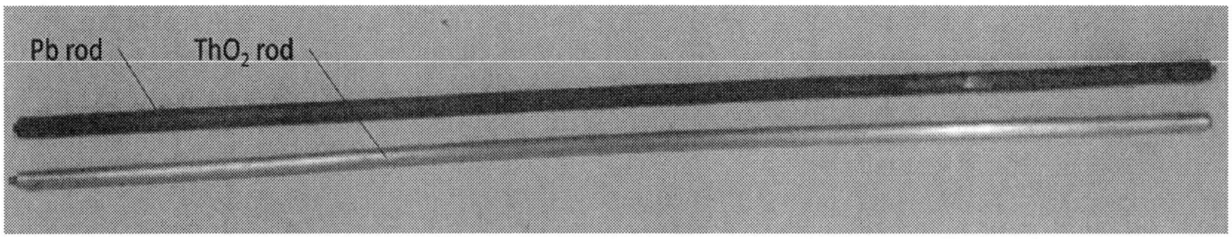

Fig. 2. Pb and ThO_2 rods.

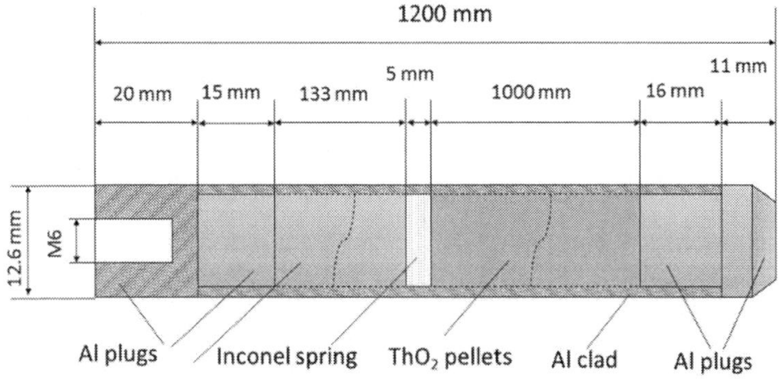

Fig. 3. Schema of ThO_2 rod.

Table 1. Characteristics of ThO_2 rod.

Mass of Th	~736 g
Mass of ThO_2	~836 g
Number of pellets	100
Pellet length	10 mm
Pellet diameter	1.064 mm
Gap thickness	0.13 mm
Clad thickness	0.85 mm
Rod diameter	12.6 mm

5 Numerical model

5. 1 Geometry

The procedure applied for the design of the Th-Pb assembly numerical model is shown in Fig. 4. Firstly, the engineering geometry has to be transferred to the computational geometry. To construct numerical geometry one needs to supply coefficients to the analytical surface equations to form a desired set of surfaces – step 1. Then using the intersection, union and complement operators it is possible to combine the surfaces into cells characterized by theirs volumes – step 2. It is also possible to define, a so called embedded geometry, which allows to build structures containing large arrays of the same elements. To do this, some cells have to be grouped together into entities called *universes*. The universes may be repeated into a square or hexagonal lattice arrangement.

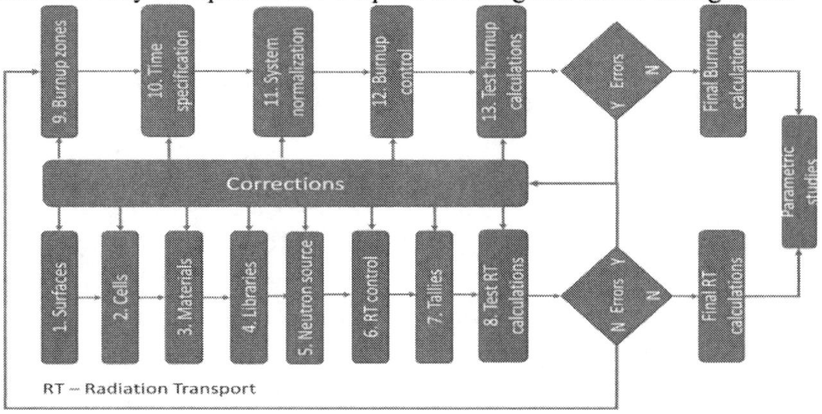

Fig. 4. Procedure applied for design of numerical models.

Fig. 5 shows the computational geometry of main structural elements of the Th-Pb assembly. The presented model is placed in the air cylinder (ϕ=1000 mm), which is surrounded by the void boundary conditions. Two main differences were introduced to the computational geometry comparing with the engineering geometry. Firstly, the wheels below the bottom plate were neglected. Secondly, the orientation of the side bars of the assembly was changed – i.e. the bars have been modeled with their faces parallel to the sides of the assembly. The numerical reconstruction of wheels and side bars rotation would demand transformation of the coordinate system and thus would complicate the numerical model and extend simulation time. Moreover, the changes are of minor importance for neutron transport and burnup simulation because the simplified elements are located outside the assembly core.

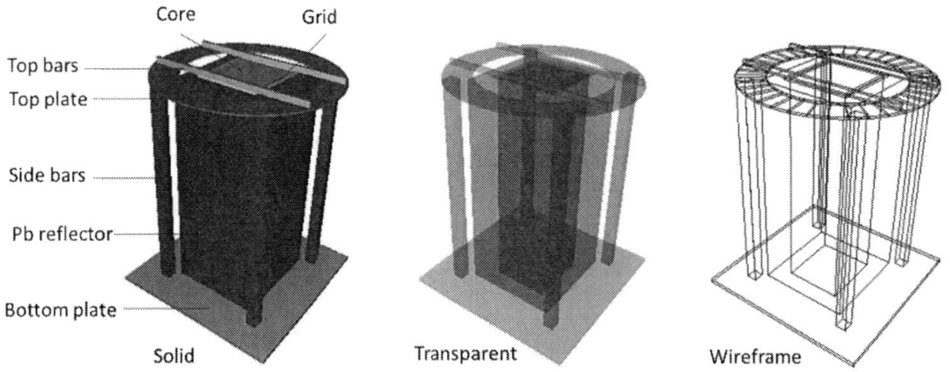

Fig. 5. Numerical model of the Th-Pb assembly.

Fig. 6 presents the cross sections of the assembly core with the Pb and ThO_2 rods and their surroundings. Both types of rods were numerically placed in the core using embedded geometry with the hexagonal lattice. Some simplification were introduced to the numerical geometry of the Th rods. The rods were divided into three axial sections namely, top of the rod, ThO_2 pellets and bottom of the rod. The upper section contains just Al filling, the hole for a M6 thread and homogenized material representing ICONEL spring (see Section 5.2). The ThO_2 in the middle section was not divided into particular fuel pellets but modeled as a one cylinder with a height of 1000 mm. The bottom section contains just compact Al plug without chamfering. The introduced simplifications will not influence the numerical results but significantly reduce the complexity of the model. The Al bars used to fix the rods, were modelled using embedded geometry but with the square lattice. The outer Pb neutron reflector was modelled as a compact cell surrounding the assembly core. The Californium neutron source was placed in the middle of the assembly – see section 5.4. The visual check of the model geometry is possible using dedicated plotting utility called Vised.

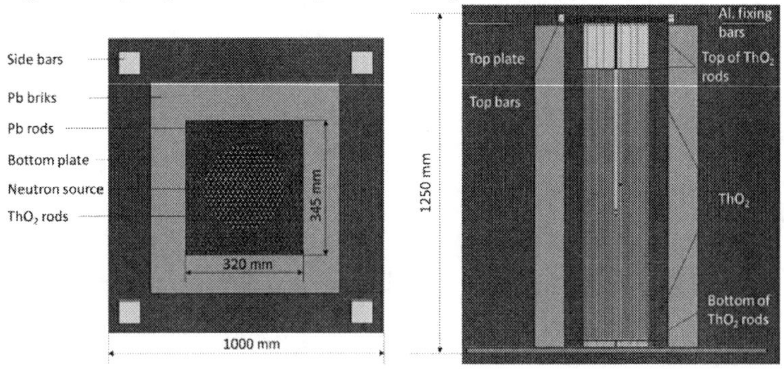

Fig. 6. Radial and axial cross-section of the assembly.

5.2 Materials, Libraries

The materials with specified densities, isotopic composition and nuclear data libraries should be assigned to each cell – step 3,4. They are represented in the input file using ZAID number where: Z means atomic number, A mass number and ID is a cross section library identifier. For instance, 90232.09c defines thorium 232 with 09c cross section library. The identifier of a library indicates the temperature, for which the cross section were evaluated, in this case 900 K. The most common nuclear data libraries available for the MCB numerical simulations are based on JEFF, JENDL and ENDF evaluations.

Usually, in the neutronic and burnup calculations one model only the most important components of the engineering geometry influencing the neutronic parameters, like neutron fluxes and spectra. Therefore, the engineering geometry does not need to be exactly transferred to the geometry of the numerical model. The less significant components are usually modelled as a homogenous mixes of various materials, which sufficiently reflects the physical behavior of the system. According to the visual analysis of the assembly the side bars, top bars and grid are empty inside, thus for the simplification they were modelled as a solid bars with the density and isotopic composition calculated for volumetric mix of inner air and surrounding steel. The numerical model is composed of eleven materials listed in Table 2.

Table 2. Materials composing numerical model [12,13].

Nr	Material	Density [g/cm^3]	Elements
1	ThO_2	9.4	Pellets
2	Cf_2O_3-Pd	12.023	Neutron source
3	Pb	11.35	Reflector bricks and rods
4	SS304L	8.0	Source capsules and holder
5	Aluminium	2.7	Fixing bars, plugs, clad
6	Steel	7.85	Plates
7	Air	0.001205	Free spaces
Nr	Homogenized materials (by volume)	Density [g/cm^3]	Elements
1	Steel 52% + Air 48%	4.08	Grid
2	Steel 31% + Air 69%	2.40	Side bars
3	Steel 13% + Air 87%	1.01	Top bars
4	Iconel 5% + Air 95%	0.4246	Spring

5.3 Neutron source, control, tallies

One of the most important issues in the numerical modeling of the Th-Pb assembly is the reconstruction of a neutron source – step 5. The source is characterized not only by the geometrical parameters but also by its intensity and energy spectrum. Moreover, the radioisotope neutron source degrades due to decay of neutron emitting isotopes and nuclear transmutation due to self-generated neutron field. All these effects should be taken into account in the numerical reconstruction of the neutron source. The Californium neutron source emitting neutrons in spontaneous fission was reconstructed numerically [14]. The source will be placed in the middle of the assembly as it is shown in Fig. 7. The intensity of the fresh source is about 10^8 n/s, which corresponds to the mass of about 43 μg ^{242}Cf. The source was modelled in the form of Cf_2O_3-Pd cermet wire embedded in two stainless steel capsules [15]. The energy spectrum of Cf source is similar to the fission spectrum, thus the Watt fission spectrum with special fitting parameters were used in the numerical modeling.

DOI: 10.1051/e3sconf/20171401047

The numerical modelling with other radioisotope neutron sources like AmBe or PuBe is also planned.

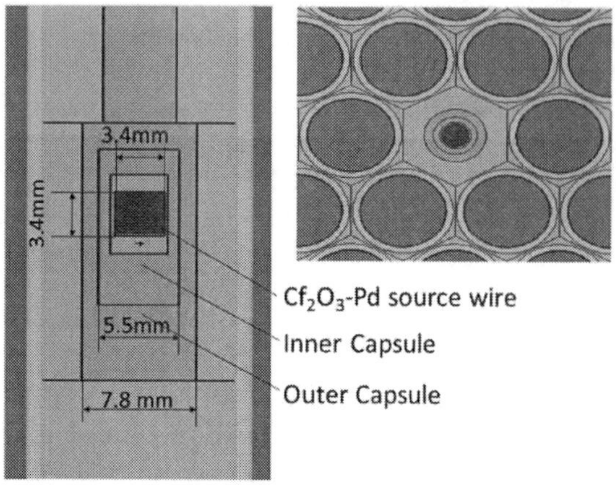

Fig. 7. Numerical model of Cf neutron source.

The main control parameters are related to the statistics of numerical simulations and thus precision of obtained results – step 6. The MCB simulation might be performed in the *kcode* mode, which provides calculation of neutron multiplication factor or in the *source* mode. The second option was chosen for the current calculations because the initial Th-Pb assembly does not contain any fissile material and is unable to induce self-sustaining fission chain reaction. The number of simulated neutron histories depends on the NPS parameter set in the input file. At the moment NPS values are quite small because the current models serves for the testing purposes. After final checkup of the model the NPS will significantly increase to the values allowing the high precision of results together with the acceptable computation time. In the MCNP/MCB modeling the additional large set of control parameters is available on request of the user [3]. These parameters are rather used for the specific types of analyses, which might be performed in additional calculations.

Recording some aspect of the average particle behavior is called tallying and tallies are the scoring functions used for tallying. Tallies provide summary information about the simulation of the physics problem. The standard type tallies can be in the input file or by the tally modification at the level of the source code. The MCB code needs particle fluxes for the estimation of the reaction rates, which in turn are necessary for calculation of the nuclide density time evolution. The numerical procedures lying behind, the so called F4 tally used to estimate the neutron flux, were implemented in the MCB code for the heating and reaction rates calculations. Thus, in the basic numerical model of Th-Pb assembly the specification of tallies in the input file is superfluous, because the quantities of interest will be automatically calculated by the MCB subroutines. However, in the future parametric analysis some tallies will be used for additional studies.

5.4 Burnup

The second stage in the design of numerical model, after test radiation transport calculations (step 8) is the implementation of the inherent MCB function for burnup calculations – step 9. The change in the fuel composition caused by the series of particle/neutron interactions and decays is described as fuel burnup. The detailed burnup analysis should predict the change in the isotopic fuel composition as the function of time and space. The control card BURN specified in the input file launches subroutines calculating the time evolution of nuclide density. The user is obliged to specify physical constraints over a transmutation system to start burnup calculations.

The geometry of the numerical model for neutron transport calculation is modified in order to divided fissile/fertile material into radial and axial burnup zones – step 9. In case of the Th-Pb assembly the thorium rods were divided into eight rings of ThO_2 rods around neutron source. Each ring in addition was divided into 20 axial zones of 50 mm length each, which gives in total 160 burnup zones. The number of burnup zones depends on the purpose of the modelling and may vary during the design of the experiments. For example, the number of rings probably will be decreased to seven, because of the use of six Pb rods around the Cf source for the intensification of neutron flux. Each burnup zone containing ThO_2 is distinguished assigning a different material number, doing so the burnup distribution of the whole assembly is obtained.

The time intervals for the reaction rates recalculations and thus the changes of material densities might be defined as a series of individual irradiation periods, or as cumulative values – step 10. For the current numerical design the irradiation time will not exceed 30 days divided into individual periods of a few days. Some additional actions related to the material processing may be activated at specified time. The system will be normalized to the source intensity of 10^8 n/s including the degradation of the neutron source due to the its decay and transmutation in every time step – step 11. The last step before burnup test calculations is related to the implementation of burnup control parameters, which determine efficiency and numerical solution of Bateman equations – step 12. For the initial calculations the standard set of burnup control parameters was used. The algorithmics used for burnup control is quite complex and it is not considered in this study [9].

6. Conclusions

The study presented the procedure developed for the design of the numerical model for the simulations of the Th-Pb fuel assembly. The design of the numerical model is the first step in order to start the integral irradiation experiments and initiation of R&D in the thorium fuel cycle at the AGH University. In the frame of this study:

- the detailed numerical model of Th-Pb assembly was designed,
- the advanced features of MCNP/MCB code were used to reconstruct real irradiation conditions,
- the model was tested in series of initial short-time and low-precision numerical simulations, to verify its credibility for modelling the irradiation experiments,
- it was shown that, designed model may serves as a basic model for a series of advanced numerical calculations and can be easily modified by the user,
- it was presented that, the development of the numerical models for the Monte Carlo simulations is a long-term task and demands detailed knowledge about engineering geometry, irradiation conditions, neutron source and other parameters influencing experimental measurements.

The future activities related to the modelling of Th-Pb assembly will focus on the modification of the numerical model for the detailed, high-precision calculations using HPC computers of ACK Cyfronet AGH. Afterwards, the numerical results of this unique will be used as a basic proof of competences of the scientific team and to boost the AGH team, towards participation in scientific projects related to thorium fuel cycle. The scientific collaboration is necessary for the knowledge exchange and useful to promote he possibility to receive funds for radiometric laboratory following to the establishment of the advanced radiometric measurement facility.

Acknowledgements The research was supported in part by PL Grid Infrastructure available at Academic Computer Centre CYFRONET AGH. In addition, partial financial support of this study through the funds of the Polish Ministry of Science and Higher Education is kindly acknowledged.

References

1. Bhabha Atomic Research Centre [online] [access: 19.09.2016], www.barc.gov.in
2. R.C. Martin, J.B. Knauer, P.A. Balo, APPL RADIAT ISOTOPES, *Production, distribution and applications of californium-252 neutron sources*, **53**, 785-792, (2000)
3. X-5 Monte Carlo Team, LANL report LA-UR-03-1987, *MCNP – A General Monte Carlo N-Particle Transport Code, Version 5*, (2005)
4. A. Wojciechowski, PROG NUCL ENERG, *Criticality of the thorium burnup in equilibrium state*, **92**, 81–90, (2016)
5. J. Cetnar, I. Królikowski, M. Oettingen, ACC Cyfronet AGH, ISBN: 978-83-61433-19-4, *Neutron transport and burnup simulations with MCB - The Monte Carlo Continuous Energy Burn-up Code*, (2016)
6. ACK Cyfronet AGH [online] [access: 19.09.2016], kdm.cyfronet.pl/portal/Prometheus
7. International Atomic Energy Agency, IAEA-TECDOC-1450, *Thorium fuel cycle – Potential benefits and challenges*, (2005)
8. International Atomic Energy Agency, IAEA Nuclear Energy Series No. NF-T-2.4, *Role of thorium to supplement fuel cycles of future nuclear energy systems*, (2012)
9. J. Cetnar, General, ANN NUCL ENERGY, *Solution of Bateman Equations for Nuclear Transmutations*, **33**, 640–645, (2006)
10. M. Oettingen, J. Cetnar, T. Mirowski, CSCI, *The MCB code for numerical modeling of fourth generation nuclear reactors*, **16**, 329–350, (2015)
11. S. I. Abdel-Khalik, P. A. Haldy, A. Kumar, FUSION SCI TECHNOL, *Blanket Design and Calculated Performance for the Lotus Fusion-Fission Hybrid Experimental Devices Test Facility*, **2**, 189-208, (1984)
12. R. J. McConn Jr, C. J. Gesh, R. T. Pagh, R. A. Rucker, R. G. Williams III, Pacific Northwest National Laboratory, PIET-43741-TM-963, PNNL-15870 Rev. 1, *Radiation portal monitor project, Compendium of Material Composition Data for Radiation Transport Modeling, Revision 1*, (2011)
13. L. R. Morss, N. M. Edelstein, J. Fuger, Springer, *The chemistry of the actinide and transactinide elements, Fourth edition*, 1, (2010)
14. R. Radev, T. McLean, Lawrence Livermore National Laboratory, LLNL-TR-664160, *Neutron Sources for Standard-Based Testing*, (2014)
15. Frontier Technology Corporation, [online], [access: 19-09-2016], www.frontier-cf252.com

E3S Web of Conferences **14**, 01048 (2017) DOI: 10.1051/e3sconf/20171401048
Energy and Fuels 2016

Coupling of high temperature nuclear reactor with chemical plant by means of steam loop with heat pump

Mariusz Kopeć[1,*]

[1]AGH-University of Science and Technology, Faculty of Energy and Fuels, Kraków, Poland

Abstract. High temperature nuclear reactors (HTR) can be used as an excellent, emission-free source of technological heat for various industrial applications. Their outlet helium temperature (700°-900°C) allows not only for heat supply to all processes below 600°C (referred to as "steam class"), but also enables development of clean nuclear-assisted hydrogen production or coal liquefaction technologies with required temperatures up to 900°C (referred to as "chemical class"). This paper presents the results of analyses done for various configurations of the steam transport loop coupled with the high-temperature heat pump designed for "chemical class" applications. The advantages and disadvantages as well as the key issues are discussed in comparison with alternative solutions, trying to answer the question whether the system with the steam loop and the high-temperature heat pump is viable and economically justified.

1 HTR as a source of process heat

According to the nomenclature proposed in the EUROPAIRS Project [1], industrial heat consumers can be grouped into the following 3 classes:

- **steam class** (~150°C up to ~600°C) where steam is used both to the heat transport and heating processes. E.g. desalination, drying, paper industry, chemical industry (soda ash, ammonia and urea), production of plastic, non-ferrous metals, refineries.
- **chemical class** (~600°C up to ~900°C) where heat is mainly consumed as the driver of chemical reactions delivering required reaction enthalpy. E.g. production of plastic, hydrogen, coal liquefaction.
- **mineral class** (above ~1000°C) where heat is used to melt solids or to drive reactions between solids. E.g. production of lime, cement, ore sintering, coke making, metallurgy.

The industrial processes belonging to these classes usually generate waste heat with temperatures enough to be used for washing, rinsing, food preparation or even district heating. In the conventional technologies the required heat is generated almost exclusively by the combustion of fossil fuels. However, the burning of fossil fuels on the global scale raises serious environmental concerns, as it is the largest source of greenhouse gases emissions, mainly CO_2. These concerns are the most important motivation for the search for alternative sources of clean process heat.

* Corresponding author: Mariusz.Kopec@agh.edu.pl

The high temperature nuclear reactor (HTR) can serve as an excellent, emission-free source of technological heat for various industrial applications, combining high efficiency with advanced safety features. The coolant temperature on HTR outlet can be as high as 900°C to 1000°C (for VHTR even 1200°C), but the temperature available for the processes is limited rather by parameters of the heat transport loop. Within the EUROPAIRS Project the following three configurations have been considered:

- steam generator and the steam based heat transport loop;
- gas-gas Intermediate Heat eXchanger and the gas based heat transport loop;
- gas-molten salt IHX and the molten salt based heat transport loop.

The use of steam transport loop is the most mature and proven technology. With existing technologies this coupling can deliver process steam at temperatures up to 600°C (or even more). All limitations are related to the resistance to creep, fatigue and corrosion of the applicable high-temperature materials, and appear virtually the same as in conventional supercritical steam systems. The near-term coupling scheme proposed in EUROPAIRS Project (with one HTR core) is shown in Fig. 1.

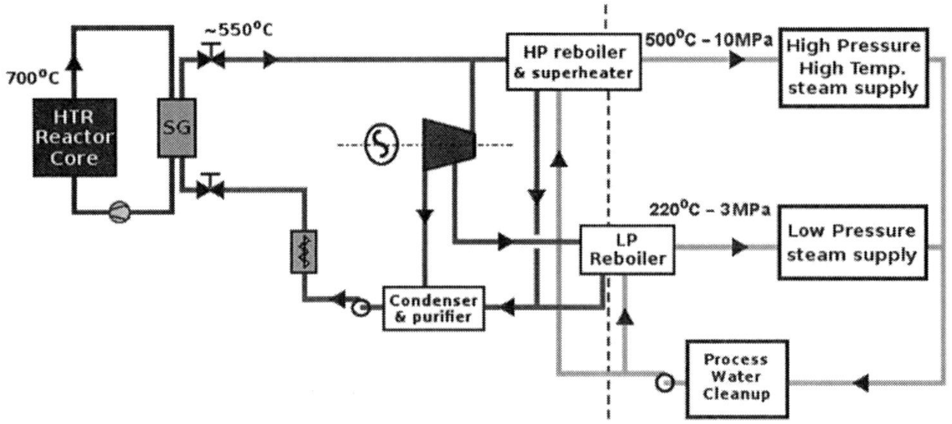

Fig. 1. System configuration proposed in EUROPAIRS Project

The gas-gas transport loop could provide heat at temperatures around 750°C to 800°C, which may be sufficient for some *chemical class* end-users. However, this technology cannot be regarded as ready to use because at these temperatures there are still challenges concerning applicable materials capable of withstanding creep damage for long lifetimes. Further development of appropriate heat exchangers is also required. Additionally, it must be taken into account that in a real situation some "safety distance" between the reactor and the chemical plant will be required. In such a case the hot helium would have to be transported over a long distance. Beside the high cost of helium, the material issues at these temperatures would require use of very expensive materials, increasing dramatically the expected installation cost.

Comparing to the gas loop, the loop on molten salts would allow for lower pressures, higher power densities and more efficient heat transfer. However, due to serious corrosion problems, high pressure drop and high risk of freezing, this configuration would require intensive R&D and probably cannot be consider even as the medium term option.

The positive long-term experience with HTRs and analysis of possible coupling schemes indicate that the current technological maturity is sufficient to deploy these reactors as an emission-free source of process heat for *steam class* end-users. Simultaneously it is clear that the HTR cannot be used as the heat source for the *mineral class* applications because the temperatures on reactor outlet are too low. But in case of

 DOI: 10.1051/e3sconf/20171401048

chemical *class applications*, the situation is more complex. Direct heating by means of gas-gas IHX and the gas heat transport loop is indeed possible as a medium-term option, but the near term solution can utilize the proven and relatively inexpensive steam loop technology with one additional step in which the temperature is raised to the level required by the end-users. At the simplest version it can be carried out by external heating (e.g. electric), but some authors suggest that higher efficiency can be achieved with a mechanical high-temperature heat pump [2].

2 Nuclear heat supply system for *chemical class* end-users

The steam-based nuclear heat supply system for *chemical class* end-users generally follows the configuration proposed in EUROPAIRS Project for the *steam class* coupling. The helium on HTR outlet has a temperature of 700°C and the steam leaves steam generator (SG) at 550°C and 10 MPa. The steam is transported over the long loop configured as a classical bottoming-cycle CHP, where the heat at ~530°C is directed through the heat exchanger towards the end-user installation, and the waste heat is used for power generation. As the temperature 530°C is too low for *chemical class* applications, in the next step it must be raised to the required level which in the presented analyzes was set to 850°C. The steam leaving the HX is expanded in a steam turbine providing mechanical work which can be used directly or converted into electrical energy, subsequently required to raise the output temperature. The latter configuration offers higher flexibility and has been assumed as the reference for the purposes of further analyzes. For simplicity, any losses due to conversion between mechanical work and electrical energy are further neglected.

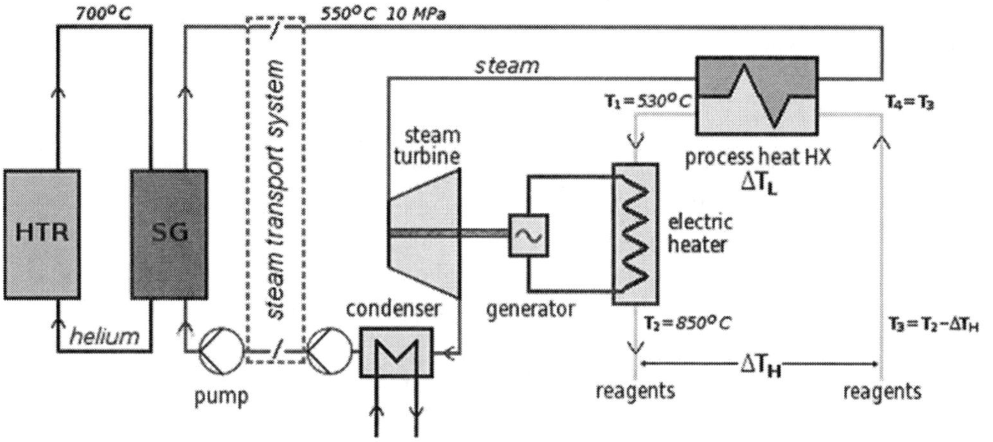

Fig. 2. Configuration of the heat transport system with an electrical heater

The additional step is necessary to increase the output temperature from 530°C to 850°C. This can be accomplished by simple electrical heating or by using the mechanical heat pump. The first configuration, shown in Fig. 2, is very simple. It is based on existing technologies and can be deployed almost immediately, but the total efficiency will be limited by efficiency of electricity production. The second one, shown in Fig. 3, will be much more complicated (and expensive), but the total efficiency can be higher than in the first case. For simplicity the same ΔT_H has been assumed for both streams in HTHX, and simultaneously the temperature drop between gas and heated medium has been set at 20°C.

 DOI: 10.1051/e3sconf/20171401048

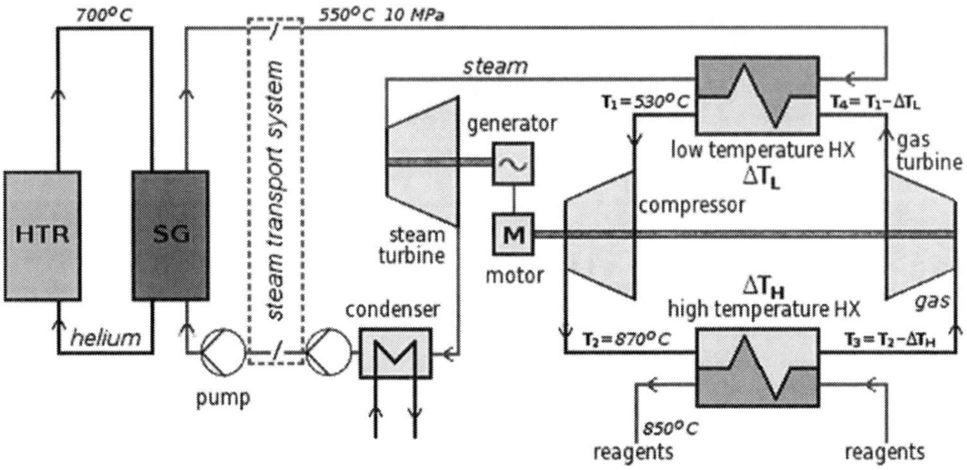

Fig. 3. Configuration of the heat transport system with a mechanical heat pump

It should be noted that regardless of the considered system configuration, most of its parameters will be determined by four basic factors:

- T_L – the temperature retrieved from the HX on steam loop – here $T_L = T_1 = 530°C$
- T_H – the gas temperature after compression – here $T_H = T_2 = 870°C$
- ΔT_H – the temperature span at which the end-user consumes the heat supplied.
- η_e – efficiency of electricity production.

In the *chemical class* the high-temperature heat is mainly used as a driver of endothermic chemical reactions, and is usually consumed at almost constant temperature. It means that this temperature span will be rather narrow, and it is safe to assume that in most situations $\Delta T_H < 200°C$.

Operating experience with the high temperature nuclear reactors shows that the efficiency of electricity production in the Rankine cycle is lower than 40% (e.g. Peach Bottom 1 – 34.8%, Fort St.Vrain – 39.2%, THTR – 39.5%) [3]. The steam parameters used in this work (550°C @ 10 MPa) are the most close to those in Peach Bottom reactor (538°C @ 10 MPa, no reheat) [4] whereas the other reactors used higher steam pressures (16.6 and 18.6 MPa respectively) as well as steam reheat. For this reason, it is justified to use in analyzes $\eta_e = 35\%$ what is very close to the efficiency of Peach Bottom reactor. In addition, it should be noted that in case of heat transport system the steam turbine will work in the bottoming cycle so the efficiency of electricity production will be lower than it would be for the fresh steam.

2.1 Key indicators of system performance

In presented analyses it is assumed that the whole energy passed to the end-user installation in form of high-temperature heat is produced by the same high temperature reactor. The high temperature heat supply systems will use the low temperature heat flux $\dot{Q}_L$ retrieved from LT HX (or process HX) and the mechanical power $\dot{W}_M$ to pass the high temperature heat flux $\dot{Q}_H$. It should be noted that the temperatures called here "low" are in fact around 530°C whereas the "high temperatures" may be well above 850°C. The energy balance is as follows:

$$Q_H = Q_L + W_M \tag{1}$$

 DOI: 10.1051/e3sconf/20171401048

Both electricity and heat come from the same source, but the former is generated with a low efficiency $\eta_e = 0.35$. Therefore the overall efficiency is defined as:

$$\eta = \frac{Q_H}{Q_L + W_M/\eta_e} \tag{2}$$

The next two indicators of interest are the share of the heat used:

$$R_{QL} = Q_L/Q_H \tag{3}$$

and the share of mechanical work used:

$$R_{WM} = W_M/Q_H \tag{4}$$

This last value determines also the overall efficiency:

$$\eta = \frac{1}{1 + R_{WM}(1/\eta_e - 1)} \tag{5}$$

From the point of view of system optimization the value of R_{WM} should be as low as possible, because the power, generated at a relatively low efficiency, is more expensive than heat. Low value of R_{WM} corresponds to a high value of R_{QL} which simply means that the heat passed from the steam loop is used effectively. However, it is important that both heats must be positive. A negative value indicates that the heat is transported in the opposite direction.

2.2 Electric heating

The small temperature drop in the chemical loop (ΔT_H) has serious consequences for the system shown in Fig. 2, because:

$$\Delta T_L = \Delta T_H + (T_1 - T_2) = \Delta T_H - 320°\text{C} \tag{6}$$

and when $T_1 = 530°C$ and $T_2 = 850°C$ the process HX would work inversely, passing heat from the chemical loop to the steam loop always when $\Delta T_H < 320°C$. This means that in practice this heat exchanger never will transfer heat to the chemical loop and therefore is not needed, as well as the whole heat transport loop. The entire high-temperature heat required by the end-user will be supplied exclusively by the electric heater, and the electric power may be generated on site of the HTR power plant. The system gets really simple and this configuration will be used as the reference one.

Substitution of $Q_L = 0$ in equations (1)-(4) gives:

- $\eta = \eta_e = 0.35$
- $R_{WM} = 1.0$
- $R_{QL} = 0.0$

In fact, this configuration does not even require the use of high-temperature reactors, since electricity can be provided by any source. However, the zero-emission requirement imposes restrictions on the method for producing electricity. The deployment of electric heating does not require intensive R&D and this solution can be used almost at once.

DOI: 10.1051/e3sconf/20171401048

2.3 High-temperature heat pump

A heat pump is a device that uses external power to transfers heat from one fluid at low temperature to another one at higher temperature. Heat pumps are commonly used at much lower temperatures, transforming worthless waste heat to useful heat of higher temperature. Performance of conventional heat pumps is described by the ratio of heat supplied to the work consumed (COP), which often exceeds 1. However, this parameter cannot be used for high-temperature heat pumps, because in their case there is no worthless heat and instead of COP the overall efficiency (2) will be used.

2.3.1 Heat pump without regeneration

The described here high-temperature heat pump consists of a gas compressor, a gas turbine and two heat exchangers (LT and HT). The compressor and the turbine are mounted on the same shaft, and are driven by an electric motor. In terms of thermodynamics, this device implements a reverse Brayton cycle, which for reversible processes is shown in Fig. 4.

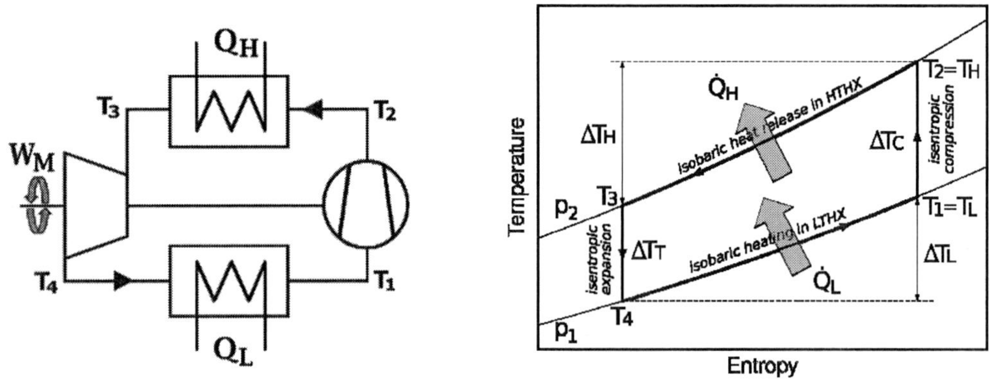

Fig. 4. Simple HT heat pump and the reverse Brayton cycle for reversible processes

It should be noted that the temperatures T_L and T_H are always set by the external conditions, and must be the same in all cycles analyzed.

Reversible processes

In case of an ideal gas there are two adiabatic processes:

$$T_2 p_2^{\left(\frac{1-\kappa}{\kappa}\right)} = T_1 p_1^{\left(\frac{1-\kappa}{\kappa}\right)} \; and \; T_3 p_2^{\left(\frac{1-\kappa}{\kappa}\right)} = T_4 p_1^{\left(\frac{1-\kappa}{\kappa}\right)} \tag{7}$$

where the appropriate temperature ratios are the same:

$$T_1/T_2 = T_4/T_3 \Rightarrow T_4 = T_3(T_1/T_2) \tag{8}$$

The work required to drive the system is the difference between the work consumed by the compressor (W_C) and the work done by the turbine (W_T):

$$W_M = W_C - W_T = m\, c_p[\Delta T_C - \Delta T_T] \tag{9}$$

The heat transferred through HTHX is:

$$Q_H = m\, c_p \Delta T_H \tag{10}$$

The share of mechanical work R_{WM} and the overall efficiency η does not depend on the temperature span ΔT_H so the results are valid also for very narrow temperature windows. For $T_1 = 530°C$, $T_2 = 870°C$, and $\eta_e = 0.35$ the calculated performance indicators are:

- $\eta = 0.64$
- $R_{WM} = 0.30$
- $R_{QL} = 0.70$

These values are really much better than the ones in case of electric heater, but are valid only for the ideal cycle. In reality the processes of compression and expansion are irreversible and the pressure drop in heat exchangers and piping cannot be neglected. On the other hand, these values indicate the limits, which in fact cannot be improved.

Irreversible processes

When the processes of compression and expansion are irreversible, the temperature ratios in eq. (8) are valid for temperatures T_{2s} and T_{4s} in equivalent isentropic processes:

$$T_1/T_{2s} = T_{4s}/T_3 \implies T_{4s} = T_3(T_1/T_{2s}) \tag{11}$$

The temperatures T_{2s} and T_{4s} are determined by internal efficiencies of compressor (η_C) and turbine (η_T) respectively:

$$T_{2s} - T_1 = \eta_C(T_2 - T_1) \quad and \quad T_4 - T_3 = \eta_T(T_{4s} - T_3) \tag{12}$$

Now, the share of mechanical work R_{WM} depends not only on the temperature span ΔT_H but also on the internal efficiencies η_C and η_T. When the temperature span ΔT_H is below the limit ΔT_H^{MIN}, the $R_{WM} > 1.0$ and the LTHX will transport heat in an opposite direction. Part of work will be lost for heating steam.

For the realistic values $\Delta T_H = 100$ and $\eta_C = \eta_T = 0.90$ the calculated performance indicators are:

- $\eta = 0.40$
- $R_{WM} = 0.81$
- $R_{QL} = 0.19$

The overall efficiency is still higher than it was for electric heating, but the gain is not very impressive. Simultaneously, the calculated $\Delta T_H^{MIN} = 74.7°C$ determines the limit below which the overall efficiency will be worse than for the simple reference case.

Irreversible processes and pressure drop

The pressure drops in cycle are characterized by the parameter β:

$$\beta = \left(\frac{p_4}{p_1}\frac{p_2}{p_3}\right)^{\frac{\kappa-1}{\kappa}} \tag{13}$$

where p_i are the pressures for the respective temperatures T_i and $\kappa = C_p/C_v$. With these pressure ratios it is possible to calculate the mechanical work W_M, its share R_{WM} and the overall efficiency η. For $\beta = 1.02$ (1% pressure drop on both LT and HT HXs), the respective performance indicators are nearly the same as for electric heating.

- $\eta = 0.36$
- $R_{WM} = 0.95$
- $R_{QL} = 0.05$

The conclusion is that in comparison with the simplest configuration, the system with the heat pump gets much more complex but with negligible performance improvement.

Considering the uncertainties regarding the realistic values of internal efficiencies, pressure drops and required width of temperature window, it is clear that the simple heat pump cannot guarantee a satisfactory results.

2.3.2 Heat pump with regeneration

Heat pump performance can be significantly improved by using the regenerative heat exchanger (Fig. 5). Although it does not enable the recovery of waste heat, but allows adjustment of the system to a narrow temperature window. The heat at temperatures lower than required by the chemical processes is used to heat up gas leaving the LT HX. The compression starts at temperature higher than the temperature on the LT HX outlet and the compressor requires less work to get the required temperature T_H.

In real situations the regenerative HX (RHX) cannot pass heat at the same temperatures for both streams and some shift ΔT_o must be taken into account. The relationships between temperatures shown in Fig. 5 explain that the temperature difference ΔT_R at the RHX cannot be arbitrary and is determined by the assumed values of T_L, T_H, ΔT_H and ΔT_o:

$$\Delta T_R = T_H - T_L - \Delta T_H - \Delta T_o \quad (14)$$

Simultaneously, the work consumed by compressor depends only on the required temperature span ΔT_H and the temperature shift ΔT_o :

$$W_C = m\, c_p(\Delta T_H + \Delta T_o) \quad (15)$$

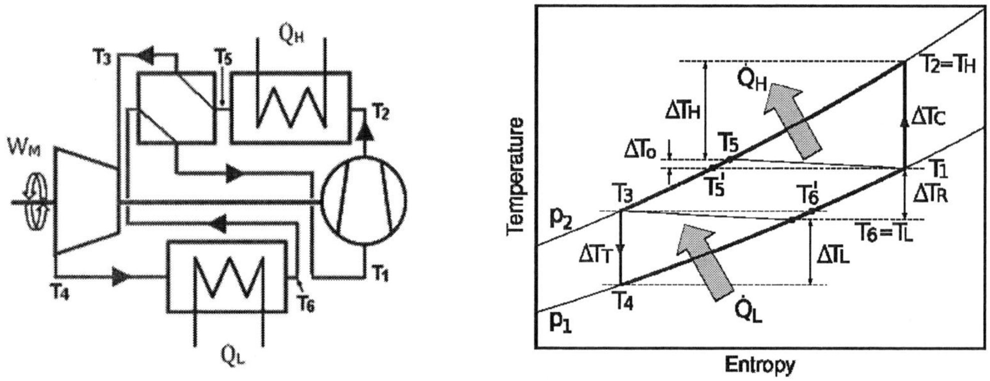

Fig. 5. HT heat pump with regenerative HX and the reverse Brayton cycle for reversible processes

The calculated performance indicators of the heat pump with a RHX are shown in Table 1 together with the results obtained in the previous case.

Table 1. Comparison of performance indicators calculated in various models.

Model	η		R_{WM}		R_{QL}	
	no RHX	RHX	no RHX	RHX	no RHX	RHX
electric heating	0.35	x	1.00	x	0.00	x
reversible processes	0.64	0.63	0.30	0.32	0.70	0.68
irreversible processes	0.40	0.54	0.81	0.46	0.19	0.54
irrevers. + pressure drop	0.36	0.48	0.95	0.59	0.05	0.41

While the regenerative heat exchanger doesn't help much in systems with reversible processes, the greatest efficiency improvement can be observed in case of models with irreversible processes (0.54 vs 0.40). When additionally the pressure drops are taken into account, the efficiency improvement is still good (0.48 vs 0.36). However, it should be remembered that the pressure drops in heat exchangers depend on the fluid velocity, which for gaseous media will be high. The dependence of overall efficiencies η on pressure drop coefficients β calculated for three selected internal efficiencies $\eta_i = \eta_C = \eta_T$ are shown in Fig. 6.

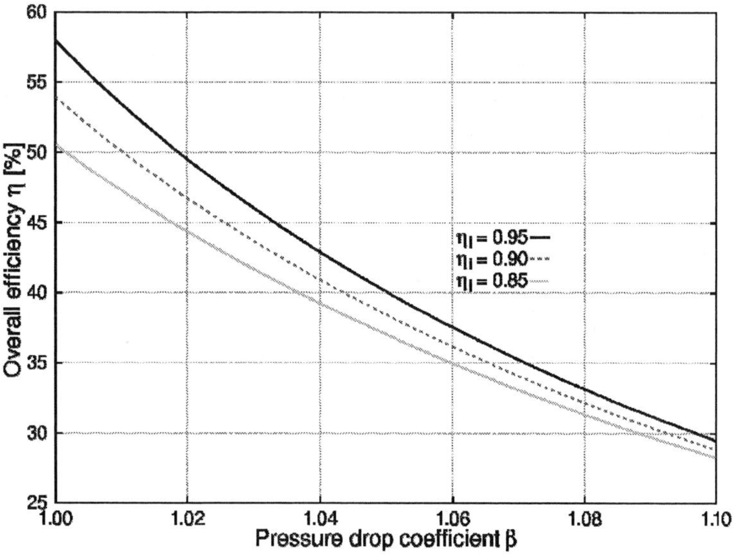

Fig. 6. Overall efficiency η as a function of pressure drop coefficient β

3 Key issues of HT heat pumps

The *chemical class* end-users expect that the nuclear heat supply system will work reliably, supplying heat at the required temperature without any interruptions. In this context, the biggest challenges are related to the design of high-temperature heat pump capable of long term operation in temperatures above 850°C. The key issues are related to:

- materials capable of long term operations in temperatures 870-880°C. The existing candidate materials are Alloy 617, Haynes 230, Hasteloy X and Alloy 800H, but only the last one is currently approved for high-temperature applications and only up to 760°C [4];
- compressor design. The compressor will work at highest temperatures (870-880°C) while the rotating parts will cause mechanical stress. Since the highest temperature is required, the compressor cannot be cooled. These conditions are a big challenge for the compressor design;
- design of compact heat exchangers. The required gas heat exchangers must be capable of transferring large thermal power in temperatures up to 880°C and, at the same time, must withstand creep damage for long lifetimes;
- turbine design – will depend on the working gas. In case of helium the internal efficiency must be improved.

Due to the above mentioned issues the design of the heat pump with the expected parameters still requires intensive R&D work, and can be considered only as a medium-term option.

4 Conclusions

Coupling of HTR reactors with *chemical class* installations by means of steam-based heat transport loop followed by a high-temperature heat pump is possible, but shouldn't be considered as a near-term solution. Simple thermodynamic calculations show that the use of heat pumps instead of electric heating could improve overall efficiency by several percentage points, but the actual numbers depend strongly on parameters of key devices. In pessimistic scenario the overall efficiency may prove to be lower than for a simple system with electric heating. But even in the best case this efficiency will be lower than 0.66 - significantly less than the result for the direct high-temperature heat transfer (1.0 when all loses omitted). At the same time it should be noted, that the biggest challenge for the construction of high-temperature heat pump will be issues with materials capable of long-term operation at temperatures of 870-880°C. In fact, the same problems prevent the successful deployment of direct coupling by means of the previously mentioned gas loop. Progress in the field of high-temperature materials is therefore essential both to construction of the high-temperature heat pump and to construction of the direct gas-based heat transport loop. The latter solution would offer significantly higher overall efficiency, being simultaneously less complicated and easier in exploitation.

Therefore it should be concluded that in *chemical-class* applications the only HTR heat supply system feasible in a near term is a steam loop with electric heating. The configuration with a high-temperature heat pump is not feasible in a near term, mainly due to material issues. The construction of such an installation is likely to be possible in future, but probably will not be profitable because of availability of an alternative solution offering much better performance.

References

1. E. Bogusch, *EUROPAIRS Final Report*, EC FP7 (2011)
2. A. Marmier, M.A. Fütterer, Nucl. Eng. Des., **238**, 2272, (2008)
3. J.M. Beck, L.F. Pincock, *High Temperature Gas-Cooled Reactors Lessons Learned Applicable to the Next Generation Nuclear Plant*, INL/EXT-10-19329 (2011)
4. G.A. Gibbs, *NGNP Project 2011 Status and Path Forward*, INL/EXT-11-23907 (2011)

DOI: 10.1051/e3sconf/20171401049

Spatial corrections for reactivity measurement in lead accelerator driven system

Paweł Gajda[1,*], and *Michał* Orliński[1]

[1]AGH University of Science and Technology, Faculty of Energy and Fuels, Department of Nuclear Energy, al. Mickiewicza 30, 30-059 Kraków, Poland

Abstract. One of the key areas of the development of Accelerator Driven Systems (ADS) are reactivity monitoring techniques. Since the measurement in the future industrial reactor have to be made in the real time applied methods should be accurate, simple and robust. Therefore methods based on point kinetic model are considered. Necessary experimental validation of selected methods was carried out within research project FP7 FREYA using VENUS-F reactor. This paper presents results obtained using the Sjöstrand method and the source multiplication method. Since ADS core behaviour differs from the point kinetics obtained reactivity value depends on the detector position in the system. From the results it is clear that measurement results strongly depend on the position of the detector in the system. For the Sjöstrand method these spatial effects can be successfully corrected using MCNP-calculated correction factors. Those correction factors do not change within the range of reactivity changes covered in the experiments. Spatial effects affecting source multiplication method are more complex and they depend also on neutron flux distribution in the core.

1 Accelerator driven systems

The Accelerator Driven System (ADS) is composed of three main components: nuclear reactor working in subcritical state, spallation neutron source and a proton accelerator. The idea of such device is relatively old and war first proposed by E. Lawrence in the early 1950's [1]. However no such industrial scale device was ever constructed. The revival of this idea is observed over last two decades, this time as one of the technologies for transmutation of spent nuclear fuel.

Growing interest in the solutions to spent nuclear fuel issue alternative to the geological storage is observed worldwide. This strategy is called partitioning and transmutation (P&T) and has advantage of reusing large portion of the spent fuel. The three main benefits of such approach are limited overall volume of waste, limited required storage time and better fuel economy. P&T technologies are partly used at the industrial scale today, mainly as reusing plutonium in form of the MOX fuel in regular light water reactors. However, the fully closed fuel cycle requires new technologies to increase the ability of the reactor to burn transuranic elements, especially minor actinides.

* Corresponding author: pgajda@agh.edu.pl

The main advantage of the ADS is fact that the reactor is working in the subcritical state which leads to enhanced safety properties by increasing the safety margin to prompt supercriticality. It is critical when dealing with minor actinides in the fuel which lead to disadvantageous changes in such safety parameters as reactivity coefficients and delayed neutron fraction. It causes limitation of minor actinides content in the fuel for critical reactors. The enhanced safety margin in ADS can compensate for disadvantageous proprieties of minor actinides and leads to more efficient transmutation. Number of dedicated actinide burners required to close the fuel cycle can be therefore reduced using ADS in compare to use of critical fast reactors [2].

Because the reactor works in the subcritical state the reactivity of the core has to be known at any time during operation of the reactor. This requirement is a relatively new issue, which did not occur in critical reactors used to date. Appropriate reactivity monitoring methods need to be developed and tested. The key requirements for utilised method are accuracy and robustness, but also simplicity since reactivity must be determined in real-time [3].

2 Experimental set-up

A set of experiments devoted to validation of the reactivity monitoring techniques was performed within European Project FP7 FREYA (Fast Reactors Experiments for Hybrid Applications). Similar experiments where previously conducted in previous projects like MUSE in FP5 and EUROTRANS in FP6, however it was the first time the measurement where performed in the lead subcritical reactor, therefore in conditions close to future industrial ADS. The experiments described in this paper were performed at VENUS-F reactor at SCK-CEN facility in Mol, Belgium by international group of scientists.

The VENUS-F reactor is a zero-power research reactor. It uses metallic uranium fuel placed in the solid lead matrix. Use of lead makes it the first mock-up of fast neutron spectre lead ADS and, as was stated before allows for neutronics parameters close to the industrial ADS. The reactor core itself is a 12x12 square matrix, that can be filled with fuel assemblies or lead bars. It is then surrounded by a lead reflector. Different loading patterns are possible, with different number of fuel assemblies used, to allow for several core configurations with different neutron multiplication. During the experiments described in this paper the reactor was in SC1 configuration with $k_{eff} \approx 0.96$. The cross section of this configuration is shown in the Fig. 1. The neutron source is placed in the centre of the core surrounded by 93 fuel assemblies (marked violet in the figure). There are also six safety rods (light blue) and two control rods (red). The rest of the core matrix is filed with lead assemblies (yellow) serving as the neutron reflector. This part is called the inner reflector while the cylindrical lead structure surrounding whole core matric called the outer reflector.

The reactor is a versatile device that can work both in critical and subcritical configuration. In the latter mode four fuel assemblies are removed from the centre of the core and replaced by D-T neutron source of the GENEPI-3C neutron generator [4]. The reactor is then driven by 14 MeV neutrons from the generator.

The GENEPI generator is able to work in three modes: pulsed, continuous and beam-trips (continuous with short, periodical interruptions of the beam). In experiments presented in this paper the generator worked in both continuous and pulsed mode (PNS).

A set of fission chambers was placed in the reactor for measuring the neutron flux. The exact positions of the fission chambers in the system are also shown in the Fig. 1.

DOI: 10.1051/e3sconf/20171401049

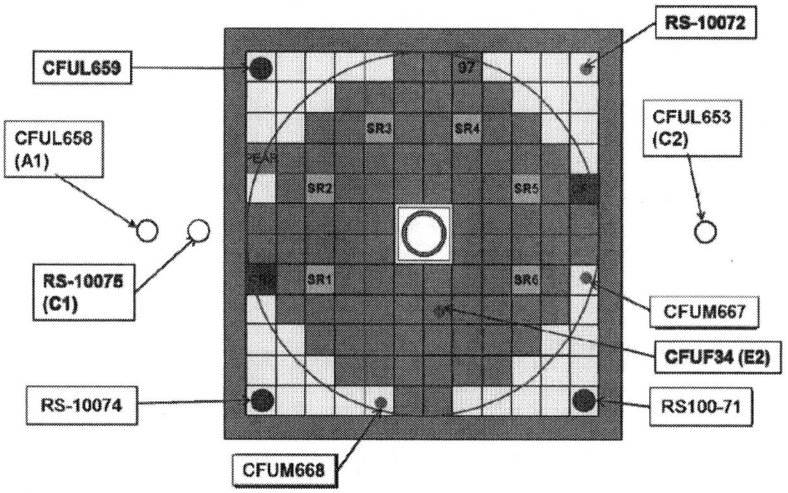

Fig. 1. VENUS-F reactor cross section and detector placement in the system [5]

3 The Sjöstrand method

In the first experiment described in this paper the generator was working in the pulsed mode. It then produce 600 ns long neutron pulses with 200 Hz frequency. The time evolution of the neutron flux in the system was then measured after every neutron pulse to assess the reactivity. The Sjöstrand method (sometimes also called the area method) was used to obtain the reactivity value [6].

The reference value of ρ = 5.28(13) \$ was used for all the experiments for core configuration SC1 with both control rods in position CR@479.3 mm. It was measured using MSM method and mentioned CR position corresponds to criticality point achieved in the critical core configuration [7, 8]. For other CR positions the reference values were obtained using MCNP simulation code to calculate reactivity difference to the reference point. MCNPX 2.5 with JEFF 3.1 data libraries was used for all calculations presented in this paper.

3.1 Preliminary results

Reactivity values for SC1 core with control rods in different positions calculated basing on neutron flux measured by every detector are shown in the Table 1. Graphic comparison of the results and reference value for selected case (CR@479.3 mm) is shown in the Fig. 2.

It is clearly visible that obtained reactivity value depends on detector placement in the reactor. Only detector placed in the active part of the core (CFUF34) is showing reactivity values consistent with the reference. The other detectors are giving values to criticality. This overestimation grows with distance from the core centre. This so called spatial effects are caused by the fact that behaviour of the system varies from the point kinetic model on which the method is based. Also the neutron flux spatial distribution in subcritical systems is more uneven than in critical reactors. The flux distribution is also changing over time after the neutron pulse from the source. This phenomena is then affecting the reactivity assessment with the Sjöstrand method. Similar effects were observed before in other subcritical systems, such as YALINA [9]. It should be noted that in case of detectors placed in the outer reflector the prompt neutron flux decay is not fully observed which can also affect the measurement.

Table 1. Reactivity values obtained with the Sjöstrand method

Detector	ρ [$]			
	CR@600mm	CR@479.3mm	CR@240mm	CR@0mm
CFUF34	-5.09(8)	-5.25(9)	-5.69(18)	-6.30(13)
CFUM667	-4.85(3)	-4.99(3)	-5.79(11)	-6.02(7)
CFUM668	-4.87(3)	-4.98(3)	-5.64(10)	-5.98(7)
RS-10071	-4.81(1)	-5.00(1)	-5.55(1)	-5.96(2)
RS-10072	-4.81(1)	-4.99(2)	-5.61(3)	-5.95(2)
RS-10074	-4.84(1)	-5.03(1)	-5.62(3)	-5.98(2)
CFUL659	-4.75(1)	-4.92(1)	-5.50(1)	-5.88(1)
RS-10075	-4.78(2)	-4.90(3)	-5.44(3)	-5.81(4)
CFUL658	-4.48(3)	-4.61(3)	-5.10(4)	-5.43(4)
CFUL653	-4.62(2)	-4.78(3)	-5.26(3)	-5.62(3)

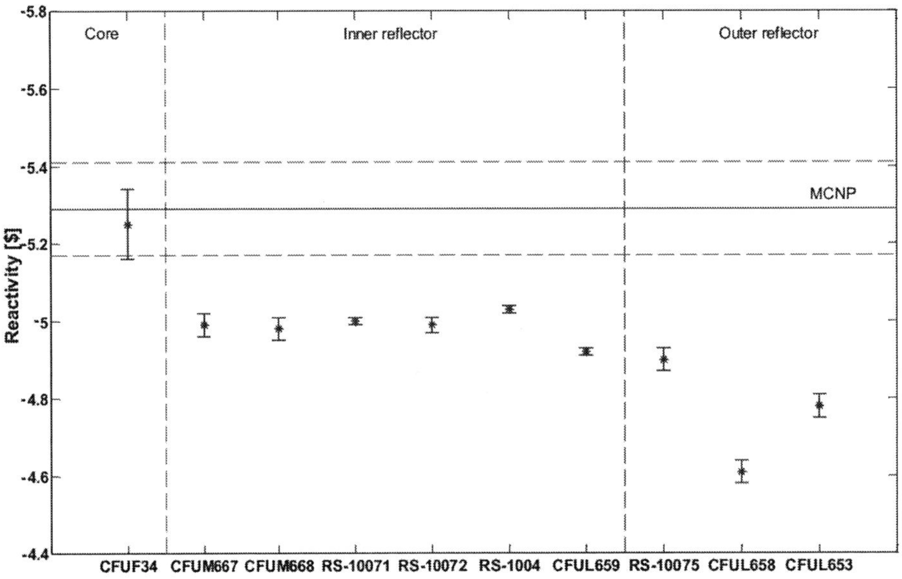

Fig. 2. Results for the Sjöstrand method - CR@479.3 mm

3.2 Spatial effects correction

For the method to be proved suitable for ADS there is a need to correct observed spatial effects. The correction method applied in this case is based on calculating the correction factors for each detector position with the MCNP simulation code [10]. The whole procedure requires calculating the neutron flux for each every detector with and without taking into account delayed neutron as well as core reactivity. It is then possible to calculate the expected reactivity value given by the detector placed in that position. Since the same relation between core reactivity and expected result from the detector should occur both in simulation and experiments, they calculated relations of those values can be then used to correct the experimental values.

The correction factors were calculated separately for different control rod positions used in the experiments. However their values do not significantly change with core reactivity and once calculated correction factor can be used at least in same range of reactivity.

The results after applying the corrections are shown in Table 2 and Fig. 3. Corrected results show good agreement both with reference values and between different detectors. Only exception are the detectors placed in the outer reflector, where despite applying the corrections obtained values are still higher than those obtained in the rest of the system. However, as it was stated before, the full prompt neutron flux decay wasn't observed in this detectors. This phenomena can fully explain the discrepancy of those results even after applying the spatial correction. Using lower frequency of the pulses to allow for full prompt decay should allow to obtain proper result. For all other detectors proposed correction method was proved suitable to use in this subcritical system. Similar results were achieved by other groups involved in the FREYA projects [11, 12].

Table 2. Spatially corrected reactivity values obtained with the Sjöstrand method

Detector	ρ [$]			
	CR@600mm	CR@479.3mm	CR@240mm	CR@0mm
CFUF34	-5.08(9)	-5.23(10)	-5.70(19)	-6.34(14)
CFUM667	-5.12(5)	-5.25(5)	-6.12(13)	-6.33(9)
CFUM668	-5.17(5)	-5.25(5)	-5.97(12)	-6.36(9)
RS-10071	-5.13(5)	-5.31(5)	-5.91(5)	-6.30(6)
RS-10072	-5.14(5)	-5.30(5)	-5.99(6)	-6.33(6)
RS-10074	-5.09(4)	-5.31(5)	-5.90(6)	-6.32(6)
CFUL659	-5.02(4)	-5.19(5)	-5.83(5)	-6.26(5)
RS-10075	-5.04(5)	-5.15(5)	-5.74(6)	-6.13(7)
CFUL658	-4.78(5)	-4.90(5)	-5.45(6)	-5.78(7)
CFUL653	-4.94(5)	-5.09(5)	-5.63(6)	-5.99(6)

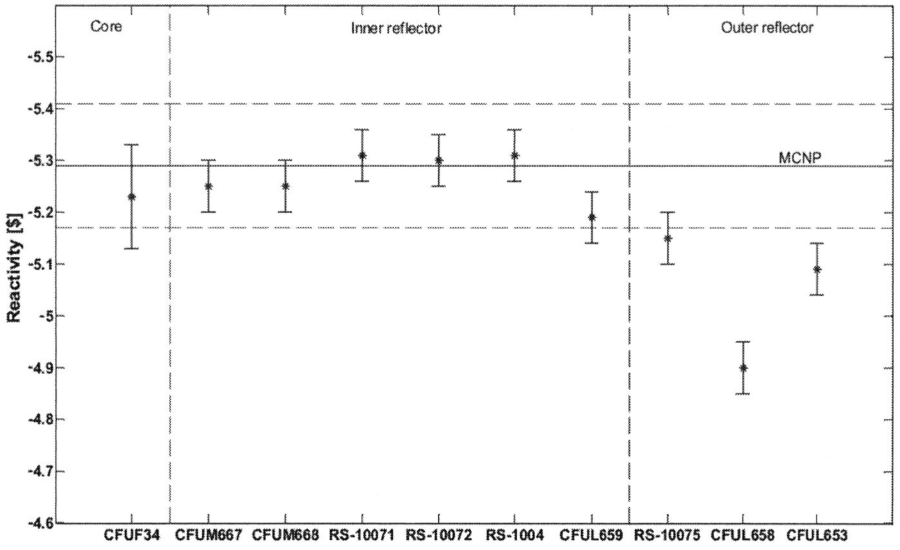

Fig. 3. Results for the Sjöstrand method after corrections - CR@479.3 mm

4 The source multiplication method

In this set of experiments the generator was operated in the continuous mode and the control rods here moved step by step between extreme positions separately or together. The source multiplication method was used to determine core reactivity [13]. In this method the change in the neutron multiplication factor is described by following formula:

$$\frac{CR_2}{CR_1} = \frac{1-k_{eff1}}{1-k_{eff2}} \tag{1}$$

where CR_1 is neutron count rate in the detector before change of the reactivity, CR_2 is the count rate after the change and k_{eff1} and k_{eff2} are neutron multiplication factors before and after the change of reactivity.

It should be noted that this method allows for measuring only relative changes in the reactivity. All values where therefore calculated in relation to the reference value for control rods position CR@479.3 mm mentioned before.

4.1 Measurement results

The results are presented in the Table 3. They are shown as measured reactivity change caused by control rods movement as measured by each detector.

It is clearly visible that once again the obtained results depends on the detector position. However in this case the relation between detector placement and measured reactivity changes is different. Because the measurement is based on static neutron flux measurement instead of its evolution in time the key factor affecting measured reactivity changes are local flux disturbances caused by the control rods which can be greater that the global change of the neutron flux in the core. This effect is visualised in the Fig. 4.

Table 3. Reactivity change due to CR insertion measured with the source multiplication method

Detector	**Δk [pcm]**		
	CR1 & CR2 0-600mm	CR1 0-600mm CR2@479.3mm	CR2 0-600mm CR1@479.3mm
CFUF34	763(36)	371(21)	401(23)
CFUM667	941(41)	542(28)	416(23)
CFUM668	838(38)	463(25)	586(30)
RS-10071	846(39)	611(31)	420(24)
RS-10072	913(40)	606(31)	422(24)
RS-10074	911(40)	404(23)	603(32)
CFUL659	848(38)	406(23)	598(31)
RS-10075	1238(61)	413(23)	641(33)
CFUL658	1296(63)	421(24)	670(34)
CFUL653	1217(61)	638(32)	428(24)
MCNP (simulation)	841(16)	408(16)	418(16)

It is clearly visible that measured reactivity change is bigger in detectors placed in proximity of the control rod used. What is important, unlike in PNS experiment and Sjöstrand method, the measured reactivity change is dependent not only on detector position but also on which control rod was used to cause the change. It is possible to correct observed discrepancies by modified source multiplication method which relies on correction factors to compensate such local changes of the neutron flux [14]. Those have to be calculated not only for each detector but also for each case of control rod movement separately. It would therefore require significant additional calculation effort. However preliminary assessment of the reactivity is possible without using the correction factors by choosing the detectors placed as far away as possible from the moving control rod [15].

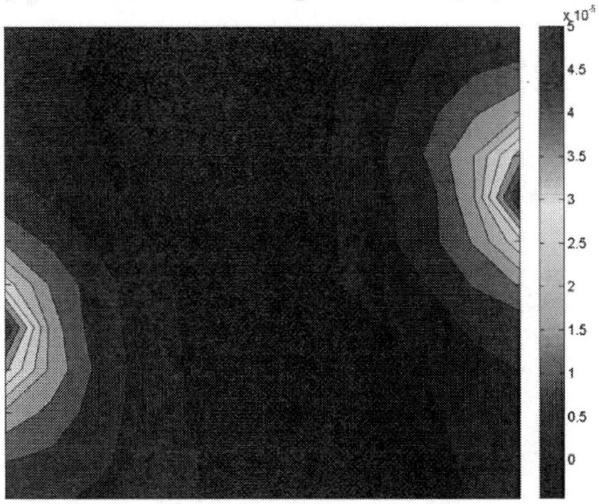

Fig. 4. Neutron flux change with both control rods fully extracted and inserted

Despite fact that uneven rod insertion wasn't used in PNS experiments described before the correction factors were calculated for such case to assess if such uneven flux distribution would affect therm. The results are shown in the Table 4. For the whole range of control rods positions that can be achieved in this core configuration the correction factors are not affected neither by change of reactivity, nor by asymmetrical rod insertion.

Table 4. Correction factors for the Sjöstrand method in the core with uneven CR insertion

Detector	Correction factor			
	CR1@600mm CR2@600mm	CR1@600mm CR2@0mm	CR1@0mm CR2@600mm	CR1@0mm CR2@0mm
CFUF34	1.001(8)	0.998(8)	1.000(8)	0.994(7)
CFUM667	0.948(7)	0.948(7)	0.949(7)	0.951(8)
CFUM668	0.942(7)	0.945(7)	0.943(9)	0.94(7)
RS-10071	0.937(7)	0.943(7)	0.941(8)	0.946(7)
RS-10072	0.936(8)	0.941(8)	0.938(9)	0.940(8)
RS-10074	0.951(7)	0.946(9)	0.948(7)	0.946(8)
CFUL659	0.946(8)	0.943(9)	0.941(8)	0.940(9)
RS-10075	0.949(9)	0.951(9)	0.948(9)	0.948(8)
CFUL658	0.938(9)	0.937(11)	0.941(9)	0.939(9)
CFUL653	0.936(10)	0.938(10)	0.935(10)	0.939(10)

5 Summary and conclusions

Methods used to determine the reactivity are derived from point kinetic model and are affected by so called spatial effects, meaning that obtained value depends on detector position within the system. Such result is consistent with previous experiments and was confirmed for fast spectre subcritical system, in conditions more similar to possible future industrial ADS than in previous experiments. It was shown that in PNS measurement the spatial effects can be corrected by calculating proper correction factors for every detector position. However, use of this method in real-time measurement would require calculating the correction factors in advance. The next step should be therefore an assessment of range of conditions that once calculated correction factor can be used and its dependence on reactivity, neutron flux distribution, etc. In the set of conditions covered by the experiments the correction factors did not change, however it has covered a reactivity range of less than 1.5 $.

Spatial effects affecting source multiplication method are more complex and therefore more difficult to correct. Because it can measure only relative changes of reactivity, in authors view it should be used only for preliminary assessment supplementary to other, more accurate method.

We appreciate the efforts and support of all the scientists and institutions involved in the FREYA project. Financial support by the European Commission (through the contracts #269665) and by the Polish Ministry of Science and Higher Education is gratefully acknowledged (contract no 11.11.210.200).

References

1. J.L. Heilbron, R.W. Seidel, B.R. Wheaton, Lawrence and His Laboratory - A historian's view of the Lawrence Year, Chapter 5: Cold War in Science (1996). Retrieved from http://www.lbl.gov/Science-Articles/Research-Review/Magazine/1981/
2. *Accelerator-driven Systems (ADS) and Fast Reactors (FR) in Advanced Nuclear Fuel Cycles: A Comparative Study* (OECD Publications, Paris, 2002)
3. P. Baeten, H. Ait Abderrahim, Progress in Nuclear Energy, **43**, 413-419 (2003)
4. A. Billebaud, P. Baeten, H. Ait Abderrahim, et al., The GUINEVERE Project for Accelerator Driven System Physics, *International Conference GLOBAL 2009 The Nuclear Fuel Cycle: Sustainable Options & Industrial Perspectives* (Paris, 2009)
5. A. Billebaud, FREYA SC1 detector configuration, FP7 FREYA internal document (LPCS Grenoble, 2012)
6. N.G. Sjöstrand, Arkiv för fysik, **11**, 13 (1956)
7. J.L. Lecouey et al., Recommended value for reactivity (SC1) and related business, FREYA WP1 Meeting 30-31 Oct 2014, Grenoble (2014)
8. J.L. Lecouey, N. Marie, G. Ban, et al., Ann. of Nucl. Energy, **83**, 65-75 (2015)
9. P. Gajda, J. Janczyszyn, W. Pohorecki, Nukleonika: Intern. J. Nucl. Res., **58**(2), 287–293 (2013)
10. V. Bécares, E. Gonzales, D. Villamarin, et al., Ann. of Nucl. Energy, **53**, 331–341 (2013)
11. N. Marie (LPCC), G. Lehaut (LPCC), J.L. Lecouey, et al., Reactivity monitoring using the area method for the subcritical VENUS-F core within the framework of the FREYA Project, arXiv:1306.1063 [physics.ins-det] (2013)
12. W. Uyttenhove, D. Lathouwers, J.-L. Kloosterman, et al., Ann. of Nucl. Energy, **72**, 286–297 (2014)
13. M. Tsuji, N. Suzuki, Y. Shimazu, J. of Nuc. Sc. and Tech., **40**(3), 158–169 (2003)
14. T. Endo, A. Yamamoto, Y. Yamane, Ann. of Nucl. Energy, **38**(11), 2417–2427 (2011)
15. J. Cetnar, G. Domańska, P. Gajda, J. Janczyszyn, Nukleonika: Intern. J. Nucl. Res., **59**(4), 137–143 (2014)

 DOI: 10.1051/e3sconf/20171401050

Comparison of statistical evaluation of criticality calculations for reactors VENUS-F and ALFRED

Jerzy Janczyszyn, *Grażyna* Domańska* and *Jerzy* Cetnar

AGH University of Science and Technology, Faculty of Energy and Fuels, Department of Nuclear Reactors, al. Mickiewicza 30, 30-0591 Krakow, Poland

Abstract. Limitations of correct evaluation of k_{eff} in Monte Carlo calculations, claimed in literature, apart from the nuclear data uncertainty, need to be addressed more thoroughly. Respective doubts concern: the proper number of discarded initial cycles, the sufficient number of neutrons in a cycle and the recognition and dealing with the k_{eff} bias. Calculations were performed to provide more information on these points with the use of the MCB code, solely for fast cores. We present applied methods and results, such as: calculation results for stability of variance, relation between standard deviation reported by MCNP and this from the dispersion of multiple independent k_{eff} values, second order standard deviations obtained from different numbers of grouped results. All obtained results for numbers of discarded initial cycles from 0 to 3000 were analysed leading for interesting conclusions.

1 Introduction

As stated by F.B. Brown [1] the limitations of correct evaluation of k_{eff} to be addressed more thoroughly are:

- Discarding a sufficient number of initial cycles (neutron generations) so that the contamination of results is negligible.
- Performing calculations of k_{eff} for sufficient number of neutrons (batch) in each cycle.
- Recognizing and dealing with the bias in the k_{eff} uncertainties.

Also Mervin and Maldonado [2] discussed the problem of under-prediction in the uncertainty of k_{eff} evaluation by MCNP. To provide some information on these points, when k_{eff} and its standard deviation are calculated for a fast subcritical core, calculations were performed with the use of the MCB code [3] for reactors VENUS-F (subcritical) [4, 5] and ALFRED [6, 7]. Methods and results are presented in the following sections.

* Corresponding author: Grazyna.Domanska@fis.agh.edu.pl

2 Calculations

The scheme of calculations with the use of KCODE was organized in such a way that the results would be convenient for applying the Hartley test of variation [8] and at the same time the calculation will not be excessively long. Respective to the Hartley tables of the test:

- the highest number of repetition in a series for which the values in table are evaluated is m = 12, and
- the highest number of results (generations) in a series is l = 61.

The idea of calculations was to assess the dependence of k_{eff} and its standard deviation (SD) on the increasing number of discarded initial cycles (DIC) and constant number of accepted ones (61) and also to compare different ways of the k_{eff} (SD) evaluation.

The assumed batch of 10000 neutrons was applied that is regarded as sufficient for producing reliable precision of evaluation. It allowed for performing the calculations for 0, 1, 2, 3, 4, 5, 10, 15, 20, 25, 50, 100, 200, 400, 800, 1500 and 3000 DIC (for 100, 200 and 800 DIC calculations were repeated 2 or 3 times). More discarded cycles resulted in excessive computation time. For reducing the uncertainty of k_{eff} and better comparing the standard deviation evaluated by MCB (MCNP), called also the apparent SD [2], with that derived from the dispersion of k_{eff}, called the real SD [2], the sets of 12 results were repeated 5 times so the total number of results for k_{eff} and its SD for each number of DIC amounted to 60..

3 Results

For each of the 210 (2 x 5 x 21) rows in Tables 1 and 2, for different number of discarded cycles, the Hartley test was applied answering the question whether all variances from series of m = 12 repetitions are statistically equal. The tested variances taken from MCB code results were based on the least squares solution of a multivariate linear regression of the cycle k_{eff} estimates, i.e. collision, absorption and track length, used in MCNP code and reported by T.J. Urbatch et al. [9]. Only in one of the 210 cases, both for VENUS and ALFRED, the test answer was negative for the confidence level α=0.01 (in the first raw for 100 discarded cycles; k_{eff} = 0.96614). This effect, can be regarded as stochastically justified. Thus, the conclusion is that the k_{eff} variance produced by MCB code is stable within the 12 repetitions, starting from 0 up to 3000 DIC.

Next, for the 60 results (5 sets x 12 repetitions) the two SD were evaluated: one, as the average value of SDs computed for each k_{eff} by the MCB code, and the other, from the dispersion of 60 values of k_{eff}, in the same way for both reactors.

Also the effect of grouping results in subsets was assessed. The subsets of the following multiplicity m can be obtained: 2, 3, 4, 5, 6, 10, 12, 15, 20, 30 and 60. For all of the subsets k_{eff} and its both SDs were calculated. The numbers of results n for each of the 3 variables were equal to: 30, 20, 15, 12, 10, 6, 5, 4, 3, 2 and 1, respectively; m x n was always equal to 60.

The following definitions of variables are used to present the obtained results, while symbols σ are used for SD produced directly by MCNP (apparent) and s for SD resulting from the dispersion of MCNP results (real):

k_i - k_{eff} from MCB for l = 61 generations of 10000 source neutrons

$k_{bar} = \overline{k}_{eff}$ - average k_{eff} from one group of results (Eqn. 1)

$$\overline{k}_{eff} = k_{bar} = \left(\sum_{i=1}^{m} k_i \right) \Big/ m \qquad (1)$$

$k_{tot} = \overline{k}_{bar}$ - average from all group averages (k_{bar}), i.e. the estimated system multiplication factor (Eqn. 2)

$$k_{tot} = \bar{k}_{bar} = \left(\sum_{j=1}^{n} k_{bar,j}\right)\Big/ n = \left(\sum_{j=1}^{n}\sum_{i=1}^{m} k_{i,j}\right) / (m \cdot n) = \sum_{i=1}^{60} k_i / 60 \tag{2}$$

σ_i - SD of k_i, from MCB, for 61 generations of 10000 source neutrons,
σ_{bar} – average of σ_i from a group of results (Eqn. 3)

$$\sigma_{bar} = \left(\sum_{i=1}^{m} \sigma_i\right)\Big/ m \tag{3}$$

σ_{tot} - average from all group averages (σ_{bar}), i.e. the estimated SD of the multiplication factor k_{bar} (Eqn. 4) and
$\sigma_{k_{tot}}$ - the estimated SD of the system multiplication factor k_{tot}, i.e. SD1 (Eqn. 5)

$$\sigma_{tot} = \left(\sum_{j=1}^{n} \sigma_{bar,j}\right)\Big/ n = \left(\sum_{j=1}^{n}\sum_{i=1}^{m} \sigma_{i,j}\right) / (m \cdot n) = \sum_{i=1}^{60} \sigma_i / 60 \tag{4}$$

$$\sigma_{k_{tot}} = \sigma_{tot} \Big/ \sqrt{m \cdot n} = \sigma_{tot} / \sqrt{60} = \sigma_{tot} / 7.746 \tag{5}$$

So, the k_{tot} , σ_{tot} and $\sigma_{k_{tot}}$ for every set of m and n is the same and equal to the one for all 60 results in one group (or 60 "groups" with single result in each).
On the way of group calculations we can also evaluate SDs from the dispersion of group averages: of k_{tot} ($s_{k_{tot}}$), i.e. the SD from dispersion of k_{bar} (Eqn. 6) and of σ_{tot} ($s_{\sigma_{tot}}$) from dispersion of σ_{bar} (Eqn. 7):

$$s_{k_{tot}} = \sqrt{\left[\sum_{j=1}^{n} \left(k_{bar,j} - k_{tot}\right)^2\right] \Big/ \left[(n-1)n\right]} \tag{6}$$

$$s_{\sigma_{tot}} = \sqrt{\left[\sum_{j=1}^{n} \left(\sigma_{bar,j} - \sigma_{tot}\right)^2\right] \Big/ \left[(n-1)n\right]} \tag{7}$$

Another SDs can be obtained from: the dispersion of k_i within a group and dispersion of this SD between groups.
They are:
s_j - SD of the single value of k_i in the j-th group (Eqn. 8)

$$s_j = \sqrt{\sum_{i=1}^{m} \left(k_i - k_{bar}\right)^2 \Big/ (m-1)} \tag{8}$$

s_{bar} - average of s_j from n single group results (Eqn. 9)

$$s_{bar} = \left(\sum_{j=1}^{n} s_j\right)\Big/ n \tag{9}$$

s_{tot} - SD of the average of sbar (Eqn. 10)

$$s_{tot} = s_{bar} / \sqrt{60} \tag{10}$$

One can further cheque the role of such values as $\bar{s}_\sigma$(average SD of s_{σ_j}, from the dispersion of σ_i in groups) (Eqn. 11) and further SD of σ_{tot} - $s^*_{\sigma_{tot}}$ (Eqn. 12).

$$\bar{s}_\sigma = \sum_{j=1}^{n} s_{\sigma_j} / n = \sum_{j=1}^{n} \sqrt{\sum_{i=1}^{m} \left(\sigma_i - \sigma_{bar}\right)^2 / (m-1)} / n \tag{11}$$

$$s^*_{\sigma_{tot}} = \bar{s}_\sigma / \sqrt{60} \tag{12}$$

 DOI: 10.1051/e3sconf/20171401050

Figs. 1 – 3, below, should help to distinguish between the different ways of evaluation of SDs values.

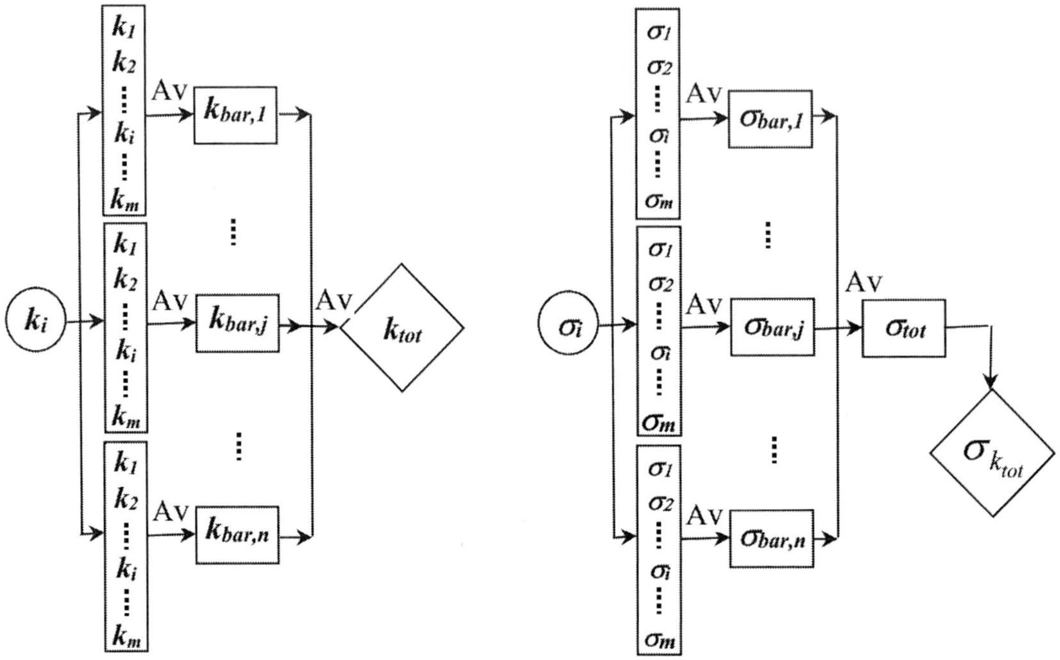

Fig. 1. Scheme of the calculations of k_{tot} (Eqn. 2) and $\sigma_{k_{tot}}$ (Eqn. 5) from the MCB data. Av – means calculation of the average.

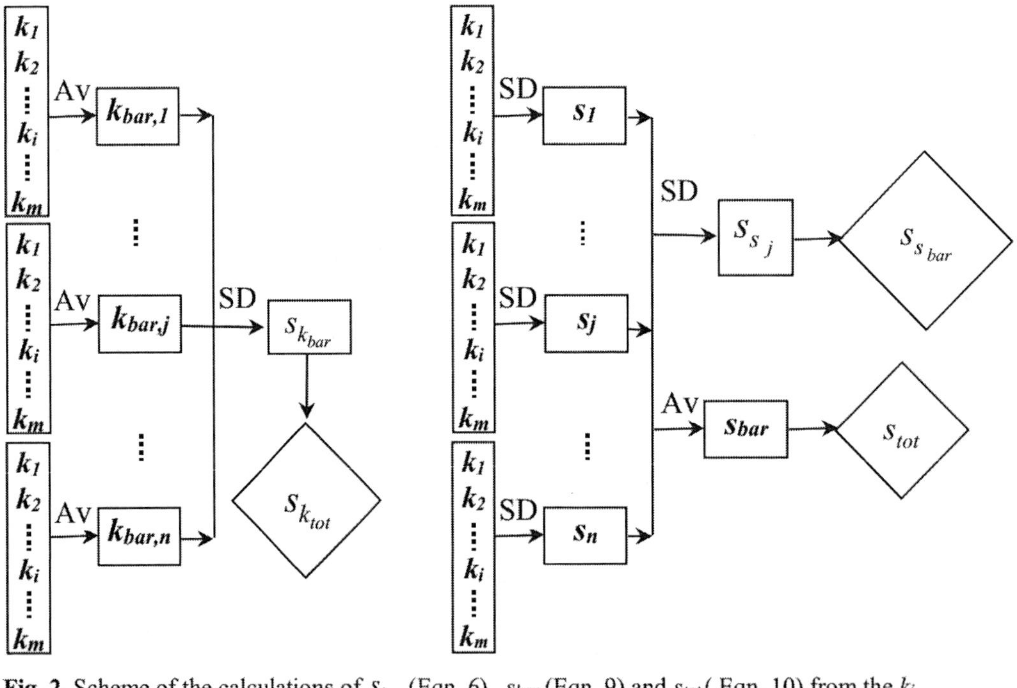

Fig. 2. Scheme of the calculations of $s_{k_{tot}}$ (Eqn. 6), s_{bar} (Eqn. 9) and s_{tot} (Eqn. 10) from the k_i dispersion. Av – means calculation of the average, and SD calculation of the standard deviation.

DOI: 10.1051/e3sconf/20171401050

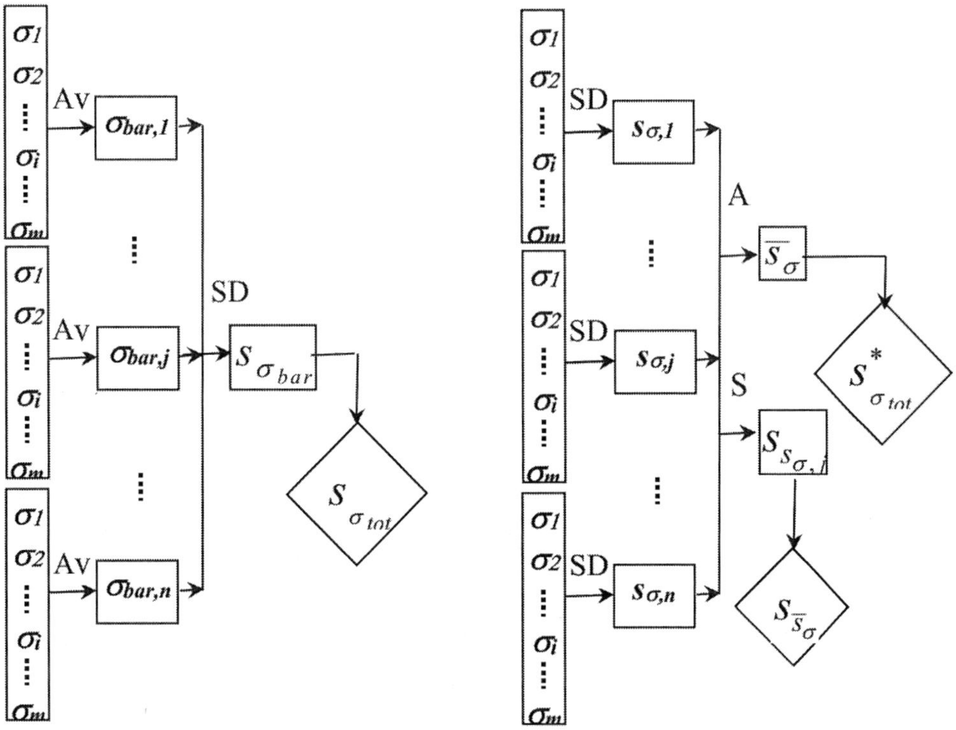

Fig. 3. Scheme of the calculations of (Eqn. 7), (Eqn. 11) and (Eqn. 12) from the σ_i dispersion. Av – means calculation of the average, and SD calculation of the standard deviation.

Exemplary results of the above described evaluations, for 50 DIC, are presented in Table 1 and on the graphs in Figs 4 and 5.

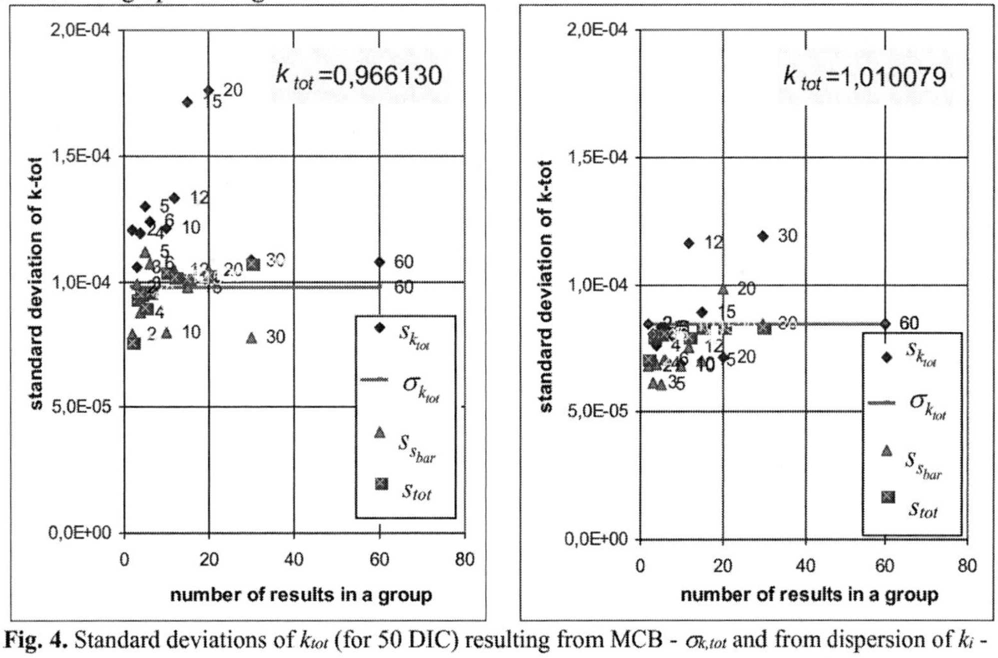

Fig. 4. Standard deviations of k_{tot} (for 50 DIC) resulting from MCB - $\sigma_{k,tot}$ and from dispersion of k_i - $s_{k,tot}$, s_{tot} and $s_{s,bar}$, (see Eqs. 5, 6, 9 and 10). Data for VENUS (left) and ALFRED (right)

 DOI: 10.1051/e3sconf/20171401050

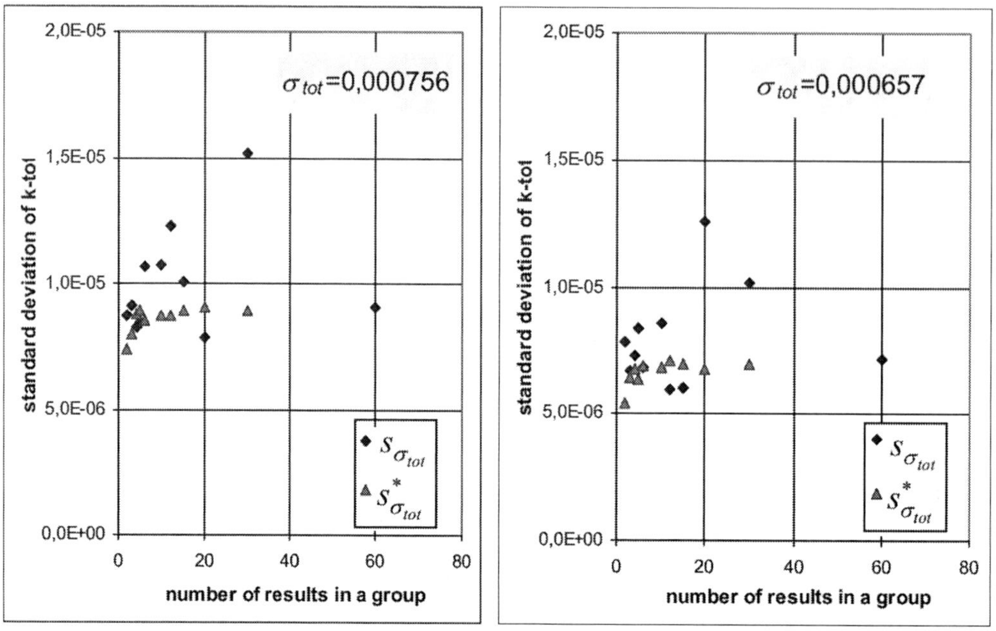

Fig. 5. Standard deviations of σ_{tot} (for 50 DIC) resulting from dispersion of σ_i - $s_{\sigma,tot}$ and $s^*_{\sigma,tot}$. See Eqs. 7 and 12. Data for VENUS (left) and ALFRED (right).

Further, influence of the number of DIC on keff SD ($s_{k_{tot}}$) was assessed, both for VENUS and ALFRED. The results are presented in Fig. 6. One can see that the SD is distinctly higher only in case when no initial cycle (generation) is discarded. The difference is more evident for higher number of groups. It is not the case for k_{eff} itself (see Fig. 8) when 3 cycles should be discarded as minimum. In Fig. 7 we present values of the same SDs as in Fig. 5, but for all numbers of DIC. This allows for better observation of the effect of grouping on the quality of obtained SD evaluation. It seems obvious from Fig. 7 that for low number of groups (n < 10) the spread of results is very high, yielding both overestimated or underestimated SDs. For more groups ($n \geq 10$) with lower number of results ($m \leq 6$) the differences are lower and usually the lowest fluctuations between different DICs are observed for 60 "groups" with single result. However, in this case we are not able to evaluate the other SDs shown in the Table 1. In Fig. 7 are results of the same SDs but obtained in opposite sequence. In case of SD*, after calculation SDs of σ in groups, the respective average value was obtained. As the result we can observe more smooth dependence but with the same tendency with regard to the number of DICs.

DOI: 10.1051/e3sconf/20171401050

Table 1. Exemplary numerical criticality statistics evaluation of reactors VENUS and ALFRED based on the results from MCB code modelling (for 50 DIC).

m=	2	3	4	5	6	10	12	15	20	30	60	
n=	30	20	15	12	10	6	5	4	3	2	1	
variable	**VENUS results**											**average**
k_{tot}	0.966130											0.966130
σ_{tot}	0.000756											0.000756
s_{bar}	0.000585	0.000721	0.000735	0.000696	0.000742	0.000804	0.000790	0.000781	0.000795	0.000830	-	0.000748
$\sigma_{k.tot}$	0.000098											0.000098
$s_{k.tot}$	0.000121	0.000106	0.000119	0.000130	0.000124	0.000121	0.000133	0.000171	0.000176	0.000108	0.000108	0.000129
$s_{s\text{-}bar}$	0.000079	0.000099	0.000088	0.000112	0.000107	0.000080	0.000105	0.000097	0.000105	0.000078	-	0.000095
s_{pt}	0.000076	0.000093	0.000095	0.000090	0.000096	0.000104	0.000102	0.000101	0.000103	0.000107	-	0.000097
$s_{\sigma\text{-}tot}$	0.000009	0.000009	0.000008	0.000008	0.000011	0.000011	0.000012	0.000010	0.000008	0.000015	0.000009	0.000010
$s^{*}_{\sigma\text{-}tot}$	0.000007	0.000008	0.000009	0.000009	0.000009	0.000009	0.000009	0.000009	0.000009	0.000009	-	0.000009
variable	**ALFRED results**											**average**
k_{tot}	1.010079											1.010079
σ_{tot}	0.000657											0.000657
s_{bar}	0.000545	0.000615	0.000624	0.000626	0.000622	0.000648	0.000616	0.000642	0.000644	0.000644	-	0.000623
$\sigma_{k.tot}$	0.000085											0.000085
$s_{k.tot}$	0.000084	0.000080	0.000076	0.000083	0.000083	0.000069	0.000116	0.000089	0.000072	0.000119	0.000085	0.000087
$s_{s\text{-}bar}$	0.000068	0.000061	0.000069	0.000061	0.000071	0.000068	0.000075	0.000070	0.000098	0.000085	-	0.000072
s_{pt}	0.000070	0.000079	0.000081	0.000081	0.000080	0.000084	0.000080	0.000083	0.000083	0.000083	-	0.000080
$s_{\sigma\text{-}tot}$	0.000008	0.000007	0.000007	0.000008	0.000007	0.000009	0.000006	0.000006	0.000013	0.000010	0.000007	0.000008
$s^{*}_{\sigma\text{-}tot}$	0.000005	0.000006	0.000007	0.000006	0.000007	0.000007	0.000007	0.000007	0.000007	0.000007	-	0.000007

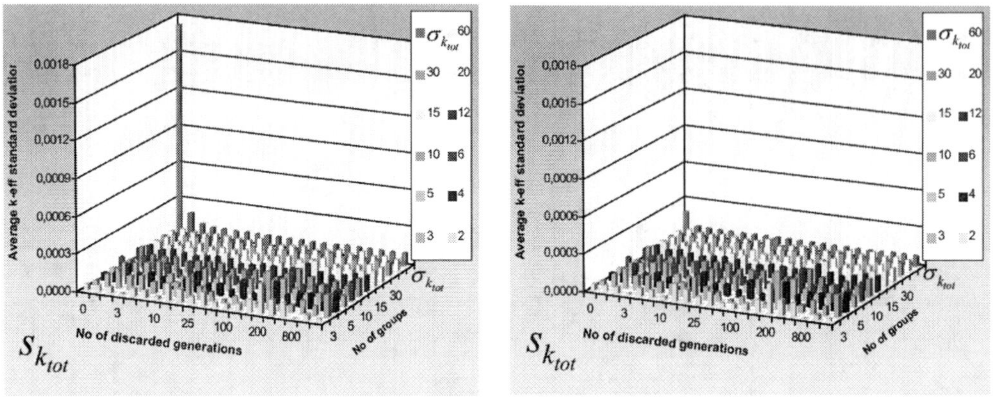

Fig. 6. Cumulated results of the k_{eff} SD ($s_{k_{tot}}$), derived from the average values of k_{eff} in groups and their dispersion between groups, and $\sigma_{k_{tot}}$ (VENUS –left, ALFRED – right)

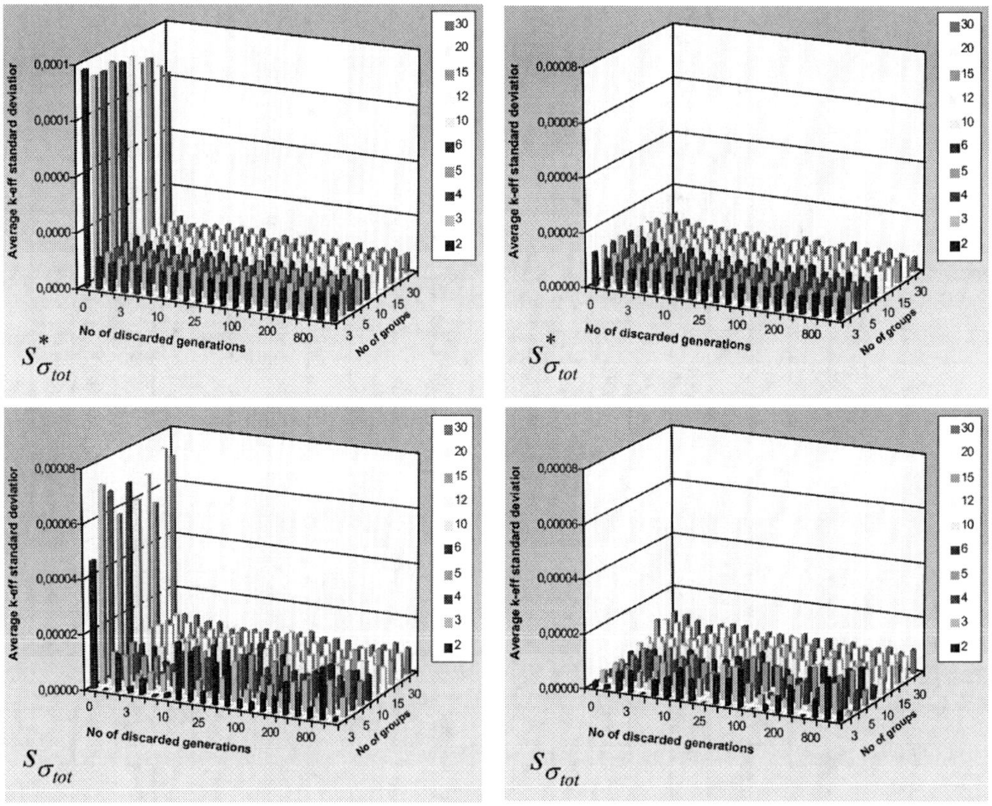

Fig. 7 Cumulated results of the σ_{tot} SD* and σ_{tot} SD, derived respectively from the values of, σ SD in groups and their average from all groups and average values of σ in groups and the averages dispersion between groups.

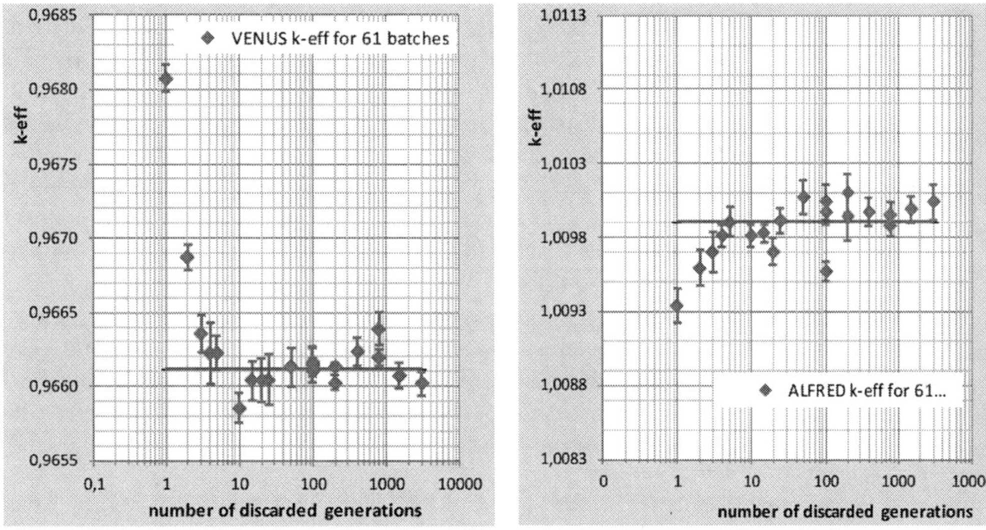

Fig. 8. Results of the dependence of average k_{eff} on the number of DIC (from 60 results).

4 Conclusions

The following conclusions are only valid for a fast reactor type similar to reactors VENUS and ALFRED used in calculations.

1. With the use of Hartley test of variation it is proven that the k_{eff} variance produced by MCB code is stable within 12 independent repetitions, starting from 0 up to 3000 discarded initial cycles. This is thus most probable that the variance will be also stable for all cases of more numerous repetitions and DICs.
2. For 50 DIC, the average SDs, of a **single** value of k_{eff}, obtained from the MCNP evaluation method, - σ_{tot} (0.000756 for VENUS and 0.000657 for ALFRED) are almost equal to: the same SDs resulting from the dispersion of k_{eff} - s_{bar} (0.000748 for VENUS and 0.000623 for ALFRED), as the average from group calculations (without 60 results group), and - $s_{k_{bar}}$ (000833 for VENUS and 0.000655 for ALFRED), from all 60 results of k_{eff}.
3. The respective SD of the **average** value of k_{eff} (from 60 results), called further the first order SD, are: for VENUS $s_{k_{tot}}$ (0.000098), s_{tot} (0.000097) and $\sigma_{k_{tot}}$ (0.000108), and respectively for ALFRED (0.000087, 0.000080 and 0.000085.
4. Standard deviations of the above SDs (second order SD) are: $s_{\sigma_{tot}}$ (VENUS – 0.000009 and ALFRED – 0.000007) for the average value of σ_{tot} and evaluated from the dispersion of 60 σ_i values, as well as: $s^{*}_{\sigma_{tot}}$ and $s_{\sigma_{tot}}$ respectively: VENUS – 0.000009 and ALFRED – 0,000007 and VENUS – 0.000010 and ALFRED – 0,000008, both on the basis of group calculations.
5. Thus the final result of the VENUS and ALFRED cores evaluation of k_{eff} (for 50 DIC) are: 0,966130 ± (0,000098 ± 0,000009) and 1,010099 ± (0,000085 ± 0,000007). This result recalculated for control rods submerging effect, from 40 cm (simulation) to 47 cm (experiment), reads 0.96505, very near to the experimental value: 0.96415±0.00086 [5].Values of the first and the second order SD are only slightly dependent on the way of calculation, whether basing on the MCB (MCNP) methodology or on the observation of dispersion of the MCB results for k_{eff} and σ.
6. The graphs in Figs. 4 and 5 show certain differences in the spread and distribution of results obtained from group evaluation of SDs: $s_{k_{tot}}$ and $s_{\sigma_{tot}}$ are more dispersed

between different number of groups then s_{tot} and $s^*_{\sigma_{tot}}$, in turn distributions of the later are more smooth. This may result from the sequence of operations in calculation. The smooth distributions are obtained when the calculation of SD precedes the operation of averaging, while the dispersed distributions appear in the case of opposite sequence. This effect is also visible for all numbers of DIC in Fig. 7.

7. Results obtained for different number of DIC show that k_{eff} SD ($s_{k_{tot}}$) is distinctly higher only in case when none of initial cycles was discarded. It is not the case for k_{eff} itself when minimum 3 cycles should be discarded. For the dependence of SD on the number of groups it was possible to see the effect of grouping on the quality of evaluated SD. It seems obvious that for low number of groups (n < 10) the spread of results is very high, yielding both overestimated or underestimated SD. For more groups ($n \geq 10$) with lower number of results ($m \leq 6$) the differences are lower and usually the lowest fluctuations between different DICs are observed for 60 "groups" with single result. However, the grouping is useful for evaluation of the second order SDs.

This work was supported by the EC in the frame of FREYA project No 269665 and by Polish Ministry of Higher Education. All calculations were carried in the Academic Computer Centre CYFRONET, AGH.

References

1. Forrest B. Brown, A Review of Monte Carlo Criticality Calculations – Convergence, Bias, Statistics,
   LA-UR-08-06558 Los Alamos Lab, and
   Int. Conf. On Mathematics, Computational Methods & Reactor Physics (M&C 2009) Saratoga Springs, New York, May 3-7, 2009, CD-ROM ANS, LaGrange Park, IL
2. Brenden T. Mervin and Guillermo I. Maldonado, Uncertainty Analyses for Localized Tallies in Monte Carlo Eigenvalue Calculations, *International Conference on Mathematics and Computational Methods Applied to Nuclear Science and Engineering* (M&C 2011) Rio de Janeiro, RJ, Brazil, May 8-12, 2011, on CD-ROM, Latin American Section (LAS) / American Nuclear Society (ANS) ISBN 978-85-63688-00-2
3. J. Cetnar, W. Gudowski and J. Wallenius, MCB: A continuous energy Monte Carlo Burnup simulation code, in *Actinide and Fission Product Partitioning and Transmutation*, **EUR 18898** EN, OECD/NEA (1999) 523
4. http://science.sckcen.be/en/Facilities/VENUS
5 A. Billebaut et all., FREYA. Current sub-critical core results, Deliverable 269665-D1.1, SCK-CEN Mol, 2013
6. http://www.world-nuclear-news.org/NN-Consortium-established-to-build-Alfred-2012134.htm 2013
7. http://www.world-nuclear.org/information-library/nuclear-fuel-cycle/nuclear-power-reactors/generation-iv-nuclear-reactors.aspx 2016
8. H. O. Hartley, The Use of Range in Analysis of Variance, Biometrika, **37**, (1950) 271-28
9. T. J. Urbatsch, R. A. Forster, R.E. Prael, and R.J. Beckman, Estimation and Interpretation of k_{eff} Confidence Intervals in MCNP, in *Proceedings of the Nuclear Criticality Technology Safety Project* R.G. Sanchez (Ed.) San Diego, California, May 16-17, 1995 pp 85-88, LA-13439-C Conference, **UC-714**, LANL Los Alamos, New Mexico 87545 (1998).

Increasing natural ventilation using solar chimney

Magdalena Nakielska[1,*], *Krzysztof* Pawłowski[1]

[1] UTP University of Science and Technology in Bydgoszcz, Al. prof. S. Kaliskiego 7, 85-796 Bydgoszcz, Poland

Abstract. At a time when much attention is paid to the problem of energy saving is looking for new technical solutions, among others in the construction industry. At present, searching new solutions is a worldwide trend in creating buildings. Those solutions enable a comfortable use of building structures in harmony with natural environment and ensure reducing energy consumption. Solar chimneys, which facilitate the gravity ventilation by using solar energy, may be a solution to the problem. In hot climates solar chimney has been applied in the process of passive cooling of a building. In temperate zones and cold is used to assist natural ventilation. The problem of the solar chimney is not popular in Poland. Construction in Bydgoszcz two research positions, enabled the analysis of the impact of solar chimneys on the thermal comfort of rooms. This article presents experimental studies on solar chimney located in Poland. The analysis of the flow of air in the room during the day have been measured using two test-models situated on the 3.1 building of UTP University of Science and Technology at 7 Kaliska street in Bydgoszcz.

1 The introduction

As the time passes, it can be observed that the price of coal, gas, petroleum and electric energy is constantly increasing. It results in conscious saving based, for example, on thermo-modernization of buildings or construction of installation systems that do not consume electric energy, yet create comfortable conditions in rooms.

The climate inside consists of such elements as: the amount of air supplied into a room that has a proper temperature, humidity and cleanness, and maintaining necessary temperature to feel thermal comfort in a room. In order to fulfill those conditions, a heating system and spaces ventilation ought to be designed correctly. Ventilation is a process of removing the internal air from spaces and making room for fresh air which is supplied from the outside [1]. This process can be natural (gravitational) or mechanical. In a gravity ventilation, the natural draught in a vertical ventilation duct causes the flow of air. The air is supplied through the windows and is exhausted through the gravity ducts. The air flow is forced by differential pressure. Air pressure inside is higher than at the outlet of a ventilation duct. The factor that enables the air exchange is density differences between internal and

* Corresponding author: magdalena.nakielska@utp.edu.pl

DOI: 10.1051/e3sconf/20171401051

external air. The density of air depends on its temperature. The higher the temperature and the chimney, the higher differential pressure [2], [3].

Unfortunately, in some seasons of the year, a gravity ventilation system does not work in a satisfactory way. High temperatures or high humidity influence on poor working of a natural ventilation in a room and cause the decrease in the quality of microclimate features inside. In order to reduce this problem, many different solutions are adopted to support the operation of a gravity ventilation. Those solutions stimulate the gravity ventilation to work better and, at the same time, improve the microclimate in a room. These are the solutions that do not generate higher costs of facility maintenance, which is of a great importance as far as economic calculation is concerned. At the same time, the facility's demand for the energy does not rise, which is, undoubtedly, an advantage, taking into consideration current legal articles. What is more, energy consumption is connected with carbon dioxide emission into the atmosphere. Currently, the reduction of carbon dioxide emission in the context of power engineering is a priority not only for Poland but also for the whole Europe. The European Union predicts the reduction of carbon dioxide emission by the year 2050 at 80-90%. The EU in its energy package places on member states of EU the duty of 20% reduction of greenhouse gases until 2020, as one of the elements of climate preservation policy.

Relating to the systems that shape the inner climate in rooms in the summer, current legal articles in "The domestic plan aiming at increasing the number of buildings with low energy consumption" mention exact solutions for eliminating or reducing the power of refrigerating system for air conditioning systems that use refrigerating units. Therefore, what should be done is to:

- limit heat gain (the reduction of solar gain by using sun protection and by the limitation of internal gains),
- adjust air streams to real load,
- use alternative cooling methods (night cooling, the use of ground energy, free cooling and passive cooling).

With reference to the use of alternative cooling methods, the aim of this publication is to present a test stand that is a solar chimney and the analysis of carried out research that confirms justified use of the solar chimney in order to increase the air exchange in a room.

2 Solar chimney

Solar chimneys that one can find all over the world differ from one another in construction and materials. The most important external factors that influence the chimneys construction are: the climate, the cardinal points direction, insolation. Despite numerous differences, the chimneys that are being tested have the same elements: air inlet and outlet, glazing and thermal mass.

2.1 The structure and the principle of operation of a solar chimney

The components of a solar chimney are presented in the Fig.1. During a day, the sun warms the test stand through the glazing. The additional heat flux supplied to the air, which is exhausted from the room, causes a temperature rise in the duct and, simultaneously, forces warmer air to flow upwards through the outlet. The rise of the difference of air density that is being exhausted between the inlet and outlet results in the enhancement of the "chimney effect". In addition, the energy stored during a day in the walls is released at night. A black colour of thermal mass increases the accumulation of the energy. The energy accumulated at night enables increased velocity of the air flow to be maintained in the duct of a solar

 DOI: 10.1051/e3sconf/20171401051

chimney. It also causes the intensification of the air exchange in the room where the solar chimney is one of the elements of the ventilation system.

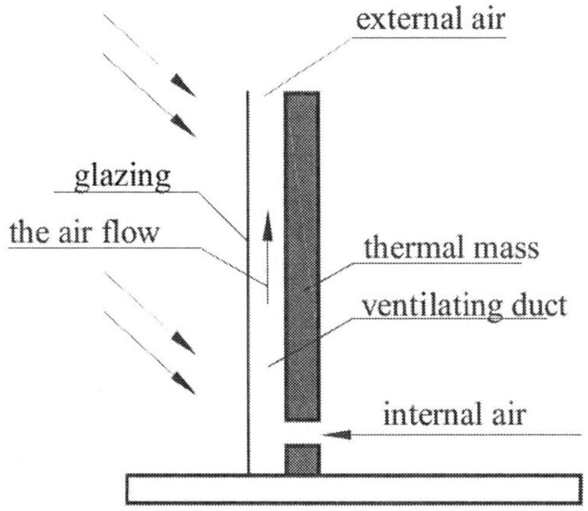

Fig.1. Schematic diagram of solar chimney

The solar chimney as one of the elements of a ventilation system has been known for many years. Along with the technique development, many people have aspired to get to know the phenomena occurring during the working of the chimney. In order to achieve this aim, a lot of seats of learning have begun the research work and used various methods such as: experimental, analytical and computer. R. Khanal and C. Lei [4] present in their article the review of some selected publications on this subject. The tests that dominate in the chosen publications are experimental ones, combined with analytical or numerical. The results of analyses, which are the most similar to real conditions, are obtained by using a test stand. It is usually an expensive and labour-consuming solution. Yet, it allows us to find an optimal solution as regard the height and sectional view of the chimney, the surface of glazing, the angle of inclination and the material that should be used to construct chimneys for specific latitudes and the intensity of solar radiation.

2.2 The examples of real models of solar chimneys

The experiment that was carried out by C. Alfonso and A. Olivera [5] makes it possible to broaden the knowledge on justified use of solar chimneys. A test model was made in Porto, in Portugal. The research was conducted by using a tracer gas and then, the results were compared with theoretical values taken from a simulation program. A test object were two identical rooms as regard their geometry. The rooms were made of concrete and were insulated outside. The same solutions were adopted connected with supplying and removing air, yet one chimney was built by means of traditional technology, while there was glazing on the southern wall of the other chimney. The chimneys were made of brick and they were 0.2 meter long, 1 meter wide and 2 meters high. There was also a 5-centimeter insulation on the solar chimney. The heated rooms and chimneys were equipped with measuring apparatus to measure the temperature, the stream of air volume and the velocity of air movement. The temperature outside, wind velocity and wind direction were read from the weather station. All the data were entered into a computer program. Several relations were checked, for example: the influence of solar radiation on ventilation efficiency, the influence of insulation and the height of a chimney on the test results. The conclusion that confirms justified use of solar chimneys is that a solar chimney had a higher efficiency at about 10-22% than a traditional chimney, which is based on average climatic data for Portugal.

At University of Technology in Brno, there were two, full-dimensional solar chimneys built. One of them is of a light structure and the other with thermal mass in order to accumulate the heat. The chimneys are 0.75 meter long, 0.20 meter wide and 2.0 meters high, and have glazing with south exposure at an angle of 30 °. P. Charvat, M. Jicha and J. Stetina [6] describe the results of research the aim of which was to test the capacity of solar chimneys in the climate conditions in the Czech Republic. It was found that the flow of air in a solar chimney between 8 a.m. and 6 p.m. was 25% larger than in a traditional chimney.

3 Testing the air flow in a solar chimney

At UTP University of Science and Technology in Bydgoszcz, al. Kaliskiego 7, there are two rooms chosen to modify the existing gravity ventilation system by adding a new element – a solar chimney. The principle of similarity was dominant in the process of choosing the spaces. Both rooms have close cubic capacity and are situated in the same building, have the same windows and similar number of ventilating ducts. There is old woodwork which made it easy to create real evaluating conditions of functioning gravity ventilation system.

Solar chimneys are superstructures built on the existing ventilating ducts. The difference between them is in materials used to create them. One superstructure, which is 0.47 meter long, 1.75 meter wide and 1.46 meter high, was made of B20 ferroconcrete reinforced every 15 centimeters with rods 10 millimeters in diameter (Fig.2.). The second test stand is a solar chimney made of solid bricks glued together. This superstructure is 0.47 meter long, 1.86 meter wide and 1.45 meter high. Three walls are made of ferroconcrete elements and solid brick. The barrier with south exposure is glazing. It is made of Pilkington Optilam 6,4 safety glass 6.33 millimeters thick, with overall solar energy transmittance of 79%, heat-transfer coefficient of 5.7 W/(m^{2}K) and a surface area of, adequately, 1.88 square meter and 2.02 square meters. The walls of solar chimneys are painted black inside and insulated outside with ESP 70-0.037 expanded polystyrene 10 centimeters thick and plastered all over.

Fig.2. Solar chimney south side

Materials used to build the walls of a solar chimney were chosen after the analysis of available literature on heat accumulation of building materials. The work [7] confirms the use of ferroconcrete and ceramic brick. This publication presents the results of laboratory research of 10 samples of different materials. It was tested how thermal energy stored in different samples was emitted during cooling down. The samples were put in the drier for three hours with a maximum temperature of 230^{0}C inside. Then, the samples were taken out of the drier to take the temperature on their surface for specific time. The results helped to calculate the ability of material to give up stored energy during specific time (9000 s). A steel sample gained the highest power - 8.874 W, a concrete sample gained lower – 5.3-6.6 W, ceramic brick – 3.72 W and the lowest was gained by cellular concrete – 1.616 W and chamotte brick – 2.408 W. It was decided that solar chimneys would be built out of solid ceramic brick and ferroconcrete because of their availability, relatively low price and their popularity in building industry.

3.1 Methodology of research

The aim of research that was carried out is the analysis of air flow in the spaces where a solar chimney is one of the elements of a gravity ventilation system. Measurements were taken during windless days in order to eliminate the influence of wind on the results. The first control measurement was taken on the 9-10 July, 2013 before constructing the test stands. The measurement had been being taken for 24 hours with a parameters reading every 2 hours.

When the solar chimneys were built, the measurements were taken in the summer on the following days: 6-7 July, 19-20 July, 9-10 August 2014.

The measurements had been being taken for 24 hours with a parameters reading every 2 hours. The following parameters were measured: the velocity, temperature and humidity of air. The C310 hytherograph was used to take a relative temperature and humidity and the velocity of air was measured by the TA 430 hot-wire anemometer.

The research was done at two testing points: on the roof, directly at the chimney outlet, and in the rooms, on the outlet grilles of the gravity ventilation system.

The following parameters were measured:

- air velocity at the level of intake grates, at 9 testing points on each of five grates located in the middle of 9 elementary fields resembling a square in shape, which the grate was divided into [8],
- air velocity at the level of the solar chimney outlet, measured at 10 testing points located in the middle of 10 elementary fields, which the grate was divided into [8],
- the temperature and humidity of air inside at specific measuring points that were located in the area where people stayed at 1.2 meter above the floor level,
- the temperature and humidity outdoors.

3.2 The research results of air flow in a solar chimney

Figures from 3 to 6 present, in a graphic way, streams of air volume in rooms and at the outlets of chimneys.

DOI: 10.1051/e3sconf/20171401051

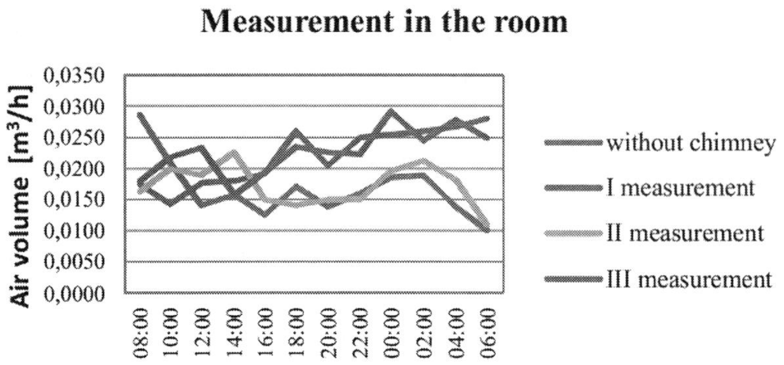

Fig.3. Measurement in the room - solar chimney ferroconcrete

The value of the stream of air volume during twenty-four hours, in each measurement cycle, is changing. These values from 8 a.m. to 4 p.m. indicate little fluctuations from 0.015 m^3/s to 0.023 m^3/s. This is the time of a direct influence of sunrays on glazing with south exposure. From 4 p.m. to 4 a.m., the values have a growing tendency and after that, they are decreasing. The results obtained from the 1st and 3rd measurement indicate a similarity to the obtained values of the stream of air volume.

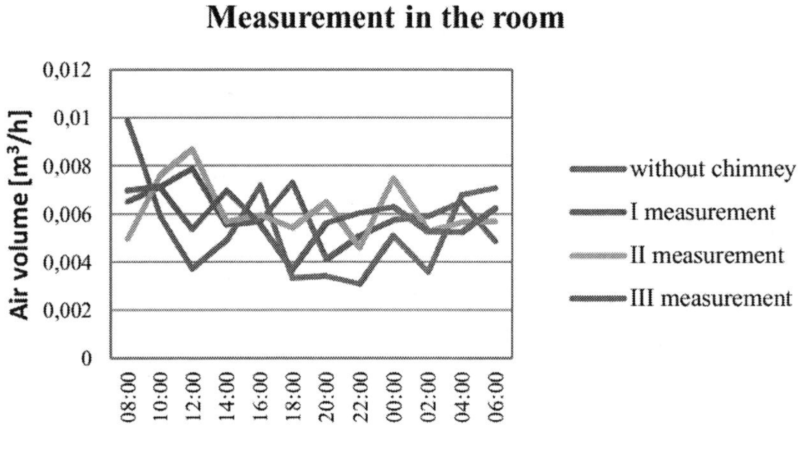

Fig.4. Measurement in the room - solar chimney made of solid bricks

The value of a volume stream during twenty-four hours, in each measurement cycle, is changing. After building a solar chimney, the values are higher, from 0.004 m^3/s to 0.008 m^3/s.

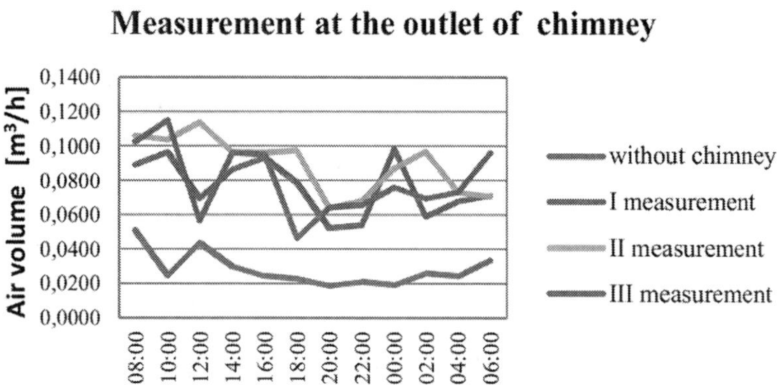

Fig.5. Measurement at the outlet of chimney - solar chimney ferroconcrete

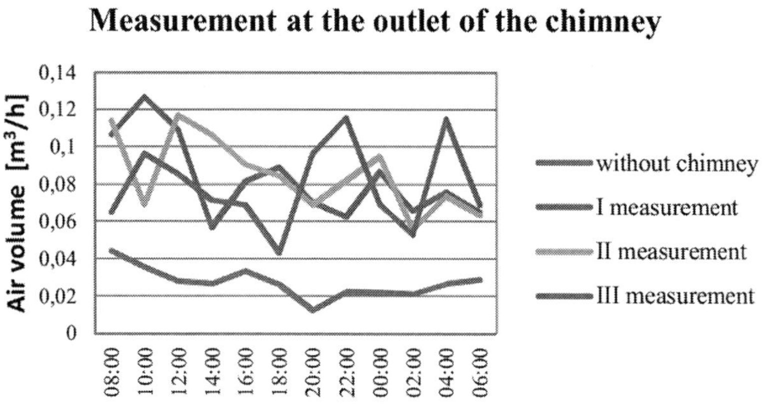

Fig.6. Measurement at the outlet of chimney - solar chimney made of solid bricks

The values of a volume stream at both test stands have clearly increased after building a solar chimney. The values indicate fluctuations from 0.05 m^3/s to 0.12 m^3/s. Higher values were obtained during a direct impact of sunrays on glazing. From 4 p.m. to 4 a.m., the values of the stream of air volume have a falling tendency.

4 Summary

On account of the necessity of limiting the amount of electric energy consumption aiming at cooling down buildings, there is a growing need for introducing alternative solutions in this range.

Solar chimneys are more and more often used elements in building engineering. They use solar energy, acquired at a low cost, in order to improve microclimate indoors. In Poland, the working of a solar chimney as one of the elements of a natural ventilation system has not been examined enough yet. The research presented in this article and carried out at two test stands, illustrates justified use of such a solution. When compared to the situation without a solar chimney, the research shows intensification of air exchange. We can observe the increase in the stream of air volume. It results from intensified air exchange in a room.

References

1. PN-B-01411:1999 Wentylacja i Klimatyzacja Terminologia.
2. S. Opaliński, S. Rabczak, *Wentylacja grawitacyjna* (Oficyna Wydawnicza Politechniki Rzeszowskiej, 2003)
3. A. Pełech, *Wentylacja i klimatyzacja: podstawy* (Oficyna Wydawnicza Politechniki Wrocławskiej, 2008)
4. R. Khanal, C. Lei, Energy and Buildings **43**, 1811-1819 (2011)
5. C. Afonso, A. Oliveira, Energy and Buildings **32**, 71-79 (2000)
6. P. Charvat, M. Jicha, J. Stetina, Solar chimney for ventilation and passive cooling. World Renewable Energy Congress VIII, Denver, Colorado USA (2004)
7. B. Zegardło, A. Halicka, 2011. Analiza właściwości cieplnych betonu z kruszywem z odpadów ceramiki sanitarnej. Budownictwo i Architektura **9**, 39-49.
8. L. Kołodziejczyk, S. Mańkowski, M. Rubik, *Pomiary w inżynierii środowiska* (Arkady Warszawa, 1980)

The cooling effect by adsorption-desorption cycles

Eliza Wolak[1,*]

[1]AGH University of Science and Technology, Faculty of Energy and Fuels, Cracow, Poland

Abstract. Adsorption appliances may turn out to be an alternative to compression-type refrigerators. The adsorption refrigeration machine may be driven by a low-grade heat source, especially solar energy. Solar adsorption cooling systems are environment-friendly and have zero ozone depletion potential. Therefore, the adsorption refrigeration is one kind of energy saving refrigeration methods. The merits of the adsorption refrigeration systems will be more significant especially when it is used in vehicles (automobiles, ships and locomotives), to preserve food and medicines and in air-conditioning. The paper presents the advantages and disadvantages as well as the evolution of the technology of adsorptive refrigeration systems. The methods of improving of adsorption refrigeration systems through improvements in adsorbents properties, use of advanced cycles and hybrid systems is also presented. Possible applications and perspectives for development of adsorption cooling systems are also analyzed. The paper describes a test stand of the adsorption-desorption refrigeration. The present investigations have been carried out utilizing the activated carbon granules as an adsorbent and methanol as an adsorbate. The paper demonstrates the measurement of temperature changes in the adsorbent bed and condenser during adsorption-desorption cycles.

1 Introduction

The energy problem has already been known since the global energy crisis broke out in the 1970s. In the process of the primary energy utilization a great amount of waste heat is discharged into the environment. The emission of the waste heat causes energy loss and also thermal pollution. The growing energy demand and the global climate change are compelling reasons to look for effective utilization of waste thermal energy and renewable energy resources. It is assumed that approximately fifteen percent of all the electricity produced in the whole world is utilized for refrigeration and air-conditioning processes of various kinds, and the energy consumption for air-conditioning systems has recently been estimated to constitute 45% of the whole households and commercial buildings [1].

In times of the energy crisis the quest for and recovery of thermal energy are particularly justified in the case of high levels of sunlight exposure and limited access to electrical energy. Utilisation of solar energy and low-grade waste heat in heat-driven refrigeration cycle is an attractive option. Solar-powered adsorption cooling systems do not have global warming

* Corresponding author: eklimows@agh.edu.pl

potential. Solar energy is clean and environmentally friendly. The adsorption refrigeration machine may be driven by a low-grade heat source. The waste heat (engine exhaust, industrial waste heat etc.) can be effectively utilized in adsorption air conditioning and cooling systems. It can be more significant especially when it is used in vehicles (ships, locomotives and automobiles).

The wide development of adsorption refrigeration systems using solid adsorbents occurred since 1970s. Since then, much theoretical and experimental work about adsorption cooling systems has been proposed and developed. The early research work studied the performance of the adsorbent – refrigerant pairs [2, 3]. They improved the basic theory of adsorption refrigeration with typical adsorbents: mainly on activated carbon, zeolite, silica gel, $CaCl_2$ and the refrigerants were mainly methanol, water, ammonia, hydrogen. New cycles, new systems and methods are proposed to improve heat and mass transfer, increase of COP (coefficient of performance) and decrease of the costs. In the last 20 years studies on the adsorption refrigeration systems have been conducted in France, USA, Japan, India, China and other countries. The main topics are enhancement of heat and mass transfer performance, new system design and its utilization. With the progress of adsorption refrigeration technology, some silica gel-water adsorption chillers have been commercialized successfully on the market.

2 Adsorption refrigeration cycles

The adsorption refrigeration cycles may be divided into intermittent cycles and continuous cycles. The wide classification of various adsorption cycles is shown in Fig. 1.

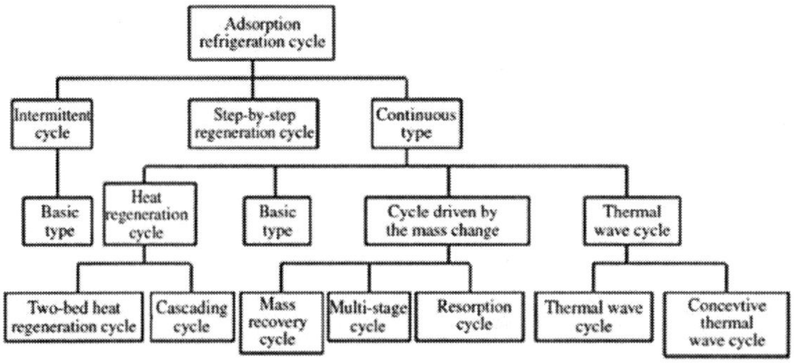

Fig.1. Classification of the adsorption refrigeration cycle [4].

The earliest cycle is the single-bed intermittent cycle. For one bed utilized in the adsorption refrigeration system on condition that only the heating and desorbing process as well as cooling and adsorbing process are involved, it is defined as the basic adsorption refrigeration cycle. This cycle is very simple and mostly used in the solar powered refrigerator. A basic intermittent adsorption refrigeration cycle consists of adsorber/regenerator with adsorbent bed, a condenser and an evaporator. The efficiency of the basic cycle is very low and the cooling is not continuous. The problem of the low COP value (sometimes less than 0.1) was mainly caused by the big temperature fluctuation of the adsorption bed [5].

For the continuous refrigeration cycles there are two or more adsorbent beds running alternately. More adsorbent beds can ensure that at any time at least one of them is at the cooling and adsorption stage, for which the cooling power can be output continually. The two-bed operating system is generally used in the cycle when the desorption time is the same as the adsorption time. When the optimum adsorption time differs significantly from the optimum desorption time, the multi-bed system may be used. In order to improve the efficiency and practicability of the adsorption refrigeration cycles, many various advanced adsorption refrigeration cycles were proposed in the literature. One can distinguish the following advanced adsorption cycles: heat recovery cycle, mass recovery cycle, thermal wave cycle, convective thermal wave cycle, cascade cycles [4, 6].

2.1 Heat recovery cycle

The COP value may also be increased by the improvement in energy use – delivered and emitted during the process. For that reason, adsorption cycles for systems consisting of at least two adsorbent beds were worked out. The heat recovery process is generally operated at the switching time for adsorption and desorption processes and may recover the sensible heat to improve the efficiency. After switching between adsorption and desorption, the cooling or heating medium immediately circulates between the hot adsorber and cold adsorber in a closed loop. Then a part of the sensible heat of the hot adsorber recovers. The refrigeration cycle's COP with heat recovery is significantly larger than that of the cycle without the heat recovery process. It can improve COP by 25% [6].

2.2 Mass recovery cycle

The mass recovery cycle may greatly improve the system performance, compared with the conventional cycle powered by low temperature heat [7]. The process involves changing the pressure in adsorbent beds. In the cold adsorbent bed connected with the evaporator the pressure is close to the evaporation pressure, while it is much lower than the condensing pressure. Meanwhile, in the hot adsorbent bed connected with the condenser the pressure is close to the condensing pressure that is much higher than evaporation pressure. In these conditions, connecting the hot bed with cold bed at the switching time may greatly increase the desorption rate of the hot bed, which will be helpful for the improvement of the adsorption quantity of the hot bed in the next half cycle for the cooling and adsorption process. The principle of the mass recovery cycle is shown in Fig. 2.

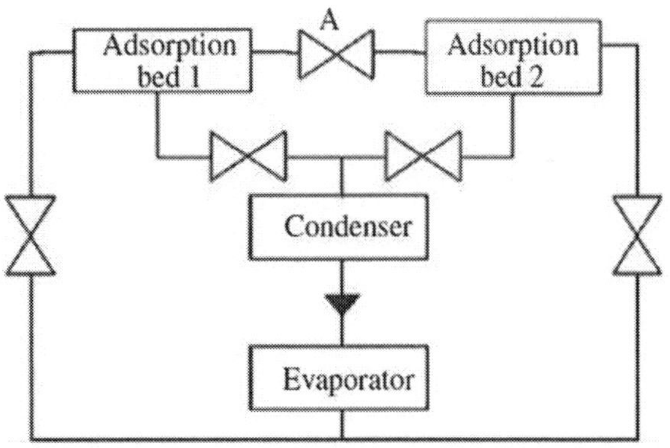

Fig. 2. Principle of the mass recovery cycle [4].

When the desorption process of adsorbent bed 1 completes, the adsorption process of adsorbent bed 2 also finishes. Then the valves connecting the adsorption beds, evaporator, and condenser, and open valve A for mass recovery between the high-pressure adsorbent bed (hot bed) and low-pressure adsorbent bed (cold bed) are closed. The mass recovery will proceed between two beds.

The heat and mass recovery cycle is currently one of the most commonly and successfully used cycles.

2.3 Thermal wave cycle

The use of the thermal wave cycle in the adsorption system was first proposed by Shelton [4]. The merits of a thermal wave cycle is that the heat can be transferred from the hot adsorbent bed to the cold adsorbent bed because the difference in temperature is large. The comparison of the continuous cycle and the thermal wave cycle is shown in Fig. 3.

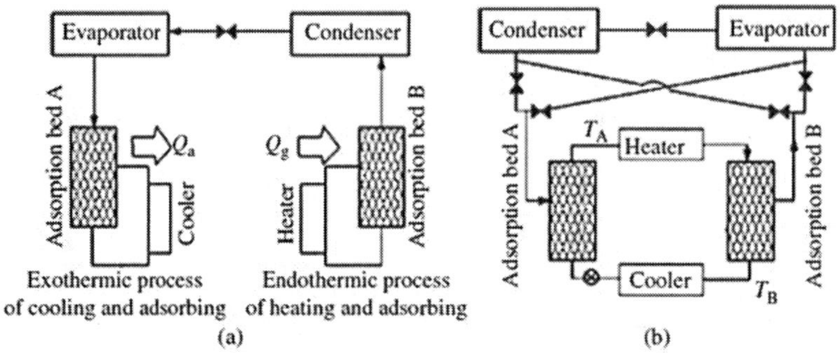

Fig. 3. Schematic of heat flow a) Basic cycle b) Thermal wave cycle [4].

In the continuous adsorption refrigeration cycle adsorption and desorption processes proceed alternatively. When one of the beds is heated by the external heat source for desorption, the second bed is also cooled by an external cooling source for adsorption. The concept of the thermal wave stipulates that heat is effectively utilized and that the external heat source, which is necessary for the desorption process, is greatly reduced. 80% of heat required in the desorption process may be provided by the heat released in the adsorption process. The basic principle of such a cycle is using single heating and cooling fluid circuit to connect two adsorbent beds, the cooler and the heater. The circuit of the fluid may transfer the released heat from the adsorption bed to the desorption bed, and recover the adsorption heat for the improvement of the energy efficiency of the system.

2.4 Convective thermal wave cycle

In the literature [9] Critoph proposed as an improvement to the thermal wave cycle a pioneering method, which is the convective thermal wave cycle. In this concept the refrigerant acts as the heat transfer fluid. Instead of heating the bed directly, the refrigerant is first heated outside the bed and then circulated through the bed to heat up the adsorbent. Such a cycle uses the forced convection between the refrigerant gas and adsorbent, that is the adsorbent is directly heated or cooled by the high pressure refrigerant vapor to obtain a higher heat density. This solution requires the use of pumps forcing the refrigerant circuit. According to Critoph, the theoretical COP value of the forced convective thermal wave cycle reaches 0.9 at the evaporating temperature of 9°C, the condensing temperature of 40°C, and desorption temperature of 200°C.

2.5 Cascade cycles

In cascading cycles, different working pairs are used in different temperatures ranges. The main characteristic of the cycle is that the heat will be transferred from the working pair with a higher equilibrium adsorption/desorption temperature to the working pair with the lower equilibrium adsorption/desorption temperature. These are applied when there exists a high temperature gradient between the heat source and the evaporator/refrigeration space. Douss and Meunier [10] conducted their studies on the basis of a cascading adsorptive heat pump. They worked on the activated carbon-methanol intermittent cycle and the two adsorber zeolite-water cycle. Driving heat was supplied by an external heat source to zeolite adsorbers while activated carbon adsorber was heated by heat recovered from zeolite adsorber under adsorption.

The above-mentioned cycles are still being investigated and developed in order to obtain the best possible effects and cooling capacities. One also attempts to combine the above-mentioned cycles with each other. This, however, results in an increase of the complexity of the systems and makes them sophisticated and this it turn affects the failure rate of adsorptive refrigerating appliances. Nevertheless, the advanced adsorption cycles are a promising direction for development of this branch of practical use of the desorption phenomenon.

3 Commercial application of adsorptive refrigerators

A significant part of adsorptive refrigerating systems described in the literature involves a prototypical appliance usually made on a small laboratory scale and theoretical models constructed on the basis of experimental data with the aid of computer programs. There are also appliances created on a bigger scale serving until today as air-conditioners for facilities, but they are still a single instance. The development of technology of adsorptive refrigeration is primarily limited by high costs of the design of the equipment and by the still low efficiency of such appliances [11].

The first instances of commercialization of adsorption refrigeration systems occurred in the United States and Japan. The data in Table 1 was collected and presents the commercial use of adsorption refrigeration systems together with a list of the main companies on the market.

Building integrated solar energy systems has the potential to provide a sustainable solution to the ever-increasing demand. They can provide cooling in the summer, heating in the winter and hot water all through the season. The basic idea of the green building is to use renewable natural energy sources (solar, wind, geothermal energy) instead of coal, oil and other fossil fuels. The green building uses the most convenient, non-polluting energy for power generation, heating and refrigeration, as well as to supply hot water to meet living needs. The solar adsorption air-conditioner was used in the Green Building, which was built by the Institute of Building Science Research in Shanghai City [19]. The authors calculated that the payback period of the system would be 5-8 years if the hot water is used for space heating and cooling requirements of the building. If hot water is consumed only for baths or showers for such buildings, the payback period would be reduced to about 2 years.

Certain developed countries initiated the "Climasol-Plan", whose aim is to promote a comprehensive approach to reduce the energy consumption of buildings and the development of passive cooling techniques. They launched certain demonstration projects of solar air-conditioners. For example, in 1999, in the Hospital of Freiburg University (Freiburg, Germany), a solar powered silica gel-water adsorption air-conditioning system was installed.

DOI: 10.1051/e3sconf/20171401052

Table 1. Commercial use of adsorption refrigeration systems.

Application	**Company/utilization**
Solar powered systems	
Adsorption chiller	a Japanese company: Nishiyodo Kuchou Manufacturing (1986) [12]
Large cooling capacity adsorption chillers above 70kW (zeolite – water)	a Japanese company: Mayekawa Use of waste heat from cogeneration systems like diesel engines, gas engines, gas turbines or fuel cells. Mayekawa has already successfully installed its adsorption chiller in a shopping mall in Japan.
Solar adsorption icemaker (ice makers and solar refrigerators)	a French company: BLM [13]
Adsorption chiller	HIJC (Heat Integrated Joint Companies) Company in the USA – a service provider
Adsorption chiller (silica-gel water)	Nischiyodo Kuchouki Co.Ltd., Japan [4]
CCHP (Cogeneration system for cooling, heat and power)	For first time installed in Maltester Hospital in Kammenz of Germany[4]
Silica-gel water adsorption refrigeration chiller	Since 2003 Macom Company, Japan [4]
Adsorption CCHP system, powered by waste heat from a 185kW diesel engine	Tokai Optical Co.Ltd. in Nagoya of Japan (2003) [4]
Solar thermal freezer	Zeotech GmbH, Germany Company (2011)
Adsorption Chiller Aggregates (water-silica gel chillers)	SorTech AG., Germany (2002) [14
Water-zeolite chillers	InvenSor GmbH is a German company. Usually employs waste heat from CHP units, solar installations and industrial production processes.
The Green Building, which was built by the Institute of Building Science Research	Shanghai City
Solar powered silica gel-water adsorption air-conditioning system	In 1999, in a Hospital of Freiburg University (Freiburg, Germany)
Solar adsorption air-conditioning system	A cosmetics company in Sarantis S.A, Greece
The locomotive air conditioner (zeolite–water)	Air conditioning prototype Shanghai Railway Bureau
Waste thermal energy	
Adsorption refrigeration system for the production of chilled water (exhaust gas from the diesel engine of a fishing boat) zeolite-water	Fishing boat [15]
Adsorption air-conditioning system – engine exhaust gas	Automobile [16]
Adsorption air-conditioner (zeolit 13X - water)	Cooling the interior of cockpit in a locomotive [17]
Adsorption refrigeration cycle (activated carbon-methanol)	Waste heat from a fuel cell-electric vehicle [18]

Another example is the solar silica gel-water adsorption air-conditioning system installed in the cosmetics company in Sarantis S.A, in Greece.

Unfortunately, there are still excessive production costs and insufficient competitiveness compared to the compressor refrigerators. Many companies that produce solar powered adsorption refrigerators have to stop production. For example, SunMark, BLM, Zeopower company (USA), EG Solar (Germany), Solaref (France), unsuccessfully attempted to commercialize intermittent solar adsorption refrigerators using zeolite water pair [20].

4 Experimental part

The paper presents the initial measurements of the new testing station. The sketch of testing station is shown in Fig.4. The control study of the newly constructed testing station of the adsorbents in the adsorption refrigeration system was conducted. The system of activated carbon – methanol was studied. The most crucial part of the system is the absorber/regenerator which is a steel tube (dimensions: diameter: 60 mm, wall thickness: 2 mm, length: 270 mm). Inside the adsorber there is a grid tube with a diameter of 35mm arranged concentrically. The tested adsorbent – activated carbon – formed from a coconut shell, in the amount of approx. 170g – was placed inside the adsorber between the steel pipe and the grid tube. Such an arrangement of adsorbent maximizes the contact of the activated carbon – methanol vapor, has a positive influence on the regeneration of the bed, facilitating heat transfer to the adsorbent bed from the outside. The evaporator and the reservoir of excess refrigerant of a thin-walled aluminum may have the volume of 150 ml. The adsorber is heated by two band heaters with a diameter of 60 mm and a length of 135 mm and the power of 900 W each. The cooling of the adsorber is carried out using a fan. The adsorber was linked with the evaporator by a 6mm polyurethane cable. A system of valves and pipes connecting the adsorber to the evaporator performed the role of a condenser.

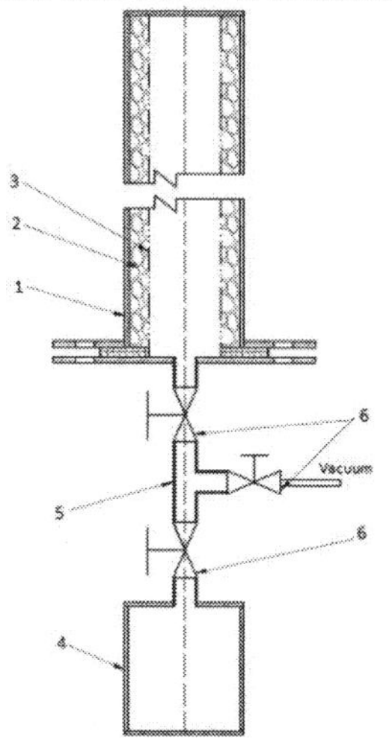

Fig.4. Sketch of testing station of adsorption cooling system
1-adsorber/generator, 2-adsorbent bed, 3-steel grid, 4-evaporator, 5-condenser, 6-valves

4.1 Experimental procedure

Control measurements were designed to check the correctness of the operation test station. A double cycle of adsorption and regeneration of the adsorbent bed was conducted. For comparison's sake, the results of a single adsorption-desorption cycle for the different temperature conditions was also provided.

Before measurements were made, the system was connected to a vacuum pump for a period of 30 minutes and the system pressure was reduced to -0,9bar (according to the indication of the pump pressure gauge). The evaporator was filled with 100cm^3 of methanol and was connected to the system.

The adsorption process was initiated by opening a valve connecting the generator to the evaporator. During the process, the temperature change in the generator and in the evaporator was registered.

After closing the valve separating the generator from the rest of the system (after completion of the adsorption process) another process was started – the process of desorption – the process of the regeneration of the adsorbent bed. The desorption process was carried out by heating the generator with a band heater. Temperature was measured electronically using the recorder, which registers changes in temperature sensors in the generator. All sensors are placed at one level in the middle of the bed height.

After completion of the desorption the valve separating the generator was closed. The heater was removed from the surface of the generator and after cooling the system to room temperature the researchers proceeded to the next cycle – the adsorption phase.

Table 2 presents the temperature changes that accompanied adsorption. During the process of adsorption of methanol vapors the heat of adsorption is released on the carbon adsorbent, resulting in an increase in temperature. At the same time heat is absorbed in the condenser therefore the temperature decreases. In the experiment there was an increase of temperature by 13.7 in the first adsorption cycle, and by 11.1 in the second adsorption cycle while the temperature decreases by 5°C in the condenser. In a single cycle of temperature an increase was registered in the adsorbent bed equal to 16.6 and the temperature in the condenser fell by 10.4.

Table 2. The measuring of temperature change in the adsorber and in the condenser

	Adsorber			Condenser		
[°C]	T_{min}	T_{max}	ΔT_a	T_{min}	T_{max}	ΔT_p
The double adsorption-desorption cycles $T_{ad} = 28°C$						
Adsorption I	28.0	41.7	13.7	27.8	22.9	4.9
Adsorption II	27.2	38.3	11.1	27.2	22.5	4.7
A single adsorption-desorption cycle $T_{ad} = 25°C$						
Adsorption	24.9	41.5	16.6	24.4	14.0	10.4

For comparison's sake, Fig. 5 presents the temperature change in the adsorbent bed and in the condenser for the two adsorption processes. The adsorption temperature for the first measurement (I) was 25° C and for the second measurement (II) it was 28°C.

 DOI: 10.1051/e3sconf/20171401052

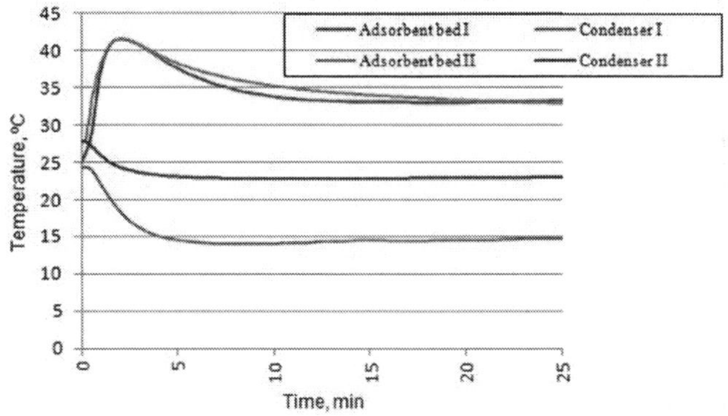

Fig. 5. Temperature profile in the adsorber and condenser during the adsorption cycle.

In accordance with Fig. 5, as a result of adsorption (I) at 25°C, a lower temperature was obtained in the condenser. For this cycle, the cooling effect is greater. The system cooled by 10.4 degrees. Fig. 6 shows the change in temperature inside the generator during desorption. It takes the system approx. 15 min to reach the 120°C temperature of desorption. The graph shows the warm-up period adsorbent bed to desorption temperature.

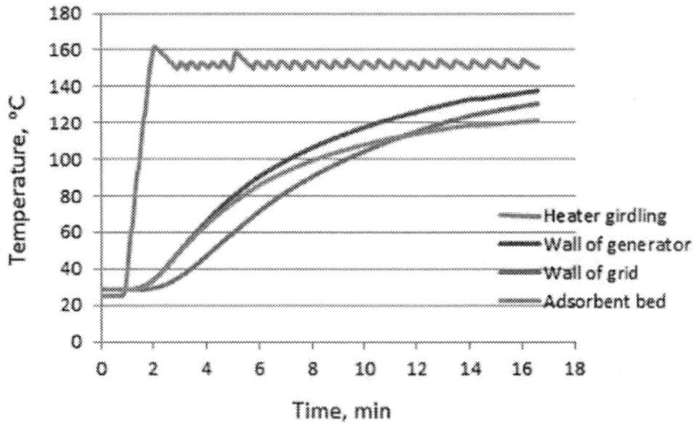

Fig. 6. Temperature changes during desorption process.

5 Conclusions

The present experiment demonstrated that it is possible to produce a low temperature by means of a simple mechanism using activated carbon bed for adsorption/desorption methanol. The presented results demonstrate that correct operation was performed at the test station.

After analyzing the initial measurements it was established that the desorption time for the system is 1h, temp. 120°C, cooling time deposits: 0.5h, adsorption time: 0.5h.

The arrangement that was built requires some refinement and improvement. First of all, one should insulate or isolate the refrigeration area and one should install the pressure sensor inside the system which will enable the researcher, among other things, to calculate the adsorbed amount of the refrigerant in the system. The presented system allows for the effect of adsorption cooling. Further work will involve the increasing of the cooling effects.

The review of commercial applications of adsorptive refrigerators demonstrated that adsorption cooling technologies are attractive alternatives. Environment-friendliness and energy usage-related benefits of this technology are also impressive. However, much effort should be done to resolve the problem of the adsorption cooling, mainly to improve heat

 DOI: 10.1051/e3sconf/20171401052

exchange, to increase COP efficiency and to reduce production costs. Therefore, further work must be done to help solve the above-mentioned problems.

The work was carried out within the research to keep the research potential AGH (11.11.210.244).

References

1. RK. SurA Das, **1**, 190 (2010)
2. R.E. Critoph, Carbon, **27**, 63 (1989)
3. N.C. Srivastava, I.W. Eames, App Therm Eng, **18**, 707 (1998)
4. R.Wang, L. Wang, J. Wu, *Adsorption refrigeration technology* (John Wiley & Sons, Singapur 2014)
5. L.W. Wang, J.Y.Wu, R.Z. Wang, App Therm Eng, **23**, 1453 (2003)
6. D. C. Wang, Y. H. Li, D. Li, Y. Z. Xia, J. P. Zhang, ", Renew and Sust Ener Rev, **14**, 344 (2010)
7. A. Akahira, K.C.A. Alam,Y. Hamamoto, A. Akisawa, T. Kashiwagi, Int J Refrig, **27**, 225 (2004)
8. S.V. Shelton, ASME J Ener Res Tech,**112**(3), 69 (1990)
9. R.E. Critoph, Heat Recov Sys & CHP, **14**(4), 343 (1994)
10. N. Douss, F. Meunier, Chem Eng Sci, **44**(2), 225 (1989)
11. E. Wolak, S. Kraszewski, E3S Web of Conferences (2016) in press
12. http://www.r718.com/news/viewprintable/3530
13. F. Meunier, Appl Therm Eng **61,** 830 (2013)
14. http://www.sortech.pl
15. R. Zhu, B. Han, M. Lin, Y.Yu, Int J Refrig, **15** (1), 31(1992)
16. M. Suzuki, Heat Recov Sys & CHP, **13**(4), 335 (1993)
17. Y. Z. Lu, R. Z. Wang, M. Zhang, S. Jiangzhou, Ener Conv and Manag, **44**, 1733(2003)
18. P. Yang, Heat Trans – Asian Res, **39** (7), 523 (2010)
19. R.Z. Wang, X.Q. Zhai, Energy **35**(11), 4407 (2010)
20. P. Goyal, P. Baredar, A. Mittal, A.R. Siddiqui, Ren and Sust En Rev, **53**, 1389 (2016)

E3S Web of Conferences **14**, 01053 (2017) DOI: 10.1051/e3sconf/20171401053
Energy and Fuels 2016

Comparison of mineral oil and esters as cooling liquids in high voltage transformer in aspect of environment protection

Grzegorz Dombek[1], *Przemyslaw* Goscinski[1], and *Zbigniew* Nadolny[1,*]

[1]Poznan University of Technology, 60965 Poznan, Poland

Abstract. The article presents the results of computer simulation of temperature field of transformer filled with mineral oil and alternatives insulating liquids which include synthetic and natural esters. The analysis was based on the results of computer simulation of temperature field of 10 MVA distribution transformer, using the program COMSOL Multiphysics.

1 Introduction

Mineral oil is one of the most commonly used insulating liquid [1]. It is used in many electrical equipment, such as power transformers. The universality of application of mineral oil results from its many positive qualities and its very good recognition over the years. However, at the end of XX century, people began to use alternative insulating liquids. Restrictive requirements for safety of electrical power equipment and environmental protection were the main reason of the use of the alternative liquids.

In the late XIX century, the idea of the use of alternative liquids in the electrical power devices was founded. At that time vegetable oils, which now are called natural esters, were used as an insulating liquid in transformers.

Long term of the use of mineral oils in electrical power equipment caused the improve of them production and treatment. Satisfied identification of their properties is main reason of their application as insulating liquids in power industry. Improving manufacturing processes and treatment of mineral oils have contributed to a significant reduction of their prices, which is one of the important criteria taken into consideration by potential users of transformers. Furthermore, affordable price means that mineral oils are still the most commonly used insulation liquids, even if alternative liquids are characterized by significantly better environmental and fire properties.

Currently, natural and synthetic esters are considered as alternatives to mineral oil [2-7]. These liquids have a number of features that are recognized for their benefits, comparing to mineral oil. These characteristics are primarily ecological properties such as biodegradability and non-toxicity, and properties conditioning operational safety which include high flash and fire point and low calorific value [2, 8, 9]. For this reason, when fire safety and environment protection play a key role, the use of transformers filled esters is justified solution.

* Corresponding author: zbigniew.nadolny@put.poznan.pl

DOI: 10.1051/e3sconf/20171401053

This applies above all electrical power equipment used in built-up areas and densely populated areas. Furthermore, companies engaged in the insuring network assets, including power transformers, are becoming increasingly aware of the fire risk associated with the use of electrical equipment insulating liquid characterized by low values of fire and flash point, which speaks in favour of esters.

On the basis of information presented above, it can be concluded that the esters, unlike mineral oil, are characterized by several positive environmental and fire properties, so that more and more willing they are used in electrical equipment, including transformers. One of the basic functions of insulating liquid is heat transport, that is generated mainly by the winding and the core [2, 10-13]. Cooling effectiveness of transformer depends on properties of used liquid, because the liquid fills significant space of the transformers [14-16]. Therefore, this paper presents a comparative analysis of the temperature field of 10 MVA transformers filled with mineral oil and esters (synthetic and natural).

2 Research object

Figure 1 shows the scheme of 10 MVA distribution transformer, which was used in computer simulation of temperature field.

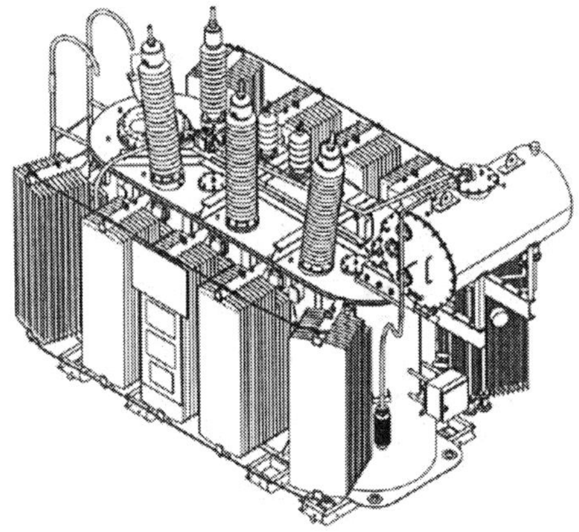

Fig. 1. 10 MVA distribution transformer model.

The transformer used ON-AN cooling type - active parts of the transformer: oil-cooled (natural flow), and the tank: air-cooled (natural convection and radiation). Furthermore, the transformer has a 11-step adjustment of the upper voltage winding in the range of +/-16% of the rated voltage. Cooling system of the transformer is aided by ten B-520FL radiators, of which the manufacture deep-drawn sheet metal were used. Each of the radiators is equipped with four ribs with a height of 1.4 m and eleven ribs with a height of 1.6 m. Simplified construction and the small number of radiators in the presented transformer helped to significantly reduce the time of computer simulation.

DOI: 10.1051/e3sconf/20171401053

3 Computational model

The analysis temperature field distribution can be made using following models: analytical models, numerical models, and real models. The paper presents the results of analysis of numerical models. Computational model was created (fig. 2) on the basis of object presented in chapter 2, which was a 10 MVA distribution transformer, using COMSOL Multiphysics software. The geometry of the model was simplified because the real object has a number of structural elements (e.g. conservator, bushings, control cabinet), which do not affect its thermal efficiency [17]. Table 1 shows main parameters of used model.

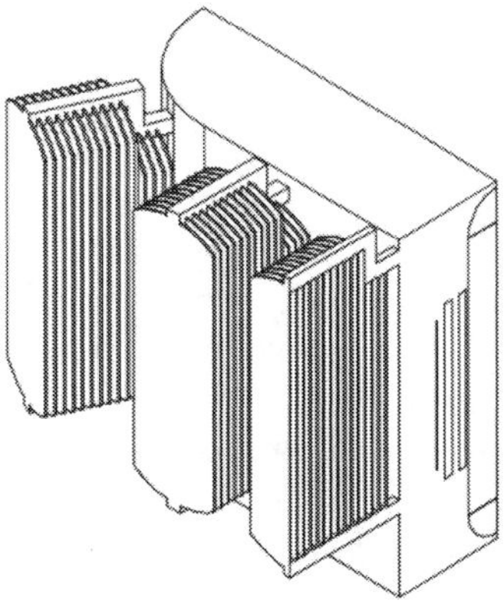

Fig. 2. 10 MVA distribution transformer computational model.

Table 1. Main parameters of the 10 MVA distribution transformer model.

Rated power	10 MVA
Rated voltages	HV: 115 kV, LV: 16,5 kV
Rated currents	HV: 50 A, LV: 350 A
Types of cooling	ON-AN
Number of phase	3
Vector group symbol	Ynd11
No-load current	0.12%
Total weight	23 800 kg
Mass of insulating oil	6 400 kg
No-load losses	6 600 W
Load losses	57 400 W

Presented in figure 2 computational model consists of tank, lower and upper voltage winding, regulating winding and radiators. Therefore, the transformer design is symmetrical and has a symmetrical spacing of radiators along the axis. So, it was decided to modelling only a quarter of the transformer. Authors resigned from the modelling stamping ribs radiators. Mesh model was built with tetragonal and pyramidal elements, whose number does not exceed 13 million. It allowed to achieve appropriate accuracy of the obtained results, which was 0.01%, and significantly reduce simulation time.

During computer simulation of insulating liquids, material parameters corresponding to the mineral oil, natural and synthetic esters were adopted. This means that in the case of mineral oil the value of heat transfer factor α equal to 125.70 $W \cdot m^{-2} \cdot K^{-1}$, natural ester 111.80 $W \cdot m^{-2} \cdot K^{-1}$, and synthetic ester 110.72 $W \cdot m^{-2} \cdot K^{-1}$ was accepted.

Heat transfer factor α is the key parameter, which describes the ability of liquid to transfer of the heat. The factor α considers many thermal properties, such as thermal conductivity, specific heat, viscosity, density and kinematic extension factor. All mentioned parameters have significant influence on heat transfer by the liquids.

The boundary conditions were used in the last step of the calculation. It means that ambient temperature and heat transfer factor α are known. It was assumed that the losses in the core and windings are heat sources. The values of these losses were determined in the result of the short-circuit attempt on presented research model, and then implemented in analysed computational model. The value of the losses was 64 kW. Moreover, according to the standard [18] it has been adopted that the ambient temperature is 20°C, and atmospheric pressure is 1 atm.

In order to determine the distribution of temperature field in the transformer in COMSOL Multiphysics, Heat Transfer module was used. The module considers thermal properties of analysed liquids and type of flow. This module considers temperature dependence of thermal properties such as thermal conductivity, viscosity, density and specific heat, which determine the heat transfer in liquids. Therefore, authors have presented the results of research on the thermal properties of the insulating liquids in previous publications, presented research results were used in the simulations [19-22].

In order to receive the dependence of time, simulations were conducted in non-stationary mode. Steady state was reached after about 14 hours for all analysed cases. During the heating trials carried out on the presented research model after about 10 hours steady state was reached. The differences in the two times result from two reasons. First one is related to the fact that in the case of the real object, in order to faster obtaining steady state under heating test voltage is initially greater than rated voltage, thereby shortening the duration of the test. On the other hand, in the case of the computer simulation, the use of the function corresponding to a gradual increase in power windings and core leads to lengthening its duration.

4 Simulation results

On the basis of presented research, object and computational model simulations of the distribution of temperature field in 10 MVA distribution transformer filled with mineral oil, synthetic ester and natural ester were carried out. Figure 3 shows the distribution of temperature field in model of a 10 MVA distribution transformer filled with one of analysed liquid – mineral oil.

Figure 4 shows the temperature distributions in the transformer, on the section of the windings to the inner surface of the tank in the case of three types of insulating liquids. After reaching the steady state of the heat flow maximum temperature in the case of mineral oil equals 65.2°C, natural ester 66.3°C and synthetic ester 66.5°C. As we can see, the temperature of the liquids in the case of natural ester is 1.1°C, and in the case of synthetic ester 1.3°C greater than mineral oil temperature. Taking into the consideration of the 5 degrees Clark right, which says that the temperature drop or rise of 5°C will result in accordingly twice lengthening or shortening life of the transformer insulation paper, these differences are important. This means that the use of natural or synthetic ester as an insulating liquid in transformer will raise the temperature. Consequently, it will contribute to shorten the life of the transformer.

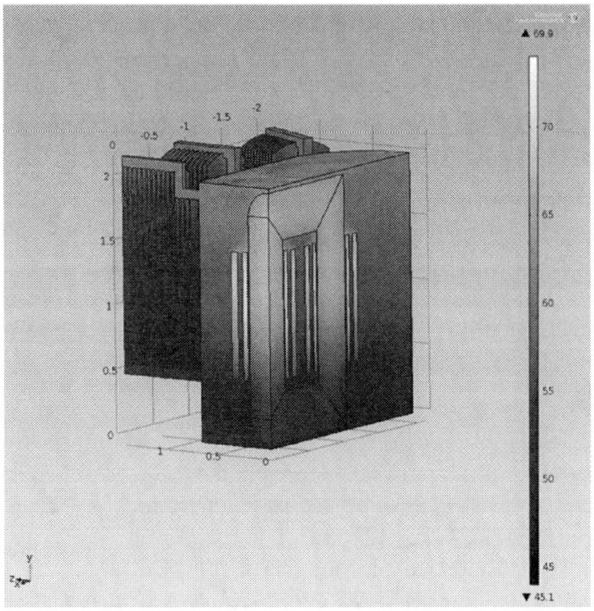

Fig. 3. The distribution of temperature field in 10 MVA distribution transformer filled with mineral oil.

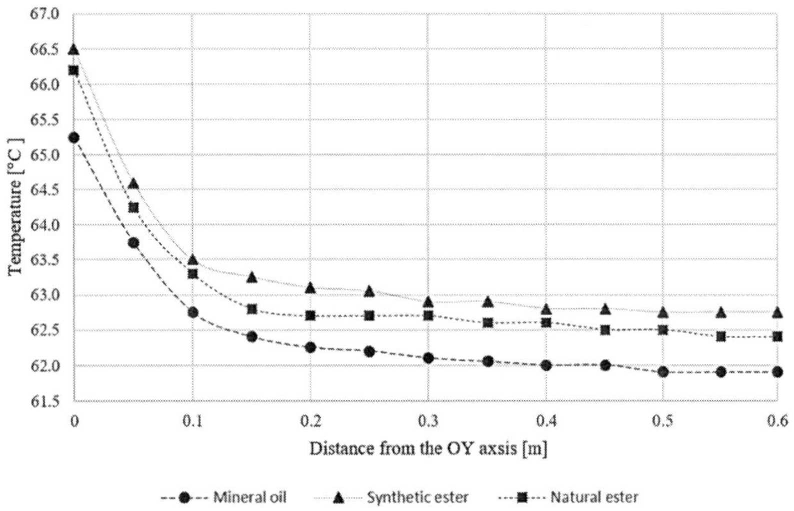

Fig. 4. The distribution of temperature field in a model of 10 MVA distribution transformer in insulating liquid on the section of the windings to the inner surface of the tank, at the height of the inlet to radiator.

5 Conclusions

On the basis of received results, we can say that the temperature of the transformer filled with esters is more than 1°C higher than the temperature of the transformer filled with mineral oil. In the case of transformer filled with synthetic or natural ester, higher temperature is caused by three times higher viscosity of natural and synthetic esters in relation to the mineral oil. Also, higher temperature will influence negatively the operation of transformers filled esters.

In the case of power transformers (e.g. 160 MVA) temperature increase in transformers filled with esters can be even greater as compared to the temperature of the transformer filled with mineral oil. Therefore, in order to use synthetic or natural ester, it should be considered the changing the geometry of the transformer (increasing its size), the reduce of load or the modification of cooling system.

The research was financed from resources of the Ministry of Science and Higher Education for statutory activities No. 04/41/DS-PB/4240, name of the task: The Influence of Various Parameters on Thermal Properties of Dielectric Liquids Used in High Voltage Transformer.

References

1. T.O. Rouse, IEEE Electr. Insul. Mag., **14**, 6-16 (1998)
2. Cigre WG A2.35, *Experiences in service with new insulating liquids. Brochure 436* (Cigre, Paris, 2010)
3. E. Gockenbach, H. Borsi, *Annual report Conference on Electrical Insulation and Dielectric* Phenomena, 521-524 (2008)
4. T.V. Oommen, IEEE Electr. Insul. Mag., **18,** 6-11 (2002)
5. T.B. Marchesan, A.J. Fanchin, *IEEE/PES Transmission and Distribution Conference and Exposition*, 329-333 (2010)
6. P. Thomas, *Proceedings of International Symposium on Electrical Insulating Materials*, 135-136 (2005)
7. N. Berger, M. Randoux, G. Ottmann, P. Vuarchex, Electra, **171**, 33-56 (1997)
8. K. Longva, *Nordic Insulation Symposium*, 293-295 (2005)
9. C.P. McShane, G.A. Gauger, J. Luksich, IEEE Transmission and Distribution Conference, **2**, 890-894 (1999)
10. K. Feser, Z. Radakovic, IEEE T. Power Deliver, **18**, 1284-1292 (2003)
11. F. Delgado, A. Ortiz, I. Fernandez, A. Arroyo, J.A. Macias, *Electrical Systems for Aircraft, Railway and Ship* Propulsion, 1-6 (2012)
12. R. Lopatkiewicz, Z. Nadolny, Przegląd Elektrotechniczny, **84**, 50-52 (2008)
13. A. Cichon, S. Borucki, T. Boczar, M. Lorenc, Eur. Phys. J-Spec. Top., **154**, 31-38 (2008)
14. R. Lopatkiewicz, Z. Nadolny, P. Przybylek, *IEEE International Conference on the Properties and Applications of Dielectric Materials*, 1-4 (2012)
15. R. Lopatkiewicz, Z. Nadolny, P. Przybylek, Przegląd Elektrotechniczny, **86**, 55-58 (2010)
16. R. Lopatkiewicz, Z. Nadolny, P. Przybylek, W. Sikorski, Przegląd Elektrotechniczny, **88**, 126-129 (2012)
17. L. Kasprzyk, A. Tomczewski, K. Bednarek, Przegląd Elektrotechniczny, **86**, 57-60 (2010)
18. IEC 60076-2:2011
19. G. Dombek, Z. Nadolny, Eskploatacja i Niezawodnosc – Maintenance and Reliability, **18**, 413-417 (2016)
20. Z. Nadolny, G. Dombek, P. Przybylek, *IEEE 2016 Conference on Electrical Insulation and Dielectric Phenomena*, 857-860 (2016)
21. Z. Nadolny, G. Dombek, P. Przybylek, D. Przadka, *IEEE 2016 Conference on Electrical Insulation and Dielectric Phenomena*, 538-541 (2016)
22. G. Dombek, Z. Nadolny, Eskploatacja i Niezawodnosc – Maintenance and Reliability, **19**, 62-67 (2017)

E3S Web of Conferences **14**, 02001 (2017) DOI: 10.1051/e3sconf/20171402001
Energy and Fuels 2016

Investigation of sewage sludge gasification with use of flue gas as a gasifying agent

Izabella Maj[1,*], *Edyta* Kudlek[2], *Piotr* Ostrowski[1] and *Marek* Pronobis[1]

[1]Silesian University of Technology, Institute of Power Engineering and Turbomachinery, Konarskiego 20, 44-100 Gliwice, Poland
[2]Silesian University of Technology, Institute of Water and Wastewater Engineering, Konarskiego 18, 44-100 Gliwice, Poland

Abstract. The paper presents results of investigation of low-temperature sewage sludge gasification with use of flue gas as a gasifying agent. Tests were conducted in a laboratory stand, equipped with a gasification reactor designed and constructed specifically for this purpose. During presented tests, gas mixture with a composition of typical flue gases was used as a gasifying agent. The measuring system ensures online measurements of syngas composition: CO, CO_2, H_2, CH_4. As a result of gasification process a syngas with combustible components has been obtained. The aim of the research was to determine the usability of sewage sludge for indirect co-firing in power boilers with the use of flue gas from the boiler as a gasifying agent and recirculating the syngas to the boiler's combustion chamber. Results of presented investigation will be used as a knowledge base for industrial-scale sewage sludge gasification process. Furthermore, toxicity of solid products of the process has been determined by the use of Microtox bioassay. Before tests, solid post-gasification residues have been ground to two particle size fractions and extracted into Milli-Q water. The response of test organisms (bioluminescent Aliivibrio fischeri bacteria) in reference to a control sample (bacteria exposed to 2% NaCl solution) was measured after 5 and 15 minutes of exposure. The obtained toxicity results proved that thermal treatment of sewage sludge by their gasification reduces their toxicity relative to water organisms.

1 Introduction

Gasification is a process of thermochemical conversion of a carbon-based feedstock into a combustible product gas with the use of a gasifying agent, such as steam, air or gas mixtures. Depending on the source, the gasification temperature is defined above 700-950 ° C, with a combustible gas as a main product of the process, what differs gasification from pyrolysis. The technology discussed in this paper is defined as a low-temperature gasification, due a temperature range lower than in typical gasification process and a combustible gas as a main product. In general, oxygen, air or steam are used as gasifying agents [1], while in discussed investigation synthetic flue gase with composition of N_2, CO_2

* Corresponding author: izabella.maj@polsl.pl

and O_2 are used. To the best of our knowledge there is very limited data regarding alternative fuels gasification with flue gases as a gasifying agent. However, increasing number of studies under coal and biomass thermal conversion are paying attention to use CO_2 and air-CO_2 mixtures as a gasifying agent. There are some experimental studies on pyrolysis of coal and biomass under N_2 and CO_2 atmosphere separately [2], however no mixture of this gases have been used.

The amount of sewage sludge has been increasing due to escalation of World's population and higher standard of living in developing countries. The disposal of waste such as sewage sludge in a pressing problem. It is necessary to implement effective and environmentally safe disposal way immediately, hence growing numbers of sewage sludge thermal utilization studies have been performed last years. Thermal processes involve converting of the organic part of the sewage sludge, resulting with final disposal of ashes only. Sewage sludge is a type of biomass fuel with a calorific value ranging up to value of a coal. In recent decades ono- and co-combustion of sewage sludge are probably the most developed thermal utilization methods, however technologies of gasification and pyrolysis are developing constantly [3, 4].

Results of presented investigation will be used as a knowledge base for industrial-scale sewage sludge gasification process. Industrial-size rotary gasification reactor using the flue gas as a gasifying agent has been designed and successfully built in a power plant in Radlin town. In field of sewage sludge gasification and pyrolysis some studies of rotary kiln and fixed bed reactors are known [5, 6] but mainly in a laboratory scale. However no studies were found in the literature about the gasification of sewage sludge with use of flue gases as gasifying agent. The main aim of presented laboratory study is to investigate an usability of sewage sludge to gasification process in an industrial-size rotary reactor prior to cofiring the process' product (syngas) in power boilers.

2 Laboratory research

Laboratory research was conducted to develop design requirements and to identify the usability of sewage sludge for low-temperature gasification in industrial-scale reactor. Tests were performed in relation to the planned construction of the industrial installation and conducted in a laboratory stand, equipped with a gasification reactor designed and constructed specifically for this purpose (Fig. 1)

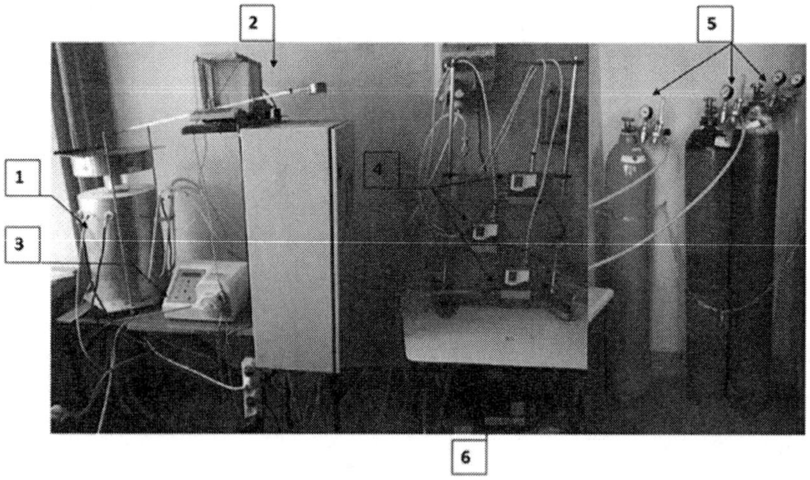

Fig. 1. Experimental gasification stand (1-reactor, 2-scales, 3-water pump, 4-gas flow meters, 5-gas tanks, 6-air compressor).

The experimental stand enables the weight loss of a sample and resulting gas composition to be recorded. The sample mass is approximately 3 g, which distinguishes the method from the classic TGA method, which is frequently used for fuel thermal conversion investigations. Moreover, in classic thermogravimetric analysis the process temperature rises with a set heating rate, while the discussed experimental setup enables to place the fuel sample in a constant temperature field and to keep the temperature stable throughout the measurement. This type of thermogravimetric analyzer may be used to determine thermal behavior of wide range of fuels during their combustion or other thermal treatment. Presented experimental setup enables to investigate gasification, pyrolysis, torrefaction and combustion, depending on agent composition and temperature requested. The station provides a possibility for the thermal degradation of an alternative fuel sample of approx. 3 g in the temperature range of 180-750°C with a gasification agent of a known composition and humidity.

The measuring system ensures online measurements and recording trends of:

- temperature
- mass loss
- composition of process gas

Before laboratory tests, properties of sewage sludge has been determined and presented in table 1.

Table 1. properties of the dried sewage sludge.

Parameter	Value	Unit
Lower heating value	10 214	kJ/kg
Ash	27,5	%
Moisture	7,2	%

3 Low-temperature gasification results

During presented tests of low-temperature sewage sludge gasification gas mixture with a composition of typical flue gas was used as a gasifying agent. On-line measurement of syngas composition revealed significant shares of combustible components (CO, H_2, CH_4). At the same time the weight loss of a fuel sample have been recorded (Fig. 2)

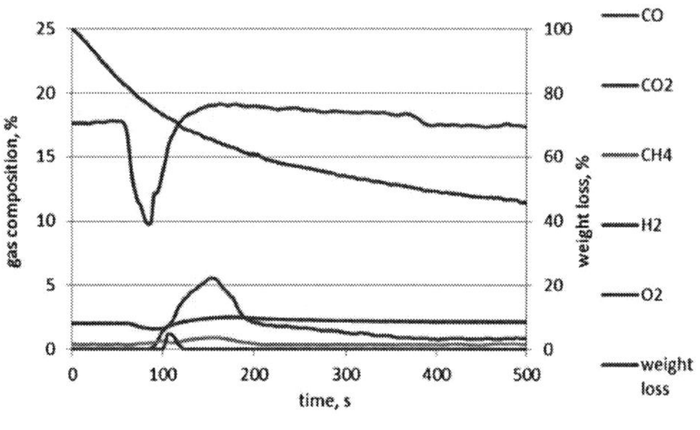

Fig. 2. Weight loss and gas composition for 550 °C sewage sludge gasification.

Tests conducted in the temperature range of 350-550°C with the application of synthetic flue gas as a converting agent identified the impact of the process temperature on the concentration of the selected combustible components (CO, CH_4 and H_2) in the process gas (Table 2).

Table 2. Time constant and maximum combustible components concentration

Temperature	T	4T	max CO	max CH_4	max H_2
°C	s	s	%	%	%
350	598	2392	2,06	0,80	0,40
450	564	2256	3,55	1,38	0,40
550	495	1980	6,07	1,01	1,14

Note: T – time constant, in field of process dynamics, defined as a time of reaching 63% of conversion. 4T defined as a time of full conversion (in inertial model process).

Depending on the process temperature, the reduction time was determined for 63% of the organic part of the fuel (Table 2). On this basis, the recommended time for the fuel to remain in the reactor was established, which was fourfold higher than the process time constant T (reduction time of 63% of organic part of the fuel).

4 Toxicity

The toxicity measurements of water samples after a 15 min and 24 h contact time with sewage sludge before and after gasification were carried out according to the Screening Test procedure of MicrotoxOmni system in Microtox analyzer Model 500. The Microtox® bioassay registers the change in the intensity of light emitted by test organisms, which are the bioluminescence bacteria *Allivibrio fisheri* exposed to toxic substances.
The rate of bioluminescence inhibition in reference to a control sample (2% NaCl solution) was measured after 5 and 15 minutes of exposure.

Absorbance measurement was performed by the use of a UV/VIS CECIL spectrophotometer model CE 1021 by Cecil Instruments Limited

The toxicological evaluation was made to assess the impact of sewage sludge before and after its gasification process on the environment. This allows for comprehensive evaluation of the feasibility of sewage sludge gasification taking into account its storage both as a substrate prior gasification and as a waste product. Figure 4 presented the change in the toxicity of water exposed to sewage sludge before and after gasification in 350, 450 and 550 °C. Dry sewage sludge of a granulation of ≤ 0.5 after a 15 min contact with water increases its toxicity according to toxicity classes [7] to the level of low toxic solutions. For the 24 h exposure the toxicological effect assume values exceeding 50%, which indicates the elution of substances form sewage sludge which have a negative impact on the test organisms and inhibit their metabolic processes. The water was classified as toxic. Water extracts of sewage sludge of a granulation greater than 0.5 mm do not exhibit a adverse effects on test organisms. Only bacteria after a 15 min contact time with solution after a 24 h extraction indicated a low toxicity of the tested samples.

DOI: 10.1051/e3sconf/20171402001

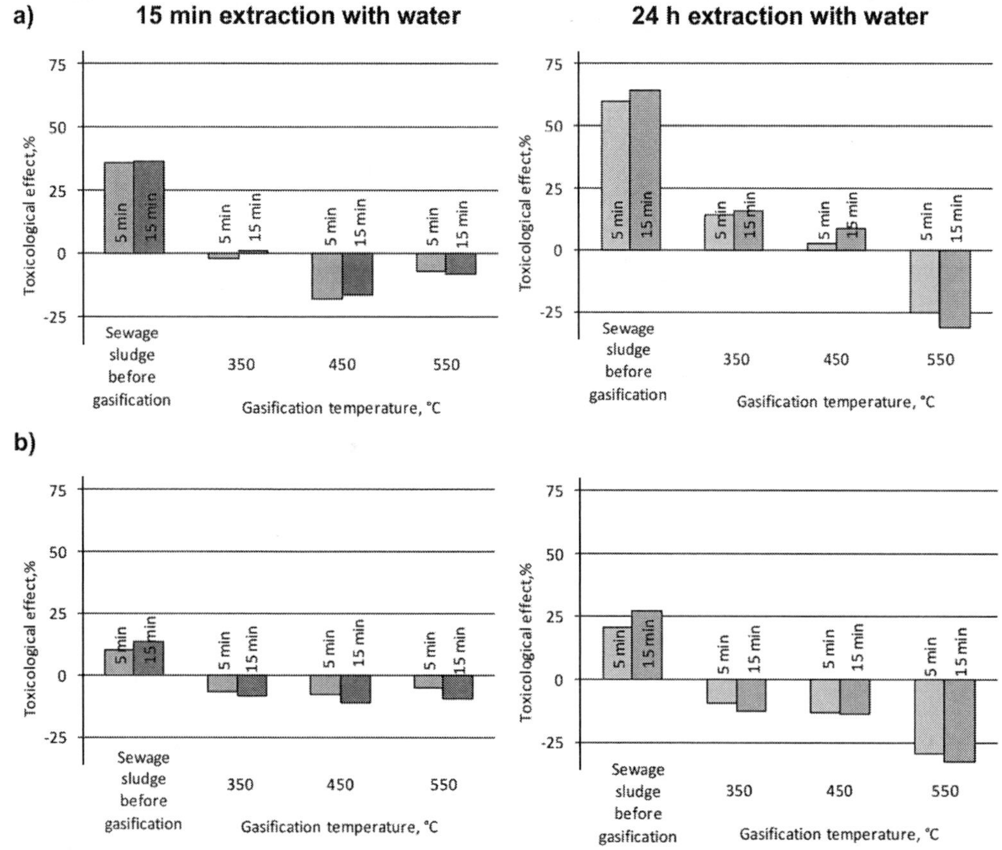

Fig. 3. Changes of toxicity of water after 15 min and 24 h extraction of sewage sludge and sewage sludge ashes with the granulation of a) ≤ 0.5 mm and b) > 0.5 mm

Water extracts of post-gasification sewage sludge ashes pointed a positive impact on living organisms. The toxic effect takes negative values. This phenomena indicated a stimulation of metabolic processes of bacteria relative to control sample. The gasification process results in the removal of the organic compounds from sewage sludge and decrease their negative properties. It should be emphasized that the organic compounds next to heavy metals are responsible for the negative impact on different ecosystems including the aquatic environment. The decrease of organic compounds has been proven by absorbance measurements at wavelength of 254 nm corresponding to the absorbance maximum for organic compounds (Fig 4.).

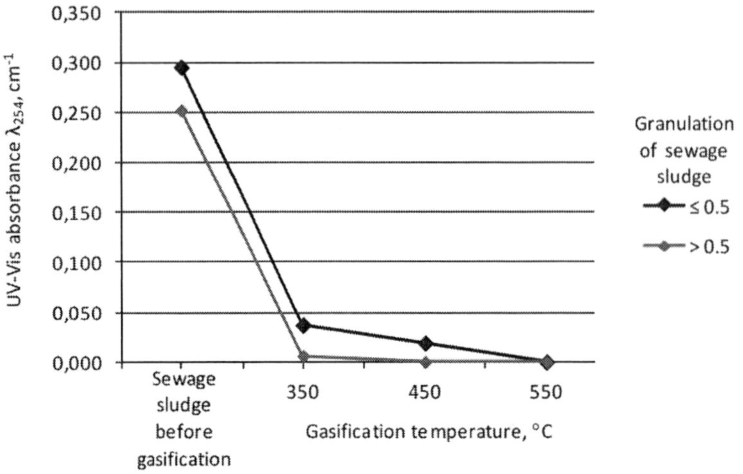

Fig. 4. Decrease of the concentration of organic substances in the gasification process expressed by the UV-Vis absorbance measurement

5 Industrial-size installation

Presented laboratory tests aimed to identify the usefulness of sewage sludge and to establish design recommendations to industrial-scale gasification process. The large-scale process can be performed in rotary gasification reactor using boiler flue gas as the converting medium (Fig 5).

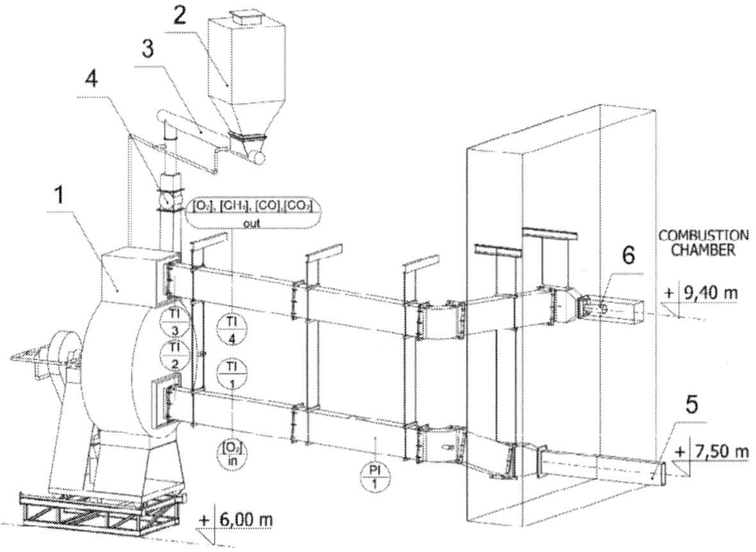

Fig. 5. Industrial-size installation concept (1-rotary reactor, 2-fuel container, 3-screw feeder, 4-rotary feeder, 5-suction duct, 6-injector)

The reactor is connected to the boiler, which implies recirculation of the product gas (syngas) into the combustion chamber. The solid products of the gasification process are transported to the container equipped with a water lock. A helical line of working vanes inside the reactor ensure continuous transport of solid products towards the chute. The construction of the reactor enables to use a wide range of alternative fuels (biomass, industrial and municipal waste). Industrial-size rotary gasification reactor using the flue gas as a gasifying agent has been designed and successfully built in Marcel power Plant in Radlin town, south of Poland. An obvious advantage of this technology is its considerable flexibility relative to the properties of the co-fired fuel (water content, particle size, composition), which is particularly important due to seasonality and the non-homogeneity of alternative fuels. Furthermore, this technology achieves high thermal shares of co-firing and separation (of solid products and ashes) from the thermal degradation installation and from the boiler, which enables their separate utilisation, e.g., as a fertiliser. The technology of using boiler flue gas as the converting medium has been developed at the Institute of Power Engineering and Turbomachinery of the Silesian University of Technology for several years.

6 Conclusions and further operation

During presented tests of low-temperature sewage sludge gasification gas mixture with a composition of typical flue gas was used as a gasifying agent. On-line measurement of syngas composition (CO, CO_2, H_2, CH_4) revealed significant shares of combustible components. At the same time a weight loss of a fuel sample have been investigated to determine the kinetics of a process and recommended time for the fuel to remain in the reactor. Results of presented investigation will be used as a knowledge base for industrial-scale sewage sludge gasification process. Industrial-size rotary gasification reactor using the flue gas as a gasifying agent has been designed and successfully built in a power plant in Radlin town, south of Poland. Several successful tests of the installation have been performed witch use of biomass feedstock (wood pellets). Low-temperature sewage sludge gasification in this industrial-size reactor will be the next step after presented laboratory research.

Moreover the toxicological tests have been performed to determine the impact of solid gasification products (sewage sludge ashes) on water organisms. The toxicological test results of water after a 15 min and 25 h contact time with sewage sludge ashes does not indicate a toxic effect on living organisms. Whereas dry sewage sludge before gasification process has a negative impact on the investigated sample. Presented tests demonstrated the impact of sewage sludge thermal treatment on their toxicity and show that even the lowest investigated process temperature (350°C) greatly reduces the toxicity of the process products, hence their further disposal would be less restricted due to the environmental issue.

This paper is the result of investigations conducted with the funds BKM/554/RIE5/2014.

References

1. J. Speight, *Gasification of Unconventional Feedstocks*, Chapter 3 – Gasifier Types, 54 (2014)
2. S. Zellagui, C. Schönnenbeck, N. Zouaoui-Mahzoul, G. Leyssens, O. Authier, E. Thunin, L. Porcheron, J.-F. Brilhac, Fuel Processing Technology, *Pyrolysis of coal and woody biomass under N2 and CO2 atmospheres using a drop tube furnace - experimental study and kinetic modeling*, **148**, 99 (2016)
3. D. Fytili, A. Zabaniotou, Renew Sust Energ Rev, *Utilization of sewage sludge in EU application of old and new methods - a review,* **12**, 116 (2008)
4. S. Werle, Ecol Chem Eng S., *Sewage sludge-to-energy management in Eastern Europe: a Polish perspective*, **22**, 459 (2015)
5. Y. Chun , S.Kim, K. Yoshikawa, Applied Energy, *Pyrolysis gasification of dried sewage sludge in a combined screw and rotary kiln gasifier*, **88**, 1105 (2011)
6. S. Werle, Energy Procedia, *Gasification of a Dried Sewage Sludge in a Laboratory Scale Fixed Bed Reactor*, **66**, 253 (2015)
7. S. Werle, M. Dudziak, Chemical Review, *Evaluation of tixicity of sawage sludge and gasification waste-products,* 92 (2013)

Simulation analysis of wastes gasification technologies

Leszek Stępień[1,*], *Marek* Ściążko[2] , *Grzegorz* Czerski[1], and *Łukasz* Kiełtyka[1]

[1] AGH w Krakowie, al. Mickiewicza 30, 30-060 Kraków
[2] Instytut Chemicznej Przeróbki Wegla, ul. Zamkowa 1, Zabrze

Abstract. Each year a significant growth in the amount of wastes generated is observed. Due to this fact technologies enabling utilization of wastes are needed. One of the ways to utilizes wastes is thermal conversion. Most widely used technology for thermal conversion is gasification that enables to produce syngas that can be either combusted or directed to further synthesis to produce methanol or liquid fuels. There are several commercially available technologies that enable to gasify wastes. The first part of this study is subjected to general description of waste gasification process. Furthermore the analysis and comparison of commercially available gasification technologies is presented, including their process arrangement, limits and capabilities. Second part of the study is dedicated to the development of thermodynamic model for waste gasification. The model includes three zones of gasification reactors: drying, gasification and eventually ash melting. Modified Gibbs minimization method is used to simulate gasification process. The model is capable of predicting final gas composition as a function of temperature or equivalence ratio. Calculations are performed for a specified average wastes composition and different equivalence ratios of air to discuss its influence on the performance of gasification (temperature of the process and gas composition). Finally the model enables to calculate total energy balance of the process as well as gasification and final gas temperature.

1 Introduction

Thermochemical conversion of wastes, represented by gasification process, has a relatively short history, but since the beginning of 1990's it became a commercially available technology. Since then, considerable growth in the amount of converted wastes is observed. As a result in 2010 over 5 mln tonnes of wastes were converted thermally worldwide. Main advantages of thermal conversion of wastes are: strong reduction of the waste in mass (about 70–80%) and in volume (about 80–90%), destruction of organic contaminants, such as halogenated hydrocarbons, utilization of recyclables from the thermal residues, such as ferrous and non-ferrous metals from bottom ash and slag, reduction of greenhouse gas emissions from anaerobic decomposition of the organic wastes, land savings due the decreased volume resulting in smaller landfills [1-3].

[*] Corresponding author: lstepien@agh.edu.pl

Gasification is a process that has been known for ages. At first only coal was used in commercial scale, but with time some other organic feedstock like biomass or different kind of wastes were also tested and used. Its main idea is to convert solid organic substance into useful gas in a presence of a gasifying agent. The producer gas in case of oxygen blown gasification consists mainly of carbon monoxide and hydrogen with the share depending on process conditions and reactor applied. In a case of air gasification, average low calorific value (LHV) of obtained gas is in the range between 4-6 MJ/m3 due to high fraction of nitrogen in a product, while for oxygen gasification it gives between 10-15 MJ/m3. Addition of steam as leads to the increased hydrogen concertation, which is desirable in case of using syngas for chemical synthesis.

Nowadays over 200 mln Mg of coal is gasified worldwide. Main products of gasification is syngas, used for the production of methanol or hydrogen which is then utilized for ammonia synthesis. Although gasification is very well known and widely used technology its implementation to waste conversion is quite recent. The main obstacle is the heterogeneity of wastes, including various chemical composition, particle sizes and heating value. That is the reason why special pretreatment process are required to operate any waste conversion facility and that needs to include drying, crushing and mixing.

On a map of various conversion technologies gasification of wastes can be placed between pyrolysis and combustion. Main differences concern the temperature and excess of air (oxygen) used in the process (Fig. 1). In a commercial practice each of these technologies is mainly used for specific feedstock, for instance pyrolysis is mainly used for obtaining oils from tires and other synthetic materials but is not necessarily recommended for municipal and commercial wastes. In case of combustion processes the feedstock is not that important as in case of pyrolysis, but amount of flue gas, which has to be cleaned, is very large (approximately 14 m^3/kg waste) what increases both the equipment size and operational costs. For this reasons gasification presents itself as a most advisable technology for waste conversion.

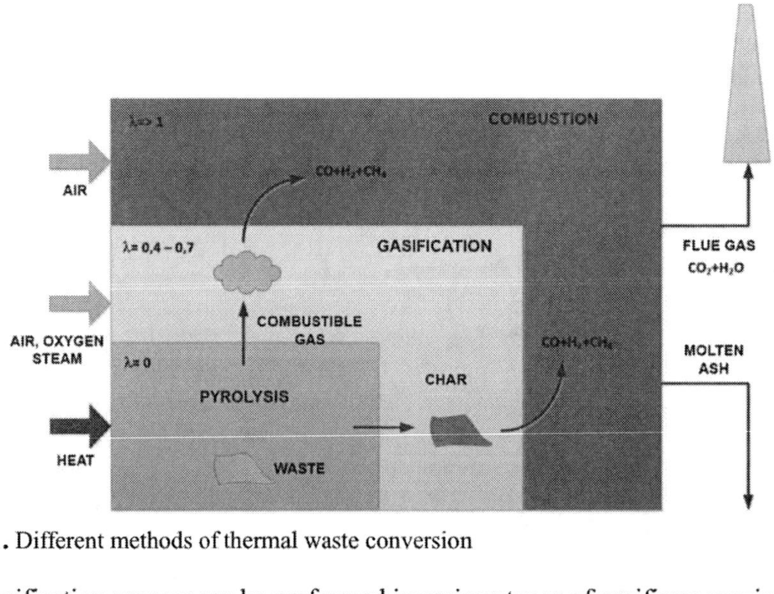

Fig. 1. Different methods of thermal waste conversion

Gasification process can be performed in various types of gasifiers: moving bed, fluidized bed or entrained, but in a case of waste gasification only the first two are applicable. There are many types of moving bed reactors, but the most common are the ones fed from the top, where the feedstock is moving downward in the counter current to gasifying agent. In the top zones the feedstock is dried, then undergoes pyrolysis and finally the char is gasified and obtained ash can be melted in the bottom zone of the reactor. Average temperature in the gasification zone can reach approximately 1100-1150 K and in the bottom part it can reach up to 1700-1900 K. This high temperature also provides the possibility to utilize hazardous, toxic and medical wastes, which is not possible in case of lower temperature reactors. To obtain such high temperature the feedstock need to be improved, which can be done by addition of met coke, that not only increases the calorific value of feedstock but also serves as a kind of grate that enables molten ash to flow downward. In that case it also necessary to use oxygen enriched air to increase the temperature. Finally ash, from this type of reactors can be collected in two ways molten or granulated after cooling and separation process.

Other reactor that can be used for waste conversion is fluid bed gasifier. Its biggest advantage is almost ideal mixing and as a result better mass and energy transfer between gas and solid phase. But on the other hand it is more difficult in operation, because it requires preprocessing of waste such as drying and grinding to specific particle size. Usually one of the particle dimensions shouldn't exceed 6mm, otherwise there is a risk of choking and defluidization. Fluid bed reactor also requires addition of sand to improve the performance of fluid bed and its stability. The temperature reached in this type of beds does not exceed 1100 K which is too low in case of processing medical wastes. The solution to that problem is to install upstream the additional equipment for a gas processing, for instance plasma reactor that will be able to increase the temperature up to at least 1500 K and reduce the amount of dioxins and other impurities in produced gas.

2 Technology review

Fig. 2 shows basic types of technological systems used in gasification facilities. They can be divided into four different groups:

a) single step gasification without any additional units,
b) gasification with built-in plasma torch,
c) gasification with the addition of coke
d) gasification with external plasma reactor.

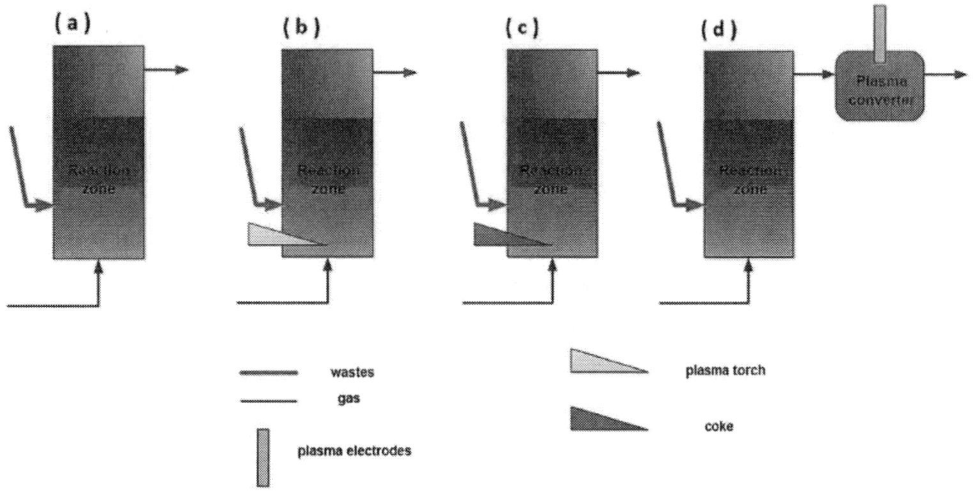

Fig. 2. Basic technological systems for waste gasification

Considering integration of technological units and systems two main groups can be distinguished, that result in a legal qualification of the whole facility what is shown in Fig. 3. First group (A) includes integrated gasification and combustion facilities, that gasify wastes and then immediately combust syngas. Produced flue gas is cleaned and released to atmosphere. Final product of this type of facility is steam that can be used to produce energy. This type of configuration is categorized as a incineration plant with all the legal consequences.

Second group (B) of facilities includes gasification of wastes resulting in cleaned syngas, that can be used either to produce energy (in-situ or transported to another facility) or for further chemical processing, for instance production of methanol, synthetic fuels or ammonia. In this situation gasification with synthesis falls in a category of waste recycling which is in considered higher in a hierarchy of waste management. Still unclear question is to qualify distant gas combustion. In case of natural gas quality it may be considered as a regular combustion process without the restriction imposed on incineration.

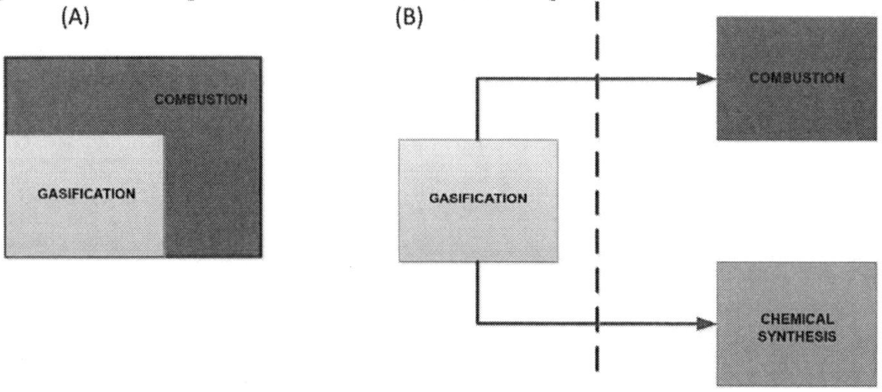

Fig. 3. Integration of technological units

One of the leading countries in the field of waste conversion is Japan, which from the beginning chose gasification as the main option suitable for waste conversion because it is able to provide satisfying performance in both, technological and environmental aspects. Two leading companies offering waste gasification systems in Japan are JFE Engineering Corporation [4], since 2003 having 10 fully operational plants in Japan and Nippon Steel & Sumikin Engineering Co.[5], having 43 operating facilities working since 1980. The scale of both technologies is fully commercial and ranges from 10 000 -230 000 Mg/a.

Both Japanese technologies use similar types of moving bed reactor, requiring addition of extra coke to increase heating value of waste and also serving as grid to assure the flow of molten slag in a bottom part of the reactor. Also both of this technologies require air and enriched air for operation. Enriched air is used in bottom part of the gasifier to increase the temperature up to 1600°C to ensure ash melting [6].

The JFE High-Temperature Gasifying and Direct Melting Furnace is one of the most compact furnaces, which is able to gasify combustibles and melt ash in a single step. Combustibles of wastes are gasified by the high temperature of the furnace. Non-combustibles of wastes are melted and turned into slag and metals which can be reused. Since a reducing atmosphere is maintained in the interior of the furnace, hazardous heavy metals may vaporized to the gaseous phase and molten ash is converted to safe slag. The molten ash from the furnace bottom is quenched in a water-granulation conveyor to form granulated slag and metals [7].

Both of the aforementioned process are able to process wide range of wastes from municipal and commercial wastes, through ash and sewage sludge up to low calorific excavated landfill wastes and even toxic, hazardous and medical wastes. An important advantage is also a fact that they do not require any pretreatment and only limit is the size of ducts delivering feedstock to the reactor.

Besides Japanese technologies also other corporations from different countries are interested in the field of waste converting, but they do not have many fully operational plants. The examples of such technologies are the ones offered by Alter NRG [8], Enerkem [9]and Leveraged Green Energy [10].

Alter NRG offers plasma gasifier with oxygen used as gasifying agent where various feedstocks can be gasified using the very high temperatures achievable with plasma. The gasifier is equipped with Westinghouse Plasma Corp. torch systems to ensure the internal temperatures in the reactor are sufficient to guarantee complete conversion of carbonaceous material to syngas and to melt all the inorganic material. Temperature inside the reactor reaches 1600°C near the plasma torch to ensure ash melting and the gas leaves this unit at temperature of about 850°C.

Enerkem offers fluid bed gasification technology for pretreated municipal wastes.

LGE offers a process called Gasplasma, which is a two-step process in which organic waste is first gasified by heating it in a low oxygen environment. As a second step, the syngas is cleaned by a plasma torch, which cracks the complex organic compounds into their constituent parts resulting in a clean synthetic gas with even higher calorific content. It is believed that this technology is able to present efficiency of about 50% when producing heat (via combustion of syngas), but it was not tested in commercial application, because it has just one operating pilot plant located in the UK.

The summary of presented technologies is shown in Table 1.

Table 1 Summary of waste gasification technologies

	JFE	Nippon	LGE	Enerkem	Alter NRG
Type of gasifier	moving bed, coke addition, ash melting unit	moving bed, coke addition, ash melting unit	Gasifier + plasma converter for syngas	Fluid bed gasification	Moving bed, plasma torch for ash melting
Oxidant	enriched air, air	enriched air, air	oxygen	Air/oxygen	Air/oxygen
Feedstock requirements	none, can gasify medical and hazardous wastes	none, can gasify medical and hazardous wastes	Not specified	post-sorted municipal solid waste (after recycling and composting)	sorted and shredded (< 15cm), can gasify medical and hazardous wastes
Product	steam	steam	syngas or energy	syngas for chemical production,	clean syngas, production of energy or chemicals
References	10	43	None (1 pilot plant)	1 under construction	6 + 2 under construction

3 Model development

Presented model is based on the Gibbs minimization method, which is well known and widely used to modeling complex processes such as gasification [11-14]. In this study

 DOI: 10.1051/e3sconf/20171402002

gasification is modelled with modified Gibbs minimization method, which uses some experimentally obtained constraints to include the deviations from thermodynamic equilibrium state caused by kinetic influences on the process. Detailed description of the model, balance equations and equilibrium modifications, are presented in author's previous study [15]. The following assumptions were applied to perform following calculations:

- gasification reactor is assumed to be a moving bed reactor, fed from the top,
- waste feed is 1000 kg/h,
- the basic amount of air fed into the reactor is calculated based on equivalence ratio typical for biomass gasification [16,17] ER=0.25,
- enriched air is assumed to contain up to 35% of oxygen,
- nitrogen from air and inorganic substance from is treated as inert and does not influence chemistry of the process,
- heat loss of the system is estimated to be equal to 15% of physical enthalpy of products.

The reactor itself can be divided into 3 zones: drying, gasification and ash melting. For this reason, aforementioned model was improved and two additional modules were included. First one is used to calculate energy consumption in drying process and the second one calculates energy consumption for ash melting.

Furthermore the calculations were performed in two different configurations presented in Figure 4. In first case, both wastes and additional coke were treated as one stream provided into the gasification reactor and they both undergo to all gasification reactions leading to a final product. Gibbs minimization is performed on a mixture of coke and waste and final gas composition is calculated. Drying of wastes results in decreasing temperature of produced syngas. It is assumed that biomass reaches temperature of 150°C in the drying zone and as a result syngas is cooled to the temperature T_{out} that can be calculated by solving the equation based on the energy balance (1) of this process.

$$\Delta h_{vap} + m_w {C_p}_{waste} (T_d - T_0) + n_{gas} {C_p}_g (T_{out} - T_0) = n_{gas} {C_p}_g (T_{proc} - T_0) \tag{1}$$

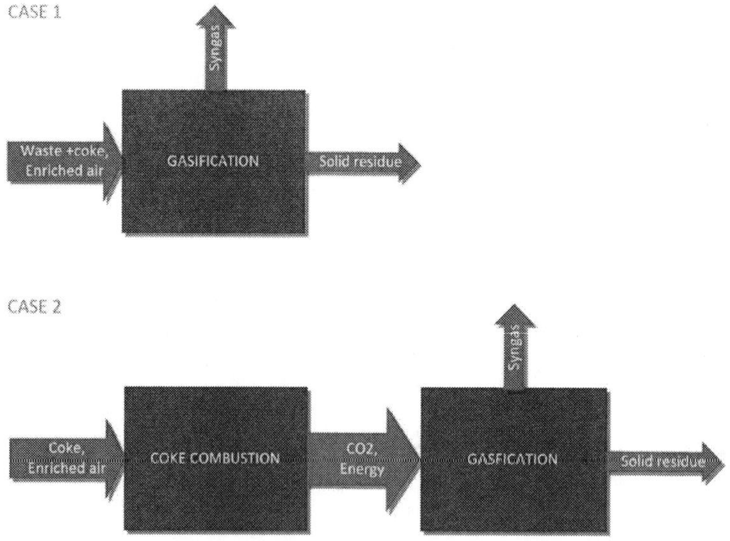

Fig. 4. Two cases of model configuration

Energy consumption needed for ash melting is included in overall energy balance of gasification process as a part of heat loss. Overall energy balance is used to calculate temperature of gasification zone. Ash melting temperature was approximated using equations proposed by Holubcik et al. [18] It was assumed that ash consist only of 4 main components SiO_2, CaO, Al_2O_3 and Fe_2O_3.

In case 2 coke and enriched air are treated separately. Coke undergoes partial combustion and produced heat is used to melt ash and deliver energy to endothermic gasification reactions. In second step Gibbs minimization is performed on wastes with air as gasifying agent and energy released during combustion is added to the overall balance of the process. In this case the excess of energy produced in melting zone is added to energy balance of gasification zone and gasification temperature is calculated. Drying zone is calculated in the same manner as in Case 1.

Chemical composition of such wastes used for the calculations is presented in Table 2 while typical morphology of municipal wastes is presented in Table 3.

Table 2 Average waste composition (as received)

	C	H	N	S	O	Moisture	Ash	Cl
% mas	50.50	3.00	1.90	0.46	21.10	21.54	0.30	1.20

Table 3 Typical morphology of municipal waste

Type	[%]
organic wastes	19.7
paper	14.6
plastics	14.1
particles <10mm	9.6
glass	8.6
particles 10-20mm	8.1
others	5.9
food waste	5.4
textiles	3.9
Multicomponent waste	3.6
neutral	3.3
metals	2.0
wood	0.6
green wastes	0.3
electrical wastes	0.27
hazardous materials	0.02
batteries	0.01
Total	**100**

4 Results and discussion

Presented calculations were performed only for technologies including addition of coke (reference to Nippon and JFE). As was previously mentioned they both present similar constructions of gasifiers and both can be treated as moving bed reactors. Because of that and the lack of detailed diagram flows of both technologies it was decided that they can be modeled by the same algorithm. Technologies using plasma torch or plasma reactor were neglected in this part of the study. According to the aforementioned assumption for Case 1 air stream was set to be 2000 m^3/h. It remained constant and only oxygen concentration was change from ambient air up to 35% of oxygen to observe the changes in gas composition and process temperature. Syngas composition is presented in Fig. 5. As can be seen the increase in carbon monoxide and carbon dioxide concentration can be observed as well as in the amount of syngas that is produced (Fig. 8). Also with the increasing amount of oxygen an increase in gasification temperature is observed which grows from 1250K form ambient air up to 1600K for 35% oxygen. Gas temperature after drying of biomass remains constant at about 600K and is does not depend on oxygen level. Heating value of produced syngas presents maximum for oxygen concentration at about 30% (Fig. 7). Fig. 8 also presents carbon conversion changes. Optimal conversion level is reached for highest oxygen concentration and highest temperature.

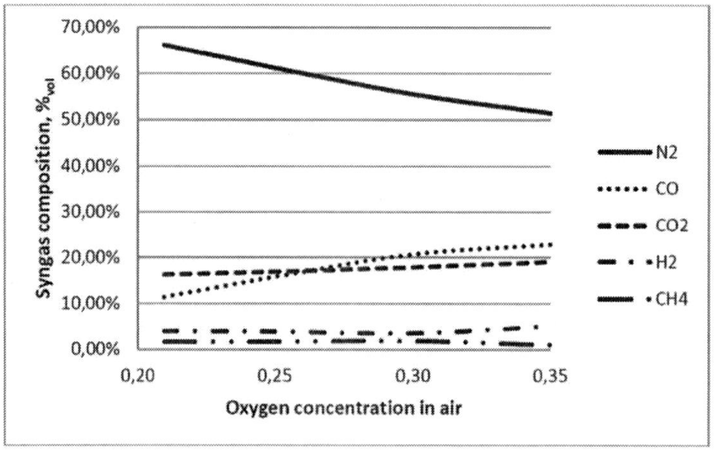

Fig. 5. Syngas composition as a function of oxygen concentration (case 1)

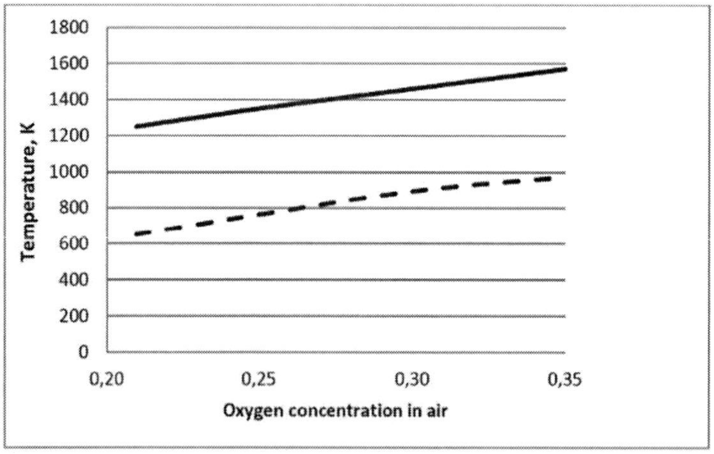

Fig. 6. Gasification and final gas temperatures as a function of oxygen concentration

 DOI: 10.1051/e3sconf/20171402002

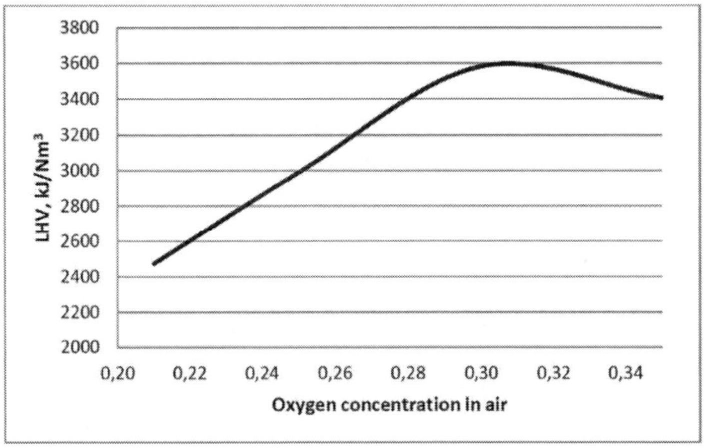

Fig. 7. Syngas LHV

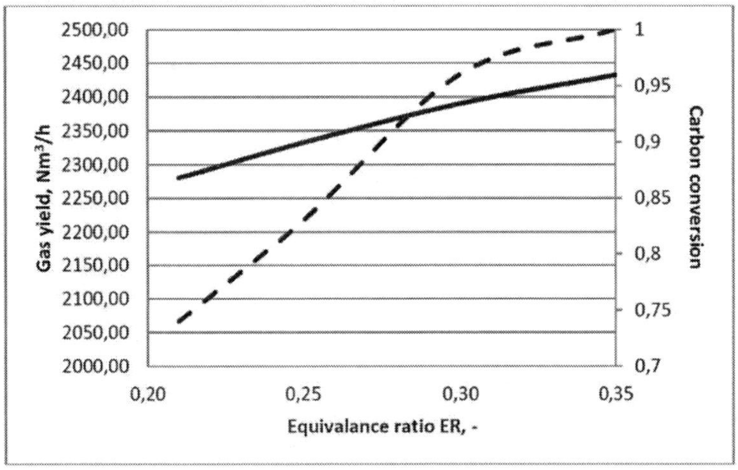

Fig. 8. Gas yield (solid) and carbon conversion efficiency (dotted)

For Case 2 additional coke was first partially burned in the atmosphere of enriched air and produced energy was partly used for ash melting and the excess of energy is added to energy balance of gasification block. Additionally carbon dioxide and nitrogen from coke combustion is added to gasification as an extra gasification agent. Basic gasification agent is air aid it's amount was calculated based on equivalence ratio which was changed from ER=0.2 up to ER=0.35. This resulted in different streams of gasification agent ranging between 1500-2600 Nm^3/h. Fig. 9 presents changes in syngas composition with the change of ER ratio. A slight increase in the concentration of carbon dioxide is observed as well as an increase in the amount of carbon monoxide. Both hydrogen and methane concentrations are on low level and do not exceed 4%. Fig. 10 shows significant increase in gas yield, but that is mainly caused by increasing stream of gasification agent and amount of oxygen that is fed to the system. Similarly to case one, carbon conversion reaches 100% for the highest stream of oxygen. Also with the increase of oxygen amount one can observe an increase in both gasification and final syngas temperature. Similarly to the previous case temperature drop is constant at about 600 K. LHV of produced syngas (Fig. 11) also shows maximum value for ER equal to approximately 0.25.

Comparing both models two significant differences can be observed regarding nitrogen concentrations and concentrations of carbon monoxide and carbon dioxide. In case 1 higher concentrations of carbon monoxide and carbon dioxide were obtained. Predicted temperatures were similar in both cases as well as the values of LHV of produced syngas, which was slightly higher for case two.

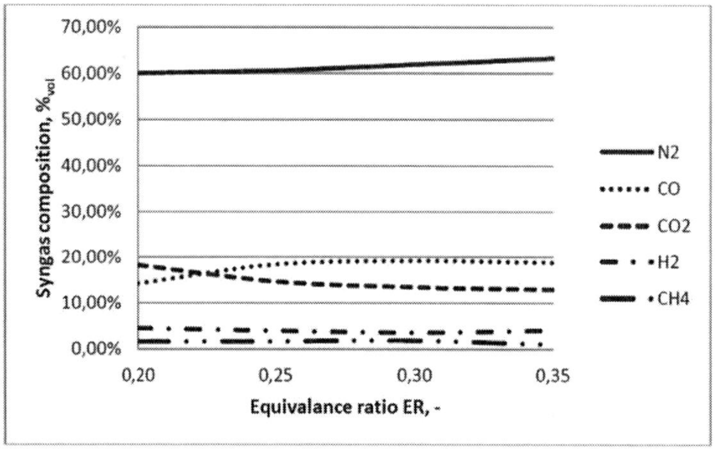

Fig. 9. Syngas composition as a function of oxygen concentration

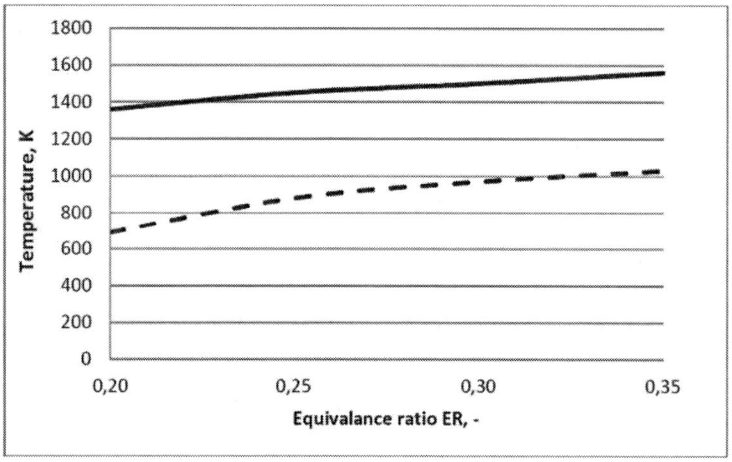

Fig. 10. Gasification and final gas temperatures as a function of oxygen concentration

 DOI: 10.1051/e3sconf/20171402002

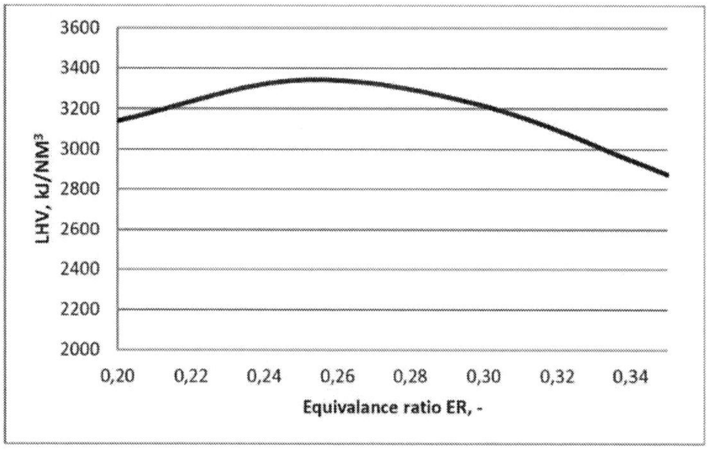

Fig. 11. Syngas LHV

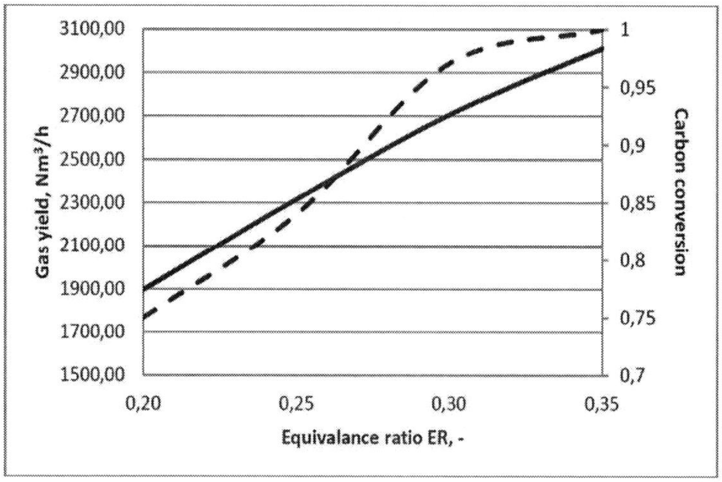

Fig. 12. Gas yield (solid) and carbon conversion efficiency (dotted)

5 Conclusions

An equilibrium model based on Gibbs minimization method was used to simulate the possible performance of two commercially available technologies in municipal waste gasification – JFE and Nippon Steel. Two different model cases with different approach to additional coke combustion, were presented, both producing similar results concerning final gas composition waste conversion and temperature.

- There is about $20\%_{vol.}$ of CO and CO_2, while concentrations of H_2 and CH_4 are much lower and do not exceed $5\%_{vol.}$,
- The maximum LHV is reached for 25-30% of oxygen in gasifying agent. Produced gas LHV is approximately 3.5 MJ/Nm^3
- Estimated gas yield ranges between 2.6-3 Nm^3/kg of waste.

Due to lack of detailed data about presented technologies, proper validation of presented values is impossible. Presented model may be useful in future assessing the technologies for perspective design of gasification facility and choosing proper feedstock for its operation.

Research was done within the framework of statutory work 11.11.210.213.

References

1. U. Arena, Waste Management **32**, 625-639 (2012)
2. C.S. Psomopoulos, C.S., Bourka, A., Themelis, N.J., 2009. Waste-to-energy: a review of the status and benefits in USA. Waste Management 29, 1718–1724.
3. S. Consonni, M. Giugliano, M. Grosso, Waste Management **25**, 123–135 (2005)
4. http://www.jfe-eng.co.jp/en/ (accesed 31.08.2016)
5. https://www.eng.nssmc.com/english/ (accesed 31.08.2016)
6. N. Tanigaki, *Proven Waste Gasification Technology for Circular Society - Key Technical Information for Deployment*, World Waste to Energy City Summit London, 2015
7. JFE High-Temperature Gasifying and Direct Melting Furnace System - Operational Results of "Fukuyama Recycle Power", JFE Technical Report, March 2011
8. http://www.alternrg.com/waste_to_energy/energy-from-waste/ (accesed 31.08.2016)
9. http://enerkem.com/about-us/technology/ (accesed 31.08.2016)
10. http://lgefund.net/technologies/ (accesed 31.08.2016)
11. X. Li, Fuel **80**, 195-207 (2001)
12. S. Shabbar., I. Janajreh Energy Conv .and Manag. **65**, 755-763 (2013)
13. E. Esmaili , N. Mahinpey, C. Jim Lim, Chem. Engi. Res. and Des. **91**, 1361-1369 (2013)
14. P. Kangas, I. Hannula, P. Koukkari, M. Hupa, Fuel **129**, 86-94 (2014)
15. M. Ściążko, L. Stępień, Chem. and Proc. Eng. **36**, 73-87 (2015)
16. A. Bhavanam, R.C.Sastry, Int. jour. Of Chem. Eng. And App. **6**, 425*-432 (2011)
17. B. Sharmina, Energies **6**, 6508-6524 (2013)
18. M. Holubcik, J. Jandacka, M. Malcho, , The holistic approach to Env. **5**, 119-125 (2015)

Simplified method for calculating SNCR system efficiency

Marek Pronobis[1,*], *Robert* Wejkowski[1], *Katarzyna* Jagodzińska[1] and *Tomasz* Kress[1]

[1]Division of Boilers and Steam Generators, Institute of Power Engineering and Turbomachinery, Silesian University of Technology, Konarskiego 20, 44-100 Gliwice, Poland

Abstract. SNCR (Selective Non-Catalytic Reduction) technology is aimed at reducing NO_x emissions. SNCR efficiency is appropriately high only for the reaction temperature range called 'the SNCR temperature window'. It is a narrow temperature range defined in various ways in the literature, which makes it difficult to evaluate the $DeNO_x$ system's efficiency. Therefore, this study attempts to approximate the relationship between SNCR system efficiency and the flue gas temperature. The approximation was performed on the basis of literature data and verified using data from an experiment. Measurements were performed in a Polish boiler with a maximum continuous rating of 230 t/h. The verified, evaluated function could be used to forecast efficiency of SNCR systems in existing units that use urea or ammonia as a reagent. The approximation results are polynomial functions that depend on flue gas temperature, which fit the literature data with the coefficient of determination $R^2 = 0.83$-0.86. Therefore, these equations could be used by the designer or operator of the boiler for preliminary determination of current SNCR system efficiency.

1 Introduction

The SNCR (Selective Non-Catalytic Reduction) system efficiency is defined by the equation:

$$\eta_{SNCR} = \frac{S_1 - S_2}{S_1}, \% \tag{1}$$

S_1 - concentration of NO_x in flue gas, without the SNCR system, %.
S_2 - concentration of NO_x in flue gas, with the SNCR system, %.

Its value is appropriately high only for a specific reaction temperature range called 'the SNCR temperature window'. Diagrams presenting the relationships between η_{SNCR} and flue gas temperature are in references [1-6]. There are also relationships between flue gas temperature and ammonia slip ; Fig. 1 shows unified charts from the literature data for urea as a reagent and Fig. 2 shows these for ammonia.

* Corresponding author: marek.pronobis@polsl.pl

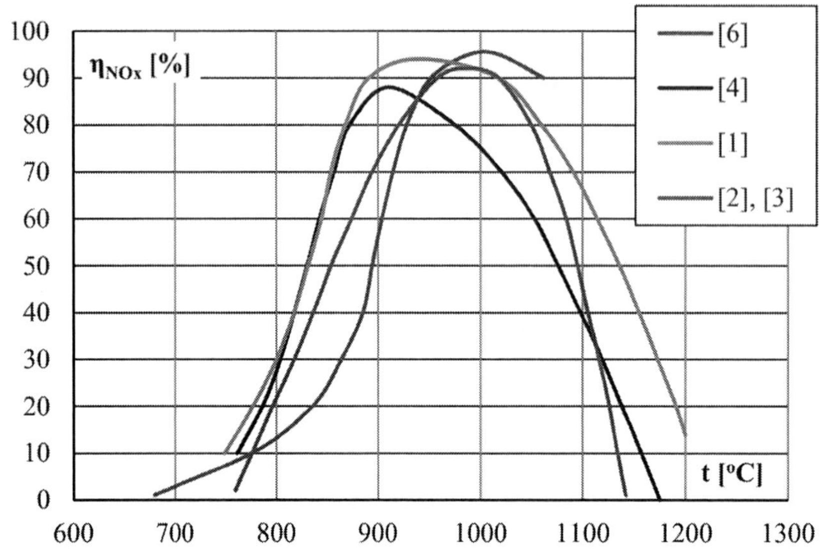

Fig. 1 The SNCR system efficiency as a function of the flue gas temperature, using urea as a reagent

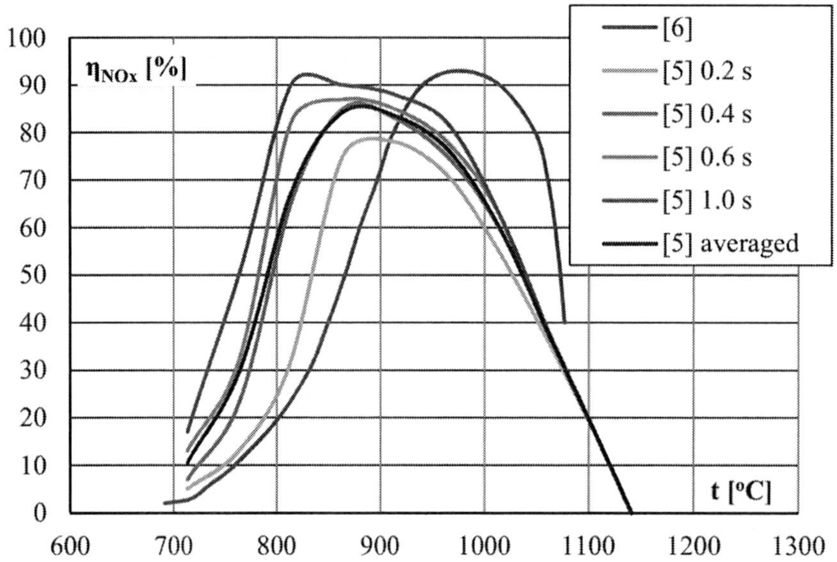

Fig. 2 The SNCR system efficiency as a function of flue gas temperature, using ammonia as a reagent

The SNCR temperature window range is relative and depends on the assumed value of the SNCR system's efficiency. Fig. 3 shows the 'old operating range' with a minimal η_{SNCR} value of 45 % ; however, if the $DeNO_x$ system efficiency had to be higher, the temperature range would be significantly reduced. In [7] the SNCR temperature window is about 850-1100 °C for urea, whereas in [2, 3] it is about 880-1080 °C for urea and for ammonia (Fig. 3).

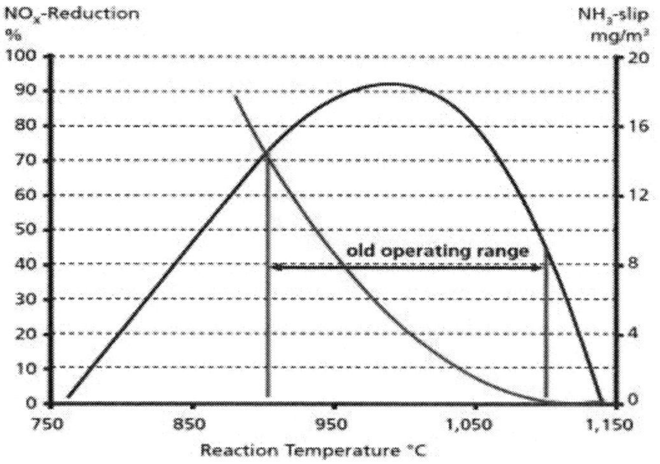

Fig. 3 The SNCR system efficiency and the ammonia slip as a function of the flue gas temperature [3]

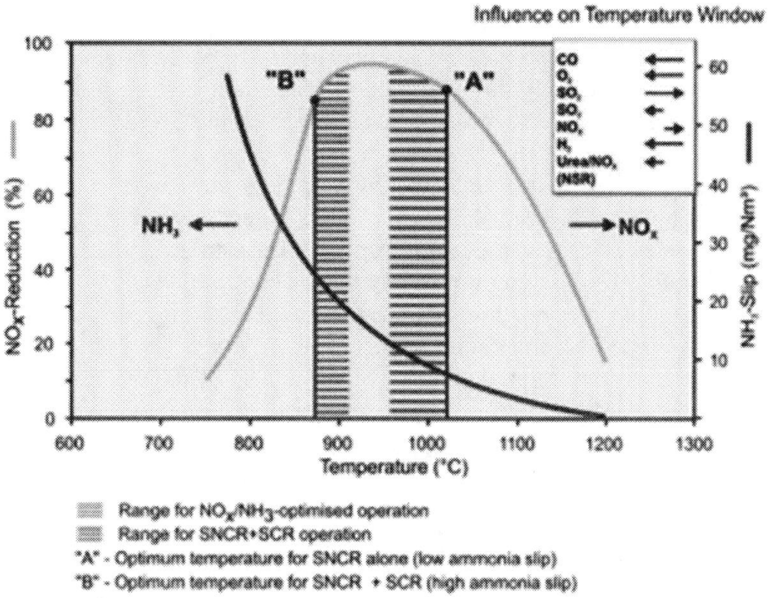

Fig. 4 The SNCR system efficiency and the ammonia slip as a function of the flue gas temperature [1]

The different course of the relationship between η_{SNCR} and the flue gas temperature is presented in [4, 6]. In [5], the SNCR system efficiency reached values above 90 % when ammonia was used as the reagent. The lab-scale tests were performed in an isolated tubular reactor in which full mixing of reagent and flue gas occurred in reaction times above 0.2 s (Fig. 2). Reaction time τ_p has a meaningful impact on NO_x non-catalytic reduction. Decreased τ_p causes a narrowing of the SNCR temperature window for which η_{SNCR} is above 60 % - from range of 780-1020 °C (τ_p = 1 s) to 830-1000 °C (τ_p = 0.2 s), corresponding to a temperature range change Δt of 240 to 170 K. Decreased τ_p also decreases the maximal value of η_{SNCR} by on average about 10 % (Fig. 2).

Reaction time significantly depends on the flue gas temperature. For temperatures above 950 °C NO_x reduction occurs in less than 0.1 s, whereas for 850 °C the requisite time is about 0.5 s [2, 9]. Changing τ_p from 0.1 to 0.5 s is related to a flue gas distance of 1-5 m, assuming the typical velocity of 10 m/s. Therefore, from the level of reactant injection to the level of appropriate reduction there must be several meters of space in which the flue gas temperature attains the SNCR temperature window.

The most important issue for NO_x reduction efficiency is an appropriate choice of reagent injection location; if it is located in an area with flue gas temperatures above the range of the SNCR temperature window, ammonia oxidation would occur, which increases the NO concentration in flue gas, whereas NH_3 slip decreases to zero. In temperatures below the SNCR temperature window, a weak reactant conversion occurs, which results in increased NH_3 slip.

The appropriate mixing of reactant with flue gas is also necessary to obtain efficient NO_x reduction. Optimal reagent droplet size is a meaningful issue because excessively large drops of reagent decrease the rate of evaporation at high flue gas temperatures and increase NH_3 slip at lower temperatures. Excessively small drops result instead in rapid reagent evaporation, which causes a deterioration of NO_x reduction.

Reactant droplets (urea solution or ammonia water) heat very slowly at the initial moment of injection because of heat removal through evaporation. Drop temperature increases by an average of 60-80 K and is unrelated to the flue gas temperature [10] - reduction of NO_x does not occur during the initial phase. Later, due to the high pressure of ammonia vapour, H_2O is evaporated straight away and nitrogen compounds (NH_2, NH_3) are released and heated immediately due to radiation. Their temperature becomes equal to the local flue gas temperature, which must be in the range of the SNCR temperature window for NO_x reduction to occur.

Urea could be injected deeper into the furnace than ammonia, because the reaction between urea and NO_x could occur only after complete evaporation of the solution drops. Using urea solution as a reagent prevents from its release in a cool area near the furnace water-walls and thus avoids generating NH_3 slip. Therefore, using urea as the reagent enables its injection into areas of higher temperatures than in the case of using ammonia, which is a significant facilitation in boiler technology.

The advantage of using ammonia water is the rapidity of its evaporation and reaction with nitrogen oxide; therefore, in the case of reagent injection between the closely spaced heating surfaces in the boiler, ammonia water is preferred as a reagent. Other advantage of ammonia is its lesser propensity to cause corrosion of a heating surfaces.

There is also a solution combining both reagents, in which urea is dissolved in ammonia water [11]. Such a system is not a part of presented approximations.

2 The results of approximation

Figs. 5 and 6 show the approximation of the relationship between the SNCR system efficiency and the flue gas temperature for urea and ammonia as reagents. They were evaluated through the regression of values from Figs. 1 and 2.

The calculation of η_{SNCR} must be performed until the trend lines reach the x-axis in the charts. It must be assumed that $\eta_{SNCR} = 0$ for the temperatures not included in the range of 740-1220 °C for urea and 725-1162 °C for ammonia.

As mentioned before, the NO_x reduction efficiency depends on the flue gas composition and various operational parameters. However, the local contents of H_2, CO or SO_3 as well as the reagent droplets' residence time are not measured in industrial practice. Generally, only the temperature distribution at the SNCR system level is measured through acoustic [7] or HVT pyrometers [12]. Therefore the flue gas temperature is the only variable in the proposed correlations.

Due to measuring only one variable the resulting functions cannot be very accurate. Nonetheless they are very useful for estimating the SNCR system efficiency when lacking specific process data.

 DOI: 10.1051/e3sconf/20171402003

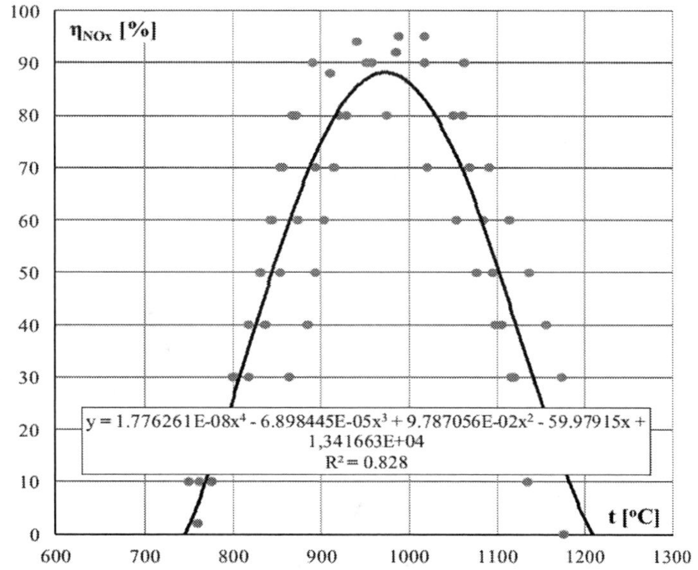

Fig. 5 The approximation of the relationship between the SNCR system efficiency and the flue gas temperature when using urea as a reagent

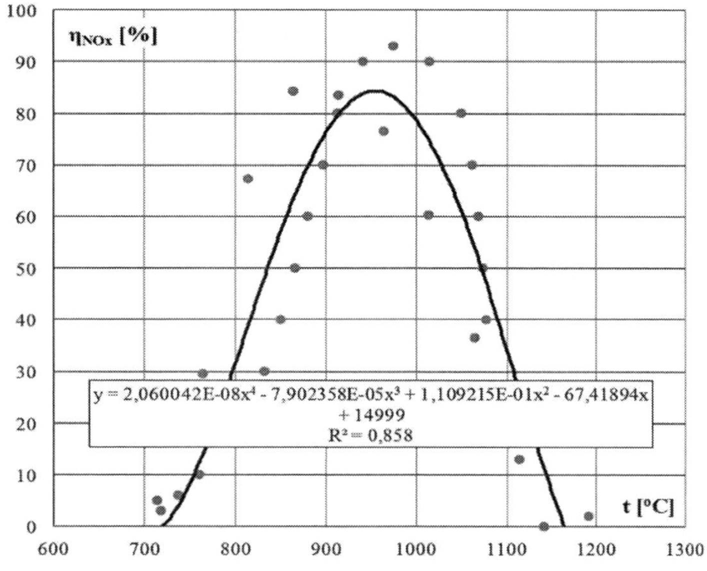

Fig. 6 The approximation of the relationship between the SNCR system efficiency and the flue gas temperature when using ammonia as a reagent

The values of the coefficient of determination R^2 (R^2 = 0.828 and 0.858) mean that more than 80 % of the η_{SNCR} depends on the flue gas temperature - the remaining part consists of the influences of other variables.

The obtained correlations denote that the SNCR temperature window values are higher when urea is used as a reactant than when ammonia is used. In the case of urea the maximal value of the SNCR system efficiency is 88 % at about 970 °C (Fig. 5), whereas for ammonia it is 84 % at about 950 °C (Fig. 6). From these developed correlations, the temperature for maximum efficiency when using urea is slightly lower than that reported in [6], while the corresponding temperature for the NH_3 is the same.

 DOI: 10.1051/e3sconf/20171402003

3 The results of measurements in an existing boiler

The SNCR system efficiency was calculated on the basis of measurements performed in an existing boiler with a maximum continuous rating of 230 t/h. Figs. 7, 8 and 9 show the locations of the SNCR reagent injection points. The system consists of 6 nozzles on the furnace front wall (marked as X, Y, Z, X', Y' and Z'), 3 nozzles on the furnace right wall (10 and 8) and 3 on the furnace left wall (36 and 9).

The points located on the furnace side walls are arranged nearly vertically:

- on the furnace right wall: ca 1010 mm from the tube row separating the furnace and the boiler convection section,
- on the furnace left wall: ca 1700 mm from the mentioned row of tubes.

Nozzles located at those points are oriented at an angle of 30° to the furnace's front wall.

The points located at the furnace front wall form an irregular band with an average height of 1533 mm above the reference level (19.25 m). The maximum difference between the heights of the ports is 770 mm. All nozzles located at the furnace's front wall are arranged horizontally.

Such nozzle locations are expedient in terms of SNCR reagent injection into the empty space at the end of the furnace. Placing them further, into the area of lower flue gas temperatures, was excluded because of the convection surfaces placed there.

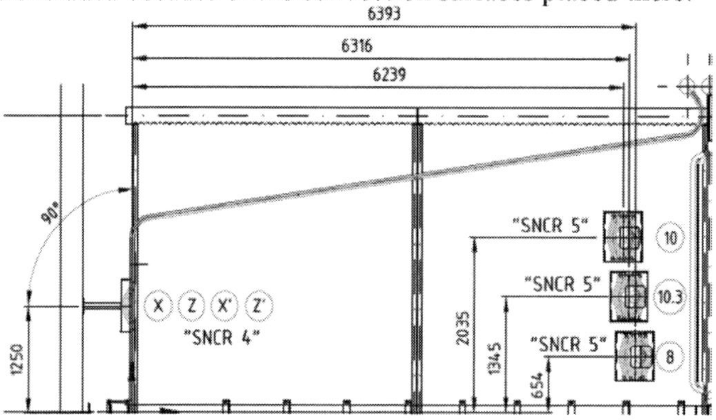

Fig. 7 The SNCR nozzle locations at the furnace's right wall

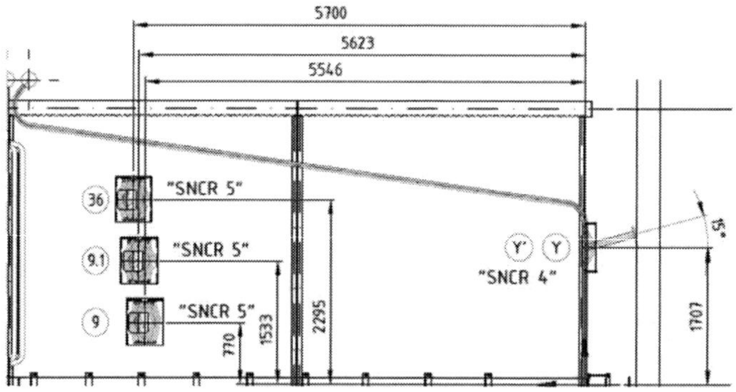

Fig. 8 The SNCR nozzle locations at the furnace's left wall

DOI: 10.1051/e3sconf/20171402003

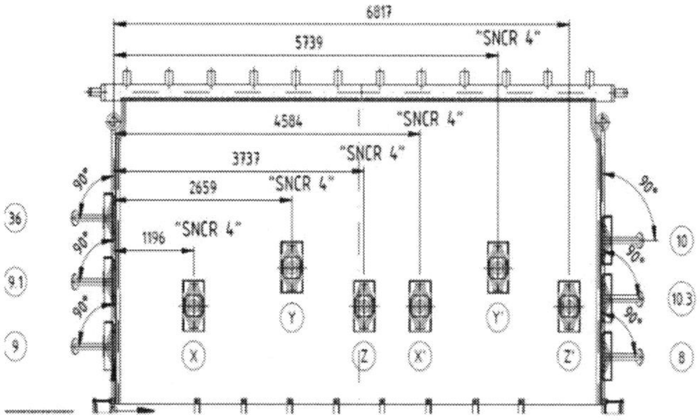

Fig. 9 The SNCR nozzle locations at the furnace's front wall

To evaluate the accuracy of the nozzle locations the measurement of flue gas temperature was performed using a bare thermocouple that was inserted into the boiler at a 1-2 m depth. Table 1 shows the obtained measurement data.

Such a measurement method is characterized by an inaccuracy resulting from the radiation of some part of the heat from the thermocouple junction to the relatively cold furnace water-walls and other heating surfaces. Thus, the measured flue gas temperatures are lower than in reality. The inaccuracy could be determined on the basis of the chart shown in Fig. 10 [13]. The measurement results refer to the case when the measured flue gas temperature T_g [K] is higher than the effective temperature of the surroundings T_∞ [K] (marked as the 'upper layer').

Using the chart shown in Fig. 10 the real values of the flue gas temperature were calculated (Table 1). T_∞ was calculated as the geometric average of the external furnace water-wall temperature as well as the flue gas temperature measured at each point. The results of the calculations confirm that the underestimation of the flue gas temperature could exceed 200 K.

The measurements were performed at a maximum distance of 2 m from the furnace walls. Because the furnace depth and width are about 7.5 m each, evaluation of the flue gas temperatures in the middle of furnace was made on the basis of the approximation shown in Fig. 11. The flue gas temperatures measured at the furnace's front wall are marked as black circles (at 2 m depth) and black triangles (at 1 m depth). The flue gas temperatures measured at the furnace's left and right walls are marked as red circles (at 2 m depth) and red triangles (at 1 m depth).

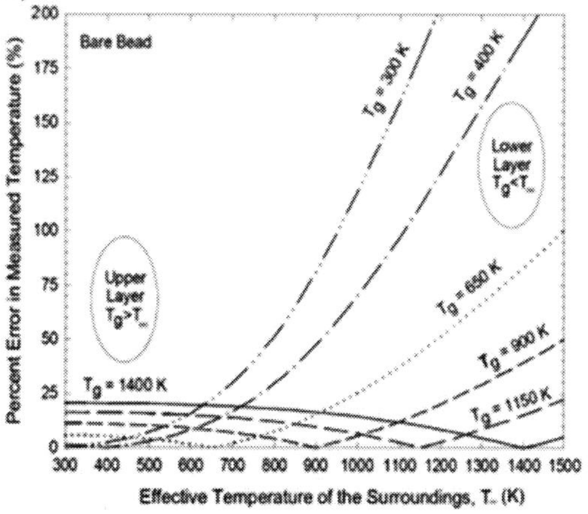

Fig. 10 The accuracy of the flue gas temperature measurement using a bare thermocouple [13]

Table 1 Measured and calculated flue gas temperature values [°C]

Measuring point	Measured values		Real values (calculated)	
	1m	2m	1m	2m
X	967	1045	1109	1203
Y	960	1041	1101	1199
Z	989	1088	1140	1279
X'	975	1101	1125	1307
Y'	953	1060	1094	1247
Z'	969	1030	1112	1186
10	1005	1040	1158	1198
8	1000	1105	1153	1312
36	1025	-	1181	-
9	1015	1027	1170	1183

Missing values were calculated on the basis of the distribution of the measured values and by averaging adjacent data. It was assumed that the distribution of the measured values at a depth of 2 m remains steady in the further part of the duct. Thus, the set of 10 temperature values was created; each represents 10 % of the duct width area. At 1 m from the furnace's front wall the flue gas temperatures have lower values than at the same distance from the furnace's side walls; therefore, it was assumed that the flue gas temperature at 1 m from the furnace's front wall is, on average 1114 °C. As a result, the flue gas temperature distribution, marked as white circles in Fig. 11, is valid for the entire duct length (*x*-coordinate) with the exception of an area of 1 m in width at the furnace's front wall.

That (or similar) simplification is necessary due to the relatively low number of measurement points. In the future, with a more accurate flue gas temperature measurement, it would be possible to improve the verification of the obtained approximations.

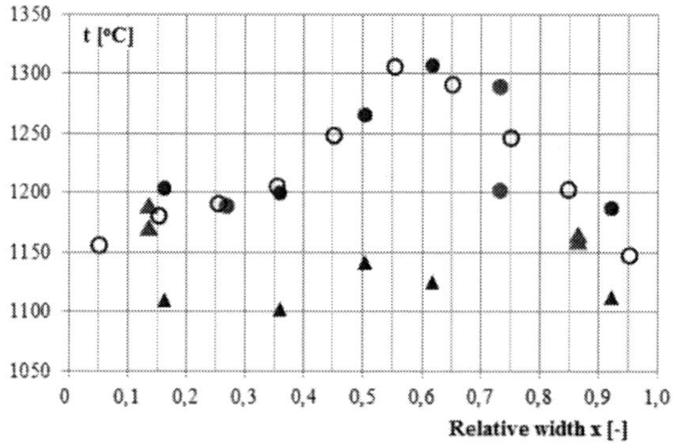

Fig. 11 The measured and the assumed flue gas temperatures

▲ Front 1 m ▲ Right and left side 1 m
● Front 2 m ● Right and left side 2 m
○ Assumed mean temperature of 10% section of flue gas duct

Calculations of the SNCR system's efficiency were performed on the basis of the NO concentration in flue gas measurements. Measurements were executed during system operation and when the system was deactivated via. Siemens ULTRAMAT 23 gas analysers and conducted by the staff of the Division of Boilers and Steam Generators of the Silesian University of Technology and the staff of the Process Analytics and Chromatography Division of the Siemens Company.

The obtained η_{SNCR} is about 8 %. It has to be noticed that this value is relatively low for industrial conditions. The gas analysers used here are characterised by an inaccuracy of about ±2 %; due to this, the inaccuracy of η_{SNCR} determination is about ±3 %.

4 The results of comparison between the measured and calculated data

The results of calculations are shown in Table 2. Calculations were performed on the basis of the approximation from Fig. 5 for the data from Fig. 11.

The average value of η_{SNCR} was calculated as the weighted arithmetic mean with the distances in the furnace to which they correspond to as weights.

Table 2 The results of η_{SNCR} calculation on the basis of the obtained approximation and measured data

Relative width	The flue gas temperature (side walls)	The SNCR system efficiency (side walls)	The flue gas temperature (front wall)	The SNCR system efficiency (front wall)	The SNCR system efficiency (average)
x [-]	**t_{fg_sw} [°C]**	**η_{SNCR_sw} [%]**	**t_{fg_fw} [°C]**	**η_{SNCR_fw} [%]**	**η_{SNCR} [%]**
0,05	1155	22	1114	44	25
0,15	1175	12	1114	44	17
0,25	1190	6	1114	44	11
0,35	1202	2	1114	44	8
0,45	1248	0	1114	44	6
0,55	1305	0	1114	44	6
0,65	1290	0	1114	44	6
0,75	1245	0	1114	44	6
0,85	1200	3	1114	44	8
0,95	1145	27	1114	44	29
				Average	**12**

As mentioned before, the estimated value of η_{SNCR} in an existing boiler is about 8 %.

To match the values of η_{SNCR} from Figs. 1 and 2 to the measured values in existing industrial facilities implementation of a correction coefficient, $\Delta\eta_{SNCR}$ is needed:

$$\eta_{SNCR} = \eta_{SNCR(calculated)} \quad \Delta\eta_{SNCR} \ [\%] \tag{2}$$

For the case described above, the $\Delta\eta_{SNCR}$ calculated from eq. (2) is 4 %.

Taking into account the inaccuracy of NO measurement (±3 %), the real value of $\Delta\eta_{SNCR}$ would be 1 or 7 %.

The ΔηSNCR cannot take the form of a multiplier, because in the case of low SNCR system efficiency values, the inaccuracy of NO measurements would give fluctuations of ΔηSNCR at the level of several tens of percent.

5 Conclusions

The approximations obtained here could be used by a boiler designer or operator for preliminary determination of the current SNCR system efficiency; however, in most cases, implementation of a correction coefficient is necessary to reduce the inaccuracy of approximations with reference to measured values. The inaccuracy of approximations is a standard issue that results from its nature. Nonetheless, such functions, representing the relationship between the SNCR system efficiency and the flue gas temperature, are very useful when one lacks specific operation data.

Future work will focus on further verification of the evaluated approximations as well as on evaluation of the correction coefficient.

The investigations presented in this paper were financed within the research program BKS/103/RIE5/2015.

References

1. B. von der Heide, *Advanced SNCR Technology for Power Plants* (POWER-GEN International, Las Vegas, 2011)
2. Z. Teuber, T. Motyczyńska, *SNCR - niekatalityczna metoda redukcji NO_x dla dużych bloków energetycznych [SNCR – the non catalytic reduction of NO_x emissions for big power plants]* (11th International Conference on Boiler Technology, Szczyrk, 2010) [in Polish]
3. W. Schüttenhelm, P. Reynolds, *SNCR for Low NO_x Emissions - Case Study of a Swedish Waste-to-Energy Plant* (on-line)
4. W. H. Sun, *Selective Non Non-Catalytic Reduction: Overview* (WPCA/Duke NO_x Seminar, North Carolina, 2005)
5. D. Wojichowski, *SNCR System Design, Installation, and Operating Experience* (DOE NETL, Pittsburgh, 2002)
6. J. K. Sorrels, D. D. Randall, C. Richardson Fry, K. S. Schaffner, *Selective Non-catalytic Reduction* (on-line)
7. W. Schüttenhelm, K. H. Huber, Z. Teuber, VGB PowerTech **12** (2013)
8. J. Kuropka, M. A. Gostomczyk, Ochrona środowiska 4 **63** (1996) [in Polish]
9. A. Aleksik, *Redukcja tlenku azotu w przemysłowym spalaniu odpadów przy zastosowaniu wtrysku mocznika [Reduction of nitrogen oxide using urea injection in an industrial waste incineration]* (PhD thesis, Wrocław University of Science and Technology, Wrocław, 1997) [in Polish]
10. J. Krüger, W. Gotthardt, K. Rieger, T. Franke, S. Krüger, VGB PowerTech **4** (2012)
11. B. von der Heide, *Möglichkeiten und Grenzen der SNCR [Possibilities and limitations connected with SNCR]* (Conference on the Waste Management and Energy, Berlin, 2012) [in German]
12. S. Kalisz, Sz. Ciukaj, K. Mroczek, T. Kress, R. Wejkowski, M. Pronobis, H. Kubiczek, *Impact of halloysite fuel additive on 650 t/h PC boiler co-firing biomass. Part I - deposition characteristics* (12th International Conference on Boiler Technology, Szczyrk, 2014)
13. L. G. Blevins, *Behavior of Bare and Aspirated Thermocouples in Compartment Fires* (33rd National Heat Transfer Conference, Albuquerque, 1999)

Artificial neural networks application for solid fuel slagging intensity predictions

Sławomir Kakietek[1,*], *Jerzy* Szymczak[1]

[1]Instytut Energetyki Zakład Procesów Cieplnych, Augustówka 36 St.,02-981 Warsaw, Poland

Abstract. Slagging issues present in pulverized steam boilers very often lead to heat transfer problems, corrosion and not planned outages of boilers which increase the cost of energy production and decrease the efficiency of energy production. Slagging especially occurs in regions with reductive atmospheres which nowadays are very common due to very strict limitations in NO_x emissions. Moreover alternative fuels like biomass which are also used in combustion systems from two decades in order to decrease CO_2 emissions also usually increase the risk of slagging. Thus the prediction of slagging properties of fuels is not the minor issue which can be neglected before purchasing or mixing of fuels. This however is rather difficult to estimate and even commonly known standard laboratory methods like fusion temperature determination or special indexers calculated on the basis of proximate and ultimate analyses, very often have no reasonable correlation to real boiler fuel behaviour. In this paper the method of determination of slagging properties of solid fuels based on laboratory investigation and artificial neural networks were presented. A fuel data base with over 40 fuels was created. Neural networks simulations were carried out in order to predict the beginning temperature and intensity of slagging. Reasonable results were obtained for some of tested neural networks, especially for hybrid feedforward networks with PCA technique. Consequently neural network model will be used in Common Intelligent Boiler Operation Platform (CIBOP) being elaborated within CERUBIS research project for two BP-1150 and BB-1150 steam boilers. The model among others enables proper fuel selection in order to minimize slagging risk.

1 Introduction

Slagging and fouling issues are a headache of boiler operators, especially from last two decades when alternative fuels became more and more important and NO_x emissions limitations left an imprint on combustion low excess air strategy. Fouling phenomena is more or less solved by ash blowers (steam, acoustic) located in the second part of the boiler which usually removes deposits from tubes. However in case of slagging in the combustion chamber and first plate heaters removing the deposit is more complicated and sometimes

* Corresponding author: slawomir.kakietek@ien.com.pl

almost impossible. Thus avoiding the slagging is rather the proper direction to deal with this phenomena.
The mechanism of slagging is very complex. Several main factors influence slagging like: amount and chemical composition of ash, physical properties of ash particles, distribution of mineral matter in coal particles and the mechanism of ash forming, temperature and excess air level in combustion chamber and aerodynamic as well. Due to such complex phenomena a number of methods of slagging intensity predictions were created as: determination of ash fusibility temperatures, slagging indexers by Rundygin [1] determined on the basis of ash analysis or Żelkowski [2] determined by laboratory investigations in small scale, pilot scale experiments [3] or more detailed experiments based on SEM or QEMSCNAN [4,5] or even real scale measurements. However whilst ash fusion temperatures for the same fuel samples can differ up to 300 C deg. depending on which laboratory they were investigated [4], the indexers are very often based on specific group of fuels and additionally they are determined on the basis of ash derived in laboratory conditions not from the pulverized flame, laboratory methods do not reflect all important aspects of pulverized flame in the boiler, even real scale measurements are limited due to e.g. number of fuels or conditions which can be investigated in real conditions. More detailed, advanced experiments (e.g. SEM) seems to be more reliable than others and rather on that basis advanced numerical CFD models of slagging should be developed [6]. Nevertheless they are more expensive and time consuming then others. These factors undermine their usability in real applications unfortunately.
Even laboratory scale experiments do not reflect all conditions in the boiler, relatively good results can be obtained in case of small scale comparative laboratory measurements where however pulverized flame is investigated. In such methods intensity of slagging is determined first for base fuel (currently utilized in the boiler) and secondly for other fuels or their blends which are supposed to be used in the boiler. This approach however has also limiting factors like: time, significant amount of fuel samples and costs. The solution of these issues may lay in smart connection of comparative laboratory investigations and artificial neural networks (ANN).
ANN were successfully utilized in slagging related issues [7-12]. These so-called "black-box model", in spite of lack of model transparency and elegance, were able to extract information from highly complex, non-linear and sometimes ill-defined issues, provided that a representative set of data is available. Most of presented applications of ANN in slagging phenomena were based on typical multilayer feedforward neural network (MFNN) and on available, selected data from either laboratory set of experiments or particular boiler. The issue arises where limited training data for ANN is available. Moreover these available data "describes" several areas of slagging phenomena like slagging of hard coals, lignite coals, their mixtures and raw or torrefied biomass or refused-derived fuels (RDF). In such spectrum of fuels with limited training data typically used MFNN may be useless.
In this paper, the authors explored the potential use of principal component analysis (PCA) technique connected with MFNN and statistical analysis to predict the slagging intensity defined by two indexers obtained from laboratory scale experiments based on limited available data. The inputs to ANN were defined as technical analysis of fuel (moisture, ash and volatiles content, and calorific value) and ash mineral analysis (eleven compounds). Levenberg-Marquardt training algorithm was used in order to optimize the training process. Finally PCA hybrid neural network was tested against not seen data and sensitivity analysis was performed to determine the most critical factors for network performance.

 DOI: 10.1051/e3sconf/20171402004

2 Experimental data

Several series of experiments were conducted in small scale 20 kW upstream drop tube reactor presented in Fig. 1. The insulated one meter length cylindrical ceramic combustion chamber **(1)** has inner diameter 100mm and 10mm thickness of walls. Flue gases leave combustion chamber **(1)** through exhaust fan **(2)**. In the bottom part of the chamber there is a low flame gas burner **(3)**. Through the axis of the burner solid fuel dust with transport air **(4)** (primary air) is supplied into the combustion chamber. The thickness of heat insulation of the chamber **(8)** made by very thin ceramic fibres is about 150 mm. Gas fuel **(5)** (propane) is supplied into the burner from the cylinder, then is mixed with secondary air **(6)** in the mixing chamber of the burner **(3)** and afterwards through stabilizing cartridge gets the chamber **(1)** where is ignited. The gas burner **(3)** generates flue gases at the specific temperature and gas composition to make the ignition of solid fuel possible. By the change of gas **(5)** or secondary air **(6)** the temperature of flue gases, excess of air, quantity of supplied heat into the chamber **(1)** and temperature at the outlet of chamber **(9)** may be controled and changed during measurements. The solid fuel is supplied through coal dust feeder **(10)** which enables changing the fuel feed rate and quantity of primary air **(7)**. About one meter from the front of the burner **(3)** there is a deposit 10 mm diameter pipe **(11)** hanged on beams of the extensometer balance **(12)**. Bending the beams as a result of slag deposit causes the resultant voltage which is the function of mass increase of deposit pipe. The temperature of flue gases is measured close to deposit pipe by thermocouple **(13)**.

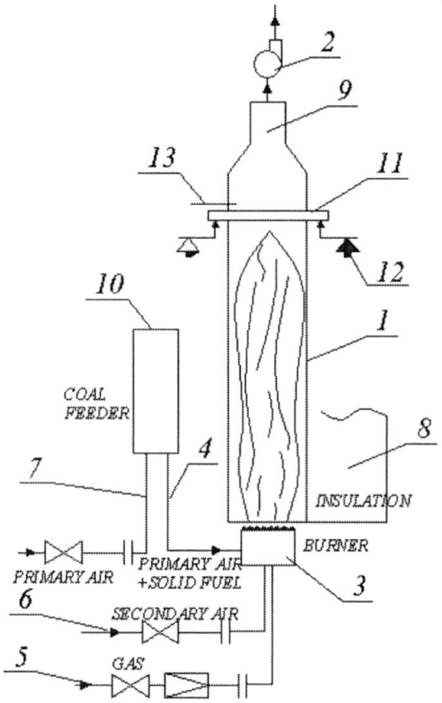

Fig. 1. Scheme of the 20 kW slagging test fasility

Investigations consist of burning pulverized samples of fuel with quantity about 0.41-0.55 g/s (1.5 - 2 kg/h) in heated in advance by gaseous fuel combustion, chamber in suitable, variable during duration of experiments conditions (excess air, temperature). In this way rate of slag deposition for various temperatures and the beginning temperature of slagging is obtained. The gaseous fuel creates the suitable conditions for experiments and is continuously used during the whole experiment. The effect of gaseous flame itself for solid fuel combustion is negligible since solid fuel is injected with higher speed into already mixed and hot flue gases similarly as in real scale boilers fuel is injected into the combustion chamber. Higher amount of water from propane combustion is compensated by more predried solid fuel in laboratory conditions as compared to real ones. T_{beg} and k_z, which are respectively, beginning temperature of slagging and index of slagging intensity are determined by below formulas:

$$B_z = \frac{M_z \cdot F_k}{M_p \cdot F_r} \tag{1}$$

$$B_z = e^{k_z \cdot \left(\frac{T_{sp}}{Tbeg} - 1 \right)} - 1 \tag{2}$$

where:
B_z – relative slagging intensity,
M_z – collected mass of slag on deposition pipe [g],
M_p – mass of ash flown through the combustion chamber of the slagging test facility [g],
F_k – area of the cross section of the combustion chamber of the slagging test facility [m^2],
F_r – active area of pipe [m^2],
T_{beg} – beginning temperature of slagging [K],
k_z – index of slagging intensity
T_{sp} – temperature of combustion [K].

More than 40 different solid fuels and their blends were investigated as hard or lignite coals, raw and torrefied biomasses, saw dusts (SDI - SDIII), sewage sludge (SS) and RDF. Tab. 1 presents technical and ash mineral analysis of investigated fuels and their blends, T_{begcal}, Slag-Index and Foul-Index calculated on commonly available formulas [13] and experimentally obtained k_z and T_{beg}. Blends were prepared on mass (SDI, SDII, SDIII, SS, RDF) or energy basis (P, B, S, TP, TB, TS). It might been seen from the very beginning that calculated slagging indexers usually do not correspond with experimentally obtained k_z and T_{beg}. This is especially visible in case of T_{begcal} which generally does not change markedly over the all fuels (including biomass) with several exceptions keeping the value around 1250 K. The opposite behaviour may be seen in case of T_{beg} which changes from 459 to 1301 K. For instance interesting behaviour might be seen in case of hard coal HC1 and raw and torrefied beech (B, TB). HC1 characterized by unusual ash mineral analysis has high k_z but simultaneously high T_{beg}. The opposite situation occurs in case of beech, raw and torrefied. After adding 20% of TB by energy to this coal T_{beg} of this blend decrease markedly. After adding 40% one will expect further decrease of T_{beg} but this is not the point. However k_z increases for this blend even for pure TB k_z was not too high and the amount of hard coal (the fuel with higher k_z) decreases when adding more TB.

Table 1. Technical and ash analysis of fuels. Calculated and laboratory obtained slagging indexers

Description	Abb.	State	Abb.	HC1	HC2	HC3	LC1	LC2	P	B	S	TP	TB	TS	HC1+ 20% S	HC1+ 20% P	HC1+ 20% B	HC1+ 20% TS
			Unit	Hard coal Wieczorek mine	Hard coal Opole PP	Hard coal Janina mine	Lignite coal Turów mine	Lignite coal Bełchatów mine	Biomass - Pine	Biomass - Beech	Biomass - Straw	Torr. Pine	Torr. Beech	Torr. Straw	Blend on energy basis	Blend on energy basis	Blend on energy basis	Blend on energy basis
Moisture	M_{ar}	as rec.	%	2.70	1.90	4.20	17.40	10.00	7.40	7.20	7.70	6.20	3.90	4.90	4.30	4.07	4.07	3.37
Ash	A_d	dry	%	3.40	28.30	11.70	7.40	17.84	0.91	1.63	5.30	0.71	1.34	5.04	4.01	2.67	2.86	3.90
Volatiles	V_d	dry	%	33.65	24.79	33.70	53.17	36.90	83.60	83.10	76.90	77.00	79.30	76.10	47.50	48.26	48.74	46.64
Calorific Value	$q_{V,net}$	as rec.	MJ/kg	30.97	22.64	24.58	20.78	14.06	17.64	16.65	15.40	19.40	18.51	17.05	25.99	27.07	26.60	26.71
Silica	SiO_2	dry	%	18.00	51.60	48.70	25.10	49.50	30.90	10.10	43.60	13.70	6.40	40.50	26.20	21.77	15.59	24.88
Iron	Fe_2O_3	dry	%	16.70	6.38	11.00	7.01	3.14	3.44	1.51	1.11	11.80	9.10	1.94	11.71	12.82	12.07	12.18
Aluminium	Al_2O_3	dry	%	11.80	24.50	26.30	20.60	19.40	2.50	6.60	1.30	2.40	0.90	1.20	8.44	9.08	10.21	8.56
Manganese	Mn_3O_4	dry	%	0.29	0.08	0.04	0.07	0.02	0.94	0.59	0.09	1.61	0.86	0.09	0.23	0.48	0.38	0.23
Titanium	TiO_2	dry	%	0.40	1.01	0.90	2.80	1.05	0.21	0.37	0.16	0.17	0.07	0.10	0.32	0.34	0.39	0.31
Calcium	CaO	dry	%	18.30	5.74	3.23	7.57	16.20	21.40	35.10	14.50	26.10	31.90	13.60	17.08	19.21	23.43	16.86
Magnesium	MgO	dry	%	11.50	2.65	1.76	11.10	0.99	4.56	3.32	2.15	5.58	4.67	2.40	8.51	9.47	9.00	8.72
Sulphur	SO_3	dry	%	17.20	4.11	3.50	14.40	9.00	1.46	0.99	2.57	2.48	1.98	2.89	12.51	12.59	12.25	12.82
Phosphorus	P_2O	dry	%	0.20	0.44	0.20	0.22	0.11	3.20	1.21	2.16	2.49	1.77	2.34	0.83	1.08	0.51	0.85
Sodium	Na_2O	dry	%	3.58	0.69	0.95	9.05	0.22	0.70	0.48	0.61	0.79	0.60	1.46	2.63	2.74	2.63	2.93
Potassium	K_2O	dry	%	0.75	2.47	2.30	1.00	0.04	12.70	10.10	18.10	10.40	12.20	22.10	6.31	4.25	3.60	7.28
Calculated beg.temp.of slagging	Tbegcal	-	K	1256	1273	1294	1224	1251	1258	1361	1235	1346	1591	1245	1247	1257	1277	1252
Calculated slagging index	Slag-Index	-	-	0.56	0.17	0.35	0.48	0.27	0.03	0.06	0.07	0.07	0.24	0.09	0.33	0.37	0.46	0.36
Calculated fouling index	Foul-Index	-	-	7.29	0.73	0.82	7.40	0.08	17.06	31.31	15.14	37.60	101.55	23.39	11.82	10.85	12.08	14.52
Laboratory obtained slagging index	Kz	-	-	**18.00**	**11.39**	**4.00**	**2.23**	**6.43**	**6.68**	**2.32**	**12.33**	-	**4.05**	**5.04**	**2.72**	**2.95**	**1.77**	**7.58**
Laboratory obtained beg.temp. of slagging	Tbeg	-	K	**1301**	**1171**	**752**	**652**	**881**	**608**	**464**	**973**	-	**731**	**658**	**543**	**743**	**459**	**1003**

Description	Abb.	State	Abb.	HC1+ 40% TS	HC1+ 20% TP	HC1+ 40% TP	HC1+ 20% TB	HC1+40 %TB	HC3+ 20% S	LC1+ 20% S	LC1+ 20% P	LC1+ 20% TP	LC1+ 20% TB	LC2+ 20% TS	LC2+ 20% TB	LC3	LC4	LC5	LC6
			Unit	Blend on energy basis	Blend on energy basis	Blend on energy basis	Blend on energy basis	Blend on energy basis	Blend on energy basis	Blend on energy basis	Blend on energy basis	Blend on energy basis	Blend on energy basis	Blend on energy basis	Blend on energy basis	Lignite coal Jóźwin mine	Lignite coal Drzewce mine	Lignite coal Kazimierz mine	Lignite coal Lubstów mine
Moisture	M_{ar}	as rec.	%	3.89	3.65	4.44	3.03	3.31	5.13	15.12	15.27	15.21	14.70	9.17	9.12	12.00	4.10	11.90	4.30
Ash	A_d	dry	%	4.29	2.67	2.06	2.83	2.36	9.99	6.91	6.02	6.09	6.19	15.75	15.46	10.60	25.90	12.50	8.40
Volatiles	V_d	dry	%	56.59	45.40	55.24	46.26	56.67	45.23	58.75	59.64	57.82	58.40	43.29	43.02	47.00	40.60	45.00	47.30
Calorific Value	$q_{V,net}$	as rec.	MJ/kg	23.45	27.83	25.21	27.53	24.68	22.13	19.51	20.11	20.51	20.33	14.55	14.70	20.76	19.23	20.67	23.73
Silica	SiO_2	dry	%	30.16	16.83	15.86	14.79	12.15	47.34	29.45	26.33	22.87	21.36	48.03	43.28	9.64	70.30	24.10	15.20
Iron	Fe_2O_3	dry	%	8.73	15.37	14.26	14.60	12.86	8.36	5.62	6.25	7.95	7.43	2.94	4.00	12.00	5.26	6.31	8.29
Aluminium	Al_2O_3	dry	%	6.07	9.25	7.12	8.79	6.30	19.63	16.06	16.75	17.04	16.66	16.43	16.73	1.69	1.09	2.28	3.72
Manganese	Mn_3O_4	dry	%	0.18	0.65	0.95	0.45	0.58	0.05	0.07	0.26	0.37	0.23	0.03	0.14	0.52	0.11	0.43	0.36
Titanium	TiO_2	dry	%	0.24	0.34	0.29	0.31	0.23	0.70	2.18	2.25	2.29	2.25	0.90	0.91	0.21	0.28	0.22	0.42
Calcium	CaO	dry	%	15.76	20.41	22.18	22.06	25.16	6.24	9.20	10.51	11.19	12.44	15.78	18.47	27.50	11.30	27.10	43.90
Magnesium	MgO	dry	%	6.58	9.89	8.55	9.61	8.05	1.86	8.99	9.71	10.02	9.81	1.22	1.52	5.20	1.13	5.23	4.02
Sulphur	SO_3	dry	%	9.47	13.21	9.87	12.99	9.52	3.25	11.61	11.65	12.07	11.92	8.00	7.99	41.50	9.55	32.60	21.70
Phosphorus	P_2O	dry	%	1.36	0.82	1.34	0.63	0.99	0.72	0.68	0.85	0.66	0.53	0.47	0.35	0.04	0.01	0.04	0.06
Sodium	Na_2O	dry	%	2.43	2.82	2.19	2.76	2.08	0.86	7.06	7.27	7.43	7.36	0.42	0.27	0.32	0.07	0.14	0.14
Potassium	K_2O	dry	%	12.29	3.37	5.56	3.91	6.53	6.52	5.02	3.49	2.84	3.24	3.64	1.79	0.11	0.03	0.08	0.09
Calculated beg.temp.of slagging	Tbegcal	-	K	1249	1271	1288	1285	1323	1254	1226	1229	1233	1237	1244	1241	1361	1265	1262	1338
Calculated slagging index	Slag-Index	-	-	0.25	0.48	0.40	0.55	0.52	0.36	0.39	0.42	0.49	0.53	0.29	0.33	12.94	0.25	3.58	2.04
Calculated fouling index	Foul-Index	-	-	18.48	12.15	17.56	14.78	25.18	2.60	9.10	8.84	9.60	10.60	1.49	0.89	1.68	0.02	0.32	0.67
Laboratory obtained slagging index	Kz	-	-	**4.83**	**16.15**	**9.61**	**4.04**	**14.02**	**6.96**	**9.17**	**11.59**	**9.03**	**4.70**	**12.88**	**9.20**	**7.12**	**7.84**	**7.97**	**8.37**
Laboratory obtained beg.temp. of slagging	Tbeg	-	K	**710**	**1098**	**1017**	**836**	**1136**	**1038**	**1083**	**1120**	**1057**	**896**	**1044**	**1075**	**896**	**1063**	**1006**	**1056**

Abb.	State	Abb.	HC4	HC4+ 5 % SS	HC4+ 10 % SS	HC4+ 15 % SS	HC4+ 20 % SS	HC4+ 5 % SDI	HC4+ 10 %SDI	HC4+ 15 %SDI	HC4 + 20 %SDI	HC4 + 10 %RDF	HC4 + 20 %RDF	HC5	HC5+ 14%SDII	HC6	HC6+ 5% SDIII
		Unit	Hard coal from RPA	Blend on mass basis	Blend on mass basis	Blend on mass basis	Blend on mass basis	Blend on mass basis	Blend on mass basis	Blend on mass basis	Blend on mass basis	Blend on mass basis	Blend on mass basis	Hard coal Dolna Odra PP	Blend on mass basis	Harc coal Stalowa Wola PP	Blend on mass basis
M_{ar}	as rec.	%	4.40	4.83	5.40	5.43	5.60	4.63	4.70	5.30	5.90	4.00	3.90	1.70	2.50	2.30	2.10
A_d	dry	%	10.40	9.36	8.10	7.71	7.10	12.43	14.70	16.15	17.60	11.10	11.10	22.60	21.60	25.20	27.30
V_d	dry	%	28.90	39.95	41.50	43.05	44.60	31.80	33.10	34.40	35.70	34.40	38.50	24.90	26.10	27.50	26.90
$q_{v,net}$	as rec.	MJ/kg	26.64	24.24	23.84	23.43	23.02	24.56	23.84	23.12	22.40	26.44	26.10	24.70	24.55	22.95	22.27
SiO_2	dry	%	38.10	34.20	30.30	29.50	28.70	37.85	37.60	37.40	37.20	35.50	34.30	50.00	50.60	53.14	57.50
Fe_2O_3	dry	%	6.10	8.30	10.50	11.05	11.60	5.85	5.60	5.45	5.30	5.65	5.44	6.44	6.35	9.51	6.00
Al_2O_3	dry	%	28.70	24.95	21.20	19.95	18.70	28.75	28.80	27.80	26.80	24.00	23.70	27.40	26.50	18.90	24.03
Mn_3O_4	dry	%	0.13	0.14	0.15	0.16	0.17	0.15	0.17	0.19	0.21	0.12	0.12	0.08	0.09	0.17	0.06
TiO_2	dry	%	1.46	1.27	1.07	1.02	0.96	1.43	1.40	1.38	1.35	1.73	1.98	1.08	1.08	0.86	1.29
CaO	dry	%	12.10	12.40	12.70	12.65	12.60	11.90	11.70	12.15	12.60	14.40	16.40	3.27	3.71	5.93	2.41
MgO	dry	%	3.42	3.29	3.16	3.13	3.09	3.64	3.85	3.68	3.51	2.84	2.75	2.72	2.84	4.02	1.97
SO_3	dry	%	5.25	5.53	5.81	5.78	5.75	5.46	5.67	6.22	6.76	10.50	11.20	3.30	3.42	3.32	1.73
P_2O	dry	%	2.04	7.22	12.40	14.15	15.90	1.95	1.86	1.92	1.98	1.84	1.79	0.65	0.66	0.40	0.60
Na_2O	dry	%	0.54	0.50	0.46	0.52	0.57	0.58	0.61	0.68	0.74	0.69	0.63	1.03	1.00	0.80	0.67
K_2O	dry	%	0.65	0.78	0.91	1.02	1.13	0.90	1.15	1.26	1.36	0.73	0.47	3.12	3.14	2.18	2.54
Tbegcal	-	K	1252	1247	1243	1242	1241	1251	1250	1249	1247	1245	1242	1282	1278	1264	1308
Slag-Index	-	-	0.55	0.32	0.37	0.36	0.35	0.32	0.34	0.36	0.38	0.35	0.35	0.16	0.16	0.22	0.11
Foul-Index	-	-	0.40	0.54	0.72	0.86	1.02	0.50	0.59	0.67	0.76	0.56	0.47	0.88	0.90	0.92	0.53
Kz	-	-	**11.89**	**9.65**	**7.61**	**9.62**	**9.64**	**4.97**	**5.34**	**5.38**	**5.56**	**6.77**	**7.81**	**6.97**	**6.35**	**9.85**	**13.86**
Tbeg	-	K	**1163**	**1096**	**1013**	**1123**	**1191**	**982**	**987**	**992**	**991**	**1034**	**1028**	**1060**	**1020**	**1100**	**1220**

This proves that slagging properties of fuel blends are sometimes not additive as one could expect on the basis of T_{begcal}, slag and fouling indexers determined on the basis of ash mineral analysis. Nevertheless this is not the topic of this paper to be discussed further.

3 ANN simulations

Principal component analysis (PCA) is a well-known, widely used statistical technique. In pattern recognition, PCA is used in various forms for optimal feature extraction and data compression. In general PCA finds an orthogonal set of directions in the input space and provides a way of finding the projections into these directions in an ordered fashion. The first principal component is the one that has the largest projection (the projection is the shadow of the data cluster in each direction). The orthogonal directions are called the eigenvectors of the correlation matrix of the input vector, and the projections the corresponding eigenvalues. Since PCA orders the projections, the dimensionality can be reduced by truncating the projections to a given order. PCA is normally done by analytically solving an eigenvalue problem of the input correlation function. However, Sanger and Oja [14, 15] demonstrated that PCA can be accomplished by a single layer linear neural network (Fig. 2) trained with a modified Hebbian learning rule. Oja's rule was find by one of the author to be more attractive than Sangers for this application [16].

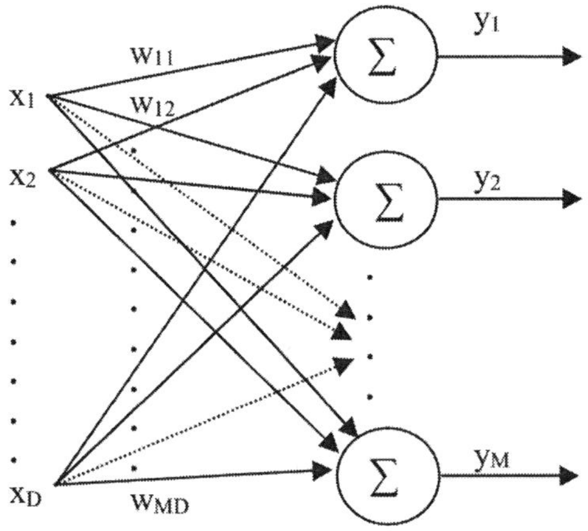

Fig. 2. A PCA network to project the data from D to M dimensions (x – inputs, w –weights, y - outputs)

What is interesting in the PCA network is computing the eigenvectors of the correlation function of the input without ever computing the correlation function. Sanger showed that this learning procedure converges to the correct solution, i.e. the weights of the PCA network approach the first m principal components of the input data matrix. The outputs are therefore related to the eigenvalues and can be used as input to another neural network. Thus PCA can be used for data reduction in conjunction with MFNN. The importance of PCA analysis is that the number of inputs for the MFNN can be significantly reduced, which positively impacts the number of required training patterns and the training time as well.

Authors proposed PCA-MFNN hybrid network with one hidden layer with statistical analysis of outputs which classifies which results are acceptable on the basis of averages and standard deviations of trained and tested twenty neural networks. Main investigated elements in ANN were : principal components number (PC) ranging from 3 to 8, hidden neurons (HN) number ranging from 6 to 10 and multiply of training patterns by random duplication of existing training data. Levenverg-Marquard training algorithm was used to train supervisied part of network – MFNN, whilst for unsupervised – PCA, Oja's rule. In MFNN tangh activation function was used both in hidden and output layer. 15 inputs were defiened : four parameters from technical analysis of fuel and eleven from ash mineral chemical analysis as presented in Table 1. 38 sets of data were choosen for training while 8 for testing.

4 Results and discussion

Figures 3-6 present the results of MFNN and PCA-MFNN simulations. Graphs presents training results, average testing results and the best test results. In case of MFNN no satisfactory results were obtained. In that case also training process was unstable, very often not reaching intended threshold. In case of hybrid PCA-MFNN for 3-4 PC elements similar behaviour was observed during training however testing results were more promising. Best results were obtained for PC ranging from 5-8. Further increase of PC does not decrease the test error and starts to increase training time. It might be seen that for some testing sets no satisfactory results were obtained. This might be caused by e.g. not enough training data sets or not proper training data sets selection for training. Presented test results are averages from 20 ANN. Results with too high standard deviations are signed as too uncertain. This is

 DOI: 10.1051/e3sconf/20171402004

especially important in final real scale application in Common Intelligent Boiler Operation Platform (CIBOP).

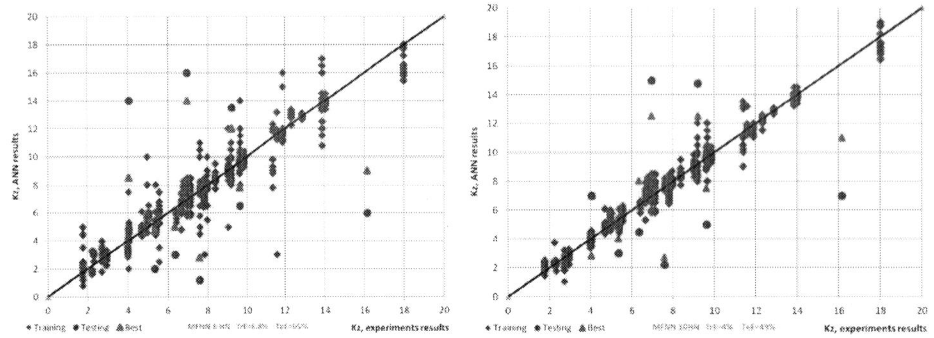

Fig. 3. MFNN simulation results for k_z predicting for 6 and 10 hidden neurons

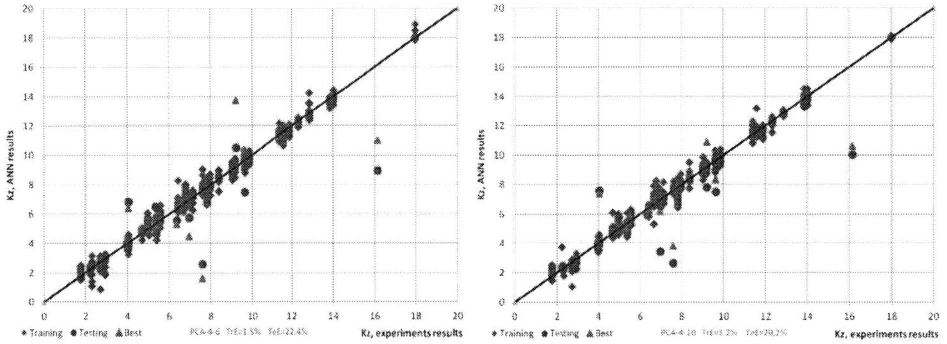

Fig. 4. PCA simulation results for k_z predicting for 4 PC elements and 6 or 10 hidden neurons

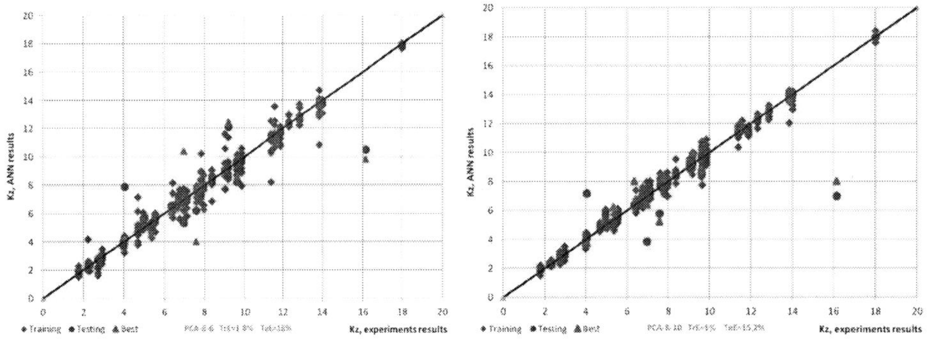

Fig. 5. PCA simulation results for k_z predicting for 8 PC elements and 6 or 10 hidden neurons

DOI: 10.1051/e3sconf/20171402004

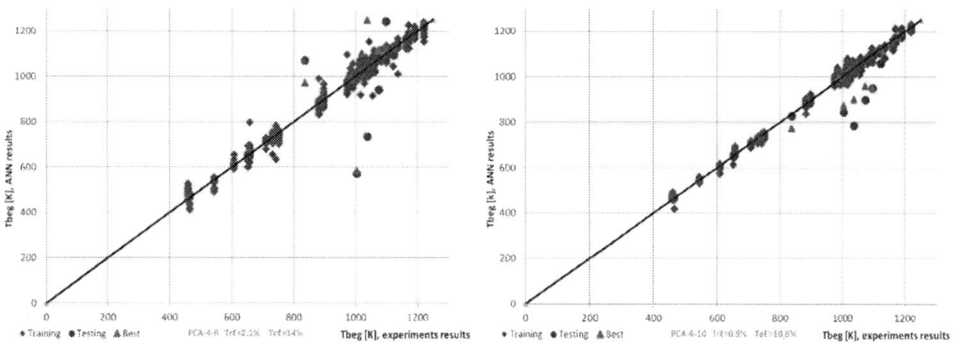

Fig. 6. PCA simulation results for T_{beg} predicting for 8 PC elements and 6 or 10 hidden neurons

Figure 7 presents the linear and average PCA-MFNN sensisivity analysis [17] of more important ash components (inputs to ANN) with k_z and T_{beg} (outputs from ANN). In general there is no linear corelations of searched parameters with ash mineral analysis. Maximum values of linear correlation coefficient can be found for silica, pottasium, aluminium and titanium in relation to T_{beg} which seems to be ''more realistic'' parameter compared to k_z intensity. In case of PCA-MFNN sensisivity for T_{beg} and k_z covers each other and reach maximum for phosphorus and potassium. However other parameters of ash are also important in PCA-MFNN performance. Figure 8 highligts the sensivity behaviour for k_z and T_{beg} for different number of PC elements (for 8 hidden neurons). In case of T_{beg} increasing the number of PC usually increases or decreases the importance of given ash mineral parameter. In case of k_z this behaviour can not be classified in such manner rather.

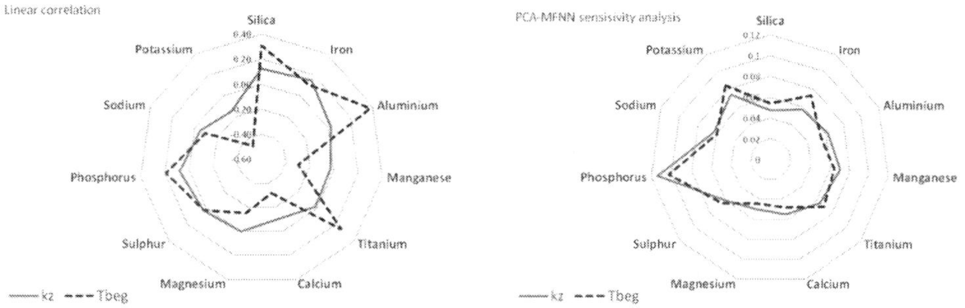

Fig. 7. Linear correlation and PCA-MFNN sensivity analysis between inputs and outputs of ANN

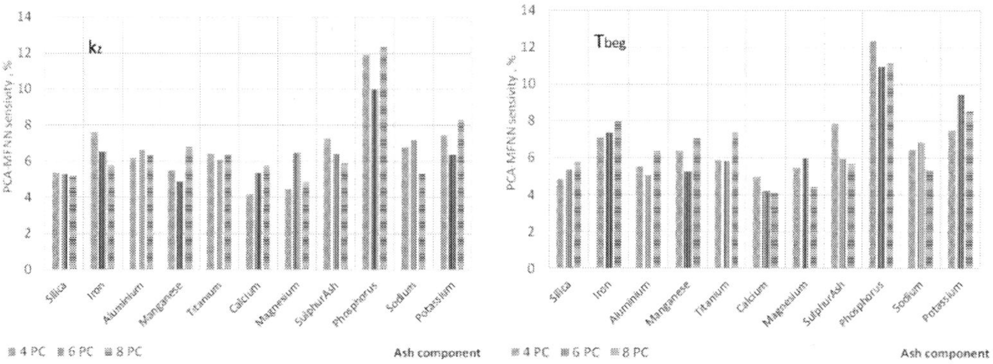

Fig. 8. Sensisivity (importance) for different number of PC elements for k_z and T_{beg}

5 Conclusions

Reasonable predictions of index of slagging intensity (k_z) and beginning temperature of slagging (T_{beg}) were obtained for PCA-MFNN hybrid network with statistical analysis of output data. It was found that more than 4 PC elements and Oja's rule instead of Sanger's rule should be applied. There are still some predictions errors especially for slagging intensity k_z which are too high or the results have too high standard deviations. Duplication of existing training data sets does not have visible effect for ANN performance.
The ANN model needs further improvement by enlarging the training database and better selection methods of training data to be used in training process. After further improvements neural network model will be used in Common Intelligent Boiler Operation Platform (CIBOP) being elaborated within CERUBIS research project for two BP-1150 and BB-1150 steam boilers.

The authors would like to acknowledge the European Commision for financial support under the European Fifth Framework Programme for Energy, Environment and Sustainable Development Programme and Research Fund for Coal and Steel, and Polish Ministry of Science and Higher Education fo co-financing of intenational projects (PowerFlam2 project Nne5-2001-00907, Cerubis project RFCR-CT-2014-00008).

References

1. J.A. Rundygina, *Nauczne Trudy*, Moscow (1983)
2. J. Żelkowski, A. Sturm. O. Mauritz, VGB Power-Tech, **10** (1998)
3. T. Heinzel, V. Siegle, H. Spliethoff, K.R.G. Hein, Fuel Proces. Techn., **54**, 109-125 (1998)
4. R.P. Gupta, T.F. Wall, I. Kajigaya, S. Miyame, Y. Tsumita, Prog. Energ. Comb. Sc., **24**, 523-543 (1998)
5. Y. Liu, R. Gupta, A. Sharma, T. Wall, A. Butcher, G. Miller, P. Gottlieb, D. French, Fuel, **84**, 1259-1267 (2005)
6. L.Y. Huang, J.S. Norman, M. Pourkashanian, A. Williams, Fuel, **75**, 271-279 (1996)
7. C. Zheng, G. Liu, Y. Huang, Chem. Eng. Min. Proces., **10**, 61-75 (2002)
8. C. Yin, Z.Y. Luo, M.J. Ni, K.F. Cen, Fuel, **77**, 1777-1782 (1998)
9. H. Salehfar, S.A. Benson, Neurocom., **23**, 195-206 (1998)
10. E. Teruel, C. Cortes, L.I. Diez, I. Arauzo, Chem. Engine. Sc., **60**, 5035-5048 (2005)
11. Y.P. Liu, M.G. Wu, J.X. Qian, Thermoch. Acta, **454**, 64-68, (2007)
12. J. Qian, X. Jiang, F. Wang, Y. Chi, J. Yan, Thermoch. Acta, **526**, 29-34, (2011)
13. M. Pronobis, *Modernizacja kotłów energetycznych* (2000)
14. T. Sanger, Neural Networks, **12**, 459-473 (1989)
15. E. Oja, Jour. Mathem. Biol., **15**, 239-245 (1982)
16. S. Kakietek, W. Rybak, Proc. from AI-21 Conf., Siedlce, Poland (2006)
17. T. Tchaban, M.J. Taylor, J.P. Griffin, Neur. Comp. and Appl., **7**, 309–317 (1998)

E3S Web of Conferences **14**, 02005 (2017) DOI: 10.1051/e3sconf/20171402005
Energy and Fuels 2016

Thermovolumetric investigations of steam gasification of coals and their chars

Stanisław Porada[1,*], *Grzegorz* Czerski[1], *Przemysław* Grzywacz[1], *Tadeusz* Dziok[1], *and Dorota* Makowska[1]

[1] AGH University of Science and Technology, Faculty of Energy and Fuels, Mickiewicza Av. 30, 30-059 Cracow, Poland

Abstract. The process of steam gasification of three coals of various rank and three chars obtained from these coals by the ex-situ method at 900 °C was compared. In the coal gasification process, the pyrolysis stage plays a very important part, which is connected with its direct impact on the kinetics of gasification of the resulting char. What is more, taking into consideration the impact of pyrolysis conditions on char properties, it should be anticipated that the gasification kinetics of coal and char, formed from it by the *ex situ* method, will be different. In order to examine and compare the process of gasification of coals and chars, an isothermal thermovolumetric method, designed by the authors, was applied. For all the examined samples the measurements were performed at three temperatures, i.e. 850, 900, and 950 °C, and at the pressure of 0.1 MPa. An evaluation of the impact of raw material on the steam gasification of the examined samples was made. The carbon conversion degree and the kinetic parameters of CO and H_2 formation reaction were calculated. It was observed that the course of gasification is different for coals and chars obtained from them and it can be concluded that coals are more reactive than chars. Values of kinetic parameters of carbon monoxide and hydrogen formation calculated for coals and corresponding chars are also different. Due to the observed differences the process of gasification of coals and of chars with steam should not be equated.

1 Introduction

Coal gasification is a clean and effective method of coal conversion to a gas which has a very broad range of applications in the synthesis of chemicals, substitutes of petroleum and natural gas as well as in power generation [1-3]. In the era of dwindling resources of primary energy carriers and increasing consumption, a revival of interest in the gasification technology can be observed, which is reflected in the recent rapid growth of the number of coal gasification plants [4] and the number of works devoted to problems of coal gasification [5-10].

In order to further develop and optimise the operation of gasification reactors it is necessary to acquire a thorough knowledge of the process and factors which influence it.

* Corresponding author: porada@agh.edu.pl

One of the fundamental problems is getting to know the impact of the applied raw material on the kinetics of the gasification process. An analysis of the course of gasification reveals that it is a complex process which involves: (1) coal pyrolysis with the formation of char and volatile products of pyrolysis, (2) gasification of the resulting char, and (3) secondary reactions in the gaseous phase. In spite of the fact that the gasification of char, limiting the course of the entire process, is the slowest stage, the coal pyrolysis stage, which directly influences char properties, should not be underestimated. Factors having an effect on the reactivity of char, apart from the properties of the applied coal, include: the final temperature of pyrolysis, the rate of its growth, pressure, and the character of the atmosphere [11-23].

The pyrolysis stage plays a very important part in the gasification process, which is connected with its direct impact on the kinetics of gasification of the resulting char. What is more, taking into consideration the impact of pyrolysis conditions on char properties, it should be anticipated that the gasification kinetics of coal and char, formed from it by the *ex situ* method, will be different.

The aim of the work was to investigate and compare the steam gasification kinetics of selected coals and chars formed from them by the *ex situ* method. In the majority of examinations presented in literature, the measurements concerning gasification are conducted for chars, leaving out the pyrolysis stage, and the authors of the paper apply coals in their works [1,7,9,10].

An isothermal thermovolumetric method, developed by the authors, which consists in an analysis of the composition of the resulting gas, was used for examining the process of steam gasification of coals and chars. Thanks to such an approach, the conditions of measurements better reflect the process of coal gasification occurring in the commercial reactors. The conducted measurements allowed for the determination of kinetic parameters of gaseous products formation. Additionally, the carbon conversion degree was calculated.

2 Material and method

2.1 Characteristics of samples and conditions of measurements

The object of this study were three Polish coals of different rank and three chars obtained from these coals by the *ex situ* method at 900 °C and under atmospheric pressure. For this purpose a Jenkner type retort was used. Mass sample of coals used amounted to 5 kg. Yield of resulting char was depending on coal type and amounted to from about 2 to 3 kg. The coals selected for the examinations comes from abundant resources and with good prospects of being put to use. For the purposes of examinations, samples were used with the particle size of 0.315-0.5 mm and the mass of 1 g. Table 1 shows the characteristics of the examined samples. The proximate and ultimate analyses were supplemented with the composition of ash, and, additionally, the alkali index, which influences the gasification kinetics [24], was calculated in accordance with the Eq. (1):

$$AI = A^a \frac{Fe_2O_3+CaO+MgO+Na_2O+K_2O}{SiO_2+Al_2O_3} \tag{1}$$

A^a – ash content in sample, -,
Fe_2O_3, CaO, MgO, Na_2O, K_2O, SiO_2, Al_2O_3 – content of particular oxides in ash, %.

From an analysis of Table 1 it can be inferred that chars are featured by greater contents of carbon and ash and lower contents of hydrogen, sulphur, moisture and volatile matter with respect to initial coals. Despite small differences in ash composition between

DOI: 10.1051/e3sconf/20171402005

coals and chars, greater values of the alkali index (AI) are observed for chars, which is connected with a higher content of ash in them.

For all the samples, the examinations were conducted for three temperatures, i.e. 850, 900 as well as 950 °C, and at the pressure of 0.1 MPa.

Table 1. Characteristics of examined coals and coal chars

Parameter	Bełchatów coal	Bełchatów char	Janina coal	Janina char	Wieczorek coal	Wieczorek char
Proximate analysis (%)						
Moisture - M^a	19.4	1.6	14.1	2.8	3.3	1.0
Ash - A^a	19.9	40.8	12.8	15.6	13.0	20.5
Volatile matter - VM^{daf}	57.0	0.7	42.1	1.2	38.2	4.9
Fixed Carbon – FC	26,1	57,2	42,3	80,6	51,7	74,7
Ultimate analysis (%)						
C^{daf}	67.1	97.5	79.6	98.5	83.8	96.5
H^{daf}	5.2	1.1	5.5	1.0	5.8	1.06
S^a_t	0.81	1.02	4.00	0.83	1.71	0.83
Ash composition (%)						
SiO_2	48.5	49.9	45.9	47.1	49.0	61.9
Al_2O_3	23.7	22.8	26.1	26.7	17.9	14.7
Fe_2O_3	3.9	3.6	10.1	7.5	10.5	8.3
MgO	0.95	0.85	1.56	1.61	2.75	2.05
CaO	11.74	11.69	2.88	3.26	5.69	3.19
K_2O	0.39	0.28	2.40	3.13	2.26	2.23
Na_2O	0.17	0.06	2.46	2.76	1.35	1.25
Alkali index AI (-)	4.73	9.25	2.53	3.86	2.83	5.42

2.2 Laboratory equipment

The investigations of steam gasification were conducted with the use of unique laboratory equipment which allows for measurements of the kinetics of gasification under isothermal conditions within a broad pressure range. The measuring device consists of three basic systems: a high pressure reactor with a heating system, a system for supplying the reactor with steam, inert gas (argon) and coal as well as a system for collecting and analyzing the resulting gas. The system for supplying the reaction zone with steam and argon consists of a micropump dispensing water, a steam generator, high pressure cylinders with argon and a set of pressure reducing valves, control and release valves, filters, manometers and a flow controller.

Inside the reactor a retort is located with the diameter of 20 mm and with an especially designed quartz grid. After stabilizing the parameters of examinations (temperature, pressure, and flow rates) a piston feeder puts a coal sample with the mass of 1g onto the grid. The resulting gas, after the evolution of water-tar condensate and drying, decompresses and is continuously analysed in respect of the contents of CO, CO_2 and CH_4. Moreover, the gas is taken into pipettes in order to be analysed in respect of the content of hydrogen with the use of a gas chromatograph equipped with a thermal conductivity detector (TCD). A diagram of the laboratory equipment is presented in Fig. 1.

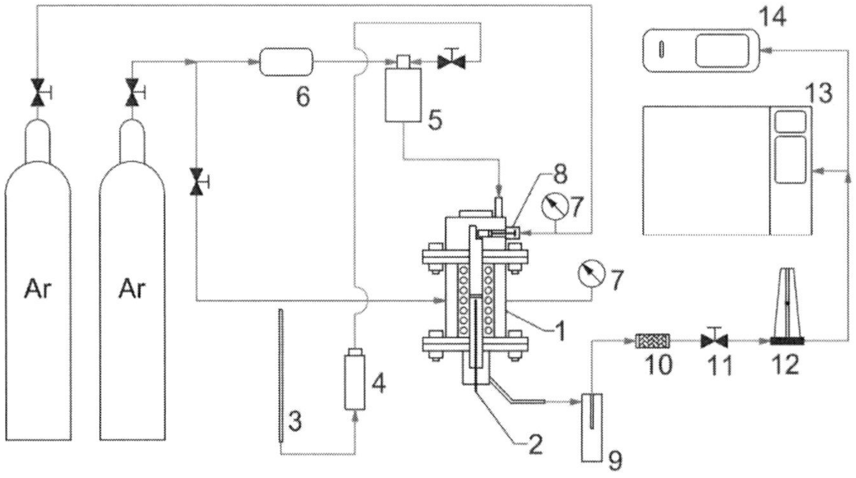

Fig. 1. The laboratory equipment for kinetic examinations of gasification process. 1 - reactor, 2 - thermocouple, 3 - water tank, 4 - water pump, 5 - steam generator, 6 - mass flowmeter, 7- pressure gauge, 8 - sample feeder, 9 - cooler, 10 - filter, 11 - pressure regulator, 12- rotameter, 13 - gas chromatograph, 14 - CO, CO_2 IR analyser

2.3 Methodology of evaluation of gasification process

In order to evaluate the gasification process on the basis of changes in the concentration of the analyzed components, their time curves of formation were determined, which enabled the determination of the yields of the products and the composition of the resulting gas. Moreover, for the most important products, i.e. carbon monoxide and hydrogen, the reaction rate constants and the kinetic parameters of their formation reaction were calculated. The work [7] describes the methodology of determining the kinetic parameters in detail. Assuming the first order of reaction, the formation rate of a product can be expressed by Eq. (2):

$$ln\frac{V_\infty}{V_\infty - V} = k \ t \quad (2)$$

k – constant reaction rate, 1/min,
V_∞ – maximal volume of gas product, cm^3,
V – volume of released gas component at time t, cm^3,
t – time, min

On the basis of Eq. (2), the formation rate constants were determined for carbon monoxide and hydrogen at three temperatures (850, 900 and 950 °C), and then, based on the

Arrhenius Equation (3), the values of the activation energy E and the pre-exponential factor A were calculated.

$$k = A \times e^{-\frac{E}{RT}} \quad (3)$$

The carbon conversion degree was estimated on the basis of the Eq. (4):

$$X = \frac{(V_{CO}+V_{CO2}+V_{CH4})\, M_C}{V_{mol}\, m\, C^{daf}} \times 100\ \% \quad (4)$$

V_{CO}, V_{CO2}, V_{CH4} – volume of released gas component, dm^3/g,
M_c – molar mass of carbon, g/mole,
m – mass of sample, g,
C^{daf} – dry ash free carbon content in sample, -,
V_{mol} – molar volume of gas at temperature of 273 K and pressure of 101325 Pa, dm^3/mole.

3 Results and discussion

3.1 Impact of raw material on gasification process

On the basis of the composition of the resulting gas the curves of formation rate of CO, CO_2, H_2 and CH_4 were developed, and are presented in Fig. 2. Generally, in the case of coals at the first stage of the process, considered equivalent to the course of pyrolysis, high rates of evolution of gaseous products are observed. The second stage is the proper gasification of the resulting char which is featured by significantly lower formation rates of the examined gaseous components.

In the case of chars, the shape of the curves is different. At the initial stage of the process, there are no peaks of gaseous products formation rates, which are characteristic for pyrolysis, and, instead, a consistent growth of the formation rates of CO, CO_2 and H_2 is observed, which, after reaching the maximum, slowly decrease. Hydrogen is characterized by the highest formation rate, both in the case of coals and chars. The curves of formation of carbon monoxide and carbon dioxide are diverse and depend on the examined material. Methane is formed only during the first minutes of the gasification of coals, which is connected with the occurring pyrolysis reactions. During the gasification of chars, the formation of methane is not observed.

The shape of the formation curves of gaseous products of gasification constitute a proof of reactivity of the examined material. It can be observed that coals of different rank differ in reactivity, and the highest reactivity is that of the Bełchatów brown coal of the lowest rank, while the lowest reactivity is that of the Wieczorek bituminous coal of the highest rank. The obtained results correspond to those reported by other researchers [19-21]. This fact may be attributed to a high content of functional groups rich in oxygen, a big share of transient pores and macropores as well as a high dispersion of the inorganic substance in coals of low rank.

Similarly to coals, chars also differ in reactivity, and the chars obtained from the Bełchatów and Janina coals of low rank are more reactive than the char from the Wieczorek coal. A comparison of the gasification rate for coals and chars obtained from them leads to the conclusion that coals are more reactive than chars obtained from them by the *ex situ* method, despite the fact that the alkali index had higher values for chars than for coals. This statement can be attributed to the impact of different conditions of char formation on its reactivity and the fact that higher values of alkali index for chars are connected with a higher content of ash in them and not with increased share of alkali and alkaline earth metals that catalyse gasification reactions.

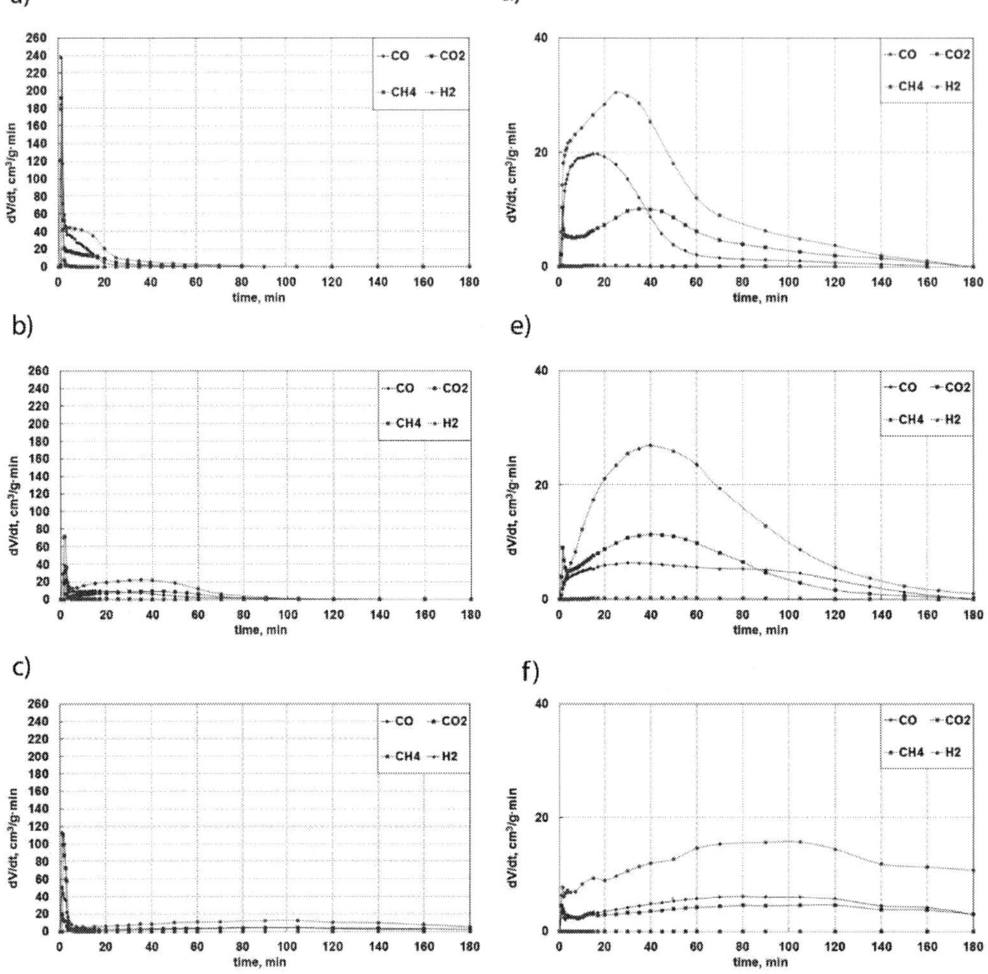

Fig. 2. Curves of evolution rate of gaseous products of gasification (t = 900 °C): *a) Bełchatów coal; b) Janina coal; c) Wieczorek coal; d) Bełchatów char; e) Janina char; f) Wieczorek char*

3.2 Analysis of carbon conversion degrees

An analysis of the curves of the carbon conversion degree for coals and chars makes it possible to notice easily differences in their reactivity, the measure of which is the slope of the curve translating into the time of reaching the maximal conversion degree. Carbon conversion degree curves for all examined samples are shown in Fig. 3.

Analysing the shape of the curves of the conversion degree for coals and chars, certain differences can be observed. Namely, in the case of coals, in contrast with chars, in the first minutes a rapid growth can be noticed of the conversion degree, after which the slope of the curve decreases. This phenomenon is caused by the reactions of coal pyrolysis, occurring in the first minutes, and by a rapid evolution of gaseous products. It can also be stated that coals, especially those of a low rank, are featured by a higher reactivity when compared to chars formed from them by the *ex situ* method.

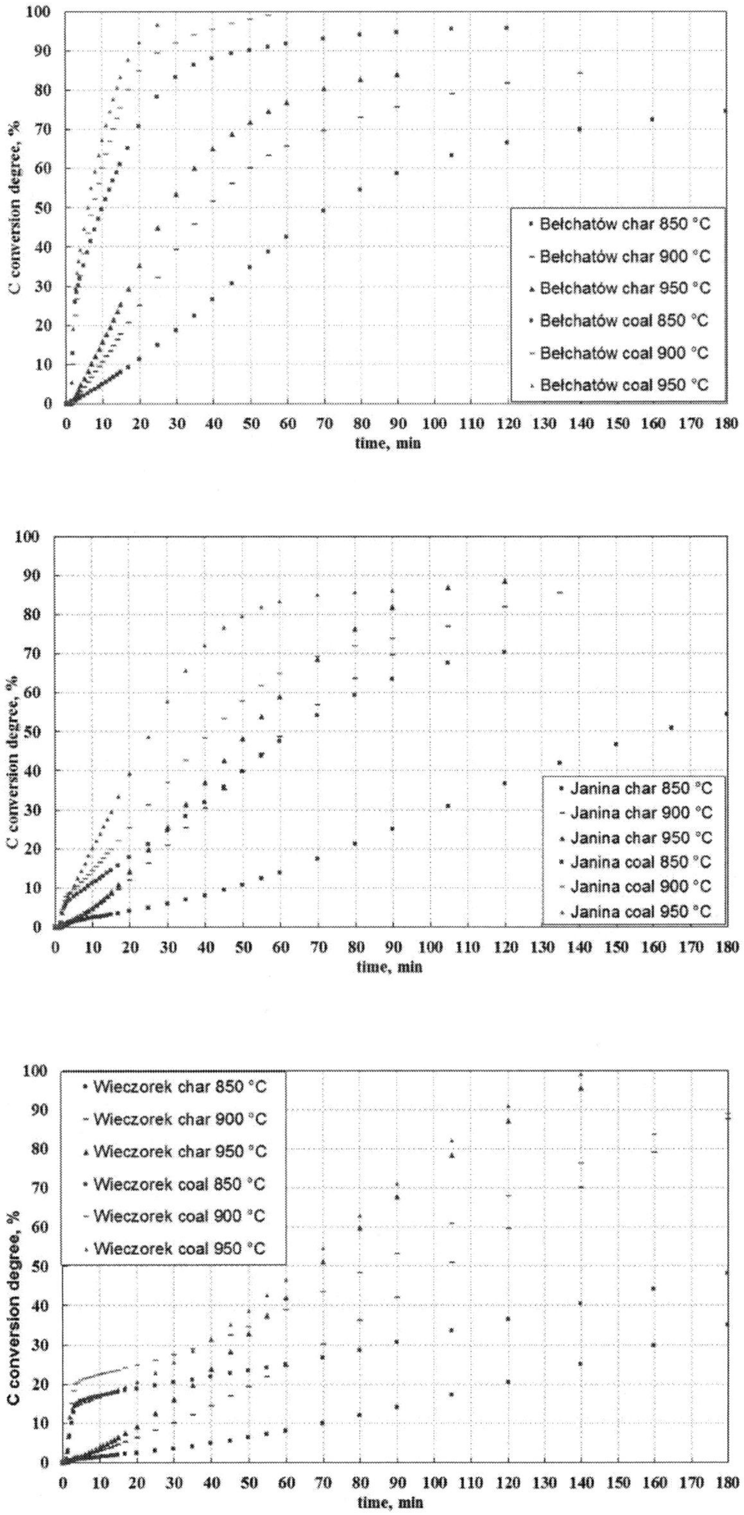

Fig. 3. Comparison of carbon conversion degree curves

3.3 Kinetic parameters of gasification process

In order to quantitatively describe the process of gasification, on the basis of the Eq. (2), the formation rate constants of carbon monoxide k_{CO} and of hydrogen k_{H2} were calculated for the three examined temperatures. Next, on the basis of the Eq. (3), kinetic parameters i.e. the activation energy E and the pre-exponential factor A of CO and H_2 formation reactions were determined. The results of the calculations are presented in Table 2 and Table 3. Comparing kinetic parameters for coals and for chars it can be seen that they are different. In the case of carbon monoxide, both energy activation and pre-exponential factor values for Bełchatów and Janina chars are higher than for corresponding coals. In the case of Wieczorek samples activation energy assumes a slightly lower value for char and its lower reactivity is associated with much lower value of pre-exponential factor. In the case of hydrogen such a clear relationship is not observed which can be explained by a more complex mechanism of its formation.

Table 2. Kinetic parameters of carbon monoxide formation reaction

Kinetic parameters of CO formation	Bełchatów coal	Janina coal	Wieczorek coal	Bełchatów char	Janina char	Wieczorek char
E [kJ/mol]	140	218	269	201	230	254
A [1/min]	2.58E+01	2.73E+04	2.07E+06	5.93E+03	3.50E+04	3.07E+05

Table 3. Kinetic parameters of hydrogen formation reaction

Kinetic parameters of H_2 formation	Bełchatów coal	Janina coal	Wieczorek coal	Bełchatów char	Janina char	Wieczorek char
E [kJ/mol]	133	181	181	106	248	242
A [1/min]	8.95E+00	7.28E+02	2.40E+02	3.25E-01	3.77E+05	9.63E+04

4 Conclusions

The process of steam gasification of three coals of different coal ranks and three chars obtained from them by the *ex situ* method at 900 °C was compared. The obtained results of the examinations lead to the following most important conclusions:

a) the courses of the gasification process of coals and chars obtained from these coals by the *ex situ* method differ from each other: in the case of coals, in the first minutes, a rapid evolution of gaseous products is observed, which is connected with the pyrolysis of the coal matter; this phenomenon is not observed during the gasification of chars,
b) generally, coals are more reactive than chars obtained from them, the reactivity of char decreases along with a growth of the parent coal rank and similarly the reactivity of coals decreases along with the growth of its rank,
c) kinetic parameters of formation reactions of the most important products i.e. carbon monoxide and hydrogen calculated for coals and chars are different; generally it can be stated that in case of gasification process both kinetic parameters should be considered i.e. activation energy and pre-exponential and not only activation energy because of their compensation effect on reactions kinetics,

d) differences in the gasification of coals and chars suggest that conditions in which the char is formed bear on its properties that is why gasification process of coal and the *ex situ* prepared char should not be equated.

This work was prepared as a part of the statutory activity of the Faculty of Energy and Fuels at the AGH University of Science and Technology, No. 11.11.210.213

References

1. G. Czerski, T. Dziok, A. Strugała, S. Porada, Przem. Chem. **93**, 1393–1400 (2014)
2. X. Yujie, Z. Guiyan, Ch. Haisheng, D. Binlin, T. Chunqing, Int. J. Hydrogen Energy **37**, 11805–11814 (2012)
3. D. Y. Young, J. L. Seung, Y. Yongseung, Koeran J. Chem. Eng. **24**, 350 (2007)
4. C. Higman, State of the Gasification Industry – the Updated Worldwide Gasification Database, Proceedings of Gasification Technologies Conference; 2013 Oct 16; Colorado Springs, USA
5. A. Strugała, G. Czerski, Przem. Chem. **91**, 2181-2184 (2012)
6. A. M. Dubinin, E. V. Cherepanova, O. A. Obozhin, Solid Fuel Chemistry **49(2)**, 31–33 (2015)
7. S. Porada, G. Czerski, T. Dziok, P. Grzywacz, D. Makowska, Fuel Process. Technol. **130**, 282–291 (2015)
8. K. Murakami, M. Sato, N. Tsubouchi, Y. Ohtsuka, K. Sugawara, Fuel Process. Technol. **129**, 91–97 (2015)
9. S. Porada, G. Czerski, P. Grzywacz, T. Dziok, D. Makowska, Przem. Chem. **93** 2059–2063 (2014)
10. S. Porada, A. Rozwadowski, Przem. Chem. **93**, 384–387 (2014)
11. M. J. G. Alonso, A. G. Borrego, D. Alvarez, J. B. Parra, R. Menendez, J. Anal. Appl. Pyrolysis. **58–59**, 887–909 (2001)
12. L. Tingting, Z. Lei, D. Li, L. Chun-Zhu, Fuel **117**, 1190–1195 (2014)
13. A. Molina, F. Mandragon, Fuel **77** 1831–1839 (1998)
14. R. C. Messenbock, D. R. Dugwell, R. Kandiyoti, Fuel **78**, 781–793 (1999)
15. C. Li, J. Zhao, Y. Fang, Y. Wang , Front. Chem. Eng. China. **4(4)**, 385–393 (2010)
16. S. Kajitani, S. Hara, H. Matsuda, Fuel **81**, 539–546 (2002)
17. M. A. Elliott editor. Chemistry of coal utilization. John Wiley & Sons Inc; 1981.
18. K. E. Benfell, G. S. Liu, D. G. Roberts, D. J. Harris, J.A. Lucas, J.G. Bailey, T.F. Wall, P. Combust. Inst. **28**, 2233–2241(2000)
19. D. P. Ye, J.B. Agnew, D.K. Zhang, Fuel **77**, 1209–1219 (1998)
20. C. Z. Li, Fuel **86**, 1664–1683 (2007)
21. M.J.G. Alonso, A.G. Borrego, D. Alvarez, R. Menendez, Fuel **78**, 1501–1513(1999)
22. Q. Yan, J. Huang, J. Zhao, Ch. Li, L. Xia, Y. Fang, J. Ther. Anal. Calor. **116**, 519–527 (2014)
23. J. Tanner, S. Bhattacharya, Chem. Eng. J. **285**, 331–340 (2015)
24. M. Sakawa, Y. Sakurai, Y. Hara, Fuel **73**, 717–720 (1982)

Assessment of the content of arsenic in solid by-products from coal combustion

Faustyna Wierońska[1,*], *Dorota* Makowska[1] and *Andrzej* Strugała[1]

[1]AGH University of Science and Technology, Faculty of Energy and Fuels, Al. A. Mickiewicza 30, 30-059 Krakow, Poland

Abstract. The coal combustion processes constitute one of the major sources of heavy metals emission into the atmosphere. From the point of view of the reduction of the emission of heavy metals and the selection of the correct exhaust gas treatment system, it is important to monitor the amount of trace elements in the solid fuels and in the solid by-products from coal combustion. One of these highly toxic elements is arsenic. The average content of arsenic in Polish hard coals and lignites is 0 ÷ 40 mg/kg [1] and 5 ÷ 15 mg/kg [2], respectively. The world average content of arsenic in hard coals and lignites, is equal to 9.0 ± 0.8 and 7.4 ± 1.4 mg/kg [3], respectively. During coal combustion processes, a significant amount of arsenic enters the atmosphere through gases and fly ashes. The proportions in which those two forms of arsenic occur in exhaust gases depend on the conditions of combustion processes [4]. The aim of the research was to determine the content of arsenic in coal blends and by-products of their combustion (slag, fly ash, gypsum, filter cakes). The determination of the arsenic quantity was performed using the Atomic Absorption Spectrometry with the electrothermal atomization.

1 Introduction

During the year in Poland the consumption of hard coal reaches about 72,8 million tons, nearly 60% of which is connected with the energy sector [5]. The growing demand for energy can increase the release of large amounts of toxic elements and their compounds into the atmosphere. A reduction in emissions of these elements and compounds into the environment is achieved in exploited installations through the use of different flue gas treatment systems (the post-combustion methods). In domestic conditions there are installations for catalytic reduction of NOx, installations for desulfurization and systems for the removal of fly ashes, such as the electrostatic precipitator and the bag filters. Apart from the application of flue gas treatment systems, modern coal combustion technologies including pulverized coal combustion, fluidized bed combustion system or supercritical and ultra-supercritical technologies also have an impact on the emission of toxic elements [6].

One of the more toxic elements is arsenic. In Poland, during the combustion processes in the sector of production and transformation of energy, more than 4.8 Mg of arsenic

* Corresponding author: wieronsk@agh.edu.pl

and its compounds are emitted annually [7]. From the point of view of the environmental impact, this element belongs to the group of the most toxic elements together with cadmium, lead, mercury and chromium. It also belongs to the highly volatile elements in respect of coal combustion processes [8]. Arsenic has a negative impact on living organisms. It causes, among others, nervous system disorders, heart functions disorder or hepatic impairment. It is also regarded by the International Agency for Research on Cancer as a carcinogenic element [9, 10].

In coal arsenic is associated primarily with pyrite As-S_{pyr}. The average content of this element in Polish hard coals and lignites is 0 ÷ 40 mg/kg [1] and 5 ÷ 15 mg/kg [2], respectively. Additionally, the world average content of arsenic in hard coals and lignites, is equal to 9.0 ± 0.8 and 7.4 ± 1.4 mg/kg, respectively [3]. During coal combustion processes, arsenic bound to pyrite As-S_{pyr} and arsenic occurring in organic compounds As_{org} enter the exhaust gases through gases and fly ashes. The proportions of those two forms of arsenic are different and depend on the conditions of combustion processes. Arsenic associated with clay remains in the bottom ash [3, 4, 11]. The relationship between the content of As in coal and the amount of ash has been well described [3]. It can be concluded that coal with a higher ash content has a higher concentration of arsenic. Moreover, it has been proven that with the reduction of the particle diameter of fly ash, the content of As grows.

The highest contents of the discussed element were recorded also in the fly ash containing large amounts of magnesium and calcium. Therefore, in order to reduce the emissions of arsenic, sorbents based on calcium oxide are considered [12]. Arsenic located in the fly ash is almost totally removed from the exhaust gases using the electrostatic precipitator and the bag filters [3, 4, 12, 13]. The concentration of As in the fly ash depends strictly on the initial content of this element in coal. The world average content of arsenic in the fly ash from hard coal is 46 ± 5 mg/kg [14], but this value can fluctuate in very wide ranges even within one coal basin. In the case of coals from the Upper Silesian Coal Basin, the As content may range from 3 to 190 mg/kg [15]. Research [15] has shown, that the highest As content is associated with the smallest fly ash particles (with diameter below 20 μm).

In view of the negative effects of arsenic on the environment, it is important to know the content of this element in the solid by-products of coal combustion, owing to their usage (fly ash, gypsum) or storage (ash-slag). The amount of As in these products depends not only on the burnt coal, but also on the type of furnace and the flue gas treatment system. The investigations carried out within the framework of this study aimed at determining the content of the discussed element in the solid by-products of coal combustion in power plants.

2 Experimental

2.1 Examined material

The samples of coal blends burnt in the boilers and solid by-products of their combustion, i.e.: the average daily samples of fly ashes, ash-slag samples and products of desulphurization (gypsum and filter cakes) were analyzed. The examined samples came from two Polish fossil fuel power plants, which use thermal hard coal blends for combustion. Both plants use the same flue gas treatment system consisting of Selective Catalytic Reduction (SCR) to reduce the level of NOx, an Electrostatic Precipitator (ESP) and the Flue Gas Desulphurization (FGD) based on the wet lime method. In both cases, for the purposes of combustion, the Pulverized Coal Fired Boilers fired by coal coming from different mines of the Upper Silesian Coal Basin, are installed. In the analyzed samples the content of the elements: Si, Al, Ca, Fe, Mg, Na and K was investigated (tables 1 and 2).

Table 1. The amount of Si, Al, Ca, K, Na, Mg and Fe in examined samples [%] – Power Plant 1.

Sample	**Content of the elements [%]**						
	Si	Al	Ca	K	Na	Mg	Fe
Coal	5.68	1.98	0.53	0.53	0.20	0.38	1.24
Ash-slag	18.31	10.10	2.71	2.24	0.41	1.65	6.27
Fly ash	20.31	11.19	2.47	2.48	0.48	1.61	4.31
Gypsum	0.275	0.061	-	0.019	0.003	0.035	0.044
Filter cake	6.18	2.27	15.88	0.76	0.11	0.80	1.47

Table 2. The amount of Si, Al, Ca, K, Na, Mg and Fe in examined samples [%] – Power Plant 2.

Sample	**Content of the elements [%]**						
	Si	Al	Ca	K	Na	Mg	Fe
Coal	4.80	1.76	0.50	0.36	0.16	0.29	1.15
Ash-slag	20.86	8.80	1.99	1.79	0.49	1.22	6.88
Fly ash	17.20	9.34	2.54	1.98	1.31	1.48	6.14
Gypsum	0.296	0.052	-	0.015	0.005	0.014	0.037
Filter cake	5.69	1.99	16.40	0.79	0.10	0.94	2.37

2.2 Analytical methods

The samples were prepared in accordance with the Polish standard PN-G-04502:2014-11. For coal samples, the determination of ash content in analytical conditions was executed in accordance with the standard PN-G-04512:1980/Az1:2002. For all samples, the determination of the chemical elements Si, Al, Ca, Fe, Mg, Na and K was carried out using Atomic Absorption Spectrometry with flame atomization (FAAS). The content of arsenic in all samples was determined with the application of Graphite Furnace Atomic Absorption Spectrometer (GFAAS) Hitachi Z-2000 using the Zeeman background correction effect. The applied method was featured by the following analytical parameters: analytical spectral line 193,7 nm; flow of argon: 200 cm^3/min (atomization – 30 cm^3/min); modifiers: Ni and Mg solution. The uncertainty of the As content ± 0.26 mg/kg, was used.

In order to perform the analyses with the use of AAS, all samples were digested in the Berghoff SpeedWave4 microwave system. Nearly 130 mg of sample was mineralized with the combination of spectrally pure nitric acid (V) and hydrofluoric acid. Subsequently, supersaturated boric acid was used as the complexing reagent.

DOI: 10.1051/e3sconf/20171402006

3 Results and discussion

The results of the investigations of the average arsenic content in samples of hard coal blends and in the solid by-products of their combustion were collected in table 3. The concentration of As in coal, in the case of both power plants, was lower than the world average content of arsenic in hard coals (9.0 ± 0.8 mg/kg). Accordingly, these coals can be classified as the ones with a low content of arsenic.

Table 3. The amount of arsenic in examined samples [mg_{As}/kg_{sample}].

Sample	Power Plant 1	Power Plant 2
Coal	5.10	2.29
Ash-slag	14.15	1.63
Fly ash	28.21	12.46
Gypsum	1.51	0.68
Filter cake	25.60	28.06

The ash content in coal derived from the first power plant was 21.7 %, and from the second power plant it was 11.1 %. It can be observed that a nearly twofold higher concentration of As in coal was determined for the sample with a higher ash content. The obtained result confirms the trend of the arsenic content towards growing along with the increasing ash content in coal. On the one hand, fly ashes and filter cakes are characterized by the highest content of this element on the other hand, the lowest concentration is displayed by the samples of gypsum (table 3). It was also observed that the ratio of the As content in fly ash and gypsum to the amount of this element in coal blends for both power plants is the same and amounts to: 5.5 for fly ash and 0.3 for gypsum. However, in the case of ash-slags and filter cakes, a similar relationship was not recorded. The content of As in filter cakes is similar. For the first and the second power plant this amount of As is 25.6 mg/kg and 28.1 mg/kg, respectively. This provides evidence, that the examined element is accumulated in those wastes. It can be connected with the easy arsenic leaching by water and the accumulation in waste filtration as well as the presence of large amounts of calcium in the analyzed filter cakes (table 1 and 2) [3, 12]. The larger arsenic content in ash-slag from the first power plant can also result from an increased calcium and magnesium content when compared to the second power plant.

4 Conclusions

The goal of this study was to examine the amount of arsenic in hard coal and solid by-products from coal combustion. Based on the investigations, the following conclusions can be drawn:

- The content of As in the analyzed samples is lower than the world average content of As in hard coals (9.0 ± 0.8 mg/kg).
- The relationship between the content of As in coal and the ash content was confirmed. Along with the increase of ash content, the amount of As in coal grows.
- Arsenic accumulates mostly in fly ash and filter cakes.
- The gypsum samples are characterized by the lowest concentration of arsenic.

Acknowledgements: This paper was prepared as a part of the statutory activity of the Faculty of Energy and Fuels at the AGH University of Science and Technology, No. 11.11.210.213.

References

1. J. Widawska-Kuśmierska, Występowanie pierwiastków śladowych w polskich węglach kamiennych, Przegląd Górniczy, nr **7-8**, s. 455-459 (1981)
2. M. Wagnar, Oznaczenie pierwiastków toksycznych i szkodliwych w węglu i jego popiele, [w:] Eksploatacja węgla brunatnego jako metoda ograniczania szkodliwego oddziaływania na środowisko pierwiastków obecnych w węglu i w produktach jego spalania (na przykaładzie KWB Bełchatów) - Monografia, pod red. M. Stryszewski, Katedra Górnictwa Odkrywkowego, Wydział Górniczy AGH, Kraków (2001)
3. Ya.E. Yudovich, M.P. Ketris, Arsenic in coal: a review, International Journal of Coal Geology **61**, p.141- 196 (2005)
4. V.C Pandey et al., Arsenic hazards in coal fly ash and its fate in Indian scenario, Resources, Conservation and Recycling Volume **55**, Issues 9–10 p.819–835 (2011)
5. GUS, Główny Urząd Statystyczny, Zużycie Paliw i Nośników Energii w 2014 roku, Warszawa (2015)
6. U. Lorenz, Skutki spalania węgla kamiennego dla środowiska przyrodniczego i możliwości ich ograniczania, Mat. Szkoły Eksploatacji Podziemnej. Sympozja I Konferencje nr 64., Instytut GSMiE PAN, Kraków, s. 97-112 (2005)
7. KOBiZE, Krajowy Bilans Emisji SO_2, NOx, CO, NH_3, NMLZO, pyłów, metali ciężkich i TZO za lata 2013 – 2014 w układzie klasyfikacji SNAP I NFR, Warszawa (2016)
8. K. Srogi, Pierwiastki śladowe w węglu, Wiadomości Górnicze **2**/2007, s. 87-96
9. IARC MONOGRAPHS, Arsenic, Metals, Fibres and Dusts, vol.**100C**, s. 41-93 (2012)
10. E. Gawęda, Arsen i jego związki w środowisku pracy – zagrożenia, ocena narażenia, Bezpieczeństwo Pracy **3**/2005, s 26-28
11. J. M. Pacyna, Lead, Mercury, Cadmium and Arsenic in the Environment: chapter 7: Atmospheric Emissions of Arsenic, Cadmium, Lead and Mercury from High Temperature Processes in Power Generation and Industry, SCOPE 31, SCOPE (1987)
12. R.M. Bolanz, J. Majzlan, L. Jurkovic, J. Göttlicher, Mineralogy, geochemistry, and arsenic speciation in coal combustion waste from Nováky, Slovakia, Fuel **94** s. 125–136 (2012)
13. H. Nalbandian, Trace element emissions from coal, IEA Clean Coal Centre (2012)
14. M.P. Ketris, Ya.E. Yudovich, Estimations of Clarkes for Carbonaceous biolithes: World averages for trace element contents in black shales and coals. International Journal of Coal Geology, **78**, pp. 135-148 (2009)
15. H.R. Parzentny, L. Róg, Potentially hazardous trace elements in ash from combusiton of coals in limnic series (Upper Carboniferous) of the Upper Silesian Coal Basin (USCB), Górnictwo i geologia, t. **2**, z. **3**, s. 81-91 (2007)

Method selection for mercury removal from hard coal

Tadeusz Dziok[1,*], *Andrzej* Strugała[1]

[1]AGH University of Science and Technology, Faculty of Energy and Fuels, Al. A. Mickiewicza 30, 30-059 Krakow, Poland

Abstract. Mercury is commonly found in coal and the coal utilization processes constitute one of the main sources of mercury emission to the environment. This issue is particularly important for Poland, because the Polish energy production sector is based on brown and hard coal. The forecasts show that this trend in energy production will continue in the coming years. At the time of the emission limits introduction, methods of reducing the mercury emission will have to be implemented in Poland. Mercury emission can be reduced as a result of using coal with a relatively low mercury content. In the case of the absence of such coals, the methods of mercury removal from coal can be implemented. The currently used and developing methods include the coal cleaning process (both the coal washing and the dry deshaling) as well as the thermal pretreatment of coal (mild pyrolysis). The effectiveness of these methods various for different coals, which is caused by the diversity of coal origin, various characteristics of coal and, especially, by the various modes of mercury occurrence in coal. It should be mentioned that the coal cleaning process allows for the removal of mercury occurring in mineral matter, mainly in pyrite. The thermal pretreatment of coal allows for the removal of mercury occurring in organic matter as well as in the inorganic constituents characterized by a low temperature of mercury release. In this paper, the guidelines for the selection of mercury removal method from hard coal were presented. The guidelines were developed taking into consideration: the effectiveness of mercury removal from coal in the process of coal cleaning and thermal pretreatment, the synergy effect resulting from the combination of these processes, the direction of coal utilization as well as the influence of these processes on coal properties.

1 Introduction

Mercury and its compounds are characterized by highly toxic properties and negatively affect the functioning of the human organisms [1]. It should be emphasized, that any amount of mercury emitted to the environment is a potential threat to life and health, which is caused by the natural biogeochemical cycle of mercury [2]. One of the main sources of mercury emission to the environment is that constituted by the coal utilization processes, in which mercury is commonly found [3-5]. This issue is particularly important for Poland, because the Polish energy production sector is based on brown and hard coal [6] and the forecasts show that this trend in energy production will continue in the coming years [7]. Moreover, it

* Corresponding author: tadeusz.dziok@agh.edu.pl

should be pointed out, that Poland is a major producer of coke in the EU [8]. At the moment of the emission limits introduction, methods of reducing mercury emission will have to be implemented in Poland.

The amount of mercury emitted in the coal utilization processes may be reduced using a number of methods. One of these methods involves using coal with a relatively low mercury content. Obviously, the availability of such coals is limited. However, in the case of their absence, methods of mercury removal from coal can be applied. The currently used and developing methods include the coal cleaning process [9, 10] as well as the thermal pretreatment of coal (mild pyrolysis) [11-13]. It should be mentioned that the effectiveness of these methods varies for different coals. This is caused by the diversity in coal origin, various characteristics of coal, and especially by the various modes of mercury occurrence in coal [9, 14-16]. The coal cleaning process allows for the removal of mercury occurring in mineral matter, mainly in pyrite. The thermal pretreatment of coal allows for the removal of mercury occurring in organic matter as well as in the inorganic constituents characterized by a low temperature of mercury release.

In this paper, guidelines for the selection of a method for mercury removal from hard coal were presented. The guidelines were developed considering the effectiveness of mercury removal from coal in the processes of coal cleaning and thermal pretreatment, the synergy effect resulting from the combination of these processes, the direction of coal utilization as well as the influence of these processes on coal properties.

2 Aim of study

The main aim of the study was to develop guidelines for the selection of a method for mercury removal from hard coal. To achieve that goal, the effectiveness of mercury removal from coal in the processes of coal cleaning and thermal pretreatment was determined. While developing the guidelines the direction of coal utilization as well as the influence of the analyzed processes on coal properties were taken into consideration.

3 Experimental

3.1 Examination of the coal cleaning process

Within the framework of the coal cleaning process investigations, the samples derived from domestic industrial plants (coal washing) as well as from demo plants (dry deshaling) were examined. Six washing plants of coking coals and two dry separators of steam coal: the air concentrating table (DD1-DD6 cases) and the X-ray optical separator (DD7 case) were analyzed. The content of mercury in raw coals and rejects derived from the coal cleaning process are given in Table 1. The method of mercury content determination was previously presented in work [10].

The effectiveness of mercury removal from coal in the process of coal cleaning was determined using the Eq. (1).

$$\eta_{cc} = \frac{Hg_{rejects} \cdot \gamma_{rejects}}{Hg_{raw}} \quad (1)$$

η_{cc} – effectiveness of mercury removal from coal in the process of coal cleaning [%],
Hg_{raw} – mercury content in raw coal [µg/kg],
$Hg_{rejests}$ – mercury content in rejects [µg/kg],
$\gamma_{rejects}$ – yield of rejects [%].

Table 1. Mercury content in raw coals and rejects derived from coal cleaning process.

Analyzed cases		**Hg_t^d [µg/kg]** Raw coal	**Hg_t^d [µg/kg]** Rejects
Coal washing	CW1	65	49
	CW2	170	36
	CW3	60	17
	CW4	83	47
	CW5	117	98
	CW6	98	50
Dry deshaling	DD1	113	319
	DD2	146	121
	DD3	88	162
	DD4	137	93
	DD5	95	78
	DD6	124	160
	DD7	86	157

3.2 Examination of the coal thermal pretreatment process (mild pyrolysis)

For the purposes of examining the coal thermal pretreatment process, due to the lack of appropriate industrial and pilot plants in Poland, the studies were conducted on bench scale. The investigations were carried out using the laboratory equipment for coal thermal pretreatment in fixed bed - Fig.1. The thermal pretreatment of coal was performed for the following process parameters:

- final temperature: 300 °C,
- heating rate: 5 °C/min,
- residence time at the final temperature: 30 min,
- purge gas: nitrogen,
- purge gas flow: 5 dm^3/min,
- sample weight: 500 g,
- coal particle size: below 10 mm.

Five hard coals were examined: two raw coking coals (CW3 and CW5) and three steam coals (SC1, SC2 and SC3). Mercury content in steam coals was as follows: 138, 37 and 91 µg/kg (dry basis). The effectiveness of mercury removal from coal in the thermal pretreatment process was determined using the Eq. (2).

$$\eta_{tp} = \frac{Hg_{before} - \gamma_{after} \cdot Hg_{after}}{Hg_{before}} \cdot 100\% \qquad (2)$$

η_{tp} – effectiveness of mercury removal from coal in the process of thermal pretreatment [%],
Hg_{before}– mercury content in coal before thermal pretreatment [µg/kg],
Hg_{after} – – mercury content in coal after thermal pretreatment [µg/kg],
γ_{after}– yield of coal after thermal pretreatment [-].

 DOI: 10.1051/e3sconf/20171402007

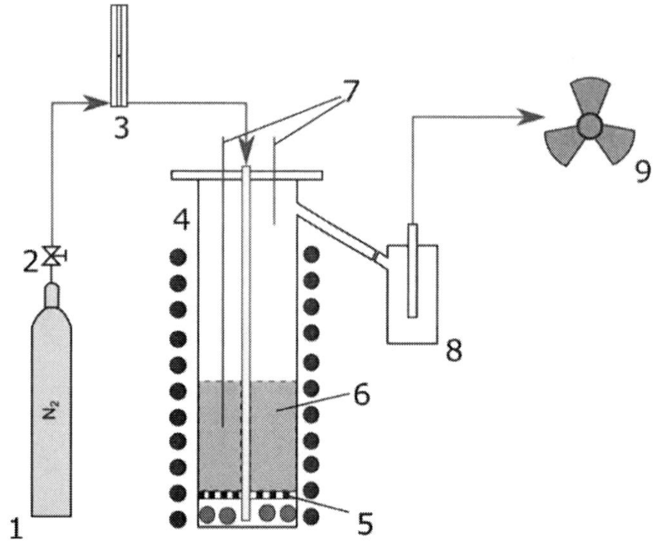

Fig. 1. Bench scale laboratory equipment for coal thermal pretreatment in fixed bed (1 - purge gas cylinder, 2 - gas reducer with control valve, 3 - rotameter, 4 - retort placed in oven, 5 - grid, 6 - coal, 7 - thermocouples, 8 – cooler, 9 - exhaust).

Additionally, the influence of the thermal pretreatment process of coal on its properties was studied. For this purpose, examinations of six coals widely used by domestic cokeries and coal fired power plants were performed (three coking coals and three steam coals). For all the selected coals, changes in calorific value and sulfur content were examined. In the case of coking coals, changes of dilatometric parameters were additionally studied. The characteristics of the examined coals are given in Table 2.

Table 2. Characteristics of hard coals used for examinations of changes in coal properties in the process of thermal pretreatment

Parameter	Symbol	Unit	Coking coals			Steam coals		
			CC1	CC2	CC3	SC1	SC2	SC3
Moisture	M^{ad}	%	2.0	1.3	1.5	9.1	3.3	2.4
Ash	A^{ad}	%	8.1	6.4	10.0	7.1	29.2	35.2
Volatile matter	VM^{daf}	%	31.0	20.5	23.8	38.1	35.9	34.7
Lower heating value	LHV^{ad}	kJ/kg	30813	32083	30753	24229	20233	18926
Sulfur	S_t^d	%	0.68	0.44	0.59	1.11	0.96	0.61
Mercury	Hg_t^d	μg/kg	60	62	106	46	100	207

3.3 Identification of the occurrence of the synergy effect resulting from the combination of both processes

For the identification of the occurrence of the synergy effect resulting from the combination of coal cleaning and thermal pretreatment process, examinations were performed of three cases of feed and clean coal produced in selected sections of the coal processing plant - Table 3. Subsequently, in accordance with the method described in paragraph 3.2, the thermal pretreatment of clean coals was conducted.

Table 3. Mercury content in feed and clean coals produced in selected sections of the coal processing plant used for the identification of the occurrence of the synergy effect resulting from the combination of coal cleaning and thermal pretreatment process

Analyzed case	Coal preparation section	Hg_t^d [μg/kg]	
		Feed coal	Clean coal
SE1	jigging	148	114
SE2	jigging	45	30
SE3	dense media separation	114	40

4 Influence of the selected method on the effectiveness of removing mercury from hard coal

4.1 Comparison of effectiveness of mercury removal from hard coal

The determined effectiveness of mercury removal from hard coal in the analyzed processes of coal cleaning and thermal pretreatment was presented in Fig.2. For each of the analyzed methods, mercury removal effectiveness was significant, although higher effectiveness was observed for the cleaning process, both washing and dry deshaling. The mercury removal effectiveness in the coal cleaning process ranged from 8 to 96% (coking coal washing: 21-76%; dry deshaling of steam coals: 8-96%), and in the coal thermal pretreatment at 300 °C it ranged from 45 to 70%. In view of the obtained results, it can be concluded that each of the analyzed methods could potentially be used for reducing mercury emissions from the coal utilization processes. Nevertheless, the analyzed methods show a wide variation in effectiveness, especially for the coal cleaning process.

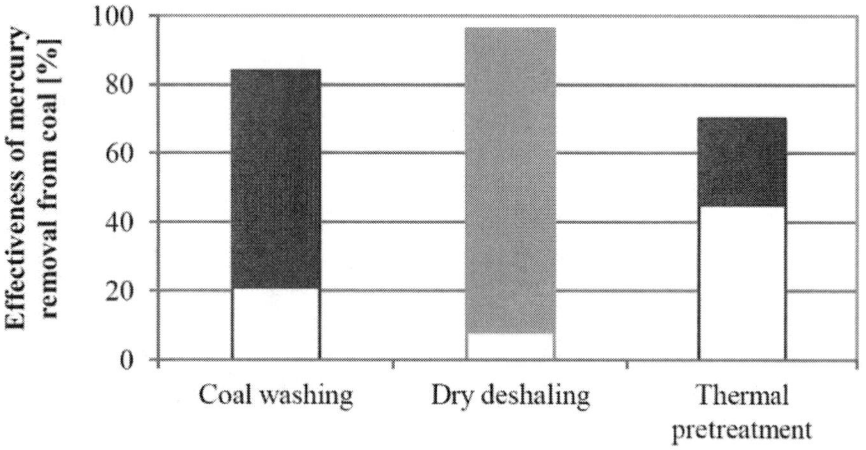

Fig. 2. Comparison of mercury removal efficiency from hard coal in processes of coal cleaning and coal thermal pretreatment

An increase in the amount of mercury removal from coal could be obtained by the combination of coal cleaning and coal thermal pretreatment. The synergy effect resulting from the combination of these methods is presented in Fig. 3. For each of the analyzed cases, the beneficial effect was obtained.

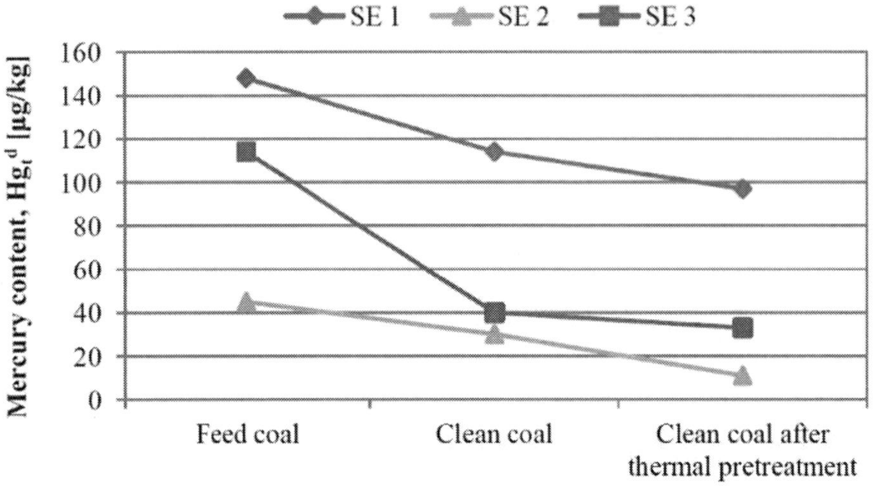

Fig. 3. Changes of mercury content in hard coal resulting from the combination of coal washing and thermal pretreatment at 300°C

4.2 The influence of the thermal pretreatment process of coal on its properties

In most of the investigated samples, the process of coal thermal pretreatment at 300°C allows for an improvement in the quality of coal, i.e.: a decrease in moisture content, an increase in the calorific value (Fig. 4) as well as partial sulfur removal (Fig. 5). Thus, the application of this process will enhance the properties of hard coal used for heat and power production.

Nevertheless, the process of thermal pretreatment (300 °C) may result in a deterioration in the coking properties, which limits the applicability of this method for coking coals. The changes of dilatometric parameters are shown in Table 4. Especially significant changes were observed for maximum dilatation (b).

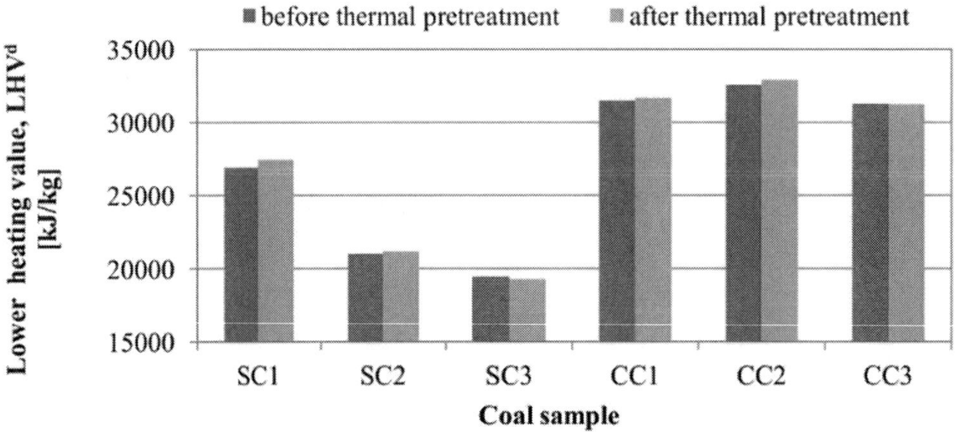

Fig. 4. Changes of lower heating value of hard coals in the process of thermal pretreatment at 300 °C.

DOI: 10.1051/e3sconf/20171402007

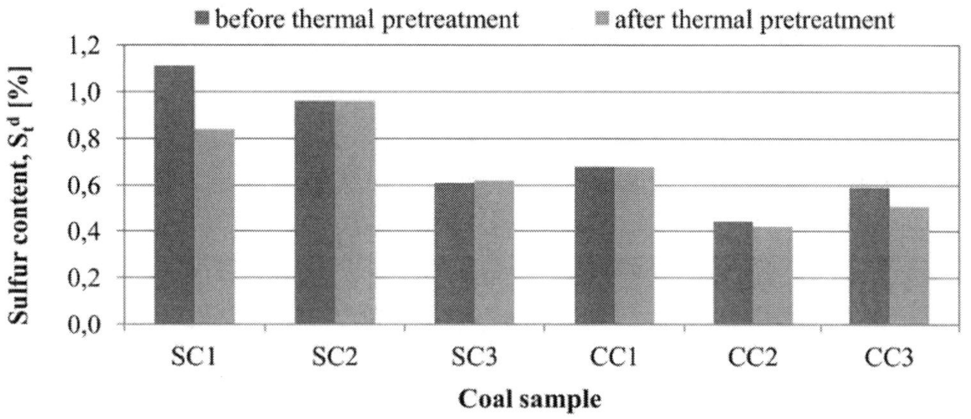

Fig. 5. Changes of sulfur content in hard coals in the process of thermal pretreatment at 300 °C.

Table 4. Changes of dilatometric parameters in coking coals in the process of thermal pretreatment at 300 °C.

Analyzed coal samples		t_I [°C]	t_{II} [°C]	t_{III} [°C]	a [%]	b [%]
CC1	before thermal pretreatment	387	436	469	29	43
	after thermal pretreatment	372	440	467	21	-5
CC2	before thermal pretreatment	408	453	485	28	16
	after thermal pretreatment	414	547	550	25	-24
CC3	before thermal pretreatment	394	443	481	30	58
	after thermal pretreatment	403	469	490	28	-27

t_I - softening temperature; t_{II} - temperature of maximum contraction, t_{II} - temperature of maximum dilatation; a - maximum contraction; b - maximum dilatation.

4.3 Guidelines for the selection of a method for mercury removal from coking coal

The recommended methods of mercury removal from coking coals are presented in Fig. 6. In the case of coking coals, mercury removal should be based on the coal washing process, which is already commonly used. Practically, only this method could ensure a coal quality which is required by the customers (the content of ash, sulfur and alkaline compounds as well as coking properties, etc.). In the case of coking coals, due to a deterioration in the coking properties of coal, the process of thermal pretreatment at the temperature of 300 °C or higher is not acceptable.

The additional possibility of mercury removal from coking coals depends on the type of coke oven battery. In the case of cokeries using gravity charging coke oven batteries, a further mercury removal is possible by applying the coal pre-drying process. The coal pre-drying process is used for increasing the bulk density of coal in the coke oven chamber. This improves the quality of the produced coke or gives an opportunity to increase the share of semi-soft coking coals in the blend instead of more expensive and less widely available hard coking coals, without a deterioration in the quality of the produced coke. The coal pre-drying process is conducted at 200 °C, at which temperature the deterioration of the coking properties of coal does not take place [17]. In this range of temperatures, there occurs a partial mercury removal from coal. Thus, the additional possibility of mercury removal from coal could be obtained without applying a separate unit, because it would take place in the coal pre-drying unit. It should be pointed out, that the implementation of the coal pre-drying process will require removing mercury from process gases using sorbents. In the case of cokeries using the coke dry quenching technology, coke dust could be used as a sorbent [18].

In the case of cokeries using stamp charging coke oven batteries, mercury could be removed from coal only in the washing process.

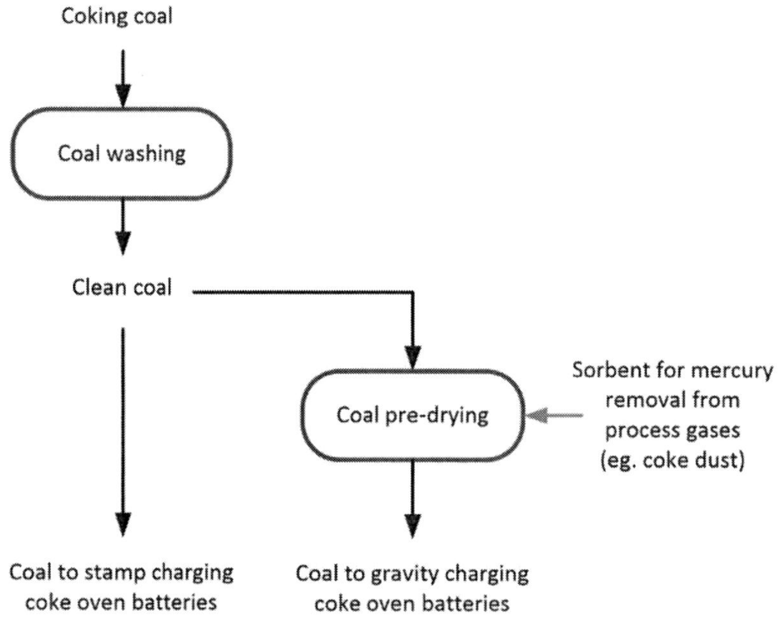

Fig. 6. Block diagram of recommended methods of mercury removal from coking coals

It should be mentioned that in the process of coal washing the middling products are also produced, which are widely used for heat and power production. In the case of a high mercury content in the middling products, its removal could be obtained in the process of thermal pretreatment.

4.4 Guidelines for the selection of a method for mercury removal from steam coal

The block diagram of the proposed methods of mercury removal from steam coals is presented in Fig. 7. For all steam coals, the dry deshaling process is recommended. This process allows for the removal of mercury as well as a decrease in ash content and thus an increase in calorific value.

The additional possibility of mercury removal from steam coal is provided by the thermal pretreatment of clean coal from the dry deshaling process. The performed analysis of the combination of the coal cleaning and thermal pretreatment process has shown the synergy effect. Nevertheless, the application of such a combination is recommended only for coals in which mercury occurs in organic matter as well as in the inorganic constituents characterized by a low temperature of mercury release. In order to determine the applicability of the combination of the coal cleaning and thermal pretreatment process, bench scale examinations are necessary. It is worth mentioning that the thermal pretreatment of coal would provide additional benefits in the form of an improvement in the quality of steam coal (e.g. a decrease in sulfur content).

The proposed methods of mercury removal from steam coal enable the production of low-mercury coal for power stations, combined heat and power plants, gasification plants, as well as low-mercury household fuel.

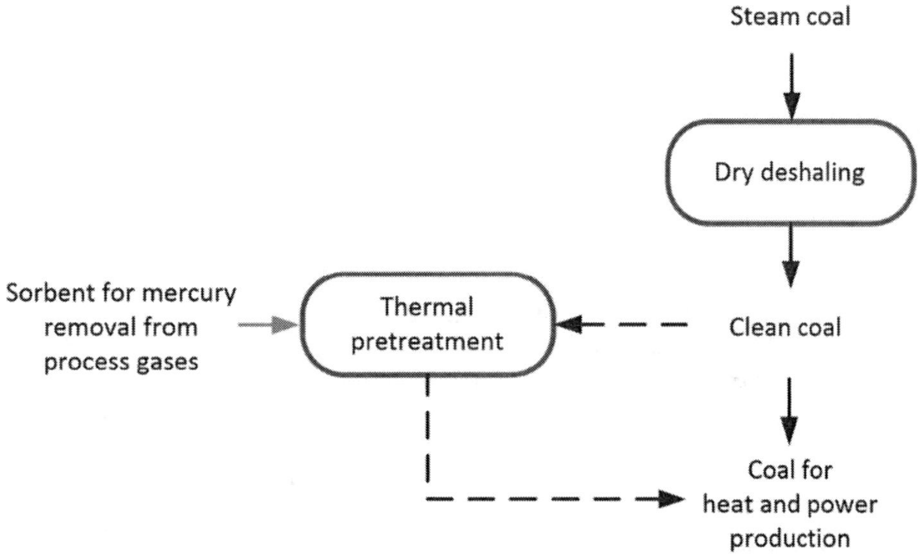

Fig. 7. Block diagram of recommended methods of mercury removal from steam coals.

It should be pointed out that, mercury removal from steam coal can also be performed in the process of coal washing. This particularly applies to coals, for which ensuring quality parameters (e.g. limits of ash content and heating value) requires the application of the coal washing process.

5 Conclusions

Each of the analyzed methods i.e. coal washing, dry deshaling, thermal pretreatment and the combination of these methods allows for the removal of mercury. However, it should be pointed out, that the selection of an appropriate method should take into account the effectiveness of mercury removal from coal, the direction of coal utilization as well as the influence of the analyzed methods on coal properties. For the removal of mercury from hard coal, depending on the direction of its utilization, the following methods are proposed:

- for coking coals used in stamp charging coke oven batteries - the coal washing process,
- for coking coals used in gravity charging coke oven batteries - the coal washing process combined with the coal pre-drying process,
- for steam coal - the dry deshaling process; in the case of coal for which ensuring the quality parameters is required - the coal washing process,
- for steam coal in which mercury occurs in organic matter as well as in the inorganic constituents characterized by low temperature of mercury release - the dry deshaling process combined with the thermal pretreatment process.

This paper was prepared as a part of the statutory activity of the Faculty of Energy and Fuels at the AGH University of Science and Technology, No. 11.11.210.213

References

1. R. Eisler, *Mercury hazards to living organisms,* (Taylor and Francis Group, Boca Raton 2006).
2. B. Gworek, J. Rateńska, *Environmental Protection and Natural Resources* **41**, 614-623 (2009) [in Polish].
3. L.L. Sloss, *Scientific and Industrial Conference*, Krakow (2015).
4. E.G. Pacyna, J.M. Pacyna, F. Steenhuisen, S. Wilson, *Atmos. Environ.* **40**, 4048-4063 (2006).
5. P. Burmistrz, K. Kogut, M. Marczak, J. Zwoździak, *Fuel Process. Technol.* **152**, 250-258 (2016).
6. PSE S.A., *Miesięczne raporty z funkcjonowania Krajowego Systemu Elektroenergetycznego i Rynku Bilansującego,* (2016) [in Polish], http://www.pse.pl/index.php?modul=8&id_rap=212.
7. Ministry of Economy of the Republic of Poland, *Prognoza zapotrzebowania na paliwa i energię do 2030 roku*, Warszawa (2009) [in Polish].
8. T. Wenecki, M. Jarno, *Koksownictwo Conference*, Karpacz (2015) [in Polish].
9. M. Mastalerz, A. Drobniak, *Int. J. Coal. Geol.* **62**, 223-236, (2005).
10. T. Dziok, S. Strugała, A. Rozwadowski, M. Macherzyński, A. Ziomber, *Mineral Resources Management*, **31(1)**, 107-122, (2015) [in Polish].
11. F.D. Guffey, E.A. Bland, *Fuel Process. Technol.* **85**, 521-531 (2004).
12. T. Chmielniak, E. Misztal, I. Mazurek, M. Sajdak, *Przem. Chem.* **94(4)**, 480-486 (2015) [in Polish].
13. T. Dziok, S. Strugała, A. Rozwadowski, A. Okońska. *Przem. Chem.* **93(12)**, 2034-2037 (2014) [in Polish].
14. S.F. Diehl, M.B. Goldhaber, J.R. Hatch, *Int. J. Coal. Geol.* **59**, 193-208 (2004).
15. A. Ohki, K. Sagayama, S. Tanamachi, *Powder Technol.* **180**, 30-34 (2008).
16. Ł. Uruski, J. Górecki, M. Macherzyński, T. Dziok, J. Gołaś, *Fuel Process. Technol.* **140**, 12-20 (2015).
17. P. Żarczyński, *Doctoral dissertation*, AGH University of Science and Technology, Faculty of Energy and Fuels, Krakow (2015) [in Polish].
18. P. Burmistrz (red.), *Fuels technology - challenges and opportunities,* Collective work, (Faculty of Energy and Fuels, AGH University, Kraków, 2016) [in Polish].

Mercury emissions from polish pulverized coal-fired boiler

Michał Wichliński[1,*], *Grzegorz* Wielgosz[1] , and *Rafał* Kobyłecki[1]

[1] Energy Engineering Department, Faculty of Infrastructure and Environment, Czestochowa University of Technology, Poland

Abstract. The current paper presents the research results carried out at one of Polish power plants at a pulverized hard coal-fired 225 MW unit. The research was carried out at full load of the boiler (100% MCR) and focused on analysis of mercury content in the input fuel and limestone sorbent for wet flue gas desulfurization (FGD) system, as well as investigation of mercury content in the combustion products, i.e. fly ash, slag, FGD product (gypsum) and FGD effluents (waste). Within the framework of the present study the concentration of mercury vapor in the exhaust gas was also investigated. The analysis was performed using Lumex RA-915+ spectrometer with an attachment (RP-91C). The measurements were carried out at three locations, i.e. before the electrostatic precipitator (ESP), downstream the ESP, and downstream the wet FGD plant. Design of the measurement system allowed to determine both forms of mercury in the flue gas (Hg^0 and Hg^{2+}) at all measurement locations.Based on the measurement results the balance of mercury for a pulverized coal (PC) boiler was calculated and the amount of mercury was assessed both in the input solids (fuel and sorbent), as well as the gaseous and solids products (flue gas, slag, ash, gypsum and FGD waste).

1 Introduction

Getting rid of pollutants, such as sulfur dioxide, nitrogen oxides or particulate matter, that are emitted to the atmosphere due to the combustion processes associated with the production of electricity and/or heat has become an important issue. The next step in the aim to protect the environment will be focused on the reduction of heavy metal emission, particularly mercury, cadmium, or lead [1].

One of the main instruments supporting the reduction of mercury emission is Minamata Convention that had been drawn up in Kumamoto, Japan, in October 2013 and signed by the Polish Government in September 2014. The Convention has become an international act with a global reach; its main aim is to protect both human health and the environment from negative affections brought about by anthropogenic emissions of mercury and its compounds. Until the end of 2016 the Minamata Convention had been ratified by 35 countries [2].

Mercury is usually determined in Earth's crust at the level of roughly 0.05 ppm. The most important mercury-containing minerals are: cinnabar (HgS), calomel (Hg_2Cl_2), and native

* Corresponding author: michal.wichlinski@gmail.com

mercury, all highly toxic, as well as the dimethylmercury that is formed in aquatic environment by microorganisms. The dimethylmercury is particularly dangerous since it is absorbed by and accumulated in living organisms [3].

New BAT (Best Available Technology) rules, which will soon be implemented for mercury emission associated with the combustion of solid fuels in power plants, very strictly point out the need to control and minimize the emission of mercury particularly that the emission from a single Polish power plant boiler may even yield roughly 50 kg/year [5]. The Hg-problem might become even more difficult for China where in 2010 the anthropogenic emission of mercury was estimated at roughly 576 Mg (i.e. over 35% of total global emission) [6].

It is estimated that the total emission of mercury yield somewhere between 5 and 15 thousand ton per year, and roughly 30-50% of that emission is of associated with human activity. From the anthropogenic emission more than half of mercury compounds emitted to atmosphere is contributed by the combustion of fossil fuels since millions tons of hard coal and lignite are being burned every year to produce electricity and heat.

High temperature conditions in the furnaces favor series of chemical reactions leading to the decomposition of mercury-containing chemical compounds. Practical result of those reactions is the presence of only gaseous metallic mercury, Hg^0,in the flue gas at temperatures above 900°C.Cooling of the flue gas and the presence of such components as NO_2, HCl, SO_2, H_2O, or Cl_2, may then favor the oxidation of mercury and/or their adsorption by fly ash particles. As a result, mercury emitted from fossil fuel combustion plants may exist in three forms, i.e. either elemental mercury (volatile metallic mercury) gas, Hg0,oxidized gaseous mercury (volatile ionic mercury), Hg^{2+}, or mercury adsorbed on dust particles, $Hg_{(p)}$ [4, 7].Due to very low mercury content in coals (usually 80-100 ng/g as in the case of Polish coals [16, 17, 18]) and its random distribution in fuel particles the exact measurement of mercury becomes quite difficult. Most common mercury analyzers are operated based on atomic adsorption spectrometry with atomic vapor generation (CV-AAS). In that method the sample is burned and mercury vapors are accumulated on a gold sorbent trap; later on then deposit is heated and mercury vapors are determined. Another way to detect mercury is the Zeeman spectroscopy method with high frequency modulated polarization of light; the method enables to determine mercury content without accumulation on the gold sorbent. The sample is introduced into a heated chamber where it is burned and the mercury vapor concentration is determined on-line. Some other convenient way is the use of mercury traps. Despite various measurement techniques applied, the results and the determination of mercury content in solid samples are comparable and in a good agreement, as confirmed e.g. in [18].

So far, the investigation of mercury emissions from fossil fuel burning power plants has been the subject of several publications. Tang and Pan [8] reported emission of mercury from two 600 MW pulverized coal fired boiler systems. One of the system was equipped with wet flue gas desulphurization (FGD) facility and electrostatic precipitator (ESP).The mercury was introduced to the system with coal and limestone and the overall mercury load was estimated as 13.62 g/h. The largest amount of mercury, 3.73 g/h, was captured by wet FGD plant; accumulation on fly ash particles was just 0.7 g/h, while the mercury in the flue gas was calculated as roughly 5.27 g/h. The other boiler system investigated by Tang and Pan [8] was additionally equipped with an SCR (selective catalytic reduction plant) to control the emission of NOx. The main source of Hg was again coal and sorbent and mercury input to that boiler system was estimated as 7 g/h. The amount of mercury emitted to atmosphere was roughly 0.17 g/h; the majority of Hg was captured by fly ash particles (3.77 g/h) and gypsum (2 g/h).

Pudasaineeet al. [9] also investigated the emission of mercury from a number of South Korean supercritical 500-600 MW pulverized hard coal fired power plants. The power generating facilities were equipped with SCRs, ESPs and wet FGDs. The authors reported that the total amount of mercury was introduced with coal and its concentration in the flue gas in the furnace was about 27.3 mg/Nm^3 and that the amount of Hg^{2+} in the flue gas downstream the SCR facility increased significantly from 3.7% to 14.7%. The measurements carried out downstream the ESP indicated that the amount of mercury in the flue gas decreased to roughly 8.5 mg/Nm^3 and over 82% of the mercury was in the form of Hg^{2+}. The measurements at the stack indicated that downstream the FGD plant the emission of mercury decreased to 2.9 mg/Nm^3, and the share of Hg^{2+} was roughly 46%. As reported by the authors, the share of various forms of mercury determined for other power generation facilities were quite similar.

Burmistrz et al. [10] investigated the emission of mercury from two (370 MW and 225 MW) Polish power plants burning hard coal. One of the facilities was equipped with an SNCR (selective non-catalytic reduction system to control the emission of NOx), was well as the ESP and wet FGD. As reported, over 10.5 g/h of mercury was introduced to the boiler system. Downstream the ESP the mercury content in the exhaust gas was reduced to roughly 4 g/h, and downstream the wet FGD it was decreased to 2.5 g/h. The reduction was mainly caused by fly ash particles which carried more than 7.7 g/h of mercury. The gypsum (product of wet FGD plant)carried out slightly more than 1.8 g/h of Hg. In case of the other facility the mercury input was estimated at roughly 7.9 g/h. Lack of the SNCR brought about that the amount of mercury captured by the ESP was lower, just roughly 4.8 g/h. The operation of wet FGD was reported quite poor since only 0.53 g/h of mercury was captured there and over 3.6 g/h was emitted to the atmosphere via the stack.

As briefly discussed above, the data and results reported by various authors may differ quite significantly. Consequently, detailed industrial data are still needed in order to assess properly current emissions of mercury. Therefore, the intention of this paper is to provide some data on emission and balance of mercury in a 225 MW pulverized hard coal fired power generation plant in Poland.

2 Research methodology

The tests were carried out at a225 MW plant burning Polish hard coal. The boiler system was equipped with SNCR, ESP, and wet FDG. The investigations were carried out at two locations, i.e. downstream the ESP and downstream the wet FGD. Unfortunately, due to technical reasons it was not possible to carry out measurements upstream the ESP. The investigations were focused on flue gas sampling from those two locations and the determination of mercury, Hg^0, and Hg^{2+}. Furthermore, the amount of mercury in coal, sorbent, fly ash, slag, and gypsum (wet FGD product) was also determined. Sampling locations are schematically shown in Fig. 1.

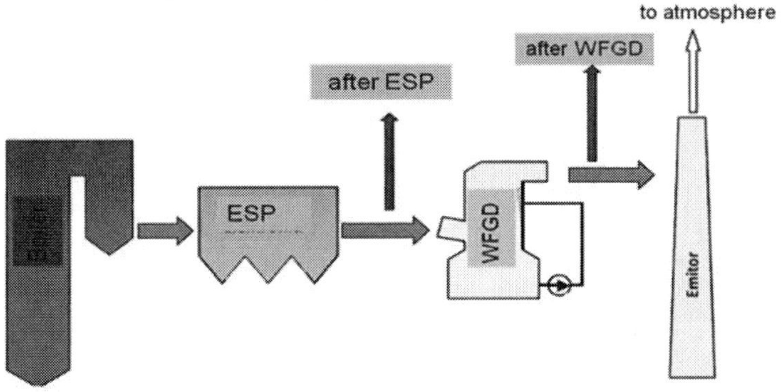

Fig. 1. Locations of mercury samplings.

The determination of mercury was carried out with the use of a mercury analyzer RA-915+ produced by Lumex corporation. The analyzer is a spectrometer operated based on the Zeeman effect (Zeeman Atomic Absorption Spectrometry with High Frequency Modulated Light Polarization, ZAAS-HFM) and allows to avoid the need for accumulation of mercury on a gold sorbent trap. The concentration of mercury in the gas phase can be determined directly by sucking in the gas to the measurement section of the spectrometer via an internal pumping device.

In order to determine mercury concentrations in solid samples the spectrometer 'Lumex' was coupled to an additional RP-91C device. The measurements could be carried out for mercury concentrations between 0.5 mg/kg and 10,000 mg/kg; the maximum error was ± 20%. The speciation of mercury was possible due to the determination of Hg^0 mercury directly by Lumex spectrometer, and the application of a part of commonly known Ontario Hydro method that allowed the investigation of Hg^{2+} mercury. The sampling device that was inserted into the measurement zone consisted of two equally long tubes connected to each other. The tubes were heated with by a ribbon heater that allowed to maintain the temperature high enough to prevent any condensation of steam. The temperature of the heating band was maintained by a temperature controller. The flue gas was parallelly sampled: via one of the tubes directly to Lumex spectrometer, while the other tube transported the flue gas to the atmosphere via a glass impinger trap. Each of the impingers in the trap contained 100 ml of KCl, allowing the dissolution of mercury Hg^{2+}. The impingers were placed in specially-designed containers where constant temperature of roughly 10°C was maintained. No filters were placed upstream the impingers and the spectrometer. The determination of Hg^{2+} mercury captured in the impingers was carried out in laboratory with the application of a cold vapor technique, where the mercury dissolved in KCl was snatched away with $SnCl_2$ (100 g/dm^3) [15]. Both sampling tubes that were inserted into the measurement zone were made of stainless steel while all other connectors were manufactured of silicone. One single measurement consisted of two 30-minute sampling times separated by a 60 minute interval. In each case the results presented in this paper are the averages of two corresponding measurement runs. The schematics of the sampling system is shown in Fig. 2.

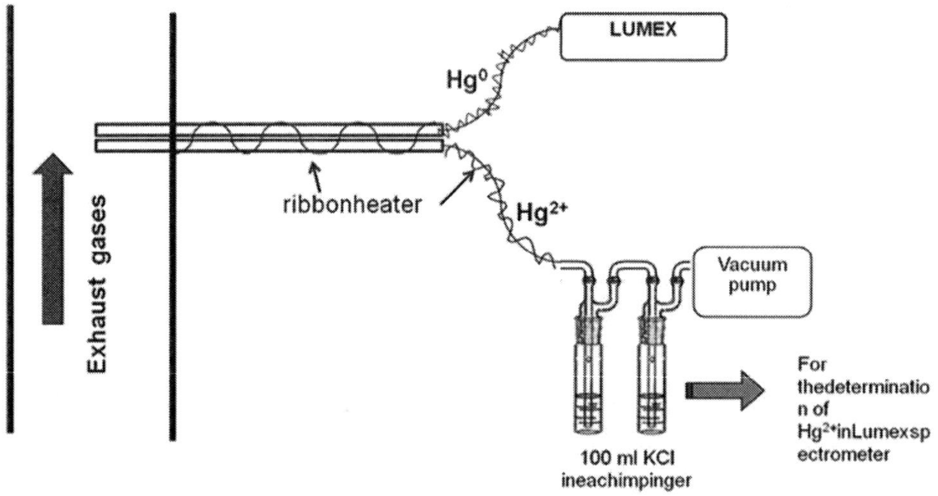

Fig. 2. Sketch of the sampling system to determine both mercury Hg0 and Hg2+.

DOI: 10.1051/e3sconf/20171402008

3 Results and discussion

During the measurement campaign at the plant the boiler was fired with a blend of hard coals from Polish mines. Ultimate and proximate analysis of the coals is shown in Table 1.

Table 1. Main parameters of coal combusted at the plant during the tests (except for transient moisture, all other values are given as 'air dry')

Parameter	Unit	Value
Transient (external) moisture, as received	%	13.6
Internal (hygroscopic) moisture	%	6.9
Ash	%	27.8
Volatile matter	%	23.2
Fixed carbon	%	42.1
Higher heating value, HHV	kJ/kg	21600
Lower heating value, LHV	kJ/kg	20700
Carbon, C	%	55.2
Total mercury, Hg	ng/g	153.6

The content of mercury in substrates and combustion products are shown in Table 2 as determined at the plant. Based on the data on mercury concentrations in various substances (cf. Table 2), as well as the information on flue gas, ash, slag, coal, and sorbent flow rates in the boiler system the flows of mercury (mg/s) at various locations were calculated and also shown in Table 2. As can be seen, the balance of mercury is in a very good agreement since the difference between mercury input and output does not exceed 11%.

Table 2. The balance of Hg for a 225 MW pulverized hard coal fired boiler

		Hg content	Flow rates: solids [kg/s]; gas [Nm^3/s]	Hg flow rate [μg/s]	Mass balance [%]
Inlet	Coal	135.6[ng/g]	22.43	2259.6	89.8
	Limestone	9.0[ng/g]	2.25	20.0	
Outlet	Fly ash	180.1[ng/g]	3.78	728.3	
	Slag	5.2[ng/g]	1.62	5.29	
	Gypsum	156.1[ng/g]	4.11	587.3	
	FGD waste	111.3[ng/g]	0.11	4.57	
	Hg^0 in flue gas	1450[ng/m^3]	187.5	272.7	
	Hg^{2+} in flue gas	2400 [ng/l]	187.5	450.0	

Figure 3 shows the amount of mercury introduced with coal and sorbent to the boiler system and the mercury that is transported with combustion products (fly ash, slag, flue gas, gypsum, and FGD waste). As it may be seen, the mercury in the combustion products was mainly determined in the fly ash, gypsum and the flue gas. In case of the flue gas roughly 60% of mercury was in the form of Hg^{2+}.

DOI: 10.1051/e3sconf/20171402008

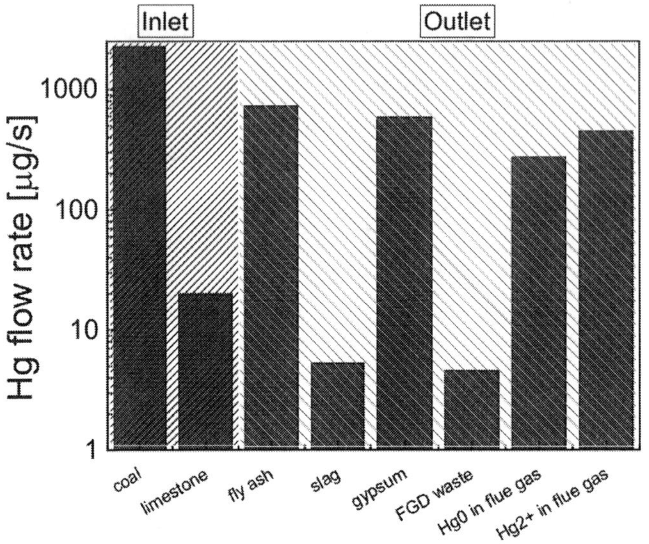

Fig. 3. Mercury mass balance for the pulverized coal fired 225 MW boiler.

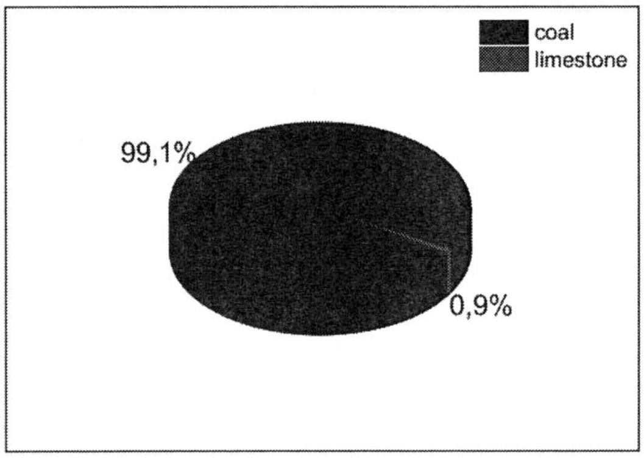

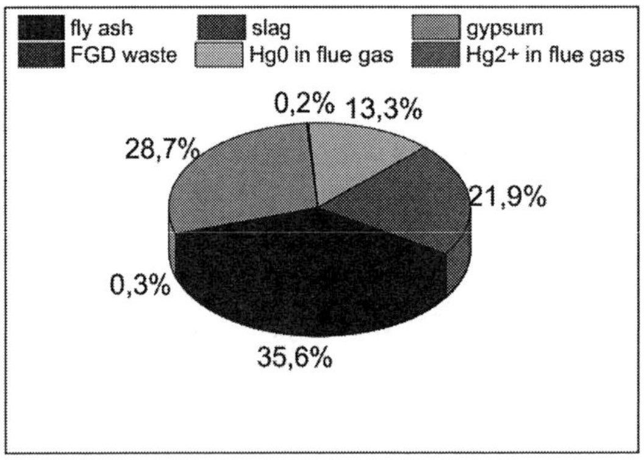

Fig 4. The shares of individual input and output components and their contribution to mercury inventory.

The shares of individual substrates and products, as well as the quantities of mercury and mercury inventory for the boiler system are shown in Fig. 4. Analysis of the results indicates that the major source of mercury (over 99%) is coal; the amount of mercury introduced with sorbent is negligible (less than 1%). As for mercury output, the largest shares, roughly 35% of total Hg, are contributed by both fly ash and flue gases. In the latter case, the majority of mercury is in the form of Hg^{2+} (22% of the total mercury input) – the contribution of Hg^0 is just 13.3% of the input mercury. The contribution of gypsum mercury is slightly lower and has been estimated as roughly 29% of total mercury input load. Since the oxidized form of mercury can be easily dissolved in water [12,13] it is commonly accepted that wet FGDs are usually very effective and can remove over 90% of Hg^{2+} mercury from the flue gas. However, in some cases, the efficiency of mercury removal in wet FGD plants may be, surprisingly, significantly lower, usually due to non-optimized operation or inappropriate chemical conditions in the scrubber [11].

4 Summary

The results presented and discussed in the current paper may be briefly summarized as follows:

1) For the boiler system investigated, the main source of mercury (over 99%) was coal containing almost 136 ng/g of Hg.

2) Roughly 36% of the total mercury was discharged with exhaust flue gases; however, comparable amounts were determined in the fly ash (35% of total Hg) and gypsum (29% of the input mercury).

3) The majority of flue gas mercury was determined as Hg^{2+} contributing to roughly 22% of the total mercury input. The share of Hg^0 was lower and estimated as roughly 13%of the total mercury input.

References

1. L. Sloss, *Legislation, standards and methods for mercury emission control*, IEA Clean Coal Centre, **CCC/195** (2012).
2. http://www.mercuryconvention.org/ (access 01.09.2016).
3. W. Seńczuk W. (Eds.): *Toksykologia. Podręcznik dla studentów, lekarzy i farmaceutów* Wydanie IV. Warszawa: Wydawnictwo Lekarskie PZWL, (2002) (in polish).
4. S. Hławiczka, *Rtęć w środowisku atmosferycznym*, Instytut Podstaw Inżynierii Środowiska Polskiej Akademii Nauk, Zabrze, (2008) (in Polish).
5. A. Glodek, J.M. Pacyna, *Mercury emission from coal-fired power plants in Poland*, Atmospheric Environment **43** (2009).
6. Zi-Jian Zhou, Xiao-Wei Liu, Bo Zhao, Zhen-Guo Chen, Hai-Zhong Shao, Le-Le Wang, Ming-Hou Xu, *Effects of existing energy saving and air pollution control devices on mercury removal in coal-fired power plants*, Fuel Processing Technology **131** 99–108 (2015).
7. K.C. Galbreath, C.J. Zygarlicke, *Mercury speciation in coal combustion and gasification flue gases*, Environmental Science & Technology **30** 2421–2426 (1996).
8. N. Tang, S. Pan, *Study on mercury emission and migration from large-scale pulverized coal fired boiler*, Journal of Fuel Chemistry and Technology, **41 (4)** (2013).
9. Deepak Pudasainee, Jeong-Hun Kim, Young-Sik Yoon, Yong-Chil Seo, *Oxidation, reemission and mass distribution of mercury in bituminous coal-fired power plants with SCR, CS-ESP and wet FGD*, Fuel **93** 312–318 (2012).

10. P. Burmistrz, K. Kogut, M. Marczak, J. Zwoździak, *Lignites and subbituminous coals combustion in Polish power plants as a source of anthropogenic mercury emission*, Fuel Processing Technology **152** 250–258 (2016).
11. S. Niksa, N. Fujiwara, *The Impact of Wet FGD Scrubbing on Hg Emissions From Coal-Fired Power Stations*, The MEGA Symposium, Washington, DC, (2004).
12. K.E. Redinger, A. Evans, R. Bailey, P. Nolan, *Mercury Emissions Control in FGD Systems, EPRI/DOE/EPA Combined Air Pollutant Control System*, Washington, DC, (1997).
13. M.S. DeVito, W.A. Rosenhoover, *Hg Flue Gas Measurements from Coal-fired Utilities Equipped with Wet Scrubbers*, 92nd Annual Meeting of the Air & Waste Management Association, St. Louis, (1999).
14. J. Górecki, A. Łoś, M. Macherzyński, J. Gołaś, P. Burmistrz, K. Borovec, *A portable, continuous system for mercury spaeciation in flue gas and process gases,* Fuel Processing Technology, Volume **154** 44-51 (2016).
15. R. Kobyłecki, M. Wichliński, G. Wielgosz, Z. Bis, *Emission of mercury from Polish large-scale utility boilers*, J. of Ecol. Eng., vol. **17/5** 128–131 (2016).
16. M. Wichliński, R. Kobyłecki, Z. Bis, *The investigation of mercury contents in Polish coal samples*, Archives of Env. Protection, vol. **39/2** 141–150 (2013).
17. J. Gołaś, A. Strugała (Eds.), *Mercury as a coal combustion pollutant*, AGH, Kraków.
18. B. Białecka, I. Pyka (Eds.), *Rtęć w polskim węglu kamiennym do celów energetycznych i w produktach jego przeróbki*, GIG, Katowice (2016), (in Polish).

Analysis of gas-phase mercury sorption with coke and lignite dust

Marta Marczak[1,*], *Piotr* Burmistrz[1], *Mateusz* Karczewski[1], *Jakub* Szczurowski[1], *Kamil* Gałkowski[1]

[1]AGH University of Science and Technology, Faculty of Energy and Fuels, al. Mickiewicza 30, 31–464 Krakow, Poland

Abstract. In recent years the problem of mercury emission became a widely discussed topic. Its high impact is caused by its toxicity and ability to accumulate in living organisms, properties that justified the United States Environmental Protection Agency (US EPA) to classify mercury as hazardous pollutant. The problem of mercury emission is crucial for countries like Poland, where the most of the emission is caused by coal-depended energy sector. Current technology of mercury removal utilizes adsorption of mercury on the surface of activated carbon. Due to high price of activated carbon, this technological approach seems to be uneconomical and calls for cheaper alternative. One possible solution can be usage of other sorptive materials obtained from thermal processes like coke production. Example of such material is coke dust obtained from dry quenching of coke. The aim of this work was to analyse the sorption potential of lignite and coke dust and determine parameters influencing mercury behaviour during combustion.

1 Introduction

Prevention of mercury atmospheric emission gained a lot of popularity in recent years. Mercury is a heavy metal generally known for its toxicity, as well as high persistence in living organisms once absorbed. In spite of its trace presence in the nature, mercury and its derivatives were classified by United States Environmental Protection Agency (US EPA) as adverse for human health [1,2].

Mercury migration in natural environment is connected not only with natural sources, but also with those of anthropogenic origin. These include mainly gold mining and coal combustion processes. Coal combustion alone is responsible for annual emission of 869 Mg of Hg on global scale [3], which creates noticeable potential for emission reduction.

Poland is one of the biggest mercury emitters in Europe with emission of 10-20 Mg per annum [3,4,5], originating mostly from combustion of coal for heating and energy production. One reason of such share in emission is the energy production profile in which up to 87% of energy is produced is from subbituminous and lignite coal [6]. Another reason

* Corresponding author: mmarczak@agh.edu.pl

is that subbituminous coals and lignites contain from 18 – 518 μg Hg kg^{-1} and 60 – 665 μg Hg kg^{-1}, respectively.

During combustion process Mercury can take several forms dependent on the conditions of combustion. Higher temperatures above the range of 90-120°C turn it into gaseous form Hg^0 which is the hardest to remove. If the temperature is lower, mercury becomes more reactive and it can either get absorbed by fly ash ($Hg_{(p)}$), or turn into ionic form Hg^{2+} that forms compounds with other flue gas constituents. These two forms are much easier to remove than the Hg^0. Thus, the removal technology should incorporate influence of temperature to promote acquisition of $Hg_{(p)}$, and Hg^{2+} [10].

Studies from various pilot installations as well as those which already were included in industrial systems, hints that oxidation of mercury to its ionic form Hg^{2+} through its intermediate form $Hg_{(p)}$ is influenced by residual char and calcium in fly ashes as well as large content of halogens and sulphur in used fuel [11,12]. These studies also indicate that fuel plays key role in emissive potential of the system. Combustion of fuel with suboptimal elemental composition (small amount of Cl, Br, Fe and large content of Ca) might promote mercury migration to flue gas, instead of partial concentration in bottom ashes. This forces usage of a dedicated system segment for mercury removal to aid already functioning flue gas clean up elements, which can, but only to some extent, remove mercury passively. According to experience from US industrial plants, the most promising active technologies seems to be based on adsorption [13-16].

Among mercury removal technologies adsorption with use of activated carbon seems to be most widely used. Activated carbon is as material with developed porous structure that has proven to capture up to 90% of flue gas mercury [17]. Its main drawback is the production cost. It is estimated that cost of 90% mercury reduction varies between 66 and 220 thousand of USD per 1kg of mercury [18]. Another drawback is that activated carbon is also not fitting well to the task of mercury removal due to highly developed porous structure (as seen in table 1) that cannot be used because of relatively low concentration of mercury in flue gas. Microporous structure of activated carbon decrease the kinetic equilibrium time of adsorption process to several hours. What is more, production of activated carbon causes potential environmental problem connected with its pretreatment process [6,12,18]. This is why the technology needs more environmentally friendly and cheaper sorbents. Lignite and coke dust seems to be very promising materials for this application. Coke dust is obtained as by-product of coke dry quenching process in coking plant. Its main advantage is mesoporous structure developed without any additional treatment, decreasing production cost by several times resulting in several times with only 2.5-5 times lower volume of pores, much better kinetics of adsorption allowing for high utilization of surface and lack of environmental impact due to lack of additional pre-treatment steps.

The aim of this work was to analyse lignite coal dust and coke dust, as cheap sorbents that could substitute expensive activated carbon in flue gas mercury removal process. 10 sample mixtures of polish lignite coal and coke from pulverized coal-fired boiler localized in power plant in southern Poland were used. The mercury content of coal mixtures used in Polish energy sector were determined and the parameters influencing mercury behaviour during combustion process were identified.

2 Experimental

10 samples of lignite and coke dust mixtures were taken from pulverized coal-fired boiler from power plant located in southern Poland. The samples were homogenized and air dried until stable sample mass was achieved. Subsequently, an ultimate and proximate analysis were carried out. The ultimate analysis was performed for determination of carbon, hydrogen and sulphur content with use of ELTRA CHS – 580 apparatus, according to PN-

G-04584:2001 polish standard. Chlorine content was evaluated as a content of chlorine anions in water solution with use of direct reading spectrophotometer (DR/2000 HACH). Chlorine present in sample was transferred into liquid solution by burning a 0.5 g sample under layer of chlorine-absorbing Eschka mixture (60-72% MgO and 30-36% Na_2CO_3, POCH) mixed with sample in 1:1 weight ratio. The remaining residue was transferred to boiling deionized water and prepared for spectrophotometric analysis with use of mercuric thiocyanate (99.99%, POCH, Poland) and addition of ferric ion solution (POCH, Poland), according to PN-ISO 587:2000 standard.

For proximate analysis, humidity, volatile matter and ash content were determined with thermogravimetric analyser (ELTRA Thermostep) and calorific values of the mixtures were assayed with Leco AC 500 calorimeter, according to PN-81/G-04513 Polish standard.

All samples were additionally characterized for their ash content and sorbent materials were analyzed for their surface area. The ash content was determined by X-ray fluorescence using EDXRF spectrometer (Panalytical Epsilon 3XLE). Surface area of dusts were determined from nitrogen vapor adsorption/desorption isotherms in the temperature of 77 K with use of gas sorption analyzer using mass spectrometer (Quantachrome Instruments AUTOSORB-1-C). Before the measurement dusts were pretreated in vacuum degasser in 473 K for 12h.

The mercury concentration in samples was determined by cold vapor atomic spectrometry method (AMA 254 mercury analyzer). The method was applied for solid samples in two detection ranges (0.5-40 ng of Hg, and 40-600 ng of Hg) and detection limit of 0.01ng of mercury, allowing mercury content determination, even in trace amounts.

Sorptive potential of chosen dust materials was evaluated on laboratory scale. The 1g of fuel blend was combusted under stable air flow of 300 cm^3/min and at 1123 K. The vessel with sorbent was placed at the outlet of the furnace with outlet pipe temperature stabilized with heating belt. The outlet temperature was set at either 363K or 393K. The temperature of sorbent was stabilized by thermocouple (Fig. 1).

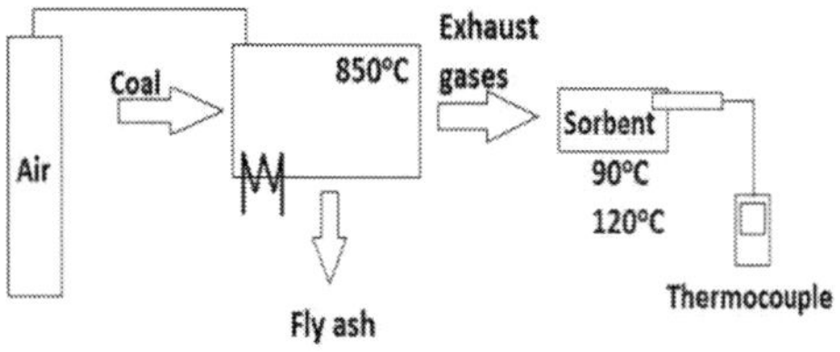

Fig. 1. Laboratory scale mercury removal measurement system for sorbent analysis

3 Results and Discussion

3.1 Sorbent comparison

Table 1 presents porous structure parameters determined from nitrogen vapor sorption/desorption isotherms in 77K. Coke dust and lignite dust are macroporous materials with lack of micropores and partial development of mesoporous structure. The total pore volume of 0.018 for coke dust is about two times smaller than that of lignite dust with its pore volume of 0.054 $cm^3 \cdot g^{-1}$, suggesting bigger sorption capacity of a lignite. Data acquired on pore sizes also suggest that micropore/mesopore/macropore share of pore volume are 33%/50%/27% and 0%/46%/53% for coke and lignite dust, respectively. Both dust have large microporous structure. Coke dust has widely developed microporous structure and low macropore structure, while lignite dust has as much of macropore volume as mesoporous but is lacking on micropore structure. Hence, their behaviour during the experiment cannot be

 DOI: 10.1051/e3sconf/20171402009

Table 2. Elemental content of combusted hard coals

Sample	M^a	V^a	A^a	Cl^a	C^a	H^a	S^a	Br^a	$q_{V,gr}^a$	Hg^a
	[% m/m]							[ppm]	[kJ·kg^{-1}]	[µg·kg^{-1}]
W5	4.6	26.28	24.0	0.31	57.1	3.94	1.14	12.3	22.8	120.4
W6	5.4	28.20	17.7	0.33	61.5	4.35	1.23	11.8	24.4	132.2

judged on pore volume alone. Nevertheless, volume of mesopores plays a key role in their suitability as a sorbent. Mesoporous structure is important factor for sorption of large molecules, like heavy metals and has deciding role in efficiency and kinetics of adsortpion in case of particles. It is also important to mention that despite lower total pore volume of coke dust it has up to 5-10 times bigger specific surface than lignite dust. Pore volume and capacity share of both dusts are also several times smaller than that of activated carbons.

Lignite and coke derived dusts are substitues of commercial activated carbons with several times lower costs and significantly smaller impact on environment on their prodcuction stage as they don't need any additional treatment steps. Due to presence of various factors that influence mercury removal, several additional analysis was performed on waste dust in order to evaluate its potential as sorbent.

Table 1. Parameters of porous structure for analyzed sorbents, based on nitrogen vapor sorption/desorption isotherms measured in 77K.

Sample	S_{BET} [m^{2}g^{-1}]	V_{DR} [cm^{3}g^{-1}]	V_{BJH} [cm^{3}g^{-1}]	V_{total} [cm^{3}g^{-1}]	SC_{Hg} [mg g^{-1}]
Coke dust (unsorted)	16,5 (16.5-32,0)	0.006	0.009	0.018	0.110-0.129
Lignite dust	3.5	0.001	0.025	0.054	0.279
Activated carbon dust	670.5	0.307	0.055	-	24.052

SC_{Hg} –mercury sorption capacity, presented as [mg/g],
S_{BET} – specific surface based on Brunauer-Emmet-Teller method,
V_{DR} – micropore volume based on Dubinin-Raduszkiewicz method,
V_{BJH} – mesopore volume based on Barret-Joyner-Halenda method,
V_{total} - total pore volume.

As coal combustion-based atmospheric mercury emission is mostly influenced by mercury content in burned fuel alongside with its elemental composition, mainly sulphur, chlorine, bromine, calcium and iron content. Table 2 presents results of elemental analysis of fuel blend samples used in the experiment, along with average concentrations of mercury and chlorine. The elemental composition shows relatively similar carbon and hydrogen content between analyzed coal blends, with average carbon content of 59% with 4% of difference between blends and average hydrogen content of 4,1% with deviation below 0.4%. Most of other basic parameters is also comparable with slight difference in ash content that ranges from 24 for W5 blend to 17.7 for W6 blend. The most important parameter – the fuel mercury content ranges between 120 and 132 µg·kg^{-1}. The average concentration of chlorine, the agent partially responsible for mercury oxidation promotion, equaled up to 0.3% with concentration of other promoting agents, bromine and sulphur of around 12ppm and 1.2%, respectively.

Low concentration of promoting agents like Cl and Br, resulted in increased production of gaseous mercury Hg^0 instead of easier to remove $Hg_{(p)}$ or oxidized mercury Hg^{2+} [15]. This confirms the need for utilization of an active method in mercury removal.

 DOI: 10.1051/e3sconf/20171402009

The ash content of fuel blends was also evaluated (Table 3). Both sample dust and blends have typical content for hard coals with high concentration of aluminium and silicon oxides and relatively low content of other compounds. Although general composition of fuel blends has low concentration of chlorine, the elevated amount of iron (<7%), and low amount of calcium (>3%), might promote oxydation of mercury along with scarce amount of chlorine, as their individual impact on the process is not fully understood.

Table 3. Metallic content of ashes for fuel and sorbents

Sample	SiO_2	Al_2O_3	Fe_2O_3	CaO	MgO	K_2O	Na_2O	P_2O_5	SO_3	MnO	TiO_2
	[%m/m]										
W5	52.29	28.71	6.85	3.16	2.32	2.79	0.86	0.41	1.45	0.08	1.01
Coke dust	42.46	28.74	9.69	6.73	2.49	2.81	1.23	1.61	2.50	0.15	1.35
Lignte dust	26.34	4.00	27.15	20.10	5.90	0.20	0.15	0.01	15.77	0.24	0.21

The characteristics of sorbents analyzed during the experiment were presented in Table 4. The best sorbent, given its purpose of atmospheric mercury emission reductor, should contain as low content of mercury and calcium, large content of chlorine, bromine, and iron. Lignite dust has mercury content of about 73.4 $\mu g \cdot kg^{-1}$. Coke dust contains around seven times less mercury that lignite dust, four times more chlorine and two times more sulphur than the lignite dust, suggesting it is more suitable for the application as a mercury adsorbent. What is more, it also contains three times less calcium than lignite dust. However, lignite dust contains noticably higher amount of iron, as shown in Table 3.

Table 4. Results of sorbent ultimate analysis

Sample			M^a	V^a	A^a	Cl^a	C^a	H^a	S^a	$q_{v,gr}^a$	Hg^a
			[%m/m]							[$kJ \cdot kg^{-1}$]	[$\mu g \cdot kg^{-1}$]
Lignite dust			8.0	49.39	5.2	0.05	58.4	5.25	0.54	22.7	73.4
Coke dust	Grain size [mm]	Y [%]	0.4	3.19	9.8	0.2	85.0	0.16	0.59	29.3	10.5
	unsorted	-									
	x>0.2	39.6	1.3	5.1	12.0	-	79.8	-	0.65	27.8	11.2
	x<0.2	34.3	2.6	6.7	17.7	-	75.1	-	1.02	26.2	12.1
	0.63>x >0.05	2.4	1.6	7.1	17.8	-	73.4	-	1.05	25.7	14.8

The scheme of dust sorbent testing system for mercury removal in combustion processes was presented in Figure 1. The system utilizes combustion of 0.5g fuel blend in controlled conditions at 1123K, constant air flow of 300 cm^3/min and exhaust gases temperature, regulated to match industrial conditions of flue gas approaching and leaving electrostatic precipitator, defined as 363K and 393K for this experiment, respectively. Results of sorbent evaluation was presented in Table 5. Efficiency of mercury removal using given sorbent was calculated according to formula (1) :

$$\eta = \frac{S_z - S_s}{C - A} \cdot 100\% \tag{1}$$

Where:

 DOI: 10.1051/e3sconf/20171402009

C - mass of mercury in fuel,
S_S - mass of mercury in sorbent before analysis,
S_Z - mass of mercury in sorbent after analysis,
A - mass of mercury in ash.

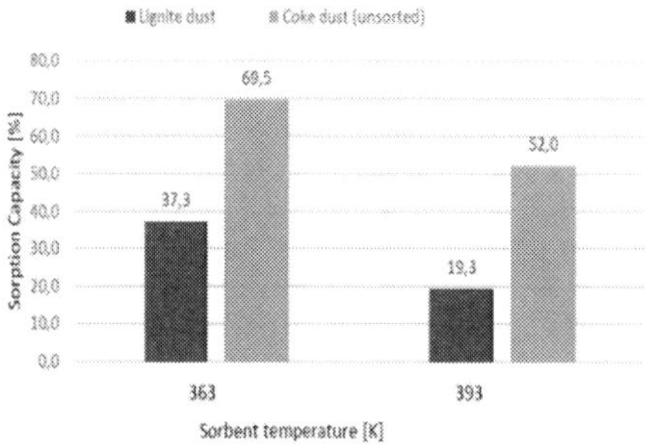

Fig. 2. Influence of sorbent temperature on Hg sorption

Based on data presented in Table 5 higher flue gas mercury sorption was achieved for coke dust, despite lignite coal having up to 2 times higher sorption capacity. Mercury take up for coke dust sorbent kept in 393K was on average around threefold higher than that of lignite coal. It may be caused by the fact that coke dust with its less developed mesoporous structure may saturate faster than lignite dust. On the other hand, lignite dust required longer time of contact with flue gas to reach its adsorption equilibrium, leading to lower overall sorption.

Both sorbents confirmed the impact of flue gas temperature on removal efficiency as much higher mercury take up was achieved in lower temperature of 363K. This phonomenon can be explained by increased Hg^0 content in higher temperature of 393K. This result suggest that mercury sorption from perspective of higher efficiency should be located after electrostatic precipitator, where flue gas temperature can be lowered to mentioned 363K. However, this setup is charged with additional cost of addition surplus gas particulate matter removal unit dedicated for sorbent dust flue gas cleanup. In case of mounting the injection system in higher temperature, the problem would be solved by presence of electrostatic precipitator after sorbent injection point. In this variant coke dust sorbent managed to remove around 52% of flue gas mercury when flue gas temperature was kept at 393K. Coke dust was also better than lignite dust, giving sorption efficiency over two times higher than lignite with sorption efficiency of around 19.3 %. Figure 3 presents simplified distribution balance of mercury during W5 fuel blend combustion. The scheme assumes that all mercury transferred from fuel to flue with exception of 0.5% that absorbed by bottom ash. During mercury sorption kept at temperature of 363K, up to 80% of mercury was emitted with flue gas to the

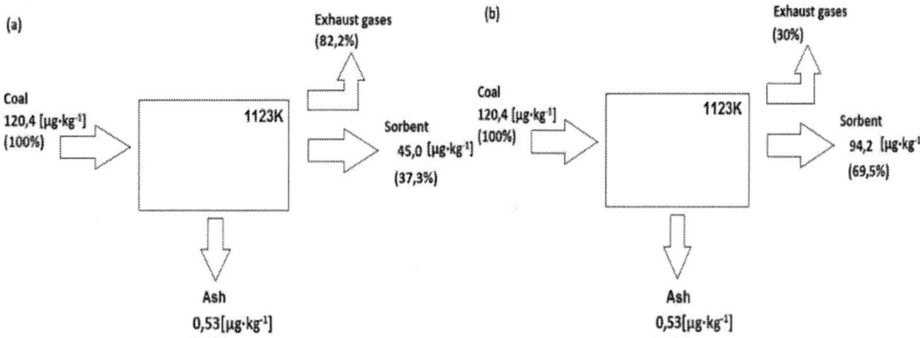

Fig. 3. Simplfied distribution balance of Hg during combustion process with (a) lignite dust (b) coke dust

atmosphere (Fig 3a). With use of coke dust sorption in the same conditions the atmospheric emission of mercury was reduced to 30% (Fig. 3b).

Table 5. Amount of mercury adsorbed by dust during coal mixtures combustion

Sample	ST [K]	Coke dust (unsorted) [μg·kg^{-1}]	Lignite Dust [μg·kg^{-1}]
W5	363	83.7	44.9
W5	393	77.9	23.3
W6	363	62.7	44.3
W6	393	58.9	20.7

3.2 Influence of grain size on sorption efficiency

Among factors that can influence efficiency of surface-based processes like adsorption, grain size seems to be one of most important, where sorption ability of given material should decrease with its grain diameter (Fig. 4). Therefore, influence of grain diameter was also tested for Coke dust, as it seemed more promising material than the lignite dust. Unsorted coke dust was ground and sieved into three fractions: above 0.2m; under 0.2mm, and between 0.063mm and 0.05mm. Dust with smallest diameter size gave the best result with 1.5 times higher efficiency than that of unsorted dust, which is in agreement with expected dependency. This fraction of dust achieved almost 100% of mercury removal efficiency. It must be noted that the producton cost of this dust increased, which was caused by higher energy output needed and material loss connected with small size dust. The efficiency of mercury removal for other fractions was considerable (Table 4).

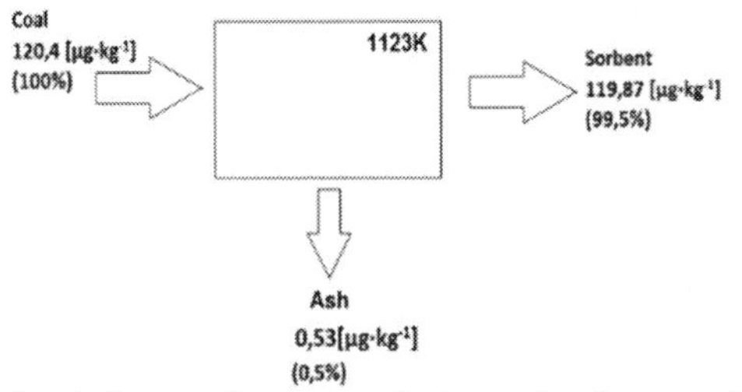

Fig. 4. Simplified distribution balance of mercury during coke dust combustion process with grain diameter between 0.065 and 0.05 mm

4 Conclusions

The aim of this study was to analyze two blends of coals combusted in polish enegetic sector. The average mercury content in these blends was 126.3 μg·kg^{-1}. Content of other elements important for mercury removal was as follows : chlorine (0.3%) bromine (12 ppm), iron (6.8%) and calcium (2.6%).

From the perspective of optimization of mercury removal, waste dusts seem to be a better choice due to their lower cost of production. Among two tested materials, lignite dust and coke dust, the latter turned out to be better in several aspects. In comparison with lignite it possess twice as much sulphur, four times more chlorine and three times less calcium. Moreover its is 33% more efficient in mercury removal. Coke dust kept at 363K adsorbed up to 83.7 μg·kg^{-1} of mercury, and up to 62.7 μg·kg-1of mercury at 393K. For lignite dust those results were much lower and equaled up to 44.9 at 363K and 23.3 μg·kg^{-1} at 393K.

Higher removal efficiency for dust in lower temperature suggest that their injection unit should be placed farther into the cleanup system, possibly after electrostatic precipitator.

However, considering economical aspects mercury removal efficiency in higher temperature is still satysfying and allows for dust capture along with other particles in flue gases without any additional investments.

The best material was also revieved for rain size mercury removal influence evaluation. Coke dust was also analyzed for influence of grain size on mercury removal efficiency. Grain fractions included : above 0.2mm, under 0.2mm and fraction between 0.063mm and 0.05mm. The coke dust with the lowest diameter managed to adsorb 99.6% of flue gas mercury which confirmed that additional material pretreatment allows for efficiency improval. In contrast, the highest diameter size fraction removed only 69.7% of mercury. Optimal size fraction (in economical terms) was fraction under 0.2mm as its removal efficiency of exceeded 80% with relatively low energy input required for receiving desired diameter size.

The important merit of waste dust fixed bed adsorption is much lower cost of cleanup process in comparison with commercial dust sorbents. The cost of mercury removal is set around 40-90 thousand USD per kg of mercury, while it should not be higher than 3 000 USD in commercial use [18]. Sorbents that were used in presented study were confirmed to be a good alternative to currently used activated carbon, and have noticeable potential for use in Polish energy sector. Results of presented laboratory scale experiment allow for estabilishing a new method for flue gas mercury removal in hard coal-based boilers.

References

1. US EPA, *Mercury study report to Congress*, EPA-452/R-97-003, (US Government PrintingOffice, Washington DC 1997).
2. US EPA, *A study of hazardous air pollutant emissions from electric utility steam generating units. Final Report to Congress*, EPA-453/R-98-004a; (US Government Printing Office, Washington DC 1998).
3. AMAP/UNEP, *Technical Background Report for the Global Mercury Assessment*, (Arctic Monitoring and Assessment Programme, Oslo, Norway/UNEP Chemicals Branch, Geneva, Switzerland, 2013).
4. KOBiZE (National Centre for Emissions Management), Poland's Informative Inventory, Report 2013.
5. KOBiZE (National Centre for Emissions Management), Poland's Informative Inventory, Report 2015.
6. P. Burmistrz, K. Kogut, M. Marczak, J. Zwoździak, Fuel Process. Technol. **152** 250-258 (2016).
7. T. Dziok, A. Strugała, A. Rozwadowski, M. Macherzyński, Fuel **159** 206-213 (2015).
8. M. Wichliński, R. Kobyłecki, Z. Bis, Arch. Environ. Protec. **39** 141–150 (2013).
9. I. Bojkowska, G. Sokolowska, *Mercury in mineral raw materials exploited in Poland as potential sources of environmental pollution* (PIG Bulletin, 2001).
10. M.A. Lopez-Anton, Y. Yuan, R. Perry, M.M. Maroto-Valer, Analysis of mercury species present during coal combustion by thermal desorption, Fuel **89** 629-634 (2010).
11. T. Gale, B. Lani, G. Offen, Fuel Process. Technol. **89** 139 (2008).
12. M. Berry, R. Semmes, T. Campbell, S. Glesmann, R. Glesmann, Mat. Power Plant Air Pollutant Control Mega Symposium, Baltimore, MD, Curran Associates, 2006.
13. S. Sjostrom, M. Durham, C.J. Bustard, C. Martin, Activated carbon injection for mercury control: Overview, Fuel **89** 1320-1322 (2010).
14. S.B. Ghorishi, R.M. Keeney, S.D. Serre, et al., Development of a Cl-impregnated activated carbon for entrained-flow capture of elemental mercury, Environ. Sci. Technol. **36** 4454-4459 (2002).

15. J.W. Graydon, X-Z. Zhang, D.W. Kirk, C.Q. Jia, Sorption and stability of mercury on activated carbon for emission control, J. Hazard. Mater. **168** (2-3) 978-982 (2009).
16. Q. Zhou, Y-F. Duan, Y-G. Hong, et al., Experimental and kinetic studies of gas phase mercury adsorption by raw and bromine modified activated carbon, Fuel Process. Technol. **134** 325-332 (2015).
17. L. Liberti, M. Notarnicola, V. Amicarelli, V. Campanaro, F. Roethel, L. Swanson, Waste Manag. Res. **16** 2 (1998).
18. W. Nowak, J. Pacyna, I. Majchrzak-Kucęba, Sorbenty z popiołu dla energetyki (2010).
19. H. Clack, Methods for reducing mercury emission from coal combustion, In: Mercury as coal combustion pollutant, Kraków (2014).

The method of determination of mercury adsorption from flue gases

Stanisław Budzyń[1], *Piotr* Burmistrz[1], *Krzysztof* Kogut[1], *Barbara* Tora[2,*]

[1]AGH University of Science and Technology, Faculty of Energy & Fuels, Department of Fuels Technology, Mickiewicz Avenue 30, 30-059 Krakow, Poland
[2]AGH University of Science and Technology, Faculty of Mining and Geoengineering, Department of Environmental Engineering and Mineral Processing, Mickiewicz Avenue 30, 30-059 Krakow, Poland

Abstract. For several recent years Faculty of Energy and Fuels of the AGH University of Science and Technology in Krakow conduct intensive studies on the occurrence of mercury contained in thermal and coking coals, as well as on the possible reduction of fossil-fuel mercury emissions. This research focuses, among others, on application of sorbents for removal of mercury from flue gases. In this paper we present the methodology for testing mercury adsorption using various types of sorbents, in laboratory conditions. Our model assumes burning a coal sample, with a specific mercury content, in a strictly determined time period and temperature conditions, oxygen or air flow rates, and the flow of flue gases through sorbent in a specific temperature. It was developed for particular projects concerning the possibilities of applying different sorbents to remove mercury from flue gases. Test stand itself is composed of a vertical pipe furnace inside which a quartz tube was mounted for sample burning purposes. At the furnace outlet, there is a heated glass vessel with a sorbent sample through which flue gases are passing. Furnace allows burning at a defined temperature. The exhaust gas flow path is heated to prevent condensation of the mercury vapor prior to contact with a sorbent. The sorbent container is positioned in the heating element, with controlled and stabilized temperature, which allows for testing mercury sorption in various temperatures. Determination of mercury content is determined before (coal and sorbent), as well as after the process (sorbent and ash). The mercury balance is calculated based on the Hg content determination results. This testing method allows to study sorbent efficiency, depending on sorption temperature, sorbent grain size, and flue-gas rates.

1 Introduction

Thermal coal combustion is one of the major sources of anthropogenic mercury emission to the atmosphere. In Poland, this is the dominant source with annual share of approximately 56% [1], which amounts to 5 700 kg of emitted mercury per year. In Polish power plants flue gases leaving boiler are treated in a following manner. Fly ash is removed using electrostatic precipitators (ESPs), fabric filters, or in a lesser extent, cyclone separators. Sulfur oxides are

* Corresponding author: tora@agh.edu.pl

removed using wet and semi-dry installations. Nitrous oxides are removed either by catalytic or by non-catalytic reduction [2]. Depending on mercury content in coal, its chemical composition, type of boiler, combustion conditions, and flue gases treatment installation, the mercury content in gases released to the atmosphere varies from 1 μg/Nm^3 to even 30 μg/Nm^3 [1, 3]. Combination of high mercury content in coal, unfavorable elemental composition (low chlorine, bromine and iron content, high calcium content), requires, in addition to passive methods, usage of additional technologies reducing mercury emission (so called 'active methods') [4, 5]. Potentially the Best Available Techniques (BATs) include: injection of pulverized activated coal into flue gases, adsorption on immobilized activated carbon bead, and usage of sulfur-impregnated sorbents. Reduction of mercury emission can be achieved by selection of coals to be combusted, its physical enrichment, mild pyrolysis, as well as combustion in fluidized bed [5].

One of the methods with highest efficiency, which is widely used in USA, is an injection of pulverized activated carbons into flue gases. These activated carbons are often modified to facilitate mercury removal [5, 6]. Although adsorptive methods are highly efficient, their cost is very high, reaching from 40 to 90 thousand USD per 1 kg of mercury removed. This cost can be decreased by usage of cheaper sorbents like selected fractions of coke dust, especially those acquired from coke dry cooling installations [7]. The first step to determine if a given sorbent is suitable for mercury removal from flue gases, is a laboratory test. This enables precise evaluation of efficiency of mercury removal from specified flue gases by a given sorbent. Such evaluation can be achieved by measuring mercury content in gases before and after sorbent treatment. However, such measurements on laboratory scale encounter several problems, which can significantly reduce reliability of the results. Because of that, the mercury balance for measuring system is more reliable method of mercury removal assessment in laboratory scale experiments.

For recent years Faculty of Energy & Fuels is performing research on mercury occurrence in coals as well as on technologies reducing mercury emission to the atmosphere. Within this research lot of effort is made to use pulverized and granulated cheap sorbents to remove mercury from gaseous phase. In this paper, several methods of laboratory scale mercury adsorption measurements were presented. Calculations of efficiency of mercury removal were based on mercury balance and not on direct measurements conducted before and after adsorption.

2 Experimental

2.1 Sorbents

Within the scope of research three types of sorbents were analyzed: mineral (A), organic- waste product from energo-chemical coal processing (B), and organic- produced during waste pyrolysis (C). Properties of these sorbents were presented in Table 1.

Table 1. Properties of studied sorbents.

Parameter	Unit	Sorbent		
		A	B	C
Carbon content	%	0.1	87.2	78.0
Sulfur content	%	< LOQ	0.59	2.64
Specific Surface area S_{BET}	m^2/g	34.5	16.5	74.7
Total pore volume (<300 nm)	cm^3/g	0.135	0.018	0.0239
Density	g/cm^3	2.16	1.99	no data

2.2 Measuring system

Figure 1 presents scheme of a test stand for measuring mercury sorption from flue gases from solid fuel combustion. It consists of tube furnace with temperature regulation, quartz combustion pipe, gas cylinder, flowmeter and sorbent containing container. Temperature of flue gases can be controlled between quartz tube outlet and combustion pipe. The fuel sample is combusted in flow of air or oxygen, and flue gases are directed through the sorbent container. Subsequently, the amount of absorbed mercury is measured. Analysis is performed in defined, controlled conditions which include: temperature and time of combustion, temperature of flue gases going through the sorbent, and flow rate. The fuel sample is inserted in a small ceramic container, which is gradually transferred into the area of highest temperature. Combustion time in this area is 10 min. Single fuel sample input is around 1 g. During single experiment it is possible to combust several portions of fuel, the amount of which depends on mercury content. The sorbent container is located inside a heating element, which allows stabilization of sorbents temperature during the measurement. This allows simulation of industrial conditions as well as analyzing mercury sorption in different temperatures. Sorbent mass, which is determined mainly by its density and granulation, varies between 1 g and 4 g. Before the procedure, mercury content is measured in both fuel and sorbent according to methods described in section 2.3. After the procedure, mercury content is analyzed in ash as well as in sorbent.

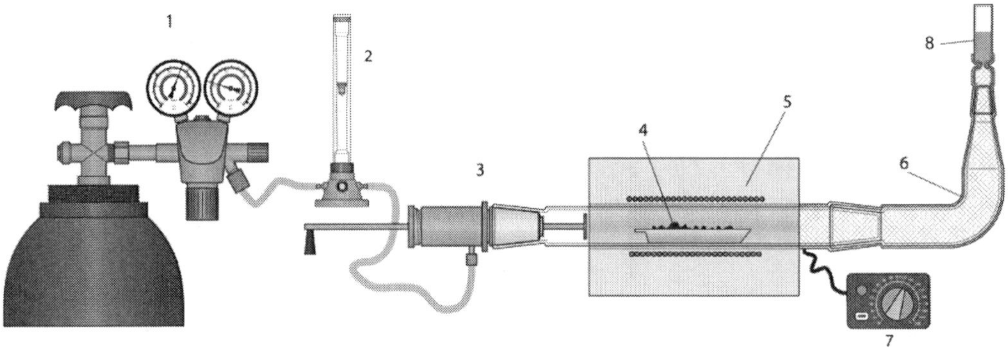

Fig. 1. Scheme of a test stand. 1- source of oxidizer, 2- rotameter, 3- oxidizer supply system, 4 – coal sample, 5 – tube furnace, 6 – heated flue gases stream, 7 – temperature regulator, 8 – sorbent bed.

In all instances experimental conditions and the coal used were the same (see table 2).

Table 2. Experimental conditions.

Parameter	Description
Fuel	lignite
Combustion temperature	850°C
Flue gases flow rate	5 cm^3/s
Combustion time	10 min
Mass of the sorbent	1 g

2.3 Mercury content analysis

The content of mercury in samples of coal, sorbent and ash was measured using atomic absorption spectrometry with cold vapor atomization (CV-AAS) method on automatic mercury analyzer MA-3000 (Nippon Instruments Corporation). Sample mass was between 50.0 and 52.5 mg, with analytical granulation of 0.2 mm. Program for subbituminous coal samples was used. Briefly, sample was heated for 2 min at 180°C, then heated for 7 minutes at 850°C. Subsequently, mercury vapors were reduced to elemental form (Hg^0), which was captured in amalgam by gold covered sand bed. After the sample was decomposed the amalgam was heated to approximately 700°C. Released mercury vapors were directed into cuvette and measured with UV light at 253.7 nm. The final result was a mean from three to five analysis for the same samples. When standard deviation exceeded 10%, additional measurements were performed.

2.4 Comparison with other methods

Other analytical methods for mercury content determination are presented in many publications [8-19]. They are based on the following detection methods:

- atomic absorption spectrometry (AAS) – this method draws upon a high vapour pressure of Hg at relatively low temperatures. A very often used technique is cold-vapour atomic absorption spectrometry,
- atomic fluorescence spectrometry (AFS) – initially, atomic fluorescence spectrometry with flame atomization was used, further modifications included electrothermal atomization and cold-vapour method,
- atomic emission spectrometry (AES) – traditional induction methods such as flame induction or arc discharge are replaced with DC-induced plasma, radio frequency-induced plasma and microwave-induced plasma,
- mass spectrometry (MS) – first applications of MS for determining the mercury content appeared as a variation of method based on spark ionization (Spark Source – SS-MS). Coupling MS with inductively-coupled-plasma (ICP-MS) found a much more applications,
- UV-visible spectrophotometry (Colorimetry) – this had been a popular method of total mercury determination until 1960s when the AAS method was introduced,
- neutron activation analysis (NAA) – main advantages of this method are short analysis time, non-destructive analysis, large sensitivity and precision,
- X-ray fluorescence spectrometry (XRD) – the main advantages are the same as in case of NAA. Additionally it allows for testing many elements simultaneously,
- electron-capture detection spectrometry (ECD) – this method is widely used for determining organic mercury compounds. Its main advantage is a possibility of methylmercury immediate determination without need of turning it into volatile form.

Table 3 presents detection limit of selected methods of the mercury determination.

Table 3. Detection limit of selected methods of the mercury determination. [9, 12, 14, 15, 17, 18]

Methods	Detection limit ng
Atomic absorption spectrometry	0.1
Atomic emission spectrometry	0.005
Mass spectrometry	0.005
Colorimetry	100
Neutron activation analysis	0.01
X-ray fluorescence spectrometry	~25
Electron-capture detection spectrometry	0.5

2.5 Calculating the balance

The above mentioned method allows for preparation of mercury balance through analysis of mercury distribution between combustion products and sorbent. The following data is required for calculating the balance:

- mass of mercury in fuel (*C*),
- mass of mercury in sorbent before analysis (S_B),
- mass of mercury in sorbent after analysis (S_A),
- mass of mercury in ash (*A*).

Thus the efficiency of mercury removal using given sorbent was calculated according to formula (1):

$$\eta = \frac{S_A - S_B}{C - A} \cdot 100\% \qquad (1)$$

3 Results and discussion

3.1 Validation of mercury content determination

Table 4 presents results of validation of method in terms of CV-AAS method accuracy. Table 5 presents summary of validation of mercury content determination method in terms of repeatability and reproducibility for coal samples.

Table 4. Results of validation of mercury content determination accuracy.

Parameter	Certified Reference Material (CRM)		
	SRM 2693	**LECO 502-680**	**SRM 2692c**
Nominal value [ppb]	37.3	110	179.0
Extended uncertainty [ppb]	7.7	11	6.9
Number of repetition CRM	6	6	6
Mean value [ppb]	38.7	112.3	185.8
Standard deviation SD [ppb]	3.5	3.3	4.7
Variation coefficient CV [%]	9.0	2.9	2.5
Range of results [ppb]	8.0	6.8	10.8
Methods accurate	YES	YES	YES

DOI: 10.1051/e3sconf/20171402010

Table 5. Summary of validation of mercury content determination method in terms of repeatability and reproducibility for coal samples (number of repetitions = 6).

Sample number	Mean [ppb]	SD_r [ppb]	CV_r [%]	SD_R [ppb]	CV_R [%]
1	28.7	3.3	11.4	4.1	14.3
2	41.0	2.1	5.1	2.7	6.6
3	49.9	2.3	4.6	2.9	5.8
4	67.2	2.9	4.3	3.7	5.5
5	106.2	7.3	6.9	9.1	8.6
6	138.8	3.6	20.6	5.2	3.7
7	153.6	8.0	5.2	10.7	7.0
8	195.5	16.6	8.5	26.6	13.6
9	330.3	11.5	3.5	15.2	4.6

Where: SD – standard deviation, CV – variation coefficient, indexes: r – repeatability, R – reproducibility.

Table 6 presents additional parameters of mercury determination method acquired during validation.

Table 6. Parameters of mercury content determination method.

Parameter	Value
Detection limit [ng]	0.002
Quantification limit [ng]	0.006
Range of application [ng]	0.006–20
Linearity (*r* coefficient)	0.9999
Extended uncertainty (relative) [%]; depending on range	3.0–10.0

The full validation performed for CV-AAS method showed that:

(i) method is accurate for coal samples with mercury content between 30 and 330 ppb,
(ii) repeatability and reproducibility of the method are acceptable,
(iii) method extended uncertainty fits between 3 and 10%, depending on range,
(iv) method quantification limit is 0.006 ng of mercury,
(v) method is highly linear.

To sum up, performed validation confirmed that the CV-AAS method using MA-3000 automated mercury analyzer is useful for determining mercury content in coal samples.

3.2 Sorbents analysis

Figure 2 presents mercury content in sorbent before and after experiment.

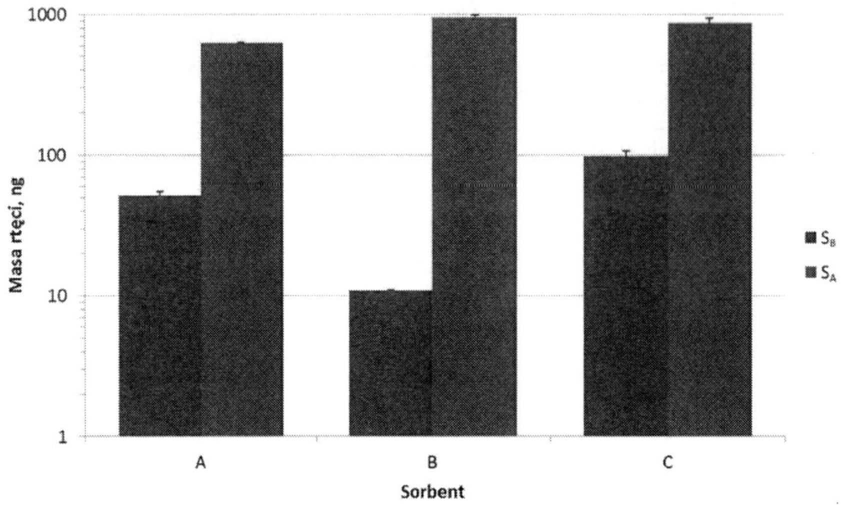

Fig. 2. Mass of mercury in sorbents before (S_B), and after (S_A) experiment

For calculations it was assumed that all mercury in coal is transferred to combustion products (ash and flue gases), and subsequently it is transferred to sorbents bed. This assumption is fulfilled because of tightness of installation and forced flow of oxidizer and flue gases. Values of mercury removal efficiency are presented in Table 7.

Table 7. Efficiency of mercury removal for tested sorbents. [%]

Sorbent	Mercury efficiency removal
A	43.5
B	85.1
C	68.3

4 Conclusions

During presented study the main challenge was determination of mercury content in analyzed samples. Most of presently used methods use different variants of atomic spectrometry, mass spectrometry, or fluorescence spectroscopy [19]. In presented system scope of analysis was reduced to mercury content in solid matrix (coal sample, ash, and sorbent). Determination of mercury content in solid phase is easier than in gaseous phase, because it lacks complications resulting from condensation during flue gases cooling, mercury speciation, and high dust level present in industrial installations.

Presented laboratory test stand for studying mercury sorption enables:

(i) analysis of mercury distribution between fuel, combustion products and ash, without need for troublesome measurements of mercury content in gaseous phase,
(ii) analysis of sorbents efficiency based on mercury balance,
(iii) analysis of influence of sorbents temperature on mercury sorption,
(iv) analysis of sorbents granulation on mercury sorption,
(v) analysis of gas flow rate on mercury sorption.

The experiments reported here were supported by Ministry of Science and Higher Education in Poland (Project AGH University of Science and Technology No 11.11.210.213 and 11.11.100.276).

References

1. P. Burmistrz, K. Kogut, M. Marczak, J. Zwoździak, Fuel Process. Technol. **152** 250-258 (2016).
2. J. Zyśk, A. Wyrwa, M. Pluta, Atmos. Environ. **45** 605-610 (2011).
3. S.X. Wang, L. Zhang, G.H. Li, Y. Wu, J.M. Hao, N. Pirrone, F. Sprovieri, M.P. Ancora, Atmos. Chem. Phys. **10** 1183-1192 (2010).
4. J.H. Pavlish, L.L. Hamre, Y. Zhuang, Fuel **89** 838-847 (2010).
5. H. Clack: *Methods for reducing mercury emission from coal combustion.* [in:] *Mercury as a coal combustion pollutant*, 105-120, (Oficyna Drukarska Jacek Chmielewski, Warszawa, 2014).
6. E.S. Olson, A. Azenkeng, J.D. Laumb, R.R. Jensen, S.A. Benson, M.R. Hoffmann, Fuel Process. Technol. **90** 1360-1363 (2009).
7. P. Burmistrz, L. Czepirski, K. Kogut, A. Strugała, Przem. Chem. **93** No. 12 (2015), 1-6.
8. B. Welz, M. Schubert-Jacobs, Frasenius J Anal. Chem. **331** 324-329 (1988).
9. A. Urba, K. Kvietkus, J. Sakalys, Z. Xiao, O. Lindqvist, Water, Air, Soil Pollut. **80** 1305-1309 (1995).
10. R.L. Lancione, D.M. Drew, Spectrochim. Acta B **40** 107-113 (1985).
11. R.I. Pereiro, C.A. Diaz, Anal. Bioanal. Chem. **372** 74-90 (2002).
12. C. Dietz, Y. Madrid, C. Camara, J. Anal. At. Spectrom. **16** 1397-1402 (2001).
13. E. Björn, T. Larsson, L. Lambertsson, U. Skyllberg, W. Frech, Ambio **36** 443-451 (2007).
14. National Institute for Minamata Disease, *Research on Mercury. Development in Mercury Analysis*, http://www.nimd.go.jp/archives/english/tenji/c_corner/c07.html (access at: 2016.09.26).
15. P. Jones, S. Hardy, J. Chromatogr. A **765** 345-352 (1997).
16. Annual Report of Institute of High Energy Physics, *Study of Chemical Species with Nuclear Analytical Techniques*, Chinese Academy of Sciences 109-122, (2001).
17. I.N. Aretaki, P. Koulouridakis, N. Kallithrakas-Kontos, Anal. Chim. Acta **562** 252-257 (2006).
18. A.M. Caricchia, G. Minervini, P. Soldati, S. Chiavarini, C. Ubaldi, R. Morabito, Microchem. J. **55** 44-55 (1997).
19. M. Macherzyński, J. Górecki, K. Styszko: *Chemical analysis of mercury and its compounds.* [in:] *Mercury as a coal combustion pollutant*, 105-120, (Oficyna Drukarska Jacek Chmielewski, Warszawa, 2014).

Analysis of green liquor influence on coal steam gasification process

Mateusz Karczewski[1,*], *Stanisław* Porada[1]

[1]AGH University of Science and Technology, Faculty of Energy and Fuels, al. Mickiewicza 30, 30-059 Cracow, Poland

Abstract. Gasification is a clean and efficient technology with a long history dating up to the 19th century. The possible application of this process ranges from gas production and chemical synthesis to the energy sector and therefore this technology holds noticeable potential for future applications. In order to advance it, a new efficient approaches for this complex process are necessary. Among possible methods, a process enhancing additives, such as alkali and alkaline earth metals seems to be a promising way of achieving such a goal, but in practice might turn to be a wasteful approach for metal economy, especially in large scale production. This paper shows alkali abundant waste material that are green liquor dregs as a viable substitute. Green liquor dregs is a waste material known for its low potential as a fuel, when used separately, due to its low organic content, but its high ash content that is also abundant in alkali and alkaline earth elements seems to make it a suitable candidate for application in coal gasification processes. The aim of this work is an evaluation of the suitability of green liquor waste to work as a potential process enhancing additive for coal steam gasification process. During the experiment, three blends of hard coal and green liquor dregs were selected, with consideration for low corrosive potential and possibly high catalytic activity. The mixtures were gasified in steam under four different temperatures. Their energies syngas yield, coal conversion degree and energies of activation were calculated with use of Random Pore Model (RPM) and Grain Model (GM) which allowed for their comparison.

1 Introduction

Among various methods of coal processing, gasification is already renowned technology with background reaching back to 19th century. Its biggest past application was town gas production from coal but was later replaced by the use of natural gas. Recently, as a clean process with extensive potential in both energy and chemical industry, the interest in this technology increased as a part of the search for sustainable and more efficient ways of energy production. To this day, a large variety of aspects were analyzed i.e. use of different operating conditions, agents and solid fuel types being the most popular [1-5].

One of more interesting aspects of this research is process enhancement by use of catalytic materials. Their addition is reported to change the kinetics of gasification process and lower

* Corresponding author: mateusz.karczewski@agh.edu.pl

its activation energy. To date, several works already stated the positive effects of various additives on process rate and/or syngas composition [6-8] with alkali and alkaline earth metals being the most popular due to lower price in comparison with other catalysts such as iron [9] or noble metals [10] and their high process efficiency enhancement [11-13].

On the contrary, a certain amount of issues may arise if such approaches were applied to industrial-scale activities. For example, wet impregnation of coal with material mixing and drying might prove to be too complex for implementation in larger installations, despite ensuring high catalyst-fuel contact [14], leaving physical mixing as only viable approach [15]. The other issue is the need of use of market alkali or alkaline compounds which might be a wasteful approach for the metal economy, especially in large scale production. The solution for this problem might be the utilization of waste materials with a high abundance of alkali and alkaline earth metals, such as green liquor dregs.

Green liquor dregs are Kraft paper mill side product with ash composition containing slightly different content that can be found in typical ashes derived from combustion processes. Its mineral composition is based mostly on potassium, magnesium, calcium, manganese abundant [16] wood ash and iron (if wood is recycled) with the highest presence of calcium originating process of caustization that uses lime for spent solvent recovery. Another source of active material in the waste is solvent itself that consist mostly of sodium sulfate for lignin breakdown [17]. As a result, green liquor dregs are a partially carbonaceous material with high mineral content rich in various alkali components. This kind of waste product can find its use as cheap catalyst material that can process together with fuel in order to utilize the catalytic potential of its mineral constituents with the additional conversion of its carbon content.

Therefore the aim of this work is an evaluation of green liquor wastes as a possible catalytic additive for coal steam gasification process as a material abundant in the potentially process enhancing components. The content of materials has been analyzed in order to determine proper waste and coal fuel blends for further research. The mixtures were steam gasified with coal in four temperatures. Their kinetic parameters were calculated basing on several kinetic models i.e. Random Pore Model (RPM) and Grain Model (GM) which allow for their comparison along with syngas yield and coal conversion degree.

2 Material and method

2.1 Material preparation

The experiment was performed with coal obtained from Szczygłowice mine and green liquor waste received from one of Polish paper mills. Both coal and green liquor were air-dried until a stable weight was obtained and then ground in ring roller mill (EKOLAB LAB-09-200). The resulting powders were the sifted for fractions under 0.2 mm in diameter.

2.2 Elemental and Technical Analysis

The measurements of elemental composition were conducted with Eltra CHS-580 analyzer, with accordance to polish standardized laboratory procedures [18, 19]. The measurement provided information on carbon and hydrogen content but also allowed for sulfur measurement.

The proximate analysis comprised of moisture, volatile matter and ash content and the lower heating value was also performed in order to obtain data for indices calculations. Moisture, volatile matter and ash analysis were performed with a thermogravimetric analyzer (Eltra Thermostep) and the lower heating value was measured with a bomb calorimeter

Table 1. Ultimate and proximate analysis of used hard coal and green liquor waste.

Proximate analysis	**Coal**	**Green Liquor**
Moisture - M^a [% wt]	1.8	4.4
Ash - A^a [% wt]	7.7	56.0
Volatile matter - VM^{daf} [% wt]	25.0	29.9
Lower Heating Value - LHV^{daf} [kJ/kg]	33247	5477
Ultimate analysis	**Coal**	**Green Liquor**
Carbon - C^{daf} [% wt]	88.8	55.4
Hydrogen - H^{daf} [% wt]	5.21	2.92
Sulfur - S^{daf} [% wt]	0.46	0.63
Oxygen&Nitrogen - $O+N^*$ [% wt]	5.54	41.09

* calculated by as difference between other constituents and 100%
^{a} – analytical state
^{daf} – dry ash-free state

(Leco AC 350), according to the polish standards [20,21]. Table 1 present all results from proximate and ultimate analysis.

Both materials were also characterized for their ash composition which was determined by X-ray fluorescence using EDXRF spectrometer (Panalytical Epsilon^{3XLE}). For later calculations, the resulting ash elemental content was assumed to be present only in the form of oxides and therefore its content was recalculated to include oxygen presence and adjusted to 100%. Resulted oxides ash concentration was shown in Table 2.

Table 2. Characteristic of green liquor and hard coal ash constituents.

Ash constituent	**Coal [wt%]**	**Green Liquor [wt%]**
MgO	17.24	3.28
Na_2O	11.72	2.77
K_2O	2.09	4.29
Fe_2O_3	3.70	7.96
SiO_2	3.14	38.69
Al_2O_3	1.33	29.30
CaO	49.65	6.53
P_2O_5	0.59	2.35
TiO_2	0.02	1.35
Cl	0.22	-

2.3 Determination of fuel blends composition

With data on fuel elemental composition, potential blend weight ratios were estimated with the corrosive potential of resulting fuel as main quality. To achieve this, several indices aimed at corrosion prediction presented elsewhere [16] were calculated for various amount of waste addition. From the results, an index with the lowest tolerance for the waste addition was chosen as the general limit for green liquor addition. Out of analyzed indices, Cl ratio and S ratio, presented below, provided the lowest waste addition limits:

$$Cl\,ratio = \frac{\%(Cl + K_2O + Na_2O)}{\%(SiO_2 + Al_2O_3)} \tag{1}$$

$$S\,ratio = \frac{\%(S + K_2O + Na_2O\,)}{\%(SiO_2 + Al_2O_3)} \tag{2}$$

Where compound name states the content by weight of the corresponding compound. The results have shown that the presence of slight slagging, achievable for Cl ratio above 1 and for S ratio above 0,5 [16] suggested potential slight slagging presence at 23% and 63% of waste addition, respectively. As the calorific value of green liquor waste is rather low, the lower amount was chosen for further blends preparations in order to keep fuel blend quality.

For further evaluation, catalytic properties were also taken into account. Therefore four blends with the addition of 5, 10, 15 and 20% of green liquor waste were prepared in order to cover the whole range of established green liquor addition margin with the addition of (from now on also named as GL5, GL10, GL15 and GL20 blends, respectively) by weight were analyzed for their content of potentially enhancing elements, as shown in Table 3. For the sake of comparison, additional sample of untreated Szczygłowice hard coal was also used (named as Coal from now on).

Table 3. Elements in ashes of analyzed fuel blends and coal used in the experiment with potential catalytic influence.

	Elemental content in fuel mixture [wt%]				
	Coal	**GL5**	**GL10**	**GL15**	**GL20**
Mg	2.61	3.07	3.53	3.99	4.46
Na	2.20	2.66	3.12	3.59	4.05
K	3.40	3.86	4.33	4.79	5.25
Fe	6.32	6.78	7.25	7.71	8.17
Ca	5.19	5.65	6.11	6.57	7.04
Mn	0.14	0.60	0.51	0.70	0.88

Table 3 shows that analyzed elements only occur in certain, relatively small quantities in resulting blends despite their large concentration in waste material. Most of the alkali materials appear in a range between 3 and 5% with more abundant in given materials elements like Fe and Ca appeared in quantities between 5 and 7%. According to recent advances in catalytic additives [23,24] the amount around 3.40% for Na and K and around 5% for Ca [24] are very favourable, therefore the fuel blends consisting of amounts up to 15%.seems to have the best catalytic potential, suggesting use of the samples with 5%, 10% and 15% waste addition for further experiment.

2.4 Steam gasification

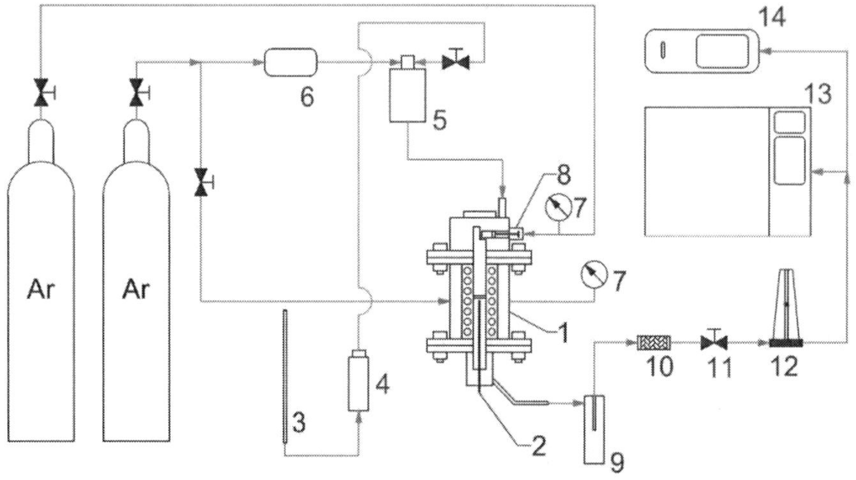

Fig. 1. The laboratory equipment for kinetics examinations of coal steam gasification process.
1 - reactor, 2 - thermocouple, 3 - water tank, 4 - water pump, 5 - steam generator, 6 - mass flowmeter, 7- pressure gauge, 8 – biomass feeder, 9 - cooler, 10 - filter, 11 - pressure regulator, 12- rotameter, 13 - gas chromatograph, 14 – CH_4, CO, CO_2 IR analyser.

Measurement of the gasification process along with kinetics determination was carried out with the use of the system shown in Figure 1. The process applied elevated pressure of 11 bars with isothermal conditions carried out at four temperatures: 850, 900, 950 and 1000°C and with controlled sample injection. This system was already successfully used in other gasification kinetic measurements [24,25]. In resulting syngas flow, the content of methane, carbon monoxide and carbon monoxide was continuously controlled by a gas analyzer (Fuji Infrared Gas Analyzer) utilizing infrared radiation adsorption for concentration determination. The content of hydrogen was analyzed using gas chromatograph (HP 5890) equipped with thermal conductivity detector (TCD). The details on methane, carbon monoxide, and carbon dioxide concentrations in flue gas allowed for calculation of changes in formation rates for measured process gasses as a function of time. Such data was used for composition and conversion degree determination by calculating the yield of each product from the area under curve dV/dt=f(t). The carbon conversion degree was also evaluated with use of the volume of carbon-based gaseous products with given equation:

$$X(t) = \frac{(V_{co}(t)+V_{CO2}(t)+V_{CH4}(t))\,M_C}{V_{mol}\,m\,C^{daf}}\ 100\ \% \qquad (3)$$

where:
$V_{CO}(t)$, $V_{CO2}(t)$, $V_{CH4}(t)$ - volume of released gas component at standard conditions as a function of time, dm^3/g
V_{mol} - volume of one mole of gas at temperature of 273 K and pressure of 101325 Pa, dm^3/mol
M_c- molar mass of carbon, g/mol
m – sample mass, g
C^{daf}- dry ash free carbon content, -

In order to determine the activation energy Grain Model (GM) and Random Pore Model (RPM) [26] were used as models suitable for gas-solid heterogeneous reactions. The GM and RPM models are described by equations (2) and (3), respectively.

 DOI: 10.1051/e3sconf/20171402011

$$\frac{dX}{dt} = k_{GM}(1 \quad X)^{\frac{2}{3}} \tag{4}$$

$$\frac{dX}{dt} = k_{RPM}(1 \quad X)\sqrt{1 \quad \Psi \ \ln(1 \quad X)} \tag{5}$$

where:
k_X- pre-exponential factor for given model, 1/s
Ψ - parameter is associated with the pore structure of the test material, due to nature of the process set at 2.

3 Results and discussion

The degree of carbon conversion, calculated according to equation (1), from gas products for every fuel blend in various temperatures was presented in Table 4. Starting from the lowest temperature of 850°C for gasification, fuel reaches the lowest conversion rates of 43.6% for pure coal sample and rises noticeably with the addition of green liquor to 67, 71.1 and 71.2% for blend GL5, GL10 and GL15, respectively. The carbon conversion increased with the rise of reaction temperature, achieving rates between 51.9 and 79.8 for 900°C, 56.7 to 90.4 for 950°C and 59.4 to 99.7 for 100°C. The lowest conversion rates in all analyzed temperatures were obtained for gasification of Coal, while GL10 blend has achieved the best results in every given temperature with a slight drop in conversion with the bigger addition of waste in the GL15 blend and even lower rates when lowest waste addition in GL5 blend was used.

Table 4. Total carbon conversion degree calculated with respect to gas products.

Carbon to syngas conversion [wt%]				
RT[°C]	**Coal**	**GL5**	**GL10**	**GL15**
1000	59.4	92.4	99.7	94.8
950	56.7	85.7	90.4	88.1
900	51.9	74.3	79.8	78.9
850	43.6	67	71.1	71.2

RT – reaction temperature [°C]

The results are complemented by syngas compounds total production, presented in Table 5. They show that addition of green liquor has a noticeable influence on every syngas constituent. Methane concentrations increased with the biggest effect in GL10 blend, where the yield of CH_4 increased from 18 for coal up to 34 cm^3/g in the case of lowest temperature and doubled from 32 to 65 cm^3/g when the highest temperature of 1000°C was used. The increase of methane content hints that catalytic properties of green liquor might enhance not only gasification process but also the pyrolysis as methane production only occurs during initial contact of fuel with heat that is later followed by proper reaction with the gasifying medium. The analogical increase occurred in hydrogen and carbon monoxide yields, where GL10 blend increased hydrogen production from 1294 to 1557 cm^3/g for 850°C and from 1693 up to 2123 for 1000°C. Analogical yield for carbon monoxide accounts for an increase from 276 up to 609 cm^3/g and from 509 up to 902 cm^3/g, respectively. The only syngas constituent not following the trend is carbon dioxide that keeps relatively constant yield between temperatures for every blend with a minor increase when the temperature is lower, but followed by decrease when a low temperature of 850°C is achieved, possibly due to the low conversion of fuel. What is more, CO_2 shows the highest yield increase for GL5 blend instead of GL10. The GL5 blend achieved a CO_2 yield in the range of 516-637 cm^3/g in comparison with 368-425 cm^3/g yield for coal. As GL5 blend seems to give the lowest increase in other syngas constituents out of all fuel mixtures, this effect might be caused by low contact of enhancing elements with coal particles.

Table 5. Total amount of gas product obtained from given fuel blend in given temperature, expressed as cm^3/g

Sample	Temperature [°C]	Syngas Yield [cm^3/g]			
		CO	H_2	CO_2	CH_4
Coal:	850	276	1294	368	18
	900	333	1490	425	30
	950	427	1574	400	33
	1000	509	1693	376	32
GL5	850	473	1438	516	23
	900	539	1598	553	31
	950	650	1629	615	30
	1000	672	2006	637	37
GL10	850	609	1557	425	34
	900	688	1712	470	48
	950	801	1942	515	49
	1000	902	2123	455	65
GL15	850	506	1502	546	22
	900	615	1671	543	34
	950	740	1882	545	45
	1000	742	2084	501	51

Based on results of syngas yields, the best sample for calorific gas production was chosen and its process of production was compared in respect of hydrogen and carbon dioxide production with that of coal (Fig. 2 and Fig. 3).

Analysis of hydrogen production during gasification process (Fig 2) revealed several changes that occurred when green liquor additive was used. The first influence can be seen during pyrolysis step where a rapid increase of production was visibly enhanced. Moreover, the drop in hydrogen rate of formation occurred with almost instant stabilization of hydrogen production after pyrolysis step ended whereas additive-free ended its pyrolysis step by a stronger drop in hydrogen rate of formation followed by rate of formation increase that marked the beginning of fuel-steam reactions. This might suggest that green liquor constituents either enhanced bond breaking inside of coal structure and steam to dissociate into hydrogen or caused gasification process to occur along with pyrolysis. The temperature also plays an important role during gasification of both coal and fuel blends as it causes much higher fuel depletion that leads to a decrease of reaction time with the increase of gasification temperature.

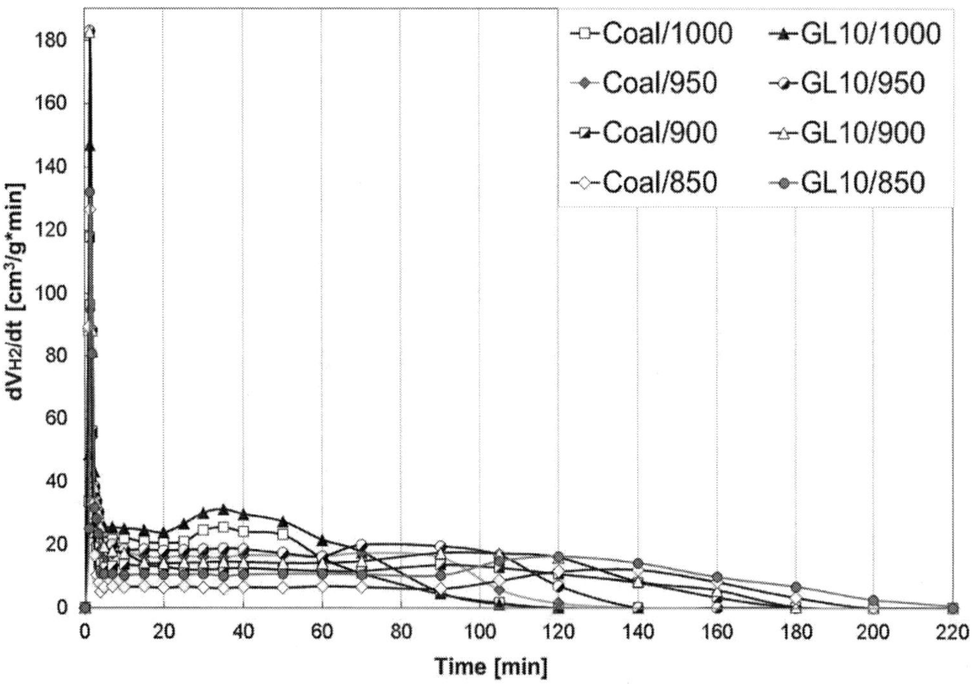

Fig. 2. Influence of green liquor addition on the kinetics of hydrogen production in comparison with untreated coal sample.

Analysis of gas production behavior for carbon monoxide (Fig 3) revealed the analogical effect of increased gas production and smoothening of pyrolytic products formation rate drop. When comparing the duration of gasification step between carbon monoxide and hydrogen it can be also noticed that carbon monoxide drop in rate of formation occurs slightly

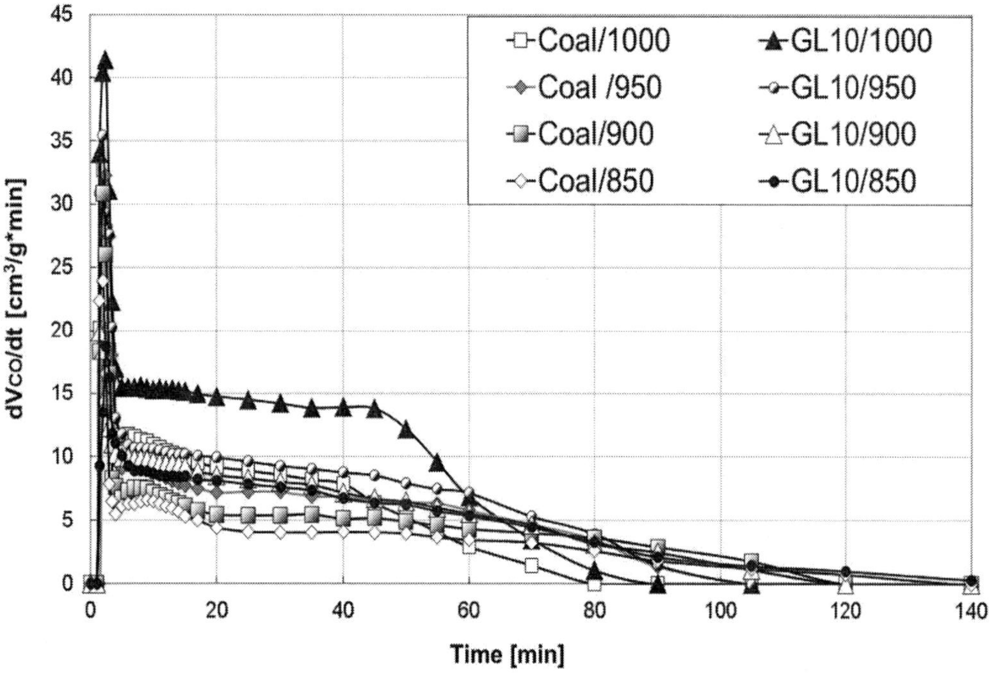

Fig. 3. Influence of green liquor addition on the kinetics of carbon monoxide production in comparison with untreated coal sample.

before hydrogen rate of formation drop and it is not followed by a visible increase of rate of formation before the drop, unlike hydrogen. The source of hydrogen increase might suggest that with depletion of sample steam began to react with produced carbon monoxide, resulting in syngas production consisting almost only of carbon dioxide and hydrogen at the end of the process. As in the case of both coal and coal with green liquor, the resulting moments where reaction ended were very similar, it can be assumed that addition of green liquor had minimal influence on this step of gasification.

Table 6. The activation energy for steam gasification process for coal and fuel blends calculated with use of Random Pore Model and Grain Model.

	Coal		GL5		GL10		GL15	
	RPM	**GM**	**RPM**	**GM**	**RPM**	**GM**	**RPM**	**GM**
E_a **[kJ/mol]**	48.1	49.7	42.8	42.5	38.5	38.2	39.6	38.8
k_0 **[1/s]**	0.109	0.104	0.0122	0.0120	0.0135	0.0136	0.0131	0.0134

Grain Model and Random Pore Model were used to calculate activation energy for every blend used in the experiment, basing on data from four different temperature measurement applied. The results were shown in Table 6. The results show that addition of green liquor to coal influenced not only syngas yield and conversion rates, but also decreased the energy of activation. The calculated values for both models ranged from 48.1 kJ/mol for RPM and 49.7 for GM obtained for untreated coal to activation energies of 42.8 to 38.5 range for RPM and 42.5 to 38.2 kJ/mol range for GM with the lowest energy of activation achieved at GL10 blend. Those values point that addition of green liquor allowed the process to utilize the presence of green liquor constituents to use the different mechanism with lower energy of activation. Moreover the results for different amounts of green liquor additives shown that such mechanism can efficiently work only for limited ash composition, as fuel with 15% of green liquor addition has shown slightly worse results than fuel blend with 5% lower waste content, suggesting that mechanism of the process is heavily influenced by contact between fuel and active components of green liquor.

4 Conclusions

The use of green liquor waste as a catalytic additive for coal gasification in steam seem to be a viable option for process enhancement. The blended mixture prepared based on information obtained from corrosion indices and existing research allowed for the determination of waste addition suitable for process enhancement with potentially low adverse effect on used installation.

The results of steam gasification have shown that the best blend out of analyzed waste:coal ratios in fuel is 10% by weight of green liquor addition. This mixture allowed for the heavy jump in the degree of carbon conversion, up to 99.7% for highest temperature from only 59.4% when untreated coal was gasified. Kinetic analysis has shown that presence of green liquor in fuel blend increases yield both during pyrolysis and during gasification phase, possibly by allowing gasification to occur along with pyrolysis.

The activation energy calculated also shows noticeable improvement, as the energy of activation decreased for fuel blends to achieve the lowest state for GL10 blend. The experiment has proven potential that green liquor holds as a cheap catalyst, but several aspects like its influence for a wider range of fuels and influence of catalytic components already residing in ashes on final process performance still needs to be addressed for a clearer image of its versatility as a fuel additive.

Acknowledgements: This paper was prepared thanks to funding from Faculty of Energy and Fuels at the AGH University of Science and Technology as part for Ph.D. Students Grants 2016, No. 15.11.210.341.

References

1. M. Mondragon, A. Mondragon, F. Mondragon, Fuel **77(15)** 1831-1839 (1998)
2. K. Jayaraman, I. Gokalp, Fuel **153**, 320-327 (2015)
3. C. Li, J. Zhao, Y.Fang, Y. Wang, Front. of Chem. Eng. in Cn. **4(4)**, 385-393 (2010)
4. D.J. Marchand, E. Schneider, B.P. Williams, Y.L. Joo, J. Kim, G.T. Kim, S.H. Kim, Fuel Process Technol. **130**, 292-298 (2015)
5. A. G. Collot, Int. J. Of Coal Geo. **65(3)**, 191-212.
6. F. Huhn, J. Klein, H. Jüntgen, Fuel **62(2)**, 196–199 (1983)
7. W.W. Li, K.Z. Li, X. Qu, R. Zhang, J.C. Bi. Fuel Proc. Tech. **126**, 504-512 (2014)
8. A. Sharma, T. Takanohashi, K. Morishita, T. Takarada, Fuel **87**, 491-497 (2008)
9. X. Qi, X. Guo, L. Xue, C. Zheng, C. J. Anal. Appl. Pyrol. **110**, 401-407 (2014)
10. K. Otto, M. Shelef, Carbon **15(5)**, 317–325 (1977)
11. J. Tang J. Wang, Fuel Process. Technol. **142**, 34-41 (2016)
12. D. W. McKee, Carbon, **17**, 419-425 (1979)
13. Z.L. Liu, H.H. Zhu, Fuel **65(10)**, 1334–1338 (1986)
14. C.L. Spiro, D.W. McKee, P.G. Kosky, E.J. Lamby, Fuel **63(5)**, 686-691 (1984)
15. J. Kopysciński, R. Rahman R. Gupta, C. Mims, J. Hill, Fuel **117**, 1181-1189 (2014)
16. M. K.Misra, K. W Ragland, A. J. Baker, Biom. and Bioen. **4(2)** (1993)
17. M. Mahmoudkhani, D. W. Keith, Int. J. of Gre. Gas Cont. **3(4)** (2009)
18. Polish Norm: PN G-04571:1998 Solid Fuels - Determination Of Carbon, Hydrogen And Nitrogen Contents Using The Automatic Analyzers - Macro Method
19. Polish Norm: PN G-04584:2001 Solid Fuels - Determination Of Total Sulphur And Of Ash Sulphur Content With The Use Of The Automatic Analysers
20. Polish Norm: PN-81-G-04513 Solid Fuels – Determination of lower and higher heat of combustion
21. PN-EN 14918:2010 Solid Biofuels - Determination of net calorific value
22. M.Karczewski, M. Marczak, P. Burmistrz, D. Makowska, Corrosive components of nutshells and their chars, E3S W. Of Conf. **10**, 00113 (2016)
23. S. Porada, A. Rozwadowski, K. Zubek, Studies of catalytic coal gasification with steam. Polish Journal of Chemical Technology, **18(3)**, 97-102 (2016).
24. K. Zubek, G. Czerski and S. Porada, E3S W. of Conf. 2017 (To be published)
25. S. Porada, A. Rozwadowski, Przem. Chem. **93,** (2014)
26. S. Porada, G. Czerski, T. Dziok, P. Grzywacz, D. Makowska, Fuel Proc. Tech. **130**, (2015)

The influence of catalytic additives on kinetics of coal gasification process

Katarzyna Zubek[1,*], *Grzegorz* Czerski[1], and *Stanisław* Porada[1]

[1]AGH University of Science and Technology, Faculty of Energy and Fuels, al. Mickiewicza 30, 30-059 Krakow, Poland

Abstract. Catalytic coal gasification is a process that has the potential to become one of the efficient industrial technology of energy production. For this reason, the subject of this study was to analyze the kinetics of catalytic gasification of 'Janina' coal with steam. Isothermal measurements were performed at 800 °C, 900 °C, 950 °C and 1000 °C at a pressure of 1 MPa using cations of sodium, potassium and calcium as catalysts. During examination the thermovolumetric method was used. This method allows to determine the formation rates of a gaseous product such as carbon monoxide, hydrogen, methane and carbon dioxide as well as their contribution to the resulting gas. Moreover, the influence of catalysts on the kinetics of CO and H_2 formation at various temperatures was determined and the kinetics parameters were calculated with the use of isoconversional model, Random Pore Model and Grain Model. The obtained results confirmed the positive effect of catalysts on the coal gasification process. The catalytic measurements were characterized by higher reaction rate and shorter duration of the process, and the calculated values of the kinetic parameters were lower than for the gasification process without the addition of catalysts.

1 Introduction

The process of coal gasification with steam is a well-known technology, the basics of which were formed during the production of town gas at the turn of the 19th and 20th centuries. The production of gas from coal has been supplanted by the use of natural gas and has become a technology with limited uses. Along with the technical advances, interest in the gasification process returned, due to environmental concerns and the search for new, sustainable ways of energy production. Gasification is a clean and effective process of gas generation that may be used to produce power, chemicals or synthetic fuels. Moreover, this technology allows decreasing greenhouse gas emissions, in particular CO_2 [1]. Because of that numerous works aimed at improving gasification process are conducted. The use a variety of gasifying agents, operating conditions, types of fuel as well as catalysts are analyzed [2-7]. This research is particularly important in the case of fluidized bed reactors operating at low temperatures, where the high reactivity of the fuel is essential to ensure the efficient course of the process. Numerous of works confirmed the influence of catalysts on increasing the process rate or

* Corresponding author: kzubek@agh.edu.pl

changing the resulting gas composition [8-10]. Currently, due to the high efficiency and low price, catalysts based on alkali and alkaline earth metals have gained the greatest popularity [11,12]. However, research is continually conducted in order to find new solutions, such as catalysts based on noble or transition metals [13,14]. In order to use catalyst on an industrial scale, beyond the high catalytic activity it should be characterized by the low price. This condition is not fulfilled in the case of noble metals. Catalysts based on transition metals are much cheaper, but at high temperatures quickly become deactivated and their regular exchange or activation significantly increase the costs of their use [15,16]. Besides the type of the catalyst, an important factor determining its effect on the reaction is a method of adding catalytically active material to the feedstock. This occurs generally in two ways: by wet impregnation and the physical mixing [17]. The most important advantage of the first method is the ability to ensure good contact between coal and catalyst [18]. However, physical mixing is characterized by simplicity and short execution time, moreover certainty that the amount of catalyst will be exactly as established [19]. The addition of catalytically active substances improves the process efficiency as a result of changes in the kinetics, resulting in the reduction of activation energy. Therefore, there is a necessity to analyze the kinetics of coal gasification with catalysts in order to determine their impact on the process. This is an important step providing information needed to design gasifiers. The complexity of the gasification, intensified by the addition of catalyst makes that kinetics of this process is still a current issue which requires further research.

The aim of this study was to analyze the kinetics of gasification of 'Janina' coal with cations of sodium, calcium and potassium, which were introduced by the wet impregnation method. Isothermal measurements were conducted at 800 °C, 900 °C, 950 °C and 1000 °C under a pressure of 1 MPa with the use of the thermovolumetric method, developed by AGH. On the basis of the obtained results the influence of catalysts on the gasification process was determined, the formation rate of the individual components of synthesis gas, as well as its composition was defined. With the use of isoconversional method the values of activation energy were calculated and these results were compared with the kinetic parameters calculated on the basis of Random Pore Model (RPM) and Grain Model (GM).

2 Material and methodology

2.1 Characteristic of material

Coal from the Polish mine 'Janina' was selected as feedstock for the research. Samples with a particle size below 0.2 mm and a weight of 1 g were used in the measurements. The results of proximate and ultimate analysis, as well as the composition of 'Janina' coal ash are summarized in Table 1.

Table 1. Characteristic of 'Janina' coal.

Parameter					
Proximate analysis	**(%wt)**	**Ultimate analysis**	**(%wt)**	**Ash composition**	**(%wt)**
Moisture - M^a	8.7	Carbon - C^{daf}	77.8	SiO_2	59.4
Ash - A^a	14.0	Hydrogen - H^{daf}	3.9	Al_2O_3	23.3
Volatile Matter - VM^{daf}	46.1	Sulfur - S^a_t	1.0	Fe_2O_3	8.2
				MgO	1.2
				CaO	2.2
				K_2O	2.9
				Na_2O	2.8

Ions of alkali and alkaline earth metals - sodium, potassium and calcium, introduced by wet impregnation method were used as catalytically active materials. Obtained samples of coal contained 3% wt. of the corresponding metal.

2.2 Methodology of measurements

Description of the laboratory equipment on which the measurements were carried out has been presented in previous publications [20,21], and its schematic diagram is shown in Figure 1.

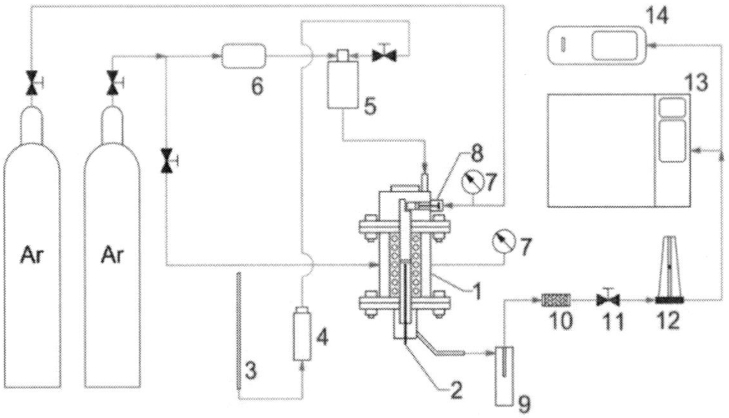

Fig. 1. The laboratory equipment for kinetic examinations of coal gasification process.
1 - reactor, 2 - thermocouple, 3 - water tank, 4 - water pump, 5 - steam generator, 6 - mass flowmeter, 7- pressure gauge, 8 - coal feeder, 9 - cooler, 10 - filter, 11 - pressure regulator, 12- rotameter, 13 - gas chromatograph, 14 - CO, CO_2 IR analyser [20,21].

The equipment consists of several basic systems: a pressure reactor with a heating system (1-2), a system for feeding the reactor with the gasifying agent - steam (3-5), carrier gas - argon (6-7), and coal (8), system for collecting and purifying the resulting gas (9-11) and the gas analysis system (12-14). In the resulting gas the content of carbon monoxide, carbon dioxide and methane was continuously controlled by analyzer operating on the principle of infrared radiation adsorption. In addition, the content of hydrogen was analyzed using gas chromatograph equipped with thermal conductivity detector (TCD).

Isothermal measurements of the kinetics gasification process were carried out at 800 °C, 900 °C, 950 °C and 1000 °C under an elevated pressure of 1 MPa. The weight of the sample fed into the reaction zone was 1 g, steam flow was 0.3 g/min, while the argon flow was 2 dm^3/min.

2.3 Methodology of kinetics parameters calculation

Based on the concentrations of the main components of the gas (CO, H_2, CO_2, CH_4) the formation rates of these gaseous products, as well as their contribution to the resulting gas were determined. During the measurements the time required to transfer gas from the reaction zone into the analyzer has been taken into account.

The reactions of formation of carbon monoxide and hydrogen may be expressed as:

$$X(t) = \frac{V_{x(t)}}{V_{x(total)}} \quad (1)$$

where: $V_{x(t)}$ is a volume of CO or H_2 formed from the beginning of the reaction to time t, dm^3/g, $V_{x(total)}$ is the total volume of CO or H_2 formed during the whole reaction, dm^3/mol.

The effect of temperature and concentration of gaseous reactant on the gasification process rate can be expressed by the equation (2):

$$\frac{dX}{dt} = k(T)G(C)f(X) \tag{2}$$

where: k and G are the rate constants, dependent on temperature T and reagent concentration C and f(X) accounts for physical or chemical changes occurring with the progress of the reaction. The apparent reaction rate constant k(T) and the concentration of the reagent G(C) are generally described in the form of an Arrhenius equation (3):

$$k(T)G(C) = Ae^{\frac{-Ea}{RT}}C^n \tag{3}$$

where: A is a pre-exponential factor, E_a is activation energy and n is an order of the reaction, taking into account the concentration of the gas reactant.

In order to analyze the kinetics of the gasification process, two approaches were used. Firstly, isoconversional method, which allows to determine the activation energy was applied. Combining and integrating equations 2 and 3, provides equation 4 which describe isoconversional method:

$$\ln(t) = \ln\left(\frac{F(X)}{C^n A}\right) + \frac{E_a}{RT} \tag{4}$$

Equation 4 allows to determine the activation energy from the slope of $ln(t)=f(1/T)$ for X = 0.25, 0.5, 0.75, even though $f(X)$ in equation 2 is unknown.

In the second approach two models, usually used to describe heterogeneous gas-solid reaction were used, i.e. Grain Model (GM) and Random Pore Model (RPM). Experimental data were analyzed in the range of CO and H_2 formation reactions degree from 0 to 80%.

The GM model assumes that the reaction occurs at the external surface of the char particle and surface area decreases non-linearly as the reaction proceeds, and RPM model assumes that the reaction surface changes as a result of two competing processes: the increase of the pore structure during gasification and its destruction due to coalescence of adjacent pores. GM and RPM models are described respectively by equations 5 and 6.

$$\frac{dX}{dt} = k_{GM}(1 - X)^{\frac{2}{3}} \tag{5}$$

$$\frac{dX}{dt} = k_{RPM}(1 - X)\sqrt{1 - \Psi \cdot \ln(1 - X)} \tag{6}$$

where: Ψ is associated with the pore structure of the material [22].

3 Results and discussion

During steam gasification process coal is transformed into gas, which is composed mainly of CO and H_2. However, as a result of a number of parallel reactions various gaseous products are also formed, among others methane and carbon dioxide. The changes in the formation rates of CO, H_2, CO_2 and CH_4 during the gasification of the examined samples at the temperature of 800 °C are shown in Figure 2.

 DOI: 10.1051/e3sconf/20171402012

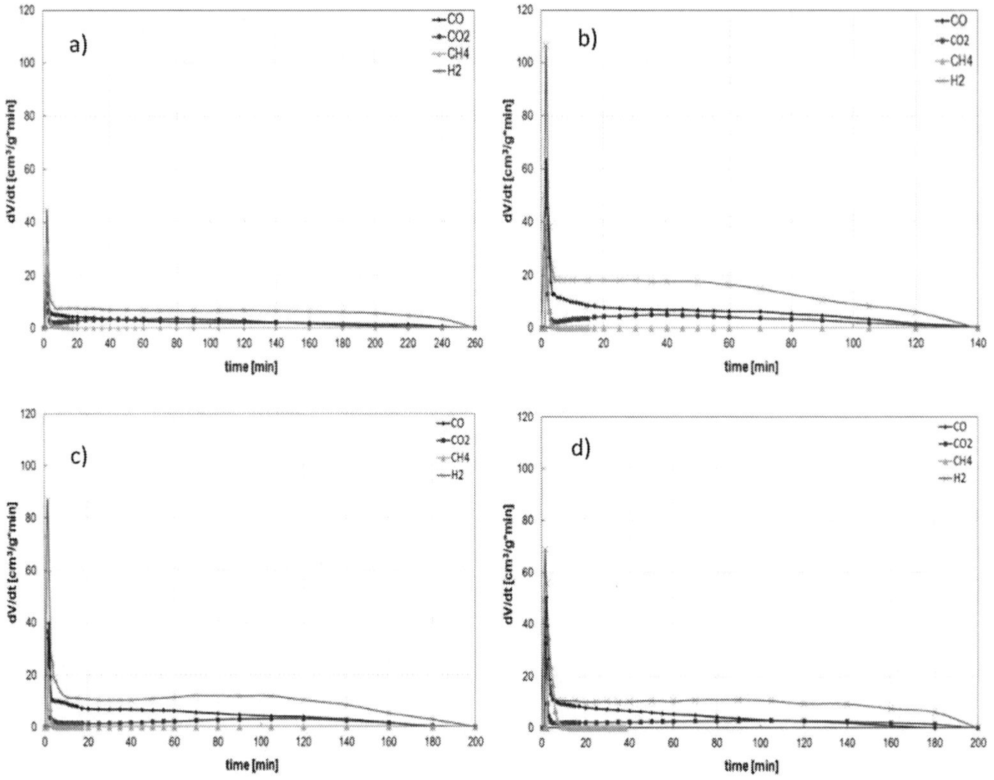

Fig. 2. The changes in the formation rates of gaseous products during the gasification of a) coal; b) coal with Na^+; c) coal with K^+; d) coal with Ca^{2+}.

The whole process can be divided into two stages: pyrolysis and gasification of previously formed char. During the pyrolysis formation rates of gaseous products are the highest, which can be observed in the first minutes of the process. Then there is much slower gasification stage.

During the whole process of coal gasification without the catalyst, the formation rate of hydrogen was the highest, higher more than twice than very similar formation rates of carbon monoxide and carbon dioxide. The last of the analyzed components - methane was formed with a relatively high rate, but its release was limited only to the pyrolysis stage. The obtained results confirmed the positive effect of catalysts on the gasification process. The addition of the cations of sodium, potassium and calcium substantially catalyzed the pyrolysis stage, resulting in a marked increase in formation rates of all the gaseous products in the initial stage of the process. During the char gasification stage increased formation rates of the basic components of gas i.e. CO and H_2 can be observed as a result of the addition of the catalysts. The average CO formation rate during the catalytic gasification was almost two times higher than in the case of coal gasification without a catalyst. However, the predominant gaseous product was still hydrogen, the formation rate of which increased, in particular in the case of using sodium ions as a catalyst. The average formation rate of hydrogen during the catalytic gasification was more than 2 times higher when the coal was impregnated with sodium ions and more than 1.5 during the gasification with Ca^{2+} - catalyst which the slightest improved the formation rate of H_2 at 800 °C compared to the non-catalytic gasification process. The addition of catalytically active materials had not a significant impact on the formation rates of carbon dioxide and methane, which proves that these catalysts mainly catalyze the reaction of gasification of element C with steam.

The composition of the gas obtained from gasification of 'Janina' coal allows to assess the impact of the individual catalysts on the process at a given temperature and it is shown in Table 2. The accuracy of CO and CO_2 measurements was ~5 %, while CH_4 and H_2 was ~1 %.

Table 2. The composition of the resulting gas.

'Janina' coal	Temperature [°C]	Share [%] CO	H_2	CO_2	CH_4
without catalyst:	800	21.2	57.8	19.4	1.6
	900	24.3	58.3	15.7	1.7
	950	28.7	56.0	13.2	2.1
with catalyst:	1000	34.8	55.4	8.1	1.7
➢ Na^+					
	800	25.2	59.7	14.0	1.1
	900	28.6	58.7	11.5	1.2
	950	33.2	55.9	9.4	1.4
	1000	30.3	58.7	9.6	1.4
➢ K^+					
	800	31.3	56.1	11.9	0.7
	900	34.0	55.5	9.0	1.5
	950	33.7	56.4	8.7	1.2
	1000	34.2	55.8	8.5	1.5
➢ Ca^{2+}					
	800	26.5	57.7	14.3	1.5
	900	30.5	57.3	10.7	1.4
	950	32.5	55.1	10.6	1.8
	1000	33.2	55.6	9.5	1.8

The contribution of hydrogen to the resulting gas from 'Janina' coal gasification was the largest and represented from 55% to 60%, depending on the analyzed case. There was no clear effect of temperature or catalyst on the share of H_2 released in the process. With rising temperature the share of carbon monoxide increased in each case at the expense of carbon dioxide, enhancing the calorific value of the resulting gas. This phenomenon was most evident during the gasification of coal without the catalysts, where the contribution of CO and CO_2 at the lowest temperature was almost identical. This may be caused by an exothermic, reversible reaction of conversion of carbon monoxide with steam, the equilibrium of which is shifted towards the substrate the more the higher is the temperature of the process, as well as by the Boudouard reaction. The addition of catalysts resulted in an increased share of the carbon monoxide in the resulting gas at a given temperature as compared with the non-catalytic coal gasification carried out under similar conditions. The contribution of CO to the resulting gas was the highest in the case of coal gasification with potassium ions added. The addition of Ca^{2+} or Na^+ resulted in gas with a similar share of CO, but lower than in the case of utilization of K^+. Exceptions were the measurements carried out at 1000 °C in which the composition of resulting gas was almost identical in each case. Neither temperature nor addition of the catalysts affected the share of methane, whose contribution to the resulting gas was negligible due to its formation only during the pyrolysis stage.

The next step was to present the progress of the reaction of main products formation, expressed as a function of released CO and H_2 at the time. Gasification, as an example of a highly endothermic reaction, proceeds faster with increasing temperature. A similar effect gave the addition of the catalysts. Effect of catalytic additives on the process, expressed

DOI: 10.1051/e3sconf/20171402012

as a function of the formation of CO at various temperatures was shown in Figure 3. At 800 °C the impact of the catalysts was the most visible. All considered catalytic additives accelerated the reaction, simultaneously reducing process time and the most active catalyst was sodium. With the increase of temperature to 950 °C catalytic gasification proceeded more efficiently compared to the non-catalytic process, and the effect of the addition of sodium and calcium ions was very similar. The least effective catalyst was potassium. At 1000 °C the addition of the catalysts was unfounded because the non-catalytic process was more efficient. This effect may be associated with the interaction of mineral matter with catalytic additives, which at high temperatures leads to the formation of complex compounds inhibiting the gasification process [23].

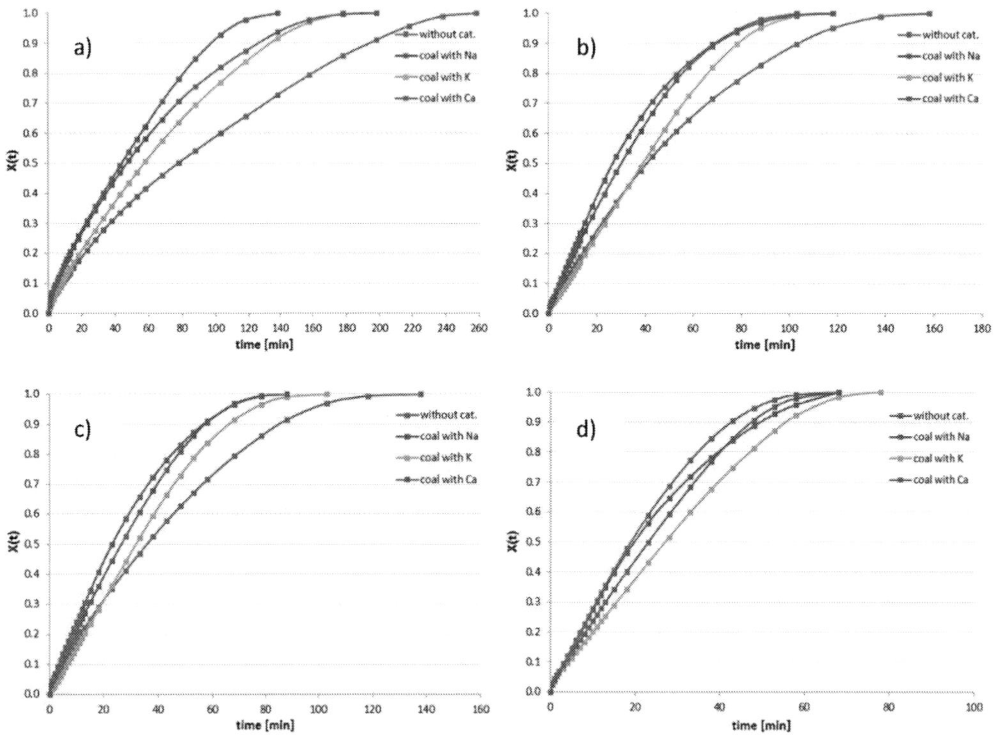

Fig. 3. Effect of catalysts on CO formation reaction at a) 800 °C; b) 900 °C c) 950 °C d) 1000 °C.

The addition of sodium most effectively influenced on the reaction of H_2 formation, while the effect of calcium and potassium was very similar in all tested temperatures, as shown in Figure 4. As already mentioned, during catalytic coal gasification at 1000 °C compounds inhibiting the process rate may be formed. This phenomenon could influence on the reaction of H2 formation at the highest analyzed temperature. Practically the whole process of non-catalytic gasification was faster than the gasification with K^+ or Ca^{2+}. However, the best of the catalysts – Na^+ increased the rate of the gasification process. Nevertheless, due to the fact that the positive effect of sodium on the progress of CO formation reaction at 1000 °C was not observed, catalytic gasification at temperatures above 950 °C is not a preferred solution.

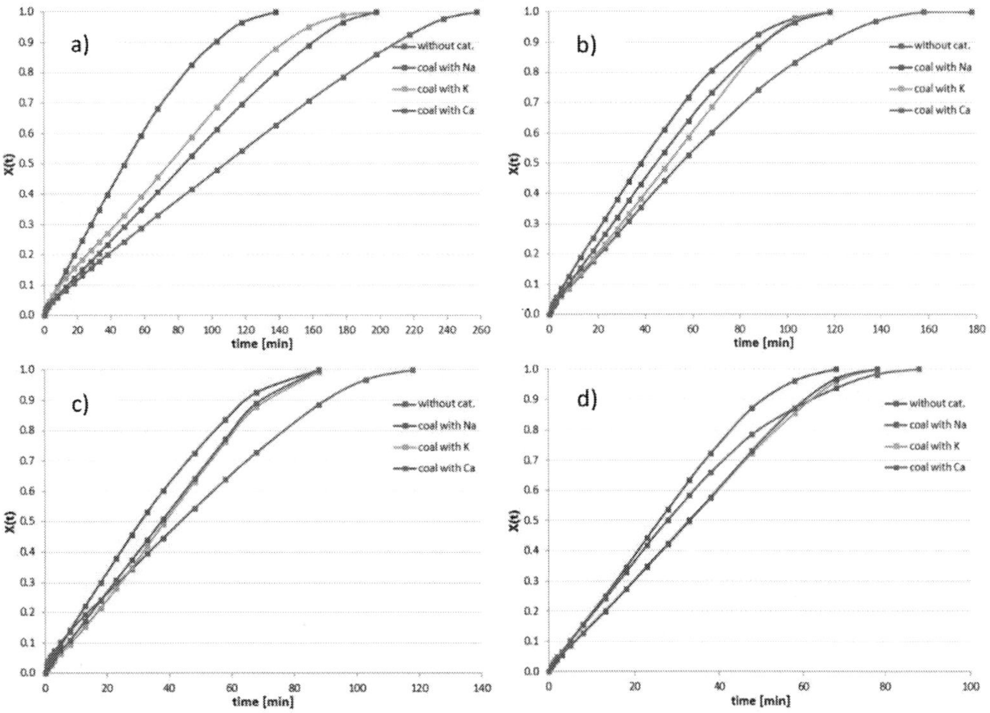

Fig. 4. Effect of catalysts on H_2 formation reaction at a) 800 °C; b) 900 °C c) 950 °C d) 1000 °C.

Another step of this work was to determine the kinetic parameters of the gasification process. With the use of isoconversional method the values of activation energy of CO and H_2 formation reactions were calculated and the obtained results are summarized in Table 3.

Table 3. Values of activation energy calculated with the use of isoconversional method.

'Janina' coal	CO	H_2
	E_a [kJ/mol]	E_a [kJ/mol]
without catalyst	72.0 ± 8.8	72.8 ± 1.4
with addition of:		
➢ Na^+	41.5 ± 4.5	34.8 ± 1.6
➢ K^+	43.6 ± 5.1	48.3 ± 1.4
➢ Ca^{2+}	40.1 ± 9.0	54.0 ± 1.6

'Janina' coal was characterized by high reactivity as evidenced by relatively low values of the calculated activation energy. Activation energies of CO and H_2 formation during non-catalytic coal gasification were the highest and simultaneously very similar. The addition of catalysts caused their significant reduction in a various way. The values of activation energy of CO formation were similar irrespective of the catalyst used and ranged between 40.1 and 43.6 kJ/mol. This lack of difference could be caused by the fact that the calculated values were vitiated by relatively large errors. Moreover, the obtained E_a of CO formation included a series of reactions that took place during the gasification process, in which the CO was created. The various effect of tested catalysts on these reactions could lead to similar values of E_a. In the case of E_a of H_2 formation differences were much bigger - from 34.8 using Na^+ as a catalyst to 54.0 using ions of calcium. This sequence could result from the fact that the sodium ions (to a lesser extent potassium ions) may catalyze the water gas shift reaction [24,25]. The results confirmed the catalytic activity of alkali and alkaline-earth metals in the gasification process, wherein the sodium more effectively catalyzed the reaction of H_2 formation than the reaction of CO formation, the effect of potassium on the formation of both gaseous product was comparable, while calcium was more effective in the reaction of CO formation. In order to obtain the values of both kinetic parameters: activation energy - E_a and pre-exponential factor - A, two additional models i.e. Random Pore Model and Grain Model were used, and the results are shown in Table 4.

Table 4. Values of kinetics parameters calculated with the use of GM and RPM.

'Janina' coal	CO		H_2	
	A [1/min]	**E_a [kJ/mol]**	**A [1/min]**	**E_a [kJ/mol]**
	Grain Model			
without catalyst	57.9 ± 5.7	80.3 ± 13.5	20.2 ± 2.5	72.4 ± 8.8
with addition of:				
➢ Na^+	2.3 ± 0.6	45.3 ± 3.0	0.6 ± 0.3	33.4 ± 4.5
➢ K^+	2.7 ± 0.8	48.8 ± 3.2	2.2 ± 0.8	48.4 ± 3.9
➢ Ca^{2+}	2.9 ± 0.9	48.5 ± 5.0	3.1 ± 1.1	52.5 ± 5.2
	Random Pore Model			
without catalyst	48.4 ± 4.8	80.2 ± 13.6	17.4 ± 2.4	72.5 ± 8.8
with addition of:				
➢ Na^+	2.0 ± 0.6	45.5 ± 2.9	0.5 ± 0.2	33.8 ± 4.4
➢ K^+	2.1 ± 0.7	48.3 ± 3.3	1.9 ± 0.7	48.6 ± 3.7
➢ Ca^{2+}	2.5 ± 0.9	48.6 ± 5.1	2.9 ± 1.3	53.1 ± 6.0

The values of E_a calculated on the basis of RPM and GM models were almost identical, and were within the margin of error of which the results calculated from isoconversional method were burdened. Calculated values of A were also similar regardless of the model used, but the differences were greater than in the case of activation energy. The highest value of A was obtained for reaction of CO formation during non-catalytic gasification. In this process the value of A, calculated for the reaction of H_2 formation was much lower, although activation energies were not significantly different. The addition of catalysts decreased the values of pre-exponential factor. The results obtained for the reaction of CO formation during the catalytic gasification process were very similar. In case of reaction of H_2 formation differences appeared, but were consistent with the changes in the values of the activation energy.

4 Conclusion

The measurements of catalytic coal gasification allowed to draw the conclusion that catalysts used in this work have a positive impact on the process. During the gasification of 'Janina' coal H_2 and CO were the main products and the rate of hydrogen formation was almost twice as high as carbon monoxide. The addition of each catalyst resulted in an increase in the formation rates of these gaseous products, particularly in the case of using sodium ions as a catalyst. These findings were confirmed by kinetics parameters calculated on the basis of all the models used. Values of kinetics parameters were considerably lower in the case of catalytic gasification than the process without a catalyst. The use of catalysts resulted in an increased share of CO in the resulting gas compared to analogous non-catalytic measurements in the temperature range from 800 to 950 °C. This effect was also observed with increasing temperature of the measurements. However, the addition of the catalyst at 1000 °C was unfounded, since did not improve the efficiency of the process. Nevertheless, as the purpose of the use of a catalyst is to improve the process rate at low temperatures, the obtained results are satisfactory and promising. These findings are particularly important in the case of gasification in fluidized bed reactors, operating at low temperatures since they indicate that the use of catalysts based on alkali and alkaline-earth metals significantly improves their work.

The work was conducted within the framework of the grant no. 15.11.210.336.

References

1. M. Pańczyk, T. Borowiecki, Adsorbenty i katalizatory. Wybrane technologie a środowisko, **2**, 275-288 (2012)
2. T. Kyotani, K. Kubota, J. Cao, H. Yamashita, A. Tomita, Fuel Process. Technol. **36**, 209-217 (1993)
3. C. Li, J. Zhao, Y. Fang, Y. Wang, Front. Chem. Eng. China. **4,** 385-393 (2010)
4. K. Jayaraman, I. Gokalp, Fuel **153**, 320-327 (2015)
5. X. Hung, F. Zhang, M. Fan, , Sustainable Catalytic Processes **7**, 179-199 (2015)
6. D.J. Marchand, E. Schneider, B.P. Williams, Y.L. Joo, J. Kim, G.T. Kim, S.H. Kim, Fuel Process Technol. **130**, 292-298 (2015)
7. C.L. Spiro, D.W. McKee, P.G. Kosky P.G, E.J. Lamby, Fuel **62(2)**, 180–184 (1983)
8. W.W. Li, K.Z. Li, X. Qu, R. Zhang, Fuel Process. Technol. **126**, 504-512 (2014)
9. A. Sharma, T. Takanohashi, K. Morishita, T. Takarada, Fuel **87**, 491-497 (2008)
10. F. Huhn, J. Klein, H. Jüntgen, Fuel **62(2)**, 196–199 (1983)
11. J. Tang J. Wang, Fuel Process. Technol. **142**, 34-41 (2016)
12. Z.L. Liu, H.H. Zhu, Fuel **65(10)**, 1334–1338 (1986)
13. X. Qi, X. Guo, L. Xue, C. Zheng, C. J. Anal. Appl. Pyrol. **110**, 401-407 (2014)
14. K. Otto, M. Shelef, Carbon **15(5)**, 317–325 (1977)
15. S. Ratchahat, S. Kodama, W. Tanthapanichakoon, H. Sekiguchi, Int. J. Hydrogen Energ. **40(35)**, 11809-11822 (2015)
16. I.G. Lee, A. Nowacka, C.H. Yuan, S.J. Park, J.B. Yang, Int. J. Hydrogen Energ. **40(36)**, 12078-12087 (2015)
17. J. Kopyscinski, R. Rahman R. Gupta, C. Mims, J. Hill, Fuel **117**, 1181-1189 (2014)
18. C.L. Spiro, D.W. McKee, P.G. Kosky, E.J. Lamby, Fuel **63(5)**, 686-691 (1984)
19. C. Phuhiran, T. Takarada, Int. J. Hydrogen Energ. **39(8),** 2649-3656 (2014)
20. S. Porada, G. Czerski, P. Grzywacz, T. Dziok, Przem. Chem. **93**, 2059-2063 (2014)
21. S. Porada, G. Czerski, T. Dziok, P. Grzywacz, D. Makowska, Fuel Process. Technol, **130**, 282-291 (2015)
22. G. De Micco, A. Nasjleti, A.E. Bohé, Fuel **95**, 537-543 (2012)
23. J. Tang, J. Wang, Fuel Process. Technol. **142**, 34-41 (2016)
24. J. Pazmino, M. Shekhar, W. Wiliams, C. Atakay, J. Miller, Catal. **286**, 279-286 (2012)
25. B. Zugic, D. Bell, M. Flytzani-Stephanopoulos, Appl. Catal.,B. **144**, 243-251 (2014)

Concentration of main elements in mineral matter of Polish coking coal and its cleaning products

Krzysztof Bytnar[1,*], *Dorota* Makowska[1]

[1]AGH University of Science and Technology, Faculty of Energy and Fuels, Mickiewicza Av. 30, 30-059 Cracow, Poland

Abstract. The chemical composition of coal mineral matter, and particularly the content of major elements (Al, Si, Fe, Ca, K, Na, Mg), plays an important role in the processing and use of bituminous coals. Mineral matter is the ballast, reducing the calorific value of the fuel. The aim of the study was to analyze the concentrations of main elements of coking coal mineral matter and products of its cleaning in heavy media, jiggers and flotation machines. Analysis of elements was performed through the technique of the atomic absorption spectrometry AAS. The results of main elements concentrations of coal mineral matter and its cleaning products are discussed in relation to each operation of mechanical treatment. As a result of the cleaning process in heavy media, the content of mineral substances decreased seven times, mainly due to the removal of components such as silicon, aluminum and iron from coal. Similar results were obtained in the case of cleaning in jiggers, leading to a significant reduction of the contents of silicates and aluminosilicates. In both cases, there was an increase in the content of sodium. Despite the significant reduction of the contents of alkali metals' compounds, their content in concentrates as converted to oxides is quite high (0,43-0,81% of dry coal). The lowest alkalis content is the displayed by a coal concentrate from the flotation process dewatered in a centrifuge.

1 Mineral matter of coal

Coals, in particular bituminous coals, due to their properties are a valuable natural resource with a variety of applications in a number of industries. In addition to coal combustion, this raw material is also used in processes like pyrolysis, gasification and liquefaction.

From the point of view of the processes during the thermochemical treatment of coal, the elemental composition of coal mineral matter plays an important role, in particular content of the so-called major elements (Al, Si, Ca, Mg, Fe, Na and K) which determine the directions of processing and properties the final products.

The mineral matter contained in the coal is a ballast, causing a decrease in the calorific value, or in the case of coking coal causes deterioration of the coal coking

* Corresponding author: bytnar@agh.edu.pl

properties. In addition, many compounds that occurs in the coal mineral matter (alkali, sulphur compounds, iron and magnesium) have a negative impact on the process of thermochemical processing of coal. The most common minerals occurring in the coal are the silicates, aluminosilicates, carbonates, sulfides, sulfates, oxides, hydroxides, halides and trace elements.

The above-mentioned alkaline compounds which mode of occurrence and negative effects have been discussed in several papers [1-3] cause, inter alia, lower the ash melting temperature. It can lead to the formation of agglomerates and closing the slats in coal boilers in the combustion process. In gasification processes alkaline compounds cause the formation of lesions in gasifiers reducing their efficiency. In the process of the pyrolysis about 40% of alkaline compounds passes to coke, being the source of the blast furnace accretions that breaking away of from the walls of the blast furnace leads to clogging of the nozzles and the blast furnace burnout. Therefore, the coke industry sets certain requirements for the content of alkali compounds in the coking coal, namely the sum of Na_2O and K_2O does not exceed 0,2 to 0,3 % wt.

A significant part of coal mineral matter is aluminum (Al) and silicon (Si). They are basic component of silicates and aluminosilicates. The share of silicate may be up to 40 to 60% of coal mineral matter (Table 1). Among the silicates, quartz and opal in 800°C transform the structure of its crystal lattice - 870°C quartz is converted to tridymite, and opal at 100°C loses water, and at high temperature it passes transformations similar to quartz.

Table 1. Examples of mineral components containing alkalis found in Polish coals [4].

Name, description			Formula
Spatial silicates	***Potassium feldspars:***		
	Orthoclase, microcline		$K[AlSi_3O_8]$
	Sodium and Potassium feldspars (plagioclases):		
	albite		$Na[AlSi_3O_8]$
	anorthite		$Ca[Al_2Si_2O_8]$
Phyllosilicates	***Micas:***		
	<u>dioctahedral structure:</u>	muscovite	$KAl_2[(OH)_2/AlSi_3O_{10}]$
	ordinary micas:	hydromuscovite	$(K,H_3O)Al_2[(OH)_2/AlSi_3O_{10}]$
		illite	$(K,H_3O)Al_2[(OH)_2/AlSi_3O_{10}]$
	hydromicas		
	<u>*trioctahedral structure:*</u>		
	ordinary micas:	biotite	$K(Mg,Fe,Mn)3(OH,F_2/AlSi_3O_{10}]$
	Montmorillonites:		
	<u>*dioctahedral structure:*</u>	montmorillonite	$(Al_{1.67}Mg_{0.33})[(OH)_2[Si_4O_{10}]Na_{0.33}(H_2O)_4$
	montmorillonites	beidelite	$Al_2[(OH)_2[Al_{0.5}Si_{3.5}O_1]\cdot(Ca, Na)_{0.5}(H_2O)_4$
Hali-des	***Anhydrous chlorides:***		
	halite		NaCl
	sylvine		KCl
Sulphates	***Anhydrous sulphates:***		
	alunite		$KAl_3(OH)_6/[(SO_4)_2]$
	natroalunite		$NaAl_3(OH)_6/[(SO_4)_2]$
	jarosite		$KFel_3(OH)_6/[(SO_4)_2]$
	natrojarosite		$NaFe_3(OH)_6/[(SO_4)_2]$
Carbon-ates	***Anhydrous carbonates:***		
	calcite		$CaCO_3$
	dolomite		$CaMg(CO_3)_2$
	ankerite		$CaFe(CO_3)_2$
Phosph-ates	***Phosphates:***		
	goyazite		$SrAl_3(PO_4)_2(OH)_5\cdot H_2O$
	crandallite		$CaAl_3(PO_4)_2(OH)_5\cdot H_2O$
	fluorine apatite		$Ca_3(PO_4)_3F$

The clay minerals in a temperature range of 600 ÷ 800°C release water, kaolinite in 800°C converted to an intermediate form $Al(AlSiO_5)_6$ to end at 1100°C in the presence of fluxes to transform $3Al_2O_3 \cdot 2SiO_2$. A halloysite transformed in a similar way, while illite at about 1250°C melts and loses potassium. Calcium (Ca) is usually in the form of carbonates and sulphates (Table 1). Carbonates in terms of incidence in the coal seams are second. The most common carbonate is the calcite and its several polymorphic variants. Calcium is also included in the double anhydrous carbonates such as dolomite or ankerite. While on the sulphate anion calcium forms anhydrous calcium sulfate - anhydrite, and its hydrated form - gypsum. The small amounts of calcium can combine with the phosphate anion of a series of minerals, so called apatites [4]. In the case of carbonates we can find a large range of thermal reaction. Siderite at a temperature of 450 to 500°C decomposes to magnetite, dolomite decomposes at 765 ÷ 895°C and calcite decomposes at 900°C. The obtained CaO reacts with SO_2, to next decomposed in higher temperatures again to CaO and SO_2. Hydrated sulfates (epsomite, gypsum) lose crystalline water at about 400°C, and the thermal dissociation of the anhydrous compounds occurs in 900°C.

Iron (Fe) is most common in combination with sulfur, mainly in the form of iron sulfides - pyrite, which is present both in free and bound form of mineral matter and can occur in its hydrothermal variety - marcasite. This was formed both by syngenetic transformation (at the same time as coal bed was formed) and as a result of epigenetic transformation (formed much latethan coal bed). As a result of the oxidation reaction of pyrite in the coal weathering, hydrated iron sulphates was formed – malanterite and jarosite, but they occur occasionally. Iron also rarely occurs in the form of iron sulphates and compounds with oxygen and a hydroxyl group (hematite and goethite). Iron sulfides decompose in 500°C giving as final products SO_2 and Fe_2O_3. According to this scheme galena and sphalerite are transformed. With increasing temperature to 1000°C formed oxides are volatilized. Iron hydroxides dehydrate at 110 ÷ 150°C giving hematite, while aluminum hydroxide loses water in 600°C converts to corundum, but in contrast to the iron oxides does not go to volatilize at higher temperatures. The iron contained in the coal as a result of the pyrolysis process enters coke and in the temperature range of 600 ÷ 1000°C due to its catalytic effect increases the coke reactivity. Whereas both silicon and aluminum act as inhibitors reducing the reactivity of the coke to the temperature 800°C.

Elemental sulfur goes into volatile at 444°C, while the oxalates decompose above 800°C precipitate the coal. Halides (halite and sylvite) are volatilize at temperatures close to 1000°C. Calcium phosphates are partially decomposed at about 800°C, and their com In British English is « sulphate »plete decomposition occurs above 1300°C. The last group of troublesome elements present in the coal are alkali: Na2O and K2O that are volatilize at a temperature of about 900°C and CaO passing under these conditions to sulfates and their subsequent transformation are similar as in the case of carbonates.

2 Experimental

The subject of the study was to analyze the content of major elements (Al, Si, Fe, Ca, K, Na, Mg) in mineral matter of coking coal and products of its beneficiation processes. The study focused on both, raw coking coal and the products of mechanical treatment from mines of the Upper Silesian Coal Basin (USCB). Taking into account all stages of coal preparation, individual products of coal cleaning was analyzed: raw coal (feed), concentrate, intermediate product and a waste from dense medium washers, jig washers and flotation. Three-product enrichment scheme implemented in the USCB mines processing plants shown in Figure 1.

The raw coal (grain size 0 to 120 mm) is headed to the processing plant, where it is screened on a screen with a mesh of 20 mm. Size grade: 20 to 120 mm is directed to

 DOI: 10.1051/e3sconf/20171402013

cleaning in a dense medium washer (a specific weight of 1.45 ÷ 1.80 g/cm^3), and the coal having a particle size of 0÷20 mm is cleaned in water pulsating jigs. The smallest coal grain class of less than 0.5 mm from the dewatering of jig wash product undergoes to flotation and the resulting concentrate is drained centrifuges (or dried). Plants have a closed water-mule loop and the waste management after the flotation process. [5].

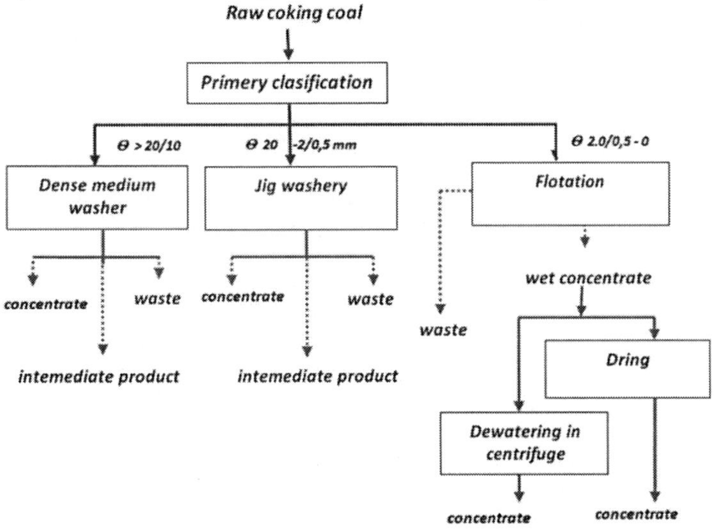

Fig. 1. The three-tier coking coal processing plants in Upper Silesian Coal Basin

Analytical samples (grain size <0.2 mm) of semi-soft coal (Polish type 34) and hard coal (Poilsh type 35), and its enrichment products, taken directly from coal processing plants of individual USCB mines, have been prepared. Samples with a weight of 0.15 ÷ 0.40 g were digested with acidic solution in microwave closed system using nitric and hydrofluoric acids. Digestion were performed in a microwave oven Berghoff Speedwave Four. The analysis of the elements Si, Al, Ca, Fe, Mg, K and Na in the prepared solutions was performed by atomic absorption spectrometry using a spectrometer Z-2000 from Hitachi with Zeeman background correction and flame atomization (F-AAS). For the determination of elements such as Fe, Mg, K, standard burner was used, fed with a mixture of acetylene and air. Due to high temperature atomization of Si, Al and Ca, for determination of this elements high-temperature burner fed a mixture of acetylene and nitrous oxide was used. As a source of radiation single element hollow cathode lamps were used. Calibration was performed using standard solutions. The analytical method was tested by measuring certified reference materials, samples from inter-laboratory comparisons and by standard addition method. Determination of ash in analytical samples were performed according to polish standard PN-G-04512: 1980, and the determination of moisture by PN-80/G-04511.

3 Results and discussion

Results of investigations of concentrations of individual elements in output coals and concentrates obtained therefrom for the three mines USCB shown in Table 2 and discussed in relation to the various methods of mechanical processing. The results are presented as a weighted average percentage of the analyzed elements in the samples of coal and products of its beneficiation. The amount of mineral matter has been estimated based on the ash content on a Parr formula [6].

Investigated coals are characterized by varying content of mineral matter. The main components are silicon and aluminum, iron and calcium and magnesium are present in relatively small amounts. In comparison with other coking coals [7], analyzed coals contains significant amounts of alkaline compounds.

Table 2. The content (% by weight, dry base) of individual elements in output coals and their concentrates obtained in the coal cleaning process.

Element	Mineral	Raw coal	Dense medium washer		Jig washer		Flotation	
			Feed	Concentrate	Feed	Concentrate	Feed	Concentrate
Si	A	12.55	16.28	1.17	7.90	1.40	3.18	1.39
	B	6.30	14.78	1.19	6.99	0.62	6.11	1.86
	C	1.97	17.44	0.49	8.43	1.10	9.05	1.40
Al	A	4.91	7.08	0.87	4.77	1.19	2.71	0.89
	B	2.58	5.69	0.57	3.38	0.51	2.70	0.78
	C	7.22	7.43	0.64	4.10	0.96	3.25	1.27
Ca	A	0.30	1.17	0.23	0.37	0.27	1.12	0.21
	B	0.97	0.88	0.29	0.30	0.17	0.61	0.44
	C	0.41	0.22	0.21	0.63	0.19	0.80	0.34
Fe	A	1.76	4.59	1.83	1.46	0.73	0.68	0.42
	B	1.46	2.18	0.83	1.34	0.56	1.09	0.60
	C	2.15	1.43	0.59	1.57	0.43	0.96	0.32
Mg	A	0.35	0.59	0.18	0.33	0.13	0.17	0.09
	B	0.38	0.51	0.12	0.25	0.08	0.07	0.12
	C	0.31	0.26	0.05	0.38	0.06	0.31	0.08
Na	A	0.44	0.17	0.09	0.19	0.09	0.19	0.12
	B	0.09	0.11	0.06	0.09	0.05	0.11	0.07
	C	0.19	0.19	0.07	0.11	0.07	0.35	0.11
K	A	1.19	1.38	0.09	0.89	0.11	0.44	0.12
	B	0.52	1.29	0.03	0.72	0.04	0.77	0.15
	C	1.61	1.65	0.05	0.90	0.10	0.87	0.22
A^d [%]	A	56.8	58.0	7.70	33.3	7.90	16.8	6.3
	B	22.2	48.7	4.6	28.9	4.0	22.2	8.9
	C	52.7	65.1	3.8	30.3	6.0	28.7	8.0
M^d [%]	A	57.8	62.7	8.6	36.3	8.8	18.1	7.1
	B	23.5	52.7	5.2	31.4	4.5	24.2	9.5
	C	57.1	70.4	5.2	33.0	6.8	31.2	8.7

Differences in the quantitative composition of mineral matter are visible at the stage of primary grain classes separation of raw coal, which can be attributed to differentiated content of mineral matter in examined coals. The feed led to the enrichment of jig washers (grain class 0÷20 mm) contains half the inorganic material than coal led to the enrichment in dense medium washers (grain class below 20 mm), which is reflected in much lower concentration of silicon, aluminum, calcium and iron. Alkali content (especially sodium) do not differ significantly from each other. Despite a lower potassium content in the grain fraction of 0 to 20 mm (the jig) calculated ratios of potassium to sodium do not differ in the case for both factions of individual mines. As a result of the coal cleaning process in dense medium washers content of inorganic matter decreases dozen times (concentration of total aluminum and silicon). Coal was purified primarily of silicates, aluminosilicates, and the iron compounds (pyrite). The contents of other elements was also reduced significantly. The least noticeable reduction can be seen in the case of sodium, which can be attributed due to crystallization of NaCl derived from circulating groundwater on grain surface and inside them. The loss of sodium is about the same regardless of type of the coal cleaning process in each grain class.

The coal cleaning process in jigs allowed to reduce the content of mineral matter several times. The substantial amounts of silicon, aluminum and potassium were removed from coal. Iron, magnesium and sodium was reduced to a lesser extent.

In the case of flotation (grain class below 0.5 mm) the feed has much lower content of mineral matter then feeds to other coal cleaning processes (more than three times lower). However, it has high content of calcium in respect of the feed to the jig washers. Despite the significant reduction of alkali, their content in coal concentrates (after conversion to oxides) is relatively high, considering main use of this type of coal - coke industry. The lowest content of alkali metals was noticed in concentrates dewatered in a centrifuge then in other processes (Table 3).

Table 3. The content of alkali oxides in coals from Upper Silesian Coal Basin mines.

Oxides in coal [%wt.dry mass]	Mineral	Raw coal	Dense medium washer		Jig washer		Flotation	
			Feed	Concentrate	Feed	Concentrate	Feed	Concentrate
K_2O	A	2,86	3,33	0,22	2,14	0,27	1,06	0,29
Na_2O		1,19	0,46	0,24	0,51	0,24	0,51	0,32
Total:		4,05	3,79	0,46	2,65	0,51	1,57	0,61
K_2O	B	1,25	3,11	0,07	1,74	0,10	1,86	0,36
Na_2O		0,24	0,30	0,16	0,51	0,13	0,30	0,19
Total:		1,49	3,41	0,23	2,25	0,23	2,16	0,55
K_2O	C	3,88	3,98	0,12	2,17	0,24	2,10	0,53
Na_2O		0,51	0,51	0,19	0,30	0,19	0,94	0,30
Total:		4,39	4,49	0,31	2,47	0,43	3,04	0,83

4 Conclusion

Knowledge of the major inorganic element content of coal is important in assessing the usefulness of the material for a specific process. The chemical composition of the mineral matter can constitute the basis for predicting the quality of solid product of coking, combustion and gasification of coal. The process of coal cleaning, in each methods, ie. dense medium washers, jiggers and flotation significantly reduces the content of mineral matter, mainly the content of silicates, aluminosilicates and iron, magnesium and potassium. The highest efficiency shows the process of cleaning in dense medium, but the best quality in terms of metallic elements content has a final product of flotation. Coal concentrates from each cleaning process are characterized by relatively high content of alkali metal compounds, which may be important from the view point of using coal in coke industry.

This paper was prepared as a part of the statutory activity of the Faculty of Energy and Fuels at the AGH University of Science and Technology, No. 11.11.210.213

References

1. A. Karcz, Karbo; R. 47 nr 10,2002
2. J. Węgiel, H. Machowska, E.Strzednicka; Koks, Smoła, Gaz, vol. XXXVI, nr 5
3. K. Bytnar, T. Dziok, P.Grzywacz, Karbo, tom. LVI, nr 4, 2011
4. A. Strugała; Gospodarka Surowcami Mineralnymi, tom 16, T.15. z.1, s. 5-30,1998
5. J. Kowalczyk, G. Strzelec, Inżynieria mineralna, vol. V, nr 2 (13), 2004
6. A. Karcz; Koksownictwo cz.1, Wydaw. AGH, Kraków 1991
7. K. Bytnar, A. Strugała, Karbo, vol. LII, nr 1, 2007

Evaluation method of economic efficiency of industrial scale research based on an example of coking blend pre-drying technology

Piotr Żarczyński[1,*], *Andrzej* Strugała[2], and *Krzysztof* Kwaśniewski[3]

[1]Arcelor Mittal Poland S.A. Zdzieszowice Unit, Poland
[2]AGH University of Science and Technology, Faculty of Energy and Fuels, Cracow, Poland
[3]AGH University of Science and Technology, Faculty of Management, Cracow, Poland

Abstract. The research on new and innovative solutions, technologies and products carried out on an industrial scale is the most reliable method of verifying the validity of their implementation. The results obtained in this research method give almost one hundred percent certainty although, at the same time, the research on an industrial scale requires the expenditure of the highest amount of money. Therefore, this method is not commonly applied in the industrial practices. In the case of the decision to implement new and innovative technologies, it is reasonable to carry out industrial research, both because of the cognitive values and its economic efficiency. Research on an industrial scale may prevent investment failure as well as lead to an improvement of technologies, which is the source of economic efficiency. In this paper, an evaluation model of economic efficiency of the industrial scale research has been presented. This model is based on the discount method and the decision tree model. A practical application of this proposed evaluation model has been presented based on an example of the coal charge pre-drying technology before coke making in a coke oven battery, which may be preceded by industrial scale research on a new type of coal charge dryer.

1 Introduction

The development and level of knowledge have a key meaning for the progress of a civilization. In particular, the development of scientific knowledge has a significant impact on the technological, social and economic progress of societies and countries. However, the availability of scientific knowledge is limited to experts, due to its fragmentation and specialization also in the scope of the same discipline. The empirical studies are particularly important in business activities. They allow for the practical application of the theoretical research results by developing methods and techniques that can be used in production processes and, consequently, for the creation of innovations [7].

According to statistics, only a small part of the initial ideas, approx. 1-2%, is suitable for further consideration and then for application in the economy. For this reason, while

* Corresponding author: piotr.zarczynski@arcelormittal.com

conducting the scientific research focused on the innovations development, it is extremely important to engage in the study both the scientific institutions and the economic ones, as this helps in their practical orientation from the very beginning. Conducting research is related to large expenditures and therefore many enterprises are reluctant to engage in research activities. Usually, they focus their strategies on the acquisition of certain technologies which are available on the market. At the same time, they lose the chance to become a pioneer of innovation in the area of their activities, and thereby also the opportunities to gain exceptional economic benefits [1].

The best way to conduct the research related to new and innovative solutions, technologies or products is to carry out the entire research cycle including the basic (theoretical), laboratory and implementation (industry) research. Especially this last stage includes the transfer of results of basic and laboratory research on an industrial scale - the transition from models and prototypes to the mass production phase. It requires the highest involvement of the entities, thus incurring the research costs and creating the need to take the risk of obtaining unfavorable results. At the same time, they give an almost 100 percent certainty of the results.

In the case of the decision to implement new and innovative technologies, it is reasonable to carry out industrial research, both because of the cognitive values and its economic efficiency. Research on an industrial scale may prevent investment failure as well as lead to technologies improvement, which is the source of economic efficiency. Further on in the paper, an evaluation model has been presented of economic efficiency of the industrial scale research based on the discount method and the decision tree model.

2 Evaluation method of economic efficiency of industrial scale research

In the case of investment decisions concerning complex investment projects, the decision tree method can be applied. In the first place, it is useful in the case of projects implemented sequentially, that is, when the decisions form a decision chain. Therefore, the risk occurs only at certain times during the investment implementation. An example of such a decision may be a decision to conduct an industrial research. Decisions taken in the next and successive moments depend on the results achieved so far. The decision tree method allows one to structure the complex problems and develop their graphical presentation in the decision tree form which contains the most important moments in a chronological order [11]. Figure 1 represents an example of a decision tree.

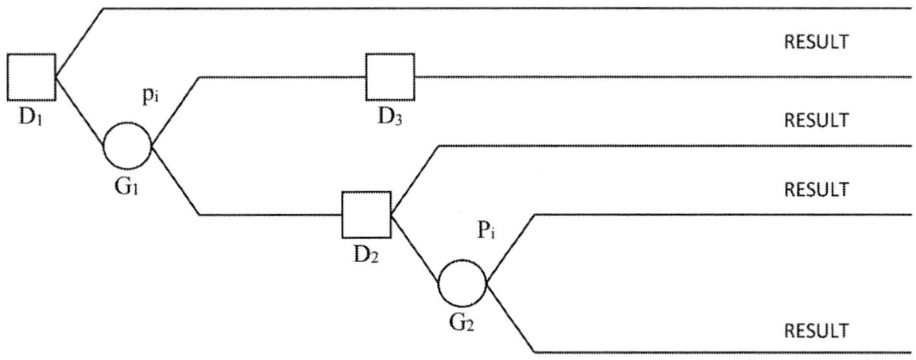

Fig. 1. Decision tree model (D_i – decision node, G_i, - chance node, p_i – probability i).

The decision tree consists of decision nodes and chance nodes. The decision nodes represent the possible ways of action at a given moment and the decision-maker makes a choice. The chance nodes are the beginning of the branches describing the consequences of decisions taken by the decision-maker which depend only on the external environment. It is necessary to determine expenditures related to potential decisions, the value and distribution of cash flow over time as well as the their probability of obtaining. It is possible to quantify the value of future decisions, the total probability and measurement of risk expressed through standard deviation. Very complex decision trees can take a vague shape and transform into over-complex trees with a significant number of decisions possible to be undertaken and with a significant number of possible results. Therefore, the application of this method for very complex projects can be problematic. The correct determination of the probability of each solution is particularly important and difficult. In the case of the decision tree elaborated for the investment project including industry research, the determination of the probability of good results can be performed by way of an cexpert analysis, using the results of the earlier research stages.

Each of the branches and decision nodes is characterized by cash flows and expenditures realted to the undertaken decisions. The determination of their value and distribution in time is necessary for evaluating the project value for the investor with the use of discount methods. In recent years, during the economic efficiency evaluation of the investment project (calculating the NPV), the Free Cash Flow approach (FCF) is applied, which takes into account in a better way the operational profit achieved by the company in relation to capital cost and investors' expectations. Free cash flows can be defined as a surplus or deficit of cash that may be a result of operating or investing activities undertaken by a company, after satisfying all the financial expectations of entities financing the activities (donors of capital). So, free cash flows are the surplus or shortage of cash generated as a result of business activities after deducting all expenses (except expenses of the debt, but after taxation), at the disposal of entities financing the company (owners and creditors) [11, 14].

The above methodology has been used to evaluate the economic efficiency of the industrial research of the coal charge pre-drying before the coking process for the investment project taking into consideration the implementation of this operation in an existing coking plant.

3 Case study

Unceasing efforts aimed at improving the economic efficiency of coke production are made in the coke-making industry. On the one hand, they have to meet increasingly higher quality requirements of the coke customers, and, on the other, they lead to the achievement of competitive advantage. The dominant cost item in the coke production is the cost of the coking coal blend. Therefore, the activities aiming at the coal blend cost reduction without the coke quality deterioration are sought. One of such activities is coal charge pre-drying implementation, which allows one to reduce the coking coal blend cost due to the possibility of increasing the share of semi-soft coals (Polish type 34) – it is the main source of economic benefits [2, 3, 6].

3.1 Characteristics of coal charge pre-drying operation

The Coking coal charge pre-drying operation is the optimized version of coal blend thermal treatment. It involves the controlled pre-drying of the blend (to the temperature cir. 85 °C), which causes a partial removal of its moisture - from cir. 9% to cir. 5 ÷ 6%. The coking

DOI: 10.1051/e3sconf/20171402014

process including the pre-drying operation includes two stages. At the first stage, which takes place before charging a blend to the oven, the blend is pre-dried in a special device designed for this purpose. At the second stage, i.e. after charging the thermally treated blend to the oven, the essential coking process takes place [2].

The implementation of the coal charge pre-dring provides a number of technological advantages, of which the most important are [2, 3, 6, 12, 16]:

- increased coal blend bulk density in the chamber after its charging,
- shorter charging time of the chambers,
- increased production capacity of a coke battery,
- higher durability of the battery's brickwork,
- improvement of the homogeneity of the bulk density distribution for the entire height and length of the oven,
- improvement of coke making properties of the coal blend
- constant moisture content in the coal blend, independent of atmospheric conditions,
- better quality of coke produced on the basis of an unchanged coal blend,
- possibility to utilise the waste heat produced in coking plant,
- improvement of the energy efficiency of the general coking process and, consequently, a reduction of the unit emissions to the environment (including CO2),
- a wider product range offered by the coke plant,
- improvement of the coking plant production flexibility by obtaining the opportunities for active management of production capacity (production capacity diversification of the coke oven battery depending on the decision to start or stop the pre-drying installation).

The technological benefits arising from the coal charge pre-drying implementation can be used in two competing directions:

- improvement in the quality of coke produced on the basis of the current coal blend,
- increase in the share of semi-soft coals (with worse coking properties) in blends without deterioration of the quality of resulting coke.

The first direction is quite commonly used in coking plants in the countries of the Far East (Japan, South Korea, China) with the use of the technology called CMC (Coal Moisture Control) [8]. In Polish conditions, this second direction is particularly interesting, due to the presence of significant quantities of cheaper semi-soft coals (Polish type 34) with weaker coking properties and the deficit of hard coals (Polish type 35) with the best coking properties - the missing amount of coals has to be imported.

The situation described above resulted in the fact that the coal charge pre-drying technology has become an object of extensive studies and analyses, and the concepts of its implementation have been considered by some Polish coking plants in recent years [5, 6, 9, 13, 15]. It also inspired the Institute for Chemical Processing of Coal in Zabrze (IChPW) to design the multi-stream dryer, which can be applied in the process of coal charge pre-drying. However, this dryer requires further research on an industrial scale. In this situation, a natural decision problem occurs whether to use a proven device (e.g. the CMC 2nd generation indirect dryer) in an industrial-scale pre-drying operation or to conduct an industrial scale research of a multi-stream dryer designed by the IChPW. So defined, the issue can be presented and solved by the application of the proposed evaluation model of economic efficiency of the industrial scale research based on the discount method and the decision tree model [10].

Table 1. Basic technical, technological and economic assumptions for considered scenarios of coal charge pre-drying investment project in existing coking plant.

Item	Unit	Coke plant without pre-drying	Coke plant with pre-drying	
			Dryer: indirect II generation CMC	Dryer: multi-stream designed by ICHPW
Technological data				
Production capacity (wet base)	ths. Mg/a	3 000	3 208	3 206
Increase of production capacity	%	-	6,92	6,88
Blend content recipe: -semi-soft coal (type 34) -hard coal (type 35)	ths. Mg/a % %	 10 90	 20 80	 25 75
Blend content utilization: *-semi-soft coal (type 34)* *-hard coal (type 35)*	ths. Mg/a *ths. Mg/a* *ths. Mg/a*	4 062 *406* *3 656*	4 346 *869* *3 477*	4 346 *1 087* *3 260*
Pre-drying installation characteristics				
Type of dryer	-	-	Indirect	Direct
Drying medium	-	-	Steam	Coke oven gas waste gases
Number of devices	pieces	-	3	5
Unit drying capacity	Mg/h	-	250	125
Number of dryers in operation	pieces	-	2	4
Preliminary moisture content of charge	%	9,5	9,5	9,5
Moisture content after pre-drying operation	%	-	6,0	6,0
Economic data				
Total Capex *- I year:* *- II year:*	Mil. € *Mil. €* *Mil. €*	-	80 *24* *56*	64 *24* *40*
Analysed lifetime period	Years	-	15	15
Loan share	%	-	68	60
Loan gross interest rate	%	-	4,5	4,5
Cost of equity	%	8,0	8,0	8,0

3.2 Application of evaluation model of economic efficiency of the industrial scale research based on the example of coal charge pre-drying

In the considered coking plant a decision is being analysed to launch a coal charge pre-drying installation for 4 top charging coke oven batteries with base production capacity equal to 3 million tonnes of wet coke per year. There are two possible scenarios of investment project implementation. The first scenario assumes an immediate decision to use the CMC 2nd generation indirect dryer. The second scenario assumes conducting the industrial scale research of the pre-drying process with the application of the research station equipped with a multi-stream dryer designed by the IChPW with the drying capacity

DOI: 10.1051/e3sconf/20171402014

equal to 10 Mg/h of wet charge. The cost of conducting industrial research equals EUR 5 million. Based on the industrial research results, a decision concerning the choice of the dryer will be undertaken: either the CMC 2nd generation indirect dryer or the multi-stream direct dryer designed by the IChPW. In both scenarios the investment period is 2 years. In the first year of investment works common for both types of dryers are performed with the capital expenditures of EUR 24 million and it does not determine the decision about the dryer choice. The industrial research is successful if it results in obtaining the capacity of the multi-stream dryer to prepare the pre-dried blend consisting of 25% of semi-soft coals and 75% of hard coals, allowing for the production of the same coke quality. On the basis of the expert analysis, the probability of success of 30% was assumed.

In the case of the CMC 2nd generation indirect dryer, the coal charge pre-drying installation consists of 3 devices with the capacity of 250 Mg/h each, two in operation and one spare. Whereas, in the case of the multi-stream direct dryer designed by the IChPW, the coal charge pre-drying installation contains 5 dryers, each with a capacity of 125 Mg/h, four of which are in operation. The basic technical, technological and economic assumptions are presented in table no. 1.

Based on the presented main and additional detailed assumptions, the economic efficiency of each investment project scenario has been evaluated by using the discounting method with the application of the Free Cash Flow to Firm approach (FCFF). In the case of the scenario involving industrial research, in the calculations, the effect connected with expenditures on that research was taken into consideration. According to the considered decision problem, a decision tree has been developed. It is presented in Figure 2. Table 2 shows the calculation of NPV for each branch of the decision tree.

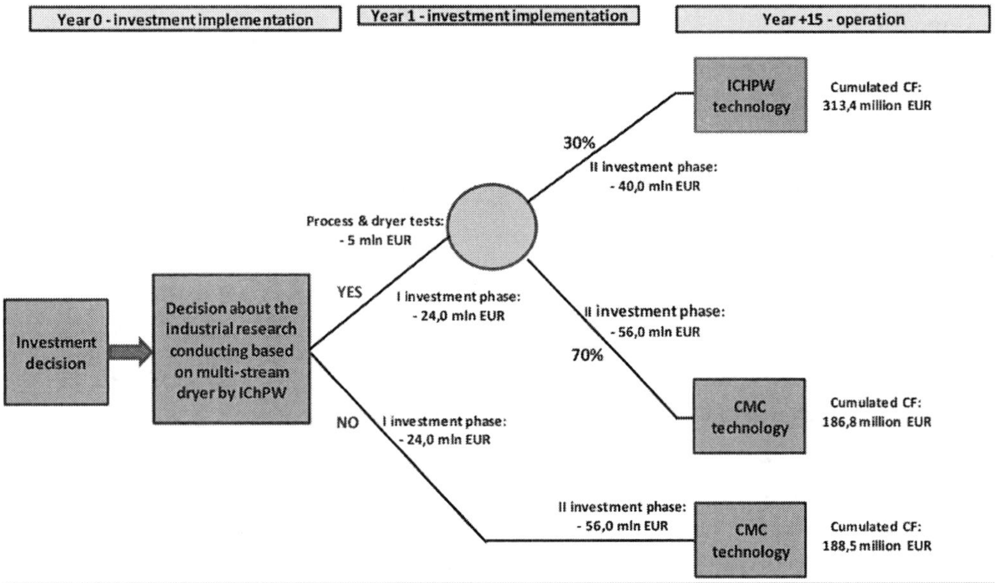

Fig. 2. Decision tree for investment project of pre-drying operation implementation in a coking plant preceded by conducting industrial research.

Table 2. NPV determination for particular branches of the decision tree for the coal charge pre-drying operation implementation in a coking plant.

[million EUR]

Item		0	1	2	3	4	5	6	7	8	9	10	11	12	13	14	15	16
		I1	I2	E1	E2	E3	E4	E5	E6	E7	E8	E9	E10	E11	E12	E13	E14	E15
Coal charge pre-drying operation with CMC II generation dryer application without industrial research	FCF	-24,0	-56,0	9,2	14,2	14,2	14,2	14,2	14,2	13,6	13,6	13,6	13,6	10,8	10,7	10,7	10,7	10,7
	dFCF	-24,0	-53,3	8,4	12,3	11,7	11,1	10,6	10,1	9,2	8,7	8,3	7,9	6,0	5,7	5,4	5,1	4,9
	NPV	**48,0**																
Coal charge pre-drying operation with CMC II generation dryer application with industrial research	FCF	-29,0	-56,0	9,2	14,2	14,2	14,2	14,2	14,2	13,5	13,5	13,5	13,5	10,5	10,5	10,5	10,5	10,5
	dFCF	-29,0	-53,2	8,3	12,2	11,6	11,0	10,5	10,0	9,0	8,6	8,1	7,7	5,7	5,4	5,2	4,9	4,7
	NPV	**40,6**																
Coal charge pre-drying operation with multi-stream ICHPW dryer application with industrial research	FCF	-29,0	-40,0	17,4	22,4	22,4	22,4	22,4	22,4	21,8	21,8	21,8	21,8	19,4	19,4	19,4	19,4	19,4
	dFCF	-29,0	-37,9	15,6	19,0	18,0	17,1	16,2	15,3	14,1	13,4	12,7	12,0	10,1	9,6	9,1	8,6	8,1
	NPV	**132,1**																

Table 3 presents the determination of the expected value E (NPV) in the case of conducting industrial research.

Table 3. NPV determination for particular branches of the decision tree for the coal charge pre-drying operation implementation in a coking plant.

Item	NPV
Expected value of Project NPV E(NPV) in the case of conducting industrial research and choice of the dryer based on it after 1 year	68,05
Project NPV in the case of immediate decision about choosing the II generation CMC dryers	47,99

The economic efficiency evaluation results show that even with a low probability of obtaining positive results (30%) of industrial research on the development of the multi-stream dryer, the decision to conduct the tests is economically justified. The expected value of E (NPV) for such a sequence of the project implementation exceeds by far the NPV for the scenario which determines the type of a dryer at the beginning of the project.

4 Conclusions

The practical application of evaluation model of economic efficiency of industrial research based on the decision tree and discounting shows that the industrial research generates a measurable value.

The presented model uses common and understandable evaluation techniques of investment projects, i.e. the decision tree model and discounting. Therefore, it is easy to use and understand by a wide group of stakeholders. Due to its versatility, it can be used in many areas.

References

1. R. Budziński, University of Szczecin, *Scientific methodology – unpublished lecture*, (2011),
2. A. Czaplicki, Karbo, *Pre-drying operation before coking process*, **special edition**, 47-57, 2007,
3. A. Czaplicki, M. Janusz, M. Kosewska, K. Wróblewska, Karbo, *Coal blend preparation methods and their influence on coke quality in top charging system,* **2**, 88-99, (2012),
4. Z. Frankl, J. Salcewicz, Coke, Tar & Gas *The heating rate as a factor in improving the coals sinterability*, **6**, (1961),
5. A. Karcz, (Reasearch project AGH) *Research of coking coals pre-drying process in terms of assessing the possibility of extending the raw materials base for the production of high-quality coke*, (2007),
6. A. Karcz, A. Strugała, Mineral Resources Management, *Increasing chances of utilizing the domestic coking coal resources through technological operations in coal blend preparation*, **24**, 5-18 (2008),
7. B. Klepacki, Annals of agriculture science, *The chosen problems connected with methodology of economic scientific* research, **96**, 38-46, (2009),
8. K. Kushioka, Y. Konno The Iron and Steel Institute of Japan, *New Moisture Control System of Coal for Coking*, (1985),
9. W. Latocha, W. Kaczmarek, A. Strugała, P. Żarczyński, Energy Policy Journal *Extending the coal base of the Polish cokemaking industry through the implementation of the coal charge pre-drying and through the use of the overseas coals*, **14**, 215–229 (2011),
10. Patent: Patent Office Bulletin No. 24 (1041) - Application No P.399258 of 21.05.2012 (2013),
11. W. Pluta, *Capital budgeting*, (2000),
12. R. Poultney, R. Willmers, The Coke oven Managers' year-book 2000, *Assessment of coal pre-treatment systems*, (2000),
13. A. Sobolewski, A. Czaplicki, P. Żarczyński, Institute for Chemical processing of Coal in Zabrze, *The conceptual design of the pilot coal charge pre-drying installatio for ArcelorMittal Poland SA Zdzieszowice Unit - unpublished*, (2013),
14. P. Szczepankowski, WSPiZ Publisher, *Company finance management* (2004),
15. A. Tramer, M. Kosewska, K Wróblewska, A. Sobolewski The 1st Polish-Czech Konference, *Opportunities to improve the quality of coke by coal charge pre-drying*, (2001),
16. P. Żarczyński, A. Strugała, A. Sobolewski, W. Kaczmarek, Mineral Resources Management, *Evaluation of potential improvements to coking process energy efficiency as a consequence of implementing a coal blend pre drying operation*, **29**, 151–165, (2013),

E3S Web of Conferences **14**, 02015 (2017) DOI: 10.1051/e3sconf/20171402015
Energy and Fuels 2016

Biomass - alternative renewable energy source to the fossil fuels

Dorota Koruba[1,*] , *Jerzy Zbigniew* Piotrowski[1], *Jolanta* Latosińska[1]

[1] Kielce University of Technology, Faculty of Civil and Environmental Engineering

Abstract. The article presents the fossil fuels combustion effects in terms of the dangers of increasing CO_2 concentration in the atmosphere. Based on the bibliography review the negative impact of increased carbon dioxide concentration on the human population is shown in the area of the external environment, particularly in terms of the air pollution and especially the impact on human health. The paper presents biomass as the renewable energy alternative source to fossil fuels which combustion gives a neutral CO_2 emissions and therefore should be the main carrier of primary energy in Poland. The paper presents the combustion heat results and humidity of selected dry wood pellets (pellets straw, energy-crop willow pellets, sawdust pellets), dried sewage sludge from two sewage treatment plants of the Holly Cross province pointing their energy potential. In connection with the results analysis of these studies the standard requirements were discussed (EN 14918:2010 "Solid bio-fuels-determination of calorific value") regarding the basic parameters determining the biomass energy value (combustion heat, humidity).

1 Introduction

As a result of human activity we can observe rapid climate changes which are particularly evident in the increase of the greenhouse gases concentration (carbon dioxide, methane and nitrous oxide) in the atmospheric air. The anthropogenic greenhouse gas emissions are caused by economic activities especially the fossil fuels burning such as coal, oil and natural gas.[4] The carbon dioxide concentration increased for the last 200 years since the beginning of the industrial revolution when the CO_2 concentration reached 280ppm and in 1958 it reached a value of 316ppm. In May of 2013 CO_2 concentration for the first time exceeded 400 ppm, in March 2016 exceeded 405ppm (Fig.1) in the graphs rapid and continual increase in the this gas concentration is shown.

* Corresponding author: dkoruba@tu.kielce.pl

 DOI: 10.1051/e3sconf/20171402015

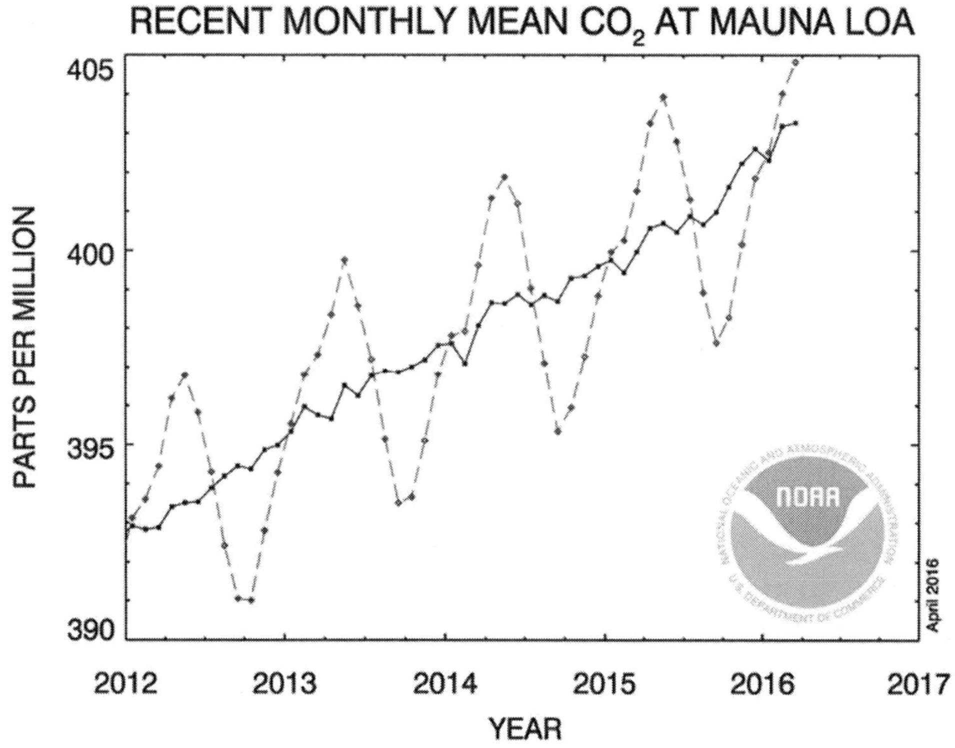

Fig. 1. The carbon dioxide concentration measured at Mauna Loa observatory in Hawaii. Source: Earth System Research Laboratory, Global Monitoring Division (http://esrl.noaa.gov/gmd/ccgg/trends).

The climate change effects observed in all natural systems, among other hydrological systems (melting snow, glaciers), which cause changes in water resources in terms of quality and quantity as well as land-based systems, eg. change in the species geographical ranges [1]. The negative climate change effects also have an impact on the human population by the presence of extreme weather conditions (high temperatures - drought, heavy rain - floods, hurricanes and fires) which also affect crops, residential and public buildings.[8] Increasing the carbon dioxide concentration results in an increased number of fungi spores (Alternaria spp., Cladospirium spp., Penicillium spp., Aspergillus spp., etc.), soil bacteria and pollen (Birch, Clone, Tilia, etc.) in the atmospheric air which cause allergies, respiratory diseases in the human population[3,4,5,9]. With the increase of carbon dioxide concentration in the atmospheric air the gas concentration in indoor air also increases[16].

The carbon dioxide high concentrations cause fatigue, headaches, dizziness, disorientation, breathlessness to the humans and in combination with carbon monoxide coming for example from fireplaces or cigarette smoke, through accumulation of the harmful effects of the last gas may in a longer term result in health deterioration. Above symptoms are the same as the symptoms of Sick Building Syndrome (SBS)[1,2,5].

In order to stop advancing climate change, in particular increase the concentration of carbon dioxide in the air in the working biomass is proposed as an alternative fossil fuels[6,13,14]. Biomass can be an essential alternative to fossil fuels due to the zero carbon dioxide emission because during the combustion - CO_2 emissions to the atmosphere is the same as the amount of CO_2 absorbed by plants during photosynthesis[10]. As a result of photosynthesis (1),

$$CO_2 + 2\,H_2O \xrightarrow{hv} O_2 + H_2O + (CH_2O) + 470kJ \qquad (1)$$

and specifically carbohydrate synthesis of 1 carbon dioxide mole accumulates chemical energy of 470kJ. During the year on the earth the biomass amount produced by photosynthesis can be 220bln tones of dry weight. Therefore the energy contained in the biomass can be a significant contribution as a renewable energy source[12].

2 The definition of biomass

Directive 2001/77/EC, of the European Parliament and the Council from 27th September 2001 on the promotion of the internal market for electricity produced from renewable energy sources has introduced a biomass definition[23].

In its meaning biomass is "the biodegradable fraction of products, waste and residues from agriculture (including plant and animal substances), forestry and related industries as well as the biodegradable fraction of industrial and municipal waste"[22-27].

On the basis of the above mentioned Directive in 2005, its interpretation of the biomass definition was also introduced by the Ministry of Economic Regulation on 19th December 2005 on detailed duties scope to obtain and submit to the origin certificates redemption, the substitute fee and purchase of electricity and heat from renewable sources energy[27]. The Regulation defines biomass as "solid or liquid plant or animal origin, which are biodegradable, derived from products, waste and residues from agriculture and forestry, and related industries, as well as parts of other wastes that are biodegradable".

Solid bio-fuels, which include biomass, have been subjected to classification by the European Committee for Standardization CN/TC335[24], and are divided into:

- products from agriculture and forestry,
- plant waste from agriculture and forestry,
- plant waste from food processing industry,
- wooden waste with the exception of wood waste which may contain halogenated compounds or heavy metals, which are a result of wood treatment and waste from construction and demolition sites.

3 Methodology of experimental studies of selected biomass pellets and dried sewage sludge

3.1. Materials and research methods

The research material: willow pellets, sawdust pellets, straw pellets, sewage sludge (Sewage sludge "S" from the sewage treatment plants in Sitkówka-Nowiny (S), subjected to the anaerobic stabilization, dewatered and dried in disk. The nominal capacity of the plant is 72 000 m^3/d at load of 275 000 RLM. Sewage sludge "Ł" from the sewage treatment plants in Olsztyn, subjected to the anaerobic stabilization, dewatered and dried in disk. The nominal capacity of the plant is 34 767 m^3/d at load of 350 000 RLM).

Research material in the form of pellets was grinded and then separated into fractions with dimensions of 0.5mm, 0.25mm, 0.125mm <0.125mm. The dried sludge was fractioned with a diameter of 0.2 mm[17-21].

Determination of moisture content, ash, and the heat of combustion were carried out on three representative samples of all pellet fraction with a molecular diameter 1g.

When testing the moisture content of the sample in a closed crucible was dried at 105°C for 30 minutes. Drying was repeated until the difference in weight between weighing does not exceed 0.5% of absolute. Samples were weighed with the accuracy 0.0001g.

Prior to the ash content determination the samples were dried to constant weight at 105°C for 30 minutes in closed crucibles. Examination of ash content was carried out at a temperature of 815°C. Samples were weighed with the accuracy 0.0001g.

3.1.1 Measuring the combustion heat of selected biofuels

Applied Technologies: calorimeter KL-12Mn, decomposition vessel (bomb), bottle with pure oxygen, PC, analytical scale, knife grinder for Biomass LMN 100, bio-fuels hydrometer BIO-1, vibrating shaker.
The methodology of research in the field of combustion heat is based on the PN-EN 14918:2010 "Solid bio-fuels - Determination of calorific value"[17], which specifies a method for solid fuel calorific calculations at a constant volume and at 25°C in a calorimeter bomb.
The heat of combustion is the amount of heat, which separates the complete combustion of the fuel mass units, when the combustion process is carried out at constant volume in an atmosphere of gaseous oxygen under the assumption that the temperature of the fuel prior to combustion temperature of the combustion products is 20°C, the products combustion CO_2 and SO_2 are in the gaseous state has not oxidized nitrogen and water content of the fuel by burning and formed during combustion due to oxidation of hydrogen is in the liquid.
The solid bio-fuels combustion heat studies were calculated on the basis of calorimetric research, according to the formula:

$$Q = K \quad (T_3 \quad T_2 \quad k)\left[\frac{J}{g}\right] \tag{2}$$

where:
Q - The heat of combustion of fuel samples [J/g]
K - Calorimeter constant
T_2 , T_3 – temperature characteristic of the balance sheet
k - correction for heat exchange with the environment calorimeter
where:

$$k = 0,5 \quad [0,2 \quad (T_2 \quad T_1) + 0,2 \quad (T_4 \quad T_3)] + 0,2 \quad (n \quad 1) \quad (T_4 \quad T_3) \tag{3}$$

where:
n - number of minutes in cycle 2
T_1, T_4 – temperature characteristic of the balance sheet

The work of the calorimeter is divided into 5 cycles:
0 - the inclusion of the calorimeter and stabilize the temperature inside the calorimeter (cycle time is approx. 2 minutes).
1- recording the temperature T_1 and measure a time interval equal to 5 minutes.
2 - recording the temperature T_2 and the ignition of the fuel sample in a bomb calorimeter, this cycle continues until reaching its maximum temperature.
3 - registration T_3 (maximum) and the measurement of the next section of time equal to 5 minutes.
4 - registration of temperature T_4 and finish the job

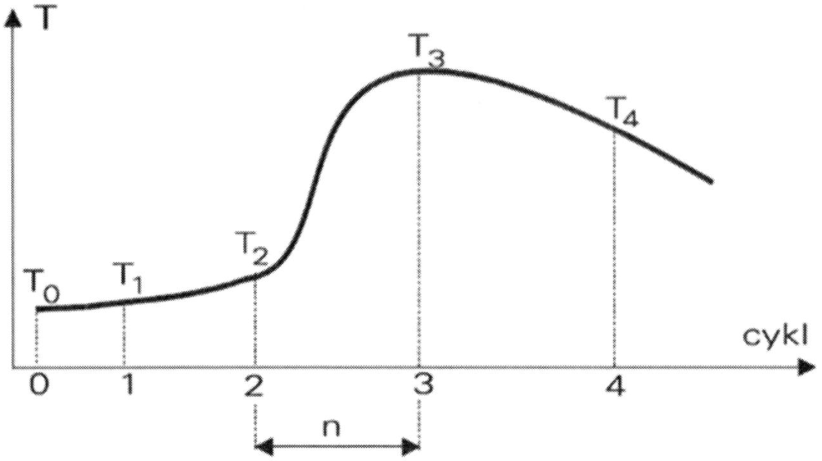

Fig. 2. Changes in temperature measurement cycle in the calorimeter.

The heat of combustion is called the upper calorific value, since it corresponds to the total heat produced from the fuel. The calorific value is a measure of the amount of heat obtained by combustion of unit of mass (or volume) of fuel.
The calorific value of organic substances containing carbon, hydrogen, sulfur and oxygen is described by the following empirical equation:

$$W_u = 33900C + 121000\left(H_2 \quad \frac{O_2}{8}\right) + 10500S \quad 2500\left(W + \frac{9}{8}O_2\right) \left[\frac{kJ}{kg}\right]$$

where: C, H_2, O_2, S are mass participation of these elements, respectively W is contents of water (humidity of material).

3.1.2. Measuring the moisture content of selected biofuels

Applied Technologies: analytical scale, laboratory oven.

The research method in determining the moisture content in the tested specimens prepared on the basis of PN-EN ISO 18134-1:2015-11, "Solid Bio-fuels - Determination of Moisture Content - Drier method - part 1: Total Moisture - Reference Method"[20]. Test methods described in this International Standards applies to all solid bio-fuels. The moisture content of solid bio-fuels (in the operating state) is always given in relation to the total weight of the test specimen (wet state).

Moisture content (W^a) of biomass in the test sample is calculated by the formula:

$$W^a = \frac{m_2 - m_3}{m_2 - m_1} \; 100\% \tag{4}$$

where:
m_1- is the mass of the empty drying container, in g;
m_2 - is the mass of the drying container and test portion before drying, in g;
m_3 - is the mass of the drying container and test portion after drying (weigh when still hot), in g;

3.1.3. The measurement of the ash content in selected biofuels

Applied Technologies: muffle furnace, analytical scale

Research methodology for the ash content determination in the tested specimens was calculated on standards: PN-ISO 1171:2002, PN-EN ISO 18122:2016-01 Solid bio-fuels - Determination of Ash Content[18,19].

The ash content was calculated from the formula:

$$A = \frac{m_3 - m_1}{m_2 - m_1} \times 100 \quad (5)$$

where:

m_1 - is the mass of the empty drying container, in g;
m_2 – is the mass of the dish plus the test sample, in g;
m_3 - is the mass of the dish plus ash, in g;

4 Test results and discussion

The combustion heat, ash and moisture content form the basis for the solid fuels technical evaluation in terms of their suitability for energy purposes. For each tested pellets fraction moisture was examined, combustion heat and ash. The combustion heat, moisture and ash content test results in the tested specimens are shown in Table 1 and figure 3,4.

The ash content results is expressed as the average weight of two measurements with an accuracy of 0.1%. The results from two tested specimens calorific determinations did not differ among themselves more than 120 J/g, the result is given as the average arithmetic number of the determinations.

Table 1. Results of combustion heat, calorific value and ash investigated pellets

The research material	**Fractions**	**Moisture [%]**	**Combustion heat [MJ/kg]**	**Calorific value [MJ/kg]**	**Ashes A [%]**
Willow pellet	0.5 mm	6.13	21	20	0.7
	0.25mm	6.09	22	21	0.9
	0.125 mm	6.20	23	23	1.1
	< 0.125 mm	6.48	22	21	1.5
Gray straw pellet	0.5mm	7.00	22	21	0.5
	0.25mm	7.23	23	22	0.7
	0.125mm	6.85	22	21	1.0
	< 0.125mm	6.44	23	22	1.4
Sawdust pellet	0.5 mm	6.41	22	21	0.6
	0.25mm	6.32	22	21	0.8
	0.125 mm	6.20	22	21	1.3
	< 0.125 mm	6.70	23	22	1.6

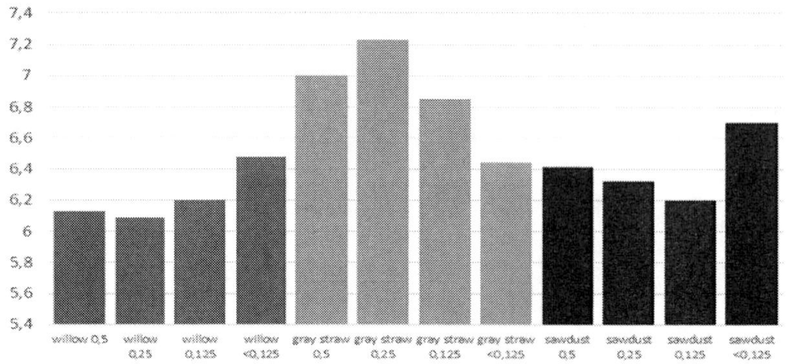

Fig. 3. Moisture particular types of pellets with the division into factions

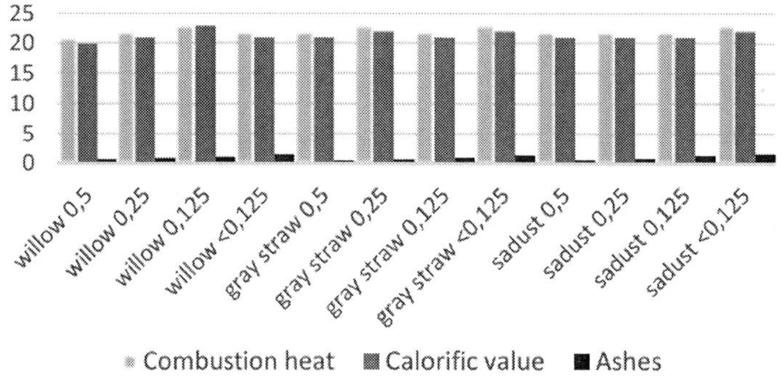

Fig. 4. Comparison of combustion heat, calorific value and ash content of pellets investigated.

The literature data and results indicate that the lower the humidity the higher the energy (eg. Gray straw fraction <0.125)[7,11]. The results (table 1, figure 4) indicates that the analyzed pellets willow, straw, sawdust has a very low ash content, compared to the sludge or coal if the ash content may be approx. 22%. Compared to the available technical data of coal and lignite, whose calorific value is in the range 24-31 MJ/kg, the results of the tested pellets of solid biomass essentially do not differ. Indicate that it could have the potential to replace fossil fuels, renewable energy source which is biomass.

The ash content in the analyzed fuels increased with decreasing fraction used to study the material. Smaller fractions of biomass have a higher content of mineral substance in the composition.

The lack of correlation between the diameter of the fraction of the pellets tested and the result of the heat of combustion.

Table 2. Results moisture combustion heat, calorific value and ash investigated sewage sludge

The research material	Fractions	Moisture [%]	Combustion heat [MJ/kg]	Calorific value [MJ/kg]	Ashes A [%]
Sewage sludge „S”	0,2 mm	7,98	18	17	65,7
Sewage sludge „L”	0,2 mm	9,43	17	16	67,1

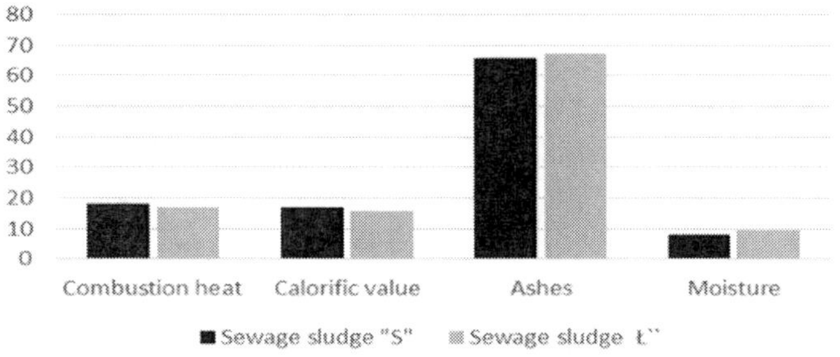

Fig. 5. Comparison moisture combustion heat, calorific value and ash content of the dried sewage sludge from two sewage.

Sewage sludge is also a "clean" biomass in terms of CO_2 emissions during the combustion. However, in the sewage sludge case there is a problem with the ash content and this problem is also reflected in research methodology (PN-EN 14918:2010)[17] of combustion heat, which assumes the performance correctness of the complete specimen combustion in a calorimeter bomb[15]. In the case of sewage sludge as well as some plants species it is not possible to achieve complete combustion in a calorimeter bomb, for the sludge specimen of "L" from 1 specimen gram the residue was 0,7094g, in the case of sewage sludge "S" 0.6995g. Residues weight after combustion is similar to the ash content in the specimen. Therefore, the research methodology work should taken which will include all types of biomass.

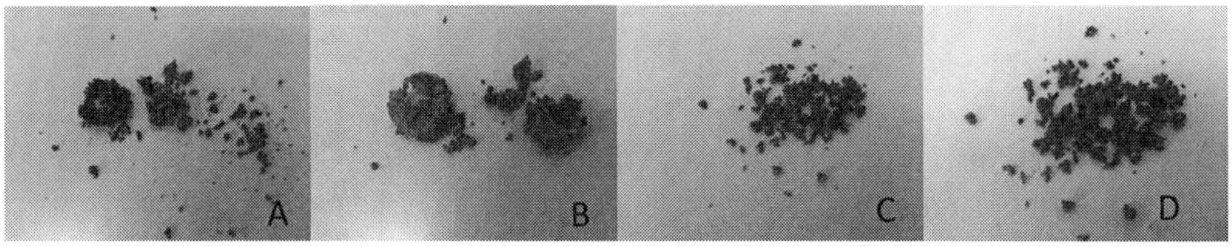

Fig. 6. Picture of the residues after combustion of 1 g tested specimen of the sewage sludge 'Ł' (pic. A,B) and 'S' (pic. C,D).

5 Conclusion

1. The intensive growth of emissions from the fossil fuels combustion caused air quality deterioration as well as the indoor air deterioration. Fossil fuels replacement with biomass can be of fundamental importance for the protection of public health.
2. Continuous increase in the carbon dioxide concentration increases the temperature of our planet resulting in long-lasting period the heat waves, increased the air pollution in terms of microbial contamination, fires, droughts and floods - all of these climate changes have a major impact on the human population health.
3. The results of the research will be the basis to develop their own research methods in the humidity, the heat of combustion and ashes.
3. The research on the research methodology development should be continued especially on the combustion heat of all biomass types.

References

1. D'Amato G. et al. Meteorological conditions, climate change, new emerging factors, and asthma and related allergic disorders. A statement of the World Allergy Organization; doi 10.1186/s40413-015-0073-0; World Allergy Organization Journal 8 (**25**): 1-52; (2015)
2. Epstein P. R., Rogers Ch.: The impacts of CO_2 and climate change on public health in the inner city - Report from the Center for Health and the Global Environment Harvard Medical School; pages:1-28; 2004
3. Lang-Yona N.: Changes in atmospheric CO_2 influence the allergen city of Aspergillus fumigatus; Global Change Biology 19, 2381–2388, doi: 10.1111/gcb.12219; (2013)
4. McMichael A.J. et al:. Climate change and human health risks and responses; WHO Library Cataloguing-in-Publication Data; ISBN 92 4 156248 X; Geneva (2003).
5. Medina Á., Rodríguez A., Sultan Y. and Magan N.: Climate change factors and *Aspergillus flavus*: effects on gene expression, growth and aflatoxin production; World Mycotoxin Journal, **8** (2): 171-179; 2015
6. Mółka J., Łapczyńska-Kordon B.: Właściwości energetyczne wybranych gatunków biomasy; Inżynieria Rolnicza; 6 (**131**); s: 141-147; (2011)
7. Niedziółka I., Szpryngiel M.,: Ocena cech jakościowych pelletów wytworzonych z biomasy roślinnej; Agricultural Engineering; Z. 2(136) T. 1; s. 267-276; (2012).
8. Pachauri, R.K and Reisinger, A., Core Writing Team (eds.): Climate Change 2007 Synthesis Report; IPCC c/o World Meteorological Organization (WMO); published by the intergovernmental panel on climate change; ISBN 92-9169-122-4; (2008)
9. Paterson R.R.M., Lima N. Further mycotoxin effects from climate change; Food Research International Elsevier Ltd, 44: 2555–2566; (2011)
10. Roszkowski A. Energia z biomasy – efektywność, sprawność i przydatność energetyczna cz.1, 2. Problems of Agricultural Engineering; z. 1 (79); s. 97–124 (2013); z. 2 (80), s. 55–68 (2013);
11. Szyszlak-Bargłowicz J., Piekarski W., Wartość opałowa biomasy łodyg ślazowca pensylwańskiego w zależności od wilgotności; Inżynieria Rolnicza 8(117) s.: 223-230 (2009).
12. Ściążko M., Zuwała J., Sobolewski A., Przewodnik metodyczny. Procedury bilansowania i rozliczania energii wytwarzanej w procesach współspalania; Wydawnictwo Towarzystwa Gospodarczego Polskie Elektrownie i Instytutu Chemicznej Przeróbki Węgla; (2007)
13. Wasielewski R., Tora B.: Bariery stosowania paliw alternatywnych w energetyce; Polityka Energetyczna 11(2), s: 129-136; (2008)
14. Wielgosiński G. Czy Biomasa jest paliwem ekologicznym? - rozdział w monografii Polska Inżynieria Środowiska pięć lat po wstąpieniu do Unii Europejskiej. Tom. I, Monografie Komitetu Inżynierii Środowiska PAN, Lublin, vol. **58**, 347-356 (2009)
15. Wielgosiński G., Nowicki M.: Badania nad termicznym rozkładem osadów ściekowych; Cieplne maszyny Przepływowe - Turbomachinery, Politechnika Łódzka, Nr **126**, 181-189 (2004)
16. Wolf J. et al.: Elevated Atmospheric Carbon Dioxide Concentrations Amplify Alternaria alternata Sporulation and Total Antigen Production; Environmental Health Perspectives; **118** (9): 1223-1228; (2010)

17. PN-EN 14918:2010: „Solid bio-fuels - Determination of calorific value"
18. PN-ISO 1171:2002: "Solid fuels - Determination of Ash Content"
19. PN-EN ISO 18122:2016-01: "Solid bio-fuels - Determination of Ash Content"
20. PN-EN ISO 18134-1:2015-11: "Solid Bio-fuels - Determination of Moisture Content - Drier method - part 1: Total Moisture - Reference Method "
21. PN-EN 14918:2010: "Solid bio-fuels - Determination of calorific value"
22. Dyrektywa Parlamentu Europejskiego i Rady 2009/28/WE z dnia 23 kwietnia 2009 r. w sprawie promowania stosowania energii ze źródeł odnawialnych
23. Dyrektywa 2001/77/WE Parlamentu Europejskiego i Rady z dnia 27 września 2001 roku w sprawie wspierania produkcji na rynku wewnętrznym energii elektrycznej wytwarzanej ze źródeł odnawialnych
24. CEN/TS 14961:2005 (E), Solid biofuels - Fuel Specification and Classes, CEN/TC 335 - Solid biofuels; (2005)
25. Ustawa z dnia 20 lutego 2015 r. o odnawialnych źródłach energii (Dz.U. 2015 poz. 478)
26. Ustawa z dnia 27 kwietnia 2001r. o odpadach (Dz.U. 2001 nr 62 poz. 628)
27. Rozporządzenie Ministra Gospodarki. w sprawie szczegółowego zakresu obowiązków uzyskania i przedstawienia do umorzenia świadectw pochodzenia, uiszczenia opłaty zastępczej oraz zakupu energii elektrycznej i ciepła wytworzonych w odnawialnych źródłach energii; z dn. 19 grudnia 2005 r., Dz.U. 2005 nr 261 poz. 2187

Technical and economic analysis of using biomass energy

Małgorzata Piaskowska-Silarska [1,*], *Stanisław* Gumuła [2], *Krzysztof* Pytel[1], *Piotr* Migo[1]

[1]The Pedagogical University, Faculty of Mathematics, Physics and Technical Science, ul. Podchorążych 2, 30-084 Kraków, Poland
[2]AGH University of Science and Technology, Faculty of Energy and Fuels, al. Mickiewicza 30, 30-059 Krakow, Poland

Abstract. In the first part of the article were presented the technical possibilities of obtaining solid biomass, biogas, landfill gas, a biogas from wastewater treatment plants, bioethanol and biodiesel. Then processes was described, allowing use of energy from biomass. As first was discussed the incineration which includes drying and degassing of the wood materials, wood gas burning at 1200^{0}C, post-combustion gas and heat transfer in the heat exchanger. Then had been described gasification, or thermochemical conversion process, occurring at high temperature. It is two-stage process. In the first chamber at deficiency of air and at relatively low temperatures (450-800^{0}C), the fuel is being degasified, resulting in creating combustible gas and a mineral residue (charcoal). In the second stage, secondary combustion chamber and at a temperature of about 1000-1200^{0}C and in the presence of excess of oxygen resultant gas is burned. A further process is pyrolysis. It consists of the steps of drying fuel to a moisture level below 10%, milling the biomass into very small particles, the pyrolysis reaction, separation of solid products, cooling and collecting bio-oil. Then discusses co-generation, which is combined production of heat and electricity. In this situation where the biomass contains too much water it can be used for energy purposes through biochemical processes. The alcoholic fermentation results in decomposition of carbohydrates taking place under anaerobic conditions, and the product is bioethanol. Another biochemical process used for the production of liquid biofuels is esterification of vegetable oils. Methane fermentation in turn causes a decomposition of macromolecular organic substances with limited oxygen available. As a result, we obtain alcohols, lower organic acids, methane, carbon dioxide and water. There was analysis of economic increasing of solid biomass energy, biogas and liquid biofuels in the following article.

1 Introduction

The use of biomass as central heating and domestic hot water preparation is nothing innovative, but only return to thermal energy generation by man for thousands of years. Using wood for heating in Europe and other parts of the world was very popular and widely used.

* Corresponding author: mpiasko@up.krakow.pl

It was not until the discovery of fossil fuels, greatly reduced the use of biomass. The main reason for this was significantly greater calorific value of fossil fuels, and better technical possibilities of their storage and transportation. If the present resources of conventional fossil fuels, ie. Coal, oil or gas did not exhaust itself at such a fast pace, and their price on the world market did not grow as dynamic and interest in biomass as an energy source would probably be marginal.

2 The analysis of technical possibilities to use a biomass as Energy

Analyzing the possibility of obtaining biomass energy in Poland, one have to start with a discussion of its resources. Considering the technical and financial costs of 739 TWh theoretical potential of biomass we get real economic potential, of more than four times smaller. If we compare it with the demand for energy, it turns out that it is more than four times smaller (Fig. 1).

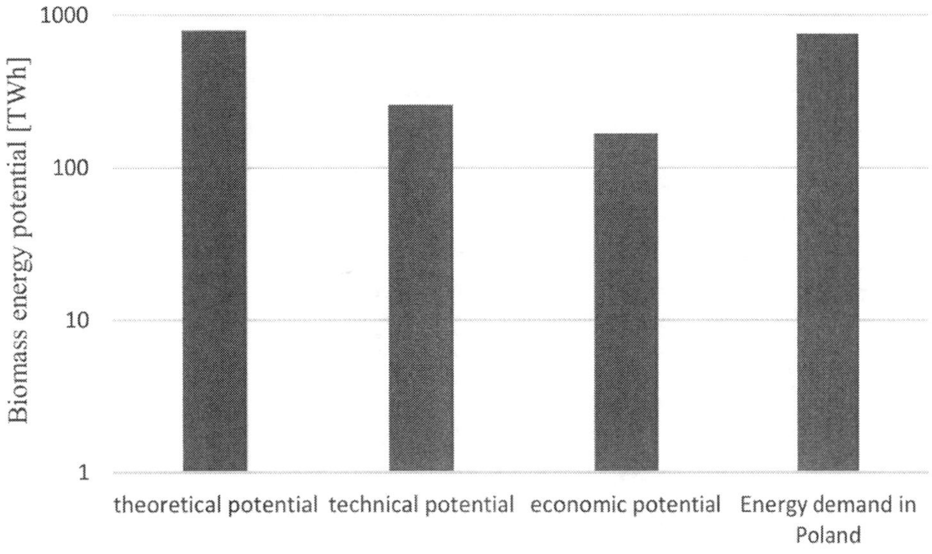

Fig. 1. Biomass resources in Poland [14].

However, as can be seen in Fig. 2, the proportion of energy from biomass in households per 1 inhabitant is comparable with conventional energy sources such as natural gas and electricity.

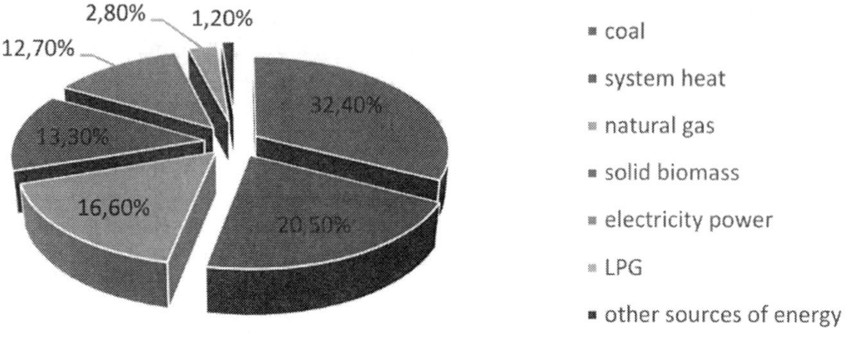

Fig. 2. Structure of energy consumption in households calculated per 1 inhabitant by specific energy sources in 2014. [13].

Comparing while obtaining energy from biomass and other renewable energy sources (Fig. 3), it is clear that it represents 88%.

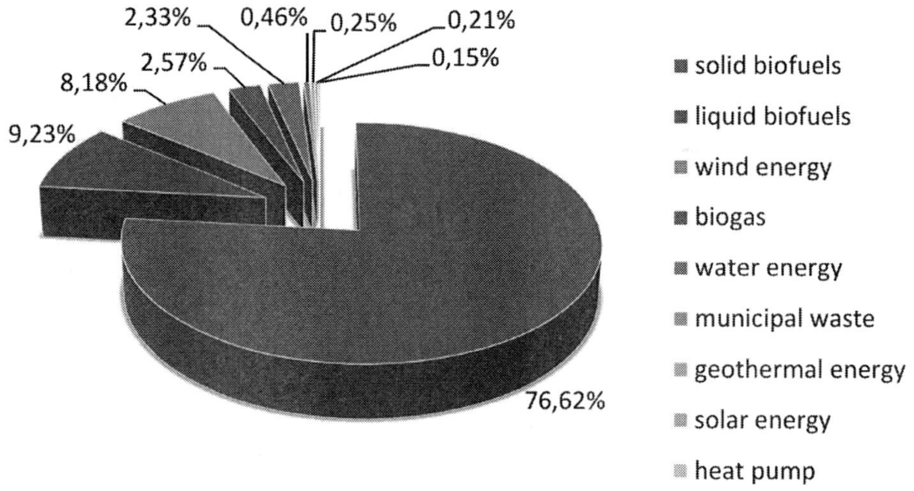

Fig. 3. Obtaining energy from renewable sources by media in 2014 in Poland [13].

If we trace the share of solid biofuels, liquid biofuels and biogas (Fig. 4) in obtaining energy, in 2010-2014, we will notice a continuous decrease in the use of solid biofuels. There is a growing of liquid fuels (except for the year 2011, when the trend was reversed) and biogas.

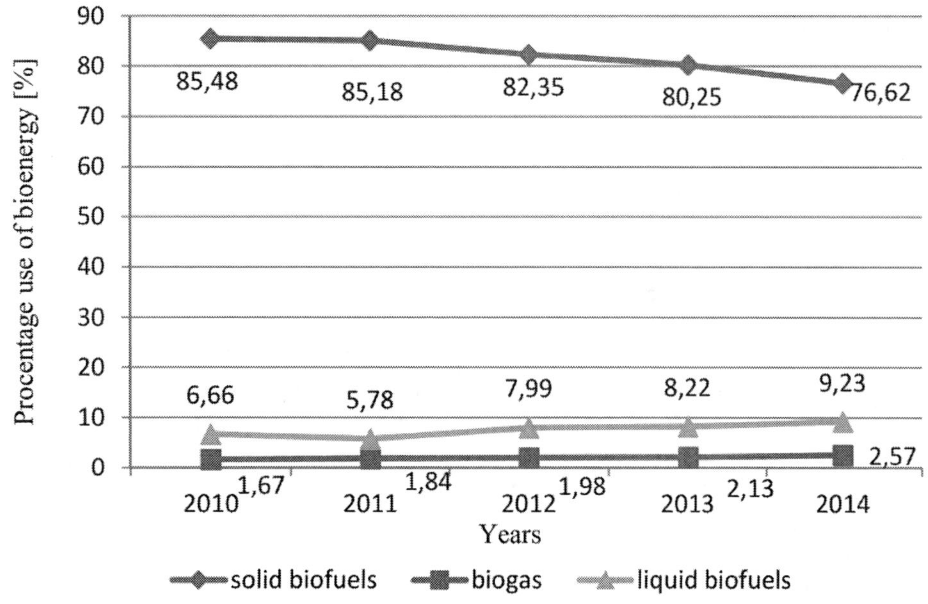

Fig. 4. The share of solid biofuels, liquid biofuels and biogas in obtaining energy in 2010-2014 [13]

In the next graph one can see how was as obtaining biogas in 2010-2014, depending on its place of production. More and more we gain biogas from wastewater treatment plants and other biogases. The share of landfill gas in obtaining energy is getting smaller (except the year 2011 when there was a temporary increase in its use) [1-4, 6, 10, 16].

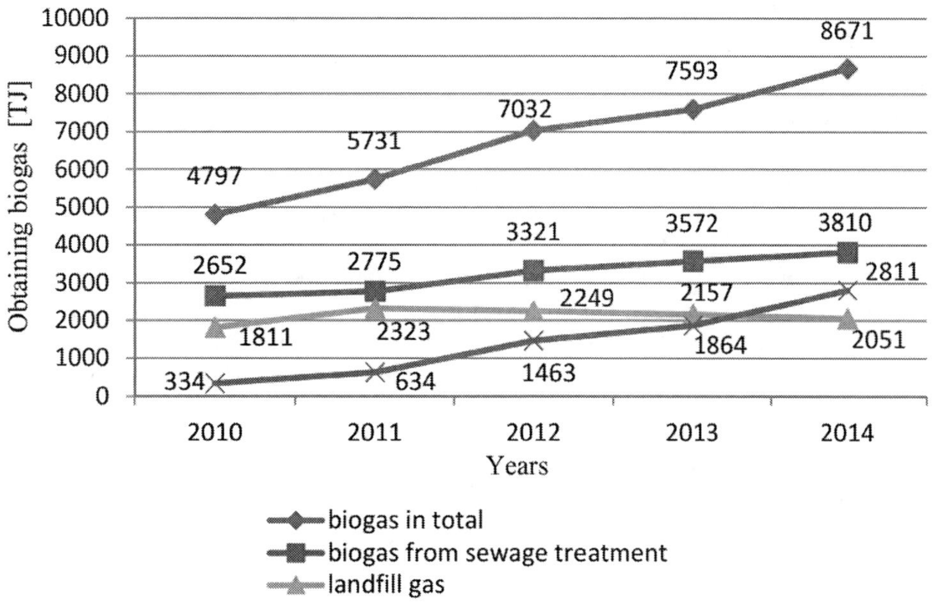

Fig. 5. Obtaining biogas in 2010-2014 [13].

3 Technology of obtaining energy from biomass

3.1 Combustion

Combustion is the most common and simplest means of obtaining heat and electricity from biomass. Combustion efficiency of wood meet environmental protection standards, should be performed in three phases:
- drying and degassing the wood material, which results in a timber gas
- burning wood gas at a temperature of 1200^{0}C,
- afterburning gas and the heat transfer in the heat exchanger.

In addition, high temperature, appropriately long burning time, with good access of oxygen can reduce emissions of carbon monoxide, hydrocarbons and polyaromatic hydrocarbons. In addition to efficient combustion of biomass, which is also reflected in the ecology, it is worth using boilers with increased heat transfer surface.
The efficiency of the combustion process affects the amount of supplied air. In modern boilers for combustion is supplied so called primary and secondary air. The primary air mixed with the fuel is used in the gasification and burning charcoal. The secondary air is used for combustion of volatiles.
Boilers for combustion of biomass have the power of several kW to several hundred MW. Typical furnace is composed with a combustion chamber lined with usually high-temperature resistant ceramic material and the grid. We can divide grates fixed, mechanical grates flat and stepped. Boilers with grates fixed or mechanical horizontal used to fuel combustion-dried. While the boilers equipped with moving grates stepped are used for wet fuels. This solution allows the evaporation of water from the fuel in the first phase and then completely burned as you move into the hearth. It is also used in boilers with fluidized bed furnace, which enables efficient combustion of biofuels poor quality while maintaining low emissions. Boilers for combustion of biomass can be equipped with automatic control and forced supply air [5, 7-9, 11, 12, 15, 17, 23, 24, 27, 32].

3.2 Gasification

Gasification is a thermochemical conversion process, occurring at high temperature, where the product is a gas:
- combusted to provide thermal energy.
- used in microwave, gas turbines, used for the production of electricity and machines, performing mechanical work.

The process of gasification of solid fuels takes place in two stages. In the first chamber at a shortage of air and relatively low temperatures (450-800C), the fuel is degasified, resulting in a combustible gas and a mineral residue (charcoal). In the second stage, the secondary combustion chamber at a temperature of about 1000-1200C, and in the presence of excess oxygen is burned of resultant gas.
An advantage of gasification is efficiency at high-level 35% and in the near future even 45-50%, while the efficiency of small and medium-sized equipment used for combustion is 15-20% [11, 12, 25, 26, 28, 29].

3.3 Pyrolysis

The product of pyrolysis is a liquid bio-oil or bio-called pyrolysis oil, which is a mixture of oxygenated hydrocarbons. The pyrolysis process includes the following stages:
- fuel drying to a moisture level below 10%,

- milling the biomass into very small particles in order to provide a quick reaction, and
- pyrolysis reaction,
- separation of solid products,
- cooling and collecting bio-oil.

During pyrolysis the biomass is converted thermally in the absence of oxygen. Depending on the course of the pyrolysis process can be distinguished: conventional rapid and instant.
Fast pyrolysis is characterized in that the small particles of biomass, low humidity, are very rapidly heated to a temperature of 450-5500C, thereby forming pyrolity oil (75%), charcoal (12%) and a mixture of combustible gas (13%). Pyrolity oil can be used directly as a fuel or used to produce other substances. The main advantage is the ease of storage and transport. Charcoal and a mixture of combustible gases are burned directly, thus providing heat for the process of pyrolysis [11, 12, 19, 21, 30].

3.4 Combined Heat and Power

Combined Heat and Power allows the combined production of heat and electricity. Thus allows to reduce fuel consumption and reduce emissions. In associated systems utilization rate of the chemical energy of the fuel is as high as 80-90%, thanks to the recovery of high-quality heat from the exhaust.
In the case of acquisition of straw as fuel, you can use a system of CHP: CHP plant from a steam turbine or steam engine (with lower values of electrical power). It would consist from a steam boiler with heating steam, steam turbine and generator of electricity. Shredded fuel would be given to the first lock fire, and then feeding screw stepped grate, which naturally occur combustion. A pair formed in the boiler passed to a steam turbine [11, 12, 22].

3.5 Biochemical processes

In a situation where the biomass contains too much water it can be used for energy purposes thanks to biochemical processes. Alcoholic fermentation causes the decay of carbohydrates, taking place under anaerobic conditions after adding yeast to the raw materials such as corn, wheat, grapes, sugar beets. Fermentation product is bioethanol. Another biochemical process used for the production of liquid biofuels is the esterification of oil. It allows you to be methyl esters: biodiesel, liquid biofuels from vegetable oils containing methanol (eg. rapeseed, soybean).Methane fermentation causes the degradation of macromolecular organic substances with limited oxygen. As a result, we obtain alcohols, lower organic acids, as well as methane, carbon dioxide and water [11, 12, 20, 31].

4. Economic analysis of biomass energy generation in Poland

The economic analysis was carried out for three types of biomass boilers, ie. For combustion fragmented wood, pellets and wood chips. For the purpose of calculating annual need for thermal energy was assumed the level 180 GJ or 50 000 kWh.
Fuel consumption for a given type of boiler was performed on the basis of the following formula:

$$B = 3600 \times Q / W_d \times \eta \times (\rho) \quad (1)$$

where:
Q – the need for heating demand [kWh/year]
W_d – the calorific value of fuel – [kJ/kg], [kJ/m^2]
η – the boiler efficiency
ρ – fuel density [kg/dm^3]

Calorific value and efficiency of the boilers are presented below.

Table 1. Calorific value and fuel efficiency of boilers for domestic purpose.

Fuel Type	Fragmented wood	Pellet	Wood chips
The calorific value [kJ/kg]	14 000	19 000	13 000
Efficiency[%]	80	85	82

The annual consumption of fragmented wood:
$B_{dk} = 3600 \times Q / W_d \times \eta = 3600 \times 50\,000 / 14\,000 \times 0{,}80 = 16\,071$ [kg]
The annual consumption of pellets:
$B_{pel} = 3600 \times Q / W_d \times \eta = 3600 \times 50\,000 / 19\,000 \times 0{,}85 = 11\,145$ [kg]
The annual consumption of wood chips:
$B_{zd} = 3600 \times Q / W_d \times \eta = 3600 \times 50\,000 / 13\,000 \times 0{,}82 = 16\,885$ [kg]
The annual cost of heating was calculated using the following formula.

$$K = B \times K_{pal} [PLN] \tag{2}$$

where:
B – the annual fuel consumption [kg/year], [m^3/year], [dm^3/year]
K_{pal} – cena paliwa [PLN/unit.]
Biomass prices are shown in the table below.

Table 2. Average Price of biomass [33,34]

Fuel Type	Fragmented wood	Pellet	Wood chips
Cost	330 PLN/t	780 PLN /t	240 PLN /t

The annual cost of heating with a fragmented wood:
$K_{dk} = B \times K_{dk} = 16{,}071 \times 330 = 5\,303{,}43$PLN]
The annual cost of heating with pellets:
$K_{pel} = B \times K_{pel} = 11{,}145 \times 780 = 8\,693{,}10$ [PLN]
The annual cost of heating with wood chips:
$K_{zd} = B \times K_{zd} = 16{,}885 \times 240 = 4\,052{,}40$ [PLN]

Coal price: Kwk = 760 PLN / t
Calorific value: Wd = 27 000 kJ / kg
The calculation of annual coal consumption:
$BWK = 3600 \times Q / Wd \times \eta = 3600 \times 50\,000/27\,000 \times 0.80 = 8\,330$ [kg / year]
The calculation of the annual operating costs for the coal boiler:
$Kwk = B \times Kwk = 8.33 \times 760 = 6\,330{,}80$ [PLN / year]
For comparison, the annual cost of heating with coal is very popular 6 330,80 [PLN].

 DOI: 10.1051/e3sconf/20171402016

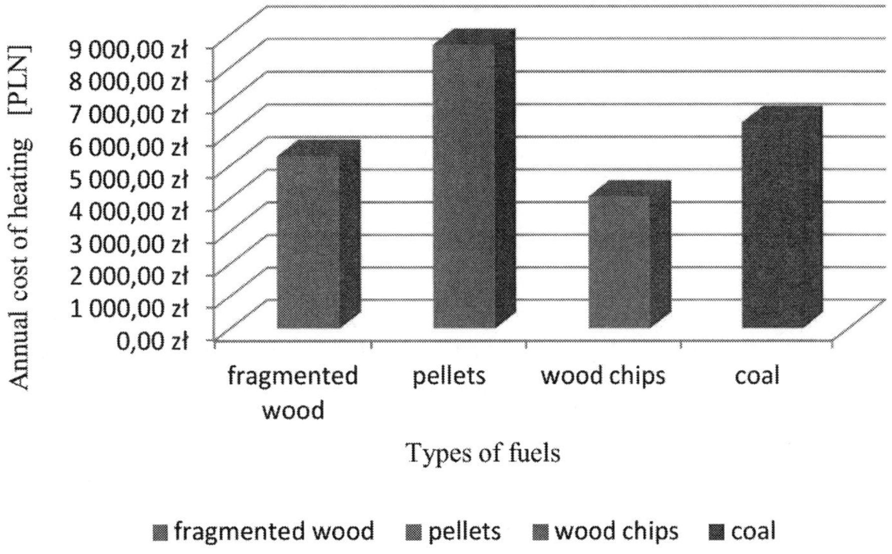

Fig. 6. Cost of heating

From the performed calculations we can state that the both heating with wood chips and pellet are cheaper than coal. An additional benefit is the low emission accompanying the combustion of biomass.

Another source of energy and at the same time a threat to the environment is landfill gas. The cost of installation use of biogas is approx. US $ 1 million. However, depending on the quantity and quality (calorific value) of landfill gas, the way of its management and technology as well as market prices of retrieved heat and electricity, the payback time ranges from 2 to 10 years. For an example the village of Malopolska is less than 5.5 years [5, 6]

From 2013 onwards, each available on the Polish market fuel has a bio additive in the form of biodiesel or bioethanol in the size of 7.1%. In addition, at some stations it is the ability to purchase pure biodiesel that replaces diesel. Its price is a little lower compared with the price of diesel fuel.

5 Summary

Analyzing the technical possibilities of biomass utilization in Poland, it is clear that these can cover less than a quarter of the demand for energy. In 2014., Its share accounted for 13.3% and was comparable with conventional energy sources such as natural gas and electricity. If we compare the energy use of biomass and other renewable sources, it turns out that it represented up to 88%. During the period considered, ie. 2010-2014, one can observe an increase share of liquid biofuels and biogas.

The growing use of biomass mainly results from its low price. Conducted economic analysis showed that both wood chips and fragmented wood is cheaper than the commonly used coal. In addition greatly can reduce emissions associated with a combustion of coal. A similar ecological and economic benefit we can get obtaining energy gas formed in landfills.

Good example of biomass use is being carried out in two municipalities of Malopolska by project From Field to Energy. It is aimed explore potential use of local biomass for NES pellets and pellets from mixed vegetable biomass. It will also determine the impact of the replacement of old boilers for automatic boilers, designed to burn pellets with reduced caloric value to reduce low-emission [35].

References

1. Energia ze źródeł odnawialnych w 2014 r., Informacje i opracowania statystyczne, GUS, Warszawa (2015)
2. Dyrektywa 2001/77/WE z dnia 27 września 2001 r. w sprawie wspierania produkcji na rynku wewnętrznym energii elektrycznej wytwarzanej ze źródeł odnawialnych (2001)
3. Dz. U. Nr 169, poz. 1199, Ustawa z dnia 25 sierpnia 2006 r. o biokomponentach i biopaliwach ciekłych (2006)
4. Dz. U. Nr 34, poz. 182, Rozporządzenie Ministra Gospodarki z dnia 23 lutego 2010 r. zmieniające rozporządzenie w sprawie szczegółowego zakresu obowiązków uzyskania i przedstawienia do umorzenia świadectw pochodzenia, uiszczenia opłaty zastępczej, zakupu energii elektrycznej i ciepła wytworzonych w odnawialnych źródłach energii oraz obowiązku potwierdzania danych dotyczących ilości energii elektrycznej wytworzonej w odnawialnym źródle energii (2010)
5. J. Dudek, P. Klimek, Polityka Energetyczna – Energy Policy Journal **11** (2008)
6. M. Piaskowska-Silarska, Polityka Energetyczna – Energy Policy Journal **16** (2013)
7. L. Borowik, P. Ptak, Przeglad Elektrotechniczny **88** (2012)
8. P. Dulian, B. Garbarz-Glos, W. Bąk, M. Antonova, C. Kajtoch, K. Wieczorek-Ciurowa, H. Noga, Ferroelectrics **497** (2016)
9. B. Garbarz-Glos, W. Bąk, H. Noga, M. Antonova, A. Kalvane, W. Śmiga, Integrated Ferroelectrics **173** (2016)
10. H. Noga, Information technologies and technical education **1** (2009)
11. http://biomasa.org
12. http://energiaodnawialna.net
13. http://stat.gov.pl
14. http://www.zielonaenergia.eco.pl
15. T. Prauzner, Przegląd Elektrotechniczny **88** (2012)
16. W. Sobczyk, K. Sternik, E.J. Sobczyk, H. Noga, Rocznik Ochrona Srodowiska **17** (2015)
17. W. Śmiga, B. Garbarz-Glos, W. Piekarczyk, H. Noga, D. Sitko, M. Karpierz, M. Livinsh, Integrated Ferroelectrics **173** (2016) .
18. I. Turekova, J. Depesova, A. Haskova, T. Bagalova, EDULEARN **15** (2015)
19. W. Hudy, K. Jaracz. K. Pytel, International Carpathian Control Conference **16** (2015)
20. K. Pytel, W. Hudy, H. Noga, W. Kulinowski, International Carpathian Control Conference **17**, 401–406, (2016)
21. K. Pytel, W. Hudy, H. Noga, W. Kulinowski, International Carpathian Control Conference **17**, 407–412, (2016).
22. K. Pytel, K. Jaracz, S. Gumula, International Carpathian Control Conference **13**, 592–595, (2012)
23. S. Gumuła, W. Hudy, M. Piaskowska-Silarska, K. Pytel, Przemysl Chemiczny, **94** (2015)
24. K. Pytel, J. Suchanicz, M. Livinsh, A. Sternberg, Phase Transitions, **87** (2014)
25. S. Gumuła, K. Pytel, Mechanics/AGH University of Science and Technology **24(4)** (2005)
26. K. Pytel, K. Jaracz, Poznan University of Technology Academic Journals, Electrical Engineering **70** (2012)
27. S. Gumuła, K. Pytel, M. Piaskowska-Silarska, Polish Journal of Environmental Studies, **23**, 2315-2320, (2014)
28. S. Gumuła, K. Pytel, M. Piaskowska-Silarska, Polish Journal of Environmental Studies, **23**, 2321-2325, (2014)

29. J. Suchanicz, B. Garbarz-Glos, G. Stopa, T. Kruzina, J. Kusz, M. Zubko, W. Hofmeister, I. Jankowska-Sumara, D. Wcisło, K. Konieczny, R. Rosiek, A. Finder, K. Pytel, M. Dambekalne, A, Sternberg, Integrated Ferroelectrics **108** (2009)
30. J. Suchanicz, G. Stopa, J. Kusz, M. Zubko, W. Hofmeister, M. Antonova, A. Kalvane, M. Dambekalne, A. Sternberg, I. Jankowska-Sumara, B. Garbarz-Glos, D. Wcisło, K. Konieczny, K. Pytel, A. Finder, Journal of Materials Science **45** (2010)
31. K. Pytel, J. Suchanicz, M. Livinsh, A. Sternberg, Condens. Matter Phys., 16 (2013)
32. J. Suchanicz, A. Finder, I. Smeltere, K. Konieczny, I. Jankowska-Sumara, B. Garbarz-Glos, M. Sokolowski, R. Bujakiewicz-Korońska, K. Pytel, M. Antonova, A. Sternberg, Integrated Ferroelectrics, **123** (2011)
33. http://www.cenypaliw.eu
34. http://ebiomasa.pl
35. http://ros.edu.pl/images/roczniki/2016/028_ROS_V18_R2016.pdf

DOI: 10.1051/e3sconf/20171402017

Estimation of energy potential of agricultural enterprise biomass

Vasyl Lypchuk[1,*], *Kateryna* Syrotiuk[2]

[1] Kielce University of Technology, Faculty of Management and Computer Modelling, 25-314, Kielce, Poland
[2] Lviv National Agrarian University, Faculty of Economics, 80381, Lviv-Dublany, Ukraine

Abstract. Bioenergetics (obtaining of energy from biomass) is one of innovative directions in energy branch of Ukraine. Correct and reliable estimation of biomass potential is essential for efficient use of it. The article reveals the issue of estimation of potential of biomass, obtained from by-products of crop production and animal breeding, which can be used for power supply of agricultural enterprises. The given analysis was carried with application of common methodological fundamentals, revealed in the estimation of production structure of agricultural enterprises, structure of land employment, efficiency of crops growing, indicators of output of main and by-products, as well as normative (standard) parameters of power output of energy raw material in relation to the chosen technology of its utilization. Results of the research prove high energy potential of by-products of crop production and animal breeding at all of the studied enterprises, which should force its practical use.

1 Introduction

Bioenergetics is an important direction for development of the sector of renewable sources of energy in Ukraine, because large potential of biomass, available for the energy production, is concentrated here.

Use of energy, produced of by-products of crop production and animal breeding, is an actual and prospective direction of economic activity, securing increase of its marketability level and economic stability of agricultural enterprises. Besides, growing of biomass for energy purposes can be a separate direction of activity at agricultural enterprises.

According to the data of energy balances of Ukraine in 2010, the whole bioenergetics substituted 1,1 bln m^3 of gas in a year [1], and the indicator made 3 bln m^3 of gas in 2014 [2]. Share of renewable sources of energy made 2,6% of the total supply of primary energy in 2014. Total supply of primary energy of biomass increased to 1,934 mln t of o.e., that equals 2,25 bln m^3 of gas, and above 3,15% more than in 2013.

In 2010-2014, increase rate made 42% annually. If the rate is kept till 2020, bioenergetics will be capable to produce 5,4 bln m^3 of gas in a year.

* Corresponding author: wlipczuk@tu.kielce.pl

Europe-integration attempts of Ukraine and reforming of its economy demonstrate that Ukraine's energy branch should be developed in the direction, which corresponds to the European standards and vectors. In the European Union, share of renewable sources makes 15% (as of 2013) in the gross ultimate energy consumption (GUE), including biomass – about 9% of GUE or 62% of all RSE. In some countries of the EU, the share of biomass among all renewable sources fluctuates between 30-40% (Luxemburg, Cyprus, Ireland), up to 80-95% (Estonia, Latvia, Lithuania, Hungary, Poland, Finland). Until 2010, share of RSE in the gross ultimate energy consumption of the EU should make 20% (obligatory target), and it can achieve at least 27% (new obligatory target, set by the European Commission in October 2014) until 2030 [3].

The European Commission developed Energy Roadmap 2050 [4], analyzing ways to achieve the set targets as to cut of emission of greenhouse gas, securing reliability and competitive capacity of energy supply systems. In the Roadmap, the EU describes five possible scenarios of power engineering development. "Scenario of RSE" is coordinated with the forecast of development prospects of the EU power engineering, launched by the European Commission on RSE (EREC) – "RE-Thinking 2015" [5].

It is known, that consumption of energy should be based on the principles of reasonability and correspond to ideas of the theory of sustainable development of an economy. Thus, rationalization of the use of biomass energy needs persistent systematic work, focused at optimization of each economic process. Considering the existing resource potential and large amount of unemployed agricultural lands, bioenergetics is a prospective direction for domestic producers.

Efficient use of the potential of agricultural enterprises biomass is possible under application of an appropriate mechanism, practically revealed in a set of managerial decisions, aimed at solution of energy tasks.

Correct and reliable estimation of agricultural biomass potential is an essential precondition for efficient use of it.

Aim of the research is to estimate possibilities of self-energy-supply of technological processes of agrarian production, depending on specialization of enterprises and basing on determination of potential of agricultural biomass, which can be used for energy purposes.

Object of the research is made by energy potential of by-products of crop production and animal breeding at agricultural enterprises with different specialization. The research was carried out on the base of agricultural enterprises in Lviv region, demonstrating rather dynamic development of bioenergetics.

2 Study of energy potential of agricultural enterprise biomass

Research on the energy potential of biomass was made with application of common methodological principles, revealed in the estimation of production structure of agricultural enterprises, structure of lands employment, efficiency of agricultural crops growing, including energy ones, indicators of output of the main and by-products, as well as normative (standard) parameters of energy output of the energy raw material according to the chosen technology of its utilization.

To estimate energy potential of by-products of crop production and animal breeding we made grouping of agricultural enterprises according to their production structure and determined four the brightest examples of typical enterprises with different structure of products (only products of crop production, with production of goods of crop production and animal breeding, taking different share in the total structure of commodity products), in particular:

LLC "Brody Agro" deals only with production of crop products, specializing in growing of cereals and industrial crops (Figure 1).

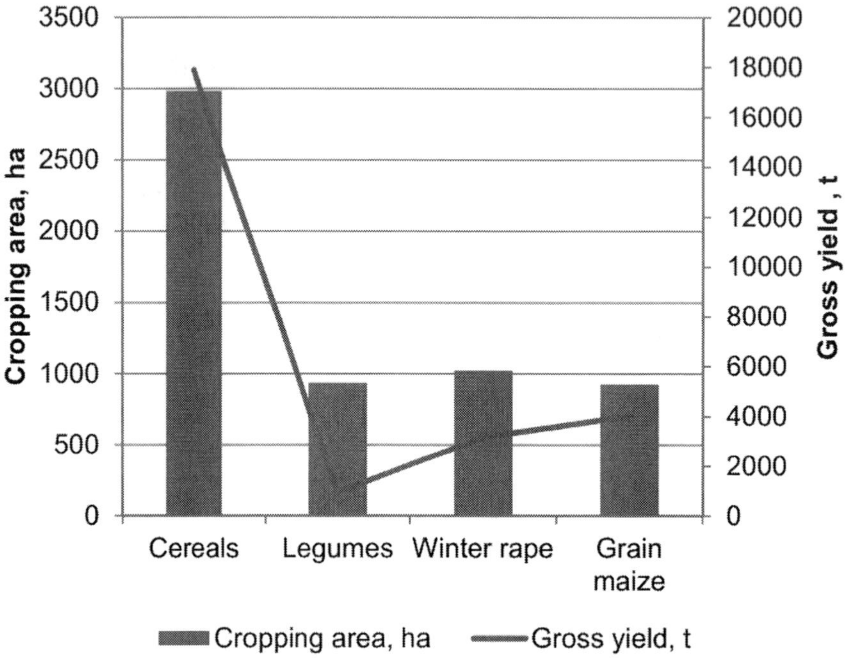

Fig.1. Cropping area and gross yield of agricultural crops at the 1st type enterprise.

PIE "Bilyi Stik" is engaged in production of crop products and animal breeding, with privilege to animal breeding. It is specialized in production of beef and dairy cattle, and in crop production – specialized in cereals growing (Figure 2).

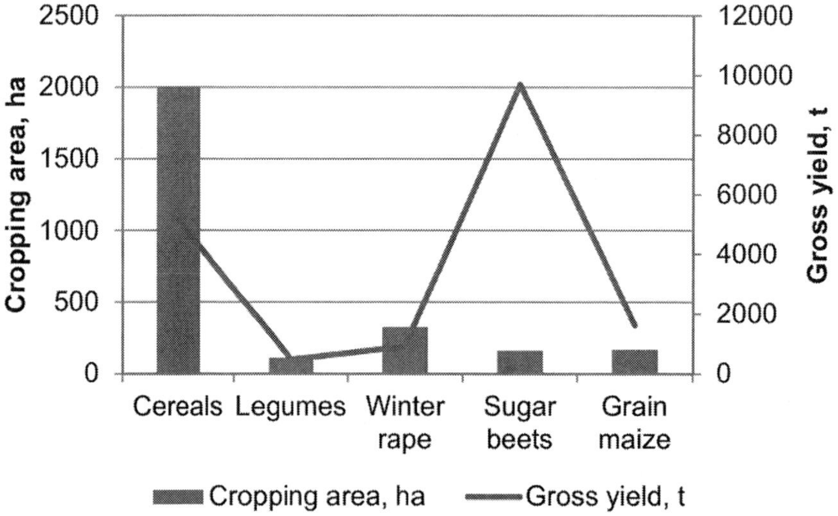

Fig. 2. Cropping area and gross yield of agricultural crops at the 2nd type enterprise.

DOI: 10.1051/e3sconf/20171402017

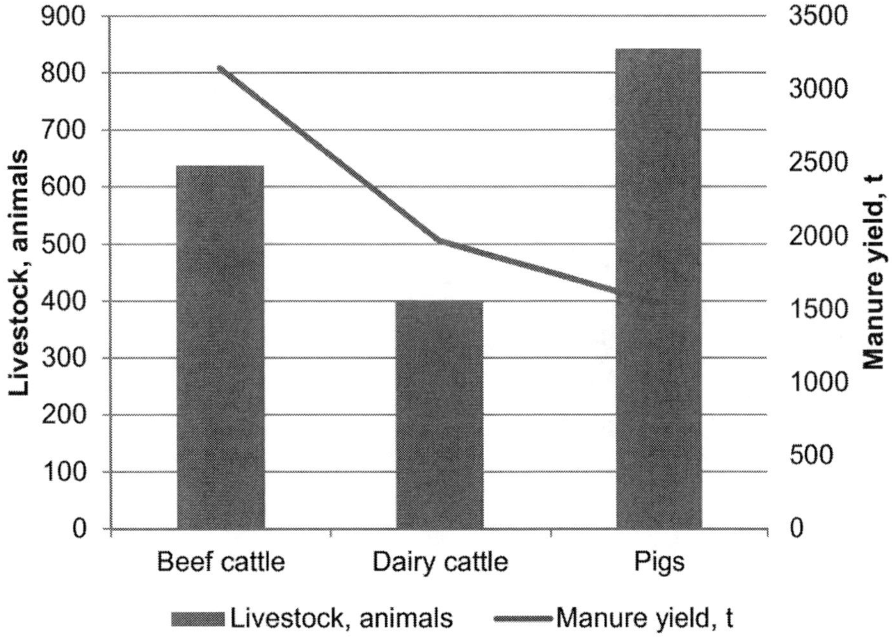

Fig. 3. Animals of cattle and amount of manure production at the 2nd group enterprise.

PE "Agro-Progres" deals with production of crop products and animal breeding, with accent to crop production. It is specialized in cereals growing and production of pig-raising products (Figure 3).

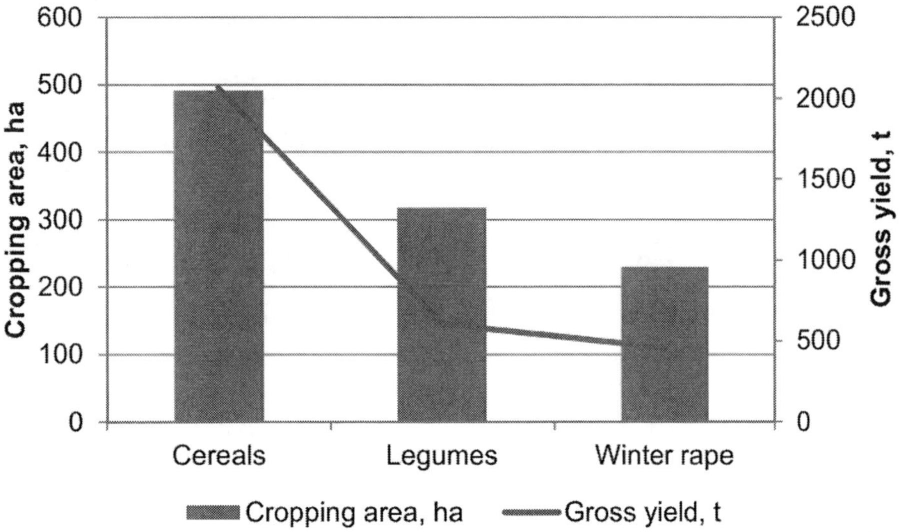

Fig. 4. Cropping area and gross yield of agricultural crops at the 3rd type enterprise.

Number of pigs makes 86 animals, supplying 156 t of manure.

LLC named after Danylo Halytskyi produces crop products and products of animal breeding in equal volume. In crop production, it is specialized in growing of cereals and rape, and in animal breeding – in beef and dairy cattle (Figure 5, 6).

DOI: 10.1051/e3sconf/20171402017

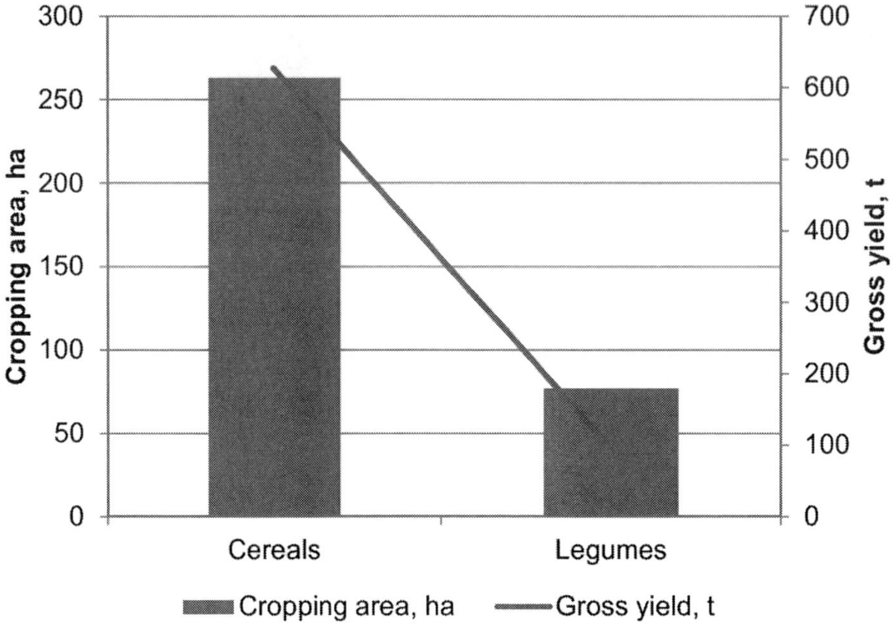

Fig. 5. Cropping area and gross yield of agricultural crops at the 4th type enterprise.

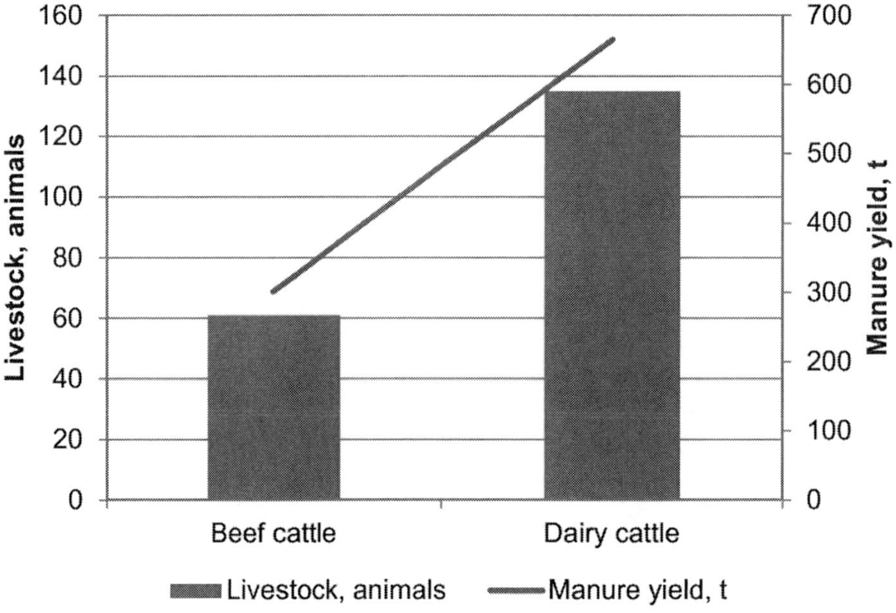

Fig. 6. Cattle livestock and amount of manure production at the 4th type enterprise.

Such differentiation of enterprises enables estimation of the amount of by-products of crop production and animal breeding for energy purposes at the enterprises with different production structure.

To find out impact of the amounts of production of agricultural products at the investigated enterprises, we make their energy estimation by consolidated indexes [6-11].

To estimate energy potential (E_{cr}) of residues from crop production, in particular in Ukraine and Poland, they use the following formula:

$$E_{cr} = V \cdot K_1 \cdot K_2 \cdot K_3 \cdot Q_{lsh}^{cr} , \quad (1)$$

where V – volume of main products, ton;

K_1 – coefficient of by-products;

K_2 – coefficient of by-products availability;

K_3 – coefficient of by-products use for technological needs;

Q_{lsh}^{cr} – low specific heat of biological by-products combustion, GJ/t.

Coefficient of by-products is made by correlation of dry matter of crop by-products of a certain crop to weight of the obtained yield.

Coefficient of by-products availability is considered as a share of total amount of crop by-products, which can be actually harvested, i.e. generally available for utilization.

Coefficient of by-products use for technological needs is calculated by deduction of the share of biomass, which can be used for other agricultural needs (silage, bedding, forage products, etc.)

To determine energy potential of by-products from animal breeding one can apply two methods [6, 7].

The first one is to determine amount of secondary raw material (manure) and biogas yield on its base with consideration of its heating value.

The second one is to determine amount of normative biogas yield per a head of cattle.

We used the first method, as it is able to give accurate description of energy characteristics of secondary raw materials. Thus, energy estimation of secondary products of animal breeding can be made with application of the following formula:

$$E_{an} = n_a \cdot K_{my} \cdot M_{my} \cdot Q_{lhv}^{bg}, \quad (2)$$

where n_a - livestock of cattle, animals;

K_{my} - manure yield from one animal of cattle, t/animal;

M_{my} - normative (standard) biogas yield from a ton of manure, m^3/t;

Q_{lhv}^{bg} - low heating value of biogas, GJ/m^3.

Basing on the above-mentioned methods it is necessary to make estimation of energy potential of by-products from crop production and animal breeding in terms of the investigated enterprises.

To make consolidated comparative estimation of specific energy potential of each of the studied enterprises we determined its value per 1 ha of arable land and agricultural land (Table 1-4).

Table 1. Amount and energy potential of crop production by-products at the 1st type enterprise.

Indicators	Cereals	Legumes	Winer rape	Grain maize
Energy by-product, t	11285	364	2657	781
Energy potential, GJ	178305	5095	38533	12498
Specific energy potential, GJ/ha of arable land	59.77	5.43	37.70	13.50
Specific energy potential, GJ/ha of agricultural land	30.09	0.86	6.50	2.11

Analysis of the table 1 shows that main energy potential is demonstrated by cereals both in total amount and by specific indicators.

Table 2. Amount and energy potential of agricultural by-products at the 2nd type enterprise.

Indicators	Cereals	Legumes	Winter rape	Sugar beets	Grain maize	Beef cattle	Dairy cattle	Pigs
Energy by-product, t (m^3)	3265	182	773	4540	315	314375	196607	99882
Energy potential, GJ	51584	2545	11213	14991	5045	7274.6	4549.5	2311.3
Specific energy potential. GJ/ha of arable land	25.73	22.52	34.40	92.54	29.85	11.4	11.4	2.7
Specific energy potential. GJ of agricultural land	20.58	1.02	4.47	5.98	2.01	2.90	1.81	0.92

Data of the Table 2 prove that enterprises. growing industrial crops have rather high specific energy potential. It is particularly true for by-products from sugar beets growing. Regardless of the considerable amount of produced manure. both specific and energy potential is insufficient. In our case. it is explained by low energy efficiency of manure utilization.

Table 3. Amount and energy potential of agricultural by-products at the 3rd type enterprise.

Indicators	Cereals	Legumes	Winter rape	Pigs
Energy by-product. t (m^3)	1302	229	383	10202
Energy potential. GJ	20572	3207	5552	236.1
Specific energy potential. GJ/ha of arable land	41.81	10.09	24.14	2.7
Specific energy potential. GJ/ha of agricultural land	8.21	1.28	2.21	0.09

Results of the Table 3 prove that energy potential is mainly revealed by grain production by-products. similar to the previous examples. Utilization of by-products of pig raising also demonstrates low energy efficiency because of small livestock.

Table 4. Amount and energy potential of agricultural by-products at the 4th type enterprise.

Indicators	Cereals	Winter rape	Beef cattle	Dairy cattle
Energy by-product. t	396	36	30058	66521
Energy potential. GJ	6252	522	695.5	1539.3
Specific energy potential. GJ/ha of arable land	23.8	6.8	11.4	11.4
Specific energy potential. GJ/ha of agricultural land	12.33	1.03	1.37	3.04

Analysis of the table 4 proves that energy efficiency of cereals growing prevails over products of animal breeding.

3 Estimation of the level of energy substitution at agricultural enterprises

Considering the fact that the investigated enterprises have different area of agricultural lands and apply different technologies to grow agricultural products. it is reasonable to determine coefficient of energy substitution. characterizing correlation of energy potential of by-products to energy needs of agricultural production. To calculate it we use the following formula:

$$K_{es} = \frac{E_{cr} + E_{an}}{N_e}. \quad (3)$$

where. N_e – are equivalent energy needs of agricultural enterprise. GJ.

Equivalent energy needs are determined according to the volume of consumed kinds of fuel and electric energy. calculated in energy unit (GJ) (Table 5).

Table 5. Estimation of energy capacity of utilized energy resources at the enterprises of different types.

Indicators	1 type		2 type		3 type		4 type	
	Liquid fuel. t	**Electric energy. kW/h**	**Liquid fuel. t**	**Electric energy. kW/h**	**Liquid fuel. t**	**Electric energy. kW/h**	**Liquid fuel. t**	**Electric energy. kW/h**
Number of units	405	91395	140	494574	67	40775	19	17442
Energy capacity. GJ	17390	329	6030	1780	2868	147	809	63

Data of the table 5 prove that transportation and field works take the main share of energy needs that is argued by large volume of purchased liquid fuel. The volume exceeds the need of electric energy in energy equivalent.

Comparative estimation of energy efficiency of agricultural production at the studied enterprises is presented in the Table 6.

Table 6. Integrated indicators of energy potential of agricultural by-products. energy needs of an enterprise and coefficient of its energy supply.

Indicators	1 type	2 type	3 type	4 type
Energy potential of crop production. GJ	234431	85378	29331	6774
Energy potential of animal breeding. GJ	0	14135	236	2235
Total energy potential. GJ	234431	99513	29567	9009
Need of energy. GJ	17719	7810	3014	872
Coefficient of energy supply	13	13	10	10

Data of the table 6 prove that. according to the coefficient of energy supply. all studied enterprises can be energy self-sustainable. It means that amount of theoretical energy potential of by-products is larger than their energy needs.

Actual level of energy supply first depends on technology and efficiency of secondary by-products utilization.

4 Conclusions

The carried research proves that amount of energy potential of crop production by-products considerably depends on yield capacity of agricultural crops. which has two constituents: the first one is based on dynamics of yield capacity rise and is forecasted to some degree; the second one – depends on changeable character of weather conditions and has scholastic character. Speaking about products of animal breeding. relatively low energy efficiency of by-products results in insufficient share of it in the structure of energy balance of an enterprise. However. it does not mean that there is no need to make manure utilization. It is connected with the fact that in the process of manure utilization the main share of efficiency is provided by ecological. agrotechnical and social component.

The carried analysis of the estimation of by-product energy potential was made on the base of the data. not reflecting dynamics of development of crop production and animal breeding at the investigated enterprises and without consideration of the impact of natural-climatic conditions on farming activity.

Considering dynamics of productive activity of the enterprises and changeable natural conditions. prospective estimation of the level of energy substitution of agricultural production should be based on the results of scholastic modelling of its energy consumption and energy supply.

References

1. *Energy balance of Ukraine. 2010.* Available at: http://www.ukrstat.gov.ua/operativ/operativ2012/energ/en_bal/arh_2012.htm
2. *Energy balance of Ukraine. 2014.* Available at: http://www.uabio.org/img/files/news/pdf/energy-balance-ukraine-2014.pdf
3. G.G. Geletuxa. T.A. Zhelyezna. P.P. Kucheruk. Ye.M. Olijny'k. *The current state and prospects of bioenergy development in Ukraine*. **9**. 32 (Policy Brief BAU 2014)
4. *Energy Roadmap 2050.* COM(2011) 885 final. 15.12.2011. Available at: https://ec.europa.eu/energy/sites/ener/files/documents/2012_energy_roadmap_2050_en_0.pdf
5. *RE-Thinking 2050.* A 100% Renewable Energy Vision for the European Union. EREC (2010)
6. G.G. Geletuxa T.A. Zheleznaya M.M. Zhovmir Yu.B. Matveyev O.I. Drozdova *Evaluation of the energy potential of biomass in Ukraine. Part 2: energy crops. liquid biofuels. biogas. Industrial heat.* **33**. 57-64 (2011)
7. V.O. Dubrovin. H.H. Heletukha. and S.H. Kudria. *Methods generalized assessment of technically achievable energy potential of biomass.* Viol_prynt. Kyiv. Ukraine. **25** (2013).
8. B. Koscik. *Bioenergy of Subcarpathian* (Jaroslaw. 2007)
9. W. Podkowka Z. Podkowka. A. Kowalczyk-Jusko. P. Pasyniuk. *Agricultural biogas renewable source of energy. Theory and practical application* (Warsaw. 2012)
10. W. Lewandowski. M. Ryms. *Biofuels. Environmentally friendly renewable sources of energy* (Warsaw. 2013)
11. A. Grzybek. P. Gradziuk. K. Kowalczyk. *Straw - energy fuel* (Warsaw. 2001)

DOI: 10.1051/e3sconf/20171402018

Co-combustion of biodiesel with oxygenated fuels in direct injection diesel engine

Wojciech Tutak[1,*], *Arkadiusz* Jamrozik[1] *and Michał* Pyrc[1]

[1]Czestochowa University of Technology, Faculty of Mechanical Engineering and Computer Science, 42-201 Czestochowa, Armii Krajowej 21 Av., Poland

Abstract. The paper presents results of experimental investigation of co-combustion process of biodiesel (B100) blended with oxygenated fuels with 20% in volume. As the alternative fuels ware used hydrated ethanol, methanol, 1-butanol and 2-propanol. It was investigated the influence of used blends on operating parameters of the test engine and exhaust emission (NO_x, CO, THC, CO_2). It is observed that used blends are characterized by different impact on engine output power and its efficiency. Using biodiesel/alcohol blend it is possible to improve engine efficiency with small drop in indicated mean effective pressure (IMEP). Due to combustion characteristic of biodiesel/alcohol obtained a slightly larger specific NO_x emission. It was also observed some differences in combustion phases due to various values of latent heat of evaporation of used alcohols and various oxygen contents. Test results confirmed that the combustion process occurring in the diesel engine powered by blend takes place in a shorter time than in the typical diesel engine.

1 Introduction

Co-combustion of fuel blends is one of the methods to utilize alternative fuels in internal combustion engines. Renewable biofuels are important energy resources for reduction of green house gases, better air quality, less dependency on oil imports, and new jobs and markets [1]. Reducing emissions of toxic compounds is the subject of many activities both engineers and legislators [2]. Co-combustion of fuels is used not only in relation to liquid or gas fuels and it is used for solid fuels as well [3, 4]. The use of biodiesel or its blends with diesel has a negative impact on cold operation of engine under 0° C and increases NO_x emissions as compared to diesel. In the first solutions of co-combustion in diesel engines gas fuel was used [5]. The gas was delivered into the engine intake system through the mixer. In order to reduce nitrogen oxides emissions especially in industrial gas engines used the two-stage combustion systems [6, 7, 8]. As an alternative fuel for compression ignition engines the so-called oxygenated fuels are used. Alcohol is a renewable fuel because it can be produced from the biomass [9]. These fuels in the structure have oxygen, which affects the combustion process. Moreover, alcohols have several important disadvantages such as low cetane numbers and high latent heat of evaporations, and cannot be directly used in diesel

* Corresponding author: tutak@imc.pcz.czest.pl

engines [1, 10, 11]. In recent years, to power diesel engines the mixtures of diesel or biodiesel with alcohols were used. Alcohols causes phase separation at low temperature. Fraction of alcohol in blend cannot be high due to low value of cetane number of alcohol which causes the increase in ignition delay. As the number of carbon in alcohols increases, they can be more easily blended with diesel and biodiesel. With the increase of carbon in alcohols, the mass percent of oxygen decreases while cetane number, density and calorific value increase [12, 13]. Moreover, because these higher alcohols have good solvent capabilities, they can easily be blended with diesel and biodiesel [12, 13]. The alternative fuels are derived from animal fats or in the cultivation process of algae [14, 15]. In practice the combustion process of more than one fuel is possible by using dual fuel technology or blend mode. In dual fuel technology are used two independent power supply systems, one is an existing direct injection system which delivers diesel fuel and second is so called fumigation system which injects an alcohol fuel into the intake port of an internal combustion engine. Into the engine cylinder the air-fuel mixture, nearly homogeneous is delivered. The ignition process is controlled by the injected dose of diesel fuel. This requires addition of an injector, along with a separate fuel tank, lines and controls [10]. The second method is to provide fuels blend to engine combustion chamber by direct injection system. This system is a simpler solution because it does not require practically any changes in the engine construction. Blend of fuels diesel/alcohol is prepared and then bring the mixture to the engine, using a typical supply system for a diesel engine [11]. The biggest difficulty is that the large percentage of alcohol do not mix with diesel fuel, hence use of diesel–alcohol blends is not feasible. Also, the blends are not stable and separate in the presence of trace amounts of water. In such a power system cannot change the ratio of diesel/alcohol [11]. In the literature is lot of works on diesel-biodiesel-alcohol or diesel-esters mixtures co-combustion in diesel engine [14]. As the oxygenated fuels is considered ethanol which can be produced by the fermentation of natural sugars, starches or cellulosic biomass. In paper [16] authors presented results of experimental investigation of co-combustion of hydrated ethanol blend with diesel fuel in direct injection diesel engine. They stated that with increase in ethanol fraction in blend increased the ignition delay and decreased combustion duration as well. In addition they stated that with increase in ethanol fraction increased NO_x emission with near to constant THC emission and increased unrepeatability of IMEP. Other alcohol taken into account as an alternative fuel is methanol. Methanol is also known as wood alcohol. Presently, methanol is produced from synthetic gas or biogas and is used as a fuel for IC engines [17]. In paper [18] are presented results of combustion analysis of methanol/diesel blends based on the cylinder pressure analysis in a compression-ignition engine. The results showed that increasing methanol mass fraction in the methanol/diesel blends resulted in the increase of the heat release rate in the premixed burning phase and shortening of the combustion duration of the diffusive burning phase [18]. The results indicated that by using methanol-blended diesel fuels, smoke opacity, CO and THC emissions reduced while CO_2 and NO_x emissions increased. Propanol is a 3-carbon structure, straight chain alcohol with higher energy density and higher cetane number which makes it a potential alternative to lighter alcohols (methanol and ethanol) as blending component with diesel [19]. In the paper [20] authors presented results of the impact of propanol as additive to diesel fuel on engine performance. Propanol reduced the smoke emissions of rapeseed oil but increased NO_x, total hydrocarbons (THC) and CO emissions significantly. A drop in peak pressure and a slight increase in ignition delay were observed with increasing propanol content in the oil [20]. Propanol fraction in blend causes improvement in performance of the engine due to the higher percentage of premixed combustion as a result of low cetane number of propanol. Butanol is one of the alcohols which have more advantages than ethanol and methanol as an alternative fuel for internal combustion engines. This alcohol has properties closer to fossil-based fuels. Butanol has a lower auto-ignition temperature than methanol and ethanol. Therefore, butanol can be ignited easier when burned in diesel engines. Author of paper [21] presented results of experimentally investigation of the performance, combustion characteristics, NO emissions, and stability of a diesel engine fuelled by diesel – biodiesel – butanol blend. They stated that the addition of butanol with a percentage of up to 20% to the biodiesel – diesel fuel blend resulted in a tolerable change in the engine performance, combustion characteristics, NO emission, and engine stability compared to the diesel fuel, which indicates that the butanol has a great potential to be used as a fuel additive in the diesel engines [21].

In this paper are presented results of comparative analysis of diesel engine powered by four blends of biodiesel with four alcohols: ethanol, methanol, 2-propanol and 1-butanol. Composition of blend consists of 80% biodiesel and 20% alcohol in volume fraction. There is presented an analysis of thermodynamic parameters and emission characteristics of engine powered by blends for the same engine settings.

2 Experimental setup

The study was conducted on the 1-cylinder direct injection natural aspired diesel engine. Detailed engine parameters are presented in Table 1. Tests conducted with a constant angle of diesel fuel injection and constant rotational speed equal to 1500 rpm. The test stand was equipped with measurement system (Fig. 1). At each mode of operation, the engine was allowed to run 5 minutes until the exhaust gas temperature and exhaust gas composition have attended steady-state values and data acquisition process was subsequently conducted. The ambient temperature was maintained at a constant level. Each experiment was conducted three times and the results were found to agree with each other with the 95% significance level.

Table 1. Main engine parameters.

Parameter	**Value**	**unit**
Displacement volume	0.573	dm^3
Bore	90	mm
Stroke	90	mm
Compression ratio	17:1	-
Rated power (at 3000 rpm)	7	kW
Crankshaft rotational speed	1500	rpm
Injection pressure	210	bar
Injection timing	17	deg bTDC

Measuring system:

- Exhaust gas analyzer: THC, CO, CO_2, O_2 – Bosch BEA 350: THC: range 0 – 9999 ppm vol, accuracy: 12 ppm vol, CO: range 0 – 10 % vol, accuracy: 0.06 % vol, CO_2: range 0 – 18 % vol, accuracy: 0.4 % vol, O_2: range 0 – 22 % vol, accuracy: 0.1 % vol, : range 0.5 – 9.999, accuracy: 0.01,
- Exhaust gas analyzer: NO_x – Radiotechnika AI9600: accuracy: 32 ppm for range 0 – 1000 ppm,
- In-cylinder pressure sensor – Kistler 6001, sensitivity: ±0.5%,
- Charge amplifiers – Kistler 5001, linearity of FS <±0,05%,
- Air/Fuel Ratio Meter LM-2, range 7.35-22.38,
- Data acquisition module – National Instruments USB-6251 – 16 bits resolution, sampling frequency 20 kHz, Resolution for the data acquisition system – 1.0 CA deg, with software [22].

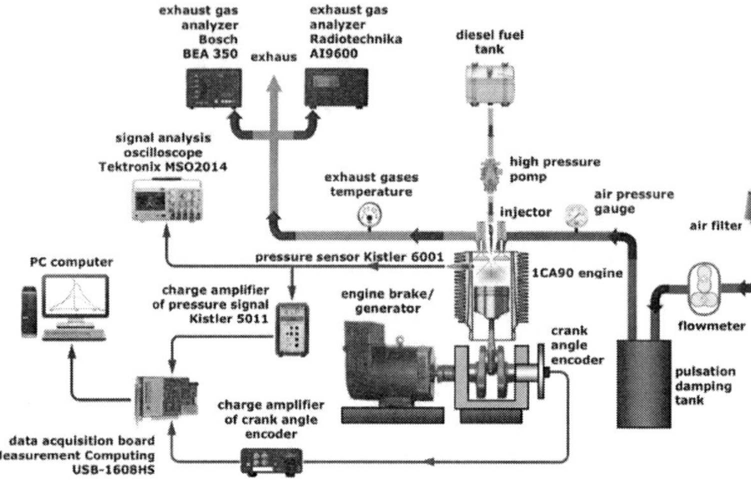

Fig. 1. The test stand.

The motivation of this study was to compare operating parameters and emission characteristics of compression ignition engine powered by blends: biodiesel-ethanol (BE), biodiesel-methanol (BM), biodiesel-propanol (BP) and biodiesel-butanol (BB). The used blends consist of 80% biodiesel (B100) and 20% alcohol in volume. The researches ware conducted under the same conditions, with unchanged settings of the test engine. With such a power system is guaranteed the same volumetric dosage of injected fuel. For all analyzed cases recorded 100 consecutive engine cycles. Each measurement was carried out three times. Parameters of engine obtained by biodiesel supply were taken as a reference for blends. In Table 2 fuels properties are presented. In Table 3 the main parameters of blends are presented.

Table 2. Fuels properties.

Formula	B100 (B)	Methanol (M)	Ethanol (E)	Propanol (P)	Butanol (B)
	$CH_3(CH_2)nCOOH_3$	CH_3OH	C_2H_5OH	C_3H_7OH	C_4H_9OH
Cetane number	56	2	11	12.0	15
LHV, MJ/kg	37	20	26.8	30.6	33.8
Latent heat of evaporation, kJ/kg	300	1162	918	727.9	626
Carbon (by wt. %)	77.1	37.48	52.14	60.0	64.82
Hydrogen (by wt. %)	12.1	12.48	13.02	13.3	13.49
Oxygen (by wt. %)	10.8	49.93	34.73	26.6	21.59
Density, kg/m^3	832	791	789	815.0	810
Auto-ignition temperature, K	534	658	698	573.0	616

It is visible that alcohols are characterized by lower LHV and cetane number in comparison with biodiesel but higher value of heat of evaporation (Fig. 2). With the increase in carbon atoms in alcohol particle increases LHV and cetane number which makes them more acceptable to the engine. Viscosity and density are two crucial properties for engine performance because it affects spray formation, spray penetration, fuel droplet size, injection timing, injection pressure, atomization and combustion characteristics [10]. Propanol is a 3-carbon and butanol 4-carbon structure with higher energy density and higher cetane number which makes these alcohols a potential alternative to lighter alcohols such as methanol and ethanol. Butanol is characterized by relatively high value of cetane number which is better than lower alcohols to help auto ignition.

Table 3. Blends properties.

	B100	BM20	BE20	BP20	BB20
LHV, MJ/kg	37	33.6	34.96	35.72	36.36
Latent heat of evaporation, kJ/kg	300	472.4	423.6	385.57	365.2
Oxygen content, % wt.	10.8	18.6	15.6	13.9	12.9

In Fig. 2 is the comparison of fuels composition. It can be indicated that hydrogen content in fuels is near the same. In alcohols with the increase in carbon in structure the oxygen atoms decreased. In case of methanol the oxygen content is near to 50% in mass. Biodiesel is characterized by the lowest oxygen content from all analyzed fuels. In case of blends the biodiesel – methanol blend has the highest oxygen content.

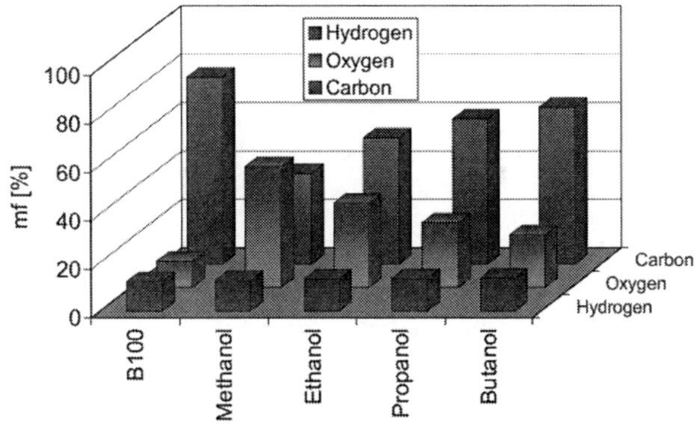

Fig. 2. Fuels mass fraction.

In Fig. 3 the comparison of LHV and heat of evaporation of sole fuels and blends is presented. The higher value of LHV has B100 and in case of blends the BB20 has the highest value of LHV.

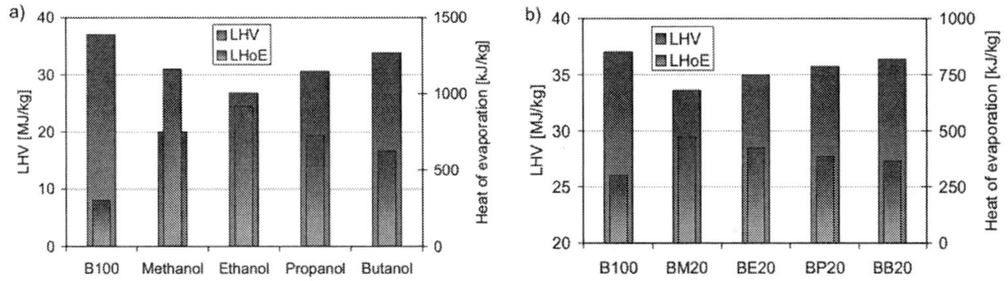

Fig. 3. Comparison LHV of fuels (a) and blends (b) used during investigation.

The dominant component in the blend is biodiesel B100 and its properties largely determine the properties of the mixture. Nevertheless, the addition of alcohol causes a change in the value of both LHV and heat of evaporation of the blend (Fig. 3b). It is visible that with the increase in carbon atoms in the structure of alcohol the LHV of blend increased as well. LHV of blends are lover than for B100.

 DOI: 10.1051/e3sconf/20171402018

3 Results and discussion

During experimental researches obtained the in-cylinder pressure traces vs. crank angle and emission of exhaust gases. On the basis of indication results thermodynamic parameters of the test engine such as: indicated mean effective pressure (IMEP), indicated thermal efficiency (ITE), specific fuel consumption (SFC), heat release and combustion phases were determined.

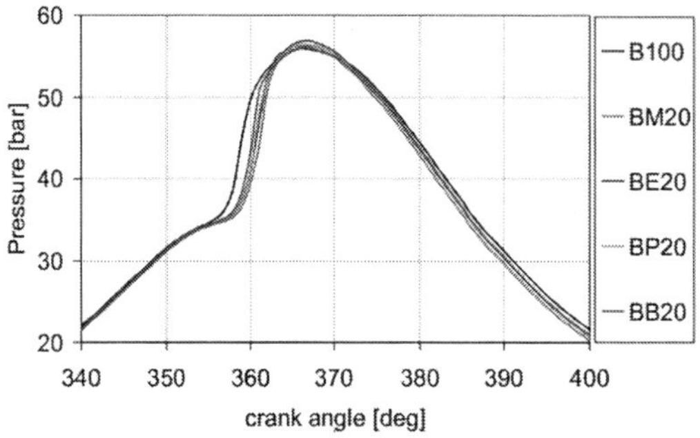

Fig. 4. In-cylinder pressure traces.

In Fig. 4 the pressure traces for engine powered by biodiesel and blends are presented. It can be stated that with alcohol participation in fuel the combustion process occurs with delay in comparison to B100. There are no significant differences in the peak pressure. Generally with increase in ignition delay during combustion process in diesel engine it is obtained higher value of peak pressure. In this case due to lower value of LHV of alcohol does not notice that increase in peak pressure.

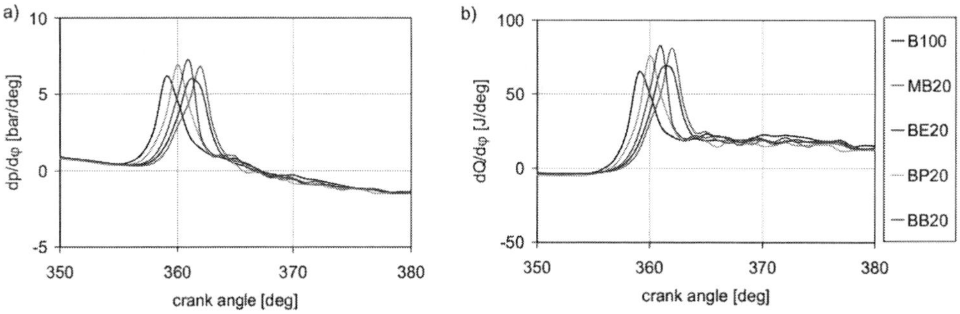

Fig. 5. The pressure rise and heat release rate.

In Fig. 5 courses of pressure rise and heat release rate for analyzed cases are presented. The peak pressure rises provide information about hardness of the engine. In all analyzed cases the maximum of dp/dφ was less than 10 bar/deg which is the limit value for a diesel engine. The higher value of dp/dφ reached for biodiesel-butanol blend and it was equal to 7.2 bar/deg and it was higher of 15% compared to B100 powering. It can be explained by the fact that butanol is characterized by relatively high LHV and the lowest value of the heat of vaporization of used alcohols. On the basis of pressure traces the heat release rate traces which are used to describe combustion process were calculated. On the basis of heat release rate courses can be stated that alcohol fraction in blend causes ignition delay and heat release occurs more rapidly. For all blends combustion received higher values of dQ/dφ compared to B100 fuelling. Near the same values of maximum dQ/dφ obtained for BB20 and BM20 blends but for BM20 it was obtained for larger angles after TDC. It was due to higher value of heat of evaporation of methanol which causes increase in ignition delay and burning after ignition occurs very rapidly. In case of biodiesel-ethanol fuelling of engine the dQ/dφ course has a slightly different shape. Combustion process of BE blend occurred with lowest rate from blends and it was the most extended in time. It can be explained among other things, the water content in ethanol which had an impact on the combustion process. Longer ignition delays as a result of lower cetane number of alcohols causes a wider fraction of the fuel to burn during the premixed combustion phase. The enhanced oxygen content of alcohol improves the combustion process, particularly during the diffusion phase [19].

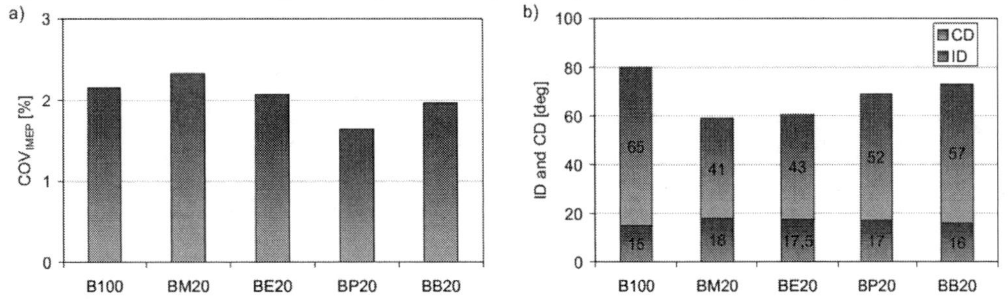

Fig. 6. Unrepeatability of IMEP (a) and combustion phases (b).

In Fig. 6a the coefficient of unrepeatability of IMEP (COV_{IMEP}) for all analyzed cases is presented. It is assumed that the engine unrepeatability of IMEP should not exceed 10%. On unrepeatability of engine cycles has the impact, among others, flow processes occurred during intake stroke [23]. In all analyzed cases COV_{IMEP} was at the level o 2%. The lowest value of COV_{IMEP} reached at BP20 burn and it was equal to 1.65% and the highest value obtained at BM20 fuelling 2.35%. In case of combustion phases the lowest value of ignition delay (ID) obtained for the engine powered by B100 and it was 15 deg of CA, the highest value of ID reached for powering by BM20 and it was equal 18 deg of CA. The participation of alcohol in the blend causes the increase in ignition delay. With the increase in carbon (C) content in the alcohol structure the impact on ID is getting smaller due to lower value of latent heat of evaporation and the rising cetane number (Tab. 1 and 2). Analyzing the combustion duration (CD) it can be stated that the presence of alcohol in blend with small carbon content in the structure (methanol and ethanol) causes a significant reduction in the CD. In case of BM20 the combustion duration was equal to 42 deg of CA but for BB20 (alcohol with 4-carbon in the structure) the CD was equal to 57 deg of CA which was reduced by 8 deg relative to the B100 burning.

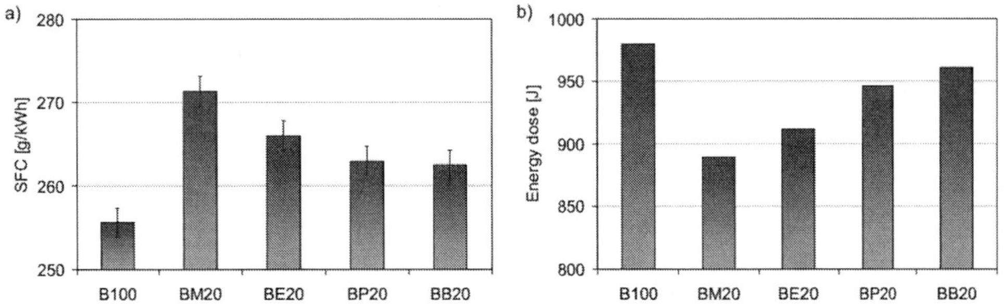

Fig. 7. Specific fuel consumption (a) and energy doses (b).

In Fig. 7 the specific fuel consumption and energy doses contained in fuel provided for engine cylinder for one engine cycle are presented. With the carbon mass fraction in alcohol the LHV increased as well (Table 1). Blend of B100 and methanol is characterized by the lowest LHV and therefore at a dose of fuel supplied to the cylinder is contained the smallest amount of energy. In BM blend in comparison with B100 it was a 10% less energy. This was reflected

 DOI: 10.1051/e3sconf/20171402018

in SFC which was the highest for BM powering and it was of 6% higher in comparison with B100 powering.

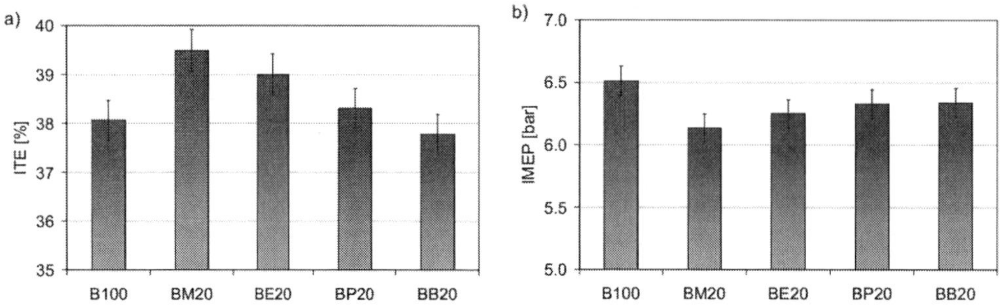

Fig. 8. Indicated thermal efficiency (a) and indicated mean effective pressure (b).

The heating value of alcohol increases with increase number of carbon atoms which indicates higher heating values for butanol from all analyzed alcohols. However, the heating value of butanol is 9.5% lower than biodisel (B100) and hence would require more amount of biodiesel/butanol blends to produce the same power output in the engine. For the lower alcohols this ratio is even higher. Higher burning velocities of alcohols lead to higher efficiency [19]. The indicated mean effective pressure depends on energy dose supplied to the cylinder of engine (Fig. 7b). This relationship confirmed obtained results (Fig. 8b). For all blends obtained lower value of IMEP in comparison with B100 fuelling (Fig. 8a). In case of indicated thermal efficiency (ITE) the highest value obtained for BM20 burning and it was 1.5% higher than for B100. In case of BB20 burning the ITE reached lowest value.

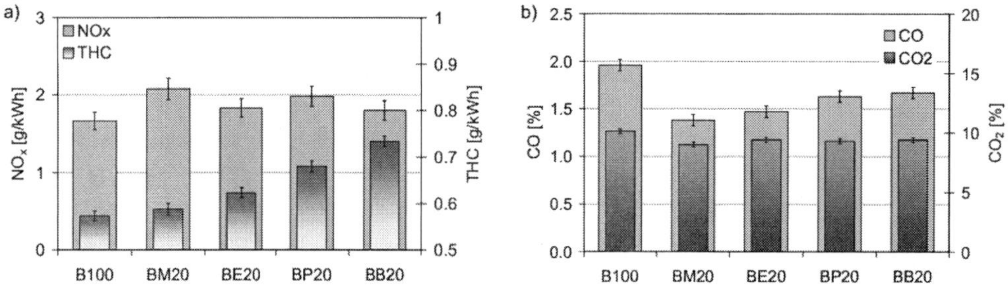

Fig. 9. Specific NO_x and THC emission (a) and emission of CO and CO_2 (b).

In Fig. 9a the specific emissions of nitrogen oxides (NO_x) and total unburned hydrocarbons (THC) are presented. The formation of NO_x in the internal combustion engine is determined by the thermal mechanism, in which the most important parameters are the temperature of combustion and oxygen content [24]. In case of oxygenated fuels, such as alcohols, the oxygen content as well as the combustion temperature could be higher which leads to the higher NO_x emissions. Co-combustion of biodiesel with oxygenated fuels increases the ignition delay of the mixture and the shortening of the combustion process which can also cause an increase in NO_x emissions. On the other hand alcohols, compared to diesel and biodiesel, characterized by a lower calorific value and higher latent heat of evaporation which can lead to decrease of temperature in the cylinder and the reduction of NO_x. On the basis of the results it can be stated that THC emission increased with the presence of alcohol content in the blend. This is due to the higher latent heat of vaporization of alcohol which causes slow evaporation, incomplete air/fuel mixing and development of a quench layer. With the increase in carbon (C) content in the alcohol structure THC emission increased as well. In case of BB20 the specific THC emission was higher of 28% in comparison with B100 fuelling. Analyzing the specific NO_x emissions can be stated that presence of alcohol in the blend causes rather slight increase in NO_x emission. The higher NO_x emission was observed for the combustion of the BM20. It can be explained by the higher oxygen content in methanol which is almost 50% weight (Fig. 2). Emission of CO and CO_2 generally are lower with alcohol content at full load of the engine due to the high in-cylinder temperature. In case of CO emission the lowest value obtained for BM20 burning and with the increase in carbon in the alcohol structure CO emission increased but does not exceed the value (1.96%) obtained by B100 burning.

4 Conclusions

In the paper are presented experimental results of comparative analysis of co-combustion of oxygenated fuels with biodiesel in the direct injection diesel engine. The test engine was powered by blends of biodiesel with four alcohols: methanol, ethanol, 2-propanol and 1-butanol. Composition of blend consists of 80% biodiesel and 20% alcohol in volume. On the basis of indication results determined thermodynamic parameters of the test engine such as: indicated mean effective pressure (IMEP), indicated thermal efficiency (ITE), specific fuel consumption (SFC), heat release and combustion phases. Conclusions regarding the thermodynamic parameters of the test engine:

- there are no significant differences in the peak pressure,
- the higher value of $dp/d\varphi$ reached for biodiesel-butanol blend and it was equal to 7.2 bar/deg and it was higher of 15% compared to B100 powering,
- in all analyzed cases COV_{IMEP} was at the level of 2%,
- the lowest value of ignition delay obtained for the engine powered by B100 and it was 15 deg of CA, the highest value of ID reached for powering by BM20 and it was equal 18 deg of CA,
- the combustion duration (CD) in presence of alcohol in blend with small carbon content in the structure (methanol and ethanol) causes a significant reduction,
- in case of BM20 the CD was equal to 42 deg of CA but for BB20 (alcohol with 4-carbon in the structure) the CD was equal to 57 deg of CA which was reduced by 8 deg relative to the B100 burning

Conclusions regarding the exhaust emission of the test engine:

- THC emission increased with the presence of alcohol content in the blend,
- with the increase in carbon (C) content in the alcohol structure THC emission increased as well. In case of BB20 the specific THC emission was higher of 28% in comparison with B100 fuelling,
- the specific NO_x emissions with presence of alcohol in the blend causes rather slight increase,
- CO and CO_2 emissions generally are lower with alcohol content at full load of the engine due to the high in-cylinder temperature.

Presented study showed that it is possible to co-combustion of alcohols (as renewable biofuels) in compression ignition engine without decreasing its performance and to the benefit of the environment. At a constant volumetric dose of injected fuel and thus a smaller amount of energy in fuel dose it can be achieved an acceptable operating parameters of the engine. The next stage of investigation should be the optimization of the engine in terms of the fuel injection angle for all used blends.

 DOI: 10.1051/e3sconf/20171402018

Acknowledgements: Research was also financed by the Ministry of Science and Higher Education of Poland from the funds dedicated to scientific research No. BS/PB 1-103-3030/2011/S.

References

1. A. Atmanli, Fuel **176**, 6 (2016)
2. G. Szendr, A. Török, Transport **29**, 5 (2014)
3. A. Kijo-Kleczkowska, K. Środa, M. Kosowska-Golachowska, T. Musiał, K. Wolski, Fuel **170**, 19 (2016)
4. A. Kijo-Kleczkowska, K. Środa, M, Kosowska-Golachowska, T. Musiał, K. Wolski, Waste Manage. **53**, 16 (2016)
5. W. Tutak, A. Jamrozik, Therm. Sci. **18**, 1 (2014)
6. A. Jamrozik, W. Tutak, Chem. Process Eng. **32**, 4, (2011)
7. A. Jamrozik, J. Mech. Sci. Techn. **29**, 5 (2015)
8. A. Jamrozik, W. Tutak, B. Pol. Acad. Sci.-Tech. Sci. **62**, 4 (2014)
9. A. Atmanli, E. Ileri, B. Yuksel, N. Yilmaz, Appl. Energ. **145**, 7 (2015)
10. W. Tutak, K. Lukács, S. Szwaja, A. Bereczky, Fuel **154**, 10 (2015)
11. W. Tutak W, Energ. Convers. Manage. **86**, 9 (2014)
12. A. Atmanli, Fuel **172**, 8 (2016)
13. B.R. Kumar, S. Saravanan, Renew Sust. Energ. Rev. **60**, 11 (2016)
14. M. Mikulski, K. Duda, S. Wierzbicki, Fuel **164**, 13 (2016)
15. M. Dębowski, M. Zieliński, A. Grala, M. Dudek, Renew. Sust. Energ. Rev. **27**, 8 (2013)
16. W. Tutak, A. Jamrozik, M, Pyrc, M. Sobiepański, Fuel Process Technol. **149**, 9 (2016)
17. S. Kumar, J.H. Cho, J. Park, I. Moon, Renew. Sust. Energ. Rev. **22**, 26 (2013)
18. X. Zhen, Y Wang, Renew. Sust. Energ. Rev. **52**, 16 (2015)
19. B. Rajesh Kumar, S. Saravanan, Renew. Sust. Energ. Rev. **60**, 31 (2016)
20. C. Mishra, P. Mishra, B. Kar, N. Katiyar, SAE Tech. Paper 2015-26-0055, (2015)
21. I. Amr, Appl. Therm. Eng. **103**, 8 (2016)
22. M. Gruca, J. Kones **11**, 6 (2004)
23. W. Tutak, A. Jamrozik, Chem. Process Eng. **32**, 3 (2011)
24. L. Zhu, C.S. Cheung, W.G. Zhang, Z. Huang, Fuel **90**, 7 (2011)

 DOI: 10.1051/e3sconf/20171402019

Experimental investigations on combustion, performance and emissions characteristics of compression ignition engine powered by B100/ethanol blend

Wojciech Tutak[1,*], *Arkadiusz* Jamrozik[1] , and *Michal* Pyrc[1]

[1]Czestochowa University of Technology Sciences, Faculty of Mechanical Engineering and Computer Science, 42-201 Czestochowa, Armii Krajowej 21 Av., Poland

Abstract. In the study are presented the results of co-combustion of biodiesel B100 with ethanol fuel as blend. The 1-cylinder direct injection compression ignition engine was used during the study. Tests were conducted at a constant angle of fuel injection and constant rotational speed equal to 1500 rpm. Results of thermal cycle parameters and emission characteristics are presented. On the basis of results stated that ethanol fuel fraction in blend causes the increase in peak heat release rate. With the increase in ethanol fuel fraction the ignition delay increased but combustion duration decreased. With the increase in ethanol fuel fraction in blend thermal efficiency increased as well. It also noticed almost constant emission of THC, the increase in NO_x emissions and decrease emissions in CO and CO_2.

1 Introduction

Diesel engines are widely used to power engineering devices for its good durability and thermal efficiency. On the other hand these engines are major contributors of air pollutant emission such as carbon monoxide (CO) and nitrogen oxides (NO_x) [1]. Biofuels are a renewable, nontoxic, eco-friendly and sustainable alternative fuel for compression ignition engines. Biodiesel has obtained much attention as an alternative fuel to reduce greenhouse and combustion emissions. There has been a wide-spread study on biodiesel-fuelled engine performance and combustion emission [2,3]. Biodiesel (B100) is considered as one of the most promising alternative fuels for internal combustion engine. Biodiesel fuel is considered as friendly fuel to co-combustion with other biofuels using blends or dual fuel systems. The use of biodiesel as the alternative diesel engine fuel is increasing rapidly. However, alcohol such as ethanol is competitive alternative fuel due to their liquid nature, high oxygen content and their production from renewable biomass. Ethanol is a form of renewable energy and it can be produced from carbon based agriculture feed stocks [4]. Ethanol fuel can be considered as a renewable fuel. It has been identified that the usage of biodiesel fuel reduces carbon monoxide, hydrocarbon and particulate matter emissions but

* Corresponding author: tutak@imc.pcz.czest.pl

DOI: 10.1051/e3sconf/20171402019

nitric oxide emissions are increased compared to diesel fuel operation [5]. Because of advantages and the increasing share of biofuels in fuel market researches on their efficient use are still needed. B100 can be burned in diesel engine and it can be mixed with other fuels such as diesel or oxygenated fuels. Authors [6] investigated the diesel engines fuelled with 20% biodiesel-diesel blended fuel. The study revealed that the maximum torque achievable with the biodiesel-diesel blended fuel was slightly lower than that achievable with neat diesel fuel at high-load conditions. Smoke was decreased by more than 20% in all modes. Authors [7] investigated operate parameters of diesel engine powered by diesel fuel, rape seed methyl ester and methanol. They stated that adding methanol to biodiesel in high loads reduced CO in 2–13% and in whole engine load range – 13–45% reduced concentration of soot. However emission rates of NO_x, HC and CO_2 were close to or higher than reference diesel fuel. Nowadays renewable fuels are increasingly used in dual fuel engines [8,9]. Ethanol is a biomass based renewable energy source, which can be produced with relatively low cost. Ethanol for a long time was considered as curious engine fuel. At first ethanol was associated with spark-ignition engines. In recent years ethanol is used as a fuel for diesel engines [10]. It is utilized in both methods: blend or dual fuel co-combustion [8-12]. There are some work considered the use of the mixtures of mineral diesel with biodiesel and it impact on operating parameters and emission. Authors [13] stated that brake specific fuel consumption increases with higher percentage of biodiesel in the blends. Brake thermal efficiency decreases with the increased percentage of biodiesel in the blends. CO and THC emissions are improved with the addition of biodiesel to diesel. NO_x emission is found to be increased with pure biodiesel compared to mineral diesel [13]. Some authors investigated the burning of mineral diesel-biodiesel-ethanol mixtures and the impact of that fuel on operating parameters and emission [14]. The experimental results showed that the performance of CI engine was improved with the use of that blend. Exhaust emissions are also improved. Tse et al. [15] presented results of the combustion characteristics and particulate emissions of a diesel engine fuelled with diesel-biodiesel-ethanol blended fuel. They stated that ethanol causes higher peak in HRR, longer ignition delay. They also stated that ethanol causes shorten the combustion duration resulting in longer premixed and shorter diffusive combustion duration when compared with biodiesel [15]. The emphasis is also on the search for sources of alternative fuels for engines and improvement of combustion systems [16-19]. Co-combustion of fuels is used not only in relation to liquid or gas fuels but it is used for solid fuels as well [1,20-22].

In recent years several experimental studies are carried out, it is evident that further studies are necessary for a better understanding on the application of ethanol–biodiesel blend in diesel engines. In this paper is investigated the various ethanol fuel (EF) fraction in blend with biodiesel (B100) on engine performance and emission. In this paper are presented results of investigation of one-cylinder direct injection air cooled diesel engine in which is made the attempt to burn a mixture of diesel-ethanol with large ethanol fuel (EF) fraction.

2 Experimental setup

The study used the 1-cylinder direct injection natural aspired compression ignition engine. Detailed engine specifications are presented in Table 1. Tests conducted at a constant angle of diesel fuel injection and constant rotational speed equal to 1500 rpm. The test stand equipped with measurement system and additional supply system with regulated pressure of fuel supplied mechanical fuel pump (Fig. 1).

Measuring system:

- Exhaust gas analyzer: THC, CO, CO_2, O_2 – Bosch BEA 350: THC: range 0 – 9999 ppm vol, accuracy: 12 ppm vol, CO: range 0 – 10 % vol, accuracy: 0.06 % vol, CO_2: range 0 –

DOI: 10.1051/e3sconf/20171402019

18 % vol, accuracy: 0.4 % vol, O_2: range 0 – 22 % vol, accuracy: 0.1 % vol, λ: range 0.5 – 9.999, accuracy: 0.01,
- Exhaust gas analyzer: NO_x – Radiotechnika AI9600: acuracy: 32 ppm for range 0 – 1000 ppm,
- In-cylinder pressure sensor – Kistler 6001, sensitivity: ±0.5%,
- Charge amplifiers – Kistler 5001, linearity of FS <±0,05%,
- Air/Fuel Ratio Meter LM-2, range 7.35-22.38,
- Resolution for the data acquisition system – 1.0 CA deg,
- Data acquisition module – National Instruments USB-6251 – 16 bits resolution, sampling frequency 20 kHz.

At each mode of operation, the engine was allowed to run a few minutes until the exhaust gas temperature and exhaust gas composition have attended steady-state values and data acquisition process was subsequently conducted. The ambient temperature was maintained at a constant level. Each experiment was conducted three times and the results were found to agree with each other with the 95% significance level.

Table 1. Main engine parameters.

Parameter	Value	unit
Displacement volume	0.573	dm^3
Bore	90	mm
Stroke	90	mm
Compression ratio	17:1	-
Rated power (at 3000 rpm)	7	kW
Crankshaft rotational speed	1500	rpm
Injection pressure	210	bar
Injection timing	17	deg bTDC

Biodiesel (B100) has physical properties similar to those of conventional diesel fuel. It is also non-toxic and biodegradable. Ethanol fuel (EF) which used to diesel engine powering it was wet ethanol which consisted 89% of C_2H_5OH and 11% of H_2O. The blend of EF and diesel fuel showed separation, which hinders the formation of usable fuel blends. However, the use of biodiesel gives close to stabile and homogenous blend

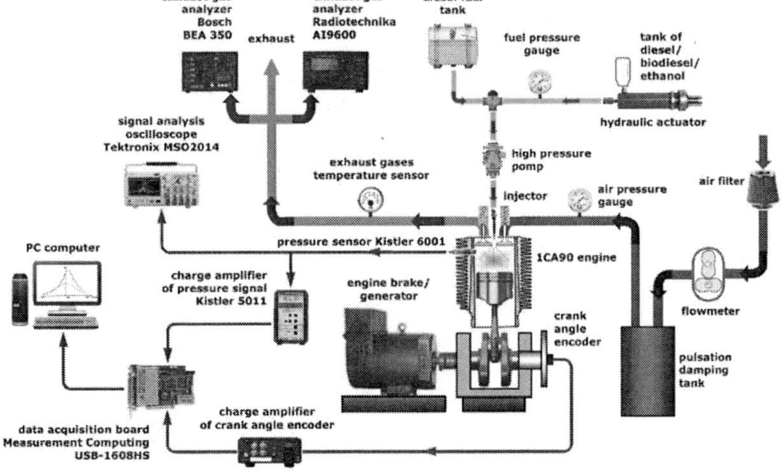

Fig. 1. Diagram of the experimental setup.

Biodiesel molecule has a polar end with affinity for ethanol, which is also polar in nature, thus, biodiesel can be used as an alternative stabilizer in preventing the separation of ethanol from diesel fuel [15]. Ethanol fuel as a fuel for compression ignition engine generally is used by blending with conventional diesel fuel or with using dual fuel technology [8,23]. Analyzing the usefulness of alcohol to co-combustion in IC engine it should be taken into account its properties such as: lower heating value (LHV) and also stoichiometric air fuel ratio (A/F_{st}) and heat of evaporation. Some of the disadvantages of alcohol fuels are low lubricity, difficulty of vaporization and high-auto-ignition temperature. Ethanol contains oxygen atom so that it can be viewed as partially oxidized hydrocarbon [4]. The main difficulties are more ethanol fuel then diesel fuel is required by mass and volume. Additionally the poor auto-ignition capability of ethanol is responsible for severe knock due to rapid burning of vaporized alcohol and combustion quenching caused by high latent heat of vaporization.

Table 2. Fuels properties.

Properties	B100	Ethanol
Molecular formula	$CH_3(CH_2)_nCOOH_3$	C_2H_5OH
Cetane number	56	~11
Lower heating value, (MJ/kg)	37.1	26.9
Density at 20°C, kg/m^3	855	789
Viscosity at 40°C, (mPa s)	4.51	1.1
Heat of evaporation, (kJ/kg)	250	840
Stoichiometric air fuel ratio	12.5	9.0
Flash point, (°C)	>101	425
Oxygen content, wt %	10.8	34.8
Carbon content, wt %	77.1	52.2
Hydrogen content, wt %	12.1	13

In Fig. 2a are presented the heat of evaporation and LHV of B100 and EF. Additionally, LHV of alcohol is lower in comparison to biodiesel fuel hence to obtain the same engine performance the higher amounts of alcohol should be provided (Fig. 2).

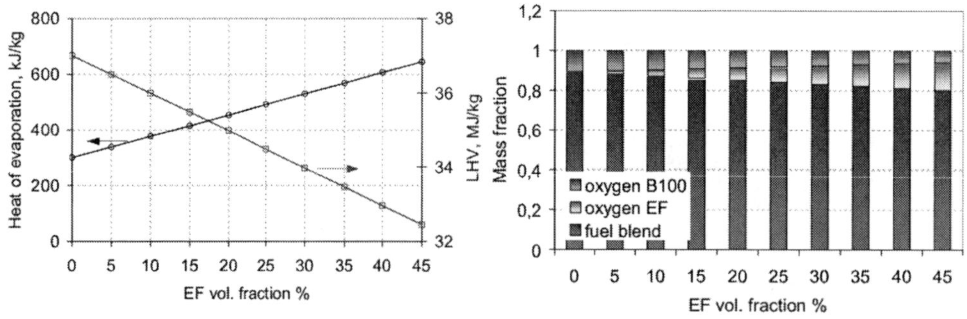

Fig. 2. Heat of evaporation and lower heating value (a) and oxygen mass fraction (b) in blend.

Biodiesel is characterized by lover value of LHV than conventional diesel fuel of near to 12% because of presence of substantial amount of oxygen in the fuel, but B100 has higher value of cetane number than diesel fuel. Biodiesel in comparison to diesel fuel has higher cetane number which could enhance the cetane number of biodiesel-ethanol blend. Whereas the diesel engine prefers fuels of high cetane number which makes auto-ignition process easily and gives small ignition delay. During the tests used BE blends in large values of ethanol fuel fraction up to 45%. As the limit of EF fraction in blend was quality of the engine work and possibility to auto-ignition of used fuels blend. As presented in Fig. 2a with the increase in EF fraction the value of heat of evaporation increased as well. It caused difficulty in vaporization of fuel dose and connected with decreased of cetane number by EF participation it worsens auto-ignition process. In case of B100 fuel the oxygen mass fraction is equal to 10.8%. In EF the oxygen mass fraction is 34.8%. As presented in Fig. 2b with the increase in EF fraction the oxygen fraction in blend increases as well.

2.1 Test method

Analysis of combustion process in the internal combustion engine usually is carried out with the rate of heat release. Heat release rate (HRR) is calculated on the basis of the measured in-cylinder pressure data and crank angle readings. The basis for determining the heat release rate is the first law of thermodynamics and the equation of state [23]. A parameter indicating the performance of the engine is the indicated mean effective pressure which is determined on the basis of the instantaneous pressure in the cylinder. On the basis of pressure courses determined the factor of unrepeatability as COV_{IMEP} [23]. The averaged value of *IMEP* of the three measurements containing 100 recorded engine cycles [bar].
After rearranging and simplifications, the heat release rate vs. crank angle is obtained in well-known form as follows:

$$HRR = \frac{1}{\kappa - 1}\left[\kappa p \frac{dV}{d\varphi} + V \frac{dp}{d\varphi}\right] \tag{1}$$

where: κ - isentropic exponent, V – cylinder volume, p – in cylinder pressure.
Instantaneous cylinder volume V is precisely described by engine geometry. Due to omitting as follows: heat transfer to walls, crevice volume, blow-by and the fuel injection effect, the resulted heat release rate is termed as the net heat release rate. The cumulative net heat released is obtained by integrating Eq. 1 over the crank angle φ.
A parameter indicating the performance of the engine is the indicated mean effective pressure which is determined on the basis of the instantaneous pressure in the cylinder. Indicated pressure is calculated as:

$$IMEP = \frac{1}{V_d}\int_0^{720} p \frac{dV}{d\varphi} \tag{2}$$

where: V_d - displacement volume [m^3],
The indicated thermal efficiency (ITE) is calculated on the basis of equation:

$$ITE = \frac{IMEP \cdot V_d \cdot 0.5 \cdot n \cdot t}{V_f \cdot \rho_f \cdot LHV} \tag{3}$$

where: *IMEP* – indicated mean effective pressure [%], V_f - volume of fuel delivered to the engine cylinder [m3], ρ_f - fuel density [kg/m^3], *LHV* - lower heating value of fuel [MJ/kg], n - rotational speed of engine, t - time consumption of the fuel delivered to the engine cylinder [min].
On the basis of pressure courses determined the factor of unrepeatability as COV_{IMEP}.

$$COV_{IMEP} = \frac{\sigma_{IMEP}}{IMEP_{mean}} \cdot 100\% \tag{4}$$

where: σ_{IMEP} – standard deviation of *IMEP*.

The standard deviation of the *IMEP*:

$$\sigma_{IMEP} = \sqrt{\frac{1}{N-1}\sum(IMEP_i - IMEP_{mean})^2} \quad (5)$$

where: $IMEP_i$ - value of *IMEP* in following cycles [bar], $IMEP_{mean}$ - averaged value of *IMEP* of the three measurements containing 100 recorded engine cycles [bar].

3 Results

The researches conducted with various fuel blends of B100/EF. It was using blends with volumetric fractions of EF from 5 to 45% (blend with 5% of EF – BE05). During experimental research recorded 100 consecutive engine operation cycles with resolution of 1 deg of CA. The results are used to evaluate the combustion process. The heat release rate are calculated on the basis of the measured in-cylinder pressure data and crank angle readings individually for each cycle and than these were averaged. On the basis of measurement data the operating parameters as *IMEP* and *ITE* of engine were determined. The study was conducted at a constant angle of the start of fuel injection equal to 17 deg before TDC. This constant injection timing of the test engine was due to the fact that such engine is still manufactured in Poland and it is used to power various devices.

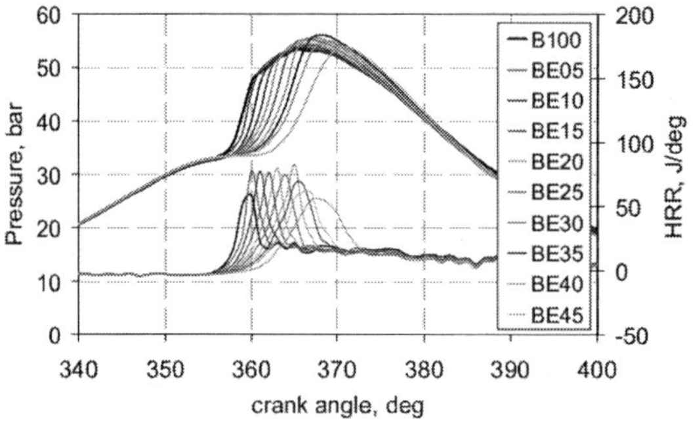

Fig. 3. Pressure and heat release rate traces of test engine.

The variation of in-cylinder pressure and heat release rate are shown in Fig. 3 for different ethanol volumetric fraction from 5 to 45, respectively. As a reference values were taken results of B100 burning in the test engine. Each pressure waveform is obtained by averaging 100 cycles of engine operation. The peak in-cylinder pressure occurs further away from the TDC (top dead center) in the expansion stroke with increase of EF fraction, which is similar to the results of Qi et al. [24].

Observing the pressure waveform can be seen that with increasing share of EF in blend the combustion process take place with higher ignition delay. In case of engine powered by biodiesel the peak pressure (p_{max}) was equal 53 bar and was achieved at 6 deg after TDC. With the increase of EF fraction in blend p_{max} reach highest values and reached for higher values of crank angles after TDC. In Fig. 3 are presented the HRR traces of co-combustion of B100 and EF as well. On the basis of these data combustion process can be characterized. It can be stated that EF fraction in blend causes the increase in peak HRR. Up to BE30 blend noticed the increase in max value of HRR. The increase in peak HRR was about 30% in comparison with B100. The peak heat release rate increases and up to BE30 is kept near on the same level. After exceeding 30% of EF fraction in blend the HRR started to decrease due to lover value of LHV of blend and relatively high value of heat of evaporation. It was du to high fraction of EF in blend which already causes significant increase in the amount of heat needed to evaporate the fuel dose. Increased peak values of HRR are directly related to thermal load. On the basis of HRR traces by integration the

MFB traces are determined. The MFB traces are used to the combustion phases determined. The ignition delay (ID) is defined as the angle from the beginning of fuel injection to 10% of MFB and combustion duration (CD) as a range of crank angle 10-90% of MFB. Ignition delay period consist of physical delay and chemical delay which occur simultaneously. In the physical delay takes place atomization, vaporization and mixing of air fuel, and in the chemical delay attributed pre-combustion reactions [25].

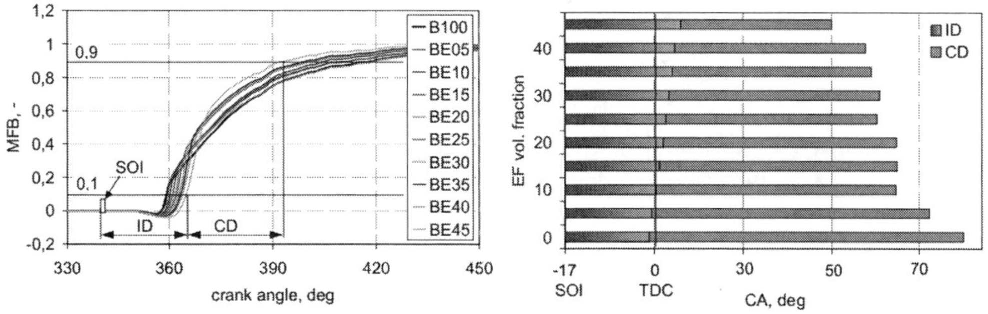

Fig. 4. Mass fraction burned (a) and combustion phases (b).

In Fig. 4a are presented courses of mass fraction burned (MFB) for combustion of all analyzed blends. There are indicated the periods of ignition delay (ID) and combustion duration (CD). In Fig. 4b are presented the combustion phases of test engine. Because ethanol has a lower cetane number and higher heat of vaporization requirements than diesel, ethanol blended fuels also decrease overall cetane number and increase ignition delay [26,27]. It can be stated that with the increase in EF fraction the ignition delay increased. Ethanol blended fuels exhibited unstable ignition characteristics. The increase in ignition delay at BE45 blend in comparison with B100 was equal to 6 deg CA. The inverse relationship was obtained in case of combustion duration. With the increase in EF fraction in blend the combustion duration decreased. In case of BE45 the combustion duration was 2-times lower in comparison with B100 fuel. For the pure biodiesel fuelling the ratio of CD to ID was 3.75 but in case of 45% of EF fraction this ratio was 1.32. Similar results are obtained by Tse et al. [15].

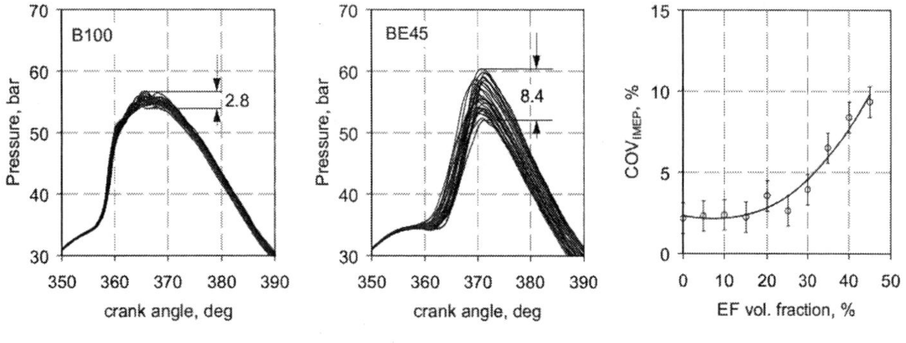

Fig. 5. Cycles variations of the engine powered by B100 and BE45 and unrepeatability of IMEP.

It was due to two characteristic properties of EF. The first it is heat of evaporation which cause the ignition delay increase by lowering the combustion temperature. The second it is oxygen content that is obviously favors the combustion process. Both of these properties counteract each other during the combustion process. Donahue et al. stated that higher oxygen content in the spray (from oxygenated fuel) reduced pyrolysis and increased oxidation, thus shortening the combustion duration [28]. As seen in the Fig. 4 all the blends give faster combustion in all analyzed cases than biodiesel which could compensate the ignition delay in start of combustion.

DOI: 10.1051/e3sconf/20171402019

The faster combustion process could lead to the increase in ITE which is visible in the Fig. 6. In Fig. 5a are presented the peak in-cylinder pressure variations for the test engine powered by B100 and blend BE45. In case of powering by pure biodiesel the peak pressure variation are equal to 2.8 bar but in case of BE45 it is 8.4 bar. The unrepeatability of successive cycles of engine operation has the impact not only on engine performance but also on the emission of toxic components of exhaust gases. With the increase in EF fraction in blend it observed the increase in unrepeatability of cycles (Fig. 5). Despite the significant unrepeatability of consecutive engine cycles ware there no cycles without combustion. In all analyzed cases the COV_{IMEP} was lower then 10%. The increase value of COV_{IMEP} is directly connected with the unstable auto-ignition characteristic of ethanol-biodiesel bend. With the increase in EF fraction the cetane number decreased at the same time, which worsens the process of auto-ignition as well.

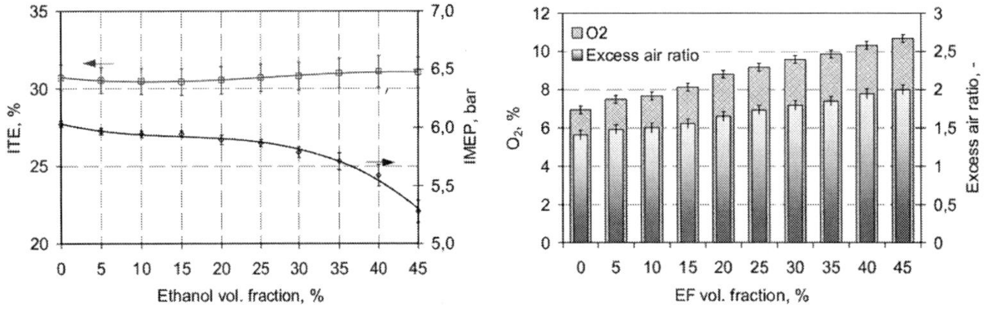

Fig. 6. Indicated thermal efficiency (ITE) and indicated mean effective pressure (IMEP) and the oxygen content in the exhaust gases.

In Fig. 6a is presented the indicated thermal effciiency and the indicated mean effective pressure for burned blends. The injection system ensures a constant volume dose of fuel that with the increase in EF fraction the energy value of the fuel dose decreased. Of course, it would then affect the value of *IMEP*. However, the decrease in the value of *IMEP* was only 0.56 bar at BE45. The more visible decrease in *IMEP* noticed after exciding 30% of EF in blend. On of the most important characteristics of thermal cycle of internal combustion engine is thermal efficiency. In analyzed cases with the EF fraction increase ITE increased as well. Similar results obtained by Zhu et al. [28]. Up to BE25 blend ITE was kept near on the same level. The measurement error of IMEP is $\delta IMEP = 2.1\%$ and error of ITE is $\delta ITE = 3.2\%$. In Fig. 6b is presented the oxygen content in the exhaust of engine powered by various blends. With the increase in EF fraction in blend while decreasing amount of hydrocarbon in the blends causes lower requirement of oxygen to combustion process. The increasing in EF fraction causes the increasing of stoichiometric air fuel ratio of the blend.

The NO_x emission increased with the EF fraction in blend (Fig. 7). The highest value of NO_x emission noticed for 45% of EF fraction equal to 4.4 g/kWh and it was higher by 30% in comparison with B100 fuelling. In case of EF fractions combustion process occurred faster with higher in-cylinder pressure and temperature as well. The NO_x formation rate strongly depends on temperature and oxygen presence. In compression ignition engine generally the excess air ratio is higher then one and additionally locally the oxygen is supplied by the fuel (Fig. 3). In the combustion chamber of the engine during relatively short premixed combustion phase are conditions to NO_x formation.

The presence of intrinsic oxygen can also cause of the slightly increased NO_x emission for the blends. Higher in-cylinder pressure could contribute to the increased NOx emissions, due to the increased peak combustion temperature at high engine loads [29].

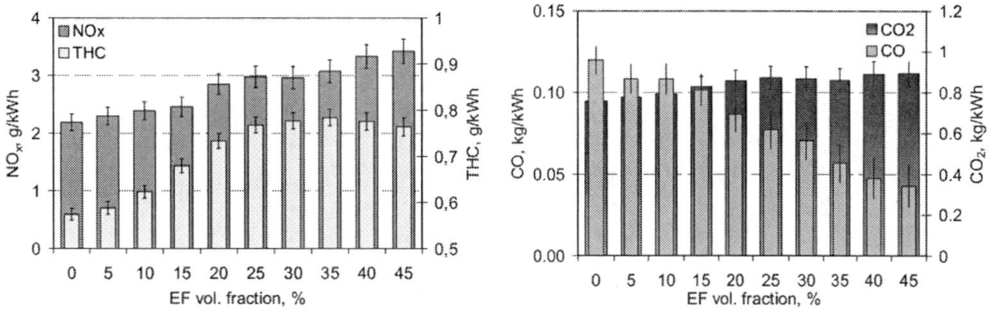

Fig. 7. Emission of THC, NO_x and CO, CO_2.

Analyzing the emissions of THC it can be stated that it was noticed slightly the increase in THC emission during the tests. The highest THC emission was indicated for 35% of EF fraction in blend and it was higher of about 30% compared to pure B100 powering. In the literature it has also been observed that the presence of alcohol in the blends contributes the increase in THC emission [25]. The THC emission is higher since temperature is lower due to the higher heat of evaporation of ethanol fuel thus combustion poor. Variations of CO and CO_2 emissions as a function of blend fractions are presented in Fig. 7. Emissions of CO depend on the air-fuel ratio relative to the stoichiometric proportion. Reach combustion process invariably produces CO and emissions increase nearly linearly with the deviation of the stoichiometric [4]. For all analyzed blends the specific emission of CO was decreased. It can be stated that combustion process occurred more complete with EF fraction. As compared to baseline biodiesel, ethanol blended fuels result in lower CO emissions in this case. Higher specific emission of CO_2 is due to lower values of IMEP.

4 Conclusions

The paper presents results of co-combustion of biodiesel B100 with ethanol fuel as a blend. In the researches used the mixture of up to 45% by volume of ethanol fuel. The analysis of the operating parameters and emission of the engine are presented in the study. The following main results are summarized:

- up to BE35 obtained the increase in peak pressure, with the increase in EF fraction in blend observed the increase in unrepeatability of cycles, up to BE30 blend noticed the increase in peak of HRR. The increase was about 30% in comparison with B100, the increase in ignition delay at BE45 blend in comparison with B100 was equal to 6 deg CA, with the increase in EF fraction in blend the combustion duration decreased. In case of BE45 the combustion duration was 2-times lower in comparison with B100 fuel, with the increase of EF the lower IMEP obtained but ITE was at the same level as in case of B100. The NOx emission increased with the EF fraction in blend up to BE35.
- emissions of THC increased with the increase in EF fraction,
- lower emission of CO noticed which indicates that combustion process occurred more complete in EF presence.

Summing up the results of research can be concluded that blend of B100 with ethanol fuel can be burned in the compression ignition engine with acceptable operating parameters and emission. Fuel B100 relatively well mixed with ethanol and form a relatively stable mixture.

References

1. G. Szendr, A. Török, Transport **29**, 5 (2014)
2. A. Ghurri, K. Jae-duk, S. Kyu-Keun, J. Jae-Youn, K. Hyung Gon, Journal of Mechanical Science and Technology **25**, 4 (2011)
3. M. Mikulski, K. Duda, S. Wierzbicki, Fuel **164**, 13 (2016)
4. B.S. Chauhan, N. Kumar, S.S. Pal, Y.D. Jun, Energy **36**, 8 (2011)
5. K. Anand, R. P. Sharma, P. S. Mehta, Biomass and bioenergy **35**, 8 (2011) 533-541.
6. S. Park, H. Kim, B. Choi, Journal of Mechanical Science and Technology **23**, 9 (2009)
7. J. Zaglinskis, K. Lukács, Á. Bereczky, Fuel **170**, 15 (2016)
8. W. Tutak, Energy Conversion and Management **86**, 8 (2014)
9. W. Tutak, K. Lukács, S. Szwaja, A. Bereczky, Fuel **154**, 10 (2015)
10. A. Murcak, C. Hasimoglu, I. Çevik, H. Kahraman, Fuel **153**, 8 (2015)
11. A. Jamrozik, W. Tutak, Bulletin of the Polish Academy of Sciences Technical Sciences **62**, 11 (2014)
12. Z. Şahina, O. N. Aksub, Renewable Energy **77**, 11 (2015)
13. A. Datta, S. Palit, B.K. Mandal, Journal of Mechanical Science and Technology **28**, 5 (2014)
14. H. Aydin, C. Ilkılıç, Applied Thermal Engineering **30**, 5 (2010)
15. H. Tse, C. W. Leung, C.S. Cheung, Energy **83**, 7 (2015)
16. A. Jamrozik, Journal of Mechanical Science and Technology **29**, 9 (2015)
17. W. Tutak, A. Jamrozik, Thermal Science **18**, 10 (2014)
18. M. Kisielewska, M. Dębowski, M. Zieliński, Bioprocess and Biosystems Engineering **38**, 10 (2015)
19. A. Jamrozik A., W. Tutak, Chemical and Process Engineering **32**, 4 (2011)
20. A. Kijo-Kleczkowska, K. Środa, M. Kosowska-Golachowska, T. Musiał, K. Wolski, Fuel **170**, 19 (2016)
21. A. Kijo-Kleczkowska, K. Środa, M, Kosowska-Golachowska, T. Musiał, K. Wolski, Waste Manage **53**, 16 (2016)
22. M. Kosowska-Golachowska, A. Kijo-Kleczkowska, A. Luckos, K. Wolski, T. Musiał, Archives of Thermodynamics **37**, 1 (2016)
23. W. Tutak, A. Jamrozik, M. Pyrc, M. Sobiepański, Fuel Processing Technology **149**, 9 (2016)
24. D.H. Qi, H. Chen, R.D. Matthews, Y.ZH. Bian, Fuel **89**, 6 (2010)
25. H. Zou, L. Wang, S. Liu, Y. Li. Front. Energy Power Eng. China **2**, 3 (2008)
26. N. Yilmaz, F. M. Vigil, A. B. Donaldson, T. Darabseh, Fuel **115**, 3 (2014)
27. S. Lennox, K. Lukács, A. Torok, A. Bereczky, M. Mbarawa, A. Penninger, A. Kolesnikov, Fuel **107**, 9 (2013)
28. L. Zhu, C.S. Cheung, W.G. Zhang, Z. Huang, Fuel **90**, 7 (2011)
29. F. Lujaji, K. Lukacs, A. Bereczky, M. Mbarawa, Fuel **90**, 5 (2011)

The influence of amount of copper of modified vermiculites on catalytic properties in SCR-NH_3

Bogdan Samojeden[1,*], *Marek* Możdżeń[1]

[1]Department of Fuels Technology, al. A. Mickiewicza 30, 30-059 Kraków, AGH University of Science and Technology

Abstract. Vermiculite was prepared by modifications with nitric acid, pillaring with Al oligocations and polymer (polyamide), impregnation with Cu. Obtained products were used as catalysts in SCR-NH_3 process. All modifications increased catalytic performance compared to base vermiculite which was used as a reference. Sample pillared with polymer and impregnated with 5%wt. Cu showed best catalytic properties in the process, compared to other polymer and Cu modified samples.

1 Introduction

Nitrogen oxides, referred as NO_x, are one of the main pollutants deriving from fuels combustion. They are connected with admixtures and impurities in fuels or may occur as a product of side reactions. The negative impact of NO_x on the environment cannot be neglected since they are the cause of acid rains and degeneration of ozone layer. The main product of fuel combustion is nitrogen monoxide (NO), which undergoes oxidation to NO_2 in contact with oxygen from the atmosphere. Subsequently, nitrogen dioxide reacts with water, hence the nitric acid is created, which is one of the components of acid rains. Moreover, NO_x in general take part in the formation of smog and also are one of the greenhouse gases [1].

Hence the need to reduce emission of such pollutants, and that is where SCR comes to the aid. There are two ways of such reduction [2]: modification of combustion process and conversion of flue gases. Nowadays SCR is the most effective and most widely used method of NO_x removal from flue gases emitted by stationary sources. It is a process where NO_x are reduced to molecular nitrogen and water, with the aid of various catalysts. NO_x reduction takes place when the gases pass through the catalyst chamber in the installation, while reductants, such as ammonia or urea, are mixed with the gases (injected into the stream) prior. Main reactions occurring in SCR are as follows [3]:

$$4NO + 4NH_3 + O_2 \Rightarrow 4N_2 + 6H_2O$$

$$4NH_3 + 3O_2 \Rightarrow 2N_2 + 6H_2O$$

$$2NO_2 + 4NH_3 + O_2 \Rightarrow 3N_2 + 6H_2O$$

SCR process utilizes reducing agents such as anhydrous ammonia, aqueous ammonia and urea. Thanks to the use of catalyst, NO reduction process can be maintained at lower temperatures (300-400°C compared to ~1000°C as in non-catalytic process).

* Corresponding author: bsamo1@agh.edu.pl

Typical industrial catalysts for this process are vanadium (V) oxide promoted with tungsten (IV) oxide on the carrier made of titanium (IV) oxide set on the monolith. The highest activity for such catalyst is at temperatures ranging from 250 to 400°C. In energetic installations, it is most commonly placed after the boiler and before the precipitator. However, such placement exposes the catalyst to harmful effects of various dusts and sulfide (IV) oxide, which reduces its lifetime [3]. Hence the need to study new catalysts with suitable properties for SCR process. So far, activated carbons and clay minerals are thoroughly studied to meet the conditions of a good SCR catalyst. Vermiculite and montmorillonite seem to be most suitable candidates thanks to their properties and ease of modification. It is also proved, that the addition of transition metals, such as Fe and Cu have great impact on catalytic properties [4]. Clay minerals, as layer materials can undergone numerous modifications, where the most important one, apart from acid activation, is pillaring with various agents, such as oligocations (Keggin ions) or polymers. Introduction of oligocations is a well known process, that leads to increase of the distance between layers. Because of such improvement, additional agents can be placed inside the material, hence giving or improving catalytic properties [5]. Since carbon (in form of activated carbon), gives very good catalytic results in SCR process [6], a polymer (polyamide) was used to introduce it into the vermiculite. The purpose of such modification was to check the impact of carbon, in other form than activated carbon which is already used as catalyst, on catalytic properties in the process.

2 Experimental and results

2.1 Sample preparation

Base material, natural vermiculite supplied by Sigma-Aldrich Company, has undergone various modifications, which purpose was to give/improve catalytic properties. All modification steps are shown in Figure 1.

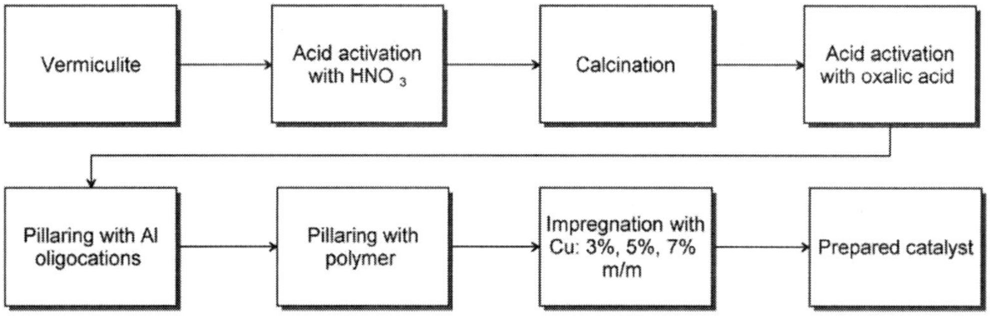

Fig.1 Preparation of samples

The first step of modification was the acid activation with 20% solution of nitric acid. Process was carried out at the acid boiling point under constant stirring for one hour. Activated material was filtrated, washed with distilled water to neutral pH. Then, vermiculite was calcined at 600°C for 4 hours. The next step was acid activation with 0.12M solution of oxalic acid, at 80°C under constant stirring for 3 hours (then filtrated, washed with distilled water and dried for 24 hours at 80°C) followed by pillaring with Al oligocations. Pillaring agent was prepared by slow addition of 0.4M NaOH to 0.4M solution of $AlCl3_3$, under constant stirring for 24 hours. OH/Al ration was equal 2.4. After that, the pillaring solution was left to age in room temperature for 72 hours. Pillaring process was carried out under the following conditions: pillaring solution was slowly added to the activated vermiculite suspension of 2%wt. in distilled water, in an amount of 12 mmol of Al per 1 g of vermiculite, for about 24 hours. Pillared vermiculite was subsequently washed with distilled water until disappearance of chloride ions and dried for 24 hours at 80°C. Furthermore, Al-modified vermiculite was additionally pillared with polymer (polyamide). 5%wt. solution was added into Al-pillared vermiculite and stirred constantly for 24 hours at room temperature. Then it was filtrated and dried for 24 hours at °C. The final step was the incipient wetness impregnation. The Cu ions precursor, $Cu(NO_2)_2$, was added as a 3, 5 and 7%wt. solution in respect to vermiculite mass. Material was dried for 24 hours at 80°C, hence the prepared catalyst was obtained.

Fig.2 Modified vermiculite before (A) and after (B) pillaring

2.2 Catalytic tests

Prepared samples were tested as catalysts in the selective catalytic reduction of NO_x with ammonia. All tests were performed on AGH University of Science and Technology, Kraków, Poland. Parameters of SCR process were as follows:

- the reaction gas - 800 ppm NO, 800 ppm NH_3, in He with 3% v/v addition of O_2;
- gas velocity - 100ml/min;
- reaction temperature - 150 - 400°C;
- mass of the catalyst - 200mg;
- gas flow controller - ABB Uras-14;
- Concentration of gases in product stream measured every 65 seconds by NDIR (nondispersive infrared sensor) (Hartmann and Braun).

NO conversion was calculated based on formula:

$$\text{NO Conversion [\%]} = \frac{NO_{in} - NO_{out}}{NO_{in}} * 100\% \quad (1)$$

Table 1 explains the preparation of samples and assigned symbols.

Table 1. Parameters of prepared samples

Sample	V	VN20	VN20g	VN20g3	VN20g5	VN20g7
Acid activation	-	HNO_3 20%	HNO_3 20%	HNO_3 20%	HNO_3 20%	HNO_3 20%
Cu impregnation	-	-	-	3%	5%	7%
Al-pillaring	-	+	+	+	+	+
Polymer pillaring	-	-	+	+	+	+

Table 2 shows the results of catalytic tests performed at 150, 200, 250, 300, 350 and 400 °C in the form of N_2O concentrations at the outlet and NO conversions.

Table 2. N_2O concentration and NO conversion at given temperatures

Parameter	Temperature [°C]	Sample					
		V	VN20	VN20g	VN20g3	VN20g5	VN20g7
N_2O Concentration [ppm]	150	12	38	155	187	323	214
	200	14	48	165	191	340	232
	250	15	68	172	200	352	255
	300	27	103	175	206	356	265
	350	17	112	176	234	393	297
	400	29	177	188	213	362	260
NO Conversion [%]	150	0.5	0.5	7.1	8.2	1.6	2.0
	200	3.0	15.2	16.1	10.0	19.6	15.7
	250	6.9	55.6	61.4	32.4	51.0	51.1
	300	8.9	90.1	93.1	66.2	80.0	82.6
	350	9.3	98.2	92.7	93.4	94.3	92.6
	400	12.4	60.5	61.5	64.4	60.3	61.2

2.3 Discussion

As can be seen in general in Figures 5 - 11, which were prepared based on Table 2, the biggest difference in N_2O concentration at the outlet, as well as NO conversion, occurs between non-modified (or only activated) vermiculites and modified ones. It is connected with enhancing catalytic properties via acid activation and pillaring. Base vermiculite (V), shows minimal, practically insignificant ability to catalyze SCR process (maximum NO conversion 12.4%), thus this sample is considered only as a reference. As a general rule for all modified vermiculites, with temperature increase, NO conversion rises, up to 350 °C, then slightly goes down at 400 °C. It means, that 350 °C is an optimal temperature for those catalysts.

Catalytic performance for polymer-modified samples with addition of Cu, varies between temperatures. Generally lower performance (compared to sample VN20) at 250 - 350 °C range, may be caused as a result of covering the active phase by polymer chains [7]. In case of montmorillonite, pillaring with the same polymer improved catalytic properties. Addition of Cu, increases the NO conversion in following orders: 7% > 5% > 3% at 250 and 300 °C; 5%, 3%, 7% at 350 °C. It is visible especially at 350 °C, where NO conversion for Cu-modified samples is highest for VN20g5 (as compared with all polymer-modified ones), which is 94.3%. These results correspond to Chmielarz et al. research [8-9], where NO conversion higher than 80% was reported. Studies for other layered materials, also show promising results, high levels of NO conversion, at various temperatures (usually ~350 °C). This parameter, obtained for PILCs (pillared interlayered clays) modified with copper goes up to almost 100% at 350 °C [10]. Polymer-modified sample (VN20g) shows best NO conversion at 250 and 300 °C, however at 350 °C it is only slightly lower (as well as catalysts with addition of Cu) than sample VN20 without further modifications.

DOI: 10.1051/e3sconf/20171402020

Table 3. S_{BET} of obtained samples

	Sample					
S_{BET}	V	VN20	VN20g	VN20g3	VN20g5	VN20g7
m^2/g	5	6	7	2	3	2

Table 3 shows specific surface area (S_{BET}) of obtained samples. Nitrogen physisorption on catalysts was performed using ASAP 2020 Micromeritics instrument after degassing at 105 °C for 24 h at 1 Pa vacuum. The adsorption–desorption isotherms of nitrogen at −196 °C were treated by the standard Brunauer–Emmett–Teller (BET) procedure. S_{BET} for p/p_0 range = 0.05–0.25 was calculated [11]. These results show slight increase in S_{BET} of samples modified only via acid activation and pillared with polymer. However, addition of copper seems to decrease specific surface area. This is cause simply by deposition of copper species on the surface, and, what is also possible, partially inside the pores. Hence the lower parameter. Additionaly, as shown in Fig 3, partial subsidence of the structure has occured, which is indicated by amorphous part in XRD diffractogram.

Figure 4, which is an UV-Vis spectra of modified samples, gives information about Cu modification. According to Chmielarz et al. [9], the absorption band centered at 225 nm, here observed for each Cu-impregnated sample, is attributed to isolated Cu^{2+} species. Additional absorption band, reported around 360 nm, and attributed to charge-transfer between Cu^{2+} and oxygen in oligonucelar $[Cu\text{-}O\text{-}Cu]_n$ species has not been observed, thus indicating that only Cu^{2+} species exist in prepared samples [9].

According to Chmielarz et al. [8], vermiculites are stable materials in tested temperatures, as compared to activated carbons, which undergo oxidation to CO_2 at elevated conditions.

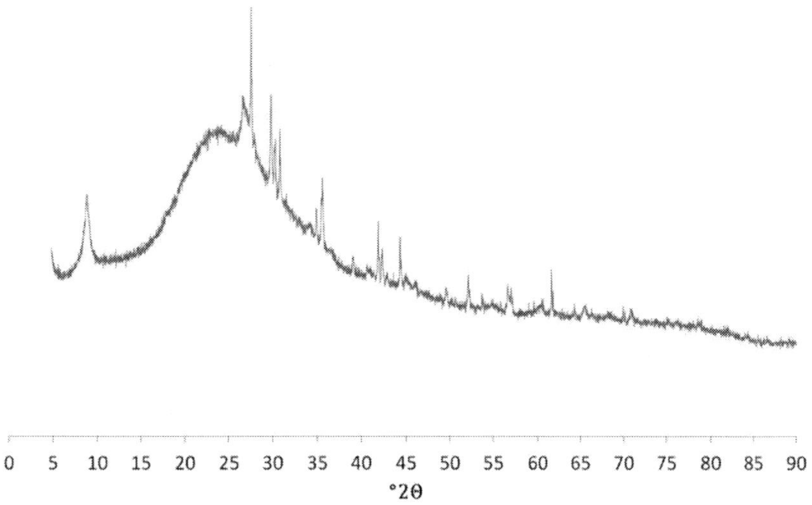

Fig. 3. XRD pattern for acid activated vermiculite (VN20)

 DOI: 10.1051/e3sconf/20171402020

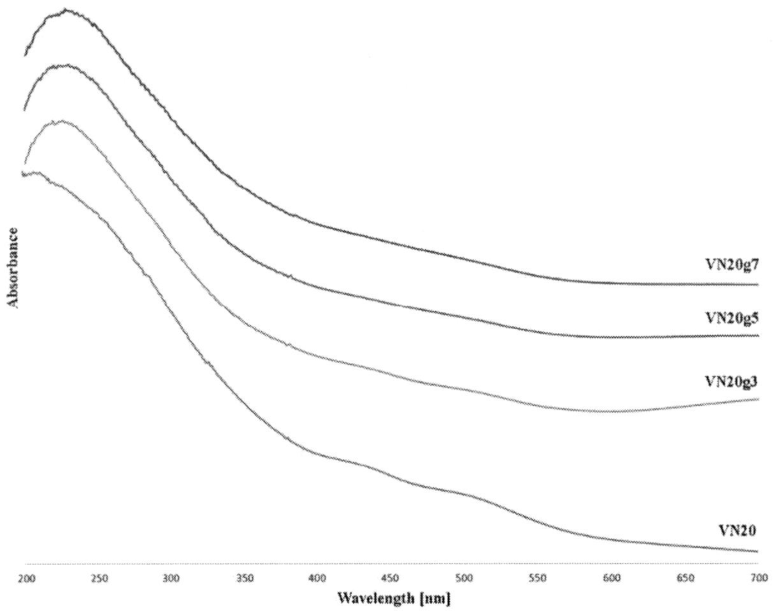

Fig. 4. UV-Vis spectra of base sample VN20 (not modified with Cu) and samples VN20g3, VN20g5 and VN20g7, impregnated with Cu, respectively 3, 5 and 7% wt.

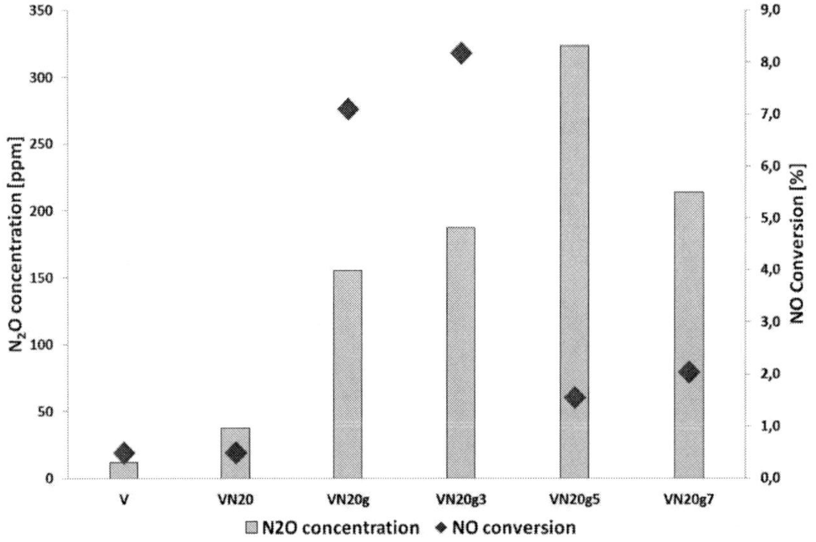

Fig. 5. N_2O concentration at the outlet and NO conversion at 150°C

 DOI: 10.1051/e3sconf/20171402020

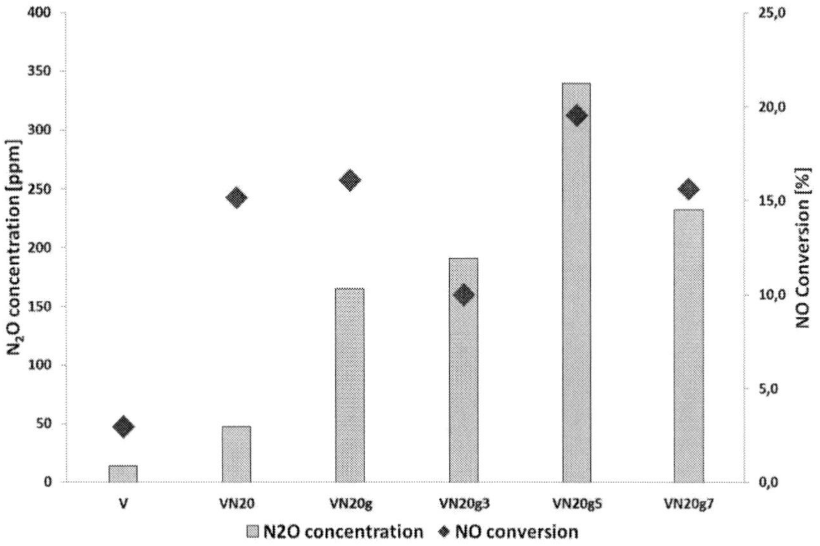

Fig. 6. N_2O concentration at the outlet and NO conversion at 200°C

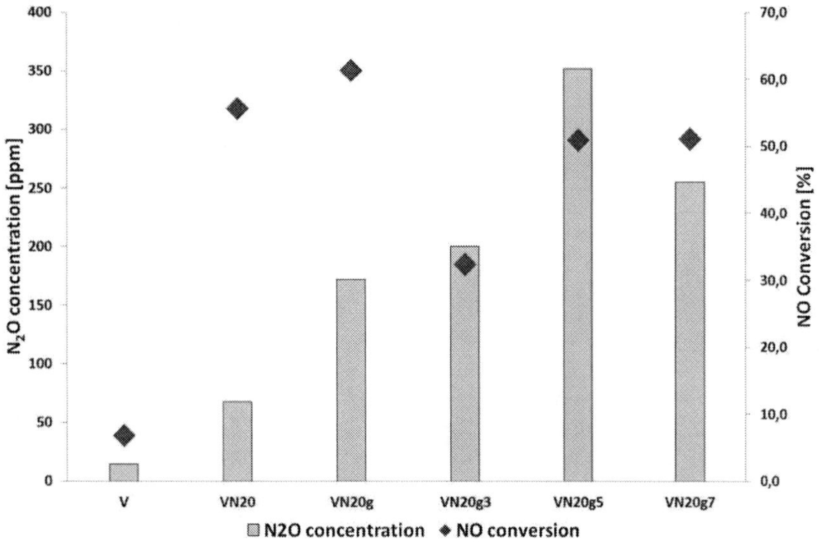

Fig. 7. N_2O concentration at the outlet and NO conversion at 250°C

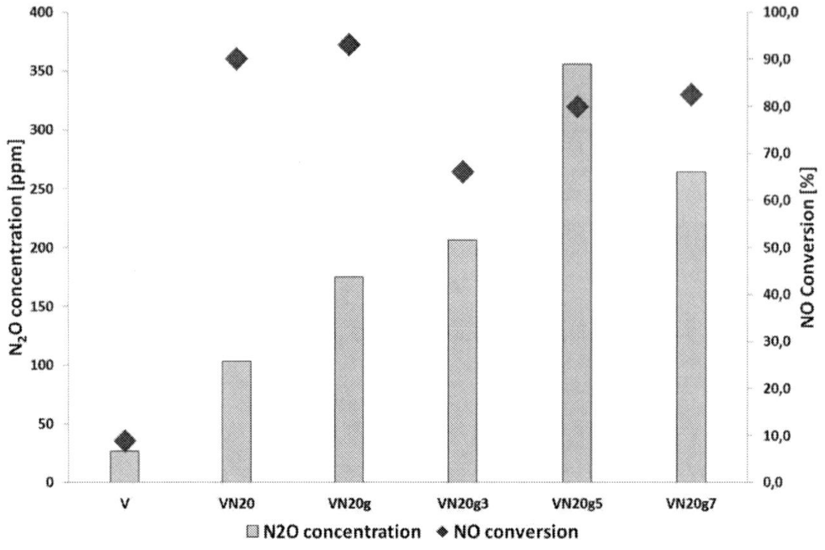

Fig. 8. N_2O concentration and at the outlet NO conversion at 300°C

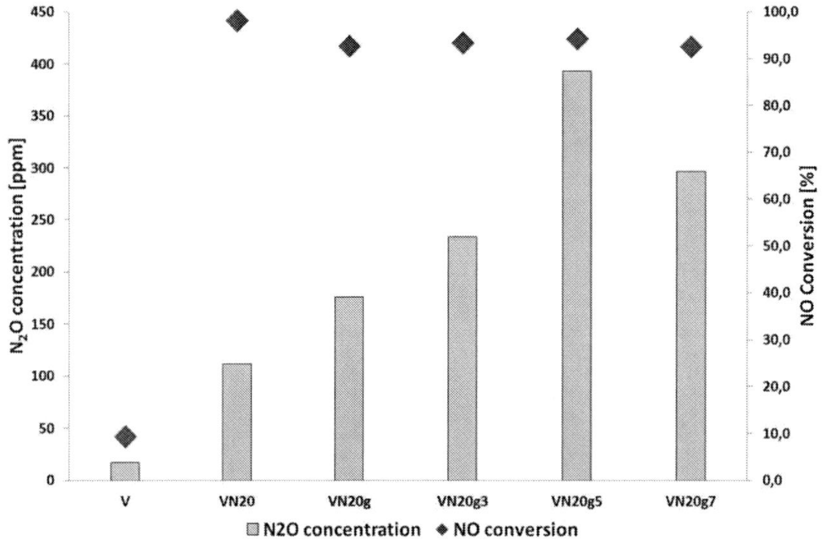

Fig. 9. N_2O concentration at the outlet and NO conversion at 350°C

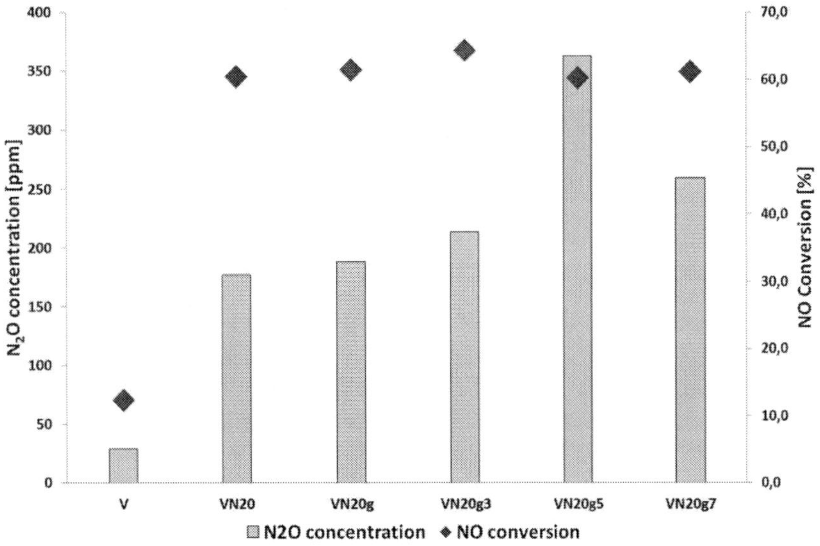

Fig. 10. N_2O concentration at the outlet and NO conversion at 400°C

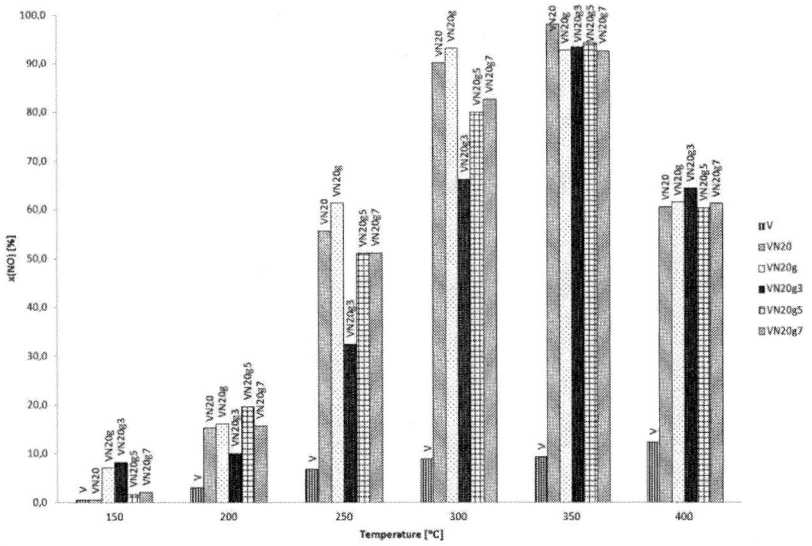

Fig. 11. NO conversion of all samples at temperatures ranging from 150 to 400 °C

3 Conclusion

The aim of this work was to study the influence of the copper amount on catalytic properties of modified vermiculites. Raw supplied vermiculite was modified by acid activation with nitric acid, pillaring with Al oligocations and polymer (polyamide), as well as impregnation with Cu.

Catalytic tests have shown promising results. Polymer, and Cu-modified samples perform well at 350 °C; NO conversions are the highest among all temperatures (94.3% for VN20g5,

92.6% for VN20g7). Modification with Cu slightly improves catalytic performance of polymer-modified vermiculites at this temperature.

Due to partial subsidence of the structure, the dependence of Cu-modification with other modification cannot be clearly described. Cu species decreased specific surface area, because of their deposition on the surface. Pillaring itself increases S_{BET}, however it may also allow the penetration of pores with active material.

The work was supported by AGH 11.11.210.213.

References

1. B. Samojeden, Paliwa i energia XXI wieku, *Wpływ modyfikacji węgli aktywnych związkami azotu na ich właściwości fizykochemiczne*, 519 – 536, (2014)
2. B. Samojeden, J. Klinik, T. Grzybek, H. Papp, *Charakterystyka modyfikowanych węgli aktywnych jako katalizatorów w reakcji DeNOx*, Gospodarka Surowcami Mineralnymi **24,** 295 - 303, 2008
3. B. Samojeden, Paliwa i energia XXI wieku, *Węgle aktywne w procesach usuwania tlenków azotu z gazów odlotowych ze źródeł stacjonarnych*, 426 - 431, (2014)
4. L. Chmielarz, A. Kowalczyk, M. Michalik, B. Dudek, Z. Piwowarska, A. Matusiewicz, App. Clay Sci. **49**, 156 - 162, 2010
5. G. Lagaly, K. Beneke, Colloid Polym Sci **269**, 1198 – 1211, 1991
6. B. Samojeden, T. Grzybek, Energy **116**, 1484 – 1491, 2016
7. M. Motak, Cat. Today **137**, 247-252, 2008
8. L. Chmielarz, P. Kuśtrowski, M. Michalik, B. Dudek, Z. Piwowarska, R. Dziembaj, Cat. Today **137**, 242 - 246, 2008
9. L. Chmielarz, P. Kuśtrowski, Z. Piwowarska, B. Dudek, B. Gil, M. Michalik, App. Cat. B: Environmental **88**, 331 - 340, 2009
10. L. Chmielarz, Z. Piwowarska, P. Kuśtrowski, B. Gil, A. Adamski, B. Dudek, M. Michalik, App Cat B: Environmental **91**, 449 - 459, 2009
11. Brunauer S, Emmett PH, Teller E, J Am Chem Soc **60**, 309–319, 1938

E3S Web of Conferences 14, 02021 (2017) DOI: 10.1051/e3sconf/20171402021
Energy and Fuels 2016

The impact of co-firing sunflower husk pellets with coal in a boiler on the chemical composition of flue gas

Monika Zajemska[1,*], *Paweł* Urbańczyk[2], *Anna* Poskart[1], *Dariusz* Urbaniak[3], *Henryk* Radomiak[1], *Dorota* Musiał[1], *Grzegorz* Golański[4] and *Tomasz* Wyleciał[1]

[1]Czestochowa University of Technology, Faculty of Production Engineering and Materials Technology, Department of Industrial Furnaces and Environmental Protection, 19 Krajowej Av., 42-201 Czestochowa, Poland
[2]Office of Technical Inspection, 8 Przybylaka Str., 41- 300 Dąbrowa Górnicza, Poland
[3]Czestochowa University of Technology, Institute of Thermal Machinery, 21 Krajowej Av., 42-201 Czestochowa, Poland
[4]Czestochowa University of Technology, Faculty of Production Engineering and Materials Technology, Institute of Materials Engineering, 19 Krajowej Av., 42-201 Czestochowa, Poland

Abstract. The aim of the article is the offer of control of environmentally undesirable products of coal and biomass co-firing with the use of numerical methods. The article analyzes the impact of co-combustion of coal and pellets made of sunflower husk in power plant boiler with a capacity of 125 MW on the composition of the flue gas. Based on the data obtained from a real object, in particular: the combustion temperature, the reagent fluxes, i.e. air, coal and biomass, as well as the elemental composition of the fuels supplied into the boiler, the initial conditions were formulated and the computer simulations were made, leading to determining the chemical composition of flue gas. Numerical calculations were performed using the latest version of CHEMKIN-PRO software. Chemical mechanism was developed by the Italian group The CRECK Modeling and it included 134 compounds and 4169 chemical reactions, including reactions of forming chlorine compounds. Additionally, the impact of selected parameters, e.g. temperature on the chemical composition changes of the combustion products was analyzed. Developed calculation model with the implemented chemical mechanism was verified by data collected from the real conditions, leading to increasing the reliability of the results.
The calculations showed that the most important influence on the composition of the flue gas from the co-firing process of coal with sunflower husk has a composition of biomass. It should be emphasized that the results of computer simulations obtained by the authors have an useful aspect and can be applied in practice, especially to the analysis of the mechanism of chloride corrosion which is possible to occur due to the chlorine content in the biomass. They may also be useful for evaluating the unburned hydrocarbons produced by combustion of rich mixtures ($\lambda < 1.0$).

* Corresponding author: zajemska@wip.pcz.pl

1 Introduction

Co-firing of biomass with coal can be considered at the moment for not only economical, but also an ecological way of its energy use. Noteworthy is the fact that the co-firing does not require costly modernization of existing energy systems. One of the most important factor is the appropriate share of biomass in the fuel mixture [1-4].

The biomass have the greatest practical importance in the power industry, where the combustion process, in spite of its complexity and the many difficulties both logistics (transport and storage of biomass) and technical (preparation of biomass to combustion, i.e. drying, grinding, supplying, etc.), is carried out on a large scale [1-3, 5]. Most of the problems with the energetic use of biomass is associated with the disadvantages of biomass (e.g. large and variable, depending on many factors, moisture content), and the considerable differences in physicochemical properties compared to coal [6-9]. This variety is especially present in the content of the element carbon (in the biomass it is almost two times less than in the coal) and element oxygen (about four times greater the content in the biomass), and the volatiles, as well as the moisture. The presence of chlorine, potassium, sodium and phosphorus is significant, because these elements can cause many operating problems during combustion process such as slagging, fouling and corrosion [9-13].

In the power engineering most commonly used is biomass from plants in the form of pellets. In the recent times the silvers, woodchips, sawdust, straw of cereal and rape as well as pellets made of sunflower husk are popular [14-17]. Their biggest advantages are the high caloric value of up to 21 MJ/kg and a relatively low chlorine content compared to other types of biomass [10, 14, 15, 18].

The process of biomass co-firing due to its diverse chemical composition has a significant impact on emissions of pollutants such as SO_2 and NO_x, as well as the chlorine compounds (e.g. HCl) [19, 20]. Therefore, it is extremely important to monitoring the chemical composition of flue gas during biomass combustion, in particular sulphur, chloride or potassium. Only the appropriate selection of the quantity and quality of biomass can contribute to the economical and ecological effects [3, 5, 10].

The results of calculations, in particular the estimation of HCl content in the exhaust gas can be extremely useful in the real conditions, namely during the assessment of the risk of corrosion being a result of co-combustion of fuels containing chlorine in their composition. Furthermore, this results can be an important part of discussion on the modernization of existing combustion chambers, as well as these newly designed.

2 The procedure of modeling

Calculations of the chemical composition of the products of coal and biomass co-firing were made using the latest version of CHEMKIN-PRO software. The chemical mechanism was taken from the website [21] of The Creck Modeling Group. The extended chemical mechanism used in calculations is adapted for burning of solid fuels, in particular biomass. The reactions set includes reactions of formation of both the NO_x and SO_2, as well as the chlorine. It should be noted that the calculations with simplified mechanisms were also performed, but the results showed significant differences in the measured and calculated values of concentrations. Whereas the best agreement with experimental results was obtained for the mechanism proposed in this article. The chemical mechanism of combustion included 134 compounds and 4169 reactions. The Perfect Stirred Reactor (PSR) was used in the calculation model.

Initial conditions were based on data from a power (steam) boiler, OP-140, operating in one of the national heating plant, namely:

 DOI: 10.1051/e3sconf/20171402021

- dimensions of the combustion chamber: a cross-section of the combustion chamber 6.055 x 6.055 m; height of the cuboid 21.75 m;
- boiler power - 112.5 MW (in steam); 128 MW (in fuel);
- the amount, power and type of built-up burners: 4 pcs. of dust burners, burner heat output of 32 MW;
- average flow of fuel, coal - 5.10 kg/s biomass - 0.53 kg/s;
- average flow of air - 37.02 m^3/s;
- temperature of the preheated air - 246°C;
- average temperature of the exhaust gas in the chamber - 1373 K;
- excess air ratio λ= 1.47;
- technical and elemental analysis of coal and biomass (Table 1);
- average flue gas composition: O_2 = 6.7%; CO_2 = 13%; CO = 29 ppm; NO = 153 ppm; SO_2 = 320 ppm; NO_x = 161 ppm.

Table 1. Technical and elemental analysis of coal and pellets made of sunflower husk.

Fuel	Q kJ/kg	M %	A %	C %	N %	S %	H %	O %	Cl %
coal	22361	12.10	17.64	55.60	0.90	0.46	3.50	9.80	-
pellet	16466	13.60	5.30	44.07	1.15	0.18	5.332	30.36	0.078

3 The results of computer simulations

As a result of the calculation the chemical composition of the flue gas at the outlet of boiler was obtained. Only selected compounds were examined, namely, NO, SO_2, CO_2, CO, O_2 and HCl. The calculated concentration values were compared with measured values (Fig. 1).

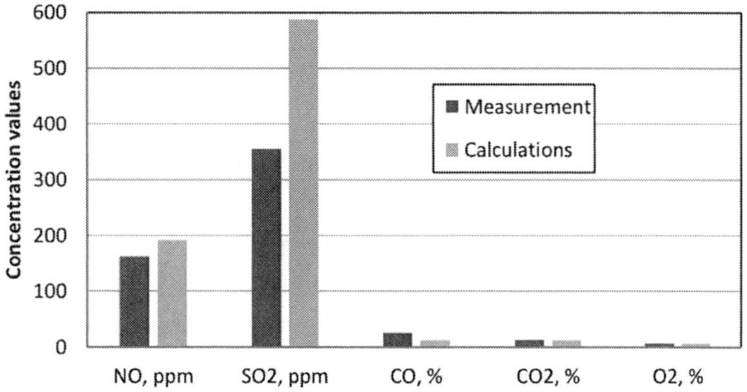

Fig. 1. Comparison of selected results of the calculated and measured values.

Fig. 1 shows that the calculated concentrations are in most cases close to the values obtained from real conditions. The biggest differences were observed for sulfur dioxide. The calculated concentration of sulfur dioxide in the flue gas was higher by about 200 ppm from measured concentrations and reached a value of 588 ppm and 355 ppm respectively. Higher levels were also observed for nitric oxide, however not so large as in the case of SO_2, namely the calculated concentration reached value of 192 ppm, and the measured was equal 162 ppm. Values of other concentrations, i.e. CO, O_2 and CO_2 were comparable.

DOI: 10.1051/e3sconf/20171402021

Occurred discrepancies can be explained primarily by the fact that the elemental composition of biomass is quite varied, which depends on its suppliers (Table 2). Much higher concentrations of SO_2 observed during calculations what is associated with the sulfur content in the pellet of sunflower husk. As is apparent from Table 2, the sulfur content in the biomass varies from 0.1 to 0.25%. For the calculations a sulfur content equal to 0.18 was applied while in fact this value in a burned fuel could be much lower.

Table 2. Technical and elementary analysis of pellets made of sunflower husk, depending on the supplier.

	Supplier 1	**Supplier 2**	**Supplier 3**
W, %	8.9	8.7	11.3
A, %	5.2	4.2	3.4
V, %	68.0	69.2	66.8
Q, MJ/kg d. m	16.9	17.3	16.3
	Content of component, %		
C	44.94	45.27	44.17
H	5.37	5.52	5.21
N	1.06	1.66	0.82
S	0.17	0.253	0.09
Cl	0.071	0.03	0.05
F	0.023	0.01	0.02
Na	0.021	0.01	0.01
K	1.402	1.02	0.89
O	34.31	34.35	34.96

Due to the fact that the combustion process in the analyzed boiler occur in conditions of varying temperatures being in the range of 1300 to 1440 K additionally the effect of temperature on the chemical composition of selected combustion products was analyzed. As an example the mole fraction of NO, SO_2 and HCl were presented in the Figs. 2-4, respectively.

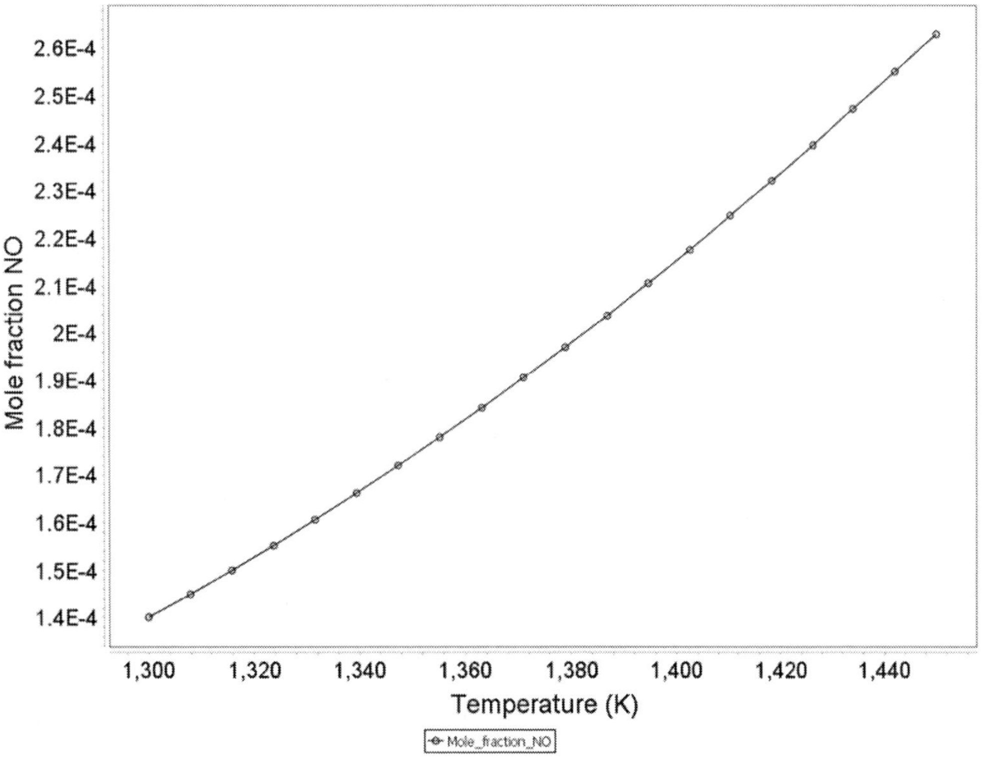

Fig. 2. The change of the molar fraction of NO as a function of the combustion temperature.

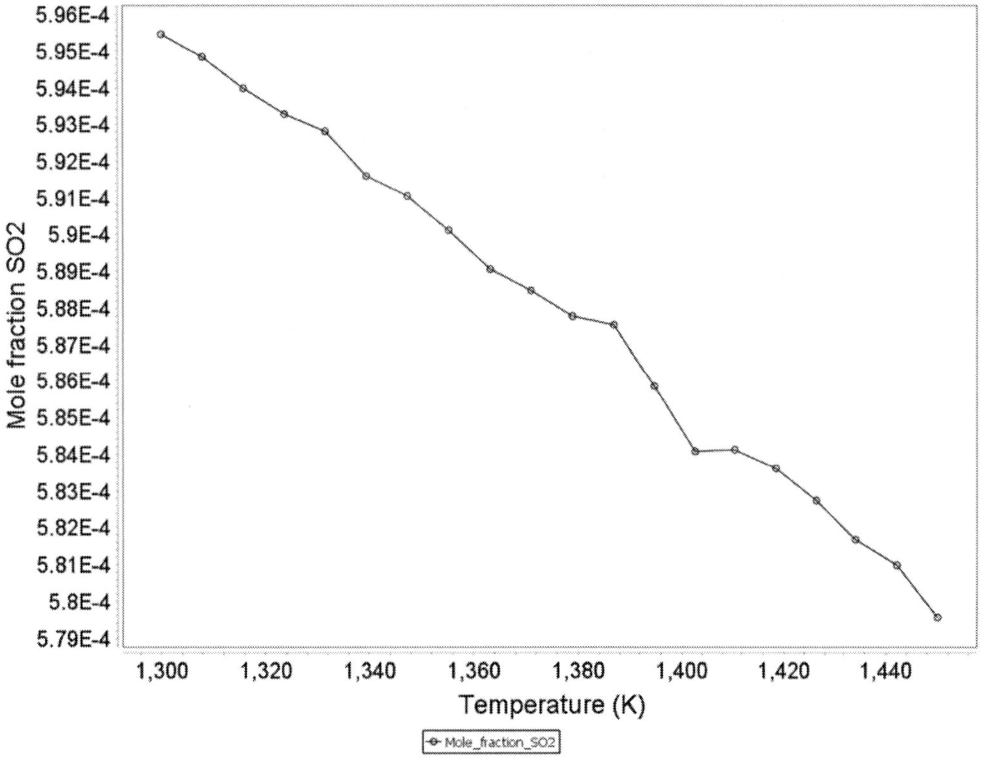

Fig. 3. The change of the molar fraction of SO_2 as a function of combustion temperature.

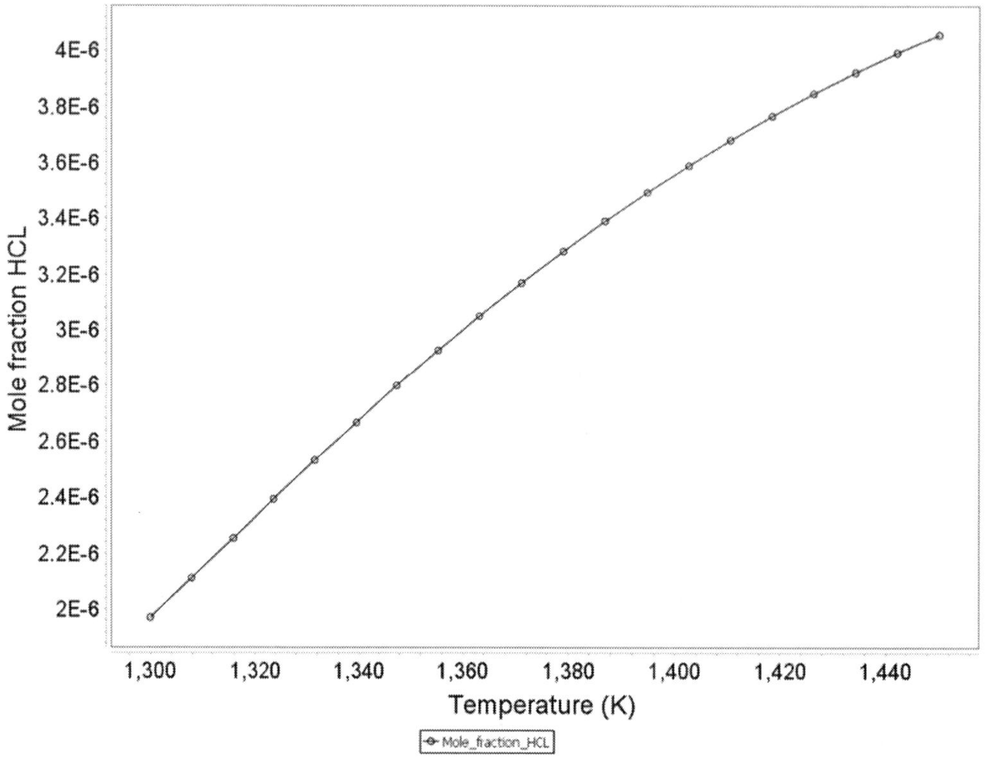

Fig. 4. The change of the molar fraction of HCl as a function of combustion temperature.

As can be seen from the Fig. 2-4 the effect of temperature is most noticeable in the case of NO. As shown in Fig. 2 the NO concentration increases with temperature and for the highest analyzed temperature, i.e. 1440 K it reaches a value of 260 ppm, while at a temperature of approx. 1300 K this value was slightly above 140 ppm. In the case of SO_2 concentration (Fig. 3) no significant difference in the concentration values as a function of temperature was observed. For all the analyzed temperature range, these values vary from 579 to 596 ppm. The situation is similar in the case of HCl (Fig. 4). The change of temperature for studied range of temperature did not significantly affect on the change in concentration of HCl. The concentration values are within the range between 2 to 4 ppm.

4 Summary

The calculations showed that a decisive influence on the composition of the flue gas from the co-firing process of coal with sunflower husk has a composition of biomass. Not without significance is also the combustion temperature, which significantly affects the concentration of nitrogen oxides in flue gas (Fig. 2).

The greatest differences in the calculated and measured concentrations was observed for sulfur dioxide in flue gases. The calculated values were by approx. 40% higher than the measured values. The reason for such a large discrepancy was an assumption in the calculations a relatively high sulfur content of the fuel in comparison to others (Table 2). The smallest differences were recorded for oxygen and carbon dioxide, which proves the correctness of the model adopted for the calculations, as well as the chemical mechanism.

In order to avoid the chloride corrosion risk a particularly important was to evaluate the content of HCl in the flue gas. It has been shown that the temperature in the analyzed range i.e. 1300-1440 K does not significantly affect HCl formation.

It should be emphasized that t he use of CHEMKIN-PRO software in the calculations can be very useful for the evaluation of biomass co-firing products at higher its content than analyzed in the article (10% mass fraction, 7.5% share of the heat). Leading combustion of biomass with greater than 10% share in the mixture, due to the enhancement of such phenomenon as corrosion, special control is required.

In addition, due to the fact that the CHEMKIN-PRO software allows for the analysis of the chemical mechanisms of the formation of compounds responsible for the corrosion, it can be used at the stage of design of boilers, in particular in regard to the proper selection of construction materials.

References

1. A. Zyga, *Współspalanie jako metoda konwersji biomasy.* Materiały konferencyjne „Nowoczesna Energetyka Europy Środkowo-Wschodniej 2015". *http://kne.itc.pw.edu.pl/attachments/article/154/Agnieszka%20Zyga.pdf*
2. T. Golec, R. Lewtak, B. Świątkowski, B. Glot, Czysta Energia **9**, 26–29 (2010)
3. K. Nowak, D. Proszak-Miąsik, S. Rabczak, JCEEA, **4**, 147–154 (2014)
4. S. Tokarski , K. Głód, M. Sciążko, J. Zuwała, Energy **92**, 24–32 (2015)
5. S. Grądziel, K. Majewski, Piece Przemysłowe & Kotły **V-VI**, 37–43 (2013)
6. K. Han, S. Niu, CH. Lu, Process Saf. Environ. **88**, 425–430 (2010)
7. B.M. Jenkins, L.L. Baxter, T.R. Miles, T.R. Miles, Combustion properties of biomass. Fuel Process. Technol. **54**, 17–46 (1998)
8. R. Saidur, E.A. Abdelaziz, A. Demirbas, M.S. Hossain, S. Mekhilef, Renew. Sust. Energ. Rev. **15**, 2262–2289 (2011)
9. J. Mółka, B. Łapczyńska – Kordon, Inżynieria rolnicza **6**, 141–147 (2011)
10. N. Bątorek-Giesa, B. Jagustyn, Ochrona Środowiska i Zasobów Naturalnych **40**, 396–401 (2009)
11. E. Cardozo, C. Erlich, L. Alejo, H. Torsten, Fuel **115**, 778–787 (2014)
12. A. Magdziarz, A.K. Dalai, J.A. Koziński, Fuel **6**, 135–145 (2016)
13. G. Golański, J. Kępa, *Nowoczesne stale dla energetyki – charakterystyka* (Wydawnictwo Politechniki Częstochowskiej, Częstochowa, 2011)
14. V.K. Verma, S. Bram, F. Delattin, P. Laha, I. Vandendael, A. Hubin, J. De Ruyck, Appl. Energ. **90**, 17–23 (2012)
15. R. Lopez, C. Fernandez, J. Fierro, J. Cara, O. Martínez, M.E. Sanchez, Energy **74**, 845–854 (2014)
16. A.A. Zabaniotou, E.K. Kantarelis, D.C. Theodoropoulos, Bioresource Technol. **99**, 3174–3181 (2008)
17. K. Sirisomboon, P. Charernporn, J. Energy Inst. **30**, 1-8 (2015)
18. S.V. Vassilev, D. Baxter, Ch.G. Vassileva, Fuel **117**, 152–183 (2014)
19. A. Magdziarz, M. Wilk, M. Zajemska, Chem. Process Eng. **32**, 423–433 (2011)
20. A. Magdziarz, M. Wilk, Rynek Energii **99**, 127–130 (2012)
21. *http://creckmodeling.chem.polimi.it/*

DOI: 10.1051/e3sconf/20171402022

Investigations of torrefied biomass grindability using a modified Hardgrove test

Mateusz Tymoszuk[1,*]

[1]Division of Boilers and Steam Generators, Institute of Power Engineering and Turbomachinery, Silesian University of Technology, Konarskiego 20, 44-100 Gliwice, Poland

Abstract This paper presents results of investigations of torrefied biomass grindability using a modified Hardgrove test. The following types of torrefied biomass were used during the testing: alder chips, palm kernel shells and willow chips. Tests were conducted in a standard Hardgrove test mill for 50 cm^3 volumetric samples (the same volume as applied in [1]). Volumetric samples of raw biomass were also tested for comparison purposes. Two ranges of the sample particle size were applied during the investigations: 0.6-1.18 mm (as in [2]) and 0.2-1.18 mm. As opposed to the standard Hardgrove test, the cumulative passing mass of the product was determined for a wider range of passing of standard sieves. The obtained results point to an increase in the grindability of torrefied fuels compared to raw biomass. They can be used to categorize torrefied biomass fuels due to their grindability, and as basic information in semi-technical scale studies on torrefied biomass comminution features.

1 Introduction

Economic and legal conditions of the Polish power industry enforce higher ($\geq$ 20 %) participation of biomass in the thermal energy mix per power unit. If biomass fuels are fired, the anticipated deterioration in operating conditions requires an appropriate output of the biomass milling plant, a proper grain size of pulverized biomass and also proper operation of the power boiler (while maintaining acceptable levels of combustible fractions in fly ash and deposits on the boiler heating surfaces). The biomass quality is crucial for all those operating parameters, which necessitates adequate categorization of biomass fuels in terms of their suitability for the power industry utilization [3, 4].

The majority of PC boilers used for energy production are equipped with coal preparation installations based on ring-ball and bowl mills. Milling and co-milling of large amounts of biomass in this type of pulverizers deteriorates the mill output and makes it very difficult to ensure the proper particle size distribution of the product fed into the combustion chamber [5]. Contrary to brittle materials, such as coal, the raw biomass anisotropic and fibrous structure makes milling by compression ineffective [6, 7]. Such plants would have to be replaced with installations based on knife or beater mills, characterized by a higher output and lower energy consumption of the biomass milling process. This would result in high investment expenditures and the mill installation higher maintenance costs. The other

* Corresponding author: mateusz.tymoszuk@polsl.pl

approach is to change comminuted biomass properties through the torrefaction process, which makes it possible to improve the pre-treated fuel grinding properties [8].

Torrefaction is a process of heat treatment of raw biomass in an inert or low-oxygen atmosphere at approximately atmospheric pressure and a temperature of about 220-350°C. The process is carried out to achieve the desired properties with minimal losses of energy, resulting mainly from the loss of volatiles. The processed biomass is characterized by higher grindability (and thus less energy is needed for fine grinding), takes hydrophobic properties and displays higher resistance to biological agents [9, 10]. The torrefaction process product grindability depends on many factors, such as the torrefaction reactor structure, the process medium, temperature and time of exposure.

A material grindability is defined as the ratio between energy consumed during the comminution process and the process effect: the material cumulative retained (passed) amount for a defined mesh size, the increased surface of the ground material or the increase in the content of fine particles (in a defined particle size range). The knowledge of mechanical properties, such as the modulus of elasticity, compressibility, strength and micro-hardness, is not sufficient to determine the material grindability. Therefore, the fuel grindability is determined by means of model testers utilizing the same grinding mechanisms as industrial mills. Grindability determination methods are based on different comminution theories, such as Kick`s, Rittinger`s and Bond`s [11].

The most commonly used grindability test for coals is the Hardgrove Grindability Index [2], which is used to predict the mill output, performance and energy requirements, and – qualitatively – also the particle size distribution after milling. The Hardgrove Grindability Index (HGI) test is based on Rittinger's theory stating that the energy needed to reduce the size is proportional to the new surface area generated in the comminution process [11]. The Hardgrove grindability test is performed using a laboratory test mill (cf. Fig. 1) consisting of a stationary grinding bowl made of hardened iron or steel, with a horizontal ball track, in which eight 1" balls run along a circular path. The balls are driven from above with a rotating ring at a speed of 20 rpm ±1 rpm. A definite pressure of 284 N ±2 N on the balls is obtained by weights, the shaft, the top grinding ring and gear. An adjustable counter is used to stop the mill automatically after reaching 60 ±0.25 revolutions. Before performing the test, it is advisable to calibrate the device following the procedure described in [2].

Fig. 1 Standard Hardgrove grindability tester

In order to determine the HGI for 50 g ±0.01 g of air-dried material, which has a limited particle size range of 0.6-1.18 mm, the sample is ground in the grindability machine. After the upper ring completes 60 ±0.25 revolutions, the grounded sample is sieved through a 75 μm passing sieve. The received cumulative passed mass G_{75} is the basis for determining the Hardgrove Grindability Index, according to the following equation:

$$HGI = 13 + 6.93 \cdot G_{75} \tag{1}$$

HGI – Hardgrove Grindability Index for the tested material,
G_{75} – cumulative mass passing through a 75 μm mesh sieve [g].

As a general rule, higher HGI values mean that the fuel is easier to grind, requiring lower power inputs and giving higher mill outputs.

As fuel, biomass is characterized by a wide range of densities, depending on the biomass type, growth conditions and pre-treatment. According to research conducted by Joshi [12] and Agus and Waters [13], mills are volumetric devices and the traditional HGI method favours denser materials with small volumes. Therefore, performing HGI tests with a fixed mass approach (50g) for fuels such as biomass can be inappropriate [14]. In order to even out grindability ratings of different types of biomass and coal, Bridgeman et al. [1] used the fixed volume approach (50 cm^3) for each sample as opposed to fixed mass (50 g).

This paper presents results of investigations of torrefied biomass grindability using a modified Hardgrove test, with the volumetric sample approach and with the test sample wider particle size range of 0.2-1.18 mm. The HGI sample standard particle size range of 0.6-1.18 mm can be insufficient for biomass samples due to the inhomogeneous character of biomass fuels. A narrow particle size range can lead to high amounts of mass being rejected from the initial sample, which could have a negative effect on the obtained results reliability.

2 Materials and Methods

The following types of torrefied biomass were used during the testing: alder chips (**T1**), palm kernel shells from two different manufacturers (raw biomass came from the same batch, **T2** and **T3**) and willow chips (**T4**). The tests were conducted in a standard Hardgrove test mill for 50 cm^3 volumetric samples (the same volume as applied in [1]). Volumetric samples of raw biomass (marked as **R1**, **R2-3**, **R4** and corresponding to respective samples marked with T denotation) were also tested for comparison purposes. Two ranges of the torrefied sample particle size were applied during the investigations: 0.6-1.18 mm (as in [2]) and 0.2-1.18 mm. For raw biomass, only the 0.6-1.18 mm particle size range was investigated. Compared to standard Hardgrove tests, the other testing parameters remained unchanged.

Biomasses T1, T2 and T4 were torrefied by the same manufacturer in the following conditions: the process temperature: 350°C in an atmosphere with low-oxygen concentrations (tor-gas) and close to atmospheric pressure (pressure slightly lower than atmospheric: 0.8-1 Pa), the torrefaction process duration: 10 min; the torrefaction reactor fumes were used as tor-gas to fuel the process.

Biomass T3 was torrefied in the following conditions: the process temperature: 310°C in an atmosphere with low-oxygen concentrations (tor-gas) and close to atmospheric pressure, the torrefaction process duration: 40 min; the torrefaction reactor fumes were used as tor-gas to fuel the process.

The Hardgrove grindability index equation is developed for coals and 50 g samples. For this reason, the Hardgrove formula application for biomass and torrefied biomass volumetric samples necessitates a modification of the standard formula. In this paper, grindability is expressed through a comparison of the cumulative passed percentage for raw and torrefied biomass, for a set of sieves with selected mesh sizes (fractions not occurring in the pre-milled sample). Additionally, the modified grindability index HGI_{modif}, corresponding to the standard HGI was determined using the following formula:

$$HGI_{modif.} = 13 + 6.93 \cdot \frac{P_{75}}{2} \quad (2)$$

HGI – Hardgrove Grindability Index for the tested material,
P_{75} – cumulative percentage passed through a 75 μm mesh sieve [%].

Representative initial samples (about 2.5-3 kg) of raw and torrefied biomasses for the particle size ranges of 0.2-1.18 (torrefied) and 0.6-1.18 (raw and torrefied) were prepared in a knife mill with adjustable rotational speed and replaceable sieves in the range of 1 to 12.5 mm, using the gradual pulverization method; each time the required grain size fraction was sifted out carefully using a vibratory sieve shaker. The desired size fractions were sifted out of the biomass initial samples, prior to comminution in the knife mill. It should be mentioned that the biomass initial samples were characterized by a high amount of fine fractions.

For each raw and torrefied biomass three 50 cm^3 samples were milled in the Hardgrove tester. Afterwards, a sieve analysis of the obtained product was performed using a pneumatic siever. The content of fractions produced during the test (and not occurring in the pre-milled sample) is determined with a set of the following sieve meshes:

- for 0.6-1.18 mm samples, mesh sizes: 75, 90, 125, 180, 200, 355, 500 μm,
- for 0.2-1.18 mm samples, mesh sizes: 75, 90, 125, 180 μm.

The results of the basic technical analysis of raw and torrefied biomasses are shown in Table 1 and Table 2, respectively.

Table 1 Raw biomass basic technical analysis

Item	Parameter	Symbol	Unit	R1	R2-3	R4
1	Moisture content (as received)	W^{as}	%	26.1	20.5	29.4
2	Moisture content (air-dried $HGI_{modif.}$ samples)	W^{HGI}	%	7.7	8.8	8.6
3	Higher heating value (dry)	HHV^d	MJ/kg	19.9	20.6	19.7
4	Ash content (dry)	A^d	%	0.6	2.54	2.5
5	Volatile fractions	V^{daf}	%	77.4	73.8	72.0
6	Weight of 50 cm^3 $HGI_{modif.}$ sample	m	g	12.2	37.6	9.3
7	Bulk density of 50 cm^3 $HGI_{modif.}$ samples	ρ_b	kg/m^3	244.4	752.0	185.7

DOI: 10.1051/e3sconf/20171402022

Table 2 Torrefied biomass basic technical analysis

Item	Parameter	Symbol	Unit	T1	T2	T3	T4
1	Moisture content (as received)	W^{as}	%	16.2	14.6	8.3	17.7
2	Moisture content (air-dried $HGI_{modif.}$ samples)	W^{HGI}	%	8.4	7.1	6.4	8.1
3	Higher heating value (dry)	HHV^{d}	MJ/kg	25.2	25.0	22.8	23.9
4	Ash content (dry)	A^{d}	%	1.8	6.6	5.4	4.3
5	Volatile fractions	V^{daf}	%	54.6	51.2	57.1	58.0
6	Weight of 50 cm^3 $HGI_{modif.}$ sample	m	g	10.6	32.6	34.9	9,6
7	Bulk density of 50 cm^3 $HGI_{modif.}$ samples	ρ_b	kg/m^3	212	652	698	192

The high contents of moisture in biomasses T1, T2 and T4 for the as-received state are the effect of water-spraying cooling of torrefied biomass leaving the reactor.

3 Test results

The results of raw and torrefied biomass grindability testing based on the modified Hardgrove test are listed in Table 3 and Table 4 for the particle size range of 0.6-1.18 mm, and in Table 5 – for the particle size range of 0.2-1.18. The product particle size distributions shown in the tables are based on averaged results of three grindability tests, for each biomass type.

Table 3 Content of fractions produced during the test in the raw biomass product for 0.6-1.18 mm samples

cumulative % passed	Unit	R1	R2-3	R4
P_{75}	[%]	6.04	2.17	5.26
P_{90}	[%]	6.48	2.60	5.99
P_{125}	[%]	7.46	3.16	7.38
P_{180}	[%]	8.44	3.62	9.78
P_{200}	[%]	8.78	3.97	10.53
P_{355}	[%]	12.30	5.50	17.03
P_{500}	[%]	18.50	8.07	23.97
$HGI_{modif.}$	[-]	33.9	20.5	31.2

Table 4 Content of fractions produced during the test in the torrefied biomass product for 0.6-1.18 mm samples

cumulative % passed	Unit	T1	T2	T3	T4
P_{75}	[%]	67.4	12.4	6.1	45.0
P_{90}	[%]	70.6	14.3	1.2	50.3
P_{125}	[%]	73.9	17.3	1.7	56.0
P_{180}	[%]	76.8	22.7	3.2	60.7
P_{200}	[%]	78.1	25.4	1.7	63.0
P_{355}	[%]	84.4	39.7	10.5	73.6
P_{500}	[%]	88.8	51.9	13.2	81.1
$HGI_{modif.}$	[-]	246.7	56.1	34.3	168.9

Table 5 Content of fractions produced during the test in the torrefied biomass product for 0.2-1.18 mm samples.

cumulative % passed	Unit	T1	T2	T3	T4
P_{75}	[%]	59.5	10.3	6.2	47.8
P_{90}	[%]	65.9	11.9	7.2	52.0
P_{125}	[%]	70.5	14.9	8.8	56.6
P_{180}	[%]	75.3	20.0	11.7	63.9
$HGI_{modif.}$	[-]	219.2	48.8	34.4	178.8

4 Discussion of results and conclusion

The comparison of the modified Hardgrove test results obtained for torrefied and raw biomass samples points to a significant increase in the biomass grindability index after the torrefaction process. The torrefied biomass cumulative percentage passed for the 75 μm mesh size is approximately ten times higher for alder chips, six and three times higher for palm kernel shell T2 and T3 samples, respectively, and nine times higher for willow chips.

The results obtained for the 0.6-1.18 and 0.2-1.18 particle size ranges both show similar tendencies in the torrefied biomass grindability categorization. Nevertheless, respective values of $HGI_{modif.}$ (based upon P_{75}) are slightly different for samples T1, T2 and T4. Only for sample T3 does $HGI_{modif.}$ remain unaffected by the change in the sample particle size range. A wider grain size range of sample particles can cause suppression of the milling effect due to the high energy demand for the fine fraction grinding (samples T1 and T2), which is not necessarily a negative factor. In most industrial milling installations, coal and biomass are fed into the mill with a relatively high amount of fine fractions. In the case of the torrefied willow sample, P75 is increased for the 0.2-1.18 mm particle size range, which may be the effect of the higher content of brittle particles in the sample wider range.

It has to be noted that torrefied biomass with high $HGI_{modif.}$ values, in spite of its high grindability, can be problematic if comminuted in industrial mills. The HGI tester is a batch device and cannot be used to predict a continuous milling circuit precisely. For milling low-density materials, such as torrefied biomass, structural changes in the industrial mill classifier could be necessary for the circuit proper operation.

The obtained results can be used to categorize torrefied biomass fuels due to their grindability and as basic information in semi-technical scale studies on torrefied biomass comminution features.

Investigations presented in this work were financed within the KIC InnoEnergy innovation Project "Biocoal for Power Generation".

References

1. T.G. Bridgeman, I. Jones, P. T. Williams, D. J. Waldron, *An investigation of the grindability of two torrefied energy crops*, Fuel **89**, 3911-3918, 2010.
2. British Standard BS 1016-112:1995.
3. Draft regulation of the Minister of Economy of 19 July 2010 (version 1.4) on detailed scope of obligations to obtain origin certificates and submit them for redemption, payment of a substitution fee, purchase of electricity and heat generated from renewable energy sources as well as on the obligation to confirm the data on the amount of electricity produced from a renewable energy source.
4. P. Basu, C. Kefa, L. Jestin, *Boilers and Burners - Design and Theory*, Springer, New York 2000.
5. K. Mroczek, *Analysis of coal mill operation*, Chemical and Process Engineering **30**, 83–98, 2009.
6. J.S. Tumuluru, C.T Wright., J.R. Hess, K.L. Kenney, *A review of biomass densification systems to develop uniform feedstock commodities for bioenergy application*, Biofuels, Bioproducts and Biorefining **5**, 683-707, 2011.
7. M. Yu, Womac, L. O. Pordesimo, *Review of Biomass Size Reduction Technology*, ASAE Annual International Meeting, Las Vegas, Nevada 2003.
8. D. T. Van Essendelft, X. Zhou, B. S. –J. Kang, *Grindability determination of torrefied biomass materials using the Hybrid Work Index*, Fuel **105**, 103–111, 2013.
9. J. Li, A. Brzdekiewicz, W. Yang, W. Błasik, *Co-firing based on biomass torrefaction in a pulverized coal boiler with aim of 100% fuel switching*, Applied Energy, **99**, 344-354, 2012.
10. K. Malińska, *Biowęgiel odpowiedzią na aktualne problemy ochrony środowiska [Biocoal as response to current environmental protection problems]*, Inżynieria i Ochrona Środowiska, **15**, No 4, 387-403, 2012.
11. A. Jankovic, H. Dundar, R. Mehta, *Relationships between comminution energy and product size for a magnetite ore*, Journal of the South African Institute of Mining and Metallurgy **110**, Issue 3, 141-146, 2010.
12. N. R. *Joshi Relative grindability of bituminous coals on volume basis*, Fuel **58**(6):477-478, 1979.
13. F. Agus, P. Waters, *Determination of the grindability of coals, shales and other minerals by a modified Hardgrove-machine method*, Fuel, **50**(4):405-431, 1971.
14. O. Williams, C. Eastwick, S. Kingman, D. Giddings, S. Lormorb, E. Lester, *Investigation into the applicability of Bond Work Index (BWI) and Hardgrove Grindability Index (HGI) tests for several biomasses compared to Colombian La Loma coal,* Fuel Volume **158**, 379–387, 2015.

Dielectric loss factor of sintered coal ash

Dorota Nowak-Woźny[1,*], *Leszek* Woźny[2], and *Laura* González Valdés[1]

[1]Faculty of Mechanical and Power Engineering, Wrocław University of Science and Technology, 27 Wybrzeże Wyspiańskiego St, 50-370 Wrocław
[2]Faculty of Electrical Engineering, Wrocław University of Science and Technology, 27 Wybrzeże Wyspiańskiego St, 50-370 Wrocław

Abstract. This work presents the study of dissipation factor (tanδ) for ash prepared from a power plant coal. Values of tanδ were measured for samples of coal ash with additive of biomass and kaolinite, sintered at various temperatures up to 1100°C. Procedure of direct tanδ measurement is described and the results of loss factor as a function of frequency are shown. Quantitative and qualitative changes of tanδ were observed for the different tested samples. The measurement data is compared with SEM microscopic images of the samples sintered at the maximum temperature. The results suggest that direct measurement of tanδ can be a useful tool for slagging and fouling risk assessment.

1 Introduction

The massive built-up of ash deposits formed on heat exchange surfaces during coal, biomass, or other fuel combustion is a very important exploiting problem because it may force to interrupt the energy conversion process in order to clean the boiler's elements. Thus, it is needed to elaborate an assessing method able to predict with precision the slagging and/or fouling tendency of a fuel.

The slagging and fouling process occurs inside the boiler when ash particles collide with the boilers surface and heating tubes, condense on cooler elements and diffuse and react with the construction materials. Properties of such formed deposits depend on the ash temperature, the temperature of the surface on which the ash gathers, and on the fuel's composition.

Sintering phenomena are associated with the bonding or welding processes of the ash particles, which is closely associated with the physico-chemical processes that take place on the ash surface [1]. Because of the complex character of the sintering processes and of the complicated physico-chemical structure of the ash, it is difficult to transfer the real slagging and fouling process into laboratory conditions for precise prediction of slagging and fouling tendency. Deposits formation mechanisms on heating parts of boilers are connected with particular ash particles interactions. The interactions model for particular ash particles takes into account such properties as ash viscosity, ash grain size, ash grains surface tension and heat conduction of the surface at which the ash is gathered.

* Corresponding author: dorota.nowak-wozny@pwr.edu.pl

The ash slagging and fouling process depends strongly on the surface tension and ash viscosity [2]. At high temperatures it appears a liquid phase on the surface of ash particles. For ash fraction of smaller grains the sintering effect starts for higher viscosity [3]. Viscosity and surface tension depend inversely on temperature. The mechanism of deposits formation is described in literature [2-10].

From physical point of view, the ash pellets used in standard tests, behave as ceramic material consisted of grains formed by different materials and with different dimensions and shapes. Consequently, the standard Ash Fusion Test (AFT) – widely used to assess slagging and fouling tendency in coal conversion process – may not represent accurately the sintering processes occurring at micro-scale of material and does not indicate exactly at which temperature the first melt/slag occurs. Contrarily, the first melt/slag point is very important in the real processes occurring in the boiler. The mechanical or pressure drop tests [11] seem to be more adequate because they are based on change of physical properties (mechanical strength and porosity) associated with the ash microstructure. Regarding to chemical reactions and phase transitions, tests based on phase equilibrium calculations [12-16] may be the way more appropriate.

Because of the dielectric properties of the ash, we suggest that the results from AC electrical test, especially dissipation factor versus frequency and sintering temperature, could be very interesting to analyse the problem.

The electrical properties of the ash depend on its microstructure and composition. The main components of the coal ash are aluminosilicates. In the ash it is common to encounter as well unburned coal particles. The existence of these particles may cause the change in electrical behaviour of ash attributed to the polarization effects on the grains surface or to ion conductivity depending on metal cations [17, 18]. In ceramic materials dielectric losses are attributed to presence of electric dipoles and loosely bound ions as well as the space charge polarization related mainly to the granular microstructure of ash [19].

The complex chemical composition of the ashes, their structure, microstructure, and numerous heterogeneities can cause various internal dielectric losses. It has to be reflected in the values of various dielectric parameters and their characteristics depending on the frequency.

The dissipation factor tanδ (also called the loss factor) is the parameter describing the internal friction for moving of coupling chargers (dipoles) and free charges (mainly ions) in dielectric materials. It can be used to determine the general dielectric properties of materials. Measurements of tanδ can be the convenient and relatively simple method for quick diagnostics of various materials with high resistance, including ashes after combustion in electro-energetic power furnaces.

For these reasons, the presented work is an attempt to apply tanδ measurement for characterization of various types of ash samples. The study does not include any other dielectric measurements. The aim was not to evaluate the dielectric properties of the materials but to assess the state of the ash by direct measurements of tanδ. Measurements were conducted for several samples of ash samples preheated at different temperatures T_s. The loss factor measurements were analysed together with scanning electron microscopy (SEM) images.

2 Samples

The ash samples used in this work were obtained from three sort of fuels:

1. C1 hard coal ash,
2. S mixture of 80 wt. % of hard coal ash with 20 wt. % of cereal pellets ash,
3. S1 mixture of 95 wt. % of hard coal ash with 5 wt. % of kaolinite.

The ashes were prepared according to the commonly applied procedure.

The fuels were ground, sieved (500 μm) and next heated in a muffle furnace at 500°C for 48 hours. Next, after re-grinding and mixing the obtained ash, samples were created in form of pellets (5 mm in diameter and about 4 mm in thickness) under a pressure of 1 MPa. Weigh percent of oxide content for tested samples is given in Table 1.

Table 1. Oxide content of samples, wt %.

Samples	C1	S	S1
SiO_2	48.9	47.7	51.2
Fe_2O_3	5.67	5.87	3.41
Al_2O_3	25.4	24.4	28.8
Mn_3O_4	1.06	1.00	0.91
TiO_2	0.08	0.08	0.05
CaO	3.65	3.65	2.24
MgO	2.57	2.86	1.60
SO3	3.57	3.57	2.13
P_2O_5	0.60	2.28	0.59
Na_2O	0.98	0.94	0.64
K_2O	2.79	3.74	2.07

3 Measuring details

Disssipation loss factor tanδ is the convenient parameter for characterization of dielectric materials. It is defined as the power lost in the form of heat in the volume of the material. Its value depend on sample temperature and on frequency and intensity of external electric field.

It can be determined as

$$\tan\delta = G/(\omega\varepsilon), \tag{1}$$

where:

- G – dielectric bulk conductivity,
- ω – pulsation of the external electric field,
- ε – permittivity of the dielectric.

Tested ash samples were sintered initially for 2 hours at fixed temperature of 500°C. After sintering the samples were quenched to room temperature (by dropping to metal plate) in order to freeze the equilibrium state of the ash microstructure and phase composition. As quick as possible the annealed samples were electrically tested. Before each test surfaces of samples were polished for improvement electrical contact. The dissipation factor (tanδ) measurements were carried out at room temperature for applied electrical field frequencies in range of 500 Hz to 1 MHz.

After electrical tests the samples were again sintered at 100 degree higher temperature and again quenched. This procedure was repeated up to 1100°C. Such thermal treatment let us measure the dissipation factor in a thermodynamic equilibrium state. Repetition of sintering of the same sample at higher temperature and annealing up to a new equilibrium state give us the information about the sintering process step by step.

The dissipation factor was measured using QuadTech 7600 Plus Precision LCR Meter. The measured samples were placed between the brass electrodes in low humidity chamber.

DOI: 10.1051/e3sconf/20171402023

The measurements setup is presented in Figure 1. After overall annealed procedure the microstructure of samples was recorded using SEM COXEM EM-30AX.

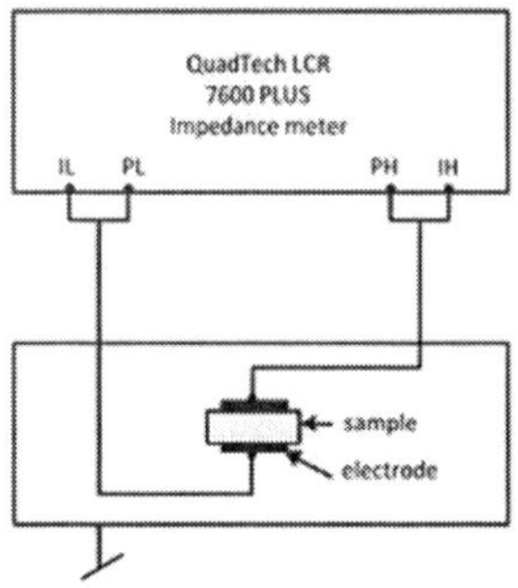

Fig. 1. Measurement setup.

4 Dissipation factor measurements

For samples C1, S and S1 the dependences of tanδ on the frequency for a fixed sintering temperature are presented on Figures 2, 3 and 4 respectively.

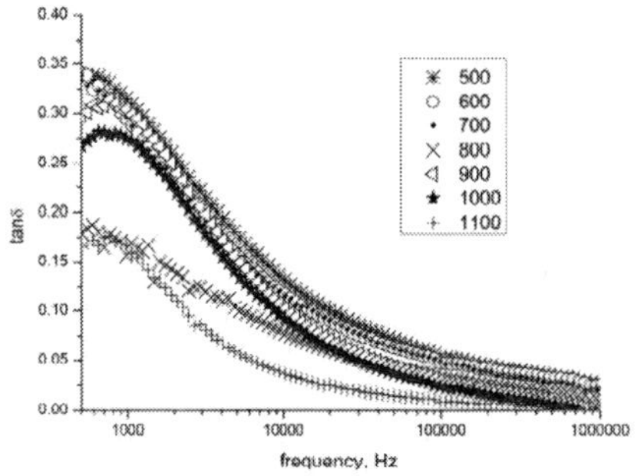

Fig. 2. Dependence of tanδ on electric field frequency for sample C1 sintered at various temperatures.

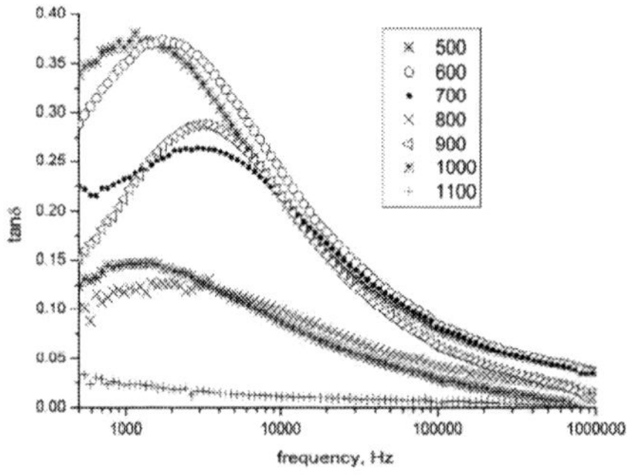

Fig. 3. Dependence of tanδ on electric field frequency for sample S sintered at various temperatures.

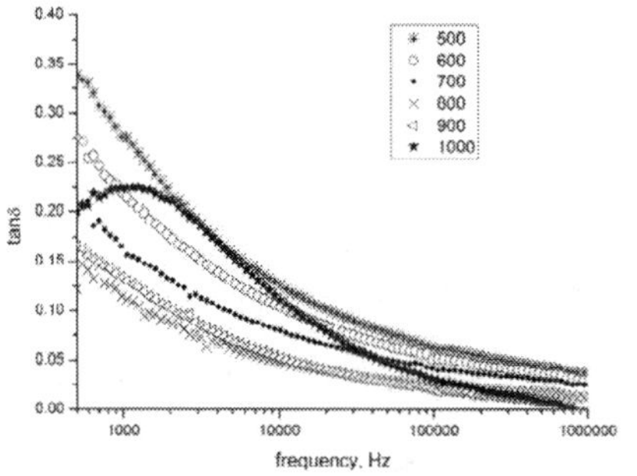

Fig. 4. Dependence of tanδ on electric field frequency for sample S1 sintered at various temperatures.

As it can be seen in Figure 2 (sample C1 – pure coal ash) the plots show a peaking behaviour for all sintering temperatures. The tangent loss characterized by the appearance of the peak suggests the presence of relaxing dipoles in the specimen [20]. It was observed a soft shifting of the peaks towards higher frequencies and a decrease of the value of dissipation factor for higher temperatures. These effects exist and are visible in the graphs until 800°C. At 900°C the tanδ peak significantly increase and for samples sintered at higher temperatures – decreases and slightly shifts to lower frequencies. In all sintering temperature region the relaxation mechanism of energy dissipation is dominant. This suggest that tanδ plot describes the same mechanism at various temperatures [21]. The frequency for maximum value of tanδ changes in the range of 500 Hz to 1 kHz and maximum dissipation factor value ranges from 0.38 to about 0.19.

For sample S (ash of coal and biomass, Figure 3) the character of losses is similar as for pure coal ash (sample C1) but observed changes are more drastic. Dependence of tanδ on frequency for sample sintered at 1100°C has different character suggesting a change of main power loss mechanism, which may be induced by a higher degree of crystallinity at that temperature [20]. Thus, the addition of biomass does not change the conduction mechanism when the samples are sintered at temperatures up to 1000°C. Sintering at 1100°C involves a change in the conduction mechanism, probably associated with the slag phase formation.

For the third sample S1 (ash of coal and kaolinite, Figure 4) there is a more interesting issue. For samples sintered at temperatures in the range of 500 to 900°C all dependences have decreasing character. It suggests the existence of conductive dielectric loss mechanism as a dominant one. Values of tanδ are less than 0.37 for all frequencies. But for the sample sintered at 1000°C a radical change in the nature of losses appears. Dependence of tanδ on frequency has peaking character, suggesting the presence of relaxing dipoles [20]. This change in the trend of the graph may indicate a change of the dominant conduction mechanism which in turn, is temperature dependent.

The detailed explanation of dissipation factor changes for tested samples is very not easy and requires further research.

5 Microstructure observations

The microstructure of the surface for all samples was studied under scanning electron microscope (SEM). The images are shown in Figures 5-7.

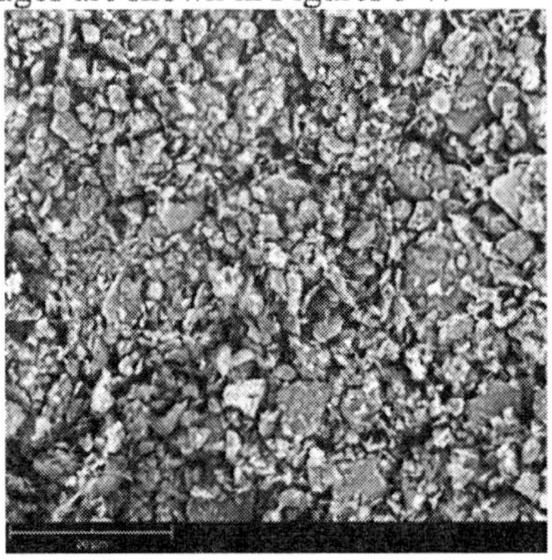

Fig. 5. Microstructure of sample C1 sintered at 1100°C.

Fig. 6. Microstructure of sample S sintered at 1100°C.

Fig. 7. Microstructure of sample S1 sintered at 1000°C.

Sample C1 sintered at 1100°C exhibits a microstructure consisting of grains of dimension up to a few μm and agglomerates with dimensions about 20 μm (Figure 5). The grains and agglomerates are distributed rather uniformly. Some of the grains have visible rounded edges but there are not observed the necks between grains that are characteristic for higher temperatures at which a greater amount of ash is present in slag form. Individual grains are weakly connected to each other. Numerous deep pores are also visible.

The microstructure of sample S is quite different (Figure 6). More homogeneous areas are adjacent to regions completely disordered. The main area of the sample is a solid material as closely interconnected grains by characteristic necks and with little or no pores [22]. Disordered regions are characterized by the existence of deep pores originating specific volume defects. In the whole sample can be seen also isolated single larger grains. These particular sintered grains are connected to each other by the necks which play a significant role in the diffusion paths for atoms and/or ions. In this way there are formed specific conduction channels.

Yet another microstructure is visible for sample S1 (Figure 7). It seems to be less homogeneous than sample C1 with a lower degree of agglomeration. More closely packed and larger individual grains are observed, with dimensions up to about 20 μm. Deep pores are also a characteristic feature of this microstructure.

6 Conclusions

On the basis of the results of studies it can be concluded that:

1. The value of dissipation factor tanδ depends on both the kind of ash and the frequency range of the measurement. Frequency dependence of tanδ is characteristic for the type of ash and the temperature at which the sample of ash was sintered. Therefore, the tanδ measurements can be convenient tool for ash characterization.
2. In the case of ash obtained from pure coal (sample C1) dependence of tanδ on frequency has a resonant character. For sintering temperature up to 800°C frequency of tanδ maximum increases from 500 Hz to about 1 kHz and the value of the maximum dissipation factor decreases. At 900°C it was observed a rapid increase of the maximum dissipation factor and slight slip to the lower frequency. Further sintering at temperatures 900, 1000 and 1100°C leads to continuous decrease in the maximum dissipation factor value. It can be caused by the changes in microstructure, chemical transitions or similar effects.
3. For sample S obtained by the combustion of pure carbon with addition of biomass the relaxation character of tanδ was observed for sintering temperatures in the range of 500-1000°C. However, conducting character of tanδ is observed for sample sintering at 1100°C. This change may be associated with appearance of liquid slag phase.
4. For ash S1 resulting from the mixture of coal ash and kaolinite the conducting character of tanδ are observed for samples sintered below 900°C. Relaxation character of losses appear at 1000°C. This change is complex to explain but is probably associated with the structure and composition transitions at high temperature for ash and kaolinite mixture.
5. The object of this work was to develop a method which could be used for laboratory studies of the sintering behaviour of ash samples as a direct and non-destructive test of slagging characteristics and slagging hazards assessment.
6. These preliminary studies are very promising. It seems that the direction of discussed research is appropriate and useful in practice. The tanδ method for characterization of fuel ashes can be a very convenient and effective instrument. Nevertheless, the interpretation of the results is rather complicated and so this method still needs further work and more detailed analysis.

References

1. D. Nowak-Woźny, W. Moroń, G. Hrycaj, W. Rybak, Archivum Combustionis **30 -3**, 177-191 (2010)
2. J. Frenkel, J. Phys. (USSR) **9**, 385-391 (1945)
3. E.J. Raask, Therm. Anal. **16**, 91-102 (1979)
4. R.W. Bryers, Prog. Energy Combust. Sci. **22 (1)**, 29-120 (1996)
5. F.J. Frandsen, Energy Fuels **23**, 3347–3378 (2009)
6. T.F. Wall, A. Lowe, L.J. Wibberley, I.McC. Stewart, Prog. Energy Combust. Sci. **5**, 1-29 (1979)
7. A.Y. Al-Otoom, L.K. Elliott, T.F Wall, B. Moghtaderi, Energy Fuels **14 (5)**, 994–1001 (2000)
8. J. Tomeczek, H. Palugniok, Fuel **81 (10)**, 1251-1258 (2002)
9. D. Nutalapati, R. Gupta, B. Moghtaderi, T.F. Wall, Fuel Process. Technol. **88 (11-12)**, 1044–1052 (2007)
10. M. Zevenhoven-Onderwater, R. Backman, B.J. Skrifvars, M. Hupa, Fuel **80 (10)**, 1503–1512 (2001)
11. A.Y. Al-Otoom, G.W. Bryant, L.K. Elliott, B.J. Skrifvars, M. Hupa, T.F. Wall, Energy Fuels **14 (1)**, 227-233 (2000)
12. B.J. Skrifvars, R. Backman, M. Hupa, Fuel Process. Technol. **56**, 55-67 (1998)
13. D. Lindberg, R. Backman, P. Chartrand, M. Hupa, Fuel Process. Technol. **105**, 129-141 (2013)
14. J.C. van Dyck, M.J. Keyser, Fuel **116**, 834-840 (2014)
15. T. Rizvi, P. Xing, M. Pourkashanian, L.I. Darvell, J.M. Jones, W. Nimmo, Fuel **141**, 275 284 (2015)
16. S. Chakravarty, A. Mohanty, A. Banerjee, R. Tripathy, G.K. Mandal, M. Raviathul Basariya, M. Sharma, Fuel **150**, 96-101 (2015)
17. W.J. McCarter, G. Starrs, T.M. Chrisp, Cem. Concr. Res. **34**, 1837–1843 (2004)
18. G.H. Zhang, K.C. Chou, Metall. Mater. Trans. B **43**, 64-72 (2012)
19. D. Maurya, J. Kumar, Shripal, J. Phys. Chem. Solids **66**, 1614-1620 (2005)
20. S. Hanjitsuwan, P. Chindaprasirt, K. Pimraksa, Int. J. Miner. Metall. Mater. **18-1**, 94-99 (2011)
21. D.K.Mahato, A. Dutta, T.P.Sinha, Phys. B **406**, 2703–2708 (2011)
22. W.D. Kingery, M. Berg, J. Appl. Phys. **26**, 1205–1212 (1955)

Extractive waste from hard coal mining in Poland – balance, status of management and environmental aspects

Justyna Pyssa[1,*]

[1]AGH University of Science and Technology, Al. Mickiewicza 30, 30-059 Krakow, Poland, Faculty of Energy and Fuels

Abstract. Extractive waste is produced at all stages of mining activities – during preparatory and operational works as well as in processes of mechanical processing of coal. In a significant part it comprises a full-value commercial product or a material for the land reclamation. Because of considerable amounts of waste generated by the coal extractive and processing industry, legal, technical, ecological and social aspects related to the possibilities of management of stored materials have been analyzed in the article.

1 Introduction

Hard coal mining is one of the biggest waste-generating industry branches in Poland. Waste is produced first of all during the coal extraction and its processing and it largely depends on the quality and condition of coal deposits, methods of breaking of coal beds and the technology of enrichment of mining product.

The processes of extraction of hard coal have been and are still accompanied by the generation of significant amounts of waste. In the past, due to limited possibilities of utilization, the larger part of waste was stored on the ground in a form of bings, dumps, dumping grounds and in mining waste landfills. Major differences in the manner of proceeding with wastes followed the introduction of payments for storage of waste in the 80's of the previous century. Towards the end of the 90's of the last century 80 percent of waste produced by the hard coal mining was economically utilized. Currently around 5 percent of wastes generated by mining is stored.

2 Coal mining in Poland

Hard coal mines in Poland are located in Upper Silesia and the Dabrowskie Basin in the Upper Silesian Coal Basin (which includes the Upper Silesian Industrial Region and the Rybnik Coal Area) and also in Lower Silesia and the Lublin region – the Lublin Coal Basin. Places of occurrence of hard coal deposits are marked in black in the figure 1.

* Corresponding author: jpyssa@agh.edu.pl

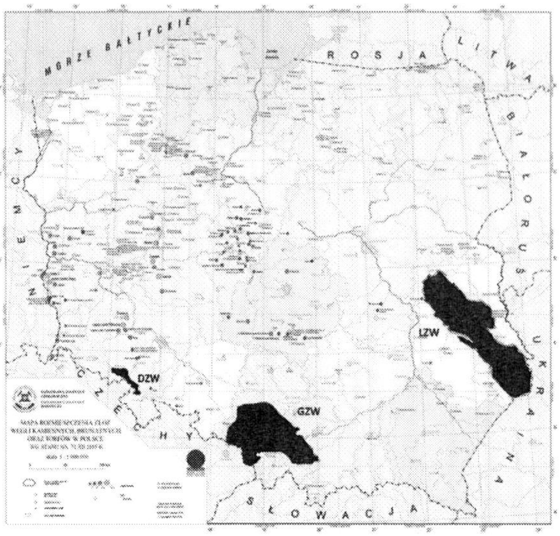

Fig. 1. Deposits of hard coal in Poland [1] DZW – the Lower Silesian Coal Basin, GZW – the Upper Silesian Coal Basin, LZW – the Lublin Coal Basin

According to the state as of the 31st of December in Poland there are 155 deposits of hard coal. Recoverable resources altogether amount to 51 960 million Mg, exploited resources come to 19 806 million Mg, whereas not exploited resources total at 28 177 million Mg [2].

The table 1 presents the condition of resources of hard coal in the year 2014 with the division by mining districts.

Table 1 Hard coal resources in 2014 [2]

Deposit resources	Geologically documented deposit					
	Number			Resources in million Mg		
	Exploited	Not exploited	No longer exploited	Exploite d	Not exploited	No longer exploited
Upper Silesian Coal Basin	50.0	44.0	42.0	19,024.0	19,213.0	3,735.0
Lublin Coal Basin	1.0	10.0	—	9,565.0	781.0	8,784.0
Lower Silesian Coal Basin	—	1.0	7.0	—	179.0	244.0

The Upper Silesian Coal Basin is the main deposit of Poland. In this region all currently operating hard coal mines are located (apart from one). Up to 78.9% of documented recoverable resources of Polish coal mines are present in GZW – the Upper Silesian Coal Basin. In the Lublin Coal Basin there is one coal mine working – Bogdanka. In the Lower Silesian Coal Basin the exploitation of hard coal was finished in the year 2000 when the extraction in the Nowa Ruda coal mine was abandoned. The reason for discontinuation of the exploitation of deposits from that basin was difficult mining and geologic conditions which made the extraction unprofitable [3].

3 Hard coal extraction in Poland

Now there are 30 hard coal mines operating in Poland which extract altogether around 65 million Mg (the data for the year 2015). In relation to the year 2014 the extraction of hard coal decreased by 1.4 percent [3]. According to the data by the Ministry of Economy, the extraction of hard coal in the period of January to June 2016 amounted to 33598.8 thousand Mg and was higher than in the same period in 2015 by 330.2 thousand Mg [4]. The extraction of steam coal in the analyzed period of time in the year 2016 increased by 0.9 percent, while the coking coal extraction increased by 1.2 %. The detailed data is presented in the figure 2.

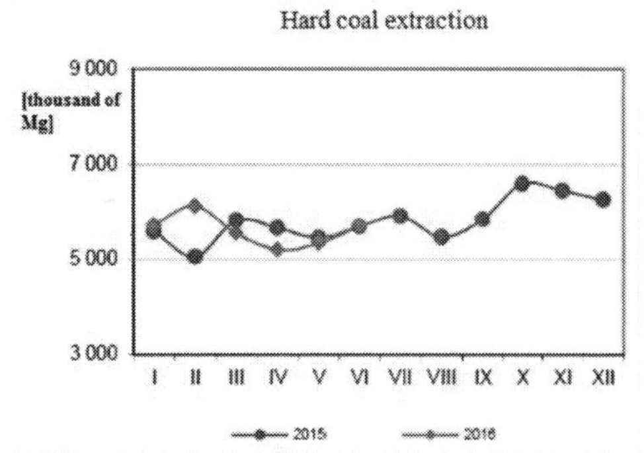

Fig. 2. Hard coal extraction in Poland in the years 2015 and 2016 [4]

4 Extractive waste from hard coal mining

The petrographic and chemical composition of waste from hard coal mining is varied and depends on geologic conditions of the exploited deposit. The waste derives from bottom and ceiling parts as well as overgrowth of coal beds which become the output during the deposit exploitation and then are separated from the coal in processes of enrichment.

The waste from hard coal mining is most often divided into mining waste and tailing waste because of the source of origin. The mining waste arises as a result of performing preparatory and enabling works. It is referred to as the gangue or waste rock [5]. It comprises around 6 percent of total mass of mining waste [6]. The tailing waste is generated on the ground and connected with the processes of hard coal enrichment – during sorting, rinsing and flotation of coal [7]. It comprises about 94% percent of total mass of hard coal mining waste [6].

4.1 Petrographic, mineral and chemical composition

The primary source of origin of the mining waste in Poland are building rocks along with coal decks of the Upper Carboniferous coal-bearing series of the Lower Silesian, Upper Silesian and Lublin coal basin. Siltstones, sandstones and conglomerates dominate in the Lower Silesian coal basin. In the Upper Silesian coal basin there are mudstones, siltstones and sandstones, while mudstones and siltstones prevail in the Lublin coal basin [7].

The mineral composition of hard coal mining waste is dependent on the share of particular carboniferous rocks in that waste. It is the loamy minerals such as: illite (28 to 82%) and kaolinite (9 to 65%) – that dominate. Quartz (3 – 37%), chlorite (less than 10 percent) and pyrite also occur. Locally sulphides of other metals – galena (Pb), sphalerite (Zn), chalcopyrite (Cu) and marcasite (Fe) – can also be present. In the waste there are also potassium feldspars, plagioclases and muscovite [7, 8].

From a chemical point of view, the hard coal waste in Poland can be characterized with the presence of silica – the content of which ranges from 34.7 to 66.9 percent – and aluminium oxide (Al_2O_3) – with the volume at about 21 percent. In the waste one can also find ferric oxides – 0.9 to12.9%, potassium oxides – 2 to 4.2 percent and calcium, sodium and titanium oxides in quantities of 1 to 2 percent. The data is presented in the table 2.

Table 2. Elements occurring in waste from hard coal mining in Poland [author's own research based on 7, 9, 10]

Element	Lublin Coal Basin	Upper Silesian Coal Basin
As, Sn, Cd, Co, Cu, Mo, Ni,	less than 100 mg/kg	more than 100 mg/kg
Ba	270 to 762 mg/kg	locally even up to 3,000 mg/kg
Cr	132	less than 100 mg/kg
Zn	to 161 mg/kg	to 332 mg/kg
Mn	—	1,360 mg/kg
Hg	less than 0.3 mg/kg	less than 0.3 mg/kg

In the hard coal mining waste slightly increased contents of radioactive isotopes of uranium, potassium and thorium occur [7, 11].

The volume of such elements as potassium, calcium, magnesium, nitrogen and phosphorus in the waste is very important in the process of reclamation with the use of waste, especially at the application of soilless method. The content of potassium is high enough [12], while other elements are present in insufficient quantities. The content of calcium is especially low [7].

The average content of carbon in the waste amounts to 9.4%. The presence of carbon may involve a potential threat of thermal activity of waste at using it for engineering and reclamation works or during storage [7, 12].

4.2 Amount of generated mining waste

The hard coal mining is intrinsically linked to producing considerable quantities of waste. In the 80's of the previous century the extraction of 1 Mg of coal was accompanied by 0.5 Mg of the extractive waste. Currently the number has decreased to 0.25 – 0.35 Mg of waste which is a result of the application of modern technologies of coal extraction while taking into consideration the coal deposits which are more and more difficult to access [13]. In line with regulation of the Minister for the Environment as of the 29th of December 2014 concerning the catalogue of waste (Journal of Laws 2014, item 1923) [14], the waste generated as a result of the hard coal extraction was classified into the group 01– wastes resulting from exploration, mining, quarrying, physical and chemical treatment of minerals. And this group includes:

- 01 01 02 – wastes from mineral non-metalliferous excavation,
- 01 04 12 – tailings and other wastes from washing and cleaning of minerals other than those mentioned in 01 04 07 and 01 04 11,
- 01 04 81 – wastes from floating enrichment of coal other than those mentioned in 01 04 80.

The figure 3 presents the data regarding the extraction of coal and the amount of extractive waste produced by the hard coal mining in the years 2010 to 2013.

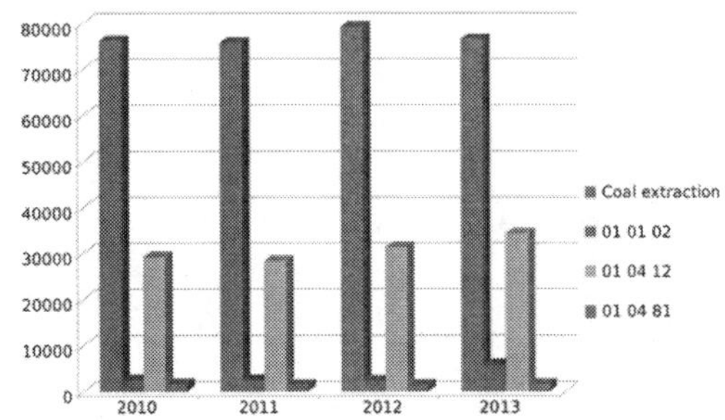

Fig. 3. Amount of generated extractive waste and coal extraction in the years 2010 to 2013 [author's own research based on [6]

In the analyzed period of time the biggest quantities of waste were produced in the group 01 04 12 (tailings and other wastes from washing and cleaning of minerals other than those mentioned in 01 04 07 and 01 04 11) – from 82.3 percent of all waste generated in the year 2013 to 89.5 percent in the year 2012. The waste from the group 01 01 02 (wastes from mineral non-metalliferous excavation) made up from 6.5% in the year 2012 to 14% in the year 2013, while the smallest amount of waste was produced in the group 01 04 81 (wastes from floating enrichment of coal other than those mentioned in 01 04 80) from 3.8% in the year 2013 to 4.8% in the year 2010.

The table 3 presents ways of proceeding with the extractive waste.

Table 3. Amount of waste in the years 2010 – 2014 in division by subgroups and along with way of managing [author's own research based on [6]

[thousand Mg]	2010	2011	2012	2013	2014[1]
010102					
Generated	2,391.0	2,385.8	2,253.4	5,827.3	5,173.3
Recovery	2,304.2	2,125.0	1,624.1	1,534.6	1,159.9
Neutralized	82.7	167.2	509.3	3,821.9	3,380.8
Stored	47.7	165.8	384.3	3,733.8	3,133.0
Temporarily warehoused	4.1	93.6	120.0	470.8	369.5
010412					
Generated	29,157.5	28,306.2	31,240.8	34,361.1	35,965.9
Recovery	26,618.1	22,520.3	26,332.9	27,792.4	14,570.2
Neutralized	2,206.1	5,150.7	4,768.9	6,397.0	5,603.8
Stored	1,001.6	5,150.7	4,768.9	6,397.0	5,603.8
Temporarily warehoused	333.3	635.2	139.0	171.7	65.1
010481					
Generated	1,596.6	1,395.2	1,429.9	1,571.0	1,672.0
Recovery	1,421.8	1,274.4	1,385.7	1,481.7	911.4
Neutralized	159.3	120.3	44.2	89.3	126.7
Stored	46.7	120.3	44.2	89.3	126.7
Temporarily warehoused	15.5	0.5	0	0	0

[1]The data for the year 2014 relate to the waste managed by the producer on their own and cannot be compared to the preceding years due to changes in the legislation.

5 Current legal status

The basic legislative act concerning the management of extractive waste is the Act as of 10th of July 2008 on extractive waste (Journal of Laws No 138, item 865) [15] which transposes the provisions of the Directive 2006/21/WE into the national legislation.
The primary objective of the act is the prevention of waste generation in the extractive industry and reduction of the adverse impact of waste on people and the environment. Those assumptions are supposed to be executed through the implementation of:

— rules of management of extractive wastes,
— rules of administration of facility for neutralizing extractive wastes,
— procedures related with obtaining permits and licenses connected with the extractive waste management,
— procedures related with the prevention of major functional failure in waste neutralization facilities of the category A.

Within the act a hierarchy of proceeding with extractive wastes was determined which obliges the producers to:

— use such manners of exploration, recognition, extraction, processing and storing which prevent the generation of extractive wastes and allow to keep their amount at the lowest possible level as well as reduce their negative influence on the environment and risk to human health and life, taking into consideration the best available techniques,
— submit extractive wastes to recovery and, if it is not possible due to technological reasons or not justified due to economic reasons, to submit it to neutralization processes in accordance with requirements for the environmental protection and the programme of extractive waste management, taking into consideration the best available techniques,
— transfer extractive wastes which cannot be submitted to recovery – for technological or economic reasons – to the nearest destinations in which it can be submitted to neutralization, taking into consideration the best available techniques and especially to facilities for the extractive waste neutralization.

The primary objective of the new act on extractive wastes – the Act of the 16th of November 2012 (Journal of Laws no 0, item 1513) [16] – is the prevention of extractive waste generation in industrial extraction and reduction its negative influence on the environment and human health and life. According to the act, the stage of extractive waste management commences at the phase of planning of mining activities – so at the stage of design of the method of mineral extraction. Environmental aspects have become the crucial issue in the extractive waste management and they are the very key condition determining the selection of recovery process. Technological and economic issues ceased to be the priority.

The implementing acts are as follows:

— The regulation of the Minister for the Environment as of the 5th of April 2011 on particular criteria for extractive waste neutralization facilities (Journal of Laws 2011 No 86, item 477).
— The regulation of the Minister for the Environment as of the 15th of July 2011 on criteria of classification of extractive waste as neutral waste (Journal of Laws 2011 No 175, item 1048).
— The regulation of the Minister for the Environment as of the 20th of June 2013 on the characteristics of extractive waste (Journal of Laws 2013, item 759).
— The regulation of the Minister for the Environment as of the 29th of May 2014 on conducting monitoring of extractive waste neutralization facilities (Journal of Laws 2014, item 875).

DOI: 10.1051/e3sconf/20171402024

— The regulation of the Minister for the Environment as of the 11th of February 2015 on financial guarantee and its equivalent amount for extractive waste neutralization facility (Journal of Laws 2015, item 311).

6 Directions of use of waste of hard coal mining

From the analysis of available data it follows that around 93 percent of waste produced during the exploitation and processing of raw minerals is economically used. Out of this quantity merely 30 percent is used industrially and nearly 70% is applied for leveling of grounds, engineering works and so-called 'earth structures' and 'landscape forms' [17, 18].

To the primary directions of the economic use of waste of hard coal mining one should include [16, 19]:

— reduction at the 'source' – at the stage of design and through the optimization of technologies applied for drilling the deposit,
— recovery – in a coal mine under the ground and on the ground,
— neutralization through storage.

One of elementary ways of limiting the generation of waste which arises during the hard coal extraction is the adjustment of machines and supports to smaller coal decks. Carbon and tailing waste is used in the coal mine for filling mine workings as a component of hydraulic and pneumatic backfill [11]. The addition of 50 percent of gangue to the hydraulic backfill influences the increase of its compressibility and flow resistance and improves the water yield and sedimentation rate [7, 20]. Ashes from the coal combustion are also added to the hydraulic backfill [11, 21-23]. The dominating technology of hydraulic backfill is based on the backfilling sand with the addition of other materials – ground fossil stone of particle size of more than 0,1 mm most frequently [24]. Very fine-grained waste coming from the flotation of coal along with binding material can also be added. The waste is used in mining technologies for sealing goafs (abandoned workings), preparing paste backfill, eliminating empties in the rock mass, producing backfilling corks and belts and in the fire safety practice [7, 11, 23].

In the Polish hard coal mining the paste backfill was used for the very first time in the „*Jastrzębie*" coal mine during the exploitation of a deck number 510/1-2. It was applied because of the geological structure and physicochemical properties of the deck – which was very low coal firmness. The content of the used paste backfill included: 40 to 50 percent of fly ashes, 50 to 60% of waste from coal floatation and 2 to 3 percent of cement [24]. The method allows to use salty mine waters as the mixing (batch) water [25].

The common practice is using the extractive waste for reclamation and leveling of land, elimination of workings after the exploitation of raw materials [7, 13, 17]. In order to achieve that the waste with codes 01 02 12 and 01 01 02 are the most frequently used [18]. Filling – in of the area is done up to the heights of adjacent non-transformed lands. In order to be used in this way the waste must meet requirements in the range of chemical composition as defined in the regulation. To be able to use the extractive waste in such way it is necessary to limit thermal phenomena connected with processes of the oxidation of carbon substance and pyrite. For that purpose one uses the waste deriving from the combustion of coal (slag, fly ashes, ash-slag mixtures, waste from desulphuring exhaust gases) which serve as screens and also reduce leaching of waste as well as limit the migration of contamination to groundwater [7]. One also uses fine-grained waste coming from floatation which disintegrate easily due to the presence of argillaceous rocks and foster the concentration of waste.

The coarse-grained and extractive waste from the processes of coal enrichment have been applied as the aggregate in road construction and engineering works. It is first of all used in earthworks by road investments for forming embankments and slopes [11, 13, 18, 26].

Mining waste – because of the high content of argillaceous rocks – has been used in the production of ceramic building materials. In the process of firing ceramics mudstone and sandstones present in the waste play a role in weakening the ceramic mass [7]. However, not all sorts of waste can be used in this way. The limitation is the occurrence of pyrite and siderite in the waste which are components responsible for obtaining products of lowered quality.

Waste containing large amounts of block shale is used in the cement industry as a clay – and silicon-bearing material in the production of portland cement [7].

The modern technology is sintering coal mud with the addition of sewage sludge the result of which is the production of high-quality aggregate [13, 19].

Coal slime – these are by-products resulting from rinsing coal dusts which were previously deposited in tailing ponds and now are used as low-energy fuel [13]. They are incinerated either immediately in power plants in boilers with fluidized bed or burnt together with the coal dust [7, 27]. The modern technology applied by Haldex SA company is the manufacturing of granulate on the basis of coal mud (from the current production and collected from settlement ponds) and calcium oxide [13]. That product can be characterized with lower moisture content and the calcium oxide included in it plays a role of sorbent material binding sulphur monoxides. There is also the generation of granulate based on grinded bottom ash. The granulate can be characterized with good energetic properties and resistance against the action of external factors at small portions of adhesive [28].

Another coal mud-based product made by Haldex company is BioCarbohumus. The addition here is- apart from the coal mud- the sewage sludge. The high content of nitrogen and phosphorus in the sewage sludge contribute to making BioCarbohumus rich in nutrients which are essential for the development of plants [13, 29]. The product has been applied in the reclamation of industrially contaminated lands and waste landfills [7, 13].

7 Impact on environment

The waste generated as a result of exploration and processing of coal affects all components of the environment. During movement, storing and technical reclamation the waste undergoes the action of external factors. Under the influence of sun and wind the surface of waste dries which causes enhanced emission of dusts to the atmosphere. The access of air makes the coal and mineral substance oxidize which results in the emission of CO, CO_2, NO_x, SO_2, polycyclic aromatic hydrocarbons (PAH) and C_xH_y. As a result of the oxidation of pyrite located in stored coal waste a strong exothermic reaction takes place which may lead to self-heating of waste and generating the critical temperature of coal ignition. This leads often to the spontaneous combustion of waste [7].

An extremely important issue is the impact of stored carbon waste onto the hydrosphere. The most important factor deciding about the scale of effect of the waste on the aquatic environment is the sort and content of soluble components in the fresh waste. It concerns especially the quantity of chlorides and primary sulphates. Soluble components, which have come into existence as an outcome of physicochemical processes (weathering) occurring after waste disposal, are of high significance. This mainly concerns the decomposition of reactive sulphides (first of all iron) as a result of which sulphates and sulphuric acid arise. Clay minerals – calcium carbonate and magnesium carbonate in particular – have neutralizing effect [7]. Another very important factor is the availability of leaching water. Both the scale and range of transportation of contaminants away from the structure of waste landfill depends on the geological structure of bedrock and hydrogeological conditions of all area. Increased amounts of sulphates, chlorides and sodium were found in aqueous extracts.

8 Summary

Hard coal mining is one of biggest producers of waste in Poland. The amount of generated extractive waste is strictly related to: volume of coal production, currently exploited part of deposit, conducted exploitation system and applied technologies of enrichment in a processing plant. The implementation of rules for managing the extractive waste and search for savings through its economic use (gaining profits and/or avoiding losses) can undoubtedly bring measurable benefits. Unfortunately, a considerable part of extractive waste is still managed by storing on bings. It results in the increase of hard coal production costs and causes permanent adverse changes in the natural environment. As it stems from the data placed in this article a significant part of extractive waste, after its proper processing, can be used in an economic manner.

The work is financed within the frames of statutory research no 11.11.210.213 at the Faculty of Energy and Fuels.

References

1. http://geoportal.pgi.gov.pl/css/surowce/images/2015/mapy/large/large_8.jpg
2. *Environment 2015* (Central Statistical Office of Poland, Warsaw, 14.12.2015). http://stat.gov.pl/en/topics/environment-energy/environment/environment-2015,1,7.html
3. *The balance of mineral resources deposit in Poland as of 31.12.2015* (PIG-PIB, Warszawa, 2016) https://www.pgi.gov.pl
4. Ministry of Energy, *Information on the Functioning of coal mining in June and in the period January-June 2016* (Warszawa, 07. 2016) http://www.me.gov.pl/Gornictwo/Realizacja+Programu+dzialalnosci+gornictwa+wegla+kamiennego+w+Polsce
5. W. Kotowski, *Utylizacja i gospodarka odpadami* (Wyd. WSEiA, Bytom, 2006)
6. Central Statistical Office of Poland, 2016 http://swaid.stat.gov.pl
7. Z. Mirkowski, J. Badera, *01 04 12 Rekultywacje wyrobisk z wykorzystaniem odpadów z wydobycia węgla kamiennego w województwie lubelskim* (Lublin, 2015)
8. Z. Bzowski, Wiadomości Górnicze, **64**, 2, 93–98 (2013) http://yadda.icm.edu.pl/yadda/element/bwmeta1.element.baztech-article-AGHL-0005-0041?q=3fb1ccd2-8ebd-45ed-8c89-fc6d0492f4e9$4&qt=IN_PAGE
9. B. Klojzy-Karczmarczyk, J. Mazurek, Zeszyty Naukowe IGSMiE PAN, **90**, 67–78 (2015); http://yadda.icm.edu.pl/yadda/element/bwmeta1.element.baztech-7b8b9097-4a20-48f4-832a-88a564b3f857
10. A. Strugała, D. Makowska, K. Bytnar, T. Rozwadowska, Polityk Energetyczna – Energy Policy Journal **17**, 4, 77–88 (2014); https://www.min-pan.krakow.pl/Wydawnictwa/PE174/(z4)-06-strugala-makowska-i-in.pdf
11. J. Pyssa, J. Rokita, Przegląd Górniczy **68/12**, 132–138 (2012) http://yadda.icm.edu.pl/yadda/element/bwmeta1.element.baztech-e36b801d-f368-442d-8733-1763ec5f83bf
12. K. Bojarska, Z. Bzowski, Górnictwo i Geologia, **7**, 2, 101–113 (2012) https://www.polsl.pl/Wydzialy/RG/Wydawnictwa/Documents/kwartal/7_2_8.pdf
13. M. Kugiel, R. Piekło, Górnictwo i Geologia, **7**, 1, 133–145 (2012) https://www.polsl.pl/Wydzialy/RG/Wydawnictwa/Documents/kwartal/7_1_12.pdf

14. Regulation of the Minister for the Environment as of the 9th of December 2014 concerning the catalogue of waste (Journal of Laws 2014, item 1923)
15. Act as of 10th of July 2008 on extractive waste (Journal of Laws No 138, item 865)
16. Act of the 16th of November 2012 (Journal of Laws no 0, item 1513)
17. S. Góralczyk, E. Żbikowska, Prace Naukowy Instytutu Górnictwa Politechniki Wrocławskiej. Studia i Materiały, **41**, 93–104 (2012). http://yadda.icm.edu.pl/baztech/element/bwmeta1.element.baztech-article-BPW6-0025-0009
18. J. Kamyk, A. Kot-Niewiadomska, Inżynieria Środowiska **38**, 23–31 (2015) http://yadda.icm.edu.pl/yadda/element/bwmeta1.element.baztech-a4ea3150-b936-45a8-b79d-3b48f444ea1a
19. S. Góralczyk, I. Baic, Polityka Energetyczna – Energy Policy Journal, **12(2/2),** 145–157 (2009) http://meeri.pl/se/pelne_teksty23/k23z_pe/k23_goralczyk_baic_z.pdf
20. F. Plewa, Z. Mysłek, *Zagospodarowanie odpadów przemysłowych w podziemnych technologiach górniczych.* (Politechnika Śląska, 2001)
21. J. Pyssa, Gospodarka Surowcami Mineralnymi – Mineral Resources Management. **21/3**, 83–93 (2005) http://www.meeri.eu/Wydawnictwa/GSM213/pyssa.pdf
22. F. Plewa, M. Popczyk, P. Piontek, Polityka Energetyczna – Energy Policy Journal **12(2/2)**, 485– 495 (2009). http://www.min-pan.krakow.pl/se/pelne_teksty23/k23z_pe/k23_plewa_popczyk_piontek_z.pdf
23. J. Pyssa, E3S Web of Conferences, **10**, 00077 (2016) http://dx.doi.org/10.1051/e3sconf/20161000077
24. F. Plewa, M. Popczyk, P. Pierzyna, Polityka Energetyczna – Energy Policy Journal **16(4)**, 257–270 (2013) http://yadda.icm.edu.pl/baztech/element/bwmeta1.element.baztech-dc62e6fd-04c9-4341-8b55-27d6b510800c/c/plewa.pdf
25. J. Palarski, F. Plewa, Z. Małysek, *Odzysk i unieszkodliwianie odpadów w górnictwie podziemnym* Wyd. Politechniki Opolskiej (Gliwice, 2012)
26. J. Pyssa, Przegląd Górniczy, **66/5**, 38–44 (2010) http://yadda.icm.edu.pl/baztech/element/bwmeta1.element.baztech-article-BGPK-2859-1183
27. I. Jelonek, Z. Mirkowski, International Journal of Coal Geology, **139**, 228–236 (2015) http://dx.doi.org/10.1016/j.coal.2015.02.005
28. A. Szymkiewicz, J.J.Hycnar, A. Fraś, R. Przystaś, T. Józefik, I. Baic, Journal of the Polish Mineral Engineering Society, **1**, 19-30 (2012) http://www.potopk.republika.pl/Full_text/im_29_r3_szymkiewicz_hycnar.pdf
29. J. Pyssa, Logistyka, **4**, 9633–9642 (2015) http://yadda.icm.edu.pl/yadda/element/bwmeta1.element.baztech-c0765b1c-2f45-4afc-9cc0-743542ece8a9

Limitation of the environmental impact of ashes from combustion of sewage sludge by the synthesis of zeolites

Jolanta Latosińska[1,*]

[1]Kielce University of Technology, Faculty of Environmental, Geomatic and Energy Engineering, Al. Tysiąclecia Państwa Polskiego 7, 25-324 Kielce, Poland

Abstract. The combustion of municipal sewage sludge allows for the recovery of renewable energy. However, it simultaneously causes the formation of ash which is not neutral for the environment. The study presents the research on the possibility of using sewage sludge ash as a raw material for the synthesis of zeolites. The synthesis of zeolites was performed with the use of an indirect fusion method and a direct hydrothermal method. The research on sewage sludge ash after zeolitization included the identification of crystalized phases, the observation of changes of ash particles surface and the measurement of cation exchange capacity (CEC). The research results proved that optimal conditions for the formation of zeolite Y were the synthesis with the indirect fusion method at the sewage sludge ash to hydroxy sodalite ratio of 1:1.8, the activation temperature of 60°C and the crystallization temperature of 90°C. Hydroxy sodalite was found in samples from both methods of zeolitization. However, in the case of direct hydrothermal method, particles of hydroxy sodalite structure were less numerous indicating a small conversion of sewage sludge ash to crystalline zeolite. CEC values comparable to commercial zeolites resulted from zeolitization with the indirect fusion method.

1 Introduction

Municipal sewage sludge, an unavoidable by-product of wastewater treatment plants, is classified as biomass [1]. Sewage sludge contains 40-85% of organic matter which is the source of renewable energy [2]. The recovery of energy from sewage sludge may be realized by mono-combustion and co-combustion. However, the combustion of sewage sludge causes the formation of sewage sludge ash, whose weight equals approximately 10% of the weight of incinerated sewage sludge [3]. There are known methods of using sewage sludge ash in the production of building materials [4-5], the stabilization of soils [6] and the recovery of phosphorus [7]. However, the dominant method of sewage sludge ash disposal is a landfilling [8]. The landfilling of sewage sludge ash poses a threat to the natural environment [9].

* Corresponding author: jlatosin@tu.kielce.pl

Zeolites are tectosilicates with a framework of $[AlO_4]^{-5}$ and $[SiO_4]^{-4}$tetrahedra, connected with one common oxygen atom. The connection of tetrahedra generates the existence of empty spaces in the structure [10]. Due to such properties as high ion exchange, high thermal, mechanical and chemical stability, zeolites are commonly used as ion exchangers, molecular sieves, catalysts and adsorbents [11-12]. Synthetic zeolites can be produced from natural raw materials such as kaolinite [13], montmorillonite [14]. Waste materials are also used for the synthesis of zeolites as the carriers of SiO_2 and Al_2O_3. Zeolites are obtained among others from coal fly ash [15], rice husk ash [16], municipal incinerator fly ash [17].

A significant similarity of the oxide composition of sewage sludge ash to the composition of coal fly ash used for the synthesis of zeolites is the basis for the beginning of research in this aspect (tab.1.).

Table 1. Chemical composition of ashes.

Component	Sewage sludge ash	Coal fly ash	
	[18]	[19]	[20]
SiO_2	35.7	34.4	38.5
Al_2O_3	6.7	17.9	15.9
Fe_2O_3	9.6	10.4	9.6
CaO	17.5	20.4	14.2
MgO	4.51	1.5	1.8
SO_3	1.2	n.d.	n.d.
K_2O	1.6	0.8	1.6
Na_2O	0.5	1.5	0.9
P_2O_5	19.0	n.d.	n.d.
TiO_2	1.0	1.0	n.d.

n.d. – no data

The synthesis of zeolites from ashes can be realized with one of the methods: a direct hydrothermal method [21], an indirect fusion followed by a hydrothermal method [22-23], a two step hydrothermal method [24] and a molten salt liquid free method [25]. The most commonly used methods are the hydrothermal and fusion methods [26-27]. The ratio of silicone to aluminum used for the synthesis of the raw material influences the type of obtained zeolite. Moreover, factors such as the concentration and type of alkaline, temperature, pressure and reaction time determine the obtainment of different zeolites from the same fly ash [28].

The aim of the study is the evaluation of the possibility of using sewage sludge ash for the synthesis of zeolites and the evaluation of the impact of the zeolitization method on the ion-exchange properties of obtained products. It is a pioneering study on zeolitization of municipal sewage sludge ash by the indirect fusion method.

2 Materials and methods

2.1 Materials

In this study sewage sludge from a municipal wastewater treatment plant in Sitkówka-Nowiny near Kielce was used. The wastewater treatment plant receives sewage from Kielce agglomeration, located in the central part of Poland.

Sewage sludge ash was obtained as a result of sewage sludge incineration in a laboratory furnace. Sewage sludge was dried in a laboratory drier at 105°C. Then, it was crushed in a mortar to a fraction < 125 µm. Sewage sludge was incinerated in a laboratory furnace at the temperature of 600°C within 11 minutes. The time of sample incineration included the period of furnace warm-up, combustion at a given maximal temperature and the furnace cooling down to the temperature of 20°C.

2.2. Methods of zeolite synthesis

The synthesis of zeolite materials was investigated with two methods: the indirect fusion and the direct hydrothermal method. Different ratios of sewage sludge ash to NaOH samples were used to explore the effect of this parameter on zeolitization. The synthesis conditions for each sample are shown in Table 2.

Table 2. Experimental conditions of sewage sludge ash zeolitization.

Sample	SSA:NaOH g/g	Activation temperature, °C	Crystallization temperature, °C	
			Indirect fusion	Direct hydrothermal
S1	1:1.4	60	60	-
S2	1:1.4	60	90	-
S3	1:1.4	90	60	-
S4	1:1.4	90	90	-
S5	1:1.4	60	-	60
S6	1:1.4	60	-	90
S7	1:1.4	90	-	60
S8	1:1.4	90	-	90
S9	1:1.8	60	60	-
S10	1:1.8	60	90	-
S11	1:1.8	90	60	-
S12	1:1.8	90	90	-
S13	1:1.8	60	-	60
S14	1:1.8	60	-	90
S15	1:1.8	90	-	60
S16	1:1.8	90	-	90

2.2.1 Indirect fusion followed by a hydrothermal method

Sewage sludge ash particles in the amount of 10 g were mixed and ground with granular NaOH to obtain a homogeneous mixture. The mixture was heated in an electric laboratory furnace at the temperature of 550°C for 1 hour. The sintered mixture was ground and mixed with distilled water (the applied concentration 3 molar NaOH), followed by an aging process with an agitation in a shaking water bath at given temperatures within 12 hours. Then the mixture was crystallized under static conditions at the given temperature within 6 hours. After the completion of the crystallization, the solid product was washed several times with distilled water until the pH of the solution reached 10.0, then dried at 105°C for 10 hours.

2.2.2 Direct hydrothermal method

Sewage sludge ash particles in the amount of 10 g were mixed with a 3 molar NaOH solution. Then the samples were subjected to activation and crystallization under respectively the same process conditions as for the indirect fusion method followed by a hydrothermal method.

2.3 Characterization methods

The chemical compositions of sewage sludge and sewage sludge ash were determined by using the X-ray fluorescence spectroscopy (XRF) method. The research on sewage sludge ash before and after the process of zeolitization included a phase analysis by the X-ray diffraction (XRD). The identification of phases was performed by the comparison of recorded diffractograms and the benchmarks in base ICDD PD-2 and PDF-4+2014.

The evaluation of grain structure change was observed with the scanning electron microscopy (SEM).

The cation exchange capacity (CEC) was determined with the use of ammonium acetate. CEC is a measurement of the number of cations in the channels by unit weight that may be replaced by other ions in solutions. The exchangeable cations of samples were replaced by NH_4^+ using 1M ammonium acetate [29]. The concentration of ammonium ions in solutions was determined with the use of ion chromatography.

3 Results and discussion

The characteristics of sewage sludge and sewage sludge ash are presented in Table 3. A significant increase of the content of certain elements in sewage sludge ash in comparison to their content in sewage sludge is caused by among others different volatility of compounds. The volatility of organic compounds is higher in comparison to the volatility of inorganic compounds. Sewage sludge ash is a high-silica ash because the ratio of SiO_2 and Al_2O_3 equals 3.94.

Table 3. Characteristics of sewage sludge and sewage sludge ash.

Parameter	Unit	Sewage sludge	Sewage sludge ash
pH	-	7.5	-
moisture	%	72.6	-
SiO_2	%	8.81	20.8
Al_2O_3	%	2.11	5.28
Fe_2O_3	%	3.74	9.45
CaO	%	5.66	14.3
MgO	%	1.42	3.72
SO_3	%	0.01	1.45
K_2O	%	0.58	1.51
Na_2O	%	0.16	0.41
P_2O_5	%	7.06	19.4
TiO_2	%	0.32	0.73
Mn_2O_3	%	0.03	0.10

SrO	%	0.02	1.45
ZnO	%	0.03	0.36
BaO	%	0.04	0.09
CuO	%	0.02	0.05
TOC	% mas.	33.69	-

Sewage sludge ash consists of: amorphous substance – 76.71% mas., quartz – 9.12% mas., calcite 4.06% mas., dolomite – 3.28% mas., muscovite – 4.42% mas., potassium feldspar – 1.72% mas., plagioclase – 0.47% mas. and other – 0.22% mas.

The XRD diffractograms were presented in figure 1 – figure 2. Diffractograms of samples S1, S2, S3, S4 and S9 indicate that zeolites were not formed while obtained phases were apatite, quartz and magnetite. Therefore for given activation and crystallization temperatures, no differences were found in formed crystalline phases for samples with the SSA:NaOH ratio of 1:1.4. Whereas, a noticeable lack of zeolite structure in sample S5 was caused by an insufficient value of the activation and crystallization temperatures.

Zeolite Y (faujasite) is presented by the diffractogram of sample S10. Depending on the silicone to aluminum ratio of the raw material, faujasite is divided into X and Y. Zeolite X is obtained as a result of the synthesis of raw materials with the ratio between 2 and 3 [29], while zeolite Y with the ratio from 3 and more [31]. Zeolite Y, obtained under the conditions of zeolitization of sample S10, proves that the synthesis at the crystallization temperature of 90°C is more favorable than at 60°C [32].

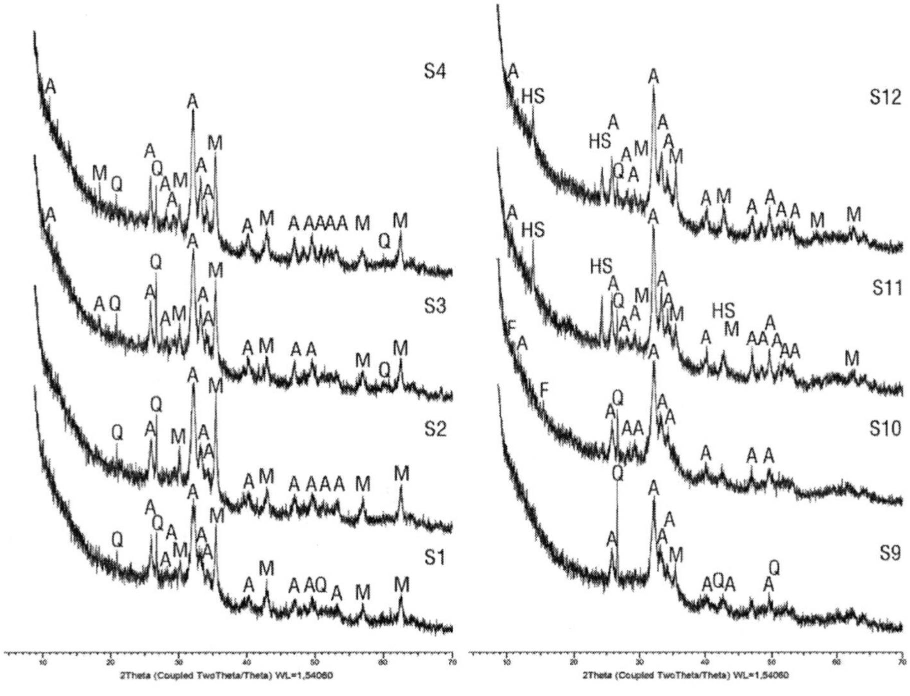

Fig. 1. XRD of the synthetic zeolites by indirect fusion method: A-apatite; F- faujasite; M- magnetite, HS-hydroxy sodalite; Q-quartz.

Diffractograms of samples S11 and S12 show that except for apatite, quartz and magnetite, hydroxy sodalite was also formed. Hydroxy sodalite has the same framework structure as sodalite. Its structure is made up of a framework of 4 – and 6– membered rings

DOI: 10.1051/e3sconf/20171402025

of SiO_4 and AlO_4 tetrahedra [33]. The pore size of hydroxy sodalite is smaller than of the zeolites with an eight-member ring aperture [34]. This feature provides many advantages, particularly the access to small molecules like ammonia and water. Therefore hydroxy sodalite is an attractive material for wide applications, such as hydrogen separation, catalysts and pigment occlusion [35-36].

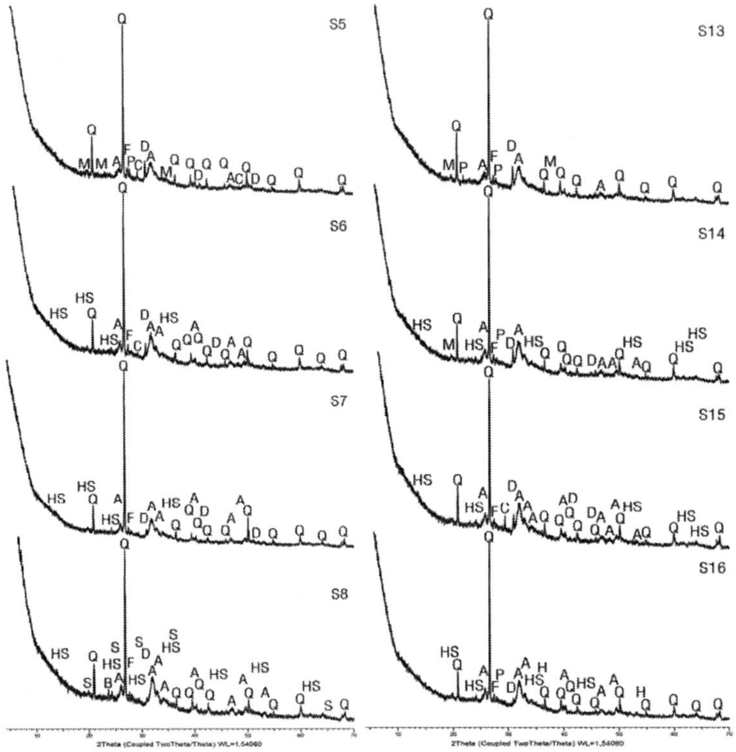

Fig.2. XRD of the synthetic zeolites by direct hydrothermal method; Q- quartz, F- potassium feldspar, A-apatite, HS-hydroxy sodalite, M- muscovite, D- dolomite, P- plagioclase, C- calcite.

Samples S5-S8 and S13-S16 obtained as a result of zeolitization with the direct hydrothermal method are characterized by a greater diversity of identified phases in comparison to samples from the indirect fusion method. All samples S5-S8 and S13-S16 except for quartz and apatite contained dolomite and potassium feldspar. The presence of dolomite and potassium feldspar in sewage sludge ash and subsequently in samples after zeolitization with the direct hydrothermal method indicates that both mentioned phases were not entirely transformed. Sample S5, except for the above mentioned phases, contained also plagioclase, calcite and muscovite which were present in sewage sludge ash. The lack of new phases in sample S5 and S13 proves that the activation and crystallization temperatures had insufficient values for the occurrence of effective zeolitization.

Diffractograms of samples S6-S8 and S14-S16 present a zeolite – hydroxy sodalite. The intensity of hydroxy sodalite peaks for these samples is smaller than for samples S11 and S12. For samples S5-S8 and S13-S16 the influence of the SSA:NaOH ratio on the formation of new phases was not found. Whereas, the greatest change in the quality of phases in comparison with the sewage sludge ash used was stated only for samples S7 and S8. Calcite, muscovite and plagioclase were not found in samples S7 and S8. Therefore, the activation temperature of 90°C is required in order to conduct the process of zeolitization. According to [37], zeolitization of sewage sludge ash with the hydrothermal method for among many a 3 molar concentration of NaOH, the crystallization temperature of 100°C

 DOI: 10.1051/e3sconf/20171402025

and the crystallization time of 10 hours also did not cause the formation of clear zeolites, i.e. Zeolite P and hydroxy sodalite. Other crystalline forms present in sewage sludge ash were also in samples after the process of zeolitization.

Figure 3 shows the SEM micrograph of the original sewage sludge ash. The SEM presents particles with a compact structure and a smooth surface because of the amorphous substance covering them.

Fig. 3.SEM of sewage sludge ash.

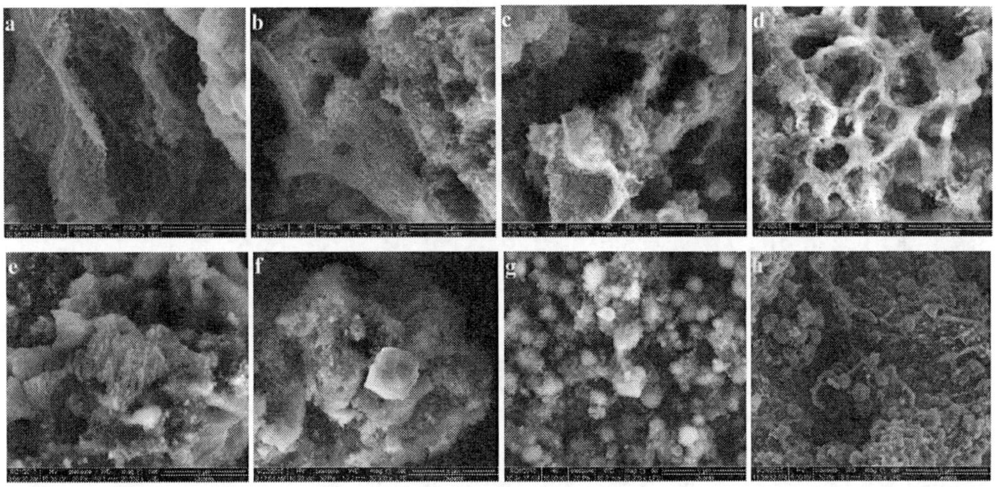

Fig. 4. SEM of the synthetic zeolites by indirect fusion method; (a) S1; (b) S2; (c) S3; (d) S4; (e) S9; (f) S10; (g) S11; (h) S12.

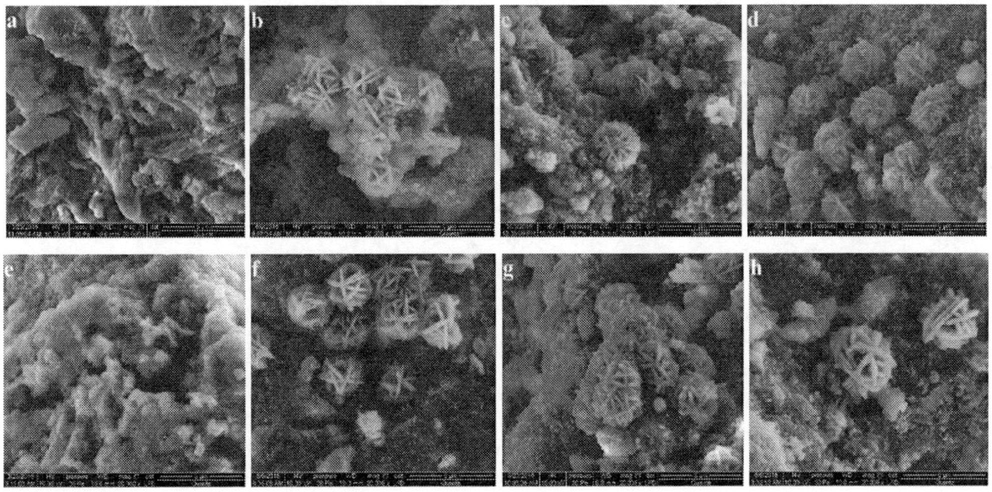

Fig. 5. SEM of the synthetic zeolites by direct hydrothermal method; (a) S5; (b) S6; (c) S7; (d) S8; (e) S13; (f) S14; (g) S15; (h) S16.

 DOI: 10.1051/e3sconf/20171402025

Figure 4 shows images of samples after zeolitization by the indirect fusion method. The morphology of samples S1, S2, S3 and S4, obtained at the lowest SSA:NaOH ratio has porous surface (fig.4.a-fig.4.d). The structure of sample S9-S12, obtained at the ratio SSA :NaOH 1 :1.8 (fig.4.e-fig.4.h) is also not homogeneous. Apart from the porous structure, there are also forms with the shape similar to long needles with uneven surface (fig.4.e). The morphology of SEM image for zeolite Y is an octahedral shape (fig.4.f). As illustrated in figures 4.g–4.h, the hydroxy sodalite has a particle size smaller than 5 μm.

Figure 5 presents SEM of samples after the process of zeolitization with the direct hydrothermal method. Samples S5 and S13 (fig.5.a, fig.5.e) are characterized by a solid structure, a lack of zeolites and clearly noticeable amorphous substance. For samples S6-S8 and S14-S16 the presence of not numerous particles of hydroxy sodalite structure on the amorphous substance indicates a small conversion of sewage sludge ash to crystalline zeolite.

SEM images of the samples formed in this study reveal the change in morphology connected with the zeolitization method.

SEM images confirm the results of the phase analysis. Peaks of high intensity in a low angle range characteristic for zeolite phases do not appear for samples S6-S7 and S14-S16. The peaks are small, often with the intensity similar to the background.

The CEC results of sewage sludge ashes and ashes after zeolitization are presented in figure 6. The CEC values of sewage sludge ashes subjected to zeolitization with the indirect fusion method are higher in comparison to the values of raw sewage sludge ash. There is a relation for this method of zeolitization, i.e. higher CEC values of samples with the SSA:NaOH ratio of 1:1.4 in comparison to samples with the SSA:NaOH ratio of 1:1.8. Whereas, there is no significant influence of the activation and crystallization temperatures on CEC values in the case of all tested samples for both methods of zeolitization.

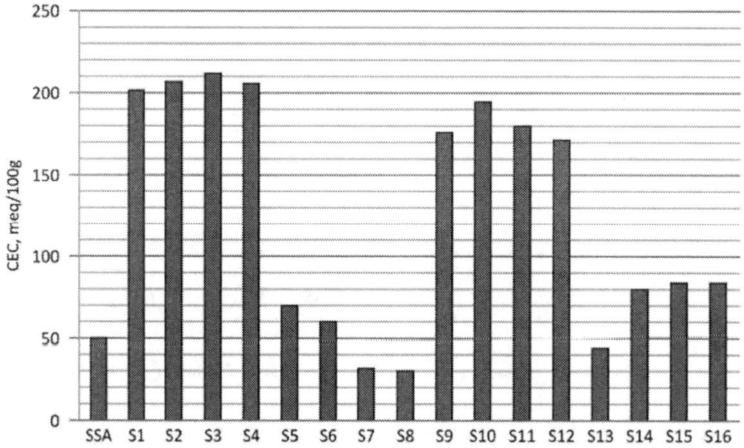

Fig. 6. CEC of sewage sludge ash (SSA) and samples after zeolitization.

The CEC values for samples S7, S8 and S13 obtained as a result of zeolitization with the direct hydrothermal method are lower than those of raw sewage sludge ash. Whereas, the CEC values for samples S5, S6, S14-S16 are higher than the values of raw sewage sludge ash but they do not reach the levels of the CEC values of samples subjected to the synthesis with the indirect fusion method. Sample S3 has the highest CEC value from all the tested samples. The CEC value is 426% higher than the value for the raw sewage sludge ash. Whereas, the lowest CEC value belongs to sample S8, obtained as a result of zeolitization with the direct hydrothermal method for the SSA:NaOH ratio of 1:1.4. The CEC value is more than 39% lower than the value for the raw sewage sludge ash.

All samples obtained as a result of the synthesis with the fusion method have significantly higher CEC values in comparison to CEC values of the remaining samples. For samples S1-S4 and S9, the increase of the CEC values is not simultaneous with the formation of zeolite phases. SEM images show that samples S1-S4 and S9-S12 have a more developed porous structure than samples obtained as a result of zeolitization with a direct hydrothermal method. Therefore, it is possible that higher CEC values are not necessarily the evidence of the ion exchange but the occurrence of a different process, for instance the sorption on the porous surface.

Sewage sludge ash subjected to zeolitization with the indirect fusion method has the CEC values at a similar level as the CEC values of natural zeolites. The CEC values of natural zeolites are in the range of 200-500 meq/100g [38-39]. There are zeolites of higher CEC values, for instance pure hydroxy sodalite has a CEC value of 920 meq/100g [40]. CEC values of the tested samples, significantly lower than CEC values of pure hydroxyl sodalite are approximately the evidence of a small degree of conversion of tested sewage sludge ash into a zeolite material. It is the approximation because other processes may occur apart from the ion exchange. Moreover, the comparison of the obtained CEC values with the literature data involves a little authoritativeness due to the diversity of methods of conducting the ion exchange process.

4 Conclusions

The study presents the test results of the synthesis of zeolites from sewage sludge ash with the indirect fusion method and the direct hydrothermal method. The results show that synthesis products of better properties can be obtained with the use of the indirect fusion method rather than the direct hydrothermal method.

From a practical point of view, the optimal conditions for the synthesis of zeolite Y are provided by the indirect fusion method at the sewage sludge ash to sodium hydroxide ratio of 1:1.8, the activation temperature of 60°C and the crystallization temperature of 90°C.

The indirect fusion method should also be used in order to obtain hydroxy sodalite from sewage sludge ash. Moreover, zeolitization with the indirect fusion method caused an increase of the cation exchange capacity value of sewage sludge ash to the level of commercial zeolites.

The combustion of municipal sewage sludge, apart from the energy recovery, allows to obtain a material which can be used as a raw material for the synthesis of zeolites. The obtained zeolites can be used for the removal of ammonium ions from sewage.

The project was funded by the National Science Centre allocated on the basis of the decision DEC-2011/03/D/ST8/04984, in Poland.

References

1. Directive 2012/27/EU of the European Parliament and of the Council of 25 October (2012)
2. M. De la Guardia, A. Morales-Rubio, Trends Anal. Chem. **8**, 311-318 (1996)
3. J. Werther, T. Ogada, Prog. Energy Combust. Sci.**25**, 1-27 (1999)
4. D.-F. Lin, C.-H. Weng, J. Environ. Engineering **6**, 922-927 (2001)
5. J. Monzó, J. Payá, M.V. Borrachero, I. Girbés, Waste Manage. **23**, 373-381 (2003)
6. L. Chen, D-F. Lin, J. Hazard. Mater. **162**, 338-343 (2009)
7. C. Adam, G. Kley , F.G. Simon, Mater. Trans. **48**, 3056-3061(2007)
8. Eurostat http://appsso.eurostat.ec.europa.eu

9. J. Latosińska, J. Gawdzik , EPE Journal, **38**, 31-44 (2012)
10. M. Król, J. Morawska, W. Mozgawa, W. Pichór, Ceram. Mater. **66**, 129-134 (2014)
11. L. Bandura, R. Panek, M. Rotko, W. Franus, Microporous and Mesoporous Mater. **223**, 1-9 (2016)
12. W. Qiu, Y. Zheng, Chem. Eng. J. **145**, 483-488 (2009)
13. M. Murat, A. Amokrane, J.P. Bastide, L. Montanaro, Clay Miner.**27**, 119-130 (1992)
14. P. Cañizares, A. Durán, F. Dorado, M. Carmona, Appl. Clay. Sci. **16**, 273-287 (2000)
15. X. Querol, N. Moreno, J.C. Umaña, A. Alastuey, E. Hernández, A. López-Soler, F. Plana, Int. J. Coal Geol. **50**, 413 - 423 (2002)
16. D. Prasetyoko, Z. Ramli, S. Endud, H. Hamdan, B. Sulikowski, Waste Manag. **26**, 1173-1179 (2006)
17. G. C. C. Yang, T. Y. Yang, J. Hazard. Mater. **62**, 75-89 (1998)
18. J. Latosińska, J. Gawdzik, Ecol. Chem. Eng. S. **21**, 465-475 (2014)
19. D. Wu, Z. Hu, X. Wang, S. He, H. Kong, Front. Environ. Sci. Engin. China. **1**, 223-220 (2007)
20. K.-M.Lee, Y.-M. Jo, J Mater Cycles Waste Manag. **12**, 212-219 (2010)
21. X. Querol, A. Alastuey, J.L. Fernandez-Turiel, A. Lopez-Soler, Fuel. **74**, 1126-1231 (1995)
22. N. Shigemoto, H. Hayashi, K. Miyara. J Material Science. **28**, 4781-4786 (1993)
23. N. Czuma, K. Zarębska, P. Baran, *Analysis of the influence of fusion synthesis parameters on the SO2 sorption properties of zeolites produced out of fly ash*, E3S Web of Conferences 10, 00010 (2016)
24. H. Tanaka, S. Furusawa, R. Hino, J Material Synthetic Processing. **10,** 143-148 (2002)
25. M. Park, C.L. Choi, W.T. Lim, M.C. Kim, J. Choi, N.H. Heo, Microporous and Mesoporous Mater. **37**, 81-89 (2000)
26. A. Molina, C. Poole, Minerals Engineering. **17**, 167-173 (2004)
27. C.A. Ríos, C.D. Williams, O.M. Castellanos, Bistua. **4**, 60-71 (2006)
28. A. Derkowski, W. Franus, E. Beran, A. Czimerova, Power Technology. **166**,47-54 (2006)
29. N. Czuma, K. Zarębska, P. Baran, W. Franus, *The process of fly ash magnetic separation impact on hydrothermal synthesis of zeolites*, E3S Web of Conferences 10, 00009, (2016)
30. H. Minato, *Standarizaton of Methods for Zeolite Speciality Determination and techniques for Zeolite resources utilization, Natural Zeolites* - SOFIA'95 282-292 (1995)
31. D.N. Stamires, Clays Clay Miner. **21**, 379-389 (1973)
32. K. Fukui, T. Nishimoto, M. Takiguchi, H. Yoshida, KONA Power and Particle Journal. **24**, 183-190 (2006)
33. T. Wajima, K. Munakata, Y Ikegami, Mater. Trans. **51**, 849-854 (2010)
34. B. Bayati, A.A. Babaluo, P. A. Namini, Iran J. Chem. Chem. Eng. **28,** 1-12 (2009)
35. J.C. Kim, M. Choi, D-S. Kim, H.J. Song, D-W. Kim, J. Ceram. Soc. Jpn. **123,** 1022-1026 (2015)
36. M.K. Naskar, D. Kumdu, M. Chatterjee, Mater. Lett. **65,** 436-438 (2011)
37. J.S. Lee, S.W. Eom, H.Y. Choi, Kor. J. Env. Hlth. **33**, 317-324 (2007)
38. K. Margeta, N. Z. Logar, M. Šiljeg, A. Farkas, *Natural Zeolites in Water Treatment – How Effective is Their Use, Water Treatment,* W. Elshorbagy (Ed.), InTech, 81-112 (2013)
39. F.A. Mumpton, *Mineralogy and Geology of Natural Zeolites*, Mineralogical Soc. America (1981)
40. T. Henmi, Soil Sci. Plant Nutr. **33**, 517-521 (1987)

E3S Web of Conferences **14**, 02026 (2017) DOI: 10.1051/e3sconf/20171402026
Energy and Fuels 2016

Safety transport of radioactive waste in the nuclear power area

Ivana Tureková[1,*], and *Linda* Makovická Osvaldová [2]

[1] UKF Nitra, Faculty of Education, Tr. A. Hlinku 1, 949 74 Nitra, Slovakia
[2] University of Žilina, Faculty of Security Engineering, Univerzitná 8215/1, 010 26 Žilina, Slovakia

Abstract. Radioactive wastes require strict rules for manipulation with them due to the hazards for the human health and environment, not excluding the hazards during their internal transport. The article deals with the transport of packing unit inside of the company and it proposes the possible alternatives so that meet the limit conditions and reduce the manipulation time with the radioactive material in the packing unit. The packing unite isolates fixated liquid waste from the environment while it also serves as protection. There are also important external radiation characteristics of package unit, which consist of measurable values of the scratch contamination surface and dose power on the surface of package unit. Thus, the paper is aimed to point out the necessity of the logistics during manipulation with the package unit in the process of internal transport so that the dose power of exposed employees would achieve the lowest possible level and meet the strict limits in a full extent.

1 Introduction

Radioactive waste (RAW) treatment includes technological processes, procedures and operations used to achieve chemical and physical form appropriate for transport, storing or long-term storing of RAW [1]. In case of liquid RAW or dispersible waste, RAW treatment also includes immobilisation (solidification) [2]. Aim of RAW treatment is to create a system consisting of fixed RAW, a fixation matrix and storage/transport file (container, barrel) meeting the following requirements [3]:

- low permeability and low leachability of wastes;
- maximum load efficiency of the system (low proportion of empty space in the container);
- chemical, mechanical, thermal and radiation stability during the time required;
- limited amount of gaseous substances ;
- resistance against chemical substances and biological organisms [4].

RAW inspection before transport confirms weather the packaging file is in compliance with radiological criteria, mechanical or chemical properties applicable to the given type of the RAW waste dump. RAW transport must be carried out in accordance with the conditions stipulated by the legislation on safe RAW transport [5, 6]:

* Corresponding author: iturekova@ukf.sk

 DOI: 10.1051/e3sconf/20171402026

- Observing the threshold limit values for transportation workers;
- Reducing the risks for transportation workers and people, in case of accidents, to an acceptable level by reducing radioactivity level of the load;
- Limiting unfixed RAW transport;
- Eliminating danger to persons or the environment even in case of major traffic accidents [7].

The goal of this article is to find the safest way of manipulation RAW during transport.

2 Fixation process of radioactive waste

Transport in question is represented by contaminated radioactive inorganic sorbent (waste) which was created during the technological process of dowtherm cleaning. Radioactive dowtherm was used as cooling for fuel elements as a substitute for "Chrompik".

Chrompik is aqueous solution of $K_2Cr_2O_7$ in a concentration of 3-5 %. Chrompik (later Dowtherm) should prevent corrosion of fuel cladding [8].

Beta and gamma activity of organic dowtherm ranges from 10^6 - 10^9 Bq.dm^{-3} and alpha activity is at 10^4 Bq.dm^{-3}. Basic physical characteristics of the substance are specified in Table 1.

Table 1. Basic characteristics of dowtherm [9].

composition	biphenyl (25.5 % of the volume) / / diphenyl (73.5 % of the volume)
colour	colourless to light yellow
solidification point	12.0 °C
boiling point (1 atm.)	257.1 °C
flashpoint	113 °C
ignition point	118 °C
auto-ignition point	599 °C
density	1056 kg.m^{-3}
critical temperature	497 °C
critical pressure	31.34 x 10^5Pa
critical volume	3.17 dm^3.kg^{-1}
average molecular weight	166 g.m^{-1}

RAW fixation is performed into organic or inorganic matrix (e.g. cement, aluminosilicate SIAL matrix).

Shielded cart with a 60 l metal barrel is placed under a fixing apparatus where active inorganic sorbent has been measured out. The amount of sorbent ensures dose rate of maximum 30 mSv/h on the surface of the barrel before fixation. Dose rate on the surface of the barrel with active sorbent has been measured in the shielded truck using gamma probe, displaying the data on the control panel. Homogenisation of the inorganic sorbent is carried out using dispersion stirrer. Homogenisation is followed by fixation into inorganic SIAL matrix. The main SIAL matrix components are natural inorganic materials containing SiO_2, Al_2O_3, CaO, MgO, Na_2O, K_2O, Fe_2O_3 and TiO_2. Stabilization of radioactive sludge and sorbents into SIAL matrix is a cold process as the processing itself takes place at ambient temperature and the final product does not emit heat which would release gases or aerosols. Stabilisation of the product takes 28 days until the fixation matrix together with the sorbent have not reached the desired properties (leachability, compressive strength) which are the conditions for further radioactive waste treatment [10].

Excellent mechanical properties of the final product - high resistance against liquids, excellent radioactive stability and low leachability are the advantages of fixation into alumina-silica matrix.

The majority of mobile equipment in the nuclear power plant (NPP) uses a packaging file for LRAW (liquid radioactive waste) fixation - 200 l barrel used for transport. Packaging file consists of several layers:

1. First layer is represented by a 60 l re-closable barrel with adjusted inner wall to increase its resistance to corrosive effects of modified and stabilized waste together with solidification matrix. Dose rate on the surface of the barrel during transport has a maximum level of 20 mSv/h.
2. Another layer is represented by an adjusted 200 l barrel with a concrete lining serving as shielding. Steel pipes of different thickness may be inserted into the system to increase shielding effect. Barrels are custom made depending on the given type of radioactive waste with the corresponding dimensions of the bore, thickness of the lid and pre-calculated coefficient of radiation shielding [10, 11, 12].

RAW transport must be carried out in accordance with the legislation on RAW transport which includes:

- Observing the threshold limit values of effective dose for workers of transport;
- Reducing the risks for drivers and population, in case of accidents, to an acceptable level by reducing radioactivity level of the load;
- Limiting unfixed RAW transport;
- Eliminating danger to persons or the environment in case of major traffic accidents [13, 14, 15].

Radioactive transport is controlled by a R - command. R - command is a written statement on compulsory technical and organizational measures ensuring the safety of employees at work in conditions of increased radiation risk determining the place, time and conditions for carrying out work, necessary measures and means in order to ensure radiation safety, composition of workers and persons responsible.

Limit values and restrictions during transport are stated in Table 2.

Table 2. Limit values and restrictions during transport [16].

Restriction		Limit value
Radiation dose limit at R-command - absorbed IED* by worker		1 mSv
Maximum dose limit of a 60 l barrel		$20\ mSv.h^{-1}$
Dose Rate (DR) for radioactive material transportation (public notice UJD 57/2006 Coll.)	On the surface of the vehicle	$2\ mSv.h^{-1}$
	On the surface of the vehicle under exclusive use	$10\ mSv.h^{-1}$
	At 2 m distance	$0.1\ mSv.h^{-1}$

**IED - individual effective dose*

Methodology of analysis (Figure 1) shows input and restrictive parameters during LRAW transport. The goal was to optimize transportation procedure of the packaging file observing the threshold limit values total absorbed dose and reduction of transportation time of the packaging file have been established as crucial parameters.

DOI: 10.1051/e3sconf/20171402026

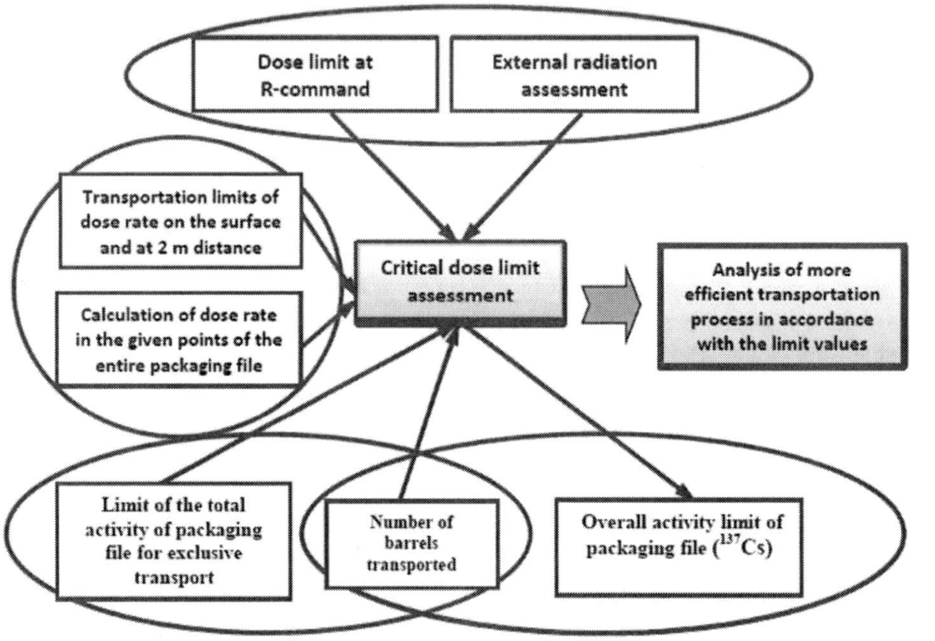

Fig. 1. Factors influencing risk analysis of packaging file transportation [17].

2.1 Present-day situation during packaging file transportation

Transport model (alternative 0) was carried out by a special transportation method with the maximum dose rate DR of 5 mSv/h on the surface of the packaging file. The dose absorbed by the workers participating in the transport is represented by the product of DR at the corresponding distance at which the worker is located during the operation and the time (t) required for the operation. Doses absorbed by the workers refer to the four shielded containers representing a packaging file with fixation product of used sorbents (4 x 60 l).

Before transportation, it is necessary to carry out operations and handling procedures which are directly related to transportation and the measurement of radioactive properties of the transported material. The initial condition is represented by a 60 l barrel placed onto a shielded transport cart under the crane on the floor level of +6 m (room A). The process, carried out by workers, was divided into 10 major steps (Figure 2).

Table 3 shows measurement output of dose rates and radiation load during transport. Trial measurement (alternative 0) was carried out with four pieces of 60 l barrels with dose rate on the surface of the transported barrels ranging between 18.3 mSv/h and 19.7 mSv/h. Increase in overall absorbed dose depending on time are shown in Figure 3.

DOI: 10.1051/e3sconf/20171402026

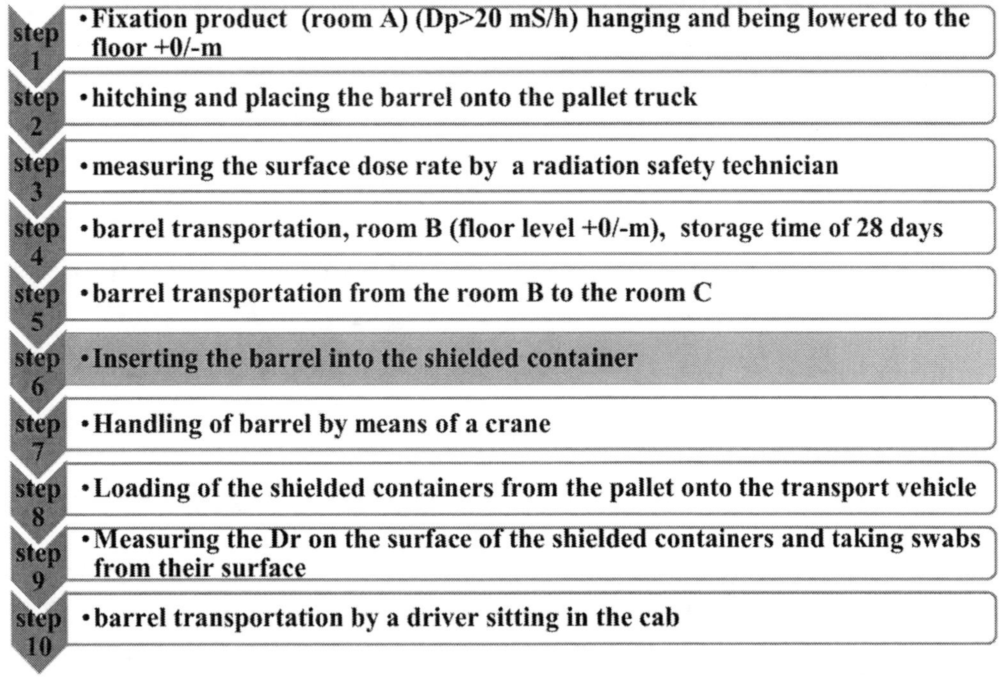

Fig. 2. Description of procedures during liquid radioactive waste transport (alternative 0).

Table 3. Calculation of radiation load during handling and transport [5].

Step	Number of employees	Distance from the source [m]	Time [min]	DR [mSv/h]	Dose [mSv]	Overall dose [mSv/h]
1.	2	0.5 -1.5	8	0,502	0.066	0.122
2.	2	1.0	6	1.433	0.143	0.286
3.	1	2.0	4	0.232	0.015	0.015
4.	2	1.5	7	0.502	0.058	0.116
5.	2	1.0	5	1.433	0.119	0.238
6.	2	0.5	2	4.57	0.152	0.304
7.	1	2.0	10	0.232	0.038	0.038
8.	4	0.5	5	0.358	0.029	0.116
9.	1	0.5	6	0.358	0.035	0.035
10.	1	2.5M + shielding of the cab	40	0.012	0.008	0.008
Sum total	17		93		0.663	1.278

Step-by-step logistics of the transport recorded different increments of absorbed dose which was related to time interval of transport, dose rate of the environment, time of the procedure and distance. Insertion of the 60 l barrels into the shielded containers seemed to be the most critical operation. The procedure, which takes 2 minutes only, is carried out by 2 employees.

From the point of view of time schedule, works were being carried out for 28 days which resulted from the technological requirements within the work program and the two different R-commands. Doses absorbed by workers transporting the fixation product of used sorbents (steps 1- 4) have been carried out at one R-command and workers when transporting or packaging four shielded barrels for transmission range (steps 5- 10) and the other tasks at the second R-command. All absorbed doses have been recorded into the central information system.

DOI: 10.1051/e3sconf/20171402026

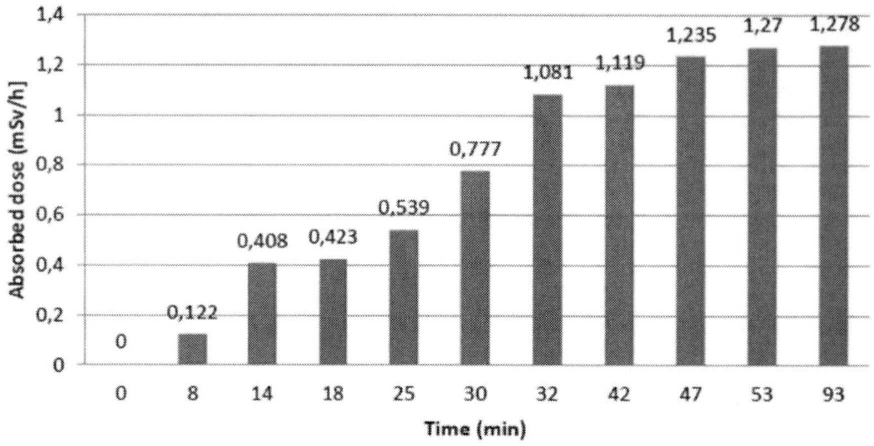

Fig. 3. Increment of overall absorbed dose for each step of the procedure depending on time (Alternative 0).

2.2 New solution for handling fixation products

Two 60 l barrels of high dose rate (maximum dose rate is 20 mSv/h) have been selected as test specimens for the experiment. Measurements have been carried out three times (see Table 4). Average values for calculating dose rate reduction coefficients depending on ionising radiation source (Table 5) according to the equation 1 were used. Dose rate reduction coefficient is defined as a ratio of maximum point dose rate measured directly on the ionising radiation source to dose rate in the given distance. In case of exclusive transport, all measurements, however, must meet one condition - dose rate at any place within the external surface of the transported unit shall not exceed 10 mGy/h.
New solution for handling of fixation products

$$\mathrm{Dp(i)} = \frac{\mathrm{Dp(k)}}{\mathrm{k(i)}} \tag{1}$$

where:
Dp(i) - is the dose rate value measured at a distance i (metres) from the barrel,
Dp(k) - is the dose rate value measured directly on the surface,
k(i) - is the dose rate reduction coefficient for the distance i.

Table 4. Dose rates results of 60 l barrels at different distances [5].

Barrel no.	Measurement	Dose Rate (DR) at a distance of [m]				
		0 (on the surface) [mSv/h]	0.5 [mSv/h]	1.0 [mSv/h]	1.5 [mSv/h]	2.0 [mSv/h]
1	1	19.7	4.58	1.52	0.44	0.21
	2	19.7	4.47	1.47	0.50	0.22
	3	19.7	4.02	1.51	0.47	0.21
2	1	18.3	4.25	1.41	0.41	0.19
	2	18.3	4.15	1.36	0.47	0.21
	3	18.3	3.73	1.407	0.44	0.20

Table 5. Dose rate reduction coefficients in the axis of ionising radiation source

Barrel No./ measurement	"c" value in "i" distance [m]						
	0	0.5	1.0	1.5	2.0	2.5	3.0
1/1	1	4.3	12.9	43.9	92.2	148.2	197.6
1/2	1	4.4	13.4	38.7	85.7	144.7	189.7
1/3	1	4.9	13	41.3	91	145.6	191.6
2/1	1	4.6	14.7	46	86.1	152.3	190.4
2/2	1	4	16.7	38.5	83.3	142.9	185.2
2/3	1	4	13	30.4	78.3	137	182.7
Average	**1**	**4.37**	**13.95**	**39.8**	**86.1**	**145.1**	**189.5**

By means of evaluation of the data obtained during the whole process of handling of the final product, the measurements - carried out under dosimetric supervision - show a considerably high radiation load for workers in one cycle of the process. One cycle includes handling and transport of four pieces of 60 l barrels, whereas 83 pieces were initially considered. The new proposals´ aim was to use spatial and technical equipment during RAW handling and transportation. Compliance with radiation load conditions for workers during each step as well as time reduction of such operations were the two conditions that had to be met.

Technical solutions and workplace layout enabled two alternative ways of handling of the final fixation product of used sorbents. The final fixation product of used sorbents is located in the shielded cart in room A, at the elevation of +6 m. This room is linked with room C (elevation of +0/-m) which serves, at the same time, as a transport corridor. Both of them are linked by a crane.

3 Results

3.1 Alternative 1

Transport vehicle is prepared in room C below the crane from the room A. Surface of the transport vehicle is covered with polyethylene foil. A pallet for four 60 l barrels is placed on the loading area at 1 m distance from the end of the loading area and 2 metres from the driver's cab. Two employees fix the transported barrel by a vacuum holder (Figure 4), lower it on the pallet placed on the loading area and loosen the vacuum holder. Radiation safety officer measures the DR on the transported barrel surface. This procedure is repeated three times and, in the end, after releasing the vacuum holder, workers cover the pallet with four 60 l barrels with a polyethylene foil. Radiation safety officer takes a swab from the packaging file surface and evaluates the measured values of DR and surface contamination taking into account the conditions and limits for transport in nuclear power plant site.

Time schedule of each step - alternative 1 (Figure 5).

DOI: 10.1051/e3sconf/20171402026

Fig. 4. Vacuum holder [5].

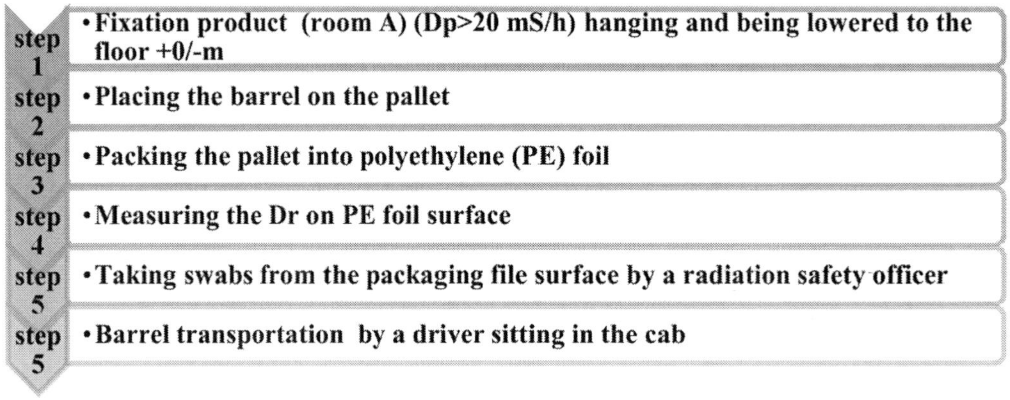

Fig. 5. Sequence of procedures during handling and transportation according to alternative 1.

Time schedule for steps 1 to 5 refers to 4 pieces of transported barrels. DR and time data obtained from the detailed description of the steps necessary for handling of the fixation product of used sorbents of alternative no. 1 are specified in Table 6.

Compared to alternative no. 0, temporary storage in room B was omitted, enabling to skip four steps of the procedure; a new step - wrapping pallets into polyethylene foil to reduce surface contamination - has been added. DR values on the surface of the packaging file (pallet with four pieces of barrels wrapped in PE foil) were 20 mSv/h. It means that the limit of 10 mSv/h - set for exclusive transport in the nuclear power plant site - has been exceeded. DR limit on the surface of the transported packaging file on the edge of the vehicle is 1.43 mSv/h. This limit for exclusive transport would then be met. If we consider a given promising alternative solution, the subject is assessed Nuclear safety agency. Significant reduction in dose absorbed by workers participating in the process can be considered as a positive result of alternative no. 1 if the same result has been achieved.

 DOI: 10.1051/e3sconf/20171402026

Table 6. Calculation of radiation load during handling and transportation, alternative 1 [5].

Step	Number of employees	Distance from the source	Time [min]	DR [mGy/h]	Dose [mSv]	Overall dose [mSv/h]
1.	2	0.5 -1.5	8	0.502	0.066	0.122
2.	2	1.0	6	1.433	0.143	0.286
3.	2	1.0	1	1.433	0.023	0.046
4.	1	2.0	4	0.232	0.015	0.015
5.	1	0.5	1	4.57	0.076	0.076
6.	1	2,0*	40	0.137	0.091	0.091
Sum total	**8**		**60**		**0.414**	**0.636**

**shielding by the cab*

3.2 Alternative 2

The procedure remains the same as with alternative no.1. Having different gear for the transport vehicle, overall reduction of DR can be achieved. Before transportation, workers place a pallet with four pieces of shielded barrels onto the vehicle. Barrels are opened then and concrete lids are placed on the loading area of the vehicle near the pallet. The pallet with four shielded barrels has been prepared on the loading area at 1 m distance from the end of the loading area and 1.6 m from the driver's cab. Two employees fix the transported barrel by a vacuum holder and lower it into the pre-prepared shielded barrel. The barrel is placed then into the opening in the shielded concrete barrel and the vacuum holder is released. To load the whole consignment, the procedure is repeated four times. When the 60 l barrels are placed into the shielded containers, workers close the shielded containers by concrete lids and secure the 200 l MEVA containers by lids and hoops. Workers leave then the loading area and stay at a distance fixed by a radiation safety technician. Radiation safety technician measures the DR on the surface of the transported shielded container by taking swabs to check the surface contamination (SC), assesses the values of DR and SC taking into consideration transportation conditions and limits in NPP site and issues the documents necessary for transport.

Time schedule of each step of alternative no. 2 are in figure 6 and the DR data and time data obtained from the detailed procedure description necessary for handling of the fixation product of used sorbents are specified in Table 7.

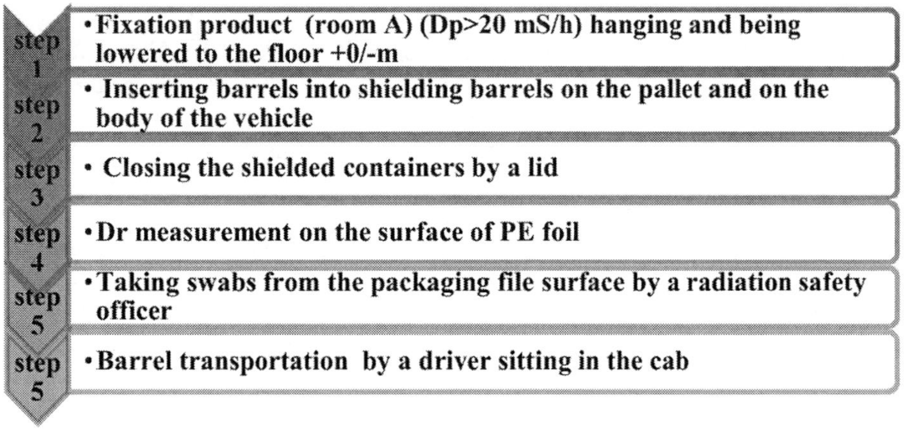

Fig. 5. Sequence of steps during handling and transportation under alternative 2.

DOI: 10.1051/e3sconf/20171402026

Table 7. Calculation of radiation load during handling and transportation, alternative 2 [5].

Step	Number of employees	Distance from the source	Time [min]	DR [mGy/h]	Dose [mSv]	Overall dose [mSv/h]
1.	2	0.5 -1.5	8	0.502	0.066	0.122
2.	2	1.0	6	1.433	0.143	0.286
3.	2	1.0	4	0.358	0.023	0.046
4.	1	2.0	4	0.058	0.004	0.004
5.	1	0.5	2	1.44	0.038	0.038
6.	1	2.0*	45	0.058	0.043	0.043
Sum total			67		**0.317**	**0.539**

**shielding by the cab*

By using shielded containers for transport, the DR on the packaging file surface decreased from 20 mSv/h to 5 mSv/h, compared to the alternative No. 1. This means that the limit for land transport in nuclear power plant site has been met. By means of DR reduction coefficient, DR value on the edge of the vehicle will be 0.454 mSv/h provided that the packaging file is placed at the distance of approx. 0.8 m from the edge. None of the measured and calculated values exceed the limit for exclusive transport. The conclusions presented above show that the procedure in alternative no.2 is applicable into practice without changing the legislation or having exceptions to the specified limits. Significant reduction in doses absorbed by workers participating in the operation with the same result - to transport the final product from the point A to point B - can be considered as a positive result of alternative no.2. However, alternative no.2 compared to alternative no.1 requires more equipment (shielded containers) what means higher expenses.

4 Discussion

All of the above mentioned procedures of handling of the final fixation product of used sorbents can be compared considering several aspects:

- observing the limits of radiation safety,
- time needed for the process,
- economic aspect,
- technological demands.

Alternative no. 0 is more time consuming, since it includes the initial re-stocking of the final product of the room A into room B and from the room C. This means the highest dose rate for workers.

Alternative no.1 and no.2 omit the process of stock transfer in order to reduce radiation load for workers participating in the process. Stock transfer and temporary storage of final product is worked into the valid work program and justified by running test for compactness of the final product which are performed on every 10th barrel containing the final product. This operation is not described in the individual steps as it is not linked directly with transport and, moreover, radiation dose for workers carrying out measurement would be the same. Alternative no.1 and 2 only describe operations required for transport of the final product. Since strength tests of each tenth barrel are one of the conditions of product storage, it means that, with the overall production of 100 pieces of 60 l barrels, the proposed procedure in alternative no.1 and 2 is applicable for 90% of the barrels from the overall production. 10% of the barrels would be handled following the valid work program.

Table 8 shows absorbed doses for workers who have taken part in transport and handling of final fixation product. The table below shows time data of individual steps of the procedure.

Table 8. Comparison of time schedule and radiation rate for each procedure

Work procedure	Number of employees	Time [min]	Absorbed dose / worker [mSv]	Overall absorbed dose [mSv/h]
According to the program	17	93	0.075	1.278
Alternative 1	8	**60**	0.079	0.636
Alternative 2	8	65	**0.067**	**0.539**

From the point of view of radiation load, alternative no. 2 is the most appropriate from the above mentioned solutions. With the overall production of 100 pieces of 60 l barrels, the whole procedure needs to be repeated 25 times, resulting in overall dose reduction for all workers present from 31.95 mSv to 13,475 mSv. Difference between estimated absorbed dose according to the valid work procedure program and suggested alternative no.2 is 16,628 mSv. At the same time, it will reduce the time required for transport of packaging file by almost one third.

Conclusion

Radiation safety is an important component and limiting condition for all the works in nuclear power plant site. On the basis of work procedures and technological methods of radioactive waste processing, it was possible to accurately measure and monitor doses absorbed by all workers. By evaluating the measured values of dose rates and time needed for each step - handling and transport - it may be possible to make an analysis of absorbed doses during each steps of the work procedure and to propose an alternative which would reduce time during packaging file transport in NPP site but, in particular, it would lead to reduction in absorbed doses for the employees involved.

Acknowledgments: This article was supported by the Grant Agency Ministry of Education SR KEGA - project no. 014UKF-4/2016.

References

1. IAEA: The Principles of Radioactive Waste Management, Safety Series No. 111-F, IAEA, Vienna (1995).
2. M. ZACHAR, V. NEČAS, Problematika nakladania s rádioaktívnymi odpadmi v Slovenskej republike. *Jaderná energetika v pracích mladé generace.* ČNS. Brno, Česká republika..ISBN 978-80-02-02116-2, p. 49-55,(2008)
3. IAEA, Predisposal Management of Low and Intermediate Level Radioactive Waste: Safety Guide No. WS-G-2.5. Vienna: IAEA (2003)
4. IAEA, Regulations for the Safe Transport of Radioactive Material: Safety Requirements No.TS-R-1. Vienna: IAEA, (2005)
5. Hlubík, M. Radiačná bezpečnosť pri transportoch rádioaktívnych odpadov v areáli JE [Diploma thesis]. MTF STU, (2013)
6. VYHLÁŠKA ÚJD SR č.57/2006 Z.z., ktorou sa ustanovujú podrobnosti o požiadavkách pri preprave rádioaktívnych materiálov (2006)

7. Vyhláška ÚJD SR č. 30/2012 Z.z., ktorou sa ustanovujú podrobnosti o požiadavkách pri nakladaní s jadrovými materiálmi, rádioaktívnymi odpadmi a vyhoretým jadrovým palivom (2012)
8. Dulanska, S., Meloun, M., Matel, L. Scaling model for prediction of radionuclide activity in cooling water using a regression triplet technique. Journal of radioanalytical and nuclear chemistry 285.3. pp.547-553. (2010)
9. Engineering and Operating Guide for DOWTHERM SR-1 and DOWTHERM 4000: Inhibited Ethylene Glycol-based Heat Transfer Fluids, www.dowtherm.com (2008)
10. IAEA, Radioactive Waste Management – Status and trends. Issues No. 4., Vienna, (2005)
11. Seidl, M., Tomek. M. *Transportation of radioactive materials.* Slovenské jadrové fórum. Ročník 6., No I, (2011)
12. SÚJB, Přeprava radioaktivních látek. Bezpečnostní návod BN-JB-1.13. https://www.sujb.cz, (2011)
13. IAEA, Regulations for the Safe Transport of Radioactive Material,Viena, (2009)
14. Tureková, I., Rybakowski, M., Szabová, Z., Dudarsku, G. Efficiency rating of vision in terms of prevention of road accidents at work. Advanced Materials Research, vol. 734-737, 2013, pp. 1613-1616 (2013)
15. Shcherba, I.D., Bekenov, L.V., Antonov, V.N., Noga, H., Uskokovic, D., Zhak, O., Kovalska, M.V., Electronic structure and X-ray spectroscopic properties of YbNi2P2, Journal of Electron Spectroscopy and Related Phenomena, Volume 212, pp. 5-10. (2016)
16. Strážovec, R. et al. Aktualizácia limít a podmienok pre prevádzku technológii. JZ TSÚ RAO. SNUS, Častá - Papiernička, (2014)
17. Hornáček, M., Nečas, V., Bezák, P. Calculation of the dose load during dismantling of large components in the process of decommissioning of nuclear installations. ICEM2013, September 8-12, Brussels, (2013)

E3S Web of Conferences **14**, 02027 (2017) DOI: 10.1051/e3sconf/20171402027
Energy and Fuels 2016

Energetic effects of adsorption of C_6-C_8 hydrocarbons on hard coal samples of different metamorphism

Agnieszka Orzechowska-Zięba[1,*] and *Adam* Nodzeński[1]

[1]AGH University of Science and Technology, Cracow, Poland

Abstract. The analysis of sorption equilibrium of aliphatic saturated and unsaturated hydrocarbons on low and medium rank coals were performed. Isosteric sorption enthalpy correlation with the degree of fill of sorption capacity was presented for the low rank coal. There was found a crucial role of primary pore structure in the coal grains in the process of sorption, especially in case of temperature anomalies occurrence. Hexane differently than other hydrocarbons of longer chains, represent the property of micropore penetration. Due to literature the presence of double bound in other molecules causes difficulties with penetration into porous structure of coal.

1 Introduction

The analysis of structure, chemical and physical characteristic of coal as well as organic reactivity of coal are basic research measurements. The correlation between chemical and physical structure and physicochemical properties of hard coal are significant for the determination of its usefulness in the coal processing into chemical and energy raw materials as well as in the research aimed at determination of gases accumulation in coal.

Hard coal seams can be treated as a specific collector of mine gases. Those gases are mainly: CO_2, CH_4 i N_2, its share in the seams is directed by geological history of the deposit. Seams posses different properties than typical reservoirs with porous fulfillment. Those differences are a result of coal ability to deform under the influence of external mechanical load, pressure of mine gases and water in the pores and channels of coal, as well as sorption capacity of coal substance [1]. These factors affect the permeability of the deposit, at the same time remain in close relation to the techniques of obtaining methane, predicting emissions of gas in the process of mining or uncontrolled outbursts of gases and rocks.

The sorption experiments of gases and vapors on coal, especially the analysis and interpretation of the equilibrium state of sorption process at constant temperature, carried out initially in the fifties and sixties of the last century, are still of scientists interest. It is an important source of information for describing the characteristics of various coal types with respect to their porosity, coal coalification degree, type and composition of macerals, chemical structure etc.

*Corresponding author: azieba@agh.edu.pl

Coal is a complex structure, in which sorption properties are significantly associated with the degree of coalification [2,3,4]. Multiple oxygen groups on the surface, control the degree of hydrophilicity of the coal surface. It is known that the structure of the coal to large extend represent apolar character. Therefore, coal is said to be a system of dual nature: hydrophobic and hydrophilic. The reactive groups in a very large degree influence sorption of polar substances and in a characteristic way regulate sorption of apolar substances such as methane or aliphatic hydrocarbons as well as aromatic hydrocarbons and carbon dioxide [5]. By the analysis of phenomena connected with the structure responsible for the process of sorption/desorption of gases/vapors, the occurrence of outburst of gases and rocks as well as accumulation and capturing of these gases can be explained [6,7,8,9]. Interactions sorbate - sorbent play an important role in the description and modeling of processes such as outburst of gases and rocks.

2 Materials and experimental methods

From the group of aliphatic saturated hydrocarbons there were selected sorbates: hexane, heptane, octane, and for the comparison two from alken group: 1-hexene, 1-heptene. Adsorbates selection was dictated by cognitive and practical aspects. What is important is that those substances differ in molecule structure, molecular dimensions, bonds present and the dipole moment value, but at the same time have similar physicochemical properties, particularly, the value of the vapor pressure of the examined temperatures. What's more are components of gas mixtures present in the mining gases of coal seams. Selected sorbates were of analytical grade purity. In standard conditions they were present in the liquid state.

Samples of coals of different coalification degree (ISO 11760 : 2007) from polish coal mines were selected as a basic research material:

-low rank coal – Brzeszcze coal mine (seam 325, level 430m) – ***B,***
-medium rank coal – Victoria coal mine (field "Witold", seam 307) – ***Vi,***
-high rank coal – Wałbrzych (field "Chrobry", seam 430) – ***W.***

Above there were presented the symbols of samples, used instead of full names, which were used in this paper. Systematic of samples was proposed for clarity and readability of the work. Samples of coal were selected in the way that carbon contend would differ and that they had different average reflectivity index R_n^0. It is claimed that the value R_n^0 indicates a degree of metamorphism of the examined coal. For the study, sample of coal of grain size ϕ 1.0-1.5mm were chosen, grains were of irregular shape. Grains of this size were large enough to minimize the effect of fractioning, at the same time small enough to reach equilibrium and allow to fulfil the condition of isotherms reversibility. In addition, it was possible to use a uniform particle size distribution for all types of performed analysis, which proves the results to be comparative. Majority of sorption research on hard coal presented in literature is performed on powder samples or on samples of granulation not higher than 0,9mm. As a result of such drastic fragmentation lost of primary pore structure can be observed as well as lost of chemical character of outer coal grains surface, which can have influence on higher values of sorbed amount of gas/vapors [20,21]. In order to characterize the research material there were analysis carried out: porosimetry and densimetry analysis, helium density measurements. Furthermore Boehm method was used to determine the surface functional groups. Ultimate and proximate (Table 1) and petrographic (Table 2) analysis were taken from the literature [10,11]. Characterisation of coal samples is shown in Tables 1-4.

Table 1. Selected parameters of ultimate and proximate analysis of experimental hard coal samples [10,11]

Coal sample	Chemical composition [wt. %]			Technical analysis[wt.%]			
	C^{daf}	H^{daf}	$(O+N)^{daf}$	S^{daf}	V^{daf}	W^a	A^a
B	79.60	5.09	14.70	0.72	40.87	2.9	15.50
Vi	87.58	4.88	6.41	---	27.9	1.25	7.78
W	92.13	2.95	3.76	---	9.46	0.77	9.10

Explanation: W – moisture; A –ash; V – volatile matter;
a – analytical state; daf – dry and ash-free state

Petrographic analysis of examined coals was performed by optical microscopy method by the Coal Geology Research Group, Faculty of Geology, Geophysics and Environmental Protection, AGH. Macerals and mineral matter contents for coals studied were presented in Table 2. The organic matter was formed basically of macerals of the vitrinite group 60.7 - 55.5% vol., liptinite 0.0 - 9.6% vol. and inertinite 27.5 - 38.4% vol. This data were taken from the work [10,11]. Sample *Vi* was characterised by a low content of liptinite group maceral with a relatively high content of inertinite group.

Table 2. Macerals and mineral matter contents for studied hard coal samples coals

Petrographic analysis [vol.%]						
Coal sample	Macerals group			Clay minerals	Pyrite	Carbonates
	Vitrinite	Liptinite	Inertinite			
B	60.7	9.6	27.5	2.1	trace amount	---
Vi	56.1	4.9	31.8	7.2	1.5	2.6
W	55.5	0.0	38.4	6.1	0.2	2.0

As a complementary to ultimate and proximate analysis and petrographic measurements, porosimetric and densimetric analysis was performed. Based on the densimetric results along with porosity and total pore volume (ε i V_{dens}) it can be stated that sample *B* was characterized by high value of total pore volume and in consequence, by high value of porosity. Densimetric measurement results were presented in Table 3. To give more detailed view on the structure of analyzed coals the pore volume distribution measured with the use of porosimetry was presented.

Analysis of porosimetric data was limited to macropores r>25 nm due to the significant compressibility of the coal in higher pressures.

As complementary analysis for characterizing the samples of coal, surface coal functional groups were determined by Boehm method [12]. This method is successfully used for adsorbents such as activated carbons or soot. Multiple measurements of functional groups for selected coals, even at ten time larger analytical samples gave differentiated results. Therefore, the paper presents the arithmetic average of this measurements. For accurate estimation of content of these groups, other methods, based on different principles should be used. The results were shown in Table 4. Based on the experimental data, in all hard coals there were quantified very small amounts of carboxylic, phenolic and lactonic groups. Coal of high degree of metamorphism was characterized by a very low content of the lactone group. With the increase of coal coalification degree, decrease in the amount of lactone groups was observed and to a lesser extend phenolic groups.

DOI: 10.1051/e3sconf/20171402027

Table 3. Comparison of high pressure porosimetric and densimetric measurements

Sample		*B*	*Vi*	*W*
d_{He}[g/cm^3]		1.38	1.41	1.45
d_{Hg}[g/cm^3]		1.23	1.30	1.32
ε[%]		10.86	7.80	8.96
$V_p \cdot 10^2$[cm^3/g]		8.83	6.00	6.79
$V_{mac} \cdot 10^2$[cm^3/g]		1.20	3.39	0.81
$(V_{mic} + V_{mes}) \cdot 10^2$[cm^3/g]		7.63	261	5.98
V_{mac}/V_p[%]		13.6	43.5	11.9
$V \cdot 10^2$ [cm^3/g] in ranges [nm]	**25 – 100**	0.39	0.7	0.26
	100 – 2500	0.72	2.21	0.53
	2500 – 7500	0.09	0.48	0.02

Where V_p, V_{mac}, V_{mes}, V_{mic}, volume of pores, volume of macro-, meso- and micropores.
Real coal density d_{He} was determined using the automated apparatus Accu Pyc 1330 (Micrometrics).
Apparent coal density d_{Hg} was determined using the apparatus Pascal 440 CE Instruments.

Table 4. The results of determination of oxygen functional groups with the use of Boehm method on the hard coal samples of different coalification degree

	Carboxylic pK=6.37	**Lactonic pK=10.75**	**Phenolic pK=15.47**	**Sum**
	[μmol/g]			
B	36.2	16.3	112.4	164.9
Vi	86.8	6.8	89.4	186.0
W	27.5	2.5	74.4	104.4

3 Results of sorption experiments

Prior to the sorption measurements hard coal samples were subjected to long-term coal degasification process until the vacuum 10^{-3} Pa was reached. In order to accelerate the degassing of samples, process of "helium flushing" took place. In this process the sample was left in the atmosphere of helium for 24 h. The helium atoms themselves do not adsorbed on the adsorbent surface, but provided the necessary kinetic energy to the molecules of adsorbed vapors and gases, in this way removed it from the surface of the sorbent [13]. After each "helium flushing" samples were again degassed. As prepared samples were subjected to sorption experiments.

Sorption properties in a coal – sorbate configuration and obtaining sorption isotherms can be considered in terms of static and dynamic methods. For this sorption studies static method–microburette liquid, using volumetric apparatus was selected [14]. Consecutive points of the isotherms were measured by dosing the sorbate vapor portion by opening the mercury valve, so producing a simultaneous increase in the pressure in the sorption space. Estimation of the amount of sorbed hydrocarbon was determined on the basis of the vapor pressure measured before dosing, after dosing and after reaching the sorption equilibrium. Measurements were performed in the temperatures of 298, 303, 308 K.

Hydrocarbons vapors may derive from ideal gas laws as a result of intermolecular interactions (even in low pressure conditions). Therefore the derivations of hydrocarbons vapors from ideal gas properties had to be included. For this purpose fugacity coefficient ϕ and fugacity f_a of analysed hydorcarbons were examined as a function of pressure. Fugacity coefficient was calculated based on Berthelot equation [16], as well as parameters of critical point, using integral formula linking compression and fugacity coefficient. Taking into account not ideality of vapors, while calculating the sorbed amount, the fugacity coefficient was included instead of pressure in the equation for a calculation.

3.1 Hard coal-hydrocarbons adsorption isotherm

Examples of sorption isotherms for the investigated systems were shown in Figures 1 and 2 (298K) and Figures 3 and 4 (308K). Sorption isotherms were presented as a relation between the amount of adsorbed hydrocarbon vapors and its relative fugacity. Low sorption capacity of hard coal and a slight decrease in liquid amount in dosing part of apparatus caused a large increase in pressure on the measuring side of apparatus, which may explain not accurately appointed equilibrium points and deviations of these points on the experimental isotherms. Because of the low vapor pressure, particularly in case of long-chain hydrocarbons, during the measurement the optimal time for reaching equilibrium was chosen, after which no equilibrium pressure change was noticed (two, in some cases tree days).

In Figures 1-4 sorption isotherms of saturated and unsaturated hydrocarbon on coals *B* and *Vi* were presented. Isotherms can be characterized by the type II isotherm by the Brunauer classification, except isotherms of octane on coal *Vi*. Type II isotherm is associated with the formation of a multimolecular adsorption layer [17]. Sorption isotherms of heptane and octane on coal *Vi* can be classified into rare type III isotherms.

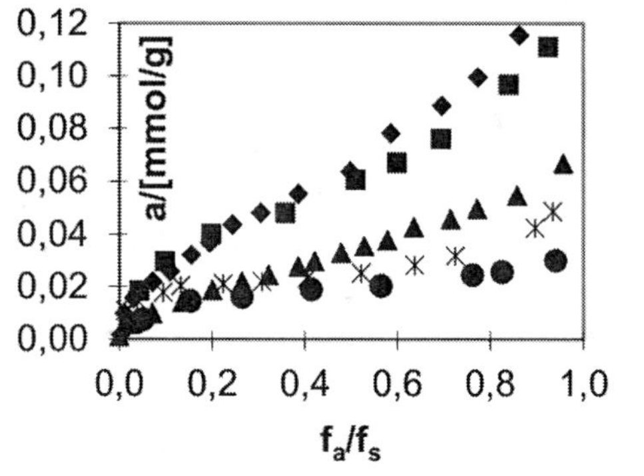

Fig. 1. Sorption isotherms of examined hydrocarbons on coal B in the temperature 298K

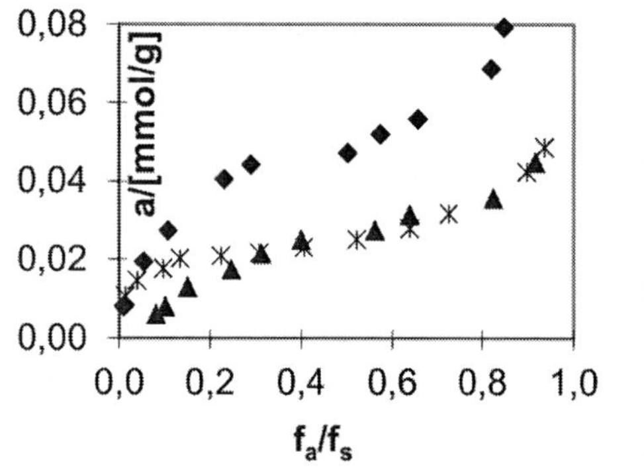

Fig. 3. Sorption isotherms of examined hydrocarbons on coal B in the temperature 308K

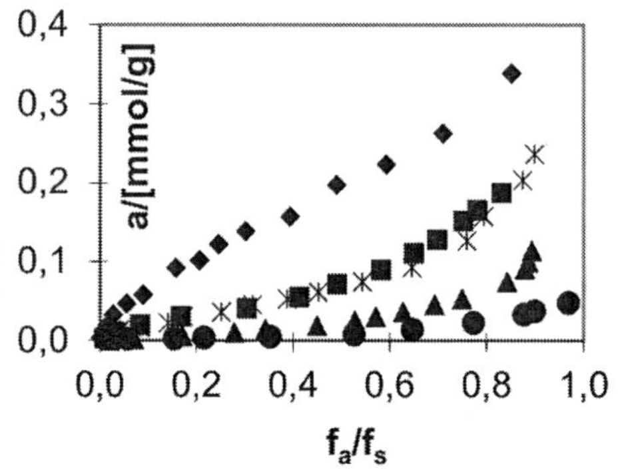

Fig. 2. Sorption isotherms of examined hydrocarbons on coal Vi in the temperature 298K

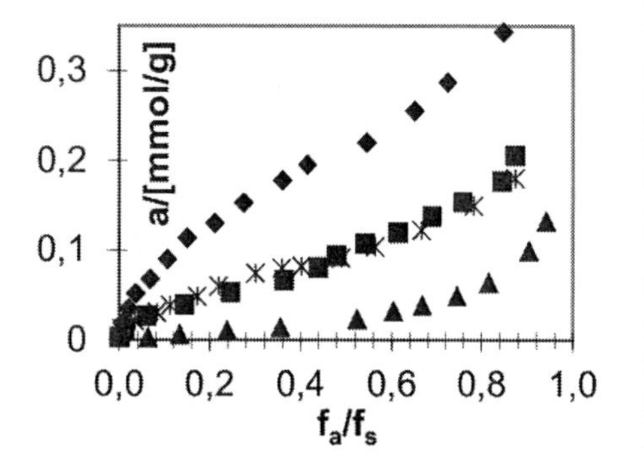

Fig. 4. Sorption isotherms of examined hydrocarbons on coal Vi in the temperature 308K

Legend: ●-n-heptane,▪-1-heptene, ▲-n-octane, ♦-1-heksene, x-n-heksane

Analyzing the adsorption isotherms of hydrocarbons on the coal of different coalification degree, it can be found that sorption of unsaturated hydrocarbons was higher than saturated ones. This was the case for both – coal *B* and coal *Vi*. Privileged polar hydrocarbon sorption on the coal samples was probably associated with the impact of the double bond (electrons

π) with polar groups on the surface of coal. Examined hydrocarbons sorption capacity of vapors was relatively low, in particular of apolar hydrocarbons. This was most probably caused by adsorption mechanism of capturing these sorbents by hard coal.

Comparing isotherms in Figure 1 it is worth noting that coal *B* isotherms of unsaturated hydrocarbons coincided practically through the entire range of relative fugacity, while for sample Vi (Fig. 2) isotherm of 1-hexene (*a* [mmol/g]-almost three times higher) was placed above the isotherm of 1-heptene and isotherms of other sorbates. Kind of regularity was observed in case of hexane, heptane and octane isotherms on sample *Vi*.

Comparison of sorption values suggests that the porous structure of the tested hard coal samples represented the properties of molecular sieve. In general, it can be stated that the low adsorption capacity of examined coal samples was the result of pores blocking by the molecules adsorbed on the surface. This effect may suggest discoid and globular groups creation on the surface [18,19] and the incomplete hydrocarbon chain penetration into pore structure of coal.

However, based on the value of the amount adsorbed on coal test samples, it can be stated that the coal *Vi* had more than twice higher sorption capacity with respect to C_6 hydrocarbons, and several times lower in relation to heptane and octane. Sorption capacity of the two samples of coal in relation to 1 - heptene was almost the same.

Sample *W*, with a high degree of metamorphism, in respect to analyzed sorbates showed indistinguishable and practically immeasurable adsorption. This suggests adsorption on the outer surface of coal grains, which surfaces along with granules diameters of approx. 1.0 - 1.5mm, is small. As a result in this paper the sorption isotherms of coal W were not presented.

This preliminary analysis of the chosen adsorption isotherms indicated that in relation to organic sorbates of relatively long chain, the mesoporosity and macroporosity was of highest importance. The adsorption increase in the initial ranges of the relative fugacity was low (the smaller the longer the chain was). Only above f_a/f_s value of 0.5, there was observed the increase in the amount of the adsorbed molecules.

The study found that the sorption isotherms of hydrocarbon vapors on hard coal in the investigated pressures ranges and temperatures were irreversible, in each of examined cases open hysteresis loop was easily visible (the predominance of sorption processes in the microporous coal structure) [20, 21].

4. Elaboration and analysis of research results

Using the experimental adsorption data up to adsorption relative fugacity, $f_a/f_s = 0.3$, for both hard coal samples *B* and *Vi*, experimental data was fitted using equations of adsorption isotherms. The attempt for a formal description of the experimental isotherms using the thermal sorption equation isotherms was performed along with preparation of calculated isotherms.

Thermal adsorption equation, which author used to describe the adsorption equilibrium, was derived from Jagiello and co-authors [22,23]. The experimental data were governed by the thermal equation of sorption based on an empirical fact that isosteric heat of sorption is in fact temperature-invariant in a wide range of temperatures.

The applied numerical method minimizes the deviations of the regression curves from all experimental points. That is why the results were correlated both in terms of pressure and temperature, which is of key importance when analyzing data obtained at different temperatures, enabling us to derive temperature relationships and to predict the shape of sorption/desorption isotherms at any temperature in the range of experimental data [24, 25].

The condition of temperature in variance Q_{st} is given by the formula

$$-R\left(\frac{\partial lnp}{\partial\left(\frac{1}{T}\right)}\right)_v = RT^2\left(\frac{\partial lnp}{\partial T}\right)_v = Q_{st}(v)\ (1)$$

Integrating this equation for v=const yields the general form of the equation

$$lp = -\frac{1}{RT} Q_{st}(v) + f(v) \qquad (2)$$

where f(v) is a function dependent on the amount of gas sorbed (v).
For the numerical calculations the equation (1a) was used in a form:

$$lnp = \frac{1}{T} \sum_{i=0}^{n} A_i v^i + \sum_{i=0}^{k} B_i v^i + zlnv \qquad (3)$$

Where $A_o...A_{ni}$ $B_o...B_k$ – best fit parameters, (n), (k) – degree of polynomials approximating the functions present in the equation, p – equilibrium pressure, v – the amount of gas adsorbed, z – parameter, which value can be 0 or 1 in accordance to the fact whether approximation data are in low fill ranges.

This equation describes well the adsorption equilibrium data of gases and vapors on different kinds of adsorbents in wide ranges of pressures and temperatures with an accuracy corresponding to the accuracy of the experiment, at a relatively small number of empirical parameters. It can be treated as the regression equation defining the value of the adsorption, depending on the pressure and temperature. It also allows for good accuracy extrapolation for other temperatures and pressures, provide crucial information about a heterogeneity of the adsorbent and about energy of adsorbate-adsorbent interactions.
The quality of fit is given by the formula

$$\sigma = \sqrt{\frac{1}{N} \sum \left[ln\left(p_i^{doś}\right) - ln\left(p_i^{obl}\right) \right]^2} \qquad (4)$$

Parameters of equation and fitting errors σ were determined. It was found that in most cases a satisfactory fit was obtained for $n = 4$, $k = 0$ and $z = 1$.

Examples of sections generated isotherms (Fig. 5-8) were presented for the tested coal samples. On the graphs, points without fill represented experimental data points. The points with color filling represented isotherms calculated according to the thermal adsorption equation. The aim of this fitting, in the relative pressures, was to determine isosteric heat of adsorption. The graphs did not show the full range of adjustment, but illustrated only the most probable course of isotherms in the low fills. Very often it differed from the experimental points shown as examples, what was caused by parameters obtained in the full fit range.

The presented isothermal sections allowed to draw only qualitative conclusions. It was visible that isotherms for coal *B* had rather regular shape and the increase in temperature lead to a reduction of sorption value. Isotherms of non-typical shape can be observed for coal *Vi*. With the increase in temperature the amount of water vapor adsorbed increased, what for adsorbent of rigid structure is unlikely to happen. Furthermore, with increasing hydrocarbon chain length, the isotherms changed its shape. The porous structure of a sample *Vi* presented the dominant share of mesopores with radious in the ranges of 100-2500 [nm] (see. Tab. 3), thus raising the temperature would lead to increase of the molecule kinetic energy and the pore diameters enlargement. This might increased penetration of the sorbate in the pore system and consequently increased sorption value.

The relatively large diameter of coal grains should also be taken into account (1.0-1.5mm), as they might contain closed pores which might "open" with rising temperature. The increase in the chain length reduced this effect to the extent that, in the case of octane isotherms, in the initial range isotherms, were virtually indistinguishable. Increase in octane sorption at a temperature of 298 K was observed at pressures corresponding to the beginning of capillary condensation process in mesopores. Similar effects were caused by the presence of a double bond. This was most likely result of the double bond reactions with the functional groups present on the surface of the macropores, hence blocking access to the interior of the particles capillaries.

DOI: 10.1051/e3sconf/20171402027

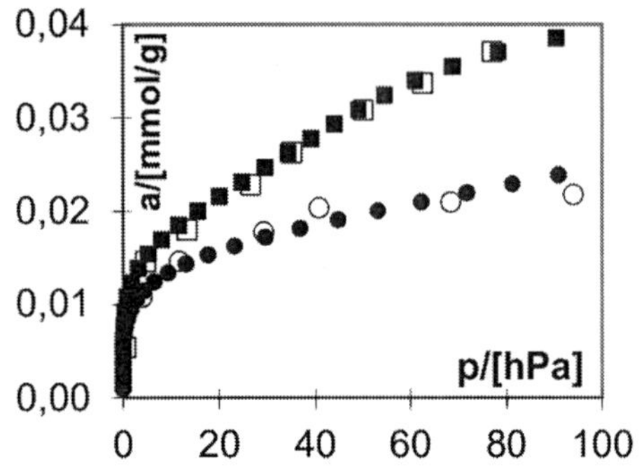

Fig. 5. Experimental and calculated sorption isotherms in B cola – n-heksane system

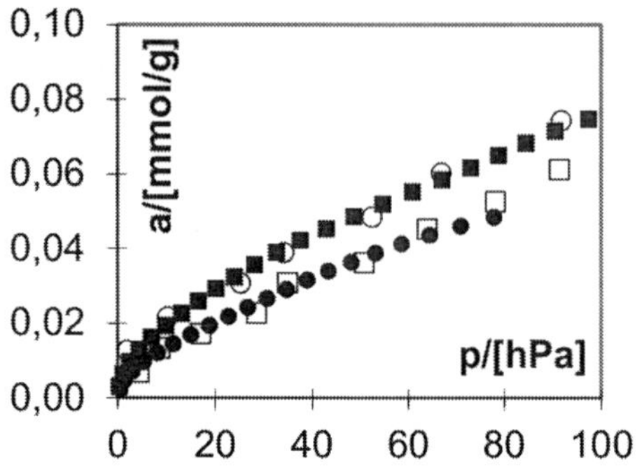

Fig. 7. Experimental and calculated sorption isotherms in Vi coal – n-heksane system

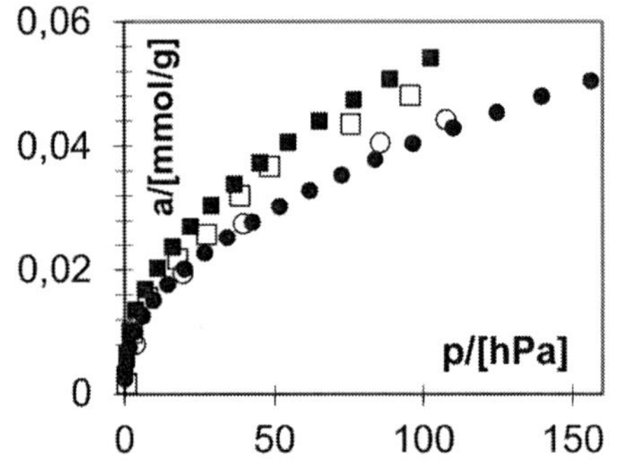

Fig. 6. Experimental and calculated sorption isotherms in B coal – 1-heksene systems

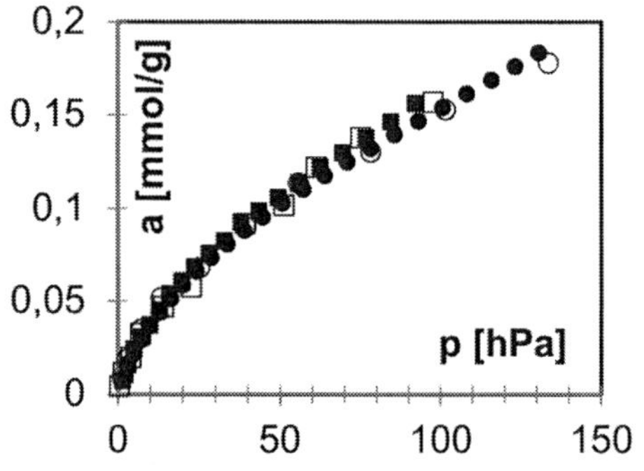

Fig. 8. Experimental and calculated sorption isotherms in Vi coal– 1-heksene systems

Legend: □ 298K; ○ 308K; ■, ● - calculated isotherms

4.1. Energetic effects of sorption process

Research on the energetic effects of the sorption process allowed for quantitative and qualitative characteristics of the binding forces between the hydrocarbon and coal surface. Isosteric heat of adsorption plays important role in the deliberations on energetic effects of adsorption process. By definition it is the thermal effect associated with a virtual moving of 1 mol of gas from volume to adsorption phase at constant temperature and pressure.

The knowledge of thermal sorption equation parameters (3) allows for numerical calculation of isosteric heats of adsorption in a function of fill due to equation (5):

$$Q_{st}(v) = -R \cdot \left(\frac{\partial \ln p}{\partial (1/T)} \right)_v = -R \sum_{i=0}^{n} A_i v^i \qquad (5)$$

A_0 constant from equation (5) allows for calculation of boundary values of isosteric heat of adsorption with $v \to 0$.

$$Q_{st}^0 = -R \cdot A_0 \qquad (6)$$

By definition isosteric heat of sorption Q_{st} is equal to sorption enthalpy ΔH_{st} with opposite sign:

$$Q_{st} = -\Delta H_{st} \qquad (7)$$

Comparing discussed above method for determining isosteric heat of adsorption with the use of the classical method based on plotting isotheres, it should be noted that the numerical method in addition to its convenience and time saving, has another advantage. It is based on the minimization of deviations from the regression curves of all experimental points, as a result data are correlated in case of pressure and temperature at the same time. This is important for the analysis of data obtained at different temperatures (families of isotherms) and allows for the preparation of temperature dependencies.

Table 5. Isosteric heat of sorption calculated with the use of (5) and (6) equations

Hydrocarbon	Q^0_{st} [kJ/mol]	Ranges of filling [dm³/kg]	Q_{st} for V_{min} [kJ/mol]	Q_{st} for V_{max} [kJ/mol]	$\Delta Q_{st}/\Delta v$ [(kJ·dm³)/mol]
		Coal B			
n-Hexane	115.9	0.291–0.515	97.1	92.8	19.19
n-Heptane	81.7	0.089–0.538	81.2	77.8	7.57
n-Octane	66.5	0.067–0.672	66.3	64.4	3.14
1-Hexene	54.5	0.045–1.367	53.7	43.5	7.71
1-Heptene	62.0	0.291–1.098	59.7	55.7	4.95

Table 5 presented the values of hydrocarbons isosteric heat of sorption for coal *B* which calculation was based on isotherms determined using thermal adsorption equation. Q^0_{st} values (determined from extrapolation) characterized direct interaction of sorbate molecules with the surface of coal. In the last column of Table 5 there was shown index characterizing the average slope of curves shown in Figure 9 in relation to the axis of fill. Index provided information how the value of isosteric heat changed with increasing by 1 mol amounts adsorbed and was a measure of the distance between isotherms determined at different temperatures along with increase in filling.

The heat of adsorption as a function of filling, calculated using equation (1) was presented by a constant line in Figure 9 for sample *B*. Presented dependencies were of declining character, with increasing values of sorption once the values of isosteric heat of sorption decrease, however values were not reaching the heat of condensation. Isosteric heat values may indicate the physical nature of the hydrocarbons molecules interaction with hard coal [22]. Isosteric heat of sorption curves as a function of filling the surface may suggest energy heterogeneity of the studied coal adsorbents.

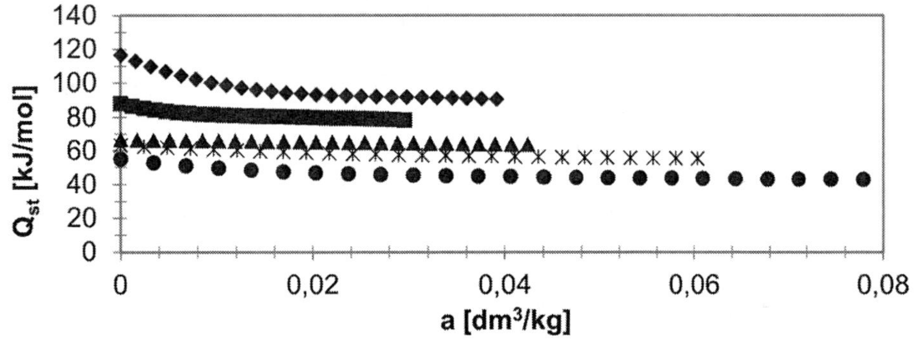

Fig. 9. Dependence of isosteric sorption heats on the amounts of sorbed hydrocarbons on the studied coal *B*((Legend: •-n-heptane,▪-1-heptene, ▲-n-octane, ♦-1-hexene, x-n-hexane)

Taking into consideration the values of isostheric heat of adsorption Q^0_{st} of sorbates, once there was no fill (for example sample *B*) it can be noted that the highers value of energy of sorption was represented by the system hexane-coal. This fact may suggest that porous structure of coal *B* was of the best availability for hexane molecules. This high energy of sorption may imply that capturing of hexane molecules took place in micropore structure. For the remaining saturated hydrocarbons examined energy of adsorption was clearly lower.

5 Conclusions

In conclusion it can be stated that the isotherms of sorption depend on the kind of vapors and the type of coal:

• It can be said that in case of examined systems sorption took place mainly on the macropores surface of hard coal;

• Samples *Vi* and *B* represented much higher sorption of polar sorbate. As the molecules dimensions (in group: saturated, unsaturated hydrocarbon) were practically uniform, the differences were most probably related to association possibilities of unsaturated hydrocarbons on the surface of macropores;

• It can be expected that in case of unsaturated hydrocarbons interconnections adsorbate-adsorbate system played more important role than of adsorbate-adsorbent. Explanation to this could be stronger interconnections of unsaturated hydrocarbons and surface polar groups of hard coals *Vi* and *B*.

The investigation of energetic effects of sorption process enabled quantitative and qualitative characteristic of bounds of hydrocarbons with the surface of coal. While considering the energy aspects of the adsorption, isosteric heat of adsorption played very important role. The knowledge of thermal adsorption equation parameters allowed for numeric calculation of the isosteric heat of sorption value in a function of filling. In case of system coal/hexane high value of boundary adsorption enthalpy was observed, which can indicate penetration of this hydrocarbon into the microporous structure. For the remaining sorbates distinguishably lower values of terminal adsorption enthalpy were fund, what proves that hydrocarbons didn't penetrate in to the micropores.

The work done within the statutory research of AGH (11.11.210.244)

References

1. G. Ceglarska – Stefańska, *Współzależność procesów sorpcyjnych i dylatometrycznych zachodzących w układach: węgle kamienne – para wodna, dwutlenek węgla, metan*, Zeszyty Naukowe AGH – Chemiaz.**16**, Kraków **(1990)**, in Polish
2. S. Helle, A. Gordon, G. Alfaro, X. Garcia, C. Ulloa,*Coal Blend combustion: link between unburnt carbon in fly ashes and maceral Composition*,Fuel Processing Technology, **80**, 209 – 223, (2003)
3. C.O. Karacan, G.D. Mitchell, *Behaviour and effect of different coal microlithotypes during gas transport for carbon dioxide sequestration into coal seams,*International Journal of Coal Geology, **53**, 201 – 217, (2003)
4. H.B. Vathaluru, R.J. Brookeb, D.K. Zhanga, H.M. Yana, *Effects of moisture and coal blending on Hardgrove Grindability Index of Western Australian coal*,Fuel Processing Technology, **81**, 1, 67 – 76, (2003)
5. M. Żyła, *Badania nad związkami między sorpcyjnymi własnościami węgli a ich chemiczną naturą*, Rozprawa doktorska, Kraków, (1963), in Polish
6. D.W. Van Krevelen, J. Schuyer, *Węgiel, chemia węgla i jego struktura*, Warszawa PWN, (1959)
7. A. Marzec, *Molecular structure of coal,*Chemia Stosowana, **25**, 381, (1981)

8. A. Nodzeński, S. Hołda, *Sorpcja i desorpcja metanu z węgli kamiennych w zakresie podwyższonych ciśnień, W.: Układ Węgiel Kamienny – Metan w aspekcie desorpcji i oddziaływania metanu z gazów kopalnianych*, Wydawnictwo AGH, Kraków, 153 – 178, (2000), in Polish
9. G. Ceglarska – Stefańska, K. Zarębska, S. Hołda, *Expansion and contraction of coal at various metamorphism under influence of CH_4 and mixture CO_2 and CH_4 sorption/desorption process, W: Theoretical and experimental studies of interfacial phenomena and their technological applications*: V Ukrainian–Polish symposium ; 4–9 September, Odessa, 25–26, (2000)
10. G. Ceglarska-Stefańska, K. Zarębska, *Sorpcja CO_2 i CH_4 na niskouwęglonych węglach z KWK Brzeszcze,* KARBO **1**, 31–34 (2006), in Polish
11. A. Nodzeński, *Wysoko ciśnieniowa desorpcja ditlenku węgla i metanu z węgla kamiennego Zagłębia Dolnośląskiego*, Rozprawy i Monografie 95, Uczelniane Wydawnictwo Naukowo – Techniczne AGH, Kraków (2000), in Polish
12. H.P. Boehm,*Chemical Identification of Surface Groups*,Advan. Catal., **16**, 179-274 (1966)
13. S. Saha, B.K. Sharma, S. Kumar, G. Sahu, Y.P. Badhe, S.S. Tambe, B.D. Kulkarni, *Density measurments of coal samples by different probe gases and their interrelation,* Fuel, **86**, 1594-1600, (2007)
14. M. Lasoń, M. Żyła, *Aparatura do wyznaczania izoterm sorpcji i desorpcji par metodą mikrobiuretek*, Chemia Analityczna, **8**, 279 – 286, (1963), in Polish
15. G. Jodłowski, M. Wójcik, *Comparative analysis of sorption of small molecule hydrocarbons and polar substances in polish hard coal*, (2013), Adsorption **19**, 813–819
16. K. Pigoń, Z. Ruziewicz, Chemia Fizyczna 1, Podstawy Fenomenologiczne, PWN,Warszawa, (2005), s. 107, in Polish
17. J. Ościk, *Adsorpcja*, PWN, Warszawa, (1983), in Polish
18. G.S. Jodłowski, *Modele matematyczne sorpcji w węglu kamiennym*, Rozprawa doktorska, AGH Kraków (1999),in Polish
19. M. Wójcik, M. Żyła,*Sorpcja par węglowodorów alifatycznych na węglach kamiennych*, Karbo**8**, 232-235, (2002), in Polish
20. A. Dudzińska, *Investigation of adsorption and desorption of acetylene on hard coal samples from Polish mines*, International Journal of Coal Geology, **128-129**, (2014), 24-31
21. A. Dudzińska, N. Howaniec, A. Smoliński, *Experimental Study on Sorption and Desorption of Propylene on Polish Hard Coals*, Energy&Fuels, **29** (8), 4850-4854, (2015)
22. J. Jagiełło, M. Lasoń, A. Nodzeński, *Thermodynamic description of the process of gas liberation from a coal* bed, Fuel **71**, 431, (1992)
23. L. Czepirski, J. Jagiełło, *Virial-type thermal equation of gas-solid adsorption,* Chem. Eng. Sci., **44**, 4. 797-801,(1989)
24. P. Baran, K. Zarębska, A. Nodzeński, *Energy aspects of CO2 sorption in the context of sequestration in coal deposits* Journal of Earth Science v **25**, (4), 719–726, (2014)
25. P. Baran, K. Zarębska, M. Bukowska, *Expansion of hard coal accompanying the sorption of methane and carbon dioxide in isothermal and non-isothermal processes*, Energy & Fuels, **29**, 3, 1899–1904. (2015)

Application of multiple sorption model to estimation of CO_2 sequestration capacity or CH_4 recovery in polish hard coals

Grzegorz Jodłowski[1,*], *Natalia* Florencka, *Elżbieta* Vogt[1], *Marta* Wójcik[1], and *Magda* Ziółkowska[1]

[1]AGH University of Science and Technology, Faculty of Fuels and Technology, al. A. Mickiewicza 30, 30-059 Kraków, Poland

Abstract. The aim of the study is to estimate the theoretical capacity of coal deposits in terms of carbon dioxide storage and methane recovery estimate during the injection of carbon dioxide. The Multiple Sorption Model was used for this purpose. The range of sorption measurement data on which the simulations were based does not exceed the critical point for both methane and carbon dioxide. The model allows to calculate absorption, adsorption, expansion and total theoretical sorption. For further consideration absorption was used as the process of the percolation of the gas in the bulk of the coal matter as well as the total theoretical sorption, the occurrence of which is possible due to the presence of fractures in the coal seam. Calculated levels of absorption and the total theoretical adsorption were applied to estimate the possible storage capacity of carbon dioxide based on the data associated with the content of coal in the mining fields of individual Polish coal mines. Moreover, MSM development for the gas mixture sorption can be used to roughly assess the recovery of methane naturally occurring in coal deposits during such a process.

1 Introduction

Mitigation of carbon dioxide pollution is the focus of various works and investigations. The most popular method is the storage of gas in the geological structures [1]. These geological methods are partially used, such as pumping in of CO_2 to exhausted oil deposits – Enhanced Oil Recovery or injection of this gas into a coal deposit with simultaneous CH_4 recovery [2-6]. An interesting alternative to carbon dioxide sequestration is the storage of this gas in the coal bed of a closed colliery or in unmineable coal seams. The latter option is appropriate to deep coal deposits whose exploration is not reasonable from the economic point of view and the depth causes the necessity of CO_2 compression up to overcritical liquid but the first option does not require such high conditions. Introduction of CO_2 to a coal bed eventually gives the possibility of to recover CH_4 and to use it as a fuel. Enhanced Coal Bed Methane (ECBM) is a commonly used approach to analyse the process of methane recovery [3,7-11].

* Corresponding author: jodlowsk@agh.edu.pl

The proposed alternative view to a rough analysis of this process on the basis of *ex situ* measurements is the application of Multiple Sorption Model (MSM). The model involves a theoretical description of the molecule of gas location in the bulk or on the surface of coal substance developed on the basis of detailed thermodynamic considerations.

2 Multiple Sorption Model

The MSM model is based on the copolymer representation of coal substance. The main components of the coal substance are:

- arene domains – semi-crystalline, rigid forms composed of plates of condensed benzene rings, building aggregates of several flakes;
- crosslinked chains – semi-rigid forms of aliphatic, alicyclic and heterocyclic chains connected with each other and with arene domains;
- non-crosslinked chains – the elastic part of coal substance consist of aliphatic, alicyclic and heterocyclic free chains;
- pores – empty spaces in coal substance;
- mineral admixtures – small aggregates dispersed in coal material whose influence on the sorption process is negligible. [12]

Arene domains and crosslinked chains constitute a limitedly rigid macromolecular phase which simultaneously constitutes the network and enables non-crosslinked chains to flow (crawl) between cells of the network. The participation and structure of the above-mentioned components of carbon structure are the input parameters of the model. Their values were obtained by pre-evaluation of the dual sorption model in the way of optimization of structure parameters with respect to the experimental data. Another set of independent parameters describes the texture of the coal adsorbent and distribution of sorbate molecules to different sorption subsystems.

Generally speaking, the coal substance is presented in the MSM model by the network of cells which have the size of methylene segments of the chain. Every object is constructed by the segments with the rule of the prohibition of occupation of the cell by multiple segments of the different objects. The exception to this rule is for pores and sorbate molecule segments.

The starting point of considerations for the model is the formula of free energy:

$$\Delta F = \Delta H + T\Delta S \tag{1}$$

which lead to the complex formulas describing: (i) absorption of molecules in the elastic part of coal material with the simultaneous creation of the space for a molecule (this space is occupied by the coal substance segments before molecule location); (ii) expansion – location of sorbate molecule in submicropores whose size is smaller than molecule size and whose dimensions should be increased; (iii) adsorption on the surface of pores whose size is greater than molecule size. Absorption and expansion is accompanied by contraction of pores (decreasing of dimensions of existing pores). Expansion and contraction results in the coal mass swelling, which is taken into account in the model calculations and is a measurable parameter. Literally speaking, the sorption system is divided into eleven subsystems containing absorption (relative pore radius $R=0$), nine expansion subsystems, giving the distribution of molecules in submicropores ($0<R<1$) and adsorption ($R>1$).

Creation of pore or its expansion is done by the forces of cohesion of coal material, and location on the surface occurs due to adhesion forces. In molecular scale the surface is presented by the set of spheres and their contacts. The number of contacts, character (cohesive/adhesive) and value of its energy determines the molecule direction to the subsystem. Therefore the molecule in the absorption subsystem has number of contacts $Z=12$ (the constant of the network) and a completely cohesive character, in the adsorption system

DOI: 10.1051/e3sconf/20171402028

1<Z<6 and completely adhesive character, in the sub-processes between above (expansion subsystems) 1<Z<12 and mixed character of interactions.
In conclusion, MSM makes it possible to determine the number of moles of sorbate present in each of the subsystems (distribution sorption capacity) and parameters of porous structure of a coal. The analysis of the results and the development of a model for the sorption of gas mixtures may determine the effects of competitive sorption.

3 Results of sorption modeling

In order to evaluate the numerical experiment involving the simulation of sorption isotherms, a set of isotherms of CO_2 and CH_4 measured in the laboratory is used. A sorption isotherm of a mixture of CH_4 + CO_2 has also been used for the numerical analysis. Parameters of coals used in this study are presented in Table 1.

Table 1. Parameters of coal samples used for the model

Sample	d_{He} [g/cm^3]	d_{Hg} [g/cm^3]	Ultimate analysis			Proximate analysis			
			A^a	W^a	V^{daf}	C^{daf}	H^{daf}	S^{daf}	$(O+N)^{daf}$
31	1.387	1.230	15.50	1.65	40.90	**79.60**	5.00	0.71	14.69
32	1.36	1.255	2.48	3.76	40.80$^{(a)}$	**80.88**	5.26	0.62$^{(a)}$	13.14
34	1.35	1.23	6.15	1.55	29.20	**86.40**	4.9	0.8	8.37
35	1.36	1.306	16.64	1.32	32.68$^{(a)}$	**86.55**	5.22	0.90$^{(a)}$	7.37
37	1.4	1.30	7.78	1.25	27.90	**87.60**	4.88	n.a.	6.41
41	1.29	1.26	2.27	b.d.	22.43	**90.08**	4.76	n.a.	4.13
42	1.43	1.357	3.70	0.81	6.09$^{(a)}$	**92.41**	3.02	0.46$^{(a)}$	4.09

d_{He}, d_{Hg} – mercury and helium density, respectively;
contents: A – ash, W – moisture, V – volatile matter; C – Carbon, H – Hydrogen, S – Sulphur, O – Oxygen, N – Nitrogen;
indexes: a – analytical state, daf – dry and ash free

Sorption isotherms of CO_2 and CH_4 measured up to 3.4 MPa were used for evaluation of the computed set of isotherms. The MSM model was described in previous articles [12-14]. The results that were obtained are presented on Figures and Tables 2 and 3.

 DOI: 10.1051/e3sconf/20171402028

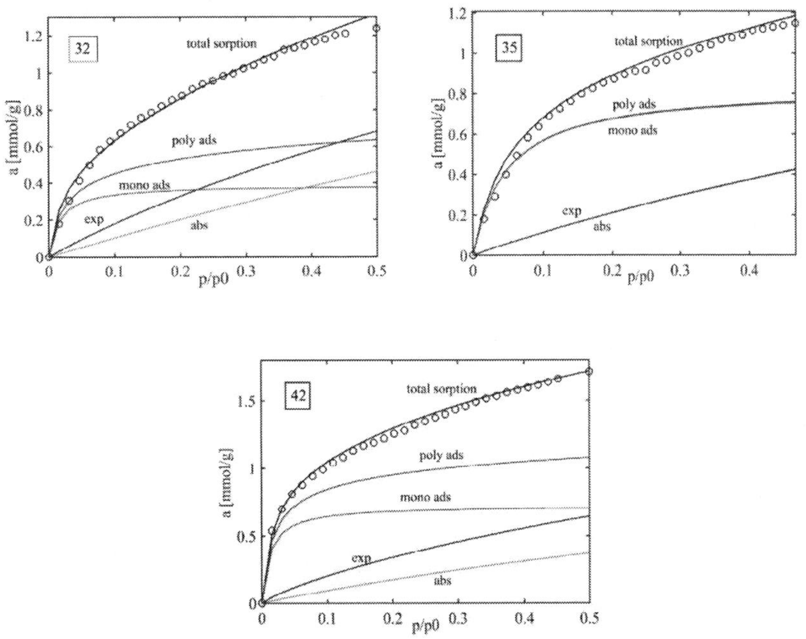

Fig. 1. Multi-sorption isotherms of **carbon dioxide** in selected samples of coal (number of samples: 32, 35, 42). Description of curves: black – theoretic total sorption, red – polylayer adsorption (poly ads), purple – monolayer adsorption (mono ads), blue – expansion (exp), green – absorption (abs), circles – experimental sorption.

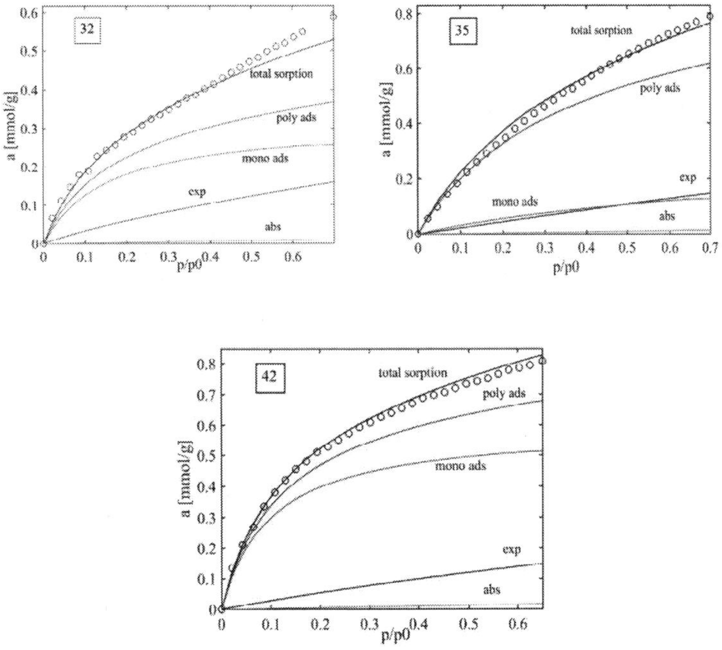

Fig. 2. Multi-sorption isotherms of **methane** on selected samples of coal (number of samples: 32, 35, 42). Curves description is the same as in Fig. 1.

DOI: 10.1051/e3sconf/20171402028

One could observe a relatively big participation of absorption and expansion in the total sorption phenomenon in Fig. 1. In the case of sample 35 the sorption process has a completely dual character – expansion is negligible. Only this sample shows such behaviour and it will be presented in Fig. 3.
Methane interacts with coal substance in a different way than carbon dioxide, and this is represented in Fig. 2 by negligible absorption. Even the expansion sub-process has smaller participation in the sorption process. Thus, methane is located mainly on the surface of pores and adsorption is taking a greater share of sorption with an increasing coal rank.

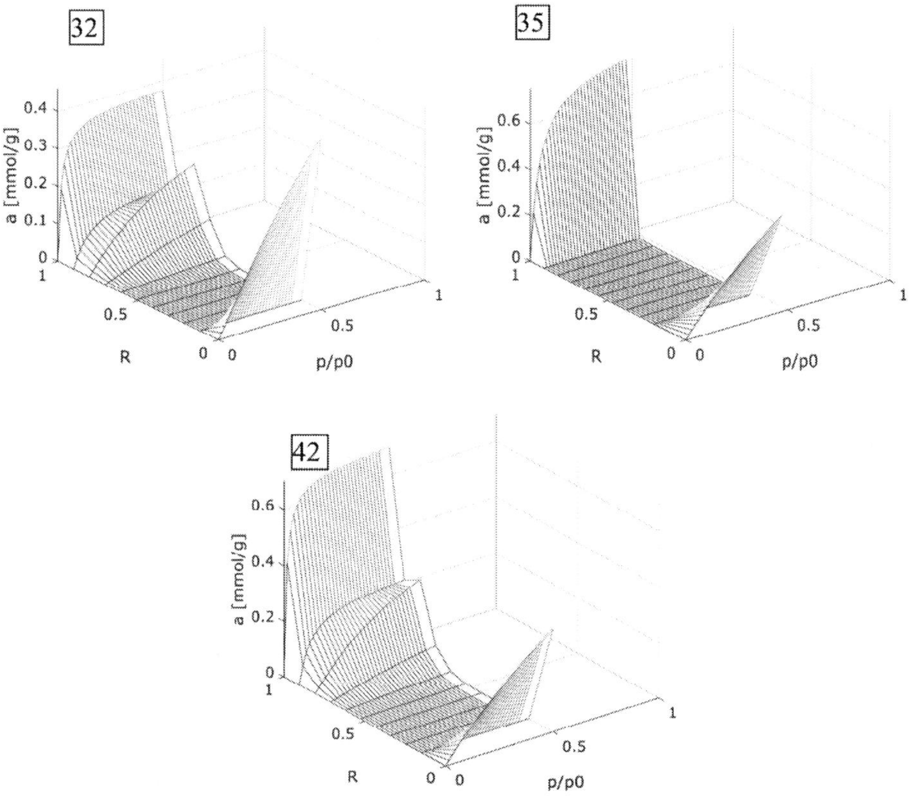

Fig. 3. Poly-sorption of **carbon dioxide** on selected samples of coal (number of samples: 32, 35, 42). Markings on the chart: a – sorption in mmol/g, R – radius of pore relative to molecule size, p/p0 – relative pressure.

Figure 3 shows the set of theoretical isotherms representing the different size of pores (sub-micropores) forming the distribution of molecules to the sub-systems connected with pore radius (R) relative to molecule size. Sample 35 shows dual character (what was stated below in Fig. 1). In this case every expansion sub-system is empty and two isotherms of adsorption and absorption summarise the total sorption capacity of this sample. The remaining samples do not exhibit similar behaviour.

 DOI: 10.1051/e3sconf/20171402028

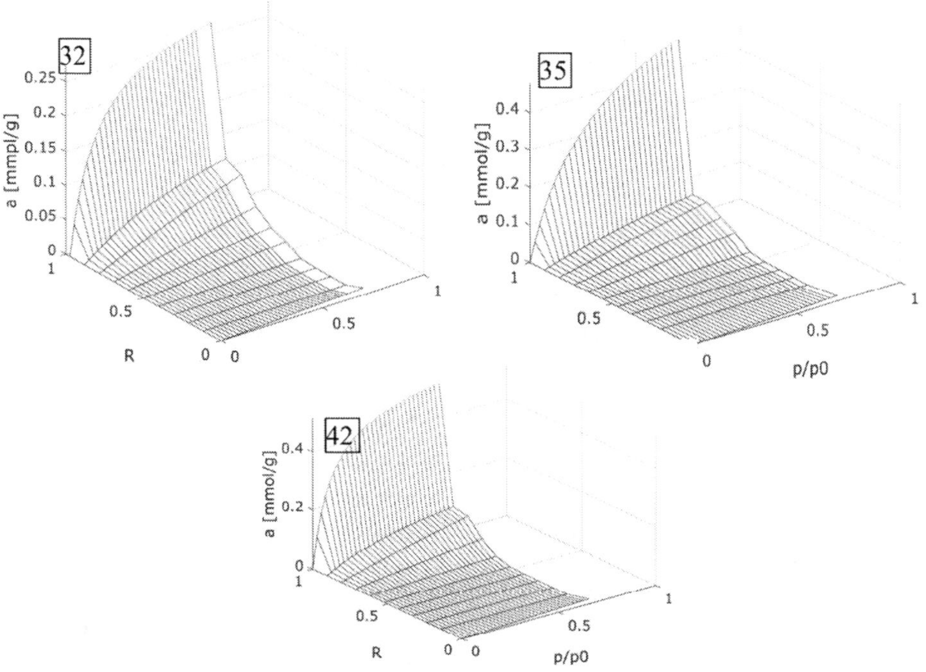

Fig. 4. Poly-sorption of **methane** on selected samples of coal (number of samples: 32, 35, 42). Markings on the chart: a – sorption in mmol/g, R – radius of pore relative to molecule size, p/p0 – relative pressure.

Isotherms of methane poly-sorption show negligible absorption (the same as in the case of multisorption presented in Fig. 2) and the small amount of sorbate molecules in expansion sub-systems. Thus the distribution of molecules capacity is relatively flat in comparison to carbon dioxide capacity distribution (Fig. 3).

Table 2. Calculated sorption capacities in subsystems using MSM for CO_2 at p=3.4 MPa

CO_2 capacity		Sample						
		31	32	34	35	37	41	42
Total sorption	mmol/g	0.747	0.763	0.930	1.180	0.875	0.830	0.800
Adsorption		0.483	0.608	0.740	0.901	0.765	0.675	0.638
Expansion		0.205	0.270	0.186	0.280	0.145	0.150	0.163
Absorption		0.090	0.148	0.178	0.133	0.140	0.148	0.163

Total sorption capacity initially increases with coal rank to achieve the maximum in the range of coals ortho-coking, then it begins to decrease slightly. It is worth noting that the absorption takes a significant part in the whole process of sorption similarly to expansion. Both sub-processes take place with the overcoming of the cohesive forces of the carbonaceous material and takes place in the bulk phase of coal deposit. Expansion and absorption give the volume tension on the coal mass what results in contraction of free pores. Eventually all these three phenomena cause the swelling of coal. Normally, in coal bed conditions, swelling is limited by the surrounding rocks and thermodynamic equilibrium shifts in favor of dissolution phenomena (absorption and expansion) in coal mass. Simultaneously, swelling results in self-sealing of the deposit after injection of CO_2 [5].

Table 3. Calculated sorption capacities in subsystems using MSM for CH_4 at 3.4 MPa

CH_4 capacity		Sample						
		31*	32	34*	35	37	41	42
Total sorption	mmol/g	0.384	0.523	0.326	0.765	n/a	n/a	0.827
Adsorption	mmol/g	0.2922	0.3656	0.2257	0.6127	n/a	n/a	0.6542
Expansion	mmol/g	0.075	0.148	0.099	0.1385	n/a	n/a	0.157
Absorption	mmol/g	0.0168	0.0094	0.0013	0.0138	n/a	n/a	0.0157

* - isotherms and computations for pressure range up to 1.8 MPa

Total sorption of methane is smaller than for carbon dioxide (see Table 3), except of anthracite. Probably it is an effect of decreasing of the participation of elastic (molecular) phase in the coal substance. This fact alone shows the differences in the interaction of methane and carbon dioxide molecule to the carbon substance. Carbon dioxide penetrates the bulk phase of coal, and methane is located on the surface of the pores, resulting in negligible absorption of methane. The shape of the methane molecule results in increased tendency to form aggregates on the surface, as opposed to carbon dioxide. In this way, also the adsorption of methane is lower than the carbon dioxide because the aggregates of methane molecules may block access to sub-micropores.
An attempt was made to develop a model for the calculation of adsorption from gas mixtures. The results of the calculations are presented in Figure 5.

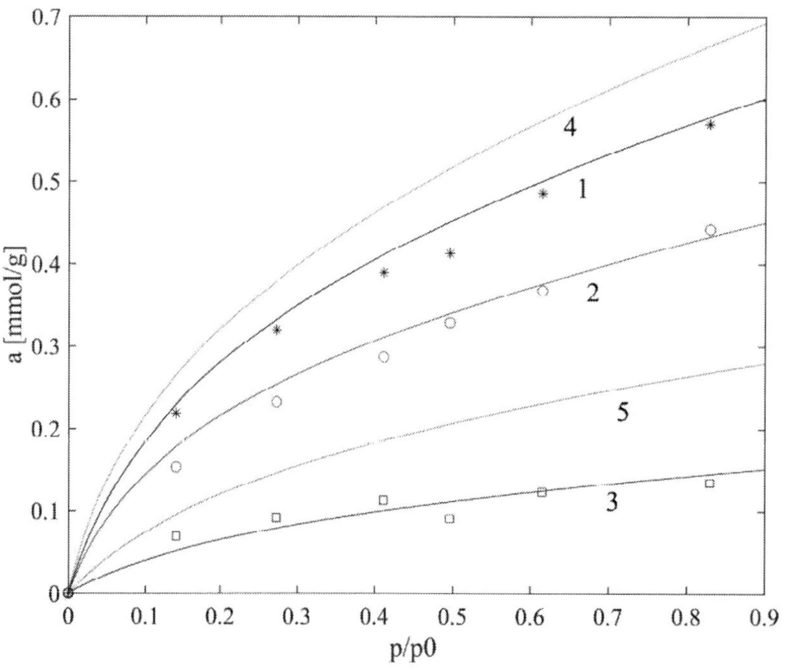

Fig. 5. Multisorption isotherm of methane and carbon dioxide mixture on a 34 coal sample. Chart description: experimental data on sorption from gas mixture – circles (CH_4), squares (CO_2) and asterisks (mix), 1 – theoretical total sorption from the mixture, 2 – theoretic total sorption of CH_4 from mixture, 3 – theoretic total sorption of CO_2 from mixture, 4 – individual sorption of CH_4, 5 – individual isotherm of CO_2.

DOI: 10.1051/e3sconf/20171402028

Sorption isotherms of methane and carbon dioxide are computed on the basis of individual sorption isotherms of these gases. The analytic formula of energy of mixing process for the mixture of gases is used for derivation of isotherms of the sorption of these gases from their mixture. However, for methane this decrease is greater in all the analyzed cases. We managed to get a satisfactory but not perfect consistency model simulations with measurements carried out in the laboratory (see fig. 5). The same procedure was carried out to the remaining sorption systems presented in this work, except samples 31 and 41 because of the lack of experimental data of methane sorption in these samples.

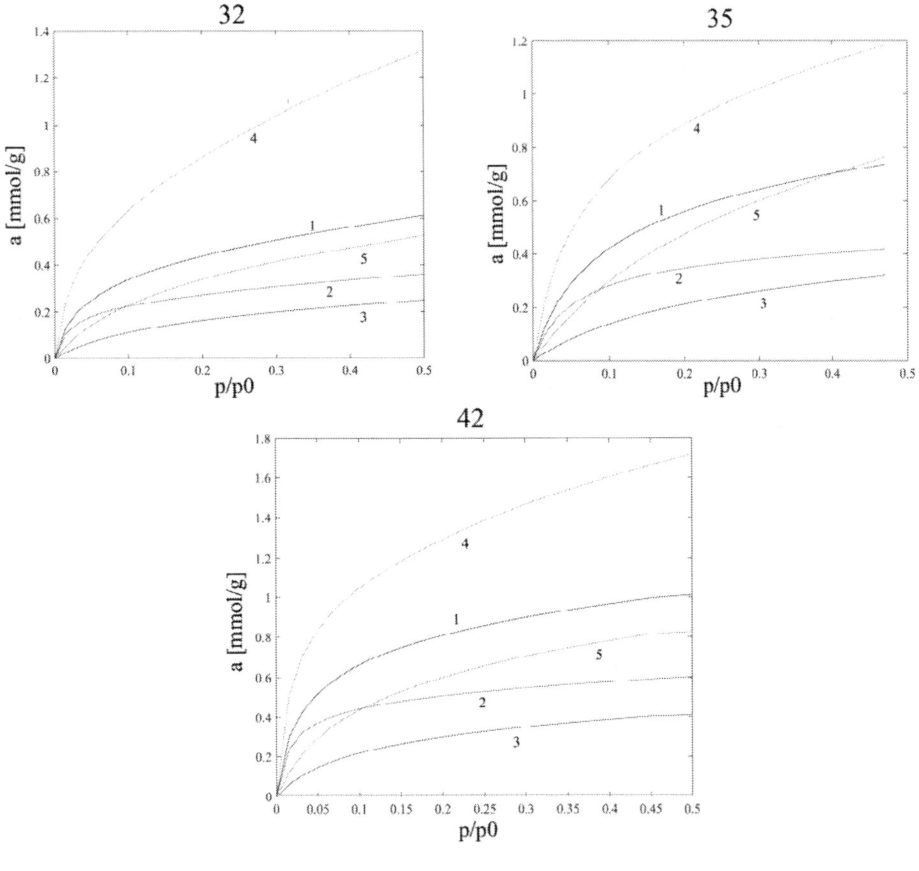

Fig. 6. Multisorption isotherm simulations of sorption from CO_2 and CH_4 mixture (number of samples: 32, 35, 42)

The decrease of methane capacity in the sorption system for the mixture of gases in comparison to sorption of an individual gas was significant in every case and it oscillated between 30 and 50%. This fact shows a big potential for methane recovery from the coal deposit after carbon dioxide injection.

4 Application of modelling results

One could use data on sorption of CO_2 and CH_4 for rough estimation of CO_2 sequestration potential and methane recovery.

The Morcinek colliery located in the Silesia region is closed. The mining field of this coal mine contains coal of types 34 and 35 accordingly to mark samples in this study. Proven coal reserves to a level of 950m of depth amount to 54 million of tonnes (operative) and 144 million of tonnes (balance) of coal [15]. The second number represents the space available for the injection of carbon dioxide. Simple calculations on the basis of modeling gives possibility of storage of 2.7 million of tonnes of CO_2 (in operative reserves) and 7.3 million of tonnes in balance reserve. This estimation involves only the expansion and absorption subsystem because probably the self-sealing of deposit appears after CO_2 injection in the butt cleat geological system [5, 8, 17].

Taking into account a typical power station of 309 MW of energy which produces 3 million of tonnes of CO_2 yearly, the capacity of the field will be enough for 3 years of sequestration with assumption of partial CO_2 capturing. Certainly, an incomplete capacity could be accessible for technical operation.

The documented volume of methane amount in the Morcinek coal field is about 2,500 million m^3. Estimation of the amount of methane desorbed from the site after injection of the above-mentioned mass of carbon dioxide gives a volume of 200-250 million Nm^3 of methane recovered during the entire operation.

A similar procedure could be applied to any mining field of colliery containing coal type identified in this study.

5 Conclusions

The proposed method of the modelling of carbon dioxide and methane gives the possibility of a rough estimation of sequestration capacity of a mining field of closed coal mines together with the possibility of estimation of methane recovery after carbon dioxide injection. Theoretical estimation shows the relatively long time operation (3 years) with the sequestration of carbon dioxide for an exemplary power station. It is obvious that not all the theoretical size can be used for storage of carbon dioxide due to technical reasons.

Another issue is the injection of carbon dioxide into the deep coal layers of over 1000 meters, where the static pressure of overburden exceeds 27 MPa [18], the carbon dioxide exists as a supercritical fluid having a density of 0.468 g/cm^3 [19]. Then sorption capacity may reach values ranging from 1.25 to 2 mmol CO_2 per gram of coal. At the same time the presence of water may reduce the value of 1.15 to 1.9 mmol/g. This means the increasing of site capacity 2 times and involves a 5 times bigger mass of coal for CO_2 sequestration. In total, a 10 times bigger storage capacity in this individual mining field.

This work was done due to financial support of the AGH-UST Statutory Research, No. 11.11.210.213

References

1. M. S. Masoudian, Journal of Rock Mechanics and Geotechnical Engineering, **8** (1), 93-112 (2016)
2. J. Q. Shi, S. Durucan, Oil & Gas Science and Technology **60** (3), 547-558 (2005)
3. T. Gentzis, International Journal of Coal Geology, **65**: 79-92 (2006)
4. R. Pini, G. Storti, M. Mazzotti, Adsorption **17**, 889-900 (2011)
5. C. S. Santarosa, D. Crandall, I. V. Haljasmaa, T-B. Hur, J. J. Fazio, R. P. Warzinski, R. Heemann, J. Marcelo, M. Ketzer, V. N. Romanov, International Journal of Coal Geology **106**, 25-34 (2013)
6. M. S. A. Perera, P. G. Ranjith, A. S. Ranathunga, A. Y. J. Koay, J. Zhao, S. K. Choi, J. Geophys. Eng. **12**, 90-107 (2015)
7. C. Ö. Karacan, F. A. Ruiz, M. Cotè, S. Phipps, International Journal of Coal Geology **86**, 121-156 (2011)
8. A. Mitra, S. Harpalani, S. Liu, Fuel **94**, 110-116 (2012)

9. P. Baran, J.Rogozińska, K. Zarębska, S. Porada, Przemysł Chemiczny, **93** (12), 2008-2012 (2014)
10. J. Zhang, K. Liu, M. B. Clennell, D. N. Dewhurst, M. Pervukhina, Fuel **160**, 309-317 (2015)
11. G. S. Jodłowski, M. Davidová, *Environmental impact and mitigation of carbon dioxide*, Technical University of Košice (2015)
12. G. S. Jodłowski, M. Wójcik, A. Orzechowska-Zięba, Adsorption. **22**, 847-854 (2016)
13. J.Milewska-Duda, J. T. Duda, G. S. Jodłowski, M. Kwiatkowski, Langmuir. **16**, 7294-7303, (2000)
14. G. S. Jodłowski, P. Baran, M. Wójcik, A. Nodzeński, S. Porada, J. Milewska-Duda, Langmuir. **13**, 5732-5735 (2013)
15. http://kwkmorcinek.pl/index.php?s=zloze (20.08.2016)
16. http://www.ppr.pl/artykul.php?id=153107 (22.06.2009)
17. C. F. Rodrigues a,*, C. Laiginhas b, M. Fernandes c, M. J. Lemos de Sousa a, d, M. A. P. Dinis, Journal of Rock Mechanics and Geotechnical Engineering **6**, 208-218 (2014)
18. S. Day, R. Sakurovs, S. Weir, International Journal of Coal Geology **74**, 203-214 (2008)
19. W. M. Haynes, *CRC Handbook of Chemistry and Physics, 95th Edition*, CRC Press (2014)

DOI: 10.1051/e3sconf/20171402029

Sorption of pharmaceuticals residues from water to char (scrap tires) impregnated with amines

Katarzyna Styszko[1,*], *Paweł* Baran[1], *Marta* Sekuła[1] and *Katarzyna* Zarębska[1]

[1]AGH University of Science and Technology, Faculty of Energy and Fuels, Al. Mickiewicza 30, 30-059 Krakow, Poland

Abstract. The study investigated the application of char activated with CO_2 and impregnated with amines solutions for removal of selected xenobiotics from aqueous solutions. The chars produced from the pyrolysis of waste tires. The solutions of monoethanolamine (MEA), diethanolamine (DEA) and polyethylenimine (PEI) were used for impregnation of char. The sorption capacity of char impregnated with amines depended on amines chemical properties. The adsorptive removal of mixture of pharmaceuticals residues by modified materials was assessed. BET surface area of materials varied from 36 $m^2 g^{-1}$ to 128 $m^2 g^{-1}$. The highest removal efficiencies up to 99 % were observed for char impregnated with PEI.

1 Introduction

The development of civilization caused the increase of the amount of wastes. Waste tires have been making greater problem in global scale. An interesting method of their disposal is to surrender to the pyrolysis process [1]. As a result of the process we obtain two products: raw pyrolysis oil and pyrolytic solid fraction – char. The liquid product is characterized by a high content of aromatic hydrocarbons and can be used as heating fuel [2,3]. Simple mixing with rapeseed oil allows to obtain heating fuel with properties required for the fuel oil. The char can be used as an inactive filler in the rubber industry. The char may be also a valuable raw material for the production of activated carbon use as an adsorbent for removing various kinds of pollutants, both liquid and gas phase [4,5]. Pyrolysis of such waste tires and their use as sorbent, possible uses of waste tires as a source of raw materials or alternative fossil fuels are an attractive alternative to disposal [6].

Pharmaceuticals and another xenobiotics residues are not sufficient removed in wastewater treatment plants [7-12]. It is important to develop and improve technologies for the removal of such compounds to ensure a sustainable healthy natural environment [13-16]. In recent years, many studies have been devoted to investigate the application of char for pollutants removal from aqueous solutions [5,17-19]. Chemical recycling of waste materials is indeed of scientific importance from the perspective of realizing effective waste management [20-31]. The porous structure of activated carbons and the presence of surface

* Corresponding author: styszko@agh.edu.pl

 DOI: 10.1051/e3sconf/20171402029

groups containing heteroatoms are the principal physical and chemical properties that will determine their performance as adsorbents. The polarity, solubility and molecular size of the adsorbate and the pH of the solution are other important factors that need to be taken into account when assessing the effectiveness of an adsorbent [1]. There are many structural varieties of pharmaceuticals residues that can be classified as either ionic or nonionic. This is one of the reasons why it is important to study the surface chemistry of the activated carbons used to eliminate them. Impregnation of selected materials is one of the possibility for preparing novel organic pollutants adsorbents. However until now little work has been carried out on the use of char (scrap tires) as adsorbent of pharmaceuticals residues [32]. According to the best knowledge of authors, char (scrap tires) impregnated with amines has been used for the first time for adsorption of pharmaceuticals residues. The objective of this study was therefore to find out if the impregnation with amines improves sorption properties of char.

2 Materials and methods

2.1 Materials

Table 1 provides the physical and chemical properties of target waste water pollutants investigated in this study. All reference standards were of >98% purity and were purchased from Sigma Aldrich (Saint Louis, USA).

Table 1. Characteristics of the compounds

Group	Compound	Molecular structure	MW	pK$_a$	Log Kow
Analgesics	Ibuprofen (IBF)	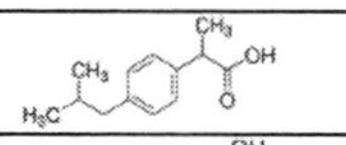	206,28	4.9	3.5-4.0
	Acetaminophen (ATP)	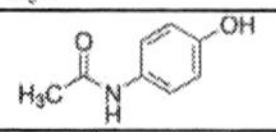	151.16	9.4	0.5-0.9
	Ketoprofen (KTP)	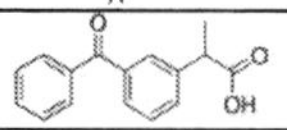	254.28	4.5	3.1-3.6
	Naproxen (NPK)	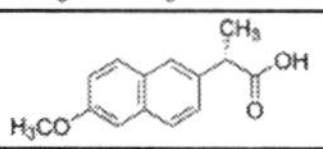	230.26	4.2	3.2-3.3
	Diclofenac (DCF)	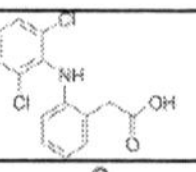	318.10	4.2	4.2-4.5
Metabolites	Salicylic acid (SAL)	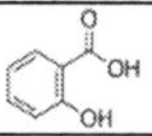	138.12	3.0	2.3-2.4
	Clofibric acid (CFA)	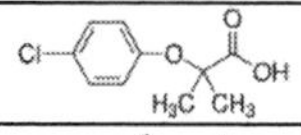	214.65	3.18	2.6
Stimulant	Caffeine (CAF)	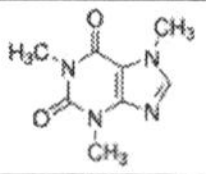	194.20	10.4	-0.07
Antiepileptic drug	Carbamazepine (CBZ)	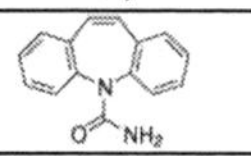	236.27	13.9	2.4-2.9
β-Blocker	Propranolol (PPN)	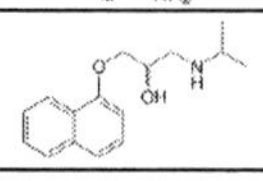	259.35	9.4	2.7-3.6

Solutions of the compounds both single ones and as mixtures were prepared in ethyl acetate and stored in the dark at 4 ^{0}C. A stock standard solution of 1000 μg mL^{-1} of each compound was used to prepare a standard mixture solution. Mixtures were used to spike

water samples. A four ring polycyclic aromatic phenol, 1-pyrenol (>99%), was used as an internal standard and was purchased from Sigma-Aldrich (Saint Louis, USA).
Derivatization reagent BSTFA + 1% TMCS was purchased from Supelco (Bellefonte, USA). Analytical grade reagents $CaCl_2$, HCl, and methanol (HPLC grade) and ethyl acetate (for GC analysis) were obtained from Avantor Performance Materials Poland S.A. (Gliwice, Poland). Deionized water (<0.07 S cm^{-1}) was used to prepare sample solvents for extraction experiments and the solid-phase extraction (SPE) was obtained from the HLP5 pure water system (Hydrolab, Gdansk, Poland). Oasis HLB 3cc extraction cartridges, (60mg) for SPE were purchased from Waters (Wexford, Ireland). Measurements of pH were carried out using a multimeter from Crison Instruments (Barcelona, Spain).

2.2 Characterisation of char and their modification

Waste tire fragments (50×50 mm) were carbonized at 673 K and activated with CO_2 at 1273 K for 15 minutes, based on procedure presented by Baran et al. [21]. Activated coal samples (CHAR) then underwent wet impregnation with nitrogen compounds solutions - monoethanolamine (MEA), diethanolamine (DEA) and polyethylenimine (PEI) (concentration was 1.5 mol L^{-1}). The impregnated samples were dried at 378 K for 1 h.

A Micromeritics Accu Pyc 1330 pycnometer was used for helium density analyses of all the samples at 296 K. To determine the influence of activation and impregnation on the specific surface area, BET method, low temperature nitrogen (N_2) adsorption isotherms for all of the samples were used. The measurements were made at liquid nitrogen temperature (77 K) using a Quantachrome Instruments Autosorb 1-C.

Helium density analysis results are presented in Table 2. The activation process resulted in increased carbon material density. Density changes may indicate the implantation of introduced substances into the samples.

Table 2. Selected properties of sorbents

Sorbents The formule of amines used to modification		Specific surface area BET [m^2 g^{-1}]	Real density [g mL^{-1}]	pH in water solution
Raw char		70.3	2.0	----
CHAR		128.9	2.1	6.78
Char impregnated with	MEA Monoethanolamine 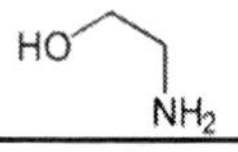	62.1	2.1	7.82
	DEA Diethanolamine 	36.3	1.9	9.26
	PEI Polyethylenimine 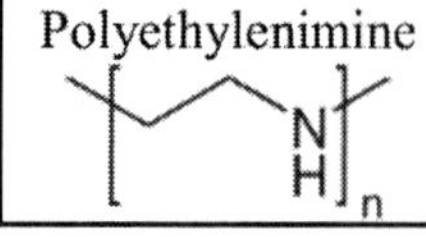	75.0	1.8	8.89

Low-temperature nitrogen-sorption-specific surface area measurements revealed relatively low values for raw char. After activation, the value increased from 70 to 128.9 m^2 g^{-1}. Analogous measurements were performed for selected samples impregnated with MEA

DEA and PEI. BET results for MEA, DEA and PEI samples indicated that the specific surface areas had decreased. In the case of sample PEI, the specific surface area is the highest among impregnated samples. Reducing specific surface area of the impregnated samples indicates that the porous structure has been blocked. Thus, it changes the chemical nature of the sorbent surface, and it may be important for the adsorption of pharmaceuticals residues.

2.3 Sorption experiments

Sorption experiments were performed according to OECD Guideline 106 [33]. Batch sorption experiments of selected compounds were conducted in 20 mL vials containing 50 mg of char in a 10 mL solution of 0.01 M $CaCl_2$ to minimize the cation exchange. Preconditioning of the char/water suspension for 21 h was conducted in $CaCl_2$ solution. Then the suspension was spiked with an appropriate working multi standard solution (50 µL, 40µg mL^{-1}) to reach 0.2 µg mL^{-1} and shaken for 24 h using a horizontal shaker at 50 rpm. Afterwards, the solutions were filtered and analyzed.

Blank and control solutions were performed in parallel with each sample. The blank and control solutions were shaken and analyzed with the same procedure as all other samples. Blank solutions included the same amount of respective sorbents as the real samples. Control samples were processed to estimate the process efficiency – the combination of the losses of tested compounds due to the adsorption on glass surface during experiments and recovery of analytes from the matrix by the samples extraction process. Control samples contained the same volume of supernatants and the mixture of standard solution but no sorbents. All experiments were performed in duplicates, each solution was analyzed two times and the obtained values were averaged.

The capacity of adsorption (µg g^{-1}) was calculated by using the following equation:

$$Q = ((C_0 - C_{eq})V)/m_s$$

where C_0 is the initial emerging contaminant concentration and C_{eq} is the emerging contaminant concentration (µg mL^{-1}) at equilibrium, V the volume of solution (mL) and m_s is the mass of the sorbent (g).

2.4 Analytical procedures

The supernatants obtained after filtration through a 0.2 µm glass fiber syringe filter were acidified to a pH of 2.0 and successively extracted by using Oasis HLB cartridges.

Analyses were carried out with the usage of Thermo Scientific GC Trace 1300 gas chromatograph coupled to the ITQ 900 ion trap mass spectrometer and a TriPlus RSH autosampler. The flow of helium through a GC column was constant and set at 1 ml min^{-1}. The programmable temperature of the vaporization injector was maintained at 250 ^{0}C, the transfer line at 280 ^{0}C and the ion source at 250 ^{0}C. The injector was operated at splitless conditions for 1 min, then turned to the split mode at the ratio of 50:1. Volume of injections was 1 µL. All the compounds separations were performed on a TG-SQC capillary column from Thermo Scientific that had 30 m x 0.25 mm inner diameter and a film thickness of 0.25 µm (5% phenyl 95% dimethylpolysiloxane). The temperatures program was as follows: 70 ^{0}C for 2 min, from 70 ^{0}C to 320 ^{0}C at 20 ^{0}C min^{-1}, and finally 5 min at 320 ^{0}C. The analyses were performed in positive mode, an electron energy 70 eV and emission current 250 µA. Helium (99.999%) was used a collision gas with flow 0.3 mL min^{-1}.

Mass spectrometry analyses were performed in the multi reaction monitoring (MRM) mode measuring the fragmentation of the precursor ions. The choice of fragmentation products for each substance was based on the most intense signal. For the acquisition of the MRM transitions the analytical run was split into time windows considering the expected retention time of the selected compounds. Data was collected, analysed and processes using Thermo Xcalibur.

Compounds were quantified by MRM, using the highest characteristic precursor ion/product ion transitions and characteristic ions used are listed in Table 2. Specific and intense product ions of each target analyte were used for quantification, and a secondary product ion was used as qualifier ion for confirmatory purposes. The absolute recoveries for water samples in each sample series were determined by comparing the peak area of the compound in each control sample, to the area obtained from detecting a standard solution prepared in the same solvent mixture. The concentrations of the compounds were corrected using the recovery percentages. The method quantification limits (MQL) for all analytes ranged from 0.6 to 8 ng L^{-1}. Depending on the compound, the absolute recoveries ranged from 67 to 147% (see Table 3).

Table 3. Retention times, MS/MS parameters for the analysis of the chosen xenobiotics by GC-MS/MS, method quantification limits, R2 and SPE recoveries.

Compound	RT (min)	Precursor-ion m/z	Product-ions Quantification /confirmation m/z	R^2	MQL ng L^{-1}	SPE absolute recovery (%)
Salicylic acid	8.27	267	209/249	0.993	0.7	87
Clofibric acid	8.69	143	99/115	0.997	3.0	108
Ibuprofen	8.96	160	145/117	0.992	3.0	122
Acetaminophen	9.04	280	206/-	0.989	0.6	91
Caffeine	10.48	194	165/109	0.986	5.0	67
Naproxen	11.49	185	170/153	0.998	0.4	136
Propranolol	11,92	72, 223 (SIM)	-	0.994	8.0	139
Ketoprofen	12.10	282	253/266	0.997	3.0	88
Carbamazepine	12.42	193	165/191	0.990	0.9	129
Diclofenac	12.49	214	179/151	0.993	1.5	147

3 Results and discussion

Fig. 1 shows the sorption efficiency of all tested pharmaceuticals residues to char and its modification impregnated of amines. The sorption of analytes onto char ranged from 36% to almost 100%. The impregnated char with polyethylenimine (PEI) exhibited the highest sorption for all analytes, except propranolol. For the removal of propranolol, char impregnated with diethanolamine (DEA) was the most sufficient, up to 79%. The highest sorption, above 85%, without regard to kind of sorbent, was observed for naproxen, diclofenac, carbamazepine, acetaminophen and ibuprofen. Not without meaning is the increasing basicity of used for impregnation amines, which is proved by the highest sorption capacity in case of use of PEI. The tested compounds are amphoteric species and consist of single or multiple charged groups, as well as polar groups (hydroxyl, carbonyl, carboxyl, amine) with aromatic rings. The sorption of salicylic acid and ketoprofen decrased on char impregnated monoethanolamine (MEA) and diethanolamine (DEA), and not changed for clofibric acid. Salicylic acid and clofibric acid characterized the lowest values of pK_a. Acidic pharmaceuticals in solutions at neutral and alkaline pH are present in the ionic forms. This might be related to repulsive interactions between functional group of compounds and on the sorbent's surface. Similar interactions, although significantly weaker, was noted for ibuprofen and diclofenac, both acidic compounds.

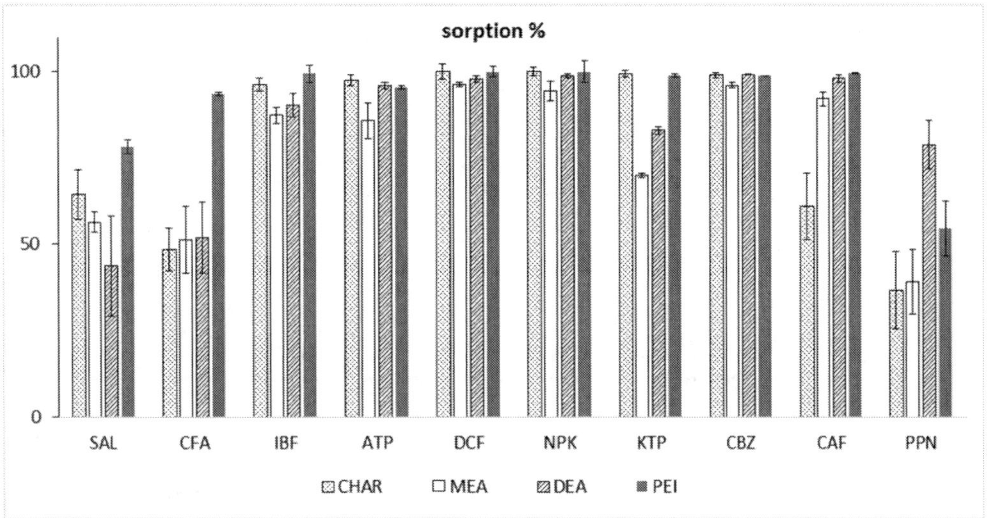

Fig. 1. Sorption of pharmaceuticals residues on char and its modifications with amines

At pH values their pK_a, the sorption affinity toward adsorbents increased significantly with the pH [13], whereas the sorption affinity dropped sharply at the pH above the pK_a values. This effect occurs in particular for char impregnated with MEA and DEA. Whereas less variation of sorption affinity due to their higher pK_a values (less variation in the ionic state from neutral to ionized form) allows carbamazepine, caffeine, acetaminophen to show strong hydrophobic interactions throughout a wide range of pH values. The octanol-water partition coeffient K_{OW} has been used to represent hydrophobicity. This is accurate if the compounds are non-ionizable, independent of pH. Under the condition of dominant hydrophobic interaction in the sorption study of higher-aromatic-containing sorbents, ionized molecules hardly attract to sorbent through this mechanism. Hovewer, the sorption studies of organic compounds in various solutions cannot be interpreted by one or two mechanisms.

Application of polar compounds of unsaturated bounds on the surface of porous material, significantly improved effectiveness of contaminations sorption. The above results demonstrated that impregnation only with amine group (PEI) allowed a higher sorption for all pharmaceuticals residues, than impregnation with amine and hydroxyl groups (MEA and DEA). PEI is a cationic polymer; the negatively charged compounds are attracted to sorbents coated in PEI. Similar observation to higher adsorption capacity toward endocrine disrupting compounds and pharmaceuticals was reported by Jung et al. [34] for N-biochar. The impregnation with compounds contains hydroxyl groups is significant for sorption of analytes with amine group in their structure as propranolol and acetaminophen. Most of applied pharmaceuticals due to the presence of hydroxyl and carboxyl groups are characterized by acidic properties, which can indicate specific mechanism of interaction or chemisorption taking place on the impregnated sorbent. As a result of the impregnation process, the pores were blocked, however decreasing the specific surface area has not influence on the sorption of analytes. The pharmaceuticals molecules are characterized by large dimensions, therefore the application of microporous adsorbents not always can be effective.

The variation in capacity of sorption of each compound to tested sorbents is given in Table 4. Sorption capacity ranged from 9.1 μg g^{-1} (propranolol) to 39.9 μg g^{-1} (diclofenac). After 24 h, 54 - 99% of the compounds sorbed to char impregnated with polyethylenimine. The high sorption capacity of sorbent impregnated with PEI at pH 8.9 is important for wastewater treatment plants, which generally operate at pH 8.

Table 4. The capacity of sorption (μg g^{-1}) of pharmaceuticals residues to char and its modifications with amines

	CHAR	*MEA*	*DEA*	*PEI*
SAL	25.8 ± 4.0	22.6 ± 1.4	20.1 ± 5.2	31.3 ± 1.1
CFA	19.4 ± 3.5	20.5 ± 4.4	20.8 ± 4.7	33.2 ± 4.9
IBF	34.8 ± 4.3	34.9 ± 1.1	36.0 ± 1.6	39.1 ± 0.7
ATP	35.7 ± 3.9	34.3 ± 2.3	38.3 ± 0.4	37.7 ± 0.6
KTP	39.1 ± 0.9	28.0 ± 2.1	33.2 ± 2.8	38.7 ± 1.0
NPK	39.5 ± 0.4	37.7 ± 1.3	39.5 ± 0.3	39.8 ± 0.1
DCF	39.6 ± 0.5	38.5 ± 0.3	39.1 ± 0.5	39.9 ± 0.1
CAF	19.3 ± 5.4	34.8 ± 1.5	39.1 ± 0.5	39.5 ± 0.4
CBZ	39.2 ± 0.5	38.3 ± 0.3	39.6 ± 0.1	38.5 ± 1.2
PPN	10.2 ± 0.8	15.7 ± 2.8	31.5 ± 3.3	21.8 ± 3.9

The current work is significant from an environmental perspective in that it provides a basis or future studies which may look at other environmental variables affecting sorption processes. The sorption data can be useful in predicting ecological risks of potential compounds from different therapeutic groups and metabolites through the understanding their transport in natural organic matter.

4 Conclusions

For the sorbents received as a result of process of activation and impregnation, the usefulness of its application for the removal of pharmaceuticals from water was investigated, based on the GC-MS analysis. The results of analysis showed, that the most effective and universal sorbent for the removal of pharmaceuticals from water among tested ones is the char impregnated with PEI, as the least compound adsorbed amount was over 50%. The smallest values of sorption capacity were found in case of non-impregnated char, which proves the hypothesis that nitrogen compounds used for the impregnation improved sorption capacity of received chars. The formation of hydrogen bound is probable between amine group of impregnating agent and hydroxyl group present in the pharmaceuticals and metabolites. Confirmation for this is the increase in sorption capacity along with the increase in sorbate acidity.

Acknowledgment: This work was supported by AGH University Grant no 11.11.210.244. Some of the measurements were performed using the scientific equipment belonging to the laboratories of AGH-UST Energy Centre Cracow, Poland.

References

1. B. Acevedo, C. Barriocanal, I. Lupul, G. Gryglewicz, Fuel **151**, (2015).
2. L. Zhang, B. Zhou, P. Duan, F. Wang, Y. Xu, Chem. Eng. J. **285**, (2016).
3. I. Hita, M. Arabiourrutia, M. Olazar, J. Bilbao, J.M. Arandes, P. Castaño, Renewable and Sustainable Energy Reviews **56**, (2016).
4. A.S. Al-Rahbi, P.T. Williams, Waste Manage. (Oxford) **49**, (2016).
5. S. Uçar, S. Karagöz, Fuel **137**, (2014).
6. M. Sienkiewicz, J. Kucinska-Lipka, H. Janik, A. Balas, Waste Manage. (Oxford) **32**, 10 (2012).
7. K. Nosek, K. Styszko, J. Gołaś, International Journal of Analytical Environmental Chemistry **94**, 10 (2014).
8. M. Petrović, S. Gonzalez, D. Barceló, TrAC, Trends Anal. Chem. **22**, 10 (2003).

9. C.I. Kosma, D.A. Lambropoulou, T.A. Albanis, J. Hazard. Mater. **179**, 1–3 (2010).
10. R.A. Hamza, O.T. Iorhemen, J.H. Tay, Environmental Technology & Innovation **5**, (2016).
11. S. Castiglioni, R. Bagnati, R. Fanelli, F. Pomati, D. Calamari, E. Zuccato, Environ. Sci. Technol. **40**, 1 (2005).
12. B. Kasprzyk-Hordern, R.M. Dinsdale, A.J. Guwy, Water Res. **43**, 2 (2009).
13. K. Styszko, K. Nosek, M. Motak, K. Bester, Comptes Rendus Chimie **18**, 10 (2015).
14. S.M. Rivera-Jimenez, M.M. Lehner, W.A. Cabrera-Lafaurie, A.J. Hernández-Maldonado, Environmental Engineering Science **28**, 3 (2011).
15. M. Grassi, G. Kaykioglu, V. Belgiorno, G. Lofrano, in: G. Lofrano (Ed.), Emerging Compounds Removal from Wastewater, Springer Netherlands, 2012, p. 15.
16. M.B. Ahmed, J.L. Zhou, H.H. Ngo, W. Guo, N.S. Thomaidis, J. Xu, J. Hazard. Mater. **323, Part A**, (2017).
17. X. Tan, Y. Liu, G. Zeng, X. Wang, X. Hu, Y. Gu, Z. Yang, Chemosphere **125**, (2015).
18. E. Kim, C. Jung, J. Han, N. Her, C.M. Park, M. Jang, A. Son, Y. Yoon, Journal of Industrial and Engineering Chemistry **36**, (2016).
19. E.M. Cuerda-Correa, J.R. DomĂnguez-Vargas, F.J. Olivares-MarĂn, J.B.n. de Heredia, J. Hazard. Mater. **177**, 1-3 (2010).
20. K. Styszko, A. Drobniak, Ochrona Środowiska **37**, 1 (2015).
21. P. Baran, M. Krzak, K. Zarębska, J. Szczurowski, W.A. Żmuda, Przem. Chem. **96**, 6 (2016).
22. M. Wdowin, M. Franus, R. Panek, L. Badura, W. Franus, Clean Technologies and Environmental Policy **16**, 6 (2014).
23. L. Bandura, W. Franus, R. Panek, M. Wdowin, Global Journal on Advances Pure and Applied Sciences **1**, (2013).
24. W. Franus, M. Wdowin, M. Franus, Environ. Monit. Assess. **186**, 9 (2014).
25. M.K. Swarcewicz, J. Sobczak, W. Pazdzioch, Water Sci. Technol. **67**, 6 (2013).
26. J. Heo, J.R.V. Flora, N. Her, Y.G. Park, Jaeweon Cho, A. Son, Y. Yoon, Sep. Purif. Technol. **90**, (2012).
27. I. Ali, M. Asim, T.A. Khan, J. Environ. Manage. **113**, (2012).
28. K. Bester, S. Banzhaf, M. Burkhardt, N. Janzen, B. Niederstrasser, T. Scheytt, Chemosphere **85**, 8 (2011).
29. J. Xu, W. Chen, L. Wu, A.C. Chang, J Environ Qual **38**, 3 (2009).
30. T.X. Bui, H. Choi, J. Hazard. Mater. **168**, 2-3 (2009).
31. Y. Zhou, L. Zhang, Z. Cheng, J. Mol. Liq. **212**, (2015).
32. R. Acosta, V. Fierro, A. Martinez de Yuso, D. Nabarlatz, A. Celzard, Chemosphere **149**, (2016).
33. OECD, *Test No. 106: Adsorption -- Desorption Using a Batch Equilibrium Method*, (OECD Publishing, 2006).
34. C. Jung, J. Park, K.H. Lim, S. Park, J. Heo, N. Her, J. Oh, S. Yun, Y. Yoon, J. Hazard. Mater. **263, Part 2**, (2013).

Development of composite calcium hydroxide sorbent in mechanical operations and evaluation of its basic sorption properties

Paweł Gara[1,*], *Ewa* Wisła-Walsh[1]

[1]AGH University of Science and Technology, Faculty of Mechanical and Robotics, al. Mickiewicza 30, 30-059 Cracow, Poland

Abstract. This article presents the results of research carried out on the possibility of obtaining composite calcium hydroxide sorbent in the process of two-step granulation, containing additional compounds of Al, Mg and Fe, and their textural and sorption studies. For this purpose, attempts were undertaken to compact commercial calcium hydroxide powder with six additives in the laboratory roll press. The resulting compacts were crushed and sieved in order to achieve the assumed sieve fraction. Based on the obtained results, basic parameters of the process of formation of composite sorbent have been determined. Both, the selected composite sorbents fractions and additives were subsequently subjected to textural studies (determination of the specific surface area and porosity) and sorption capacity performance. In addition, for the better interpretation of the results, thermogravimetric studies were carried out both for the additives and composite sorbents, as well as the grain size distribution of the additives. The results of the physicochemical tests of the obtained composite sorbents were compared with analogic results from the study on fine-grained hydroxide sorbent without additives and carbonate sorbent. The presented results showed that in a two-step granulation process it is possible to obtain the granular $Ca(OH)_2$ sorbent, as well as composite sorbents possessing better SO_2 sorption capacity in comparison to the powder $Ca(OH)_2$ and/or to the calcium carbonate sorbent. This can be attributed to the combination of capability of the sorbent to appropriate thermal decomposition and the formation of a group of pores in the range of 0.07-0.3 microns.

1 Introduction

Practically since the 1960s, to reduce emissions of acid pollutants (SO_2, HCl, HF) generated during the combustion of fossil fuels, many methods and technologies are applied in which the sorbent is mainly a high-quality limestone ($CaCO_3$ content > 85%) of required grain size [1-3].

The change in the emission limits for sulfur dioxide, expected in the near future, and the obligation to co-fire biomass caused intensification of the search for more efficient sorbents

* Corresponding author: pgara@agh.edu.pl

that reduce emissions of SO_2 and increase sorption of other acidic pollutants such as NO_x and CO_2 [4-7].

In the fluidized bed boilers, which burn fossil fuels (coal and lignite), the sorbent reducing SO_2 emissions is mainly limestone ($CaCO_3$) with a high content of calcium carbonate (over 85%) [4,8].

Increasing the efficiency of reducing sulfur dioxide emissions from fluidized beds, as in the case of dry methods (FSI, LIMB), is possible first of all by increasing the amount of sorbent introduced to the furnace, or by replacing the calcium carbonate sorbent with a more efficient one, for example the composite hydroxide sorbent [4,7,9-10].

The research conducted so far has shown [10] that, during specific mechanical operations, development of the texture of the obtained material occurs. The BET specific surface area (S_{BET}) of the granular (compacted and crushed) sorbent is more than doubled compared to the value of the parameter for the parent hydroxide in powdery (< 0.09 mm) form. It was also found that the increase in the value of the specific surface area is a consequence of a network of pores with specific diameters and distribution forming during mechanical operations used.

The article presents results of research on the possibility of obtaining a composite hydroxide sorbent in a two-step granulation, containing additives of compounds of aluminum, magnesium and iron, added to increase the specific surface area of the resulting sorbent; the article also presents textural and sorption studies conducted to determine the effects of these additives on the degree of sorption of SO_2.

The study consisted in compacting in the LPW450 press powdery calcium hydroxide with six additives (aluminum hydroxide, magnesite, natural goethite, olivine, dolomite waste) and binder (water), the amount of which was determined in previous studies [10, 11]. The compact obtained in the process of compacting was crushed immediately after production and sieved to separate the granular fraction assumed in the studies.

The separated fractions of composite sorbents and the introduced additives were subjected to tests, which consisted in determining their texture by measuring the specific surface in the raw state, after the thermal decomposition, and after the process of sorption of SO_2. Also their sorption capacity with respect to SO_2 was established using the CI index specifying the amount of sulfur dioxide, in grams, adsorbed by 1000 g of the tested sorbent. [8,12].

For a more accurate interpretation of the results, thermogravimetric tests were carried out for the additives and the obtained composite sorbents, and a study of grain composition of the additives was conducted.

The results of physicochemical tests and sorption capacity tests of the obtained composite sorbents were compared with the corresponding results obtained for the calcium hydroxide sorbent (with no additives) and calcim carbonate sorbent.

2 The experiment

Preparation of calcium hydroxide for compacting involved adding an appropriate amount of water (20% of wt.) and uniformly distributing it in a laboratory Z-type mixer.

Samples of composite sorbents were obtained by preparing calcium hydroxide mixture with additives: aluminum hydroxide in quantities 10% and 20%, magnesite 10%, natural goethite 10%, olivine 10% and dolomite waste 10% of the dry weight of calcium hydroxide. The mixing time was 20-30 minutes, which allowed to obtain homogeneity. The materials were metered into the compaction zone of the laboratory roll press LPW 450 equipped with a gravitational feeder and smooth forming rings.

The compacting process was conducted at a peripheral speed of 0.1 m/s, while maintaining a 3 mm gap between the rolls. During the attempts to compact, changes in the

DOI: 10.1051/e3sconf/20171402030

value of three parameters were recorded: the width of the gap between the forming rolls, the unit pressure, and a torque to the shaft; their sample time characteristics are shown in Figure 1.

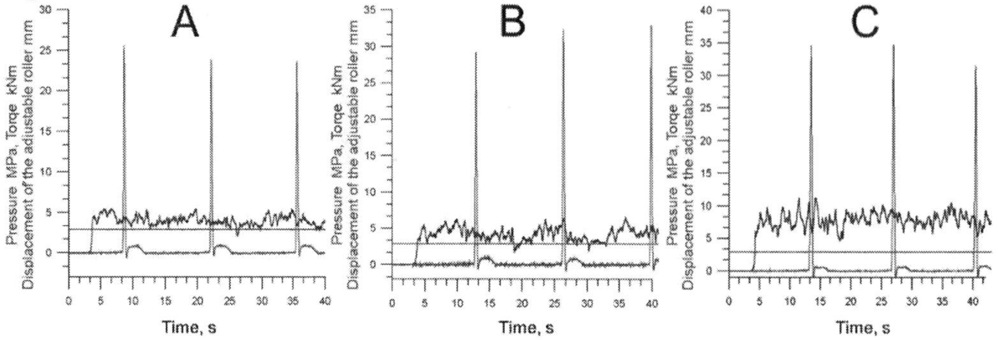

Fig. 1. Time characteristics of the unit pressure and the torque obtained while compacting samples of calcium sorbent (A), a composite sorbent with 10% (B) and 20% (C) addition of selected materials.

Immediately after the compaction process the compacts were subjected to crushing and sieving, releasing the assumed fraction. Tests of the crushing process of the compacted material were carried out using a hammer crusher. Independent parameters in the process of crushing were grid spacing value and impact velocity. Based on the results of preliminary research the adopted values were grid spacing of 3 mm and impact velocity of 8.5 m/s.

For the obtained fractions of 0.125-0.250 mm of the calcium hydroxide sorbent (with no additives), six composite sorbents, and the calcium carbonate sorbent, as well as the additives themselves, a series of tests were conducted. The analysis of the grain-size composition of the parent powdery calcium hydroxide and for the crushed and sieved samples of the additives was conducted using a MasterSizer 2000 (MalvernInstr).

The thermal analysis (TG and DTG) in the range of 25-1000°C of the tested powdery calcium hydroxide, calcium hydroxide sorbent (with no additives), and six composite sorbents, as well as of the calcium carbonate sorbent and the samples of additives, was conducted using an apparatus Netzsch STA 449 F3.

Measurement of specific surface areas, (S_{BET}), for both parent (untreated) sorbent samples, samples after the thermal decomposition at 850°C, and samples after the sorption of sulfur dioxide was conducted with the use of Nova 1200 (QuantachromeInst).

The total pore volume, (V_{CAL}) together with their distribution as a function of the diameter of sorbent samples after thermal decomposition, was determined using Poremaster 60 (Quantachrome Inst).

The SO_2 sorption index (CI) for the additives decomposed at 850°C, for the calcium hydroxide sorbent (no additives), six composite sorbents, and calcium carbonate sorbent, was determined, using laboratory equipment. The material for these tests was a fraction of the sorbent, with a grain size in the range of 0.125-0.250 mm and weight of 1.5 g. After calcination at 850°C for half an hour, the samples were contacted at this temperature (850°C) with gas containing 1780 ppm SO_2, 3% O_2 and 16% CO_2. On the basis of the obtained results, the SO_2 sorption index, CI, was calculated, specifying the amount of sulfur dioxide, in grams, adsorbed by 1000 g of the tested sorbent.

3 Results and discussion

The particle size of the parent calcium hydroxide and additives selected for the study are gathered in Table 1. The presented data show that the tested additives have a similar grain size distribution in relation to the starting calcium hydroxide, with the sample of dolomite characterized by the smallest grain size distribution, and with goethite (Ge) having the largest grain size distribution which is very well illustrated by the values of parameters S_{CUM} calculated for them, amounting to 1.94 m^2/g and 0.71 m^2/g, respectively.

Table 1. Grain size distribution - statistical parameters of additives along with their symbols.

Sample / Parameter	Tpp $Ca(OH)_2$	Al $Al(OH)_3$	Br $MgCO_3$	Ge $FeOFe_2O_3$	Ol Mg,$FeSiO_4$	Że Ca,$MgCO_3$
Mediana-D_{50} [μm]	12.94	22.47	15.94	35.45	8.17	6.72
Mode [μm]	26.13	25.27	34.11	48.18	14.57	11.79
D_{10} [μm]	3.50	4.29	2.04	4.50	2.00	1.08
D_{90} [μm]	69.58	50.50	92.99	92.75	35.74	29.99
S_{CUM} [m^2/g]	1.05	0.75	1.12	0.71	1.36	1.94

Mediana-D_{50} – The equivalent diameter of particle which has 50% of the grains found in the distribution (middle value)
Mode – The equivalent diameter of particle most commonly found in the distribution (value that occurs most often)
S_{CUM} – Cumulative surface area

In turn, comparing particle size distribution of calcium hydroxide (Tpp) with the particle size distribution of aluminum hydroxide (Al), the impact of the mode on this parameter can be noticed. S_{CUM} for aluminum hydroxide is lower by 28.57% than the value of this parameter for the calcium hydroxide, with almost comparable values of other statistical parameters of grain size distribution.

Table 2 shows the weight losses on heating of additives samples and starting calcium hydroxide, obtained from TG curves.

Table 2. Weight losses during the heating of the samples of starting calcium hydroxide and additives (TG curve).

Sample	Tpp $Ca(OH)_2$	Al $Al(OH)_3$	Br $MgCO_3$	Ge $FeOFe_2O_3$	Ol Mg,$FeSiO_4$	Że Ca,$MgCO_3$
TG Weight loss [%]						
Range 25-130°C	0.41	0.04	0.09	3.2	0.2	0.05
Range 25-600°C	19.9	33.6	10.11	32.6	1.7	1.91
Range 25-850°C	26.2	34.5	43.73	37.8	2.7	30.26

Tabulated data shown in Table 2 lead to the conclusion that heating the tested samples of additives up to 110°C is accompanied by a gradual, monotonic decrease in mass as a result of release of the physically bound water, while during heating up to about 500°C the loss is associated with the dehydroxylation of hydroxides. After exceeding this temperature, the process of decarbonisation of dolomites and carbonates begins. Similarly, Table 3 shows the weight loss on heating for the composite sorbents based on powder calcium hydroxide and for calcium carbonate sorbent (WT). The presented results of thermogravimetric (TG) tests for the obtained composite sorbents and compacted sorbent are similar to each other except for the data for the calcium carbonate sorbent. Weight loss at 850°C in the case of this sorbent is more than 43%, which is characteristic of calcium carbonates.

Table 3. Weight losses on heating for the composite sorbents sample, and for the calcium carbonate sorbent (TG), together with their symbols.

Sample	WT	Tp	TAl	TAl 20	TBr	TGe	TOl	TŻe
TG Weight loss [%]								
Range 25-130°C	0.01	0.15	0.72	2.86	0.06	0.35	0.23	0.09
Range 25-600°C	0.36	19.97	21.63	25.31	19.85	18.40	18.65	18.40
Range 25-850°C	43.88	26.16	25.91	28.04	26.32	25.41	23.67	24.68

All other sorbents are characterized by losses in mass in the range of 23-28%, which corresponds to the percentage of water formed during the dehydroxylation of the calcium hydroxide.

Table 4 shows specific surface values, (S_{BET}) for the starting calcium hydroxide and additives samples used to prepare the composite sorbents.

The specific surface areas were determined for thermally unmodified samples (S_{BET_NR}) and, in the case of the additives, also after thermal tretment at 850°C for half an hour (S_{BET_R}). For comparison, Table 4 also provides the S_{CUM} values, determined in the grain size analysis shown in Table 1.

Table 4. The values of specific surface areas (S_{BET}) and geometric surface (S_{CUM}) for the parent materials.

Sample Parameter	Tpp $Ca(OH)_2$	Al $Al(OH)_3$	Br $MgCO_3$	Ge $FeOFe2O_3$	Ol $Mg,FeSiO_4$	Że $Ca,MgCO_3$
S_{BET_NR} [m^2/g]	15.06	<0.1	12.90	69.9	7.5	4.64
S_{BET_R} [m^2/g]	(–)	92.2	21.97	14.0	1.9	16.08
S_{BET_RS} [m^2/g]	(–)	(–)	8.65	(–)	(–)	(–)
S_{CUM} [m^2/g]	1.05	0.75	1.12	0.71	1.36	1.94

Data presented in Table 4 shows that the selected addives in the raw state have varying values of specific surface areas compared to conventional sorbents derived from limestone with even much lower particle size. The commercial aluminum hydroxide has the lowest specific surface area, with the proviso that the value of the parameter obtained from the measurements of the low-temperature nitrogen adsorption agrees with the value of the geometric surface area of the sample obtained from the granulometric analysis of particle size, indicating that it is practically non-porous. In contrast, all other additives are of primary porosity, with the sample of natural goethite having the most developed porosity.

After the process of thermal decomposition (850°C-0.5 h) the size of the specific surface for calcium hydroxide and goethite decreases, despite the fact that – except for samples of olivine – the process is associated with significant weight losses resulting from dehydration, dehydroxylation and/or decarbonisation, as discussed above. This indicates sintering of pores of smaller diameter to form larger pores; in the remaining samples, the specific surface increases to varying degrees, which can be explained by the development of the texture of the decomposition products (calcium, magnesium and aluminum oxides), most likely associated with the formation of a group of small-diameter pores.

The following table (Table 5) puts together the specific surface area values of the examined composite sorbents, the calcium hydroxide sorbent (without additives) and the calcium carbonate sorbent, for undecomposed samples, S_{BET_NR}, samples decomposed at 850°C for half an hour, S_{BET_R}, and samples after sulfur dioxide sorption conducted at the same temperature, also for 0,5 h (S_{BET_RS}).

Table 5. The values of specific surface (S_{BET}) and surface (S_{POR_R}) of composite sorbents and calcium carbonate sorbent.

Sample / Parameter	WT	Tp	TAl	TAl 20	TBr	TGe	TOl	TŻe
S_{BET_NR} [m²/g]	0.48	28.4	22.35	27.73	21.35	24.2	19.72	25.33
S_{BET_R} [m²/g]	12.6	5.3	12.7	17.2	6.6	5.0	5.5	5.4
S_{POR_R} [m²/g]	11.04	4.6	13.1	23.5	7.5	5.9	4.0	4.2
S_{BET_RS} [m²/g]	(–)	1.9	3.5	7.3	2.8	2.1	2.0	2.1

S_{BET_NR} – Specific surface area values of undecomposed sorbents
S_{BET_R} – Specific surface area values of thermally treated sorbents, (850°C, 0.5 h)
S_{BET_RS} – Specific surface area values of of sorbents measured after SO_2 sorption
S_{POR_R} – Cumulative surface area (porosimetry) of thermally treated sorbents

Table 5 also includes accumulated surface area values, S_{POR_R}, for samples after thermal decomposition; the values designated by mercury porosimetry.

As shown in the Table 5, the value of the surface of undecomposed calcium carbonate sorbent (0.48 m²/g) is typical of this kind of minerals of this particular grain size (0.125-0.250 mm), and indicates that it is practically a non-porous mineral.

The specific surface area of compacted calcium hydroxide (Tp) with the same grain size is higher, compared to the value of this parameter measured for calcium carbonate. It is also almost twice higher than the specific surface area of the starting calcium hydroxide, which suggests expansion of the texture of this material during its production [7,10,11].

When the process of calcium hydroxide compaction occurs with the addition of additives in quantities of 10%, then, for samples of composite sorbents in the raw state (not thermally decomposed), a several percent decrease in the specific surface area of the obtained sorbents is observed. The results obtained for the thermally unmodified composite sorbents samples have confirmed the assumption that the grain size composition of the merged mixtures strongly influences the development of texture during the production of composite sorbents based on calcium hydroxide [10,13].

After a 0.5h process of thermal decomposition at 850°C the specific surface area of the carbonate sorbent increases as a result of its decarbonisation, which leads to the development of grain texture of the forming calcium oxide. On the contrary, in the case of thermal decomposition of compacted hydroxide sorbent, its surface decreases as a result of the „reorganization" of the pore structure [10,16].

In turn, analyzing the effects of additives on the surface area values for composite sorbents samples, it can be concluded that addition of aluminum hydroxide results in a noticeable increase in this parameter (2.4 times) at the 10% share. Increasing the amount of aluminum hydroxide additive to 20% entails a further increase in S_{BET} (3.2 times), with the

DOI: 10.1051/e3sconf/20171402030

proviso that this increase is not proportional. Introducing other additives to the obtained sorbents hardly affects the size of the specific surface area of the obtained composite sorbents.

Table 5 also shows obtained from the mercury porosimetry measurement the values of surface area, S_{POR_R} for the thermally decomposed sorbents, S_{POR_R}, and after the sorption of sulfur dioxide, $S_{BET\ RS}$, respectively.

The presented data suggest that, for the measurement of the specific surface area of decomposed samples, both of the applied measurement methods (low-temperature nitrogen sorption and porosimetry) yield consistent results, and the observed differences are due to different diameter-ranges in the capillary system tested by these methods.

In turn, specific surface values measured after SO_2 sorption suggest that the sorption of sulfur dioxide is accompanied by a similar decrease in the value of this parameter (~50%) as a result of the formation of chemisorbed calcium sulfate inside the sorbent pores [13-18] with the exception of samples of composite sorbent with added aluminum hydroxide, the process of closing the pores is accompanied by a smaller (25%) decrease in the value of the surface area in that case. It is somewhat less pronounced than for the composite hydroxide sorbent, which may indicate a relatively small contribution of these additives to the SO_2 – bonding process. In the case of aluminum hydroxide addition, the presented results may indicate that after the 0.5 h-long sorption process a part of the surface is still available; however, as will be shown further in this paper, it does not participate in the process of sorption probably because of too small pore diameters. This is well illustrated in Figure 2, which compares differential curves of pore volume distribution as a function of their diameter, for the calcium hydroxide sorbent without additives before thermal decomposition (T P NR), after thermal decomposition (T P R), and after SO_2 sorption (T P S).

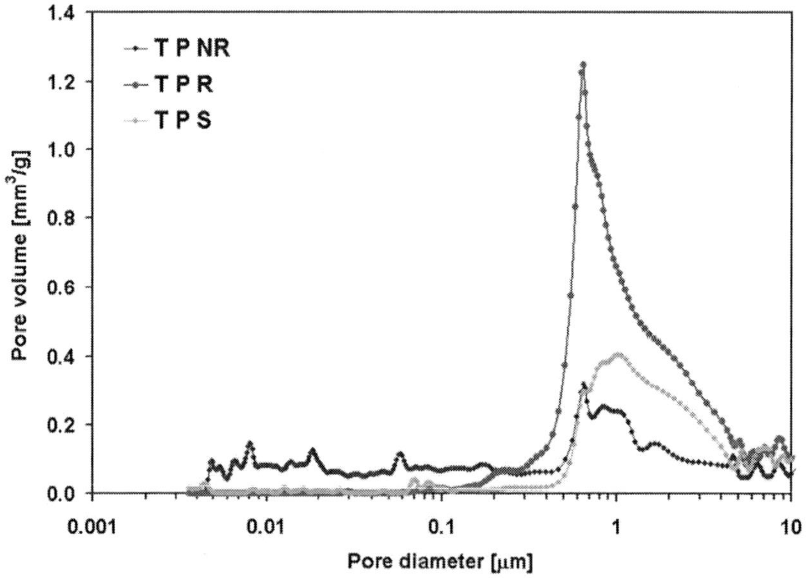

Fig. 2. Differential curves of pore volume distribution for the composite sorbent prior to heating (T P NR), after thermal decomposition (T P R), and after SO_2 sorption (T P S).

Table 6 compares the values of total pores volume present in the tested composite sorbents and in the calcium carbonate sorbent, obtained for the samples after thermal decomposition together with the mean pore diameter. Figure 3, in turn, presents differential curves of the pore volume distribution as a function of their diameters.

Table 6. The values of total pore volume (V_{CAL}) and the mean pore diameters $D_{ŚR}$ of the composite sorbents and the calcium carbonate sorbent as a function of their diameters.

Sample / Parameter	WT	Tp	TAl	TAl 20	TBr	TGe	TOl	TŻe
V_{CAL} [cm^3/g]	0.575	0.768	0.622	0.585	0.668	0.625	0.483	0.600
$D_{ŚR}$ [µm]	0.208	0.668	0.190	0.010	0.356	0.424	0.483	0.571

V_{CAL} – Total pore volume
$D_{ŚR}$ – The mean pore diameters $D_{ŚR}$

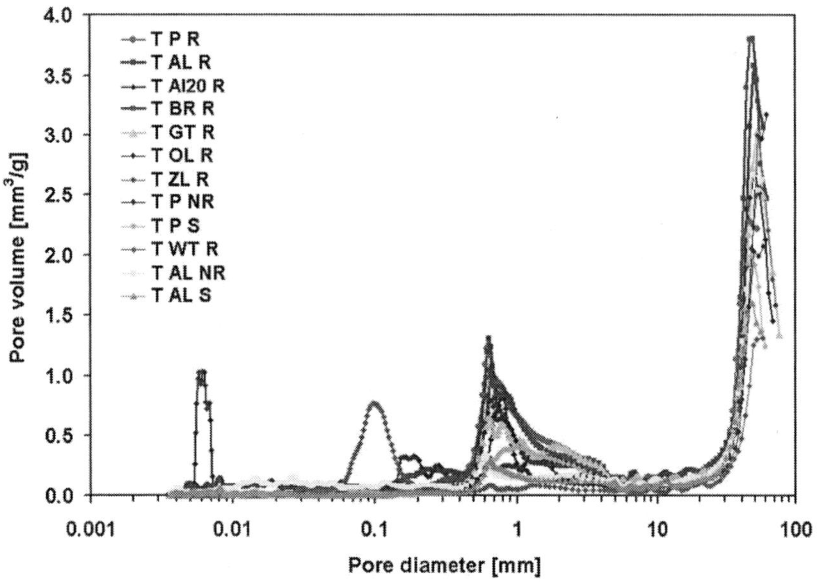

Fig. 3. Differential curves of pore volume distribution as a function of their diameters.

As mentioned above, the data in Table 6 and the differential curves of pore volume distribution as a function of their diameters (Fig.3) for the tested decomposed composite sorbents indicate that there is no correlation between the pore volume and the specific surface area; however, on the example of sample (WT) and (TAl), it can be stated (Fig. 2 and 3) that there is a correlation between the presence of pores with diameters ca. 0.07-0.3 µm and this parameter. These correlations and calcium content in the tested samples mutually affect the sorption capacity of these materials, as is shown by test of sorption index CI, results of which are presented in the Tables 7 and 8 below.

Table 7 collects the values of the SO_2 sorption index CI, calculated on the basis of measurements carried out in accordance with the above described method.

The values of these parameters obtained for the studied additives, with the exception of the additive (Że), allow us to conclude that they do not show significant sorption capacity with respect to SO_2 (Table 8).

Table 7. Sorption index CI for the tested sorbent additives.

Sample / Index	Al $Al(OH)_3$	Br $MgCO_3$	Ge $FeOFe_2O_3$	Ol $Mg,FeSiO_4$	Że $Ca,MgCO_3$
CI [g/1000g]	0	17.5	10	0	138

Table 8. Sorption index CI for the tested sorbents.

Sample Index	WT	Tp	TAl	TAl 20	TBr	TGe	TOl	TŻe
CI [g/1000g]	160	181	229	236	130	157	126	150

The values of the CI index for the tested sorbents, which are presented in Table 7 and 8, allow us to conclude that both the calcium carbonate sorbent (WT) and calcium hydroxide sorbent (Tp) (no additives) can be assessed as „excellent" [7, 8, 11] in accordance with the currently accepted criterion; however, it must be noted that, in the conditions tested, the hydroxide sorbent is characterized by higher (by 11%) reactivity in comparison to the carbonate sorbent (WT).

Introduction of 10-20% of added aluminum hydroxide causes the CI index obtained for the composite aluminum hydroxide sorbents to increase by about 30% compared to the parent (no additives) calcim hydroxide sorbent.

Other additives allow obtaining composite sorbents that, according to the above mentioned criteria, can be evaluated as „good" and „very good" sorbents.

4 Conclusion

Based on the analysis of the presented results it was found that in a two-step granulation process it is possible to obtain composite sorbents for general flue gas cleaning from fluidized bed boilers with a simultaneous increase in the melting temperature of the inert layers.

Using industrial $Ca(OH)_2$ with better technical characteristics (higher content of CaO, narrower grain size distribution) than calcium hydroxide used in the earlier studies [10], a sorbent with 22.3% smaller specific surface area was obtained.

In the tested conditions, despite the smaller specific surface area, SO_2 sorption capacity of the obtained granular hydroxide sorbent (without additives) is better than the sorption capacity of the calcium carbonate sorbent.

From among the additives introduced to the calcium hydroxide sorbent, only aluminum hydroxide increases the sorption capacity of the composite sorbent in comparison to hydroxide sorbent, yet increasing the amount of added $Al(OH)_3$ from 10% to 20% does not result in a proportional increase in the sorption of SO_2.

The introduction of additives such as olivine, dolomite, or magnesite reduces the sorption capacity of the tested composite sorbents in comparison to the calcium hydroxide sorbent. However, these additives do not undergo sintering up to 1000°C, therefore their presence may increase the melting point of the inert layer in the fluidized bed boilers and help decrease chlorine corrosion.

For the evaluation of the tested sorbents on a technical scale, the hydroxide sorbent with no additives and/or the composite sorbent should be prepared on the basis of calcium hydroxide with a broader grain size distribution.

This article has been prepared within the scope of statutory activity, Contract No. 11.11.130.957.

References

1. E. Wisla-Walsh, II National Conference on Science and Technology, "Sorbents for flue gas desulfurization", Kamien Slaski, (1997), (In Polish)
2. R.K. Srivastava, W. Jozewicz, Controlling SO_2 Emissions: An Analysis of Technologies. EPA/600/SR-00/093, (November 2000)
3. R.K. Srivastava, W. Jozewicz: J. of the Air & Waste Management Association, **51**, (2001)
4. B. Leckner, Prog. Energy Combust. Sci., **24**, (1998)
5. Yan Liu, T.M. Bisson, Hongqun Yang, Zhenghe Xu, Fuel Processing Technology, **91**, (2010)
6. T.D. Wheelock, L.K. Doraiswamy, K. Constant, Semi-Annual Technical Progress Report, April 1, 1997 - September 30, 1997
7. N. Lysek, E. Wisła-Walsh, J. Małolepszy, Proceeding of the 2nd European Technical Conference on Lime and Limestone Products. Indispensable Chemicals for Environmental Protection, Produced in Environmentally Compatible Processes, 23/24 October, 1996, Koln.
8. A. Szymanek, Limestone as an SO_2 sorbent. Part. I-III. *Air protection and waste issues* No. 1, 4, 6, (2005), (in Polish)
9. P. Maina, Chem Sci Trans, **2**, 1 (2013)
10. P. Gara, M. Hryniewicz, E. Wisła-Walsh, Polish Journal of Environmental Studies, **17**, 3A, (2008)
11. P. Gara, *Research into selection of compaction and grinding parameters of calcium hydroxide aiming at the development of the specific sorbent,* Ph D thesis, AGH Kraków, 2005 (in Polish)
12. N. Lysek, *Sorbents for flue gas desulfurization. Production and use* (Opolwap S.A., Tarnów Opolski 1997)
13. H.G. Shin, H. Kim, Y.N. Kim, H.S. Lee, J. Mater. Sci. Technol., **25,** 3 (2009)
14. J. Adanez, F. Garcia-Labiano, V. Fierro, Chemical Engineering Science **55** (2000)
15. E.J. Anthony, D.L. Granatstein, Progress in Energy and Combustion Science, **27**, (2001)
16. W. Jozewicz, B.K. Gullet, Zement-Kalk-Gips, **47,** 1 (1994)
17. R.H. Borgwardt, AIChE J, **32**, 1986
18. I. Ávila, P.M. Crnkovic, F.E. Milioli, Kai H. Luo, Applied Surface Science, **258** (2012)

Synthesis on research results of FGD gypsum briquetting

Bogdan Kosturkiewicz[1*], *Andrzej* Janewicz[1], *Marek* Hryniewicz[1], and *Łukasz* Kosturkiewicz[2]

[1]AGH University of Science and Technology, Faculty of Mechanical and Robotics, al. Mickiewicza 30, 30-059 Cracow, Poland
[2]Trzuskawica S.A., Sitkówka 24, 26-052 Nowiny

Abstract. FGD gypsum products can be characterized by significant solubility in water and dusting in dry state. These characteristics can cause a considerable pollution of air, water and soil. Among many approaches of preparing utilization of this waste, the process of compaction using briquetting has proved to be very effective. Using FGD gypsum products a new material of fertilizers characteristics has been acquired and this material is resistant to the conditions of transportation. This paper presents results of experimental briquetting of flue gas desulphurisation products in a roll press. The experiments were conducted in a laboratory roll presses LPW 450 and LPW 1100 equipped with two interchangeable forming rings that form material into saddle-shaped briquettes with volume 6,5 cm^3 and 85 cm^3. The experiments were conducted with various percentage amounts of FGD gypsum moisture. The results provided information regarding influence of moisture and roll press configuration on quality of briquettes. On the basis of obtained results, technological process and a general outline of technological line for FGD gypsum were developed. Two roll presses of own construction with different outputs were identified as appropriate for this purpose. A range of necessary works related to their adaptation for the FGD gypsum briquetting were pointed out.

1 Introduction

Due to increasingly strict environmental standards set out in the Directive [1], regarding industrial emissions 2010/75/UE (the so-called Directive IED), power plants are forced to reduce the amount of sulfur dioxide emission. The overall amount of sulfur dioxide emitted by Polish industry decreased from 1040 thousand Mg. in 2000 to 401 thousand Mg. in 2014 [2]. The most commonly used method of SO_x disposal is the wet process of desulphurization used widely in conventional power plants (about 78% of all installations worldwide [3]). It is based on the absorption of sulfur dioxide in water-limestone slurry and its reaction with calcium carbonate. The product of this reaction is synthetic gypsum in the form of calcium sulfate dihydrate crystals, referred as FGD gypsum (Flue Gas Desulphurisation). Technical and commercial usage of FGD gypsum is currently limited

* Corresponding author: kostur@agh.edu.pl

due to the surplus of demand over supply, furthermore due to transportation difficulties there is an idea to compact FGD gypsum. The subject was undertaken at the Department of Manufacturing Systems, AGH University of Science and Technology in Cracow. The results are presented in this article.

2 Experimental

The study focused on two kinds of FGD gypsum from two power plants – one from Połaniec and second from Bełchatów. FGD gypsum was homogenized in laboratory mixer with heated water jacket i.e. moisture and granulation was nearly the same in each sample. The chemical composition of mixture shown in table 1.

Table 1. Chemical composition of FGD gypsum.

Ingredient	**Value, %**	
	Połaniec power plants	**Bełchatów power plants**
$CaSO_4 \cdot 2H_2O$	~90	~90.8
Na_2O	< 0.06	< 0.05
K_2O	< 0.07	< 0.11
Al_2O_3	< 0.30	< 0.22
Fe_2O_3	< 0.15	< 0.21
SiO_2	< 0.50	< 0.48
$CaCO_3$+Mg CO_3	< 1.50	< 1.40
$CaSO_3 \cdot 2H_2O$	< 0.50	< 0.51
chlorides	< 0.01	< 0.01
H_2O	< 10	< 10

The chemical analysis of FGD gypsum was performed by using a spectrometer MiniPal 4 type: PW4025/00 (by PANanlitycal). Mixing and drying time was dependent on the moisture level of the material. Moisture content of gypsum determines the weight method. Briquetted material had moisture content of 2.3-30.6% and bulk density of 0.69-0.91 g/cm^3 (the lower humidity, the lesser bulk density). Briquetting of FGD gypsum was performed in a laboratory roll press LPW 450 with roller diameter of 450 mm, equipped with two types of interchangeable molding rings capable of producing briquettes in a saddle-shape with volume of 4 cm^3 (variant I) or 6.5 cm^3 (variant II) and laboratory roll press LPW1100 (variant II) with possibility of forming saddle-shaped briquettes with volume of 85 cm^3. Graphical visualization of forming rings are shown in Fig. 1.

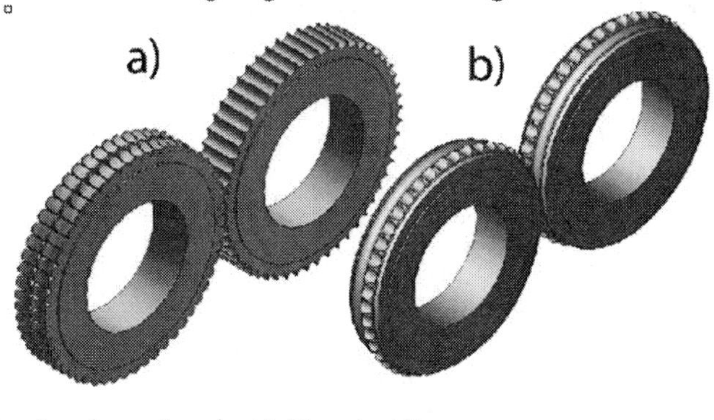

Fig.1. Models of forming rings: a) variant I, b) variant II.

More detailed description of the testing installation located in Laboratory of Departament of Manufacturing Systems AGH was featured in previous article [4].
During the evaluation process of FGD gypsum compactibility, the first determined parameters were moisture content and compacting characteristics of the material. Compaction degree was determined by dividing the volume of sampled material before merging to the volume of briquettes at certain measurement point for a given unit pressure. Characteristics of compaction were digressive with compaction degree increasing with moisture – Fig. 2.

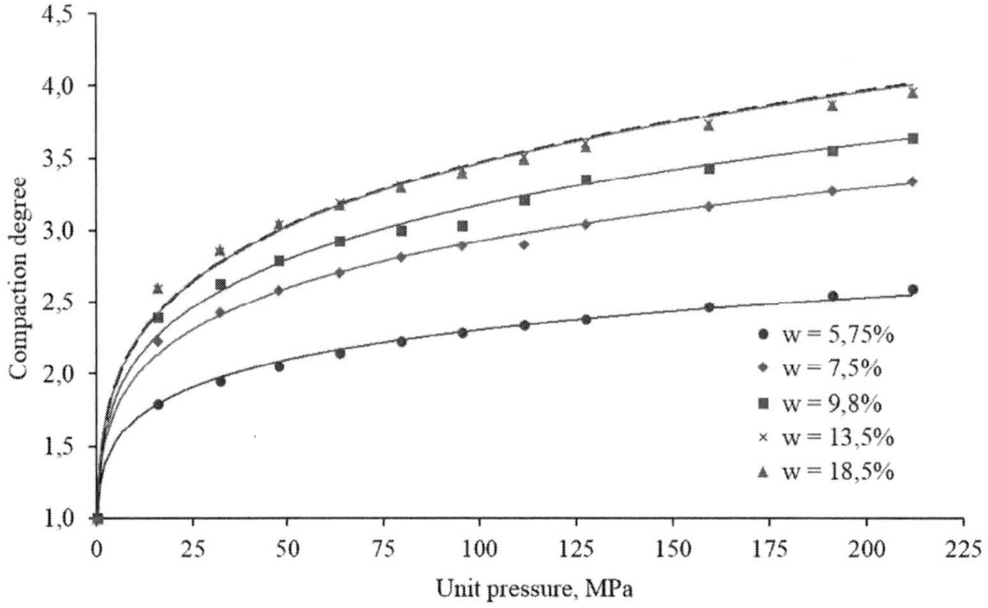

Fig.2. Compression characteristics of FGD gypsum from Połaniec power plants.

High value measured during the tests pointed need for a roll press equipped with asymmetrical compaction layout. Amount of material used for a single compaction was 9-11kg. Composition of the test materials are given in Table 2.

Table 2. Characteristic of trials.

Symbol of trial	Peripheral speed of rolls, m/s	Model of forming rings	Briquette volume, cm³	Plant of origin
A	0.1	I	4.0	Połaniec power plants
B	0.1	II	6.5	Połaniec power plants
C	0.3	II	6.5	Połaniec power plants
D	0.3	II	6.5	Bełchatów power plants
E	0.3	II	85.0	Bełchatów power plants

Briquettes were formed with a peripheral speed of rolls v_w=0.1 and 0.3 m/s, which corresponds to rotation velocity n_w=4.24 and 12.73 rpm. During compacting the initial gap between the rollers was approx. 1,0 mm. Depending on its moisture content material passed through the rollers once or twice during the process. FGD gypsum with moisture of 2.3-9.1% passed through the rolls once, while moisture content of 21.5-30.6% passed twice. Briquettes in shape of molding cavities (saddle-shaped briquettes) were formed from the material with moisture content between 2.3-25.0% (Fig. 3).

Fig. 3. FGD gypsum from Połaniec power plants briquettes (volume-6,5 cm^3) produced in roll press.

Material with moisture exceeding 25.0% failed to form briquettes. During the merging operation following parameters were measured: a torque to the shaft of the roll press (M_t) and the specific pressure in the forming cavity (p_f).
Torque value was recorded by using a measuring circuit composed of Telemetry kit MT 2555 and preamplifier EV 2510 Hottinger Baldwin Messtechnik.
After leaving the cavities, the briquettes were stored in a container from which samples were taken at random (in batches 10 pcs. each) in order to determine parameters characterising their mechanical properties. The following properties were determined for the obtained briquettes: the density (ρ_b), strength to drop right after the test (K_0), drop strength after 168 h (K_{168}) and compressive strength after 168 h (D_{168}) (for FGD gypsum with moisture content of 21.5-30.6%).
Considering the loads character the briquettes are endangered to during their reloading, transportation and thermal management, it was decided that indexes that reflect their resistance to destruction at best shall be: compression strength and resistance to drop. Compressive strength tests of briquettes were performed at the measurement stand equipped with ZWICK 1120 press of pressure range 0 ÷ 2000 N. The briquette was crushed slowly between two flat, parallel metal plates with facial areas greater than the projected area of the briquette. The speed rate of the upper plate was 0,001 m/s. The test for compressive strength, the crushing load (in N) was measured on 10 briquettes from each batch and the average value converted to compressive strength (in MPa) according equation [5,6]:

$$D = \frac{Crushing \quad load}{Cross - \sec tional \quad area \quad of \quad plane \quad of \quad fracture}, \text{ MPa} \qquad (1)$$

The strength to drop tests were performed after 168 h seasoning in order to minimize the influence of atmospheric conditions (humidity, temperature) on the value of obtained results. Durability of briquettes is determined by ASTM Method D440-86 (Drop Shatter for

 DOI: 10.1051/e3sconf/20171402031

Coal). The briquettes were dropped from a height of 2 m onto a stell plate, 60 mm thicks. This test was performed three times, each time sieving the broken mass at a sieve of mesh 18 x18 mm. Size of sieve was determined to 2/3 of average calculated from two maximum briquette dimensions measured in mutually perpendicular directions.

3 Results and discussion

Test results were processed statistically according to procedures specified for the estimation carried out by evaluation in points.

Based on the results it was possible to indicate the quality of produced briquettes. Crucial parameter for briquettes obtained from FGD gypsum is their strength to drop which determines availability for product transport. The test results of FGD gypsum briquetting in roll-press are shown in fig. 4,5,6 and 7.

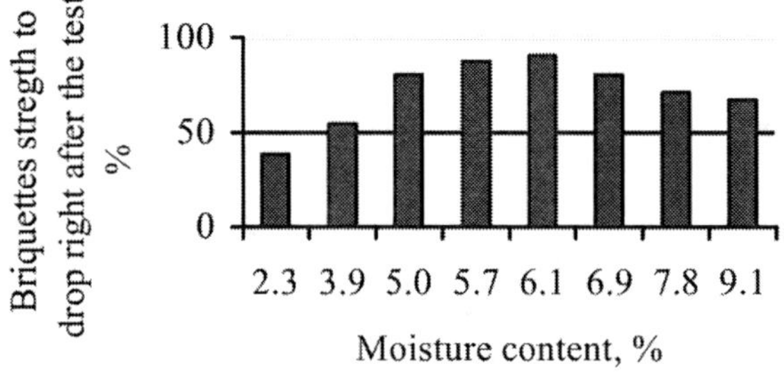

Fig.4. Effect of FGD gypsum from Połaniec power plants humidity on strength of briquettes to drop right after the test, trial C.

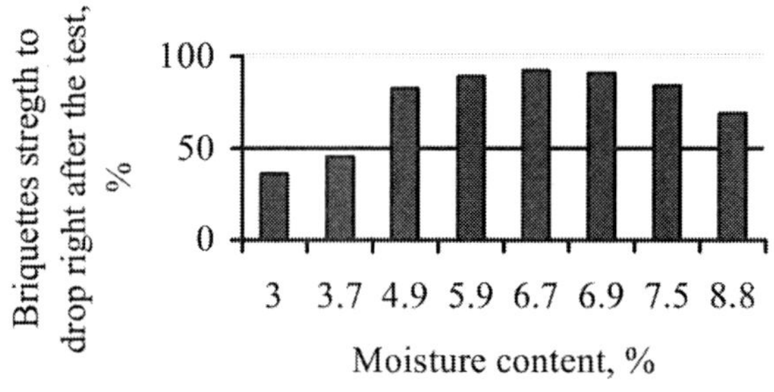

Fig. 5. Effect of FGD gypsum from Bełchatów power plants humidity on strength of briquettes to drop right after the test, trial D.

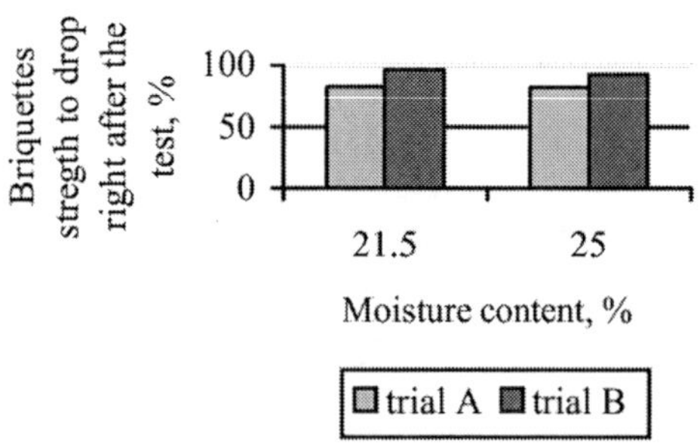

Fig. 6. Effect of FGD gypsum from Połaniec power plants humidity on strength of briquettes to drop right after the test.

DOI: 10.1051/e3sconf/20171402031

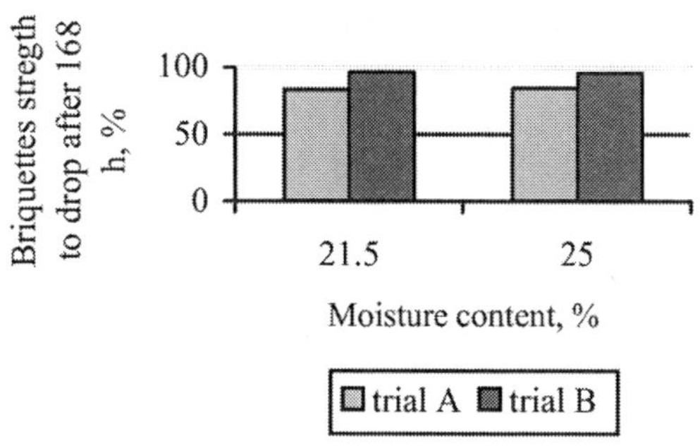

Fig. 7. Effect of FGD gypsum from Połaniec power plants humidity on strength of briquettes to drop after 168 h.

Initial strength to drop of briquettes measured directly after merging was 38.1-90.4% in case of FGD gypsum with moisture of 2.3-9.1% and 80.0-96.4% for FGD gypsum with moisture of 21.5-30.6%. After a brief seasoning of 168 h, strength to drop of briquettes has increased slightly.

Unitary energy consumption during the briquetting process in a laboratory press equipped with gravity feeder amounted to 17.89-30.96 kWh/Mg. Those calculations were carried out based on torque values on the shaft of the press.

Power consumed during briquetting process N_b was determined by the relation:

$$N_b = 2 \cdot (M_t \cdot n_r)/9.55, \text{ kW} \tag{2}$$

where: M_t - torque moment on a roll press, kNm; n_r - roll press rotational speed, rpm.

Example of time characteristic for the torque was shown in Fig. 8.

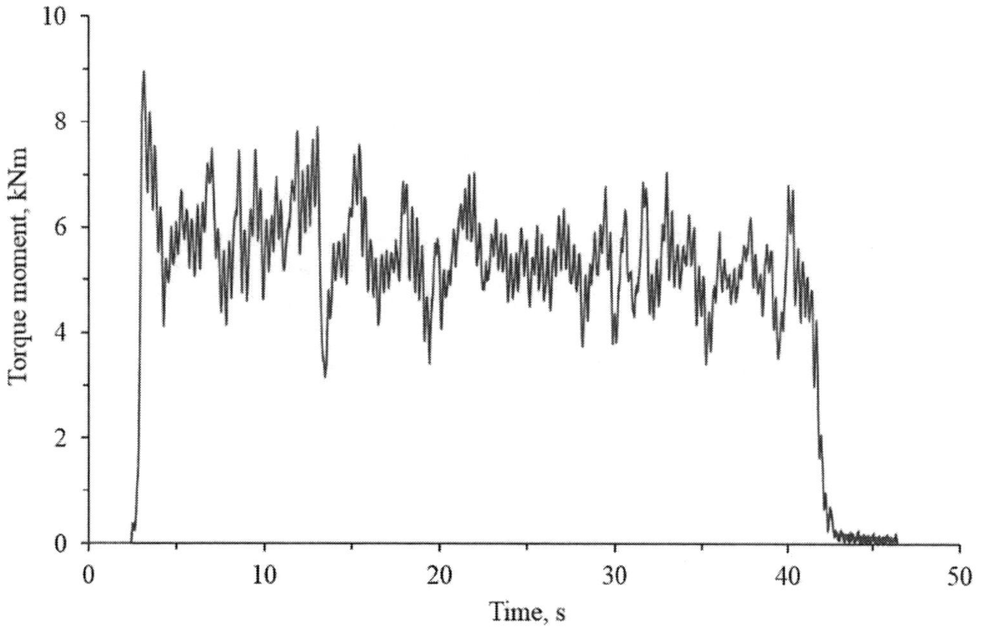

Fig. 8. Time characteristics of the torque moment on a roll press, measured at $v_r = 0.1$ m/s.

Mass performance of roll press equipped with saddle-shaped forming cavity rings W_b was determined by the relation:

$$W_b = 60 \cdot V_b \, i_b \, n_r \, \rho_b, \text{ Mg/h}, \tag{3}$$

where: V_b - briquette volume, m^3; i_b - cavities quantity, i_b= 90, -; n_r - roll press rotational speed, rpm; ρ_b- briquette density, Mg/m^3.
Unit energy demand of briquetting process (Z_e) was determined by the equation:

$$Z_b = N_b/W_b, \text{ kWh/Mg}, \tag{4}$$

where: N_b - power consumed during briquetting process, kW; W_b - mass yield of the roll press, Mg/h.
Passing the material twice through the rolls of press caused unit pressure in the cavity forming to reach up to 246 MPa (Fig. 9).

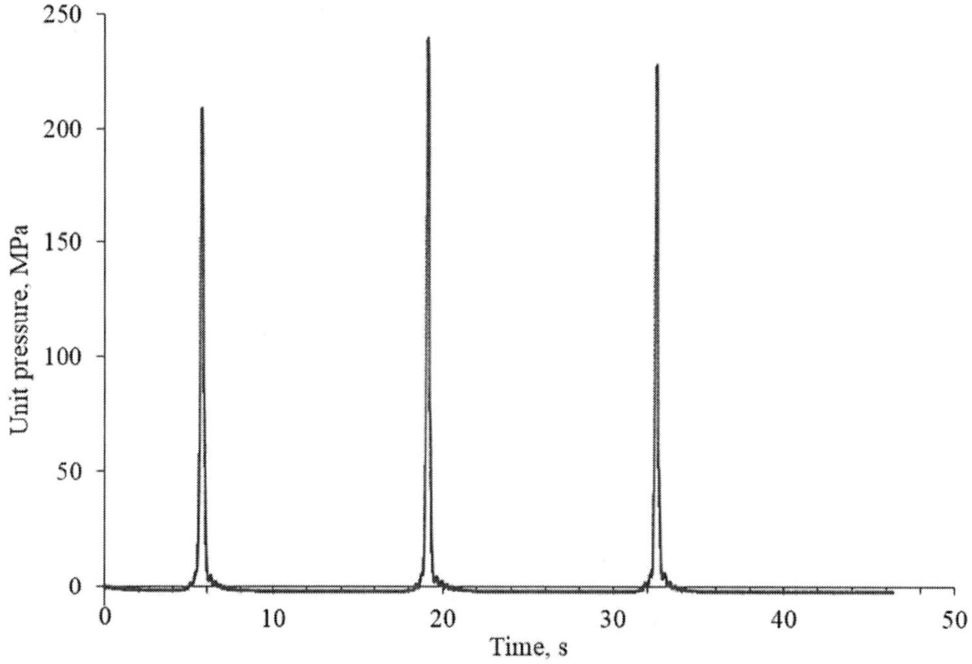

Fig. 9. Time characteristics of unit pressure in a forming cavities, measured at v_r = 0.1 m/s.

This information may be crucial in the preparation of guidelines for the design of industrial lines of merging FGD gypsum without drying the material. Compressive strength of briquettes after 168 h was 0,472-0,536 MPa while their density was contained in the range of 1.83-1.96 g/cm^3.
The study was then performed on the second roll-press LPW 1100 with rolls diameter d=1100 mm, with a compaction system allowing production of briquettes in the shape of a saddle with a volume of approx. 85 cm^3. FGD gypsum from Bełchatów power plants briquettes produced in roll press LPW 450 and LPW 1100 is presented in Fig. 10.

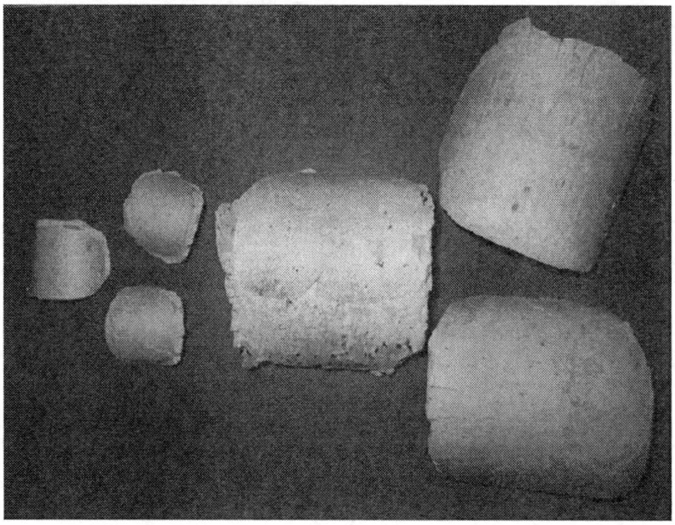

Fig.10. FGD gypsum from Bełchatów power plants briquettes produced in roll press LPW 450 (volume-6,5 cm^3) and LPW 1100 (volume-85 cm^3).

Briquetting was cast on FGD gypsum with various humidity. After compacting briquettes were left to age for a period of 72 hours in a covered area at a temperature of 18-20 °C. Figure 11 shows the influence of humidity on the strength of gypsum waste dump briquettes after aging.

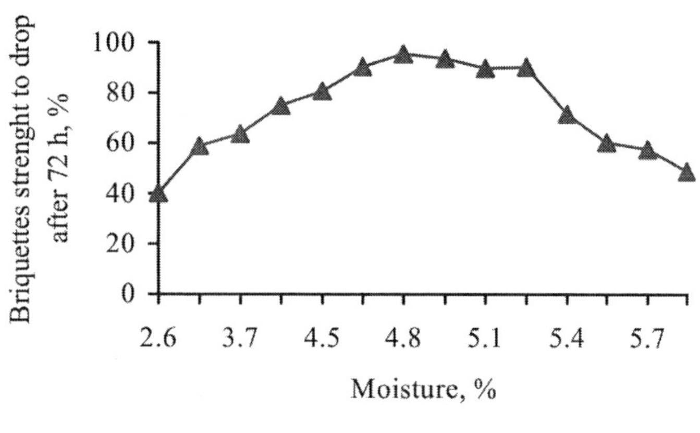

Fig.11. Relationship between briquettes' discharge strength and moisture of FDG gypsum briquetted in LPW 1100 roll press obtained from Bełchatów power plant.

Granted that the minimal discharge strength of the briquettes shouldn't be lower than 90%, it was estimated that the moisture level between 4.6 to 5.2% is the most beneficial for the FDG gypsum.

It should be noted that briquettes received in the LPW 1100 press are nearly 12 times higher in volume than those received in LPW 450. In such case lower humidity of briquetting material is conductive to obtain material with higher unitary pressure.

Positive results of laboratory tests of FGD gypsum from Bełchatów power plant were the inspiration for trial briquetting in industrial conditions. For this purpose, the third roll-press was chosen – PW 500. It was designed by a team of the University employees. Chosen parameters of the industrial press PW 500 are shown as below:

- volumetric efficiency Q= 3 m^3/h,

- pitch diameter for rollers D_p = 500 mm,
- rotational speed for rollers n_w = 11.5 rpm,
- molding cavities shape – saddle,
- briquette volume V_b = 13 cm^3,
- material hopper – gravity feed.

Tests carried out on an industrial scale confirmed the results. In order to obtain positive results it was necessary to ensure adequate moisture of the FGD gypsum. Best quality briquettes were obtained in a narrow range of 5.2-6.5% moisture. Performance of roll press PW 500 reached approx. 6.5 Mg/h briquetted material. The amount of undersize, which was successfully returned to the briquetting ranged from 10-15%. Based on the industrial tests results a concept of the transmission line of FGD gypsum from Bełchatów power plant was developed – Fig. 12.

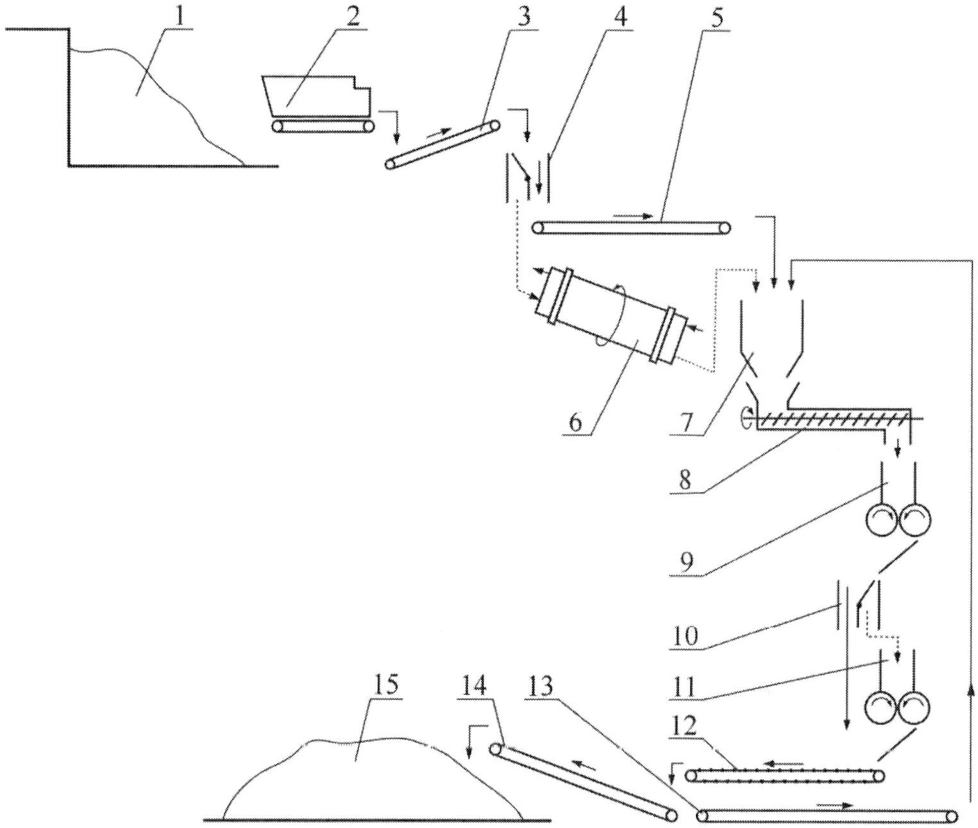

Fig. 12. Concept of the transmission line of FGD gypsum

1 - open landfill gypsum from flue gas desulphurization, 2 – box feeder, 3 - belt conveyor, 4 - two-way-distributor, 5 - belt conveyor, 6 - drum drier, 7 - one-dump bunker, 8 - screw feeder, 9 – roll press, 10 - two-way-distributor, 11 – roll press, 12 – transfer conveyor, 13,14 - belt conveyor, 15 – landfill of briquettes.

4 Conclusion

FGD gypsum from Połaniec power plants and Bełchatów power plants may be used as a material for briquetting in a roll-press equipped with a merging system capable of producing briquetted in a saddle shape. The quality of briquettes obtained from FGD gypsum depended on the moisture content and the configuration of the roll-press. Humidity of Połaniec power plants. FGD gypsum range of 5.0-6.9% formed by single pass through roll-press LPW 450, rotating at a peripheral speed of 0.3 m/s allowed to obtain briquettes with strength to drop more than 80%. Same strength to drop can be obtained by using Bełchatów power plants FGD gypsum with moisture content of 4.9-7.5%. With higher humidity content 21.5-25.0% material should be passed through roll-press twice. The highest strength to drop was obtained by briquettes formed in a roll press equipped with a second variant of forming rings. In case of gypsum with moisture content exceeding 25.0% briquettes were not obtained as the material adhered to the forming rings. The test results can be used to determine the initial design assumptions for the briquetting line of FGD gypsum. Such work should focus on the moisture content of the material, configuration of the roll-press, peripheral speed of the rollers and the chemical composition of compacted material. Briquettes made of FGD gypsum obtain sufficient strength to drop during aging process. The time needed depends on the conditions of disposal. FGD gypsum products as a new material may be used in agriculture [7] and building industry [8].

Acknowledgements This article has been prepared within the scope of statutory activity, Contract No. 11.11.130.957.

References

1. Directive 2010/75/EU of the European Parliament and of the Council (OJ L 334, 24.11.2010, p. 17)
2. Statistical Yearbook of Industry 2014, 2015, Warsaw
3. T. Pająk, M. Jurczyk, M. Cyranka, *Przem. Chem.*, **94,** 1544-1547 (2014)
4. A. Magdziarz, M. Wilk, B. Kosturkiewicz, *Chem. Proc. Eng.*, **32**, 299-309 (2011)
5. S.R. Richards, *Fuel Process. Technol.*, **25**, 89-100, (1990)
6. S.J. Mangena, V.M. du Cann, *Int. J. Coal Geol.*, **71**, 303–312, (2007)
7. A. Kaczor, J. Zuzańska, *Chem.-Did.-Ecol.-Metrol.*, **14,** 69-78 (2009)
8. J.J. Hycnar, G. Borowski, T. Józefiak, A. Malec, *Ecol. Eng.*, **45,** 51-58 (2015)

The influence of fly ash zeolites fusion synthesis parameters on the received products

Natalia Czuma[1,*], *Katarzyna* Zarębska[1], and *Paweł* Baran[1]

[1]AGH University of Science and Technology, Kraków, Poland

Abstract. Fly ash is one of possible raw material for the production of zeolites. Synthesis of zeolites with the use of fusion method enables the possibility of receiving materials of high yield of zeolites. This process may be used for efficient recycling of coal waste from power energy sector, which due to legislative indications shall no longer be directed into the landfills. The transformation of this waste into useful material of high application possibilities is possible. As received zeolites may be used eg. in purification of flue gases, which due to growing concern about the environment issues is needed. In the performed experiments the influence of a series of parameters was examined in correlation to received zeolitic material. Experimentally chosen proposition of the most beneficial process steps was presented.

1 Introduction

Polish energy sector is mainly based on the use of coal. One of by-products of the coal burning is fly ash, which due to its properties may become valuable substrate for zeolite synthesis [1,2,3]. The synthesis of zeolites with the use of fusion method is one of production options, distinguished by possibility of receiving high yields of zeolites in the synthesized material [4,5]. The zeolites obtained form a fly ash can be used in many differentiated sectors. Interesting application examples of its use are as: oil spills sorbents [6,7], additives for asphalt production [8], removal of benzene, toluene and xylene [9], ammonia ions [3] carbon dioxide [3,10] or sulphur dioxide [11,12].

In the experiments presented in this article the influence of changing the selected parameters of the performed synthesis in relation to received material was analyzed (implementation of ultrasound energy to the process, the extension of hydrothermal solution aging process, additional raw material and product of fusion grinding, solution aging for several months time and elimination of hydrothermal aging).

In the research the experiments were aimed at determining the influence of proposed process parameters change on the received zeolite material in relation to the type of received phases type and to the yield of synthesis.

[*] Corresponding author: nczuma@agh.edu.pl

DOI: 10.1051/e3sconf/20171402032

2 Experimental

With the use of fly ash, selected form one of the polish power plants, there were performed 5 fusion synthesis. The synthesis were differentiated in case of the temperature of fusion process, in each of temperatures 2-4 samples were prepared characterized by different amount of the used fly ash and alkali, the preparation of material to the process and treatment after the process, implementation of additional process step or its elimination (Table 1). As received samples were subjected to fusion process lasting 1 hour, in the muffle furnace, subsequently distillated water was added in the proportions 1g sample, 33g liquid, than shaken for 6 hours and heated in the temperature of 90^{0}C for 6 hours. Received materials were rinsed with water up to pH~10 and dried.

In Table 1 the comparison of samples were presented along with the description (in comments) of changed parameter used.

Table 1. Comparison of preformed synthesis

Process identifier	Fly ash to alkali ratio	Sample identifier	Comment
F650	2	F650-2	In this experiment the amount of used during fusion alkali was decreased so that the fly ash/ alkali ratio would change from value 2 to 3, the heating process after the fusion was performed in the temperature 90^{0}C for 6 hours
	3	F650-3	
F550	0,83	F550-0,83	In this experiment the amount of used during fusion alkali was increased so that the fly ash/alkali ratio would change from value 0,83 to 0,7, the heating process after the fusion was performed in the temperature 90^{0}C for 6 hours
	0,7	F550-0,7	
F700	0,83	F700-0,83-6H	In this experiment the amount of used during fusion alkali was increased so that the fly ash/alkali ratio would change from value 0,83 to 0,6 the heating process after the fusion was performed in the temperature 90^{0}C for 6 hours
	0,6	F700-0,6-6H	
	0,83	F700-0,83-24H	In this experiment the amount of used during fusion alkali was increased so that the fly ash/alkali ratio would change from value 0,83 to 0,6 the heating process after the fusion was performed in the temperature 90^{0}C for 24 hours
	0,6	F700-0,6-24H	
F500	1	F500-1-UA	In this experiment after the fusion the solution was subjected to the process of ultrasound disintegration, in the following step the material was rinsed with distilled water and dried
	1	F500-1	In the sample after the fusion process, shaking for 6 hours took place and heating in the temperature 90^{0}C for 6 hours
	1	F500-1-UA-G	In this experiment additional step of ultrasound disintegration was added before heating in the temperature 90^{0}C for 6 hours
	1	F500-1-3M	In this experiment after the fusion shaking for 6 hours took place, sample was left in ambient temperature for 3 months, the heating step was omitted
F600	1	F600- 1	In this experiment after the fusion, shaking for 6 hours took place, the solution was heated in the temperature 90^{0}C for 6 hours, material before fusion was mixed, after the process bigger agglomerated were fragmented, no grinding took place
	1	F600- 1-ROZ	The material before the process of fusion was mixed and ground in mortar, after the fusion process material was also ground in mortar

DOI: 10.1051/e3sconf/20171402032

Figure 1 presents sample F650-2 removed from the furnace after the fusion process.

Fig. 1. Example of sample after the fusion process

After addition of distillated water to the material solutions of tint presented on Fig. 2 were received.

Fig. 2. Exapme of sample with destillated water added prepared for the shaking process

Solutions shaking and increased temperature lead to colour change to brownish yellow. In the final stage while the process of filtration took place the solution was practically colourless. It is suspected that the colour change is related to the oxidation state of chrome compounds present in the solution.

3 Results

As a result of performed synthesis material, analized with the use of XRD method in order to define the outcome of process, was recieved.

Table 2. Comparison of synthesis results in relation to the type of zeolites recieved

Process identifier	Fly ash to alkali ratio	Sample identifier	Zeolite type
F650	2	F650-2	X, A
	3	F650-3	sodalite
F550	0,83	F550-0,83	X, A
	0,7	F550-0,7	X, A
F700	0,83	F700-0,83-6H	X, A
	0,6	F700-0,6-6H	X, A
	0,83	F700-0,83-24H	sodalite
	0,6	F700-0,6-24H	sodalite
F500	1	F500-1-UA	no zeolite
	1	F500-1	X
	1	F500-1-UA-G	X, P, sodalite
	1	F500-1-3M	no zeolite
F600	1	F600- 1	X, A, sodalite
	1	F600- 1-ROZ	X, A, sodalite

Figures 3-7 represent the comparison of results of XRD analysis on the synthesized samples, diffractograms were presented in series in which results were analyzed.

3.1 Data analysis for series of samples F650

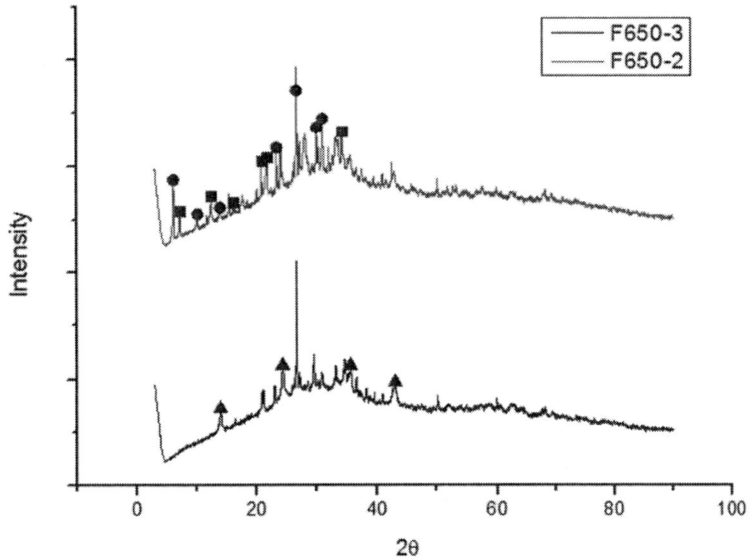

Fig. 3. Diffractograms comparison for the samples syntehsised in the fusion method series identified as F650, zeolite phases symbols: •-zeolite X, ■-zeolite A,▲-sodalite

In the fusion method experiment (F650^{0}C) the decrease of alkali amount influence on the synthesis product was tested. In preliminary experiments [11] it was fund that the use of fly ash to alkali ratio of value 2 in most cases did not result in obtaining zeolite as a product. The only case of zeolite crystallization in this fly ash/alkali ratio was the one performed in the temperature 650^{0}C, which was the criteria of selecting this temperature as a synthesis parameter for testing increase of fly ash/alkali ratio up to value 3. In the preformed

DOI: 10.1051/e3sconf/20171402032

experiment in comparison to other synthesis the hydrothermal reaction time was also increased to 24 hours (which was directed by receiving zeolite products in this time frames with fly ash/alkali ratio 2).

As a result of this way planned experiment, the reflections characteristic for zeolites X and A were obtained for fly ash/alkali ratio 2, while for fly ash/alkali ratio 3 the presence of sodalite was confirmed. Results of performed experiments indicate that it is possible to obtain zeolites of higher potential of application with the use of fly ash/alkali ratio 2. Related research [11] performed with the use of fusion method indicated that the use of higher fly ash/alkali ratio is not efficient in relation to zeolite crystallization.

3.2 Data analysis for series of samples F550

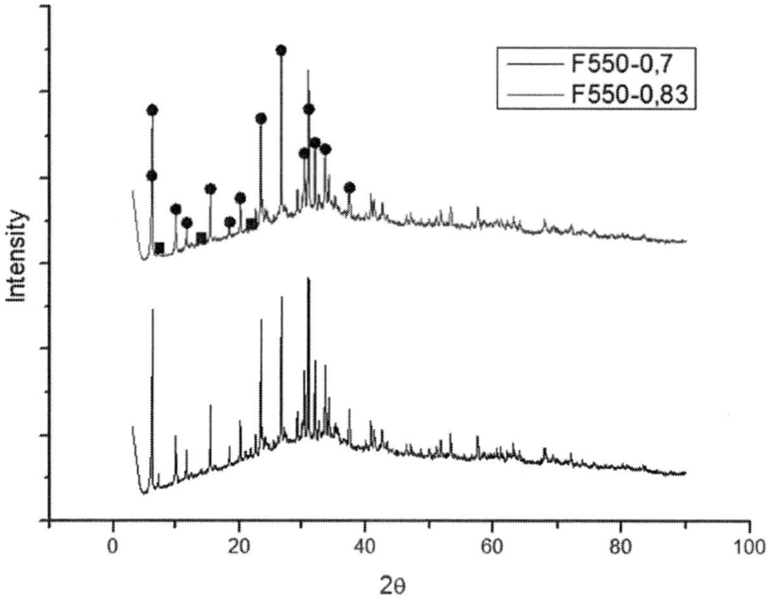

Fig. 4. Diffractograms comparison for the samples syntehsised in the fusion method series identified as F550, zeolite phases symbols: •-zeolite X, ■-zeolite A

In the experiment performed in the temperature 550⁰C it was tested how does the change in fly ash/alkali ratio will influence the synthesis product. In the previous experiments [11], in analogical conditions the synthesis was performed with the use of fly ash/alkali ratio 2, 1, 0.83. The most efficient outcome was found for fly ash/alkali ratio 0,83. Based on recieved results it is objective to find out wheather the change of proportion to value 0,7 (the value selected within the range 0,83-0,6 according to compartision options with other experimental data) will significantly affect the process and what will be the trend for observed change.

During the comparison analysis of samples diffractograms F550-0,83 and F550-0,7, the same zeolite phases were observed. However, it is worth noticing that, zeolite A presence is negligible and will not be taken into consideration. Comapring the area under the most intensive reflections characteristic for appropriate zeolite phase, it was found that F550-0,7 sampe had slightly lower efficiency in case of zeolite formation than F550-0,83 (less than 10%). This experiemnt was to verify the previous ones and indicated that the efficiency in the research is increasing in case of used fly ash/alkali ratio as follows: 2<1<0.7<0.83.

DOI: 10.1051/e3sconf/20171402032

3.3 Data analysis for series of samples F700

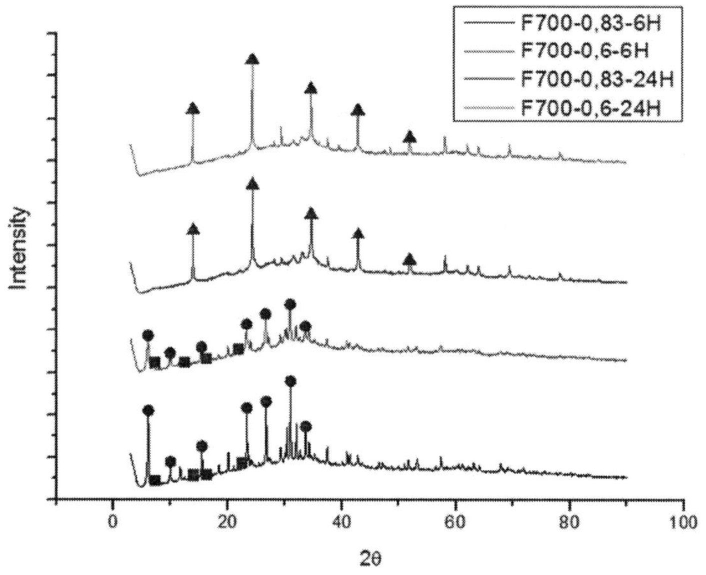

Fig. 5. Diffractograms comparison for the samples syntehsised in the fusion method series identified as F700, zeolite phases symbols: •-zeolite X, ■-zeolite A,▲-sodalite

Accoring to the received results connected with the lower efficiency with the use of lower amount of alkali, it was decided to test the influence of increasing amount of alkali added in the fusion process in the temperature of 700^{0}C so that the fly ash/alkali ratio is 0,6. The data was analysed in series by two samples synthesised in hydrotermal reaction after the fusion process which was accordingly 6 and 24 hours.

The data analysis was devided into two parts: in the first part the analysis od fly ash/alkali ratio decrease for values 0,83 to 0,6 was investigated for the hydrotermal process after the fusion, equal 6 hours. It was found that the amount of cristalizing zeolite X with the use of decreasing fly ash/alkali ratio is decreasing. It is worth noticing that in the same sample the amount of zeolite A was higher. The statement, pointing out that in given conditions fly ash/alkali ratio was more favourable for the synthesis of zeolite A, seems to be justified. In the second part outcome of analysis of samples for which the hydrotermal process after the fusion was 24 hours was presented. In this case it was found that fly ash/base ratio decrease lead to slight decrease of the amount of sodalite in comparison to ratio 0,83.

 DOI: 10.1051/e3sconf/20171402032

3.4 Data analysis for series of samples F500

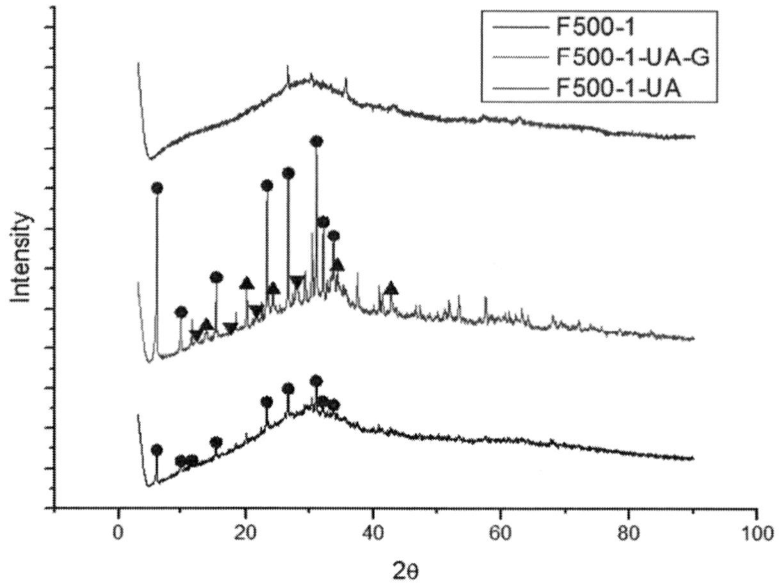

Fig. 6. Diffractograms comparison for the samples syntehsised in the fusion method series identified as F500, zeolite phases symbols: •-zeolite X, ■-zeolite A,▲-sodalit,▼-zeolite P

In the proposed experiments the influence of the use of ultrasonic homogenizer after the fusion process was tested in relation to zeolite phases obtained. It was decided to verify wheather very precisely disintegrated fly ash particles will positively influence the process. Four synthesis were proposed and tested: the basic one (F500-1) – material after the fusion process was dissolved in water, shaken for 6 hours and was subjected to heating in temperature 90⁰C for the period of 6 hours, synthesis (F500-1-UA) during which after the fusion, after the ultarsound homogenisation process (the time of homogenisation was 2 minuts), material was rinsed with distillated water and dried, and synthesis (F500-1-UA-G) during which the material was homogenised for 2 minutes, shaken and heated in temperature 90⁰C for the period of 6 hours. Additionally one of the samples (F500-1-3M) after the fusion process, after disolution and shaking was left in beaker for period of 3 months, according to [13].

The XRD diffractograms results analysis indicated that zeolite X was recieved in the sample F500-1. No zeolite cristalized in F500-1-UA as well as in F500-1-3M. The most interesting results were found for sample F500-1-UA-G, the zeolite X was recieved with efficiency around three times higher than for sample F500-1, what is more the presence of zeolite P1 and sodalite was confirmed.

3.5 Data analysis for series of samples F600

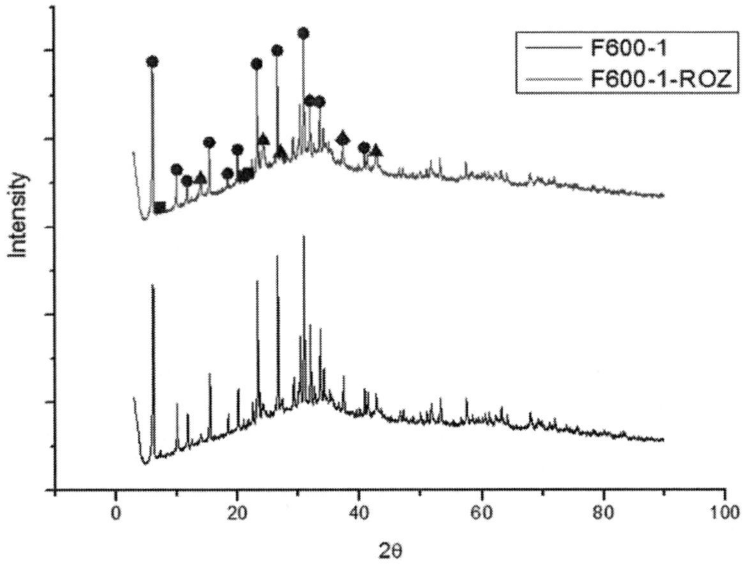

Fig. 7. Diffractograms comparison for the samples syntehsised in the fusion method series identified as F650, zeolite phases symbols: •-zeolite X, ■-zeolite A,▲-sodalite

In the performed synthesis with the use of fusion method in temperature 600^0C, the estimation of influence of fly ash and alkali manual grinding in mortar was presented in relation to the products and its synthesis efficiency. Material was ground before and after the fusion process, the results were compared with the ones recieved with the use of ultrasound energy used for disintegration. In case of other presented in the work materials the gringing process was intentionally skipped as the raw material used (fly ash and alkali) is fine-grained itself. The prepared samples were only mixed. Similar approach was presented in case of fusion product. Although the material was dense while taking if out of ceramic crucible it was breaking up into small pieces, which were dissolving in distillated water added. In connection to literature date, stating that the material has to be additionally ground [5], the influence of the grinding process on the amount and the efficiency of zeolites produced were tested.

The diffractograms analysis of samples F600-1 and F600- 1-ROZ indicated that in both cases zeolite X, sodalite and zeolite A was present. In case of zeolite A only trace reflections were found and its presence in the following analysis will be omitted. Estimating proportions of areas under the most intense reflections, it was found that zeolite X represented higher share in non ground sample. This conclusion was not expected, as due to beginning assumptions smaller particles more easily will undergo fusion process, which should have effect on higher efficiencies of zeolites in products. The explanation to this unexpected result may be fact that estimated amount of sodalite in ground sample (F600- 1-ROZ) is higher than in the one where there was no grinding (F600-1). Particles grinding resulted in obtaining material which was more favorable substrate for the crystallization of sodalite, using the building material needed for the crystallization of zeolite X. In this case in order to obtain zeolite X most probably the synthesis parameters should be changed eg. decrease of temperature or time.

4 Conclusions

Based on the performed experiments it was found that, the increase of fly ash/alkali ratio up to value 3 lead to obtaining sodalite as a product-zeolite of relativelly lower application posibilities. Based on the results comparison with previously done research it can be concluded that there is a negative impact on the formation of zeolite material once the decrease of the amount of alkali used takes place.

In order to examine tha impact of using lower fly ash/alkali ratio, the experiments with the use of decreasing fly ash/alkali ratio from 0,83 to 0,7 were done. The analysis of the results indictae that slightly lower efficiencies were obtained with the use of fly ash/alkali ratio of value 0,7. This allows to hypothesize that the ratio of lower value is less advantageous in relation to synthesis efficiency. To confirm propriety of this hypothesis the synthesis of fly ash/alkali ratio of value 0,6 was performed.

The analysis was done in two variants: with the use of hydrothermal reaction time 6 and 24 hours. It was found that for the hydrothermal reaction time 6 hours, with changed fly ash/alkali ratio form 0,83 to 0,6 the decrease of the amount of zeolite X is observed, at the same time the increase of zeolite A amonut was found for fly ash/alkali ratio 0,6. For the hydrothermal reaction time 24 hours the decrease of the amount of sodalite was observed for the fly ash/base ratio 0,6. Out of recieved results it can be concluded that the use of fly ash/base ratio 0,6 is less favourable in relation to synthesis of zeolite X and sodalite, at the same time fly ash/alkali ratio 0,6 is benefical for zeolite A synthesis in given synthesis conditions.

Comparing all of the tested ratios of fly ash and alkali it can be stated that the most efficient is the one of value 0,83.

In the series of samples the analysis of the influence of implication of utrasound energy was performed. The reference sample obtained without the ultrasound energy contained zeolite X, the use of ultrasound energy after the fusion process did not lead to zeolite crystalization (F500-1-UA). Most probably it was related to lack of time and conditions for cristals growth. The most interesting results were noted for sample for which the ultrasound disintegration was carried on before the hydrothermal aging (F500-1-UA-G). The estimated zeolite efficiency is around three times higher than in reference sample F500-1. Additionally there were present zeolite P1 phase and sodalite. In the sample F500-1-3M no zeolites were present. Analysis of this sample allows to state that despite the literature information saying that it is possible to recieve zeolite in proposed way, it was not this case. Most probably the problem might occured because of high solution evaporation rates and need for water addition. In the performed research it was also decided to check on how the disintegration of material after and before the process will influence the final product. The estimation of proportions of areas under the most intense reflections characteristic for given phases, indicates for higher amiount of zeolite X in non ground sample. Authors believe that grinding was more beneficail for cristalization of sodalite at the cost of zeolite X building material (the decrease in zeolite X efficiency in the samples which undergone the grinding process).

Sumarizing recieved experiments results it can be stated that the efficiency of fusion synthesis is influenced even by slight process parametras change. In this work the selection of positively influencing the zeolites of approptiate type was proposed.

The authors are grateful to AGH University of Science and Technology (Project 15.11.210.337) for its financial support of this work.

References

1. T. Suchecki, *Zeolity z popiołów lotnych. Otrzymywanie i aplikacje w inż. Środowiska* (Ossolineum, Wrocław, 2005)
2. I. Majchrzak-Kucęba, *Mikroporowate i mezoporowate materiały z popiołów lotnych* (Wydawnictwo Politechniki Śląskiej, Częstochowa, 2011)
3. M. Franus, M. Wdowin, L. Bandura, W. Franus, Fresen. Environ. Bull. **24**, 3a, 854-866 (2015)
4. S. S. Bukhari, J. Behin, H. Kazemian, S. Rohani, Fuel **140**, 250–266 (2015)
5. N. Shigemoto, H. Hayashi, K. Miyaura, J. Mater. Sci. **28**, 17, 4781-4786 (1993)
6. L. Bandura, M. Franus, R. Panek, A. Woszuk, W. Franus, Przem. Chem. **3**, 94, 323-327 (2015)
7. L. Bandura, M. Franus , G. Józefaciuk , W. Franus, Fuel **147**, 100-107 (2015)
8. A. Woszuk, W. Franus , Constr. Build. Mater. **114**, 556-563 (2016)
9. L. Bandura, R. Panek, M. Rotko, W. Franus, Micropor. Mesopor. Mat. **223**, 1-9 (2016)
10. Kyung-Mi Lee, Young-Min Jo, J. Mater. Cycles Waste Manag. **12**, 212–219 (2010)
11. N. Czuma, K. Zarębska, P. Baran, E3S Web of Conferences **10** (2016)
12. T. T. Suchecki1, T. Wałek, M. Banasik, Pol. J. Environ. Stud. **13**, 6, 723-727 (2004)
13. A. Derkowski, W. Franus, H. Waniak-Nowicka, A. Czímerová, Int. J. Miner. Process. **82**, 57–68 (2007)

Chemical and hydrophobic modification of activated WD-extra carbon

Eliza Wolak[1,*], *Elżbieta* Vogt[1], and *Jakub* Szczurowski[1]

[1]AGH University of Science and Technology, Faculty of Energy and Fuels, Cracow, Poland

Abstract. In this paper commercially available activated WD-extra carbon (Gryfskand) which is applied for water treatment was used. Activted carbon was modified by the following chemical agents: H_2O_2, HNO_3 and HCl. The textural characteristics of the samples were determined by the analysis of physical adsorption isotherms of nitrogen vapor at 77 K. The Boehm titration general procedure was used to determine the distribution of the surface functional groups. Chemical modifications significantly affect the chemical, structural and surface properties of activated carbons. Hydrophobization with ethereal stearic acid was performed on the raw material and samples after chemical modification. Hydrophobic properties of the samples were specified. The relationship of the chemical modification agents with hydrophobization degree was indicated.

1 Introduction

With the increase of population and the development of the industry and energy sector the amount of pollution increases as well. The energy sector is one of the main sources of emissions of coal combustion by-products. Furnace wastes are hazardous to the environment. During storage large amounts of saline waters of sedimentary and leachates contaminated with arsenic and other harmful elements that can contaminate groundwater may be produced [1]. Technological processes in power plants involve the formation of cooling water that goes into surface waters, causing changes in physical, chemical and biological properties, increasing pollution and reducing the amount of oxygen [2]. Oil and organic solvent pollutants are also very harmful.

The variety of types of pollution generated in the energy sector results in the ongoing research in new ways to reduce their emissions or eliminate the effects of their pollution. For many years activated carbons have been used in water treatment processes.

Activated carbons are known to be excellent adsorbents and are therefore used to remove a broad spectrum of dissolved organic and inorganic species from both the gas phase and liquid phase. Their effectiveness in the purification processes primarily depends on structural properties and chemical properties of the surface. Techniques to modify and characterize the surface chemical properties of activated carbons constituted the subject of interest for many scientists [3].

* Corresponding author: eklimows@agh.du.pl

The chemical modifications may be carried out by various agents, the most common of which include the following: a strong oxidizing nitric acid [4], inorganic acids: hydrochloric acid (HCl) and hydrofluoric acid (HF) [5]. Operation of the carbon-oxidizing agents causes the introduction of the activated carbon surfaces into a variety of oxygen-containing surface groups (carboxyl, lactone, phenol, ketone, quinone and ethereal groups). Many properties of carbon adsorbents are definitely influenced by chemisorbed oxygen, which can be bound in the form of various functional groups [6]. The modifying surface carbon may change the nature of hydrophobic to hydrophilic.

The most widely used treatment-enhancing properties of carbon sorbents are appropriate modifications, which result in the changing of their structural and surface parameters [7, 8]. The change of the hydrophobic properties of the carbon materials, which is applied during production filters for removing oil and organic pollutants [9] or in the purification processes used frying oils, is also possible [10].

The kind of material which is useful for the specific application results directly from its properties. The activated carbons used for the removal of pollution from solutions are characterized by high water absorption. It may be unfavorable during the regeneration process. Due to the heat regeneration of activated carbon, regardless of the type, reactors require drying the wet or moist activated carbon and an increase of the temperature to around 1000 K. Absorbed water-related negatively affects the process of regeneration of waste sorbents. Therefore in this application an important feature of carbons may be there resistance to the water. Water resistance of carbons can be changed by means of the hydrophobization process [11-14].

2 Experimental part

2.1 Sample preparation

During researches a commercially available WD-extra activated carbon (Gryfskand) which is applied for water treatment was used. It was obtained from coal dust and binder and it was used as a raw material (WD). Activted carbon (WD) was modified by the following chemical agents: HCl (1M), HNO_3 (1:1) and H_2O_2 (30%). The 10g sample of adsorbent was poured with 300cm^3 of the oxidizing agent and this led to boiling. Modification was carried out at the boiling point during 3 hours. Subsequently the samples were filtered, washed with distilled water to a constant pH and dried. The modified active carbon was obtained which is denoted by WD(HCl), WD(HNO_3) and WD(H_2O_2), respectively. Determination of pH of water extract of activated carbons were done based on PN-85 / C-97555/10 standard. The values of pH of water extract of carbons were the following: WD – 9.6, WD(HCl) – 5.2, WD(HNO_3) – 4.7, WD(H_2O_2) – 7.8. The modification of active carbon with chemical agents exerted a strong influence on the chemical character of its surface. The alkaline raw material after modification with strong HCl and HNO_3 becomes an acidic adsorbent. The modification of the alkaline adsorbent by H_2O_2 was obtained but the values of pH from 9.6 to 7.8 was decreased.

2.2 Structural properties

The standard parameters of the porous structure were calculated by means of low-temperature (77 K) nitrogen adsorption isotherms. Porous texture parameters were determined by means of an apparatus for precise measurement of physical adsorption and chemisorption – AUTOSORB-1-C (Quantachrome Instruments, USA). Before measurement, the samples were degassed at 200°C for 12 hours using a vacuum degasser system equipped with a turbomolecular pump.

The surface areas (S_{BET}) were obtained from physical adsorption isotherms data using the BET equation. The total volume pores (V_t) were read from the isotherms for the relative pressure equal to 0.995. The micropore volume (W_o), adsorption energy (E_o) and micropore surface area (S_{mic}) were calculated using the Dubinin-Radushkevich (DR) equation. The mesopore volumes (V_{BJH}), surface area of mesopore (S_{BJH}) were determined using the standard Barreta, Joynera, Halendy method (BJH). All structural parameters were summarized in Table 1.

Table 1. Structural parameters for active carbon samples.

Adsorbent	S_{BET} m^2g^{-1}	V_t cm^3g^{-1}	W_o cm^3g^{-1}	E_o $kJ \cdot mol^{-1}$	S_{mic} m^2g^{-1}	V_{BJH} cm^3g^{-1}	S_{BJH} m^2g^{-1}
WD	1583	0.845	0.591	18.37	1662	0.302	289
WD(HCl)	1235	0.627	0.448	20.96	1259	0.162	126
WD(HNO_3)	652	0.299	0.204	21.87	575	0.098	91
WD(H_2O_2)	1129	0.574	0.406	21.50	1142	0.159	138

2.3 Chemical surface characterization

The chemical properties of the surface were estimated by Boehm's method which is a general procedure used to determine the distribution of the surface functional groups [15]. The obtained results are given in Table 2.

Table 2. Concentration of oxygen functional groups determined by Boehm's method.

Adsorbent	Basic groups $mmol \cdot g^{-1}$	Acidic groups, $mmol \cdot g^{-1}$			
		Carboxyl	Lactone	Phenol	Total
WD	2.210	0.125	0.124	0.359	0.608
WD(HCl)	0.246	0.373	0.371	1.594	2.338
WD(HNO_3)	0.123	2.455	2.984	4.745	10.184
WD(H_2O_2)	2.094	0.123	0.125	0.374	0.622

The Boehm titration method gives both qualitative and quantitative information only about basic and acidic groups (in the form of carboxyl, lactone and phenol). These groups differ in their acidities and can be distinguished by neutralization with different solutions: HCl (for basic groups) and $NaHCO_3$, Na_2CO_3 and NaOH (for acidic groups). The total numbers of the groups of basic character were determined by titration of the NaOH solution, the excess of unreacted HCl. The amount and the type of surface oxygen groups by Boehm titration was examined. Figure 1 contains summarized total basic and acidic groups for all study samples.

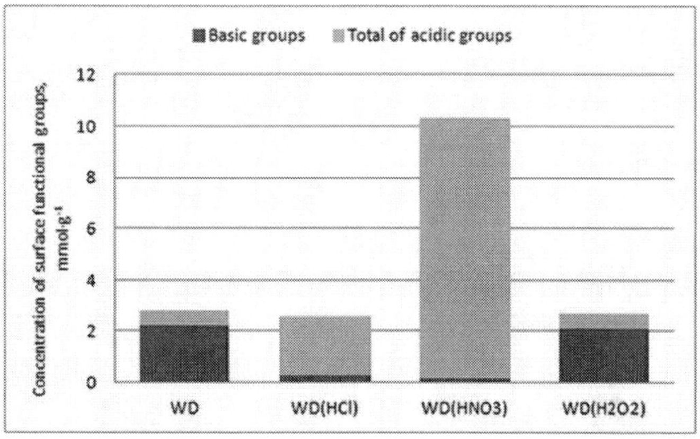

Fig. 1. Concentration of surface functional groups.

2.4 Hydrophobization process

During the hydrophobization process a commercial modifier – stearic acid – was used. The method of carbon hydrophobization with the use of stearic acid was carried out with the use of ether stearic acid solution. The concentration of the ether solution was chosen in such a way that after evaporation of the solvent, the content of stearic acid in a solid material was equal to 1%. Hydrophobization was carried out by mixing stearic acid solution with the carbon in a laboratory evaporator, at an increased temperature. The detailed descriptions of the conducted hydrophobization processes can be found in the author's earlier works [16].

2.4.1 Evaluation of the hydrophobization degree by "floating on water" test

In this work a relative evaluation of the hydrophobization degree of the analysed materials was based on a simple experiment – the so-called "floating on water" test [17]. The experiment consisted in placing a small amount of material on the surface of water in a beaker. The hydrophobization degree of materials modified in the work was evaluated on the basis of the material amount floating on the water surface for a specified period of time. Fig. 2 presents the photographs of the "floating on water" test for the materials, obtained immediately after placing the samples on the water surface.

DOI: 10.1051/e3sconf/20171402033

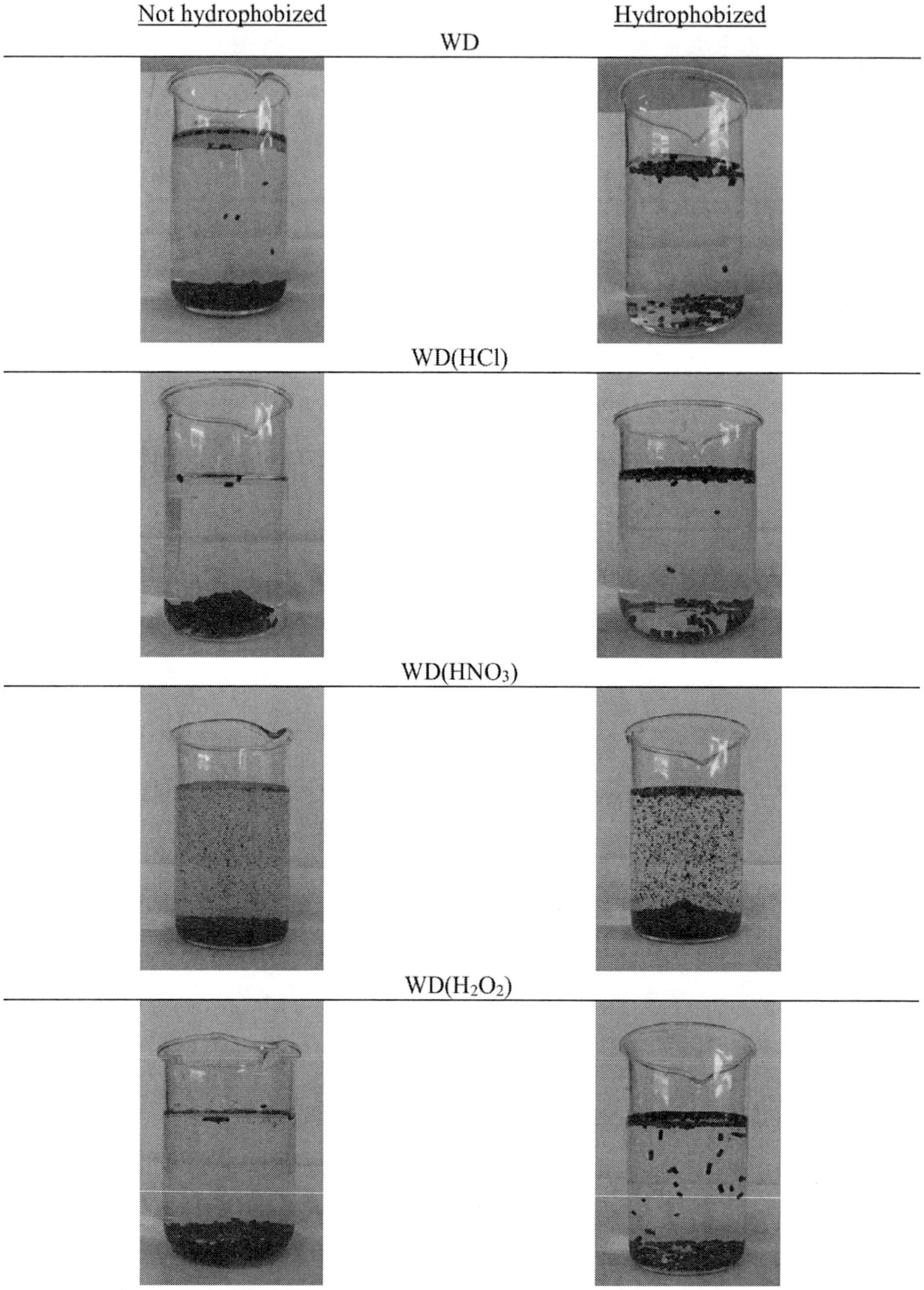

Fig. 2. The photographs of the "floating on water" test for: WD, WD(HCl), WD(HNO_3), WD(H_2O_2).

2.4.2 Evaluation of the water absorption

Water absorption was determined by means of a modified method based on the EN 1097-6:2013 standard [18]. The test portion of material was placed in a beaker and completely immersed in the water. After that the test portion was placed in the wire basket. The sample remained in the basket for a certain time until the water has stopped dripping. In our test absorbent clothes have not been used for draining excess water.

The amount of water determined during a test carried out in this manner is adequate to the amount of water retained in the bed of activated carbon which will be generated in industrial conditions. That is the amount of water, that related negatively affects the process of regeneration of waste sorbents.

The mass m_1 of the saturated test portion of material was obtained. The water absorption (WA) was calculated from the formula:

$$WA = \frac{m_1 - m_2}{m_2} \cdot 100\% \tag{1}$$

Where:
m_1 – mass of the saturated test portion [g]
m_2 – mass of the oven-dried at 105°C, test portion [g].
The obtained results are presented in Table 3.

Table 3. Water absorption.

	Type of materials modification				
		WD	WD(HCl)	WD(HNO_3)	WD(H_2O_2)
WA	not hydrophobized	164	211	292	167
	hydrophobized	142	124	206	164

3 Results and discussion

The proposed method of the modification of WD-extra active carbon influenced its surface and structural properties. The greatest impact on the structure of the porous active carbon was particularly distinct when use was made of nitric acid at the boiling point. As a result of its modification, it was observed that the resulting material was fragmented. It was demonstrated that the modifying substance (WD(HNO_3)) reduced surface area by almost 60% compared to the raw material, resulting from the partial oxidation of the carbon by a strong oxidizing agent, which is HNO_3. Both the micropore volume (0.204 cm^3/g) and particularly the mesopore volume (0.098 cm^3/g) is reduced. Nitric acid causes destruction of pore structure due to oxidation. Modification with non-oxidant hydrochloric acid and hydrogen peroxide also influenced the pore structure of the carbon, but they were not quite as big.

The modifying substances applied accounted for an increase in the concentration of the acid functional groups (Table 2 and Fig. 1). The raw carbon contained predominantly the groups of basic character and a negligible amount of acidic groups. The greatest increase of the acidic groups was observed in the case of nitric acid, which is consistent with the literature [19, 20]. The total concentration of oxygen surface functional groups goes up from 0.608mmol/g for raw active carbon (WD) to 10.184 mmol/g for WD(HNO_3). The total concentration of basic groups goes down from 2.210 to 0.123 mmol/g for appropriate carbons. Also the increase of concentration of acidic groups was particularly distinct when HCl was used. The total amount of acidic groups was 2.338 mmol/g where 68% is phenol groups and the rest carboxyl and lactone. The use of H_2O_2 had the poorest effect on the nature of the surface. The WD(H_2O_2) modifying material remained as the only alkaline carbon.

On the basis of the results obtained from the "floating on water" test (immediately after placing the samples on the water surface) it was stated (Fig. 2) that the hydrophobization process changed the water resistance of carbons. Non-hydrophobized materials became immediately moist and fell onto the beaker bottom. In the case of hydrophobized materials raw (WD) or HCl modified coals WD(HCl) almost the complete portion of the materials remained on the water surface. In the case of hydrophobized and H_2O_2 modified materials WD(H_2O_2) more than half of the sample fell to the bottom of the beaker. The worst hydrophobic properties were obtained by carbon modified with nitric acid WD(HNO_3) – almost the whole quantity of the material fell to the bottom of the beaker. The material on the water surface was observed for another 2 hours. After that only hydrophobized and HCl-modified WD(HCl)carbon remained on the water surface. This material has the best hydrophobized properties. It floated on the water also on the next day as the only one of all the modified materials.

The water absorption measurements (Table 3) confirm the above-mentioned conclusions. In the case of hydrophobized and HCl modified carbon WD(HCl) the water absorption was the smallest from the values obtained for all samples. The obtained values of water absorption (Table 3) enable us not only to compare the hydrophobic properties of hydrophobized carbons but also tell us how much of water is absorbed. In the case of hydrophobized and HCl-modified WD(HCl) coal this value is almost less than half the value for non-hydrophobized material. This information may be important during the designing of the adsorbent regeneration process. We can state that the types of chemical modification of carbons that are used affect the hydrophobization ability of the material.

The work was carried out within the research to keep the research potential AGH (11.11.210.244).

References

1. S.A. Ntim, S. Mitra, Journal of Colloid and Interface Science, **375**, p.154 (2012)
2. C. Namasivaya, D. Sangeetha, R. Gunasekaran, Trans IChemE, Part B, Process Safety and Environmental Protection , **85**(B2), p.181 (2007)
3. J.A. Menendez, J. Phillips, B. Xia, L.R. Radovic, Langmuir, **12**, p.4404 (1996)
4. A. Abdel-Nasser, El-Hendawy, Carbon, **41**, 4 (2003)
5. A. Bagreev, T.J. Bandosz, D.C. Locke, Carbon, **39**, 13 (2001)
6. A.M. Kalijadis, M.M. Vukcevic, Z.M. Jovanovic, Z.V. Lausevic, M.D. Lausevic, J. Serb. Chem. Soc.**76**, 5 (2011)
7. M. Repelewicz, K. Jedynak, J. Choma, Ochrona środowska, **31**, 45 (2009)
8. B. Buczek, E. Wolak, Adsorption, **14**, 283 (2008)
9. C.H. Lee, N. Johnson, J. Drelich, Y.K. Yap, Carbon, **49**, p. 669 (2011)
10. B. Buczek, W. Chwiałkowski, Żywność. Nauka. Technologia. Jakość, **4**, 45 (2005)
11. D. Duong Do, *Adsorption analysis: Equilibria and kinetics* (Imperial College Press 1998)
12. W.J. Thomas, B. Crittenden, *Adsorption Technology and Design* (Elsevier Science &

Technology Books, 1998)
13. R.T. Yang, *Adsorbents: Fundamentals and application* (John Wiley & Sons 2003)
14. F.G. Kerry, *Industrial gas handbook: gas separation and purification* (Taylor & Francis Group, LLC, 2006)
15. H. P. Boehm, Carbon, **40**, 145 (2002)
16. E. Vogt, Polish Journal of Chemical Technology, **10,** 1 (2008)
17. E. Vogt, Cement Lime Concrete, **3**, p. 160 (2012)
18. EN 1097-6:2013 standard
19. M. Repelewicz, K. Jedynak, J. Choma, Ochrona środowiska, **31**, 3 (2009)
20. B.K. Pradhan, N.K. Sandle, Carbon, **37**, 8 (1999)

DOI: 10.1051/e3sconf/20171402034

The new method of modifying the hydrophobic properties of expanded perlite

Elżbieta Vogt[1,*], *Łukasz* Płachta[1]

[1]University of Science and Technology, Energy and Fuel Department, Cracow, Poland

Abstract. The progressive industrialization and development of the automotive industry is the cause of the increasing demand for chemical products, especially oil products. Unfortunately, during processing, transportation or storage of these products, they get very often into the environment causing pollution. The removal of the results of accidents is still a current problem. The techniques which employ various types of sorbents deserve special attention among the several methods of eliminating the effects of pollutions. Moreover, expanded hydrophobic perlite is an interesting material among sorbents which are used on a large scale. The new method of modifying the hydrophobic properties of expanded perlite, with the use of solutions of stearic acid in organic solvents, was presented. The perlite that was used in research was produced by the PerliPol registered partnership in Bełchatów. Hydrophobic properties of the obtained materials were determined on the basis of the results achieved due to the modified film flotation method, "floating on water" test and on the basis of the value of water retention for individual samples. All grain fractions of perlite obtained hydrophobic properties which were better than or comparable to the hydrophobic properties of the HydroPerl (PerlPol) commercial material used to remove petroleum product pollution. The hydrophobization process significantly improved the adsorption capacity of modified perlite to petroleum product pollution.

1 Introduction

The progressive industrialization and development of the automotive industry is the cause of the increasing demand for chemical products, especially petroleum products. Unfortunately, during processing, transportation or storage of these products, they very frequently get into the environment causing pollution. KGPSP (The National Headquarters Of The State Fire Service) statistics (in Poland) report that the number of dangerous incidents involving petroleum hydrocarbons in 2001-2011 was between 2000 and 3000 per year [1]. The stable level of hazard preserved through the years results from the increase of safety conditions by using oil and petroleum products, probably associated with the use of modern technologies. However, accidents involving oil and its products spills, especially from tankers, transport tankers and industrial plants are still particularly dangerous for the environment [2-3]. Particularly worrying are the reports of environmental disasters in the

* Corresponding author: vogt@agh.edu.pl

 DOI: 10.1051/e3sconf/20171402034

areas of large bodies of water such as: the Erika oil tanker disaster off the coast of France, the spill in November 1999 (30 thousand tons of diesel), the Prestige tanker disaster off the coast of Spain in 2002 (50 thousand tons), the explosion of the Deepwater Horizon oil rig in 2010 (about 666 thousand tons of oil) and other disasters [2-4].

The disaster recovery is still a current problem – it is costly and difficult. The techniques which employ various types of sorbents deserve special attention among the several methods of eliminating the consequences of pollution. Different types of sorbents are used: natural and processed ones. Among the natural sorbents we may mention the following: peat, sawdust, bark and wood, cellulose waste from the production of paper and cotton products, unprocessed crushed minerals, natural zeolites. Synthetic organic sorbents such as propylene or polyurethane and processed mineral sorbents such as expanded perlite or modified zeolites are examples of processed applied sorbents [5-6].

Among these sorbents expanded perlite is an interesting material which is used on a large scale. Perlite is an alumina-silicate rock of volcanic origin. This material is extracted mainly in Turkey, Greece, USA, Italy, Japan and Hungary. This rock can be subjected to a process of expansion by soaking it in a high temperature (approx. 650-870°C) [7-9].

Many manufacturers producing sorbents on the basis of perlite are known. The main sellers on the Polish market are the following companies: PerliPol [10], Perlit-Polska [11], Zębiec [12] the Czech company PERLIT [13]. They offer raw expanded perlite and hydrophobized material. Hydrophobic perlite obtained by modification of the raw perlite has especially good sorption properties. Tests prove that hydrophobic properties significantly improve the absorption capacity of various kinds of sorbents used for the removal of pollutions, especially petroleum products. [10, 14-15] The technologies of perlite hydrophobic manufacturing that are known are based mainly on silicone compounds [10, 16].

2 Materials and methods

The perlite that was used in research, of varying grain size, was produced by the PerliPol registered partnership in Bełchatów [10]. According to the manufacturer data, the tested material has the following chemical composition: SiO_2 (65-75%), Al_2O_3 (10-18%), K_2O+ Na_2O (6-9%), MgO + CaO (2-6%), Fe_2O_3 (1-5%) [17]. Granulometric composition of grain fractions used in the research is provided in Table 1. Hydrophobized expanded HydroPerl perlite, the commercially available product of the PerliPol partnership, was used as a comparative material. HydroPerl is obtained by the hydrophobization of EP 150 raw perlite with the use of silicone compounds. The hydrophobization procedure is the manufacturer's secret [10]. The hydrophobic material, according to the manufacturer's data, is characterized by the following granulometric composition: 1,00 – 4,00 mm (20 - 70%), < 1,00 mm (30 - 80%), < 0,315 mm (max 40%).The HydroPerl absorption of petroleum products that was determined according to the manufacturer's method is less than 350 l/m^3 [17].

Table 1. Perlite type [17].

Perlite type	EP 180	EP 150	EP 100
Granulation [mm]	[%]	[%]	[%]
+ 1.00	min 60	max 20	max 20
+ 0,50	0-100	15-60	10-80
+ 0,025	max 10	min 60	min 60

 DOI: 10.1051/e3sconf/20171402034

Stearic acid as a modifier of hydrophobic properties of the raw perlite (EP 180, EP 150, EP 100) was used in research. The modification process was carried out with the use of solutions of stearic acid in organic solvents. Ethanol and petroleum ether stearic acid solutions were used. The concentration of the solution was selected in such a way that after evaporation of the solvent, the content of stearic acid in a solid material was 8% (w/w). Hydrophobization was carried out by mixing stearic acid solution with the raw perlite in a laboratory evaporator, at an increased temperature. The detailed descriptions of the conducted hydrophobization processes may be found in the author's earlier works [18-19].

2.1 Measurement methods

Hydrophobic properties of the obtained materials were determined on the basis of the results obtained by the modified film flotation method – the "floating on water" test and on the basis of the value of water retention (absorption) for individual samples [19-22]. The study of petroleum adsorption was carried out on basis of the instructions available on the website of the manufacturer, PerliPol [10, 17]

2.1.1 Evaluation of the hydrophobization degree by the "floating on water" test

In this work a relative evaluation of the hydrophobization degree of the analysed materials was based on a simple experiment, the so-called "floating on water" test [19-21]. The experiment consisted in the placing of a small amount of material on the surface of water in a beaker. The hydrophobization degree of the materials that were modified in the work was evaluated on the basis of the material amount floating on the liquid surface for a specified period of time. Fig. 1 presents the example photographs of the "floating on water" test for the EP 100 perlite. The photographs were obtained immediately after placing the samples on the water surface.

Fig. 2 presents the example photographs of the "floating on water" test for the EP 150 perlite.

Fig. 1. Photographs of the "floating on water" test for the EP 100 perlite. From the left: raw material, samples modified by: ethanol stearic acid solution, petroleum ether stearic acid solution.

Fig. 2. Photographs of the "floating on water" test for the EP 150 perlite. From the left: raw material, samples modified by: ethanol stearic acid solution, petroleum ether stearic acid solution, HydroPerl.

2.1.2 Evaluation of the water absorption

Water absorption was determined with the use of a modified method based on the [22] standard. The test portion of material was placed in the beaker and completely immersed in the water. After that the test portion was placed in a wire basket. The sample remained in the basket for a certain amount of time until the water has stopped dripping. In our test absorbent clothes have not been used for draining excess water. The amount of water determined during a test which was conducted in such a way is adequate to the amount of water retained in the layer of perlite, during the removal of pollutions.

The mass m_1 of the saturated test portion of material was obtained. The water absorption (WA) was calculated from the formula 1:

$$WA = \frac{m_1 - m_2}{m_2} \tag{1}$$

Where: m_1 – mass of the saturated test portion [g] m_2 – mass of the oven-dried at 105°C, test portion [g].

The obtained results are presented in Table 2.

Table 2. Water absorption.

Perlite type	EP 180	EP 150	EP 100	HydroPerl
WA [$g_{water}/g_{perlite}$]	Raw material			4.5
	4.9	9.3	5.6	
	Samples modified by ethanol solution			
	4.6	6.5	3.9	
	Samples modified by petroleum ether solution			
	5.0	7.4	2.6	

2.1.3 Evaluation of the petroleum products absorption

The study of petroleum adsorption was carried out on basis of the instructions available on the website of the manufacturer, PerliPol [10, 17]. The petroleum products absorption (Ads) was calculated from the formula 2:

$$Ads = \frac{15 - V_{petroleum}}{25} \cdot 1000 \qquad (2)$$

Where: Ads – petroleum products adsorption [l/m^3] $V_{petroleum}$ – volume of non-absorbed petroleum [cm^3].

The obtained results are presented in Table 3.

Table 3. Petroleum absorption.

Perlite type	EP 180	EP 150	EP 100	HydroPerl
Ads [l/m^3]	Raw material			360
	160	220	80	
	Samples modified by ethanol solution			
	280	420	480	
	Samples modified by petroleum ether solution			
	200	400	520	

3 Discussion and conclusions

It is noticed that the different fractions of the raw perlite have different initial hydrophobic properties. In the photographs (Figures 1 and 2) we may see that the whole portion of the EP 100 raw perlite falls to the bottom of the beaker. The EP 150 raw material wetted up but still floats on the water surface. The sample EP 180 raw perlite also floated on the surface of the water.

Hydrophobic properties of all materials after the modification process have improved significantly. All modified pearlite grain fractions, as well as the HydroPerk commercial product did not wet up and they float on the water surface. This can be seen in the photographs (Figs. 1, 2). All materials were left on the surface for a period of 24 hours. At this time, no change was observed in the degree of wetted up perlite.

Water absorption values (WA) listed in Table 2 confirm good hydrophobic properties of perlite after modification. Moreover, it can be seen that the amount of absorbed water depends on the composition of the grain fraction perlite. In the case of EP 180 fractions it may be assumed that the material after the modification absorbs a comparable amount of water as a raw material. Water impregnability of the EP 100 and EP 150 materials after the hydrophobization process significantly decreased. Particularly good results are obtained for the hydrophobic EP 100 material, because the obtained water absorption value is lower than in the case of the HydroPerl commercial material.

The values of petroleum product adsorption (Ads) listed in Table 2 are higher for the hydrophobized EP 150 and EP 100 material than for HydroPerl commercial material, and this is what demonstrates good sorption properties of the modified perlite. Only hydrophobized EP 180 material is characterized by lower petroleum product absorption value.The study that was conducted confirms the information from the literature that hydrophobic properties significantly improve the absorption capacity of sorbents used for the removal petroleum products [9-10, 14].

In conclusion, the raw materials (expanded perlite) are characterized by natural hydrophobic properties but upon contact with water raw expanded perlite gets wet. Better hydrophobic properties are required from the expanded perlite which is used to remove petroleum pollutions. Methods of the hydrophobization of perlite with the use of silicon compounds are known [10-13, 15]. The presented study demonstrated that one may change the hydrophobic properties of the expanded perlite with the use of stearic acid solutions in organic solvents. Almost all of the grain fractions of perlite modified during the work obtained hydrophobic properties comparable to those of hydrophobic properties of the HydroPerl commercial product (PerlPol). The hydrophobization process that is described significantly improved adsorption capacity of modified perlite to petroleum product pollutions.

Acknowledgements. The author is grateful to the AGH University of Science and Technology (Project 11.11.210.244) for financial support.

References

1. J. Rakowska, K. Radwan, Z. Ślosorz, E. Pietraszek, M. Łudzik, P. Suchorab, Scientific and Research Centre for Fire Protection, J. Tuliszkowski, National Research Institute Józefów (2012)
2. A. Jernelöv, Ambio, **39**, 5-6 (2010)
3. M. Blackburn, C.A.S. Mazzacano, C. Fallon, S.H. Black, Oil in our OSEANS. A Review of the Impacts of Oil Spills on Marine Invertebrates, Portland, OR: The Xerces Society for Invertebrate Conservation (2014)
4. M. M. Auffret, S. Duchemin, I. Rousseau, A. BoutetTanguy, D. Moraga, A. Marhic. Aquat. Living Resour. **1**, 3 (2004)
5. L. Bandura, M. Franus, G. Józefaciuk, W. Franus, Fuel **147** (2015)
6. S. Suni, A.L. Kosunen, M. Hautala, A. Pasila, M. Romantschuk, Mar. Pollut. Bull. **49** (2004)
7. M. Arifuzzaman, H. Sung Kim, Constr. Build. Mater. **93** (2015)
8. A. Rózycka, W. Pichór, Constr. Build. Mater. **120** (2016)
9. M. Gursoy, M. Karaman, Chem. Eng. J. **284** (2016)
10. PerliPol, http://www.perlipol.com.pl/pl [July 2016]
11. Perlit-Polska, http://www.perlit-polska.pl [July 2016]
12. Zębiec, http://www.zebiec.pl [July 2016]
13. PERLIT, http://www.perlit.cz [July 2016]
14. Smart Green Technology, http://www.smart-green.eu/download/2012-08-31%20Inf0ormacja%20o%20produktach.pdf [July 2016]
15. M, Giroux, E. Sahadeo, R. Libera, A. Maurizi, I. Giles, A. Marteel-Parrish, Polyhedron **114** (2016)
16. M. E. Wilson, Hydrophobic expanded perlite compositions and process for preparing the same, United States Patent, 4,889,747 (89)
17. http://www.perlipol.com.pl/dokumenty/karta_techniczna_perlitu_ekspandowanego_2010.pdf [July 2016]
18. E. Vogt, PJCT **10**, 1, (2008)
19. E. Vogt, Cem. Lime Concrete **3** (2012)
20. D.W. Fuerstenau, M.C. Williams, Colloid. Surface. **22** (1987)
21. E. Vogt, I. Opaliński, ICheaP-9 Conference, Chem. Eng. Trans. **17**, (Editor Sauro Pierucci, Rome 2009)
22. EN 1097-6:2013 standard, Tests for mechanical and physical properties of aggregate

E3S Web of Conferences **14**, 02035 (2017) DOI: 10.1051/e3sconf/20171402035
Energy and Fuels 2016

Analysis of microalgae pellets combustion in a circulating fluidized-bed

Monika Kosowska-Golachowska[1,*], *Tomasz* Musiał[1], *Dariusz* Urbaniak[1] and *Henryk* Otwinowski[1]

[1]Czestochowa University of Technology, Institute of Thermal Machinery, Armii Krajowej 21 Av., 42-201 Czestochowa, Poland

Abstract. Microalgae are expected to become an important source of high-value products with several applications in a large number of areas of biotechnology and, especially, in biofuels production. The increasing interest in microalgae as a source of biofuel (so-called third generation biofuel) is due to the several advantages. The objective of this study was to investigate combustion characteristics of microalgae (*Oscillatoria* sp.) pellets burnt in a circulating fluidized-bed (CFB) in terms of sample temperature profiles, ignition time, ignition temperature, devolatilization time and the burnout time. Spherical 10-mm microalgae pellets were tested at temperature of 850°C in a 12-kW bench-scale CFB combustor.

1 Introduction

Fossil fuels provide around 79% of world energy supply [1]. In countries such as China, Australia and Poland, coal is the primary source of power generation. However, coal combustion results in emission of CO_2 to atmosphere which is probably the main cause of global warming. Therefore, other biomass or alternative fuel solutions are needed, including microalgae. There is a growing number of works devoted to combustion [2-6], pyrolysis [7-9] and gasification [10-12] of biofuels.

Microalgae are expected to become an important source of high-value products with several applications in a large number of areas of biotechnology (such as cosmetics, pharmacy and food) and, especially, in biofuels production. The increasing interest in microalgae as a source of biofuel (so-called third generation biofuel) is due to the several advantages that it offers over terrestrial oil crops. These advantages include: high growth rate, an all-the-year-round production, high efficiency in CO_2 capture, the use of wastewater as a source of nutrients, the elimination of the need for herbicides or pesticides, and their possible cultivation in brackish water or non-arable land, resulting in a minimisation of the associated environmental impact. Microalgae can be farmed in fresh water or marine water and thus are not in competition with production of food crops [6-8, 13, 14].

Direct combustion is a thermochemical technique used to burn biomass in the presence of excess air. In the process, photosynthetically stored chemical energy in the biomass is converted into hot gases [13]. Typically, combustion takes place inside a boiler, furnace or

* Corresponding author: kosowska@imc.pcz.czest.pl

in a steam turbine at a temperature around 850 °C. The combustion process accepts various types of biomass, but the moisture content should be less than 50%. The heat produced from the combustion process does not have suitable options for storage; hence it is best utilized immediately. The cost of energy production from direct combustion is slightly higher as the biomass requires pretreatment, such as dehydrating, cutting and crushing, prior to the process. One of the simplest approaches in the literature for microalgae biomass utilization is combined heat and power production [13].

Circulating fluidized-bed (CFB) boilers are ideal for efficient power generation. They are capable of firing a broad variety of solid biomass fuels in small combined heat and power plants and large utility power plants. CFB combustion technology is increasingly becoming the market-leading technology used in the large-scale utility power sector firing a broad range of solid biomass fuels, due to its well-known benefits such as fuel flexibility, high efficiency, availability, and reliability. However, detailed knowledge of biomass specifications and full understanding of their variability of supply are paramount to design boilers with the highest efficiency and availability, and to operate them in the most economical way [15].

Microalgae biomass conversion technologies are classified in two different types such as biochemical conversion and thermochemical conversion (Figure 1). Combustion, gasification and pyrolysis are widely used thermal processes which can be used to convert dried algae to energy and fuels [16]. Most of the thermochemical processing studies on algae are based on thermogravimetric experiments on pyrolysis [7,9], gasification [7,11] and combustion [6,7].

Once you have produced the algae in many processes it's dried to a powder before it can be used in the processes to produce liquid fuels. In some cases this will not be done on site, therefore the algae powder will need to be transported. This is where small scale pellet plants come into play. To reduce the transportation costs of algae it can be pelleted. Algae pellets will also flow much better through the process hoppers etc than powder which can bridge and lead to a halt in the liquid fuel production process [17].

However, little is known about the behaviour of microalgae in the combustion process in CFB boilers. The objective of this study was to investigate combustion characteristics of microalgae pellets burnt in a circulating fluidized-bed.

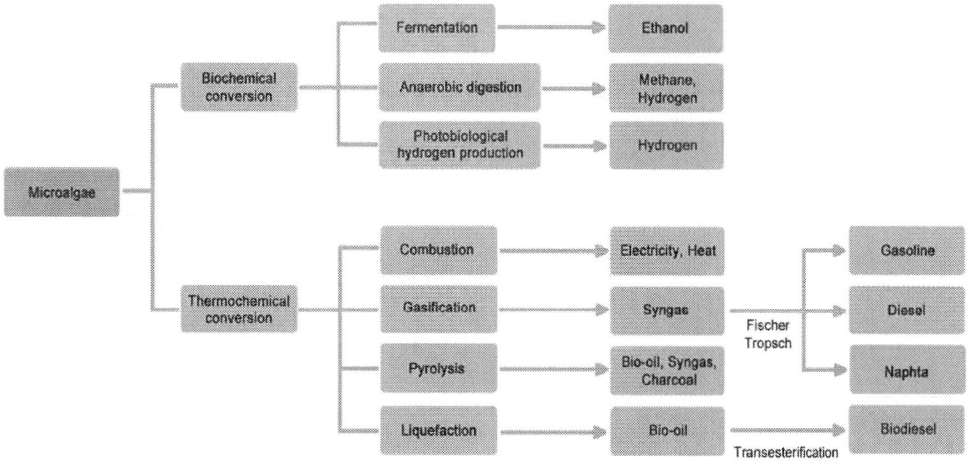

Fig. 1. Microalgae biomass conversion processes.

2 Experimental

2.1 Tested microalgae

Properties of microalgae (*Oscillatoria* sp.), including proximate and elemental analyses and higher heating value (HHV) are listed in Table 1. The proximate analysis showed volatiles (mainly organic) of 72.3 wt% in microalgae. The HHV of microalgae was 15.86 MJ/kg. The elemental analysis showed that the tested biomass mainly consisted of carbon and oxygen.

Table 1. Proximate and ultimate analyses of microalgae.

Proximate analysis (dry basis)	
Moisture content , wt %	4.4
Volatiles content, wt %	72.3
Ash content, wt %	8.9
Fixed carbon, wt %	14.4
HHV, MJ/kg	15.86
Ultimate analysis (dry basis)	
Carbon, wt %	37.7
Hydrogen, wt %	5.5
Nitrogen, wt %	4.7
Sulphur, wt %	0.46
Oxygen*, wt %	38.34

*calculated by difference

2.2 Laboratory method of algae pelletization

Figure 2 displays a flow diagram that demonstrates how the microalgae pellets were produced. The first stage involved drying algae feedstock. In the second stage, the fuel was milled in a laboratory mill. In the third stage, the milled fuel was sifted by passing it through a series of standard sieves with sizes decreasing to 0.1 mm. The sifted fuel was mixed with potato starch as a binder (approximately 8% by weight) and water in the fourth stage.

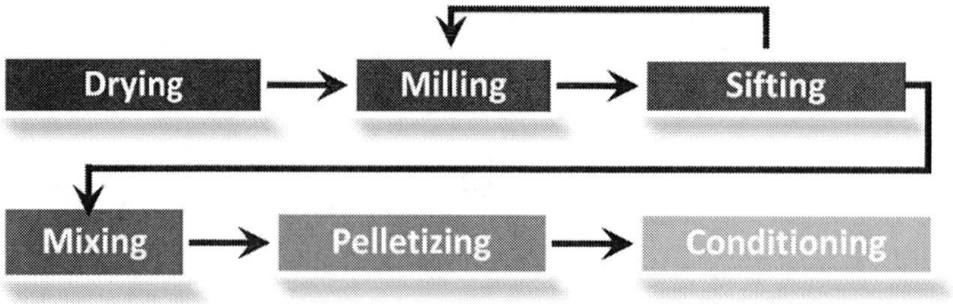

Fig. 2. Flow diagram of biomass pellet production [3].

 DOI: 10.1051/e3sconf/20171402035

In the fifth stage, the mixture was compacted using a hydraulic stamping press, which gave the pellet its spherical shape. Figure 3 illustrates system for forming pellets. The last stage involved conditioning the pellets to remove moisture. Figure 4 shows the physical appearance of the microalgae pellet.

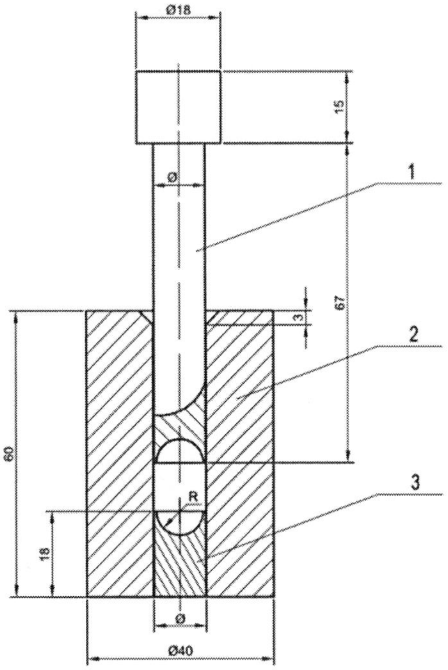

Fig. 3. Scheme of construction unit for biomass pellet formation: 1-upper punch, 2-matrix, and 3-lower punch [3].

Fig. 4. Spherical microalgae pellet (d = 10 mm).

2.3 CFB combustor and experimental procedures

Microalgae pellets combustion tests were conducted in a 12-kW bench-scale CFB combustor shown schematically in Figure 5. The facility consists of a combustion chamber (1), a cyclone (2), a downcomer (3) and a loop seal (4). The electrically-heated rectangular combustion chamber (riser), 680×75×35 mm, is the main component of the unit. Silica sand (particles smaller than 400 μm) to a mass of 0.3 kg constituted the inert bed. A rotameter (15) controlled the supply of air to the preheater (8). During combustion tests, the superficial gas velocity was kept at a constant level of about 5 m/s. The temperature was held at 850 °C by means of a microprocessor controller (11). S-type thermocouples (T1–T3) measured the temperature at three different levels inside the combustion chamber with an accuracy of ±2 °C.

A single microalgae pellet (5) was introduced into the combustion chamber and positioned stationary in the bed. To measure the temperatures in the centre and at the surface of the biomass sample a special stand was constructed. It provided a support for two S-type thermocouples.

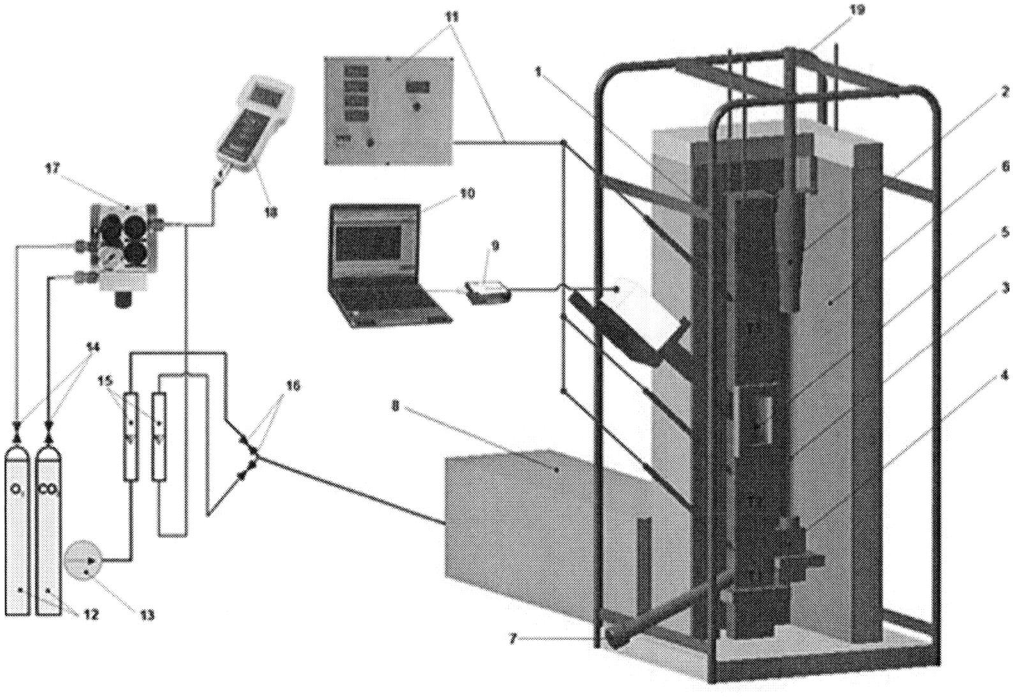

Fig,.5. Schematic diagram of the experimental apparatus [4]

1-combustion chamber, 2-cyclone, 3-downcomer, 4-loop seal, 5-biomass particle, 6-insulation, 7-drain valve, 8-preheater, 9- card, 10-computer, 11-temperature control system, 12-gas cylinders, 13-air compressor, 14-pressure regulators, 15-rotameters, 16-valves, 17-mixer, 18-gas analyser, 19-ventilation duct, T1–T3-S-type thermocouples

The tip of the first thermocouple was located inside the particle, while the second thermocouple measured the surface temperature and served as a basket in which the microalgae pellet was placed. The thermocouples were connected via a card (9) to a computer (10) in order to record the temperature measurements.

Ignition time and devolatilization time were measured by stopwatch with an accuracy of 0.1 s. The intraparticle temperature, the surface temperature, ignition time and devolatilization time were measured simultaneously.

Video and digital cameras were used to record the combustion process of microalgae pellets.

Spherical 10-mm microalgae pellets were tested at temperature of 850°C in the air atmosphere.

3 Results and discussion

Each biomass sample introduced to the combustion chamber undergoes several characteristic stages namely:

a) heating and drying,
b) ignition of volatiles,
c) combustion of volatiles,
d) combustion of remaining char.

DOI: 10.1051/e3sconf/20171402035

Almost all solid fuels experience these processes but the duration of each process depends on fuel type and its composition (moisture and volatile matter contents, total carbon content), temperature in the combustion chamber, heating rate and oxidizing atmosphere [4].

Figure 6 shows pictures of a microalgae pellet burning in a circulating fluidized bed. The first stage is heating and drying (Fig. 6a). After rapid heating, the ignition of volatiles follows. The ignition time (τ_i) was 2 s for microalgae pellets. Burning volatiles form a distinctive long flame (Fig. 6c). The average volatiles combustion time was 32 s. The last stage is combustion of char (Fig. 6d).

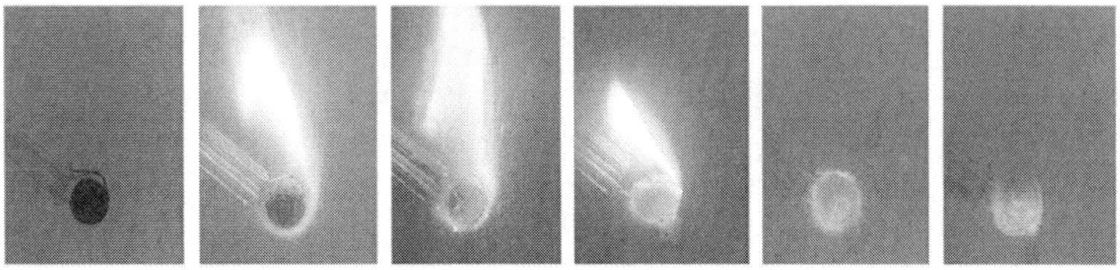

Fig. 6. Visualisation of microalgae pellet in CFBC in air.

Figure 7 shows temperatures measured at the surface and in the centre of microalgae pellet burned at 850 °C in CFB. The average ignition temperature of microalgae pellets was determined to be 308 °C in the air atmosphere. When the flame approaches its point of extinction, the surface temperature reaches its maximum value. The maximum surface temperature was approximately 1100 °C. In the next stage, i.e. char combustion, the centre temperature was higher than the surface temperature. The maximum centre temperature was approximately 1000 °C. The burnout time (τ_b) of microalgae pellet was approximately 138 s.

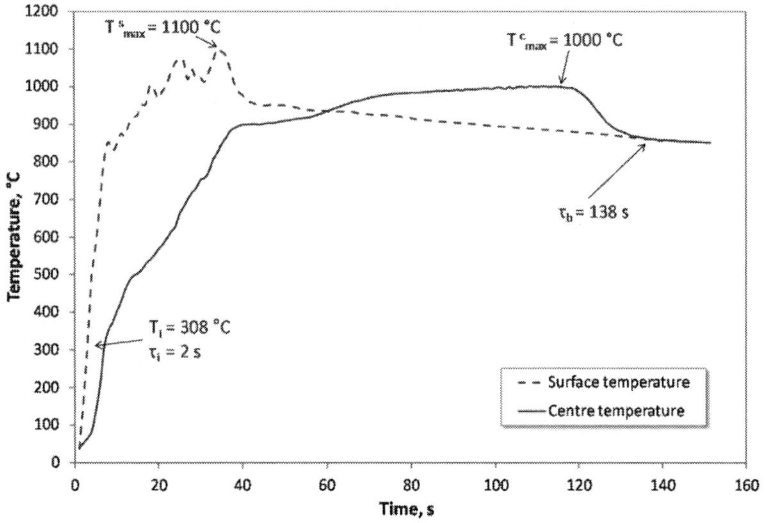

Fig. 7. Temperature profiles for microalgae pellet during combustion in CFB.

The comparison of temperature profiles for 10-mm microalgae pellet and wheat straw pellet is shown in Figure 8. In both cases, after an initial delay, the centre temperature exceeds the surface temperature and is approximately 100°C higher during the course of combustion. The temperature profiles for both pellets are similar. However, the combustion process of wheat straw pellet proceeded at slightly higher temperatures and was shorter in time compared to combustion of microalgae pellet. The burnout time of wheat straw pellet is approximately 42% shorter than that for microalgae pellet.

The average parameters during 10-mm pellets combustion in a circulating fluidized-bed are shown in Table 2.

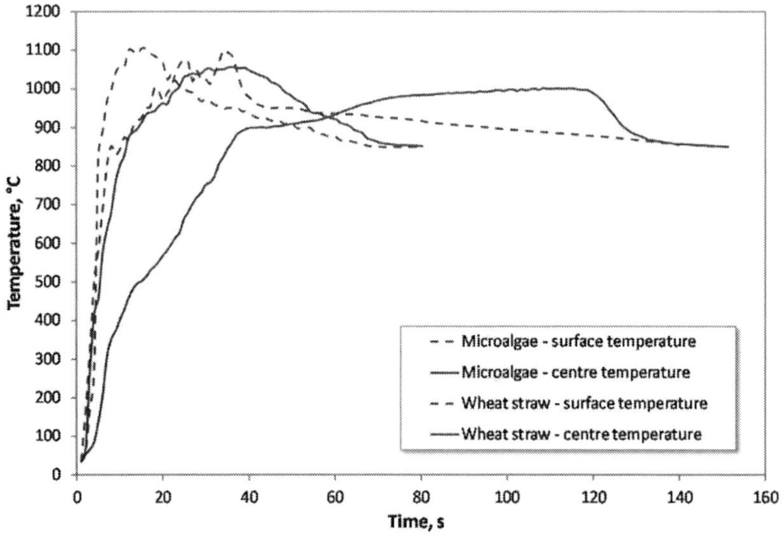

Fig. 8. Comparison of temperature profiles for microalgae pellet and wheat straw pellet during combustion in CFB.

Table 2. Average parameters during 10-mm pellets combustion in CFB at 850 °C.

Fuel	**Microalgae**	**Wheat straw** [19]	**Salix viminalis** [19]	**Bituminous coal** [19]
Ignition time, τ_i, s	2	< 1	< 1	3
Ignition temperature, T_i, °C	308	260	280	370
Maximum surface temperature, T^s_{max}, °C	1100	1150	1140	1070
Maximum centre temperature, T^c_{max}, °C	1000	1020	1030	1080
Volatiles combustion time, s	32	18	24	50
Burnout time, s	140	80	90	580

Generally, the ignition temperature decreases with an increase in volatile matter content. The comparison of the volatiles combustion time and burnout time is shown in Figure 9 and 10, respectively. The volatiles combustion time depends on the volatiles content in fuel and the density of pellet. The burnout time depends on the carbon content in fuel and the density of pellet.

DOI: 10.1051/e3sconf/20171402035

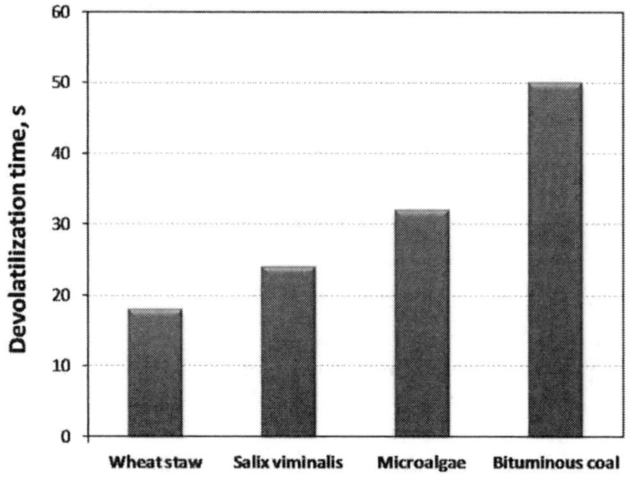

Fig. 9. Comparison of devolatilization time for 10-mm pellets in CFBC

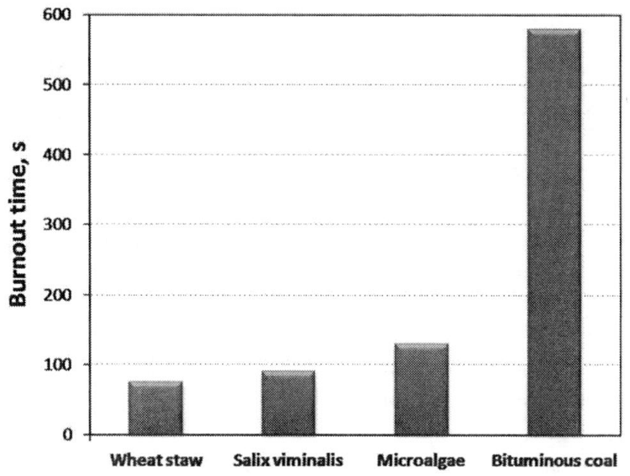

Fig. 10. Comparison of burnout time for 10-mm pellets in a circulating fluidized bed.

4 Conclusions

Spherical microalgae pellets were burned in a laboratory-scale CFB combustor at 850 °C. The higher heating value of microalgae (*Oscillatoria* sp.) was 15.86 MJ/kg. The average ignition temperature and ignition time of microalgae pellets were 308 °C and 2 s, respectively. The average volatiles combustion time and burnout time of microalgae pellet were 32 s and 138 s, respectively. The results of this experimental research show that the microalgae pellets can be successfully burned as an alternative fuel in circulating fluidized bed boilers.

Acknowledgments This work was financially supported by the Ministry of Science and Higher Education (Poland) under the statutory funds (BS-1-103-3020/2016). The support is gratefully acknowledged.

References

1. Renewables Global Status Report (2016)
2. W. Tutak, A. Jamrozik, M. Pyrc, M. Sobiepański, Fuel Processing Technology, **149**, 86-95 (2016)
3. A. Kijo-Kleczkowska, K. Środa, M. Kosowska-Golachowska, T. Musiał, K. Wolski, Fuel, **170** 141-160 (2016)
4. M. Kosowska-Golachowska, A. Kijo-Kleczkowska, A. Luckos, K. Wolski, T. Musiał, Archives of Thermod., **37**, 1, 17-30 (2016)
5. A. Kijo-Kleczkowska, K. Środa, M. Kosowska-Golachowska, T. Musiał, K. Wolski, Waste Manage. **53**, 165-181 (2016)
6. Y.T. Tang, X.Q. Ma, Z.Y. Lai, Bioresource Techn. **102**, 1879-1885 (2011)
7. L. Sanchez-Silva, D. López-González, A.M. Garcia-Minguillan, J.L. Valverde, Bioresource Technology, 130, 321-331 (2013)
8. D. Beneroso, J.M. Bermúdez, A. Arenillas, J.A. Menéndez, Bioresource Techn., **144** 240-246 (2013)
9. K. Kirtania, S. Bhattacharya, Biomass and Bioenergy **55**, 291-298 (2013)
10. G. Xue, M. Kwapinska, W. Kwapinski, K.M. Czajka, J. Kennedy, J.J. Leahy. Fuel, **121**, 189-197 (2014)
11. C.E. Figueira, P.F. Moreira Jr., R.Giudici, Bioresource Techn. **198**, 717-724 (2015)
12. J. Wang, C.L. Weller, D.D. Jones, M.A. Hanna, Biomass and Bioenergy **32**, 7, 573-581 (2008)
13. Raheem, W.A.K.G. Azlina, Y.H. Yap, M. K. Danquah, R. Harun, Renewable and Sustainable Energy Reviews, **49**, 990–999 (2015)
14. K. Szychalski, Musiał T., Kosowska-Golachowska M., Proceedings of the Students Conference 2015, Poland, Czestochowa, 83-92 (2015)
15. T. Eriksson M., V. Barišić K. Nuortimo T. Jäntti PowerGen Europe (2015)
16. D.J. Lane, P.J. Eyk, P.J. Ashman, Ch.W. Kwong, R. Nys, D.A. Roberts, A.J. Cole, D.M. Lewis, Energy&Fuels, **29**, 4, 2542-2554 (2015)
17. http://www.pelheat.com/algae_biomass_fuel.html
18. A. Magdziarz, M. Kosowska-Golachowska, A. Kijo-Kleczkowska, K. Środa, K. Wolski, D. Richter, T. Musiał, Analysis of sewage sludge ashes from air and oxy-fuel combustion in a circulating fluidized-bed, E3S Web of Conferences **10**, 00054 (2016)
19. M. Kosowska-Golachowska, K. Wolski, W. Gajewski, A. Kijo-Kleczkowska, T. Musiał, K. Środa, Rynek Energii, **3**, 124, 90-99 (2016)

The activity of selected gamma radionuclides in the Tatra National Park

Barbara Kubica[1*], *Marcin* Stobiński[1], *Grzegorz* Szaciłowski[1], and *Katarzyna* Szarłowicz[1]

[1]AGH University of Science and Technology al. A. Mickiewicza 30-059 Cracow, Poland

Abstract. Cesium is naturally occurring active metal, represented by one stable isotope ^{133}Cs, and number of artificial, unstable, isotopes. The most common artificial isotope of cesium is ^{137}Cs, present in the environment of the Tatra Mountains due to nuclear weapon testing in late `50, and nuclear reactor accident in Chernobyl in 1986. However in recent years the growth in biofuels use for power generation can be the next source of ^{137}Cs emission. Burning wood cultivated on grounds contaminated by ^{137}Cs can introduce secondary emission of this isotope to the atmosphere.This paper presents the results of determination of gamma emitting radionuclides artificial ^{137}Cs and natural ^{40}K in soil samples from the Tatra Mountains. Results show some differences in the vertical distribution of examined radionuclides. It was found that the change of activity of ^{137}Cs in the soil samples depends mostly on the soil density and on the concentration of organic material. The state of "zero" ^{137}Cs activity was developed in the form of maps.

1 Introduction

Radioactive substances are considered as one of the numerous types of pollution of the natural environment. However, various radionuclides are abundant in the environment from the Earth formation. The average content of radioactive isotopes in the Earth's crust is approx. 0.1% of which the most important are natural uranium and thorium. These elements are widespread in rocks, soil, and seas, lakes and rivers [1-5]. An example is also the potassium ^{40}K isotope, the abundance of the natural potassium is equal to 0.0117%, and the half-life $T_{½} = 1.28 \cdot 10^9$ years [6]. In contrast, the origin of artificial radioactive elements is a major concern in respect to the natural environment with special emphasis put on the protected natural areas. The main source of their presence are: nuclear weapon tests conducted in the atmosphere, catastrophes of nuclear reactors (including the largest disaster in Europe – Chernobyl), controlled effluents from the nuclear fuel reprocessing plans [1-3, 7].

The contamination of large areas of Europe after the accident of the Chernobyl nuclear reactor in 1986 has become a challenge for many research centers. Especially important artificial radionuclide is ^{137}Cs for the reason of its rather long half-life (30,08 years) and

* Corresponding author: bkubica@agh.edu.pl

large dispersion in the environment. Between the years 2000 – 2015 there were made extensive studies of anthropogenic ^{137}Cs and natural ^{40}K activity levels in soil [8-18].

This paper presents the research result concerned on the area of the Tatra Mountains. Mountain ecosystems, due to their enormous sensitivity to disruption of homeostasis, are ideal places to observe radiochemical contamination. The aim of this study was to determine the spatial distribution of gamma radionuclides: artificial ^{137}Cs and for comparison natural ^{40}K in the soil samples taken from the whole region of the Tatra Mountains with special attention to the National Park areas.

2 Materials and Methods

2.1 Study area

The Tatra Mountains lie in the southern Poland and northern Slowakia, they consist of two mountain chains: The High Tatra Mountains and The Western Tatra Mountains. Despite the fact that they are the highest mountains in the Carpathian Mountains, their surface area is relatively small. Approximately their size corresponds to the average size of a single Alpine valley [24-27]. The entire Tatra Mountains area is legally protected by establishing of national parks. Due to its unique landscape and nature, the best preserved forest fragments of the Lower and Upper mountains, UNESCO acknowledged the Tatra Mountains as the World Biosphere Reserve.

2.2 Field sampling

The soil samples were taken from 245 sampling points (60 on the Polish and 185 on the Slovak side of the mountain range) situated in all main valleys (Fig. 1 and 2). The material was collected from June to October of 2000 – 2013. Soil samples were taken from the top layer using a cylindrical probe (diameter and height of 10 cm). The soil core was divided into three equal parts namely: "a" (0-3 cm), "b" (3-6 cm) and "c" (6-9 cm). The part "a" relates to the top layer of the soil, and "c" was the deepest layer. The exact location (latitude and longitude coordinates and altitude) of the sampling points was determined using a satellite navigation system (GPS) and Garmin 76CS maps.

2.3 Laboratory works

The soil samples were dried at 105°C, weighted (total mass), deprived of stones and large fragments of plants, and then grounded and sieved through the 2 mm sieve. The samples were placed into test vessels, weighed, and their bulk density was determined. The results were presented as activity [$Bq\cdot kg^{-1}$] and as radioactive fallout [$Bq\cdot m^{-2}$]

Measurements of activity of ^{137}Cs and ^{40}K were performed using semiconductor gamma-ray spectrometers equipped with coaxial, high purity germanium (HPGe) detectors of 15% relative efficiency. The efficiency calibration was performed by the use of the Reference Materials obtained from the International Atomic Energy Agency (IAEA-154 of certified activity of ^{137}Cs – 3749 $Bq\cdot kg^{-1}$, and ^{40}K – 1575 $Bq\cdot kg^{-1}$, and IAEA-375, the certified activities were: ^{137}Cs – 5280 $Bq\cdot kg^{-1}$, and ^{40}K – 424 $Bq\cdot kg^{-1}$). In order to maintain similar properties of the samples and the reference materials (such as matrix, density and sample geometry) further correction factors were included in the efficiency calibration. The most important correction factor was the self-absorption coefficient. This is correlated with the sample density, and refers to partial absorption of emitted gamma rays by the sample itself [17-20]. All samples were measured for 72 hours.

Studies in the Tatra Mountains were performed during 15 years [8-18, 20-23]. Therefore in order to keep results comparable, the measured activities of ^{137}Cs are recalculated to the date of the 1st September 2000.

3 Results and discussion

The cesium activity concentrations vary substantially, the highest content of this radionuclide was observed in the High Tatra Mountains and on the ridge connecting Grześ Mt and Wołowiec Mt in the Western Tatraa Mountains (Fig. 1 and 2). There is also a trend indicating the competitiveness of potassium and cesium sorption in soil. In places where the soil was depleted in potassium the activity of cesium was higher. Analyses of ^{137}Cs activity concentration in the Tatra soils show a great diversity of this radionuclide which ranged from 55.8 Bq·kg^{-1} (417.8 Bq·m^{-2}) for Tomanowa Pass (1685 m a.s.l.) to 5111 Bq·kg^{-1} (8400 Bq·m^{-2}) for Pass Krzyżne (2112 m a.s.l.). After deposition cesium is readily bound by organic matter in the soil. Soil composition and properties in the Tatra Mountains, as in other mountainous systems, are closely related to climate, complexity of the vertical structure of vegetation, etc.

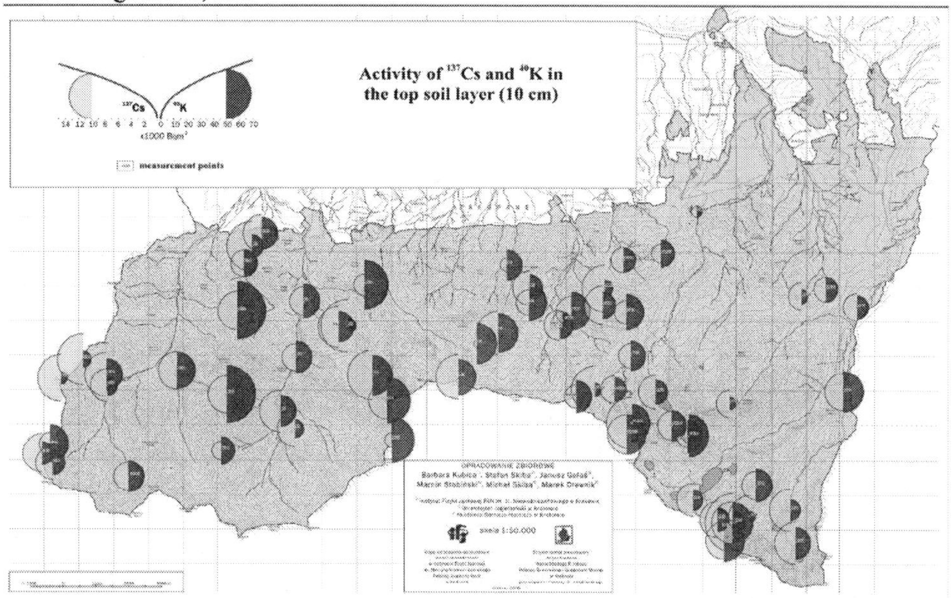

Fig. 1. The concentration of gamma radionuclides: artificial ^{137}Cs and natural ^{40}K expressed in Bq· m^{-2} in soil samples taken from the Polish part of the Tatra Mountains.

With increasing altitude the humus level thickness increases which reveal a high sorption and a high ion exchange capacity. With increasing soil density which is concomitant with decrease of organic matter content in soil (what corresponds with decreased sampling point altitude) the cesium content is markedly decreased. Potassium follows the invers proportional tendency: with increasing density of the soil, natural potassium content also increases, because the mineral components of the soil, for example potassium feldspar, mica, clay minerals, either contain or adsorb it.

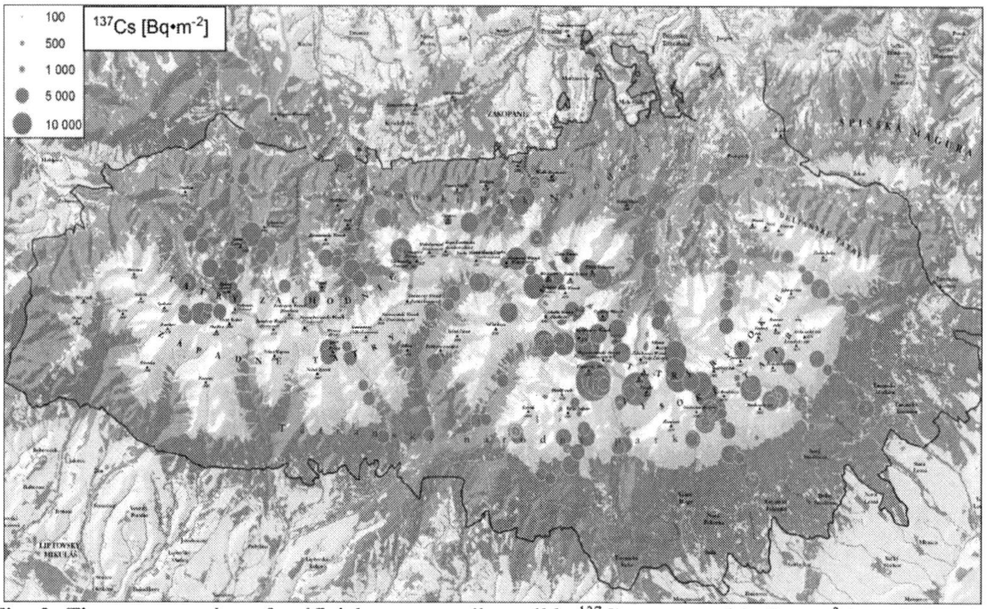

Fig. 2. The concentration of artificial gamma radionuclide ^{137}Cs expressed in Bq· m^{-2} in soil samples taken from the Tatra Mountains.

It can be seen from Fig. 2 that there is also a very large diversity of the ^{137}Cs levels in the considered region. The highest ^{137}Cs activity was observed in the area of the High Tatra Mountains on both side of the Polish and Slovakia border.

After analysis of determined ^{137}Cs activities, eleven points were selected for annual monitoring in order to assess the change in the cesium activity and migration mechanism of this gamma radionuclide in the Tatra Mountains area. The measuring points were placed in at least two places in five main valleys in the Polish part of the Tatra Mountains. In the same measurement points activity of natural radioisotope ^{40}K was verified. The changes in ^{137}Cs activity since 2000 to 2012 are presented in Fig. 3. In some locations the activity levels exceeds several times the average values reported for Poland, which equals **2.41 kBq·m^{-2}** [28].

The activity concentrations of ^{137}Cs in many places of the Tatra Mountains showed decreasing tendency which may be explained both by radioactive decay and by migration processes occurring in the environment. Mainly, the process of cesium penetration into deeper soil layers with the organic forms and its sorption on aluminosilicates should be taken into consideration.

The activity of cesium in the year 2012 changed from 10.04 kBq·m^{-2} for soil samples collected in the area of Staszicowe Stawki Lakes to 8.07 kBq·m^{-2} for the samples situated in the vicinity of Morskie Oko Lake. This is almost 4 times higher comparing to the national average.

Ordinarily, for the years 2000-2012 the decreasing tendency of cesium content in the surface layer of the soil was observed. This is obviously a result of ^{137}Cs movement downwards in the soil profile.

a b

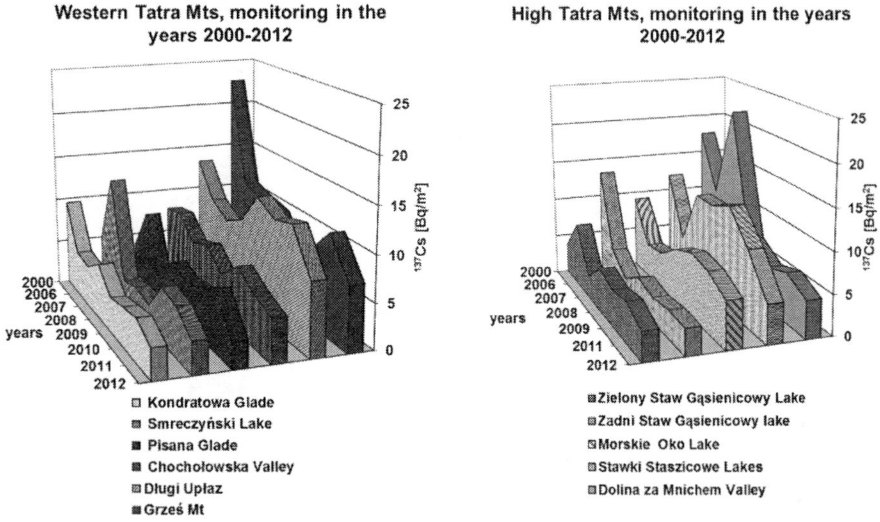

Fig. 3. The radioactivity of ^{137}Cs in the soil samples collected in the Western (a) and in the High Tatra Mountains (b) – monitoring in years 2006-2012.

The increase of ^{137}Cs content in the region of Morskie Oko Lake in the High Tatra Mountains (Fig. 3b) and on the ridge connecting Grzes Mt in the Western Tatra Mountains (Fig. 3a) can be explained by the higher rainfall rate. This may cause the ^{137}Cs introducing into environment from the upper parts of the atmosphere. It is supposed that in the higher parts of atmosphere the presence of cesium is an effect of Chernobyl accident.
Proving this thesis was taken up by the installation of probes containing a cesium selective composite sorbent with magnetic elements (hereinafter referred to as NiNCF) [29-30]. Supposing that the average rainfall in these regions amounted approximately 1700 mm per year, the calculated activity of cesium should be in the range from 5.61 mBq to 37.4 mBq. The results (Table 1) point to higher activity of this gamma radionuclide (from 33.3 mBq to 98.4 mBq). This fact can confirm the deduction that between 2007-2012 ^{137}Cs was introduced to the Tatra Mountains area along with atmospheric precipitation.

Table 1. The amounts of ^{137}Cs collected on composite magnetic sorbent (NiNCF) in different places in the Polish part of the Tatra Mountains.

Sampling points	**2008** ^{137}**Cs** [mBq·kg^{-1}]	**2009** ^{137}**Cs** [mBq·kg^{-1}]	**2010** ^{137}**Cs** [mBq·kg^{-1}]	**2011** ^{137}**Cs** [mBq·kg^{-1}]	**2012** ^{137}**Cs** [mBq·kg^{-1}]
Długi Upłaz Mt 1603 m a.s.l.	**65±31**	**83±16**	**46±27**	**98±43**	**78±49**
Grześ Mt 1639 m a.s.l.	**53±15**	**61±15**	**67±35**	**56±31**	**54±30**
Staszicowe Stawki Lakes 1893 m a.s.l.	**65±14**	**85±19**	**61±22**	**86±37**	**65±22**

4 Conclusions

The following conclusions can be drawn:

- the activity of ^{137}Cs in the soil samples from the Tatra Mountains varies significantly from 55.8 Bq·kg^{-1} (417.8 Bq·m^{-2}) for Tomanowa Pass (1685 m a.s.l) to 5111 Bq·kg^{-1} (8400 Bq·m^{-2}) for Krzyżne Pass (2112 m a.s.l). In most locations, the values are not high, as well, they are lower than the average cesium concentrations established for Poland area;
- variety of the ^{137}Cs activity in the soil samples taken from the Tatra Mountains depend mostly on the soil volume density and on the content of organic matter;
- the activity of ^{40}K increases with the depth of soil core while the ^{137}Cs level declines, in the soil surface layer (first 10 cm) because of different origin of these two isotopes;
- annual research of cesium activity indicate a downward migration trend of this radionuclide.

This work was partially supported by the Grant of AGH University of Science and Technology No. 11.11.210.244. Some of the measurements were performed using the scientific equipment belonging to the laboratories of AGH-UST Energy Centre Cracow, Poland.

References

1. R. Broda, Acta Physica Polonica B **18**, 935-951, (1987)
2. R. Broda, B. Kubica, Z. Szeglowski, K. Zuber, Radiochim. Acta **48**, 89-96, (1989)
3. P. Jaracz, E. Piasecki, S. Mirowski, Z. Wilhelmi, J. Radioanal. Nucl. Chem. **141**, 243-259, (1991)
4. The team of the President of the National Atomic Energy Agency for the construction of a nuclear power plant in Żarnowiec: Report on the consequences of the disaster in Chernobyl (with attachments), Warsaw, PAA, (1991)
5. H. Bem, *The radioactivity in the environment*, Łódź, PAN, (2005)
6. A.Z. Hrynkiewicz, Man and ionizing radiation, (PWN, Warsaw, 2001)
7. W.P. Krasnov, Wyd. Wolyn, Zhytomyr, 112, (1998)

8. B. Kubica, J.W. Mietelski, J. Gołaś, S. Skiba, E. Tomankiewicz, P. Gaca, M. Jasińska, M. Tuteja-Krysa, Polish J. Environ. Studies **11,** 537-545 (2002)
9. B. Kubica, S. Skiba, J.W. Mietelski, J. Gołaś, M. Kubica, M. Stobiński, M. Tuteja-Krysa, E. Tomankiewicz, P. Gaca, Z. Krzan, Polish J. Environ. Studies **13**, 153-159, (2004)
10. B. Kubica, M. Skiba, S. Skiba, J. Gołaś, M. Kubica, M. Stobiński, M. Tuteja- Krysa, J. Radioanal. Nucl. Chem. **266**, 3-9, (2005)
11. P. Gaca, E. Tomankiewicz, J. W. Mietelski, S. Grabowska, B. Kubica, J. Radioanal. Nucl. Chem. **267**, 443-448, (2006)
12. B. Kubica, W.M. Kwiatek, M. Stobiński, S. Skiba, M. Skiba, J. Gołaś, M. Kubica, M. Tuteja-Krysa, A. Wrona, R. Misiak, Polish J. Environ. Studies **16**, 723-729, (2007)
13. J.W. Mietelski, B. Kubica, P.Gaca, E. Tomankiewicz, S. Błażej, M.Tuteja-Krysa, J. Radioanal. Nucl. Chem., **275**, 523-533, (2008)
14. B. Kubica, S.Skiba, M. Drewnik, M. Stobiński, M. Kubica, J. Gołaś, R. Misiak, Nukleonika **55**, 377-387, (2010)
15. M. Kłosowski, K. Kozak, J. Mazur, P. Olko, B. Kubica, P. Skawiński, A. Cebulska-Wasilewska, Rapid Diagnosis in Populations at Risk from Radiation and Chemicals **73**, 351-361, (2010)
16. B. Kubica, M. Stobiński, S. Skiba, J. Gołaś, W. Reczyński, K. Szarłowicz, P. Fleischer, R. Misiak, W. Kwiatek, Science and Industry Research Spectroscopic in practice, a monograph, 533-542, (2011)
17. B. Kubica, An application of sparingly soluble hexacyanoferrates of divalent transition metal as sorbents (Habilitation thesis, Scientific Publishing PARAGRAPH, 2008)
18. M. Stobiński, An environmental risk assessment of radioactive cesium-137 and selected heavy metals accumulation in the Tatra National Park (doctoral thesis, 2011)
19. M. Mostajaboddavati, S. Hassanzadeh, H. Faghihian, M.R. Abdi, M. Kamali, J. Radioanal. Nucl. Chem. **268**, 539-544, (2006)
20. P. Jodłowski, Nukleonika **51,** 21-25, (2006)
21. J.P. Bolivar, M. Garcia-Leon, R. Garcia-Tenorio, Appl. Radiat. Isot. **48**, 1125-1126, (1997)
22. L.S. Quindos, C. Sainz, I. Fuente, J. Nicolas, L. Quindos, J. Arteche, J. Radioanal. Nucl. Chem. **270**, 339-343, (2006)
23. B. Kubica, M. Stobiński, R. Misiak, K. Szarłowicz, *Science and industry - spectroscopic methods in practice, new challenges and opportunities*, 808-814, (2014)
24. Z. Mirek, *Nature of the Tatra National Park, Tatra National Park*, 595-617, (1996)
25. W. Siarzewski, *Tatra National Park*, (2005)
26. J. Nyka, *Polish Tatras Mts – guide* (Publishing Traverse 2011)
27. J. Nyka, *Slovak Tatras Mts – guide* (Publishing Traverse 2015)
28. M. Biernacka, K. Isajenko, P. Lipiński, M. Kuczbajska, *Monitoring of the concentration of ^{137}Cs in soil- annual report for 2007*, (Chief Inspectorate of Environmental Protection)
29. B. Kubica, Z. Szeglowski, M. Tuteja-Krysa, R. Fiałkowski, Patent No. 196361 issued 14.12.2007 "Composite magnetic ion exchanger"
30. B. Kubica, Z. Hubicki, M. Stobiński, K. Szarłowicz, M. Bartyzel, Patent No 217 761 issued 14.08.2014 "The method of producing the composite sorbent based on NiFN"

The influence of the modification of acidic montmorillonites with polyacrylamide and copper deposition on SCR-NH_3 catalytic performance

Katarzyna Świrk[1,*], *Monika* Motak[1], *Teresa* Grzybek[1], *Cezary* Czosnek[1], and *Paweł* Kniaź[1]

[1]AGH University of Science and Technology, Faculty of Energy and Fuels, Department of Fuel Technology, Mickiewicza Av. 30, 30-059 Cracow, Poland

Abstract. The aim of this work was to study the influence of the modification of montmorillonites by carbonaceous and Al species as well as copper deposition. Commercial acid-treated montmorillonites, K5 and K30 (Sigma-Aldrich GmBH), were modified and used as catalysts in Selective Catalytic Reduction process of nitrogen oxides by ammonia. The characterisation was carried out by low-temperature nitrogen sorption, Fourier-transform-infrared spectroscopy and X-ray diffraction. Catalytic performance in SCR-NH_3 of so-modified montmorillonites was compared under the following conditions: mass of catalyst: 200 g, flow 100 cm^3/min, reaction mixture: 800 ppm NO, 800 ppm NH_3, 3 % O_2, and He. The modification with copper and polyacrylamide led to the increase in NO conversion. The studied catalysts showed low N_2O formation.

1 Introduction

Directive 2010/75/EU introduced substantial reduction of NOx emissions from combustion of solid and liquid fossil fuels from 1st of January 2016 [1]. The problem of a low-cost, efficient and environmentally friendly catalyst allowing to reduce NOx emissions has not yet been fully solved. Although the process of Selective Catalytic Reduction (SCR) used in industry to remove nitrogen oxides from the power plants is a well-established reaction, nevertheless, it requires the optimization of a number of aspects. The main problem is the currently applied industrial catalyst: $V_2O_5/WO_3/TiO_2$. Firstly, it is expensive. Secondly, it operates in the range of 300 – 400/450°C and thus is most often placed directly downstream of the SCR boiler, but upstream of the electrostatic precipitator and the desulfurization installation. This results in the catalyst deactivation caused by deposition of fly ash on its surface. At the same time co-firing of coal and biomass - the method used to reduce CO_2 emissions - has a negative effect on the operation of the SCR catalyst placed in the front of dust collector because of alkali and alkaline earth elements present in biomass and, consequently, in the fly ash. It leads to deterioration of the catalyst by poisoning reaction centres [2]. Placing the catalyst downstream of the dust collector would therefore allow for

* Corresponding author: swirk@agh.edu.pl

more efficient operation and longer catalysts life. However, this would require preheating of the flue gas to approximately 400/450°C. Thus new catalysts, more efficient at lower temperatures, are needed. The development of such catalyst may allow the SCR installation to be placed downstream of ESP, or directly before the stack, eliminating the problem of catalyst deactivation caused by fly ash.

In this study commercial acid-treated montmorillonites K5 and K30, supplied by Sigma-Aldrich GmBH, were modified and tested as catalysts in Selective Catalytic Reduction process of nitrogen oxides by ammonia.

2 Experimental

2.1 Catalyst preparation

Six samples of K5, and seven samples of K30 were modified with: alumina oligocations, followed by polymer treatment, with optional carbonization, and promoted with copper (5 wt. %).

Table 1. Modification procedures of the studied samples

Sample name	Copper deposit	Alumina pillaring	Polymer addition	Carbonization
K5	-	-	-	-
K5Cu	+	-	-	-
K5Al	-	+	-	-
K5AlCu	+*	+	-	-
K5AlGc	-	+	+	+
K5AlGcCu	+*	+	+	+
K30	-	-	-	-
K30Cu	+	-	-	-
K30Al	-	+	-	-
K30AlCu	+*	+	-	-
K30AlG	-	+	+	-
K30AlGc	-	+	+	+
K30AlGcCu	+*	+	+	+

*after alumina pillaring

The samples, derived from K5, were pillared with Al-polycations by chlorhydrol treatment according to the procedure performed by Vaughan [3] and described previously [4]. The modification of K30 materials was carried out according to Chmielarz et al. [5], with minor changes. K30-based catalysts were modified with a mixture of NaOH (0.4 M) and $AlCl_3$ (0.4 M) with the molar ratio of OH/Al= 2.4. After five days of aging at 65°C this solution was added to the 3 wt. % suspension of K30, and the ratio between Al ions and K30 was set at 12 mmol ions per 1 g of clay. The next step was filtration with Büchner funnel and washing with distilled water until Cl^- ions were removed from the filtrate (test with $AgNO_3$). Then the catalysts were dried for 24 hours at 100°C, and subsequently calcined for 30 minutes at 500°C.

Incipient wetness impregnation was performed for both K5 and K30-based samples with an aqueous solution of $Cu(NO_3)_2$ (5 wt. % of Cu). The samples were treated either after pillaring (K30AlCu) or polymer treatment followed by carbonization (K30AlGcCu). In order to compare the influence of copper deposition, the raw material was also impregnated with copper (K30Cu).

The polymer treatment included mixing of K5 [4] and K30 with 3% of polymer polyacryloamide using commercial product (Gigtar delivered by Zakłady Azotowe at Tarnów, Poland). After 72 hours of mixing montmorillonite samples were dried at 100°C for 12 hours. Next, carbonization was performed for 30 minutes at 500°C in argon atmosphere as proposed previously [6].

The samples designation and their modification procedures are summarized in Table 1.

2.2 Catalyst characterisation

Low-temperature sorption of nitrogen (-196°C) was used to determine textural parameters of the prepared catalysts. The specific surface areas S_{BET} for K5-based materials were examined by ASAP 2000 (Micromeritics), while for K30 series the measurements were performed by using Gemini VII 2390 (Micromeritics). Before the measurements the samples were degassed at 110°C.

XRD measurements of the modified montmorillonities were carried out by Philips PV 3020 X' Pert (for K5), and PANalytical-Empyrean (for K30) diffractometer. The structure of the samples were determined with CuKα (λ=1.5406 Å) radiation source. The measurement range of 2θ = 3-90°C was set.

2.3 Catalytic performance

The catalytic tests of NO reduction with NH_3 (SCR) were performed for 200 mg of catalyst placed in a fixed-bed microreactor. A thermocouple was used to control the temperature inside the catalytic bed. The composition of the inlet gas mixture was as follows: 800 ppm NO, 800 ppm NH_3, 3 vol. % O_2 and He to balance. The total flow was 100 cm^3/min. In order to analyze the outlet concentrations of NO and N_2O, an IR detector (ABB 2020 A0 series) was used. Before each SCR reaction, the catalysts were kept at 500°C for 30 minutes in helium atmosphere. The NO conversion [%] was calculated using the following formula (1):

$$NO_{conversion} = (NO_{in} - NO_{out}) / NO_{in} \cdot 100\% \quad (1)$$

where NO_{in} is the nitric oxide concentration at the reactor inlet [ppm], and NO_{out} at the outlet [ppm].

3 Results and discussion

3.1. Low-temperature sorption of nitrogen

Figure 1 presents an example of nitrogen sorption isotherm for K30AlGcCu. According to the IUPAC classification it represents isotherm type IV, indicating that montmorillonite-based materials have mesoporous structure. The shape of the hysteresis loop may be classified as H4 type which corresponds to the presence of homogeneous slit-shaped pores [7]. The isotherms recorded for other montmorillonites were similar to the one shown in Fig. 1.

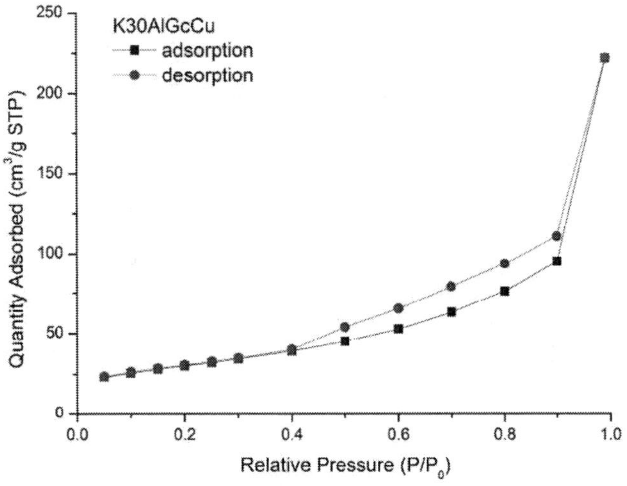

Fig. 1. Nitrogen sorption isotherm for K30AlGcCu

Table 2. Specific surface area S_{BET}, mesopore volume V_{mes}, micropore volume V_{mic}, and dominant pore size

Catalyst	S_{BET} [m²/g]	V_{mes} [cm³/g]	V_{mic} [cm³/g]	Pore size [nm]
K30	212	1.10	0.008	15.30
K30Cu	158	0.32	n/d	5.99
K30Al	151	0.21	0.007	5.02
K30AlCu	153	0.27	0.009	6.19
K30AlG	142	0.24	0.009	5.57
K30AlGc	130	0.25	0.003	6.11
K30AlGcCu	107	0.53	0.005	11.1

Table 2 shows textural parameters for the studied samples e.g. specific surface area, volume and size of dominant micro- and mesopores for K30 series. For K5 series, only specific surface areas were measured, and were equal to 164 m^2/g, 152 m^2/g, 128 m^2/g, 78 m^2/g, 98 m^2/g for K5, K5Cu, K5Al, K5AlCu, and KAlGcCu samples, respectively [4]. The decrease in specific surface area after Al-pillaring was observed. This observation is rather hard to explain, since the XRD measurements resulted in the presence of (001) reflection; it will be discussed later in the text. However, it should be mentioned that similar effect was observed by Motak [4] for Milowice montmorillonite pillared by Zr or Al oligociations.

No increase in specific surface area was observed for K30 series after alumina intercalation. Therefore, the collapse of the structure for K30Al series most likely occurred. This assumption is supported by XRD results explained further in the text and is related to disappearance of the (001) reflection for Al-pillared K30 samples. One could speculate that instead of the intercalation, the deposition of alumina in the porous structure occurred, as suggested by pore volume decrease of K30Al as compared to that of K30. Polymer modification, followed by carbonization, decreased the specific surface area further from 152 to 130 m^2/g, which is in agreement with the previous observations [4]. Dry impregnation with copper solution reduced the specific surface area of K30AlGcCu. In contrast, similar effect was not observed after the introduction of Cu^{2+} ions into K30Al.

3.2. X-ray diffraction

Fig. 2. shows diffraction patterns for K30-based materials. (001) reflection was observed only for the starting material (K30), at $2\theta = 6.2^\circ$ with $d_{001} = 1.4$ nm. d_{001} calculated from the first low-angle diffraction maximum gives information about the interlayer distances in clays. Thus, the absence of this reflection may be interpreted as the partial collapse of the structure of K30Al-based samples, as previously observed for clay materials by Mahboub et al. [7].

d_{001} of 1.6 nm was registered for K5Al, K5AlG, and K5AlGCu, while for K5 it was equal to 1.3 nm. This suggests at least partial intercalation for K5Al-based samples. The subsequent modifications of these samples, such as introduction of polymer, carbonization, and the deposition with copper, did not affect their structures [4].

The most intense reflection, which appears in all diffraction patterns, is associated with quartz (SiO_2), and it is a typical impurity for clay material [8, 9]. Copper oxide (CuO) was observed for K30Cu which was impregnated with Cu^{2+}, and calcined at 500°C. Copper oxide reflections are assigned to angles $2\theta = 35.5^\circ$, 38.7°. In the XRD patterns of other copper-modified catalysts, CuO presence was not observed. Perhaps CuO was in non-crystalline form and/or its crystallites were very small, which may suggest high dispersion of this metal oxide.

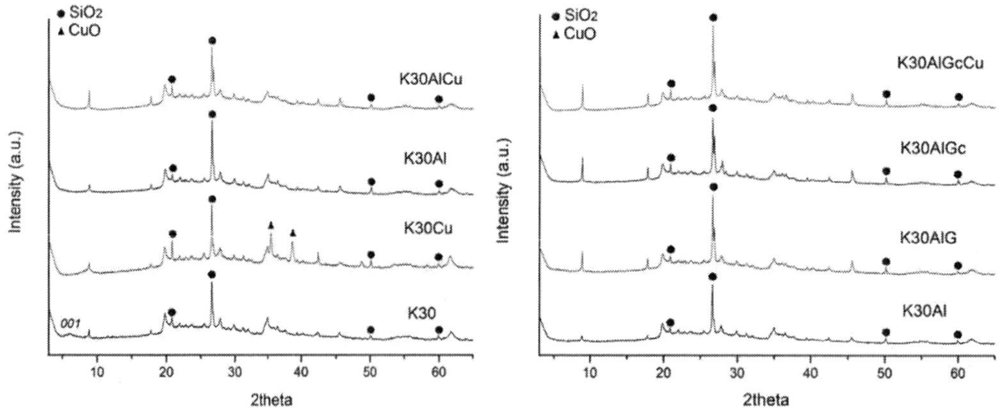

Fig. 2. XRD patterns for K30-based materials

The comparison of XRD patterns for the selected samples is presented in Fig. 3. There is a noticeable increase of reflections at $2\theta = 26.86^{\circ}$ for samples which were carbonized (K30AlGc and K30AlGcCu). Graphite reflection occurs at ca. 27° of 2θ, thus it may suggest that carbon deposition on the surface of carbonised samples took place.

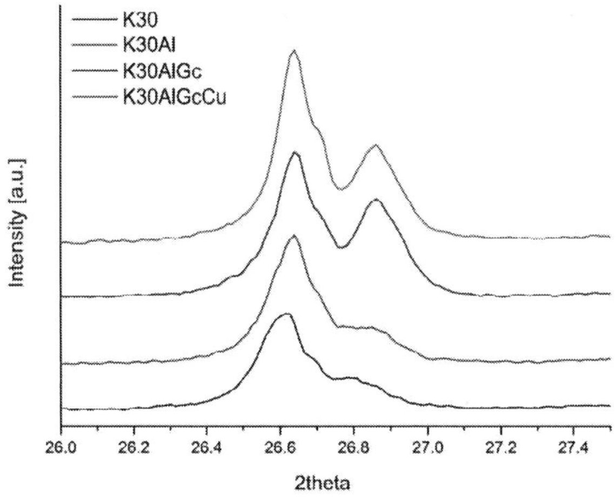

Fig. 3. XRD reflections for K30, K30Al, K30AlGc, K30AlGcCu in the range of $2\theta = 26–27.5^{\circ}$

3.3. Catalytic tests

Fig. 4. shows NO conversion versus temperature for montmorillonite-derived samples. For K5 series, NO conversion formed a sequence: K5AlGcCu> K5AlCu> K5Cu> K5AlGc> K5Al. The same sequence was observed for K30-based carriers (K30AlGc> K30Al).

As may be derived from Fig. 4., at the low temperature region (200–250°C) carbon deposition increased NO conversion, which was almost two times higher in comparison to the samples without carbonization step. Possibly carbon played a role as a second reducing agent, as suggested by Motak [4]. The catalysts K5AlGcCu and K30AlGcCu maintained NO conversion above 80% at 250 and 300°C. The activity of the catalysts based on K30 at 250°C followed a sequence: K30AlGcCu> K30Cu> K30AlCu.

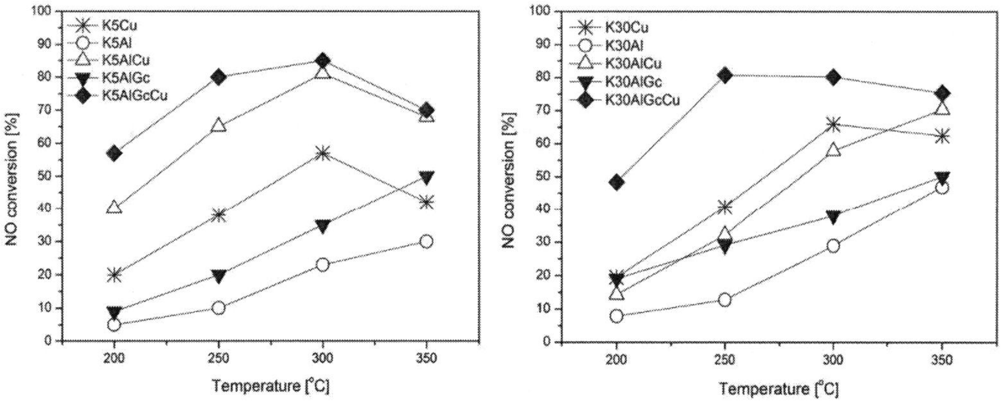

Fig. 4. NO conversion versus temperature for K5-, and K30-based materials.

Fig. 5. shows N_2O formation during catalytic tests. It should be mentioned that 20 ppm are below detection limit of the method used, therefore only values above 20 ppm should be considered. The N_2O formation was low for carriers and catalysts up to 350°C, and it did not exceed 65 ppm. K30-based samples showed slightly lower amount of N_2O in the post-reaction mixture.

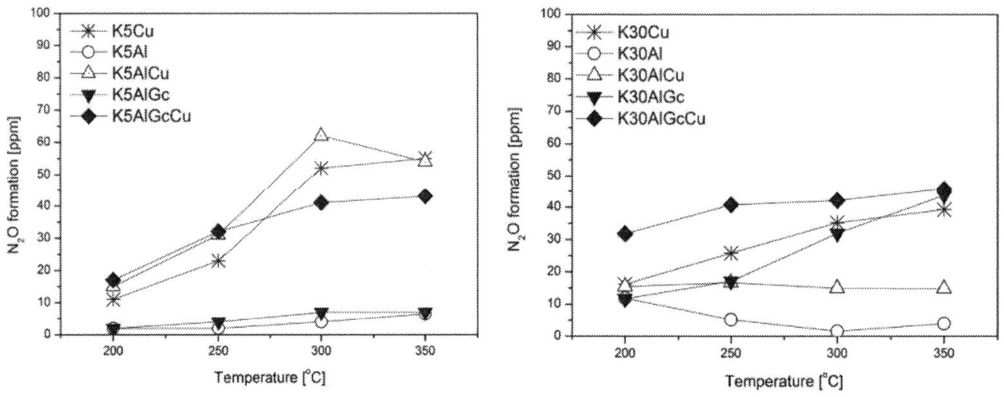

Fig. 5. N_2O formation versus temperature for K5-, and K30-based materials.

In case of K5-based materials the introduction of copper led to an increase of N_2O amount. This was not observed for K30AlCu, for which N_2O formation remained stable at very low level (less than 20 ppm). This indicates that N_2O formation is strongly dependent on the sample preparation procedure, since modification by Al oligocations was carried out in a different way for K30 than for K5 series.

4 Conclusions

Two series of catalysts, based on the commercial montmorillonite K5 and K30, modified with alumina pillaring, carbon deposition and copper impregnation were characterised by low-temperature nitrogen adsorption and X-ray diffraction. The activity and selectivity of the catalysts in the SCR-NH_3 process were tested for the mixture of 800 ppm NO, 800 ppm NH_3, and 3% O_2 in He.

Based on the low-temperature nitrogen sorption studies a decrease in surface area was observed after introduction of alumina oligocations into montmorillonite. The polyacrylamide modification followed by carbonization, and copper deposition for montmorillonites catalysts decreased the specific surface area even further.

X-ray diffraction showed d_{001} of 1.6 nm for K5Al, K5AlG, and K5AlGCu, while for K5 it was equal to 1.3 nm, indicating that pillaring was successful and the modification with polymer and copper oxide did not further influence the structure of studied clays. In the case of K30-series, the reflection (001) was observed only for the starting material at $2\theta = 6.168°$, corresponding to d_{001} of 1.4 nm. The absence of (001) reflection in diffraction patterns of other K30-based samples may suggest a collapse of the clay structure.

The catalytic tests exhibited high activity of the modified montmorillonite samples in the process of NH_3-SCR. Better results were observed for the catalysts modified with polymer and carbonaceous deposit as compared to the samples modified with only Al-oligocations. Additional copper deposition on carbonised samples resulted in a significant increase in catalytic activity. The modification with Cu improved the catalytic performance of each catalyst treated by this metal. The quantity of N_2O produced for all samples was low and did not exceed 65 ppm.

On the whole, the introduction of copper oxides, Al oligocations treatment, as well as treatment with polyacrylamide, and carbonization positively affected catalytic activity and did not decrease selectivity to nitrogen.

Acknowledgements AGH 11.11.210.213

References

1. Directive 2010/75/EU of the European parliament and of the Council, Offical Journal of the European Union (2010)
2. M. Kiełtyka, A.P. Doares Dias, H. Kubiczek, B. Sarapata, T. Grzybek, CR Chimie, **18,** 1036 (2015)
3. D. E. W. Vaughan, R. J. Lussier, in: W. Rees (Ed.), Proc. 5th Int. Conf. Zeolites, **94** (1980)
4. M. Motak, Catal Today, **137**, 247 (2008)
5. L.Chmielarz, P. Kuśtrowski, M. Zbroja, B. Gil-Knap, J. Datka, R. Dziembaj, Appl Catal B-Environmental **53**, 46 (2004)
6. T. Grzybek J. Klinik. M. Motak, H. Papp, M. Żyła, J. Colloid Interface Sci. **227**, 291 (2000)
7. R. Mahboub, Y. El Mouzdahir, A. Elmchaouri, A. Carvalho, M. Pinto, J. Pires, Colloids Surf. A **280**, 81 (2006)
8. L.Chmielarz, Naturalne krzemiany warstwowe jako materiały do syntezy katalizatorów dla procesu DeNOx, in: Adsorbenty i Katalizatory, *Wybrane technologie a środowisko*, 7 (2012)
9. V. Tabernero, C. Camejo, P. Terreros , M. Dolores Alba, T. Cuenca, Materials **3**, 1015 (2010)

Tri-reforming as a process of CO_2 utilization and a novel concept of energy storage in chemical products

Katarzyna Świrk[1,*], *Teresa* Grzybek[1], and *Monika* Motak[1]

[1]Department of Fuel Technology, AGH University of Science and Technology, Mickiewicza Av. 30, 30-059 Cracow, Poland

Abstract. Methane tri-reforming process is a combination of three reactions: steam methane reforming (SRM), dry methane reforming (DRM), and partial oxidation of methane (POM). The first two reactions have endothermic character, while the POM is the exothermic reaction, which allows to obtain the energy necessary for the other two processes in situ. CH_4 reacts with H_2O, CO_2, O_2, and synthesis gas is produced with desired H_2/CO ratios (1.5 – 2.0). The presence of H_2O and O_2 in the reaction environment can positively influence the stability of the used catalyst, by the inhibition of coke formation.

In this paper two scenarios for methane tri-reforming implementation are discussed: (i) Tri-reforming as a effective way for chemical CO_2 utilization, without the separation of carbon dioxide from flue gases from fossil fuel-fired power stations, and (ii) dry reforming of methane improved by the addition of water and oxygen, which may be applied as a chemical energy storage process. The literature on the subject of tri-reforming is shortly reviewed, including thermodynamics of the process, the possible conversions of methane and carbon dioxide, and proposed catalysts, both studied in tri-reforming, and in single processes (SRM, DRM and POM).

1 Introduction

Carbon dioxide is a gaseous compound which is also considered to be one of the main greenhouse effect causes. In 2014, among the total anthropogenic greenhouse gas (GHG) emissions, the two-thirds were represented by energy sector. In this share over 80% is assigned to the CO_2 presence. The early estimations performed by statistical office of the European Union (Eurostat) showed that CO_2 emissions had increased in 2015 compared with the previous year (change 2014/2015 was at level +0.7%) [1] in contrast to tendency which is observed from 2011 to 2014 with total decrease of CO_2 emissions to -9.6%,

* Corresponding author: swirk@agh.edu.pl

respectively at 2011/12 it was drop of -2.1%, at 2012/13 of -2.5%, and at 2013/14 it was observed CO_2 emission decrease of -5.0% [2-4].

The regulations which aim at declining concentration of greenhouse gases in the atmosphere were established [5]. Kyoto Protocol and European Union regulate the concentration of GHG emissions in the atmosphere (CO_2, CH_4, N_2O, HFCs, PFCs, SF_6) via long-time strategies (until 2050). In 2015, it took place the latest United Nations Conference on Climate Change (COP21, Paris) which assumes GHG reduction in order to preserve less than two-degree temperature growth above pre-industrial levels [5-7].

To facilitate decreasing and/or avoiding high CO_2 emissions related to energy production and use, the following methods are proposed: an increase in the efficiency of energy production, carbon capture and utilization processes (CCU), carbon capture and storage processes (CCS), and/or the application of renewable energy sources.

Particular attention is paid to CCU processes, where CO_2 can be used as feedstock in syntheses of desirable chemicals, as well as the CO_2 reputation may be improved by treating it as a valuable resource with almost zero costs of production [8]. Currently CO_2 is used in industry only on a limited scale. Urea synthesis is the best known process of CO_2 usage. The urea serves as a source of ammonia in order to reduce NOx emission from the flue gases (in selective catalytic reduction process). The other known carbon dioxide syntheses are leading to the following products: salicylic (2-hydroxybenzoic) acid which is used as raw materials for the production of drugs; cyclic carbonates - used as solvents, and components of electrolytes in Li-batteries; polycarbonates from which are manufactured optical, automotive, and electronic components [9].

Another process which may utilize CO_2 is tri-reforming of methane (TRM), delivering synthesis gas (H_2/CO) in the presence of suitable catalyst.

In this paper two scenarios for methane tri-reforming implementation are shortly discussed: (i) TRM as a effective way for chemical CO_2 utilization, without the necessity of the separation of carbon dioxide from flue gases from fossil fuels fired power stations, and (ii) dry reforming of methane (DRM) improved by the addition of water and oxygen, which may applied as a chemical energy storage process.

2 Tri-reforming of methane

Tri-methane reforming is a combination of three main reactions which take place in the single reactor: steam methane reforming (SMR), dry methane reforming (DMR), and partial oxidation of methane (POM) [10]. The two first reactions have endothermic character, while the POM is the exothermic reaction, which allows to obtain the energy necessary for the other two processes in situ (see Fig. 1).

- **Steam methane reforming (SMR):**
  $$CH_4 + \mathbf{H_2O} \leftrightarrow CO + 3H_2$$
  $H_2/CO = 3$, $\Delta H^0 (25°C) = +207$ kJ/mol
- **Dry methane reforming (DMR):**
  $$CH_4 + \mathbf{CO_2} \leftrightarrow 2CO + 2H_2$$
  $H_2/CO = 1$, $\Delta H^0 (25°C) = +248$ kJ/mol
- **Partial oxidation of methane (POM):**
  $$CH_4 + \tfrac{1}{2}\mathbf{O_2} \leftrightarrow CO + 2H_2$$
  $H_2/CO = 2$, $\Delta H^0 (25°C) = -36$ kJ/mol

TRI-REFORMING
OF METHANE (TRM)

- CO_2 utilization
- High CH_4 and CO_2 conversions
- Synthesis gas production ($H_2/CO = 1,5 - 2$)
- Improvement of catalyst stability due to the H_2O and O_2 presence

Fig. 1. Tri-reforming of methane with its main features

Thermodynamic tri-reforming calculations showed that high CH_4 and CO_2 conversions are possible. Song and Pan [10] reported respectively, over 95, and ca. 80 % at 800 – 850°C, under the following conditions: CO_2/H_2O molar ratio = 1.0, O_2/H_2O = 0.1, and the ratio of $(CO_2+H_2O+O_2)/CH_4=1.05$.

Moreover, the biggest advantage of the process is the production of synthesis gas which may be converted into methanol (MeOH) and liquid fuels (via Fischer-Tropsch synthesis). The syngas obtained in methane tri-reforming process has molar ratios in the range of 1.5 – 2.0 [10].

Additionally, the presence of H_2O and O_2 can positively influence the stability of the used catalyst, by the inhibition of catalytic coking [9,10], thus avoiding the stability problem common for dry reforming process, and improving lifetime of the catalyst. The formation and destruction of carbon deposits is the result of side reactions, depicted in Fig. 2: methane decomposition and Boudouard reaction leading to coke creation, and the latter reacting with CO_2, H_2O, and O_2.

Reactions for coke formation :

Methane decomposition:	$CH_4 \leftrightarrow C + 2\,H_2$	$\Delta H^0(25°C)$ = **+ 75 kJ/mol**
Boudouard reaction:	$2\,CO \leftrightarrow C + CO_2$	$\Delta H^0(25°C)$ = **– 172 kJ/mol**

Reactions for coke destruction :

Reaction with CO_2:	$C + CO_2 \leftrightarrow 2\,CO$	$\Delta H^0(25°C)$ = **– 172 kJ/mol**
Oxidation reaction:	$C + O_2 \rightarrow CO_2$	$\Delta H^0(25°C)$ = **– 394 kJ/mol**
Reaction with H_2O:	$C + H_2O \leftrightarrow CO + H_2$	$\Delta H^0(25°C)$ = **+ 131 kJ/mol**

Fig. 2. Reactions of coke formation and destruction

2.1 Implementation of tri-reforming of methane

2.1.1 Chemical CO_2 utilization via TRM

CO_2 may be treated as a source of carbon for chemicals and fuels feedstock. Their production may be possible via tri-reforming of methane, which utilizes CO_2 in one of its reforming reaction.

The scenario applies to fossil fuel-fired power plants, where carbon dioxide is one of the components of flue gas. Fig. 3 shows the visualization of the concept where flue gases with natural gas added are directed to the tri-reformer. As a result, synthesis gas, raw material of chemical syntheses, is obtained. It should be stressed that direct CO_2 usage is assumed without its pre-separation, thus resulting in energy savings, since conventional method, such as CO_2-MEA absorption, requires high energy for absorbent regeneration [9, 12].

DOI: 10.1051/e3sconf/20171402038

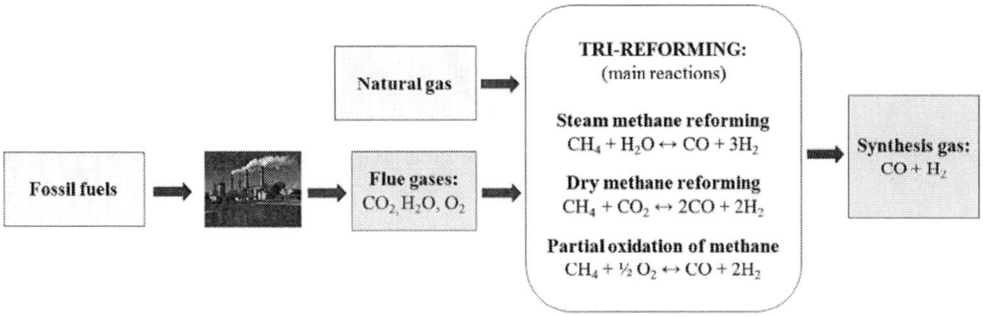

Fig. 3. Scheme of chemical CO_2 utilization by tri-reforming process implementation in coal- and gas-fired power stations.

As an example, for a typical coal-fired power station, ca. 20% of the produced energy (100 MW of 500 MW) should be used for the CO_2 capture via alkanolamines methods [10]. Tri-reforming would be also attractive for oxyfuel combustion due to high CO_2 content in flue gas.

Halmann and Steinfield [12] reported the possible CO_2 emission avoidance, as well CH_4 savings which could be gained, when syngas was produced via tri-reforming (instead of traditional steam reforming) with its following application to obtain typical bulk chemicals (methanol, ammonia, etc.), as illustrated by Fig. 4. The highest CO_2 emission avoidance (50 %) occurs for methanol production. The largest fuel savings, however, are possible for hydrogen production.

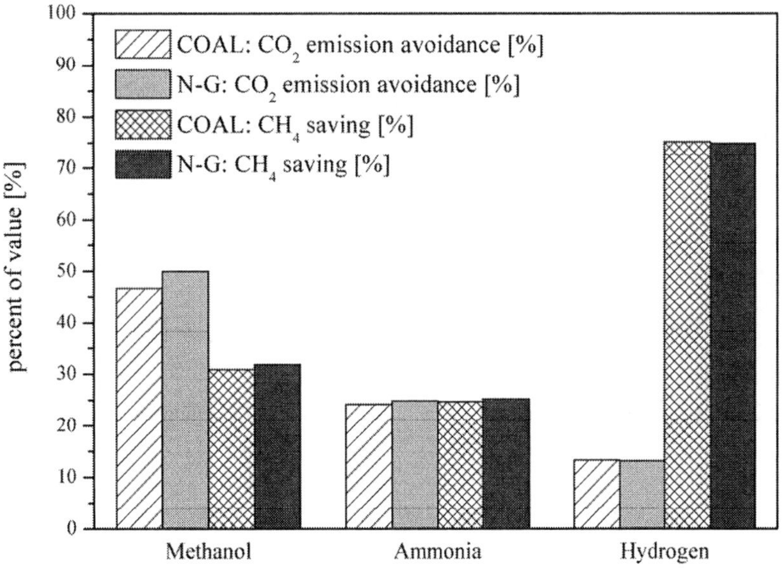

Fig. 4. Carbon dioxide emission avoidance and methane savings when methanol, hydrogen or ammonia are produced, with H_2 or syngas obtained via implementation of tri-reforming process in coal- and natural gas-fired power stations, instead of current industrial processes. The numerical data taken from [12]

DOI: 10.1051/e3sconf/20171402038

2.1.2 Energy storage via TMR

Dry reforming of methane (DRM) is an endothermic reaction which may be used in the energy storage. The surplus energy (from renewable sources or off-peak) may be stored in either chemical products (e.g. methane, via methanation reaction) or used to power an endothermic process (such as e.g. DRM) and released via reversible exothermic reaction. This concept was widely described in literature [e.g. 6, 13, 14]. DRM, however, has certain disadvantages, of which the lack of stable catalyst is the most serious. DRM catalysts are prone to catalytic coking and thus shortened lifetime [6, 9]. In this respect tri-reforming would be of advantage, because the presence/addition of the H_2O and O_2, as schematically shown in Fig. 5, would inhibit coke formation, and result in extended catalytic lifetime.

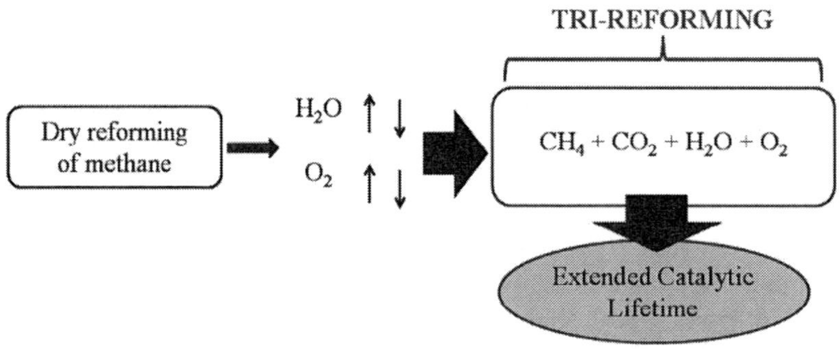

Fig. 5. Dry reforming of methane optimized by the addition of water and oxygen (tri-reforming process)

3 Catalysts for methane reforming process

The tri-reforming process requires an efficient catalyst. The required properties of the catalysts may be derived from mechanistic considerations.

Djéga-Mariadassou [15] considers the process as follows (* abbreviation for free radical):

- CH_4 is chemisorbed dissociatively on reduced metal Ni^0 (dry and steam reforming, partial oxidation of methane),
- CO_2 is activated via dissociative chemisorption; the used catalyst requires alkaline properties (dry reforming),
- H_2O activation with the formation of radical O and gaseous hydrogen (steam reforming),
- surface reactions are: (i) carbon radical *C reacts with oxygen radical *O (or hydroxyl radical *OH) in order to form CO of syngas, and (ii) two hydrogen radicals form hydrogen gas (2 *H → H_2) via associative desorption (dry and steam reforming),

In case of partial oxidation of methane (POM), it has been shown by Djéga-Mariadassou [15] and Nishimoto et al. [16] that total CH_4 oxidation takes place on reduced metal: metal oxide MO (thanks to oxygen sites O□), followed by dry reforming, and steam reforming (thanks to Ni^0). DRM proceeds satisfactorily either on precious metals (Pt, Rh, Ru, Ir) or metallic Ni, the latter preferred being much cheaper and more available [9]. SRM, the only process of the three reactions mentioned currently used on industrial scale, uses Ni based catalysts. The summary of TRM reactions and the required properties of possible catalysts, together with the proposed chemical component are presented schematically in Fig. 6.

DOI: 10.1051/e3sconf/20171402038

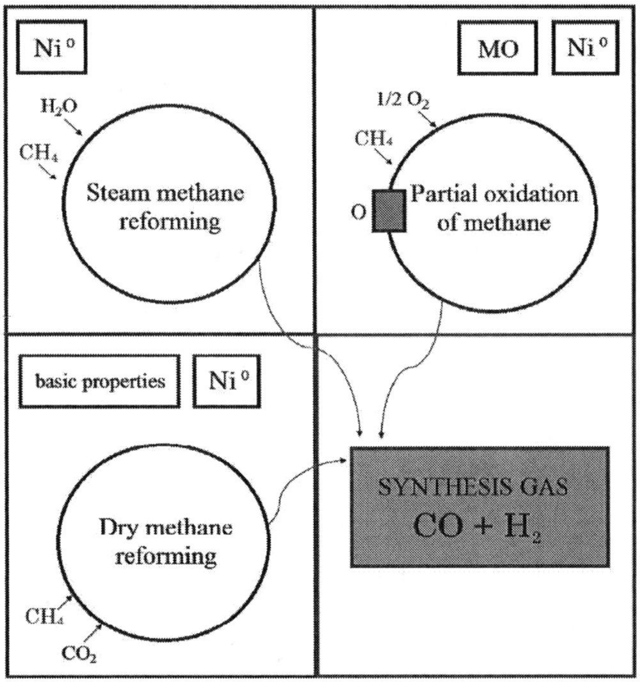

Fig. 6. Major components of TRM catalyst, based on: [15, 16]

3.1 Catalysts for single processes (SMR, DRM and POM)

Among three main tri-reforming reactions only the steam reforming of natural gas is the process on industrial scale. This reformation is focused on hydrogen production, and operates in the 700–1000°C, in the presence of nickel catalyst supported on alumina.

The dry methane reforming has not been yet implemented. The same as SRM, it shows very good properties with Ni as the active material. The newest DMR studies are mainly focused on stability increase due to the coke formation problem. Wide review was made by Dębek [6], as well as by Taniewski et. al [9]. The studied DMR catalysts were mainly Ni supported on Al_2O_3, MgO, CeO_2, ZrO_2, SiO_2, ordered mesoporous silica – SBA–15, La_2O_3, TiO_2, mixed oxides (MgO–Al_2O_3, CeO_2–ZrO_2, CeO_2–Al_2O_3), zeolites (zeolite Y, zeolite A, zeolite X, ZSM-5), clays (clinoptolite, diatomite, vermiculite, montmorlonite), and carbon-based materials (carbon nanotubes, activated carbon). Several different promoters were tested: (i) alkali, or alkaline earth metals – K, Li, Mg, Ca, Ba, (ii) rare earth metals – Ce, Zr, La, and (iii) other metals – Au, Ag, Sn, Bi, As, Pb, Cu [6, 9]. Hydrotalcites (double layered hydroxides LDHs) were found to be especially interesting for DRM. These materials may contain Ni, Mg and Al in their structure, thus fulfilling the requirements for appropriate redox and basic properties. Dębek et al [17-22] showed that high CO_2 conversions may be obtained even at relatively low temperatures (around 600°C) with appropriate tailoring of Ni:Mg:Al composition and/or the addition of other structural elements, such as Zr, or promoters (Ce). The application of La as a promoter to Ni/Mg/Al hydrotalcites was also proven advantageous [23].

The partial oxidation of methane was also shown to be efficient for Ni-based catalysts, although the optimal support is still searched. The following oxides, or mixed oxides, were investigated as carriers [24]: Al_2O_3, MgO-Al_2O_3, MgO, CaO-Al_2O_3, CaO, BaO, SiO_2, CeO_2, ZrO_2, CeO_2–ZrO_2, SrO–La_2O_3, TiO_2, Mn_3O_4, perovskite oxides, SiC, ThO_2, Y_2O_3, Yb_2O_3, alumina pillared bentonite, H-Y zeolite, ZSM-5 zeolite, active carbon, oxidized diamond, and more. In the literature of POM, also different promoters of Ni-based catalysts were considered: (i) noble metals – Pt, Pd, Ru, Ir, (ii) alkali, earth alkali metals – Na, K, B, Ca, Mg, Li, and (iii) rare earth metals – Ce, Zr, La, Y, Sm.

3.2 Catalysts for methane tri-reforming process

They are only scarce literature reports on catalysts investigated for TRM. Table 1 represents the examples of such catalysts, where the active component (Ni) was based on various supports: Al_2O_3 [25], ZrO_2 [26-28], SiO_2 spheres [11], β–SiC [29], CeO_2 [30], with or without the addition of promoter.

Table 1. Ni-based catalysts in tri-reforming of methane process

Support	Promoter	Results	Lit.
Al_2O_3	$Ce\text{-}ZrO_2$ (1	Ce improved catalyst stability by increasing the mobility of oxygen ions on the surface	[25]
ZrO_2	none	Strong interaction between $Ni\text{-}ZrO_2$ = high activity; the best results for 4.8% Ni-loaded catalyst at 800°C	[26]
ZrO_2	MgO (2	Mg improved CO_2 adsorption = high activity; and increase of thermal stability	[27]
ZrO_2	CaO (2	Samples with calcination at 700°C, and co-precipitated at pH of 10–12 gave the best catalytic results	[28]
SiO_2 spheres	none	11% Ni/SiO_2 showed stable activity at 750°C (4 hours) without deactivation. Presence of oxygen impacted coke formation and improved catalytic performance	[11]
β-SiC	Mg (1	Adding Mg to the Ni/β-SiC (molar ratio Mg/Ni = 1/1) decreased coke formation, increased basicity, and improved Ni-Mg interaction. The sequence of Ni–Mg impregnation was important	[29]
CeO_2	La (3	Adding La (10 at.%) resulted in strong interactions between Ni–La; Ce provided surface oxygen vacancies	[30]

1) impregnation, 2) co-precipitation, 3) combustion synthesis

Catalysts prepared by Lee et al. [25] contained Ni promoted with cerium and zirconium oxide on the alumina support. The major role was played weak acidic sites, basic sites, and redox abilities of Ce–ZrO_2. The catalyst displayed very good behavior by inhibition of the coke formation on the surface of catalyst, as well as on the reactor walls. The oxidant (steam or oxygen) had positive influence on the inhibition of carbon deposition, while the addition of cerium increased the mobility of oxygen ions on the catalyst surface.

Singha et al. [26] examined Ni supported on ZrO_2 without any additional promoter. It was noted that the rising Ni loading resulted in increasing Ni particle size. 4.8 wt.% Ni was reported as optimal for the used reaction conditions. High metal dispersion with strong metal support interaction (thanks to Ni–ZrO_2 presence) were found to be the most important factors in achieving high activity and stability.

Similarly observations were reported by Lai-Zhi et al. [27]. The catalyst with Ni-Mg-ZrO_2 exhibited strong metal–support interactions. The addition of Mg improved the alkaline character of the catalyst, and as a consequence improved CO_2 chemisorption. High temperature also has positive impact on the conversion of CH_4 and CO_2.

Other authors focused on the preparation conditions of the catalysts, such as pH of co-precipitation of catalyst and calcination temperature as factors which influence the catalytic performance. Lan-jie et al. [28] observed that after thermal treatment in the air at 700°C and the used preparation pH value of 10–12 it is possible to obtain proper surface area, nano-sized Ni particles, and very good Ni–ZrO_2 interaction which resulted in the more than 70% conversion of CH_4 (measurement at 700°C, 1 atm).

Majewski et al. [11] tested silica spheres which were covered by three–dimensional nickel. This resulted in shell structure which allowed the catalyst to encapsulate and protect nanoparticles via prevention of their migration during the catalytic reactions. The material was then called as nickel–silica core shell catalyst. In the catalytic tests an increase in oxygen content resulted in coke reduction and CH_4 conversion improvement (to 90%). The catalyst containing 11% Ni showed stable activity at 750°C (4 hours) without deactivation.

García-Vargas et al. [29] studied the influence of the impregnation order (of nickel and magnesium on silicon carbide SiC) on catalyst properties. Materials which had Mg impregnated in the first step were more stable and showed good catalytic performance. When Ni impregnation was the first step, weaker interaction between Ni and Mg was observed which, according to the authours, could have explained the blockage of Ni particles by Mg, the reduction of the number of Ni active sites and inferior catalytic results. It was additionally observed that the Mg/Ni molar ratio of 1/1 resulted in less coke formation, higher basicity, and better Ni-Mg interaction.

Pino et al. [30] studied Ni–CeO_2 materials loaded additionally with lanthanium (10 at.%). These catalysts showed strong interactions between Ni and La, and better Ni dispersion. Thanks to an appropriate amount of lantanium loading, the CH_4 conversion increased from 93% to 96%, as well as CO_2 conversion 83% – 86.5%.

4 Conclusions

Tri-reforming of methane is the process consisting of stream reforming of methane, dry reforming of methane, and partial oxidation of methane. It may be considered as a method for the CO_2 emissions reduction, and as a process for synthesis gas production. The TRM might be implemented in the fossil fuels fired power stations, where natural gas together with pre-cleaned flue gases (SO_2, NOx, and fly ash removed) are directed to the tri-reformer. Obtained syngas may be applied to produce chemicals and liquid fuels. The TRM application is a promising method due to high equilibrium (thermodynamic), possibility of converting CO_2 without its expensive pre-separation from flue gases, appropriate H_2/CO

DOI: 10.1051/e3sconf/20171402038

molar ratios obtained, etc. TRM may be also applied as energy storage concept. Thanks to the H_2O and O_2 addition to the dry reforming system, it is possible to prevent coke formation on the surface of the catalyst. In addition, oxygen presence allows the production of heat in situ, thereby increasing process efficiency. However, it should be stressed that for the future implementation of tri-reforming on industrial scale, the active, stable (free from coke formation), and selective catalyst is a key requirement.

Nickel-based materials are widely used as catalysts for methane reforming processes due to availability and low price of this metal. They are an acceptable alternative for noble metals (Pt, Rh, Ru, Ir, Au) which are very expensive. Literature review for TRM shows that the reduced metal (Ni^0) facilitates CH_4 dissociative chemisorption, therefore its presence is required in each of the reforming reaction. CO_2 chemisorption, on the other hand, requires basic sites, so the addition of elements of basic nature (e.g. Mg) is of the importance. Metal oxides (as support or promoters) play also an important role in case they offer oxygen sites allowing the oxidation of intermediates. Due to their structure and composition hydrotalcites (double layered hydroxides) containing Ni, Mg, Al and appropriate promoters may prove interesting materials for TRM.

Several types of single or mixed oxides were reported in the literature as possible supports for TRM catalysts: Al_2O_3, MgO, CeO_2, ZrO_2, SiO_2, La_2O_3, TiO_2. As promoters metals such as: Pt, Rh, Ru, Ir, Au, K, Li, Mg, Ca, Ce, Zr, La were reported to positively modify the catalyst structure. Their influence concerned: inhibition of carbon deposition, resistance to sintering of active phase, impovement the oxygen tolerance and its ions mobility on catalyst surface.

Acknowledgements to AGH 11.11.210.213

References

1. V. Bourgeais, M. Goll, *Eurostat news release*, **89** (2016), Available online on 23.08.2016: http://ec.europa.eu/eurostat/documents/2995521/5176346/8-07052014-AP-EN.PDF/318b117b-2065-4a61-9efb-c382f901a8ea\
2. V. Bourgeais, M. Goll, *Eurostat news release*, **105** (2015), Available online on 23.08.2016: http://ec.europa.eu/eurostat/documents/2995521/6875491/8-15062015-BP-EN.pdf/8adf74de-e79b-4778-905f-823c42c6e1b1
3. L. Corselli-Nordblad, M. Goll, *Eurostat news release*, **74** (2014), Available online on 23.08.2016: http://ec.europa.eu/eurostat/documents/2995521/5176346/8-07052014-AP-EN.PDF/318b117b-2065-4a61-9efb-c382f901a8ea
4. J. Urhausen, M. Goll, *Eurostat news release*, **80** (2013), Available online on 23.08.2016: http://ec.europa.eu/eurostat/documents/2995521/5166066/8-29052013-AP-EN.PDF/38dda41b-d760-4909-b8bd-77e651c401a3?version=1.0
5. *Energy & climate change report, Ed.* OECD/IEA (2015)
6. R. Dębek, *Novel catalysts for chemical CO_2 utilization - Doctoral thesis* (2016)
7. Citizens' summary EU plan for a competitive low-carbon economy by 2050, Available online on 28.08.2016: http://ec.europa.eu/clima/citizens/summary/docs/roadmap_2050_en.pdf
8. G. Centi, S. Perathoner S., Catal Today **148**, 191 (2009)

9. M. Taniewski, T. Grzybek, J. Cetnar, Viable CO_2 chemical sequestration applications, in *Development of coal, biomass and wastes gasification technologies with particular interest in chemical sequestration of CO_2:* a monograph **99** (2015)
10. C. Song, W. Pan, Catal Today **98**, 463 (2004)
11. A. J. Majewski, J. Wood, Int J Hydrogen Energ **39**, 12578 (2014)
12. M. Halmann, A. Steinfeld, Catal Today **115**, 170 (2006)
13. J. H. McCrary, G.E. McCrary, T.A. Chubb, Sol Energy **29**, 141 (1982)
14. T.A. Chubb, Sol Energy **24**, 341 (1980)
15. G. Djéga-Mariadassou , *Catalysis for environment: depollution, renewable energy and clean fuels international annual meeting: proceedings*, 1 (2009)
16. H.-A. Nishimoto, K. Nakagawa, N.-O. Ikenaga, T.Suzuki, Catal Lett **82**, 161 (2002)
17. R. Dębek, A.Gramatyka, M.Motak, T.Grzyebk, P.Da Costa, Przemysł Chemiczny **93**,2026 (2014)
18. R. Dębek, K. Zubek, M. Motak, P. Da Costa, T. Grzybek, Research on Chemical Intermediates **41**,9485 (2015)
19. R. Dębek, M. Radlik, M. Motak, M.E. Galvez, W. Turek, P. Da Costa, T. Grzybek, Catalysis Today 257, **59** (2015)
20. R.Dębek, K.Zubek, M.Motak, M.E.Galvez, P.Da Costa, T.Grzybek, Comptes Rendus Chimie **18**, 1205 (2015)
21. R.Dębek, M.E.Galvez, F.Launay, M.Motak, T.Grzybek, P.Da Costa, International Journal of Hydrogen Energy **41**, 11616 (2016)
22. R.Dębek, M.Motak, D.Duraczyńska, F.Launay, M.E.Galvez, T.Grzybek, P.Da Costa, Catalysis Science and Technology, **5**, 6705 (2016)
23. H. Liu, D.Wierzbicki, R.Dębek, M.Motak, T.Grzybek, P.Da Costa, M.E.Galvez, Fuel **182**, 9 (2016)
24. B. C. Enger, R. Lødeng, A. Holmen, Appl Catal A-General **346**, 1 (2008)
25. S.-H. Lee, W. Cho, W.-S. Ju, B.-H. Cho, Y.-C. Lee, Y.-S. Baek, Catal Today **87**, 133 (2003)
26. R. K. Singha, A. Shukla, A. Yadav, S. Adak, Z. Iqbal, N. Siddiqui, R. Bal, Appl Energ **178**, 110 (2016)
27. S. Lai-zhi, T. Yi-sheng, Z.Qing-de, X. Hong-juan, H. Yi-zhuo, J Fuel Chem Technol **40**, 831 (2012)
28. S. Lan-jie, W. Chang-zhen, S. Nan-nan, W. Xia, Z. Ning, X. Fu-kui, W. Wei, S. Yu-han, J Fuel Chem Technol **40**, 210 (2012)
29. J. M. García-Vargas, J. L.Valverde, J. Díez, P.Sánchez, F. Dorado, Appl Catal B-Environ **164**, 316 (2015)
30. L. Pino, A. Vita, F. Cipiti, M. Lagana, V. Recupero, Appl Catal B-Environ **104**, 64 (2011)

Ceria promotion over Ni-containing hydrotalcite-derived catalysts for CO_2 methane reforming

Radosław Dębek[1,2*], *Monika* Motak[1] , *Maria E.* Galvez-Parruca[2], *Teresa* Grzybek[1], *Patrick* Da Costa[2], and *Ludwik* Pieńkowski[1]

[1]AGH University of Science and Technology, Faculty of Energy and Fuels, Al. A.Mickiewicza 30, 30-059, Kraków, Poland
[2]Sorbonne Universités, UPMC, Univ. Paris 6, CNRS, UMR 7190, Institut Jean Le Rond D'Alembert, 2 Place de la Gare de Ceinture, 78210 Saint-Cyr-L'Ecole, France

Abstract. The catalytic activity in dry methane reforming of hydrotalcite-derived catalysts with ceria and/or nickel species introduced into hydrotalcite interlayer spaces was examined. The prepared materials were characterized (XRF, XRD, FT-IR, H_2-TPR and N_2 sorption) and subsequently tested in CO_2 methane reforming at 550 °C. The obtained results showed that the incorporation of nickel species between hydrotalcite layers resulted in active catalyst with no sign of carbon deposition. Additionally, a beneficial effect of ceria promotion was observed. Ceria-promoted sample exhibited higher activity, stability and selectivity towards DRM, which may be explained by the formation of small Ni crystallites and prevention of the formation of inactive $NiAl_2O_4$ spinel phase.

1 Introduction

The growing emissions of carbon dioxide forced implementation of various CO_2 emission reduction strategies, which may be divided into two main groups: (i) carbon capture and storage (CCS) and (ii) carbon capture and utilization (CCU) technologies [1]. The latter approach allows to convert CO_2 into added-value products, creating in this way a carbon cycle. One of the processes that converts CO_2 into added-value products is dry reforming of methane (DRM). The process leads to an industrially important mixture of H_2 and CO and may be considered an attractive route for syngas production, alternative to the currently applied steam reforming reaction [2]. It may be also applied in Chemical Energy Transmission and/or Storage systems (CETS) for transport and storage of energy [3]. It should be mentioned, however, that DRM process has not yet been commercialized due to the low price of CO_2 emissions, high endothermicity of the reaction and lack of cheap, active and stable catalysts [4].

The investigation for a stable catalyst for DRM process is currently focused on nickel-based materials, as they show similar activity to noble metals but are cheaper and more

* Corresponding author: debek@agh.edu.pl

abundant [5, 6]. However, they still undergo deactivation due to (i) the occurrence of C-forming reactions, which procced easily on large Ni crystallites, and (ii) sintering of nickel active phase. The problem of improving stability of nickel-based catalysts may be solved by increasing nickel-support interactions, which results in the formation of small Ni aggregates and prevents sintering. This may be achieved by the application of appropriate supports. Magnesia and alumina were reported to provide beneficial effects on catalytic performance in DRM due to the formation of NiO-MgO solid solution and $NiAl_2O_4$ spinel phase, respectively [7, 8]. The application of hydrotalcites (layered double hydroxides, HTs) as catalysts precursors for DRM turned out to be of advantage, as all required components i.e. Ni, magnesia and alumina may be introduced into hydrotalcite brucite-like layers [9-11]. In our previous research we showed that hydrotalcite-derived catalysts with nickel introduced into brucite-like layers were promising materials for DRM [12-14].

The other commonly used component of DRM catalyst is ceria. Due to its redox properties, promotion by cerium species may help to oxidize carbon deposits through Boudouard reaction [15, 16]. Moreover, as shown in our previous work, the addition of ceria increases the concentration of intermediate and strong basic sites, which also enhance catalysts stability [17]. Tsyganok et al. [18, 19] proposed an alternative way of introducing nickel species into hydrotalcite structure by co-precipitation in the solution of $[Ni(EDTA)]^{2-}$ complexes.

The work presented in this study was aiming at the comparison of catalytic performance of hydrotalcite-derived materials into which cerium and/or nickel species were introduced into interlayer spaces by co-precipitation in the solution of $[Ce(EDTA)]^{-}$ and/or $[Ni(EDTA)]^{2-}$.

2 Experimental

2.1 Catalyst preparation

The catalysts precursors, hydrotalcite-like materials, were synthesized by co-precipitation method at constant pH from an aqueous solution of appropriate nitrates ($Mg(NO_3)_2 \cdot 6H_2O$, $Al(NO)_3 \cdot 9H_2O$ – Sigma Aldrich) of total cations concentration equal to 1M. The precipitating agent was 1M solution of NaOH (POCH). These two solutions were added simultaneously to the 5wt.% solution of $[Ni(EDTA)^{2-}]$ or mixture of 5 wt.% $[Ni(EDTA)^{2-}]$ and 3 wt.% $[Ce(EDTA)^{-}]$, resulting in samples HTNi and HTNi-Ce, respectively. An additional reference material was synthesized by co-precipitation in 0.05 M solution of Na_2CO_3 (sample HTMgAl). The flows of various solutions were adjusted in order to maintain constant pH equal to 10. The co-precipitation was carried out under constant stirring at 65 °C. The obtained mineral suspension was aged at 65 °C for 1h and subsequently washed with warm distilled water (ca. 50-60 °C) and filtered until neutral pH of the filtrate was registered. The synthesized materials were dried in air overnight at 80 °C and ground into fine powder. Finally, prepared hydrotalcites were calcined in the stream of air (10 cm^3/min) at 550 °C for 4 h.

2.2 Physico-chemical characterization

The chemical composition of prepared catalysts was examined by X-ray fluorescence (XRF). The measurements were carried out in an energy dispersive XEPOS spectrometer from Spectro Ametek. X-ray diffraction (XRD) patterns were recorded on an Empyrean diffractometer from PANalytyical, in the 2Θ ° using a Cu X-ray source (λ=1.5046 Å). Textural properties were evaluated from the N_2 adsorption isotherms acquired at -196 °C in

 DOI: 10.1051/e3sconf/20171402039

a Belsorp Mini II equipment. Prior to each measurement, the samples were outgassed at 120 °C for 3 h. Temperature-programmed reduction (TPR) profiles were recorded with a BELCAT-M apparatus from BEL Japan, equipped with a thermal conductivity detector (TCD). The samples (ca. 50 mg) were first degassed at 100 °C for 2 h, then reduced at 7.5 °C/min using 5 % H_2 in Ar. Fourier Transform Infrared Spectroscopy (FT-IR) measurements were carried out on Frontier spectrometer from PerkinElmer. The sample was diluted in KBr in relation 1:10 and subsequently DRIFT spectra were recorded in the range of 4000-450 cm^{-1}.

2.3 Catalytic tests

The activity towards DRM of the synthesized hydrotalcite-derived materials was tested in a fixed-bed reactor. Prior to the experiments, each catalyst was reduced with hydrogen (5 % H_2 in Ar) at 900 °C. The catalytic tests were carried out at 550 °C for 5 h with the substrate molar ratio $CH_4/CO_2/Ar=1/1/8$ and GHSV equal to 20,000 h^{-1}. The total gas flow was equal to 100 cm^3/min. The products of the reaction were analysed by gas chromatography (490 Varian Micro-GC).

3 Results and discussion

3.1 Physicochemical characterization of the prepared materials

Designation of the prepared catalysts and their chemical composition are presented in Table 1. The coprecipitation of hydrotalcites in the solution of [Ni(EDTA)$^{2-}$] complexes resulted in incorporation of ca. 7 wt.% of nickel and for HTNi-Ce sample around 3 wt.% of cerium. The calculated values of M^{2+}/M^{3+} molar ratios for prepared materials were very close to the nominal ones (Table 1), pointing to the successful synthesis of HTs materials with a desired composition.

Table 1 The designation of the prepared catalysts and their chemical composition determined by XRF.

Sample	Cations in brucite-like layers	Species present in interlayer spaces	Ni (wt.%)	Mg (wt.%)	Al (wt.%)	Ce (wt.%)	M^{2+}/M^{3+}
HTNi	Mg^{2+}, Al^{3+}	Ni^{2+}	7.7	18.4	7.1	-	2.9 (3)*
HTNi-Ce	Mg^{2+}, Al^{3+}	Ni^{2+}, Ce^{3+}	7.2	17.9	7.5	3.2	3.1 (3)*
HTMgAl	Mg^{2+}, Al^{3+}	-	-	37.4	14.15	-	2.9 (3)*

* in the parenthesis are given nominal values

The XRD diffractograms recorded for the prepared catalysts are depicted in Fig. 1. The freshly prepared materials (Fig. 1A) exhibited reflections characteristic for the 3R rhombohedral layered structure of hydrotalcite (ICOD 00-014-0191). The reference sample (HTMgAl) showed reflections at 2Θ equal to ca. 11, 23 and 35 °, arising from the diffraction of X-rays on (003), (006) and (009) planes [20]. In case of samples HTNi and HTNi-Ce the shift of the first reflections to lower values of 2Θ was observed (reflections at ca. 6, 12, 18, 24 °). The position of the first reflections in hydrotalcites XRD patterns is dependent on the basal spacing between layers, and thus the type of anions present between hydrotalcite layers. The shift of the first reflections observed for samples HTNi and HTNi-Ce may be thus explained by incorporation of [Ce(EDTA)$^-$] and/or [Ni(EDTA)$^{2-}$] anions into interlayer spaces. The similar shift of the first reflection position to 2Θ ca. 6 ° was observed by Tsyganok et al. [18].

All samples exhibited rhombohedral symmetry, thus parameter *c* of unit cell may be calculated from the position of the first reflection or by averaging the positions of the first three reflections, and unit cell parameter *a* from the position of the first reflection from the doublet present at 2Θ equal to ca. 61° [21]. The calculated parameters are presented in Table 2. The parameter *a*, which describes average cation-cation distances in the brucite-like layers, had the same value for all samples, indicating that the incorporation of cerium and/or nickel species into interlayer spaces did not affect layered structure. The parameter *c*' (c'=*c*/3), which describes basal spacing in hydrotalcite structure, increased for HTNi and HTNi-Ce samples with respect to the reference HTMgAl material, pointing to the successful incorporation of cerium and/or nickel species into hydrotalcite structure.

Materials after calcination (Fig. 1B) exhibited reflections (2Θ ca. 43 and 63 °) characteristic for periclase-like mixed oxides, which is typical for the products of HTs thermal decomposition [22]. Additionally, HTNi-Ce sample showed the presence of CeO_2 phase (reflection at 2Θ ca. 30 °).

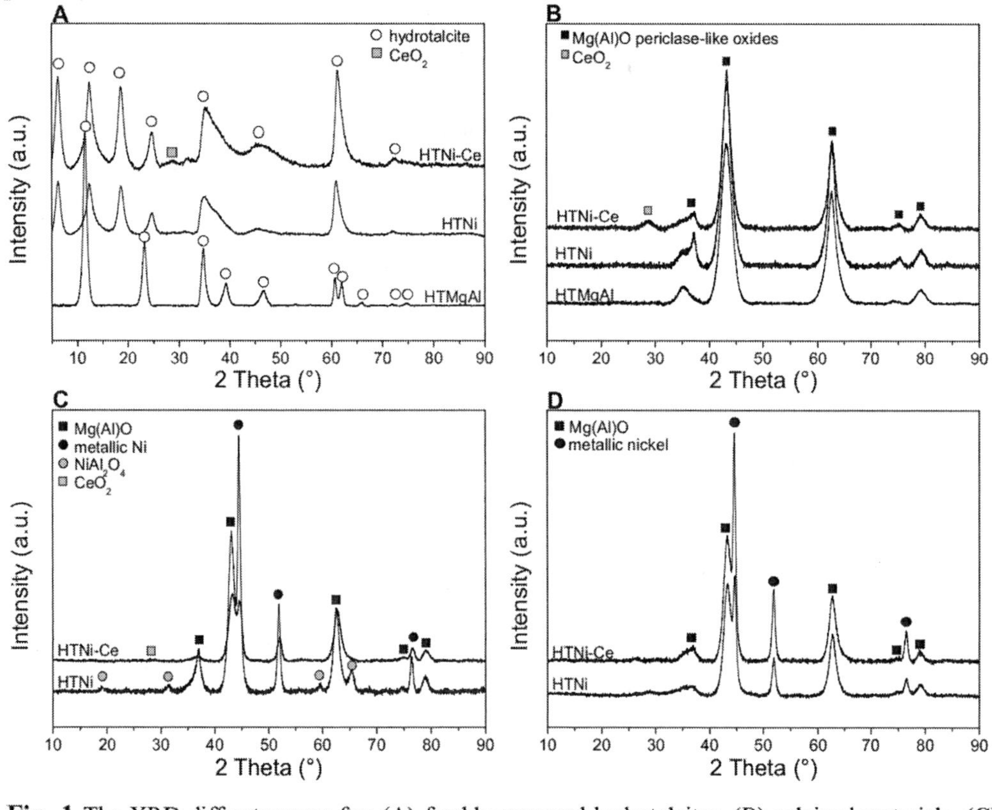

Fig. 1 The XRD diffractograms for: (A) freshly prepared hydrotalcites, (B) calcined materials, (C) reduced samples, (D) catalysts after 5 h DRM catalytic tests at 550 °C

The FT-IR spectra recorded for freshly prepared hydrotalcites are presented in Fig. 2A. All samples exhibited a number of common absorption bands observed at 3550, 3000, 1650, 1370 and 870 cm^{-1}. The broad band at 3550 cm^{-1} arises from the stretching vibration of structural OH^- groups in the metal hydroxide layer [20]. A small shoulder at ca. 3000 cm^{-1} may be attributed to a second type of OH^- stretching vibrations of interlayer hydrogen bonding with carbonate groups. The weak band at 1650 cm^{-1} was ascribed to interlayer water molecules. The strong band at 1370 cm^{-1} and weak band at 870 cm^{-1}, were attributed to the stretching vibration and out-of-plane bending vibration of CO_3^{2-} [23]. The bands in the range of 450-700 cm^{-1} are typical of the frameworks of HTs, and indicate M-OM stretching, which further confirms that the catalysts precursors are ordered hydrotalcites [24]. The presence of a weak band at ca. 1376 cm^{-1} may be indicating the presence of NO_3^- anions [25], and is in good agreement with the value of *c'* parameter calculated from XRD measurements for sample HTMgAl (Table 2).

Table 2 The unit cell parameters, anions present in the interlayer spaces and textural properties of synthesized hydrotalcites (S_{BET} – specific surface area, V_{tot} – total volume of pores; d_r – mean pore diameter).

Sample	Unit cell parameter		Anions in interlayer spaces	S_{BET} (m^2/g)	V_{tot} (cm^3/g)	d_r (nm)
	a (Å)	*c'*=c/3 (Å)				
HTNi	3.06	14.12	[$Ni(EDTA)^{2-}$]	231	0.36	6
HTNi-Ce	3.06	14.23	[$Ni(EDTA)^{2-}$] and [$Ce(EDTA)^-$]	229	0.38	6
HTMgAl	3.06	7.84	CO_3^{2-} and NO_3^-	95	0.87	35

Samples HTNi and HTNi-Ce exhibited additionally absorption bands associated with EDTA molecules. The band at 2930 and 2850 cm^{-1} are ascribed to asymmetric and symmetric C-H stretching of methylene groups. The absorption of infrared radiation by asymmetric stretching of coordinated and uncoordinated COO^- may be attributed to the bands at 2850 and 1600 cm^{-1}, respectively, while band at 1110 cm^{-1} to the vibration of C-N single bonds [18, 24]. These results agree with the XRD measurements which confirmed the presence of nickel-EDTA and Ce-EDTA chelates between brucite-like layers for samples HTNi and HTNi-Ce.

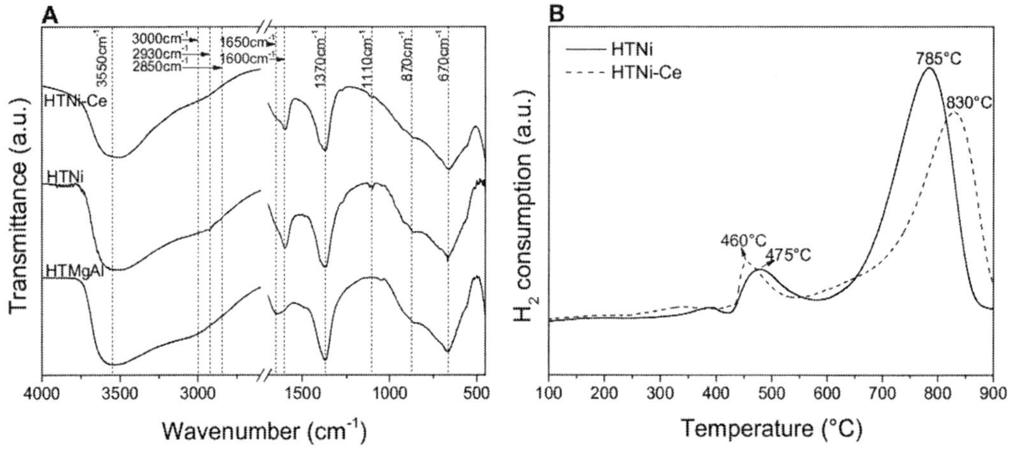

Fig. 2 FT-IR spectra (A) and H_2-TPR profiles (B) for the prepared materials.

H_2-TPR profiles recorded for Ni-containing samples are depicted in Fig. 2B. Both samples exhibited the presence of two types of Ni species, suggesting various types of interactions between NiO species and the mesoporous Mg(Al)O support. According to literature, pure NiO phase is reduced at the temperature range 220-420 °C [26-28]. Thus the reduction peaks centered at 460 and 475 °C for HTNi-Ce and HTNi, respectively, might be attributed to the reduction of NiO weakly bonded with support surface. The most intense reduction peak was observed for both catalysts between 600 and 900 °C, indicating strong interactions between nickel species and support. These peaks may be thus attributed to the reduction of Ni^{2+} in NiO-MgO solid solution or $NiAl_2O_4$ spinel phase, which were formed upon calcination [13, 24].

DOI: 10.1051/e3sconf/20171402039

The introduction of ceria species between HTs brucite-like layers modified H_2-TPR profile. The position of the reduction peaks was shifted and additionally small reduction peak centred at ca. 340 °C appeared. The explanation of the reduction process of sample HTNi-Ce is quite complicated, as the reduction profiles of NiO and ceria may be overlapping. As described in literature, the reduction of CeO_2 occurs in three stages: (i) reduction of surface oxygen or oxygen capping species (around 300 °C) [29], (ii) reduction of surface lattice oxygen (450-600 °C) [15] and (iii) total bulk reduction (up to 900 °C) [9]. Therefore, the peak observed at 340 °C and the shift of the peak in the 400-500 °C region may be attributed to the overlapping reduction profiles of NiO and CeO_2. The most intense peak was shifted to higher temperatures (from 785 to 830 °C), indicating that the addition of ceria decreased reducibility of nickel species and suggesting increased metal-support interactions for the Ce-promoted catalyst. This results stay in line with the estimated Ni crystallite sizes for the reduced samples basing on XRD results (Table 3, Fig. 1C).

The values of textural parameters calculated for the prepared catalysts basing on N_2 sorption experiments are presented in Table 2, and corresponding sorption isotherms and distribution of pores are depicted in Fig. 3. The samples exhibited specific surface areas between 100-230 m^2/g, in agreement with the data reported in literature for similar hydrotalcite-derived materials [18, 20, 24]. Samples HTNi and HTNi-Ce showed much higher values of S_{BET} with respect to the reference material (HTMgAl), which was caused by the substitution of carbonate anions by [Ce(EDTA)$^-$] and/or [Ni(EDTA)$^{2-}$] complexes. This resulted in the increase of basal spacing between hydrotalcite layers, as confirmed by XRD. The shape of N_2 adsorption isotherm and the distribution of pores for HTNi and HTNi-Ce samples also differed. Narrow pores and the shape of adsorption/desorption isotherm suggest the formation of microporous materials. Moreover the observed hysteresis loop was of type H4, pointing to the uniform distribution of slit shaped pores [30]. This clearly indicates that the type of anions present in the interlayer spaces of hydrotalcite has a strong influence on textural properties of the studied catalysts. However, the introduction of cerium species did not affect textural properties of prepared material with respect to the HTNi sample.

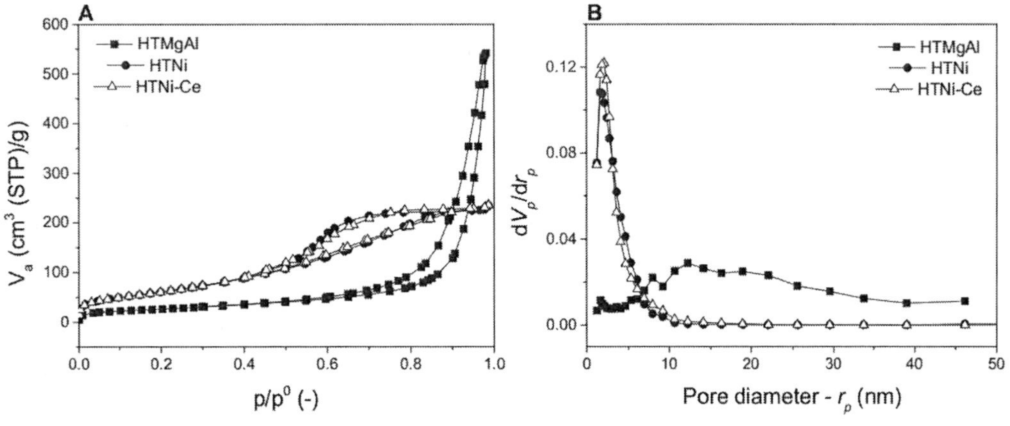

Fig. 3 N_2-adsorption/desorption isotherms (A) and pore diameter distribution (B) of the studied materials.

3.2 Low temperature CO_2 reforming of methane

The synthesized materials were tested in low temperature CO_2 methane reforming reaction at 550 °C (Fig. 4). At temperature as low as 550 °C, the Ni-containing hydrotalcite-derived materials showed good catalytic performance. The catalytic activity towards both CH_4 and

CO_2 was higher for ceria-promoted sample and reached values of ca. 30 and 37 %, respectively. Both catalysts showed higher conversions of CO_2 than those of CH_4, indicating the occurrence of reverse water gas shift reaction (RWGS). This is directly reflected in the distribution of obtained products, as an excess of carbon monoxide was observed for both catalysts (Fig. 4B). The CO_2 conversions were decreasing with the time on stream (TOS), pointing to deactivation of the catalysts. On the other hand, the opposite trend was registered for CH_4 conversions.

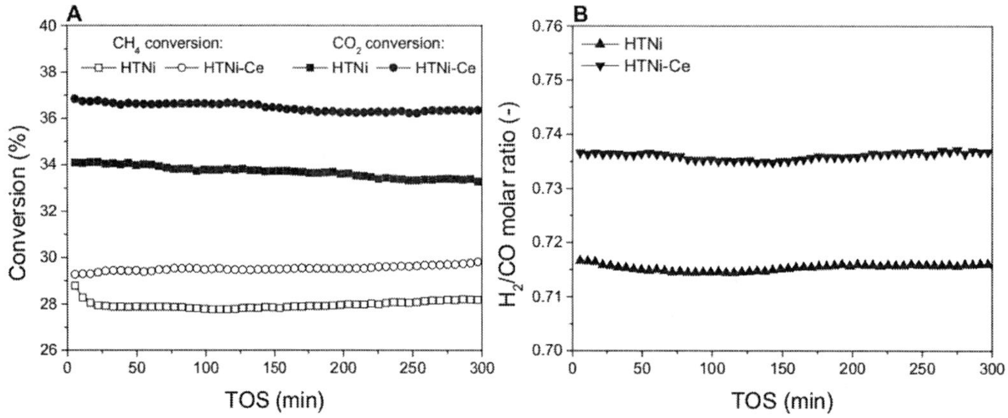

Fig. 4 The results of catalytic tests: (A) CH_4 and CO_2 conversions (B) H_2/CO molar ratio; temperature 550 °C; $CH_4/CO_2/Ar=1/1/8$; GHSV=20,000 h^{-1}; total flow 100 cm^3/min;

The beneficial effect of ceria addition on catalysts activity may be explained by the fact that Ce-promoted sample had smaller Ni crystallites with an average size of ca. 8 nm (Table 3). Small Ni crystallites are desired in dry reforming process, as it allows to avoid C-forming side reactions which proceed more easily on large Ni crystallites [31]. Therefore, sample HTNi-Ce was more selective towards DRM which resulted in higher activity and higher selectivity.

Another possible explanation of the beneficial effect of ceria addition is that sample HTNi exhibited formation of $NiAl_2O_4$ spinel phase after reduction in H_2 (Fig. 1C). As reported in literature, the formation of spinel phase has a negative effect on catalysts activity towards DRM, as nickel species in $NiAl_2O_4$ are too strongly bonded with the support and thus cannot be reduced during catalyst activation. As a result, lower amount of metallic nickel species is formed [31, 32]. Therefore the addition of Ce species, by modifying reducibility of NiO, prevents the formation of inactive spinel phase.

Table 3 The Ni crystallite size estimated for the reduced and spent catalysts after DRM at 550 °C.

Sample	Ni crystallite size* (nm)	
	Reduced samples	Catalyst after DRM tests
HTNi	12	13
HTNi-Ce	8	8

*estimated by Scherrer equation

The catalysts after reaction were analysed by XRD experiments (Fig. 1D). Both catalysts did not exhibit any graphitic carbon (reflection at 2Θ equal to ca. 26°). On the other hand, such type of carbon deposit was observed for Ni-containing hydrotalcite-derived catalyst with nickel species introduced into HTs brucite-like layers. It originated from direct CH_4 decomposition [14]. These results indicate that the incorporation of Ni species between brucite-like layers results in the catalyst resistant to the methane decomposition. This explanation is in good agreement with the obtained values of H_2/CO

DOI: 10.1051/e3sconf/20171402039

molar ratio (Fig. 4B). However, the formation of amorphous carbon on HTNi and HTNi-Ce catalysts cannot be excluded and further characterization is required for further confirmation. Basing on XRD experiments for the reduced and spent catalysts Ni crystallite sizes were estimated (Table 3). No sintering of Ni active phase was observed after 5h TOS. The Ce-promoted catalyst exhibited smaller Ni crystallites, which may explain more stable performance with respect to HTNi catalyst.

4 Conclusions

The hydrotalcite-derived materials were synthesized and tested in DRM reaction. Nickel species were introduced into HTs-based catalytic system at the co-precipitation stage, resulting in the incorporation of Ce and/or Ni species into hydrotalcite interlayer spaces. The obtained results showed that the method of nickel introduction influenced material properties, especially their texture. The prepared catalysts turned out to be active in low temperature dry reforming. Higher activity, stability and selectivity towards DRM was exhibited by ceria-promoted sample, indicating the beneficial effects of ceria addition. the formation of small Ni crystallites upon catalyst reduction in H_2. Additionally, Ce-promoted sample did not exhibit the formation of inactive $NiAl_2O_4$ spinel phase, which enhanced its catalytic performance.

The obtained results showed that the method of nickel and ceria introduction into HTs-based catalytic systems strongly influenced their catalytic properties. As reported in literature, the introduction of ceria and nickel onto hydrotalcite surface and into brucite-like layers, respectively, resulted in increased reducibility of NiO. However, it did not affect the size of Ni crystal size [9, 15, 33]. Therefore, the application of the appropriate method of catalysts synthesis may allow tailoring their catalytic properties in dry reforming reaction.

The work was financed within NCBiR strategic research project 'Technologies supporting development of safe nuclear energy', research task no. 1 'Development of high temperature reactors for industrial applications', agreement no. SP/J/166183/12, step task no. 15 'Preparation and physicochemical characterization of catalysts for dry reforming of methane'.
R. Dębek and M. Motak would like to acknowledge AGH grant 11.11.210.213.

References

1. A.J. Hunt, E.H.K. Sin, R. Marriott, J.H. Clark, Generation, Capture, and Utilization of Industrial Carbon Dioxide, Chemsuschem, 3 (2010) 306-322.
2. J.-M. Lavoie, Review on dry reforming of methane, a potentially more environmentally-friendly approach to the increasing natural gas exploitation, Frontiers in Chemistry, 2 (2014).
3. J.T. Richardson, S.A. Paripatyadar, Carbon dioxide reforming of methane with supported rhodium, Applied Catalysis, 61 (1990) 293-309.
4. N. Thybaud, Lebain, D., , Panorama des vois de valorisation du CO_2, French Agence de l'Environnement et de la Maitrise de l'Energie, Angers, France, 2010.
5. M.S. Fan, A.Z. Abdullah, S. Bhatia, Catalytic Technology for Carbon Dioxide Reforming of Methane to Synthesis Gas, Chemcatchem, 1 (2009) 192-208.
6. M. Usman, W.M.A. Wan Daud, H.F. Abbas, Dry reforming of methane: Influence of process parameters—A review, Renewable and Sustainable Energy Reviews, 45 (2015) 710-744.
7. A. Becerra, M. Dimitrijewits, C. Arciprete, A. Castro Luna, Stable Ni/Al_2O_3 catalysts for methane dry reforming, Granular Matter, 3 (2001) 79-81.

8. R. Zanganeh, M. Rezaei, A. Zamaniyan, Dry reforming of methane to synthesis gas on NiO–MgO nanocrystalline solid solution catalysts, Int J Hydrogen Energ, 38 (2013) 3012-3018.
9. C.E. Daza, J. Gallego, J.A. Moreno, F. Mondragón, S. Moreno, R. Molina, CO_2 reforming of methane over Ni/Mg/Al/Ce mixed oxides, Catal Today, 133–135 (2008) 357-366.
10. A.R. Gonzalez, Y.J.O. Asencios, E.M. Assaf, J.M. Assaf, Dry reforming of methane on Ni-Mg-Al nano-spheroid oxide catalysts prepared by the sol-gel method from hydrotalcite-like precursors, Appl Surf Sci, 280 (2013) 876-887.
11. O.W. Perez-Lopez, A. Senger, N.R. Marcilio, M.A. Lansarin, Effect of composition and thermal pretreatment on properties of Ni–Mg–Al catalysts for CO_2 reforming of methane, Applied Catalysis A: General, 303 (2006) 234-244.
12. R. Dębek, K. Zubek, M. Motak, P. Da Costa, T. Grzybek, Effect of nickel incorporation into hydrotalcite-based catalyst systems for dry reforming of methane, Res Chem Intermediat, 41 (2015) 9485-9495.
13. R. Dębek, K. Zubek, M. Motak, M.E. Galvez, P. Da Costa, T. Grzybek, Ni–Al hydrotalcite-like material as the catalyst precursors for the dry reforming of methane at low temperature, Comptes Rendus Chimie, 18 (2015) 1205-1210.
14. R. Dębek, M. Motak, D. Duraczyska, F. Launay, M.E. Galvez, T. Grzybek, P. Da Costa, Methane dry reforming over hydrotalcite-derived Ni-Mg-Al mixed oxides: the influence of Ni content on catalytic activity, selectivity and stability, Catalysis Science & Technology, 6 (2016) 6705-6715.
15. C.E. Daza, S. Moreno, R. Molina, Co-precipitated Ni–Mg–Al catalysts containing Ce for CO_2 reforming of methane, Int J Hydrogen Energ, 36 (2011) 3886-3894.
16. C.E. Daza, C.R. Cabrera, S. Moreno, R. Molina, Syngas production from CO_2 reforming of methane using Ce-doped Ni-catalysts obtained from hydrotalcites by reconstruction method, Applied Catalysis A: General, 378 (2010) 125-133.
17. R. Dębek, M. Radlik, M. Motak, M.E. Galvez, W. Turek, P. Da Costa, T. Grzybek, Ni-containing Ce-promoted hydrotalcite derived materials as catalysts for methane reforming with carbon dioxide at low temperature - On the effect of basicity, Catal Today, 257 (2015) 59-65.
18. A.I. Tsyganok, K. Suzuki, S. Hamakawa, K. Takehira, T. Hayakawa, Mg–Al Layered Double Hydroxide Intercalated with $[Ni(edta)]^{2-}$ Chelate as a Precursor for an Efficient Catalyst of Methane Reforming with Carbon Dioxide, Catal Lett, 77 (2001) 75-86.
19. A.I. Tsyganok, T. Tsunoda, S. Hamakawa, K. Suzuki, K. Takehira, T. Hayakawa, Dry reforming of methane over catalysts derived from nickel-containing Mg-Al layered double hydroxides, J Catal, 213 (2003) 191-203.
20. F. Cavani, F. Trifirò, A. Vaccari, Hydrotalcite-type anionic clays: Preparation, properties and applications, Catal Today, 11 (1991) 173-301.
21. V. Rives, Characterisation of layered double hydroxides and their decomposition products, Materials Chemistry and Physics, 75 (2002) 19-25.
22. D. Tichit, B. Coq, Catalysis by Hydrotalcites and Related Materials, CATTECH, 7 (2003) 206-217.
23. S. Kannan, A. Dubey, H. Knozinger, Synthesis and characterization of CuMgAl ternary hydrotalcites as catalysts for the hydroxylation of phenol, J Catal, 231 (2005) 381-392.
24. P. Tan, Z. Gao, C. Shen, Y. Du, X. Li, W. Huang, Ni-Mg-Al solid basic layered double oxide catalysts prepared using surfactant-assisted coprecipitation method for CO_2 reforming of CH_4, Chinese Journal of Catalysis, 35 (2014) 1955-1971.

DOI: 10.1051/e3sconf/20171402039

25. E. Lopez-Salinas, Y. Ono, Intercalation chemistry of a Mg□Al layered double hydroxide ion-exchanged with complex MCl2−4 (M □ Ni, Co) ions from organic media, Microporous Materials, 1 (1993) 33-42.
26. C. Li, Y.-W. Chen, Temperature-programmed-reduction studies of nickel oxide/alumina catalysts: effects of the preparation method, Thermochimica Acta, 256 (1995) 457-465.
27. B. Mile, D. Stirling, M.A. Zammitt, A. Lovell, M. Webb, The location of nickel oxide and nickel in silica-supported catalysts: Two forms of "NiO" and the assignment of temperature-programmed reduction profiles, J Catal, 114 (1988) 217-229.
28. A. Kadkhodayan, A. Brenner, Temperature-programmed reduction and oxidation of metals supported on γ-alumina, J Catal, 117 (1989) 311-321.
29. K.Y. Koo, S.-h. Lee, U.H. Jung, H.-S. Roh, W.L. Yoon, Syngas production via combined steam and carbon dioxide reforming of methane over Ni–Ce/$MgAl_2O_4$ catalysts with enhanced coke resistance, Fuel Processing Technology, 119 (2014) 151-157.
30. G. Leofanti, M. Padovan, G. Tozzola, B. Venturelli, Surface area and pore texture of catalysts, Catal Today, 41 (1998) 207-219.
31. Y.H. Hu, E. Ruckenstein, Catalytic Conversion of Methane to Synthesis Gas by Partial Oxidation and CO_2 Reforming, Advances in Catalysis, Academic Press2004, pp. 297-345.
32. Y.-g. Chen, J. Ren, Conversion of methane and carbon dioxide into synthesis gas over alumina-supported nickel catalysts. Effect of Ni-Al_2O_3 interactions, Catal Lett, 29 (1994) 39-48.
33. H.-P. Ren, Y.-H. Song, W. Wang, J.-G. Chen, J. Cheng, J. Jiang, Z.-T. Liu, Z.-W. Liu, Z. Hao, J. Lu, Insights into CeO_2-modified Ni–Mg–Al oxides for pressurized carbon dioxide reforming of methane, Chemical Engineering Journal, 259 (2015) 581-593.

A thermodynamic analysis of chloride corrosion in biomass-fired boilers for Fe-O-Cl-S system

Robert Kaczmarczyk[1,*], *Agata* Mlonka-Mędrala[1], and Sebastian Gurgul[2]

[1]AGH University of Science and Technology, Faculty of Energy and Fuel, 30059 Krakow Mickiewicza 30, Poland
[2] MTU Aero Engines Polska Sp. z o.o., 36002 Jasionka Tajecina 108, Poland

Abstract. The paper presents a thermodynamic analysis of chloride-induced corrosion in the Fe-O-Cl-S system. The influence of steam concentration in the gas phase on chloride-induced corrosion process was presented. Based on the parametric equations the equilibrium concentration of the gas phase was determined. The effect of alkali metals chlorides in gas phase (Na,K)Cl on formation of (Na,K)FeO_2 in the passive oxide scale layer ($FeO/Fe_3O_4/Fe_2O_3$) was analysed. Condensation of (Na,K)Cl vapors, formation of low-melting eutectic mixtures in deposits and consequences of this process on corrosion process were examined. Additionally, the role of SO_2 in chlorination and oxidation process of steel in melted ash deposits was discussed as well. The results were correlated with available in the literature laboratory experimental data and industrial corrosion process observations. Presented thermodynamic analysis was compared with assumptions of an "active oxidation" model. The results may be used for experimental research planning and a corrosion process prevention in the industry.

1 Chloride-induced high temperature corrosion process of steel

The emission limits for fossil fuel combusting units affects significantly the final product cost. [1,2]. Direct biomass combustion is the best alternative technology for fossil fuel combustion. The development of renewable energy sources is one of the priorities of Energy Policy until 2030. However, a limited material strength and corrosion resistance of boiler heat exchanging surfaces at higher temperatures is a major problem associated with biomass combustion and waste incineration. Flue gas and fly ash composition, combustion process parameters (especially flue gas and tube wall temperature), thermal gradient in the deposits and corrosive agent concentration in a wall boundary layer affect the corrosion process mechanism and kinetics. [3,4,5]. Corrosive species concentration in biomass firing units presents as follows: 25-1000 ppm HCl, 0-50 ppm (K,Na)Cl, 0-70 ppm SO_2 [3]. Moisture content in the combustion gas is around 10-20%. A passive oxide scale formed during boiler operation is composed of iron oxides. The rate of oxide scale formation in Fe-O system is rather slow and it depends on a steel grade. In the gas phase, the oxygen concentration does not affect the kinetics of the process. [6,7]. The acceleration of steel degradation process by gaseous HCl in the exhaust gas was observed in numerous laboratory experiments and field

* Corresponding author: robertk@agh.edu.pl

measurements. During biomass combustion chlorine is present mainly as HCl [8]. A so-called active oxidation model describes the steel oxidation process accelerated by chlorine in the exhaust gas. The model is widely described in the literature [9,10]. The fundamental assumptions of the „active oxidation model" are presented in [9]. The high temperature corrosion experiments in a lab-scale ware carried out in simulated atmosphere He-[O_2-HCl-SO_2] and with fly ash deposits (K_2SO_4, $CaSO_4$, (K,Na)Cl). The diffusion of Cl_2 through the pores and cracks of the metal oxide scale, formation of $FeCl_2$(g), which penetrates the scale backward to flue gas and final oxidation form a closed cycle of reactions. Such repeatability causes a severe oxide scale destruction and steel protective properties against chloride corrosion are lost. (Fig. 1)

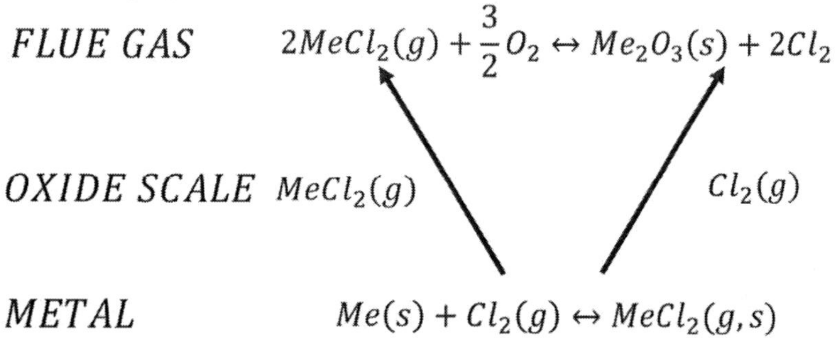

Fig. 1 Cl-induced corrosion cycle described in [9].

A steel oxidation process accelerated by chlorine presence in the flue gas called active oxidation is a possible mechanism of chlorine-induced corrosion process. However, the selectivity of oxide scale layer Fe_3O_4/Fe_2O_3 in diffusion process of gas phase components and neglecting of the other gas phase components influence (e.g. H_2O) on corrosion process rises some doubts among scientists. At high temperature, in reducing atmosphere rich in HCl, the reaction of iron and hydrochloric acid in ash deposits layer was described in [11]. In [6,12-15] the high temperature corrosion process was observed in oxidizing conditions (5% O_2) with 500-3000 ppm HCl and the metal chlorides were formed according to an active oxidation mechanism. In [5] the corrosion process was analysed under solid chlorides mixtures: KCl-$ZnCl_2$ in an oxidizing atmosphere (Ar-O_2). A so-called hot corrosion process was examined in [16-19]. The influence of sulfur dioxide on the corrosion process was analysed in [3,9,20-25]. The reaction of sulfur dioxide with alkali chlorides deposited on heat exchanging surfaces is an additional source of molecular chlorine. [9] The melting temperature of the sulfates (K_2SO_4 -1347K, Na_2SO_4 -1157K) is lower than the heat exchanger wall temperature. Therefore, the alkali sulfates in the fly ash deposits are less corrosive than the chlorides. For molar ratio S/Cl >2-2,2 the chlorine-induced corrosion is no longer a problem [20]. In [24,25] the free alkali index was determined based on elementary analysis of the selected fuels. Free alkali metals (not bounded by the Cl and S in the fuel) will form alkali metal hydroxides. [24] KOH and NaOH seems to be more corrosive than other compounds (sulfates and chlorides).

2 Chloride-induced corrosion process of iron - a thermodynamic analysis

Equilibrium calculations were performed for the Fe-[O-Cl-S-(Na,K)] system based on:
- thermodynamic properties of pure components (i) taking part in the reaction and described by chemical potential $\Delta\mu_i^0 = a + bT + cTln(T)$, the relation was determined based on molar heat capacity function (C_p),

- the Gibbs free energy change $\Delta G_T^0 = -RT\ ln\ K$,
- parametric equations, which determine composition change from initial to equilibrium $x_i^0 \rightarrow x_i$, the gas phase composition changes along a linear function of parameter $\tau \in R$, a vector parallel to the function: $x_i = x_i^0 + \tau\cos(\alpha_i)$ *(i=1,...,s*) determines a transformation of the reactants (i) from initial to final (equilibrium) state [26, 27, 28, 29],
- conditions for thermodynamic equilibrium and diffusive equilibrium,
- thermodynamic databases [30,31] and a numerical computing environment - MATLAB.

The corrosion process model reactions (j) used in the thermodynamic analysis of [O-Cl-S-(Na,K)] system are presented in Tab. 1.

Table 1. Gas phase corrosion process of iron model reactions in [O-Cl-S-(Na,K)] system.

j	reaction	ln K
1	$2HCl + 0.5O_2 \leftrightarrow Cl_2 + H_2O$	0.6225
2	$2NaCl_{(g)} + Fe_2O_3 + 0.5O_2 \leftrightarrow Na_2Fe_2O_4 + Cl_2$	1.4096
3	$2KCl_{(s)} + Fe_2O_3 + 0.5O_2 \leftrightarrow K_2Fe_2O_4 + Cl_2$	122
4	$2NaCl_{(g)} + SO_2 + O_2 \leftrightarrow Na_2SO_{4(s)} + Cl_2$	43.864
5	$2KCl_{(g)} + SO_2 + O_2 \leftrightarrow K_2SO_{4(s)} + Cl_2$	44.501
6	$2NaOH_{(c)} + Fe_2O_3 = Na_2Fe_2O_4 + H_2O$	8.544
7	$2NaCl_{(g)} + H_2O + 0.5O_2 = 2NaOH_{(c)} + Cl_2,$	-10,2072
8	$2NaCl_{(g)} + H_2O + 0.5O_2 = 2NaOH_{(g)} + Cl_2,$	-42.642

Phase stability of iron chlorides, oxides and pure iron depends on chlorine and oxygen concentration in the gas phase. A phase stability diagram for [Fe-O-Cl] system at T=800K is presented in Fig. 2. Low partial pressures of chlorine and oxygen ($P_{Cl_2} = 10^{-16}$ [atm.], $P_{O_2} = 10^{-27}$ [atm.]) protect iron against chloride and oxide corrosion. At partial pressures of oxygen $P_{O_2} = 10^{-8}$ [atm.] and chlorine $P_{Cl_2} = 10^{-3}$ [atm.], formation of chloride phase in contact with pure hematite ($FeCl_2/Fe_2O_3$) might be observed. According to reaction (1), steam and chlorine formation in the gas phase is determined by HCl and oxygen concentration in the exhaust gas. The actual composition of the gas phase was defined by a set of parametric equations [26-29]. The calculations were carried out for initial concentrations of oxygen ($x_{O_2}^0 = 5 \cdot 10^{-2}$), hydrochloric acid ($x_{HCl}^0 = 3 \cdot 10^{-3}$) and nitrogen($x_{N_2}^0 = 0.947$), which correspond to the experimental conditions in [3]. Total pressure of the system was assumed to be 1 atm. (P_{tot}=1 atm). Assuming constant water vapor concentration P_{H2O}=const., the equilibrium concentrations Cl_2=f(O_2) tend to gaseous chloride corrosion with a decrease of moisture in the flue gas. According to an active oxidation model, an activation of chlorine can be observed. [3]

In a full-scale biomass firing units and in laboratory experiments carried out in the conditions similar to the biomass combustion, the oxygen content is around 5% and HCl is 500-3000 ppm. In such conditions, iron (III) oxide - Fe_2O_3 is the stable phase and the oxide corrosion is the main process.

 DOI: 10.1051/e3sconf/20171402040

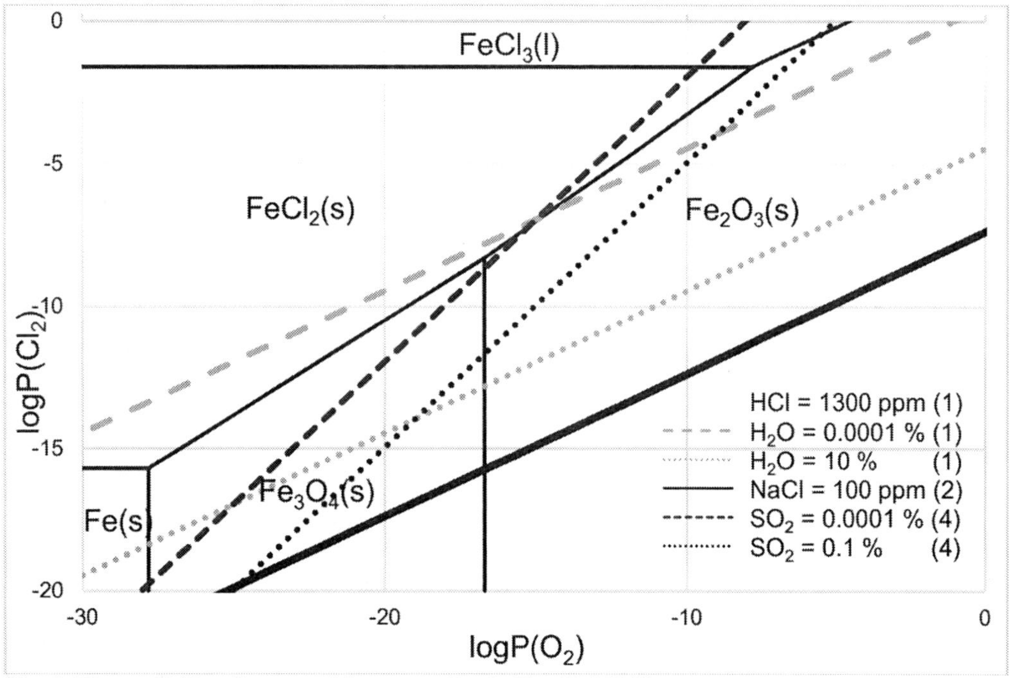

Fig. 2. Phase stability diagram of the system Fe-O-Cl at T=800 K

2.1 The actual composition of the gas phase

The equilibrium composition x_i of the gas phase depends on oxygen initial concentration ($x^0_{O_2}$) according to reaction (1). At 800 K, the equilibrium compositions x_i are shown in Tab.2 and in Fig.3. When the initial oxygen concentration in the combustion gas is reduced, moisture content, oxygen and chlorine equilibrium concentrations decrease. Under reducing conditions the gas phase composition tends to the chloride corrosion process.

Table 2. Deacon reaction gas phase compositions at T=800 K, according to parametric equation calculations [26-29].

x_i^o		lgx_i			
HCl	O_2	O_2	Cl_2	H_2O	HCl
0.003	0.05	-1,3045	-3,0732	-3,0732	-2,8822
	0.01	-2,0151	-3,1599	-3,1599	-2,7913
	0.001	-3,1158	-3,3292	-3,3292	-2,6854
	0.0001	-5,1447	-3,7312	-3,7312	-2,5802
	0.00001	-9,2218	-4,6990	-4,6990	-2,5287
	0.000001	-13,2427	-5,6990	-5,6990	-2,5235

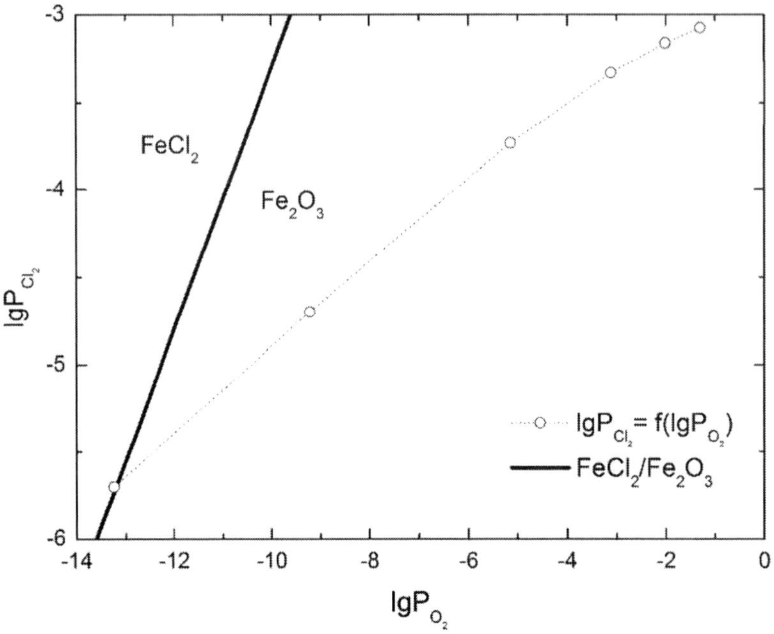

Fig. 3. Phase stability diagram of the system Fe-O-Cl at T=800 K with the equilibrium compositions of the Deacon reaction ($x^0_{HCl} = 0.003$ i $x^0_{O_2} = 0.05 - 10^{-6}$)

2.2 The role of alkali metal chlorides in corrosion process of iron

When $NaCl_{(g)}$ is added under oxidizing conditions, a new [Fe-O-Cl-Na] system is introduced. According to reaction (2), hematite reacts with $NaCl_{(g)}$ and solid $2NaFeO_2$ is formed. [9] Assuming, the activity of solid phase components equal to unity ($a_{2NaFeO_2} = a_{Fe_2O_3} = 1$), for P_{NaCl}=100 ppm, a relation $\lg P_{Cl_2} = f(\lg P_{O_2})$ is presented in Fig. 2. In these conditions, the corrosion proceed within the oxide corrosion range. Solid $2NaFeO_2$ formed on the hematite surface can be destructive for passive oxide layer (Fe_2O_3). However, according to reaction (1) and (2), $2NaFeO_2$ formation is limited by water vapor concentration in the combustion gas. At 800 K, 45 atm. of H_2O in the gas phase will be required for formation of separate phases $2NaFeO_2$ - Fe_2O_3. Although, the mechanism of $2KFeO_2$ formation is different. First, $KCl_{(g)}$ condensate on the oxide scale and next $KCl_{(s)}$ reacts with hematite (3). The steam partial pressure needed for the reaction to proceed is 10^{-50} [atm] only. Therefore, $2KFeO_2$ formation in biomass combustion atmosphere is not limited by water vapor concentration. According to (9), the decrease of hematite activity will intensify the oxide corrosion process. The activity of Fe_2O_3will decrease with water vapor concentration increase and dissolution of $2(Na,K)FeO_2$ in low melting eutectic mixtures:

$$a_{Fe_2O_3} = \frac{K_1}{K_{(2,3)}} \left(\frac{p_{HCl}}{p_{NaCl}}\right)^2 \frac{a_{2(Na,K)FeO_2}}{p_{H_2O}} \tag{9}$$

The reaction mechanism is different when fly ash rich in chlorides ($11PbCl_2$ – $50KCl$ – $39ZnCl_2$ wt.%, T_m=821 K) is deposited on heat exchanging surfaces. The oxide corrosion process is intensified by steam concentration and anticorrosive properties of iron and low-alloy steels are decreased [15].

The corrosion process of iron in atmosphere rich in chlorine and water vapor has a two way mechanism. The increase of steam concentration and low melting ash components formation in the deposits enhance the oxide corrosion attack. When the steam concentration decreases, a corrosive agent – Cl_2 is activated and chloride corrosion rate increases. The dissolution of

2(Na,K)FeO_2 in low melting eutectic mixtures will decrease the activity ($a_{2(Na,K)FeO_2} < 1$) and the oxide corrosion process will be intensified. In biomass-fired boilers, low melting eutectic mixtures are formed in the reaction of $NaCl_{(g)}$ and $KCl_{(g)}$ condensed on the oxide scale with other fly ash components.

2.3 Condensation of alkali metal chlorides

At equilibrium, in multiphase multicomponent system any two phases of the same compound in contact with each other must have the same chemical potential. Therefore, at equilibrium for NaCl and KCl in different phases the relation $\mu_{i(s)} = \mu_{i(g)}$ $(i= NaCl, KCl)$ is true. At T = 800K the vapor pressures of pure (Na, K)Cl in equilibrium with their solid phase are presented in Table 3.

$$RTlnP_i^0 = \mu_{i(s)}^0 - \mu_{i(g)}^0 \qquad (10)$$

where: R=8.314 [J/mol K], T – temperature in [K], P_i^0 equilibrium vapour pressure of „i" component in $[atm]$ (i=NaCl, KCl), $\mu_{i(s)}^0, \mu_{i(g)}^0 [\frac{J}{mol}]$- chemical potentials of „i" component in solid and gas phase.

Table 3. The equilibrium vapor pressures of NaCl and KCl at 800K.

i	T=800 K	
	lnP_i^0	$P_i^0[ppm]$
NaCl	-15.115	0.273
KCl	-15.234	0.242

In biomass-firing units and waste incineration power plants, potassium and sodium chlorides emission varies from 5 to 120 ppm. At T = 800K, the KCl/NaCl equilibrium vapor pressures (Tab. 3) are lower than the partial pressure of the chlorides in the combustion gases. Therefore the alkali metal chlorides condensate on heat exchanging surfaces and can form low melting eutectic mixtures with other fly ash components.

2.4 The role of sodium hydroxide in corrosion process of iron

The theory of corrosion attack by condensed $NaOH_{(c)}$ is not is thermodynamically favored. At T =800K and in the combustion atmosphere $P_{HCl} = 1000\,ppm, P_{NaCl} = 100\,ppm, P_{H_2O} = 10\,\%$, partial pressure of $NaOH_{(g)}$ from reaction (8) and (1) ($lgP_{NaOH_{(g)}} = -11.122$) is lower than equilibrium pressure of the evaporation process $NaOH_{(c)} \rightarrow NaOH_{(g)}$ ($lgP_{NaOH_{(g)}}^0 = -7.043$). Therefore, the $NaOH_{(g)}$ vapors will not condensate on the heat exchanging surfaces. Formation of $NaOH_{(c)}$ from reactions (1) and (7) is limited by moisture content in the flue gas. Assuming liquid oxide phase as an ideal mixture ($a_i \cong x_i$, $i= NaOH,\ Fe_2O_3,\ 2NaFeO_2$), the $NaOH_{(c)}$ concentration . $x_{NaOH_{(c)}}$is 1/10 of the equilibrium concentration form the reaction (6).

2.5 The role of sulfur dioxide in corrosion process of iron

When sulfur dioxide is present in the combustion atmosphere, alkali sulfates are formed according to reactions (4 and 5). Assuming constant conditions: P_{NaCl}=100 ppm., $P_{SO_2} = const.$ and $a_{Na_2SO_4} = 1$, the gas phase compositions $lgP_{Cl_2} = f(lgP_{O_2})$ proceed in the oxide corrosion range (Fig. 2). With an increase of SO_2 concentration, gas phase equilibrium concentrations Cl_2=f(O_2) tends to gaseous chloride corrosion. Therefore, SO_2 presence in gas phase leads to additional chlorine generation and activation of chloride corrosion process according to an active oxidation model.

Sulfur dioxide in the combustion gas affect also the condensation of NaCl and KCl vapors. The condensed alkali metal chlorides can react with other fly ash components and form eutectics of low melting point: NaCl-KCl-$MeCl_2$. The equilibrium pressure of the chloride $NaCl_{(g)}$ or $KCl_{(g)}$ depends on the gas phase equilibrium composition, according to reactions (1) and (4,5):

$$\ln P_{(Na,K)Cl} = \ln P_{HCl} + 0.5(\ln K_1 - \ln K_{(4,5)} - \ln P_{SO_2} - \ln P_{H_2O}) - 0.25 \ln P_{O_2} \quad (11)$$

In a steady-state combustion process conditions: $P_{HCl} = 1300$ ppm, $P_{SO_2} = 100\,ppm$, $P_{H_2O} = 10^{-1} atm$, the equilibrium pressure of the chlorides $NaCl_{(g)}$ and $KCl_{(g)}$ (T=800 K) depends on oxygen partial pressure, according to relations :

$$lgP_{NaCl} = -9.7763 - 0.1086\ lgP_{O_2}$$
$$lgP_{KCl} = -9.9144 - 0.1086\ lgP_{O_2} \quad (12)$$

In oxidizing combustion conditions, alkali sulfates formation according to (4,5) decrease (Na.K)Cl vapor pressure below the equilibrium value needed for the condensation process $(Na,K)Cl_{(g)} \rightarrow (Na,K)Cl_{(s)}$ (Fig. 4).

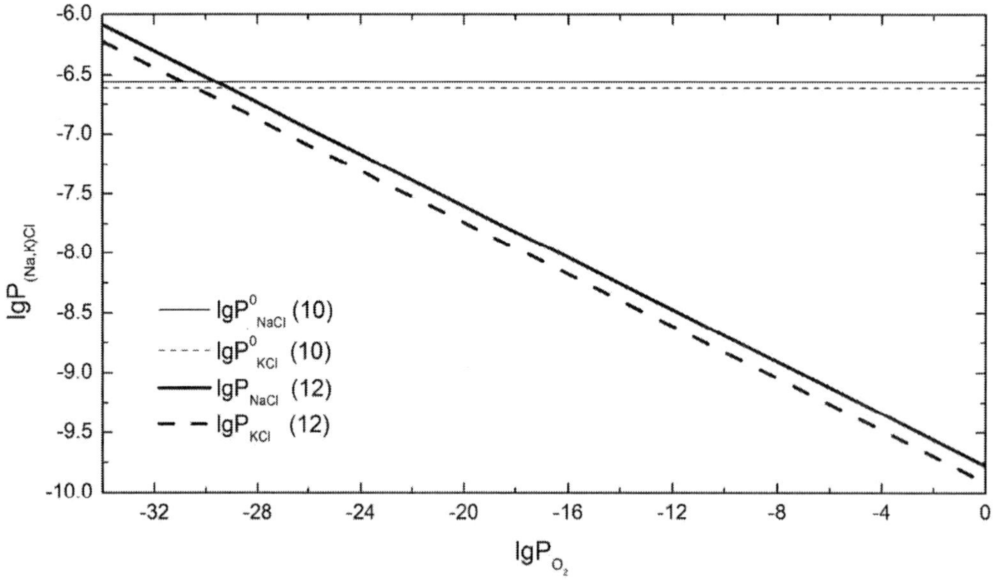

Fig. 4. Vapor pressures of alkali chlorides: NaCl and KCl in the combustion gas, at T=800 K.

Sulfur dioxide inhibits the condensation of alkali chlorides in the combustion gases. Therefore, the risk of low melting ash components formation and ash deposits fusion or partial fusion is decreased. The more stable sulfates are formed in the deposits. The solubility of hematite in fused chlorides is limited and the oxide corrosion rate is reduced. However, in a certain temperature range alkali sulfates and other complex salts mixtures can accelerate corrosion attack by a so-called „hot corrosion" process.

3 Conclusions

Several conclusions come from the presented analysis of high-temperature corrosion in Fe-[O-Cl-S-(Na,K)] system. At T=800K in a oxidizing atmosphere (O_2 - 5%, Cl_2 <1300 ppm), chloride corrosion process mechanism is not thermodynamically favored. Concentration of HCl and oxygen determine water vapor formation. In steady-state conditions, equilibrium concentrations $Cl_2=f(O_2)$ tend to gaseous chloride corrosion with a decrease of moisture in the gas phase and the activation of chlorine according to an active oxidation model can be observed. When deposits rich in chlorides (NaCl/ KCl/$MeCl_2$) are introduced to the system, the influence of the water vapor concentration on chloride corrosion process is minor and the steam accelerates the rate of the oxide corrosion. When chlorine compounds and water vapor are present in the combustion gas the corrosion process of iron has a two-way mechanism. The increase of the water vapor content and formation of low melting components in the deposits accelerates the corrosion attack in the oxide area. A decrease of H_2O partial pressure in the flue gas activates chlorine towards gaseous chloride corrosion. In reducing atmosphere the water vapor concentration in the system decreases. Low steam concentrations promotes gaseous chloride corrosion. Condensation of alkali chlorides is one of the mechanisms of low melting ash deposits components formation on heat exchanging surfaces. At T=800K, partial pressures of potassium and sodium chlorides are higher than vapor pressures of gaseous chlorides in equilibrium with the pure chlorides in the solid phase. Condensed alkali chlorides can form eutectics of low melting point and the oxide corrosion process will be accelerated ($a_{Fe_2O_3} < 1$).

The theory of corrosion attack by condensed $NaOH_{(c)}$ is not is thermodynamically favored. Sulfur dioxide in gas phase leads to additional chlorine generation and activation of chloride corrosion process according to an active oxidation model. In oxidizing combustion conditions, alkali sulfate formation decreases the alkali chlorides vapor pressure below the equilibrium value needed for the condensation process $(Na,K)Cl_{(g)} \rightarrow (Na,K)Cl_{(s)}$. Therefore, sulfur dioxide inhibits the condensation of alkali chlorides in the combustion gas and the risk of low melting ash components formation and ash deposits fusion or partial fusion is decreased. The more stable sulfates are formed in the deposits. The solubility of hematite in fused chlorides is limited and the oxide corrosion rate is reduced. In a certain temperature range alkali sulfates and other complex salts mixtures can accelerate corrosion attack by a so-called „hot corrosion" process.

Presented conclusions can be used for combustion processes optimization as well as in interpretation of steel chloride-induced corrosion experiments.

The present work was supported by the Polish Ministry of Science (Grant AGH No. 11.11.210.312)

References

1. European Commission, *Renewable energy progress report* (Brussels, 2015)
2. *Final meeting of the technical working group (TWG), for the review of the BAT reference document for large combustion plants (LCP BREF), final meeting, background paper (BP)* (Seville, 2015)
3. P.Viklund, *Superheater corrosion in biomass and waste fired boilers. Characterization, causes and prevention of chlorine-induced corrosion* (Sweden, Division of Surface and Corrosion Science, Department of Chemistry, School of Chemical Science and Engineering, 2013)
4. M. Pronobis, Modernizacja kotłów energetycznych (Wydawnictwo Naukowo-Techniczne, Warszawa 2002)
5. T. Hardy, W, Kordylewski, K. Mościcki, *Arch. Spal.*, **9**, 181-195 (2009)

6. M. Spiegel, Beeinflussung der Korrosionsgeschwindigkeit von den Verhaltnissen im rohrwandnahen Bereich bei der HT-Chlor-Korrosion. Vortag VDIWissensforum „Belage und Korrosion“, 14.06.2005 in *Dampferzeugerkorrosion*, M. Born (Freiberg, SAXONIA Standortentwicklungs-und-Verwaltungsgesellschaft mbH, 2005)
7. J. R. Nicholls, R. Newton, N. J. Simms, J. F Norton, *Mater. High Temp.,***20 (2)**, 93-108 (2003)
8. A. Zahs, M. Spiegel, H. J. Grabke, *Mater. Sci. Eng.*, **50**, 561-578 (1999)
9. H. J. Grabke, E. Reese, M. Spiegel, *Corro. Sci.*, **37**, 7, 1023-1043 (1995)
10. M.A. Uusitalo, P.M.J. Vouoristo, T.A. Mäntylä, *Corro. Sci.*, **46**, 1311-1331 (2004)
11. K. Salmenoja, M. Hupa, R. Backman, Laboratory Studies on the Influence of Gaseous HCl on Fireside Corrosion of Superheaters in *Impact of Mineral Impurities in Solid Fuel Combustion*, edited by R. P. Gupta, T. F. Wall, L. Baxter (Springer US, New York, 1999)
12. D. Bramhoff u.a.: Einfluss von HCl und Cl_2 auf die Hochtemperaturkorrosion des 2,25Cr1Mo-Stahls in Atmospharen mit hohen Sauerstoffdrucken. Werkstoffe und Korrosion in *Dampferzeugerkorrosion*, M. Born (Freiberg, SAXONIA Standortentwicklungs-und-Verwaltungsgesellschaft mbH, 2005)
13. M. Spiegel,, *Mater. Corro.*, **50**, 373-393 (1999)
14. M.A. Uusitalo, P.M.J. Vuoristo, T.A. Mantyla, *Mater. Sci. Eng.*, **346**, 168-177 (2003)
15. M. Sanchez Pasten, M. Spiegel, *Mater. Corro.*, **57**, 2, 192-192 (2006)
16. K. Zhang, Y. Niu, C. Zeng, W. Wu, *J. Mater. Sci. Technol.*, **20**, 2, 213-216 (2004)
17. B. J. Skrifvars, M. Westén-Karlsson, M. Hupa, K. Salmenoja, *Corro. Sci.*, **52**, 1011–1019 (2010)
18. B. J. Skrifvars, R. Backman, M. Hupa, K. Salmenoja, E. Vakkilainen, *Corro. Sci.*, **50**, 1274–1282 (2008)
19. Y. S. Li, Y. Niu, W.T. Wu, *Mater. Sci. Eng.*, **A345**, 64, 64-71 (2003)
20. J. Waltl, N. Rechberger, *VGB PowerTech*, **3**, 48-52 (2006)
21. M. Aho, E. Ferrer, *Fuel,* **84**, 201-212 (2005)
22. R.W. Bryers, Factors critically affecting fireside deposits in steam generators in *Impact of Mineral Impurities in Solid Fuel Combustion*, edited by R. P. Gupta, T. F. Wall, L. Baxter (Springer US, New York, 1999)
23. S.C. Srivastava, K.M. Godiwalla, M.K. Banerjee, *J. Mater Sci*, **32**, 835-849 (1997)
24. T. Blomberg, *Mater. Corro.*, 57, 170-175 (2006)
25. M. Born, *VGB PowerTech,* **5**, 107-111 (2005)
26. W. Ptak, M. Sukiennik, *Bulletin de l'Academie Polonaise des Sciences. Serie des sciences techniques*, **17**, 21-25 (1969)
27. W. Ptak, M. Sukienni, R. Olesinski, R. Kaczmarczyk, *Arch. Metall.*, **32**, 355-362 (1987)
28. R. Kaczmarczyk, S. Gurgul, *Arch. Metall. Mater.*, **59**, 4, 1379-1383 (2014)
29. R. Kaczmarczyk, S. Gurgul S, *Arch. Metall. Mater.*, **59**, 1, 145-148 (2014)
30. O. Knacke, O. Kubaschewski, K. Hesselmann, *Thermochemical Properties of Inorganic Substances* (Berlin, Springer, 1991)
31. Outokumpu HSC Chemistry, *Chemical Reaction and Equilibrium Software*, (Outokumpu Research Oy, Finland, 2002) ISBN 952-9507-08-9.

E3S Web of Conferences **14**, 02041 (2017) DOI: 10.1051/e3sconf/20171402041
Energy and Fuels 2016

Methanation of carbon dioxide by hydrogen reduction – a thermodynamic analysis

Robert Kaczmarczyk[1,*], *Agata* Mlonka-Mędrala[1] , and *Sebastian* Gurgul[2]

[1]AGH University of Science and Technology, Faculty of Energy and Fuel, 30-059 Krakow Mickiewicza 30, Poland
[2] MTU Aero Engines Polska Sp. z o.o., 36-002 Jasionka Tajecina 108, Poland

Abstract. The paper presents a thermodynamic analysis of the methanation of carbon dioxide by hydrogen reduction. Equilibrium gas phase composition was determined by means of parametric equations. Calculations were performed for the temperature range T = 500 - 700 K and the initial composition $x_{CO_2}^0 \in (0-1)$.. A crucial parameter for catalytic processes, carbon precipitation range $[C] = f(T, x_{CO_2}^0)$was presented as a function of temperature and an initial gas phase CO_2-H_2 composition. The CO_2 conversion efficiency and the methane yield in the process was determined. Obtained results may be used for experimental research prediction and in industrial pragmatic.

1 Introduction

A catalytic reduction of carbon oxides by hydrogen is commonly used in syngas treatment. Methane is the main product of the process. In a so-called methanation process, the overall content of carbon oxides in the syngas is reduced from 0.2-0.5%v to less than 2 ppmv [1]. The syngas treatment is crucial in the synthesis of ammonia. [2] Carbon monoxide appears to be destructive for the catalysts in the process. The methanation process can be used to reduce the CO content in the synthesis gas. The most commonly used catalyst in the methanation technologies is a nickel oxide on alumina support [3]. However, ruthenium catalysts possess higher activity than nickel oxide. The most common catalyst supports are: Al_2O_3, MgO, $MgAl_2O_4$, SiO_2, FeO [4,5] and activated carbon [6]. In the power industry, methanation process may be used for waste CO_2 utilization from fossil-fuel combusting units (CCU). In the coal-gasification process, synthesis gas and hydrocarbons from further conversion (eg. in Fischer-Tropsch synthesis) are produced [7-19]. However, these technologies generates high CO_2 emissions and affect the final product cost. [20, 21]. The innovative technologies for hydrocarbons generation from a mixture of carbon dioxide and hydrogen tie a valuable product generation with a waste product utilization. Thermodynamics and kinetics of the hydrocarbon synthesis form carbon dioxide and hydrogen mixture ware widely described in [22, 23]. However, only mixtures of the stoichiometric composition under standard pressure were analyzed. The synthesis of hydrocarbons was considered as a kinetically controlled polymerization process of the - $[CH_2]$ – groups on the catalyst surface.

* Corresponding author: robertk@agh.edu.pl

Under favorable conditions the final composition of the synthesis product will be correlated with a so-called ASF distribution (Anderson-Schulz-Flory) with the growth probability of hydrocarbons (α) [24,25]. The ASF equation is used for description of the hydrocarbon synthesis reaction form CO-H_2 and CO_2-H_2 mixtures. In [26-28] the detailed mechanisms of the carbon dioxide hydrogenation are presented.

2 A thermodynamic analysis of carbon dioxide methanation process by means of parametric equations method

A characteristic of carbon dioxide methanation process was accomplished based on thermodynamic fundamental principles, thermodynamic properties of pure substances taking part in the reaction and properties of solutions formed in the gas phase. For a given chemical reaction, the Gibbs free energy change ΔG_T^0 was determined:

$$\Delta G_T^0 = \sum_{i=1}^{m_1} k_{B_i} \cdot \mu_{B_i}^0 - \sum_{i=1}^{m_2} k_{A_i} \cdot \mu_{A_i}^0 \quad (1)$$

Where: A_i, $i = 1, \ldots m_1$ and B_i, $i = 1, \ldots m_2$ are reactants and products respectively, k_{Ai} and k_{Bi} are stoichiometric coefficients, $\mu_{A_i}^0$ and $\mu_{B_i}^0$ represents standard chemical potential of pure substances. The relation between Gibbs free energy change and thermodynamic equilibrium constant K is presented below:

$$\Delta G_T^0 = -RT \ln K \quad (2)$$

where: K represents the tendency of the process and is determined by reagents activities a_i.

$$K = \frac{\prod_{i=1}^{s} a_{B_i}^{k_{B_i}}}{\prod_{i=1}^{s} a_{A_i}^{k_{A_i}}} \quad (3)$$

The activity of pure components in solid or liquid phase is equal to unity. The activities of chemical species in the gas phase can be determined by the components partial pressures $P_i = x_i P_{tot}$ (P_{tot} – the total system pressure). In isothermal-isobaric conditions, the composition of the gas phase changes linearly and a conversion from initial to final (equilibrium) state can be described by a vector parallel to the linear function [29-32]:

$$x_i = x_i^0 + \tau \frac{k_i - x_i^0 \sum_{i=1}^{s} k_i}{\sqrt{\sum_{i=1}^{s} (k_i - x_i^0 \sum_{i=1}^{s} k_i)^2}} \quad (4)$$

where x_i, x_i^0 – equilibrium and initial molar fraction of „i" component, $\tau \in R$ – straight line parameter, k_i – „i" reagent stoichiometric coefficient. Positive stoichiometric coefficients (k_i) are defined for products and negative for reactants. The direction cosines are time independent and they are functions of initial composition and stoichiometric coefficients only. In the reaction($\sum_{i=1}^{s} k_i \neq 0$) the phase composition change proceed linearly and the lines are intersecting at a so-called "characteristic point" ($\overline{x_i}$). In case of the reaction ($\sum_{i=1}^{s} k_i = 0$), the direction cosine depends on stoichiometric coefficients only. Initial composition does not affect the direction cosine and a change of reagents concentration proceed linearly and the lines are parallel to each other. The presented computational process analysis was carried out based on thermodynamic data [33] and a numerical computing environment - MATLAB.

The catalytic carbon dioxide methanation process over a nikel-based catalyst may be described as follows:

$$(CO_2 - H_2) \xrightarrow[Ni]{T} (CH_4, H_2O, H_2, CO_2, CO) \tag{5}$$

The main products the methanation process are methane, water vapor and other contaminations like: unreacted carbon dioxide, hydrogen and carbon monoxide. The model reactions of the methanation process are presented in Tab. 1.

Table 1. Model reactions (j=3) for methanation CO_2-H_2 process.

j	Reaction	$\sum_{i=1}^{5} k_i$
1	$CO_{2(g)} + 4H_{2(g)} \leftrightarrow CH_{4(g)} + 2H_2O_{(g)}$	-2
2	$CH_{4(g)} + H_2O_{(g)} \leftrightarrow CO_{(g)} + 3H_{2(g)}$	2
3	$CO_{(g)} + H_2O_{(g)} \leftrightarrow CO_{2(g)} + H_{2(g)}$	0

Table 2. The parametric equations for equilibrium concentrations calculations of the methanation process

i	$x_i = \begin{cases} dla \sum_{i=1}^{5} k_i \neq 0 \Rightarrow x_i^0 + (\bar{x}_i - x_i^0)\tau, \quad gdzie \quad \bar{x}_i = \frac{k_i}{\sum_{i=1}^{5} k_i} \\ dla \sum_{i=1}^{5} k_i = 0 \Rightarrow x_i^0 + \tau \frac{k_i}{\sqrt{\sum_{i=1}^{5}(k_i)^2}} \end{cases}$		
	j=1	j=2	j=3
CH_4	$x_{CH_4}^o + (-0.5 - x_{CH_4}^0)\tau$	$x_{CH_4}^o + (-0.5 - x_{CH_4}^0)\tau$	$x_{CH_4}^o$
H_2O	$x_{H_2O}^o + (-1 - x_{H_2O}^0)\tau$	$x_{H_2O}^o + (-0.5 - x_{H_2O}^0)\tau$	$x_{H_2O}^o - 0.5\tau$
H_2,	$x_{H_2}^o + (2 - x_{H_2}^0)\tau$	$x_{H_2}^o + (1.5 - x_{H_2}^0)\tau$	$x_{H_2}^o + 0.5\tau$
CO_2,	$x_{CO_2}^o + (0.5 - x_{CO_2}^0)\tau$	$x_{CO_2}^o - x_{CO_2}^0\tau$	$x_{CO_2}^o + 0.5\tau$
CO	$x_{CO}^o - x_{CO}^0\tau$	$x_{CO}^o + (0.5 - x_{CO}^0)\tau$	$x_{CO}^o - 0.5\tau$

The parametric equations for all model reactions and compounds $(CH_4, H_2O, H_2, CO_2, CO)$ are presented in Tab. 2. For a given initial composition x_i^0 (i=CO_2, H_2) equilibrium concentration x_i for the first model reaction was determined. The τ-parameter was calculated by substitution of the corresponding parametric equations to the equilibrium constant K formula. For the model reaction (j=1) equilibrium constant is defined as:

$$K_1 = \frac{p_{CH_4} p_{H_2O}^2}{p_{CO_2} p_{H_2}^4} = \frac{x_{CH_4} x_{H_2O}^2}{x_{CO_2} x_{H_2}^4} P^{-2} \;\; ; \; where \; P = 1\,[atm] \tag{6}$$

$$K_1 = \frac{\left[x^o_{CH_4}+(-0.5-x^0_{CH_4})\tau\right]\cdot\left[x^o_{H_2O}+(-1-x^0_{H_2O})\tau\right]^2}{\left[x^o_{CO_2}+(0.5-x^0_{CO_2})\tau\right]\cdot\left[x^o_{H_2}+(2-x^0_{H_2})\tau\right]^4} \quad (7)$$

For a given initial gas composition (e.g.: $x^o_{CO_2} = x^o_{H_2} = 0.5,\ x^o_{CH_4} = x^o_{H_2O} = 0,$) a function is written as follows:

$$5.0625K_1\tau^4 + (6.75K_1{+}1)\tau^3 + 3.375K_1\tau^2 + 0.75K_1\tau + 0.625K_1 = 0 \quad (8)$$

The function has 4 roots $(\tau_1, \tau_2, \tau_3, \tau_4)$, equilibrium composition is calculated for each root by means of parametric equations for all components of the gas phase. A given τ-parameter has a physical meaning when: $0 \le x_i \le 1$ and $\sum_{i=1}^{5} x_i = 1$. The equilibrium composition of the first reaction determines the initial composition of the second one and so on. Calculations were carried out in a loop, till the final composition fulfills all equilibrium constants for all the model reactions. In the methanation process, carbon precipitate is formed according to reactions in Tab. 3. The formation of pure carbon precipitate is limited by gas phase components concentration ratios(α, β, γ). $K_\alpha, K_\beta, K_\gamma$ are the reactions equilibrium constants. The activity of the formed carbon precipitate is equal to unity ($a_{[C]}$=1).

Table 3. An interpretation of the carbon precipitation process in the gas phase $(CH_4, H_2O, H_2, CO_2, CO)$.

rection	α	β	γ	$a_{[C]}$=1
$CH_4 \leftrightarrow [C] + 2H_2$	$x^2_{H_2}/x_{CH_4}$			$[C] = \begin{cases} M \le K_\alpha \\ B \ge K_\beta \\ R \le K_\gamma \end{cases}$
$[C] + CO_2 \leftrightarrow 2CO$		x^2_{CO}/x_{CO_2}		
$CO + H_2 \leftrightarrow [C] + H_2O$			$x_{H_2O}/x_{CO}x_{H_2}$	

The analysis of methanation process in the thermodynamic equilibrium conditions was carried out for a temperature range 500 – 700 K and for initial concentrations $x^0_{CO_2} = 0 \div 1$. For the obtained results, the carbon precipitation range as a function of the temperature and the initial gas phase composition was presented in Fig. 1.

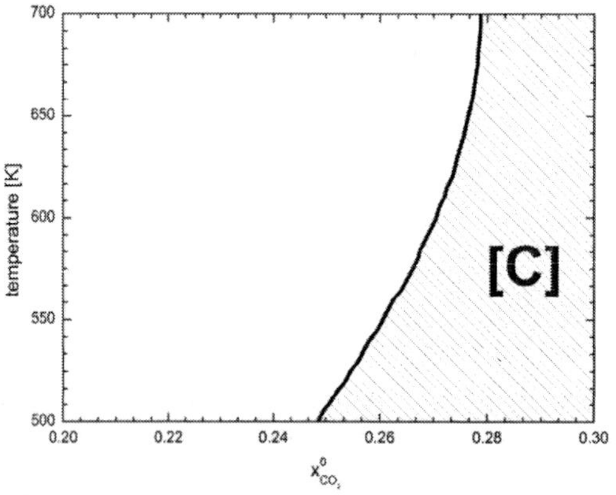

Fig. 1. The carbon precipitation range $[C] = f\left(T, x^0_{CO_2}\right)$ as a function of temperature and an initial gas phase CO_2/H_2 composition.

 DOI: 10.1051/e3sconf/20171402041

In the analyzed temperature range, carbon precipitation starts in the initial conditions: $x^0_{CO_2} > 0.248$ at 500 K and $x^0_{CO_2} > 0.279$ at 700 K. Further analysis was limited to carbon precipitation free range. The gas phase equilibrium concentrations in the analyzed process for initial compositions $x^0_{CO_2} = 0 \div 0.3$ are shown in Fig. 2. At T = 500K , when the CO_2 initial concentration increases, the equilibrium concentrations of CH_4 and H_2O increase as well, until the maximum concentration is achieved $(CH_4 = 32.38\ \%\ ;\ H_2O = 64.77\ \%\ )$. It represents a so-called stoichiometric methanation process $\left(x^0_{CO_2} = 0.2\,;\ x^0_{H_2} = 0.8\right)$. Hydrogen equilibrium concentration in under-stoichiometric range of concentrations $\left(x^0_{CO_2} < 0.2\right)$ decreases from 80 % to 2.3 % $\left(x^0_{CO_2} = 0.2\right)$. In these conditions, there is no CO and CO_2 in the equilibrium gas phase. In the over-stoichiometric range of concentrations $\left(x^0_{CO_2} > 0.2\right)$ concentration of CH_4 and H_2O is decreased, concentration of H_2 is stable and concentration of CO_2 is increased. The CO content in the gas phase changes around 1-11 ppm $(x^0_{CO_2} = 0.2 - 0.24)$. The maximal value of the equilibrium concentration functions $x_{(CH_4\ ,H_2O)} = f\left(x^0_{CO_2}\right)$ decreases with the temperature increase. At T=700 K the maximal concentrations of methane and water vapor are lower than the stoichiometric concentrations at 500 K and are equal to $CH_4 = 23.81\ \%\ ;\ H_2O = 47.89\ \%$. The increase of equilibrium concentrations of hydrogen, carbon dioxide and carbon monoxide $(H_2 = 22.59\ \%\ ;\ CO_2 = 5.45\ \%, CO = 0.27\ \%)$ is observed in the initial conditions $x^0_{CO_2} \geq 0.2$. Therefore, the methanation process proceed more effectively at lower temperatures. At 500K an increase of the CO_2 conversation efficiency (η_{CO_2}) and methane yield (y_{CH_4}) might be observed (Fig. 3).

$$\eta_{CO_2} = \frac{n^0_{CO_2} - n_{CO_2}}{n^0_{CO_2}} \tag{9}$$

$$y_{CH_4} = \eta_{CO_2} \frac{n_{CH_4}}{n_{CH_4} + n_{CO}} \tag{10}$$

where: n^0, n – methanation process products and reactants number of moles

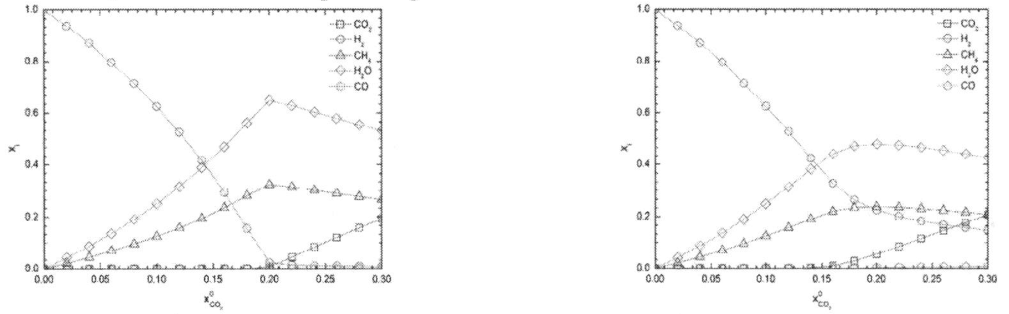

a.) $T = 500\ K$ b.) $T = 700\ K$

Fig. 2. Equilibrium concentrations $x_i = f\left(x^0_{CO_2}\right)$ of the gas phase (i=$CH_4, H_2O, H_2, CO_2, CO$) for the initial concentrations $x^0_{CO_2} = 0 \div 0.3$

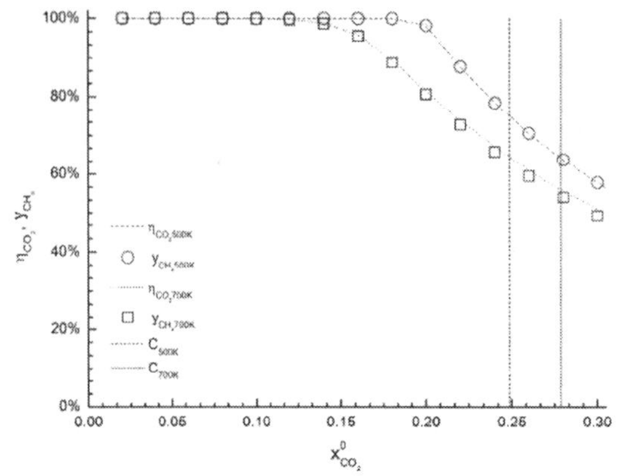

Fig. 3. CO_2 conversion efficiency, methane yield and carbon precipitation range for methanation CO_2/H_2 process within the temperature range T=500 -700 K, $x^0_{CO_2} = 0 \div 3$.

The best results in methanation process of carbon dioxide by hydrogen are observed for a so-called stoichiometric initial gas phase composition $\left(x^0_{CO_2} = 0.2\,;\, x^0_{H_2} = 0.8\right)$. The stoichiometric composition is beyond the carbon precipitation range. Furthermore, the maximal methane concentration, a minimal CO, CO_2 and H_2 concentration and 98 % of CO_2 conversion efficiency is obtained. The temperature increase of stoichiometric methanation CO_2/H_2 process decreases the methane concentration, methane yield and conversion efficiency of the process. However an increase of CO, CO_2 and H_2 concentration in the process gases is observed. The characteristics of the stoichiometric methanation process is presented in Fig. 4.

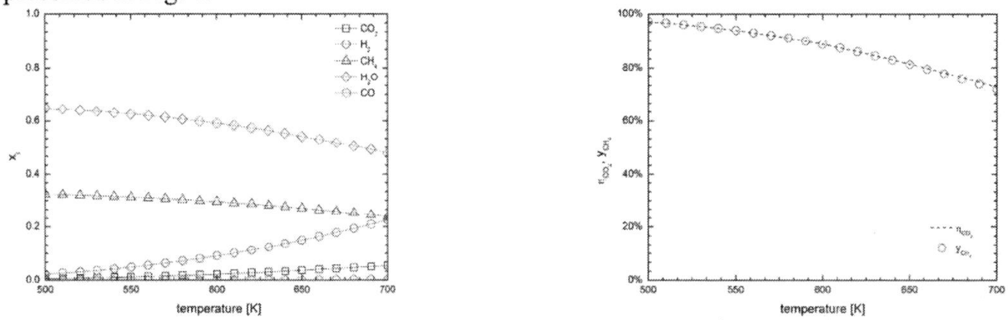

$a.)\; x_i = f(T)\,, (i = CH_4, H_2O, H_2, CO_2, CO)$ b.) $\left(\eta_{CO_2}, y_{CH_4}\right) = f(T)$

Fig. 4. The stoichiometric $\left(x^0_{CO_2} = 0.2\,;\, x^0_{H_2} = 0.8\right)$ methanation CO_2/H_2 process characteristic in a temperature function.

3 Conclusions

The maximal methane concentration in the reaction products is observed for the initial concentrations $\left(x^0_{CO_2} = 0.2\,;\, x^0_{H_2} = 0.8\right)$ coming from the stoichiometry of the reaction (j=1). Above the initial stoichiometric concentration, a decrease of an equilibrium concentration of methane, methane yield and CO_2 conversion efficiency is observed. The temperature decreases the efficiency of methanation process. The increase of temperature and initial concentration $x^0_{CO_2}$ promotes CO formation in the process gases. According to thermodynamics, in the analyzed temperature range, the carbon precipitation appears to be a problem for the initial concentration $x^0_{CO_2} > 0.248$ at 500 K and $x^0_{CO_2} > 0.279$ at 700 K.

The presented thermodynamic analysis in equilibrium conditions shows the direction and the limit of the gas phase composition, the real technological process will tend to.

The present work was supported by the Polish Ministry of Science (Grant AGH No. 11.11.210.312)

References

1. A. Gołębiowski, K. Stołecki, *Przem. Chem.* **80/11** (2001)
2. W. Bobrownicki, S. Pawlikowski, *Technologia Związków Azotowych* (WNT, Warszawa, 1974)
3. A.L. Kohl, F.C. Riesenfeld, *Oczyszczanie gazu* (WNT, Warszawa, 1965)
4. Z. Kowalczyk, K. Stołecki, W. Raróg-Pilecka, E. Miśkiewicz, E. Wilczkowska, Z. Karpiński, *Appl. Catal., A*, **342**, 35-39 (2008)
5. W.K. Jóźwiak, T.P. Maniecki, *Przem. Chem.*, **82/8-9** (2003)
6. Z. Kowalczyk, K. Stołecki, W. Raróg-Pilecka, E. Miśkiewicz, E. Wilczkowska, Z. Karpiński, *Pol. J. Chem.*, **82**, 607-612 (2008)
7. C.S. Song, W. Pan, *Catal. Today*, **98**, 463-484 (2004)
8. B.H. Davis, *Catal. Today,* **141**, 25-33 (2009)
9. B.H. Davis, *Fuel Process. Technol.*, **71**, 157-166 (2001)
10. B.H. Davis, M.L. Ocelli, *Fischer-Tropsch Synthesis, Catalysts and Catalysis* (Studies in Surface Science and Catalysis, Vol. 163, The Netherlands, 2006)
11. B.H. Davis, *Top. Catal.,* **32**, 143-162 (2005)
12. K. Dreszer, L. Więcła-Solny, *Energ. Policy,* **10**, 189-201 (2007)
13. J. Hong, P.A. Chernavskii, A.Y. Khodakov, W. Chu, *Catal. Today,* **140**, 136-141 (2009)
14. A.Y. Krylova, E.A. Kozyukov, *Solid Fuel Chem.*, **41**, 335- 341 (2007)
15. W. Liu, J. Hu, Y. Wang, *Catal. Today,* **140**, 142-148 (2009)
16. Z-T. Liu, J-L. Zhou, B-J. Zhang, *J. Mol. Catal.*, **94**, 255-261 (1994)
17. M. Luo, H. Hamdeh, B.H. Davis, *Catal. Today* **140**, 127-134 (2009)
18. Y. Ohtsuka, T. Arai, S. Takasaki, N. Tsobouchi: *Energ. Fuel.,* **17**, 804-809 (2003)
19. J. Patzlaff, Y. Liu, C. Graffmann, J. Gaube, *Appl. Catal.,* **A 186**, 109-119 (1999)
20. J. Wozniak, *Energ. Policy*, **15**, 139-148 (2012)
21. H. Majchrzak, W. Krawczyk, *Energetyka,* **652** (2008)
22. X. Xiaoding, E.B.M Doesburg, J.J.F. Scholten, *Catal. Today,* **2**, 125-170 (1987)
23. Fujimoto K., Shikada T, *Appl. Catal,* **31**, 13-23 (1987)
24. G. R. Couch, *Coal to liquids.* (IEA Clean Coal Centre, nr CCC/132, 2008)
25. M. Stelmachowski M., Nowicki L.: Applied Energy **74**, 85-93 (2003)
26. Skrzyńska E., Ogonowski J.: Nafta-Gaz, **8**, 561-570 (2009)
27. Skrzyńska E., Ogonowski J.: Nafta-Gaz, **8**, 601-608 (2009)
28. Skrzyńska E., Ogonowski J.: Nafta-Gaz, **1**, 34-42 (2010)
29. W. Ptak, M. Sukiennik, *Bulletin de l'Academie Polonaise des Sciences. Serie des sciences techniques*, **17**, 21-25 (1969)
30. W. Ptak, M. Sukiennik, R. Olesinski, R. Kaczmarczyk, *Arch. Metall.*, **32**, 355-362 (1987)
31. R. Kaczmarczyk, S. Gurgul, *Arch. Metall. Mater.*, **59**, 4, 1379-1383 (2014)
32. R. Kaczmarczyk, S. Gurgul S, *Arch. Metall. Mater.*, **59**, 1, 145-148 (2014)
33. O. Knacke, O. Kubaschewski, K. Hesselmann, *Thermochemical Properties of Inorganic Substances* (Berlin, Springer, 1991)

Modification of lithium titanate spinel by d-electron metals

Danuta Olszewska[1,*] and *Anna* Rutkowska[1]

[1]AGH University of Science and Technology, Faculty of Energy and Fuels, al. A. Mickiewicza 30, 30-059 Cracow, Poland

Abstract. A simply and cheap solid state synthesis was used to produce powders of spinel phase $Li_{4-x}Cu_xTi_5O_{12}$, where $0 \leq x \leq 0.2$, with crystallite size in ~890 nanometers range. The as-prepared samples were verified by scanning electron microscopy and X-ray diffraction. The electrochemical performance of these samples was examined by galvanostatic and voltamperometric tests. To determine transport properties an impedance spectroscopy tests were obtained. These measurements showed excellent high-rate performance and remark- ably good cyclability of the fabricated powders. Capacity retention in $Li_{3.85}Cu_{0.15}Ti_5O_{12}$ has 77% theoretical capacity after raise the discharge current from 1C to 10C and there is less than 2% of capacity loss after 50 charge/discharge cycles at 1C current rate.

1. Introduction

The world population is constantly growing, particularly in undeveloped countries, they usually consume energy in improper way. It is caused by having an inefficient energy systems and outdated devices. However, well developed countries are equally guilty. Modern society buy more and more electronic devices, drive power electric vehicles, industry is based on electricity instead of human work. All of this operations cause that world's energy sources are still decreasing. Moreover, some of the most important issues are global warming and air pollution, which are connected with combustion of traditional fuels [1]. The key to replacement of fossil fuels and traditional energy sources is to create effective systems of energy storage from renewable sources [2, 3]. Energy storage technologies - such as various types of batteries, electrochemical capacitors, pumped hydro, compressed air energy storage, flywheels etc., are very useful for many applications: backup power, energy management, load levelling, frequency regulation, voltage support, and grid stabilization [4, 5]. Unfortunately, not every type of storage is suitable for every type of application because of its special characteristics. An energy storage system should be safe, reliable, as cheap as possible and acceptable to industry. Lithium-ion batteries seem to be an attractive candidate to this goal [6]. Li-ion batteries have high gravimetric and volumetric energy, high power density and long life [7, 8]. Lithium titanate spinel - $Li_4Ti_5O_{12}$ (sign. LTO), has very good electrochemical properties and it is an attractive anode material in Li-ion batteries [6, 9]. LTO has high working potential (~1.55V vs. Li/Li^+) due to the redox couple of Ti^{3+}/Ti^{4+}[5, 10]. It

* Corresponding author: dolszew@agh.edu.pl

avoids the formation of SEI, contrary to graphite. Lithium titanate is environmentally friendly, inexpensive, of long cycle life, overcharge resistance and wide temperature work range [11]. The most crucial issue is the fact that LTO is a zero-strain material, what means that during the charge/discharge process it doesn't show any change of lattice parameter at all [12]. According to the reaction:

$$3Li + Li_4Ti_5O_{12} = Li_7Ti_5O_{12} \quad (1)$$

theoretical nondoped LTO capacity is around 175 mAh·g^{-1}. Despite all of these advantages $Li_4Ti_5O_{12}$ is an insulator with low electronic conductivity (10^{-13} Scm^{-1}) [13]. Furthermore, spinel $Li_4Ti_5O_{12}$ shows a very small volume variation during the charge/discharge process, consequently resulting in excellent cycle stability. Due to this fact, numerous researchers have been trying to improve its transport properties, by going to nanoscale [7, 14–20] or/and doping [21–27]. The second way is to dope with cations/anions in Li, Ti or O site; modification with d-electron metal could enhance electronic conductivity and reduce the electrode polarization. In this paper spinel was doped $Li_4Ti_5O_{12}$ with Cu^{2+} ions.

2. Experimental

2.1 Synthesis of main electrode materials

All samples were prepared using a one-step solid-state reaction method. In all of synthesis processes Li_2CO_3 (Aldrich, 99.8%), and TiO_2-anatase (Acros, 99+%) powders as Li^+ and Ti^{4+} sources were used. Moreover, in some powders CuO (Acros, 99+%) as a Cu^{2+} source was used. The mixed reactants were ball-milled for 20 min with a propanol (POCH, pure) added. After drying, powders were calcinated at 900°C for 4 h in a tube furnace in air atmosphere. In order to eliminate TiO_2 phase in samples, we used 105% of calculated stoichiometric Li_2CO_3 weight.

2.2 Structure and morphology

The phase characteristics and structural studies of the materials were determined by X-ray diffraction in 10-110 deg range with CuKα radiation (λ=1.5406Å), using Panalytical Empyrean diffractometer. Rietveld analysis of XRD data were done using GSAS/EXPGUI set of software [28–30]. Morphology and structure of samples were analyzed by NanoSEM 200 FEI scanning electron microscope equipped with low vacuum detector (LVD).

2.3 Preparation of lithium-ion batteries

The electrochemical tests were prepared using CR2032 coin-type cells. The working electrodes were prepared by pasting a slurry containing 70 wt.% active material, 15 wt.% graphite, 10 wt.% carbon black and 5 wt.% polyvinylidene fluoride (PVDF) dissolved in N-methyl-2-pyrrolidone onto an aluminum foil. After coating, the slurry was dried in a vacuum dryer in 70°C for 1h. Batteries were assembled in the glove-box (UNILAB, M. Braun) under argon atmosphere with controlled oxygen and water vapor pressure (<0.1 ppm) using a metal lithium foil as a counter electrode and 1M $LiPF_6$ solution in the 1:1 mixture of ethylene carbonate (EC) and dimethyl carbonate (DEC) as a electrolyte.

2.4 Electrochemical and transport measurements

The electrochemical characterizations were obtained using galvanostatic charge/discharge cycles at different rates over a voltage range of 1.3-2.2 V. The C rate was calculated based on tested-electrode weight and theoretical capacity of LTO. Cells were tested at computer-controlled galvanostat KEST 32k multichannel and ATLAS measurement. All of electrochemical tests were carried out at room temperature. Cyclic voltammetry studies were carried out using ATLAS measurement with scanning rates in the range of 0.01–5 mV s^{-1} and in 1.3–2.0 V voltage range.
Electrical conductivity of the synthesized materials was measured on disk-shaped sinters by impedance spectroscopy method in a frequency range of 10^{-1} to 10^{6} Hz, using Solartron SI 1260 Frequency Response Analyzer. The obtained data were analyzed using ZView 2.0 software (Scribner Associates).

3. Results and discussion

3.1 Structure

Names of samples with their theoretical calculated capacity and volume of cells are shown in **Table 1**. In this Table synthesis conditions are presented as well. **Fig. 1.** presents the diffraction patterns of the tested powders. **Fig. 2** presents SEM images of sample $Li_{3.85}Cu_{0.15}Ti_5O_{12}$: magnification · 10000 (left) and · 50000 (right) respectively.

Table 1. Specification of $Li_4Ti_5O_{12}$ samples doped with copper. Chemical formula, preparation conditions, structure parameters (space group, cell volume) and theoretical capacity.

	$Li_4Ti_5O_{12}$	$Li_{3.99}Cu_{0.01}Ti_5O_{12}$	$Li_{3.95}Cu_{0.05}Ti_5O_{12}$	$Li_{3.9}Cu_{0.1}Ti_5O_{12}$	$Li_{3.85}Cu_{0.15}Ti_5O_{12}$	$Li_{3.8}Cu_{0.2}Ti_5O_{12}$
Cu content	0	0.01	0.05	0.1	0.15	0.2
Theoretical capacity [mAh/g]	175.17	174.93	174.07	173.01	171.95	170.93
Cell volume [Å^3]	584.35	584.54	584.53	581.67	584.79	584.74
Preparation conditions						
Substrates	$Li_2CO_3 + TiO_2$	$Li_2CO_3 + TiO_2 + CuO$				
Milling time [min]	20					
Temperature [°C]	900					
Heating time [h]	4					

In order to detect phase composition in the sintered material, powders were examined in room-temperature by X-ray diffraction. The study demonstrates that the samples of $Li_{4-x}Cu_xTi_5O_{12}$ (x=0-0.2) have similar phase composition. We can observe the most intensive peak at the 2θ of 18.4°, 43.3°, 35.6° and 62.8°, which can be well assigned to the characteristic diffraction peaks of the cubic $Li_4Ti_5O_{12}$ spinel-type phase with the *Fd-3m* space group, with the planes at (111), (220), (311), (222), (400), (331), (422), (333), (511), (440), (531) and (444) – ICDD 04-007-9267. The XRD patterns show several impurity peaks at 2θ of 18.5°, 20.5°, 43.6° and 43.9° which can correspond with the diffraction peak of the monoclinic-type Li_2TiO_3 with the C2/c space group, listed in the ICDD 04-009-2812. The maximum amount of Li_2TiO_3-impurities were obtained in $Li_{3.95}Cu_{0.05}Ti_5O_{12}$ samples and it was at 12 wt.% range. The highest purity is in the $Li_{3.85}Cu_{0.15}Ti_5O_{12}$ powder, where there is 6.5 wt.% of Li_2TiO_3 in powder. Other samples have similar amount of impurities from 8 wt.% to 11 wt.%. In the as-prepared powders there is no TiO_2 (rutile) phase, which was observed in the our previous work. The lack of TiO_2 is caused by over-stoichiometric amount of Li_2CO_3 in mix of the precursors. All of this XDR pattern analyzes were obtained using a mathematical Rietveld method with very satisfying value of χ^2=5±1.5. According to the literature [28], the phase transition from spinel phase to the structure of rock salt causes only a slight shrinkage of the lattice parameter, of 8.3595 Å - 8.3538 Å, with only a 0.2% change in cell volume. The cell parameters of the prepared powders fall within the range 8.3605 - 8.3625 [Å]. The elementary cell volume is in the range of 584.35-584.58 [Å^3]. This confirms that the structural parameters correspond with the literature.

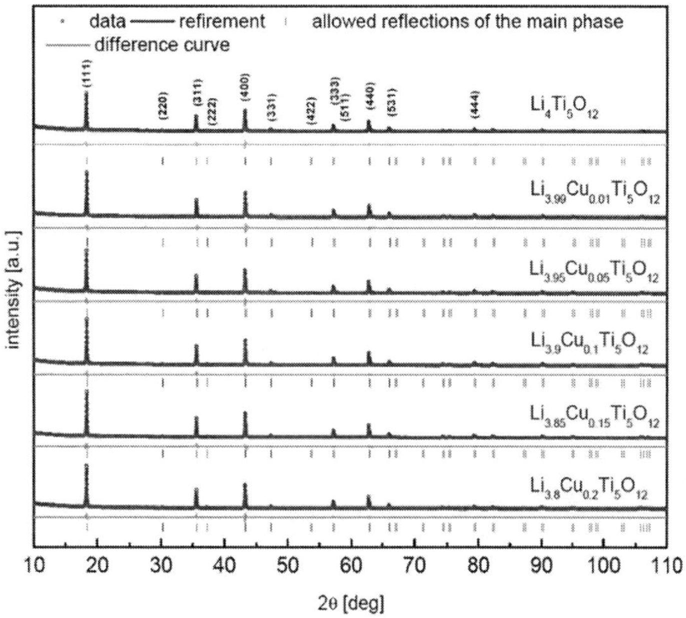

Fig. 1. X-ray diffraction patterns of samples $Li_4Ti_5O_{12}$, $Li_{3.99}Cu_{0.01}Ti_5O_{12}$, $Li_{3.95}Cu_{0.05}Ti_5O_{12}$, $Li_{3.9}Cu_{0.1}Ti_5O_{12}$, $Li_{3.85}Cu_{0.15}Ti_5O_{12}$ and $Li_{3.8}Cu_{0.2}Ti_5O_{12}$ with refinement and peak position of main phase $Li_4Ti_5O_{12}$.

Fig. 2. SEM images of sample $Li_{3.85}Cu_{0.15}Ti_5O_{12}$: magnification · 10000 (left) and · 50000 (right).

In order to get more structural information about the as-synthesized samples, they were also examined by SEM. **Fig. 2.** shows the typical SEM images of the purest $Li_{3.85}Cu_{0.15}Ti_5O_{12}$ powder. The morphology of this powder indicates a disordered morphology composed of agglomerated sintered particles with wide size distribution, in microscale. Some of the crystallites were aggregated in much bigger grains than average. The sintering temperature in this work is probably a little bit too high. Low sintering temperature is also crucial to prevent the grains from excessive growth. Moreover, too high sintering temperature leads to decreasing the Li-ion transport paths. Morphology of commercial powder $Li_4Ti_5O_{12}$ is similar to **Fig. 2.** images and was observed by Wang at al. [31].

3.2. Electrochemical and transport properties

3.2.1 Charge/discharge performance

Fig. 3 presents a specific capacity during discharge process for samples: a) $Li_4Ti_5O_{12}$, $Li_{3.99}Cu_{0.01}Ti_5O_{12}$ and $Li_{3.95}Cu_{0.05}Ti_5O_{12}$ (discharge/charge current rate: C/30, C/15 and C/6) and b) $Li_{3.9}Cu_{0.1}Ti_5O_{12}$, $Li_{3.85}Cu_{0.15}Ti_5O_{12}$ and $Li_{3.8}Cu_{0.2}Ti_5O_{12}$ (discharge/charge current rate: C/6, 2C/3 and 5C/3).

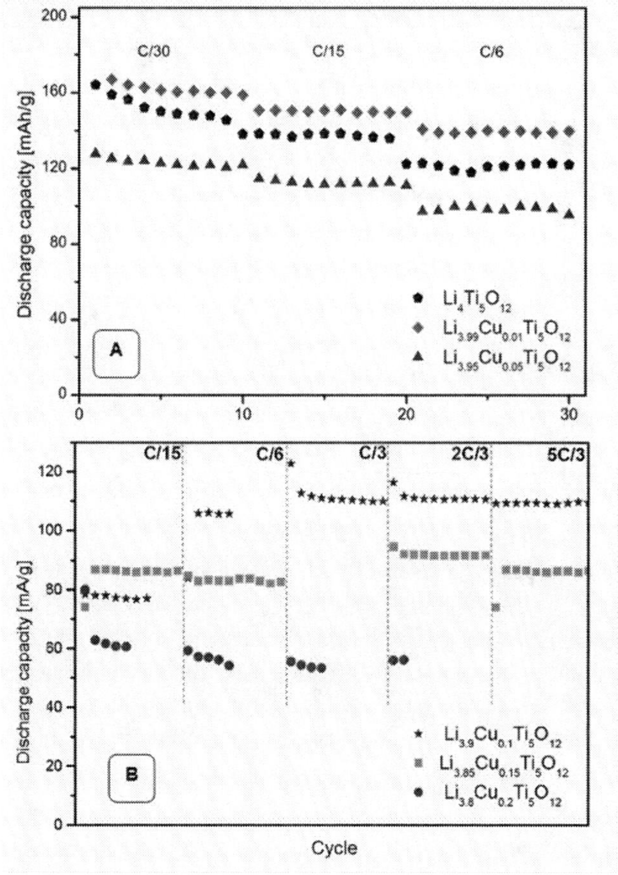

Fig. 3. Specific capacity as a function of cycle number and current rate: a) $Li_4Ti_5O_{12}$, $Li_{3.99}Cu_{0.01}Ti_5O_{12}$ and $Li_{3.95}Cu_{0.05}Ti_5O_{12}$ and b) $Li_{3.9}Cu_{0.1}Ti_5O_{12}$, $Li_{3.85}Cu_{0.15}Ti_5O_{12}$ and $Li_{3.8}Cu_{0.2}Ti_5O_{12}$.

Fig. 3a shows the specific capacity from the first of 30 cycles for samples: $Li_4Ti_5O_{12}$, $Li_{3.99}Cu_{0.01}Ti_5O_{12}$ and $Li_{3.95}Cu_{0.05}Ti_5O_{12}$, at C/30, C/15 and C/6 rates. For each sample, specific capacity significantly decreases with each next rate. Non-doped LTO has specific capacity at rates C/30, C/15 and C/6 at ~150 mAh·g^{-1}, 140 mAh·g^{-1} and 125 mAh·g^{-1} respectively. The $Li_{3.95}Cu_{0.05}Ti_5O_{12}$ sample has 40 mAh·g^{-1} smaller specific capacity than the first two samples. Next three samples were tested with C/15, C/6, C/3, 2C/3 and 5C/3, current rates achieve 110 mAh·g^{-1} in the best case. The $Li_{3.85}Cu_{0.15}Ti_5O_{12}$ and $Li_{3.8}Cu_{0.2}Ti_5O_{12}$ samples have very stable capacity characteristics. The best and well known sample is $Li_{3.85}Cu_{0.15}Ti_5O_{12}$. The cells assembled with anode based on this material show 100 % of reversibility. Moreover, specific capacity measured at the first 10 cycles and after all tests, has the same value, which responds to 77 % of theoretical capacity for 1C rate.

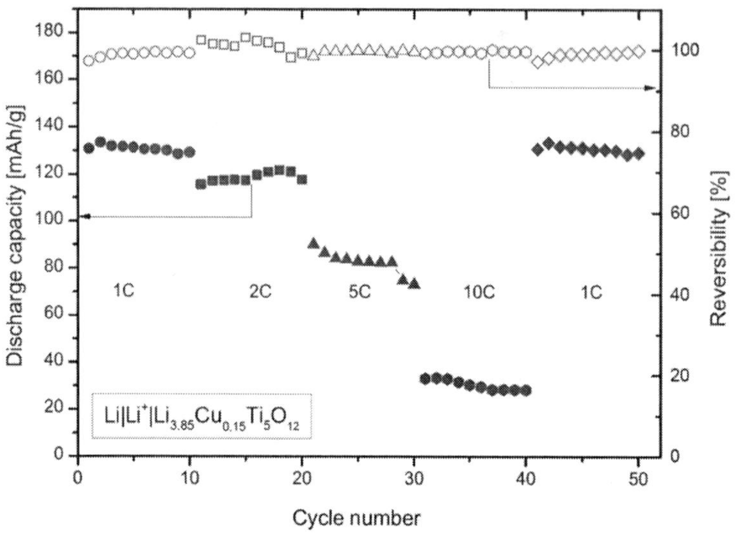

Fig. 4. Specific capacity and reversibility as a function of cycle number for$Li_{3.85}Cu_{0.15}Ti_5O_{12}$.

In **Fig. 4** we can observe change discharge capacity of $Li_{3.85}Cu_{0.15}Ti_5O_{12}$ sample as a function of cycle number and charge rate. With increased the current rate, specific capacity was decreased; the value of specific capacity for 1C rate was ~133 $mAh \cdot g^{-1}$, and for 10C rate was ~30 $mAh \cdot g^{-1}$. This sample was characterized by excellent reversibility of charge/discharge process.

3.2.2 Cyclic voltammetry

Cyclic voltammograms and integral d*Q*/d*E* shown in **Fig. 5** a and b, respectively, were obtained with $Li_{3.85}Cu_{0.15}Ti_5O_{12}$-based anode in Li-ion cell, at the different scan rates. In order to check the stability and reversibility of a process, the cell was tested at three different scan rates, each of these rates was repeated three times. As a result perfectly reversible cycles were obtained.

DOI: 10.1051/e3sconf/20171402042

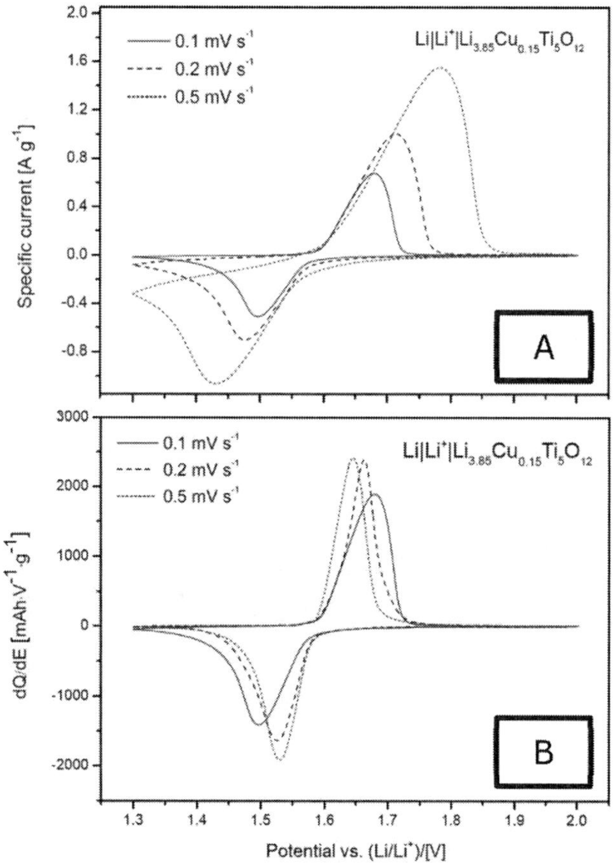

Fig. 5. Cyclic voltammograms (a) and integral d*Q*/d*E*) (b) of 3rd cycles of $Li|Li^{+}|$ $Li_{3.85}Cu_{0.15}Ti_5O_{12}$ at different scan rates.

The three cyclic voltammograms of $Li|Li^{+}|$ $Li_{3.85}Cu_{0.15}Ti_5O_{12}$ cell are shown in **Fig. 5b.** Data for the first cycle at $0.1 mVs^{-1}$ scan rates indicate small polarization. However, the next cycles are very similar. At the third cycle in each scan rates, which are presented in **Fig. 5.**, the oxidation (Li deintercalation) and reduction (Li intercalation) peaks are located at 1.65 V and 1.53 V, respectively, with the average value being 1.59 V. It corresponds to the voltage of Ti^{3+}/Ti^{4+} redox couple (1.60 V) for $Li_4Ti_5O_{12}$ material. The observed difference between potentials of anodic and cathodic peaks equals 60 mV. As can be seen in **Fig. 5.**, with the increasing scan rate, potentials of anodic and cathodic peak shift slightly from each other.

 DOI: 10.1051/e3sconf/20171402042

3.2.3 Conductivity

The impedance spectroscopy was used to study the transport properties of the samples. (**Fig. 6 a** and **b**).

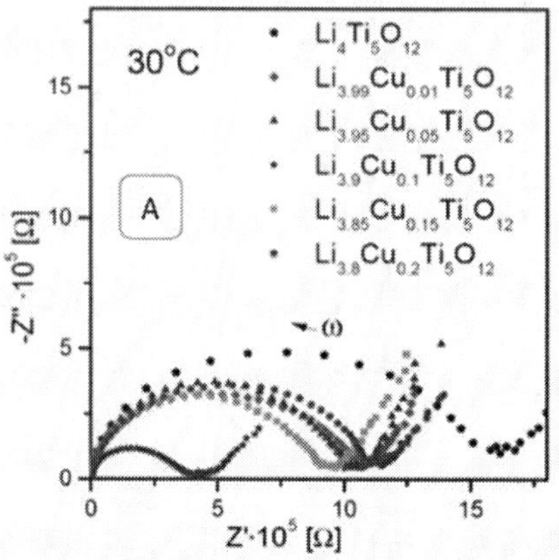

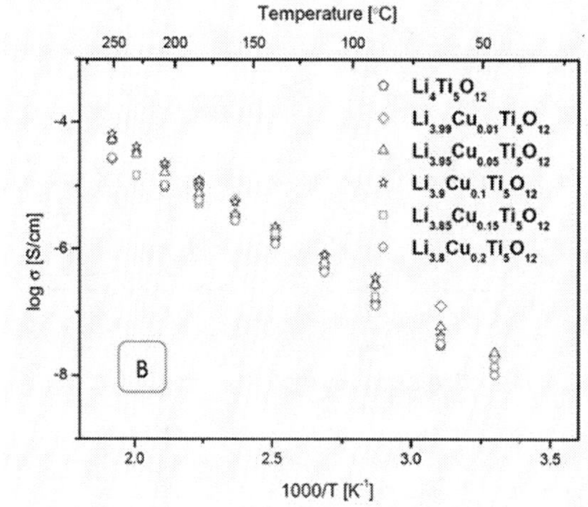

E_a (eV)	R^2
0.76	0.9981
0.46	0.9785
0.48	0.9985
0.53	0.9937
0.47	0.9916
0.50	0.9976

Fig. 6. Complex impedance plots of samples $Li_{4-x}Cu_xTi_5O_{12}$, where $0 \leq x \leq 0.2$ at 30°C temperature (a); temperature dependence of total conductivity for all studied samples $Li_{4-x}Cu_xTi_5O_{12}$, where $0 \leq x \leq 0.2$ Cu with activation energy of studied samples calculated by the standard Arrhenius equation (b).

Fig. 6b. shows the relationship between log σ and 1000/T. The activation energy of this process was presented on the right from curves. E_a of the pellets were calculated with Arrhenius equation. R^2 of calculated values are quite proper and equal 0.988±0.01 averagely. Conductivity increases with the increase of temperature, however conductivity values are similar to each other. Undoped sample has the biggest activation energy (0.76 eV compared to ~0.5eV for the others) and the biggest resistance at the same time. The pellet has resistance in the range of $15 \cdot 10^5$ [Ω], this value is 50% higher from the others. The $Li_{3.85}Cu_{0.15}Ti_5O_{12}$ sample presents the smallest energy activation.

4. Conclusion

The sample of $Li_{4-x}Cu_xTi_5O_{12}$, where $0 \leq x \leq 0.2$ were prepared using a solid-state method. In the SEM images we can observe crystallites in width spectrum of size, due to high temperature of sintering of powders. Substitution of Cu in LTO slightly changes electrochemical performance of the material. The best electrochemical properties and reversibility during intercalation/deintercalation are presented in LTO with 0.15 Cu molar ratio. The difference between cathodic and anodic peak in cyclic voltammograms suggests single-electron process of lithiation.

The financial support of the author Danuta Olszewska - the AGH University of Science and Technology, research project No.11.11.210.911 and Anna Rutkowska - the AGH University of Science and Technology, research project No.15.11.210.338.

References

1. M. Wakihara, Mater. Sci. Eng. Rep. **33**, 109 (2001)
2. X. Guo, C. Wang, M. Chen, J. Wang, J. Zheng, J. Power Sources **214**, 107 (2012)
3. Pan H, Hu Y-S, Chen L Energy Environ Sci **6**, 2338–2360 (2013)
4. U.S. Department of Energy, Grid Energy Storage, 67 (2013)
5. J. Liu, X. Li, J. Yang, D. Geng, Y. Li, D. Wang, R. Li, X. Sun, M. Cai, M.W. Verbrugge, Electrochim. Acta. **63**, 100 (2012)
6. C. Lin, M.O. Lai, L. Lu, H. Zhou, Y. Xin, J. Power Sources **244**, 272 (2013)
7. S. Goriparti, E. Miele, F. De Angelis, E. Di Fabrizio, R. Proietti Zaccaria, C. Capiglia, J. Power Sources **257**, 421 (2014)
8. T. Yuan, K. Wang, R. Cai, R. Ran, Z. Shao, J. Alloys Compd **477**, 665 (2009)
9. D. Wu, Y. Cheng, Ionics (Kiel). **19**, 395 (2013)
10. J. Liu, X. Li, M. Cai, R. Li, X. Sun, Electrochim. Acta **93**, 195 (2013) 195–201.
11. E. Pohjalainen, S. Räsänen, M. Jokinen, K. Yliniemi, D.A. Worsley, J. Kuusivaara, J. Juurikivi, R. Ekqvist, T. Kallio, M. Karppinen, J. Power Sources **226**, 134 (2013)
12. X.-D. Zheng, C.-C. Dong, B. Huang, M. Lu, Ionics (Kiel). **19**, 385 (2012)
13. C. Lai, Z. Wu, Y. Zhu, Q. Wu, L. Li, C. Wang, J. Power Sources **226,** 71 (2013)
14. Y. Zhang, L. Wu, J. Zhao, W. Yu, Mater. Chem. Phys. **166**, 182 (2015)
15. S.H. Yu, A. Pucci, T. Herntrich, M.G. Willinger, S.H. Baek, Y.E. Sung, N. Pinna, J. Mater. Chem. **21**, 806 (2011)
16. Z. Liu, N. Zhang, Z. Wang, K. Sun, J. Power Sources **205**, 479 (2012)
17. M. Li, J. Liang, X. Wang, K. Yu, J. Alloys Compd **681**, 471 (2016)
18. Z. Zhang, G. Li, H. Peng, K. Chen, J. Mater. Chem. A. **1**, 15429 (2013)
19. W. Fang, X. Cheng, P. Zuo, Y. Ma, L. Liao, G. Yin, Solid State Ionics **244**, 52 (2013)
20. Y. Xu, G. Jian, M.R. Zachariah, C. Wang, J. Mater. Chem. A. **1**, 15486 (2013)
21. Q. Zhang, X. Li, Int. J. Electrochem. Sci. **8**, 6449 (2013)
22. E.D. Jeong, H.J. Han, O.S. Jung, M.G. Ha, C.H. Doh, M.J. Hwang, H.-S. Yang, K.S.S. Hong, Mater. Res. Bull. **47, 2847** (2012)
23. C. Lin, X. Fan, Y. Xin, F. Cheng, M.O. Lai, H. Zhou, L. Lu, J. Mater. Chem. A **2**, 9982 (2014)
24. Z. Zhao, Y. Xu, M. Ji, H. Zhang, Electrochim. Acta **109**, 645 (2013)
25. S. Ji, J. Zhang, W. Wang, Y. Huang, Z. Feng, Z. Zhang, Z. Tang, Mater. Chem. Phys. **123**, 510 (2010)
26. C.-C. Yang, H.-C. Hu, S.J.J. Lin, W.-C. Chien, J. Power Sources **258**, 424 (2014)
27. C. Lin, B. Ding, Y. Xin, F. Cheng, M.O. Lai, L. Lu, H. Zhou, J. Power Sources **248**, 1034 (2014)
28. C. Larson, R.B. Von Dreele, Structure **748**, 86 (2004)
29. B.H. Toby, Powder Diffr. **21**, 67 (2006)
30. B.H. Toby, J. Appl. Crystallogr. **34**, 210 (2001)
31. W.Wang, D. Choi, Z. Yang, Metal. Mat. Transactions **A 44,** supp. 1, 21 (2013)

Numerical prediction of the chemical composition of gas products at biomass combustion and co-combustion in a domestic boiler

Henryk Radomiak[1], *Agnieszka* Bala-Litwiniak[1,*], *Monika* Zajemska[1], and *Dorota* Musiał[1]

[1]Czestochowa University of Technology, Faculty of Production Engineering and Materials Technology, Department of Industrial Furnaces and Environmental Protection, Al. A. Krajowej 19, 42-201 Czestochowa, Poland

Abstract. In recent years the numerical modelling of biomass combustion has been successfully applied to determine the combustion mechanism and predict its products. In this study the influence of the addition of waste glycerin in biomass wood pellets on the chemical composition of exhaust gases has been investigated. The pellets have been prepared from spruce- and pine wood sawdust without and with addition of waste glycerin. The waste glycerol is a undesirable by-product of biodiesel transesterification at oil manufacturing. The produced pellets were being burned in the 10 kW domestic boiler adapted to wood pellets combustion. The possibilities of pollutants generation (CO_2, CO, NO_x SO_x and compounds containing chlorine) in the exhaust gases coming from the boiler were numerically calculated using the latest version of CHEMKIN-PRO software, introduced by the American company Reaction Design. The results of the calculations correspond to the data obtained on a real object, in particular: combustion temperature, gas pressure, residence time of fuel in the burner, air flow, fuel consumption, as well as elementary composition of fuel supplied into the boiler. The proposed method of predicting the chemical composition of exhaust gases allows proper control of the combustion process and can be considered as an important step in reducing the pollutants (lower emission of NO_x, SO_x and CO_2 neutral) and thus to contribute to the improvement of the environmental quality. In addition, knowledge of the amounts of Cl-based compounds produced in combustion process (under given conditions), can serve as an important hint in terms of corrosion prevention of boiler- and chimney steels.

1 Introduction

Since decades, the industry production in many European countries has been resting on fossil fuels. Combustion of such fuels as coal, crude oil and natural gas results in substantial

* Corresponding author: abala@wip.pcz.pl

emission of pollutants which, due to increasingly high environmental standards, requires continuous limitation [1-3].

There are many alternative new and renewable energy sources which can be used instead of fossil and conventional fuels. The most common form of renewable energy, widely used in the world is biomass [4-6]. One of the numerous advantages for the energy use of biomass is the reduction in the emissions of greenhouse gases, namely carbon dioxide, and, at the same time, toxic combustion gas components - sulphur dioxide, nitrogen oxides, carbon monoxide and chlorine compounds. It is often assumed that, in the overall balance, the carbon dioxide emission is zero [7]. This results from the fact that in the biomass growth process the plant absorbs a CO_2 gas, amount which is equal to that emitted during biomass combustion. However, considering the energy needed for biomass transportation and its preparation (breaking up), it is estimated that the amount of greenhouse gases formed in the biomass combustion process is smaller by over 90% compared to the combustion of coal [2].

Therefore, in recent years increasingly popular and environmentally friendly method of heating is use of the installation of biomass boilers that utilize wood pellets. Biomass boilers are becoming more and more modern, allow for acceptable combustion efficiency and are automated. In addition, fuels such as wood and biomass are broadly available and relatively cheap [8-10]. Because for the production of pellets water is normally used as binder, there is reasonable to consider the waste glycerin instead of water which can contribute to a higher calorific value of pellets. The waste (crude) glycerin is formed as a by-product in process of transesterification of biodiesel oil manufacturing [11-13]. Due to the relatively high calorific value of the waste glycerin (16.1–22.6 MJ/kg depending on the raw material used to biodiesel production [14-16]), a reasonable solution seems to be its utilization as a heating fuel [15-19]. Unfortunately, direct combustion of the waste glycerin is unprofitable due to its high viscosity and relatively high flash-point which requires a costly modifications of heating equipment [15,17-19]. The simplest and the cheapest method of the utilization of the waste can be its co-combustion with other fuels [20-22]. Glycerin co-combustion with the wood pellets can be a significant solution, because, besides the benefits of waste management, it also contributes to the reduction of harmful combustion products. It could be a cost-effective alternative without necessity of substantial modifications of the heating devices.

This paper presents modelling of wood pellets (WP) without- and with 7% addition of waste glycerin (WGP), during their combustion in a domestic MiniBio boiler adapted for this kind of combustion. The pellets contained glycerin additions have been compared in terms of applicable quality standards with these of glycerin free pellets. Effect of waste glycerin addition has also been examined in view of concentration of harmful combustion products.

2 Material and methods

2.1 Materials - specimens preparation

The waste glycerin used in the study has been delivered from Trzebinia Refinery (RT S.A.), Poland. The minimum content of glycerin in the waste was 80 mass%, ash content - below 5 mass%, MONG (Matter Organic Non Glycerol) - below 6 mass% and the balance - being water. Calorific value of the waste was 22.6 MJ/kg [15]. As the solid component, a spruce and pine wood sawdust with fragmentation of 2–4 mm and a bulk density of 180 kg/m^3 has been used. Cylindrical, 5–15 mm long pellets, 6 ±1 mm in diameter, have been machined with the use of MGL-200 pellet device, applying the wood sawdust without- and with 7 mass% of waste glycerol additions. In further parts of the paper, the abbreviation "WP" (wood pellet) and "WGP" (wood/glycerin pellet) is applied to name the two kinds of composite pellets tested. The appearance of the tested pellets, without- and with glycerol addition, is visually shown in Fig. 1.

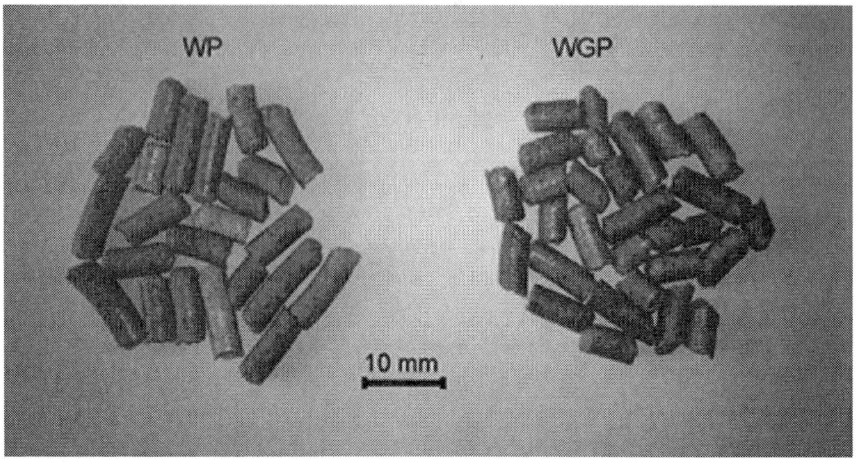

Fig. 1. Visual appearance of as prepared fuel wood pellets: without (WP) and with 7% of glycerin addition (WGP).

For the tested fuels the physicochemical characterization such as: calorific value, bulk density as well as moisture and ash contents have been determined. For simulations purposes also the elemental mass percentages of carbon [C], hydrogen [H], nitrogen [N], chlorine [Cl], phosphorus [P] and sulfur [S]. The detailed WP and WGP pellets characterization is presented in Table 1.

Table 1. Comparison of chosen physicochemical parameters and the elemental composition of the tested pellets. The numbers in brackets denote the confidential range of individual quantities.

Analyzed parameters	WP	WGP
Calorific value, MJ/kg (± 0.1 MJ/kg)	18.1	18.4
Bulk density, kg/m^3 (± 10 kg/m^3)	590	500
Moisture, % (± 0.1%)	4.2	3.4
Ash, % (± 0.2%)	0.8	0.8
C, % (±0.7%)	49.8	49.2
H, % (±0.32%)	5.85	6.24
N, % (±0.1%)	0.19	0.10
Cl, % (±0.04%)	0.20	0.02
S, % (±0.03%)	0.03	0.02
P, % (±0.01%)	0.01	0.01
O, % (bal.)	38.92	40.21

The resulting pellet parameters (diameter, bulk density, moisture- and ash content, and calorific value) are well matched with the German (DIN51731) and Swedish (SS187120) standards. To evaluate the permitted emissions for the analyzed fuels the EN 303-5 standard has been adopted.

2.2 Test stand description and modelling procedure

The 10 kW domestic biomass boiler, Mini Bio type, Kostrzewa (Poland) has been employed for pellets burning. The boiler has been equipped with a burner adapted for combustion of wood pellets having a diameter from 6 to 8 mm. The fuel feed rate ensuring the 7.5 kW working boiler output was equal to 0.000414 kg/s for the WP, and 0.000404 kg/s for the WGP while the air flow was 0.00315 and 0.00308 m^3/s for WP and WGP, respectively. The chamber's internal diameter was 0.35 m and its total length, 0.5 m. The temperature distribution inside of the combustion chamber (573 K to 1123 K, depending on the distance from the burner) was measured with the use of an NiCr-Ni thermocouple. The boiler water reservoir was installed around the combustion chamber and was equipped with a thermostat for keeping water in given temperature range. The temperature set point of the thermostat during pellets combustion was fixed on 333 K. The boiler parameters were employed for modelling purposes.

Numerical calculations of the chemical composition of products formed during combustion of the pellets were carried out with the use of CHEMKIN-PRO licensed software (*The American company Reaction Design*) . For the calculations a simplified chemical mechanism involving 62 chemical elements and compounds and 255 chemical reactions has been applied. The mechanism has been developed using the databases of the universities of Leeds [23] and Milano [24] by the Gas Research Institute, University of California (GRI - Mech version 3.0) [25]. The selection of the chemical reactor and the combustion mechanism was based on the experience of other authors, who have modelled the processes of biomass combustion and co-combustion using older versions of the Chemkin application [2, 26-29]. For example, in paper [2], the processes of co-combustion of selected grades of biomass with natural gas have been investigated. In cited papers [2,3,29], the research comprised both laboratory experiments and numerical simulations with the use of the Chemkin-Pro. The comparison of chemical analyses with numerical calculations, shows that the values obtained from calculations were slightly lower than these coming from the experiment. Very close results have been obtained in Ref. [29] where numerical calculations were done for a wood biomass sample combusted in the heat chamber with a "*Perfectly Stirred Reactor*". Based on the above-cited [23-28] and our own experiences in the numerical modelling of the biomass co-combustion, the calculations for the WP and WGP have been carried out for the PSR (Perfectly Stirred Reactor) variant. In Fig.2 a simplified scheme of the variant is presented.

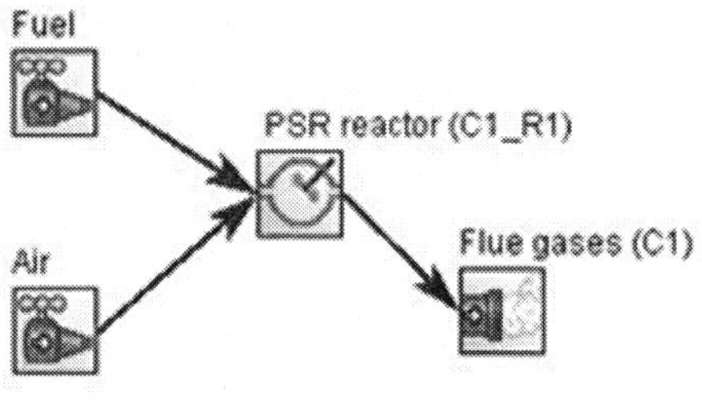

Fig. 2. Scheme of the PSR reactor used in the modelling calculations.

3 Analysis and discussion of the results

In modelling process, as maximum temperature the T = 1123 K has been assumed (the maximum temperature measured in reaction chamber during burning of the tested pellets), 1 atm as an external pressure and some different times of reactants spending in the chamber: 0.0001 to 1 s.

In the calculations, 62 products of combustion have been designated but a detailed analysis included the selected compounds. Among nitrogen oxides, the NO was the main burning product (concentrations of NO_2 and N_2O were relatively small, usually lower than 1

 DOI: 10.1051/e3sconf/20171402043

ppm). On the other hand, only traces of sulphur compounds were formed (the sum of H_2S, SO_2 and SO_3 concentrations was less than 0.001 ppm). Because of very small quantities of sulphur compounds in burning products, only [NO] and [CO] (Fig.3) as well as [CO_2] and [O_2] (Fig.4) have been presented on graphs illustrating reactant residence time in reaction chamber. It should be noted that some chlorine-rich compounds, like HCl, HOCl or Cl_2 are also formed during WP and WGP burning (Fig.5). These compounds can be a source of aggressive chloride ions that are responsible for accelerated corrosion of metallic elements of the chamber and boiler [30].

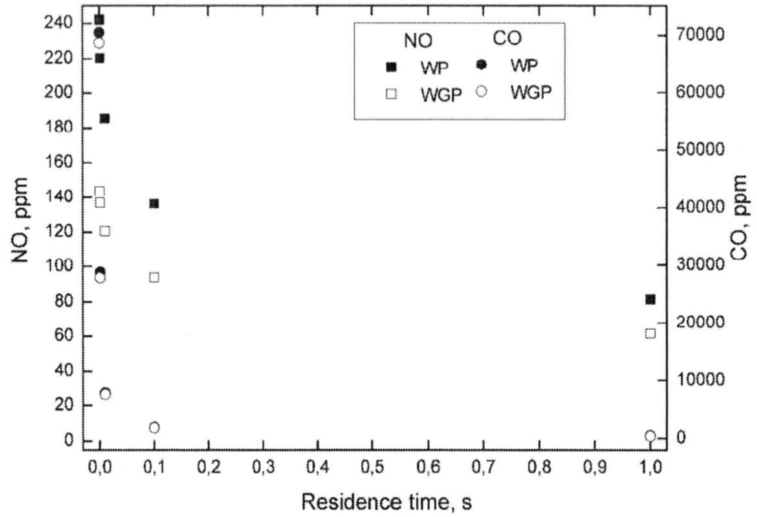

Fig. 3. Effect of waste glycerin addition on NO and CO concentration changes in burning products versus time of reactants residence in a reaction chamber.

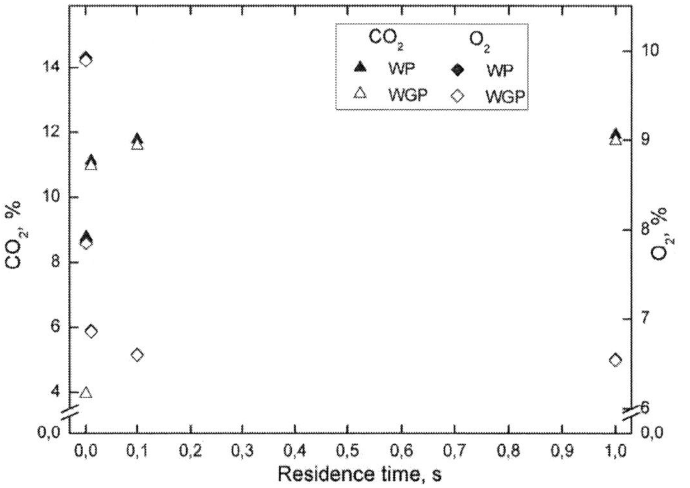

Fig. 4. Effect of waste glycerin addition on CO_2 and O_2 concentration changes in burning products versus time of reactants residence in a reaction chamber.

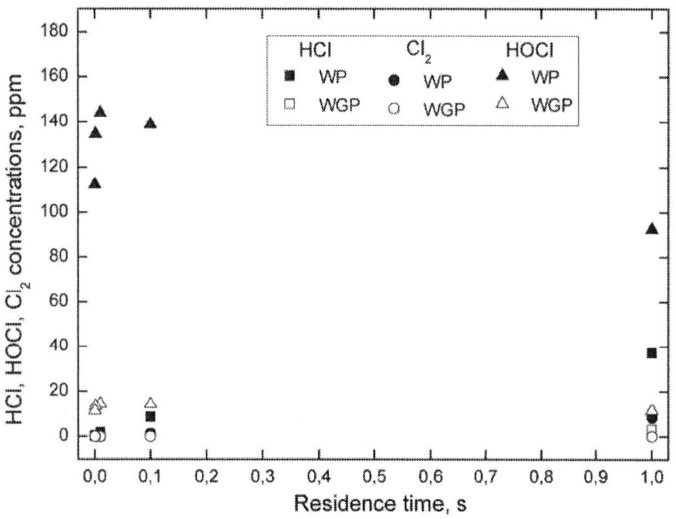

Fig. 5. Effect of waste glycerin addition on HCl, HOCl and Cl_2 concentration changes in burning products versus time of reactants residence in a reaction chamber.

It is seen from Figs 3 and 4 that NO and CO concentrations in the burning products are the greater the shorter residence time in the burner, irrespectively of presence of glycerin in the pellets. This behavior is strictly associated with limited O_2 concentration in flue gas. Concentrations of CO_2 is the greater the longer residence time in the burner due to carbon from wood oxidation (and, thus, O_2 disappearance in reaction products). The CO concentration after 0.0001 s is nearly 70000 ppm and does not practically depend on presence of glycerin in the pellets, which can be attributed to the similar carbon contents in both WP and WGP pellets (see Table 1). The situation is changed in the case of NO and results from twofold greater concentration of nitrogen in glycerin-free pellets. In fact, for the shortest residence times the NO concentration in flue gas is nearly 2 times greater for WP pellets. After 0.1 s the [NO] is still greater (by about 40 per cent) for the glycerin-free pellets. Evidently, glycerin reacts with nitrogen containing compounds present in wood with final formation of solid ionic deposits (e.g. nitrates). This most likely limits nitrogen oxidation to form NO.

Both HCl and Cl_2 concentrations increase a little with longer residence time in the chamber but they scarcely depend on glycerin presence in the pellets. However, in case of WP after time of 1s the HCl concentration is as large as nearly 40 ppm. For other cases it does not exceed several ppm. The situation becomes clearly unfavorable in view of emission of hypochlorous acid (HOCl). It should be stressed that the simulation indicates HOCl as a compound of the greatest concentration among possible chlorine agents in exhaust fume. Obviously, large HOCl emission is predicted for glycerin-free wood pellets only. Such prediction is an evident consequence of tenfold lower chlorine concentration registered analytically for the glycerin containing pellets. Possibly glycerin reacts with chlorine compounds at the stage of pellet formation and part of reaction products is not detectable by chemical analysis (Mohr method titration). In spite of ambiguity of chemical analysis relating appearance of chlorine in WGP, the present results confirm our earlier examinations [30] indicating decrease of chloride ions concentration in flue gas condensate being a result of glycerin addition to the wood pellets. The chlorides, even in small concentrations, accumulate during long-lasting boiler exploitation and cause local corrosion of steel boiler elements. This way, glycerin presence in the fuel ultimately inhibits corrosion of carbon- and stainless steel elements of the boiler.

The mechanism of possible reaction between glycerin and chlorine-rich agents present in wood is still an open question but the advantageous action of glycerin towards limitation of chlorine content in flue gas deserves further, systematic investigations.

4 Conclusions

Numerical modelling enables the evaluation of hazardous gaseous pollutants forming during the biomass combustion. Concentration of burning products depends on both the fuel chemical composition and parameters of burning process.

Co-combustion of wood pellets with the waste glycerin does not practically affect CO and CO_2 concentrations in flue gas. However, the addition of glycerin satisfactorily reduces the share of chlorine- and nitrogen-rich compounds that enter the fume. Consequently, glycerin indirectly limits concentration of both HOCl and NO in exhaust fumes.

Because of documented protection of boiler steel elements against corrosion and evident limitation of air pollution, the use of waste glycerin as a binder (instead of water) for the production of wood pellets can serve as a cost-effective solution of its use.

Acknowledgements: The work was supported by Ministry of Science and Higher Education of Poland - grant No BS/MN–205-302/2016 is acknowledged.

References

1. A. Magdziarz, S. Werle, Waste Manage. **34**, 174–179 (2014)
2. M. Zajemska, D. Musiał, H. Radomiak, A. Poskart, T. Wyleciał, D. Urbaniak, Pol. J. Environ. Stud. **23**, 1445–1448 (2014)
3. M. Zajemska, A. Poskart, Przem. Chem. **92**, 357–361 (2013)
4. P. Mc Kendry, Bioresource Technol. **83**, 37–46 (2002)
5. A. Demirbas, Energy Educ Sci Tech. **5**, 21–45 (2000)
6. S.R. Wan Alwi, Z.A. Manan, J.J. Klemeš, D. Huisingh, J. Clean. Prod. **71**, 1–10 (2014)
7. F. Cherubini, G.P. Peters, T. Berntsen, A.H. Strømann, E. Hertwich, Global Change Biol. **3**, 413–426 (2011)
8. C Serrano, H. Portero, E. Monedero, Fuel **111**, 564–73 (2013)
9. H. Sefidari, N. Razmjoo, M. Strand, Fuel **135**, 120–9 (2014)
10. H. Liu, J. Chaney, J. Li, C. Sun, Fuel **103**, 792–8 (2013)
11. A. Demirbas, *Biodiesel: A Realistic Fuel* ***Alternative for Diesel Engines*** (Springer, 2008**)**
12. B. Moser, In Vitro Cell. Dev. Biol.-Plant **45**, 229–266 (2009)
13. J. Park, D. Kim, Z. Wang, J. Lee, Appl Biochem Biotech. **154**, 246–252 (2009)
14. T. Thamsiriroj, J.D. Murphy, Energ. Fuel. **24**, 1720–30 (2010)
15. A. Bala, H. Radomiak, COW **9**, 365–369 (2012)
16. C.A.G. Quispe, C.J.R. Coronado, J.A. Carvalho, Renew. Sust. Energ. Rev. **27**, 475–93, (2013)
17. M. Gupta, N. Kumar, Renew. Sust. Energ. Rev. **16**, 4551–6 (2012)
18. S.A. Steinmetz, J.S. Herrington, Ch.K. Winterrowd, W.L. Roberts, J.O.L Wendt, W.P. Linak, P. Combust. Inst. **34**, 2749–2757 (2013)
19. P.R. Stack, *Emissions Evaluation: Combustion of Crude Glycerin and Yellow Grease in an Industrial Fire Tube Boiler* (USA: Agricultural Utilization Research Institute Marshall, 2007)
20. P. Queirós, M. Costa, R.H. Carvalho, P. Combust. Inst. **34**, 2759–2767 (2013)
21. A. Bala, H. Radomiak, Energ. Source. (to be published ID:UESO-2015-0467.R1)

DOI: 10.1051/e3sconf/20171402043

22. C. Sakkampang, T. Wongwuttanasatian, Fuel **115**, 186–199 (2014)
23. *http://www.chem.leeds.ac.uk*
24. *http://creckmodeling.chem.polimi.it*
25. *http://www.me.berkeley.edu/gri_mech*
26. A.T. Wijayanta, M.D. Alam, K. Nakaso, J. Fukai, M. Shimizu, Bioresource Technol. **110**, 645–651 (2012)
27. E.M. Fitzpatrick, K.D. Bartle, M.L. Kubacki, J.M. Jones, M. Pourkashanian, A.B. Ross, A. Williams, K. Kubica, Fuel 88, 2409–2417 (2009)
28. C. Dupont, G. Boissonnet, J.M. Seiler, P. Gauthier, D. Schweich, Fuel **86**, 32–40 (2007)
29. A. Magdziarz, M. Wilk, M. Zajemska, Chem. Process. Eng. **4**, 423–433 (2011)
30. A. Bala-Litwiniak, K. Kamieniak, Ochrona przed Korozją (Corrosion Protection) **57**, 179–183 (2014)

DOI: 10.1051/e3sconf/20171402044

Plastic waste depolymerization as a source of energetic heating oils

Marta Wołosiewicz-Głąb[1,*], *Paulina* Pięta[1] , *Sebastian* Sas[2], *Łukasz* Grabowski[1]

[1] AGH - University of Science and Technology, Faculty of Mining and Geoengineering, Department of Environmental Engineering and Mineral Processing, 30 A. Mickiewicz Ave., 30-059 Cracow, Poland
[2]AGH - University of Science and Technology, Faculty of Mining and Geoengineering, Department of Underground Mining, 30 A. Mickiewicz Ave., 30-059 Cracow, Poland

Abstract. In the past years there has been an increase in production and consumption of plastics, which are widely used in many areas of life. Waste generated from this material are a challenge for the whole of society, regardless of awareness of sustainable development and its technological progress. Still the method of disposal of plastic waste are focused mainly on their storage and incineration, not using energy contained there. In this paper technology for plastic waste depolymerization with characteristics of fuel oil resulting in the process, as an alternative to traditional energy carriers such as: coal, fine coal or coke used in households will be presented. Oil has a high calorific value and no doubt could replace traditional solutions which use conventional energy sources. Furthermore, the fuel resulting from this process is sulfur-free and chemically pure. The paper presents the installation for plastics waste depolymerization used in selected Polish Institute of Plastics Processing, along with the ability to use the main thermocatalytic transformation product.

1 Plastic waste's market analysis

Over the last several years we have seen a steady increase in the production of plastics in the world, while in Europe, their production is stable and has remained at about 58 million Mg (Fig. 1). The largest producers of plastics are China, the US and Europe, and their share in world production was respectively 26%, 19% and 20% in 2014.

The data indicate that the demand for plastics is increasing all the time. In Europe alone, in 2014 this value amounted to 47.8 million tonnes. Five European countries accounts for 2/3 of the consumption of plastic, which are: Germany, Italy, France, the United Kingdom and Spain. These numbers are reflected in the extensive use of plastics in many areas:

* Corresponding author: wolosiewiczm.@gmail.com

 DOI: 10.1051/e3sconf/20171402044

as packaging, construction, agriculture, automotive and electronics industry, and many products like furniture, etc. [1].

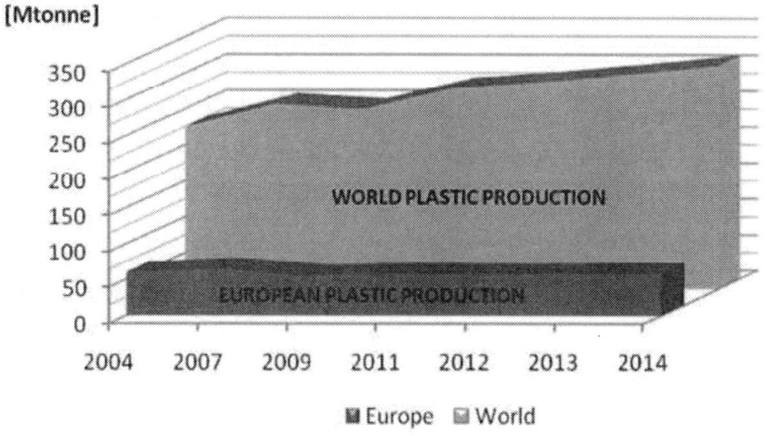

Figure 1. Plastic waste production in Europe and in the world for years 2004–2014. Source: [1]

The growing production of plastics affects the increase in the amount of waste generated and the need for their further development. An important assess direction of the plastic recycling process is the division due to their structural construction. The internal polymer structure, which are used for plastics production, determines their division into three groups: thermoplastics, thermosets and elastomers [2]. There are various types of plastics and their variations allow proper material selection for a particular application. The most commonly used plastics materials include polypropylene (PP), polyethylene (PE-LE and PE-HD), polyvinyl chloride (PVC), polystyrene (PS, EPS) and polyethylene terephthalate (PET), representing 72.8% of total demand in Europe [1]. In addition, in wastes we also find plastics listed in the table below.

Table 1. The types of plastics which can be found in waste. Source: [22]

<table>
<tr><th>Packaging waste</th><th>Construction waste</th><th>Agriculture waste</th><th>Waste removed from service</th><th>Used electrical and electronic equipment</th><th>Mixed municipal waste</th></tr>
<tr><td colspan="3">LDPE</td><td colspan="2">ABS, PP, PA, PC/ABS</td><td rowspan="3">All kinds of plastics can be found</td></tr>
<tr><td rowspan="2">PET, PP, HDPE, PS, PET-G, SAN, SMMA, ASA</td><td colspan="2">HDPE</td><td rowspan="2">PC, PMMA, POM, PBT, PVC</td><td rowspan="2">PS, PBT, PC, POM, PMMA</td></tr>
<tr><td>EPS, XPS, PUR, MDPE, C-PCW, PVDF, PPS, PPSU, PSU, PC, PMMA, ASA/PC, ASA/PVC, PVC</td><td>LLDPE, PP</td></tr>
</table>

Disposal of waste from plastics is a challenge for society as a whole, regardless of awareness of sustainable development and technological progress. In 2014, post-consumer waste contained 25.8 million of used plastic, which 29.7% were recycled, the 39.5% were used for energy recovery, while the remaining 30.8% remained in landfills, which is equivalent to Europe losing by about 8 million tonnes of valuable raw material. However, when comparing the data for the years 2006-2014 for the disposal of plastic waste, trends are encouraging, because it significantly reduced the share of waste deposited at Europe's landfill sites (Fig. 2). European leaders in the recovery of waste plastics are Switzerland, Austria, the Netherlands, Germany, Sweden, Norway, Luxembourg, Denmark and Belgium. In these countries the storage prohibition was introduced, which means that landfill gets no more than 10% of such wastes. In Germany, Switzerland and the Netherlands is stored even less - only approx. 1% [1].

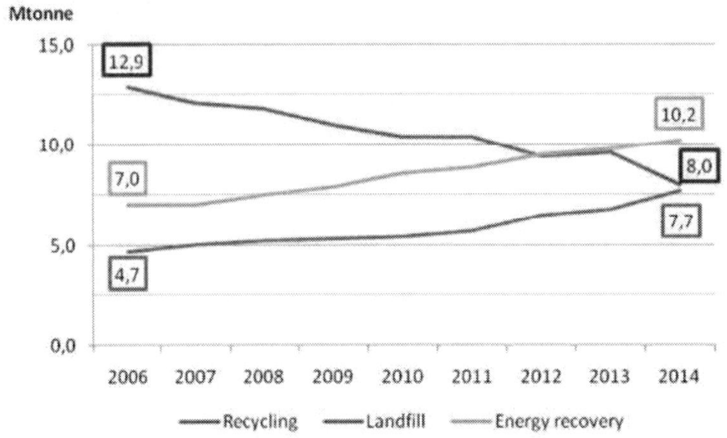

Figure 2. Total plastics waste recycling and recovery in the years 2006-2014. Source: [1]

Still, storage is present in many EU countries as the main way of plastics waste processing. Among the countries which stockpile of more than 50% post-consumer plastics waste are Hungary, Romania, Latvia, Croatia, Bulgaria, Greece, Cyprus, Malta and Poland.

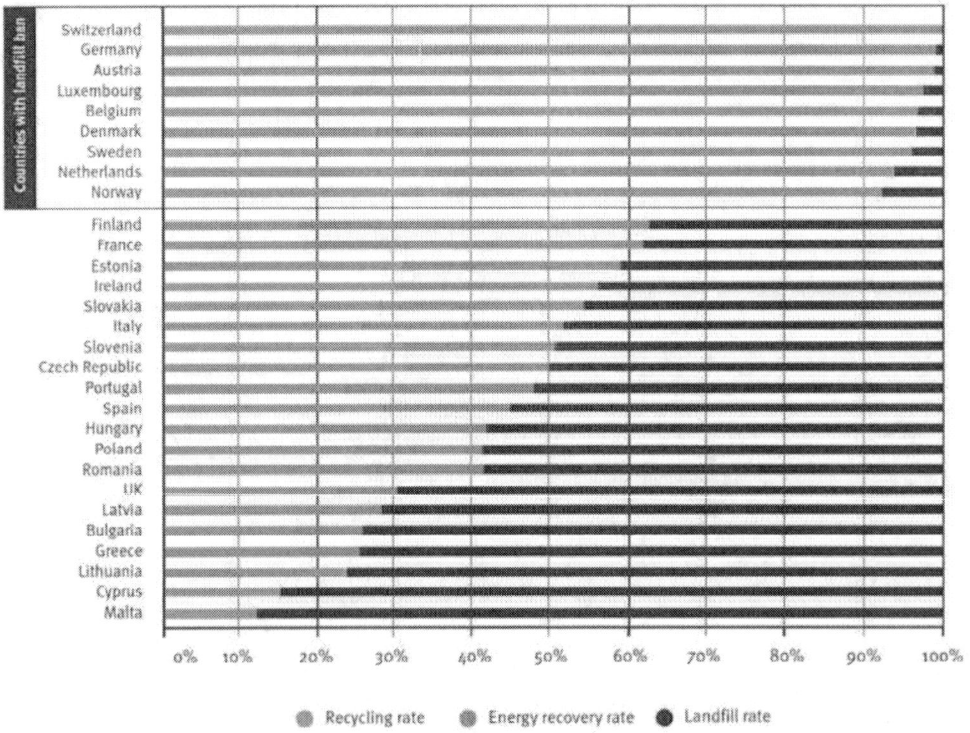

Figure 3. Treatment of post-consumer plastic waste in EU27 + 2 Source: [1]

The economy of plastics waste in Poland in 2014 was as follows: 56% was landfilled, 26% was recycled, and from 24% the energy was recovered [1]. A real opportunity to reduce the amount of plastic stored within large areas and degrading the environment and the landscape is a ban on the collection of plastic waste, as has been done in mentioned European countries. This initiative was introduced by the European Commission and described in the pages of the "Green Paper on plastics waste in the environment." According to the proposal the level of recycling of packaging waste in 2030 should increase to 80% and in the case of municipal waste - up to 70%. However, in the longer term, EU countries should strive for the complete elimination of landfill of such waste [13].

2 Plastics as an alternative energy source

Alternative fuels can be both solid and liquid waste, municipal and industrial used as a replacement for conventional fuels. According to the definition, alternative fuel is such fuel which is obtained in the process of recovery of waste with a calorific value. Table 2 summarizes the most commonly used alternative fuels. Comparing of the mean values of the individual types of alternative fuel, it can be seen that lower calorific values are obtained for both the particulate solid fuels (18-23 MJ / kg), and the impregnated (8-11 MJ / kg). In contrast, a much higher calorific value is characterized by alternative liquid fuels (30-36 MJ / kg). Post-consumer plastics are among the waste with a high calorific value. Particularly interesting polymers in terms of calorific value are polyolefins, which can have up to 47 MJ / kg, so that they can successfully compete with coal, coke or pulverized coal [3, 4].

Table 2. The calorific value of waste used as an alternative fuel. Source: [4]

Alternative fuel	Calorific value [MJ/kg]
Used tires	29,2
Plastics	40-46
Wastepaper	about 11
Used wood	about 18
Used solvents	about 25
Used rubber	30
Used oils	40
Municipal waste	7-10
Carbon slates	12-18
Coal sludge	12-18
Tar	21

Table 3. The calorific value of different fuels. Source: [5]

Fuel	Calorific value [MJ/kg]
Coal	25-28
Wood	7-14
Heating oil	40-48
Polyethylene	44

Two trends dominate regarding to the disposal of plastic waste. The first is the landfill, and the second combustion. None of these typical solutions allows for complete utilization of the energy contained in this material. The production of solid alternative fuels from mixed waste is energy intensive process which makes it very expensive. In turn, the processing of waste in cement kilns does not allow full and swift recovery of their energy, and what is more, causing huge emissions dioxin, furans, nitrogen oxides and other substances. Harmful emissions into the atmosphere from the combustion of municipal waste causes huge degradation of the environment and constitutes a threat to the flora and fauna. In this situation, the best solution may be installations for depolymerization of plastic waste, enabling purification of the natural environment, maximum recovery and utilization of the energy contained in plastics.

3 Polyolefins – a high-energy feed for depolymerization process

Polyolefins (PE, PP) are the main group of synthetic plastics. These wastes due to mass scale production and their relatively uniform chemical composition are very attractive material for a variety of chemical transformation, undertaken in order to reduce the amount of them in landfills. The data provided by the Central Statistical Office indicate that polyethylene and polypropylene are the approx. 7% of municipal waste. In Poland, approx. 860 000 Mg polyolefins per year go to landfills.

PE waste has a high calorific value (Table 2) and can be used as fuel to produce electricity and heat. However, their burning faces some technical difficulties. Starting with a special dispenser, by adjusting the furnace combustion to change of state of the raw material. A major complication may also be mechanical impurities and heterogeneity of materials, which results in the need to introduce full automation of the operation of the furnace / burner [5, 6].

An interesting feature of polyolefins is their relatively easy breakdown (cracking) at elevated temperature. At a temperature of 400 (450)°C–800°C there is a rupture of chemical bonds and the formation of polymers of low molecular weight products which can be divided into 4 groups: gaseous products, liquid hydrocarbons, solid paraffin products and coke. Depending on the applied temperature for the cracking process the share of individual product groups changes (Fig. 4). Research has shown that as a result of the catalytic cracking at a temperature of 470°C it can be approx. 85% of the liquid product, 10% of non-condensable gaseous products and 5% of coke [5, 6].

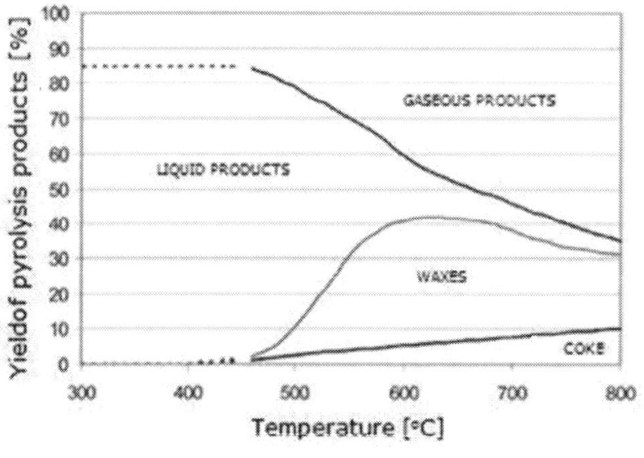

Figure 4. Approximate changes of polyolefins pyrolysis products depending on process temperature. Source: [5]

4 Plastics depolymerization technology

The most common method in the case of polyolefins' waste is the thermal cracking. It consists in subjecting the waste to high temperatures, typically in the range of 500–800°C. The decomposition products include: solid carbonaceous residue and liquid hydrocarbons and the high calorific gas [9, 10]. BASF, Fuji Tech. Conrad., Hamburg

DOI: 10.1051/e3sconf/20171402044

Pyrolysis and BP have developed pyrolysis technology on fluidized layer [8, 12]. The disadvantage of this method is on the one hand the need for a high temperature and a large amount of waste products that require further processing steps, so that they can be used for e.g. as components for fuel. The advantage of pyrolysis of waste is that it is simple process from the technical point of view.

Competitive method compared to pyrolysis is catalytic cracking. It allows thanks to the selection of suitable catalysts to control, within certain limits, both performance and distribution of the products of degradation of polymers, as well as lower the temperature of the process, which is a big advantage. Zeolites are used as catalysts in the form of particulate matter. Catalytic cracking of plastics waste on an industrial scale has been implemented by Nippon Steel Corporation (Japan) [9, 11]. The diagram of the plastic depolymerization system is presented in Fig. 5.

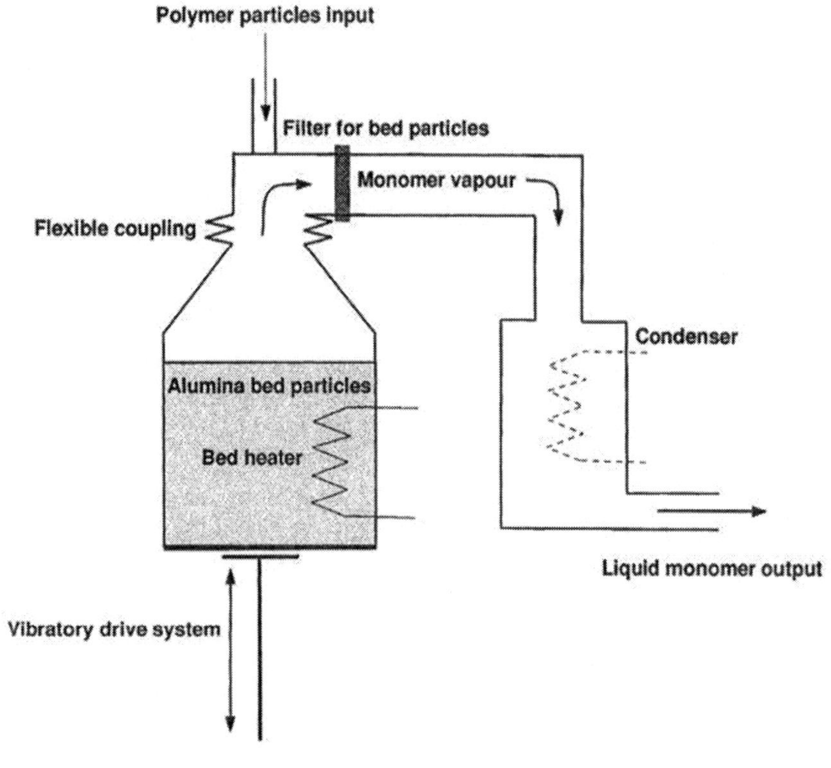

Figure 5. Schematic representation of a mechanically fluidised depolymerisation process. Source: [21]

During the catalytic cracking of a polyolefin heating is carried out at temperatures of 380-480°C, which results in progressive cracking of the polymer chains to hydrocarbon molecules formed by a low boiling point. The vapour-gas phase consists of gaseous hydrocarbons (methane, ethane, ethylene, propane, isopropylene, etc.). Products condensing in room conditions are liquids, including gasoline and oil fraction. With these fractions the paraffins are also distilled off which, although can not be separated, they have the tendency to slowly crystallize and precipitate as a sediment. The fusion of the starting waste is not easy, since polyolefins have a low coefficient of thermal conductivity, and after melting initially have the form of a very viscous liquid [5, 14]. For supporting the endothermic cracking reaction catalysts may be used provided in the system discussed in the next section. The use of cracking catalysts improves the heat transfer from the heated waste and lowers the cracking temperature and reaction products achieve greater uniformity of quality. Most catalysts are mixtures of zeolite, kaolin, or quartz. Studies [6] showed that the course of the depolymerization is influenced by the method of catalysts application. Additionally, what it is very important, catalytic process of polyolefins thermodegradation occurs with 100% relative efficiency and hence of 100% pure yield.

Plastics processing is a rapidly developing field of science because of the importance and the global incidence of these materials. Numerous studies and publications, and continually introduced new solutions confirms the dynamic development of this area of knowledge. Interesting technological solutions have been presented in the work for providing heat to the PE waste in liquid metal reactors [18], fluidized layer reactors [19, 20], or by hot gas, which is a heat transfer medium [16]. Difficulty in giving heat to melted and heated polymer and difficulties in swaging the melted polymer due to its high viscosity and the possibility of solidification, inspired research to develop techniques of cracking in an aqueous medium [17]. Heat supply supporting the pyrolysis reaction using infrared radiant heaters is proposed in [19]. Another heating technique is the use of electromagnetic waves with a frequency of 2450MHz and an ultrasound for the reaction of cracking of various waste, including polymers. A potential pre-irradiation of waste polyethylene with gamma irradiation to induce a preliminary degradation of chemical chains facilitate the following thermal degradation [5].

5 Characteristics of the technological system

The work describes technology of processing of plastics waste both polyethylene and polypropylene in the hydrocarbon mixture of paraffins and olefins in the way non-pressure catalytic cracking on the example of Polish installations of the Plastics Processing [14]. The task of the processing line is to process plastics to obtain fuel component. In the case of the depolymerization process the raw materials of PP and PE can be mixed in any proportions. Impurities are acceptable, such as: sand and dust, printed labels on the packaging. Plastics may have a different colour. Importantly, the waste must meet certain quality requirements. It is necessary that it is uniform in terms of the type of materials and can not include organic admixtures containing Cl, N, S or carbonyl groups, as well as wood, stone, and metal. The maximum moisture content should not exceed 10%, and the pollution can not have a larger share than 5% of the total charge weight. Processing capacity of installation is set at 1,460 Mg / year in cyclical work.

In the first cycle of the process of transformation the polyolefin recyclable materials are shredded into few millimetres fraction. However, research carried out at the Warsaw University of Technology showed that optimal yields of liquid fractions are obtained by granulation of the feed material of 0,063-0,071 mm or 0,08-0,1 mm [6]. The prepared material is collected in the retention tank and mixed with the catalyst (Al/Al2O3). A conveyor belt continuously supplies a material to a reactor equipped with suitable systems for monitoring and controlling. Catalytic cracking process takes place at a temperature of 420°C (± 10°C) with separation of hydrocarbon fraction from C1 to C34. Heat essential to the process is supplied from the combustion of light fuel oil in a burner with a capacity of approx. 100 kW and the burner from the combustion of an uncondensed gaseous fraction. Because of the continued supply of polyolefins in the reaction chamber reaction mass is formed, which is mixed with the anchor stirrers. The next step is to direct the hydrocarbon mixture to the distillation system, where the separation of the ingredients occurs by evaporation and condensation in the cooler with air cooling and an aqueous medium. Gas fractions (C_1 to C_4) flow into the intermediate tank, from which are collected by the gas burners of the reactor to heat the next batch of feed material. While the liquid fraction (C_5 to C_{34}) at approx. 20°C flows to the storage tank which is heated by the heat received from the coolant in the distillation system.

The use of high-energy gaseous products of polyolefins depolymerization and cooling system significantly improves the economics of raw material thermocatalysis. The applied technology was not equipped with an automatic cleaning system, and therefore impurities are removed periodically [14]. The system is equipped with two reactors, while the second one of them is cleaned from unprocessed impurities, the first works. Schematic of this circuit is shown in Fig. 6.

The final product of the processing of raw material is a mixture of saturated hydrocarbons. The resulting combination of hydrocarbons C_5-C_{34} due to the applied technology of evaporation is chemically pure. The resulting liquid fraction can be used as a raw material in the refining and petrochemical industry, or may be treated as a starting material in other branches of chemical industries for the production of chemical products from PE and PP. Another application is to use the mixture as a synthetic fuel.

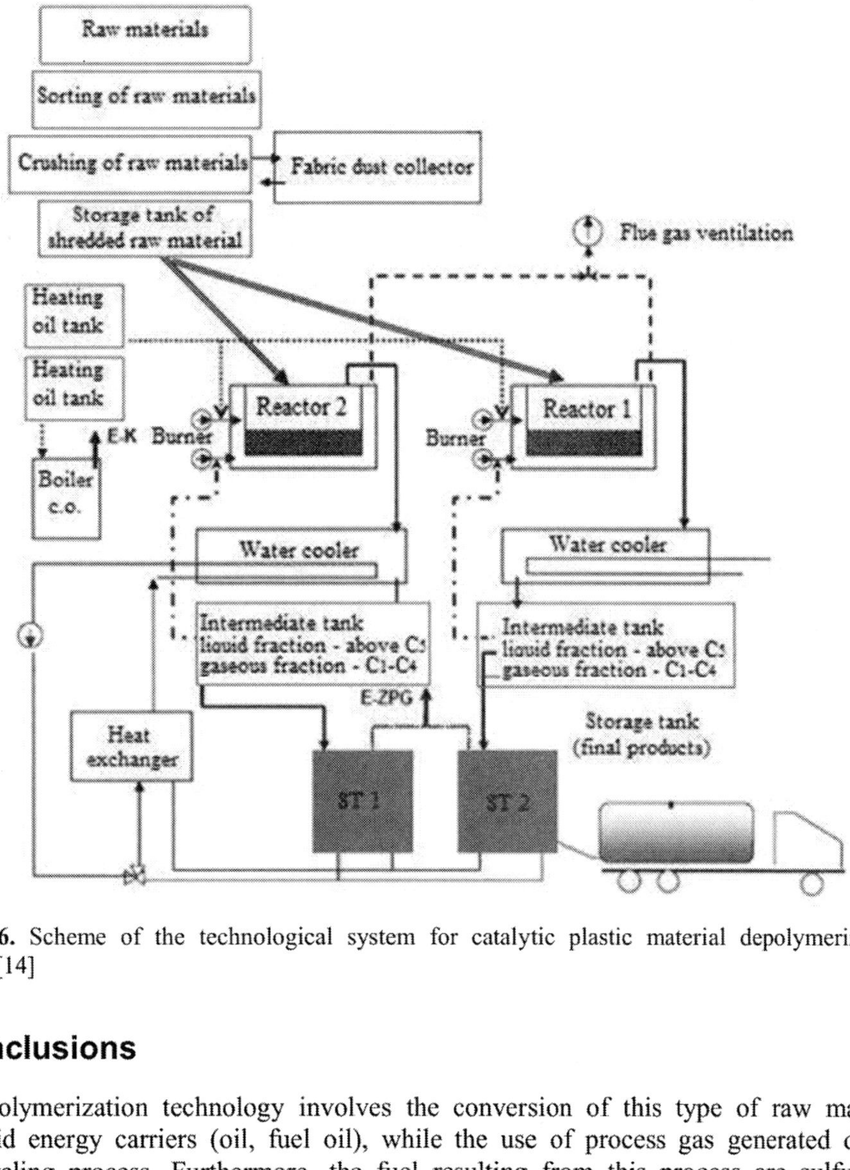

Figure 6. Scheme of the technological system for catalytic plastic material depolymerization. Source: [14]

6 Conclusions

Depolymerization technology involves the conversion of this type of raw material for liquid energy carriers (oil, fuel oil), while the use of process gas generated during the recycling process. Furthermore, the fuel resulting from this process are sulfur-free and chemically pure. Characterized in that they are ready-to-use and capable of storage and transport. At the same time the performance of the plastics depolymerization module can easily be adapted to the size and type of supplies. In 80-85%, these installations allow the recovery of the energy contained in the plastic waste. Approximately 11% is the process gas, which is a carrier of energy for operation, and the waste of coke breeze (4-9%), depending on the amount of impurities contained in the device with the target material. Depolymerization products of 1 Mg of mixed plastics are energy carrier between 4.5 and 5.9 MWe net and up to 5 MWt net. Process energy consumption is small - 0.8-1 kWh/kg, and more importantly this energy can be directly supplied from products resulting from the operation of the system. Admittedly, it can be concluded that the technology for the conversion of plastic into alternative fuels is characterized by high efficiency and is environmentally friendly.

To sum up, the technology of plastics depolymerization is an environment-friendly process, creating the possibility for safe and effective reduction of a part of the waste from the group of olefins by converting them into products of commercial applications. Used during the process PE and PP products do not meet the requirements that would allow their recycling, thus subjecting them to thermocatalytic conversion in energy products is the best ultimatum for the collection of these substances in landfills, which has more negative impact on the environment than the mentioned process of depolymerization. The biggest advantage of plastics processing plant are final products which can be widely used in various industries (refinery and petrochemical industry, production of household chemicals, energy industry for the production of gaseous and liquid). In addition, thanks to the process of depolymerization, we can limit environmental degradation as a result of depositing waste, proceeding in accordance with the waste hierarchy referred to in the Act on Waste, and closing the life cycle of waste by recycling them. In addition, a significant part of the work of the high temperature thermal cracking is the process temperature which is qualified as a medium temperature heat source. The energy in the process can be used as a source of energy (by using a recuperation process) to carry out other processes on the premises, or be used as an energy source for the process water or hot water installation for the social purposes. In case of described process, the ecological effect has been achieved simultaneously with economic effect, which is the greatest achievement of this technology.

Acknowledgement

The paper is the effect of the scientific project no. 15.11.100.021.

References

1. Plastics Europe. Plastics - the Facts 2014, An analysis of European latest plastics production, demand and waste data (2015)
2. Mróż J., Recykling i utylizacja materiałów odpadowych w agregatach metalurgicznych. Wyd. Pol. Cz., Częstochowa (2006)
3. Sorek A., Borecki M., Ostrowska-Popielska P., Selected plastic waste as a source of alternative fuels in the metallurgical industry, IMŻ, vol. 4, pp. 47-57 (2012)
4. Walendziewski J., Kałużyński M., Surma A., Sieć Naukowo-Gospodarcza "Energia" Określenie potencjału odpadów i ich rodzaju do produkcji stałych paliw alternatywnych, projekt Z/2.02/II/2.6/06/05, Wrocław (2007)
5. Grzybowski P., Transformation of the poliethylene wastes into fuel products, Chemik, 66, 7, 725-734 (2012)
6. Sokołowski J., Rokicki G., Marczewski M., Szewczyk K., Thermalcatalytic recycling of polyolefins and polystyrene. CzT Ch, 16, 2, 311-321 (2008)

7. Sokołowski J., Sposób termiczno-katalitycznego recyklingu składników polimerowych obejmujących poliolefiny i polistyren oraz przepracowanych olejów silnikowych w obecności ditlenku węgla oraz instalacja do realizacji sposobu. Patent PL 216345 B1 (2013)
8. Belgiorno V., De Feo G., Della Rocca C, Napoli R.M.A. Waste Manag., 23, 1 (2003)
9. Aguado J., Serrano D.P., Feedstock Recycling of Plastic Wastes (Ed. J.H.Clarc), The Royal Society of Chemistry, Cambridge (1999)
10. Tukker A., Plastic waste Feedstock Recycling, Chemical recycling and Incineration, Rapra Review Reports, 3(4), Report 148, Rapra Technology Ltd., Shropshire, United Kingdom (2002)
11. EPA SITE Technology Capsule, Texaco Gasification Process, Environmental Protection Agency, Office of Research and development, Cincinnati, OH45268EPA540/R94/515a[on line] [access: 25.08.2016] At the following website: http://www.epa.gov/ORD/SITE/reports/540r94514/540r945a.pdf (1995)
12. Metecan H., Ozkan A.R., Iskr R., Yanik J., Saglam M., Yuskel M., Fuel, 84, 30 619 (2005)
13. Saj E., Unia szykuje całkowity zakaz składowania tworzyw sztucznych [online],[Access: 25.08.2016] At the following website: http://www.srodowisko.abc.com.pl/czytaj/-/artykul/unia-szykuje-calkowity-zakaz-skladowania-tworzyw-sztucznych (2015)
14. Pracownia B-B Projektowo-Usługowa Zdzisław Budych. Raport o oddziaływaniu na środowisko planowanego przedsięwzięcia - Zakładu Przetwórstwa Tworzyw Sztucznych w Szwecji k. Wałcza (2012)
15. Nolte M. J., Zięba A.: Sposób i układ do prowadzenia krakingu odpadów poliolefinowych. Patent PL 351703 (2002)
16. Charles W. et al.: Cracking Process. Patent US 3,985,820, (1976).
17. Hughes W.L. et al.: Method of converting organic material into useful productsand disposable waste. Patent US 4,657,681 (1987)
18. Tatsumi T., et al.: Process for the thermal decomposition of thermoplastic resinswith a heat transfer medium. Patent US 3,974,206 (1976)
19. Kirkwood K.C., et al.: Polymer cracking. Patent US 5,364,995 (1994)
20. Fan L. et al.: Pyrolitic conversion of carbonaceous solids to fuel gas in quartzsand fluidized beds. Patent US 4,448,589 (1984)
21. Newborough M., Highgate D., Vaughan P.,Thermal depolymerisation of scrap polymers, Appl. Therm. Eng., 22, 17, 1875–1883 92002)
22. Borkowski K.., The recycling of plastics, PlasticsEurope Fundation, Poland [on line], [access: 26.08.2016] At the following website: www.pie.pl/materialy/_upload/KONF_OPAK/PlasticsEurope_KBorkowski.pdf (2015)

AUTHOR INDEX

AUTHOR INDEX

AUTHOR INDEX